The Periodic Table of Elements (long form).

Atomic weights are shown above the symbols; atomic numbers, below.

METALS

NONMETALS

TRANSITION METALS

INNER TRANSITION METALS

PERIODS	1 IA	2 IIA	3 IIIB	4 IVB	5 VB	6 VIB	7 VIIB	8	9 VIIIB	10	11 IB	12 IIB	13 IIIA	14 IVA	15 VA	16 VIA	17 VIIA	18 VIIIA
1	1.0079 H 1																1.0079 H 1	4.00260 He 2
2	6.941 Li 3	9.01218 Be 4											10.81 B 5	12.011 C 6	14.0067 N 7	15.9994 O 8	18.998403 F 9	20.179 Ne 10
3	22.98977 Na 11	24.305 Mg 12											26.98154 Al 13	28.0855 Si 14	30.97376 P 15	32.06 S 16	35.453 Cl 17	39.948 Ar 18
4	39.0983 K 19	40.08 Ca 20	44.9559 Sc 21	47.88 Ti 22	50.9415 V 23	51.996 Cr 24	54.9380 Mn 25	55.847 Fe 26	58.9332 Co 27	58.69 Ni 28	63.546 Cu 29	65.39 Zn 30	69.72 Ga 31	72.59 Ge 32	74.9216 As 33	78.96 Se 34	79.904 Br 35	83.80 Kr 36
5	85.4678 Rb 37	87.62 Sr 38	88.9059 Y 39	91.224 Zr 40	92.9064 Nb 41	95.94 Mo 42	(98) Tc 43	101.07 Ru 44	102.9055 Rh 45	106.42 Pd 46	107.868 Ag 47	112.41 Cd 48	114.82 In 49	118.71 Sn 50	121.75 Sb 51	127.60 Te 52	126.9045 I 53	131.29 Xe 54
6	132.9054 Cs 55	137.33 Ba 56	* [57-71]	178.49 Hf 72	180.9479 Ta 73	183.85 W 74	186.207 Re 75	190.2 Os 76	192.22 Ir 77	195.08 Pt 78	196.9665 Au 79	200.59 Hg 80	204.383 Tl 81	207.2 Pb 82	208.9804 Bi 83	(209) Po 84	(210) At 85	(222) Rn 86
7	(223) Fr 87	226.0254 Ra 88	† [89-103]	(261) Unq 104	(262) Unp 105	(263) Unh 106	(262) Uns 107	(265) Uno 108	(266) Une 109									

*LANTHANIDE SERIES

138.9055 La 57	140.12 Ce 58	140.9077 Pr 59	144.24 Nd 60	(145) Pm 61	150.36 Sm 62	151.96 Eu 63	157.25 Gd 64	158.9254 Tb 65	162.50 Dy 66	164.9304 Ho 67	167.26 Er 68	168.9342 Tm 69

†ACTINIDE SERIES

227.0278 Ac 89	232.0381 Th 90	231.0359 Pa 91	238.029 U 92	237.0482 Np 93	(244) Pu 94	(243) Am 95	(247) Cm 96	(247) Bk 97	(251) Cf 98	(252) Es 99	(257) Fm 100	(258) Md 101

COLLEGE CHEMISTRY
With Qualitative Analysis

with what?

Strabismus?

EIGHTH EDITION

College Chemistry
With Qualitative Analysis

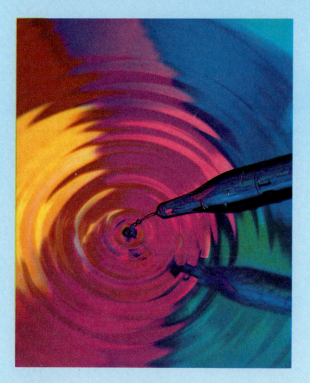

Henry F. Holtzclaw, Jr.
University of Nebraska—Lincoln

William R. Robinson
Purdue University

D. C. HEATH AND COMPANY

Lexington, Massachusetts   Toronto

Cover photo: E. R. Degginger
Additional photo credits on page A25

Editors: Mary Le Quesne and Carmen Wheatcroft

Production Supervisor: Marret McCorkle

Production Editor: Antoinette Schleyer

Designer: Victor Curran

Production Coordinator: Mike O'Dea

Photo Researcher: Martha Shethar

10 9 8 7 6 5 4 3 2 1

Published simultaneously in Canada.

Printed in the United States of America.

International Standard Book Number: 0-669-12862-7

Library of Congress Catalog Card Number: 87-80410

Preface

Chemistry is the study of the behavior of matter and of the models used to describe that behavior. Through words, photographs, diagrams, chemical equations, and mathematical equations, *College Chemistry, Eighth Edition*, introduces students to chemistry and teaches them to describe and predict chemical behavior both quantitatively and qualitatively. It offers the foundations that allow students not only to understand the chemistry presented but also to extend this understanding to new situations.

Chemistry is central to a description of the behavior of matter in many disciplines. However, we believe that it is essential that the concepts and models of the chemical behavior of the elements be presented clearly before they are applied to other systems in which additional considerations may mask the chemical principles involved. Thus we have paid careful attention to the introduction and development of chemical principles before extending them to different situations. This philosophy is also carried out in the exercises at the end of the chapters; many practice problems are provided in addition to more complicated or applied exercises.

This eighth edition of *College Chemistry* has evolved from a line of very successful general chemistry textbooks: Over one million general chemistry students have benefited from its previous editions. Two years of careful revision draw on our work with these earlier editions; our experience as teachers; the suggestions and ideas of our many users, colleagues, and collaborators; and the results of extensive market research. Throughout the revision process we have tried to make the text more interesting and readable. At the same time we have retained the clear discussion and concern for the student that has characterized the previous editions.

Important Learning Aids within the Text
This book provides several helpful tools to students and instructors.

- Over 250 examples proceed from simple, straightforward problems to those that are more complex. Many of these examples illustrate the relevance of chemistry to many other disciplines.

- Over 2100 exercises, many of which have multiple parts, are grouped by topic for easy location of problems of a particular type. Simple drill problems, with no distracting frills, help reinforce the introduction to a topic. Additional problems, including those that apply chemistry to a wide variety of other fields, will challenge the student.

- End-of-chapter material includes Summaries, which provide quick reviews of each chapter, and lists of Key Terms and Concepts. The references in Key Terms and Concepts identify the text section in which a concept is introduced.

- The glossary, a new addition to the text, enables students to review terms or look up unfamiliar terms.
- An extensive index of over 5500 entries allows the user to find topics easily. In addition, when the text introduces a new concept in terms of ideas developed earlier, there is a cross-reference to those earlier sections.

Changes in this Edition

- This edition is richly illustrated with many full color photographs and drawings selected for their pedagogical value. The art program shows chemical systems, illustrates techniques and equipment, demonstrates the behavior of chemical reactions, and elaborates on and reinforces concepts. The new photographs convincingly demonstrate that chemistry is not an isolated science.
- The text has been extensively revised. We have rewritten or replaced many chapters in order to streamline the presentation of topics and to combine similar ideas. This reorganization is particularly apparent in the chapters dealing with descriptive chemistry. We have also rewritten many sections to complement the accompanying photographs.
- The introduction to thermochemistry now appears in Chapter 4. (The more advanced concepts of thermodynamics have been left for later in the course.) With this early introduction, we can discuss energy-related topics such as ionization energy; electron affinity; lattice energy; and heats of fusion, sublimation, vaporization, and solution as specific examples of a more general concept rather than as individual, unrelated phenomena. For example, with the early coverage of thermochemistry, we can introduce bond energies as a logical component of the coverage of covalent bonding.
- The revised chapter, "Condensed Matter," now emphasizes that many of the properties of solids and liquids can be described by similar models. For example, application of the kinetic-molecular theory explains the formation of both a liquid and solid by condensation of a gas as well as evaporation of both liquids and solids.
- The discussion of the dissolution of gases introduces the analytical form of Henry's law.
- The separation of kinetics and the introduction to equilibrium into two chapters allows greater flexibility. We have added an expanded discussion of the solution of equilibrium problems by successive approximation and by use of the quadratic equation.
- The revised section on buffers more fully addresses the uses, selection, and function of this important class of chemical systems.
- Acid-base and solubility product chapters employ the revised value of K_2 for the ionization of H_2S.
- A clearer distinction is made between enthalpy, entropy, and free energy changes under standard state conditions and nonstandard state conditions.
- Combining the introduction to electrolysis of several solutions into a single section emphasizes the common aspects of the process.

Our Approach to Descriptive Chemistry

The descriptive chemistry of the elements has been extensively reorganized to present this material as concisely as possible without decreasing the overall descriptive chemistry content. It is presented in several segments that can be combined in a variety of ways.

- **Chapter 9   An introduction to general chemical behavior based on the Periodic Table.** This chapter introduces types of compounds (acids, bases, salts, electrolytes, and nonelectrolytes) and chemical reactions (addition, decomposition, acid-base, oxidation-reduction, etc.) as well as the general behavior of metals and nonmetals. It discusses prediction of reaction products based on the Periodic Table, oxidation numbers, and the activity series. This first chapter on descriptive chemistry reflects our philosophy that the basics of descriptive chemistry should not be strictly a matter of memorization.

- **Chapter 13   The chemistry of the active metals.** Chapter 13 describes the chemical behavior of aluminum and the active metals of Groups IA and IIA, elements that are strong reducing agents and that exhibit only one oxidation state in their (essentially ionic) compounds.

- **Chapter 16   A description and explanation of the behavior of acids and bases.** This chapter presents a qualitative description of these important classes of compounds.

- **Chapter 20   An introduction to the chemistry of the nonmetals.** Chapter 20 presents a general overview of the behavior of the nonmetals based on the similarity in their behaviors and the differences from the behavior of the active metals.

- **Chapters 22 and 23   The specifics of the behavior of the individual nonmetals.** These chapters provide the instructor with a selection of material that may be used to explore the behavior of specific nonmetals in greater detail.

- **Chapters 25, 27, and 28   The chemistry of the semi-metals, the transition metals, and the post-transition metals, respectively.** Again we discuss the general behavior common to each set of elements before examining individual elements. Thus the instructor is free to explore the behavior of these elements at varying levels. Chapter 27 includes new material on recent developments in superconductivity and its high-technology applications.

- **Chapter 29   The discussion of the atmosphere and natural waters.** The discussion of these systems has been incorporated into a single chapter that also deals with the challenges facing society.

- **Chapters 30 and 31   An introduction to organic and biochemistry.** A separate chapter is devoted to organic compounds. The biochemistry has been substantially revised and updated.

We are particularly proud of our revised qualitative analysis chapters. The qualitative analysis scheme is described in a new chapter that presents the chemistry involved in the separation and identification of the cations in the scheme and the principles upon which the scheme is based. This is followed by chapters with our well-tested qualitative analysis procedures to which we have added a useful and unique set of color photographs of the precipitates, solutions, and tests present in most steps of the scheme.

Supplements

The **Instructor's Guide** by Norman E. Griswold of Nebraska Wesleyan University provides instructors with chapter-by-chapter teaching aids for introductory chemistry course planning.

A **Study Guide,** also by Norman E. Griswold, is designed to strengthen students' knowledge of the facts and principles of chemistry. The *Study Guide* helps students to recognize important chemical concepts and understand how selected topics relate to each other. The *Study Guide* begins with a section about general study methods, including brief directions for solving problems and taking examinations. The main body of the *Guide* is divided into chapters that correspond to chapters in the text.

The supplements package also includes the **Solutions Guide** by John H. Meiser and Frederick K. Ault, both of Ball State University. This *Guide* consists of fully worked-out solutions to approximately half the problems from the text.

New to this edition is the **Complete Solutions Guide** by Norman E. Griswold, John H. Meiser, and Frederick K. Ault. This valuable supplement contains complete solutions to all text exercises, both numerical and discussion.

Basic Laboratory Studies in College Chemistry with Semimicro Qualitative Analysis by Grace R. Hered, City Colleges of Chicago, parallels the style and sequence of the material in *College Chemistry*. *Basic Laboratory Studies* can be used for a one- or two-semester course in either general chemistry and qualitative analysis or in chemistry for the health sciences. The manual emphasizes the use of descriptive chemistry and encourages students to think independently and sharpen their problem-solving skills in the lab. Bold red type is used effectively to reinforce caution and safety notes.

The **Instructor's Guide for Basic Laboratory Studies in College Chemistry,** also by Grace R. Hered, is a valuable new supplement featuring a wealth of resources for the laboratory instructor. Helpful suggestions for instructors include lists of possible chemicals and laboratory equipment, special notes for each lab exercise, answers to the prelab questions, and a list of locker supplies.

Two **Computerized Testing** programs are available to adopters in both Apple and IBM versions. *Heathtest* (licensed from ips Publishing, Inc.) is a powerful algorithmic testing program that enables instructors to produce multiple versions of chapter tests, mid-terms, and final exams quickly and easily. It has full graphics capability, enabling it to display a wide variety of structures, including Lewis structures, benzene rings, and carbon chains. *Archive* provides an extensive multiple choice program.

Printed **Test Item Files** for both *Heathtest* and *Archive* are also available. Three tests for each chapter are included in the *Heathtest Test Item File*. The entire data base is included in the *Archive Test Item File*.

By special arrangement with COMPress, Inc., a division of Wadsworth, Inc., D. C. Heath is able to offer adopters of *College Chemistry* a special discount on a variety of **software** programs.

New to this edition of the text is a collection of over 80 full-color **transparencies,** which are an excellent teaching aid. The transparencies have been taken from artwork in the text and have been carefully prepared to ensure clarity and ease of reading.

Also new to this edition are a set of **videotapes** prepared by Paul Kelter of the University of Wisconsin, Oshkosh. Fifty chemistry lecture demonstrations, chosen for their chemical significance and learning value, show real students interacting with their instructor as he performs a wide variety of interesting experiments. Designed to be both informative and motivational, the demonstrations consistently emphasize laboratory safety and depict the practical application of chemical principles.

Acknowledgments

No authors can take the entire credit for the success of a text; many others play an important role in its development. We are particularly indebted to the following persons who read all or part of the manuscript and made many thoughtful suggestions:

> J. M. Bellama, University of Maryland
> Toby F. Block, Georgia Institute of Technology
> James Finholt, Carleton College
> Patrick Garvey, Des Moines Area Community College
> Charles Greenlief, Emporia State University
> James F. Hall, Northeastern University
> David Hennings, Doane College
> Jerry L. Mills, Texas Tech University

In addition to John Meiser, F. Keith Ault, Norman Griswold, and Grace Hered, many present users of the text submitted many helpful ideas. They include Genevieve W. Adams, Mars Hill College; J. Keith Addy, Wagner College; John F. Albrecht, University of Wisconsin Center—Richland; C. J. Alexander, Des Moines Area Community College; Harrison C. Allison, Marion Military Institute; Wesley E. Bentz, Alfred University; Marshall Bishop, Southwestern Michigan College; Kenneth Borden, University of Indianapolis; Robert W. Braun, Milwaukee School of Engineering; John P. Carlson, Grand View College; James D. Carr, University of Nebraska—Lincoln; Joseph H. Cecil, Kent State University, Tuscarawas Campus; Wallace Coker, Gadsen State Junior College; Alan D. Cooper, Worcester State College; Alden L. Crittenden, University of Washington; David A. Darnall, Shelby State Community College; Ronald DeLorenzo, Middle Georgia College; Ronald D'Orazio, Ellsworth Community College; W. L. Ellerbrook, Clarendon College; Walter C. Emken, Central Washington University; Sandra Y. Etheridge, Gulf Coast Community College; Donald Evers, Iowa Central Community College; Henry S. Falkowski, Potomac State College of West Virginia University; Larry Ferren, Olivet Nazarene University; Tom Fico, North Florida Junior College; Walter C. Flanders, State University of New York at Morrisville; Rosemary Fowler, Cottey College; Aline B. Frappier, Community College of Rhode Island; D. B. Fraser, Essex County College; Charles Freidline, Union College; Mark Freilich, Memphis State University; Roy Gold, Colorado Northwestern Community College; Edwin S. Gould, Kent State University; Mary T. Guy, Southwest Mississippi Junior College; Robert L. Hade, Carthage College; Robert J. Hall, Hannibal—LaGrange College; Paul V. Hansen, Jr., Carthage College; John Hanson, Olivet Nazarene College; Ron Head, Okaloosa Walton Junior College; Donald P. Hoster, Community College of Baltimore; John Ide, Mitchell Community College; Robert E. L. Ingram, Arizona Western College; Victor N. Kingery, Garland County Community College; John D. Konitzer, McHenry County College; Laurence G. Ladwig, Black Hawk College; Kenneth W. Larson, Gogebic Community College; Bobbie M. Liggett, Andrew College; Eugene C. Lindblad, Dana College; William Lindquist, Worthington Community College; Clarence W. Linsey, Mid-America Nazarene College; Eugene R. Magnuson, Milwaukee School of Engineering; Garry McGlaun, Gainesville Junior College; Kenneth E. Miller, Milwaukee Area Technical College; Stephen Monts, Kankakee Community College; Terry L. Morris, Southwest Virginia Community College; Melvyn W. Mosher, Missouri Southern State College; Paul D. Neumann, Nashville State Technical Institute; William C. Nickels, Schoolcraft Community College; G. A. Nyssen, Trevecca Nazarene College; Clifford Owens, Rutgers University—Camden Campus; John F. Penrose, Jefferson Community

College; John Peslak, Jr., Hardin—Simmons University; William F. Pfeiffer, Utica College of Syracuse University; J. R. Pipal, Alfred University; Leroy A. Purchatzke, University of Wisconsin Center—Manitowoc County; James W. Rhoades, Crowder College; Mary E. Richards, Portland Community College; Samuel Rieger, Mattatuck Community College; Gary F. Riley, St. Louis College of Pharmacy; Bobby Roberson, Brewer State Junior College; Emeric Schultz, Bloomsburg University; Joel Shelton, Bakersfield College; Jesse G. Spencer, Valdosta State College; Charlene Steinberg, formerly of University of Wisconsin Center—Sheboygen; Joseph C. Stickler, Valley City State University; Art Struempler, Chadron State College; C. Larry Sullivan, Avila College; Ken Tasa, Brazosport College; Suzanne Tourtellotte, Albertus Magnus College; Michael C. VanDerveer, Indiana University, Purdue University at Indianapolis; James G. Vogel, College of Saint Teresa; Ray M. Ward, Bakersfield College; Galen L. Weick, Cloud County Community College; Laverne Weidler, Black Hawk College—East Campus; Maurice Weitlauf, Oakton Community College; Robert E. Whaley, Shelby State Community College; George Williams, State Technical Institution of Memphis; Charles H. Willits, Rutgers University—Camden Campus; Alan Wilson, Kaskaskia College; Aleen W. Wilson, Patrick Henry Community College; William H. Zuber, Jr., Memphis State University.

We appreciate the valuable contributions of Jan Dunker, who checked all examples in the chapters and the numerical exercises, and Susan Greer, who typed large portions of the manuscript. Finally, we wish to acknowledge and thank our colleagues on the editorial, design, and production staffs at D. C. Heath; their thoughtful assistance has helped us produce a better book.

Henry F. Holtzclaw, Jr.
University of Nebraska—Lincoln

William R. Robinson
Purdue University

About the Authors

Henry F. Holtzclaw, Jr.

Henry F. Holtzclaw, Jr., is Foundation Regents Professor of Chemistry at the University of Nebraska—Lincoln. He received the A.B. degree at the University of Kansas and M.S. and Ph.D. degrees in inorganic chemistry at the University of Illinois. He was Dean for Graduate Studies at the University of Nebraska for nine years (1976–85) and was Interim Chairman of the Chemistry Department in 1985–86.

His research is in synthesis, stereochemistry, and bonding of metal chelates, including metal chelates of 1,3-diketones and nitrogen and sulfur substituted 1,3-diketones. He has also worked with metal chelate polymers of various dihydroxyquinoid ligands.

Professor Holtzclaw has served as a member of the National Committee of Examiners (Advanced Chemistry Test) for the Graduate Record Examination and as a member of the Graduate Record Examination Board. He has also served on the TOEFL Policy Committee (Test of English as a Foreign Language) and on its Executive Committee and Research Committee, including a term as Chairman of the Research Committee. In the American Chemical Society, he is a Councilor and has served on the Publications Committee, the Committee on Committees, and the Nominations and Elections Committee.

He has just completed a term as President of *Inorganic Syntheses* and was Editor-in-Chief of Volume VIII of that series. He is a member of the American Chemical Society and Sigma Xi and is a Fellow in the American Association for the Advancement of Science.

William R. Robinson

William R. Robinson is a Professor of Chemistry at Purdue University. He received his B.S. and M.S. degrees in chemistry from Texas Technological College (now Texas Tech University) and the Ph.D. degree from the Massachusetts Institute of Technology. In 1973 he spent six months as an Adjunct Associate Professor in the Department of Earth and Space Sciences at the State University of New York at Stony Brook.

Professor Robinson has been active in the General Chemistry program at Purdue since joining the faculty in 1967. He has served as Director of General Chemistry and has assisted with preliminary and developmental reviews of freshman texts. He has published in *The Journal of Chemical Education.* He is a member of the American Chemical Society General Chemistry Exam Committee.

Professor Robinson's other interests include the structure, properties, and reactivity of transition metal compounds. His research activities have included thermal studies of classical coordination compounds of cobalt and chromium, synthetic and structural studies of heavy transition metal compounds containing metal-metal bonds, synthetic and structural studies of organometallic compounds, and x-ray diffraction studies of aqueous solutions. At present he is engaged in the study of the solid state chemistry and structure of transition metal oxides, sulfides, and phosphates. He is associate editor of *The Journal of Solid State Chemistry,* and a member of the American Association for the Advancement of Science, the American Chemical Society, the American Crystallographic Association, and Sigma Xi.

William H. Nebergall

William H. Nebergall (deceased) received the B.Ed. degree at Western Illinois State Teachers College, the M.S. degree at the University of Illinois, and the Ph.D. degree in inorganic chemistry at the University of Minnesota. He taught high school in Illinois and then taught at Tennessee State Teachers College, the University of Kentucky, the Wisconsin State Teachers College at Superior, and the University of Minnesota before joining the faculty at Indiana University. He served as Guest Professor for a year at Brunswick Tech University.

His research was in organometallic, organosilicon, and fluoride chemistry, and in inorganic phosphates and fluorides. He formulated the phosphate medium for Crest toothpaste that made the fluoride paste effective in dental caries control and held the patent on the toothpaste formula.

Brief Contents

Semimicro Qualitative Analysis

Contents

9 Chemical Reactions and the Periodic Table
203

10 The Gaseous State and the Kinetic-Molecular Theory
236

11 Condensed Matter: Liquids and Solids
274

12   Solutions; Colloids   313

13   The Active Metals   357

18 The Solubility Product Principle   529

19 Chemical Thermodynamics   557

20 A Survey of the Nonmetals   589

21

Electrochemistry and Oxidation-Reduction

609

22

The Nonmetals, Part 1: Hydrogen, Oxygen, Sulfur, and the Halogens

652

29 The Atmosphere and Natural Waters

850

30 Organic Chemistry

867

31 Biochemistry

892

Semimicro Qualitative Analysis

32 Chemistry of the Qualitative Analysis Scheme 927

33 General Laboratory Directions 949

34 The Analysis of Group I 955

35 The Analysis of Group II 959

40 The Analysis of Solid Materials 1013

40.1 Dissolution of Nonmetallic Solids   1013 • **40.2** Dissolution of Metals and Alloys   1015

Appendixes A1

A Chemical Arithmetic   A1 • **B** Units and Conversion Factors   A5 • **C** General Physical Constants   A6 • **D** Solubility Products   A7 • **E** Formation Constants for Complex Ions   A8 • **F** Ionization Constants of Weak Acids   A9 • **G** Ionization Constants of Weak Bases   A9 • **H** Standard Electrode (Reduction) Potentials   A10 • **I** Standard Molar Enthalpies of Formation, Standard Molar Free Energies of Formation, and Absolute Standard Entropies [298.15 K (25°C), 1 atm]   A11 • **J** Composition of Commercial Acids and Bases   A19 • **K** Half-Life Times for Several Radioactive Isotopes   A19 • **L** Apparatus for Qualitative Analysis (one student)   A20 • **M** Reagents for Cation and Anion Analysis   A20 • **N** Preparation of Solutions of Cations   A21 • **O** Laboratory Assignments   A23

Photo Credits A25

Glossary A27

Index A37

Some Fundamental Concepts

1

Gold, an element

Although some of you have chosen to be chemistry majors, many have not. In fact, few of you taking chemistry are volunteers; most of you were drafted for one reason or another. It is obvious why those of you who plan to be chemistry majors are here, but why are you others in a chemistry course? What is there about chemistry that leads so many different disciplines to require chemistry of their students?

Some of you are in the course because you need the factual content of a chemistry course in your major. Others need to use the skills of a chemist. The training of biologists, geoscientists, students of the health sciences, soil scientists, food scientists, ecologists, chemical engineers, and many others requires that students in these areas be adept at using the tools of the chemist. Most students need to be introduced to the language chemists use to describe the world around us, because it is a language used by

Extraction of metals from minerals such as this sample of wulfenite is a chemical process.

many disciplines. Students in all disciplines benefit from practicing the analytic thought processes used to interpret results of experiments that can be described in terms of a world of atoms and molecules too small to be seen.

There is, however, a more important reason for you to become familiar with the principles and vocabulary of chemistry. Much of our society is based on science and technology. First-class medical care, sufficient and varied food supplies, comfortable housing, convenient transportation, rapid communication, and overall personal comfort result directly from scientific and technological developments in the past century. Chemistry has played an important role in these developments.

For example, engineering materials are prepared, analyzed, and fabricated using processes based on chemical technology. The manufacture of L-dopa, a drug used for the treatment of Parkinson's disease, takes advantage of chemical properties of hydrogen discovered by chemists in academic laboratories. Research on recombinant DNA technology led to the bacterial synthesis of human insulin for treatment of diabetes. Enzyme-mapping techniques, developed by chemists and biologists in the late 1960s, were essential to the design and synthesis of diphenyl ether herbicides. The study of certain compounds of phosphorus led to the discovery of organophosphorus pesticides, which, along with other pesticides, reduced crop losses enough to feed an estimated 500 million more people. Even everyday items such as rubber bands, toothpaste, and the dye that gives blue jeans their color reflect the input of chemists in product development.

Nevertheless, within the last two decades, it has become apparent that the application of science and technology must be approached thoughtfully. Some of its contributions to our lives have had undesirable side effects. For example, even though energy is the key to our technological society, its production has resulted in destruction of natural beauty by strip mining, pollution of natural waters by oil spills, and the possibility of radioactive contamination from accidents in nuclear plants, as well as the alarmingly rapid depletion of natural sources of energy. The fertilizers and insecticides that have increased food production have polluted our air and water. Automobiles and jet airplanes have revolutionized transportation but have also contaminated the air. Some medicines that have saved countless lives can have toxic side effects.

Chemists deserve credit for many scientific and technological advances, but they also bear responsibility for helping to solve the problems created by some of these same advances. The responsibility cannot be assigned to chemists alone, however. It is the *joint* responsibility of physicists, sociologists, biologists, theologians, political scientists, humanists, *and* chemists. Together we must work to solve the problems of a world that sometimes seems bent on destroying itself with its own technological progress.

No single discipline offers the total perspective or the expertise necessary to accomplish this task. That will require the knowledge of many disciplines. Because few of us can master the wide variety of knowledge necessary, we must depend on others. Before we can draw on the experience of others, however, we must appreciate the strengths and limitations of their areas. Chemists must know their own field extremely well and other fields well enough to ensure the wise use of their discoveries. People in other fields must know enough chemistry and other sciences to apply their own specialized knowledge to problems involving science. The questions of why a person should study chemistry or any other science, regardless of the intended main field of endeavor, and why a scientist should study other fields in addition to science have never had a more demanding or obvious answer.

Introduction

1.1 Chemistry

Oil and nylon, gasoline and water, salt and sugar, and iron and gold are all forms of matter that differ strikingly from each other in many ways. These differences are due to differences in the composition and structure of these substances.

All forms of matter can be converted into other forms. For example, iron is converted into rust when it combines with oxygen and water; gasoline is converted into water and other substances when it burns; and oil can be converted into nylon by a series of chemical reactions. Our very existence depends on changes that occur in matter. Plants convert simple forms of matter into more complex forms, which serve as food. The chemical changes that take place when this food is digested and assimilated by the body are essential to life processes.

During our study of chemistry, we will examine many different changes in the composition and structure of matter, the causes that produce these changes, the changes in energy that accompany them, and the principles and laws involved in these changes. In brief, then, the science of **chemistry may be defined as the study of the composition, structure, and properties of matter and of the reactions by which one form of matter may be produced from or converted into other forms.**

Knowledge about chemistry has increased so much during the last century that chemists usually specialize in one of the field's several principal branches. **Analytical chemistry** is concerned with the identification, separation, and quantitative determination of the composition of different substances. **Physical chemistry** is primarily concerned with the structure of matter, energy changes, and the laws, principles, and theories that explain the transformation of one form of matter into another. **Organic chemistry** is the branch dealing with reactions of the compounds of carbon. **Inorganic chemistry** is concerned with the chemistry of elements other than carbon and of their compounds. **Biochemistry** is the chemistry of the substances comprising living organisms. The boundaries between the branches of chemistry are arbitrary, however, and much work in chemistry cuts across them.

Copper and nitric acid react, forming copper nitrate and gaseous brown nitrogen dioxide.

1.2 Matter and Energy

Matter, by definition, is anything that occupies space and has mass. All the objects in the universe occupy space and have mass; they are matter. The property of occupying space is often easily perceived by our senses of sight and touch. The **mass** of an object pertains to the quantity of matter that the object contains. The force required to change the speed at which an object moves, or to change the direction in which it moves, is a measure of its mass.

Matter can exist in three different states: solid, liquid, and gas (Fig. 1.1). A **solid** is rigid, possesses a definite shape, and has a volume that is very nearly independent of changes in temperature and pressure. A **liquid** flows and thus takes the shape of its container, except that it assumes a horizontal upper surface. Liquids, like solids, are only slightly compressible and so, for practical purposes, have definite volumes. A **gas** takes both the shape and the volume of its container. Gases are readily compressible and capable of infinite expansion.

Figure 1.1
The three states of matter as illustrated by water; ice is solid water, the ocean is liquid water, and the clouds form when gaseous water (which is invisible) condenses to form very small drops of liquid water.

Energy can be defined as the capacity for doing work, where **work** is simply the process of causing matter to move against an opposing force. For example, when we pump up a bicycle tire, we are doing work—we are moving matter (the air in the pump) against the opposing force of the air already in the tire.

Like matter, energy exists in different forms. The energy for running the pump to carry out the work of pumping up a tire could be chemical energy released in our muscles as we use a hand pump, electrical energy used to drive an electric motor on a compressor, or heat used in a steam engine running a compressor. Light is also a form of energy. Solar cells can convert light into electrical energy, which could run our compressor or do other work.

Energy can be further classified as either potential energy or kinetic energy. Since one definition of energy is the capacity for doing work, **potential energy** is the potential for doing work. A piece of matter is said to possess potential energy by virtue of its position, condition, or composition. Water at the top of a waterfall possesses potential energy because of its position; if it falls in a hydroelectric plant, it does work that leads to the production of electricity. A compressed spring, because of its condition, possesses potential energy and can do work such as making a clock run. Natural gas possesses potential energy: Because of its composition, it will burn, producing heat, another kind of energy. When a body is in motion, it also has energy, or the capacity for doing work. The energy that a body possesses because of its motion is called **kinetic energy.** As water falls from the top of a waterfall, its potential energy becomes kinetic energy.

1.3 Law of Conservation of Matter and Energy

When a piece of calcium metal is exposed to dry air, it unites with oxygen in the air (Fig. 1.2). If the product (calcium oxide) is collected and weighed, it is found to have a mass greater than that of the original piece of metal. If, however, the mass of the oxygen that combined with the metal is taken into consideration, the final mass is found to be equal to the sum of the masses of the calcium metal and the oxygen. This behavior of matter is in accord with the **law of conservation of matter: During an ordinary chemical change, there is no detectable increase or decrease in the total quantity of matter.**

The conversion of one type of matter into another (chemical change) is always accompanied by the conversion of one form of energy into another. Usually heat is evolved or absorbed, but sometimes the conversion involves light or electrical energy instead of, or in addition to, heat. When calcium metal combines with oxygen, potential energy is converted into heat. Under certain conditions the calcium may actually burn, producing light as well. Many transformations of energy, of course, do not involve chemical changes. Electrical energy can be changed into light, heat, or potential energy without any chemical change. Potential and kinetic energy can each be converted into the other. Many other conversions are possible, but all of the energy involved in any change always exists in some form after the change is completed. This fact is expressed in the **law of conservation of energy: During an ordinary chemical change, energy can be neither created nor destroyed, although it can be changed in form.**

When applied to chemical changes, the laws of conservation of matter and energy hold very well. During nuclear changes, however, these laws are violated. Both conversion of matter into energy and conversion of energy into matter can occur in nuclear reactions. So, although the laws of conservation of matter and of energy both hold in ordinary chemical reactions, to include both chemical and nuclear changes these two laws are combined into one statement: **The total quantity of matter and energy available in the universe is fixed.** Matter and energy are not distinct, but are two forms of a single entity.

Since energy can be neither created nor destroyed in a chemical reaction, why is there concern about a shortage of energy? The answer to this question is that we are not running out of energy itself, but we are running out of useful forms of energy. The energy contained in a fossil fuel such as oil can be converted, in part, into work when the fuel is burned; the rest is generally converted into waste heat, a form of energy that cannot be used. Furthermore, millions of years are required for nature to produce oil reserves to replace those used. If we are not to run out of *available* energy, we must be willing to devote money and time to research so as to develop more efficient methods of using the energy we have and to find new methods of using other energy present in the universe.

1.4 Chemical and Physical Properties

The characteristics that enable us to distinguish one substance from another are known as properties. **Chemical properties** refer to how one kind of matter transforms into another kind. Iron exhibits a chemical property when it combines with oxygen and water to form the reddish-brown iron oxide we call rust. A chemical property of

Total weight
183.233 g

(a)

Total weight
183.233 g

(b)

Figure 1.2

(a) A closed 1-liter flask filled with air and containing 0.646 grams of calcium. (b) The same flask after the calcium has reacted with 0.258 grams of oxygen from the air in the flask and formed 0.904 grams of calcium oxide. The total weight of the flask and its contents has not changed.

Figure 1.3
You can distinguish iron and chromium because they exhibit different chemical properties. Iron rusts, but chromium does not.

chromium is that it does not rust (Fig. 1.3). The **physical properties** of a particular kind of matter are those characteristics that do not involve a change in the chemical identity of the matter. Familiar physical properties include color, hardness, physical state, melting temperature, boiling temperature, and electrical conductivity (Fig. 1.4). Iron, for example, melts at a temperature of 1535°C; chromium melts at a different temperature, 190°C. As they melt, the metals change from solids to liquids but are still iron and chromium, since no change in chemical identity has occurred.

The properties of various kinds of matter enable us to distinguish one kind from another. We can distinguish between iron wire and chromium wire by their physical properties (their colors, for example) or by their chemical properties (for example, iron rusts, but chromium does not).

In order to identify a chemical property, we look for a chemical change. A **chemical change** always produces one or more different kinds of matter from those that were present before the change occurred. Iron rust contains oxygen as well as iron, and it is therefore a different kind of matter from the iron, air, and water from which it formed. Thus the formation of iron rust is a chemical change. When milk sours, the sugar (lactose) in the milk is converted into lactic acid by a chemical change, and the composition and the properties of the acid differ from those of the sugar.

The absence of a chemical change is also a chemical property. Since chromium does not rust, one of its chemical properties is that it does not undergo this chemical change.

Figure 1.4
You can distinguish these elements by their physical properties. Mercury is the silvery liquid. Bromine is the dark orange liquid that forms an orange gas. Chlorine is the pale yellow-green gas. Copper is the coppery-colored solid. Aluminum is the silvery solid.

A **physical change** is one that does not involve a change of one kind of matter into another. The melting of iron, the freezing of water, the conversion of liquid water to steam, and the condensation of steam to liquid water are all examples of physical changes. In each of these there is a change in one or more properties, but there is no alteration of the chemical composition of the substance involved. Water, whether in the solid, liquid, or gaseous state, maintains the same chemical composition. Iron, whether molten or solid, is the same kind of matter. Some physical properties can be observed only when matter undergoes a physical change. The freezing temperature of water, for example, is determined by measuring the temperature as water changes from a liquid to a solid. Other physical properties can be measured without a physical change. For example, the color of a sample can be determined without an accompanying physical change.

1.5 The Classification of Matter

All samples of matter can be classified as either pure substances or mixtures (Fig. 1.5). A **mixture** is composed of two or more kinds of matter that can be separated

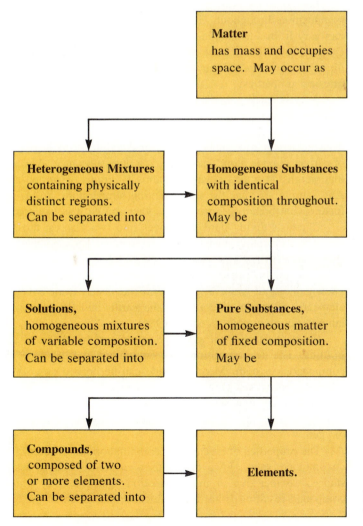

Figure 1.5
The classification of matter.

Figure 1.6
Chocolate chip ice cream is a heterogeneous mixture; chocolate syrup is a homogeneous mixture.

by physical means. Moreover, the composition of a mixture can be varied continuously. Blood consists of varying amounts of proteins, sugar, salt, oxygen, carbon dioxide, and other components mixed in water. It can be recognized as a mixture because it can be separated into solids and plasma by centrifugation, a physical method of separation. Other mixtures include granite, which is a mixture of quartz, feldspar, and mica; sugar syrup, which is a mixture of sugar and water; and air, which is a mixture of oxygen, nitrogen, carbon dioxide, water vapor, and small amounts of other gases.

When sugar dissolves in water it forms a **homogeneous mixture,** a mixture with a composition that is uniform throughout. In a homogeneous mixture of sugar dissolved in water, the amount of sugar is always the same from place to place within the mixture; so is the amount of water. Homogeneous mixtures are often called **solutions.** Other examples of homogeneous mixtures include syrup, air, soft drinks, gasoline, and a solution of salt in water. A mixture whose composition differs from point to point is a **heterogeneous mixture.** Chocolate chip ice cream is an example of a heterogeneous mixture (Fig. 1.6). At some points it consists of chocolate chips; at other points, vanilla ice cream. Other heterogeneous mixtures include granite (you can often see the separate bits of mica, quartz, and feldspar), sand in water, blood, and concrete.

Pure substances are similar to homogeneous mixtures in that both are of uniform composition throughout. However, while the composition of a homogeneous mixture can vary from one sample to another, a specific pure substance always has the same composition. Thus a **pure substance** is defined as a homogeneous sample of matter, all specimens of which have identical compositions as well as identical chemical and physical properties.

A carbonated soft drink is a homogeneous mixture, typically of water, sugar, coloring and flavoring agents, and carbon dioxide. It is not a pure substance by our definition, because its composition can vary. The amount of carbon dioxide in the drink decreases after the bottle is opened and the drink begins to lose its fizz. A sample of carbon dioxide alone, however, is a pure substance. Any sample of carbon dioxide that weighs 44 grams always contains 12 grams of carbon and 32 grams of oxygen. Any sample also has the same melting temperature, color, and other properties, no matter what brand of beverage or other source it is isolated from.

Chemists divide pure substances into two classes: those that cannot be decomposed by a chemical change and those that can. Pure substances that cannot be decomposed by a chemical change are called **elements.** Familiar examples are iron, silver, gold, aluminum, sulfur, oxygen, and carbon. At the present time, 109 elements are known; a list of these is printed on the inside front cover of this book. Of these elements, 88 occur naturally on the earth, and the other 21 have been created in laboratories (the most recent in 1984).

Pure substances that can be decomposed by chemical changes are called **compounds.** The decomposition of a compound may produce either elements or other compounds, or both. Mercury oxide can be broken down by heat into the elements mercury and oxygen (Fig. 1.7). If heated sufficiently in the absence of air, the compound sugar decomposes to the element carbon and the compound water. Water can be decomposed by an electric current into its two constituent elements, hydrogen and oxygen.

The properties of elements in combination are different from those in the free, or uncombined, state. However, the term *element* is used to designate an elemental substance, whether free or in combination. For example, white crystalline sugar is a compound resulting from the chemical combination of the element carbon, which is

Figure 1.7
Red mercury oxide is a compound. When heated it decomposes into silvery drops of mercury and oxygen gas (which is not visible).

usually a black solid when free, and the two elements hydrogen and oxygen, which are colorless gases when uncombined. Free sodium, an element that is a soft shiny metallic solid, and free chlorine, an element that is a yellow-green gas, combine to form sodium chloride, a white brittle solid.

Although there are only 109 known elements, several million chemical compounds result from different combinations of these elements. Each compound possesses definite chemical and physical properties by which chemists can distinguish it from all other compounds.

Eleven of the 88 naturally occurring elements make up about 99% of the earth's crust and the atmosphere (Table 1.1). Oxygen constitutes nearly one-half and silicon about one-fourth of the total quantity of the elements in the atmosphere and the earth's crust together. Only about one-fourth of the elements are found on the earth in the free state; the others occur only in chemical combinations with other elements.

Sugar decomposes into black elemental carbon and water when heated.

Table 1.1   Percentages of elements in the atmosphere and the earth's crust by mass

Element	%	Element	%
Oxygen	49.20%	Chlorine	0.19%
Silicon	25.67	Phosphorus	0.11
Aluminum	7.50	Manganese	0.09
Iron	4.71	Carbon	0.08
Calcium	3.39	Sulfur	0.06
Sodium	2.63	Barium	0.04
Potassium	2.40	Nitrogen	0.03
Magnesium	1.93	Fluorine	0.03
Hydrogen	0.87	Strontium	0.02
Titanium	0.58	All others	0.47

1.6 Atoms and Molecules

An atom is the smallest particle of an element that can enter into a chemical combination. Gold is an element. If we could take a gold ring and divide it into smaller and smaller pieces, eventually we would have a single atom of gold. This atom would no longer be gold if it were divided. Such a single atom of gold is incredibly small, about 0.0000000113 inch in diameter. It would require about 88 million gold atoms lying side by side to make a line 1 inch long.

The first suggestion that matter is composed of atoms is attributed to early Greek philosophers, notably Democritus. These philosophers argued that matter is composed of small indivisible particles. Indeed, the word *atom* comes from the Greek word *atomos,* which means indivisible. However, it was not until the early nineteenth century that John Dalton (1766–1844), an English chemist and physicist, presented an atomic theory *and* supported it with quantitative measurements. Since that time, repeated experiments have verified his theory, which is still used with only minor revisions.

Only a few elements, such as the gases helium, neon, and argon, consist of a collection of individual atoms that move about independently of one another (Fig. 1.8). Other elements, such as the gases nitrogen, oxygen, and chlorine, consist of pairs of atoms, each pair moving as a single unit. The element phosphorus consists of units

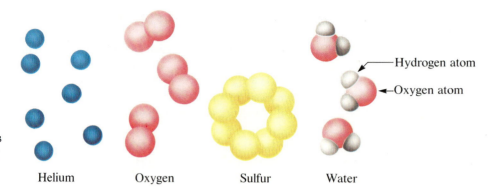

Hydrogen atom

Oxygen atom

Helium   Oxygen   Sulfur   Water

Figure 1.8

Representations of molecules of the elements helium, oxygen, and sulfur and of the compound water.

composed of four phosphorus atoms; sulfur, of units composed of eight sulfur atoms. These units are called **molecules.**

A molecule is the smallest particle of an element or compound that can have a stable, independent existence. A molecule may consist of a single atom, as in helium, of two or more identical atoms, as in nitrogen and sulfur, or of two or more different atoms, as in water. Water has a definite composition and a set of chemical properties that enable us to recognize it as a distinct substance. One might ask, To what extent can a drop of water be subdivided and still be water? The limit to which such a subdivision can be carried is the water molecule. Each water molecule is a unit that contains two hydrogen atoms and one oxygen atom (Fig. 1.8). Subdivision of a water molecule results in the formation of the gases hydrogen and oxygen, each of which has properties quite different from those of water and those of each other. A molecule of the sugar glucose contains 6 carbon atoms, 12 hydrogen atoms, and 6 oxygen atoms.

Atoms and molecules are too small to be seen even with very powerful optical microscopes, and only relatively large molecules can be seen even with electron microscopes. An appreciation of the minute size of molecules can be gained from the fact that if a glass of water were enlarged to the size of the earth, the water molecules would be about the size of golf balls.

1.7 The Scientific Method

Your study of chemistry will be concerned with the observations, theories, and laws that give this science its foundation and that form a framework into which information fits to make an integrated area of knowledge. This framework develops from the pursuit of answers to many questions, each of which can be subjected to experimental investigation by an approach often called the **scientific method.**

Scientists identify problems or questions through their own observations and experiments or from the observations, experiments, and conclusions of others. (Problems are often identified when experiments do not go as expected.) The first step in applying the scientific method to a problem involves carefully planning experiments to gather facts and obtain information about all phases of the problem. The results are examined for general relationships that will unify the observations. Sometimes a wide variety of observations can be summarized in a general verbal statement or mathematical equation known as a **law.** One example is the law of conservation of matter (Section 1.3), which summarizes the results of thousands of experimental observations. More often, however, a tentative explanation is suggested. Such a proposal is called a **hypothesis.** For instance, Dalton attempted to explain why mass is conserved in a chemical reaction when he first presented his ideas, which were really hypotheses, of the atomic nature of matter. A hypothesis is tested by further experiments, and, if it is capable of explaining the large body of experimental data, it is dignified by the name **theory.** Dalton's ideas have been so extensively tested that we now refer to the atomic theory. Theories themselves can prompt new questions or suggest new directions in which additional information can be sought.

Measurement in Chemistry

1.8 Units of Measurement

The hypotheses, theories, and laws that describe the behavior of matter and energy are usually based on quantitative measurements. When properly reported these measurements convey three ideas: the size or magnitude of the property being measured (a number), an indication of the possible error in the measurement, and the units of the measurement.

Units provide a standard of comparison for a measurement. A well-known sandwich is prepared from a quarter pound of hamburger meat (0.250 pound). The mass of the meat has a magnitude of 0.250 times the mass of the arbitrary standard of 1 pound. If the unit of reference is changed, the property being measured (in this case the mass) does not change. The same mass of hamburger meat can be reported as being a quarter pound, 4.00 ounces, or 0.000125 ton (a 0.000125-tonner?); the actual amount of meat is the same, no matter what the unit of measurement.

The results of scientific measurements are usually reported in the metric system, originally devised in France in 1799. The original metric units for length, mass, and time were the meter, the gram, and the second. Various combinations of these units are still in use. One such system that has been used frequently by chemists is the **centimeter-gram-second, or cgs, system.** More recently scientists have begun to use an updated version of the metric system, in which the units for length, mass, and time are the meter, the kilogram, and the second: the **International System of Units (SI units).** These units have been used by the U.S. National Bureau of Standards since 1964. The SI system consists of the following seven base units, from which other units of weight and measure can be derived.

Units of degrees Celsius, degrees Fahrenheit, or kelvins are used to measure the temperature of this molten metal.

Physical property	Name of unit	Symbol
Length	meter	m
Mass	kilogram	kg
Time	second	s
Electric current	ampere	A
Temperature	kelvin	K
Luminous intensity	candela	cd
Amount of substance	mole	mol

In both the SI and other metric systems, measurements may be reported as fractions or multiples of ten times a base unit by using the appropriate prefix with the name of the base unit. For example, a length can be reported in units of meters, kilometers (10^3 meters), millimeters (10^{-3} meter), or picometers (10^{-12} meter). The prefixes and their symbols denoting the powers to which 10 is raised are given in Table 1.2.

Table 1.2   Common prefixes used in the metric system

Prefix	Symbol	Factor	Example
pico	p	10^{-12}	1 picometer (pm) = 1×10^{-12} m (0.000000000001 m)
nano	n	10^{-9}	1 nanogram (ng) = 1×10^{-9} g (0.000000001 g)
micro	μ*	10^{-6}	1 microliter (μL) = 1×10^{-6} L (0.000001 L)
milli	m	10^{-3}	2 milliseconds (ms) = 2×10^{-3} s (0.002 s)
centi	c	10^{-2}	5 centimeters (cm) = 5×10^{-2} cm (0.05 m)
deci	d	10^{-1}	1 deciliter (dL) = 1×10^{-1} L (0.1 L)
kilo	k	10^{3}	1 kilometer (km) = 1×10^{3} m (1000 m)
mega	M	10^{6}	3 megagrams (Mg) = 3×10^{6} g (3,000,000 g)
giga	G	10^{9}	5 gigameters (Gm) = 5×10^{9} m (5,000,000,000 m)
tera	T	10^{12}	1 teraliter (TL) = 1×10^{12} L (1,000,000,000,000 L)

*Greek letter mu.

Since both SI and cgs units are currently in use by scientists, it is necessary to be familiar with both systems. Consequently, both systems of units will be used in this text.

1.9 Conversion of Units; Dimensional Analysis

In addition to the two systems of units in use within the scientific community, many other units are employed by other disciplines. For example, units of mass range from grams and kilograms in science, to pounds and tons in engineering, to grains and drams in medicine. The relationships between some common units are given in Table 1.3. A more complete table is given in Appendix B.

Table 1.3   Common conversion factors (to four significant figures)

Length	
1 meter = 1.094 yards	1 inch = 2.54 centimeters*
Volume	
1 liter = 1.057 quarts	1 cubic foot = 28.316 liters
Mass	
1 kilogram = 2.205 pounds	1 ounce = 28.35 grams

*Exact conversion.

Consider the conversion of a length from meters to yards. As given in Table 1.3, 1 meter = 1.094 yards. This may be read as indicating that there is exactly 1 meter per 1.094 yards or, alternatively, 1.094 yards per 1 meter. Very often a slash (/) or a division line is used instead of the word *per*. This gives the following unit conversion factors:

$$1 \text{ meter}/1.094 \text{ yards}, \quad \text{or} \quad \frac{1 \text{ meter}}{1.094 \text{ yards}}$$

$$1.094 \text{ yards}/1 \text{ meter}, \quad \text{or} \quad \frac{1.094 \text{ yards}}{1 \text{ meter}}$$

A **unit conversion factor** is used to convert a quantity in one system of units to the corresponding quantity in another system of units. For example, the second factor above can be used to convert a length in meters to the corresponding length in yards. We simply multiply the quantity in meters by the conversion factor.

How long in yards is a run of 100.0 meters?   **EXAMPLE 1.1**

From the relationship 1 meter = 1.094 yards, we can get the unit conversion factor 1.094 yards/1 meter. Multiplication gives

$$100.0 \text{ meters} \times \frac{1.094 \text{ yards}}{1 \text{ meter}} = 109.4 \text{ yards}$$

Keep in mind that the distance covered by the runner does not change when we switch from units of meters to units of yards. Multiplying by a unit conversion factor is equivalent to multiplying by 1.

Note that the unit of meters in 100.0 meters and in the conversion factor cancel, as indicated by the cancellation lines in color. Since the answer is in the correct unit, it is highly likely that the conversion has been set up correctly. If an incorrect unit conversion factor had been used, an incorrect unit would have been obtained. Consider the following conversion:

$$100.0 \text{ meters} \times \frac{1 \text{ meter}}{1.094 \text{ yards}} = 91.41 \frac{\text{meters}^2}{\text{yard}}$$

The final unit, meters^2/yard, indicates that the conversion does not convert meters into yards, since the units do not cancel to give yards.

A technique called **dimensional analysis** has been used in Example 1.1. Units are treated like algebraic quantities and carried through the calculation. Like algebraic quantities one can be multiplied by another, one divided into the other, or two or more cancelled. The final units resulting from such a treatment will provide valuable assistance in determining whether or not a calculation has been set up correctly.

1.10 Uncertainty in Measurements and Significant Figures

Suppose you weigh an object on an inexpensive balance and find, as best you can determine, that its mass is closer to 13.4 grams than to either 13.3 or 13.5 grams. You would report the mass as 13.4 grams. The uncertainty in this measurement is almost 0.1 gram, or 1 part in 134 (about 0.75%), because the true mass of the object could lie anywhere between 13.35 and 13.45 grams. If the same object were weighed on an analytical balance in your laboratory, your best efforts might show that its mass is 13.384 grams; that is, the mass is closer to 13.384 grams than to either 13.383 or 13.385 grams. The uncertainty in this case would be about 0.001 gram (1 part in 13,384, or about 0.0075%), since the true mass could lie anywhere between 13.3835 and 13.3845 grams. Any experimental measurement, no matter how carefully made,

Any measurement of the mass of silver metal deposited on the copper wire will have some degree of uncertainty.

may contain this type of uncertainty. **All measurements may be taken to have an uncertainty of at least one unit in the last digit of the measured quantity.**

The mass 13.4 grams has an uncertainty of 0.1 gram; 13.384 grams, an uncertainty of 0.001 gram; a measured volume of 25 milliliters, an uncertainty of 1 milliliter; and a measured length of 0.001378 meter, an uncertainty of 0.000001 meter. All of the measured digits in the determination are called **significant figures.**

Results calculated from measurements are as uncertain as the measurement itself. Suppose a sample weighs 78.7 grams and exactly one-third of it is iron. In order to calculate how much iron is in the sample, we need only multiply 78.7 grams by one-third. On an electronic calculator, this comes out to 26.233333 grams. This is not the correct answer, however. Because the mass of the sample is uncertain to 1 part in 787, or about 0.1%, the mass of iron in the sample must have the same uncertainty. To report the mass of iron as 26.233333 grams indicates an uncertainty of 0.000001 gram, or about 0.000004% (1 part in 26,233,333). The mass of iron should be reported as 26.2 grams (three significant figures). In calculations involving experimental quantities, the significant figures in these quantities must be taken into account in order not to overestimate or underestimate the uncertainty in the calculated result.

The following rules may be used to determine the number of significant figures in a measured quantity:

1. *Find the first nonzero digit on the left and count to the right, including this first nonzero digit and all remaining digits. Unless the last digit is a zero lying immediately to the left of the decimal point, this is the number of significant figures in the number.*

first nonzero digit on the left

0.00120530

six significant figures

first nonzero digit on the left

10.00120530

ten significant figures

first nonzero digit on the left

126.7309

seven significant figures

first nonzero digit on the left

97

two significant figures

The first three zeros on the left in the first example above, 0.00120530, are not significant. They merely tell us where the decimal point is located. We could equally well express this number in exponential notation as 1.20530×10^{-3}. In this case the number 1.20530 contains all six significant figures, and 10^{-3} serves to locate the decimal.

2. *When a number ends in zeros that are to the left of a decimal point, the trailing zeros may or may not be significant.* A mass given as 1300 grams may indicate a mass closer to 1300 grams than to 1200 or 1400 grams. In this case the zeros serve only to locate the decimal point and are not significant. If the mass is closer to 1300 grams than to 1290 or 1310 grams, then the zero to the right of the 3 is significant. If the mass is closer to 1300 grams than to 1299 or 1301 grams, both zeros are significant. The ambiguity can be avoided by using exponential notation: 1.3×10^3 (two significant figures), 1.30×10^3 (three significant figures), 1.300×10^3 (four significant figures). For convenience in this text, we will assume that when a number ends in zeros that are to the left of a decimal point, these trailing zeros are significant unless otherwise specified.

The two rules given above apply to measured quantities. If we count items rather than measure them, the result is exact, with no uncertainty. If we count eggs in a carton, we know, without any uncertainty, exactly how many eggs the carton contains. Defined quantities are also exact. By definition, 1 foot is exactly 12 inches, 1 kilogram is exactly 1000 grams, and 1 inch equals exactly 2.54 centimeters. These exact numbers, numbers with no uncertainty, can be considered to have an infinite number of significant figures.

The correct use of significant figures in a calculation carries the uncertainty of measured quantities into the result. The following rules govern the number of significant figures that should be in an answer:

1. *When adding or subtracting, report the results with the same number of decimal places as that of the number with the least number of decimal places.*

(a) Add 4.383 g and 1.0023 g.   (b) Subtract 421 g from 486.39 g.   **EXAMPLE 1.2**

(a)   4.383 g   (b)   486.39 g
      +1.0023 g   −421   g
       5.385 g   65   g

---

2. *When multiplying or dividing, report the product or quotient with no more digits than the least number of significant figures in the numbers involved in the computation.*

Multiply 0.6238 cm by 6.6.   **EXAMPLE 1.3**

$$0.6238 \text{ cm} \times 6.6 = 4.1 \text{ cm}$$

In rounding numbers off at a certain point, simply drop the digits that follow if the first of them is less than 5; to two significant figures, 8.7235 rounds off to 8.7. If the first digit to be dropped is greater than 5 or if it is 5 followed by other nonzero digits, increase the preceding digit by 1; to three significant figures, 3.8689 rounds off to 3.87; 23.3501, to 23.4. If the digit to be dropped is 5 or 5 followed only by zeros, a common practice is to increase the preceding digit by 1 if it is odd and to leave it unchanged if it is even. Dropping a 5 always leaves an even number. Thus 3.425 rounds off to 3.42, and 7.53500 rounds off to 7.54.

estimate

⊦ note

1.11 Length and Volume

The standard unit of length in both metric and SI units is the **meter (m).** In 1899 the meter was defined by international agreement to be the length of a certain platinum-iridium alloy bar kept in France (Fig. 1.9). In 1960 the meter was redefined as the length equal to exactly 1,650,763.73 wavelengths, in vacuum, of the orange line in the emission spectrum of ^{86}Kr, a particular isotope of krypton (spectra and isotopes will be discussed in subsequent chapters). Within the error of the measurement, the length of the meter was not changed; only the standard used to define it was changed.

In more familiar units of reference, a meter is about 3 inches longer than a yard; 1 meter is approximately 1.094 yards; 1 meter equals 100 centimeters or 1000 millimeters (Fig. 1.10).

Figure 1.9

The standard meter, as defined in 1899, was the distance between two lines inscribed on this x-shaped platinum-iridium bar. The kilogram was defined to be equal to the mass of the platinum-iridium cylinder.

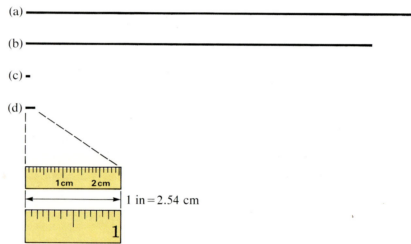

Figure 1.10

Relative lengths of (a) 1 meter, (b) 1 yard, (c) 1 centimeter, and (d) 1 inch (not actual size) and a comparison of 1 inch and 2.54 centimeters (actual size). A meter is a little longer than 1 yard and exactly 100 times longer than 1 centimeter.

Volumes are defined in terms of the standard length. The standard SI volume, a **cubic meter (m^3),** is the volume of a cube that is exactly 1 meter on an edge. A more convenient unit is the volume of a cube with an edge length of exactly 10 centimeters (cm), or 1 decimeter (10 cm = 1 dm). This volume of 1000 cm^3, or 1 dm^3, is often called a **liter (L)** and is 0.001 times the volume of a cubic meter; 1 liter is about 1.06 quarts. A **milliliter (mL)** is the volume of a cube with an edge length of exactly 1 centimeter (Fig. 1.11) and thus is equal in volume to 1 cubic centimeter (cm^3); 1 liter contains 1000 milliliters.

Figure 1.11

Comparison between a cubic-centimeter block and a dime. A cubic centimeter is equal to 1 milliliter; 1 cubic centimeter of water has a mass of 1 gram at 4°C.

Dime   1 cubic centimeter
= 1 milliliter

EXAMPLE 1.4

What is the volume in liters of 1.000 oz, given that 1 L = 1.06 qt and that 32 oz = 1 qt?

From the two relationships it is possible to determine the exact unit conversion factor 1 qt/32 oz and the inexact unit conversion factor 1 L/1.06 qt. First convert ounces to quarts.

$$1.000 \text{ oz} \times \frac{1 \text{ qt}}{32 \text{ oz}} = 0.03125 \text{ qt}$$

Then convert quarts to liters.

$$0.03125 \text{ qt} \times \frac{1 \text{ L}}{1.06 \text{ qt}} = 2.95 \times 10^{-2} \text{ L}$$

Thus

$$1.000 \text{ oz} = 2.95 \times 10^{-2} \text{ L}$$

These two steps could have been combined into one operation as follows:

$$1.000 \text{ oz} \times \frac{1 \text{ qt}}{32 \text{ oz}} \times \frac{1 \text{ L}}{1.06 \text{ qt}} = 2.95 \times 10^{-2} \text{ L}$$

The number of significant figures in the result is determined by the inexact experimental relationship that exactly 1 liter is equal to 1.06 quarts.

1.12 Mass and Weight

We said in Section 1.2 that the mass of an object is the quantity of matter that it contains. The mass of a body of matter is an invariable quantity. On the other hand, the weight of a body is the force that gravity exerts on the body; it is variable, since the attraction depends on the distance from a planet's center. If this book were taken up in an airplane, it would weigh less than it does at sea level. Far out in space, its weight would be negligible. It is well known that astronauts experience weightlessness while in outer space. However, neither the mass of the book nor that of the astronauts changes as the distance from the center of the earth changes. Scientists measure quantities of matter in terms of mass rather than weight because the mass of a body remains constant, whereas its weight is an accident of its environment. It should be noted, however, that the term *weight* is often used loosely for the mass of a substance.

The mass of an object is found by comparing it to other objects of known mass. The instrument used to make this comparison is called a **balance.** Figure 1.12 shows a two-pan balance. The object to be weighed is placed in one pan, and weights of known mass are added to the other unit until the two pans balance and the pointer comes to the center of the scale. At this point, both pans contain equal masses, and the mass of the object equals the total mass of the weights added. Weighing on a balance makes use of the fact that the gravitational attraction for objects of equal mass is the same; i.e., their weights are equal. A modern analytical balance is shown in Fig. 1.13.

The standard object to which all SI and other metric units of mass are referred is a cylinder of platinum-iridium alloy, which is also kept in France (Fig. 1.9). The unit of mass, the **kilogram (kg),** is defined as the mass of this cylinder; 1 kilogram is about

Figure 1.12

A traditional two-pan balance. Weights of known mass are added to the right pan until they balance the object in the left pan. At this point both pans contain equal masses.

Figure 1.13

A modern analytical balance. Inside the upper part of the balance is a beam with a counterweight at the back and movable weights at the front (the end of the beam with the balance pan). When an object is placed in the pan, the front end of the beam becomes heavier than the back. The dials are used to *remove* weights until balance is reestablished. The mass of the object is equal to the mass removed to restore the balance.

2.2 pounds. The **gram (g)** is equal to exactly 0.001 times the mass of the standard kilogram (1×10^{-3} kg) and is very nearly equal to the weight of 1 cubic centimeter of water at 4°C, the temperature at which water has its maximum density. A dime has a mass of about 2.3 grams.

1.13 Density and Specific Gravity

One of the physical properties of a solid, a liquid, or a gas is its density. **Density is defined as mass per unit volume.** This may be expressed mathematically as

$$\text{Density} = \frac{\text{mass}}{\text{volume}} \quad \text{or} \quad D = \frac{M}{V}$$

Substances can often be distinguished by measuring their densities since any two substances usually have different densities (Table 1.4).

Table 1.4	Densities of common materials at 25°C and 1 atm
Air	1.29 g/L
Helium gas	0.179 g/L
Water	0.997 g/cm^3
Glycerin	1.26 g/cm^3
Mercury	13.6 g/cm^3
Salt	2.17 g/cm^3
Iron	7.86 g/cm^3
Silver	10.5 g/cm^3

The term **specific gravity** denotes the ratio of the density of a substance to the density of a reference substance. The reference substance for solids and liquids is usually water. Common reference substances used in specifying the specific gravities for gases are air and hydrogen.

$$\text{Specific gravity of substance} = \frac{\text{density of substance}}{\text{density of reference substance}}$$

Note that specific gravity, being the ratio of two densities, has no units. The units in the numerator and denominator cancel each other.

When measured in the metric system of units, the density of any substance has practically the same numerical value as its specific gravity referred to water as the reference substance. For example, the density of glycerin is 1.26 g/cm^3, whereas its specific gravity referred to water (0.997 g/cm^3) as the reference substance is 1.26 (no units).

EXAMPLE 1.5

Calculate the density and specific gravity of a body that has a mass of 321 g and a volume of 45.0 cm^3 at 25°C.

STEP 1 $$\text{Density} = \frac{\text{mass}}{\text{volume}} = \frac{321 \text{ g}}{45.0 \text{ cm}^3} = 7.13 \text{ g/cm}^3$$

water's specific density .997

STEP 2

$$\text{Specific gravity} = \frac{\text{density of sample}}{\text{density of water}}$$

The density of water at 25°C is 0.997 g/cm^3.

$$\text{Specific gravity} = \frac{7.13 \text{ g/cm}^3}{0.997 \text{ g/cm}^3} = 7.15$$

What is the mass in kilograms of 10.5 gal (39.7 L) of gasoline with a specific gravity of 0.82?   **EXAMPLE 1.6**

Because gasoline has a specific gravity of 0.82, the density of 1 mL of gasoline is 0.82 times that of 1 mL of water.

$$\text{Density} = \text{specific gravity} \times \text{density of water}$$
$$= 0.82 \times 0.997 \text{ g/cm}^3$$
$$= 0.82 \text{ g/cm}^3 \text{ (or 0.82 g/mL)}$$

Conversion of 39.7 L to mL followed by multiplication by the mass of 1 mL of gasoline gives the total mass.

$$\text{Volume} = 39.7 \text{ L} \times \frac{1000 \text{ mL}}{1 \text{ L}}$$
$$= 39,700 \text{ mL}$$
$$\text{Mass} = \text{density} \times \text{volume}$$
$$= 0.82 \text{ g/mL} \times 39,700 \text{ mL}$$
$$= 33,000 \text{ g} = 33 \text{ kg}$$

What is the density in pounds per liquid quart and in grams per milliliter for the common antifreeze ethylene glycol? A 5.600-qt sample of ethylene glycol weighs 12.953 lb.   **EXAMPLE 1.7**

$$\text{Density} = \frac{\text{mass}}{\text{volume}} = \frac{12.953 \text{ lb}}{5.600 \text{ qt}}$$
$$= 2.313 \text{ lb/qt}$$

To determine the density in grams per milliliter, we need the mass in grams and the volume in milliliters. The necessary conversion factors are given in Appendix B: 1 qt = 0.9463 L; 1 lb = 0.453592 kg.

$$12.953 \text{ lb} \times \frac{0.453592 \text{ kg}}{1 \text{ lb}} \times \frac{1000 \text{ g}}{1 \text{ kg}} = 5875.4 \text{ g}$$

$$5.600 \text{ qt} \times \frac{0.9463 \text{ L}}{1 \text{ qt}} \times \frac{1000 \text{ mL}}{1 \text{ L}} = 5299 \text{ mL}$$

$$\text{Density} = \frac{5875.4 \text{ g}}{5299 \text{ mL}} = 1.109 \text{ g/mL}$$

Alternatively, the calculation could be set up in one equation as follows:

$$\frac{12.953 \text{ lb}}{5.600 \text{ qt}} \times \frac{1 \text{ qt}}{0.9463 \text{ L}} \times \frac{1 \text{ L}}{1000 \text{ mL}} \times \frac{0.453592 \text{ kg}}{1 \text{ lb}} \times \frac{1000 \text{ g}}{1 \text{ kg}} = 1.109 \text{ g/mL}$$

1.14 Temperature and Its Measurement

The word **temperature** refers to the hotness or coldness of a body of matter. In order to measure a change in temperature, some physical property of a substance that varies with temperature must be used. Practically all substances expand when temperature increases and contract when it decreases. This property is the basis for the common glass thermometer. The mercury or alcohol in the tube rises when the temperature increases because the volume of the mercury or alcohol expands more than that of the glass container.

To agree on a set of temperature values, we need **standard reference temperatures** that can be readily determined. Two such temperatures are the freezing and boiling points of water at a pressure of 1 atmosphere. On the **Celsius** scale the freezing point of water is taken at 0° and the boiling point as 100°. The heights of the mercury column at these two reference temperatures determine the 0° and 100° points on a thermometer. The space between these two points is divided into 100 equal intervals, or degrees. On the **Fahrenheit** scale the freezing point of water is taken as 32° and the boiling point as 212°. The space between these two points on a thermometer is divided into 180 equal parts, or degrees. Thus, a degree Fahrenheit is exactly $\frac{100}{180}$, or $\frac{5}{9}$, of a degree Celsius (Fig. 1.14). The relationships are shown by the equations

$$\frac{°F - 32}{180} = \frac{°C}{100}, \qquad °C = \frac{5}{9}(°F - 32), \qquad °F = \frac{9}{5}°C + 32$$

The readings below 0° on either scale are treated as negative.

The SI unit of temperature is called the **kelvin (K),** named after Lord Kelvin, a British physicist. The zero point on the Kelvin scale corresponds to $-273.15°C$. The size of the degree on the Kelvin scale is the same as that on the Celsius scale. The freezing temperature of water on the Kelvin scale is 273.15 K (0°C), and the boiling

Figure 1.14

The relationships among the Kelvin, Celsius, and Fahrenheit temperature scales.

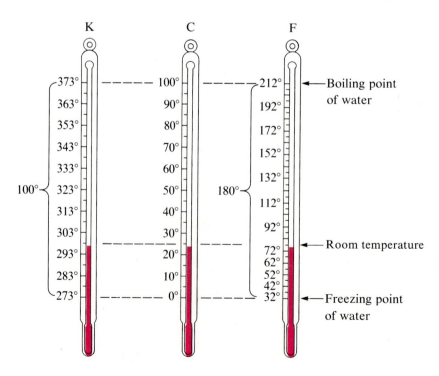

temperature is 373.15 K (100°C). A temperature on the Celsius scale is converted to its equivalent on the Kelvin scale by adding 273.15 to the Celsius reading; a temperature on the Kelvin scale is converted to its equivalent on the Celsius scale by subtracting 273.15 from the Kelvin reading. (Note that temperatures on the Kelvin scale are, by convention, reported without the degree sign. The units of temperature on the Kelvin scale are kelvins; K is read ''kelvins.''

$$K = °C + 273.15, \qquad °C = K - 273.15$$

Figure 1.14 shows the relationships among the three temperature scales. Temperatures in this book are in Celsius unless otherwise specified. The following examples demonstrate conversions between the three scales.

The boiling temperature of ethyl alcohol is 78.5°C at 1 atmosphere of pressure. What is its boiling point on the Fahrenheit scale and on the Kelvin scale?

EXAMPLE 1.8

$$°F = \frac{9}{5}°C + 32 = \left(\frac{9}{5} \times 78.5\right) + 32 = 141 + 32 = 173°F$$

$$K = °C + 273.15 = 78.5 + 273.15 = 351.6 \text{ K}$$

Convert 50°F to the Celsius scale and the Kelvin scale.

EXAMPLE 1.9

$$°C = \frac{5}{9}(°F - 32) = \frac{5}{9}(50 - 32) = \frac{5}{9} \times 18 = 10°C$$

$$K = °C + 273.15 = 10 + 273.15 = 283 \text{ K}$$

For Review

Summary

Since chemistry deals with the composition, structure, and properties of matter and with the reactions by which various forms of matter may be interconverted, it occupies a central place in the study of science and technology.

Matter is anything that occupies space and has **mass.** The basic building block of matter is the **atom,** the smallest unit of an element that can combine with other elements. In many substances, atoms are combined into **molecules.** Matter exists in three states: **solids,** of fixed shape and essentially incompressible; **liquids,** of variable shape but essentially incompressible; and **gases,** of variable shape and compressible or expansible. Most matter is a **mixture** of substances. These mixtures can be **heterogeneous** or **homogeneous.** Homogeneous mixtures are called **solutions.** Pure substances are also homogeneous. A **pure substance** may be either an **element** composed of only one type of atom or a **compound** consisting of two or more types of atoms. A pure substance exhibits characteristic **chemical and physical properties** by which it can be identified.

Chemistry is a science based on laws and theories that are derived from observations by the **scientific method.** Quantitative measurements utilize the metric system, usually with **SI units.** Common units in use include the **gram** and **kilogram** for mass, the **centimeter** and **meter** for length, the **liter** (cubic decimeter) and **milliliter** (cubic centimeter) for volume, and **degrees Celsius** or **kelvins** for temperature. Experimental measurements have an associated error that can be expressed by attention to significant figures. When using quantities in calculations, you must keep track of the units involved.

Key Terms and Concepts

Note: Section numbers are given for each term.

atom (1.6)	International System of Units	molecule (1.6)
chemical change (1.4)	(SI units) (1.8)	physical change (1.4)
chemical properties (1.4)	kelvins (1.14)	physical properties (1.4)
compound (1.5)	kilogram (1.12)	scientific method (1.7)
density (1.13)	liquid (1.2)	significant figures (1.10)
element (1.5)	liter (1.11)	solid (1.2)
energy (1.2)	mass (1.2)	solutions (1.5)
gas (1.2)	matter (1.2)	specific gravity (1.13)
gram (1.12)	meter (1.11)	temperature (1.14)
	mixture (1.5)	unit conversion factor (1.9)

Exercises

Note: Throughout the text, if an exercise number is in color the solution to that exercise is worked out in the manual prepared by J.H. Meiser and F.K. Ault, titled *Solutions Guide for General Chemistry and College Chemistry, 8th editions, by Holtzclaw and Robinson.*

You should make a special effort to develop good habits in expressing answers to exercises to the correct number of significant figures. To this end, we have paid careful attention to significant figures in the answers provided. Our answers were obtained by carrying all figures in the calculator and rounding the final answer. If you choose to do the exercise in steps, rounding after each step, you may get answers that differ from ours by one or two units in the least significant figure. These differences do not indicate an error; they simply reflect the two equally acceptable, but different, ways of rounding.

1. With what is the science of chemistry concerned?
2. What properties distinguish solids from liquids? Solids from gases? Liquids from gases?
3. Describe a chemical change that illustrates the law of conservation of matter.
4. Why is an object's mass rather than its weight used as a measure of the amount of matter it contains?
5. Describe how a scientist would proceed from the formulation of a question to the establishment of a theory.

Classification of Matter

6. How could you distinguish between an element, a compound, and a mixture.
7. In what ways do heterogeneous mixtures differ from homogeneous mixtures? In what ways are they alike?

8. Classify each of the following as a heterogeneous mixture or a homogeneous mixture: blood, ocean water, air, blueberry pancakes, pancake syrup, gasoline, a milkshake, concrete, a bowl of vegetable soup.
9. In what ways does a pure substance differ from a homogeneous mixture? In what ways are they alike?
10. Classify each of the following homogeneous materials as a solution, an element, or a compound: iron, water, shampoo, carbon dioxide, oxygen, sulfur, glucose, blood plasma.
11. Classify each of the following as a physical or chemical change: condensation of steam, burning of gasoline, souring of milk, dissolving sugar in water, melting of gold, leaves ''turning'' in color, explosion of a firecracker, magnetizing a screwdriver, melting of ice.
12. In what ways does an atom differ from a molecule? In what ways are they alike?
13. An atom and a molecule of helium are identical, but an atom and a molecule of sulfur are not. What is the difference? Figure 1.8 may prove helpful.
14. How do molecules of elements and molecules of compounds differ?

Significant Figures

15. The subtraction of two numbers, each of which contains three significant figures, can give an answer that may have one, two, or three significant figures. Explain how each case can arise and provide an example of each.
16. Indicate whether each of the following can be determined exactly or must be measured with some degree of uncertainty.

(a) The number of apples in a bag.

(b) The mass of a watermelon.

(c) The number of gallons of gasoline necessary to fill an automobile gas tank.

(d) The number of meters in exactly 5 km.

(e) The mass of this textbook.

(f) The number of minutes in one year.

(g) The time required to drive from Dallas to Chicago at an average speed of 53 miles per hour.

17. In a 1950 handbook, 1 inch was given as equal to 2.540005 cm. Since that time, 1 in has been defined as exactly 2.54 cm. Are there more, less, or the same number of significant figures in the newer value as compared to the older one?

18. How many significant figures are contained in each of the following numbers: 113; 207.033; 0.0820; 0.04109; 3.2×10^8; 9.74150×10^{-4}; 17.0?

Ans. 3; 6; 3; 4; 2; 6; 3

19. Round off each of the following numbers to two significant figures: 236; 38.1; 8.497×10^8; 7.0003; 135; 0.445.

Ans. 240; 38; 8.5×10^8; 7.0; 140; 0.44

20. Express the following numbers in exponential notation: 711.0; 0.239; 90743; 134.2; 0.05499; 10000.0; 0.000000738592.

Ans. 7.110×10^2; 2.39×10^{-1}; 9.0743×10^4; 1.342×10^2; 5.499×10^{-2}; 1.00000×10^4; 7.38592×10^{-7}

21. Perform the following calculations and give the answer with the correct number of significant figures.

(a) 1228×3442 *Ans. 4,227,000 (or 4.227×10^6)*

(b) $5.6 \times 10^2 \times 7.41 \times 10^3$ *Ans. 4.1×10^6*

(c) 2.734/28   *Ans. 0.098*

(d) 812×0.000023 *Ans. 0.019*

(e) 1.4 + 7.340 + 4.7593   *Ans. 13.5*

(f) (43 × 2.59)/28.44445   *Ans. 3.9*

22. It is necessary to determine the density of a liquid to four significant figures. The volume of solution can be measured to the nearest 0.01 cm^3.

(a) What is the minimum volume of sample that can be used for the measurement.   *Ans. 10.00 cm^3*

(b) Assuming the minimum-volume sample determined in (a), how accurately must the sample be weighed (to the nearest 0.1 g, 0.01 g, . . .), if the density of the solution is greater than 1.00 g/cm^3?

Ans. Nearest 0.01 g

Metric System; SI Units

23. Much of Chapter 1 deals with measurement. Why is measurement so important in chemistry?

24. What is the difference between mass and weight?

25. Even though a body "weighs" less on the moon than on the earth, why would an analytical balance, such as the one described in Section 1.12, indicate that the mass of an object is the same on the moon and on the earth?

26. Indicate the SI base units appropriate to express the following.

(a) The length of a 500-mile race.

(b) The mass of an elephant.

(c) The volume of a swimming pool.

(d) The speed of an automobile.

(e) The density of the metal platinum.

(f) The area of a football field.

(g) The maximum temperature at the South Pole on April 1, 1913.

27. What prefixes used with SI units indicate multiplication by the following exact quantities: 10^6; 10^{-1}; 0.01; 10^3; 0.001; 0.000001?

28. Using exponential notation, express the following quantities in terms of SI base units. (See Section 1.8.)

(a) 0.13 g   *Ans. 1.3×10^{-4} kg*

(b) 232 Gg   *Ans. 2.32×10^8 kg*

(c) 5.23 pm   *Ans. 5.23×10^{-12} m*

(d) 86.3 mg   *Ans. 8.63×10^{-5} kg*

(e) 37.6 cm   *Ans. 3.76×10^{-1} m*

(f) 54 Mm   *Ans. 5.4×10^7 m*

(g) 1 Ts   *Ans. 1×10^{12} s*

(h) 27 ps   *Ans. 2.7×10^{-11} s*

(i) 0.15 mK   *Ans. 1.5×10^{-4} K*

29. Many medical laboratory tests are run using 5.0 μL of blood serum. What is this volume in liters and in milliliters?   *Ans. 5.0×10^{-6} L; 5.0×10^{-3} mL*

30. Complete the following conversions.

(a) 13 g = _____ kg   *Ans. 0.013 kg*

(b) 17.43 m = _____ mm   *Ans. 1.743×10^4 mm*

(c) 3451 mg = _____ g   *Ans. 3.451 g*

(d) 1344 mL = _____ L   *Ans. 1.344 L*

(e) 4.18 g = _____ mg   *Ans. 4.18×10^3 mg*

(f) 27.8 m^3 = _____ L   *Ans. 2.78×10^4 L*

(g) 123 mL = _____ cm^3 *Ans. 123 cm^3*

(h) 17.38 km = _____ cm   *Ans. 1.738×10^6 cm*

Conversion of Units

A table of useful conversion factors can be found in Appendix B.

31. The mass of a competition Frisbee is 4.41 oz. What is the mass in grams?   *Ans. 125 g*

32. Calculate the length of a soccer field (110.0 yd) in meters and kilometers.   *Ans. 100.6 m; 1.006×10^{-1} km*

33. Soccer is played with a round ball between 69 and 71 cm in circumference and weighing between 400 and 450 g (two significant figures). What are these specifications in inches and ounces?

Ans. Between 27 and 28 in; between 14 and 16 oz

34. How many liters of a soft drink are contained in a 12.0-oz can?   *Ans. 0.355 L*

35. A barrel of oil is exactly 42 gallons. How many liters of oil are in a barrel?   *Ans. 159.0 L*

36. A 170-cm tall track-and-field athlete weighs 49.8 kg. Is she more likely to be a distance runner or a shot putter?

Ans. At 5 ft 7 in tall and 110 lb, a distance runner.

37. The diameter of a red blood cell is about 8×10^{-4} cm. What is the diameter in inches?   *Ans. 3×10^{-4} in*

38. Is a 197-lb weight lifter light enough to compete in a class limited to those weighing 90 kg or less?
    Ans. Yes, weight = 89.4 kg

39. A very good 197-lb weight lifter lifted 152 kg in the snatch and 192 kg in the clean and jerk. What were the masses of the weights lifted, in pounds?   *Ans. 335 lb; 423 lb*

40. Gasoline is sometimes sold by the liter. How many liters are required to fill a 10.0-gallon (liquid, U.S.) gas tank?
    Ans. 37.9 L

41. In a recent Belgian Grand Prix, the leader turned a lap with an average speed of 113.61 miles per hour. What was his speed in kilometers per hour and in meters per second?
    Ans. 182.84 km/h; 50.789 m/s

42. In a recent year 220,000,000 kg (assume three significant figures) of rayon was manufactured in the United States for use in carpets, automobile tires, and fabrics. What is this mass in pounds?   *Ans. 4.85×10^8 lb*

43. It has been estimated that if the Western Antarctic ice sheet and the Arctic Sea ice melted due to a climatic change, the level of the oceans could increase by about 6 m. How much would sea level increase, in feet?
    Ans. 20 ft (one significant figure)

44. Convert the following into the metric units indicated.
    (a) A record long jump, 8.90 m, to feet and inches.
    Ans. 29.2 ft; 3.50×10^2 in
    (b) The height of Mt. Kilimanjaro, 5963 m (the highest mountain in Africa), to feet.
    Ans. 19,570 (1.957×10^4) ft
    (c) The greatest depth of the ocean, about 1.0×10^4 m, to feet.   *Ans. 33,000 (3.3×10^4) ft*
    (d) The area of the state of Oregon, 96,981 mi^2, to square kilometers.   *Ans. 2.5118×10^5 km^2*
    (e) The volume of one gill, exactly 4 oz, to milliliters.
    Ans. 118.3 mL
    (f) The displacement of an automobile engine, 316 in^3, to liters.   *Ans. 5.18 L*
    (g) The estimated volume of the oceans, 330,000,000 mi^3, to km^3.   *Ans. 1.4×10^9 km^3*
    (h) The volume of exactly 1 L of milk to gallons (liquid, U.S.).   *Ans. 0.2642 gal*
    (i) The mass of a bushel of rye, 32.0 lb, to kilograms.
    Ans. 14.5 kg
    (j) The mass of a 65-g egg to ounces.   *Ans. 2.3 oz*
    (k) The mass of 5.00-grain aspirin to milligrams (1 grain = 0.00229 oz).   *Ans. 325 mg*
    (l) The mass of a 3525-lb car to kg.   *Ans. 1599 kg*

45. Solutions in chemistry laboratories are commonly prepared in 500-mL flasks. What is the volume of such a flask in fluid ounces? In pints?   *Ans. 16.9 oz; 1.06 pt*

46. The distance between the centers of the two oxygen atoms in an oxygen molecule is 1.21 Å. What is this distance in centimeters and in inches?
    Ans. 1.21×10^{-8} cm; 4.76×10^{-9} in

47. If a line of 1.0×10^8 water molecules is 1.00 in long, what is the average diameter of a water molecule in centimeters, angstrom units, and nanometers?
    Ans. 2.5×10^{-8} cm; 2.5 Å; 0.25 nm

48. The density of liquid mercury is 13.59 g/cm^3. What is the mass of 100 mL of mercury? What is the volume of 25 g of mercury?   *Ans. 1.36×10^3 g; 1.8 cm^3*

49. Calculate the density of aluminum if 13.8 cm^3 has a mass of 37.3 g.   *Ans. 2.70 g/cm^3*

50. Osmium is the densest element known. What is its density if 2.72 g has a volume of 0.121 cm^3?
    Ans. 22.5 g/cm^3

51. What is the mass of each of the following?
    (a) 6.00 cm^3 of bromine; density = 2.928 g/cm^3
    Ans. 17.6 g
    (b) 1000 mL of octane: density = 0.702 g/cm^3
    Ans. 702 g
    (c) 4.00 cm^3 of sodium; density = 0.97 g/cm^3
    Ans. 3.9 g
    (d) 250 mL of gaseous chlorine; density = 3.16 g/L
    Ans. 0.790 g

52. What is the volume of each of the following?
    (a) 25 g of iodine; density = 4.93 g/cm^3
    Ans. 5.1 cm^3
    (b) 2.00 g of gaseous hydrogen; density = 0.089 g/L
    Ans. 22 L
    (c) 226 g of carbon; density = 2.25 g/cm^3
    Ans. 100 cm^3

53. What is the specific gravity of uranium at 25°C and 1 atm pressure relative to water if 37.4 g of uranium has a volume of 2.00 cm^3?   *Ans. 18.8*

54. What is the specific gravity of natural gas (density 0.893 g/L) with respect to air (density 1.2929 g/L)? With respect to hydrogen (density 0.08987 g/L)?
    Ans. 0.691; 9.94

Temperature

55. Normal body temperature is 98.6°F. What is the temperature in degrees Celsius and in kelvins?
    Ans. 37.0°C; 310.2 K

56. The Voyager 1 flyby of Saturn revealed the surface temperature of the moon Titan to be 93 K. What is the surface temperature in degrees Celsius and degrees Fahrenheit?
    Ans. −180°C; −292°F

57. Zero kelvin is −273.15°C. What is 0 K on the Fahrenheit scale?   *Ans. −459.67°F*

58. Convert the following temperatures as indicated:
    (a) Temperature of the interior of a rare roast, 125°F, to degrees Celsius.   *Ans. 52°C*
    (b) Temperature of scalding water, 54°C, to degrees Fahrenheit and kelvins.   *Ans. 129°F; 327 K*
    (c) Coldest area in a freezer, −10°F, to degrees Celsius and kelvins.   *Ans. −23°C; 250 K*
    (d) Temperature of dry ice, −77°C, to degrees Fahrenheit and kelvins.   *Ans. −107°F; 196 K*

(e) The melting temperature of gold, 1336 K, to degrees Celsius and Fahrenheit.   *Ans. 1063°C; 1945°F*

(f) The boiling temperature of gold, 2966°C, in degrees Fahrenheit and kelvins.   *Ans. 5371°F; 3239 K*

(g) The boiling temperature of liquid ammonia, −28.1°F, in degrees Celsius and kelvins.

Ans. −33.4°C; 239.8 K

Additional Exercises

59. How do the densities of most substances change as a result of an increase in temperature?

60. Explain how you could determine if the outside temperature is warmer or cooler than 0°C, without using a thermometer.

61. In a recent year, a quantity of 18.4 billion pounds of phosphoric acid was manufactured in the United States for use in fertilizer. The cost of this acid averaged $318.00 per ton (1 ton = 2000 lb). What was the total value of the phosphoric acid produced? What was its cost per kilogram?
Ans. $2.93 × 10⁹; $0.351/kg

62. Given that one bushel is defined as 32 dry quarts, what is the mass in kilograms of a liter of each of the following types of grain?

(a) Wheat, 60.0 lb/bu   *Ans. 0.772 kg/L*
(b) Oats, 32.0 lb/bu   *Ans. 0.412 kg/L*
(c) Corn, 56.0 lb/bu   *Ans. 0.721 kg/L*
(d) Barley, 48 lb/bu   *Ans. 0.62 kg/L*

63. In the United States, there are about 2.5 million farms. These farms annually consume $2.4 × 10^{10}$ L of gasoline and diesel fuel, $4.90 × 10^9$ m³ of natural gas, $5.7 × 10^9$ L of liquified petroleum gas (LPG), and $1.16 × 10^{14}$ kilojoules of electricity. How many gallons of gasoline and diesel fuel, cubic feet of natural gas, gallons of LPG, and kilowatt hours of electricity were consumed? (1 kWh = $3.6 × 10^6$ J.)

Ans. 6.3 × 10⁹ gal; 1.73 × 10¹¹ ft³;
1.5 × 10⁹ gal; 3.22 × 10¹⁰ kWh

64. Moon rock is estimated to contain about 0.1% water. What mass of moon rock (in pounds) would a moon base have to process to recover 1 L of water (density 1.0 g/cm³)?
Ans. 2 × 10⁴ lb

65. Solids will float in a liquid with a density greater than the solid. A pound of butter forms a block about $6.5 × 6.5 × 12.0$ cm. Will this butter float in water (density 1.0 g/cm³)?   *Ans. Yes; its density is 0.89 g/cm³*

66. A 5.0-pint bottle contains 9.0 lb of sulfuric acid. What is the density of sulfuric acid (g/cm³)?   *Ans. 1.7 g/cm³*

67. Commercial solvents are commonly sold in 55-gallon drums. What is the weight in pounds of the carbon tetrachloride (density = 1.5942 g/cm³) in a 55-gal drum of this solvent?   *Ans. 7.3 × 10² lb*

68. A European recipe calls for the use of 1.50 kg of milk. An American who wants to make the recipe has no balance and only volume-measuring containers marked in ounces. Assuming the density of milk is 1.03 g/cm³, what volume of milk in ounces is needed?   *Ans. 49.2 oz*

69. Which of the following amounts of aluminum (density = 2.70 g/cm³) will occupy the greatest volume: 1.0 lb, 0.50 kg, 0.050 L?   *Ans. 0.50 kg*

70. What is the density of sugar (sucrose) in g/cm³ if 100.0 lb occupies 1.008 ft³?   *Ans. 1.589 g/cm³*

71. What mass in kilograms of concentrated hydrochloric acid is contained in a standard 5.0-pint container? The specific gravity of concentrated hydrochloric acid is 1.21.

Ans. 2.9 kg

72. Mercury is a very heavy metal, commonly used in thermometers. At 0°C, its density is 13.5955 g/mL; at 10°C, 13.5708 g/mL; at 25°C, 13.5340 g/mL; at 60°C, 13.4486 g/mL; and at 100°C, 13.3522 g/mL.

(a) Convert the five temperatures to degrees Fahrenheit.

(b) What is the volume of 5.152 g of mercury at each of the five temperatures?

(c) Make a graph, plotting temperature against volume.

(d) Does the plot suggest a property of mercury that makes it a suitable choice for use in thermometers? If so, what property?

(e) Can the same graph be adapted to both Celsius and Fahrenheit temperatures? If so, revise the plot to show both.

(f) Calculate the density of mercury at 60°C in pounds per cubic inch.

Ans. (a) 0°C, 32°F; 10°C, 50°F; 25°C, 77°F; 60°C,
140°F; 100°C, 212°F (b) 0°C, 0.3789 mL; 10°C,
0.3796 mL; 25°C, 0.3807 mL; 60°C, 0.3831 mL;
100°C, 0.3859 mL (d) Yes; approximately linear
relationship between temperature and volume. (e) Yes,
simply by labelling temperature axis in terms of both
temperature scales (f) 0.4859 lb/in³

Symbols, Formulas, and Equations; Elementary Stoichiometry

2

Crystals of vitamin C,
ascorbic acid

W hen speaking and writing about matter and the changes it undergoes, chemists use symbols, formulas, and equations to indicate what elements are present, the relative amounts of each, and how the combinations of elements change during a chemical reaction. They can use this information, along with the masses of the substances involved, to determine how much product will form during the course of a reaction or how much of the starting materials will be consumed.

Measurement of changes in mass as a chemical change occurs and the interpretation of the meaning of these changes have played an important role in the development of the science. One of the most fundamental laws of chemistry, the law of conservation of matter (Section 1.3), is based on such observations. There was little real progress in the development of chemistry until the French chemist Lavoisier (1743–1794) put this science on a quantitative basis. His simple observation—that when metals are burned (oxidized), the product has a larger mass than the initial mass of metal—had a profound effect upon the chemical theories of the time and provided the basis for the later work of John Dalton and his contemporaries.

2.1 Symbols and Formulas

A **symbol** is an abbreviation a chemist uses to indicate an element or an atom of the element. Symbols for several common elements are listed in Table 2.1. Some are derived from the common name of the element; others are abbreviations of the Latin name of the element. The symbols for all known elements are given on the inside front cover of this book.

To avoid confusion with other notations, the second letter of a two-letter symbol is never capitalized—Co is the symbol for the element cobalt; CO is the notation for the compound carbon monoxide, which contains the elements carbon (C) and oxygen (O).

A **formula** is an abbreviation a chemist uses to indicate a compound or a molecule of the compound. For example, NaCl is the abbreviation for sodium chloride (common table salt). A formula also contains information about the compound; it shows its composition, using the symbols of the elements present in the substance. Thus the formula NaCl identifies the elements sodium (Na) and chlorine (Cl) as the constituents of sodium chloride. Subscripts are used to indicate the relative numbers of atoms of

The bottle in the top photograph contains the gaseous element chlorine, Cl_2. The solid element sodium, Na, is shown in the middle. These two elements react vigorously (bottom) forming sodium chloride, NaCl, a compound containing sodium and chlorine.

Table 2.1	Some common elements and their symbols
Aluminum	Al
Bromine	Br
Calcium	Ca
Carbon	C
Chlorine	Cl
Chromium	Cr
Cobalt	Co
Copper	Cu (from *cuprum*)
Fluorine	F
Gold	Au (from *aurum*)
Helium	He
Hydrogen	H
Iodine	I
Iron	Fe (from *ferrum*)
Lead	Pb (from *plumbum*)
Magnesium	Mg
Mercury	Hg (from *hydrargyrum*)
Nitrogen	N
Oxygen	O
Potassium	K (from *kalium*)
Silicon	Si
Silver	Ag (from *argentum*)
Sodium	Na (from *natrium*)
Sulfur	S
Tin	Sn (from *stannum*)
Zinc	Zn

each type in the compound, but only if more than one atom of a given element is present. For example, the formula for water, H_2O, indicates that each molecule contains two atoms of hydrogen and one atom of oxygen. The formula for sodium chloride, NaCl, indicates the presence of equal numbers of atoms of the elements sodium and chlorine.

Parentheses in a formula indicate a group of atoms that may behave as a unit. The formula for aluminum sulfate, $Al_2(SO_4)_3$, indicates that there are two atoms of aluminum (Al) for each three sulfate (SO_4) groups. In most reactions of $Al_2(SO_4)_3$, the SO_4 units remain combined as a discrete unit consisting of one S (sulfur) atom and four O (oxygen) atoms. The formula shows a total of two atoms of aluminum, three atoms of sulfur, and twelve atoms of oxygen. The ratios obtained from such formulas are exact to any number of significant figures.

A **molecular formula** gives the actual numbers of atoms of each element in a molecule. An **empirical formula** (sometimes called a **simplest formula**) gives the *simplest* whole-number ratio of atoms. For example, the molecular formula of benzene, C_6H_6, identifies the benzene molecule as composed of exactly six carbon atoms and six hydrogen atoms. The simplest ratio of these atoms is one to one. Thus the empirical formula of benzene is CH. The molecular formula H_2O indicates that water has exactly two atoms of hydrogen and one atom of oxygen in one molecule (2 atoms H/1 atom O). The simplest ratio of atoms is 2 to 1, so the empirical formula is also H_2O.

Some molecules contain a single type of atom; their formulas contain only a single symbol. A molecule of the most common form of sulfur consists of eight atoms (see Fig. 1.8 in Chapter 1); the molecular formula is S_8. The molecular formulas for elemental hydrogen and oxygen (diatomic molecules) are H_2 and O_2, while the formula for neon (a monatomic molecule) is Ne.

Note that H_2 and 2H do not mean the same thing. The formula H_2 represents a molecule of hydrogen consisting of two atoms of the element, chemically combined. The expression 2H, on the other hand, indicates that the two hydrogen atoms are not in combination as a unit but are separate particles (Fig. 2.1).

In Chapter 6, we will see that many compounds do not contain discrete molecules but instead are composed of particles called **ions.** Ions are atoms or groups of atoms that are electrically charged (Section 6.1). Thus the composition of compounds such as NaCl, $Al_2(SO_4)_3$, and $CuSO_4$ cannot be identified in terms of the number of atoms of each type in a molecule, but only in terms of a *formula unit,* since such compounds do not contain physically distinct and electrically neutral molecules. The formula units of such compounds are empirical formulas because they give only the simplest whole-number ratios of elements in the compounds. The formula for copper sulfate, $CuSO_4$, indicates that there is one atom of copper for each atom of sulfur and each four atoms of oxygen in a sample of the compound, but it is an empirical formula since copper sulfate does not contain distinct molecules. The total numbers of atoms of the three elements in any sample of copper sulfate will be proportional to these numbers.

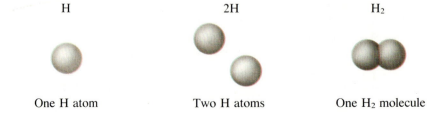

Figure 2.1

Illustration of the difference in meaning between the symbols H, 2H, and H_2.

EXAMPLE 2.1

A molecule of blood sugar, glucose, contains 6 carbon atoms, 12 hydrogen atoms, and 6 oxygen atoms. What are the empirical and molecular formulas of glucose?

The molecular formula is $C_6H_{12}O_6$, since one molecule actually contains 6C, 12H, and 6O atoms. The simplest whole-number ratio of C to H to O atoms in glucose is $1:2:1$, so the empirical formula is CH_2O.

EXAMPLE 2.2

How many oxygen atoms are present in an aspirin tablet, which contains 1.08×10^{21} aspirin molecules with the formula $C_9H_8O_4$?

The molecular formula of aspirin, $C_9H_8O_4$, indicates that there are exactly 4 atoms of oxygen in 1 $C_9H_8O_4$ molecule (4 atoms O/1 molecule $C_9H_8O_4$). This ratio can be used as a unit conversion factor to convert the number of molecules of aspirin in the tablet to the number of oxygen atoms in the tablet.

$$1.08 \times 10^{21} \text{ molecules } C_9H_8O_4 \times \frac{4 \text{ atoms O}}{1 \text{ molecule } C_9H_8O_4} = 4.32 \times 10^{21} \text{ atoms O}$$

2.2 Moles of Atoms and Atomic Weights

A chemical formula tells a chemist *how many* atoms of each type must be assembled to make a particular compound. However, we do not have to count out the actual number of atoms involved in the chemical reaction used to produce that compound. We need only get the correct ratios of the numbers of atoms. To make carbon tetrachloride, CCl_4, simply requires that four times as many chlorine atoms as carbon atoms be combined.

In the laboratory, numbers of atoms are conveniently measured in units of 6.022×10^{23} atoms. This is the unit that chemists call **one mole of atoms (1 mol).** The number of atoms in one mole of atoms of any element (1 mol atoms = 6.022×10^{23} atoms) is also called **Avogadro's number** in honor of the Italian professor of physics Amedeo Avogadro (1776–1856).

We could make a sample of carbon tetrachloride by combining 1 mole of carbon atoms (1 mol C atoms = 6.022×10^{23} C atoms) with 4 moles of chlorine atoms ($4 \times 6.022 \times 10^{23}$ Cl atoms). This would produce 6.022×10^{23} molecules of CCl_4. We could make a smaller sample of CCl_4 by using fewer moles of carbon and chlorine atoms (smaller numbers of C and Cl atoms), as long as there were four times as many chlorine atoms as carbon atoms.

EXAMPLE 2.3

How many CCl_4 molecules can be made by the combination of 0.050 mol of carbon atoms with chlorine atoms?

The formula CCl_4 tells us that one CCl_4 molecule contains one carbon atom (1 molecule CCl_4/1 atom C). If we can determine the number of carbon atoms present, we can determine the number of CCl_4 molecules that can be made. One mole of C atoms (1 mol C atoms) contains 6.022×10^{23} C atoms; 0.050 mole of C atoms contains

$$0.050 \text{ mol C atoms} \times \frac{6.022 \times 10^{23} \text{ C atoms}}{1 \text{ mol C atoms}} = 3.0 \times 10^{22} \text{ C atoms}$$

A single CCl_4 molecule contains 1 atom of carbon; therefore, we can make 1 molecule of CCl_4 for each carbon atom. Because we have 3.0×10^{22} atoms of carbon we can make

$$3.0 \times 10^{22}\ \cancel{\text{C atoms}} \times \frac{1\ \text{CCl}_4\ \text{molecule}}{1\ \cancel{\text{C atom}}} = 3.0 \times 10^{22}\ \text{CCl}_4\ \text{molecules}$$

EXAMPLE 2.4 How many moles of chlorine atoms are required to react with 0.050 mol of carbon atoms to convert all of the carbon atoms to CCl_4?

The formula CCl_4 indicates that there are 4 Cl atoms for each C atom in a molecule; thus we must take four times as many Cl atoms as C atoms, or

$$0.050\ \cancel{\text{mol C atoms}} \times \frac{6.022 \times 10^{23}\ \text{C atoms}}{1\ \cancel{\text{mol C atoms}}} = 3.0 \times 10^{22}\ \text{C atoms}$$

$$3.0 \times 10^{22}\ \cancel{\text{C atoms}} \times \frac{4\ \text{Cl atoms}}{1\ \cancel{\text{C atom}}} = 1.2 \times 10^{23}\ \text{Cl atoms}$$

We can calculate the number of moles of Cl atoms from the definition of a mole, 1 mol Cl atoms = 6.022×10^{23} Cl atoms (1 mol Cl atoms/6.022×10^{23} Cl atoms).

$$1.2 \times 10^{23}\ \cancel{\text{Cl atoms}} \times \frac{1\ \text{mol Cl atoms}}{6.022 \times 10^{23}\ \cancel{\text{Cl atoms}}} = 0.20\ \text{mol Cl atoms}$$

The number of atoms in a mole of atoms is so large that it is difficult to appreciate exactly how big it really is. As a guide to the size of the number, consider that if the entire population of the United States (approximately 225,000,000 people) spent 12 hours a day, 365 days a year, counting atoms at the rate of one atom per second, it would require about 170 million years to count the atoms in one mole of hydrogen atoms.

A mole of atoms is defined as the number of atoms contained in exactly 12 grams of ^{12}C, a particular carbon isotope. **Isotopes** are atoms that exhibit identical chemical properties but differ in mass (see Section 5.6). The actual number of atoms in one mole of atoms (6.022×10^{23} in one mole of any type of atoms) had to be determined by experiment.

One mole will always equal the same # for any atom. The mass for a mole varies with the different masses of each atom

The mass of a single atom can be calculated from the mass of one mole of atoms and the number of atoms in a mole of atoms.

$$\text{Mass of one atom} = \frac{\text{mass of mole}}{\text{Avogadro's number}}$$

The mass of a mole of oxygen atoms is 15.999 grams. The mass of a single atom is

$$\frac{15.999\ \text{g}}{1\ \cancel{\text{mol}}} \times \frac{1\ \cancel{\text{mol}}}{6.022 \times 10^{23}\ \text{atoms}} = 2.657 \times 10^{-23}\ \text{g/atom}$$

→ Mass of one atom x Avogadro's # = mass of mole

The mass of an atom is very small. Rather than describe such a small mass in grams, chemists use another arbitrary unit called either the **atomic mass unit (amu)** or the **dalton** (after John Dalton). By international agreement, an atomic mass unit (or dalton) is taken to be exactly $\frac{1}{12}$ of the mass of a ^{12}C atom; 1 amu = 1.6606×10^{-24} g. On the average, the mass of a hydrogen atom is 1.0079 amu (or daltons) on this arbitrary scale. The average mass of a fluorine atom is 18.998 amu, and that of phosphorus is 30.974 amu. Average masses are given here because many elements consist of mixtures of different isotopes. For example, 98.89% of naturally occurring carbon atoms are ^{12}C atoms with a mass of exactly 12 amu (by definition); the remainder have different masses, with 1.11% possessing a mass of 13.00335 amu, and less than 10^{-8}% having a mass of 14.0032 amu. The average mass of all of the carbon atoms in the naturally occurring mixture is 12.011 amu.

The **atomic weight** of an element is a number, without units, that is numerically equal to the average mass of an atom of the element (in amu). Thus the atomic weight of hydrogen is 1.0079, that of carbon is 12.011, that of fluorine is 18.998, and that of phosphorus is 30.974. A table of atomic weights is given inside the front cover of this book.

Let us now look at the relationship between masses of atoms and numbers of atoms. A mass of 18.998 amu of fluorine contains one atom of fluorine; a mass of 30.974 amu of phosphorus contains one atom of phosphorus. A mass of 75.992 amu (4 × 18.998) of fluorine contains four fluorine atoms. Similarly, a mass of 123.90 amu (4 × 30.974) of phosphorus contains four phosphorus atoms. The ratio of the mass of the sample of fluorine to the mass of the sample of phosphorus (75.992 amu to 123.90 amu) is the same as the ratio of their atomic weights (18.998 to 30.974), and the samples contain the same number of atoms—four. This is true of any pair of elements. **If the masses of samples of two elements have the same ratio as the ratio of their atomic weights, the samples will contain identical numbers of atoms.**

Balances in chemical laboratories measure in units of grams, not atomic mass units, but this is of little consequence in weighing out relative numbers of atoms. Just as 75.992 amu of fluorine and 123.90 amu of phosphorus contain the same number of atoms, 75.992 grams of fluorine and 123.90 grams of phosphorus also contain identical numbers of atoms, because the ratio of the masses is still 18.998 to 30.974.

The **mass of a mole of atoms** in grams is numerically equal to the atomic weight of the atom, and a mole of atoms of any element can be measured out by weighing out a mass of the element, in grams, equal to its atomic weight. Such a mass is sometimes called a **gram-atomic weight** of atoms. A sample of 30.974 grams of phosphorus contains one mole of phosphorus atoms (one gram-atomic weight of phosphorus). Similarly, 18.998 grams of fluorine contains one mole of fluorine atoms, and 12.011 grams of carbon contains one mole of carbon atoms. All of these quantities contain the same number of atoms. **A mole of atoms of any element contains the same number of atoms as a mole of atoms of any other element.**

Now we can see that it is easy to "count out" relative numbers of atoms in the laboratory (Fig. 2.2). Suppose, for example, that we have 32.1 grams of sulfur atoms and want an equal number of zinc atoms. From the atomic weight of sulfur, 32.1, we can recognize that 32.1 grams of sulfur atoms is one mole of sulfur atoms. To get one mole of zinc atoms, we simply look up the atomic weight of zinc (65.4) and then weigh out one mole of zinc atoms—65.4 grams. Finding fractional amounts of moles simply requires an extra arithmetical step.

The reaction of sodium with water. Each atom of sodium that reacts releases one atom of hydrogen. Because the ratio of the atomic weights of sodium to hydrogen is 23 to 1, 1.0 gram of sodium releases 0.043 gram of hydrogen.

Figure 2.2

Each sample contains 6.02 × 10^{23} atoms—1.00 mole of atoms: 12.0 g of carbon (at. wt 12.0), 65.4 g of zinc (at. wt 65.4), 201 g of mercury (at. wt 201), 32.1 g of sulfur (at. wt 32.1), 24.3 g of magnesium (at. wt 24.3), 63.5 g of copper (at. wt 63.5), 28.1 g of silicon (at. wt 28.1), and 207 g of lead (at. wt 207). (All values are given to three significant figures.)

EXAMPLE 2.5   How many moles of nitrogen atoms are contained in 9.34 g of nitrogen?

The atomic weights on the inside front cover indicate that the atomic weight of nitrogen is 14.0067; hence 1 mol of nitrogen atoms contains 14.0067 g, or there is 1 mol of nitrogen atoms per 14.0067 g (1 mol N atoms/14.0067 g N atoms). The quantity of nitrogen in 9.34 g can be converted to moles of nitrogen atoms as follows:

$$9.34 \text{ g N atoms} \times \frac{1 \text{ mol N atoms}}{14.0067 \text{ g N atoms}} = 0.667 \text{ mol N atoms}$$

Notice that the gram units cancel, indicating that the correct conversion factor has been used. Three significant figures in the answer are justified by the mass of nitrogen given (9.34 g).

EXAMPLE 2.6   How much sodium contains the same number of atoms as are in 18.29 g of chlorine?

From the table of atomic weights, we find that 1 mol of sodium atoms equals 22.990 g (22.990 g Na atoms/1 mol Na atoms) and 1 mol of chlorine atoms equals 35.453 g (1 mol Cl atoms/35.453 g). First find the number of moles of chlorine atoms present.

$$18.29 \text{ g Cl atoms} \times \frac{1 \text{ mol Cl atoms}}{35.453 \text{ g Cl atoms}} = 0.5159 \text{ mol Cl atoms}$$

A quantity of 0.5159 mol of Cl atoms contains the same number of atoms as are in 0.5159 mol of Na atoms. The next step, then, is to find the mass of 0.5159 mol of Na atoms.

$$0.5159 \text{ mol Na atoms} \times \frac{22.990 \text{ g}}{1 \text{ mol Na atoms}} = 11.86 \text{ g}$$

EXAMPLE 2.7   How many chlorine atoms are contained in 18.29 g of chlorine?

In Example 2.6 it was shown that 18.29 g of chlorine contains 0.5159 mol of chlorine atoms. Since 1 mol contains 6.022 × 10^{23} atoms, the conversion factor is 6.022 × 10^{23} Cl atoms/1 mol Cl atoms.

$$0.5159 \text{ mol Cl atoms} \times \frac{6.022 \times 10^{23} \text{ Cl atoms}}{1 \text{ mol Cl atoms}} = 3.107 \times 10^{23} \text{ Cl atoms}$$

2.3 Molecular Weights, Empirical Weights, and Moles of Molecules

The **molecular weight** of a molecule is the sum of the atomic weights of all the atoms in the molecule. For example, the molecular weight of chloroform, $CHCl_3$, equals the sum of the atomic weights of one carbon atom, one hydrogen atom, and three chlorine atoms; the molecular weight of aspirin, $C_9H_8O_4$, equals the sum of the atomic weights of nine carbon atoms, eight hydrogen atoms, and four oxygen atoms.

For $CHCl_3$:
$$1C = 1 \times 12.011 = 12.011$$
$$1H = 1 \times 1.0079 = 1.0079$$
$$3Cl = 3 \times 35.453 = 106.359$$
$$\text{mol. wt} = 119.378$$

For $C_9H_8O_4$:
$$9C = 9 \times 12.011 = 108.099$$
$$8H = 8 \times 1.0079 = 8.0632$$
$$4O = 4 \times 15.9994 = 63.9976$$
$$\text{mol. wt} = 180.160$$

Because the average mass of an atom in amu is numerically equal to its atomic weight, the average mass of a molecule in amu is numerically equal to its molecular weight.

As we pointed out earlier, ionic compounds such as $NaCl$, $Al_2(SO_4)_3$, and $CuSO_4$ do not contain discrete molecules and cannot be characterized properly by a molecular weight. For such compounds, we can calculate the **formula weight,** or **empirical weight,** which is the sum of the atomic weights of the atoms found in one unit, as indicated by the empirical formula.

For $Al_2(SO_4)_3$:
$$2Al = 2 \times 26.981 = 53.962$$
$$3S = 3 \times 32.06 = 96.18$$
$$12O = 12 \times 15.999 = 191.988$$
$$\text{formula wt} = 342.13$$

Ionic compounds are calculated to formula weight

Although many chemists refer loosely to the formula weight of an ionic compound as its molecular weight, there is a difference.

A **mole of molecules** contains the same number of molecules as there are atoms in exactly 12 grams of ^{12}C. Thus a mole of molecules contains 6.022×10^{23} molecules. To weigh out a mole of molecules, weigh out a mass in grams numerically equal to the molecular weight of the molecule. This mass is sometimes called the **molar mass** or the **gram-molecular weight** of the substance. Thus 119.378 grams of chloroform, the molar mass of chloroform, contains one mole of $CHCl_3$ molecules; 180.160 grams of aspirin, the molar mass of aspirin, contains one mole of $C_9H_8O_4$ molecules (Fig. 2.3). Because a mole of molecules contains Avogadro's number of molecules, 119.378 g of $CHCl_3$ and 180.160 g of $C_9H_8O_4$ each contain 6.022×10^{23} molecules.

It is also possible to weigh out a mole of an ionic compound such as $NaCl$ or KOH. A mole of an ionic compound contains 6.022×10^{23} of the units described by the formula. A mole of $NaCl$ would, therefore, contain Avogadro's number of $NaCl$ units, a unit being composed of one sodium atom and one chlorine atom. The mass of a mole of an ionic compound in grams is equal numerically to the formula weight and is called the **molar mass** or the **gram-formula weight.**

We said earlier that a chemical formula indicates the number of atoms in one molecular or empirical unit of that compound (Section 2.1). **A chemical formula also indicates the number of moles of atoms in one mole of the compound.** Ethanol (C_2H_5OH, often called ethyl alcohol) contains two carbon atoms in one molecule. It also contains two moles of carbon atoms in one mole of ethanol molecules ($2 \times 6.022 \times 10^{23}$ C atoms in 6.022×10^{23} C_2H_5OH molecules).

Figure 2.3

Each sample contains 6.02×10^{23} molecules or formula units—1.00 mole of the compound. From left to right: 58.5 g of NaCl (sodium chloride, formula wt 58.5), 18.0 g of H_2O (water, mol. wt 18.0), 74.1 g of C_4H_9OH (butyl alcohol, mol. wt 74.1), 342 g of $C_{12}H_{22}O_{11}$ (sucrose, or common sugar, mol. wt 342), and 180 g of $C_9H_8O_4$ (aspirin, mol. wt 180). One mole of carbon atoms is also shown. (All values are given to three significant figures.)

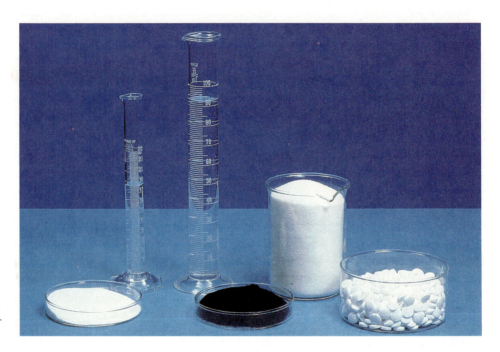

The following examples illustrate the use of the concepts of chemical formulas, atomic weights, molecular weights, and moles of atoms and molecules.

EXAMPLE 2.8

How many moles of sulfur atoms are contained in 80.3 g of sulfur?

The atomic weight of sulfur is 32.06, so 1 mol S atoms = 32.06 g sulfur (1 mol S atoms/32.06 g sulfur).

$$80.3 \text{ g sulfur} \times \frac{1 \text{ mol S atoms}}{32.06 \text{ g sulfur}} = 2.50 \text{ mol S atoms}$$

EXAMPLE 2.9

How many moles of sulfur molecules are contained in 80.3 g of sulfur if the molecular formula is S_8?

The molecular weight of S_8 is $8 \times 32.06 = 256.5$; hence 1 mol of S_8 = 256.5 g sulfur (1 mol S_8/256.5 g sulfur).

$$80.3 \text{ g sulfur} \times \frac{1 \text{ mol } S_8}{256.5 \text{ g sulfur}} = 0.313 \text{ mol } S_8$$

EXAMPLE 2.10

The recommended minimum daily dietary allowance of vitamin C, $C_6H_8O_6$, for a young woman of average weight is 4.6×10^{-4} mol. What is this allowance in grams?

The molecular weight of $C_6H_8O_6$ is 176.1, to one decimal place. Hence 1 mol $C_6H_8O_6$ = 176.1 g (176.1 g $C_6H_8O_6$/1 mol $C_6H_8O_6$).

$$4.6 \times 10^{-4} \text{ mol } C_6H_8O_6 \times \frac{176.1 \text{ g } C_6H_8O_6}{1 \text{ mol } C_6H_8O_6} = 0.081 \text{ g } C_6H_8O_6$$

EXAMPLE 2.11

How many moles of hydrogen atoms are contained in 0.500 g (500 mg) of vitamin C?

First calculate the number of moles of vitamin C, $C_6H_8O_6$; then use the information from the chemical formula, which indicates exactly 8 mol of H atoms per 1 mol of $C_6H_8O_6$.

$$0.500 \text{ g } C_6H_8O_6 \times \frac{1 \text{ mol } C_6H_8O_6}{176.1 \text{ g } C_6H_8O_6} = 2.84 \times 10^{-3} \text{ mol } C_6H_8O_6$$

$$2.84 \times 10^{-3} \text{ mol } C_6H_8O_6 \times \frac{8 \text{ mol H atoms}}{1 \text{ mol } C_6H_8O_6} = 2.27 \times 10^{-2} \text{ mol H atoms}$$

2.4 Percent Composition from Formulas

All samples of a pure compound contain the same elements in the same proportion by mass. For example, 11% of the mass of any sample of pure water is always hydrogen; 89% of the mass is always oxygen. This **law of definite proportion,** or **law of definite composition,** helped convince Dalton of the atomic nature of matter and led him to outline his atomic theory.

KNOWN AS

Since all samples of a pure compound contain the same relative amounts of elements by mass, the proportion of each element present in a compound can be used to identify the compound. The proportion commonly used is the **percent composition,** that is, the percent by mass of the element in the sample, or the fraction of the total mass of the sample due to the element, multiplied by 100.

$$\text{Percent by mass of } X = \frac{\text{mass of } X}{\text{mass of sample}} \times 100$$

EXAMPLE 2.12

Calculate the percent by mass of hydrogen and of oxygen in the compound water, H_2O.

To calculate the percent hydrogen and percent oxygen in water, we need to know the masses of hydrogen and oxygen in a sample of water with a known mass. A convenient quantity of water is one mole of water, which has a mass of 18.0 g. This mass of water contains 1 mol of oxygen and 2 mol of hydrogen. The percentages of H and O can be calculated as shown.

$$\text{Percent hydrogen by mass in } H_2O = \frac{2 \times \text{atomic weight of H}}{\text{molecular weight of } H_2O} \times 100$$

$$= \frac{2 \times 1.01}{(2 \times 1.01) + (1 \times 16.0)} \times 100 = 11.2\%$$

$$\text{Percent oxygen by mass in } H_2O = \frac{1 \times \text{atomic weight of O}}{\text{molecular weight of } H_2O} \times 100$$

$$= \frac{1 \times 16.0}{(2 \times 1.01) + (1 \times 16.0)} \times 100 = 88.8\%$$

2.5 Derivation of Formulas

An empirical formula shows the relative numbers of moles of atoms in a compound; thus, in order to write the empirical formula of a compound, we must know the relative numbers of moles of atoms in a sample of it. In the following example we are given the number of moles of C and H atoms in a sample of methane and asked to determine the empirical formula.

EXAMPLE 2.13   A sample of the principal component of natural gas, methane, contains 0.090 mol of carbon and 0.36 mol of hydrogen. What is the empirical formula of this compound?

For every 0.090 mol of C in the compound, there is 0.36 mol of H; thus the formula might be written as $C_{0.090}H_{0.36}$. But chemical formulas are customarily written in terms of whole-number ratios, so this formula must be reduced. The ratio of C to H is 1:4, so the *empirical* formula of the compound must be CH_4.

Note: To reduce two nonintegers to integers, divide each by the smaller noninteger. For this example,

$$\frac{0.090}{0.090} = 1; \qquad \frac{0.36}{0.090} = 4$$

Information about the composition of a compound is usually given in terms of either the masses of the elements in a sample or the percents by mass of the elements. If the masses of the elements are given, convert these to moles and reduce the mole ratio to the simplest whole numbers to find the empirical formula. If percents by mass are given, find the mass of each element present in a specific mass of sample (100 g is a convenient mass), convert to moles, and reduce the mole ratios to the simplest whole numbers.

EXAMPLE 2.14   A sample of the gaseous compound formed during bacterial fermentation of grain is found to consist of 27.29% carbon and 72.71% oxygen. What is the empirical formula of the compound?

Percent composition of an element in a compound is understood to indicate percent by mass. The mass of an element in a 100.0-g sample of a compound is equal in grams to the percent of that element in the sample; hence 100.0 g of this sample would contain 27.29 g of carbon and 72.71 g of oxygen.

$$100.0 \text{ g sample} \times \frac{27.29 \text{ g C}}{100 \text{ g sample}} = 27.29 \text{ g C}$$

$$100.0 \text{ g sample} \times \frac{72.71 \text{ g O}}{100 \text{ g sample}} = 72.71 \text{ g O}$$

The relative number of moles of carbon and oxygen atoms in the compound can be obtained by converting grams to moles, as shown in the following table.

For every 2.272 mol of carbon, there are 4.544 mol of oxygen. Reduced to the smallest whole numbers, the number of moles of oxygen is twice the number of moles of carbon. Hence, the number of oxygen *atoms* in the compound is also twice the number of carbon *atoms*. The simplest formula for the gaseous compound produced during fermentation must therefore be CO_2.

Element	Mass of element in 100.0 g of sample	Relative number of moles	Divide by the smaller number	Smallest integral number of moles
Carbon	27.29 g	$27.29 \text{ g C} \times \dfrac{1 \text{ mol C}}{12.01 \text{ g C}} = 2.272 \text{ mol C}$	$\dfrac{2.272}{2.272} = 1.000$	1
Oxygen	72.71 g	$72.71 \text{ g O} \times \dfrac{1 \text{ mol O}}{16.00 \text{ g O}} = 4.544 \text{ mol O}$	$\dfrac{4.544}{2.272} = 2.000$	2

The empirical formula CO_2 indicates a formula weight of 44. It can be shown by experiment that the gas actually has a molecular weight of 44. Thus the empirical formula and the molecular formula of CO_2 are the same. If the molecular weight for the gas were 88, then the molecular formula would be C_2O_4, indicating twice as many atoms as in the simplest formula, CO_2.

EXAMPLE 2.15

A sample of the mineral hematite, an oxide of iron found in iron ores and shown in the photograph, contains 34.97 g of iron and 15.03 g of oxygen. What is the empirical formula of hematite?

The steps in the solution of this problem are outlined in the following table.

A sample of black hematite and white quartz crystals.

Element	Mass of element	Relative number of moles	Divide by the smaller number	Smallest integral number of moles
Iron	34.97 g	$34.97 \text{ g Fe} \times \dfrac{1 \text{ mol Fe}}{55.85 \text{ g Fe}} = 0.6261 \text{ mol Fe}$	$\dfrac{0.6261}{0.6261} = 1.00$	$2 \times 1.00 = 2$
Oxygen	15.03 g	$15.03 \text{ g O} \times \dfrac{1 \text{ mol O}}{16.00 \text{ g O}} = 0.9394 \text{ mol O}$	$\dfrac{0.9394}{0.6261} = 1.50$	$2 \times 1.50 = 3$

Here, division by the smaller number of moles does not give two integers, so a third step is necessary. We must multiply by the smallest whole number that will give whole numbers for the relative numbers of moles, and hence of atoms, of each element: $2 \times 1.00 = 2$ and $2 \times 1.50 = 3$. The simplest whole-number mole ratio (and also whole-number atom ratio) is 2:3, and the empirical formula of hematite is Fe_2O_3.

EXAMPLE 2.16

Pure oxygen is sometimes prepared in a general chemistry laboratory by heating a compound containing potassium, chlorine, and oxygen. What is the empirical formula of this compound if a 3.22-g sample decomposes to give gaseous oxygen and 1.96 g of KCl?

The mass of oxygen produced, and therefore the mass of oxygen atoms in the 3.22-g sample, is 3.22 g − 1.96 g = 1.26 g. We can determine the number of moles of oxygen atoms by converting 1.26 g of oxygen atoms to moles of oxygen atoms. The moles of potassium atoms and of chlorine atoms can be determined from the number of moles of KCl, since there is 1 mol K and 1 mol Cl per 1 mol KCl. The steps in the solution of this problem are outlined in the following table:

Atom	Number of moles	Divide by smaller number	Smallest integral number of moles
O	$1.26 \text{ g O} \times \dfrac{1 \text{ mol O}}{16.0 \text{ g O}} = 0.0788 \text{ mol O}$	$\dfrac{0.0788}{0.0263} = 3.00$	3
K	$1.96 \text{ g KCl} \times \dfrac{1 \text{ mol KCl}}{74.6 \text{ g KCl}}$		
	$\times \dfrac{1 \text{ mol K}}{1 \text{ mol KCl}} = 0.0263 \text{ mol K}$	$\dfrac{0.0263}{0.0263} = 1.00$	1
Cl	$1.96 \text{ g KCl} \times \dfrac{1 \text{ mol KCl}}{74.6 \text{ g KCl}}$		
	$\times \dfrac{1 \text{ mol Cl}}{1 \text{ mol KCl}} = 0.0263 \text{ mol Cl}$	$\dfrac{0.0263}{0.0263} = 1.00$	1

Since the ratio of K to Cl to O is 1:1:3, the simplest formula of the compound is $KClO_3$.

2.6 Chemical Equations

When atoms, molecules, or ions regroup to form other substances, chemists use a shorthand type of expression called a **chemical equation** to describe the chemical change. Formulas indicate all substances involved and their compositions. For example, the equation for the reaction of methane with oxygen to form carbon dioxide and water (Fig. 2.4) is

$$CH_4 + 2O_2 \longrightarrow CO_2 + 2H_2O$$

The formulas for the reactants are written to the left of the arrow, and the formulas for the products are written to the right. The arrow is read as "gives," "produces," "yields," or "forms." A plus sign on the left side of an equation means "reacts with," and one on the right side, "and." Since matter cannot be created or destroyed in a chemical reaction, a chemical equation must have the same number of atoms in the products as there are in the reactants. In Fig. 2.4 we can see that one carbon atom, four hydrogen atoms, and four oxygen atoms are present in both the reactants and the products.

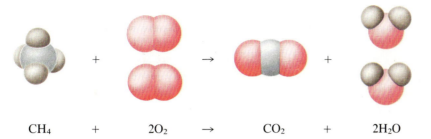

$$CH_4 \quad + \quad 2O_2 \quad \rightarrow \quad CO_2 \quad + \quad 2H_2O$$

Figure 2.4

The chemical equation for the reaction of one molecule of methane with two molecules of oxygen to give one molecule of carbon dioxide plus two molecules of water. The drawing shows how the atoms are redistributed during the chemical change.

The statement, "when water is decomposed by an electric current, hydrogen and oxygen are formed," (Fig. 2.5) can be expressed by the chemical equation

$$H_2O \longrightarrow H_2 + O_2 \tag{1}$$

However, this equation is incomplete because it does not contain the same number of atoms on both sides of the arrow. One molecule of water, which contains only one atom of oxygen, cannot be rearranged into a product that contains two atoms of oxygen. Two molecules of water, however, can provide two atoms of oxygen. We can correct the equation by balancing it—placing the proper coefficients on each side of the arrow.

$$2H_2O \longrightarrow 2H_2 + O_2 \tag{2}$$

The subscripts in a formula cannot be changed to make an equation balance; subscripts indicate a compound's atomic composition.

Most simple chemical equations can be balanced by examination. This involves looking at the equation and adjusting the coefficients so that equal numbers of atoms of each type are present on both sides of the arrow. Let's consider the reaction of aluminum, Al, with hydrogen chloride, HCl, producing aluminum chloride, $AlCl_3$, and hydrogen, H_2. First we write the unbalanced equation:

$$Al + HCl \longrightarrow AlCl_3 + H_2$$

Now we determine the number of atoms of each type on both sides. In this equation there are one Al, one H, and one Cl atom on the left and one Al, two H, and three Cl atoms on the right. The equation is not balanced because the numbers of atoms are not the same on both sides. To find the coefficients, it is usually best to consider the molecule with the most atoms first, which in this case is $AlCl_3$. This molecule contains three Cl atoms, which must all come from HCl. Therefore, we place a coefficient of 3 in front of HCl. At this stage we have

$$Al + 3HCl \longrightarrow AlCl_3 + H_2$$

Although the Cl and Al atoms are now balanced, the H atoms are not. Three HCl molecules, which collectively contain three hydrogen atoms, would give $1\frac{1}{2}$ H_2 molecules. Thus we have

$$Al + 3HCl \longrightarrow AlCl_3 + 1\frac{1}{2} H_2$$

Although now balanced, this equation is not in its most conventional form, because it indicates that half of a molecule of hydrogen is formed, and half-molecules do not exist. Therefore, we multiply the equation by 2 to obtain a balanced equation with the simplest whole-number coefficients.

$$2Al + 6HCl \longrightarrow 2AlCl_3 + 3H_2$$

There are two Al, six H, and six Cl atoms on both sides.

Figure 2.5

When an electric current is passed through water, 2 volumes of hydrogen are produced for each 1 volume of oxygen.

The reaction of aluminum with iodine produces aluminum iodide. The purple vapor is iodine gas, which has been produced by the heat of the reaction.

Balancing equations by inspection is a trial-and-error process; you may have to try several sets of coefficients before you find the correct ones.

In addition to identifying reactants and products, chemical equations often give other information, such as the state of the reactants and products. For example, the formula for a gaseous reactant or product is followed by (g); a liquid by (l); a solid by (s); and a reactant or product that is dissolved in water by (aq). The reaction of solid sodium, Na, with liquid water to give hydrogen gas, H_2, and a solution of sodium hydroxide, NaOH, in water would be indicated by this equation:

$$2Na(s) + 2H_2O(l) \longrightarrow H_2(g) + 2NaOH(aq)$$

Special conditions, such as temperature, the presence of a catalyst, or any other special circumstances that characterize the reaction, may be written above or below the arrow. The electrolysis of water described in Equation (2) is sometimes written as follows:

$$2H_2O(l) \xrightarrow{\text{Elect.}} 2H_2(g) + O_2(g)$$

where the abbreviation *Elect.* over the arrow indicates that the reaction occurs by means of an electric current. A reaction carried out by heating may be indicated with a triangle over the arrow:

$$CaCO_3(s) \xrightarrow{\triangle} CaO(s) + CO_2(g)$$

2.7 Mole Relationships Based on Equations

The chemical reaction

$$2H_2O \longrightarrow 2H_2 + O_2$$

indicates that exactly two molecules of H_2O decompose to give two molecules of H_2 and one molecule of O_2. But the relative proportion of molecules, atoms, or ions is identical to the relative proportion of *moles* of molecules, atoms, or ions. Thus the equation also indicates relative numbers of moles.

$2H_2O$	$\longrightarrow$	$2H_2$	$+$	O_2
2 molecules		2 molecules		1 molecule
$2 \times 6.022 \times 10^{23}$ molecules		$2 \times 6.022 \times 10^{23}$ molecules		6.022×10^{23} molecules
2 moles		2 moles		1 mole

The coefficients in a chemical equation indicate the relative numbers of moles of atoms or moles of molecules that react or are formed. The equation indicates that 2 moles of H_2 are formed for each 2 moles of H_2O decomposed (2 mol H_2/2 mol H_2O), that 1 mole of O_2 is formed for each 2 moles of H_2O decomposed (1 mol O_2/2 mol H_2O), and that 1 mole of O_2 forms for each 2 moles of H_2 formed (1 mol O_2/2 mol H_2). These unit conversion factors are exact to any number of significant figures.

It is therefore quite straightforward to determine, from the information provided by a chemical equation, the numbers of atoms and molecules, or the numbers of moles of these atoms and molecules, that react or are produced in a chemical process.

EXAMPLE 2.17   As shown in the photograph, aluminum reacts with iodine. How many moles of Al_2I_6 are produced by the reaction of 4.0 mol of aluminum according to the following equation?

$$2Al(s) + 3I_2(s) \longrightarrow Al_2I_6(s)$$

The balanced equation indicates that exactly 1 mol of Al_2I_6 is produced per 2 mol of Al consumed (1 mol Al_2I_6/2 mol Al). This relationship can be used to convert the number of moles of Al reacting to the number of moles of Al_2I_6 produced.

$$4.0 \text{ mol Al} \times \frac{1 \text{ mol } Al_2I_6}{2 \text{ mol Al}} = 2.0 \text{ mol } Al_2I_6$$

The unwanted units cancel, so the correct factor was used. The number of significant figures in the answer was determined by the amount of Al, since conversion factors determined from equations are exact.

How many moles of I_2 are required to react exactly with 0.429 mol of aluminum?   **EXAMPLE 2.18**

The balanced equation indicates that exactly 3 mol of I_2 reacts with 2 mol of Al. This gives us the conversion factor: 3 mol I_2/2 mol Al.

$$0.429 \text{ mol Al} \times \frac{3 \text{ mol } I_2}{2 \text{ mol Al}} = 0.644 \text{ mol } I_2$$

2.8 Calculations Based on Equations

In this section we will introduce a method of chemical calculation that relates the quantities of substances involved in chemical reactions. All such calculations proceed from the facts that (1) balanced equations give the ratios of moles of reactants and products involved in chemical reactions, and (2) the masses of the reactants and products can be determined from the numbers of moles involved, using the atomic, molecular, or formula weights of the substances participating in the reaction. The following examples illustrate the method.

Calculate the moles of oxygen produced during the thermal decomposition of 100.0 g of potassium chlorate to form potassium chloride and oxygen according to the following reaction: **EXAMPLE 2.19**

$$2KClO_3(s) \xrightarrow{\Delta} 2KCl(s) + 3O_2(g)$$

We are asked to calculate the number of moles of O_2 produced from 100.0 g of $KClO_3$, and we are given the equation for the reaction by which O_2 forms from $KClO_3$. If we can determine the number of moles of $KClO_3$ that react, we can also determine the number of moles of O_2 produced, since the equation indicates that exactly 3 mol of O_2 is produced for each 2 mol of $KClO_3$ that reacts (3 mol O_2/2 mol $KClO_3$). The number of moles of $KClO_3$ used can be found from the mass of $KClO_3$ and the relationship 1 mol $KClO_3$ = 122.5 g $KClO_3$ (Section 2.3).

$$100.0 \text{ g } KClO_3 \times \frac{1 \text{ mol } KClO_3}{122.5 \text{ g } KClO_3} = 0.8163 \text{ mol } KClO_3$$

$$0.8163 \text{ mol } KClO_3 \times \frac{3 \text{ mol } O_2}{2 \text{ mol } KClO_3} = 1.224 \text{ mol } O_2$$

Thus 100.0 g of $KClO_3$ is 0.8163 mol $KClO_3$, and 0.8163 mol $KClO_3$ will produce 1.224 mol of O_2. In this case, the chain of calculations involves the following unit conversions:

$$\boxed{\text{Mass of } KClO_3} \longrightarrow \boxed{\text{Moles of } KClO_3} \longrightarrow \boxed{\text{Moles of } O_2}$$

Note how dimensional analysis with the resulting cancellation of units is helpful in checking to see if the correct conversion factors have been used.

It is not necessary to write each step of the calculations separately. Although the logic of the problem is perhaps best approached in two steps, the calculations could have been written in one step, as follows:

$$100.0 \text{ g } KClO_3 \times \frac{1 \text{ mol } KClO_3}{122.5 \text{ g } KClO_3} \times \frac{3 \text{ mol } O_2}{2 \text{ mol } KClO_3} = 1.224 \text{ mol } O_2$$

EXAMPLE 2.20 Calculate the mass of oxygen gas, O_2, required for the combustion of 702 g of octane, C_8H_{18}.

$$2C_8H_{18} + 25O_2 \longrightarrow 16CO_2 + 18H_2O$$

In this example we are asked to calculate the mass of O_2 that reacts with 702 g of octane, C_8H_{18}, in a reaction indicated by the balanced equation. We can calculate the number of grams of O_2 required (by using the molecular weight of O_2) if we know the number of moles of O_2 required. The equation indicates the number of moles of O_2 and the number of moles of C_8H_{18} that react with each other, so the number of moles of O_2 can be determined if we can determine the number of moles of C_8H_{18} present. We can determine the number of moles of C_8H_{18} from the mass of C_8H_{18} and its molecular weight. Thus, the chain of calculations requires the following conversions:

$$\boxed{\text{Mass of } C_8H_{18}} \longrightarrow \boxed{\text{Moles of } C_8H_{18}} \longrightarrow \boxed{\text{Moles of } O_2} \longrightarrow \boxed{\text{Mass of } O_2}$$

From the molecular weight of C_8H_{18}, 1 mol C_8H_{18} = 114.2 g C_8H_{18} [(8 × 12.011) + (18 × 1.0079), rounded to one decimal place].

$$702 \text{ g } C_8H_{18} \times \frac{1 \text{ mol } C_8H_{18}}{114.2 \text{ g } C_8H_{18}} = 6.147 \text{ mol } C_8H_{18}$$

From the balanced chemical equation, 25 mol of O_2 reacts for each 2 mol of C_8H_{18} (25 mol O_2/2 mol C_8H_{18}), so

$$6.147 \text{ mol } C_8H_{18} \times \frac{25 \text{ mol } O_2}{2 \text{ mol } C_8H_{18}} = 76.84 \text{ mol } O_2$$

From the molecular weight of O_2, 1 mol O_2 = 32.00 g O_2. Hence

$$76.84 \text{ mol } O_2 \times \frac{32.00 \text{ g } O_2}{1 \text{ mol } O_2} = 2459 \text{ g } O_2 = 2.46 \times 10^3 \text{ g } O_2$$

In this example one more digit than is justified by the data was carried through the calculations, and the final answer was rounded off to three significant figures. Although the logic of the problem involves three steps, the calculations could have been handled in one step, as follows:

$$702 \text{ g } C_8H_{18} \times \frac{1 \text{ mol } C_8H_{18}}{114.2 \text{ g } C_8H_{18}} \times \frac{25 \text{ mol } O_2}{2 \text{ mol } C_8H_{18}} \times \frac{32.00 \text{ g } O_2}{1 \text{ mol } O_2} = 2.46 \times 10^3 \text{ g } O_2$$

EXAMPLE 2.21

What mass of sodium hydroxide, NaOH, would be required to produce 16 g of the antacid milk of magnesia [magnesium hydroxide, $Mg(OH)_2$] by the reaction of magnesium chloride, $MgCl_2$, with NaOH?

$$MgCl_2(aq) + 2NaOH(aq) \longrightarrow Mg(OH)_2(s) + 2NaCl(aq)$$

The equation tells us that 2 mol of NaOH are required to form 1 mol of $Mg(OH)_2$. If we calculate the number of moles of $Mg(OH)_2$ in 16 g $Mg(OH)_2$, we can determine the moles of NaOH necessary and then the mass of NaOH required.

$$\boxed{\text{Mass of } Mg(OH)_2} \longrightarrow \boxed{\text{Moles of } Mg(OH)_2} \longrightarrow \boxed{\text{Moles of } NaOH} \longrightarrow \boxed{\text{Mass of } NaOH}$$

From the formula weight of $Mg(OH)_2$, 58.3 g $Mg(OH)_2$ = 1 mol $Mg(OH)_2$. Therefore

$$16 \text{ g } Mg(OH)_2 \times \frac{1 \text{ mol } Mg(OH)_2}{58.3 \text{ g } Mg(OH)_2} = 0.274 \text{ mol } Mg(OH)_2$$

From the chemical equation, 2 mol of NaOH reacts to give 1 mol of $Mg(OH)_2$ [2 mol NaOH/1 mol $Mg(OH)_2$]. Hence

$$0.274 \text{ mol } Mg(OH)_2 \times \frac{2 \text{ mol } NaOH}{1 \text{ mol } Mg(OH)_2} = 0.548 \text{ mol } NaOH$$

From the formula weight of NaOH, 40.0 g NaOH = 1 mol NaOH. Thus

$$0.548 \text{ mol } NaOH \times \frac{40.0 \text{ g } NaOH}{1 \text{ mol } NaOH} = 22 \text{ g } NaOH$$

The mass, 16 g $Mg(OH)_2$, requires two significant figures for the answer.

The calculations here could have been handled in one step, as demonstrated in previous examples.

2.9 Limiting Reagents

Under certain conditions, all of the reactants in a chemical reaction may not be completely consumed. Consider, for example, the reaction of silver nitrate with copper metal (Fig. 2.6).

$$2AgNO_3 + Cu \longrightarrow 2Ag + Cu(NO_3)_2$$

The equation indicates that exactly 2 moles of $AgNO_3$ react per mole of Cu. If the ratio of reactants actually used in this reaction differs from 2:1, then one of the reactants will be present in excess and not all of it will be consumed. If, for example, we have 2 moles of $AgNO_3$ and 2 moles of Cu, 1 mole of Cu must remain unreacted at the end of the reaction because 2 moles of $AgNO_3$ can react with only 1 mole of Cu. The reactant

Figure 2.6
The photograph on the left shows a copper wire shortly after it is placed in a solution of silver nitrate. On the right the reaction is complete. Silver crystals cover the wire and the blue color of copper nitrate is evident in the solution. Silver nitrate is the limiting reagent in this reaction since some copper wire remains unconsumed.

LOOK AT GLOSSARY

that will be completely consumed ($AgNO_3$ in this case) is called the **limiting reagent.** This reactant limits the amount of product that can be formed and determines the theoretical yield of the reaction. Other reactants are said to be present in excess.

To determine which reagent is the limiting reagent, we calculate the amount of product expected from each reactant. The reactant that gives the smallest amount of product is the limiting reagent.

EXAMPLE 2.22

A mixture of 5.0 g of $H_2(g)$ and 10.0 g of $O_2(g)$ is ignited. Water forms according to the following equation:

$$2H_2(g) + O_2(g) \longrightarrow 2H_2O(g)$$

Which reactant is limiting? How much water will be produced by the reaction?

First, assume that all of the H_2 will react and calculate the theoretical yield of water (to the two significant figures justified), using the following steps:

$$\boxed{\text{Mass of } H_2} \longrightarrow \boxed{\text{Moles of } H_2} \longrightarrow \boxed{\text{Moles of } H_2O}$$

$$5.0 \text{ g } H_2 \times \frac{1 \text{ mol } H_2}{2.02 \text{ g } H_2} = 2.48 \text{ mol } H_2$$

$$2.48 \text{ mol } H_2 \times \frac{2 \text{ mol } H_2O}{2 \text{ mol } H_2} = 2.5 \text{ mol } H_2O$$

Now, assume that all of the O_2 will react, and calculate the theoretical yield of water (to the three significant numbers justified), using the following steps:

$$\boxed{\text{Mass of } O_2} \longrightarrow \boxed{\text{Moles of } O_2} \longrightarrow \boxed{\text{Moles of } H_2O}$$

$$10.0 \text{ g } O_2 \times \frac{1 \text{ mol } O_2}{32.00 \text{ g } O_2} = 0.3125 \text{ mol } O_2$$

$$0.3125 \text{ mol } O_2 \times \frac{2 \text{ mol } H_2O}{1 \text{ mol } O_2} = 0.625 \text{ mol } H_2O$$

At this point, it should be apparent that O_2 is the limiting reactant, since 10.0 g of O_2 gives 0.625 mol of H_2O, whereas 5.0 g of H_2 would give 2.5 mol of H_2O if the hydrogen were all converted to water. The following expression gives the mass of water equal to 0.625 mol of water:

$$0.625 \text{ mol } H_2O \times \frac{18.02 \text{ g } H_2O}{1 \text{ mol } H_2O} = 11.3 \text{ g } H_2O$$

2.10 Theoretical Yield, Actual Yield, and Percent Yield

In the preceding sections we have shown how to calculate the mass of product produced by a chemical reaction. These calculations are based on the assumptions that the reaction is the only one involved, that all of the reactant is converted into product, and that all of the product can be collected. The calculated amount of product based on these assumptions is called the **theoretical yield.** We rarely find these conditions satisfied either in the laboratory or in industrial production; the actual mass of product isolated (the **actual yield**) from a reaction is usually less than the theoretical yield and is used in calculating the **percent yield** of a reaction from the following relationship:

$$\text{Percent yield} = \frac{\text{actual yield}}{\text{theoretical yield}} \times 100$$

EXAMPLE 2.23

A general chemistry student, preparing copper metal by the reaction of 1.274 g of copper sulfate with zinc metal, isolated a yield of 0.392 g of copper. What was the percent yield?

$$CuSO_4(aq) + Zn(s) \longrightarrow Cu(s) + ZnSO_4(aq)$$

To calculate percent yield we need both the theoretical yield of copper and the actual yield. The actual yield is 0.392 g; the theoretical yield is the amount of copper that would have been obtained if all of the $CuSO_4$ had been converted into Cu and recovered. The theoretical yield can be calculated by the following steps:

$$\boxed{\begin{array}{c}\text{Mass of}\\CuSO_4\end{array}} \longrightarrow \boxed{\begin{array}{c}\text{Moles of}\\CuSO_4\end{array}} \longrightarrow \boxed{\begin{array}{c}\text{Moles of}\\Cu\end{array}} \longrightarrow \boxed{\begin{array}{c}\text{Mass of}\\Cu\end{array}}$$

$$1.274 \text{ g } CuSO_4 \times \frac{1 \text{ mol } CuSO_4}{159.6 \text{ g } CuSO_4} = 7.982 \times 10^{-3} \text{ mol } CuSO_4$$

$$7.982 \times 10^{-3} \text{ mol } CuSO_4 \times \frac{1 \text{ mol } Cu}{1 \text{ mol } CuSO_4} = 7.982 \times 10^{-3} \text{ mol } Cu$$

$$7.982 \times 10^{-3} \text{ mol } Cu \times \frac{63.546 \text{ g } Cu}{1 \text{ mol } Cu} = 0.5072 \text{ g } Cu$$

The theoretical yield of Cu is 0.5072 g. The percent yield may now be calculated.

$$\text{Percent yield} = \frac{\text{actual yield}}{\text{theoretical yield}} \times 100 = \frac{0.392 \text{ g}}{0.5072 \text{ g}} \times 100 = 77.3\%$$

For Review

Summary

A chemical **formula** identifies the elements in a substance using **symbols,** one- or two-letter abbreviations for the elements. Subscripts indicate the relative numbers of the different atoms. A **molecular formula** indicates the exact number of each type of atom in a molecule. An **empirical formula** gives the simplest whole number ratio of atoms in a substance.

The mass of an atom is usually expressed in **atomic mass units, amu.** An amu is defined as exactly $\frac{1}{12}$ the mass of a ^{12}C atom. The **atomic weight** of an element is a number without units and is numerically equal to the mass of the element (in amu). A **mole of atoms** of any given substance contains the same number of atoms as are found in exactly 12 grams of ^{12}C. A mole of any type of atoms contains **Avogadro's number** of atoms, 6.022×10^{23} atoms. The mass, in grams, of a mole of atoms is numerically equal to the atomic weight of the atom.

The mass of a molecule, in amu, is equal to the sum of the masses of its constituent atoms. The **molecular weight** of a compound is equal to the sum of the atomic weights of its constituent atoms. One **mole of molecules** contains 6.022×10^{23} molecules, and the mass of a mole of molecules, in grams, is numerically equal to the molecular weight. The subscripts in a molecular formula indicate the number of moles of each type of atom in a mole of molecules. Compounds that do not consist of individual molecules are characterized by a **formula weight,** which is equal to the sum of the atomic weights of the atoms in the empirical formula. A mole of such a compound is 6.022×10^{23} of the units described by the empirical formula.

The amount of a substance, in moles, may be determined from its mass, or the mass may be determined from the number of moles. Thus a given quantity of a substance, in moles, may be obtained by weighing. The empirical formula of a substance can be determined from the relative numbers of moles of the various types of atoms that make up the substance. If the **percent composition** is known, then the relative numbers of moles can be calculated for a sample of any specific weight of the compound.

Chemical equations indicate the regrouping of atoms that accompanies chemical reactions. Because matter can be neither created nor destroyed in a chemical reaction, the same number of atoms of each type must appear in both the reactants and the products of a reaction. The coefficients in a balanced chemical equation indicate the relative numbers of moles of reactants and products involved in a reaction. Balanced equations may be used to identify how many moles of product should result from a given quantity of reactant.

The amount of product produced by complete conversion of a reactant in a reaction is called the **theoretical yield** of the reaction. Calculation of the theoretical yield of a reaction requires a knowledge of which reagent acts as the **limiting reagent.** The amount of product that is actually isolated from a reaction, either in the laboratory or in an industrial process, is called the **actual yield** of the reaction. The **percent yield** can be calculated from the actual yield and the theoretical yield.

Key Terms and Concepts

actual yield (2.10)	Avogadro's number (2.2)	empirical weight (2.3)
atomic mass unit, amu (2.2)	chemical equation (2.6)	formula (2.1)
atomic weight (2.2)	empirical formula (2.1)	formula weight (2.3)

gram-atom (2.2)
gram-atomic weight (2.2)
gram-formula weight (2.3)
gram-molecular weight (2.3)
law of definite proportion (2.4)
limiting reagent (2.9)

molar mass (2.3)
mole of atoms (2.2)
mole of molecules (2.3)
molecular formula (2.1)
molecular weight (2.3)

percent composition (2.4)
percent yield (2.10)
symbols (2.1)
theoretical yield (2.10)
yield (2.10)

Exercises

Note: Students sometimes get answers that differ in the last decimal place from those given. This generally reflects different ways of rounding off intermediate steps in the calculation. The answers given have been determined by rounding off to the correct number of significant figures at the end of the calculation. Molecular weights have been calculated using all significant figures in the atomic weights of the constituent atoms.

Symbols, Formulas, and Chemical Equations

1. Name the following elements found in your body: C, H, O, N, P, S, I, Na, Cl, K, Ca, Fe, Br, Cu, Zn, Co, and Mg.
2. Explain why the symbol for the element sulfur and the formula for a molecule of sulfur differ.
3. Determine the empirical formulas of the following compounds: (a) dinitrogen tetraoxide, N_2O_4; (b) hydrazine, N_2H_4; (c) phosphoric acid, H_3PO_4; (d) the anesthetic cyclopropane, C_3H_6; (e) fructose, $C_{12}H_{22}O_{11}$; (f) novocain, $C_{13}H_{21}N_2O_2Cl$; (g) niacin, $C_6H_5NO_2$.
4. Write a balanced equation that describes each of the following chemical reactions: (a) Acetylene gas, C_2H_2; burns in air forming gaseous carbon dioxide, CO_2, and water. (b) When heated, pure aluminum reacts with air to give Al_2O_3. (c) Gypsum, $CaSO_4 \cdot 2H_2O$, decomposes when heated, giving calcium sulfate, $CaSO_4$, and water. (d) Pure iron reacts with sulfuric acid, $H_2SO_4(aq)$, to form a solution of iron sulfate, $Fe_2(SO_4)_3$, in water and gaseous hydrogen, H_2. (e) During photosynthesis in plants, carbon dioxide and water are converted into glucose, $C_6H_{12}O_6$, and oxygen, O_2. (f) Acid rain, a solution of sulfurous acid, $H_2SO_3(aq)$, forms when sulfur dioxide, SO_2, reacts with water. (g) Water vapor reacts with sodium metal to produce gaseous hydrogen, H_2, and solid sodium hydroxide, NaOH. (h) Gaseous water and hot carbon react to form gaseous hydrogen, H_2, and gaseous carbon monoxide, CO.
5. Balance the following equations:
   (a) $Pt + Cl_2 \longrightarrow PtCl_4$
   (b) $H_2 + S_8 \longrightarrow H_2S$
   (c) $S + O_2 \longrightarrow SO_2$
   (d) $Sc_2O_3 + SO_3 \longrightarrow Sc_2(SO_4)_3$
   (e) $Al + H_2SO_4 \longrightarrow Al_2(SO_4)_3 + H_2$
   (f) $KClO_3 \longrightarrow KCl + O_2$
   (g) $CuCO_3 + HCl \longrightarrow CuCl_2 + H_2O + CO_2$
   (h) $PCl_5 + H_2O \longrightarrow POCl_3 + HCl$
6. Balance the following equations:
   (a) $FeCl_2 + Cl_2 \longrightarrow FeCl_3$
   (b) $ZrCl_4 + H_2O \longrightarrow ZrO_2 + HCl$
   (c) $Fe + H_2O \longrightarrow Fe_3O_4 + H_2$
   (d) $P_4 + O_2 \longrightarrow P_4O_{10}$
   (e) $Pb + H_2O + O_2 \longrightarrow Pb(OH)_2$
   (f) $Ag + H_2S + O_2 \longrightarrow Ag_2S + H_2O$
   (g) $Ca_3(PO_4)_2 + C \xrightarrow{\triangle} Ca_3P_2 + CO$
   (h) $Ca_3(PO_4)_2 + H_3PO_4 \longrightarrow Ca(H_2PO_4)_2$
7. Determine both the formula and the number of moles of the missing substances in the following equations:
   (a) $Zn + CuSO_4 \longrightarrow \underline{\hspace{1.5cm}} + ZnSO_4$
   (b) $N_2 + \underline{\hspace{1.5cm}} \longrightarrow 2NH_3$
   (c) $2K + 2H_2O \longrightarrow 2KOH + \underline{\hspace{1.5cm}}$
   (d) $2C_6H_6 + \underline{\hspace{1.5cm}} \longrightarrow 12CO_2 + 6H_2O$
   (e) $2MnO_2 + 4KOH + \underline{\hspace{1.5cm}} \longrightarrow$
   $ 2H_2O + 2K_2MnO_4$
   (f) $2AgNO_3 + 2\underline{\hspace{1.5cm}} \longrightarrow$
   $ Ag_2O + H_2O + 2NaNO_3$

Atomic and Molecular Weights, Moles

8. If the mass of a ^{12}C atom were redefined to be exactly 18 nmu (new mass units), what would the mass of an average oxygen atom be in nmu?   *Ans. 24 nmu*
9. A sample of CsCl contains 0.55418 g of Cs and 0.14783 g of Cl. From this information, calculate the atomic weight of Cs, using 35.453 as the atomic weight of Cl.   *Ans. 132.90*
10. Mercury bromide, $HgBr_2$, contains 55.658% Hg and 44.342% Br. The atomic weight of Br is 79.904. Calculate the atomic weight of mercury from these data.   *Ans. 200.59*
11. Determine the molecular weight of each of the following compounds:
    (a) Hydrogen bromide, HBr   *Ans. 80.912*
    (b) Methane, CH_4 *Ans. 16.043*
    (c) Sulfuric acid, H_2SO_4 *Ans. 98.07*
    (d) Sodium hydroxide, NaOH   *Ans. 39.9971*
    (e) Hydrogen peroxide, H_2O_2 *Ans. 34.0146*
    (f) Propane, C_3H_8 *Ans. 44.096*
    (g) Aspirin, $C_6H_4(CO_2H)(CO_2CH_3)$ *Ans. 180.160*

12. Calculate the mass, in grams, of one mole of each of the following compounds:
    (a) Hydrogen fluoride, HF   *Ans. 20.0063 g*
    (b) Ammonia, NH_3 *Ans. 17.0304 g*
    (c) Nitric acid, HNO_3 *Ans. 63.0128 g*
    (d) Silver sulfate, Ag_2SO_4 *Ans. 311.79 g*
    (e) Acetic acid, CH_3CO_2H *Ans. 60.052 g*
    (f) Boric acid, $B(OH)_3$ *Ans. 61.83 g*
    (g) Caffeine, $C_8H_{10}N_4O_2$ *Ans. 194.193 g*

13. Determine the number of moles in each of the following:
    (a) 4.26 g of ammonia, NH_3 *Ans. 0.250 mol*
    (b) 113.5 g of potassium cyanide, KCN
          Ans. 1.743 mol
    (c) 30.0 g of ethylene, C_2H_4 *Ans. 1.07 mol*
    (d) 1.95×10^{-2} g of the amino acid glycine, $CH_2(NH_2)CO_2H$ *Ans. 2.60×10^{-2} mol*
    (e) 14.9 g of the insecticide Paris green, $Cu_3(AsO_3)_2 \cdot Cu(CH_3CO_2)_2$ *Ans. 0.0241 mol*

14. Determine the mass, in grams, of the following:
    (a) 1.78×10^2 mol of potassium bromide, KBr
          Ans. 2.12×10^4 g
    (b) 2.35 mol of calcium carbonate, $CaCO_3$
          Ans. 235 g
    (c) 1.518×10^{-3} mol of phosphoric acid, H_3PO_4
          Ans. 0.1488 g
    (d) 1.07 mol of acetylene, C_2H_2 *Ans. 27.9 g*
    (e) 0.22930 mol of aluminum sulfate, $Al_2(SO_4)_3$
          Ans. 78.452 g
    (f) 6.437×10^{-4} mol of caffeine, $C_8H_{10}N_4O_2$
          Ans. 0.1250 g

15. The approximate minimum daily nutritional requirement of the amino acid leucine, $C_6H_{13}NO_2$, is 8.4 millimoles. What is this requirement in grams?   *Ans. 1.1 g*

16. Up to 0.0100 percent by mass of copper(I) iodide, CuI, may be added to table salt as a dietary source of iodine. How many moles of CuI could be contained in 1.00 lb (454 g) of table salt?   *Ans. 2.38×10^{-4} mol*

17. The herbicide Treflan, $C_{13}H_{16}N_2O_4F_3$, is applied at the rate of 450 g (two significant figures) per acre to control weeds in corn, cantaloupes, cotton, and other crops. What is this application in moles per acre? In molecules per cm^2? (1 acre = 4.047×10^7 cm^2)
          Ans. 1.4 mol per acre; 2.1×10^{16} molecules per cm^2

18. Determine the mass in grams of the following:
    (a) 0.600 mol of oxygen atoms   *Ans. 9.60 g*
    (b) 0.600 mol of oxygen molecules, O_2 *Ans. 19.2 g*
    (c) 0.600 mol of ozone molecules, O_3 *Ans. 28.8 g*

19. The approximate minimum daily human nutritional requirement of the amino acid lysine, $C_6H_{14}N_2O_2$, is 8.0 g. What is this requirement in millimoles?
          Ans. 55 mmol

20. An average 55 kg woman contains 7.5×10^{-3} mole of hemoglobin (molecular weight: 64,456) in her blood. What is this quantity in grams? In pounds?
          Ans. 480 g; 1.1 lb

21. Determine the moles of hydrogen atoms, the moles of uranium atoms, the moles of water molecules, and the moles of oxygen atoms in 0.100 mol of the mineral carnotite, $K_2(UO_2)_2(VO_4)_2 \cdot 3H_2O$.
          Ans. 0.600 mol H atoms; 0.200 mol U atoms; 0.300 mol H_2O molecules; 1.5 mol O atoms

22. Determine the grams of zirconium and of oxygen in 0.3384 mol of the mineral zircon, $ZrSiO_4$.
          Ans. 30.87 g Zr; 21.66 g O

23. Determine the number of
    (a) grams of sulfur in 0.16 mol SO_3 *Ans. 5.1 g*
    (b) grams of magnesium in 7.52 mol MgS   *Ans. 183 g*
    (c) grams of carbon in 0.01008 mol of novocain, $C_{13}H_{21}N_2O_2Cl$ *Ans. 1.574 g*
    (d) moles of hydrochloric acid in 785.4 g of hydrochloric acid, HCl   *Ans. 21.54 mol*
    (e) moles of oxygen atoms in 51.7 g of gypsum, $MgSO_4 \cdot 2H_2O$ *Ans. 1.98 mol*
    (f) moles of carbon atoms in 0.6163 g of niacin, $C_6H_5NO_2$ *Ans. 3.004×10^{-2} mol*

24. Determine which of the following contains the greatest mass of hydrogen: 1 mol of CH_4; 0.4 mol of C_3H_8; 1.5 mol of H_2.   *Ans. 1 mol CH_4*

25. Determine which of the following contains the greatest mass of iron: 27 g of Fe; 122 g of $FePO_4$; 266 g of $FeCl_3$; 225 g of FeS_2.   *Ans. 225 g FeS_2*

26. Determine which of the following contains the largest mass of carbon atoms: 0.10 mol of glucose, $C_6H_{12}O_6$; 3.0 g of ethane, C_2H_6; a 1.0-g diamond (diamond is pure carbon).   *Ans. 0.10 mol $C_6H_{12}O_6$*

27. Determine
    (a) which has the greatest mass in one mole of atoms: hydrogen; carbon; nitrogen; fluorine.   *Ans. Fluorine*
    (b) which has the greatest mass in one mole of molecules: N_2; O_3; P_2; Ne.   *Ans. P_2*

28. Determine which of the following contains the greatest total numbers of atoms: 15.0 g of aluminum metal; 1.25×10^{23} molecules of carbon monoxide, CO; 0.10 mol of carbon tetrachloride, CCl_4; 5 g of sulfur dioxide, SO_2.
          Ans. 15.0 g Al

Empirical Formulas and Percent Composition

29. Calculate the percent composition of each of the following compounds to three significant figures.
    (a) Potassium chloride, KCl   *Ans. 52.4% K; 47.6% Cl*
    (b) Nitrogen trifluoride, NF_3 *Ans. 19.7% N; 80.3% F*
    (c) Chromium(III) oxide, Cr_2O_3
          Ans. 68.4% Cr; 31.6% O
    (d) Formaldehyde, CH_2O
          Ans. 40.0% C; 6.71% H; 53.3% O
    (e) Cryolite, Na_3AlF_6
          Ans. 32.9% Na; 12.9% Al; 54.3% F
    (f) Nickel diphosphate, $Ni_2P_2O_7$
          Ans. 40.3% Ni; 21.3% P; 38.4% O

30. Determine the percent composition of each of the following to three significant figures:
    (a) Ethyl alcohol, CH_3CH_2OH
        Ans. 52.1% C; 13.1% H; 34.7% O
    (b) Vitamin C, ascorbic acid, $C_6H_8O_6$
        Ans. 40.9% C; 54.5% O; 4.58% H
    (c) Trinitrotoluene, TNT, $C_6H_2(CH_3)(NO_2)_3$
        Ans. 37.0% C; 2.22% H; 42.3% O; 18.5% N
    (d) Codeine, $C_{18}H_{21}NO_3$
        Ans. 72.2% C; 7.07% H; 4.68% N; 16.0% O
    (e) Aspirin, $C_6H_4(CO_2H)(CO_2CH_3)$
        Ans. 60.0% C; 4.48% H; 35.5% O

31. Which of the following fertilizers contains the highest percentage of nitrogen: urea, N_2H_4CO; ammonium nitrate, NH_4NO_3; ammonium sulfate, $(NH_4)_2SO_4$?   *Ans. Urea*

32. Determine each of the following to three significant figures:
    (a) The percent of water in plaster of Paris, $(CaSO_4)_2 \cdot H_2O$ *Ans. 6.21%*
    (b) The percent of sulfate ion, SO_4^{2-}, in Ag_2SO_4
        Ans. 30.8%
    (c) The percent of phosphate ion, PO_4^{3-}, in $Ca_3(PO_4)_2$
        Ans. 61.2%

33. The most common flavoring agent added to food contains 39.3% Na and 60.7% Cl. Determine the empirical formula of this compound.   *Ans. NaCl (salt)*

34. The light-emitting diode used in some calculator displays contains a compound composed of 69.24% Ga and 30.76% P. What is the empirical formula of this compound?   *Ans. GaP*

35. A gaseous compound that contains 27% C and 73% O is a byproduct of winemaking. What is the empirical formula of this gas?   *Ans. CO_2*

36. Determine the empirical formula of sodium nitrite, a compound containing 33.32% Na, 20.30% N, and 46.38% O, which is used to preserve bacon and other meat.
    Ans. $NaNO_2$

37. The spines of some sea urchins are composed of a mineral that contains 40.0% Ca, 12.0% C, and 48.0% O. What is this mineral?   *Ans. $CaCO_3$*

38. Most polymers are very large molecules composed of simple units repeated many times. Thus they often have relatively simple empirical formulas. Determine the empirical formulas of the following polymers:
    (a) Polyethylene; 86% C, 14% H   *Ans. CH_2*
    (b) Polystyrene; 92.3% C, 7.7% H   *Ans. CH*
    (c) Saran; 24.8% C, 2.0% H, 73.1% Cl   *Ans. CHCl*
    (d) Orlon; 67.9% C, 5.70% H, 26.4% N   *Ans. C_3H_3N*
    (e) Lucite (plexiglas); 59.9% C, 8.06% H, 32.0% O
        Ans. $C_5H_8O_2$
    (f) PVC (polyvinylchloride); 38.5% C, 56.7% Cl; 4.8% H   *Ans. C_2H_3Cl*

39. Determine the empirical and molecular formulas of a compound with a molecular weight of 30 that contains 93.3% N and 6.7% H.   *Ans. NH; N_2H_2*

40. Dichloroethane, a compound containing carbon, hydrogen, and chlorine, is often used for dry cleaning. The molecular weight of dichloroethane is 99. A sample of dichloroethane is found to contain 24.3% carbon and 71.6% chlorine. What are its empirical and molecular formulas?   *Ans. CH_2Cl; $C_2H_4Cl_2$*

41. Determine the empirical and molecular formulas of a compound with a molecular weight of 70 that contains 85.6% carbon and 14.4% hydrogen?   *Ans. CH_2; C_5H_{10}*

42. Plaster of Paris contains 6.2% water and 93.8% $CaSO_4$. What is the empirical formula of Plaster of Paris?
    Ans. $(CaSO_4)_2 \cdot H_2O$ or $Ca_2(SO_4)_2 \cdot H_2O$

43. Nicotine contains 74.9% carbon, 8.7% hydrogen, and 17.3% nitrogen. It is known that this compound contains two nitrogen atoms per molecule. What are the empirical and molecular formulas of nicotine?
    Ans. C_5H_7N; $C_{10}H_{14}N_2$

44. A 27.5-g sample of a compound containing carbon and hydrogen contains 5.5 g of hydrogen. What is the empirical formula of this compound?   *Ans. CH_3*

45. Determine the empirical formula of gold chloride if a 126-mg sample contains 106 mg of gold.   *Ans. AuCl*

46. A 1.728-g magnesium carbonate sample decomposes upon heating, giving gaseous CO_2 and a residue of 0.821 g of MgO. Determine the empirical formula of this compound.   *Ans. $MgCO_3$*

47. Hemoglobin (molecular weight: 64,456) contains 0.35% iron by mass. How many iron atoms are present in a hemoglobin molecule?   *Ans. 4*

48. What is the product formed when 0.709 g of iron reacts with $Cl_2(g)$ to give 1.610 g of a sample containing iron and chlorine? Identify the product and write the balanced equation for its formation.   *Ans. $Fe + Cl_2 \longrightarrow FeCl_2$*

49. The 1.610-g sample of iron(II) chloride described in exercise 48 reacts with 0.450 g of chlorine to give a different iron chloride. Determine the empirical formula of this new iron chloride.   *Ans. $FeCl_3$*

Chemical Calculations Involving Equations

50. If 28.1 g of Si reacts with N_2, giving 0.333 mole of Si_3N_4 according to the equation

    $$3Si + 2N_2 \longrightarrow Si_3N_4$$

    which of the following can be determined by using only the information given in this exercise: (a) the moles of Si reacting, (b) the moles of N_2 reacting, (c) the atomic weight of N, (d) the atomic weight of Si, (e) the mass of Si_3N_4 produced?

51. (a) How many moles of gallium chloride are formed by the reaction of 1.5 mol of HCl according to the following equation:

    $$2Ga + 6HCl \longrightarrow 2GaCl_3 + 3H_2$$

    1.5 mol

    Ans. 0.50 mol

(b) What mass of $GaCl_3$, in grams, is produced?

Ans. 88 g

52. (a) How many moles of H_2 are produced by the reaction of 1.24 mol of H_3PO_4 in the following reaction?

$$2Cr + 2H_3PO_4 \longrightarrow 3H_2 + 2CrPO_4$$

Ans. 1.86 mol

(b) What mass of H_2, in grams, is produced?

Ans. 3.75 g

53. (a) How many molecules of I_2 are produced by the reaction of 0.3600 mol of $CuCl_2$ in the following reaction?

$$2CuCl_2 + 4KI \longrightarrow 2CuI + 4KCl + I_2$$

Ans. 1.084×10^{23} I_2 molecules

(b) What mass of I_2 is produced?   *Ans. 45.69 g*

54. Thin films of silicon, Si, used for fabrication of electronic components, may be prepared by the decomposition of silane, SiH_4:

$$SiH_4(g) \xrightarrow{\triangle} Si(s) + H_2(g)$$

What mass of SiH_4 is required to prepare 0.2173 g of Si by this technique?   *Ans. 0.2485 g*

55. Silicon carbide, SiC, a very hard material used as an abrasive on sandpaper and in other applications, is prepared by the reaction of pure sand, SiO_2, with carbon at high temperature. Carbon monoxide, CO, is the other product of this reaction. Write the balanced equation for the reaction and calculate how much SiO_2 is required to produce 1.00 kg of SiC.

Ans. $SiO_2 + 3C \longrightarrow SiC + 2CO$; 1.50 kg

56. Calcium chloride 6-hydrate, $CaCl_2 \cdot 6H_2O$, is a solid that is used to melt ice and snow. What mass, in kg, of $CaCl_2 \cdot 6H_2O$ can be prepared from 100 lb (45.4 kg) of $CaCO_3$ (limestone)?

$$CaCO_3 + 2HCl + 5H_2O \longrightarrow CaCl_2 \cdot 6H_2O + CO_2$$

Ans. 99.4 kg

57. Tooth enamel consists of hydroxyapatite, $Ca_5(PO_4)_3(OH)$. This substance is converted to the more decay-resistant fluorapatite, $Ca_5(PO_4)_3F$, by treatment with tin(II) fluoride, SnF_2 (commonly referred to as stannous fluoride). Products of this reaction are SnO and water. What mass of hydroxyapatite can be converted to fluorapatite by reaction with 0.100 g of SnF_2?   *Ans. 0.644 g*

58. Silver is often extracted from ores as $KAg(CN)_2$ and then recovered by the reaction

$$2KAg(CN)_2(aq) + Zn(s) \longrightarrow$$
$$2Ag(s) + Zn(CN)_2(aq) + 2KCN(aq)$$

What mass of zinc, in grams, is required to produce one ounce (28.35 g) of silver?   *Ans. 8.592 g*

59. Silver sulfadiazine burn cream creates a barrier against bacterial invasion and releases antimicrobial agents directly into a wound. What mass of silver oxide would be required to prepare 225 g of silver sulfadiazine, $AgC_{10}H_9N_4SO_2$, from sulfadiazine, $C_{10}H_{10}N_4SO_2$?

$$Ag_2O + 2C_{10}H_{10}N_4SO_2 \longrightarrow 2AgC_{10}H_9N_4SO_2 + H_2O$$

Ans. 73.0 g

Limiting Reagents

60. What is the limiting reagent when 0.25 mol of P_4 and 0.25 mol of O_2 react according to the following equation?

$$P_4 + 5O_2 \longrightarrow P_4O_{10}$$

Ans. O_2

61. What is the limiting reagent when 0.25 mol of Cr and 0.50 mol of H_3PO_4 react according to the following reaction?

$$2Cr + 2H_3PO_4 \longrightarrow 2CrPO_4 + 3H_2$$

Ans. Cr

62. What is the limiting reagent when 1.00 g of Si and 1.00 g of N combine according to the following reaction?

$$3Si + 2N_2 \longrightarrow Si_3N_4$$

Ans. Si

63. What is the limiting reagent when 10.0 g of propane, C_3H_8, is burned with 25 g of oxygen?

$$C_3H_8(g) + 5O_2(g) \longrightarrow 3CO_2(g) + 4H_2O(g)$$

Ans. O_2

64. What mass of $[Cr(H_2O)_6]Cl_3$ is produced when a solution containing 2.5×10^{-3} mol of $H_2Cr_2O_7$ is added to a solution containing 7.5×10^{-3} mol of H_2FeCl_4? The two react according to the following equation:

$$H_2Cr_2O_7 + 6H_2FeCl_4 + 5H_2O \longrightarrow$$
$$2[Cr(H_2O)_6]Cl_3 + 6FeCl_3$$

Ans. 0.67 g ·

Percent Yield

65. A student spilled his ethanol preparation and consequently isolated only 15 g instead of the 47 g theoretically possible. What was his percent yield?   *Ans. 32%*

66. A sample of calcium oxide, CaO, weighing 0.69 g was prepared by heating 1.31 g of calcium carbonate. What was the percent yield of the reaction?

$$CaCO_3(s) \xrightarrow{\triangle} CaO(s) + CO_2(g)$$

Ans. 94%

67. Toluene, $C_6H_5CH_3$, is oxidized by air under carefully controlled conditions to benzoic acid, $C_6H_5CO_2H$, used to prepare the food preservative sodium benzoate, $C_6H_5CO_2Na$. What is the yield of a reaction that converts 1.000 kg of toluene to 1.21 kg of benzoic acid?

$$2C_6H_5CH_3 + 3O_2 \longrightarrow 2C_6H_5CO_2H + 2H_2O$$

Ans. 91.3%

68. Freon-12, CCl_2F_2, is prepared from CCl_4 by reaction with HF. The other product of this reaction is HCl. What is the percent yield of a reaction that produces 12.5 g of CCl_2F_2 from 32.9 g of CCl_4?   *Ans. 48.3%*

69. Ether, $(C_2H_5)_2O$, for anesthetic use is prepared by the reaction of ethanol with sulfuric acid.

$$2C_2H_5OH + H_2SO_4 \longrightarrow (C_2H_5)_2O + H_2SO_4 \cdot H_2O$$

What is the percent yield if 1.00 kg of C_2H_5OH yields 0.766 kg of $(C_2H_5)_2O$?   *Ans. 95.2%*

70. Citric acid, $C_6H_8O_7$, a component of jams, jellies, and fruity soft drinks, is prepared industrially by fermentation of sucrose by the mold *Aspergillus niger*. The overall reaction is

$$C_{12}H_{22}O_{11} + H_2O + 3O_2 \longrightarrow 2C_6H_8O_7 + 4H_2O$$

What is the amount, in kilograms, of citric acid produced from 1 metric ton (1.000×10^3 kg) of sucrose in a reaction if the yield is 92.3%?   *Ans. 1.036×10^3 kg*

71. If 0.20 mol of $CrPO_4$ is recovered from the reaction described in exercise 61, what is the percent yield?
Ans. 80%

72. The reaction of 3.0 mol of H_2 with 2.0 mol of I_2 produces 1.0 mol of HI ($H_2 + I_2 \longrightarrow 2HI$). Determine the theoretical yield in moles and the percent yield for this reaction.   *Ans. 4.0 mol; 25%*

73. If the yield of Si_3N_4 recovered from the reaction described in exercise 62 is 98%, what mass of Si_3N_4 is recovered?
Ans. 1.6 g

74. Addition of 0.403 g of sodium oxalate, $Na_2C_2O_4$, to a solution containing 1.48 g of uranyl nitrate, $UO_2(NO_3)_2$, yields 1.073 g of solid $UO_2(C_2O_4) \cdot 3H_2O$.

$$UO_2(NO_3)_2(aq) + Na_2C_2O_4(aq) + 3H_2O \longrightarrow$$
$$UO_2(C_2O_4) \cdot 3H_2O(s) + 2NaNO_3(aq)$$

Determine the limiting reagent and percent yield of this reaction.   *Ans. $Na_2C_2O_4$; 86.6%*

75. What is the percent yield of $NaClO_2$ if 106 g is isolated from the reaction of 202.3 g of ClO_2 with a solution containing 3.22 mol of NaOH?

$$2ClO_2(g) + 2NaOH(aq) \longrightarrow$$
$$NaClO_2(aq) + NaClO_3(aq) + H_2O(l)$$
Ans. 78.2%

Additional Exercises

76. Determine the simplest formula of indium bromide if a sample of indium bromide weighing 0.100 g reacts with silver nitrate, $AgNO_3$, giving indium nitrate and 0.159 g of AgBr.   *Ans. $InBr_3$*

77. The principal component of mothballs is naphthalene, a compound with a molecular weight of about 130, containing only carbon and hydrogen. A 3.000-mg sample of naphthalene burns to give 10.3 mg of CO_2. Determine its empirical and molecular formula.   *Ans. C_5H_4; $C_{10}H_8$*

78. The equation for the preparation of phosphorus, P_4, in an electric furnace is

$$2Ca_3(PO_4)_2 + 6SiO_2 + 10C \longrightarrow 6CaSiO_3 + 10CO + P_4$$

What mass of SiO_2 is required to prepare 1.000 kg of P_4?   *Ans. 2.910 kg*

79. The empirical formula for an ion-exchange resin used in water softeners is $C_8H_7SO_3Na$. The resin softens water by removing calcium ion, Ca^{2+}, by the reaction

$$Ca^{2+} + 2C_8H_7SO_3Na \longrightarrow (C_8H_7SO_3)_2Ca + 2Na^+$$

What mass of ion-exchange resin, in grams, is required to soften 1000 gal of water with a Ca^{2+} content of 2.50 ppm (2.50 g Ca^{2+} per 1,000,000 g water)? One gallon of water has a mass of 8.0 lb.   *Ans. 93 g*

80. Sulfur may be removed from coal by washing powdered coal with NaOH(aq). The reaction may be represented as

$$R_2S(s) + 2NaOH(aq) \longrightarrow$$
$$R_2O(s) + Na_2S(aq) + H_2O(l)$$

where R_2S represents the organic sulfur present in the coal. What mass of NaOH(s), in kilograms, is required to react with the sulfur in 1.00 metric ton (1000 kg) of coal that contains 1.6% S by mass?   *Ans. 40 kg*

81. Concentrated sulfuric acid with a density of 1.84 g/cm^3 contains 97.5% H_2SO_4 by mass. How many liters of this acid is required to manufacture 1000 kg of phosphate fertilizer, $Ca(H_2PO_4)_2$, according to the reaction

$$2Ca_5(PO_4)_3F + 7H_2SO_4 + 14H_2O \longrightarrow$$
$$2Ca(H_2PO_4)_2 + 7(CaSO_4 \cdot 2H_2O) + 2HF$$
Ans. 817 L

82. Sodium bicarbonate, $NaHCO_3$, baking soda, can be purified by dissolving it in hot water (60°C), filtering to remove insoluble impurities, cooling to 0°C to precipitate solid $NaHCO_3$, and then filtering to remove the solid $NaHCO_3$, leaving soluble impurities in solution. Any $NaHCO_3$ that remains in solution is not recovered. The solubility of $NaHCO_3$ in hot water at 60°C is 164 g/L. The solubility in cold water at 0°C is 69 g/L. What is the percent yield of $NaHCO_3$ when purified by this method?   *Ans. 58%*

83. A volume of 2.27 mL of $SiCl_4$ (density = 1.483 g/mL) reacts with an excess of $H_2S(g)$, giving $HSSiCl_3$ according to the reaction

$$SiCl_4(l) + H_2S(g) \longrightarrow HSSiCl_3(l) + HCl(g)$$

If the HCl produced reacts with 8.267×10^{-3} mol of NaOH according to the reaction

$$HCl + NaOH \longrightarrow NaCl + H_2O$$

what is the percent yield of $HSSiCl_3$?   *Ans. 41.7%*

Applications of Chemical Stoichiometry

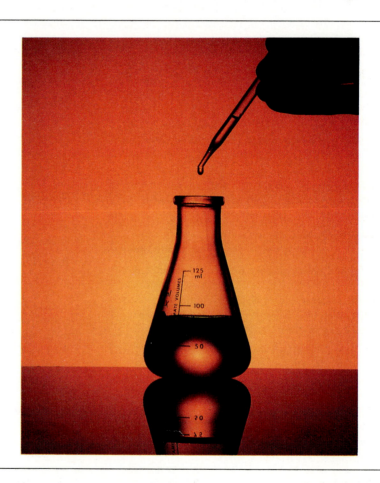

An indicator is added before titrating.

3

Chemistry is a quantitative science involving a great deal of calculation. However, the purpose of these calculations is not merely to get a number, but also to provide quantitative predictions or descriptions of how matter behaves. A correct calculation suggests that the theory used in the calculation is correct. For example, the ability to calculate accurate volumes of gas samples is one reason for believing that the gas laws (Chapter 10) are correct. Correctly calculating the amount of product that is produced in a chemical reaction shows that we understand something about what is going on in the reaction, and it allows us to use the reaction to produce a desired amount of a product. Calculations are tools that chemists use in understanding and using the behavior of matter.

One category of chemical calculations involves relationships between quantities of reactants and products based upon the mole relations given in the balanced equation for their reaction. These are calculations of the type introduced in Chapter 2. Determination of the masses of reactants required for a reaction, choosing the amount of reactant necessary to give a desired amount of product, calculation of the concentration of a solution, determination of percent composition as a check for purity, and identification of the empirical and molecular formula of an unknown product all fall into this category, which is generally called **chemical stoichiometry.** In this chapter, we will be concerned with a detailed treatment of chemical stoichiometry.

3.1 Solutions

In Chapter 2 we used chemical equations to relate moles of reactants to moles of products, and we saw how to measure out moles of substances by weighing. In the following two sections we will describe how chemical equations can be used to relate quantities of materials present in solution.

When sugar is stirred into water, the sugar dissolves and a clear mixture, or **solution,** of sugar in water is formed. This solution consists of the **solute** (the dissolved sugar) and the **solvent** (the water). A sugar solution that contains only a small amount of sugar in comparison with the amount of water is said to be **dilute;** the addition of more sugar makes the solution more **concentrated.** The concentration limits of a solution depend on the nature of the substance dissolved (the solute) and of the medium in which the solute is dissolved (the solvent). The solute or the solvent can be a gas, a liquid, or a solid, but in this chapter we will limit our attention to solutes dissolved in liquids. Solutions will be discussed in greater detail in Chapter 12.

The relative amounts of solute and solvent present in a solution, that is, its **concentration,** can be expressed as the molarity of the solution. **The molarity, M, of a solution is the number of moles of solute per exactly 1 liter (or dm^3) of solution.** To calculate molarity, divide the number of moles of solute in a given volume of solution by the volume in liters.

$$\text{Molarity} = \frac{\text{moles of solute}}{\text{liters of solution}} \qquad (1)$$

If 342 g of sucrose, 1.00 mol $C_{12}H_{22}O_{11}$, is placed in a volumetric flask and enough water is added to make 1.00 L of solution, a 1.00 M solution results.

EXAMPLE 3.1

The acid secreted by the cells of the stomach lining is a hydrochloric acid solution that typically contains 0.00774 mol of HCl per 0.0500 L of solution. What is the concentration of this acid?

The concentration of the acid can be calculated using Equation (1). Divide the moles of HCl by the volume in liters.

$$\text{Molarity} = \frac{\text{moles of HCl}}{\text{liters of solution}}$$

$$= \frac{0.00774 \text{ mol}}{0.0500 \text{ L}}$$

$$= 0.155 \text{ M, or } 0.155 \text{ mol/L, or } 0.155 \text{ mol/dm}^3$$

For calculations it may be convenient to write this molarity as the unit conversion factor 0.155 mol HCl/1 L.

This example presented the volume in liters, but many experimental measurements are in milliliters and must be converted to liters before being used to find molarities.

EXAMPLE 3.2   How many moles of sulfuric acid, H_2SO_4, are contained in 0.80 L of a 0.050 M solution of sulfuric acid?

Equation (1) can be rearranged to solve for moles of solute:

$$\text{Molarity} \times \text{liters of solution} = \text{moles of solute}$$

$$\frac{0.050 \text{ mol } H_2SO_4}{1 \text{ L}} \times 0.80 \text{ L} = 0.040 \text{ mol } H_2SO_4$$

EXAMPLE 3.3   What mass of iodine is required to make 5.00×10^2 mL of a 0.250 M solution of I_2 with chloroform, $CHCl_3$, as the solvent?

The moles of I_2 necessary to make a 0.250 M solution can be found by rearranging Equation (1) as shown in Example 3.2. The weight of I_2 required can then be calculated. The following steps are involved in the calculation:

| Volume of solution | → | Moles of I_2 | → | Mass of I_2 |

Note that the volume is given in milliliters and must be converted to liters.

$$5.00 \times 10^2 \text{ mL} \times \frac{1 \text{ L}}{1000 \text{ mL}} = 0.500 \text{ L}$$

$$\frac{0.250 \text{ mol } I_2}{1 \text{ L}} \times 0.500 \text{ L} = 0.125 \text{ mol } I_2$$

$$0.125 \text{ mol } I_2 \times \frac{253.8 \text{ g } I_2}{1 \text{ mol } I_2} = 31.7 \text{ g } I_2$$

Although the logic of this problem is perhaps best approached in three steps, the calculations could have been written in one step, as follows:

$$5.00 \times 10^2 \text{ mL} \times \frac{1 \text{ L}}{1000 \text{ mL}} \times \frac{0.250 \text{ mol } I_2}{1 \text{ L}} \times \frac{253.8 \text{ g } I_2}{1 \text{ mol } I_2} = 31.7 \text{ g } I_2$$

All of the examples in this chapter have the calculations written as separate steps to help show the logic involved in solving them. However, all of these calculations could be written as a single step. When writing single-step calculations, it is particularly important to check that the units cancel in order to avoid inversion of a term.

3.2 Titration

A known volume of a solution is usually measured with a pipet or a buret. A **pipet** is a tube that will deliver a known fixed volume of a liquid when filled to a reference line (Fig. 3.1). A **buret** is a cylinder (Fig. 3.2) that is graduated in fractions of a milliliter so that the volume of a liquid delivered from it may be accurately measured.

Burets and pipets are used to determine the concentrations of solutions (to standardize solutions) by **titration** (Fig. 3.3). During a titration, a buret is used to add a solution of a reactant to a solution of a sample. When the exact amount of reactant necessary to react completely with the sample has been added, an indicator changes color, marking the **end point** of the titration. At this point the titration is stopped, and the amount of solution added is determined by reading the volume of solution delivered from the buret. This volume is used in determining the concentration of the sample solution.

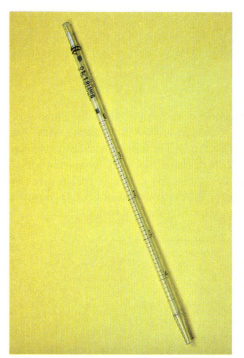

Figure 3.1 (left)
This 5.00 mL pipet will deliver precise volumes if it is filled to a reference line and allowed to drain.

Figure 3.2 (right)
A buret is used for accurately measuring the volume of a solution added during a titration.

Figure 3.3
The end point of the titration of an acid with a base using phenolphthalein as the indicator. (left) As the end point is approached, streaks of color are observed that disappear with mixing. (right) At the end point, one drop from the buret produces a permanent color change.

Just as chemical equations relate the masses of reactants and products of chemical reactions, they also relate concentrations or volumes of solutions of reactants and products. A chemical equation gives the ratios of moles of reactants and products entering into a chemical reaction. If these reactants or products are dissolved, Equation (1) in Section 3.1 can be used to relate the molarity and volume of the solution to the number of moles present.

EXAMPLE 3.4

A solution of HCl in water is standardized by titrating a solution of a 0.5015-g sample of pure, dry sodium carbonate, Na_2CO_3. What is the concentration of the HCl solution if 48.47 mL is added from a buret to reach the end point?

$$Na_2CO_3 + 2HCl \longrightarrow 2NaCl + CO_2 + H_2O$$

To determine the concentration of the HCl solution, we need to know the number of moles of HCl in some volume of the solution. At the end of the titration, we know that we have added 48.47 mL of the solution and that the HCl in it has reacted exactly with 0.5015 g of Na_2CO_3. The balanced equation tells us that 2 mol of HCl has been added for every 1 mol of Na_2CO_3 present. Thus the number of moles of HCl present in 48.47 mL of solution can be determined from the number of moles of Na_2CO_3 titrated. The concentration of the solution can be determined by the following steps:

$$\boxed{\text{Mass of } Na_2CO_3} \longrightarrow \boxed{\text{Moles of } Na_2CO_3} \longrightarrow \boxed{\text{Moles of } HCl} \longrightarrow \boxed{\text{Molarity of solution}}$$

$$0.5015 \text{ g } Na_2CO_3 \times \frac{1 \text{ mol } Na_2CO_3}{105.99 \text{ g } Na_2CO_3} = 4.7316 \times 10^{-3} \text{ mol } Na_2CO_3$$

$$4.7316 \times 10^{-3} \text{ mol } Na_2CO_3 \times \frac{2 \text{ mol } HCl}{1 \text{ mol } Na_2CO_3} = 9.4632 \times 10^{-3} \text{ mol } HCl$$

$$\frac{9.4632 \times 10^{-3} \text{ mol } HCl}{0.04847 \text{ L}} = 0.1952 \text{ M}$$

The concentration of the HCl solution is 0.1952 M.

EXAMPLE 3.5

An acidic solution of oxalic acid, $H_2C_2O_4$, was added to a flask using a 20.00-mL pipet, and the sample was titrated with a 0.09113 M solution of potassium permanganate, $KMnO_4$. A volume of 23.24 mL was required to reach the end point. What was the concentration of the $H_2C_2O_4$ solution?

$$5H_2C_2O_4 + 2KMnO_4 + 3H_2SO_4 \longrightarrow 2MnSO_4 + K_2SO_4 + 10CO_2 + 8H_2O$$

To determine the concentration of $H_2C_2O_4$ in the solution, we need to know the number of moles of $H_2C_2O_4$ in 0.02000 L (20.00 mL) of the solution, the amount delivered by the pipet. The equation tells us that for every 5 mol of $H_2C_2O_4$ present, 2 mol of $KMnO_4$ reacts. The moles of $KMnO_4$ reacting can be determined from the volume and molarity of the $KMnO_4$ solution added during the titration.

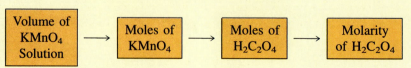

$$0.02324 \text{ L KMnO}_4 \times \frac{0.09113 \text{ mol KMnO}_4}{1 \text{ L KMnO}_4} = 2.1179 \times 10^{-3} \text{ mol KMnO}_4$$

$$2.1179 \times 10^{-3} \text{ mol KMnO}_4 \times \frac{5 \text{ mol H}_2\text{C}_2\text{O}_4}{2 \text{ mol KMnO}_4} = 5.2948 \times 10^{-3} \text{ mol H}_2\text{C}_2\text{O}_4$$

$$\frac{5.2948 \times 10^{-3} \text{ mol H}_2\text{C}_2\text{O}_4}{0.02000 \text{ L}} = 0.2647 \text{ M}$$

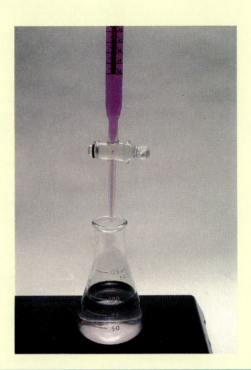

The titration of $H_2C_2O_4$ by addition of $KMnO_4$. As the purple solution is added to the colorless solution, purple $KMnO_4$ is converted to essentially colorless $MnSO_4$ by the reaction with colorless $H_2C_2O_4$. At the instant all of the $H_2C_2O_4$ has been consumed, the addition of more $KMnO_4$ will turn the solution in the flask purple; this indicates the end point of this titration.

3.3 Chemical Calculations Involving Stoichiometry

Stoichiometry is the calculation of both material balances and energy balances in a chemical system. We will discuss material balances in this chapter and energy balances in future chapters. Most stoichiometry problems involve the basic transformations, or conversions, illustrated below. The arrows indicate that one quantity is converted to the next.

Quantity A $\longrightarrow$ Moles A $\longrightarrow$ Moles B $\longrightarrow$ Quantity B

The overall conversion from Quantity A to Quantity B requires three steps (each of which may require one or more unit conversion factors of the types we have previously discussed):

1. conversion of a quantity measured in units like grams or liters into a quantity in moles.

2. determination of how many moles of a second substance are equivalent to the number of moles of the first
3. conversion of the number of moles of the second substance into a mass of that substance or a volume of a solution of a certain molarity

EXAMPLE 3.6   What volume of a 0.750 M solution of hydrochloric acid can be prepared from the HCl produced by the reaction of 25.0 g of NaCl with an excess of sulfuric acid?

$$NaCl(s) + H_2SO_4(l) \longrightarrow HCl(g) + NaHSO_4(s)$$

In order to determine the volume of 0.750 M solution that we can prepare, we need the number of moles of $HCl(g)$ that are available to make the solution. The chemical equation tells us the relationship between the number of moles of gaseous HCl produced and the number of moles of NaCl reacting. The number of moles of NaCl can be determined from the mass of NaCl used and its formula weight. Solution of the problem requires the following steps, each of which uses a single unit conversion factor:

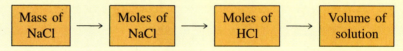

Convert the quantity of NaCl in grams to moles of NaCl. As described in Section 2.3, this conversion requires calculating the formula weight of NaCl, 58.44; thus 1 mol NaCl = 58.44 g NaCl. The conversion is then

$$25.0 \text{ g NaCl} \times \frac{1 \text{ mol NaCl}}{58.44 \text{ g NaCl}} = 0.4278 \text{ mol NaCl}$$

Convert moles of NaCl to moles of HCl. The chemical equation indicates that 1 mol of HCl is produced for every 1 mol of NaCl that reacts. The unit conversion factor is 1 mol HCl/1 mol NaCl (Section 2.7).

$$0.4278 \text{ mol NaCl} \times \frac{1 \text{ mol HCl}}{1 \text{ mol NaCl}} = 0.4278 \text{ mol HCl}$$

Convert moles of HCl to the volume of a 0.750 M solution of HCl. As shown in Section 3.1,

$$\text{Molarity} = \frac{\text{moles of solute}}{\text{liters of solution}}$$

This can be rearranged to solve for the volume of solution:

$$\text{Liters of solution} = \text{moles of solute} \times \frac{1}{\text{molarity}}$$

Since a 0.750 M solution contains 0.750 mol HCl per exactly 1 L (0.750 mol HCl/ 1 L), the reciprocal of the molarity (1/M) is expressed as follows:

$$\frac{1}{\text{molarity}} = \frac{1 \text{ L}}{0.750 \text{ mol HCl}}$$

Thus

$$0.4278 \text{ mol HCl} \times \frac{1 \text{ L}}{0.750 \text{ mol HCl}} = 0.570 \text{ L}$$

This example and those that follow illustrate a basic concept of stoichiometry. **The relative numbers of moles of the various species participating in a reaction are provided by the chemical equation that describes the system** (Section 2.7).

Although Example 3.6 illustrates the basic flow of almost any stoichiometry problem, such calculations often involve several additional steps. These may result from the nature of the material (a mixture rather than a pure compound, for example) used in a reaction or from the various ways of measuring it (measuring a volume rather than a mass, for example). In such cases the additional steps usually involve relating moles to measured quantities.

3.4 Common Operations in Stoichiometry Calculations

Conversion of a measured quantity of a substance to moles of the substance, or the reverse conversion, may involve one or more of the unit conversion factors discussed below. Careful attention to units and their cancellations will be a great help in correctly setting up the calculations.

1. INTERCONVERSION BETWEEN MASS AND MOLES. The number of moles of a substance, A, in a given mass of pure A can be determined by use of a conversion factor, as described in Section 2.3.

$$\text{Mol A} = \text{mass A} \times \frac{1 \text{ mol A}}{\text{molar mass A}}$$

Rearrangement of this expression gives an equation that can be used to find the mass of A that corresponds to a given amount of A in moles.

$$\text{Mass A} = \text{mol A} \times \frac{\text{molar mass A}}{1 \text{ mol A}}$$

2. INTERCONVERSION BETWEEN MASS OF A MIXTURE AND MASS OF A COMPONENT. The percent by mass of a component, A, in a sample of a mixture or a compound is given by the expression

$$\text{Percent A} = \frac{\text{mass A}}{\text{mass sample}} \times 100$$

where 100 is an exact multiplier. When using percent by mass in problems, it is often convenient to use the grams of component A per 100 g of sample (g A/100 g sample); this is numerically equal to the percent of A by mass. For example, a sample of coal that contains 0.23% sulfur by mass contains 0.23 g of sulfur per 100 g of coal.

The mass of A and the mass of the sample can be determined from the following expressions:

$$\text{Mass A} = \text{mass sample} \times \frac{\text{g A}}{100 \text{ g sample}}$$

$$\text{Mass sample} = \text{mass A} \times \frac{100 \text{ g sample}}{\text{g A}}$$

3. INTERCONVERSION BETWEEN VOLUME AND MASS USING DENSITY. In Section 1.13 density was defined as the mass of a sample divided by its volume

(density = mass/volume). The equation may be rearranged to find the mass of a given volume of a substance of known density:

$$\text{Mass} = \text{density} \times \text{volume}$$

or to find the volume of a given mass of a substance of known density:

$$\text{Volume} = \frac{\text{mass}}{\text{density}}$$

EXAMPLE 3.7   The reaction of chlorobenzene, C_6H_5Cl, with a solution of ammonia containing 28.2% NH_3 by mass is used in the preparation of aniline, $C_6H_5NH_2$, a key ingredient in the preparation of dyes for fabrics. What mass of NH_3 is required to prepare 125 mL of the ammonia solution with a density of 0.899 g/cm^3?

Two steps are required to solve this problem: a conversion from volume and density to mass of a mixture (the solution of NH_3), followed by a conversion from the mass of a mixture (the solution of NH_3) to the mass of a pure substance (NH_3).

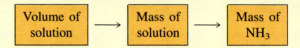

Find the mass of 125 mL of NH_3 solution (interconversion between volume and mass using density). Since 1 mL is the same as 1 cm^3, we express the volume in cm^3.

$$125 \ \text{cm}^3 \ \text{solution} \times \frac{0.899 \ \text{g solution}}{1 \ \text{cm}^3 \ \text{solution}} = 112.4 \ \text{g solution}$$

Now we find the mass of NH_3 in the solution (interconversion between mass of a mixture and mass of a component). A 28.2% NH_3 solution by mass contains 28.2 g NH_3 per 100 g solution.

$$112.4 \ \text{g solution} \times \frac{28.2 \ \text{g} \ NH_3}{100 \ \text{g solution}} = 31.7 \ \text{g} \ NH_3$$

4. INTERCONVERSION BETWEEN VOLUME OF SOLUTION AND MOLES OF SOLUTE USING MOLARITY. The number of moles of a solute, A, in a given volume of a solution of known concentration; the concentration of the solution; and the volume of the solution containing a given number of moles of the solute, A, can each be determined as described in Sections 3.1 and 3.2.

$$\text{Molarity} = \frac{\text{moles A}}{\text{volume of solution}}$$

$$\text{Moles A} = \text{molarity} \times \text{volume of solution}$$

3.5 Applications of Chemical Stoichiometry

Many of the stoichiometry problems that a student sees at the beginning of a chemistry course appear formidable. However, the majority of these problems involve nothing more complicated than applying the unit conversions given above. In most of these problems the first step or steps use unit conversions to transform a quantity of a substance to moles of the substance. The numbers of moles of additional substances that are equivalent to the first are then usually found, and, finally, the amount of one of these additional substances in some unit other than moles is determined.

The difficulty with stoichiometry problems is usually in finding the order in which to do the conversions. Unfortunately, no single technique can be used to string together these conversions for all types of problems. However, there are some useful guidelines. Many problems require that we determine the quantity of B that is equivalent to a given amount of A. In most of these cases, we will need a mole-to-mole conversion (to convert moles of A to moles of B). First, we identify the substances (A and B) to be interconverted. We then find the information that tells how much of substance A is present and the information that tells how to report the amount of substance B. After the problem is broken down into these large steps, we work on the conversions necessary to carry them out. For example, the conversion of a quantity of A to moles of A may require only one conversion, or it may require several.

One final hint: Be certain that you understand how to do each of the various types of single conversions before you combine them in complicated problems.

EXAMPLE 3.8

As shown in the photograph, potassium iodide, KI, reacts with copper nitrate, $Cu(NO_3)_2$, to form a mixture of copper iodide, CuI, and iodine, I_2. What volume of a 0.2089 M solution of KI will contain enough KI to react exactly with the $Cu(NO_3)_2$ in 43.88 mL of a 0.3842 M $Cu(NO_3)_2$ solution according to the following equation?

$$2Cu(NO_3)_2(aq) + 4KI(aq) \longrightarrow 2CuI(s) + I_2(s) + 4KNO_3(aq)$$

This problem asks us to use the volume of a $Cu(NO_3)_2$ solution of known concentration to calculate the volume of a KI solution of known concentration. The overall conversion is

$$\boxed{\text{Volume of } Cu(NO_3)_2 \text{ solution}} \longrightarrow \boxed{\text{Volume of KI solution}}$$

The addition of a solution of KI to a solution of $Cu(NO_3)_2$.

To determine the necessary volume of the KI solution, we need to know the moles of KI required to react. This can be determined from the moles of $Cu(NO_3)_2$ that react [the moles of $Cu(NO_3)_2$ present in 43.88 mL of the $Cu(NO_3)_2$ solution], which we can find by using the conversion factor obtained from the equation. The calculation involves the following steps:

$$\boxed{\text{Volume of } Cu(NO_3)_2 \text{ solution}} \xrightarrow{4} \boxed{\text{Moles of } Cu(NO_3)_2} \longrightarrow \boxed{\text{Moles of KI}} \xrightarrow{4} \boxed{\text{Volume of KI solution}}$$

Note that the volumes must be expressed in liters. (The number above an arrow indicates the paragraph in Section 3.4 that relates to the conversion.) The calculations for each step follow.

$$\frac{0.3842 \text{ mol } Cu(NO_3)_2}{1 \text{ L}} \times 0.04388 \text{ L} = 1.686 \times 10^{-2} \text{ mol } Cu(NO_3)_2$$

$$1.686 \times 10^{-2} \text{ mol } Cu(NO_3)_2 \times \frac{4 \text{ mol KI}}{2 \text{ mol } Cu(NO_3)_2} = 3.372 \times 10^{-2} \text{ mol KI}$$

$$3.372 \times 10^{-2} \text{ mol KI} \times \frac{1 \text{ L}}{0.2089 \text{ mol KI}} = 0.1614 \text{ L}$$

The reaction requires 0.1614 L, or 161.4 mL, of KI solution.

EXAMPLE 3.9   The toxic pigment called white lead, $Pb_3(OH)_2(CO_3)_2$, has been replaced by rutile, TiO_2, in white paints. How much rutile can be prepared from 379 g of an ore containing ilmenite, $FeTiO_3$, if the ore is 88.3% ilmenite by mass? TiO_2 is prepared by the reaction

$$2FeTiO_3 + 4HCl + Cl_2 \longrightarrow 2FeCl_3 + 2TiO_2 + 2H_2O$$

This problem asks how much rutile, TiO_2, can be prepared from an ore containing ilmenite, $FeTiO_3$. The overall conversion is

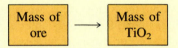

Since most conversions of this type involve mole-to-mole conversions, the chain of conversions for this problem probably includes one, too. The problem states that the $FeTiO_3$ in an ore is converted into TiO_2, so the chain of conversions should include a conversion of moles of $FeTiO_3$ to moles of TiO_2.

| Mass of ore | → | Moles of FeTiO₃ | → | Moles of TiO₂ | → | Mass of TiO₂ |

Now it only remains to convert the mass of ore to the mass of $FeTiO_3$ in the ore. The final calculation involves the following unit conversions:

| Mass of ore | →² | Mass of FeTiO₃ | →¹ | Moles of FeTiO₃ | → | Moles of TiO₂ | →¹ | Mass of TiO₂ |

(The number above an arrow indicates the paragraph in Section 3.4 that relates to the conversion.) The calculations for each step follow.

$$379 \text{ g ore} \times \frac{88.3 \text{ g FeTiO}_3}{100 \text{ g ore}} = 334.7 \text{ g FeTiO}_3$$

$$334.7 \text{ g FeTiO}_3 \times \frac{1 \text{ mol FeTiO}_3}{151.7 \text{ g FeTiO}_3} = 2.206 \text{ mol FeTiO}_3$$

$$2.206 \text{ mol FeTiO}_3 \times \frac{2 \text{ mol TiO}_2}{2 \text{ mol FeTiO}_3} = 2.206 \text{ mol TiO}_2$$

$$2.206 \text{ mol TiO}_2 \times \frac{79.90 \text{ g TiO}_2}{1 \text{ mol TiO}_2} = 176 \text{ g TiO}_2$$

Thus 176 g of TiO_2 can be prepared from the 379-g sample of ore.

EXAMPLE 3.10   Calcium acetate, $Ca(CH_3CO_2)_2$, is used as a mordant during the dyeing of fabric. The preparation of calcium acetate proceeds according to the equation

$$Ca(OH)_2 + 2CH_3CO_2H \longrightarrow Ca(CH_3CO_2)_2 + 2H_2O$$

How many grams of a sample containing 75.0% calcium hydroxide, $Ca(OH)_2$, by mass is required to react with the acetic acid, CH_3CO_2H, in 25.0 mL of a solution having a density of 1.065 g/mL and containing 58.0% acetic acid by mass?

Overall, this problem asks us to find the mass of a $Ca(OH)_2$ sample that will react with a given volume of CH_3CO_2H solution.

$$\boxed{\begin{array}{c}\text{Volume of}\\ CH_3CO_2H\\ \text{solution}\end{array}} \longrightarrow \boxed{\begin{array}{c}\text{Mass of}\\ Ca(OH)_2\\ \text{sample}\end{array}}$$

Since the solution contains CH_3CO_2H that reacts with the $Ca(OH)_2$ in the sample, we need a mole-to-mole conversion.

$$\boxed{\begin{array}{c}\text{Volume of}\\ CH_3CO_2H\\ \text{solution}\end{array}} \longrightarrow \boxed{\begin{array}{c}\text{Moles of}\\ CH_3CO_2H\end{array}} \longrightarrow \boxed{\begin{array}{c}\text{Moles of}\\ Ca(OH)_2\end{array}} \longrightarrow \boxed{\begin{array}{c}\text{Mass of}\\ Ca(OH)_2\\ \text{sample}\end{array}}$$

The density of the solution and the percent by mass of acetic acid in the solution are given, so we can calculate the mass of the solution and from that the mass of acetic acid in the solution. This will give us the steps necessary to convert from volume of solution to moles of acetic acid. To convert from moles of $Ca(OH)_2$ to mass of sample, we need to use the percent by mass of $Ca(OH)_2$ in the sample. The overall string of conversions is

$$\boxed{\begin{array}{c}\text{Volume of}\\ CH_3CO_2H\\ \text{solution}\end{array}} \xrightarrow{3} \boxed{\begin{array}{c}\text{Mass of}\\ CH_3CO_2H\\ \text{solution}\end{array}} \xrightarrow{2} \boxed{\begin{array}{c}\text{Mass of}\\ CH_3CO_2H\\ \text{solute}\end{array}} \xrightarrow{1} \boxed{\begin{array}{c}\text{Moles of}\\ CH_3CO_2H\end{array}}$$

$$\longrightarrow \boxed{\begin{array}{c}\text{Moles of}\\ Ca(OH)_2\end{array}} \xrightarrow{1} \boxed{\begin{array}{c}\text{Mass of}\\ Ca(OH)_2\end{array}} \xrightarrow{2} \boxed{\begin{array}{c}\text{Mass of}\\ \text{sample}\end{array}}$$

(The number above an arrow indicates the paragraph in Section 3.4 that relates to these conversions.) The calculations for each step follow.

$$25.0 \text{ mL solution} \times \frac{1.065 \text{ g solution}}{1 \text{ mL solution}} = 26.62 \text{ g solution}$$

$$26.62 \text{ g solution} \times \frac{58.0 \text{ g } CH_3CO_2H}{100 \text{ g solution}} = 15.44 \text{ g } CH_3CO_2H$$

$$15.44 \text{ g } CH_3CO_2H \times \frac{1 \text{ mol } CH_3CO_2H}{60.05 \text{ g } CH_3CO_2H} = 0.2571 \text{ mol } CH_3CO_2H$$

$$0.2571 \text{ mol } CH_3CO_2H \times \frac{1 \text{ mol } Ca(OH)_2}{2 \text{ mol } CH_3CO_2H} = 0.1286 \text{ mol } Ca(OH)_2$$

$$0.1286 \text{ mol } Ca(OH)_2 \times \frac{74.09 \text{ g } Ca(OH)_2}{1 \text{ mol } Ca(OH)_2} = 9.528 \text{ g } Ca(OH)_2$$

$$9.528 \text{ g } Ca(OH)_2 \times \frac{100 \text{ g sample}}{75.0 \text{ g } Ca(OH)_2} = 12.7 \text{ g sample}$$

EXAMPLE 3.11   What is the molar concentration of acetic acid in the solution described in Example 3.10?

This problem asks that we find the molarity of the acetic acid solution described in the preceding problem. Thus we must proceed from the volume of that solution to its molarity.

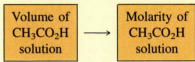

A reaction is not involved, so we do not need to do a mole-to-mole conversion. We should remember, however, that if we know the moles of acetic acid in a given volume of solution, we can calculate the molarity of the solution. The problem gives the volume of acetic acid solution (25.0 mL), so moles of acetic acid enter the chain of calculations.

$$
\boxed{\begin{array}{c}\text{Volume of}\\ \text{CH}_3\text{CO}_2\text{H}\\ \text{solution}\end{array}} \longrightarrow \boxed{\begin{array}{c}\text{Moles of}\\ \text{CH}_3\text{CO}_2\text{H}\end{array}} \longrightarrow \boxed{\begin{array}{c}\text{Molarity of}\\ \text{CH}_3\text{CO}_2\text{H solution}\end{array}}
$$

The first steps in this series of conversions have been determined in Example 3.10. Insertion of these steps gives

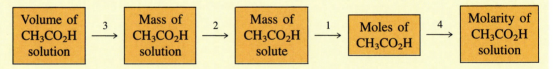

(The number above an arrow indicates the paragraph in Section 3.4 that relates to these conversions.) Steps 1 through 3 have been worked in Example 3.10, which shows that the sample contains 0.2571 mol of CH_3CO_2H.

$$
\begin{aligned}
\text{Molarity} &= \frac{\text{moles of CH}_3\text{CO}_2\text{H}}{\text{liters of solution}}\\[6pt]
&= \frac{0.2571 \text{ mol}}{0.0250 \text{ L}} = 10.3 \text{ mol/L} = 10.3 \text{ M}
\end{aligned}
$$

Note that molarity is expressed in moles per liter; hence the volume of the solution must be expressed in liters.

EXAMPLE 3.12   Small amounts of sulfuric acid, H_2SO_4, are prepared industrially by the oxidation of metal sulfides. The following sequence of reactions illustrates the manufacture of sulfuric acid from pyrites, FeS_2:

$$4FeS_2(s) + 11O_2(g) \longrightarrow 2Fe_2O_3(s) + 8SO_2(g) \tag{1}$$

$$2SO_2(g) + O_2(g) \longrightarrow 2SO_3(g) \tag{2}$$

$$SO_3(g) + H_2O(l) \longrightarrow H_2SO_4(l) \tag{3}$$

What mass of FeS_2 is required to prepare 1.00 L of H_2SO_4 (density = 1.85 g/mL)?

The problem is solved by converting the volume of H_2SO_4 into the mass of FeS_2 required to produce it. The following conversions are required:

Volume of H_2SO_4	$\xrightarrow{3}$	Mass of H_2SO_4	$\xrightarrow{1}$	Moles of H_2SO_4	$\longrightarrow$	Moles of SO_3

	Moles of SO_2	$\longrightarrow$	Moles of FeS_2	$\xrightarrow{1}$	Mass of FeS_2

(The number above an arrow indicates the paragraph in Section 3.4 that relates to these conversions.)

$$1.00 \text{ L } H_2SO_4 \times \frac{1000 \text{ mL}}{1 \text{ L}} \times \frac{1.85 \text{ g}}{1 \text{ mL}} = 1850 \text{ g } H_2SO_4$$

$$1850 \text{ g } H_2SO_4 \times \frac{1 \text{ mol } H_2SO_4}{98.07 \text{ g } H_2SO_4} = 18.86 \text{ mol } H_2SO_4$$

From Equation (3):

$$18.86 \text{ mol } H_2SO_4 \times \frac{1 \text{ mol } SO_3}{1 \text{ mol } H_2SO_4} = 18.86 \text{ mol } SO_3$$

From Equation (2):

$$18.86 \text{ mol } SO_3 \times \frac{2 \text{ mol } SO_2}{2 \text{ mol } SO_3} = 18.86 \text{ mol } SO_2$$

From Equation (1):

$$18.86 \text{ mol } SO_2 \times \frac{4 \text{ mol } FeS_2}{8 \text{ mol } SO_2} = 9.430 \text{ mol } FeS_2$$

$$9.430 \text{ mol } FeS_2 \times \frac{120.0 \text{ g } FeS_2}{1 \text{ mol } FeS_2} = 1.13 \times 10^3 \text{ g } FeS_2$$

For Review

Summary

If a substance is dissolved, that is, if it is in **solution,** a given quantity of the substance, in moles, can be obtained by measuring out a known volume of the solution, provided the molarity of the solution is also known. The **molarity** of a solution is equal to the number of moles of solute dissolved in exactly 1 liter of the solution.

Stoichiometry involves calculation of the amounts of materials in chemical reactions. Such calculations reflect the fact that the relative numbers of moles of reactants and products involved in a chemical reaction are given by the balanced chemical equation describing the reaction. Thus a balanced equation can be used to determine unit conversion factors relating moles of reactants and/or products. The quantity of a substance can be given as the mass of a pure sample of the substance, as the mass of a mixture containing a certain percent by mass of the substance, as the volume of a substance with its density, or as the volume of a solution with a certain molarity for the substance.

Volumes are often measured by use of a **pipet,** which delivers a fixed volume of a liquid, or by a **buret,** which delivers a variable volume that can be measured. Burets are used in **titrations.**

Key Terms and Concepts

buret (3.2)
concentration (3.1)
end point (3.2)

molarity (3.1)
pipet (3.2)
solution (3.1)

stoichiometry (3.3)
titration (3.2)

Exercises

Molarity and Solutions

1. Calculate the concentration in moles per liter for each of the following solutions:
   (a) 98.1 g of sulfuric acid, H_2SO_4, in 1.00 L of solution
      Ans. 1.00 M
   (b) 24.5 g of sodium cyanide, NaCN, in 2.000 L of solution
      Ans. 0.250 M
   (c) 90.0 g of acetic acid, CH_3CO_2H, in 0.750 L of solution
      Ans. 2.00 M
   (d) 2.12 g of potassium bromide, KBr, in 458 mL of solution
      Ans. 3.89×10^{-2} M
   (e) 0.1374 g of copper sulfate, $CuSO_4$, in 13 mL of solution
      Ans. 6.6×10^{-2} M

2. Determine the mass of solute present in each of the following solutions:
   (a) 0.450 L of 1.00 M NaCl solution   *Ans. 26.3 g*
   (b) 2.0 L of 0.480 M $MgCl_2$ solution   *Ans. 91 g*
   (c) 455 mL of 3.75 M HCl solution   *Ans. 62.2 g*
   (d) 2.50 mL of 0.1812 M $KMnO_4$ solution
      Ans. 7.16×10^{-2} g
   (e) 25.38 mL of 9.721×10^{-2} M KOH solution
      Ans. 0.1384 g

3. Determine the moles of solute required to make the indicated amount of solution:
   (a) 1.00 L of 1.00 M $LiNO_3$ solution   *Ans. 1.00 mol*
   (b) 4.2 L of 2.45 M C_2H_5OH solution   *Ans. 10 mol*
   (c) 275 mL of 0.5151 M $KClO_4$ solution
      Ans. 0.142 mol
   (d) 25 mL of 0.1881 M H_2O_2 solution
      Ans. 4.7×10^{-3} mol
   (e) 1856 mL of 0.1475 M H_3PO_4 solution
      Ans. 0.2738 mol

4. The lowest limit of $MgSO_4$ that can be detected by taste in drinking water is about 3.32×10^{-3} M. What is the limit in grams per liter?   *Ans. 0.400 g/L*

5. Cow's milk contains an average of 4.5 g of the sugar lactose, $C_{12}H_{22}O_{11}$, per 0.100 L of milk. What is the molarity of lactose in this milk?   *Ans. 0.13 M*

6. For reasons of taste, the maximum chloride ion content,

Cl^-, of domestic water supplies has been set at about 7.0×10^{-3} M. What is the maximum content of Cl^- in grams per liter?   *Ans. 0.25 g/L*

7. An iron content of 0.1 mg/L in drinking water can usually be detected by taste. Will the iron in a water sample with an $FeCO_3$ concentration of 5.25×10^{-7} M be detectable by taste?
   Ans. No. The solution contains 2.93×10^{-2} mg Fe/L.

8. How many moles of hydrochloric acid, HCl, are required to react with the sodium hydroxide in 25.0 mL of a 0.100 M NaOH solution?

$$NaOH(aq) + HCl(aq) \longrightarrow NaCl(aq) + H_2O(l)$$
   Ans. 2.50×10^{-3} mol

9. How many moles of mercury nitrate, $Hg(NO_3)_2$, are required to react with the calcium chloride in 14.96 mL of a 2.244 M solution of $CaCl_2$?

$$Hg(NO_3)_2(aq) + CaCl_2(aq) \longrightarrow$$
$$HgCl_2(aq) + Ca(NO_3)_2(aq)$$
   Ans. 3.357×10^{-2} mol

10. What mass of calcium carbonate is required to react with the sulfuric acid in 137.8 mL of a 0.6943 M solution?

$$CaCO_3(s) + H_2SO_4(aq) \longrightarrow$$
$$CaSO_4(s) + H_2O(l) + CO_2(g)$$
   Ans. 9.576 g

11. What mass of $PbCrO_4$, the pigment "chrome yellow," often used by artists, can be produced by addition of excess Na_2CrO_4 to 1.00 L of a 0.493 M solution of $Pb(NO_3)_2$?

$$Na_2CrO_4(aq) + Pb(NO_3)_2(aq) \longrightarrow$$
$$PbCrO_4(s) + 2NaNO_3(aq)$$
   Ans. 159 g

12. What mass of the active metal zinc is required to react exactly with the hydrochloric acid in a 125.0-mL sample of a 0.2110 M solution of HCl?

$$Zn(s) + 2HCl(aq) \longrightarrow ZnCl_2(aq) + H_2(g)$$
   Ans. 0.8622 g

13. An excess of silver nitrate, $AgNO_3$, reacted with 25.00 mL of a solution of $CaCl_2$ producing $Ca(NO_3)_2$ and 4.498 g of AgCl. What is the molarity of the $CaCl_2$ solution?   *Ans. 0.6277 M*

14. A 5.00-mL sample of the sulfuric acid solution from an automobile battery required addition of 17.48 mL of a 1.95 M solution of NaOH to react with the acid. What was the concentration of the battery acid?

$$H_2SO_4(aq) + 2NaOH(aq) \longrightarrow Na_2SO_4(aq) + 2H_2O(l)$$
Ans. 3.41 M

15. What is the concentration of calcium hydroxide in a solution formed by the reaction of 0.1000 g of calcium with enough water to give 400.0 mL of solution?

$$Ca(s) + 2H_2O(l) \longrightarrow Ca(OH)_2(aq) + H_2(g)$$
Ans. 6.238×10^{-3} M

16. The lead nitrate, $Pb(NO_3)_2$, in 47.29 mL of a 0.1001 M solution reacts with all the aluminum sulfate, $Al_2(SO_4)_3$, in 25.00 mL of solution. What is the molar concentration of the $Al_2(SO_4)_3$ in the original $Al_2(SO_4)_3$ solution?

$$3Pb(NO_3)_2(aq) + Al_2(SO_4)_3(aq) \longrightarrow$$
$$3PbSO_4(s) + 2Al(NO_3)_3(aq)$$
Ans. 6.31×10^{-2} M

Titration

17. What is the concentration of NaCl in a solution if titration of 10.00 mL of the solution with 0.250 M $AgNO_3$ requires 17.05 mL of the $AgNO_3$ solution to reach the end point?

$$AgNO_3(aq) + NaCl(aq) \longrightarrow AgCl(s) + NaNO_3(aq)$$
Ans. 0.426 M

18. Titration of 10.00 mL of a 0.444 M solution of HCl with a solution of LiSH required 23.2 mL of the LiSH solution to reach the end point. What is the molar concentration of LiSH?

$$HCl(aq) + LiSH(aq) \longrightarrow LiCl(aq) + H_2S(aq)$$
Ans. 0.191 M

19. What is the H_2SO_4 concentration in a solution of H_2SO_4 that requires 31.91 mL to titrate 2.474 g of K_2CO_3?

$$H_2SO_4(aq) + K_2CO_3(s) \longrightarrow$$
$$K_2SO_4(aq) + H_2O(l) + CO_2(g)$$
Ans. 0.5610 M

20. Potatoes may be peeled commercially by soaking them in a 3 to 6 M solution of sodium hydroxide, then removing the loosened skins by spraying with water. Does a sodium hydroxide solution have a suitable concentration if titration of 10.00 mL of the solution requires 25.3 mL of 1.87 M HCl to reach the end point?

$$NaOH + HCl \longrightarrow NaCl + H_2O$$
Ans. Yes; the concentration is 4.73 M.

21. In a common medical laboratory determination of the concentration of free chloride ion, Cl^-, in blood serum, a serum sample is titrated with a $Hg(NO_3)_2$ solution.

$$2Cl^- + Hg(NO_3)_2 \longrightarrow HgCl_2 + 2NO_3^-$$

What is the Cl^- concentration in a 0.15 mL sample of normal serum that requires 1.62 mL of 4.96×10^{-3} M $Hg(NO_3)_2$ to reach the end point?   *Ans. 0.11 M*

22. Titration of a 20.0-mL sample of a particularly acidic rain required 1.7 mL of 0.0811 M NaOH to reach the end point. If we assume that the acidity of the rain is due to the presence of sulfuric acid, H_2SO_4, what was the concentration of this sulfuric acid solution?

$$H_2SO_4(aq) + 2NaOH(aq) \longrightarrow$$
$$Na_2SO_4(aq) + 2H_2O(l)$$
Ans. 3.4×10^{-3} M

23. What volume of 0.0125 M HBr solution is required to titrate 125 mL of a 0.0100 M $Ca(OH)_2$ solution?

$$Ca(OH)_2 + 2HBr \longrightarrow Ca(Br)_2 + 2H_2O$$
Ans. 2.00×10^2 mL

24. Crystalline potassium hydrogen phthalate, $KHC_8H_4O_4$, is often used as a standard acid for standardizing basic solutions because it is easy to purify and to weigh. If 1.5428 g of this salt is titrated with a solution of $Ba(OH)_2$, the reaction is complete when 22.51 mL of the solution has been added. What is the concentration of the $Ba(OH)_2$ solution?

$$2KHC_8H_4O_4 + Ba(OH)_2 \longrightarrow$$
$$BaK_2(C_8H_4O_4)_2 + 2H_2O$$
Ans. 0.1678 M

25. A common oven cleaner contains a very caustic metal hydroxide, which we can represent as MOH. Titration of 0.7134 g of MOH with 1.000 M HNO_3 requires 17.85 mL to reach the end point.

$$MOH + HNO_3 \longrightarrow MNO_3 + H_2O$$

What is the element M and what is the formula of the hydroxide?   *Ans. Na; NaOH*

26. (a) A 5.00-mL sample of vinegar, a solution of acetic acid (CH_3CO_2H), was titrated with a 0.240 M NaOH solution. If 16.96 mL was required to reach the end point, what was the molar concentration of the acetic acid in the vinegar?

$$CH_3CO_2H + NaOH \longrightarrow CH_3CO_2Na + H_2O$$
Ans. 0.814 M

(b) If the density of the vinegar is 1.005 g/cm³, what is the percent of acetic acid by mass in the vinegar?
Ans. 4.86%

Conversion between Quantities

27. Begin with the expression for each of the following conversions, and derive the expression for the reverse conversion:

(a) Grams A to moles A

(b) Volume of a liquid of known density to mass

(c) Moles of solute to molar concentration (given the volume of solution)

(d) Percent A and mass of A to mass of a sample containing A

28. Derive a one-step conversion factor for each of the following:
    (a) Moles A to mass of a mixture containing a known percentage of A
    (b) Volume of pure A and its density to volume of a mixture of known density containing a known percentage of A
    (c) Volume of a solution of known density with a known molarity of A to percentage A in the solution

29. Determine the moles of $Ca_3(PO_4)_2$ and the mass of Ca contained in a 374-g sample of phosphate rock that contains 89.2% $Ca_3(PO_4)_2$.   *Ans. 1.08 mol; 129 g*

30. What mass of PCl_3 and how many moles of PCl_3 are contained in 10.00 mL of PCl_3, a liquid with a density of 1.574 g/cm^3?   *Ans. 15.74 g; 0.1146 mol*

31. What is the molar concentration of isotonic saline used for intravenous injections, which has a density of 1.007 g/mL and contains 0.95% by mass of NaCl?   *Ans. 0.16 M*

32. Several insects are able to withstand exposures to low temperatures because their body fluids contain large amounts of glycerol, $C_3H_5(OH)_3$. The glycerol acts as an antifreeze and reduces the freezing point of the fluids. What is the molar concentration of glycerol in a wasp's body fluid (density 1.068 g/cm^3) that contains 28.0% glycerol by mass?   *Ans. 3.25 M*

33. Sulfuric acid for laboratory use is supplied in 2.5-L bottles that contain 98.0% H_2SO_4 by mass, with a density of 1.92 g/mL. What mass of pure H_2SO_4 in kg is contained in such a bottle?   *Ans. 4.7 kg*

34. What volume of 0.250 M potassium hydroxide solution can be prepared from 174 g of KOH that contains 12.8% water by mass?   *Ans. 10.8 L*

35. What volume of 0.200 M acetic acid solution can be prepared from one pint (0.473 L) of 99.7% (by mass) acetic acid, CH_3CO_2H, with a density of 0.960 g/cm^3?   *Ans. 37.7 L*

36. What is the HCl concentration (molarity) in a commercial hydrochloric acid solution with a density of 1.198 g/cm^3 that contains 36.50% HCl?   *Ans. 11.99 M*

37. Concentrated sulfuric acid as produced industrially often contains 98.0% H_2SO_4 by mass and has a density of 1.92 g/mL. What is the concentration of H_2SO_4 in this acid solution?   *Ans. 19.2 M*

38. Ammonia for laboratory use is supplied as a solution containing 28.2% NH_3 (by mass) with a density of 0.8990 g/mL. What volume of this solution is required to prepare 500 mL of 0.500 M $NH_3(aq)$ for a general chemistry laboratory?   *Ans. 16.8 mL*

39. What volume of concentrated nitric acid, a 69.5% solution of HNO_3 with a density of 1.42 g/cm^3, is required to prepare 10.0 L of a 0.150 M solution of HNO_3?   *Ans. 0.0958 L*

40. What mass of sodium hydroxide, NaOH, containing 22.80% water (by mass) is required to prepare 0.100 L of 0.2733 M NaOH solution?   *Ans. 1.42 g*

41. What volume of gaseous carbon dioxide (density 1.964 g/L) is required to carbonate 12 ounces (355 mL) of a beverage with a density of 1.067 g/cm^3 that contains 0.66% CO_2 by mass?   *Ans. 1.3 L*

42. The limit for occupational exposure to ammonia, NH_3, in the air is set at 2.8×10^{-3} % by mass. What mass of NH_3, in grams, per cubic meter of air (density, 1.20 g/L) is required to reach this limit?   *Ans. 0.034 g*

43. A 5.0-mL sample of human plasma with a density of 1.0299 g/mL was found to contain 0.478 g of dissolved protein and 0.077 g of dissolved nonprotein material. What are the percents of protein, of nonprotein materials, and of all dissolved solids in the plasma?   *Ans. 9.3%; 1.5%; 10.8%*

44. How many grams of an insect-repellent solution containing 17.0% of N,N-dimethyl-metatoluamide, $C_{10}H_{13}NO$, can be prepared from 1.00 mol of $C_{10}H_{13}NO$?   *Ans. 960 g*

Chemical Stoichiometry

45. How many grams of CaO are required for reaction with the HCl in 27.5 mL of a 0.523 M HCl solution? The equation for the reaction is

$$CaO + 2HCl \longrightarrow CaCl_2 + H_2O$$

    Ans. 0.403 g

46. Limestone is almost pure calcium carbonate, $CaCO_3$. How much quicklime, CaO, can be prepared by roasting (heating) a metric ton, 1.000×10^3 kg, of limestone that contains 94.6% $CaCO_3$? $CaCO_3$ decomposes into CaO and CO_2 during the roasting process.   *Ans. 5.30×10^2 kg*

47. Aspirin, $C_6H_4(CO_2H)(CO_2CH_3)$, can be prepared in the chemistry laboratory by the reaction of salicylic acid, $C_6H_4(CO_2H)(OH)$, with acetic anhydride, $(CH_3CO)_2O$.

$$2C_6H_4(CO_2H)(OH) + (CH_3CO)_2O \longrightarrow$$
$$2C_6H_4(CO_2H)(CO_2CH_3) + H_2O$$

What volume of acetic anhydride (density, 1.0820 g/cm^3) is required to produce 1.00 kg of aspirin, assuming a 100% yield?   *Ans. 262 mL*

48. Elemental phosphorus, P_4, is prepared by the following reaction:

$$4Ca_5(PO_4)_3F + 18SiO_2 + 30C \longrightarrow$$
$$3P_4 + 2CaF_2 + 18CaSiO_3 + 30CO$$

What mass of phosphorus can be prepared from a 1.000-kg sample of an ore that is 86.72% $Ca_5(PO_4)_3F$, assuming a 100% yield of P_4?   *Ans. 159.8 g*

49. Automotive air bags inflate when a sample of NaN_3 is very rapidly decomposed.

$$2NaN_3(s) \longrightarrow 2Na(s) + 3N_2(g)$$

What mass of NaN_3 is required to produce 13.0 cubic feet (368 L) of nitrogen gas with a density of 1.25 g/L?   *Ans. 712 g*

50. What volume in L of air (density of 1.20 g/L and 21.0% O_2 by mass) is required to burn 1.00 gallon of gasoline (density, 0.780 g/cm^3), which can be represented as C_8H_{16}? Assume a 100% yield and complete consumption of the O_2.

$$C_8H_{16}(l) + 12O_2(g) \longrightarrow 8CO_2(g) + 8H_2O(g)$$

Ans. 4.01 × 10^4 L

51. Reaction of rhenium metal with Re_2O_7 gives a solid of metallic appearance that conducts electricity almost as well as copper. A 0.414-g sample of this material, which contains only rhenium and oxygen, was oxidized in an acidic solution of hydrogen peroxide. Addition of an excess of KOH gave 0.511 g of $KReO_4$. Determine the empirical formula of the metallic solid and write the equation for its formation.   *Ans. Re + 3Re$_2$O$_7$ ⟶ 7ReO$_3$*

52. Bronzes used in bearings are often alloys (solid solutions) of copper and aluminum. A 1.953-g sample of one such bronze was analyzed for its copper content by the sequence of reactions given below. The aluminum also dissolves, but we need not consider it because aluminum sulfate, $Al_2(SO_4)_3$, does not react with KI.

$$Cu + 2H_2SO_4 \longrightarrow CuSO_4 + 2H_2O + SO_2$$
$$2CuSO_4 + 4KI \longrightarrow 2CuI + I_2 + 2K_2SO_4$$
$$I_2 + 2Na_2S_2O_3 \longrightarrow Na_2S_4O_6 + 2NaI$$

If 35.06 mL of 0.837 M $Na_2S_2O_3$ is required to react with the I_2 formed, what is the percent Cu in the sample?

Ans. 95.5%

53. A compound found in cast iron is cementite, which contains iron and carbon. A 1.724-g sample of cementite was dissolved in a dilute solution of sulfuric acid, giving a solution of $FeSO_4$ and various carbon-containing compounds, which may be neglected because they do not react in the subsequent analysis. A 0.2753 M solution of K_2CrO_4 was added to the solution, and a volume of 34.88 mL was required for the reaction with the $FeSO_4$ according to the equation

$$6FeSO_4 + 2K_2CrO_4 + 8H_2SO_4 \longrightarrow$$
$$3Fe_2(SO_4)_3 + 2K_2SO_4 + Cr_2(SO_4)_3 + 8H_2O$$

What is the empirical formula of cementite? *Ans. Fe$_3$C*

54. Glauber's salt, $Na_2SO_4 \cdot 10H_2O$, is an important industrial chemical that is isolated from naturally occurring brines in New Mexico. A 0.3440-g sample of this material was allowed to react with an excess of $Ba(NO_3)_2$, and 0.2398 g of $BaSO_4$ was isolated. What is the percent of $Na_2SO_2 \cdot 10H_2O$ in the sample analyzed?

$$Na_2SO_4 \cdot 10H_2O(aq) + Ba(NO_3)_2(aq) \longrightarrow$$
$$BaSO_4(s) + 2NaNO_3(aq) + 10H_2O$$

Ans. 96.23%

55. Hydrogen peroxide may be prepared by the following reactions:

$$2NH_4HSO_4 \xrightarrow{\text{electric current}} H_2 + (NH_4)_2S_2O_8$$
$$(NH_4)_2S_2O_8 + 2H_2O \longrightarrow 2NH_4HSO_4 + H_2O_2$$

What mass of NH_4HSO_4 is initially required to prepare 1.00 g of H_2O_2? What mass of H_2O is required?

Ans. 6.77 g; 1.06 g

56. Potassium perchlorate, $KClO_4$, may be prepared from KOH and Cl_2 by the following series of reactions:

$$2KOH + Cl_2 \longrightarrow KCl + KClO + H_2O$$
$$3KClO \longrightarrow 2KCl + KClO_3$$
$$4KClO_3 \longrightarrow 3KClO_4 + KCl$$

What mass of $KClO_4$ can be prepared from 50.0 g of KOH?   *Ans. 15.4 g*

57. Sodium thiosulfate, $Na_2S_2O_3$, photographic hypo, may be prepared from Glauber's salt, $Na_2SO_4 \cdot 10H_2O$, by the following series of reactions:

$$Na_2SO_4 \cdot 10H_2O(s) \xrightarrow{\Delta} Na_2SO_4(s) + 10H_2O(g)$$
$$Na_2SO_4(s) + 4C(s) \xrightarrow{\Delta} Na_2S(s) + 4CO(g)$$
$$2Na_2S(aq) + Na_2CO_3(aq) + 4SO_2(aq) \longrightarrow$$
$$3Na_2S_2O_3(aq) + CO_2(g)$$

What mass of $Na_2SO_4 \cdot 10H_2O$ is required to prepare 50.0 g of $Na_2S_2O_3$?   *Ans. 67.9 g*

Additional Exercises

58. The sugar-free soft drink Tab contains 7.9 mg of the artificial sweetener saccharin, $C_7H_5SNO_3$, per 1.0 ounce. What is the molar concentration of saccharin in this drink?

Ans. 1.5 × 10^{-3} M

59. The average plasma volume in an adult is 39 mL per kilogram of weight. The average concentration of sodium ion, Na^+, is 0.142 M. What is the mass of sodium ion present in the serum of a 75-kg (165-lb) adult? What mass of NaCl, which is composed of Na^+ and Cl^- ions, would be required to provide this much sodium?

Ans. 9.5 g Na$^+$; 25 g NaCl

60. The concentrations of sodium ion, potassium ion, chloride ion, and dihydrogen phosphate ion in Gatorade Thirst Quencher are Na^+, 21.0 mM; K^+, 2.5 mM; Cl^-, 17.0 mM; $H_2PO_4^-$, 6.8 mM. These, incidentally, match the relative concentrations of these ions lost through perspiration. What mass of Na^+, K^+, Cl^- and $H_2PO_4^-$ is contained in one 8.00-oz glass of Gatorade? (mM = millimolar, 10^{-3} M)

Ans. 0.114 g Na$^+$; 0.023 g K$^+$;
0.143 g Cl$^-$; 0.16 g H$_2$PO$_4^-$

61. A copper ion, Cu^{2+}, content of 1.0 mg/L causes a bitter taste in drinking water. What mass of $CuSO_4$ will give a copper content of 1.0 mg/L and what is the $CuSO_4$ concentration in mol/L?   *Ans. 2.5 mg; 1.6 × 10^{-5} M*

62. A mixture of morphine ($C_{17}H_{19}NO_3$) and an inert solid is analyzed by combustion with O_2. The unbalanced equation for the reaction of morphine with O_2 is

$$C_{17}H_{19}NO_3 + O_2 \longrightarrow CO_2 + H_2O + NO_2$$

The inert solid does not react with O_2. If 0.0400 g of the mixture yields 0.0872 g of CO_2, calculate the percent morphine by mass in the mixture.   *Ans. 83.1%*

63. What mass of a sample that is 98.0% sulfur would be required in the production of 1.00 kg of H_2SO_4 by the following reaction sequence?

$$S_8 + 8O_2 \longrightarrow 8SO_2$$
$$2SO_2 + O_2 \longrightarrow 2SO_3$$
$$SO_3 + H_2O \longrightarrow H_2SO_4$$

Ans. 334 g

64. For many years, the standard medical laboratory technique for determination of the concentration of calcium (Ca^{2+}) in blood serum used the following reaction sequence involving ammonium oxalate, $(NH_4)_2C_2O_4$, and potassium permanganate, $KMnO_4$:

$$Ca^{2+} + (NH_4)_2C_2O_4 \longrightarrow CaC_2O_4(s) + 2NH_4^+$$
$$CaC_2O_4(s) + H_2SO_4 \longrightarrow H_2C_2O_4 + CaSO_4$$
$$2KMnO_4 + 5H_2C_2O_4 + 3H_2SO_4 \longrightarrow$$
$$K_2SO_4 + 2MnSO_4 + 10CO_2 + 8H_2O$$

What is the molar concentration of calcium in a 2.0-mL serum sample if 0.68 mL of 2.44×10^{-3} M $KMnO_4$ is required for the final reaction? How many milligrams of calcium is contained in 1.00×10^2 mL of the serum?   *Ans. 0.0021 M; 8.4 mg*

65. The amounts of active ingredient per tablet of several antacids, and equations for how these ingredients react with stomach acid (HCl), are given below. Assuming that the stomach acid is undiluted (0.155 M), calculate what volume of acid in mL will react with a pure sample of the active ingredients of each tablet.

(a) Phillip's Tablets, 0.311 g $Mg(OH)_2$

$$Mg(OH)_2 + 2HCl \longrightarrow MgCl_2 + 2H_2O$$

Ans. 68.8 mL

(b) Tums, 0.500 g $CaCO_3$

$$CaCO_3 + 2HCl \longrightarrow CaCl_2 + H_2O + CO_2$$

Ans. 64.5 mL

(c) Rolaids, 0.334 g $NaAl(CO_3)(OH)_2$

$$NaAl(CO_3)(OH)_2 + 4HCl \longrightarrow$$
$$NaCl + AlCl_3 + 3H_2O + CO_2$$

Ans. 59.9 mL

(d) Gelusil, 0.500 g $Mg_2Si_3O_8$, 0.075 g $Mg(OH)_2$

$$Mg_2Si_3O_8 + 4HCl \longrightarrow 2MgCl_2 + 3SiO_2 + 2H_2O$$
$$Mg(OH)_2 + 2HCl \longrightarrow MgCl_2 + 2H_2O$$

Ans. 66 mL

66. Hach Chemical Company of Loveland, Colorado, offers a test kit for determination of chloride ion in domestic water supplies. The kit contains a silver nitrate solution, which is added drop by drop to a 23.0-mL water sample to which an indicator has been added. When sufficient silver nitrate has been added to convert the chloride ion completely to AgCl, the solid produced turns orange. The concentration of the silver nitrate solution is such that each drop used to reach the color change corresponds to 12.5 mg of Cl^- per liter of water tested.

(a) What mass of chloride ion (mg/L) is contained in a water sample that requires 17 drops of test solution to reach the color change?   *Ans. 2.1×10^2 mg/L*

(b) What is the molar concentration of Cl^- in the sample tested in (a)?   *Ans. 6.0×10^{-3} M*

(c) If the sample size used is 5.75 mL instead of 23.0 mL, to what chloride content, in mg/L, does one drop of the test solution correspond?   *Ans. 50.0 mg/L*

(d) This test kit can also be used for bromide and iodide testing. If the reading obtained is multiplied by 2.25, the correct content of bromide in mg/L is obtained. What is the factor necessary to give iodide content in mg/L?   *Ans. 3.58*

(e) If 20 drops of the silver nitrate test solution equals 1.00 mL, what is the molar concentration of $AgNO_3$ in the test solution?   *Ans. 0.162 M*

67. Calculate the mass of sodium nitrate required to produce 5.00 L of O_2 (density = 1.43 g/L) according to the reaction

$$2NaNO_3 \xrightarrow{\triangle} 2NaNO_2 + O_2$$

if the percent yield of the reaction is 78.4%.

Ans. 48.4 g

68. On average, protein contains 15.5% nitrogen by mass. This nitrogen can be converted to gaseous ammonia, $NH_3(g)$, and analyzed by the Kjeldahl technique. The ammonia reacts with an excess of boric acid, H_3BO_3

$$NH_3(g) + H_3BO_3(aq) \longrightarrow (NH_4)H_2BO_3(aq)$$

and the product titrated with standardized hydrochloric acid.

$$(NH_4)H_2BO_3 + HCl \longrightarrow NH_4Cl + H_3BO_3$$

What is the percent by mass of protein in a sample of lobster meat if titration of the $(NH_4)H_2BO_3$ formed in a Kjeldahl analysis of a 4.95-g meat sample requires 34.05 mL of 0.2011 M HCl solution?   *Ans. 12.5%*

69. (a) Champagne can be prepared by fermentation of grape juices in the bottle. This fermentation converts sugar into enthanol and carbon dioxide.

$$C_6H_{12}O_6 \longrightarrow 2C_2H_5OH + 2CO_2$$

The carbon dioxide cannot escape and so produces a carbonated beverage. What is the percent by mass of CO_2 in a bottle containing 750 mL of champagne with a density of 0.98 g/cm^3 that contains 12.0% ethanol by mass?   *Ans. 11%*

(b) If exactly one-half of the CO_2 escapes from solution as $CO_2(g)$ when the bottle is opened, what will be the volume of the CO_2 bubbles be, assuming that $CO_2(g)$ has a density of 1.96 g/L?   *Ans. 21 L*

70. Uranium may be isolated from the mineral pitchblende, which contains U_3O_8. The pitchblende in a 6.235-g sam-

ple of an ore containing uranium was subjected to the following sequence of reactions:

$$2U_3O_8 + O_2 + 12HNO_3 \longrightarrow 6(UO_2)(NO_3)_2 + 6H_2O$$

$$(UO_2)(NO_3)_2 + 4H_2O + H_3PO_4 \longrightarrow$$
$$(UO_2)HPO_4 \cdot 4H_2O + 2HNO_3$$

$$2[(UO_2)HPO_4 \cdot 4H_2O] \xrightarrow{\triangle} (UO_2)_2P_2O_7 + 9H_2O$$

What is the percentage of U_3O_8 in the ore by mass if 1.607 g of $(UO_2)_2P_2O_7$ is isolated?   *Ans. 20.26%*

71. A 25-mL volume of solution of triethylaluminum, $Al(C_2H_5)_3$, a substance used in the production of the plastic polyethylene, was allowed to react with 25.0 mL of a 0.103 M HCl solution.

$$Al(C_2H_5)_3 + 3HCl \longrightarrow AlCl_3 + 3C_2H_6$$

The $Al(C_2H_5)_3$ was the limiting reagent, so some HCl remained unreacted. The unreacted HCl was titrated with a 0.142 M solution of NaOH; 16.75 mL was required.

$$HCl + NaOH \longrightarrow NaCl + H_2O$$

What was the concentration of $Al(C_2H_5)_3$ in the solution?
Ans. 2.6×10^{-3} M

72. What is the limiting reagent when 5.0×10^{-2} moles of HNO_3 react with 225 mL of 0.10 M $Ca(OH)_2$ solution?

$$Ca(OH)_2(aq) + 2HNO_3(aq) \longrightarrow$$
$$Ca(NO_3)_2(aq) + 2H_2O(l)$$
Ans. $Ca(OH)_2$

73. What mass of $PbSO_4$ is produced when 25.0 mL of 0.2338 M $Pb(NO_3)_2$ solution is added to 25.0 mL of 0.0971 M $Al_2(SO_4)_3$ solution, forming $PbSO_4(s)$ and $Al(NO_3)_3(aq)$?   *Ans. 1.77 g*

Thermochemistry

4

The brilliant effects of the aurora result when chemical reactions in the upper atmosphere release energy as light.

H eat is either released or absorbed during the course of almost all chemical and physical changes. Your body temperature is 37°C because many of its chemical reactions, such as the metabolic oxidation of sugar to carbon dioxide and water, produce heat, which keeps your body warm. Heat is also released when charcoal, which essentially is pure carbon, burns and forms carbon dioxide. Ice melts when it absorbs heat. Heat must be added to convert limestone into quicklime.

In the preceding two chapters relationships between the amounts of reactants and products in chemical reactions were discussed. For example, we found out how to calculate the mass of carbon dioxide produced when one gram of sugar is oxidized. In this chapter we will focus on the relationships between the amount of heat released or absorbed by a reaction and the amounts of reactants and products present. We will discuss how we could, for example, determine the amount of heat produced when one gram of sugar is oxidized. Other forms of energy will be considered later.

This chapter introduces **thermochemistry,** the determination and study of the heat absorbed or evolved in the formation and dissociation of compounds in chemical reactions and in changes of phase (changes of a liquid to a gas, a solid to a liquid, etc.). It discusses how to measure the amount of heat a chemical reaction produces or absorbs and how to use this information to determine the heat changes associated with other related reactions.

4.1 Heat and Its Measurement

Heat is the form of energy that moves spontaneously from a warmer object to a cooler object. However, heat can be *forced* to move from a cooler to a warmer object.

A chemical reaction or a physical change that produces heat is called an **exothermic** process. The burning of charcoal, for example, is an exothermic process. A reaction or change that occurs with the absorption of heat is an **endothermic** process. Melting of ice is an endothermic process, because heat is absorbed when ice melts.

The temperature of an object can be used to determine whether heat has moved into or out of it. In the absence of a phase change, we can increase the temperature of an object by adding heat, or some other form of energy, to it (Fig. 4.1). Removing heat, or some other form of energy, will reduce the temperature. For example, when hot coffee is poured into a mug, the coffee becomes cooler and the mug warmer. Heat moves from the coffee, decreasing its temperature, into the mug, increasing its temperature.

One measure of heat and other forms of energy is the calorie. A **calorie (cal)** is the amount of heat or other energy necessary to raise the temperature of 1 gram of water by 1 degree Celsius (Fig. 4.2). The SI unit of heat is the **joule, J.** There are 4.184 joules in one calorie*. A **kilojoule, kJ,** is 1×10^3 joules.

Specific heat is a physical property of a substance that may be used to measure quantities of heat. The **specific heat** is the quantity of heat required to raise the temperature of 1 gram of a substance 1 degree Celsius (or kelvin). Every substance has its own specific heat. The specific heat of water, which is one of the largest, is 4.184 joules per gram per degree (4.184 J/g °C). The specific heat of copper is 0.38 joule per gram per degree, that of aluminum is 0.88 joule per gram per degree, and that of zinc is 0.39 joule per gram per degree. Thus if a 1-gram sample of aluminum picks up enough heat to raise its temperature by 1 degree, it must have absorbed 0.88 joule.

The specific heat of a substance is a property of 1 gram of the substance. The heat capacity of a given quantity of matter describes a property of the entire quantity of matter. The **heat capacity** of a body of matter is the quantity of heat required to increase its temperature by 1 degree Celsius (or kelvin).

$$\text{Heat capacity} = \text{specific heat (J/g °C)} \times \text{mass (g)}$$

For example, 10 grams of aluminum (specific heat = 0.88 J/g °C) has a heat capacity of 8.8 J/°C; 25 grams of aluminum, a heat capacity of 22 J/°C. The greater the mass of a substance, the greater its heat capacity.

---

*Although a calorie is the same amount of energy as 4.184 joules, the joule is not defined this way in the SI system. In that system, a joule is defined as the amount of heat or other energy equal to the kinetic energy ($\frac{1}{2} Mv^2$) of an object with a mass M of exactly 2 kilograms moving with a velocity v of exactly 1 meter per second. One joule is equivalent to 1 kg m^2 s^{-2}, which is 1 newton-meter.

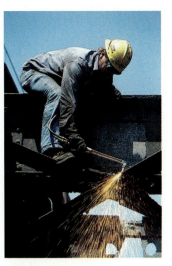

Figure 4.1

An oxyacetylene torch. Heat produced by the combustion of acetylene in oxygen enters the metal, heats it, and melts it. The sparks result when the molten metal is blown away by the gases in the flame.

Figure 4.2

Heat from the hot flame of the Bunsen burner enters the cooler water, and its temperature increases. If 800 grams of water is present, its temperature will increase by 1°C for each 800 calories it absorbs.

If the mass of a substance and its specific heat are known, the amount of heat, q, entering or leaving the substance can be determined by measuring the temperature change before and after the heat is gained or lost.

$$q = cm\ \Delta T$$

where c is the specific heat of the substance, m is its mass, and ΔT (called "delta t") is the temperature change, $T_{final} - T_{initial}$. If heat enters a substance, its temperature increases and its final temperature is higher than its initial temperature; $T_{final} - T_{initial}$ has a positive value; and the value of q is positive. If heat flows out of a substance, the final temperature is less than the initial temperature, $T_{final} - T_{initial}$ has a negative value, and q is negative. The following example shows how to calculate the amount of heat that enters a sample of water.

EXAMPLE 4.1   A flask containing 800 g of water is heated over a Bunsen burner as shown in Fig. 4.2. If the temperature of the water increases from 20°C to 85°C, how much heat (in calories) did the water absorb?

The specific heat of water is 4.184 J/g °C, or 1.000 cal/g °C. Thus 1 cal of heat will increase the temperature of 1 g of water by 1°C, and 800 cal will increase the temperature of 800 g of water by 1°C. To increase the temperature of 800 g of water by 65° will require 800 × 65 or 52,000 cal.

$$q = cm\ \Delta T = cm(T_{final} - T_{initial})$$
$$= 1.000\ cal/g\ °C \times 800\ g \times (85 - 20)°C$$
$$= 52{,}000\ cal\ (=52\ kcal)$$

Since the temperature increased, $T_{final} - T_{initial}$ is greater than zero and q is positive, indicating that heat was absorbed by the water.

EXAMPLE 4.2   Calculate the temperature change that results when 1625 J is removed from 75 g of ethyl alcohol initially at 25.0°C. The specific heat of ethyl alcohol is 2.4 J/g °C.

When heat is removed from a substance, q is negative, and the temperature of the substance decreases. Here $q = -1625$ J. The relationship between q and the temperature change is

$$q = cm\ \Delta T$$

We can rearrange this equation to solve for the temperature change, ΔT, which is the quantity necessary to determine the final temperature.

$$\text{Temperature change} = \Delta T = \frac{q}{cm}$$

$$= \frac{-1625\ J}{2.4\ J/g\ °C \times 75\ g}$$

$$= -9.0°C$$

The new temperature is not −9.0°C; this is the change in temperature, ΔT. We can calculate the final temperature as follows:

$$\Delta T = T_{final} - T_{initial}$$
$$-9° = T_{final} - 25°$$
$$T_{final} = 25° - 9° = 16°$$

4.2 Calorimetry

The process of measuring the amount of heat involved in a chemical or physical change is called **calorimetry.** There are several forms of calorimetry; the simplest is based on measuring the change in temperature as the heat involved in a chemical change is transferred into or out of a known quantity of water. The amount of heat can be determined from the temperature change of the water, as shown in Example 4.1. Of course, one must be careful that no heat other than that produced by the reaction enters the water and that no heat is lost before the reaction is complete.

Two polystyrene cups can be used to construct a **calorimeter,** a device used to measure heats of reactions. Such "coffee cup calorimeters" (Fig. 4.3) are often used in general chemistry laboratories. Because the cups are good insulators, they prevent the loss or gain of heat from the surroundings. When an exothermic reaction occurs in solution in such a calorimeter, the heat produced by the reaction is trapped in the solution and increases its temperature. When an endothermic reaction occurs, the heat required is absorbed from the solution and decreases its temperature. The change in temperature can be used to calculate the amount of heat involved in either case.

If the cups, thermometer, and stirrer used to construct the calorimeter had no heat capacity, the amount of heat produced in the reaction, $q_{reaction}$, would equal the amount of heat absorbed by the solution, q_{soln}. Because energy can be neither created nor destroyed in a chemical reaction

$$q_{reaction} + q_{soln} = 0$$

and

$$q_{reaction} = -q_{soln}$$

In some cases the amount of heat absorbed by the calorimeter is small enough to be neglected, and this equation is used to calculate the approximate amount of heat involved in a reaction. The following example shows how to determine the heat produced by the reaction of 0.0500 mole of HCl(aq) with 0.0500 mole of NaOH(aq).

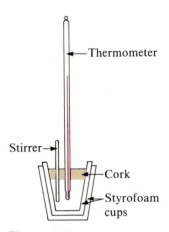

Figure 4.3

A calorimeter constructed from two polystyrene cups. A thermometer and stirrer extend through the cover into the reaction mixture.

EXAMPLE 4.3

When 50.0 mL (50 g) of 1.00 M HCl at 22.00°C is added to 50.0 mL (50 g) of 1.00 M NaOH at 22.00°C in a coffee cup calorimeter, the temperature increases to 28.87°C. Ignore the specific heat of the calorimeter and calculate the approximate amount of heat produced by the reaction between the HCl and NaOH.

$$HCl(aq) + NaOH(aq) \longrightarrow NaCl(aq) + H_2O(l)$$

The specific heat of the solution produced is 4.18 J/g °C.

At the instant of mixing, we have 100 g of a mixture of 0.500 M HCl and 0.500 M NaOH at 22.00°C. The mixture undergoes an immediate change as the HCl reacts with NaOH. This exothermic reaction produces the heat, which is trapped in the solution by the calorimeter and which raises its temperature to 28.87°C. The amount of heat, q_{soln}, can be calculated from the increase in temperature of the solution. Now let us look at the calculation.

$$q_{reaction} + q_{soln} = 0$$

and

$$q_{soln} = cm\ \Delta T$$

therefore

$$q_{reaction} = -q_{soln} = -cm\ \Delta T$$

The specific heat of the solution, c, is given as 4.18 J/g °C. The mass of the solution is 100 g. ΔT is 28.87°C − 22.00°C.

$$q_{reaction} = -q_{soln} = -cm\,\Delta T = -[cm(T_{final} - T_{initial})]$$
$$= -[4.18 \text{ J/g °C} \times 100 \text{ g} \times (28.87 - 22.00)°C]$$
$$= -2870 \text{ J}$$

This reaction produces 2870 J of heat.

If the amount of heat absorbed by a calorimeter is not small or if we want more accurate results, then we must take into account the heat absorbed both by the solution, q_{soln}, and by the calorimeter, q_{cal}. Thus

$$q_{reaction} + q_{soln} + q_{cal} = 0$$

and

$$q_{reaction} = -[q_{soln} + q_{cal}]$$

where q_{soln} is determined as shown in Example 4.3 and

$$q_{cal} = \text{heat capacity of calorimeter} \times \Delta T = C_{cal} \times \Delta T$$

The heat capacity of a calorimeter is commonly called its **calorimeter constant, C_{cal}**, and must be determined experimentally for each calorimeter. A calorimeter constant can be determined by running a reaction that produces a known amount of heat in the calorimeter or by mixing two volumes of water at different temperatures, as shown in the following example.

EXAMPLE 4.4   A 60.0-g sample of water at 84.3°C is added to 50.0 g of water at 23.2°C in a coffee cup calorimeter. Calculate the calorimeter constant if the final temperature of the mixture is 55.7°C.

In order to reach the final temperature, 60.0 g of water lost heat and cooled from 84.3°C to 55.7°C. We will call the amount of this heat q_{60}. At the same time, 50.0 g of water absorbed heat (q_{50}), the calorimeter absorbed heat (q_{cal}), and each increased its temperature from 23.2°C to 55.7°C. We need to determine q_{cal} so that we can calculate C_{cal} ($= q_{cal}/\Delta T$).

Because energy can be neither created nor destroyed, the heat lost by the 60 g of hot water must equal the heat gained by the 50 g of cooler water and by the calorimeter:

$$q_{lost} + q_{gained} = 0$$

so

$$q_{60} + q_{50} + q_{cal} = 0$$

After we calculate q_{60} and q_{50} using the equation $q = cm\,\Delta T$, we can determine q_{cal}.

$$q_{60} = 4.184 \text{ J/g °C} \times 60.0 \text{ g} \times (55.7 - 84.3)°C = -7180 \text{ J}$$
$$q_{50} = 4.184 \text{ J/g °C} \times 50.0 \text{ g} \times (55.7 - 23.2)°C = 6799 \text{ J}$$

Now we must find q_{cal}

$$q_{60} + q_{50} + q_{cal} = 0$$
$$q_{cal} = -q_{60} - q_{50}$$
$$= -(-7180 \text{ J}) - 6799 \text{ J} = 381 \text{ J}$$

During a temperature change of 32.5°C, the calorimeter absorbed 381 J. The calorimeter constant, C_{cal} (which is the amount of heat it absorbs when it changes temperature by 1 degree), is

$$C_{cal} = \frac{q_{cal}}{\Delta T} = \frac{381 \text{ J}}{32.5°C} = 11.7 \text{ J/°C}$$

why this heat, not the one for 60g

As may be seen in the next example, not all changes release heat.

EXAMPLE 4.5

The temperature of 50.0 g of water at 25.00°C contained in the calorimeter described in Example 4.4 ($C_{cal} = 11.7$ J/°C) decreases to 24.39°C when 1.00 g of solid $KClO_3$ is added and the mixture is stirred until all of the solid is dissolved. Calculate the heat involved in this change if the specific heat of the resulting solution is 4.18 J/g °C.

In this case, we start with solid $KClO_3$ and liquid water at 25.00°C. When we run the reaction in a calorimeter, no heat can enter the system from the surroundings so the heat necessary to dissolve the $KClO_3$ is extracted from the solution and the calorimeter, and the temperature of the solution and calorimeter is lowered. From the specific heat of the solution and the calorimeter constant, we can calculate the amount of heat that was removed from the solution and the calorimeter.

Be careful to note that the mass of the final solution is 51.0 g, the mass of the added $KClO_3$ plus that of the water.

$q_{reaction} + q_{soln} + q_{cal} = 0$

$q_{reaction} = -cm \, \Delta T - C_{cal} \, \Delta T$

$q_{reaction} = -cm(T_{final} - T_{initial}) - C_{cal}(T_{final} - T_{initial})$

$\qquad = -[4.18 \text{ J/g °C} \times 51.0 \text{ g} \times (24.39 - 25.00)°C] - [11.7 \text{ J/°C} \times (24.39 - 25.00)°C]$

$\qquad = -[4.18 \text{ J/g °C} \times 51.0 \text{ g} \times (-0.61°C)] - [11.7 \text{ J/°C} \times (-0.61°C)]$

$\qquad = -(-130 \text{ J}) - (-7 \text{ J}) = 137 \text{ J}$

$\qquad = 140 \text{ J} \qquad \text{(2 significant figures justified by the data)}$

A total of 140 J was absorbed by the reaction.

Not all calorimeters trap the heat of a reaction in a solution. If a reaction were run in the steel container of the calorimeter shown in Figure 4.4, the heat of the reaction

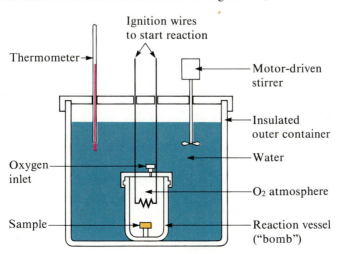

Figure 4.4

A bomb calorimeter of the type used to measure the heats of reaction involving one or more gases. The gases are contained in the pressure-proof bomb, and the reaction is set off by brief electric heating of a fuse wire in the bomb. The heat produced by the reaction is absorbed by the surroundings: the bomb, the stirrer, the thermometer, and the water. The heat of the reaction can be determined if the heat capacity of the surroundings has been determined.

would be trapped in the water surrounding the container. Such calorimeters are used to measure the heats of reactions involving gases. A modification of such a calorimeter was used to measure the heat produced by a living person. A man lived for four days inside a small room surrounded by water. From the increase in water temperature, it was found that he produced about 2,400,000 calories per day. The amount of heat produced or consumed by metabolic processes is usually reported in nutritional Calories. One Calorie equals 1000 thermochemical calories, so the heat produced by his metabolism was about 2400 nutritional Calories per day.

4.3 Thermochemistry and Thermodynamics

Thermochemistry is a branch of **chemical thermodynamics,** the chemical science that deals with the relationships between heat and other forms of energy known as work. Although we will concentrate on thermochemistry in this chapter and will postpone our broader consideration of thermodynamics until Chapter 19, we do need to consider some thermodynamic ideas that are widely used.

Substances act as reservoirs of energy; energy can be added to them or removed from them. The molecules in liquids and gases move around, and because of this motion they have kinetic energy (Section 1.2). They can store added energy by increasing the speed of their motion, thus increasing the total amount of kinetic energy present in the substance. When energy is removed, the total amount of kinetic energy is reduced. Adding or removing energy can also change the amount of other kinds of energy in a substance, such as that associated with the vibrations and rotations of molecules and with the forces that hold molecules together. The total of all possible kinds of energy present in a substance, or in a collection of substances, is called the **internal energy.** We will be interested in the changes in internal energy in a variety of substances.

The substance (or substances) we choose to study is called a **system.** Everything else is called the **surroundings.** As a system undergoes a change, its internal energy can change, and energy can be transferred from the system to the surroundings or from the surroundings to the system.

As an example of a system let us consider 1 gram of liquid water at 0°C in a container open to the atmosphere (Fig. 4.5). The water is the system. The container and the atmosphere are part of the surroundings. When this system loses 334 joules of heat to the surroundings, it changes to ice at 0°C.

The law of conservation of energy (Section 1.3) tells us that energy can be neither created nor destroyed. Because this system lost energy to the surroundings as heat during the change, the total energy in the system is less after the change than it was before. Loss of heat is one way for a system to reduce its internal energy.

The system also loses energy for a second less obvious reason. As water freezes, it expands and pushes back the atmosphere. Pushing back the atmosphere is work, just like pumping air into a bicycle tire is work. It requires energy to do work; as the system freezes, it uses some of its internal (stored) energy to do the work of pushing back the atmosphere.

As a second example of a system we can take 1 mole of carbon atoms, 4 moles of hydrogen atoms, and 4 moles of oxygen atoms combined to make 1 mole of methane, CH_4, and 2 moles of oxygen molecules, O_2. Under 1 atmosphere of pressure and at 25°C, this system would have a volume of about 73 liters. Methane, the principal

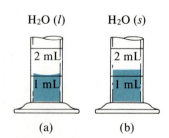

H_2O (*l*) H_2O (*s*)

(a)   (b)

Figure 4.5

When a system (a) consisting of 1 gram of liquid water at 0°C changes to (b) 1 gram of ice at 0°C, 334 joules are given up to the surroundings, and the system does work on the surroundings as it expands.

component of natural gas, burns in oxygen according to the following equation

$$CH_4(g) + 2O_2(g) \longrightarrow CO_2(g) + 2H_2O(l)$$

If the methane and oxygen in our system combine, the components rearrange themselves to give 1 mole of carbon dioxide gas, CO_2, and 2 moles of liquid water, H_2O, under 1 atmosphere pressure and at 25°C. In addition, the system gives up 890,000 joules of heat to the surroundings, reducing its internal energy by 890,000 joules.

The volume of the system after the methane burns is about 24.5 liters. The surroundings must do work on the system to push it from 73 liters to this smaller volume. Some of the energy present in the surroundings is used to do the work, and because this energy cannot disappear (remember the law of conservation of energy), it appears in the system. This adds to the total energy present in the system. So, although the system loses 890,000 joules of energy to the surroundings as heat when the methane burns, the change in internal energy of the system is less than this. The surroundings do about 4900 joules of work on the system to compress it, and the energy added by this work compensates, to a small extent, for the energy lost as heat.

We will consider how to determine the amount of work involved in a change when we discuss chemical thermodynamics more fully in Chapter 19. For now you need only recognize that work occurs when a system pushes back the surroundings against a restraining pressure or when the surroundings compress the system against a pressure. Energy is transferred into a system when it absorbs heat from the surroundings or when the surroundings do work on the system. Energy is transferred out of a system when heat is lost from the system or when the system does work on the surroundings.

4.4 Enthalpy Changes

If a change occurs under a constant pressure, and the only work done is due to expansion or contraction of the system under this constant pressure, the heat lost or absorbed by the system during the change is called the **enthalpy change** of the reaction, or the **heat of reaction,** and given the symbol **ΔH.** The heat of a reaction measured in a coffee cup calorimeter is directly proportional to the enthalpy change of the reaction, because the reaction occurs at the essentially constant pressure of the atmosphere. The heat of a reaction measured in a bomb calorimeter may not be directly proportional to ΔH, because the pressure in the bomb can change as the reaction proceeds.

The value of the enthalpy change of a reaction is given as a ΔH value following the equation for the reaction. The ΔH value indicates the amount of heat associated with the reaction of the number of moles of reactants shown in the chemical equation. The equation

$$H_2(g) + \frac{1}{2} O_2(g) \longrightarrow H_2O(l) \qquad \Delta H = -285 \text{ kJ}$$

indicates that when a system consisting of 1 mole of hydrogen gas and $\frac{1}{2}$ mole of oxygen gas at some temperature and pressure changes to 1 mole of liquid water at the same temperature and pressure, 285 kJ (285,000 J) of heat are given up to the surroundings. The negative value of ΔH indicates an exothermic reaction; a positive value of ΔH would indicate an endothermic reaction. Sometimes the amount of heat is written as if it were a reactant or a product. For this reaction we could also write

$$H_2(g) + \frac{1}{2} O_2(g) \longrightarrow H_2O(l) + 285,000 \text{ J}$$

A burning magnesium ribbon. The heat released when one mole of Mg burns in air is ΔH for the reaction.

EXAMPLE 4.6

When 0.0500 mol of HCl(aq) reacts with 0.0500 mol of NaOH(aq) to form 0.0500 mol of NaCl(aq), 2870 J of heat are produced ($q = -2870$ J, Example 4.3). What is ΔH, the enthalpy change for the reaction

$$HCl(aq) + NaOH(aq) \longrightarrow NaCl(aq) + H_2O(l)$$

run under the conditions described in Example 4.3?

ΔH for the reaction is the heat produced when 1 mol of HCl(aq) reacts with 1 mol of NaOH(aq) at constant pressure. For the reaction of 0.0500-mol amounts, $q = -2870$ J. The heat resulting from the reaction of 1-mol quantities is 20.0 times greater, 20.0×-2870 J or $-57,400$ J (-57.4 kJ). Thus the enthalpy change for the reaction is -57.4 kJ, which we write

$$HCl(aq) + NaOH(aq) \longrightarrow NaCl(aq) + H_2O(l) \qquad \Delta H = -57.4 \text{ kJ}$$

The National Bureau of Standards has been particularly active in measuring and tabulating enthalpy changes of reactions. Data are reported for a specific set of conditions called a **standard state.** Since 1981 the bureau has used a standard state of 298.15 K (25°C) and a pressure of 100 kilopascals (100 kPa, 0.987 atm). Another common standard state, the one used by the bureau before 1981, is 298.15 K and 1 atmosphere of pressure. Because ΔH of a reaction changes very little with such small changes in pressure, ΔH values (except for the most precisely measured values) are the same under either set of standard conditions. We will use the standard state of 298.15 K and 1 atmosphere and will use the symbol ΔH°_{298} to indicate an enthalpy change for these conditions. (The symbol ΔH is used to indicate an enthalpy change for a reaction where the conditions are not specified.) The enthalpy change of a reaction also depends on the state of the reactants and products (solid, liquid, or gas), so these must also be specified.

One common table of enthalpy changes contains heats of combustion. A **heat of combustion,** or **enthalpy of combustion,** is the enthalpy change for the combustion

Table 4.1   Heat of combustion for one mole of a substance under standard state conditions

Substance	Combustion reaction	Heat of combustion, ΔH°_{298} (kJ mol^{-1})
Carbon	$C(s) + \frac{1}{2}O_2(g) \longrightarrow CO(g)$	-111
	$C(s) + O_2(g) \longrightarrow CO_2(g)$	-394
Hydrogen	$H_2(g) + \frac{1}{2}O_2(g) \longrightarrow H_2O(g)$	-242
	$H_2(g) + \frac{1}{2}O_2(g) \longrightarrow H_2O(l)$	-286
Magnesium	$Mg(s) + \frac{1}{2}O_2(g) \longrightarrow MgO(s)$	-602
Nitrogen	$N_2(g) + O_2(g) \longrightarrow 2NO(g)$	$+180$
Silver	$Ag(s) + \frac{1}{4}O_2(g) \longrightarrow \frac{1}{2}Ag_2O(s)$	-15.5
Sulfur	$S(s) + O_2(g) \longrightarrow SO_2(g)$	-297
Carbon monoxide	$CO(g) + \frac{1}{2}O_2(g) \longrightarrow CO_2(g)$	-283
Methane	$CH_4(g) + 2O_2(g) \longrightarrow CO_2(g) + 2H_2O(g)$	-802
Acetylene	$C_2H_2(g) + \frac{5}{2}O_2(g) \longrightarrow 2CO_2(g) + H_2O(g)$	-1256
Methanol	$CH_3OH(l) + \frac{3}{2}O_2(g) \longrightarrow CO_2(g) + 2H_2O(g)$	-638
Isooctane	$C_8H_{18}(l) + \frac{25}{2}O_2(g) \longrightarrow 8CO_2(g) + 9H_2O(g)$	-5460

negative means heat is produced, positive is heat is absorbed

(combination with oxygen) of 1 mole of a substance under standard state conditions. Heats of combustion for several substances have been measured and are listed in Table 4.1. Most of these combustion reactions evolve heat. In fact, hydrogen, carbon (as coal or charcoal), methane, acetylene, methanol, and isooctane (a component of gasoline) are used as fuels. When silver combines with oxygen, only a little heat is produced. The reaction of nitrogen with oxygen requires heat to proceed.

The importance of the phases of the reactants and products on the value of the enthalpy change for a reaction is illustrated by the table. When liquid water, $H_2O(l)$, is formed by combustion of 1 mole of hydrogen, 286 kilocalories of heat is produced. When water vapor, $H_2O(g)$, is formed from 1 mole of hydrogen, only 242 kilocalories is produced.

EXAMPLE 4.7

As the photograph shows, the combustion of gasoline is an exothermic process. Let us assume that the heat of combustion of gasoline is the same as that of isooctane and calculate the amount of heat produced by the combustion of 1.0 L. We need to answer the question, How much heat (in kJ) can be produced by burning 1.0 L of isooctane under standard state conditions? The density of isooctane is 0.692 g/mL.

The heat of combustion of isooctane (Table 4.1) is -5460 kJ mol^{-1}; burning 1 mol of isooctane under standard conditions produces 5460 kJ of heat. If we determine how many moles of isooctane are contained in 1.0 L of isooctane and multiply that by the amount of heat produced by combustion of 1 mol of isooctane, we have answered the question. The number of moles can be determined from the molecular weight of isooctane, 114, and the mass of isooctane in 1.0 L. The mass can be determined from the volume and density of isooctane.

$$\text{Mass } C_8H_{18} = 1.0 \text{ L} \times \frac{1000 \text{ mL}}{1 \text{ L}} \times \frac{0.692 \text{ g}}{1 \text{ mL}} = 692 \text{ g } C_8H_{18}$$

$$\text{Mol } C_8H_{18} = 692 \text{ g } C_8H_{18} \times \frac{1 \text{ mol } C_8H_{18}}{114 \text{ g } C_8H_{18}} = 6.07 \text{ mol } C_8H_{18}$$

$$6.07 \text{ mol } C_8H_{18} \times \frac{-5460 \text{ kJ}}{1 \text{ mol } C_8H_{18}} = -33,000 \text{ kJ}$$

Combustion of 1.0 L of isooctane produces 33,000 kJ of heat.

The uncontrolled combustion of gasoline.

Probably the most useful tabulation of enthalpy changes is for a particular type of chemical reaction in which *1 mole* of a pure substance is formed from the free elements in their most stable states under standard state conditions. This enthalpy change is referred to as the **standard molar enthalpy of formation** of the substance formed and is designated by ΔH_f°. (Note the subscript letter f, which identifies an enthalpy of *formation under standard state conditions*.) The standard molar enthalpy of formation of $CO_2(g)$ is -394 kilojoules per mole, which is the enthalpy change for the reaction

$$C(s) + O_2(g) \longrightarrow CO_2(g) \qquad \Delta H_f^\circ = \Delta H_{298}^\circ = -394 \text{ kJ mol}^{-1}$$

with the reactants and products at a pressure of one atmosphere and 25°C and with the carbon present as graphite, the most stable form of carbon under these conditions. For silver oxide, Ag_2O, ΔH_f° is -31 kilojoules per mole; that is, it is equal to ΔH_{298}° for the reaction

$$2Ag(s) + \frac{1}{2} O_2(g) \longrightarrow Ag_2O(s)$$

Table 4.2   Standard molar enthalpies of formation.
(See Appendix I for additional values.)

Substance	ΔH_f°, kJ mol^{-1}	Substance	ΔH_f°, kJ mol^{-1}
Carbon		Hydrogen	
C(s) (graphite)	0	H$_2$(g)	0
C(g)	716.68	H$_2$O(g)	−241.8
CO(g)	−111	H$_2$O(l)	−285.8
CO$_2$(g)	−394	HCl(g)	−92.31
CH$_4$(g)	−74.8	H$_2$S(g)	−20.6
Chlorine		Oxygen	
Cl$_2$(g)	0	O$_2$(g)	0
Cl(g)	121.7	Silver	
Copper		Ag$_2$O(s)	−31
Cu(s)	0	Ag$_2$S(s)	−32.6
CuS(s)	−53.1		

The reaction of $\frac{1}{2}$ mole of O_2 and 2 moles of Ag to produce 1 mole of Ag_2O is the correct one to use for this purpose, since the enthalpy of formation refers to 1 mole of product. By convention, the standard molar enthalpy of formation of an element in its most stable form is equal to zero (see Table 4.2).

Standard molar enthalpies of formation of some common substances are given in Table 4.2 and in Appendix I. The unit of heat in these tables is kilojoules, but many older tables have calories (1 cal = 4.184 J) and kilocalories. If you have occasion to use other tables, be sure to check the units.

EXAMPLE 4.8   Ozone, O_3(g), forms from oxygen, O_2(g), by an endothermic process. Assuming that both the reactants and the products of the reaction are in their standard states, determine the standard molar enthalpy of formation, ΔH_f°, of ozone from the following information.

$$3O_2(g) \longrightarrow 2O_3(g) \qquad \Delta H_{298}^\circ = 286 \text{ kJ}$$

ΔH_f° is the enthalpy change for the formation of one mole of a substance in its standard state from the elements in their standard states. Thus ΔH_f° for O_3(g) is the enthalpy change for the reaction

$$\frac{3}{2} O_2(g) \longrightarrow O_3(g)$$

Since 286 kJ of heat are absorbed when 2 mol of O_3(g) are formed, to form 1 mol of O_3(g) will require half as much heat, or 143 kJ. Thus the enthalpy change is 143 kJ, and

$$\Delta H_f^\circ = 143 \text{ kJ mol}^{-1}$$

EXAMPLE 4.9   Hydrogen gas, H_2, reacts explosively with gaseous chlorine, Cl_2, to form hydrogen chloride, HCl, which is also a gas. What is the enthalpy change for the reaction of

1 mol of $H_2(g)$ with 1 mol of $Cl_2(g)$ if both the reactants and products are at standard state conditions?

$$H_2(g) + Cl_2(g) \longrightarrow 2HCl(g)$$

Because the standard molar enthalpies of formation of $H_2(g)$ and $Cl_2(g)$ are 0 (Table 4.2), these are the most stable states of the elements under standard state conditions. A pure substance, $HCl(g)$, is being formed under standard state conditions from elements in their most stable states under these conditions, so ΔH°_{298} must be proportional to the standard molar enthalpy of formation of $HCl(g)$. In this example, 2 mol of $HCl(g)$ are being formed; therefore,

$$\Delta H^\circ_{298} = 2 \, \Delta H^\circ_f$$

From Table 4.2 $\Delta H^\circ_f = -92.31$ kJ mol^{-1}

Therefore, $\Delta H^\circ_{298} = 2 \text{ mol HCl} \times (-92.31 \text{ kJ mol}^{-1}) = -184.6$ kJ

Thus the reaction evolves 184.6 kJ of heat.

$$H_2(g) + Cl_2(g) \longrightarrow 2HCl(g) \qquad \Delta H^\circ_{298} = -184.6 \text{ kJ}$$

Note the difference in units between ΔH (as well as ΔH°_{298}) and ΔH°_f. The enthalpy change for a reaction, ΔH (or ΔH°_{298}), gives the amount of heat produced by the reaction as it is written and has units of kilojoules or, more rarely, joules. The magnitudes of ΔH and ΔH°_{298} change as the amounts of reactant change. For example, combustion of 1 mole of $CO(g)$ produces 283 kilojoules ($\Delta H = -283$ kJ) while combustion of 2 moles of $CO(g)$ produces 566 kilojoules ($\Delta H = -566$ kJ). The standard molar heat of formation of a compound, ΔH°_f, identifies the enthalpy change associated with formation of 1 mole of the compound and has units of kilojoules per mole (kJ mol^{-1}).

4.5 Hess's Law

Some chemical changes can occur in two or more steps. For example, carbon can react with oxygen to form carbon dioxide in a two-step process. If carbon is burned in a limited amount of oxygen, carbon monoxide, CO, is formed.

$$C(s) + \frac{1}{2} O_2(g) \longrightarrow CO(g)$$

Carbon monoxide will burn in additional oxygen and form carbon dioxide, CO_2.

$$CO(g) + \frac{1}{2} O_2(g) \longrightarrow CO_2(g)$$

The equation describing the overall change of C to CO_2 is the sum of these two chemical changes:

STEP 1 $C(s) + \frac{1}{2} O_2(g) \longrightarrow CO(g)$

STEP 2 $CO(g) + \frac{1}{2} O_2(g) \longrightarrow CO_2(g)$

SUM $C(s) + \cancel{CO(g)} + O_2(g) \longrightarrow \cancel{CO(g)} + CO_2(g)$

Because the same number of moles of CO appear on both sides of the sum, there is no net chemical change in the number of moles of CO during the reaction, and the CO cancels. Another way of looking at this is that the CO produced in Step 1 is consumed in Step 2. The net change is

$$C(s) + O_2(g) \longrightarrow CO_2(g)$$

If a chemical reaction can be written as a sum of steps, the enthalpy change of the reaction is equal to the sum of the enthalpy changes of the steps. From the experimental heats of combustion (Table 4.1),

$$C(s) + \frac{1}{2} O_2(g) \longrightarrow CO(g) \qquad \Delta H^\circ_{298} = -111 \text{ kJ}$$

$$CO(g) + \frac{1}{2} O_2(g) \longrightarrow CO_2(g) \qquad \Delta H^\circ_{298} = -283 \text{ kJ}$$

$$\overline{C(s) + O_2(g) \longrightarrow \ + CO_2(g)} \qquad \overline{\Delta H^\circ_{298} = -394 \text{ kJ}}$$

You can see that ΔH for the bottom reaction is the sum of the ΔH values for the two reactions above it. This is an example of the principle stated by **Hess's law: If a process can be written as the sum of several stepwise processes, the enthalpy change of the total process equals the sum of the enthalpy changes of the various steps.** This law applies even if the steps are hypothetical.

Hess's law lets us use standard molar enthalpies of formation to calculate enthalpy changes of other reactions. However, before we use this law, let us look briefly at two important features of ΔH for a reaction.

1. ΔH for a reaction in one direction is equal in magnitude and opposite in sign to ΔH for the reaction in the reverse direction.
2. ΔH is directly proportional to the quantities of reactants or products.

The first statement tells us that if we know ΔH for the reaction

$$H_2O(s) \longrightarrow H_2O(l)$$

is 6.0 kilojoules, then ΔH for the reverse reaction,

$$H_2O(l) \longrightarrow H_2O(s)$$

is simply the negative of that for the forward reaction, or -6.0 kilojoules. The second statement indicates that if the heat of fusion of 1 mole of water is 6.0 kilojoules,

$$H_2O(s) \longrightarrow H_2O(l) \qquad \Delta H = 6.0 \text{ kJ}$$

then the heat of fusion of 2 moles of water is twice as great,

$$2H_2O(s) \longrightarrow 2H_2O(l) \qquad \Delta H = 12 \text{ kJ}$$

EXAMPLE 4.10

The standard molar enthalpy of formation of $FeCl_2(s)$ is -341.8 kJ mol^{-1}. $FeCl_3(s)$ forms when 1 mol of $FeCl_2(s)$ reacts with $\frac{1}{2}$ mol of $Cl_2(g)$ under standard state conditions; the enthalpy change of the reaction, ΔH°_{298} is -57.7 kJ. What is the standard molar enthalpy of formation of $FeCl_3(s)$?

The magnitude of the standard molar enthalpy of formation of $FeCl_3(s)$ is equal to ΔH°_{298} for the reaction that forms 1 mol of $FeCl_3(s)$ from the elements under standard state conditions:

$$Fe(s) + \frac{3}{2} Cl_2(g) \longrightarrow FeCl_3(s)$$

This reaction can be written as the sum of two others.

STEP 1 $\qquad\qquad\qquad Fe(s) + Cl_2(g) \longrightarrow FeCl_2(s)$

STEP 2 $\qquad\qquad FeCl_2(s) + \frac{1}{2} Cl_2(g) \longrightarrow FeCl_3(s)$

SUM $\qquad\qquad\qquad Fe(s) + \frac{3}{2} Cl_2(g) \longrightarrow FeCl_3(s)$

Since 1 mol of $FeCl_2(s)$ is produced by Step 1, the magnitude of ΔH°_{298} for this step is equal to the magnitude of the standard molar enthalpy of formation of $FeCl_2(s)$. ΔH°_{298} for Step 2 is given. The sum of the enthalpy changes of these two steps is equal to the magnitude of ΔH°_f for $FeCl_3(s)$.

$$Fe(s) + Cl_2(g) \longrightarrow FeCl_2(s) \qquad \Delta H^\circ_{298} = -341.8 \text{ kJ}$$

$$FeCl_2(s) + \frac{1}{2} Cl_2(g) \longrightarrow FeCl_3(s) \qquad \Delta H^\circ_{298} = -57.7 \text{ kJ}$$

$$Fe(s) + \frac{3}{2} Cl_2(g) \longrightarrow FeCl_3(s) \qquad \Delta H^\circ_{298} = -399.5 \text{ kJ}$$

Under standard state conditions, the enthalpy change of the reaction which forms 1 mol of $FeCl_3(s)$ from the elements in their most stable states is -399.5 kJ. Therefore, the standard molar enthalpy of formation of $FeCl_3(s)$ is -399.5 kJ/mol.

Hess's law can be used to determine the enthalpy change, under standard state conditions, of any reaction if the standard molar enthalpies of formation of the reactants and products are available. The procedures used are (1) decompositions of the reactants into their component elements, for which the enthalpy changes are proportional to the negative of the heats of formation of the reactants, followed by (2) recombinations of the elements to give the products, for which the enthalpy changes are proportional to the enthalpies of formation of the products. This is one reason that a table of enthalpies of formation is so useful. The following examples illustrate the process.

EXAMPLE 4.11

Phosphoric acid for use in cola soft drinks and other food products is prepared by the reaction of phosphorus(V) oxide, P_4O_{10}, with water.

$$P_4O_{10}(s) + 6H_2O(l) \longrightarrow 4H_3PO_4(l)$$

How much heat is produced by the reaction of exactly 1 mol of $P_4O_{10}(s)$ with liquid water at constant pressure under standard state conditions?

The amount of heat, q, produced by the reaction of exactly 1 mol of $P_4O_{10}(s)$ at constant pressure under standard conditions is equal to the standard enthalpy change, ΔH°_{298}, for the reaction. The reaction of $P_4O_{10}(s)$ with water can be written as the sum of three steps, two of which involve decomposition of P_4O_{10} and H_2O to their component elements, and a third which involves formation of H_3PO_4 from these elements.

With the standard molar heats of formation of $P_4O_{10}(s)$, $H_2O(l)$, and $H_3PO_4(l)$ (-2984.0, -285.83, and -1266.9 kJ mol^{-1}, respectively) from Appendix I, we have all of the information necessary to determine ΔH°_{298}.

STEP 1 $P_4O_{10}(s) \longrightarrow 4P(s) + 5O_2(g)$ $\Delta H^\circ_1 = -\left(1 \text{ mol } P_4O_{10}(s) \times \dfrac{-2984.0 \text{ kJ}}{1 \text{ mol } P_4O_{10}(s)}\right)$

$$= -(-2984.0 \text{ kJ}) = 2984.0 \text{ kJ}$$

STEP 2 $6H_2O(l) \longrightarrow 6H_2(g) + 3O_2(g)$ $\Delta H^\circ_2 = -\left(6 \text{ mol } H_2O(l) \times \dfrac{-285.83 \text{ kJ}}{1 \text{ mol } H_2O(l)}\right)$

$$= -(-1715.0 \text{ kJ}) = 1715.0 \text{ kJ}$$

STEP 3 $6H_2(g) + 8O_2(g) + 4P(s) \longrightarrow 4H_3PO_4(l)$

$$\Delta H^\circ_3 = 4 \text{ mol } H_3PO_4(l) \times \dfrac{-1266.9 \text{ kJ}}{1 \text{ mol } H_3PO_4(l)}$$

$$= -5067.6 \text{ kJ}$$

SUM $P_4O_{10}(s) + 6H_2O(l) \longrightarrow 4H_3PO_4(l)$ $\Delta H^\circ_{298} = \Delta H^\circ_1 + \Delta H^\circ_2 + \Delta H^\circ_3$

$$= (2984.0 + 1715.0 - 5067.6) \text{ kJ}$$

$$= -368.6 \text{ kJ}$$

The reaction produces 368.6 kilojoules of heat.

EXAMPLE 4.12   Calculate the heat of combustion of methane, ΔH°_{298} for the following reaction.

$$CH_4(g) + 2O_2(g) \longrightarrow CO_2(g) + 2H_2O(l)$$

This reaction can be written as the sum of three steps: (1) decomposition of CH_4 to $C(s)$ and $H_2(g)$, (2) reaction of H_2 and O_2 to form H_2O, and (3) reaction of C and O_2 to form CO_2. The enthalpy changes of these steps are proportional to the standard molar enthalpies of formation found in Appendix I.

STEP 1 $CH_4(g) \longrightarrow C(s) + 2H_2(g)$ $\Delta H^\circ_1 = -\Delta H^\circ_{f_{CH_4(g)}}$

STEP 2 $2H_2(g) + O_2(g) \longrightarrow 2H_2O(l)$ $\Delta H^\circ_2 = 2\Delta H^\circ_{f_{H_2O(l)}}$

STEP 3 $C(s) + O_2(g) \longrightarrow CO_2(g)$ $\Delta H^\circ_3 = \Delta H^\circ_{f_{CO_2(g)}}$

SUM $CH_4(g) + 2O_2(g) \longrightarrow CO_2(g) + 2H_2O(l)$ $\Delta H^\circ_{298} = \Delta H^\circ_1 + \Delta H^\circ_2 + \Delta H^\circ_3$

$\Delta H^\circ_{298} = \Delta H^\circ_1 + \Delta H^\circ_2 + \Delta H^\circ_3$

$$= -\left(1 \text{ mol } CH_4(g) \times \dfrac{-74.81 \text{ kJ}}{1 \text{ mol } CH_4(g)}\right) + \left(2 \text{ mol } H_2O(l) \times \dfrac{-285.83 \text{ kJ}}{1 \text{ mol } H_2O(l)}\right)$$

$$+ 1 \text{ mol } CO_2(g) \times \dfrac{-393.51 \text{ kJ}}{1 \text{ mol } CO_2(g)}$$

$$= -890.36 \text{ kJ}$$

The combustion of 1 mol of methane produces 890.36 kJ when the water produced is a liquid. The value of -802 kJ given in Table 4.1 for the heat of combustion of methane is for the formation of *gaseous* water.

We will consider another form of Hess's law in Chapter 19, when we explore the subject of thermodynamics more fully.

4.6 Fuel and Food

The most common chemical reaction used to produce heat is **combustion**, the rapid combination of a fuel with oxygen, which is normally accompanied by a flame. Combustion is usually started by heating a fuel in oxygen or air, but after the reaction has begun it provides the necessary heat to keep itself going.

The major fuels are fossil fuels, all of which are being depleted much more rapidly than they are being formed. They consist primarily of **hydrocarbons**, compounds composed only of hydrogen and carbon. **Natural gas** is composed of low molecular-weight hydrocarbons, principally methane (CH_4) with small amounts of ethane (C_2H_6), propane (C_3H_8), and butane (C_4H_{10}). These hydrocarbons are almost odorless, so small amounts of foul-smelling compounds called mercaptans are added to the gas to aid in detecting leaks. Some natural gas contains hydrogen sulfide, a contaminant that is removed before the gas is sold for commercial or residential use. **Petroleum**, from which oil and gasoline are produced, is a liquid that varies in its composition, depending on the location of the wells from which it is pumped. It is a mixture that contains hundreds of different compounds, primarily hydrocarbons ranging from methane to compounds containing 50 carbon atoms. **Coal** is a solid fuel composed mainly of hydrocarbons with high molecular weights. Compounds containing oxygen, nitrogen, and sulfur are also found in petroleum and coal.

In many parts of the world, wood and dried animal dung are used as residential fuels for cooking and heating.

A synthetic fuel of great interest is **hydrogen**, H_2, one of the fuels used in the space shuttle. When hydrogen burns, it forms water, a compound with no negative environmental impact. Free hydrogen does not occur naturally. It is a byproduct of refining petroleum, and it can also be prepared from water by an endothermic reaction that requires energy.

$$2H_2O(l) \longrightarrow 2H_2(g) + O_2(g) \qquad \Delta H^\circ_{298} = +572 \text{ kJ}$$

As long as this energy must be obtained by combustion of fossil fuels, or as long as hydrogen is obtained from fossil fuels, it is unlikely to become a common fuel. However, if nuclear or solar energy technology becomes suitable for the production of hydrogen, this element could serve as a way of transporting energy from production site to use site, much as natural gas and oil do now.

The amount of heat that a fuel can produce can be determined from its heat of combustion (Section 4.4) or by use of Hess's law (Section 4.5, Example 4.12).

Which will produce more heat, combustion of 1.0 g of hydrogen, H_2, or 1.0 g of methane, CH_4, assuming that the water produced is in the gas phase. Heats of combustion may be found in Table 4.1.

EXAMPLE 4.13

At first glance it might appear that methane is the choice; the heat of combustion of methane is -802 kJ/mol, and that of H_2 is -242 kJ/mol. However, we need to divide these values by the number of grams in 1 mol of each of the fuels in order to find the heat of combustion per gram.

$$\text{For } H_2: \qquad \frac{-242 \text{ kJ}}{1 \text{ mol}} \times \frac{1 \text{ mol}}{2.0158 \text{ g}} = \frac{-120 \text{ kJ}}{1 \text{ g}}$$

$$\text{For } CH_4: \qquad \frac{-802 \text{ kJ}}{1 \text{ mol}} \times \frac{1 \text{ mol}}{16.043 \text{ g}} = \frac{-50.0 \text{ kJ}}{1 \text{ g}}$$

Hydrogen provides 120 kJ of heat per gram of fuel; methane, 50.0 kJ per gram.

Food is the body's fuel. In addition, it provides the nutrients necessary to maintain the physiological functions of the body. Each day, a normally active, healthy adult requires about 130 kJ of energy from food for each kilogram of body weight. This corresponds to about 14 kilocalories or 14 Calories per pound of body weight. (Remember that the Calorie used in nutrition is what chemists call a kilocalorie; 1 Calorie = 1 kcal = 4.184 kJ.)

Most of the energy used by our bodies comes from carbohydrates. These are broken down in the stomach to glucose (blood sugar), which is transported to the cells by the blood. In the cells glucose reacts with O_2 in a series of steps, eventually producing carbon dioxide, liquid water, and energy.

$$C_6H_{12}O_6(s) + 6O_2(g) \longrightarrow 6CO_2(g) + 6H_2O(l) \qquad \Delta H^\circ_{298} = -2870 \text{ kJ}$$

The amount of heat that can be produced in the body is the same as the amount that would be produced in a calorimeter burning the same amount of glucose under constant pressure.

The average heat of combustion per gram of carbohydrate is -17 kJ (-4 kcal/g), assuming liquid water is the product of the oxidation. Carbohydrates break down rapidly in the body, so this energy becomes available quickly.

Fats also produce carbon dioxide and water when metabolized in the body. Palmitic acid, a typical fat, reacts as follows:

$$C_{15}H_{31}CO_2H(s) + 23O_2(g) \longrightarrow 16CO_2(g) + 16H_2O(l) \qquad \Delta H^\circ_{298} = -9977 \text{ kJ}$$

Fats produce significantly more energy per gram than either carbohydrates or proteins; their heats of combustion average -38 kJ per gram (-9.1 kcal/g). This high heat of combustion per gram makes fats ideal for storage of energy that is not needed to

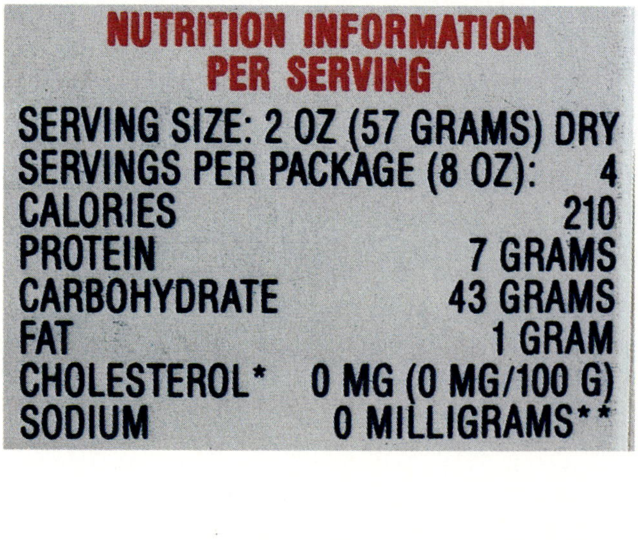

Figure 4.6

The calorie content on this label is determined from the masses of carbohydrate, fat, and protein in a serving.

NUTRITION INFORMATION
PER SERVING

SERVING SIZE: 2 OZ (57 GRAMS) DRY
SERVINGS PER PACKAGE (8 OZ):   4
CALORIES   210
PROTEIN   7 GRAMS
CARBOHYDRATE   43 GRAMS
FAT   1 GRAM
CHOLESTEROL*   0 MG (0 MG/100 G)
SODIUM   0 MILLIGRAMS**

maintain the body's function. The body does not waste energy; excess energy from overeating is stored by production of fat.

Proteins are not generally oxidized in the body; instead, their components are used to build body proteins. However, in cases where proteins are used to provide energy for the body, they produce about -17 kJ per gram (-4 kcal/g).

The Calorie content in a serving of a food (Fig. 4.6) is the sum of the Calories contained in the carbohydrates, fats, and proteins present in the serving.

For Review

Summary

Thermochemistry is the determination and study of the heat absorbed or evolved during chemical changes. **Heat, q,** is the form of energy that moves spontaneously from a hotter object to a cooler object. Heat, whether measured in units of **calories** or **joules,** is measured in a **calorimeter,** usually by determining the temperature change of water or of a solution of known **specific heat.**

As a system loses heat by an **exothermic** process, its internal energy is decreased. Heat is added to a system by an **endothermic** process, which also increases the internal energy of the system. When a system does **work,** the internal energy of the system is decreased. When work is done on a system, its internal energy increases.

If a chemical change is carried out at constant pressure and the only work done is due to expansion or contraction, q for the change is called the **enthalpy change** or **heat of reaction,** denoted by the symbol ΔH, or ΔH°_{298} if the reaction occurs under standard state conditions. Examples of enthalpy changes include **heat of combustion** and **standard molar enthalpy of formation.** The standard molar enthalpy of formation, ΔH°_f, is the enthalpy change accompanying the formation of 1 mole of a substance from the elements in their most stable states at 298.15 K and 1 atm (a **standard state**). If the enthalpies of formation are available for the reactants and products of a reaction, its enthalpy change can be calculated using **Hess's law:** if a process can be written as the sum of several stepwise processes, the enthalpy change of the total process equals the sum of the enthalpy changes of the various steps. The ΔH for a reaction in one direction is equal in magnitude, but opposite in sign, to ΔH of the reaction in the opposite direction, and ΔH is directly proportional to the quantity of reactants.

Heat is produced from fuels by combustion reactions. Common fuels include the fossil fuels—natural gas, petroleum, and coal. Food is the fuel used by the body.

Key Terms and Concepts

calorimetry (4.2)
calorie (4.1)
chemical thermodynamics (4.3)
combustion (4.6)
endothermic reaction (4.1)
enthalpy change, ΔH (4.4)
enthalpy change at standard state
    conditions, ΔH°_{298} (4.4)
exothermic reaction (4.1)

food (4.6)
fuel (4.6)
heat (4.1)
heat capacity (4.1)
heat of combustion (4.4)
heat of reaction (4.4)
hydrocarbon (4.6)
internal energy (4.3)
joule (4.1)

measurement of heat (4.1)
specific heat (4.1)
standard molar enthalpy of forma-
    tion, ΔH°_f (4.4)
standard state (4.4)
surroundings (4.3)
system (4.3)
work (4.3)

Exercises

Heat

1. A burning match and a bonfire may have the same temperature, yet we stay warm sitting around a bonfire, but not around a burning match, on a fall evening. Why not?

2. If the temperature of the 800 g of water in Fig. 4.2 increases by 25.0°C, how much heat in joules was added to the water? How much heat in calories?
   Ans. 8.37×10^4 J; 2.00×10^4 cal

3. (a) If 14.5 kJ of heat was added to the 800 g of water in Fig. 4.2, how much would its temperature increase?   *Ans. 4.33°C*

   (b) If 29 kcal of heat was added to the 800 g of water in Fig. 4.2, how much would its temperature increase?   *Ans. 36°C*

4. Explain the difference between *heat capacity* and *specific heat* of a substance.

5. How much heat, in joules and in calories, must be added to a 75.0-g iron block with a specific heat of 0.451 J/g °C to increase its temperature from 25°C to its melting temperature of 1535°C?   *Ans. 5.11×10^4 J; 1.22×10^4 cal*

6. The specific heat of ice is 1.95 J/g °C. How much heat, in joules and in calories, is required to heat a 28.4-g (1-oz) ice cube from −23.0°C to −1.0°C?
   Ans. 1.22×10^3 J; 291 cal

7. Calculate the heat capacity, in joules and in calories, of the following

   (a) 28.4 g (1 oz) of water (specific heat 4.184 J/g °C)
   Ans. 119 J/°C; 28.4 cal/°C

   (b) 28.4 g (1 oz) of lead (specific heat 0.129 J/g °C)
   Ans. 3.66 J/°C; 0.876 cal/°C

   (c) 45.8 g of nitrogen gas (specific heat 1.04 J/g°C)
   Ans. 47.6 J/°C; 11.4 cal/°C

8. Calculate the specific heat of water in the units Calories/lb °F (1 Calorie is the nutritional calorie = 1000 calories). The specific heat of water is 4.184 J/g°C
   Ans. 0.2520 Calories/lb°F

9. How many mL of water at 23°C with a density of 1.00 g/mL must be mixed with 180 mL (about 6 oz) of coffee at 95°C so that the resulting combination will have a temperature of 60°C? Assume that coffee and water have the same density and the same specific heat.   *Ans. 170 mL*

10. How much will the temperature of 180 mL of coffee at 95°C be reduced when a 45-g silver spoon (specific heat 0.24 J/g °C) at 25°C is placed in the coffee and the two are allowed to reach the same temperature? Assume that the coffee has the same density and specific heat as water.
    Ans. 1°C

Calorimetry

11. If 1.506 kJ of heat are added to 30.0 g of water at 26.5°C in a coffee cup calorimeter like that in Fig. 4.3, what is the resulting temperature of the water?   *Ans. 38.5°C*

12. A 70.0-g piece of metal at 80.0°C is placed in 100 g of water at 22.0°C contained in a coffee cup calorimeter like that shown in Fig. 4.3. The metal and water come to the same temperature at 26.4°C. How much heat did the metal give up to the water? What is the specific heat of the metal?   *Ans. 1.8 kJ; 0.49 J/g °C*

13. A 0.500-g sample of KCl is added to 50.0 g of water in a coffee cup calorimeter (Fig. 4.3). If the temperature decreases by 1.05°C, what amount of heat is involved in the dissolution of the KCl, assuming the heat capacity of the resulting solution is 4.18 J/g °C?   *Ans. 222 J*

14. When 50.0 g of 0.200 M NaCl(aq) at 24.10°C is added to 50.0 g of 0.200 M AgNO₃(aq) at 24.10°C in a coffee cup calorimeter, the temperature increases to 25.67°C as AgCl(s) forms. Assuming the specific heat of the solution and products is 4.20 J/g °C, how much heat in joules is released in the calorimeter?   *Ans. 659 J*

15. When 1.0 g of fructose, $C_6H_{12}O_6(s)$, a sugar commonly found in fruits, is burned in oxygen in a bomb calorimeter (Fig. 4.4), the temperature of the calorimeter increases by 1.58°C. If the heat capacity of the calorimeter and its contents is 9.90 kJ/°C, what is q for this combustion?   *Ans. −15.6 kJ*

16. A sample of 0.562 g of carbon in the form of graphite is burned in oxygen in a bomb calorimeter (Fig. 4.4). The temperature of the calorimeter increases from 26.74°C to 27.63°C. If the specific heat of the calorimeter and its contents is 20.7 kJ/°C, how much heat was released by this reaction?   *Ans. 18 kJ*

Enthalpy Changes

17. Which of the heats of combustion in Table 4.1 are also molar heats of formation?

18. Does $\Delta H^\circ_{f_{H_2O(g)}}$ differ from ΔH°_{298} for the reaction $2H_2(g) + O_2(g) \longrightarrow 2H_2O(g)$? If so, how?

19. For the conversion of graphite to diamond
    $$C(s, graphite) \longrightarrow C(s, diamond)$$
    $$\Delta H^\circ_{298} = 1.90 \text{ kJ mol}^{-1}$$

    Is graphite or diamond the more stable form of carbon under standard conditions?

20. When 2.50 g of methane burns in oxygen, 125 kJ of heat is produced. What is the molar heat of combustion of methane under these conditions?   *Ans. −802 kJ mol⁻¹*

21. In 1774 oxygen was prepared by Joseph Priestley by heating red mercury(II) oxide with sunlight focused through a lens. How much heat is required to decompose 1 mol of HgO(s) to Hg(l) and O₂(g) under standard conditions?   *Ans. 90.83 kJ*

22. How many kilojoules of heat will be liberated when 1 mol of manganese, Mn, is burned to form $Mn_3O_4(s)$ at standard state conditions? ΔH°_f of Mn_3O_4 is equal to −1388 kJ mol⁻¹.   *Ans. 462.7 kJ*

23. The reaction of graphite with oxygen is described by the equation

$$C(s, graphite) + O_2(g) \longrightarrow CO_2(g)$$

Assume that the reaction was run at constant pressure and calculate the molar enthalpy of formation of $CO_2(g)$ from the data in exercise 16. What conditions would be required for this to be a standard molar enthalpy of formation?
Ans. -3.9×10^2 kJ mol^{-1}

24. Calculate the molar heat of solution (ΔH for the dissolution) of KCl under the conditions described in exercise 13.
Ans. 33.1 kJ mol^{-1}

25. Assume that the solutions have densities of 1.00 g/mL and calculate the approximate value of ΔH for the reaction

$$NaCl(aq) + AgNO_3(aq) \longrightarrow AgCl(s) + NaNO_3(aq)$$

using the data in exercise 14. (The result is only approximate because the density of the solutions is assumed to be 1.00 g/mL.)
Ans. -65.9 kJ mol^{-1}

26. Assume that the combustion of fructose described in exercise 15 was carried out at constant pressure and is described by the equation

$$C_6H_{12}O_6(s) + 6O_2(g) \longrightarrow 6CO_2(g) + 6H_2O(l)$$

Calculate the molar heat of combustion of fructose under these conditions. *Ans. -2.8×10^3 kJ mol^{-1}*

Hess's Law

27. For the conversion of graphite to diamond
$$C(s, graphite) \longrightarrow C(s, diamond)$$

$$\Delta H^\circ_{298} = 1.90 \text{ kJ mol}^{-1}$$

Do the heats of combustion of graphite and carbon differ? That is, are the enthalpy changes for the following reactions the same or different?

$$C(s, graphite) + O_2(g) \longrightarrow CO_2(g)$$
$$C(s, diamond) + O_2(g) \longrightarrow CO_2(g)$$

28. Calculate ΔH for the process

$$Os(s) + 2O_2(g) \longrightarrow OsO_4(g)$$

from the following information:

$$Os(s) + 2O_2(g) \longrightarrow OsO_4(s) \qquad \Delta H = -391 \text{ kJ}$$
$$OsO_4(s) \longrightarrow OsO_4(g) \qquad \Delta H = 56.4 \text{ kJ}$$
Ans. -335 kJ

29. Calculate ΔH°_{298} for the process

$$Sb(s) + \frac{5}{2} Cl_2(g) \longrightarrow SbCl_5(g)$$

from the following information:

$$Sb(s) + \frac{3}{2} Cl_2(g) \longrightarrow SbCl_3(g) \qquad \Delta H^\circ_{298} = -314 \text{ kJ}$$

$$SbCl_3(g) + Cl_2(g) \longrightarrow SbCl_5(g) \qquad \Delta H^\circ_{298} = -80 \text{ kJ}$$
Ans. -394 kJ

30. Calculate ΔH°_{298} for the process

$$Zn(s) + S(s) + 2O_2(g) \longrightarrow ZnSO_4(s)$$

from the following information:

$$Zn(s) + S(s) \longrightarrow ZnS(s) \qquad \Delta H^\circ_{298} = -206.0 \text{ kJ}$$
$$ZnS(s) + 2O_2(g) \longrightarrow ZnSO_4(s) \qquad \Delta H^\circ_{298} = -776.8 \text{ kJ}$$
Ans. -982.8 kJ

31. Calculate ΔH for the process

$$2Hg(l) + Cl_2(g) \longrightarrow Hg_2Cl_2(s)$$

from the following information:

$$Hg(l) + Cl_2(g) \longrightarrow HgCl_2(s) \qquad \Delta H = -224 \text{ kJ}$$
$$Hg(l) + HgCl_2(s) \longrightarrow Hg_2Cl_2(s) \qquad \Delta H = -41.2 \text{ kJ}$$
Ans. -265 kJ

32. Calculate ΔH°_{298} for the process

$$3Co(s) + 2O_2(g) \longrightarrow Co_3O_4(s)$$

from the following information:

$$Co(s) + \frac{1}{2} O_2(g) \longrightarrow CoO(s)$$

$$\Delta H^\circ_{298} = -237.9 \text{ kJ}$$

$$3CoO(s) + \frac{1}{2} O_2(g) \longrightarrow Co_3O_4(s)$$

$$\Delta H^\circ_{298} = -177.5 \text{ kJ}$$
Ans. -891.2 kJ

33. Calculate the standard molar enthalpy of formation of $NO(g)$ from the following data:

$$N_2(g) + 2O_2(g) \longrightarrow 2NO_2(g) \qquad \Delta H^\circ_{298} = 66.4 \text{ kJ}$$
$$2NO(g) + O_2(g) \longrightarrow 2NO_2(g) \qquad \Delta H^\circ_{298} = -114.1 \text{ kJ}$$
Ans. 90.2 kJ mol^{-1}

34. From the molar heats of formation in Appendix I, determine how much heat is required to evaporate one mol water: $H_2O(l) \longrightarrow H_2O(g)$. *Ans. 44.01 kJ*

35. The decomposition of hydrogen peroxide, H_2O_2, has been used to provide thrust in the control jets of various space vehicles. Using the data in Appendix I, determine how much heat is produced by the decomposition of exactly 1 mol of H_2O_2 under standard conditions.

$$2H_2O_2(l) \longrightarrow 2H_2O(g) + O_2(g)$$

Ans. 54.0 kJ

36. (a) Calculate the molar heat of combustion of propane, $C_3H_8(g)$, for the formation of $H_2O(l)$ and $CO_2(g)$. The enthalpy of formation of propane is -104 kJ/mol.
Ans. -2220 kJ mol^{-1}

(b) Calculate the molar heat of combustion of butane, $C_4H_{10}(g)$, for the formation of $H_2O(l)$ and $CO_2(g)$. The enthalpy of formation of butane is -126 kJ mol^{-1}.
Ans. -2880 kJ mol^{-1}

(c) Both propane and butane are used as gaseous fuels. Which one has the higher heat of combustion per gram? *Ans. C_3H_8*

Fuel and Food

37. From the data in Table 4.1, determine which of the following fuels produces the greatest amount of heat per gram when burned under standard conditions: $CO(g)$, $CH_4(g)$, or $C_2H_2(g)$?   *Ans. CH_4*

38. The heat of combustion of hard coal averages -35 kJ g^{-1}, that of gasoline, $-33,800$ kJ L^{-1}. How many kilograms of hard coal provide the same amount of heat as is available from 1.0 L of gasoline?   *Ans. 0.97 kg*

39. Ethanol, C_2H_5OH, is used as a fuel for motor vehicles, particularly in Brazil. (a) Write the balanced equation for the combustion of ethanol to $CO_2(g)$ and $H_2O(g)$, and, using the data in Appendix I, calculate the heat of combustion of 1 mol of ethanol. (b) The density of ethanol is 0.7893 g mL^{-1}. Calculate the heat of combustion of exactly 1 mL of ethanol. (c) Assuming that the mileage an automobile gets is directly proportional to the heat of combustion of the fuel, calculate how many times farther an automobile could be expected to go on 1 gal of gasoline than on 1 gal of ethanol. Assume that gasoline has the heat of combustion and the density of n-octane, C_8H_{18} ($\Delta H_f^\circ = -208.4$ kJ mol^{-1}; density $= 0.7025$ g mL^{-1}).
Ans. (a) $C_2H_5OH(l) + 3O_2(g) \longrightarrow 2CO_2(g) + 3H_2O(g)$; -1234.8 kJ; (b) -21.16 kJ; (c) 1.487 times farther

40. A teaspoon of the carbohydrate sucrose (common sugar) is reputed to contain 16 nutritional Calories (16 kcal). What is the mass of one teaspoon of sucrose?   *Ans. 4 g*

41. What is the maximum mass of carbohydrate in a 12-oz can of Diet Coke that contains less than 1 (nutritional) Calorie per can?   *Ans. 0.25 g*

42. What mass of fat, in grams and ounces, must be produced in the body to store an extra 100 nutritional Calories?
Ans. 11 g; 0.39 oz

43. The oxidation of the sugar glucose, $C_6H_{12}O_6$, is described by the following equation.

$$C_6H_{12}O_6(s) + 6O_2(g) \longrightarrow 6CO_2(g) + 6H_2O(l) \quad \Delta H = -2816 \text{ kJ}$$

Metabolism of glucose gives the same products, although the glucose reacts with oxygen in a series of steps in the body. (a) How much heat in kilojoules is produced by the metabolism of 1.0 g of glucose? (b) How many nutritional Calories (1 cal = 4.184 J; 1 nutritional Cal = 1000 cal) are produced by the metabolism of 1.0 g of glucose?
Ans. 16 kJ; 3.7 Calories

Additional Exercises

44. A sample of $WO_2(s)$ with a mass of 0.9745 g was "burned" in oxygen in a bomb calorimeter like that shown in Fig. 4.4, and the temperature increased by 1.310°C. (a) Calculate the heat released if the heat capacity of the

calorimeter and product is 872.3 J/°C. (b) The product of the reaction is $WO_3(s)$. Assume that the heat produced is proportional to the enthalpy change of the reaction in the calorimeter and calculate the enthalpy change of the reaction per mole of product. (c) The enthalpy of formation of $WO_3(s)$ under these conditions is -842.91 kJ mol^{-1}. From these data, calculate the enthalpy of formation of $WO_2(s)$.
Ans. 1.143 kJ; -253.1 kJ mol^{-1}; -589.8 kJ mol^{-1}

45. Water gas, a mixture of H_2 and CO, is an important industrial fuel produced by the reaction of steam with red-hot coke, essentially pure carbon.

$$C(s) + H_2O(g) \longrightarrow CO(g) + H_2(g)$$

(a) Assuming that coke has the same enthalpy of formation as graphite, calculate ΔH_{298}° for this reaction.
Ans. 131.30 kJ

(b) Methanol, a liquid fuel that could possibly replace gasoline, can be prepared from water gas and additional hydrogen at high temperature and pressure in the presence of a suitable catalyst.

$$2H_2(g) + CO(g) \longrightarrow CH_3OH(g)$$

Under the conditions of the reaction, methanol forms as a gas. Calculate ΔH_{298}° for this reaction and for the condensation of gaseous methanol to liquid methanol.
Ans. -90.2 kJ mol^{-1}; -38.0 kJ mol^{-1}

(c) Calculate the heat of combustion of 1 mol of liquid methanol to $H_2O(g)$ and $CO_2(g)$.
Ans. -638.4 kJ

46. (a) Using the data in Appendix I, calculate the standard enthalpy change for each of the following reactions:
(1) $N_2(g) + O_2(g) \longrightarrow 2NO(g)$
Ans. 180.5
(2) $Fe_2O_3(s) + 3H_2(g) \longrightarrow 2Fe(s) + 3H_2O(l)$
Ans. -33.3 kJ
(3) $2LiOH(s) + CO_2(g) \longrightarrow Li_2CO_3(s) + H_2O(g)$
Ans. -89.4 kJ
(4) $CH_4(g) + N_2(g) \longrightarrow HCN(g) + NH_3(g)$
Ans. 164 kJ
(5) $CS_2(g) + 3Cl_2(g) \longrightarrow CCl_4(g) + S_2Cl_2(g)$
Ans. -279.7 kJ

(b) Which of these reactions are exothermic?
Ans. (2); (3); (5)

47. The white pigment TiO_2 is prepared by the hydrolysis of titanium tetrachloride, $TiCl_4$, in the gas phase.

$$TiCl_4(g) + 2H_2O(g) \longrightarrow TiO_2(s) + 4HCl(g)$$

How much heat is evolved in the production of exactly 1 mol of $TiO_2(s)$ under standard state conditions?
Ans. 67.1 kJ

Structure of the Atom and the Periodic Law

5

Light from excited krypton atoms is separated into a line spectrum.

A s John Dalton examined the experimental observations that led to his theory that matter was composed of atoms, he found no evidence that atoms were composed of smaller particles. Thus Dalton assumed that they were simple indivisible bodies. However, a series of discoveries beginning in the later part of the nineteenth century showed that atoms are complex systems composed of a number of smaller particles.

Many different experiments have contributed to our ideas about the structure of atoms, yet there are a few which have been particularly important in the development of these ideas. In this chapter, we will examine some of the key experiments. Then we will discuss the details of atomic structure itself and describe how several kinds of chemical behavior follow from this structure. In fact, observed regularities in chemical behavior, as shown by the Periodic Table, provide additional evidence for our view of the structure of the atom.

As will be seen in later chapters, the chemical behavior of atoms and molecules is primarily determined by the arrangement of their electrons. Thus this study of the structure of atoms is the first step in developing a model of matter that will help you to understand why substances exhibit particular kinds of chemical properties and to recall these properties systematically.

The Structure of the Atom

5.1 Electrons

Figure 5.1

Two types of discharge tubes. (a) In the absence of a charge on the plates, electrons flow from the cathode to the anode (black path). When the plates are charged, the electron stream curves toward the positive plate (red path). (b) A shadow being cast by an object in a stream of electrons. The tube glows where there is no shadow. (c) A photograph of the tube illustrated in (b).

Much of our knowledge about the structure of atoms resulted from experiments involving the passage of electricity through different gases at low pressures. The apparatus used for experiments of this type, called a **discharge tube,** or **Crookes tube,** was developed by Sir William Crookes (1832–1919). Crookes passed an electric current through a gas-filled glass tube with electrodes sealed into both ends [Fig. 5.1(a)]. As gas was pumped out of the tube, a pressure was reached at which the remaining gas glowed. With more pumping, the gas ceased to glow. However, close observation revealed that the glass at one end of the tube was glowing. Crookes suggested that this glow was produced by negative particles, which he called **cathode rays,** passing from

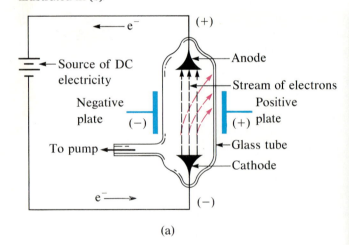

(a)

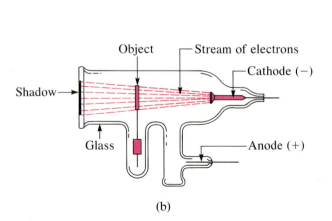

(b)

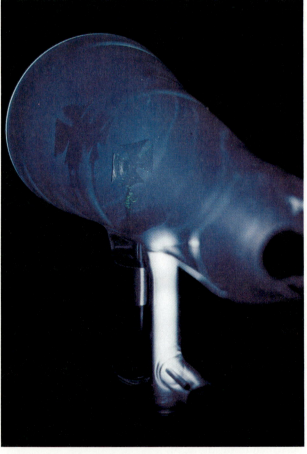

(c)

the negative electrode (**cathode**) and moving toward the positive electrode (**anode**). Those particles that missed the anode hit the glass and made it glow.

Cathode rays were shown to have the following characteristics:

1. They travel from the cathode in straight lines, as indicated by the fact that an object placed in their path casts a sharp shadow on the end of the tube [Fig. 5.1(b)].
2. They cause a piece of metal foil to become hot by striking it for a period of time.
3. They are deflected by magnetic and electrical fields [Fig. 5.1(a)], and the direction of the deflection is the same as that shown by negatively charged particles passing through such fields (a charged object will attract an object of opposite charge, whereas objects having charges of the same sign repel each other).

These characteristics are best explained by assuming that cathode rays are streams of small negatively charged particles. These particles are called **electrons.**

In 1897, the English physicist Sir J. J. Thomson measured the deflection of cathode-ray particles in both a magnetic field and an electric field. From the strengths of the fields, he determined the ratio of the charge, *e*, to the mass, *m*, of the cathode-ray particles. He found *e/m* to be identical for all of these particles, irrespective of the metal the electrodes were made of or the kind of gas in the tube. The charge was measured by the American physicist R. A. Millikan in 1909 (Fig. 5.2). From the values of *e/m* and *e*, he found the mass of the electron to be 1/1837 of that of the hydrogen atom, or 0.00055 amu. This quantitative observation that all electrons are identical, regardless of their source, along with the observation that they can be liberated from any kind of atom, proves quite conclusively that electrons are parts of all types of matter.

Electrons are emitted by some metals when the metals are heated to high temperatures. This process is called **thermal emission** and is the source of electrons in X-ray tubes and cathode-ray tubes such as television picture tubes. Electrons are also emitted from metals, most readily from active metals such as cesium, sodium, and potassium, when they are exposed to light (Fig. 5.3). This type of emission is known as the

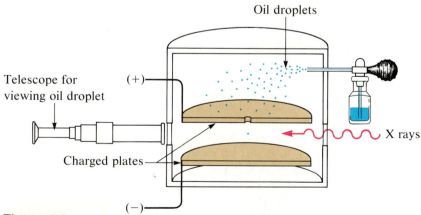

Figure 5.2

The apparatus used by Millikan. A fine drop of oil drifts from the top of the apparatus through the hole into the region between the two plates. The X rays cause the drop to pick up a negative charge, and the drop is attracted toward the positive plate. At one particular voltage, the electrical and gravitational forces are balanced, and the drop remains stationary. From the charge on the plates and the mass of the drop, the charge on the drop can be calculated.

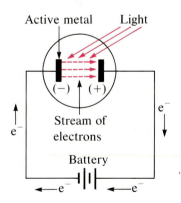

Figure 5.3

A photoelectric cell in an electric circuit. Light striking the active metal electrode causes electrons to be emitted and a current to flow.

Figure 5.5

The scattering of α particles by a gold atom. The α parti-cles, with a + 2 charge, are deflected back toward their source only when they col-lide with the much heavier, positively charged gold nu-cleus, with a + 79 charge. (Re-call that particles with a charge of like sign repel one another.) Since the nucleus is very small, compared to the size of an atom, very few α particles collide with the nu-cleus. A few other α particles pass close to the gold nu-cleus and are deflected by the repulsion, but most (shown in blue) pass through the relatively large region occupied by electrons, which are too light to deflect the rapidly moving α particles. If the nucleus were drawn to scale in this figure, it would be invisible.

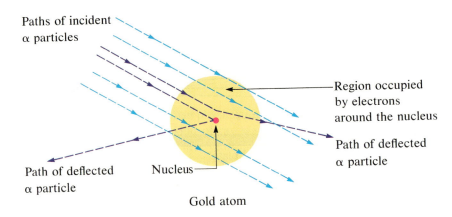

Paths of incident α particles

Region occupied by electrons around the nucleus

Path of deflected α particle

Path of deflected α particle   Nucleus

Gold atom

electrons necessary to produce an electrically neutral atom (Fig. 5.5). Rutherford's nuclear theory of the atom, proposed in 1911, is the model we still use.

Rutherford determined the diameter of the nucleus to be at least 100,000 times smaller than the diameter of the atom. (The diameter of an atom is of the order of 10^{-8} cm.) Thus a nucleus is almost unbelievably small. If the nucleus of an atom were as large as the period at the end of this sentence, the atom would have a diameter of about 40 yards.

From this same series of experiments, Rutherford found that, for many of the lighter elements, the number of positive charges in the nucleus (and thus the number of protons in the nucleus) is approximately equal to half of the atomic weight of the element. In 1914 another English physicist, Henry Moseley, reported a method for determining the number of protons in the nucleus of any atom, and hence the number of positive charges in the nucleus. The number of protons in the nucleus is called the **atomic number** of the element. Since a neutral atom contains the same number of protons as electrons, the atomic number represents the number of electrons *for a neutral atom* as well.

Moseley measured the X rays produced by the elements. A modern X-ray tube (Fig. 5.6) is a modified cathode-ray tube in which electrons are produced by thermal emission from a filament heated by an electric current. When a solid target is placed in the beam of electrons, very penetrating rays, called X rays, are produced by the ele-ments in the target. Using a series of different elements as targets, Moseley showed that the energy of the X rays produced by an element depends on its atomic number. This fact was used to determine the atomic numbers of the heavier elements.

5.6 Isotopes

At about the same time that Moseley was carrying out his experiments, A. J. Dempster and F. W. Aston, working separately, showed that some elements consist of atoms with different masses. Magnesium, for example, was shown to consist of three different types of atoms: approximately 79% have a mass of about 24 amu; 10%, about 25 amu; and 11%, about 26 amu.

Both Dempster and Aston used early versions of a **mass spectrometer** for their work. In the mass spectrometer (Fig. 5.7), gaseous atoms or molecules are bombarded by high-energy electrons, which knock off electrons and produce positively charged ions. These ions are then directed through a magnetic or electric field, which deflects their paths to an extent dependent on their mass-to-charge ratios (m/e). Ions of any mass-to-charge ratio can be made to strike the detector by varying the field. Since most

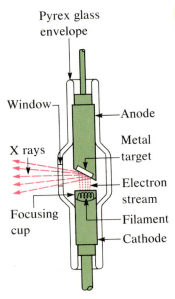

Pyrex glass envelope

Window

X rays

Focusing cup

Anode

Metal target

Electron stream

Filament

Cathode

Figure 5.6

In an X-ray tube, electrons striking a target such as tungsten cause the produc-tion of X rays. In Moseley's experiment, various elements were used as targets.

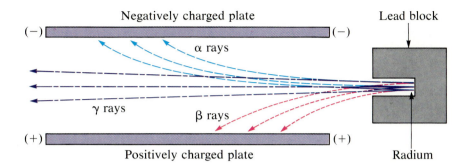

Figure 5.4
The effect of an electric field on radiation from the radioactive substance radium. Beta (β) rays are electrons. Alpha (α) rays are nuclei of helium atoms. Gamma (γ) rays are similar to X rays.

have a mass of approximately 4 amu and a charge of +2. Alpha particles are helium ions—helium atoms that have lost two electrons each.

2. The second type of ray curves toward the positive part of the field, with a deflection much greater than that of the alpha particles. This behavior indicates that these particles, called **beta (β) particles,** are negatively charged and have much less mass than alpha particles. Beta particles have been shown to be electrons.

3. The third type of ray is not deflected at all when it passes through the electric field, indicating that the particles are neither negatively nor positively charged. These rays are similar to X rays but have higher energies than X rays (and higher energies than α particles and β particles as well). These high-energy particles are called **gamma (γ) rays.**

Several other types of rays (radiation) are also known and will be discussed in Chapter 24.

Early experimenters used particles from radioactive materials to investigate the structure of atoms. For example, proof that atoms contain protons was obtained in England by Ernest Rutherford in 1919. Using high-velocity α particles from radium as projectiles, he bombarded atoms such as nitrogen and aluminum and found that protons were ejected from the atoms as a result of these collisions. These experiments indicated quite definitely that the proton is a unit of the atom.

5.5 The Nuclear Atom

Although the experiments described above indicated that atoms can be characterized as assemblies of electrons, protons, and neutrons, they gave no clue as to how these particles are arranged. The first insight into the architecture of atoms resulted from α-particle scattering experiments by Ernest Rutherford.

When Rutherford projected a beam of α particles from a radioactive source onto very thin gold foil (Fig. 5.5), he found that most of the particles passed through the foil without deflection. However, a few were diverted from their paths, and a very few were deflected back toward their source. From the results of a series of such experiments, Rutherford concluded that (1) the volume occupied by an atom must be largely empty space, because most of the α particles pass through the foil undeflected, and (2) each atom must contain a heavy, positively charged body (the nucleus) because of the abrupt change in path of a few α particles. A fast-moving, positively charged α particle can change path only when it hits or closely approaches another body (the nucleus) with a highly concentrated, positive charge.

The atom, then, was presumed to consist of a very small, positively charged **nucleus,** in which most of the mass of the atom is concentrated, surrounded by the

Figure 5.5

The scattering of α particles by a gold atom. The α particles, with a + 2 charge, are deflected back toward their source only when they collide with the much heavier, positively charged gold nucleus, with a + 79 charge. (Recall that particles with a charge of like sign repel one another.) Since the nucleus is very small, compared to the size of an atom, very few α particles collide with the nucleus. A few other α particles pass close to the gold nucleus and are deflected by the repulsion, but most (shown in blue) pass through the relatively large region occupied by electrons, which are too light to deflect the rapidly moving α particles. If the nucleus were drawn to scale in this figure, it would be invisible.

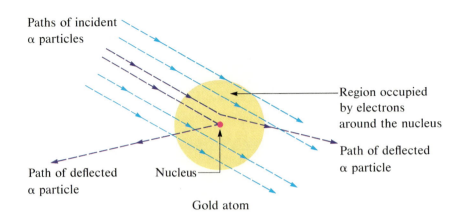

Paths of incident α particles

Region occupied by electrons around the nucleus

Path of deflected α particle

Path of deflected α particle

Nucleus

Gold atom

electrons necessary to produce an electrically neutral atom (Fig. 5.5). Rutherford's nuclear theory of the atom, proposed in 1911, is the model we still use.

Rutherford determined the diameter of the nucleus to be at least 100,000 times smaller than the diameter of the atom. (The diameter of an atom is of the order of 10^{-8} cm.) Thus a nucleus is almost unbelievably small. If the nucleus of an atom were as large as the period at the end of this sentence, the atom would have a diameter of about 40 yards.

From this same series of experiments, Rutherford found that, for many of the lighter elements, the number of positive charges in the nucleus (and thus the number of protons in the nucleus) is approximately equal to half of the atomic weight of the element. In 1914 another English physicist, Henry Moseley, reported a method for determining the number of protons in the nucleus of any atom, and hence the number of positive charges in the nucleus. The number of protons in the nucleus is called the **atomic number** of the element. Since a neutral atom contains the same number of protons as electrons, the atomic number represents the number of electrons *for a neutral atom* as well.

Moseley measured the X rays produced by the elements. A modern X-ray tube (Fig. 5.6) is a modified cathode-ray tube in which electrons are produced by thermal emission from a filament heated by an electric current. When a solid target is placed in the beam of electrons, very penetrating rays, called X rays, are produced by the elements in the target. Using a series of different elements as targets, Moseley showed that the energy of the X rays produced by an element depends on its atomic number. This fact was used to determine the atomic numbers of the heavier elements.

5.6 Isotopes

At about the same time that Moseley was carrying out his experiments, A. J. Dempster and F. W. Aston, working separately, showed that some elements consist of atoms with different masses. Magnesium, for example, was shown to consist of three different types of atoms: approximately 79% have a mass of about 24 amu; 10%, about 25 amu; and 11%, about 26 amu.

Both Dempster and Aston used early versions of a **mass spectrometer** for their work. In the mass spectrometer (Fig. 5.7), gaseous atoms or molecules are bombarded by high-energy electrons, which knock off electrons and produce positively charged ions. These ions are then directed through a magnetic or electric field, which deflects their paths to an extent dependent on their mass-to-charge ratios (*m/e*). Ions of any mass-to-charge ratio can be made to strike the detector by varying the field. Since most

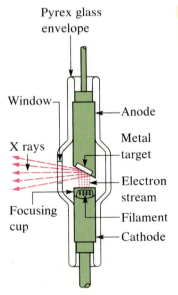

Pyrex glass envelope

Window

X rays

Focusing cup

Anode

Metal target

Electron stream

Filament

Cathode

Figure 5.6

In an X-ray tube, electrons striking a target such as tungsten cause the production of X rays. In Moseley's experiment, various elements were used as targets.

the negative electrode **(cathode)** and moving toward the positive electrode **(anode).** Those particles that missed the anode hit the glass and made it glow.

Cathode rays were shown to have the following characteristics:

1. They travel from the cathode in straight lines, as indicated by the fact that an object placed in their path casts a sharp shadow on the end of the tube [Fig. 5.1(b)].
2. They cause a piece of metal foil to become hot by striking it for a period of time.
3. They are deflected by magnetic and electrical fields [Fig. 5.1(a)], and the direction of the deflection is the same as that shown by negatively charged particles passing through such fields (a charged object will attract an object of opposite charge, whereas objects having charges of the same sign repel each other).

These characteristics are best explained by assuming that cathode rays are streams of small negatively charged particles. These particles are called **electrons.**

In 1897, the English physicist Sir J. J. Thomson measured the deflection of cathode-ray particles in both a magnetic field and an electric field. From the strengths of the fields, he determined the ratio of the charge, e, to the mass, m, of the cathode-ray particles. He found e/m to be identical for all of these particles, irrespective of the metal the electrodes were made of or the kind of gas in the tube. The charge was measured by the American physicist R. A. Millikan in 1909 (Fig. 5.2). From the values of e/m and e, he found the mass of the electron to be 1/1837 of that of the hydrogen atom, or 0.00055 amu. This quantitative observation that all electrons are identical, regardless of their source, along with the observation that they can be liberated from any kind of atom, proves quite conclusively that electrons are parts of all types of matter.

Electrons are emitted by some metals when the metals are heated to high temperatures. This process is called **thermal emission** and is the source of electrons in X-ray tubes and cathode-ray tubes such as television picture tubes. Electrons are also emitted from metals, most readily from active metals such as cesium, sodium, and potassium, when they are exposed to light (Fig. 5.3). This type of emission is known as the

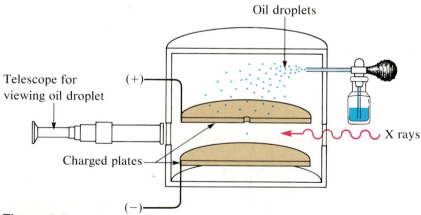

Figure 5.2

The apparatus used by Millikan. A fine drop of oil drifts from the top of the apparatus through the hole into the region between the two plates. The X rays cause the drop to pick up a negative charge, and the drop is attracted toward the positive plate. At one particular voltage, the electrical and gravitational forces are balanced, and the drop remains stationary. From the charge on the plates and the mass of the drop, the charge on the drop can be calculated.

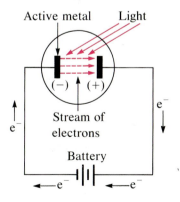

Figure 5.3

A photoelectric cell in an electric circuit. Light striking the active metal electrode causes electrons to be emitted and a current to flow.

photoelectric effect and is the basis of the electric eye, or photoelectric cell, used in some automatic door openers.

5.2 Protons

In 1886 Eugen Goldstein, a German physicist, using a Crookes tube with holes in the cathode, observed that another kind of ray was emitted from the anode and passed through the holes in the cathode. Three years later, Wilhelm Wien showed these rays to be positively charged particles. Their charge-to-mass ratio was found to be much smaller than that for electrons, and it varied with the kind of gas in the tube. Since the ratio varied, it follows that either the charge varied or the mass varied or both. Actually, both vary. Measurements showed the charge on each particle to be a positive charge, equal to that on the electron, or to some whole-number multiple of that charge, but opposite in sign. The mass of the positive particle was found to be least when hydrogen was used as the gas in the discharge tube. From the values of e and e/m for the positive particles when hydrogen was used, the value of m was calculated as 1.0073 amu. These particles are called **protons** and are one of the basic units of structure of all atoms.

The formation of positively charged particles, called **ions,** from the molecules of gas in a Crookes tube is caused by the loss of electrons from the molecules when they are struck by high-speed cathode rays. Since any neutral atom or molecule can be made to form positive ions by the loss of one or more electrons, it follows that every atom or molecule must contain one or more positive units (or protons). The simplest atom is that of hydrogen—it contains one proton and one electron.

5.3 Neutrons

The existence of the **neutron,** a particle with a mass close to that of the proton but with no charge, had been postulated for more than ten years before the English physicist James Chadwick detected this third basic unit of the atom in 1932. He showed that uncharged particles (neutrons) are emitted when atoms of beryllium (and other elements) are bombarded with high-velocity helium atoms with all electrons removed (alpha particles). Neutrons were determined to have a mass of 1.0087 amu. They are unstable outside of an atom and slowly disintegrate to form a proton and an electron.

The basic particles present in an atom are summarized in Table 5.1.

Table 5.1	Basic particles in an atom	
Particle	Mass, amu	Charge
Electron	0.00055	−1
Proton	1.0073	+1
Neutron	1.0087	0

5.4 Radioactivity and Atomic Structure

In 1896 additional evidence that atoms are complex rather than indivisible came with the discovery of radioactivity by the French physicist Antoine Becquerel. **Radioactivity** is the spontaneous decomposition of atoms of certain elements, such as radium and uranium, into other elements, with the simultaneous production of so-called rays. Three types of rays were characterized by passing them through an electric field (Fig. 5.4):

1. One type of ray curves toward the negative part of the electric field and so must consist of positively charged particles. These are called **alpha (α) particles,** which

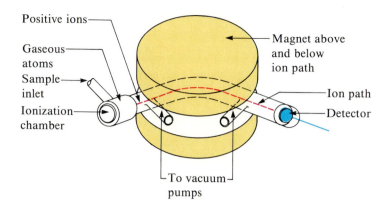

Positive ions

Gaseous atoms

Sample inlet

Ionization chamber

Magnet above and below ion path

Ion path

Detector

To vacuum pumps

Figure 5.7

A diagram of a mass spectrometer. Ions leave the ionization chamber and move into the magnetic field. Light ions experience a large deflection and collide with the inner wall of the tube. Heavy ions experience less deflection and collide with the other wall. Only those ions with a specific value of m/e pass through the field into the detector.

of the ions are singly charged ($e = 1$), the separation is primarily on the basis of mass. Taking into account the magnetic field and other characteristics of the instrument, the mass of each fragment can be determined.

A mass spectrum is determined by varying the magnetic or electric field so the ions with progressively higher masses strike the detector and plotting the relative intensities (proportional to the numbers of ions of particular masses), as indicated by the detector, versus their mass-to-charge ratios. A mass spectrum showing the three types of magnesium atoms is shown in Fig. 5.8.

These mass spectrometric observations explained why the atomic weights of many elements are not integral numbers and suggested that atoms contain uncharged particles with a mass of 1 amu. Magnesium atoms, for example, were known to have an atomic number of 12 and thus to contain 12 protons and 12 electrons with a combined mass of about 12 amu. The additional mass in these atoms (having total masses of about 24, 25, and 26 amu) must then be due to particles with masses of about 1 amu but with no charge. If they had a charge, the number of positive charges and negative charges within the atoms would not be the same, and the atoms would not be electrically neutral. The three types of magnesium atoms, therefore, must contain 12, 13, and 14 of these neutral particles, respectively. The existence of such a neutral particle with a mass of 1 amu (the neutron) was subsequently confirmed by Chadwick (Section 5.3).

We now know that the nuclei of atoms contain both protons and neutrons (except for the nucleus of the common hydrogen atom, which contains only a proton). For a given element, the number of protons does not vary, but the number of neutrons may vary within a limited range. Atoms with the same atomic number and different numbers of neutrons are called **isotopes**. Isotopes are identified with the atomic number as a subscript and the **mass number** (the sum of the number of protons and neutrons) as a superscript to the left of the symbol for the element. The naturally occurring isotopes of magnesium would be indicated as $^{24}_{12}Mg$, $^{25}_{12}Mg$, and $^{26}_{12}Mg$. The compositions of the nuclei of the naturally occurring elements with atomic numbers 1 through 10 are given in Table 5.2.

Since each proton and each neutron contributes approximately one unit to the atomic mass of an atom and each electron contributes far less, the mass of a single atom is approximately equal to its mass number (a whole number). However, many atomic weights (the average mass of a great many atoms of an element) are not close to whole numbers, because most elements exist as mixtures of two or more isotopes. The atomic weights are the weighted averages of the masses of each isotope present in a sample of the element; that is, they are equal to the sum of the masses of each isotope, each mass being multiplied by the fraction of that isotope present in the sample of the element.

are atoms of same element with different masses due to a variation in number of neutrons in nucleus

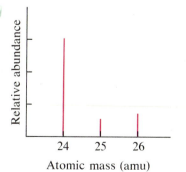

Relative abundance

24   25   26

Atomic mass (amu)

Figure 5.8

The mass spectrum of magnesium, showing three peaks due to the three naturally occurring isotopes of magnesium.

Table 5.2 Nuclear compositions of atoms of the very light elements

	Symbol	Atomic number	Number of protons	Number of neutrons	Mass, amu	% Natural abundance
Hydrogen	$_1^1\text{H}$	1	1	0	1.0078	99.985
	$_1^2\text{D}$	1	1	1	2.0141	0.015
Helium	$_2^3\text{He}$	2	2	1	3.0160	0.00013
	$_2^4\text{He}$	2	2	2	4.0026	100
Lithium	$_3^6\text{Li}$	3	3	3	6.0151	7.42
	$_3^7\text{Li}$	3	3	4	7.0160	92.58
Beryllium	$_4^9\text{Be}$	4	4	5	9.0122	100
Boron	$_5^{10}\text{B}$	5	5	5	10.0129	19.6
	$_5^{11}\text{B}$	5	5	6	11.0093	80.4
Carbon	$_6^{12}\text{C}$	6	6	6	12.0000*	98.89
	$_6^{13}\text{C}$	6	6	7	13.0033	1.11
	$_6^{14}\text{C}$	6	6	8	14.0032	$<10^{-8}$
Nitrogen	$_7^{14}\text{N}$	7	7	7	14.0031	99.63
	$_7^{15}\text{N}$	7	7	8	15.0001	0.37
Oxygen	$_8^{16}\text{O}$	8	8	8	15.9949	99.759
	$_8^{17}\text{O}$	8	8	9	16.9991	0.037
	$_8^{18}\text{O}$	8	8	10	17.9992	0.204
Fluorine	$_9^{19}\text{F}$	9	9	10	18.9984	100
Neon	$_{10}^{20}\text{Ne}$	10	10	10	19.9924	90.92
	$_{10}^{21}\text{Ne}$	10	10	11	20.9940	0.257
	$_{10}^{22}\text{Ne}$	10	10	12	21.9914	8.82

*Mass assigned as exactly 12 by international agreement.

EXAMPLE 5.1   Dempster found that magnesium contains 78.70% $_{12}^{24}\text{Mg}$ atoms (mass = 23.98 amu), 10.13% $_{12}^{25}\text{Mg}$ atoms (mass = 24.99 amu), and 11.17% $_{12}^{26}\text{Mg}$ atoms (mass = 25.98 amu). Calculate the weighted average mass of a Mg atom.

$$\text{Weighted average mass} = \frac{78.70\ _{12}^{24}\text{Mg atoms}}{100\ \text{Mg atoms}} \times \frac{23.98\ \text{amu}}{1\ _{12}^{24}\text{Mg atom}}$$

$$+ \frac{10.13\ _{12}^{25}\text{Mg atoms}}{100\ \text{Mg atoms}} \times \frac{24.99\ \text{amu}}{1\ _{12}^{25}\text{Mg atom}} + \frac{11.17\ _{12}^{26}\text{Mg atoms}}{100\ \text{Mg atoms}} \times \frac{25.98\ \text{amu}}{1\ _{12}^{26}\text{Mg atom}}$$

$$= 24.31\ \text{amu/Mg atom}$$

More precise measurements have shown this value to be 24.305 amu.

5.7 The Bohr Model of the Atom

Rutherford's nuclear model of the atom, with its very small, massive nucleus surrounded by lightweight electrons, accounted nicely for the properties of the atom as revealed by the discharge-tube experiments of Crookes and others, by the α-particle scattering experiments, and by the varying weights of the elements due to the presence of isotopes. However, one serious problem remained. According to the physical principles known at the time, such an atom should not be stable. Since there is an attractive force between the positively charged nucleus and the negatively charged electrons, the electrons would be expected to fall into the nucleus. If it were assumed that the electrons moved around the nucleus in circular orbits, centrifugal force acting on the electrons could counterbalance the force of attraction, and the electrons would stay in their orbits. However, according to classical physics, an electron moving in such a circular orbit would radiate energy continuously; since this would mean that it was continuously losing energy, it should move in smaller and smaller orbits and finally fall into the nucleus.

In 1913 Niels Bohr, a Danish scientist, proposed a solution to the problem by suggesting a new theory for the behavior of matter. He proposed that the energy of an electron in an atom cannot vary continuously, but is **quantized;** that is, it is restricted to discrete, or individual, values. The success of this theory in explaining the spectra of the hydrogen atom (Section 5.8) and of hydrogenlike ions, (which contain only one electron moving about a nucleus) led to its general acceptance.

The **Bohr model of a hydrogen atom** or a hydrogenlike ion assumed that the single electron moved about the nucleus in a circular orbit and that the centrifugal force due to this motion counterbalanced the electrostatic attraction between the nucleus and the electron. The energy of the electron was *assumed* to be restricted to certain values, each of which correspond to an orbit with a different radius. Each of these orbits could be characterized by an integer, n. Bohr showed that the energy of an electron in one of these orbits is given by the equation

$$E = \frac{-kZ^2}{n^2}$$

where k is a constant, Z is the atomic number (the number of units of positive charge on the nucleus), and n is the integer characteristic of the orbit ($n = 1, 2, 3, 4, \ldots$). For a hydrogen atom $Z = 1$ and $k = 2.179 \times 10^{-18}$ J, which gives the energy in joules. Other units of energy commonly used include electron-volts (1 eV = 1.602×10^{-19} J) and ergs (1 erg = 10^{-7} J).

A spark promotes the electron in a hydrogen atom into an orbit with $n = 2$. What is the energy, in joules, of an electron with $n = 2$?

EXAMPLE 5.2

The energy of the electron is given by the equation

$$E = \frac{-kZ^2}{n^2}$$

The atomic number, Z, of hydrogen is 1; $k = 2.179 \times 10^{-18}$ J; and the electron is characterized by an n value of 2. Thus

$$E = \frac{-2.179 \times 10^{-18} \text{ J} \times 1^2}{2^2} = -5.448 \times 10^{-19} \text{ J}$$

The distance of the electron from the nucleus is also related to the value of n.

$$\text{Radius of orbit} = \frac{n^2 a_0}{Z}$$

where a_0 is the radius of the orbit in the hydrogen atom for which $n = 1$ (0.529 Å,

Table 5.3   Calculated energies and radii for a hydrogen atom

n	Shell	Energy, eV	Distance from nucleus, Å	Distance from nucleus, pm
1	K	−13.595	0.529	52.9
2	L	− 3.399	2.116	211.6
3	M	− 1.511	4.761	476.1
4	N	− 0.850	8.464	846.4
5	O	− 0.544	13.225	1322.5
∞	—	0	∞	∞

where 1 Å$=10^{-10}$ m or 100 pm). According to this model, the closest an electron in a hydrogen atom can get to the nucleus is 0.529 Å (in an orbit with $n = 1$). Thus, the electron cannot fall into the nucleus. Because we must add energy to an electron to move it into an orbit with a larger n value, as the electron moves away from the nucleus, its energy becomes higher (Table 5.3), reaching zero at an infinite distance ($n = \infty$). As the electron moves closer to the nucleus, it loses energy, and its energy becomes lower (more negative).

Thus the Bohr model of the hydrogen atom (and of hydrogenlike ions with only one electron) postulates a single electron that moves in circular orbits about the nucleus (Fig. 5.9). The electron usually moves in the orbit for which $n = 1$, the orbit in which it has the lowest energy. When the electron is in this lowest-energy orbit, the atom is said to be in its **ground state.** If the atom picks up energy from an outside source, it changes to an **excited state** as the electron moves to one of the higher-energy orbits.

The motion of the electron in a *circular* orbit was not the most important concept of the Bohr model. Indeed, that part of the model is not correct. What was revolutionary was the idea that the electron is restricted to discrete energies that are identified by n.

The integer n is called a **quantum number.** The properties of an electron in a Bohr atom are often identified by giving its quantum number. If we know the quantum number of the electron in a hydrogen atom, we can easily evaluate the energy of the electron or the size of the orbit that it occupies. For example, an electron with the quantum number $n = 1$ resides in the first orbit with an energy of −13.595 eV and a radius of 0.529 Å.

Instead of quantum numbers, letters are sometimes used to distinguish electrons (Table 5.3). When we speak of a K electron, we mean an electron with $n = 1$. An L electron is an electron with $n = 2$; an M electron, with $n = 3$; etc. In Section 5.10, we will see that there are in fact three additional quantum numbers that also describe various properties of an electron.

$n = 6$ $r = 36a_0$

$n = 5$ $r = 25a_0$

$n = 4$ $r = 16a_0$

$n = 3$ $r = 9a_0$

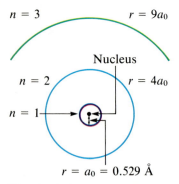

Nucleus

$n = 2$ $r = 4a_0$

$n = 1$

$r = a_0 = 0.529$ Å

Figure 5.9
A sketch of the hypothetical circular orbits of the Bohr model of the hydrogen atom drawn to scale. If the nucleus were drawn to scale, it would be invisible.

5.8 Atomic Spectra and Atomic Structure

The Bohr model of the hydrogen atom would be merely an interesting intellectual curiosity if it could not be checked against experimental data. In fact, it explains the spectrum of hydrogen atoms very well, and this observation tends to substantiate the new assumption in the model. Before we look at the agreement of the model with experiment, however, a brief introduction to light and to spectra is required.

Visible light is one form of **electromagnetic radiation.** (Radio waves, ultraviolet light, X rays, and γ rays are other examples.) All forms of electromagnetic radiation exhibit wavelike behavior. They all can be characterized by a wavelength, λ (Fig. 5.10), and a frequency, ν. The frequency is the rate at which equivalent points on a wave pass a given point. The product $\lambda\nu$ is equal to the speed, c, with which all forms of electromagnetic radiation move (2.998×10^8 m/s in a vacuum), so the wavelength and frequency are inversely proportional ($\nu = c/\lambda$). Different forms of electromagnetic radiation have different wavelengths and frequencies, but they all move with the same speed in a vacuum.

Electromagnetic radiation also has properties associated with particles called **photons**. The light emitted when one electron moves from an orbit with a larger n value to one with a smaller n value is a photon. The energy of a photon may be determined from the expression

$$E = h\nu = \frac{hc}{\lambda}$$

where h is Planck's constant (6.626×10^{-34} J s), c is the velocity, λ is the wavelength, and ν is the frequency of the radiation.

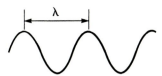

Figure 5.10

Characteristics of a wave. The distance between identical points, such as consecutive peaks, on the wave is the wavelength, λ (lambda). The number of times per second that identical points on the wave pass a given point is the frequency, ν (nu).

An experimental iodine laser emits light with a wavelength of 1.315 μm. What is the frequency of this light, and what is the energy per photon?

EXAMPLE 5.3

As noted above, frequency, ν, is inversely proportional to wavelength, λ.

$$\nu = \frac{c}{\lambda}$$

Since 1 μm = 10^{-6} m, the wavelength given, 1.315 μm, equals 1.315×10^{-6} m. Hence,

$$\nu = \frac{2.998 \times 10^8 \text{ m/s}}{1.315 \times 10^{-6} \text{ m}}$$

$$= 2.280 \times 10^{14} \text{ s}^{-1}$$

The relationship of the energy of a photon to its wavelength is

$$E = h\frac{c}{\lambda}$$

$$= \frac{6.626 \times 10^{-34} \text{ J s} \times (2.998 \times 10^8 \text{ m/s})}{1.315 \times 10^{-6} \text{ m}} = 1.511 \times 10^{-19} \text{ J}$$

Thus the frequency of the laser light is 2.280×10^{14} s^{-1} (read "per second") with an energy of 1.511×10^{-19} J per photon.

Sir Isaac Newton separated sunlight into its component colors (a spectrum) by passing it through a glass prism and showed that it contains all wavelengths (and thus all energies) of visible light. Sunlight gives a continuous **spectrum** (Fig. 5.11) such as the one we see in the rainbow and in the "rainbows" sometimes produced when a cut-glass object sits in a sunny window. Sunlight also contains ultraviolet light (very short wavelengths) and infrared light (very long wavelengths), which can be detected and recorded with instruments but are invisible to the human eye. Incandescent solids such as the filament in a light bulb also give continuous spectra. However, when an electric current is passed through a gas at low pressure, the gas gives off light that shows a spectrum made up of a number of bright lines (a **line spectrum**) when passed through a prism (Fig. 5.12). Each of these lines corresponds to a single wavelength of light, so the light emitted by such a gas consists of light of discrete energies.

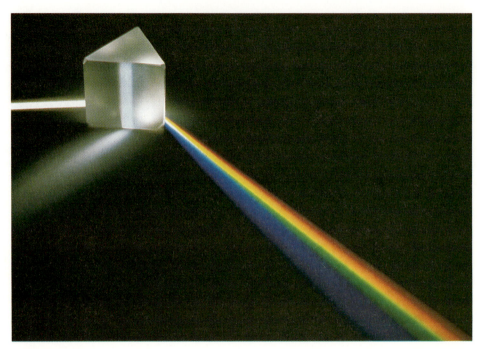

Figure 5.11
Sunlight passing through a prism is separated into its component colors, resulting in a continuous spectrum.

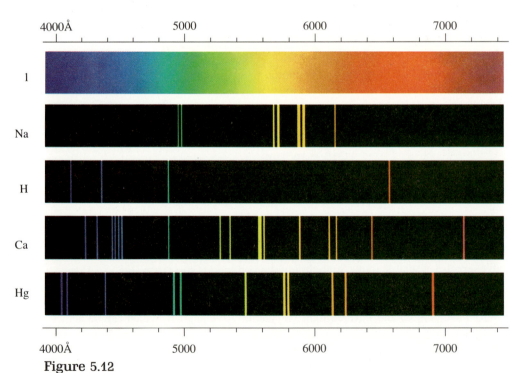

Figure 5.12
A comparison of the continuous spectrum of white light (spectrum 1) and the line spectra of the light from excited sodium, hydrogen, calcium, and mercury atoms.

If a tube containing hydrogen at low pressure is subjected to an electric discharge, light of a blue-pink hue is produced. Passage of this light through a prism produces a line spectrum, indicating that it is composed of light of several energies. J. R. Rydberg measured the frequencies of the lines in the visible spectrum and found that they could be related by an equation that gives the energies, in joules, of the photons in each line as follows:

$$E = h\nu = 2.179 \times 10^{-18} \left(\frac{1}{n_1^2} - \frac{1}{n_2^2} \right)$$

where n_1 and n_2 are integers, with n_1 smaller than n_2. This equation is empirical; that is, it is derived from observation rather than from theory.

Bohr suggested that a hydrogen atom radiates energy as light when its electron suddenly changes from a higher-energy orbit to one that has a lower energy (Fig. 5.13). Because of conservation of energy, the energy lost by an electron when it moves to a lower-energy orbit must appear somewhere; it usually appears as light. If a hydrogen atom is excited with an electric discharge, the electron gains energy and moves to an orbit of higher energy (and higher n). As the atom relaxes, the electron moves from this higher-energy level to one in which its energy is lower and emits a photon with an energy equal to the difference in energy between the two orbits. According to Bohr's theory (Section 5.7), the electron in a hydrogen atom ($Z = 1$) can have only those *discrete* energies, E_n, permitted by the equation

$$E_n \text{ (in joules)} = \frac{-k}{n^2} = \frac{-2.179 \times 10^{-18}}{n^2}$$

When the electron in a hydrogen atom falls from a higher-energy outer orbit characterized by n_2 to a lower-energy inner orbit characterized by n_1 (n_1 is less than n_2), the difference in energy is the absolute value of the difference in energy between the two orbits, $E_{n_1} - E_{n_2}$. This is the energy, E, emitted as a photon, or

$$E \text{ (in joules)} = \left| E_{n_1} - E_{n_2} \right| = 2.179 \times 10^{-18} \left(\frac{1}{n_1^2} - \frac{1}{n_2^2} \right)$$

This theoretical expression is exactly the same as the one Rydberg found. The agreement of the two equations—one experimental, one theoretical—provides powerful and indispensable evidence for the validity of the Bohr concept of atomic structure involving discrete energy levels for electrons.

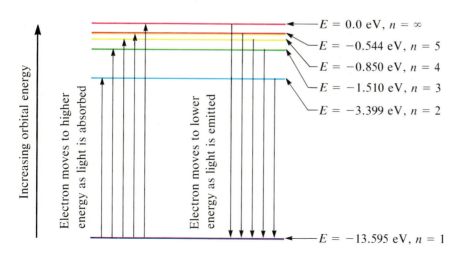

Figure 5.13

Relative energies of some of the circular orbits in the Bohr model of the hydrogen atom and electronic transitions that give rise to its atomic spectrum. Note the decreasing energy difference between levels as n increases.

EXAMPLE 5.4   What is the energy (in joules) and the wavelength (in meters) of the photon radiated when an electron, previously excited by an input of energy, moves from the orbit with $n = 4$ to the orbit with $n = 2$ in a hydrogen atom?

In this case the electron starts out with $n = 4$, so $n_2 = 4$. It finishes with $n = 2$, so $n_1 = 2$. The difference in energy between these two states is given by the expression

$$E = |E_{n_1} - E_{n_2}| = 2.179 \times 10^{-18} \left(\frac{1}{n_1{}^2} - \frac{1}{n_2{}^2} \right) \text{ J}$$

$$= 2.179 \times 10^{-18} \left(\frac{1}{2^2} - \frac{1}{4^2} \right) \text{ J}$$

$$= 2.179 \times 10^{-18} \left(\frac{1}{4} - \frac{1}{16} \right) \text{ J}$$

$$= 4.086 \times 10^{-19} \text{ J}$$

The energy of the photon emitted is equal to the difference in energy between the two orbits, 4.086×10^{-19} J. The wavelength of a photon with this energy is found from the expression $E = hc/\lambda$. Rearrangement gives

$$\lambda = \frac{hc}{E}$$

$$= \frac{6.626 \times 10^{-34} \text{ J s} \times 2.998 \times 10^8 \text{ m s}^{-1}}{4.086 \times 10^{-19} \text{ J}}$$

$$= 4.862 \times 10^{-7} \text{ m (or 486.2 nm)}$$

Even though the idea of discrete energy levels applies to all atoms, Bohr's equation works only for atoms or ions with one electron. Attempts to explain the spectra and other physical properties of more complicated atoms require the use of a more complex model, the quantum mechanical model of the atom.

5.9 The Quantum Mechanical Model of the Atom

It is now known that the laws of ordinary mechanics do not adequately explain the properties of small particles such as electrons. As a consequence, Bohr's theory of an electron moving in a circular orbit about a nucleus must be considered only a first approximation of the structure of the atom. It can explain the spectrum of an atom or ion with a single electron, but not the spectra of atoms with more. In addition, it fails to provide a satisfactory picture of chemical bonding. Thus the Bohr model has been supplanted by the more mathematical theories of quantum mechanics.

The **quantum mechanical model of the hydrogen atom** consists of one proton (as a nucleus) and one electron moving about the proton. Larger atoms have several electrons moving about a positive nucleus. Although the electrons are known to move about the nucleus, the exact path they take cannot be determined. The German physicist Werner Heisenberg expressed this in a form that has come to be known as the **Heisenberg uncertainty principle: It is impossible to determine accurately both**

the momentum and the position of a particle simultaneously. (The momentum is the mass of the particle multiplied by its velocity.) The more accurately we measure the momentum of a moving electron (or any other particle), the less accurately we can determine its position (and conversely). An experiment designed to measure the exact position of an electron will alter its momentum, and an experiment designed to measure the exact momentum of an electron will unavoidably change its position. If position and momentum are measured at the same time, the values are inexact for one or the other or both. Although we cannot pinpoint an electron's position or path, we *can* calculate the *probability* of finding an electron at a given location within an atom.

In spite of the limitations described by the uncertainty principle, the behavior of electrons can be determined in a useful way, using the mathematical tools of quantum mechanics. Both the energy of an electron in an atom and the region of space in which it may be found can be determined.

The results of a quantum mechanical treatment of the problem developed by another German physicist, Erwin Schrödinger, show that the electron may be visualized as being in rapid motion within one of several regions of space located around the nucleus. Each of these regions is called an **orbital, or atomic orbital.** Although the electron may be located anywhere within an orbital at any instant in time, it spends most of its time in certain high-probability regions. For example, in an isolated hydrogen atom in its ground state, the single electron effectively occupies all the space within about 1 Å of the nucleus. This gives the hydrogen atom a spherical shape. Within this spherical orbital the electron has the greatest probability of being at a distance of 0.529 Å from the nucleus. Note that a Bohr orbit and a quantum mechanical orbital are very different. An orbit is a circular path; an orbital is a three-dimensional region of space.

Some chemists describe the occupancy of an orbital in terms of **electron density.** The electron density is high in those regions of the orbital where the probability of finding an electron is relatively high, and low in those regions of the orbital where the probability is low.

Each electron in an atom can be described by a mathematical expression called a **wave function,** which is given the symbol ψ. The shape of an orbital that the electron occupies, the energy of the electron in the orbital (sometimes called the energy of the orbital), and the probability of finding the electron in some region within the orbital can be determined from ψ. For example, the square of the wave function, ψ^2, is a measure of the probability of finding the electron at a given point at a distance r from the nucleus; and $4\pi r^2\psi^2$ (the radial probability density) is a measure of the probability of finding the electron within the volume of a thin spherical shell (somewhat like a layer of an onion) of radius r and thickness dr (where dr is a very small fraction of r).

Figure 5.14 illustrates the electron density in the occupied orbital of a hydrogen atom in the ground state. The electron occupies the orbital in which it will have the lowest possible energy (the lowest-energy orbital). The probability of finding the electron in a thin shell very close to the nucleus is practically zero, but increases rapidly just beyond the nucleus and becomes highest in a thin shell at a distance of 0.529 Å from the nucleus. The probability then decreases rapidly as the distance of the thin shell from the nucleus increases and becomes exceedingly small at a distance greater than about 1 Å. Most of the time, but not always, the electron will be located within a sphere with a radius of 1 Å; that is, the probability of finding the electron inside this sphere is high. Figure 5.15 shows a plot of the probability of finding the electron in a thin shell as a function of r, the distance of the shell from the nucleus for three of the lowest-energy orbitals in a hydrogen atom.

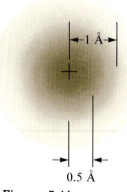

0.5 Å

Figure 5.14

Electron density for an electron of lowest possible energy in a hydrogen atom. The electron occupies the 1s orbital.

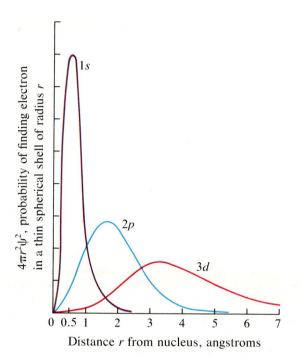

Figure 5.15

The probability of finding the electron at a given distance r from the nucleus for three of the lowest-energy orbitals in a hydrogen atom. The energy of an electron in these orbitals increases as n increases. Note the overlap of orbitals.

5.10 Results of the Quantum Mechanical Model of the Atom

The general results of the quantum mechanical model of the atom may be summarized as follows:

1. The location of an electron cannot be determined exactly. All that can be identified is the region, or volume, of space where there is a relatively high probability of finding the electron—that is, the orbital occupied by the electron.

2. Orbitals are characterized by **n, the principal quantum number,** which may take on any integral value: $n = 1, 2, 3, 4, 5, \ldots$. As n increases, the orbitals extend farther from the nucleus, and the average position of an electron in these orbitals is farther from the nucleus (Fig. 5.15). The energies of the orbitals increase as the value of n increases. For values of n greater than 1, there are several different orbitals with the same value of n. A **shell** contains orbitals with the same value of n. Shells may be identified by either the values of n or the letters K, L, M, N, O, $\ldots$, respectively (an $n = 2$ shell may be called an L shell, for example).

Shell:	K	L	M	N	O	P	Q
n:	1	2	3	4	5	6	7

3. Orbitals with the same value of n may have different shapes. The different shapes of orbitals are distinguished by a second quantum number, **l,** called the **azimuthal,** or **subsidiary, quantum number.** In atoms with two or more electrons,

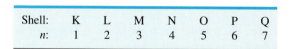

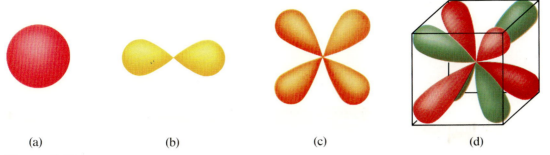

(a)   (b)   (c)   (d)

Figure 5.16

Shapes of the orbitals for (a) an *s* subshell (b) a *p* subshell (c) a *d* subshell (d) an *f* subshell. To improve the perspective, the *f* orbital is shown within a cube, with lobes of different colors.

electrons with different *l* values (electrons that occupy differently shaped orbitals) will have different energies. An electron with *l* = 0 occupies a spherical orbital [Fig. 5.16(a)]. Such an orbital is called an *s* orbital and the electron occupying it an *s* electron. An electron with *l* = 1 occupies a dumbbell-shaped region of space [Fig. 5.16(b)] and is called a *p* electron; a *d* electron, with *l* = 2, occupies a volume of space usually drawn with four lobes [Fig. 5.16(c)]; and an *f* electron, with *l* = 3, occupies a rather complex-looking volume of space usually shown with eight lobes (Fig. 5.19). Letters corresponding to the values of *l* continue alphabetically following *f*. Although *l* values higher than 3 are possible, electrons with *l* greater than 3 are not found unless they have been excited by absorption of energy by an atom.

l:	0	1	2	3	4	5	6
Letter designation:	*s*	*p*	*d*	*f*	*g*	*h*	*i*

In an atom with two or more electrons, the energies of electrons in the orbitals within a given shell increase in the following order:

$$s \text{ electrons} < p \text{ electrons} < d \text{ electrons} < f \text{ electrons}$$

The mathematics of quantum mechanics tells us that there are limits on the values that *l* can have: An electron for which the principal quantum number has a value of *n* may have an integral *l* value ranging from 0 to $(n - 1)$. These limits have been verified experimentally from the spectra of atoms.

Shell	*n* Value	Possible *l* values	Types of orbitals
K	1	0,	1*s*
L	2	0, 1	2*s*, 2*p*
M	3	0, 1, 2	3*s*, 3*p*, 3*d*
N	4	0, 1, 2, 3	4*s*, 4*p*, 4*d*, 4*f*
O	5	0, 1, 2, 3, 4	5*s*, 5*p*, 5*d*, 5*f*, 5*g*

Thus an electron in a K shell, for which $n = 1$, can only have an l value of zero. An electron in a shell with $n = 3$ can have an l value of zero (an s electron), an l value of 1 (a p electron), or an l value of 2 (a d electron), but cannot have an l value higher than 2.

    4. For l values larger than zero, orbitals with the same value of l in any given shell have the same shape and the same energy but differ in their orientation. There are $(2l + 1)$ orbitals with the same l quantum number, which differ only in their orientation about the nucleus. Each orientation is characterized by a third quantum number, **_m_,** the **magnetic quantum number,** which may have any integral value from $-l$ through zero to $+l$. A sphere may have only one orientation in space, and thus there is only one type of spherical, or s, orbital. This orbital is characterized by an l value of zero; hence m here can only be zero. There are three possible values of m (-1, 0, $+1$) for an orbital with $l = 1$ (a p orbital) and thus three possible orientations. Figure 5.17 shows how the dumbbell shape of a p orbital can be oriented in three ways along the x, y, and z axes of an xyz coordinate system. The three p orbitals are designated, as shown in the figure, p_x, p_y, and p_z to indicate their directional character. Figure 5.18 shows five orientations

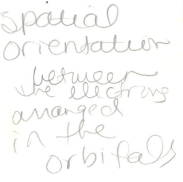

spatial orientation between the electrons arranged in the orbitals

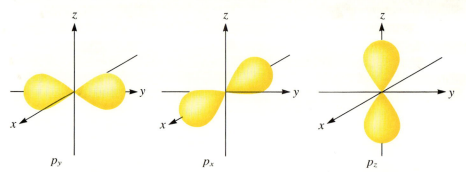

Figure 5.17

The different orientations of the three equivalent atomic p orbitals.

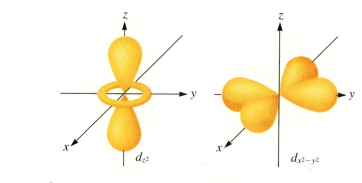

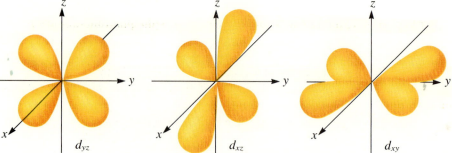

Figure 5.18

The different orientations of the five equivalent atomic d orbitals. The lobes of the d_{z^2} and $d_{x^2-y^2}$ orbitals lie along the axes, whereas the lobes of the d_{xz}, d_{yz}, and d_{xy} orbitals lie between the axes.

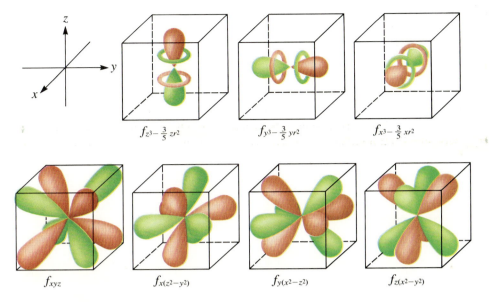

$f_{z^3-\frac{3}{5}zr^2}$ $f_{y^3-\frac{3}{5}yr^2}$ $f_{x^3-\frac{3}{5}xr^2}$

f_{xyz} $f_{x(z^2-y^2)}$ $f_{y(x^2-z^2)}$ $f_{z(x^2-y^2)}$

Figure 5.19

The different orientations of the seven equivalent atomic f orbitals. To improve the perspective, each of the seven orbitals is shown within a cube, with lobes of different colors. Only one f orbital is shown in each cube.

of the d orbitals ($l = 2$; $m = -2, -1, 0, +1, +2$). Figure 5.19 shows seven orientations of the f orbitals ($l = 3$; $m = -3, -2, -1, 0, +1, +2, +3$). There is no general agreement on how best to represent the seven f orbitals. Figure 5.19 shows one way.

In a free atom, the p orbitals in the same shell have the same energy and are referred to as **degenerate.** Degenerate orbitals are orbitals with the same energy. Likewise, the five d orbitals in a given shell are degenerate, as are the seven f orbitals. Each set of degenerate orbitals, that is, orbitals with the same l value in a given shell, is referred to as a **subshell.**

Indicate the number of subshells, the number of orbitals in each subshell, and the values of n, l, and m for the orbitals in the N shell of an atom.

EXAMPLE 5.5

The N shell is the shell with $n = 4$, so it contains subshells with $l = 0, 1, 2,$ and 3. Hence there are four subshells.

For the subshell with $l = 0$ (the s subshell), m can only be equal to zero. Thus there is only one $4s$ orbital. For $l = 1$ (the p subshell), $m = -1, 0,$ or $+1$ so we find three $4p$ orbitals; for $l = 2$ (the d subshell), $m = -2, -1, 0, +1,$ or $+2$, giving five $4d$ orbitals; and for $l = 3$ (the f subshell), $m = -3, -2, -1, 0, +1, +2,$ or $+3$, giving seven $4f$ orbitals.

Thus there are 16 orbitals, with the values of n, l, and m indicated above, in the 4 subshells of the N shell of an atom.

---

5. Electrons have a fourth quantum number, s, the **spin quantum number.** Electrons behave as if each were spinning about its own axis. The spin quantum number specifies the direction of spin of an electron about *its own axis*. The spin can be either counterclockwise or clockwise and is designated arbitrarily by either $s = +\frac{1}{2}$ or $s = -\frac{1}{2}$. Electrons with the same spin quantum number are said to have **parallel spins.**

For every possible combination of n, l, and m, there can be two electrons differing only in the direction of spin about their own axes. Each orbital can contain a maximum of two electrons that have identical n, l, and m values but differ in their s values.

Table 5.4   Summary of allowed values for each quantum number

Shell	n	l	m	s	Maximum number of electrons in subshell	Maximum number of electrons in shell
K	1	0 (s)	0	$+\frac{1}{2}$	$\left.\begin{array}{c}1\\1\end{array}\right\}2$	2
	1	0	0	$-\frac{1}{2}$		
L	2	0 (s)	0	$+\frac{1}{2}, -\frac{1}{2}$	2	8
	2	1 (p)	-1	$+\frac{1}{2}, -\frac{1}{2}$	$\left.\begin{array}{c}2\\2\\2\end{array}\right\}6$	
	2	1	0	$+\frac{1}{2}, -\frac{1}{2}$		
	2	1	$+1$	$+\frac{1}{2}, -\frac{1}{2}$		
M	3	0 (s)	0	$+\frac{1}{2}, -\frac{1}{2}$	2	18
		1 (p)	$-1, 0, +1$	$\pm\frac{1}{2}$ for each value of m	6	
		2 (d)	$-2, -1, 0, +1, +2$	$\pm\frac{1}{2}$ for each value of m	10	
N	4	0 (s)	0	$\pm\frac{1}{2}$ for each value of m	2	32
		1 (p)	$-1, 0, +1$	$\pm\frac{1}{2}$ for each value of m	6	
		2 (d)	$-2, -1, 0, +1, +2$	$\pm\frac{1}{2}$ for each value of m	10	
		3 (f)	$-3, -2, -1, 0, +1, +2, +3$	$\pm\frac{1}{2}$ for each value of m	14	
O	5	0 (s)	0	$\pm\frac{1}{2}$ for each value of m	2	50[a]
		1 (p)	$-1, 0, +1$	$\pm\frac{1}{2}$ for each value of m	6	
		2 (d)	$-2, -1, 0, +1, +2$	$\pm\frac{1}{2}$ for each value of m	10	
		3 (f)	$-3, -2, -1, 0, +1, +2, +3$	$\pm\frac{1}{2}$ for each value of m	14	
		4 (g)	$-4, -3, -2, -1, 0, +1, +2, +3, +4$	$\pm\frac{1}{2}$ for each value of m	18[a]	

[a]The total number of 50 electrons for the O shell, including the 18 electrons for the g subshell, is the number of electrons theoretically possible. No element presently known contains more than 32 electrons in the O shell.

6. The energy of an electron in an atom is limited to discrete values. The mathematics is far more complicated than in the Bohr model, but the energy can be determined from the wave function that describes the behavior of the electron.

7. The maximum number of electrons that may be found in a shell with a principal quantum number of n is $2n^2$. Since each atomic orbital can hold a maximum of two electrons, the number of atomic orbitals in a shell is given by n^2. Table 5.4 summarizes all of the allowed combinations of quantum numbers that may describe an electron in the first five shells of an atom.

5.11 Orbital Energies and Atomic Structure

The energies of atomic orbitals increase as the principal quantum number, n, increases. If an atom contains two or more electrons, the energies of the orbitals increase within a shell as the azimuthal quantum number, l, increases. Although the relative energies of certain orbitals vary from atom to atom, the increasing order of energy is roughly that shown in Fig. 5.20. The energy of an electron in an orbital is indicated by the vertical coordinate in the figure, the orbital with electrons of lowest energy being the $1s$ orbital at the bottom of the diagram. Orbitals that have about the same energy are indicated by the braces at the right of the figure. For atomic numbers greater than 20, the relative energies of the orbitals may differ slightly from the order shown. The actual energies of the electrons in these orbitals vary from atom to atom. For example, the energies of the electrons in $1s$ orbitals become lower and lower as the atomic numbers of the atoms increase.

The arrangement of electrons in the orbitals of an atom (that is, the **electron configuration** of the atom) is described by a number that designates the number of the principal shell, a letter that designates the subshell, and a superscript that designates the number of electrons in that particular subshell. For example, the notation $2p^4$ (read "two-p-four") indicates four electrons in the p subshell of the shell for which $n = 2$; the notation $3d^8$ ("three-d-eight") indicates eight electrons in the d subshell of the shell for which $n = 3$.

In general, electrons fill orbitals from the bottom to the top of Fig. 5.20. Each set of orbitals is filled before electrons occupy the next set immediately above. Figure 5.21 gives this order in a format that is easy to remember.

The electron configuration of an atom in its ground state can be predicted using the following guidelines:

1. Electrons in an atom occupy the lowest possible energy levels, or orbitals. The first electron that is placed in a set of atomic orbitals goes into the $1s$ orbital. When the $1s$ orbital is filled, the next electron goes into the $2s$ orbital, and so on. The order of orbital energies is given in Figs. 5.20 and 5.21.

2. The maximum number of electrons in an orbital is limited to two, according to the **Pauli exclusion principle: No two electrons in the same atom can have the same set of four quantum numbers.** Thus a $1s$ orbital is filled when it contains two elec-

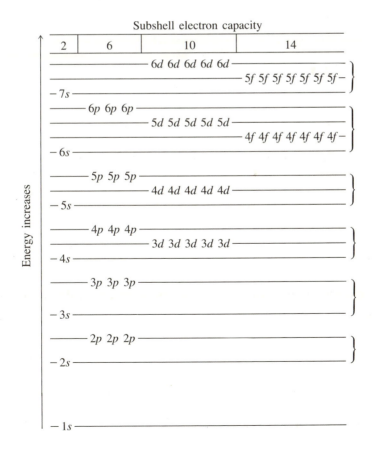

Figure 5.20

Generalized energy-level diagram for atomic orbitals in an atom with two or more electrons (not to scale). Orbitals of about the same energy are indicated by braces.

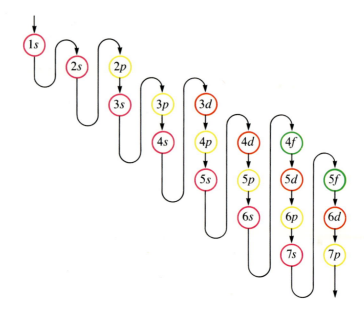

Figure 5.21

Order of occupancy of atomic orbitals. The orbitals fill in the order indicated by the connecting lines.

trons: one with $n = 1$, $l = 0$, $m = 0$, and $s = +\frac{1}{2}$; the other with $n = 1$, $l = 0$, $m = 0$, and $s = -\frac{1}{2}$. Since no other combination of quantum numbers is possible for an electron in the $1s$ orbital (Table 5.4), a third electron must occupy the $2s$ orbital. The $2s$ orbital, in turn, is filled when it contains two electrons, so that the fifth electron is forced into one of the three $2p$ orbitals, and so forth.

3. Subshells containing more than one orbital are filled as described by **Hund's rule: Every orbital in a subshell is singly occupied (filled) with one electron before any one orbital is doubly occupied, and all electrons in singly occupied orbitals have the same spin;** that is, their spin quantum numbers are the same. For example, a $3p^2$ electron configuration has one unpaired electron in each of two different $3p$ orbitals. Both electrons have the same spin and the same spin quantum number. A d^6 electron configuration has a pair of electrons having different spins in one d orbital and four unpaired electrons having the same spin in the remaining four d orbitals.

5.12 The Aufbau Process

In order to illustrate the systematic variations in the electronic structures of the various elements, chemists "build" them in atomic order. Beginning with hydrogen, they add one proton to the nucleus and one electron to the proper subshell at a time until they have described the electron configurations of all the elements. This process is called the **aufbau process** from the German word *aufbau* (building up). Each added electron occupies the subshell of lowest energy available, according to the rules given in Section 5.11. Electrons enter higher-energy subshells only after lower-energy subshells have been filled to capacity.

A single hydrogen atom consists of one proton and one electron. The single electron is found in the $1s$ orbital around the proton. The electron configuration of a hydrogen atom is customarily represented as $1s^1$ (read "one-s-one").

After hydrogen, the next simplest atom is that of the noble gas helium, with an atomic number of 2 and an atomic weight of approximately 4. The helium atom contains two protons and two neutrons in its nucleus and two electrons in the $1s$ orbital, which is completely filled by these two electrons. The electron configuration of helium is, therefore, $1s^2$ (read "one-s-two"). Note that in this atom the shell of lowest energy, the K shell, is completely filled.

The atom next in complexity is lithium, with an atomic number of 3, an atomic weight of approximately 7, and three protons, four neutrons, and three electrons. Two electrons fill the $1s$ orbital, so the remaining electron must occupy the orbital of next lowest energy, the $2s$ orbital. Thus the electron configuration of lithium is $1s^2 2s^1$.

An atom of the metal beryllium, with an atomic number of 4 and an atomic weight of approximately 9, contains four protons and five neutrons in the nucleus, and four electrons. The fourth electron fills the $2s$ orbital, making the electron configuration $1s^2 2s^2$.

An atom of boron (atomic number = 5) contains five electrons: two in the K shell ($n = 1$), which is filled, and three in the L shell ($n = 2$). Since the s subshell in the L shell can contain only two electrons, the fifth electron must occupy the higher-energy $2p$ subshell. The electron configuration of boron, therefore, is $1s^2 2s^2 2p^1$.

Carbon (atomic number = 6) has six electrons. Four of them fill the $1s$ and $2s$ subshells. The remaining two electrons occupy the $2p$ subshell, giving a $1s^2 2s^2 2p^2$ configuration. To describe the distribution of the electrons in the $2p$ subshell, we use Hund's rule. These electrons occupy different $2p$ orbitals but have the same spin and hence the same spin quantum number.

Nitrogen (atomic number = 7) has a $1s^2 2s^2 2p^3$ configuration, with one electron in each of the three $2p$ orbitals, in accordance with Hund's rule. These three electrons have parallel spins. Oxygen atoms (atomic number = 8) have the configuration $1s^2 2s^2 2p^4$, with a pair of electrons in one of the $2p$ orbitals and a single electron in each of the other two. The two unpaired electrons in separate $2p$ orbitals have the same spin. Fluorine atoms (atomic number = 9) have the configuration $1s^2 2s^2 2p^5$, with only one $2p$ orbital containing an unpaired electron. All of the electrons in the neon atom (atomic number = 10) are paired, with a configuration $1s^2 2s^2 2p^6$. Both the K and L shells of the noble gas neon are filled (two electrons in the K shell and eight in the L shell).

The sodium atom (atomic number = 11) has eleven electrons, one more than the neon atom. This electron must go into the lowest-energy subshell available, the $3s$ orbital, giving a $1s^2 2s^2 2p^6 3s^1$ configuration. We can abbreviate this as [Ne]$3s^1$. The symbol [Ne] represents the configuration of the two filled shells in Ne (neon), $1s^2 2s^2 2p^6$, which are identical to the two filled inner shells in a sodium atom. Similarly, the configuration of lithium may be represented as [He]$2s^1$, where [He] represents the configuration of the helium atom, which is identical to that of the filled inner shell of lithium. Writing the configurations in this way emphasizes the similarity of the configurations of sodium and lithium. Both atoms have only one electron in an s subshell outside a filled set of inner shells.

The magnesium atom (atomic number = 12), with its 12 electrons in a [Ne]$3s^2$ configuration, is analogous to beryllium, [He]$2s^2$. Both atoms have a filled s subshell outside their filled inner shells. Aluminum (atomic number = 13), with 13 electrons and the electron configuration [Ne]$3s^2 3p^1$, is analogous to boron, [He]$2s^2 2p^1$. The electron configurations of silicon (14 electrons), phosphorus (15 electrons), sulfur (16 electrons), chlorine (17 electrons), and argon (18 electrons) are analogous in the compositions of their outer shells to carbon, nitrogen, oxygen, fluorine, and neon, respectively. See Table 5.5 for the configurations of these elements. The table lists the lowest-energy, or ground-state, electron configurations for atoms of each of the known elements.

Table 5.5   Electron configurations of the elements

Atomic number	Symbol	Electron configuration	Atomic number	Symbol	Electron configuration	Atomic number	Symbol	Electron configuration
1	H	$1s^1$				73	Ta	$[Xe]6s^24f^{14}5d^3$
2	He	$1s^2$	37	Rb	$[Kr]5s^1$	74	W	$[Xe]6s^24f^{14}5d^4$
			38	Sr	$[Kr]5s^2$	75	Re	$[Xe]6s^24f^{14}5d^5$
3	Li	$1s^22s^1 = [He]2s^1$	39	Y	$[Kr]5s^24d^1$	76	Os	$[Xe]6s^24f^{14}5d^6$
4	Be	$[He]2s^2$	40	Zr	$[Kr]5s^24d^2$	77	Ir	$[Xe]6s^24f^{14}5d^7$
5	B	$[He]2s^22p^1$	41	Nb	$[Kr]5s^14d^4$	78	Pt	$[Xe]6s^14f^{14}5d^9$
6	C	$[He]2s^22p^2$	42	Mo	$[Kr]5s^14d^5$	79	Au	$[Xe]6s^14f^{14}5d^{10}$
7	N	$[He]2s^22p^3$	43	Tc	$[Kr]5s^14d^6$	80	Hg	$[Xe]6s^24f^{14}5d^{10}$
8	O	$[He]2s^22p^4$	44	Ru	$[Kr]5s^14d^7$	81	Tl	$[Xe]6s^24f^{14}5d^{10}6p^1$
9	F	$[He]2s^22p^5$	45	Rh	$[Kr]5s^14d^8$	82	Pb	$[Xe]6s^24f^{14}5d^{10}6p^2$
10	Ne	$[He]2s^22p^6$	46	Pd	$[Kr]4d^{10}$	83	Bi	$[Xe]6s^24f^{14}5d^{10}6p^3$
			47	Ag	$[Kr]5s^14d^{10}$	84	Po	$[Xe]6s^24f^{14}5d^{10}6p^4$
11	Na	$[Ne]3s^1$	48	Cd	$[Kr]5s^24d^{10}$	85	At	$[Xe]6s^24f^{14}5d^{10}6p^5$
12	Mg	$[Ne]3s^2$	49	In	$[Kr]5s^24d^{10}5p^1$	86	Rn	$[Xe]6s^24f^{14}5d^{10}6p^6$
13	Al	$[Ne]3s^23p^1$	50	Sn	$[Kr]5s^24d^{10}5p^2$			
14	Si	$[Ne]3s^23p^2$	51	Sb	$[Kr]5s^24d^{10}5p^3$	87	Fr	$[Rn]7s^1$
15	P	$[Ne]3s^23p^3$	52	Te	$[Kr]5s^24d^{10}5p^4$	88	Ra	$[Rn]7s^2$
16	S	$[Ne]3s^23p^4$	53	I	$[Kr]5s^24d^{10}5p^5$	89	Ac	$[Rn]7s^26d^1$
17	Cl	$[Ne]3s^23p^5$	54	Xe	$[Kr]5s^24d^{10}5p^6$	90	Th	$[Rn]7s^26d^2$
18	Ar	$[Ne]3s^23p^6$				91	Pa	$[Rn]7s^25f^26d^1$
			55	Cs	$[Xe]6s^1$	92	U	$[Rn]7s^25f^36d^1$
19	K	$[Ar]4s^1$	56	Ba	$[Xe]6s^2$	93	Np	$[Rn]7s^25f^46d^1$
20	Ca	$[Ar]4s^2$	57	La	$[Xe]6s^25d^1$	94	Pu	$[Rn]7s^25f^6$
21	Sc	$[Ar]4s^23d^1$	58	Ce	$[Xe]6s^24f^2$	95	Am	$[Rn]7s^25f^7$
22	Ti	$[Ar]4s^23d^2$	59	Pr	$[Xe]6s^24f^3$	96	Cm	$[Rn]7s^25f^76d^1$
23	V	$[Ar]4s^23d^3$	60	Nd	$[Xe]6s^24f^4$	97	Bk	$[Rn]7s^25f^86d^1$
24	Cr	$[Ar]4s^13d^5$	61	Pm	$[Xe]6s^24f^5$	98	Cf	$[Rn]7s^25f^{10}$
25	Mn	$[Ar]4s^23d^5$	62	Sm	$[Xe]6s^24f^6$	99	Es	$[Rn]7s^25f^{11}$
26	Fe	$[Ar]4s^23d^6$	63	Eu	$[Xe]6s^24f^7$	100	Fm	$[Rn]7s^25f^{12}$
27	Co	$[Ar]4s^23d^7$	64	Gd	$[Xe]6s^24f^75d^1$	101	Md	$[Rn]7s^25f^{13}$
28	Ni	$[Ar]4s^23d^8$	65	Tb	$[Xe]6s^24f^9$	102	No	$[Rn]7s^25f^{14}$
29	Cu	$[Ar]4s^13d^{10}$	66	Dy	$[Xe]6s^24f^{10}$	103	Lr	$[Rn]7s^25f^{14}6d^1$
30	Zn	$[Ar]4s^23d^{10}$	67	Ho	$[Xe]6s^24f^{11}$	104	Unq	$[Rn]7s^25f^{14}6d^2$
31	Ga	$[Ar]4s^23d^{10}4p^1$	68	Er	$[Xe]6s^24f^{12}$	105	Unp	$[Rn]7s^25f^{14}6d^3$
32	Ge	$[Ar]4s^23d^{10}4p^2$	69	Tm	$[Xe]6s^24f^{13}$	106	Unh	$[Rn]7s^25f^{14}6d^4$
33	As	$[Ar]4s^23d^{10}4p^3$	70	Yb	$[Xe]6s^24f^{14}$	107	Uns	$[Rn]7s^25f^{14}6d^5$
34	Se	$[Ar]4s^23d^{10}4p^4$	71	Lu	$[Xe]6s^24f^{14}5d^1$	108	Uno	$[Rn]7s^25f^{14}6d^6$
35	Br	$[Ar]4s^23d^{10}4p^5$	72	Hf	$[Xe]6s^24f^{14}5d^2$	109	Une	$[Rn]7s^25f^{14}6d^7$
36	Kr	$[Ar]4s^23d^{10}4p^6$						

Potassium (atomic number = 19) and calcium (atomic number = 20) have one and two electrons, respectively, in the N shell ($n = 4$). Hence potassium corresponds to lithium and sodium in outer shell configuration, whereas calcium corresponds to beryllium and magnesium.

Beginning with scandium (atomic number = 21; see Table 5.5), additional electrons are added successively to the $3d$ subshell after two electrons have already occupied the $4s$ subshell. After the $3d$ subshell is filled to its capacity with 10 electrons, the $4p$ subshell fills. Note that for three series of elements, scandium (Sc) through copper (Cu), yttrium (Y) through silver (Ag), and lutetium (Lu) through gold (Au), a total of ten d electrons are successively added to the $n - 1$ shell next to the outer n shell to bring that $n - 1$ shell from 8 to 18 electrons. For two series, lanthanum (La) through lutetium (Lu) and actinium (Ac) through lawrencium (Lr), fourteen f electrons are successively added to the third shell from the outside (the $n - 2$ shell) to bring that shell from 18 to 32 electrons.

EXAMPLE 5.6

Gaseous cesium (Cs) atoms have been used in an experimental laser. These atoms emit light when their outermost electron is excited to a $7p$ orbital and then falls back into the $6s$ orbital. Without reference to Table 5.5, write the electron configurations of cesium (a) in its ground state and (b) when the outermost electron is excited to a p orbital.

From the order of filling presented in Fig. 5.21, when Cs is in its ground state, we can fill the orbitals in the following order with 55 electrons:

$$1s^2 2s^2 2p^6 3s^2 3p^6 4s^2 3d^{10} 4p^6 5s^2 4d^{10} 5p^6 6s^1$$

However, note that many chemists write the orbitals in increasing order of the quantum numbers:

$$1s^2 2s^2 2p^6 3s^2 3p^6 3d^{10} 4s^2 4p^6 4d^{10} 5s^2 5p^6 6s^1$$

Either notation is acceptable. Since Cs contains one more electron than Xe, these configurations can both be abbreviated as $[Xe]6s^1$.

The outermost Cs electron is the electron in the $6s$ orbital. When this electron is excited to a higher-energy p orbital, the configuration changes to $[Xe]7p^1$. It should be noted that this configuration is not the ground-state configuration, the stable configuration. It cannot be predicted from the rules presented in Section 5.11. It occurs only when sufficient energy is put into the atomic system to move the electron from the ground-state $6s$ orbital out to the higher-energy $7p$ orbital.

Students are sometimes troubled by exceptions to the order for the filling of orbitals, which is shown in Fig. 5.21. For instance, the electron configurations of chromium (at. no. = 24), copper (at. no. = 29), and lanthanum (at. no. = 57), among others, are not those expected from the figure. In general, such exceptions involve subshells with very similar energies, and small effects lead to changes in the order of filling.

Half-filled and completely filled subshells represent conditions of preferred stability. This stability is such that an electron shifts from the $4s$ into the $3d$ orbitals in order to gain the extra stability of a half-filled $3d$ subshell (in chromium) or a filled $3d$ subshell (in copper).

In other atoms, certain combinations of repulsions between electrons lead to minor exceptions in the expected order of filling, because the magnitude of the repulsions is greater than the small differences in energy between subshells. From Fig. 5.21, we would expect the electron configuration of lanthanum (at. no. 57) to be $[Xe]6s^2 4f^1$. Instead, as shown in Table 5.5, La has no $4f$ electron but has one $5d$ electron: $[Xe]6s^2 5d^1$. This exception reflects the fact that the $4f$ and $5d$ subshells have very similar energies (see Fig. 5.20) and that electrons change from one sublevel to the other easily. A close inspection of Table 5.5 will disclose several such exceptions.

If you realize that exceptions in the expected order of filling of orbitals result from the similar energies of various subshells, the ability to predict the electron configuration for each element, based upon the expected order of addition of electrons, is sufficient for most purposes.

Group IVA: Carbon (in the form of diamond)

Group IA: Potassium

Group VIIA: Chlorine

Some Elements of the Periodic Table

Group VIA: Sulfur

Group VA: Bismuth

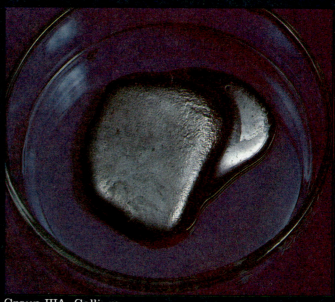

Group IIIA: Gallium

Group IIB: Mercury

Group IIA: Calcium

Group IVA: Tin

The Periodic Law

5.13 The Periodic Table

It became evident early in the development of chemistry that certain elements could be grouped together by reason of their similar properties. The members of one such grouping are lithium, sodium, and potassium. These elements all look like metals, all conduct electricity well, all react with chlorine, forming white, water-soluble compounds with one chlorine atom per metal atom, and all react with water, giving hydrogen gas and metal hydroxides with one hydroxide (OH) group per metal. If M is used to represent a metal atom, the formulas of the chlorides and hydroxides would be MCl and MOH, respectively. A second grouping includes calcium, strontium, and barium, which also look like metals and conduct electricity well. However, when they react with chlorine, they give white solids that contain two chlorine atoms per metal atom, or MCl_2. When they react with water, they give hydrogen gas and a hydroxide with two hydroxide groups per metal atom, or $M(OH)_2$. Fluorine, chlorine, bromine, and iodine also exhibit similar properties to each other. They do not conduct electricity and are nonmetallic. They react with hydrogen to form compounds that contain one hydrogen atom and one halogen atom: HF, HCl, HBr, and HI. When they react with sodium, they form compounds that contain one sodium atom per halogen atom: NaF, NaCl, NaBr, and NaI.

Dimitri Mendeleev in Russia and Lothar Meyer in Germany, working independently, observed that the properties of the elements are periodic functions of their atomic weights. As an understanding of the structure of the atom and electronic configuration developed, however, it became apparent that the properties of atoms were actually periodic functions of their atomic numbers rather than their atomic weights. The modern statement of this periodic relationship, **the periodic law,** is: **The properties of the elements are periodic functions of their atomic numbers.**

The Periodic Table is an arrangement of the atoms in increasing order of their atomic numbers that collects atoms with similar properties in vertical columns. Figure 5.22 gives one common form of the Periodic Table. A copy of the Periodic Table is also presented inside the front cover of this text.

As mentioned, lithium (Li), sodium (Na), and potassium (K) exhibit similar chemical properties. These elements are found in the leftmost vertical column in the table, along with rubidium (Rb), cesium (Cs), and francium (Fr). These six elements comprise a group known as the **alkali metals,** all of which have similar chemical properties. [Hydrogen (H) is listed in both column IA and column VIIA because it possesses a small number of similarities to elements in those columns. However, it is considered to be neither.] The second vertical column of elements from the left in the Periodic Table includes calcium (Ca), strontium (Sr), and barium (Ba), along with beryllium (Be), magnesium (Mg), and radium (Ra), and thus it should be expected (and it is observed) that their chemical behavior is similar. These six elements are referred to as the **alkaline earth metals.** The group of elements including fluorine (F), chlorine (Cl), bromine (Br), iodine (I), and astatine (At) is located in the second vertical column from the right in the table and is known as the **halogens.**

There are changes in the properties of the elements in a vertical column, so the chemical behavior of these elements, while similar, is not identical. Adjacent elements in each horizontal row differ decidedly in both chemical and physical properties, but these properties change in a regular way across each row. The horizontal and vertical

The Periodic Table of Elements (long form).

PERIODS	1 IA	2 IIA	3 IIIB	4 IVB	5 VB	6 VIB	7 VIIB	8	9 VIIIB	10	11 IB	12 IIB	13 IIIA	14 IVA	15 VA	16 VIA	17 VIIA	18 VIIIA
1	1.0079 H 1																	4.00260 He 2
2	6.941 Li 3	9.01218 Be 4											10.81 B 5	12.011 C 6	14.0067 N 7	15.9994 O 8	18.998403 F 9	20.179 Ne 10
3	22.98977 Na 11	24.305 Mg 12											26.98154 Al 13	28.0855 Si 14	30.97376 P 15	32.06 S 16	35.453 Cl 17	39.948 Ar 18
4	39.0983 K 19	40.08 Ca 20	44.9559 Sc 21	47.88 Ti 22	50.9415 V 23	51.996 Cr 24	54.9380 Mn 25	55.847 Fe 26	58.9332 Co 27	58.69 Ni 28	63.546 Cu 29	65.39 Zn 30	69.72 Ga 31	72.59 Ge 32	74.9216 As 33	78.96 Se 34	79.904 Br 35	83.80 Kr 36
5	85.4678 Rb 37	87.62 Sr 38	88.9059 Y 39	91.224 Zr 40	92.9064 Nb 41	95.94 Mo 42	(98) Tc 43	101.07 Ru 44	102.9055 Rh 45	106.42 Pd 46	107.868 Ag 47	112.41 Cd 48	114.82 In 49	118.71 Sn 50	121.75 Sb 51	127.60 Te 52	126.9045 I 53	131.29 Xe 54
6	132.9054 Cs 55	137.33 Ba 56	* [57-71]	178.49 Hf 72	180.9479 Ta 73	183.85 W 74	186.207 Re 75	190.2 Os 76	192.22 Ir 77	195.08 Pt 78	196.9665 Au 79	200.59 Hg 80	204.383 Tl 81	207.2 Pb 82	208.9804 Bi 83	(209) Po 84	(210) At 85	(222) Rn 86
7	(223) Fr 87	226.0254 Ra 88	† [89-103]	(261) Unq 104	(262) Unp 105	(263) Unh 106	(262) Uns 107	(265) Uno 108	(266) Une 109									

METALS — TRANSITION METALS

NONMETALS

INNER TRANSITION METALS

*LANTHANIDE SERIES

138.9055 La 57	140.12 Ce 58	140.9077 Pr 59	144.24 Nd 60	(145) Pm 61	150.36 Sm 62	151.96 Eu 63	157.25 Gd 64	158.9254 Tb 65	162.50 Dy 66	164.9304 Ho 67	167.26 Er 68	168.9342 Tm 69	173.04 Yb 70	174.967 Lu 71

†ACTINIDE SERIES

227.0278 Ac 89	232.0381 Th 90	231.0359 Pa 91	238.029 U 92	237.0482 Np 93	244) Pu 94	(243) Am 95	(247) Cm 96	(247) Bk 97	(251) Cf 98	(252) Es 99	(257) Fm 100	(258) Md 101	(259) No 102	(260) Lr 103

Figure 5.22

The Periodic Table of Elements (long form). Atomic weights are shown above the symbols; atomic numbers, below.

variations in behavior will be discussed more fully in the following sections and in Chapter 9.

The modern Periodic Table (Fig. 5.22) consists of 7 horizontal rows of elements (often referred to as **periods** or **series**) and 32 vertical columns of elements (referred to as **families** or **groups**). In order to fit the table on a single page, some of the elements have been written below the main body of the table. The **lanthanide series** fits between elements 56 (barium, Ba) and 72 (hafnium, Hf). The **actinide series** fits between elements 88 (radium, Ra) and 104 (unnilquadium, Unq).

The **first short period** of the table contains the elements hydrogen (H) and helium (He). The **second short period** contains eight elements, beginning with lithium (Li) and ending with neon (Ne). The **third short period** also contains eight elements, beginning with sodium (Na) and ending with argon (Ar).

The **fourth period** is the first of two long periods, each of which contains 18 elements. The fourth period includes the elements from potassium (K) through krypton (Kr). Within this period are the elements from scandium (Sc) through copper (Cu), which are known as the **first transition series**. The **fifth period** begins with rubidium (Rb) and ends with xenon (Xe). Within this period are the elements yttrium (Y) through silver (Ag), which comprise the **second transition series.**

The **sixth period,** beginning with cesium (Cs) and ending with radon (Rn), contains 32 elements. The **third transition series,** made up of lanthanum (La) and the elements hafnium (Hf) through gold (Au), is included in the sixth period. The third transition series is split; between lanthanum and hafnium is a series of 14 elements, cerium (Ce) through lutetium (Lu), called the **first inner transition series,** or the **lanthanide series** or the **rare earth elements.** Lanthanum behaves very much like the elements cerium through lutetium. Hence lanthanum is often included in the lanthanide series, even though in terms of electronic structure it is more properly considered to be the first element of the third transition series.

The **seventh period** extends from francium through element number 118. However, no elements after element 109 have been characterized. The known elements in this period include a part of the **fourth transition series** (actinium, and elements 104 through 109). The unreported elements 110 and 111 would also be members of this series. The elements between actinium (Ac) and element 104, namely the elements thorium (Th) through lawrencium (Lr), make up a second inner transition series referred to as the **actinide series.** Actinium is sometimes included with the actinide series because of its similarity in properties to those elements.

The discoverer of a new element names the element. However, until the name is recognized by the International Union of Pure and Applied Chemistry (IUPAC), the recommended name of the new element will be based on the Latin words for its atomic number. These names will also be used for elements, such as element 110, which have not yet been prepared or discovered. An international dispute about credit for the discovery of element 104 led to this new set of systematic names for the heavier elements. In 1964, Soviet scientists reported element 104, which they named kurchatovium, after the leader of their nuclear research program. Five years later, physicists at the University of California suggested that the Soviets were wrong and that, in fact, *they* had prepared element 104. They named it rutherfordium, after the British scientist. Neither name was formally adopted, although each is used in its country of origin. In 1979 IUPAC recommended using the name unnilquadium (abbreviated Unq) for element 104 until the question of the original discoverer could be settled. In addition, they recommended unnilpentium (Unp) for element 105, unnilhexium (Unh) for element 106, unnilseptium (Uns) for element 107, unniloctium (Uno) for element 108, and unnilennium (Une) for element 109. These names are based on the following Latin words for the atomic numbers: *nil* = 0, *un* = 1, *bi* = 2, *tri* = 3, *quad* = 4, *pent* = 5, *hex* = 6, *sept* = 7, *oct* = 8, and *enn* = 9.

Many of the groups (or families) in the Periodic Table are identified by Roman numerals and letters at the top of the vertical columns. The assignment of the letters A or B to the groups is arbitrary, and some Periodic Tables, particularly European ones, use a different system from the one that appears in the text. While this book was being produced, there was discussion about an IUPAC recommendation that the groups be numbered, from left to right, from one to eighteen. We have included this scheme as well in Fig. 5.22.

5.14 Electronic Structure and the Periodic Law

When arranged in order of increasing atomic number, the elements with similar chemical properties recur at definite intervals, i.e., periodically. This behavior reflects the periodic recurrence of similar electron configurations in the outer shells of these elements. Figure 5.23 shows the electron configurations of the elements with atomic numbers 1 through 18, inclusive, arranged as these atoms are arranged in the Periodic Table. Note the periodicity with respect to the number of **valence electrons; that is, electrons in the outermost shell.**

Elements in any one group have the same number of electrons in their outermost shell: lithium and sodium have only one, beryllium and magnesium have two, fluorine and chlorine have seven electrons in the **valence shell**—the outermost shell (or shells in certain heavier atoms) containing the valence electrons. The similarity in chemical properties among elements of the same group occurs because they have the same numbers of valence electrons, and the number of electrons in the valence shell of an atom determines its chemical properties. **It is the loss, gain, or sharing of valence electrons that determines how elements react.** Thus the Periodic Table is simply an arrangement of atoms that puts elements with the same number of valence electrons in the same group. This arrangement is emphasized in Fig. 5.24, which shows in Periodic Table form the electron configuration of the last subshell to be filled by the aufbau process.

IA							VIIIA
H $1s^1$	**IIA**	**IIIA**	**IVA**	**VA**	**VIA**	**VIIA**	**He** $1s^2$
Li [He]$2s^1$	**Be** [He]$2s^2$	**B** [He]$2s^22p^1$	**C** [He]$2s^22p^2$	**N** [He]$2s^22p^3$	**O** [He]$2s^22p^4$	**F** [He]$2s^22p^5$	**Ne** [He]$2s^22p^6$
Na [Ne]$3s^1$	**Mg** [Ne]$3s^2$	**Al** [Ne]$3s^23p^1$	**Si** [Ne]$3s^23p^2$	**P** [Ne]$3s^23p^3$	**S** [Ne]$3s^23p^4$	**Cl** [Ne]$3s^23p^5$	**Ar** [Ne]$3s^23p^6$

Figure 5.23

The electron configurations of the first 18 elements arranged in the format of the Periodic Table.

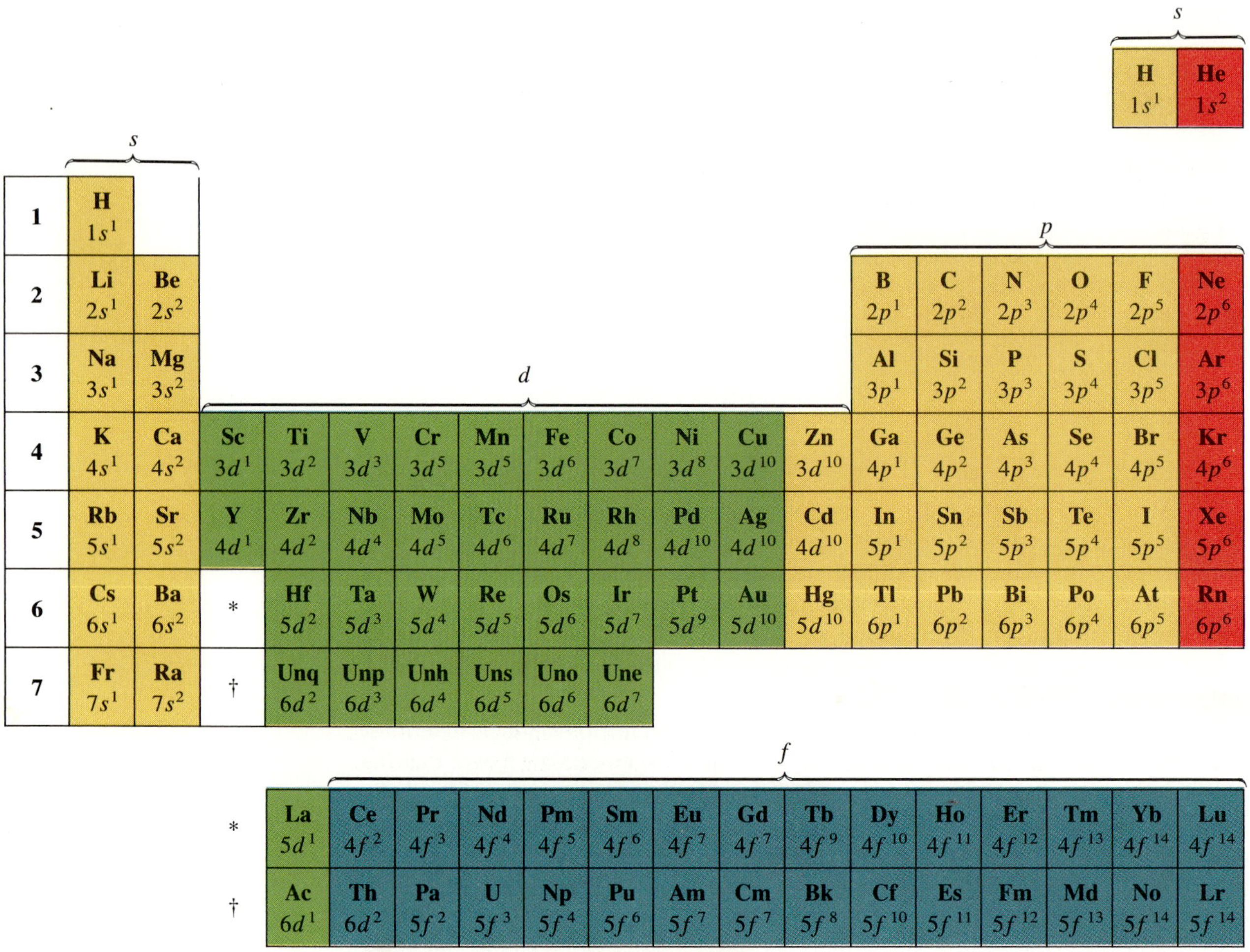

s

		H $1s^1$	He $1s^2$

s

p

			d												*p*					
1	H $1s^1$																			
2	Li $2s^1$	Be $2s^2$												B $2p^1$	C $2p^2$	N $2p^3$	O $2p^4$	F $2p^5$	Ne $2p^6$
3	Na $3s^1$	Mg $3s^2$												Al $3p^1$	Si $3p^2$	P $3p^3$	S $3p^4$	Cl $3p^5$	Ar $3p^6$
4	K $4s^1$	Ca $4s^2$	Sc $3d^1$	Ti $3d^2$	V $3d^3$	Cr $3d^5$	Mn $3d^5$	Fe $3d^6$	Co $3d^7$	Ni $3d^8$	Cu $3d^{10}$	Zn $3d^{10}$	Ga $4p^1$	Ge $4p^2$	As $4p^3$	Se $4p^4$	Br $4p^5$	Kr $4p^6$
5	Rb $5s^1$	Sr $5s^2$	Y $4d^1$	Zr $4d^2$	Nb $4d^4$	Mo $4d^5$	Tc $4d^6$	Ru $4d^7$	Rh $4d^8$	Pd $4d^{10}$	Ag $4d^{10}$	Cd $4d^{10}$	In $5p^1$	Sn $5p^2$	Sb $5p^3$	Te $5p^4$	I $5p^5$	Xe $5p^6$
6	Cs $6s^1$	Ba $6s^2$	*	Hf $5d^2$	Ta $5d^3$	W $5d^4$	Re $5d^5$	Os $5d^6$	Ir $5d^7$	Pt $5d^9$	Au $5d^{10}$	Hg $5d^{10}$	Tl $6p^1$	Pb $6p^2$	Bi $6p^3$	Po $6p^4$	At $6p^5$	Rn $6p^6$
7	Fr $7s^1$	Ra $7s^2$	†	Unq $6d^2$	Unp $6d^3$	Unh $6d^4$	Uns $6d^5$	Uno $6d^6$	Une $6d^7$									

f

	La $5d^1$	Ce $4f^2$	Pr $4f^3$	Nd $4f^4$	Pm $4f^5$	Sm $4f^6$	Eu $4f^7$	Gd $4f^7$	Tb $4f^9$	Dy $4f^{10}$	Ho $4f^{11}$	Er $4f^{12}$	Tm $4f^{13}$	Yb $4f^{14}$	Lu $4f^{14}$
*	La $5d^1$	Ce $4f^2$	Pr $4f^3$	Nd $4f^4$	Pm $4f^5$	Sm $4f^6$	Eu $4f^7$	Gd $4f^7$	Tb $4f^9$	Dy $4f^{10}$	Ho $4f^{11}$	Er $4f^{12}$	Tm $4f^{13}$	Yb $4f^{14}$	Lu $4f^{14}$
†	Ac $6d^1$	Th $6d^2$	Pa $5f^2$	U $5f^3$	Np $5f^4$	Pu $5f^6$	Am $5f^7$	Cm $5f^7$	Bk $5f^8$	Cf $5f^{10}$	Es $5f^{11}$	Fm $5f^{12}$	Md $5f^{13}$	No $5f^{14}$	Lr $5f^{14}$

Figure 5.24

The order of occupancy of atomic orbitals in the Periodic Table. The electron configuration of the last subshell to be occupied as the atoms are built up by the aufbau process is shown.

It is convenient to classify the elements in the Periodic Table into four categories according to their atomic structures.

1. *Noble gases*. Elements in which the *s* and any *p* orbitals in the outer shell are completely filled. The noble gases are the members of Group VIIIA (shown in red in Fig. 5.24).

2. *Representative elements*. Elements in which the last electron added enters an *s* or a *p* orbital in the outermost shell but in which this shell is incomplete. The outermost shell for these elements is the valence shell. The representative elements are those in Groups IA, IIA, IIIA, IVA, VA, VIA, VIIA and IIB of the Periodic Table (shown in yellow).

3. *Transition elements*. Elements in which the second shell, counting from the outside, is building from 8 to 18 electrons as its *d* orbitals fill. The outermost *s* subshell, the *ns* subshell, and the *d* subshell of the next shell in, the $(n - 1)d$ subshell, contain the valence electrons in these elements. Thus the $(n - 1)d$ and *ns* subshells are the valence shells in the transition elements. There are four transition series (shown in green).

(a) First transition series: scandium (Sc) through copper (Cu); $3d$ subshell filling.

(b) Second transition series: yttrium (Y) through silver (Ag); $4d$ subshell filling.

(c) Third transition series: lanthanum (La), plus hafnium (Hf) through gold (Au); $5d$ subshell filling.

(d) Fourth transition series (incomplete): actinium (Ac), plus elements 104 through 109; $6d$ subshell filling. (If elements 110 and 111 are discovered, this will complete the series.)

(Although the aufbau principle predicts that zinc, cadmium, and mercury would be transition elements, in fact, they are representative elements.)

4. *Inner transition elements*. Elements in which the third shell, counting in from the outside, is building from 18 to 32 electrons as its f orbitals fill. The valence shells of the inner transition elements consist of the $(n - 2)f$, $(n - 1)d$, and ns subshells. There are two inner transition series (shown in blue).

(a) First inner transition series: cerium (Ce) through lutetium (Lu); $4f$ subshell filling.

(b) Second inner transition series: thorium (Th) through lawrencium (Lr); $5f$ subshell filling.

(Lanthanum and actinium, because of their similarities to the other members of the series, are sometimes included as the first elements of the first and second inner transition series, respectively.)

5.15 Variation of Properties Within Periods and Groups

Elements within a group have identical numbers and generally identical distributions of electrons in their valence shells. Thus we expect them to exhibit very similar chemical behavior. Across a period, we might expect a smoothly varying change in chemical behavior, since each element differs from the preceding element by one electron. However, sometimes the similarities or differences within a group or across a period are not as regular as we might expect. This is because the loss, gain, or sharing of valence electrons depends on several factors, including (1) the number of valence electrons, (2) the magnitude of the nuclear charge and the total number of electrons surrounding the nucleus, (3) the number of filled shells lying between the nucleus and the valence shell, and (4) the distances of the electrons in the various shells from each other and from the nucleus.

Examples of these effects on the periodic variation of some physical properties will be considered in the following paragraphs.

1. VARIATION IN COVALENT RADII. (See table inside the back cover.) There are several ways to define the radii of atoms and thus to determine their relative sizes. We will use the **covalent radius,** which is defined as half the distance between the nuclei of two identical atoms when they are joined by a single covalent bond. A covalent bond will be discussed in Chapter 6; for now, we will simply use the covalent radius as one measure of the size of an atom. In general, from left to right across a period of the Periodic Table, each element has a smaller covalent radius than that of the one preceding it (Table 5.6). This change in size can be attributed to the increasing nuclear charge across the period, with the added electrons going into partially occupied shells. Each element in the Periodic Table has one more electron and a nuclear charge that is one higher than the preceding element. Within a period, however, the number of shells is constant. In general, within a given period, the larger nuclear charge results in a larger force of electrostatic attraction between the nucleus and the electrons, because the

Table 5.6

Atom	Covalent radius, Å	Nuclear charge	Electron configuration
Na	1.86	+11	$1s^2 2s^2 2p^6 3s^1$
Mg	1.60	+12	$3s^2$
Al	1.43	+13	$3s^2 3p^1$
Si	1.17	+14	$3s^2 3p^2$
P	1.10	+15	$3s^2 3p^3$
S	1.04	+16	$3s^2 3p^4$
Cl	0.99	+17	$3s^2 3p^5$

additional electrons are in the same shell. This causes the decrease in covalent radii across the period.

Our covalent radii are based on interatomic distances between two identical atoms held together by chemical bonds. Noble gases, however, do not bond this way. Thus covalent radii are not available for the noble gases.

Proceeding down a group of the Periodic Table, succeeding elements have larger covalent radii as a result of greater numbers of electron shells (Table 5.7), a factor that more than offsets the effect of the larger nuclear charge of each succeeding element. To be more specific, the increasing nuclear charge might lead us to expect that electrons would be held more tightly and pulled closer to the nucleus. However, the total number of shells increases down a group, and shells with larger principal quantum numbers have larger radii. The larger size of the shells coupled with repulsions between the increasing numbers of electrons overcome the increased nuclear attraction, so that the atoms increase in size down a group.

Table 5.7

Atom	Covalent radius, Å	Nuclear charge	Number of electrons in each shell
F	0.64	+ 9	2, 7
Cl	0.99	+17	2, 8, 7
Br	1.14	+35	2, 8, 18, 7
I	1.33	+53	2, 8, 18, 18, 7
At	1.4	+85	2, 8, 18, 32, 18, 7

2. VARIATION IN IONIC RADII. As shown in the table inside the back cover, the radius of a positive ion is less than the covalent radius of its parent atom. A positive ion forms when one or more than one electron is removed from an atom. Usually, the representative elements form positive ions by loss of all of their valence electrons. The loss of all electrons from the outermost shell results in a smaller radius, because the remaining electrons occupy shells with smaller principal quantum numbers (and smaller radii). In fact, even the radii of these remaining filled electron shells decrease (relative to their size in the neutral atom), because of the decrease in the total number of electron–electron repulsions within the atom. The decreasing repulsions give rise to a greater average attraction of the nucleus per remaining electron, an effect spoken of as an increase in the *effective* nuclear charge. Thus the covalent radius of a sodium atom ($1s^2 2s^2 2p^6 3s^1$) is 1.86 Å, whereas the ionic radius of a sodium ion ($1s^2 2s^2 2p^6$) is

0.95 Å. Proceeding down the groups of the Periodic Table, positive ions of succeeding elements have larger radii, corresponding to greater numbers of shells.

A simple negative ion is formed by the addition of one or more than one electron to the valence shell of an atom. This results in a greater force of repulsion among the electrons and a decrease in the effective nuclear charge per electron. Both effects cause the radius of a negative ion to be greater than that of the parent atom. For example, a chlorine atom ($[Ne]3s^23p^5$) has a covalent radius of 0.99 Å, whereas the ionic radius of a chloride ion ($[Ne]3s^23p^6$) is 1.81 Å. For succeeding elements proceeding down the groups, negative ions have more electron shells and larger radii. (These effects are apparent in the table inside the back cover.)

Ions and atoms that have the same electron configuration, such as those in the series N^{3-}, O^{2-}, F^-, Ne, Na^+, Mg^{2+}, and Al^{3+} and those in the series P^{3-}, S^{2-}, Cl^-, Ar, K^+, Ca^{2+}, and Sc^{3+}, are termed **isoelectric.** The greater the nuclear charge, the smaller the ionic radius in a series of isoelectronic ions and atoms. This trend is illustrated in Table 5.8.

Table 5.8

Species:	N^{3-}	O^{2-}	F^-	Ne	Na^+	Mg^{2+}	Al^{3+}
Radius, Å:	1.71	1.40	1.36	1.12	0.95	0.65	0.50

Electron configuration: $1s^22s^22p^6$

Species:	P^{3-}	S^{2-}	Cl^-	Ar	K^+	Ca^{2+}	Sc^{3+}
Radius, Å:	2.12	1.84	1.81	1.54	1.33	0.99	0.81

Electron configuration: $1s^22s^22p^63s^23p^6$

As we shall see later, many properties of ions can be explained in terms of their sizes and charges.

3. VARIATION IN IONIZATION ENERGIES. The amount of energy required to remove *the most loosely bound* electron from a gaseous atom is called its **first ionization energy.** This change may be represented for any element X by the equation

$$X(g) + energy \longrightarrow X^+(g) + e^-$$

The energy required to remove the second most loosely bound electron is called the **second ionization energy;** to remove the third, the **third ionization energy;** and so forth. First ionization energies increase in an irregular way from left to right across a period (Table 5.9). This overall increase may be attributed to the fact that the electrons lost come from the same shell, while the nuclear charge increases.

Figure 5.25 (p. 129) shows the relationship between the first ionization energies and atomic numbers of several elements. The values of the first ionization energies are provided in Table 5.9. Note that the ionization energy of boron is less than that of beryllium. This is explained by differences in the attraction of the positive nucleus for electrons in different subshells. On the average, an *s* electron is attracted to the nucleus more than a *p* electron on the same principal shell. This means that an *s* electron will be harder to remove from an atom than a *p* electron in the same shell. The electron removed during the ionization of beryllium ($[He]2s^2$) is an *s* electron, whereas the electron removed during the ionization of boron ($[He]2s^22p^1$)is a *p* electron; this results in a smaller first ionization energy for boron even though its nuclear charge is greater by one unit. The two additional nuclear charges in carbon are sufficient to make its first ionization energy larger than that of beryllium. The first ionization energy for oxygen is slightly less than that for nitrogen because of the repulsion between the two electrons

Table 5.9   First ionization energies of some of the elements. (Energy in electron-volts per atom; 1 eV per atom corresponds to 96.49 kJ per mole of atoms.)

1	2	3	4	5	6	7	8	9	10	11	12	13	14	15	16	17	18
1 **H** 13.6																	2 **He** 24.6
3 **Li** 5.4	4 **Be** 9.3											5 **B** 8.3	6 **C** 11.3	7 **N** 14.5	8 **O** 13.6	9 **F** 17.4	10 **Ne** 21.6
11 **Na** 5.1	12 **Mg** 7.6											13 **Al** 6.0	14 **Si** 8.1	15 **P** 11.0	16 **S** 10.4	17 **Cl** 13.0	18 **Ar** 15.8
19 **K** 4.3	20 **Ca** 6.1	21 **Sc** 6.5	22 **Ti** 6.8	23 **V** 6.7	24 **Cr** 6.8	25 **Mn** 7.4	26 **Fe** 7.9	27 **Co** 7.9	28 **Ni** 7.6	29 **Cu** 7.7	30 **Zn** 9.4	31 **Ga** 6.0	32 **Ge** 8.1	33 **As** 10	34 **Se** 9.8	35 **Br** 11.8	36 **Kr** 14.0
37 **Rb** 4.2	38 **Sr** 5.7	39 **Y** 6.4	40 **Zr** 6.8	41 **Nb** 6.9	42 **Mo** 7.1	43 **Tc** 7.3	44 **Ru** 7.4	45 **Rh** 7.5	46 **Pd** 8.3	47 **Ag** 7.6	48 **Cd** 9.0	49 **In** 5.8	50 **Sn** 7.3	51 **Sb** 8.6	52 **Te** 9.0	53 **I** 10.5	54 **Xe** 12.1
55 **Cs** 3.9	56 **Ba** 5.2	[57-71] *	72 **Hf** 7	73 **Ta** 7.9	74 **W** 8.0	75 **Re** 7.9	76 **Os** 8.7	77 **Ir** 9.2	78 **Pt** 9.0	79 **Au** 9.2	80 **Hg** 10.4	81 **Tl** 6.1	82 **Pb** 7.4	83 **Bi** 8	84 **Po** 8.4	85 **At** ...	86 **Rn** 10.7
87 **Fr** ...	88 **Ra** 5.3	[89-103] †	104 **Unq** ...	105 **Unp** ...	106 **Unh** ...	107 **Uns** ...	108 **Uno** ...	109 **Une** ...									

*LANTHANIDE SERIES	57 **La** 5.6	58 **Ce** 6.9	59 **Pr** 5.8	60 **Nd** 6.3	61 **Pm** ...	62 **Sm** 5.6	63 **Eu** 5.7	64 **Gd** 6.2	65 **Tb** 6.7	66 **Dy** 6.8	67 **Ho** ...	68 **Er** ...	69 **Tm** ...	70 **Yb** 6.2	71 **Lu** 5.0
†ACTINIDE SERIES	89 **Ac** 6.9	90 **Th** ...	91 **Pa** ...	92 **U** 4	93 **Np** ...	94 **Pu** ...	95 **Am** ...	96 **Cm** ...	97 **Bk** ...	98 **Cf** ...	99 **Es** ...	100 **Fm** ...	101 **Md** ...	102 **No** ...	103 **Lr** ...

occupying the same $2p$ orbital in the oxygen atom. Since these two electrons occupy the same region of space, their repulsion overcomes the additional nuclear charge of the oxygen nucleus. Analogous changes occur in succeeding periods.

The attractive force exerted on the valence electrons by the positively charged nucleus is partially counterbalanced by the repulsive forces between electrons in inner shells and the valence electrons. An electron being removed from an atom is thus shielded from the nucleus by these inner shells. This shielding and the increasing distance of the outer electron from the nucleus provide an explanation of the fact that, proceeding down the groups, succeeding elements generally have smaller first ionization energies. As a rule, the elements in a group have the same outer electron configuration and the same number of valence electrons.

4. VARIATION IN ELECTRON AFFINITIES. The **electron affinity** is a measure of the energy involved when an electron is added to a gaseous atom to form a negative

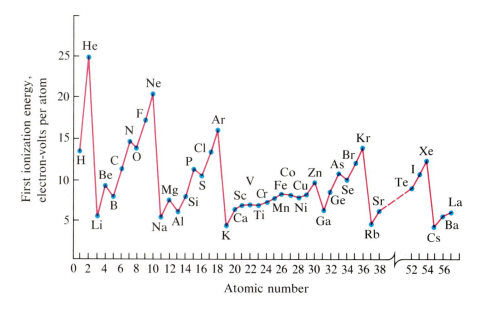

Figure 5.25

A graphic illustration of the periodic relationships between first ionization energy and atomic number for some of the elements.

ion. A positive value for electron affinity indicates that energy is produced when an electron is added to an atom. The change is expressed for any element X by the equation

$$X(g) + e^- \longrightarrow X^-(g) + energy$$

A negative value for electron affinity indicates that energy must be added to force the electron onto the atom, as represented in the equation

$$Energy + X + e^- \longrightarrow X^-$$

Elements to the left within a period have little tendency to form negative ions; thus their electron affinities tend to be small or negative (Table 5.10). Elements to the right within a period tend to have larger electron affinities.

Table 5.10 Electron affinities of some elements.
(Electron affinities in electron-volts per atom;
1 eV per atom = 96.49 kJ per mole.)

IA							VIIIA
H 0.75	IIA	IIIA	IVA	VA	VIA	VIIA	**He** −0.2[a]
Li 0.62	**Be** −2.5[a]	**B** 0.24	**C** 1.27	**N** 0.0	**O** 1.46	**F** 3.34	**Ne** −0.3
Na 0.55	**Mg** −2.4[a]	**Al** 0.46	**Si** 1.24	**P** 0.77	**S** 2.08	**Cl** 3.61	**Ar** −0.36[a]
K 0.50	**Ca** −1.6[a]	**Ga** 0.4[a]	**Ge** 1.20	**As** 0.80	**Se** 2.02	**Br** 3.36	**Kr** −0.4[a]
Rb 0.48	**Sr** −1.7[a]	**In** 0.4[a]	**Sn** 1.25	**Sb** 1.05	**Te** 1.97	**I** 3.06	**Xe** −0.4[a]
Cs 0.47	**Ba** −0.5[a]	**Tl** 0.5	**Pb** 1.05	**Bi** 1.05	**Po** 1.8[a]	**At** 2.8[a]	**Rn** −0.4[a]

[a]Calculated value.

Although succeeding elements across the periods of the Periodic Table generally have progressively greater electron affinities, exceptions are found among the elements of Group IIA, Group VA, and Group VIIIA. These groups have filled *ns* subshells, half-filled *np* subshells, and all subshells filled, respectively. In each case, the completely filled or half-filled subshells represent relatively stable configurations.

For Review

Summary

A number of experiments have shown that atoms are composed of protons, neutrons, and electrons. **Protons** are relatively heavy particles with a charge of $+1$ and a mass of 1.0073 amu. **Neutrons** are relatively heavy particles with no charge and with a mass of 1.0087 amu, very close to that of protons. **Electrons** are light particles with a mass of 0.00055 amu (1/1837 of that of a hydrogen atom) and a charge of -1. Rutherford's α-particle scattering experiments showed that the atom contains a small, massive, positively charged **nucleus** surrounded by electrons. The size of the nucleus is at least 100,000 times smaller than the size of the atom. It was subsequently shown that the nucleus consists of protons and neutrons and that the number of protons in the nucleus, the **atomic number,** determines the type of atom, i.e., the element. Isotopes are atoms with the same atomic number but different **mass numbers.** An element is usually composed of several isotopes, and its atomic weight is the weighted average of the masses of the isotopes involved.

Bohr described the hydrogen atom in terms of an electron moving in a circular orbit about a nucleus. In order to account for the stability of the hydrogen atom, he postulated that the electron was restricted to certain discrete energies characterized by a **quantum number, *n*,** which could have only integer values ($n = 1, 2, 3, \ldots$). Thus the energy of the electron was assumed to be **quantized.** From this model we can calculate the energy E, of an electron in a hydrogenlike system:

$$E = \frac{-kZ^2}{n^2}$$

where k is a constant and Z is the atomic number. The radius of the orbit can also be calculated as follows:

$$r = \frac{n^2 a_0}{Z}$$

where a_0 is the radius of the orbit in the hydrogen atom for which $n = 1$. Bohr suggested that an atom emits energy, in the form of a photon, only when an electron falls from a higher-energy orbit to a lower-energy orbit. Because the Bohr model accounted for the line spectra produced by samples of hydrogen gas in a discharge tube, his postulation of quantized energies for the electrons in atoms was widely accepted.

Light and other forms of electromagnetic radiation move through a vacuum with a speed, c, of 2.998×10^8 m/s. This radiation shows wavelike behavior, which can be characterized by a frequency, v, and a wavelength, λ, such that $c = \lambda v$. Electromagnetic radiation also has the properties of particles called **photons.** The energy of a photon is related to the frequency of the radiation: $E = hv$, where h is Planck's constant. The line spectrum of hydrogen is obtained by passing the light from a discharge tube through a prism. From a measurement of the wavelength of each line, the energy of each can be determined. These energies correspond to the values predicted when an electron in a hydrogen atom falls from a higher-energy to a lower-energy orbit.

Atoms with two or more electrons cannot be described satisfactorily by the Bohr model, and it has been replaced by the **quantum mechanical model.** This model identifies the behavior of an electron in terms of a **wave function, ψ,** which identifies the **orbital,** or region of space, occupied by the electron. It is possible to calculate the energy of the electron from ψ or to determine the probability of finding the electron in a given location in the orbital. The distribution of an electron within an orbital may be described in terms of **electron density.**

An orbital is characterized by three quantum numbers. The **principal quantum number, n,** may be any positive integer. The energy of the orbital and its average distance from the nucleus are related to n. The **azimuthal quantum number, l,** can have any integral value from 0 to $(n - 1)$. The shape of the orbital is indicated by l, and within a multielectron atom the energies of orbitals with the same value of n increase as l increases. The **magnetic quantum number, m,** with values ranging from $-l$ to $+l$, describes the orientation of an orbital. A fourth quantum number, the **spin quantum number, s** $(s = \pm\frac{1}{2})$, describes the spin of an electron about its own axis. Orbitals with the same value of n occupy the same **shell.** Orbitals with the same values of n and l occupy the same **subshell.**

n	Shell	Subshells
1	K	$1s$ $(l = 0)$
2	L	$2s$ $(l = 0)$, $2p$ $(l = 1)$
3	M	$3s$ $(l = 0)$, $3p$ $(l = 1)$, $3d$ $(l = 2)$
4	N	$4s$ $(l = 0)$, $4p$ $(l = 1)$, $4d$ $(l = 2)$, $4f$ $(l = 3)$

There are $(2l + 1)$ orbitals in the l subshell of any shell.

By adding electrons one by one to atomic orbitals in the order $1s$, $2s$, $2p$, $3s$, $3p$, $4s$, $3d$, $4p$, etc., and following both the **Pauli exclusion principle** (no two electrons can have the same set of four quantum numbers) and **Hund's rule** (whenever possible, electrons remain unpaired in degenerate orbitals), we can predict the **electron configurations** of most atoms. The periodic recurrence of similar electron configurations led to the statement of the **periodic law** and the construction of a **Periodic Table,** in which elements with similar electron configurations (and, consequently, similar chemical behavior) are found in the same **group,** or **family. Valence electrons** in the outermost s and p orbitals are responsible for the chemical behavior of the **representative elements:** Group IA (the alkali metals, s^1), Group IIA (the alkaline earth metals, s^2), Group IIIA (s^2p^1), Group IVA (s^2p^2), Group VA (s^2p^3), Group VIA (the chalcogens, s^2p^4), Group VIIA (the halogens, s^2p^5), and Group IIB (s^2). **Transition metals** may have valence electrons in the outer shell and first inner shell, the ns and $(n - 1)d$ orbitals, respectively. The **inner transition metals** may have valence electrons in three shells: the ns, $(n - 1)d$, and $(n - 2)f$ orbitals. The **atomic radii** and **ionic radii** of elements, their **ionization energies,** and their **electron affinities** vary in a periodic way with the nature of the valence orbitals involved.

Key Terms and Concepts

alpha particles (5.4)
atomic number (5.5)
aufbau process (5.12)
azimuthal quantum number, l
    (5.10)
beta particles (5.4)
Bohr model of the atom (5.7)
cathode rays (5.1)
covalent radius (5.15)
degenerate orbitals (5.10)
electron affinity (5.15)
electron configuration (5.12)
electron density (5.9)
electrons (5.1)
excited state (5.7)
family of elements (5.13)
first ionization energy (5.15)
gamma rays (5.4)

ground state (5.7)
Heisenberg uncertainty principle
    (5.9)
Hund's rule (5.11)
ionic radius (5.15)
ionization energy (5.15)
ions (5.2)
isoelectronic species (5.15)
isotopes (5.6)
magnetic quantum number, m
    (5.10)
mass number (5.6)
neutrons (5.3)
noble gases (5.13)
nucleus (5.5)
orbital (5.9)
Pauli exclusion principle (5.11)
periodic law (5.13)

Periodic Table (5.13)
periods of elements (5.13)
photoelectric effect (5.1)
photons (5.8)
principal quantum number, n
    (5.10)
protons (5.2)
quantized (5.7)
quantum mechanical model of the
    atom (5.9)
radioactivity (5.4)
shell (5.10)
spectrum (5.8)
spin quantum number, s (5.10)
subshell (5.10)
wave function (5.9)
valence electrons (5.14)
valence shell (5.14)

Exercises

Atomic Structure

1. What is the evidence that the cathode rays in a discharge tube are negatively charged particles? What is their source?
2. How are cathode rays and β particles similar?
3. How are positive rays and α particles similar? How do they differ?
4. Describe and interpret the experiment that shows that an atom contains a small, positively charged nucleus of relatively large mass.
5. Must the numbers of protons, electrons, and neutrons be the same or different in each of the following: (a) two different isotopes of the same element, (b) two atoms of the same element with the same mass, (c) two atoms of different elements with the same mass number?
6. In what way is the lightest hydrogen nucleus unique?
7. Using the data in Table 5.2, sketch the mass spectrum expected for a sample of neon atoms.
8. From the data given in Table 5.2, calculate the atomic weight of naturally occurring neon.   *Ans. 20.17*
9. Explain why atomic weights of elements often differ markedly from whole numbers, although protons and neutrons have masses that are very nearly 1 amu.
10. Determine the number of protons and neutrons in the nuclei of each of the following elements.
    (a) ^7_3Li, (b) $^{14}_7N$, (c) $^{27}_{13}Al$, (d) $^{51}_{23}V$, (e) $^{106}_{46}Pd$.
11. Identify the following elements:
    (a) ^9_4X, (b) $^{28}_{14}X$, (c) $^{63}_{29}X$, (d) $^{182}_{74}X$, (e) $^{202}_{80}X$.

The Bohr Model of the Atom

12. How are the Bohr model and the Rutherford model of the atom similar? How do they differ?
13. What does it mean to say that the energies of the electrons in hydrogen atoms are quantized?
14. Figure 5.13 gives the energies of an electron in a hydrogen atom for values of n from 1 to 5. What is the energy, in eV, of an electron with $n = 6$ in a hydrogen atom?
          Ans. −0.3778 eV
15. What is the lowest possible energy, in eV, for an electron in a He$^+$ ion?   *Ans. −54.40 eV*
16. Excited H atoms with electrons in very high-energy levels have been detected. What is the radius (in centimeters) of a H atom with an electron characterized by an n value of 110? How many times larger is that than the radius of the H atom in its ground state?
          Ans. 6.40 × 10^{-5} cm; 12,100 times larger
17. Calculate the radius of a Li^{2+} ion using the Bohr model of the electronic structure of this ion.   *Ans. 0.176 Å*
18. Which is larger, a hydrogen atom with an electron in an orbit with $n = 4$ or a He$^+$ ion with an electron in an orbit with $n = 5$?   *Ans. H, with n = 4*

Orbital Energies and Spectra

19. (a) From the data given in Fig. 5.13, determine the energy, in eV, of the photon emitted when an electron in an excited hydrogen atom moves from an orbit with $n = 4$ to one with $n = 1$.   *Ans. 12.745 eV*

(b) From the data given in Fig. 5.13, determine the energy, in eV, of the photon required to move an electron in a hydrogen atom from an orbit with $n = 1$ to one with $n = 4$.   *Ans. 12.745 eV*

(c) Using the Bohr model, calculate the energy, in eV, of the photon emitted when an electron in an excited hydrogen atom moves from an orbit with $n = 4$ to one with $n = 1$.   *Ans. 12.75 eV*

(d) Using the Bohr model, calculate the energy, in eV, of the photon required to move an electron in a hydrogen atom from an orbit with $n = 1$ to one with $n = 4$.   *Ans. 12.75 eV*

20. (a) From the data given in Fig. 5.13, calculate the energy, in eV, required to ionize a hydrogen atom. Calculate the energy in J per atom.
    Ans. 13.595 eV; 2.1782 × 10^{-18} J per atom

    (b) Using the Bohr model, calculate the energy, in eV, required to ionize a hydrogen atom. Calculate the energy in J per atom.
    Ans. 13.60 eV; 2.179 × 10^{-18} J per atom

21. When heated, lithium atoms emit photons of red light with a wavelength of 6708 Å. What is the energy in J and the frequency of this light?
    Ans. 2.961 × 10^{-19} J; 4.469 × 10^{14} s^{-1}

22. FM-104, an FM radio station, broadcasts at a frequency of 1.039×10^8 s^{-1} (103.9 megahertz). What is the wavelength of these radio waves, in meters?
    Ans. 2.885 m

23. Using the Bohr model, calculate the energy, in eV and J, of the light emitted by a transition from the $n = 2$ (L) to the $n = 1$ (K) shell in He$^+$.
    Ans. 40.80 eV; 6.537 × 10^{-18} J

24. Consider a collection of hydrogen atoms with electrons randomly distributed in the $n = 1, 2, 3, 4,$ and 5 shells. How many different wavelengths of light will be emitted by these atoms as the electrons fall into the lower-energy states?   *Ans. 10*

25. Calculate the lowest and highest energies, in electron volts, of the light produced by the transitions described in exercise 24.
    Ans. From n = 5 to n = 4, 0.3060 eV; from n = 5 to n = 1, 13.06 eV

26. Calculate the frequencies and wavelengths of the light produced by the transitions described in exercise 25.
    Ans. = 7.399 × 10^{13}; 3.158 × 10^{15} s^{-1}; 4.052 × 10^{-6} m; 9.494 × 10^{-8} m

27. Light that looks blue has a wavelength of 4800 Å. What is the frequency of this light? What is the energy of a photon of this blue light, in joules?
    Ans. 6.246 × 10^{14} s^{-1}; 4.138 × 10^{-19} J

28. Does a photon of the blue light described in exercise 27 have enough energy to excite the electron in a hydrogen atom from the $n = 1$ shell to the shell with $n = 2$?
    Ans. No; 1.6336 × 10^{-18} J is required.

29. X rays are produced when the electron stream in an X-ray tube knocks an electron out of a low-lying shell of an atom in the target, and an electron from a higher shell falls into the lower-lying shell. The X ray is the photon given off as the electron falls into the lower shell. The most intense X rays produced by an X-ray tube with a copper target have wavelengths of 1.542 Å and 1.392 Å. These X rays are produced when an electron from the L or M shell falls into the K shell of a copper atom. Calculate the energy separation in eV of the K, L, and M shells in copper.
    Ans. M, K: 8.907 × 10^3 eV; M, L: 8.66 × 10^2 eV;
    L, K: 8.041 × 10^3 eV

The Quantum Mechanical Model of the Atom

30. How are the Bohr model and the quantum mechanical model of the atom similar? How do they differ?

31. How do electron shells, subshells, and orbitals differ?

32. Discuss the following: electron density, atomic orbital.

33. What information about an electron in an atom is available from the wave function that describes the electron?

34. What are the four quantum numbers used to describe an electron in the quantum mechanical model of an atom? What does each describe?

35. Which of the following orbitals are degenerate in an atom with two or more electrons: $3d_{xy}, 4s, 4p_z, 4d_{xy}, 4p_x$?

36. Consider the atomic orbitals, (i), (ii), and (iii), shown below in outline.

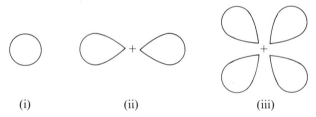

(i)   (ii)   (iii)

(a) What is the maximum number of electrons that can be contained in atomic orbital (iii)?   *Ans. 2*

(b) How many orbitals with the same value of l as orbital (i) can be found in the shell $n = 4$? How many as orbital (ii)? How many as orbital (iii)?
    Ans. 1; 3; 5

(c) What is the smallest n value possible for an electron in an orbital of type (iii)? Of type (ii)? Of type (i)?
    Ans. 3; 2; 1

(d) What are the l values that characterize each of these three orbitals?   *Ans. i, 0; ii, 1; iii, 2*

(e) Arrange these orbitals in order of increasing energy in the M shell. Is this order different in other shells?
    Ans. i < ii < iii; no

37. Identify the subshell in which electrons with the following quantum numbers are found:
    (a) $n = 2, l = 1$ *Ans. 2p*
    (b) $n = 3, l = 0$ *Ans. 3s*

(c) $n = 4$, $l = 2$ *Ans. 4d*
(d) $n = 7$, $l = 1$ *Ans. 7p*
(e) $n = 4$, $l = 3$ *Ans. 4f*

38. What type of orbital is occupied by an electron with the quantum numbers $n = 3$, $l = 2$? How many degenerate orbitals of this type are found in a multielectron atom?
                                                        Ans. 3d; 5

39. Write the quantum numbers of the six electrons in a carbon atom. For example, the quantum numbers for one of the $2s$ electrons will be $n = 2$, $l = 0$, $m = 0$, $s = +\frac{1}{2}$.

40. The quantum numbers that describe the electron in the $2s$ orbital of a lithium atom are $n = 2$, $l = 0$, $m = 0$, $s = +\frac{1}{2}$. Excitation of the electron can promote it to energy levels described by other sets of quantum numbers. Which of the following sets of quantum numbers cannot exist in an excited lithium atom (or any other atom)?
    (a) $n = 2$, $l = 1$, $m = -1$, $s = +\frac{3}{2}$
    (b) $n = 3$, $l = 2$, $m = 0$, $s = +\frac{1}{2}$
    (c) $n = 3$, $l = 3$, $m = -2$, $s = -\frac{1}{2}$
    (d) $n = 4$, $l = 1$, $m = -2$, $s = +\frac{1}{2}$
    (e) $n = 27$, $l = 14$, $m = -8$, $s = -\frac{1}{2}$ *Ans. a, c*

Electronic Configurations

41. Using complete subshell notation ($1s^2 2s^2 3p^6$, etc.), predict the electronic configuration of each of the following atoms: (a) $^{16}_{8}O$, (b) $^{27}_{13}Al$, (c) $^{32}_{16}S$, (d) $^{40}_{18}Ar$, (e) $^{40}_{20}Ca$, (f) $^{48}_{22}Ti$, (g) $^{55}_{25}Mn$, (h) $^{75}_{33}As$, (i) $^{119}_{50}Sn$, (j) $^{175}_{71}Lu$, (k) $^{222}_{86}Rn$

42. Using complete subshell notation ($1s^2 2s^2 2p^6$, etc.), predict the electronic configuration of each of the following ions: (a) N^{3-}, (b) Mg^{2+}, (c) Al^{3+}, (d) Cl^-, (e) Sc^{3+}, (f) Pb^{2+}, (g) Ce^{4+}.

43. Identify the atoms whose electronic configurations are given below.
    (a) $1s^2 2s^2 2p^6$ (b) $1s^2 2s^2 2p^6 3s^2 3p^6 4s^2 3d^3$
    (c) $[Ar]4s^2 3d^7$ (d) $[Ar]4s^2 3d^{10} 4p^6$
    (e) $[Kr]5s^2 4d^{10} 5p^2$ (f) $[Xe]6s^2 4f^4$

44. The $4s$ orbitals fill before the $3d$ orbitals when building up electronic structures by the aufbau process. However, electrons are lost from the $4s$ orbital before they are lost from the $3d$ orbitals when transition elements are ionized. The ionization of copper ($[Ar]3d^{10}4s^1$) gives Cu^+ ($[Ar]3d^{10}$), for example. The electronic structures of a number of transition metal ions are given below; identify the transition metals.
    (a) M^{3+}, $[Ar]3d^3$
    (b) M^{2+}, $[Ar]3d^2$
    (c) M^{3+}, $[Ar]3d^4$
    (d) M^{3+}, $[Ar]3d^2$
    (e) M^{3+}, $[Kr]4d^6$
    (f) M^{2+}, $[Kr]4d^9$

45. Using subshell notation ($1s^2 2s^2 2p^6$, etc.), predict the electronic structures of the following ions: (a) V^{4+}, (b) Co^{3+},

(c) Cr^{2+}, (d) Pb^{2+}, (e) Ag^+. Read the introduction to exercise 44 before attempting this exercise.

46. Which of the following electronic configurations describe an atom in its ground state and which describe an atom in an excited state?
    (a) $[He]2s^1$
    (b) $[Ar]5s^1 4d^3$
    (c) $[Ar]4s^2 3d^{10} 4p^4$
    (d) $[Ar]4s^2 3d^9 4p^3$
    (e) $[Kr]6s^2 6p^1$
    (f) $[Xe]6s^2 4f^{14} 5d^{10} 6p^2$

47. Which of the following sets of quantum numbers describes the most easily removed electron in a boron atom in its ground state? Which of the electrons described is most difficult to remove?
    (a) $n = 1$, $l = 0$, $m = 0$, $s = -\frac{1}{2}$
    (b) $n = 2$, $l = 1$, $m = 0$, $s = -\frac{1}{2}$
    (c) $n = 2$, $l = 0$, $m = 0$, $s = \frac{1}{2}$
    (d) $n = 3$, $l = 1$, $m = 1$, $s = -\frac{1}{2}$
    (e) $n = 4$, $l = 1$, $m = 1$, $s = \frac{1}{2}$ *Ans. b; a*

48. $N^+(g)$ can be produced from $N(g)$ by removing one electron from any of the occupied orbitals of the nitrogen atom. Several of these processes are:
    (a) $1s^2 2s^2 2p^3 \longrightarrow 1s^1 2s^2 2p^3$
    (b) $1s^2 2s^2 2p^3 \longrightarrow 1s^2 2s^1 2p^3$
    (c) $1s^2 2s^2 2p^3 \longrightarrow 1s^2 2s^2 2p^2$
    The first ionization potential of N is the energy of which of these processes? Which process will require the most energy?   *Ans. c; a*

The Periodic Table and Periodic Properties

49. Describe the noble gases, representative elements, inner transition elements, and transition elements in terms of filling of s, p, d, and f subshells.

50. State the periodic law. How does the periodic law correlate with electronic structure?

51. In terms of electronic structure, explain why there are 2 elements in the first period of the Periodic Table, 8 elements in the second and third periods, 18 elements in the fourth and fifth periods, and 32 elements in the sixth period.

52. Which groups have the following electronic structures in their valence shells? (n represents the principal quantum number.)
    (a) ns^2
    (b) $ns^2 np^1$
    (c) $ns^2 np^3$
    (d) $ns^2 (n-1)d^6$
    (e) $ns^2 (n-1)d^2$

53. In terms of the electronic configuration of the H atom, explain why H could be included in either Group IA or Group VIIA of the Periodic Table.

54. The formula for silicon chloride is $SiCl_4$. What is the chemical formula for the chlorides of C and Ge, two other members of Group IVA?

55. Most representative metals that form positive ions in chemical compounds do so by the loss of all of their valence electrons. Which group of representative elements in the Periodic Table forms tripositive ions, M^{3+}?

56. Representative nonmetals that form negative ions in chemical compounds do so by filling their valence shells with electrons. Which group in the Periodic Table forms dinegative ions, X^{2-}?

57. Why is the radius of an atom larger than the radius of a positive ion formed from it?

58. Why is the radius of an atom smaller than the radius of a negative ion formed from it?

59. Explain the decreasing radius of the isoelectronic ions Na^+, Mg^{2+}, Al^{3+}.

60. Explain the increasing radius of the isoelectronic ions F^-, O^{2-}, and N^{3-}.

61. Explain the decreasing radii for the atoms Na, Mg, Al, Si.

62. Which of the following would have the higher ionization energy: (a) K or Ca? (b) S or Cl? (c) O or S? (d) Sr or Ba? (e) P or O? Explain why.

63. Arrange each of the atoms or ions in the following groups in order of increasing size, based on their location in the Periodic Table.
    (a) Cs, K, Li, Na, Rb
    (b) B, Be, C, F, Li
    (c) Br^-, Ca^{2+}, K^+, Se^{2-}
    (d) Al^{3+}, F^-, Na^+, Mg^{2+}, O^{2-}
    (e) As^{3-}, Ca^{2+}, Cl^-, K^+

Additional Exercises

64. Consider one of the isotopes of uranium used in nuclear fission, $^{233}_{92}U$. (a) How many protons, neutrons, and electrons are contained in this atom? (b) How may protons, neutrons, and electrons are contained in a $^{233}_{92}U$ ion with a charge of +3?

65. Atoms of the isotope $^{16}_{8}O$ readily form ions with a charge of -2. How many protons, neutrons, and electrons are contained in such an ion?

66. How would the results of the Rutherford α-particle scattering experiment differ if the atom were only twice as large as its nucleus?

67. (a) Are the charges on the nuclei of atoms quantized?
    (b) Are the masses of individual atoms quantized?
    (c) Is the atomic weight of an element that is a mixture of isotopes quantized?

68. Using the Bohr model, what is the energy, in eV, of an electron with $n = 3$ in the ion Be^{3+}? What is the radius of the circular orbit?   *Ans. 24.18 eV; 1.19 Å*

69. The bright yellow light emitted by a sodium vapor lamp, or by heated compounds containing sodium, has wavelengths of 5896 and 5890 Å. These lines result from the relaxation of electrons from one or the other of two closely spaced energy levels to a common lower level. What is the energy separation, in eV, of the two closely spaced levels?   *Ans. 0.002 eV*

70. In Chapter 1, the meter was defined as 1,650,763.73 times the wavelength of a certain Kr line. What is the frequency of this radiation (to four significant figures)?
    Ans. 4.949×10^{14} s^{-1}

71. Draw an energy-level diagram like that in Fig. 5.20, showing the energies of the atomic orbitals in a hydrogen atom.

72. Figure 5.15 shows the radial probability density of three of the lowest-energy orbitals in a hydrogen atom. Which other orbitals are degenerate with each of these three?

73. Write the electronic configurations of the following elements, using complete subshell notation: B, Ne, Na, Ni, Se, Y, Ge.

74. Sketch the shape of an orbital with $n = 3$, $l = 0$; of an orbital with $n = 3$, $l = 1$.

75. Using [NG] to represent the electronic configuration of a noble gas, write the general electronic configurations for the alkali metals, the alkaline earths, the halogens, and the elements of Groups VIA, IIIB, and VIIB. The general electronic configuration for the elements of Group IIIA, for example, would be $[NG]ns^2np^1$.

76. The energy required to remove an electron from a Si^{3+} ion is 45.08 eV. To remove an electron from an Al^{3+} ion requires 120.2 eV. Explain the large difference.

77. Assume that, in another universe, the values of the quantum number m are limited to zero or positive integers up to the value of l for a particular subshell; thus for $l = 2$, m could be 0, 1, or 2. If, with this restriction, all other quantum numbers behaved as described in this chapter, describe the electronic configuration of nitrogen (at. no. 7) and the configuration of fluorine (at. no. 9) How many unpaired electrons would these atoms contain?
    Ans. N, $1s^22s^22p^3$; F, $1s^22s^22p^43s^1$; each with 1 unpaired electron

78. Write a periodic table for the first 14 elements in another universe if the quantum numbers of these elements behave as described in exercise 77.

79. A gaseous ion, X^+, in its most stable state, has three unpaired electrons. If X is a representative element, to which group of the Periodic Table does it belong?

80. The gaseous ion X^{2+} has no unpaired electrons in its most stable state. If X is a representative element, it could be a member of one of two possible groups. Which ones are they?

81. The ion X^- has three unpaired electrons. If X is a representative element, to which group does it belong?

82. One wavelength of light emitted by a sample of heated Li atoms is found at 2084 Å. Could this light be emitted by a Li^{2+} ion?   *Ans. Yes*

83. Which of the following electron configurations is found in the largest neutral atom? In the smallest neutral atom?
    (a) $1s^2 2s^2$
    (b) $1s^2 2s^2 2p^6 3s^2$
    (c) $1s^2 2s^2 2p^6 3s^2 3p^6 4s^2$
    (d) $1s^2 2s^2 2p^6 3s^2 3p^6 4s^2 3d^{10}$

84. Which of the following electron configurations is found in the largest positive ion with one positive charge? In the smallest?
    (a) $1s^2$
    (b) $1s^2 2s^2 2p^6$
    (c) $1s^2 2s^2 2p^6 3s^2 3p^6$
    (d) $1s^2 2s^2 2p^6 3s^2 3p^6 3d^{10}$

85. Each of the following electron configurations is found in a negative ion. Arrange them in increasing order of their size. The atomic number is given for each.
    (a) $(Z = 7)$ [Ne]
    (b) $(Z = 8)$ [Ne]
    (c) $(Z = 9)$ [Ne]
    (d) $(Z = 34)$ [Kr]
    (e) $(Z = 35)$ [Kr]

86. Identify two different atoms that lose electrons with the quantum numbers $n = 3$, $l = 0$, $m = 0$, $s = +\frac{1}{2}$ during the first ionization of each.

87. What are the n and l values of the electrons lost when an Mn atom forms a Mn^{3+} ion?

88. The value of one of the quantum numbers describing the electron that is lost when a silver atom forms a Ag^+ ion cannot be predicted. Which quantum number is it?

Chemical Bonding, Part 1: General Concepts

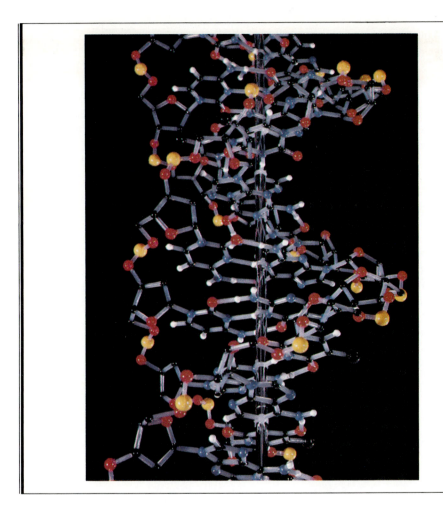

6

Covalent bonds are present in this complex DNA molecule as well as in simple molecules.

Although it is important to understand the electron configurations of isolated atoms, most chemists do not deal with isolated atoms. They usually study groups of two or more atoms and the attractive forces, called **chemical bonds,** that hold them together. When elements react and form a compound, chemical bonds form between the different atoms in the compound. When compounds undergo chemical reactions, the bonds between atoms are rearranged. When the atoms separate, the bonds are destroyed, and the compounds no longer exist.

In this chapter we will examine two types of chemical bonds and determine how to predict their formation from a consideration of the electron configurations of the atoms involved. We will see that the forces that hold atoms together are electrical in nature and that formation of chemical bonds between atoms involves changes in their electron configurations. We will also discover that loss of energy often accompanies formation or rearrangement of chemical bonds in chemical reactions.

6.1 Ionic Bonds; Chemical Bonding by Electron Transfer

When an element that loses electrons easily reacts with an element that gains electrons easily, one or more electrons are completely transferred, and **ions** are produced. A compound formed by this transfer is stabilized by **ionic bonds,** which involve strong electrostatic attractions between the ions of opposite charge present in the compound.

As an atom loses one or more than one electrons, it is left with a positive charge and becomes a **positive ion,** or **cation.** For example, when a sodium atom, which loses electrons easily, combines with an element that gains electrons easily, the sodium atom gives up its one valence electron and is left with a positive charge of 1. When a calcium atom loses its two valence electrons, a positive charge of 2 remains.

$$Na \cdot \longrightarrow Na^+ + e^- \qquad Ca: \longrightarrow Ca^{2+} + 2e^-$$

| Sodium | Sodium | Calcium | Calcium |
| atom | cation | atom | cation |

The dot notation in these equations is often used to describe changes in the electron configurations of representative elements. Each of these electron-dot formulas, or **Lewis symbols,** consists of the symbol for the element with one dot added for each valence electron present in the atom or ion. When two dots are written adjacent to each other, as in $Ca:$, they represent two electrons paired (with opposite spins) in the same orbital. The Lewis symbols for the atoms of the third period are given in Table 6.1.

Table 6.1	Lewis symbols for the atoms of the third period	
Atom	Electron configuration	Lewis symbol
Sodium	$[Ne]3s^1$	Na $\cdot$
Magnesium	$[Ne]3s^2$	Mg :
Aluminum	$[Ne]3s^23p^1$	$\dot{Al}$:
Silicon	$[Ne]3s^23p^2$	$\cdot \dot{Si} \cdot$
Phosphorus	$[Ne]3s^23p^3$	$\cdot \ddot{P} \cdot$
Sulfur	$[Ne]3s^23p^4$	$: \ddot{S} \cdot$
Chlorine	$[Ne]3s^23p^5$	$: \ddot{Cl} \cdot$
Argon	$[Ne]3s^23p^6$	$: \ddot{Ar} :$

For clarity, different symbols (· , × , ∘) are sometimes used as a convenience to distinguish electrons according to their sources. You should remember, however, that all electrons are identical, regardless of origin.

meaning it doesn't take much to get rid of electrons on the right

Atoms that easily form cations have relatively low ionization energies. As was noted in Section 5.15, atoms with low first ionization energies tend to be those lying to the left in a period or those lying toward the bottom of a group in the Periodic Table. These elements are metals.

Atoms that readily pick up electrons have relatively high electron affinities and lie to the right in the Periodic Table (Section 5.15). These elements, the nonmetals, are only a few electrons shy of having a filled valence shell and can pick up the electrons lost by metals, thereby filling the valence shell. A neutral atom, as it fills its valence shell by gaining one or more electrons, becomes a **negative ion,** or **anion.** Nonmetals such as F, Cl, Br, I, O, and S form anions.

$$:\overset{..}{\underset{..}{Cl}}\cdot\ +\ e^-\ \longrightarrow\ :\overset{..}{\underset{..}{Cl}}:^-\qquad :\overset{..}{\underset{.}{S}}\cdot\ +\ 2e^-\ \longrightarrow\ :\overset{..}{\underset{..}{S}}:^{2-}$$

Chlorine   Chloride   Sulfur   Sulfide
atom   anion   atom   anion

Thus, when metals combine with nonmetals, electrons are transferred from the metals to the nonmetals and ionic compounds (compounds that contain ions) are formed. The use of Lewis symbols to show the transfer of electrons during formation of ionic compounds is illustrated by the following examples:

Metal		*Nonmetal*		*Ionic Compound*	
$Na^{\times}$	+	$:\overset{..}{\underset{..}{Cl}}\cdot$	$\longrightarrow$	$Na^+[:\overset{..}{\underset{..}{Cl}}\overset{\times}{:}^-]$	(1)
Sodium atom		Chlorine atom		Sodium chloride (sodium ion and chloride ion)	
$Mg^{\times}_{\times}$	+	$:\overset{..}{\underset{.}{O}}\cdot$	$\longrightarrow$	$Mg^{2+}[:\overset{..}{\underset{\times}{O}}\overset{\times}{:}^{2-}]$	(2)
Magnesium atom		Oxygen atom		Magnesium oxide (magnesium ion and oxide ion)	
$Ca^{\times}_{\times}$	+	$2:\overset{..}{\underset{..}{F}}\cdot$	$\longrightarrow$	$Ca^{2+}[:\overset{..}{\underset{..}{F}}\overset{\times}{:}^-]_2$	(3)
Calcium atom		Fluorine atoms		Calcium fluoride (calcium ion and two fluoride ions)	

The actual reactions do not occur exactly as written, since the nonmetals involved exist as the diatomic molecules Cl_2, O_2, and F_2. However, the transfer of electrons from the metals to the nonmetals is an important feature of the reaction, and the products have the electron configurations shown.

A crystal of sodium chloride consists of sodium ions and chloride ions (Fig. 6.1). The regular three-dimensional arrangement arises because each anion and cation pulls the maximum number of oppositely charged ions around itself. Each sodium ion is surrounded by six chloride ions, and each chloride ion is surrounded by six sodium ions. The force that holds the ions together in the crystal is the electrostatic attraction between ions of opposite charge. This electrostatic attraction is very strong in ionic compounds. For example, it requires 769 kilojoules of energy to separate the Na^+ and Cl^- ions in a mole of NaCl.

Any given ion in a crystal of sodium chloride exerts the same electrostatic force on each of its six immediate neighbors of opposite charge; therefore, it is impossible to identify any single pair of sodium and chloride ions as being a molecule of sodium chloride. In ionic compounds no molecules are present; a crystal of an ionic compound is an aggregation of ions. The formula of an ionic compound represents the relative

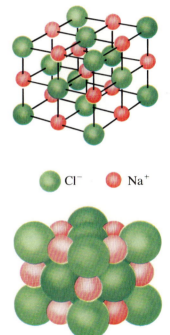

Figure 6.1

The arrangement of sodium and chloride ions in a crystal of sodium chloride (common salt). The smaller red spheres represent sodium ions; the larger green ones, chloride ions. The upper structure is an "expanded" view that shows the geometry more clearly.

Cl⁻   Na⁺

numbers of ions necessary to balance the ionic charges. A crystal of sodium chloride is electrically neutral, so it must contain the same numbers of Na^+ and Cl^- ions. Its *empirical* formula is NaCl. A crystal of sodium oxide contains twice as many Na^+ as O^{2-} ions. Its empirical formula is Na_2O. It follows that the term *molecular weight* has no significance in connection with ionic compounds. The formula NaCl represents one *formula unit* of sodium chloride; it cannot be said to represent a molecule, for there are no molecules of sodium chloride.

6.2 The Electronic Structures of Ions

With a few exceptions found among the heavy atoms at the bottom of Groups IIB, IIIA, IVA, and VA, *representative elements* (Section 5.14) tend to lose all valence electrons when forming cations, making it easy to remember the charges on the positive ions. The charge is equal to the group number of the element forming the ion, because the group number is the same as the number of electrons in the outermost shell of a representative element. Members of Group IA have ionic charges of $+1$; members of Groups IIA and IIB, $+2$; etc. The exceptions to this rule are Hg_2^{2+} (a diatomic ion that contains a covalent bond between the mercury atoms), Tl^+, Sn^{2+}, Pb^{2+}, and Bi^{3+}, which form in addition to the expected ions Hg^{2+}, Tl^{3+}, Sn^{4+}, Pb^{4+}, and Bi^{5+}.

The positive ion produced by the loss of all valence electrons from a representative metal will have either the stable **noble gas electron configuration** (the same electron configuration as a noble gas), in which the outermost shell in the ion has the configuration ns^2np^6 (or $1s^2$ for cations of the second period), or a **pseudo noble gas electron configuration,** in which the outermost shell has the configuration $ns^2np^6nd^{10}$. An example of the latter is the zinc ion, Zn^{2+}, with the electron configuration $1s^22s^22p^63s^23p^63d^{10}$, which forms from a zinc atom, with the configuration $1s^22s^22p^63s^23p^63d^{10}4s^2$.

EXAMPLE 6.1   Write the electron configurations and indicate the charges of the ions formed from the elements K, Ca, Ga, and Cd.

As members of Groups IA, IIA, IIIA, and IIB, respectively, these elements form the ions K^+, Ca^{2+}, Ga^{3+}, and Cd^{2+}. These ions are formed by the loss of all of the valence electrons from the atoms, giving noble gas electron configurations for K^+ and Ca^{2+} and pseudo noble gas electron configurations for Ga^{3+} and Cd^{2+}, as shown in the following table:

	Atom		Ion
K	$1s^22s^22p^63s^23p^64s^1$	K^+	$1s^22s^22p^63s^23p^6$
Ca	$1s^22s^22p^63s^23p^64s^2$	Ca^{2+}	$1s^22s^22p^63s^23p^6$
Ga	$1s^22s^22p^63s^23p^63d^{10}4s^24p^1$	Ga^{3+}	$1s^22s^22p^63s^23p^63d^{10}$
Cd	$1s^22s^22p^63s^23p^63d^{10}4s^24p^64d^{10}5s^2$	Cd^{2+}	$1s^22s^22p^63s^23p^63d^{10}4s^24p^64d^{10}$

Transition and inner transition metal elements behave differently from representative elements. Most transition metal cations have $+2$ or $+3$ charges resulting from loss of their outermost s electrons (or electron), sometimes followed by loss of one or two d electrons from the next-to-outermost shell. For example, copper $(1s^22s^22p^63s^23p^63d^{10}4s^1)$ forms the ions Cu^+ $(1s^22s^22p^63s^23p^63d^{10})$ and Cu^{2+}

$(1s^22s^22p^63s^23p^63d^9)$. Although the d orbitals of the transition elements are the last to fill when building up electron configurations by the aufbau principle, the outermost s electrons are the first to be lost when these atoms ionize. When the inner transition metals form ions, they usually have a $+3$ charge resulting from the loss of their outermost s electrons and a d or f electron.

The formation of monatomic anions by representative elements always involves formation of ions with noble gas electron configurations. Thus negative ions such as N^{3-}, S^{2-}, I^-, and O^{2-} form when a neutral atom picks up enough electrons to fill its valence shell completely. Keeping this in mind makes it simple to determine the charge on any monatomic negative ion; the charge is equal to the number of electrons necessary to fill the valence orbitals of the parent atom. Oxygen, for example, has the electron configuration $1s^22s^22p^4$, whereas the oxygen anion (oxide ion) has the noble gas electron configuration $1s^22s^22p^6$. Since two additional electrons are required to fill the valence shell of the oxygen atom, the oxide ion has a charge of -2.

EXAMPLE 6.2

Write the electron configurations of a phosphorus atom and its negative ion, and give the charge on the anion.

Phosphorus is a member of Group VA. A neutral phosphorus atom has the electron configuration $1s^22s^22p^63s^23p^3$, or $[Ne]3s^23p^3$. A phosphorus atom requires three additional electrons to fill its valence shell, so the charge of the phosphorus anion is -3. The electron configuration of P^{3-} is $[Ne]3s^23p^6$.

The electronic differences between an atom and its ion give them very different physical and chemical properties. Sodium *atoms* form sodium metal, a soft, silvery-white metal that burns vigorously in air and reacts rapidly with water. Chlorine *atoms* form chlorine gas, Cl_2, a greenish-yellow gas that is extremely corrosive to most metals and very poisonous to animals and plants. The vigorous reaction between sodium and chlorine forms the compound sodium chloride, common table salt, which contains *ions* of sodium and chlorine. The ions exhibit properties entirely different from those of the elements sodium and chlorine. Chlorine is poisonous, but sodium chloride is essential to life; sodium atoms react vigorously with water, but sodium chloride is stable in water.

6.3 Covalent Bonds; Chemical Bonding by Electron Sharing

Many compounds do not contain ions but instead consist of atoms bonded tightly together in molecules. The bonds holding such atoms together form when pairs of electrons are shared between atoms—that is, when the electrons occupy orbitals of both atoms involved in the bond. Such a bond is called a **covalent bond.** Electron pairs are shared and covalent bonds are formed when two atoms have about the same tendency to give up or to pick up electrons.

The simplest substance in which atoms are covalently bonded is the hydrogen molecule, H_2. A hydrogen atom has one electron in its $1s$ shell. The two electrons (an electron pair) from the two hydrogen atoms in a hydrogen molecule are shared by the two nuclei (Fig. 6.2). These shared electrons spend most of their time in the region between the two nuclei, and the electrostatic attraction between the two positively charged nuclei and the two negatively charged electrons holds the molecule together.

$$H\cdot \ + \ \cdot H \qquad H:H$$
Hydrogen atoms $\longrightarrow$ Hydrogen
molecule

Figure 6.2

Combination of two hydrogen atoms to form a hydrogen molecule, H_2, through covalent bonding.

The bond resulting from this attraction is very strong, as is evidenced by the large amount of energy required to break the covalent bonds in 1 mole of hydrogen molecules—436 kilojoules.

$$H_2(g) \longrightarrow 2H(g) \qquad \Delta H = +436 \text{ kJ}$$

Conversely, this same quantity of energy is evolved when a mole of hydrogen molecules is formed from hydrogen atoms.

$$2H(g) \longrightarrow H_2(g) \qquad \Delta H = -436 \text{ kJ}$$

In forming ionic compounds, many metals and nonmetals tend to assume noble gas electron configurations. Atoms forming covalent molecules by electron sharing have the same tendency. Each electron of the pair in the covalent bond in a hydrogen molecule spends an equal amount of time near each nucleus, so the $1s$ orbital of each hydrogen atom in the molecule is occupied by both electrons. In effect, each atom has the electronic structure of a helium atom.

The sharing of one pair of electrons by two atoms in a molecule of chlorine gives each atom the electronic structure of an atom of the noble gas argon, since a free atom of chlorine has seven electrons in its outer shell, one less than an atom of argon.

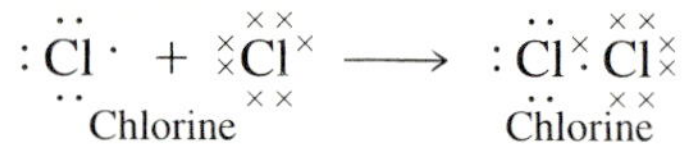

$$:\!\overset{..}{\underset{..}{Cl}}\cdot \; + \; \overset{\times\times}{\underset{\times\times}{\times}}\!\overset{}{Cl}\!\times \longrightarrow \; :\!\overset{..}{\underset{..}{Cl}}\!\overset{\times\times}{\underset{\times\times}{\times}}\!Cl\!\times$$

Chlorine atoms   Chlorine molecule

The **Lewis structure** shown above for Cl_2 indicates that each Cl atom has three pairs of electrons that are not used in bonding (called **unshared pairs,** or **lone pairs**) and one shared pair of electrons (written between the atoms). A dash is sometimes used to indicate a shared pair of electrons.

$$H\!-\!H \qquad\qquad :\!\overset{..}{\underset{..}{Cl}}\!-\!\overset{..}{\underset{..}{Cl}}\!:$$

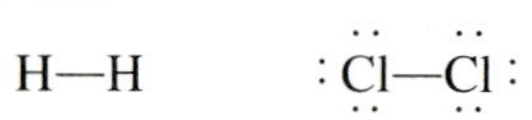

A single shared pair of electrons is called a **single bond.**

The bonding in the molecules of the other halogens (the molecules F_2, Br_2, I_2, and At_2) is like that in the chlorine molecule, with one single bond between atoms and three unshared pairs per atom. However, the electron configurations of the atoms in these molecules are like those of the noble gases Ne, Kr, Xe, and Rn, respectively.

The number of covalent bonds (shared electron pairs) that an atom can form can often be predicted from the number of electrons needed to fill its valence shell. Each atom of a Group IVA element has four electrons in its valence shell and can accept four more electrons to achieve a noble gas electron configuration. These four electrons can be gained by forming four covalent bonds, as illustrated for carbon in CCl_4 (carbon tetrachloride) and silicon in SiH_4 (silane).

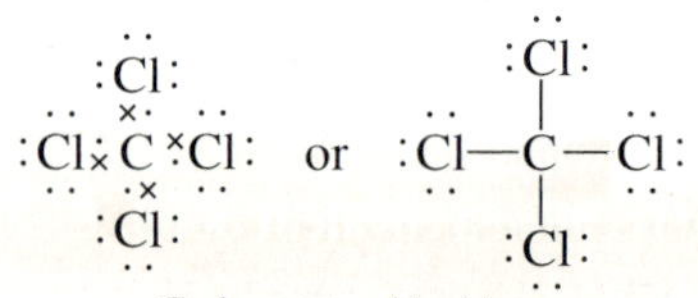

:Cl:
:Cl×C×Cl: or :Cl—C—Cl:
:Cl:
Carbon tetrachloride 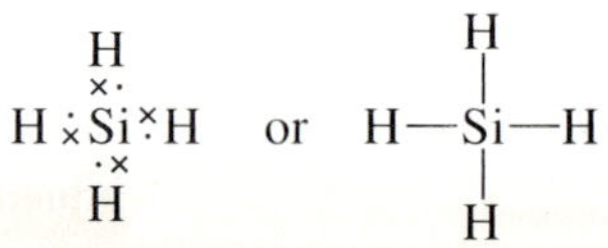

H
H×Si×H or H—Si—H
H
Silane

Note that the Lewis structure of a molecule in general does not indicate its three-dimensional shape. As will be discussed in Chapter 8, carbon tetrachloride and silane are not flat molecules.

Nitrogen and the other elements of Group VA need three additional electrons to achieve a noble gas configuration. These three electrons can be gained by formation of three single covalent bonds, as in NH_3 (ammonia). Oxygen and the other atoms in

Group VIA need only two electrons to fill their valence shell; thus they can form two single covalent bonds. The elements in Group VIIA, such as fluorine, need to form only one single covalent bond in order to fill their valence shell.

$$\underset{\underset{\text{H}}{|}}{\overset{\underset{\text{H}}{|}}{:\!N\!-\!H}} \qquad :\!\overset{\underset{\text{H}}{|}}{\underset{\cdot\cdot}{O}}\!-\!H \qquad H\!-\!\underset{\cdot\cdot}{\overset{\cdot\cdot}{F}}:$$

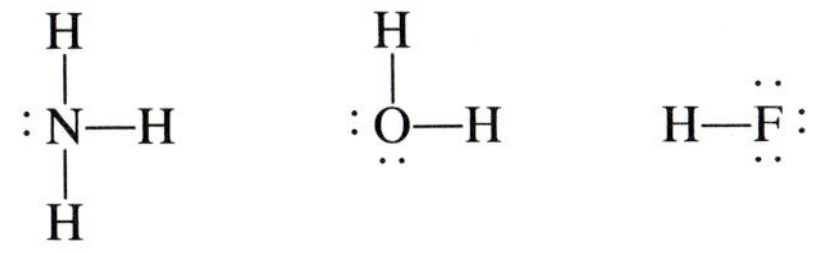

A pair of atoms may share more than one pair of electrons. For example, two pairs of electrons are shared between the carbon and oxygen atoms in CH_2O (formaldehyde) and between the two carbon atoms in C_2H_4 (ethylene), giving rise in both cases to a **double bond.**

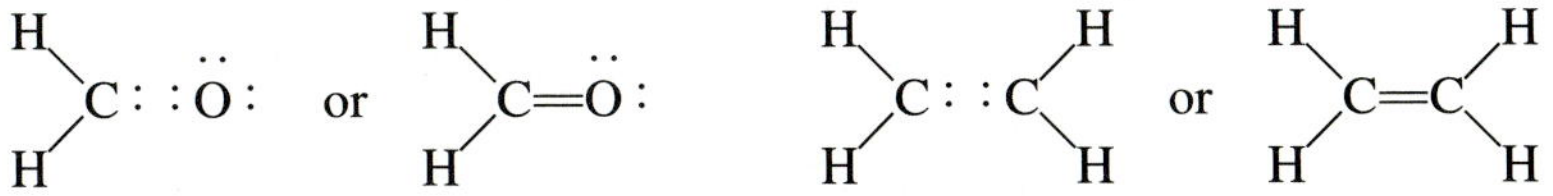

When three electron pairs are shared by two atoms, as in CO (carbon monoxide) and N_2 (nitrogen molecule), a **triple bond** is formed.

$$:C:::O: \quad \text{or} \quad :C\!\equiv\!O: \qquad\qquad :N:::N: \quad \text{or} \quad :N\!\equiv\!N:$$

Under normal conditions, the ability to form double or triple bonds is limited almost exclusively to bonds between carbon, nitrogen, and oxygen atoms. For example, the element nitrogen forms the N_2 molecule, which contains a triple bond, whereas the element phosphorus (also in Group VA) forms the P_4 molecule, which contains only single bonds (Fig. 6.3). Phosphorus, sulfur, and selenium sometimes form double bonds with carbon, nitrogen, and oxygen.

The molecules of CO and N_2 are said to be **isoelectronic;** they have the same arrangement of unshared pairs and bonding pairs of electrons. Other examples of pairs of isoelectronic species are HCl and OH^-, and NH_3 and H_3O^+.

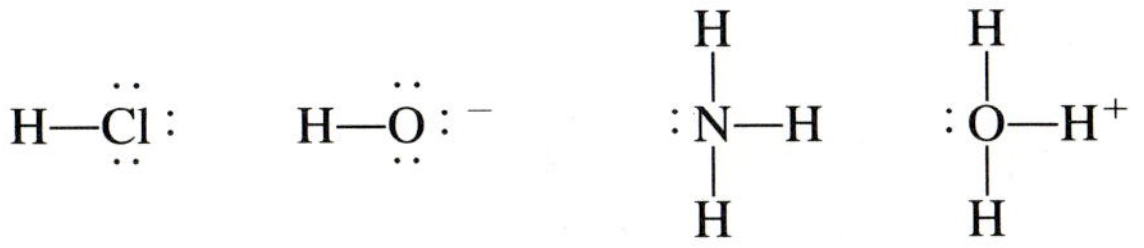

$$H\!-\!\underset{\cdot\cdot}{\overset{\cdot\cdot}{Cl}}: \qquad H\!-\!\underset{\cdot\cdot}{\overset{\cdot\cdot}{O}}:^{-} \qquad \underset{\underset{\text{H}}{|}}{\overset{\underset{\text{H}}{|}}{:\!N\!-\!H}} \qquad \underset{\underset{\text{H}}{|}}{\overset{\underset{\text{H}}{|}}{:\!O\!-\!H^+}}$$

Ions such as OH^- and H_3O^+ that are composed of more than one atom are called **polyatomic ions** (see Section 6.12). The atoms in a polyatomic ion are held together by covalent bonds. Thus compounds containing polyatomic ions are stabilized by both covalent bonds and ionic bonds. For example, potassium nitrate, KNO_3, contains the K^+ cation and the polyatomic NO_3^- anion. It has an ionic bond resulting from the electrostatic attraction between the ions K^+ and NO_3^- and covalent bonds between the nitrogen and oxygen atoms in NO_3^-.

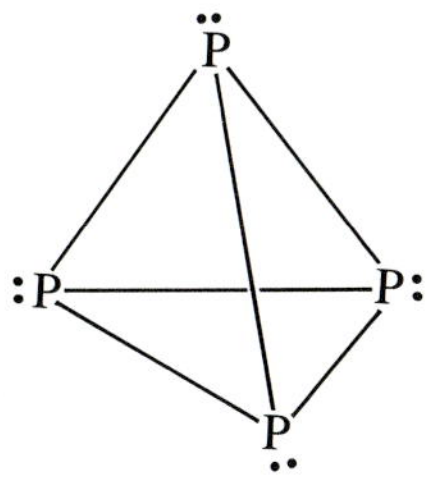

Figure 6.3
The Lewis structure of P_4.

6.4 Covalently Bonded Atoms Without a Noble Gas Configuration

Elements in the second row of the Periodic Table never form compounds in which they have more than eight electrons in their valence shell. However, larger atoms, in which the outermost electron shell is one where $n = 3$ or greater, can share more than four pairs of electrons with other atoms. For example, the phosphorus atom in PCl_5 shares five pairs of electrons (a total of 10 electrons). Sulfur shares six electron pairs (12 electrons) in the SF_6 molecule; and iodine shares seven electron pairs (14 electrons) in IF_7. In some molecules, such as IF_5 and XeF_4, the number of electrons in the outer shell of the central atom exceeds eight, even though some of its electron pairs are not shared.

(The three unshared pairs usually found on each Cl and F atom in these Lewis structures have been omitted for clarity.)

A few molecules contain atoms that do not have the noble gas configuration since they have fewer than eight electrons in their valence shell. For example, boron, with three valence electrons, shares electron pairs with three chlorine atoms in BCl_3 (boron trichloride).

In this molecule, each chlorine atom has an argon configuration. However, boron, with only six electrons in its outer shell, does not have a noble gas configuration; one of its valence shell orbitals is vacant.

An atom like the boron atom in BCl_3, which does not have a filled valence shell, is very reactive. It readily combines with a molecule containing an atom with an unshared pair of electrons. The unshared pair is shared by both atoms when the bond forms. For example, NH_3 reacts with BCl_3 to form the compound BCl_3NH_3, because the unshared pair on nitrogen can be shared with the boron atom.

A bond like the one between the boron and nitrogen atoms, formed when one of the atoms provides both bonding electrons, is called a **coordinate covalent,** or **dative, bond.** Coordinate covalent bonds can also occur within ions, as when a water molecule combines with a hydrogen ion to form a hydronium ion or when an ammonia molecule combines with a hydrogen ion to form an ammonium ion.

| Water molecule | Hydronium ion, H_3O^+ | Ammonia molecule | Ammonium ion, NH_4^+ |

The formation of a coordinate covalent bond is only possible between an atom or ion with an unshared pair of electrons in its valence shell and an atom or ion that needs a pair of electrons to acquire a more stable electron configuration. The difference between the coordinate covalent bond and the normal covalent bond is in the mode of formation—that is, whether each atom contributes one electron or one atom contributes both. Once established, covalent bonds are indistinguishable from one another, since electrons are identical regardless of their source. All of the O—H bonds in H_3O^+ are the same, as are the N—H bonds in NH_4^+.

6.5 Electronegativity of the Elements

Electrons in a covalent bond between identical atoms are shared equally. However, the electron pair in a covalent bond between unlike atoms may not be shared equally; the electrons may spend more time closer to one atom than to the other. For example, a hydrogen atom is covalently bonded to a chlorine atom in a molecule of hydrogen chloride:

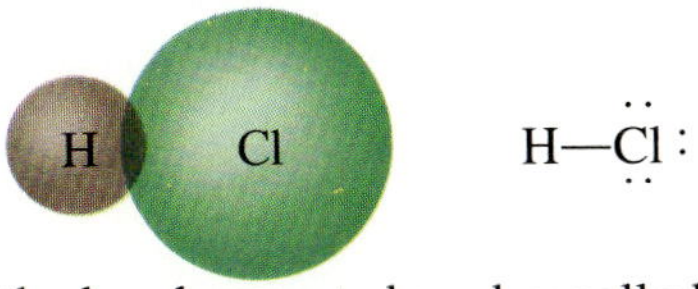

But the electrons forming the bond are not shared equally by the atoms, as they are in H_2 and Cl_2. A chlorine atom attracts electrons more strongly than a hydrogen atom, so the electrons of the shared pair are more closely associated with the chlorine atom. Thus the hydrogen atom develops a small positive charge (sometimes called a partial positive charge), and the chlorine atom develops a partial negative charge. This does not mean that the hydrogen atom has lost its electron; it means that the electrons of the shared pair spend more time *on the average* in the vicinity of the chlorine atom than they do near the hydrogen atom.

The extent of attraction of an atom for a shared pair of electrons is known as its electronegativity. **Electronegativity** is a measure of the attraction of an atom for electrons *in a covalent bond*. The larger the electronegativity of an element, the more strongly an atom of that element attracts electrons. Electronegativities are given in Table 6.2. Because the electronegativity of a given element varies a little depending on the compound in which the element is found, the entries are average values.

Table 6.2 Approximate electronegativities of some of the elements, shown in a periodic table arrangement

NONMETALS

H 2.1																H 2.1	He . . .
Li 1.0	Be 1.5			METALS								B 2.0	C 2.5	N 3.0	O 3.5	F 4.1	Ne . . .
Na 1.0	Mg 1.2			TRANSITION METALS								Al 1.5	Si 1.7	P 2.1	S 2.4	Cl 2.8	Ar . . .
K 0.9	Ca 1.0	Sc 1.2	Ti 1.3	V 1.4	Cr 1.6	Mn 1.6	Fe 1.6	Co 1.7	Ni 1.8	Cu 1.8	Zn 1.7	Ga 1.8	Ge 2.0	As 2.2	Se 2.5	Br 2.7	Kr . . .
Rb 0.9	Sr 1.0	Y 1.1	Zr 1.2	Nb 1.2	Mo 1.3	Tc 1.4	Ru 1.4	Rh 1.4	Pd 1.4	Ag 1.4	Cd 1.5	In 1.5	Sn 1.7	Sb 1.8	Te 2.0	I 2.2	Xe . . .
Cs 0.9	Ba 1.0	La-Lu 1.1-1.2	Hf 1.2	Ta 1.3	W 1.4	Re 1.5	Os 1.5	Ir 1.6	Pt 1.4	Au 1.4	Hg 1.4	Tl 1.4	Pb 1.6	Bi 1.7	Po 1.8	At 2.1	Rn . . .
Fr 0.9	Ra 1.0	Ac-Lr 1.1-	Unq . . .	Unp . . .	Unh . . .	Uns . . .	Uno . . .	Une . . .									

In general, for the representative elements, electronegativity increases going across a period from left to right and going up a group. Thus the nonmetals, which lie toward the right in the Periodic Table, tend to have higher electronegativities than the metals, although there are some irregularities in the center of the table. This is another example of the periodic variation of properties among the elements.

Fluorine, the most chemically active nonmetal, has the highest electronegativity (4.1), and cesium and francium, the most chemically active metals, share the lowest electronegativity (0.9). Because the metals have relatively low electronegativities and tend to assume positive charges in compounds, they are often spoken of as being **electropositive;** conversely, nonmetals are said to be **electronegative.**

6.6 Polar Covalent Bonds and Polar Molecules

A **polar covalent bond** has a positively charged end and a negatively charged end. Such a bond results when a covalent bond is formed between atoms of different electronegativities, because the shared pair of electrons stays closer to the more electronegative atom. As mentioned, the chlorine atom in a hydrogen chloride molecule attracts the pair of electrons of the covalent bond more strongly than does the hydrogen atom (Section 6.5). The hydrogen-chlorine bond is polar, with the chlorine atom somewhat negative and the hydrogen atom somewhat positive.

The greater the difference between the electronegativities of the two atoms involved in a bond, the greater is the polarity of the bond. Thus, the polarity of the bond in the hydrogen halides increases in the order HI < HBr < HCl < HF, which corresponds to the increase in electronegativity of the halogens: I (2.2) < Br (2.7) < Cl (2.8) < F (4.1). If the difference in electronegativity between two atoms is sufficiently large, the shared electron pair will spend all of its time on the more electronegative atom, resulting in ionic bonding rather than covalent bonding. This is the case with salts such as LiF, NaCl, K_2O, Li_3N, and CsBr, where the electronegativity difference is greater than about 2. The other extreme is achieved when identical atoms share a pair of electrons, as in H—H, where the bond is covalent with no polarity. Bonds between atoms with small differences in electronegativities are primarily covalent in character. This is the case with CO_2, CCl_4, I_2O_5, NI_3, ICl, and NO. It follows that bonds that are primarily ionic are formed when elements at the extreme left of the Periodic Table react with elements at the extreme right, whereas bonds that are primarily covalent are formed when elements close together in the table react with one another.

There is no sharp dividing line between compounds in which the bonding is covalent and those in which the bonding is ionic. In intermediate cases, the molecules have bonds that possess some of the nature of both covalent and ionic bonds, or **covalent bonds with partial ionic character.**

If a molecule has a positive end and a negative end, it is said to be **polar** and to possess a **dipole.** Polar molecules contain polar covalent bonds. In a molecule of hydrogen chloride, for example, the hydrogen atom is at the positively charged end of the molecule, and the chlorine atom is at the negatively charged end. Polar molecules tend to align when placed in an electric field, with the positive end of the molecule oriented toward the negative plate and the negative end toward the positive plate (Fig. 6.4). Polar molecules are attracted to an electrically charged object, nonpolar molecules are not (Fig. 6.5).

Polar molecules randomly oriented in the absence of an electric field

Positive plate   Negative plate

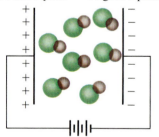

Polar molecules tending to line up in an electric field

Figure 6.4

Polar molecules, such as hydrogen chloride, tend to align in an electric field, with the positive ends oriented toward the negative plate and the negative ends toward the positive plate.

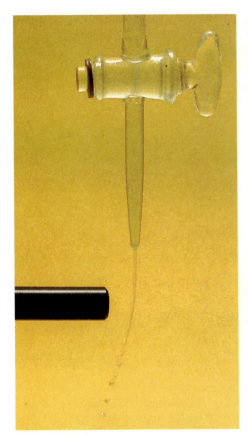

Figure 6.5
The stream of water (left) is deflected toward the electrically charged rod because the polar water molecules, H_2O, are attracted by the charge; the nonpolar carbon tetrachloride molecules, CCl_4, are not attracted and a stream of carbon tetrachloride (right) is unaffected.

Polar covalent bonds are sometimes found in nonpolar molecules. Such molecules contain several polar covalent bonds directed in such a way as to give symmetrical molecules, thus they are nonpolar. This is illustrated by $HgCl_2$, in which each of the covalent bonds is polar but the molecule as a whole is nonpolar. Each chlorine atom is negative with respect to the positive mercury atom, so each mercury-chlorine bond is polar. However, these bond polarities cancel each other because they point in opposite directions, giving an electrically symmetrical molecule.

$$Cl^- \!-\!^+Hg^+\!-\!^-Cl$$

6.7 Writing Lewis Structures

When we write the Lewis structure for a molecule or ion that we can consider to be formed from individual atoms, we form single or multiple bonds by pairing up the unpaired electrons on the reacting atoms.

$$H\cdot \ + \ :\overset{..}{Br}\cdot \ \longrightarrow \ H\!:\!\overset{..}{Br}\!:$$

$$2H\cdot \ + \ :\overset{..}{S}\cdot \ \longrightarrow \ H\!:\!\overset{..}{\underset{..}{S}}\!:$$
$$\phantom{2H\cdot \ + \ :\overset{..}{S}\cdot \ \longrightarrow \ H\!:\!}H$$

$$\cdot\overset{..}{N}\cdot \ + \ \cdot\overset{..}{N}\cdot \ \longrightarrow \ :N\!:\!:\!:N\!:$$

Sometimes, however, a molecule or ion is more complicated, and then it is helpful to follow the general procedure outlined below. Let us determine the Lewis structures of $PO_2F_2^-$ and CO as examples in following this procedure.

STEP 1. Draw a skeleton structure of the molecule or ion showing the arrangement of atoms and connect each atom to another with a single (one electron pair) bond.

$$F-\overset{\overset{\displaystyle F}{|}}{\underset{\underset{\displaystyle O}{|}}{P}}-O^- \qquad C-O$$

When several arrangements of atoms are possible, as for $PO_2F_2^-$, we must use experimental evidence to choose the correct one. However, as a rule, the less electronegative element is often the central atom (as in $PO_2F_2^-$, SF_6, ClO_4^-, and PCl_5), with the exception that hydrogen is not found as a central atom.

STEP 2. Determine the total number of valence electrons in the molecule or ion.

A. For a *molecule,* this is equal to the sum of the number of valence electrons on each atom.
B. For a *positive ion,* this is equal to the sum of the number of valence electrons minus the number of positive charges on the ion (one electron is lost for each single positive charge).
C. For a *negative ion,* this is equal to the sum of the number of valence electrons plus the number of negative charges on the ion.

$PO_2F_2^-$: Number of valence electrons =

$$5(\text{P atom}) + 6(\text{O atom})$$
$$+ 6(\text{O atom}) + 7(\text{F atom})$$
$$+ 7(\text{F atom}) + 1(\text{negative charge})$$
$$= 32$$

CO: Number of valence electrons = 4(C atom) + 6(O atom) = 10

STEP 3. Deduct the two valence electrons that are used in each of the bonds written in Step 1. Distribute the remaining electrons as unshared pairs, so that each atom (except hydrogen) has eight electrons if possible. If there are too few electrons to give each atom eight electrons, convert single bonds to multiple bonds, where possible. Remember, the ability to form multiple bonds is limited almost exclusively to bonds between carbon, nitrogen, and oxygen, although phosphorus, sulfur, and selenium will sometimes form double bonds with carbon, nitrogen, and oxygen. Thus some elements may have fewer than eight electrons in their valence shells when they function as central atoms. For example, see the Lewis formula for BCl_3 in Section 6.4.

In the ion $PO_2F_2^-$, the total of 32 valence electrons can be distributed as 4 electron pairs (eight electrons) in bonds and 3 unshared pairs around each of the fluorine and oxygen atoms. The 10 valence electrons in CO could be distributed as one bonding pair and 4 unshared pairs, but this would leave at least one atom with fewer than 8 electrons. Filled valence shells about the carbon and oxygen atoms can only be obtained if each atom has 3 shared pairs in a triple bond and 1 unshared pair.

$$:\!\overset{\overset{\displaystyle \cdots}{..}}{\underset{}{F}}\!:$$
$$:\!\overset{..}{\underset{..}{F}}\!-\!\overset{\overset{\displaystyle |}{|}}{\underset{\underset{\displaystyle :\!\overset{..}{\underset{..}{O}}\!:}{|}}{P}}\!-\!\overset{..}{\underset{..}{O}}\!:^{-} \qquad :C\!\equiv\!O:$$

STEP 4. In those molecules in which there are too many electrons to have only eight electrons around each atom, the central atom may have more than eight electrons in its valence shell (see, for example, the Lewis formulas of PCl_5, SF_6, IF_5, and XeF_4 in Section 6.4). However, note that the outer atoms contain a maximum of eight electrons in their valence shells.

Although heavier atoms can contain more than eight electrons, a Lewis formula with more than eight electrons for an atom of the first or second period is almost certainly incorrect and should be reexamined.

EXAMPLE 6.3

Jupiter's atmosphere contains small amounts of the compounds methane, CH_4, acetylene, HCCH, and phosphine, PH_3. What are the Lewis structures of these molecules?

Since hydrogen cannot be a central atom, the skeleton structure (Step 1) of CH_4 must be

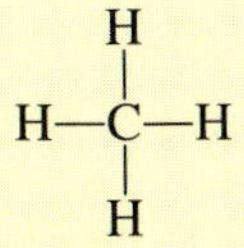

$$\begin{array}{c} H \\ | \\ H-C-H \\ | \\ H \end{array}$$

CH_4 contains eight valence electrons, four from the carbon atom and one from each of four hydrogen atoms (Step 2). All of the valence electrons are used in writing single bonds in the skeleton structure. Since none remain to be distributed, the Lewis structure is the skeleton structure.

The molecule HCCH has the skeleton structure

$$H-C-C-H$$

The molecule contains ten valence electrons, six of which are used in the single bonds in the skeleton structure, leaving four electrons to be distributed. The attempts to distribute these electrons as unshared pairs, as illustrated below, do not give the carbon atoms eight electrons each.

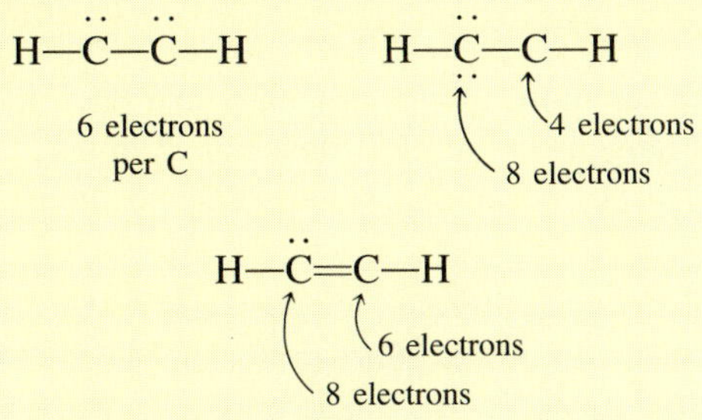

$$H-\overset{..}{\underset{}{C}}-\overset{..}{\underset{}{C}}-H \qquad H-\overset{..}{\underset{..}{C}}-C-H$$

6 electrons per C   4 electrons
                           8 electrons

$$H-\overset{..}{C}=C-H$$

6 electrons
8 electrons

Only with a triple bond will both carbon atoms have eight electrons.

$$H-C\equiv C-H$$

The skeleton structure of PH_3 is

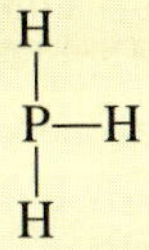

$$\begin{array}{c} H \\ | \\ P-H \\ | \\ H \end{array}$$

Six of the eight valence electrons in the molecule (five from the phosphorus atom, three from the three hydrogen atoms) are used to form single P—H bonds in the skeleton structure. The two remaining electrons must appear as an unshared pair on the phosphorus atom in order to give it eight electrons.

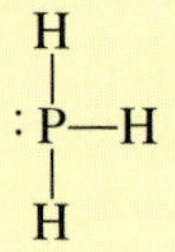

$$\overset{\displaystyle H}{\underset{\displaystyle H}{\,\vert\,}}$$
$$:P—H$$

EXAMPLE 6.4   Write the Lewis structure of the compound ClF_3.

Chlorine is the central atom in this molecule since it is the less electronegative element. The skeleton structure (Step 1) is

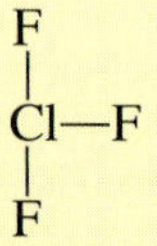

$$\overset{\displaystyle F}{\underset{\displaystyle F}{\,Cl—F\,}}$$

ClF_3 contains 28 valence electrons, 7 from the chlorine atom and 21 from the three fluorine atoms. Of these, 6 are used in the single bonds, leaving 22 to be placed as unshared pairs. Three unshared pairs can be placed on each fluorine atom, leaving two pairs of electrons. Since the valence shell of chlorine has an n value greater than 2 ($n = 3$), it can hold the two remaining electron pairs. The Lewis structure is

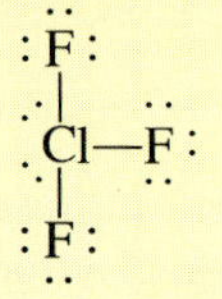

$$\overset{\displaystyle :\!F\!:}{\underset{\displaystyle :\!F\!:}{\,\overset{..}{Cl}—\overset{..}{F}:\,}}$$

6.8 Resonance

Sometimes two or more Lewis structures can have the same arrangement of atoms but different arrangements of electrons. For example, both Lewis structures of SO_2, sulfur dioxide, have the atoms in the same positions, but some electrons in different positions.

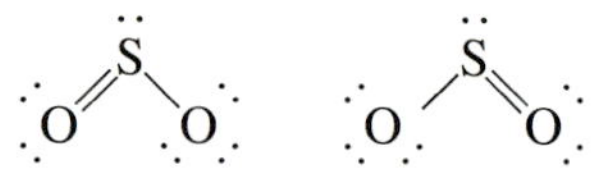

A double bond between two given atoms is shorter than a single bond between the same two atoms. However, experiments show that both sulfur-oxygen bonds in SO_2 have the same length and are identical in all other properties. Since it is not possible to write a single Lewis structure for sulfur dioxide in which both bonds are equivalent, we must apply the concept of **resonance.** If two or more Lewis structures with the same arrangement of atoms can be written for a molecule or ion, the actual distribution of electrons is an *average* of that shown by the various Lewis structures. The actual distribution of electrons in each of the sulfur-oxygen bonds in SO_2 is an average of that shown in the two Lewis structures—that is, an average of a double bond and a single bond. The individual Lewis structures are called **resonance forms.** The actual electronic structure of the molecule (the average of the resonance forms) is called a **resonance hybrid** of the individual resonance forms. A double-headed arrow is used between Lewis structures to indicate that they are resonance forms and that *the true distribution of electrons is an average of the individual resonance forms.*

The distribution of electrons in N_2O, nitrous oxide, may be represented by two resonance forms as follows:

$$: \ddot{N}\!\!=\!\!N\!\!=\!\!\ddot{O}: \quad \longleftrightarrow \quad :N\!\!\equiv\!\!N\!\!-\!\!\ddot{\underset{..}{O}}:$$

Since the electron distribution in N_2O is an average of these two resonance forms, the distribution of electrons in the bond between the two nitrogen atoms may be seen to be greater than that in an ordinary double bond but less than that in an ordinary triple bond. The distribution of electrons in the bond between nitrogen and oxygen is between that in a single bond and that in a double bond.

It must be emphasized that a molecule described as a resonance hybrid *never* possesses an electronic structure described by a single resonance form. The actual electronic structure is *always* an average of that shown by all resonance forms. A simple analogy to a resonance hybrid is the mule, which is the hybrid offspring of a male donkey and a female horse. Just as the characteristics of a mule are fixed, so the properties of a resonance hybrid are fixed. The mule is not a donkey part of the time and a horse part of the time. It is always a mule. Correspondingly, a molecule with an electronic structure described by a resonance hybrid does not exhibit one electronic structure part of the time and another electronic structure the rest of the time. Instead, the material is always in the form of the intermediate resonance hybrid, which cannot be written with a single Lewis structure.

6.9 The Strengths of Covalent Bonds

Stable molecules exist because strong covalent bonds hold the atoms together; energy must be added to break the bonds and separate the atoms. The strength of a covalent bond between two atoms is measured by the energy required to break it—that is, the energy necessary to separate the bonded atoms. The stronger a bond, the more energy is required to break it.

The energy required to break a covalent bond in a gaseous substance is called the **bond energy** of the bond. The bond energy for a diatomic molecule, D_{X-Y}, is defined as the standard enthalpy change for the reaction.

$$XY(g) \longrightarrow X(g) + Y(g) \qquad D_{X-Y} = \Delta H^\circ_{298}$$

For example, the bond energy of the H—H bond, D_{H-H}, is 436 kJ per mole of H—H bonds broken:

$$H_2(g) \longrightarrow 2H(g) \qquad D_{H-H} = \Delta H^\circ_{298} = 436 \text{ kJ}$$

Bond energies for commonly occurring diatomic molecules range from 946 kJ per mole for N_2 (triple bond) to 150 kJ per mole for I_2 (single bond), and from 569 kJ per mole for HF to 295 kJ per mole for HI.

Molecules with three or more atoms necessarily have two or more bonds. The sum of all of the bond energies in such a molecule is equal to the standard enthalpy change for the reaction that breaks all of the bonds in the molecule. For example, the sum of the four C—H bond energies in CH_4, 1660 kJ, is equal to the standard enthalpy change of the reaction

$$\begin{array}{c} H \\ | \\ H\!-\!C\!-\!H(g) \\ | \\ H \end{array} \longrightarrow C(g) + 4H(g) \qquad \Delta H^\circ_{298} = 1660 \text{ kJ}$$

The *average* C—H bond energy, $D_{C—H}$, is 1660/4 = 415 kJ per mole because there are four moles of C—H bonds broken in the reaction.

The strength of a bond between the same two atoms increases as the number of shared electron pairs in the bond increases. Thus double or triple bonds between two atoms are generally stronger than single bonds between the same two atoms. Usually, however, a double bond is not quite twice as strong as a single bond between the same two atoms, and a triple bond is not quite three times as strong as a single bond. Average bond energies for some common bonds appear in Table 6.3.

Table 6.3   Some average bond energies (kJ mol^{-1})

					Single bonds						
H	C	N	O	F	Si	P	S	Cl	Br	I	
436	415	390	464	569	395	320	340	432	370	295	H
	345	290	350	439	360	265	260	330	275	240	C
		160	200	270	—	210	—	200	245	—	N
			140	185	370	350	—	205	—	200	O
				160	540	489	285	255	235	—	F
					230	215	225	359	290	215	Si
						215	230	330	270	215	P
							215	250	215	—	S
								243	220	210	Cl
									190	180	Br
										150	I

		Multiple bonds		
C=C, 611	C=N, 615	C=O, 741	N=N, 418	O=O, 498
C≡C, 837	C≡N, 891	C≡O, 1080	N≡N, 946	

Tabulated enthalpy changes (Appendix I) can be used to obtain bond energies.

EXAMPLE 6.5   Evaluate the bond energy of a Cl—Cl single bond to one decimal from the data in Appendix I.

The energy of a mole of single bonds between chlorine atoms, $D_{Cl—Cl}$, is equal to the standard enthalpy of the reaction

$$Cl_2(g) \longrightarrow 2Cl(g) \qquad D_{Cl—Cl} = \Delta H^\circ_{298}$$

This is the reaction for the formation of 2 mol of Cl atoms. Referring to Appendix I, we find that the enthalpy of formation of 1 mol of chlorine atoms, $\Delta H^\circ_{f_{Cl(g)}}$, is 121.7 kJ. The enthalpy of formation of 2 mol of Cl atoms is 2 × 121.7 kJ = 243.4 kJ; hence $D_{Cl—Cl}$ = 243.4 kJ, and 243.4 kJ is required to break the Cl—Cl bonds in one mol of Cl_2 molecules. The rounded value is given in Table 6.3.

EXAMPLE 6.6   Evaluate the bond energy of the C≡O bond in CO(g) from the data in Appendix I.

The energy of 1 mol of $C \equiv O$ bonds, $D_{C \equiv O}$, is equal to the standard enthalpy of the reaction

$$CO(g) \longrightarrow C(g) + O(g) \qquad D_{C \equiv O} = \Delta H^\circ_{298}$$

This reaction can be written as three steps, each of which involves an enthalpy of formation (Hess's law, Section 4.5),

$$CO(g) \longrightarrow C(s) + \tfrac{1}{2}O_2(g) \qquad \Delta H^\circ_1 = -\Delta H^\circ_{f_{CO(g)}}$$
$$C(s) \longrightarrow C(g) \qquad \Delta H^\circ_2 = \Delta H^\circ_{f_{C(g)}}$$
$$\tfrac{1}{2}O_2(g) \longrightarrow O(g) \qquad \Delta H^\circ_3 = \Delta H^\circ_{f_{O(g)}}$$
$$\overline{CO(g) \longrightarrow C(g) + O(g) \qquad \Delta H^\circ_{298} = \Delta H^\circ_1 + \Delta H^\circ_2 + \Delta H^\circ_3}$$

$$
\begin{aligned}
D_{C \equiv O} = \Delta H^\circ_{298} &= -\Delta H^\circ_{f_{CO(g)}} + \Delta H^\circ_{f_{C(g)}} + \Delta H^\circ_{f_{O(g)}} \\
&= -1 \text{ mol } CO(g) \times (-110.5 \text{ kJ mol}^{-1}) + 1 \text{ mol } C(g) \times \\
&\quad (716.68 \text{ kJ mol}^{-1}) + 1 \text{ mol } O(g) \times (249.2 \text{ kJ mol}^{-1}) \\
&= 1076.4 \text{ kJ}
\end{aligned}
$$

Thus 1076.4 kJ is required to break the $C \equiv O$ bond in $CO(g)$. The rounded value is reported in Table 6.3.

A knowledge of bond energies is helpful in understanding why some reactions are exothermic and others are endothermic. An exothermic reaction (ΔH negative) results when the bonds in the products of a reaction are stronger than the bonds in the reactants. An endothermic reaction (ΔH positive) results when the bonds in the products are weaker than those in the reactants.

Consider the following reaction:

$$H_2(g) + Cl_2(g) \longrightarrow 2HCl(g)$$

To form the two moles of HCl, one mole of H—H bonds and one mole of Cl—Cl bonds must be broken. This requires the input of 436 kJ + 243 kJ = 679 kJ (the bond energies of the H—H and Cl—Cl bonds, respectively, Table 6.3). Upon formation of two moles of HCl, two moles of H—Cl bonds are formed (bond energy = 432 kJ per mole, Table 6.3), releasing 2 × 432 kJ = 864 kJ. Because the bonds in the products are stronger than those in the reactants by 185 kJ, we should expect the reaction to release 185 kJ more energy than it consumes. This excess energy would be released as heat, so the reaction should be exothermic with a ΔH value of −185 kJ, as observed.

The following example illustrates how bond energies can be used to estimate approximate enthalpy changes for reactions. Chemists seldom calculate ΔH values this way, because the bond energies involved are averages and standard molar enthalpies of formation give more accurate values. However, if the necessary enthalpy changes are not available, bond energies can be used to get a useful approximation.

EXAMPLE 6.7

Methanol, CH_3OH, is manufactured by the reaction

$$CO(g) + 2H_2(g) \longrightarrow CH_3OH(g)$$

Calculate the approximate enthalpy change, ΔH, of the reaction from the bond energies in Table 6.3.

The enthalpy change for this reaction is approximately equal to the sum of the enthalpy changes of three reactions, each of which involves bond breaking or bond formation:

$$C\equiv O(g) \longrightarrow C(g) + O(g) \qquad\qquad \Delta H = D_{C\equiv O}$$

$$2H-H(g) \longrightarrow 4H(g) \qquad\qquad\qquad \Delta H = 2D_{H-H}$$

$$C(g) + O(g) + 4H(g) \longrightarrow H-\underset{\underset{H}{|}}{\overset{\overset{H}{|}}{C}}-O-H \qquad \Delta H = -3D_{C-H} - D_{C-O}$$
$$-D_{O-H}$$

$$CO(g) + H_2(g) \longrightarrow CH_3OH(g) \qquad \Delta H = D_{C\equiv O} + 2D_{H-H}$$
$$-3D_{C-H} - D_{C-O} - D_{O-H}$$

In Table 6.3, we find that

$$D_{C\equiv O} = 1080 \text{ kJ mol}^{-1}, \quad D_{H-H} = 436 \text{ kJ mol}^{-1},$$
$$D_{C-H} = 415 \text{ kJ mol}^{-1}, \quad D_{C-O} = 350 \text{ kJ mol}^{-1},$$
$$\text{and } D_{O-H} = 464 \text{ kJ mol}^{-1}.$$

Thus

$$\Delta H = [1080 + (2 \times 436) - (3 \times 415) - 350 - 464] \text{ kJ} = -107 \text{ kJ}$$

6.10 Oxidation Numbers

The oxidation number of an atom, sometimes called the oxidation state, is a useful bookkeeping concept. Oxidation numbers are of value in writing chemical formulas, naming compounds, and keeping track of the redistribution of electrons during chemical reactions. The **oxidation number** can be determined by application of the following rules:

1. The oxidation number of an atom of any element in its elemental form is considered to be zero. The atoms of Na, N_2, P_4, and S_8 all have oxidation numbers of zero.

2. The oxidation number of a monatomic ion is equal to the charge on the ion. The ions Li^+, Co^{2+}, Cl^-, and N^{3-} have oxidation numbers of $+1$, $+2$, -1, and -3, respectively.

3. The oxidation number of fluorine in a compound is always -1.

4. The elements of Group IA (except hydrogen) in compounds have an oxidation number of $+1$.

5. The elements of Group IIA in compounds have an oxidation number of $+2$.

6. The elements of Group VIIA have an oxidation number of -1 when they are combined in compounds with less electronegative elements. Chlorine has an oxidation number of -1 in NaCl, CCl_4, PCl_3, and HCl, for example.

7. Oxygen has an oxidation number of -2 with three exceptions:

   (a) In compounds with fluorine, it has a positive oxidation number (fluorine is more electronegative than oxygen). For example, oxygen has an oxidation number of $+2$ in oxygen difluoride, OF_2.

   (b) In peroxides (compounds that contain an O—O single bond), it has an oxidation number of -1. An example is hydrogen peroxide, H_2O_2.

   (c) In superoxides (compounds that contain O_2^-), it has an oxidation number of $-\frac{1}{2}$. An example is potassium superoxide, KO_2.

8. Hydrogen has an oxidation number of -1 in compounds with less electronegative elements and an oxidation number of $+1$ in compounds with more electronegative elements.

9. The sum of the oxidation numbers of all atoms in a neutral compound is zero. The sum of the oxidation numbers of all atoms in an ion is equal to the charge on the ion.

If the oxidation numbers are known for all but one kind of atom in a compound, its oxidation number can be calculated.

Calculate the oxidation number of sulfur in the compounds Na_2SO_4, Na_2SO_3, and H_2S.   **EXAMPLE 6.8**

The oxidation number for sulfur in Na_2SO_4 can be calculated from the oxidation numbers for sodium and oxygen. The two sodium atoms, each with an oxidation number of $+1$ (Group IA, rule 4), total $+2$; the four oxygen atoms, each with an oxidation number of -2 (rule 7), total -8. For the sum of the oxidation numbers to be zero, sulfur must have an oxidation number of $+6$. A similar calculation shows the oxidation number of sulfur in Na_2SO_3 to be $+4$.

In H_2S, hydrogen is bonded to a more electronegative element and so has an oxidation number of $+1$ (rule 8). For the sum of the oxidation numbers of all three atoms to be zero, sulfur must have the oxidation number -2.

As demonstrated in this example, an element may have more than one oxidation number, depending on the compound in which it is found. Other examples include iron, with an oxidation number of $+2$ in $FeCl_2$ and $+3$ in $FeCl_3$, and tin, with oxidation numbers of $+2$ and $+4$ in $SnCl_2$ and $SnCl_4$, respectively. The oxidation number of chlorine in each of these examples is -1. Chlorine, however, exhibits an oxidation number of $+1$ in $NaClO$, $+3$ in $NaClO_2$, $+5$ in $NaClO_3$, and $+7$ in $NaClO_4$.

It should be emphasized that, although oxidation numbers are convenient when writing formulas and balancing oxidation-reduction equations, they are quite arbitrary. For some compounds, calculating the oxidation number of a member atom gives a fraction.

Calculate the oxidation number of Fe in the mineral lodestone, a magnetic oxide of iron with the formula Fe_3O_4.   **EXAMPLE 6.9**

In Fe_3O_4, oxygen is combined with a less electronegative element and is assigned an oxidation number of -2. The four oxygen atoms contribute a total of -8. For the sum of the oxidation numbers in the compound to be zero, the oxidation numbers of the three iron atoms must also total $+8$. Hence the oxidation number for iron in lodestone is $+\frac{8}{3}$ or $+2\frac{2}{3}$. (The oxidation number always refers to one atom of an element.)

6.11 Applying Oxidation Numbers to Writing Formulas

The formula of a compound may be written using the oxidation numbers of the atoms involved because the sum of the oxidation numbers of the atoms in the compound must equal zero.

EXAMPLE 6.10   Write the formula for magnesium chloride using the oxidation numbers of its constituent elements.

Magnesium is a member of Group IIA and so has an oxidation number of $+2$. Chlorine, a member of Group VIIA, is combined with a less electronegative element and so has an oxidation number of -1.

The formula of magnesium chloride cannot be $MgCl$, because $+2$ and -1 do not add up to 0. For the total of the oxidation numbers for the compound to be zero, the atoms must be in a ratio of one magnesium ion to two chloride ions, or $MgCl_2$.

EXAMPLE 6.11   Write the formula for aluminum oxide.

Aluminum, the less electronegative element, is a member of Group IIIA and will lose three electrons, giving Al^{3+} with an oxidation number of $+3$. Oxygen, a member of Group VIA, needs two electrons to fill its valence shell, giving the ion O^{2-} with an oxidation number of -2. Since $+3$ and -2 do not add up to 0, AlO is not the correct formula for aluminum oxide. By inspection, it is readily seen that two atoms of aluminum will give a total of six positive units of oxidation number, that three atoms of oxygen will give six negative units of oxidation number, and that the algebraic sum of the oxidation numbers will thus be zero. The correct simplest formula for aluminum oxide is, therefore, Al_2O_3.

6.12 Polyatomic Ions

Polyatomic ions contain more than one atom (Section 6.3). Several examples, selected from many such ions, are given in Table 6.4.

Table 6.4   Some common polyatomic ions

Ammonium	NH_4^+	Carbonate	CO_3^{2-}	Phosphate	PO_4^{3-}
Acetate	$CH_3CO_2^-$	Sulfate	SO_4^{2-}	Diphosphate	$P_2O_7^{4-}$
Nitrate	NO_3^-	Sulfite	SO_3^{2-}	Arsenate	AsO_4^{3-}
Nitrite	NO_2^-	Thiosulfate	$S_2O_3^{2-}$	Arsenite	AsO_3^{3-}
Hydroxide	OH^-	Peroxide	O_2^{2-}		
Hypochlorite	ClO^-	Chromate	CrO_4^{2-}		
Chlorite	ClO_2^-	Dichromate	$Cr_2O_7^{2-}$		
Chlorate	ClO_3^-	Silicate	SiO_4^{4-}		
Perchlorate	ClO_4^-				
Permanganate	MnO_4^-				

When writing formulas of compounds that include more than one polyatomic ion, we enclose the formula of the ion in parentheses and indicate the number of such ions with a subscript. Examples include $(NH_4)_2CO_3$ and $Al_2(SO_4)_3$. In $(NH_4)_2CO_3$, two ammonium ions, each with a $+1$ ionic charge, balance the -2 ionic charge of the carbonate ion. In $Al_2(SO_4)_3$, two aluminum ions, each with a charge of $+3$, and three sulfate ions, each with a charge of -2, are present and the charges are balanced. It should be noted that the sum of the oxidation numbers of the various atoms, as well as the sum of the total charges of the ions, equals zero for each compound.

The following examples illustrate the process of writing formulas for compounds containing polyatomic ions using ionic charges.

EXAMPLE 6.12

Write the formula for iron perchlorate, given that the iron is present as the +3 ion.

The iron ion here is Fe^{3+}. With charges of +3 for the iron ion and -1 for the perchlorate ion (see Table 6.4), we can see that the formula $FeClO_4$ is incorrect. By combining three perchlorate ions and one iron ion, we can make the sum of the positive and negative charges be zero; the correct formula is $Fe(ClO_4)_3$.

EXAMPLE 6.13

Write the formula for calcium phosphate, which is one component of a phosphate rock used in fertilizer production.

Since the charge for the calcium ion is +2 (Ca is a member of Group IIA) and the charge on the phosphate ion is -3, the formula cannot be $CaPO_4$. The sum of the charges on the ions must be zero for the compound. By including three calcium ions and two phosphate ions, the sum becomes zero. Thus $Ca_3(PO_4)_2$ is the correct formula for calcium phosphate.

6.13 The Names of Compounds

1. BINARY COMPOUNDS. **Binary compounds** contain two different elements. The name of a binary compound consists of the name of the more electropositive element followed by the name of the more electronegative element with its ending replaced by the suffix $-ide$. Some examples are as follows:

NaCl, sodium chloride	Na_2O, sodium oxide
KBr, potassium bromide	CdS, cadmium sulfide
CaI_2, calcium iodide	Mg_3N_2, magnesium nitride
AgF, silver fluoride	Ca_3P_2, calcium phosphide
HCl, hydrogen chloride	Al_4C_3, aluminum carbide
LiH, lithium hydride	Mg_2Si, magnesium silicide

A few polyatomic ions have special names and are treated as if they were single atoms in naming their compounds. Thus NaOH is called sodium hydroxide; HCN, hydrogen cyanide; and NH_4Cl, ammonium chloride.

If a binary hydrogen compound is an acid (Section 9.2) when it is dissolved in water, the prefix *hydro-* is used instead of the word *hydrogen* and the suffix *-ic* replaces the suffix *-ide* when we are referring to the aqueous solution.

$HCl(g)$, hydrogen chloride	$HCl(aq)$, hydrochloric acid
$HBr(g)$, hydrogen bromide	$HBr(aq)$, hydrobromic acid
$H_2S(g)$, hydrogen sulfide	$H_2S(aq)$, hydrosulfuric acid
$HCN(g)$, hydrogen cyanide	$HCN(aq)$, hydrocyanic acid

If a nonmetallic element with variable oxidation number can unite with another nonmetallic element to form more than one compound, the compounds can be distinguished by the Greek prefixes *mono-* (meaning one), *di-* (two), *tri-* (three), *tetra-* (four, *penta-* (five), *hexa-* (six), *hepta-* (seven), and *octa-* (eight). The prefixes precede the name of the constituent to which they refer.

CO, carbon monoxide	N_2O_5, dinitrogen pentaoxide
CO_2, carbon dioxide	SO_2, sulfur dioxide
NO_2, nitrogen dioxide	SO_3, sulfur trioxide
N_2O_4, dinitrogen tetraoxide	BCl_3, boron trichloride

A second method of naming different binary compounds that contain the same elements uses Roman numerals placed in parentheses after the name of the more electropositive element to indicate its oxidation number. This method of naming binary compounds is usually applied to those in which the electropositive element is a metal, but is occasionally applied to other compounds as well.

$FeCl_2$, iron(II) chloride	SO_2, sulfur(IV) oxide
$FeCl_3$, iron(III) chloride	SO_3, sulfur(VI) oxide
Hg_2O, mercury(I) oxide	NO, nitrogen(II) oxide
HgO, mercury(II) oxide	NO_2, nitrogen(IV) oxide

The system of nomenclature used in this book was formulated by a committee of the International Union of Pure and Applied Chemistry. However, there is another, older system that you may encounter elsewhere. According to the older system, when the more electropositive element is a metal, the lower oxidation number of the metal is indicated by using the suffix *-ous* on the name of the metal. The higher oxidation number is designated by the suffix *-ic*. Sometimes the stem of the Latin name for the metal is used. Thus $FeCl_2$ is ferrous chloride, and $FeCl_3$ is ferric chloride; Hg_2O is mercurous oxide, and HgO is mercuric oxide.

2. TERNARY COMPOUNDS. **Ternary compounds** contain three different elements. As pointed out earlier, a few, such as NH_4Cl, KOH, and HCN, are named as if they were binary compounds. Chlorine, nitrogen, sulfur, phosphorus, and several other elements form oxyacids (ternary compounds with hydrogen and oxygen) that usually differ from each other in oxygen content. The most common acid of a series usually bears the name of the acid-forming element, ending with the suffix *-ic,* as in chloric acid ($HClO_3$), sulfuric acid (H_2SO_4), nitric acid (HNO_3), and phosphoric acid (H_3PO_4). If the "central" element (Cl, S, etc.) of a related acid has a higher oxidation number than it does in the most common form, the suffix *-ic* is retained and the prefix *per-* is added. The name perchloric acid for $HClO_4$ illustrates this rule. If the "central" element has a lower oxidation number than it does in the most common acid, the suffix *-ic* is replaced with the suffix *-ous*. Examples are chlorous acid ($HClO_2$), sulfurous acid (H_2SO_3), nitrous acid (HNO_2), and phosphorous acid (H_3PO_3). If the same central element in two acids has lower oxidation numbers than it does in its *-ous* acid, the acid having the lower of the two oxidation numbers is named by adding the prefix *hypo-* and retaining the ending *-ous*. Thus HClO is hypochlorous acid, and H_3PO_2 is hypophosphorous acid.

Metal salts of the oxyacids (compounds in which a metal replaces the hydrogen of the acid) are named by identifying the metal and then the negative acid ion. For these salts the ending *-ic* of the oxyacid name is changed to *-ate,* and the ending *-ous* is changed to *-ite*. The salts of perchloric acid are perchlorates, those of sulfuric acid are sulfates, those of nitrous acid are nitrites, and those of hypophosphorous acid are hypophosphites. This system of naming applies to all inorganic oxyacids and their salts. The names of the oxyacids for chlorine and their corresponding sodium salts are given in Table 6.5.

Table 6.5 Names of oxyacids of chlorine and the corresponding sodium salts

Cl Oxidation number	Acids	Salts
+1	$HClO$, hypochlorous acid	$NaClO$, sodium hypochlorite
+3	$HClO_2$, chlorous acid	$NaClO_2$, sodium chlorite
+5	$HClO_3$, chloric acid	$NaClO_3$, sodium chlorate
+7	$HClO_4$, perchloric acid	$NaClO_4$, sodium perchlorate

Oxyacids that contain more than one replaceable hydrogen atom can form anions that contain hydrogen; H_2SO_4 gives HSO_4^- (hydrogen sulfate), and H_3PO_4 gives $H_2PO_4^-$ (dihydrogen phosphate) and HPO_4^{2-} (hydrogen phosphate). Metal salts of these ions are named like other salts of oxyacids; KH_2PO_4, for example, is potassium dihydrogen phosphate.

The system of nomenclature for a class of compounds known as coordination compounds will be described in Chapter 26.

For Review

Summary

When atoms react, their electronic structures change, giving chemical bonds. An **ionic bond** results when electrons are transferred from one atom to another. The resulting compound consists of regular three-dimensional arrangements of ions held together by the strong electrostatic attractions between ions of opposite charge. The charges of ions of the representative metals may be determined readily since, with few exceptions, these ions have a **noble gas electron configuration** or a **pseudo noble gas electron configuration.**

Covalent bonds result when electrons are shared as electron pairs between atoms, with the electrons attracted by the nuclei of both atoms. When only one atom furnishes both electrons of an electron-pair bond, the bond is called a **coordinate covalent bond.** A **single bond** results when one pair of electrons is shared between two atoms. The sharing of two or three pairs of electrons between two atoms constitutes a **double bond** or **triple bond,** respectively. The distribution of electrons in bonds and **unshared pairs** in a molecule can be indicated with a **Lewis structure.** Most Lewis structures can be written by inspection or by starting with a skeleton structure consisting of the constituent atoms connected by single bonds and then distributing the remaining electrons as unshared pairs or in multiple bonds to give all atoms filled valence shells, if possible. In some instances, a central atom may have an empty valence orbital, or if it is an atom for which n is greater than or equal to 3 for the valence shell, it may have more than eight electrons in its valence shell.

If two or more Lewis structures with an identical arrangement of atoms but a different distribution of electrons can be written, the actual distribution of electrons in the molecule cannot be described by a single Lewis structure but is an average of the distributions indicated by the individual Lewis structures. The actual electron distribution is called a **resonance hybrid** of the individual Lewis structures (called **resonance forms**).

Even though pairs of electrons are often shared between different kinds of atoms, they may not be shared equally. The electrons of a shared pair may spend more time near one atom than near the other, giving a **polar covalent bond.** The measure of the ability of an atom to attract a pair of electrons is called its **electronegativity.** The greater the difference in the electronegativities of two atoms in a covalent bond, the more polar the bond. Some molecules that contain polar bonds also possess a **dipole;** that is, these molecules have a positive end and a negative end.

The strength of a covalent bond is measured by its **bond energy,** the enthalpy change required to break that particular bond in a mole of molecules. The bond energies of triple bonds are generally larger than those of double bonds, and double bonds larger than single bonds.

Oxidation numbers are assigned to atoms in molecules using a set of rules, as described in this chapter. The electronegativities of the elements figure prominently in these rules. The chemical formula of a compound may be written using the oxidation numbers of the atoms in the compound. A knowledge of the chemical formula is necessary to name a compound. Compounds and ions may be named systematically, using accepted nomenclature.

Key Terms and Concepts

anion (6.1)
binary compounds (6.13)
bond energies (6.9)
cation (6.1)
coordinate covalent bond (6.4)
covalent bond (6.3)
dipole (6.6)
double bond (6.3)

electronegativity (6.5)
ionic bonds (6.1)
ionic compounds (6.1)
ions (6.1)
isoelectronic (6.3)
Lewis structure (6.3)
Lewis symbols (6.1)
oxidation number (6.10)

polar covalent bond (6.6)
polyatomic ions (6.3)
resonance forms (6.8)
resonance hybrid (6.8)
single bond (6.3)
ternary compounds (6.13)
triple bond (6.3)
unshared pairs (6.3)

Exercises

Ions and Ionic Bonding

1. Why does a cation have a positive charge?
2. Why does an anion have a negative charge?
3. Select the atoms you would expect to form positive ions and the ones you would expect to form negative ions from the following list: As, Br, Ca, Cd, F, Ga, Li, N, S, Sn, Zn.
4. Predict the charge on the monatomic ions formed from the following elements in ionic compounds containing these elements: (a) As, (b) Br, (c) Ca, (d) Cs, (e) F, (f) Ga, (g) N, (h) S, (i) O, (j) Mg, (k) I, (l) K
5. Identify the following ions as having (i) a noble gas electron configuration, (ii) a pseudo noble gas electron configuration, or (iii) neither (i) nor (ii): (a) P^{3-}, (b) Sr^{2+}, (c) Cu^{2+}, (d) Cu^+, (e) Hg^{2+}, (f) Cr^{2+}, (g) Sn^{2+}, (h) Se^{2-}, (i) Sm^{3+}
6. The elements that form the greatest concentration of monatomic ions in sea water are Cl, Na, Mg, Ca, K, Br, Sr, and F. Write the Lewis symbols for the ions formed from these elements.

7. Write the electron configuration and the Lewis symbol for each of the following atoms and for the monatomic ion found in ionic compounds containing the element: (a) Al, (b) Cl, (c) Mg, (d) Na, (e) P, (f) S.
8. Write the formulas of each of the following ionic compounds using Lewis symbols: (a) NaBr, (b) $MgCl_2$, (c) Li_2O, (d) AlF_3, (e) Ga_2O_3, (f) CaS.
9. M and X in the Lewis structures listed below represent elements in the third period of the Periodic Table. Write the formula of each compound using the chemical symbols of the elements.
    (a) $[M^{2+}][X^{2-}]$ (b) $[M^+]_3[X^{3-}]$
    (c) $[M^{3+}][X^-]_3$ (d) $[M^{2+}]_3[X^{3-}]_2$
10. Why is it incorrect to speak of the molecular weight of NaCl?

Covalent Bonding

11. How does a covalent bond differ from an ionic bond?
12. What are the characteristics of two atoms that will form a covalent bond?

13. When should you expect a compound to contain a covalent bond? To contain an ionic bond?
14. Predict which of the following compounds are ionic and which are covalent, based on the location of their constituent elements in the Periodic Table.
    (a) SO_2 (b) Na_2S
    (c) ClF_3 (d) $CuCl_2$
    (e) MgO (f) PCl_3
    (g) H_2O (h) TiO_2
    (i) $CaBr_2$ (j) Cl_2CO
    (k) NH_3
15. How are single, double, and triple bonds similar? How do they differ?
16. Write Lewis structures for the following: (a) HF, (b) AsF_3, (c) $SeCl_2$, (d) H_3O^+, (e) $GeCl_4$, (f) BH_4^-, (g) PCl_4^+
17. Write Lewis structures for the following: (a) O_2, (b) CO_2, (c) H_2CO, (d) FNO, (e) H_2CCH_2.
18. Write Lewis structures for the following: (a) CN^-, (b) HCN, (c) HCCH, (d) NO^+, (e) C_2^{2+}
19. Write the Lewis structure for the diatomic sulfur molecule, S_2, which is found in sulfur vapor.
20. Methanol H_3COH, is used as a fuel in some race cars. What is the Lewis structure of methanol?
21. The skeleton structures of a number of biologically active molecules are given below. Complete the Lewis structures of these molecules.
    (a) The amino acid alanine
    (b) Urea
    (c) Succinic acid
    (d) Carbonic acid
    (e) Histidine

(f) Vitamin C (ascorbic acid)

22. Write the Lewis structure for phosphoric acid, H_3PO_4, which has one oxygen atom and three OH groups bonded to the phosphorus.
23. The atmosphere of Titan, the largest moon of Saturn, contains methane (CH_4) and traces of ethylene (C_2H_4), ethane (C_2H_6), hydrogen cyanide (HCN), propyne (H_3CCCH), and diacetylene (HCCCCH). Write the Lewis structures of these molecules.
24. The following species do not satisfy the noble gas structure rule. Write the Lewis structure for each. (a) BCl_3, (b) ClF_5, (c) AsF_5, (d) ICl_4^-, (e) XeF_4, (f) S_2F_{10} (contains an S—S bond), (g) PCl_6^-.

Resonance

25. Which of the following pairs represent resonance forms of the same species?
26. Draw the resonance forms for the following: (a) selenium dioxide, OSeO; (b) the nitrite ion, NO_2^-; (c) the carbonate ion, CO_3^{2-}; (d) the acetate ion,
27. Ozone, O_3, is the component of the upper atmosphere that protects the earth from ultraviolet radiation. Write the resonance forms that describe the electronic structure of this molecule.
28. Explain why nitric acid, $HONO_2$, contains two distinct types of nitrogen-oxygen bonds instead of three types. (*Hint:* consider the resonance hydrid of nitric acid.)

29. Write the resonance forms for the nitrate ion, NO_3^-, and show that this ion is isoelectronic with the carbonate ion, CO_3^{2-}.
30. Sodium oxalate, $Na_2C_2O_4$, contains the oxalate anion. Write the resonance forms of this anion, which contains a carbon-carbon bond (O_2CCO_2).
31. The skeleton structure of benzene, C_6H_6, is given below. Write the resonance forms of benzene.

32. Bromstyrol, C_8H_7Br, has a strong hyacinth-like odor and is used in lilac and hyacinth scents. Write the resonance forms of bromstyrol, the skeleton structure of which is given below.

Electronegativity and Polar Bonds

33. What is meant by the electronegativity of an element?
34. How does the electronegativity difference of the atoms in a covalent bond affect the polarity of the bond?
35. How is the polarity of a covalent bond related to the polarity of a molecule containing the bond?
36. Explain why some molecules that contain polar bonds are not polar.
37. Give examples of polar covalent molecules and nonpolar covalent molecules that contain polar covalent bonds.
38. Which of the following molecules or ions contain polar bonds? O_3, P_4, S_2^{2-}, NO_2^-, ClO_3^-, H_2S, BH_4^-, HCN
39. Which of these molecules or ions are polar? NO_2^+ (a linear ion); CN^-; O_3 (a nonlinear molecule); BrCl; H_2S (a nonlinear molecule); Br_2; BrCl; BF_3 (a planar molecule).
40. One of the most important polar molecules is water, H_2O, a bent molecule with an angle of 105° between the two O—H bonds. Which end of the O—H bond in water is positive and which is negative? Sketch a figure of the H_2O molecule and indicate the positive and negative ends of the molecule.

Bond Energies

41. Using the data in Appendix I, calculate the bond energies of F_2, Cl_2, and FCl. All are gases in their most stable form at standard state conditions.

> *Ans. F—F = 158.0 kJ per mole of bonds;*
> *Cl—Cl = 243.36 kJ per mole of bonds;*
> *F—Cl = 255.15 kJ per mole of bonds*

42. Using the data in Appendix I, calculate the bond energies of N_2, O_2, and NO. All are gases in their most stable form at standard state conditions.

> *Ans. N≡N = 945.408 kJ per mole of bonds;*
> *O=O = 498.34 kJ per mole of bonds;*
> *N=O = 631.62 kJ per mole of bonds*

43. Using the data in Appendix I, calculate the Ti—Cl single bond energy in $TiCl_4$. *Ans. 430.0 kJ*
44. (a) Using the bond energies given in Table 6.3, determine the approximate enthalpy change for the formation of ethylene from ethane.

$$C_2H_6(g) \longrightarrow C_2H_4(g) + H_2(g)$$

(b) Compare this with the standard state enthalpy change. *Ans. 128 kJ; 136.94 kJ*

45. The enthalpy of formation of $AsF_5(g)$ has been determined to be -16.46 kJ g^{-1} of arsenic using the reaction $2As(s) + 5F_2(g) \rightarrow 2AsF_5(g)$. Using this information and the data in Appendix I, calculate the As—F single bond energy in AsF_5. *Ans. 386 kJ*
46. Using the bond energies in Table 6.3, calculate the approximate enthalpy change for each of the following reactions.

(a) $H_2(g) + F_2(g) \longrightarrow 2HF(g)$ *Ans. 542 kJ*
(b) $2CO(g) + O_2(g) \longrightarrow 2CO_2(g)$ *Ans. -306 kJ*
(c) $CH_4(g) + Cl_2(g) \longrightarrow CH_3Cl(g) + HCl(g)$
     Ans. -104 kJ
(d) $H_2C=CH_2(g) + H_2(g) \longrightarrow H_3CCH_3(g)$
     Ans. -128 kJ
(e) $CH_3CH_2CH_3(g) \longrightarrow CH_3CH=CH_2(g) + H_2(g)$
     Ans. 128 kJ

Oxidation Numbers

47. What are the oxidation numbers of the following elements in their compounds? (a) F, (b) Mg, (c) Rb, (d) Na, (e) K, (f) Ca
48. Chlorine forms binary compounds with each of the following elements: Mg, Co, F, N, Na, O, S, Zn. In which of the compounds will chlorine have a positive oxidation number?
49. What are the oxidation numbers of hydrogen in the binary compounds it forms with Al, Cl, Mg, N, Na, O, and S?
50. Determine the oxidation number of each of the following elements in the compound indicated:

(a) B in B_2O_3 (b) C in CF_4
(c) Cl in HCl   (d) P in P_4O_{10}
(e) Co in $LiCoO_2$ (f) N in HNO_3 ($HONO_2$)
(g) S in SO_2Cl_2 (h) P in Na_3PO_4

51. Determine the oxidation number of N in each of the following compounds: Na_2NH, NH_4F, NO, N_2H_4, HNO_3, $Ca(NO_2)_2$.
52. Determine the oxidation number of Mn in each of the following compounds: $LiMnO_2$, K_2MnCl_4, $ZnMn_2O_4$, Mn_2O_7, $K_5Mn(CN)_6$ (CN is an ion with a charge of -1).
53. Sodium thiosulfate, $Na_2S_2O_3$, is used as a fixing agent in

photography. What is the oxidation number of each of the elements in $Na_2S_2O_3$?

54. Ammonium perchlorate, $(NH_4^+)(ClO_4^-)$, and aluminum powder are the primary components of the 2 million lb of solid fuel in the booster rockets used to launch the space shuttle. These substances react, giving $AlCl_3$, Al_2O_3, H_2O, and N_2 as the principal products. What are the oxidation numbers of each element in the fuel and in the products of the reaction?

Names of Compounds

55. Do the suffixes *-ous* and *-ic* correspond to any particular oxidation state? How are they used?
56. Name the following binary compounds: (a) NaF, (b) CuS, (c) HgI_2, (d) AgBr, (e) KH, (f) $AlCl_3$, (g) $MgCl_2$, (h) LiCl, (i) Na_2O, (j) K_3P, (k) Mg_3N_2.
57. Name the following compounds, each of which contains polyatomic ions: (a) KOH, (b) LiCN, (c) NH_4Br, (d) $Mg(OH)_2$, (e) $Al(CN)_3$, (f) NH_4F, (g) $Ca(CN)_2$.
58. Name the following compounds, each of which contains polyatomic ions: (a) $(NH_4)_2CO_3$, (b) $KClO_4$, (c) $BaSO_4$, (d) $NaNO_3$, (e) $CuCrO_4$, (f) $KMnO_4$, (g) $Mg_3(PO_4)_2$, (h) $Al(CH_3CO_2)_3$.
59. Name the following compounds. (Since the metal atom in each can exhibit two or more oxidation states, it will be necessary to identify the oxidation number of the metal as you name the compound.) (a) $FeCl_3$, (b) Cu_2S, (c) $MnBr_2$, (d) $Pb(OH)_2$, (e) $TlNO_3$, (f) $Cr(ClO_4)_3$, (g) $SnCl_4$, (h) $CuSO_4$
60. Name the following compounds, using Greek prefixes for each element as needed: (a) NO_2, (b) N_2O_3, (c) N_2O_4, (d) ClO_2, (e) Cl_2O_7, (f) PCl_5.

Formulas of Compounds

61. Why must a sample of calcium chloride, $CaCl_2$, contain two chloride ions for each calcium ion?
62. Using oxidation numbers as a guide, write the formula of each of the following compounds: (a) lithium chloride, (b) magnesium oxide, (c) sodium sulfide, (d) calcium chloride, (e) hydrogen iodide, (f) aluminum bromide.
63. Using the ionic charges as a guide, write the formula of each of the following compounds: (a) magnesium perchlorate, (b) sodium nitrate, (c) calcium sulfate, (d) potassium hydroxide, (e) ammonium phosphate, (f) aluminum nitrate, (g) sodium cyanide.
64. Using the indicated oxidation state as a guide, write the formula of each of the following compounds: (a) cobalt(II) bromide, (b) lead(II) sulfide, (c) mercury(II) oxide, (d) iron(II) nitrate, (e) copper(I) sulfide, (f) chromium(II) sulfate, (g) tin(II) perchlorate.
65. Write the formula of each of the following compounds: (a) sulfur dioxide, (b) carbon tetrafluoride, (c) dinitrogen pentaoxide, (d) sulfur hexafluoride, (e) phosphorus trichloride, (f) lead dioxide.
66. The following compounds have been approved by the FDA as food additives and may be used in foods as neu-

tralizing agents, sequestering agents, sources of minerals, etc. What is the chemical formula of each compound?
(a) Potassium iodide
(b) Sulfur dioxide
(c) Calcium oxide
(d) Calcium chloride
(e) Hydrogen chloride
(f) Sodium hydroxide
(g) Magnesium carbonate
(h) Ammonium carbonate
(i) Carbon dioxide
(j) Copper(I) iodide
(k) Sulfuric acid
(l) Potassium sulfate
(m) Sodium hydrogen carbonate
(n) Ammonium aluminum sulfate
(o) Sodium dihydrogen phosphate
(p) Aluminum sulfate
(q) Iron(III) phosphate
(r) Potassium dihydrogen phosphate
(s) Iron(II) sulfate
(t) Potassium hydrogen sulfite
(u) Sodium sulfite

Additional Exercises

67. Complete and balance the following equations:
(a) $K + I_2 \longrightarrow$
(b) $Cs + P_4 \longrightarrow$
(c) $Mg + S_8 \longrightarrow$
(d) $H^+ + NH_3 \longrightarrow$
(e) $Al + O_2 \longrightarrow$
(f) $Zn + Br_2$
68. How does the electronic structure of an isolated nitrogen atom differ from that of a nitrogen atom in a molecule of nitrogen, N_2?
69. X may indicate a different representative element in each of the following Lewis formulas. To which group does X belong in each case?

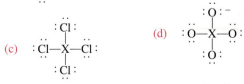

70. The number of covalent bonds between two atoms is called bond order. Bond order may be an integer, as with $:N \equiv N:$ (bond order = 3), or a fraction, as with SO_2 (bond order = $1\frac{1}{2}$ due to the resonance $O = S - O \longleftrightarrow O - S = O$). Arrange the molecules and ions in each of the following groups in increasing bond order for the bond indicated. Unshared pairs have been omitted for clarity.

(a) C—O   bond   order: $C \equiv O$; $O = C = O$; $CH_3 - O - O - CH_3$; HCO_2^-

$$\left(\quad H - C \overset{\displaystyle O}{\underset{\displaystyle O^-}{\Big\backslash}} \quad \longleftrightarrow \quad H - C \overset{\displaystyle O^-}{\underset{\displaystyle O}{\Big\backslash}} \quad \right)$$

(b) N—N   bond   order: $CH_3 - N = N - CH_3$; N_2; $H_2N - NH_2$; N_2O ($N = N = O \longleftrightarrow N \equiv N - O$).

(c) C—N bond order: $H_3C - NH_2$; CH_3CONH^-

$$\left(\quad (CH_3 - C \overset{\displaystyle O}{\underset{\displaystyle NH^-}{\Big\backslash}} \quad \longleftrightarrow \quad CH_3 - C \overset{\displaystyle O^-}{\underset{\displaystyle NH}{\Big\backslash}} \quad \right)$$

$H_2C = NH$; $H_3C - C \equiv N$.

(d) C—C bond order: $H_3C - CH_3$; $HC \equiv CH$; $H_2C = CH_2$; C_6H_6

71. Arrange the following ion and molecules in increasing order of their C—O bond order (see exercise 70): CO; CO_3^{2-} (C is the central atom, since it is less electronegative); H_2CO (C is the central atom); CH_3OH (with a CO bond, three CH bonds, and one OH bond).

72. Why is the bond order (exercise 70) of a bond between H and another atom never greater than 1?

73. As the bond order (exercise 70) between two atoms of the same type increases, the distance between the two atoms, the bond distance, decreases. How would you expect the bond distance in NO_2^- to compare to that in NO_3^-?

74. Separate the following species into three different groups of isoelectronic molecules and ions: NCN^{2-}, CH_3^-, NNO, H_3O^+, NO_2^- (ONO$^-$), O_3 (OOO), NH_3, CO_2 (OCO), ClNO.

75. A compound with a molecular weight of about 45 contains 52.2% C, 13.1% H, and 34.7% O. Write two possible Lewis structures for this compound.

76. Which of the following represents the molecular structure of hydroxylamine?

(a)   (b)

(Hint: Using bond energies, determine if the conversion of structure (a) to structure (b) is endothermic or exothermic.)   Ans. (b)

Chemical Bonding, Part 2: Molecular Orbitals

7

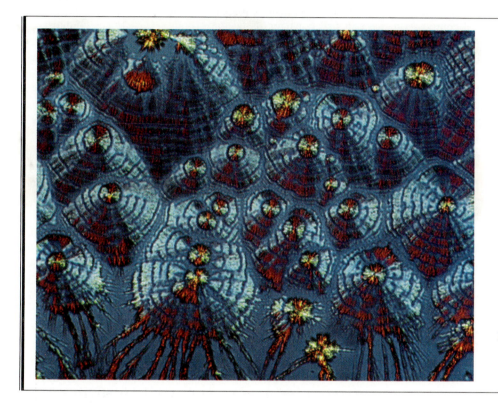

Photomicrograph of Na_2CO_3, which contains ionic and covalent bonds.

In Chapter 6 we considered bonding from the viewpoint that the electrons partici-pating in bonding are located in the atomic orbitals of the bonded atoms. The distribution of valence electrons in a covalent molecule was treated simply in terms of shared and unshared pairs of electrons. Except for noting that shared pairs of electrons spend much of their time between the atoms they connect, little was said about the distribution of the electrons in a molecule, and nothing was said about the relative energies of electrons in molecules. Molecular orbital theory provides a model for describing the distribution of electrons throughout the molecule and the energies of those electrons.

7.1 Molecular Orbital Theory

Molecular orbital theory treats the distribution of electrons in molecules in much the same way as the distribution of electrons in atomic orbitals is treated for atoms. Using the techniques of quantum mechanics, the behavior of an electron in a molecule is described by a wave function, ψ, which may be used to determine the energy of the electron and the shape of the region of space within which it moves. As in an atom, an electron in a molecule is limited to discrete (quantized) energies, and the region of space in which the electron is likely to be found is called an *orbital*. However, because the orbital extends over all of the atoms in the molecule and an electron might be found near the nucleus of any of the atoms in the molecule, the orbital is called a **molecular orbital.** Like an atomic orbital, a molecular orbital is full when it contains two electrons (with opposite spin).

Because the types of orbitals found in molecules vary with the molecular shape and composition, we will first consider molecular orbitals in molecules composed of two identical atoms (H_2 or Cl_2, for example). Such molecules are called **homonuclear diatomic molecules.** The types of molecular orbitals commonly observed in these diatomic molecules, **sigma (σ) orbitals** and **pi (π) orbitals,** are shown in Fig. 7.1. In three of these orbitals, called σ_s, σ_p, and π_p orbitals (read as ''sigma-s,'' ''sigma-p,'' and ''pi-p''), the electrons are more or less between the nuclei. Electrons in these orbitals are attracted by both nuclei at the same time and are of lower energy than they would be in the isolated atoms. Thus electrons in these orbitals stabilize the molecule; the orbitals are called **bonding orbitals.** Electrons in the orbitals called σ_s^*, σ_p^*, and π_p^* (read as ''sigma-s-star,'' ''sigma-p-star,'' and ''pi-p-star'') are located well away from the region between the two nuclei. Electrons in these latter orbitals are of higher energy than they would be in the isolated atoms, so they destabilize the molecule. The orbitals are called **antibonding orbitals.**

The exact wave functions that describe the behavior of an electron in molecular orbitals are very difficult to determine. Thus approximate wave functions are usually used instead. One approximation involves using the sum or the difference of the wave functions of the constituent atoms to describe the behavior of the electron in the molecule. The wave functions of any number of atomic orbitals may be added together, or subtracted from one another, to give the wave function of a molecular orbital. At this stage, however, we will use only one atomic orbital on each of the two atoms of a diatomic molecule to approximate a molecular orbital in the molecule.

In a diatomic molecule two molecular orbitals result from each two atomic orbitals that combine. The two molecular orbitals can be thought of as being described by the addition or the subtraction of the parts of the two atomic orbitals that **overlap**—that is,

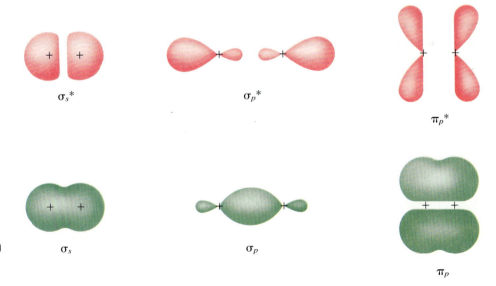

Figure 7.1

Molecular orbitals found in homonuclear diatomic molecules. Bonding molecular orbitals are shown in green; antibonding molecular orbitals in red. The plus signs (+) indicate the locations of nuclei.

σ_s^* σ_p^* π_p^*

σ_s σ_p π_p

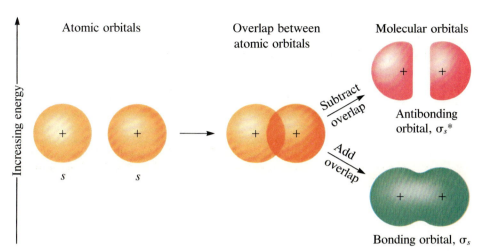

Figure 7.2

A representation of the formation of sigma (σ) molecular orbitals by the combination of two s atomic orbitals. The bonding molecular orbital is shown in green; the antibonding molecular orbital in red. The plus signs (+) indicate the locations of nuclei.

share the same region of space. The molecular orbital resulting from the addition of the overlapping parts of two atomic orbitals has a relatively large density between the two nuclei, indicating that electrons occupying such an orbital spend a large amount of time between the nuclei (Fig. 7.2, bottom right). This is a bonding situation, and the molecular orbital is a bonding orbital. A second molecular orbital, arrived at by subtraction of the overlapping parts of the atomic orbitals, has a lower density between the nuclei, indicating that the electrons spend little time between them (Fig. 7.2, top right). Such a situation is unstable, and the molecular orbital is an antibonding orbital.

Figure 7.2 illustrates the two types of molecular orbitals that can be formed from two s atomic orbitals on adjacent atoms: a σ_s (bonding) orbital, formed by addition of the two s orbitals on adjacent atoms, and a σ_s^* (antibonding) orbital, formed by subtraction of the s orbitals. (The subscript letters designate the type of atomic orbitals added or subtracted to give a molecular orbital.) The bonding molecular orbital is of lower energy than either of the atomic orbitals that combine to produce it. The antibonding orbital is of higher energy than either of the two atomic orbitals. Electrons fill the lower-energy bonding molecular orbital before the higher-energy antibonding molecular orbital, just as they fill atomic orbitals of lower energy before filling those of higher energy. Electrons favor the lower-energy bonding orbital because a given electron is attracted by both nuclei in the bonding orbital but usually by only one nucleus in the antibonding orbital. Notice in Fig. 7.2 that an electron in an antibonding orbital has a low probability of being found in the space between the nuclei.

Figure 7.3 illustrates the four kinds of molecular orbitals that can be formed from end-to-end or side-by-side combinations of two p atomic orbitals on adjacent atoms: σ_p, σ_p^*, π_p, and π_p^*. When combined end-to-end [Fig. 7.3(a)], two p atomic orbitals form two σ molecular orbitals (as is the case with atomic s orbitals). However, when two p atomic orbitals are combined side-by-side [Fig. 7.3(b)], the two resulting molecular orbitals are π orbitals.

Each atom contains three p atomic orbitals: p_x, p_y, and p_z (Section 5.10). One of these, say the p_x orbital, combines end-to-end with a corresponding p_x atomic orbital of another atom to form two σ molecular orbitals, σ_{p_x} and $\sigma_{p_x}^*$. Each of the other two p atomic orbitals, p_y and p_z, combines side-by-side with a corresponding p atomic orbital of another atom, giving rise to two sets of π bonding and π^* antibonding molecular orbitals. The set of π and π^* orbitals from the p_y atomic orbitals is oriented at right

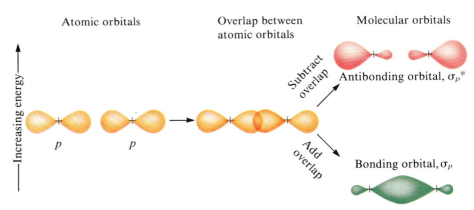

(a) The bonding and antibonding molecular orbitals formed by the end-to-end overlap of two p atomic orbitals.

Figure 7.3

A representation of the formation of sigma (σ) and pi (π) molecular orbitals by the combination of p orbitals. Bonding molecular orbitals are shown in green; antibonding molecular orbitals in red. The plus signs (+) indicate the locations of nuclei.

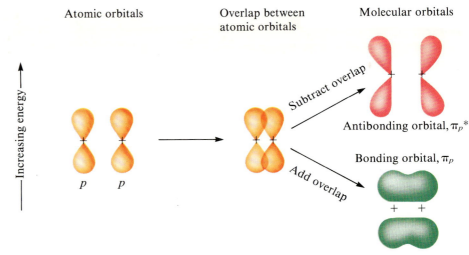

(b) The bonding and antibonding molecular orbitals formed by the side-to-side overlap of two p atomic orbitals.

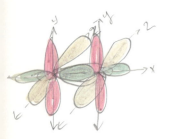

angles to the set from the p_z orbitals. It is as if one set of π and π^* molecular orbitals were oriented along the y axis of a set of coordinates and the other set were oriented along the z axis. The notations π_{p_y} and π_{p_z} (commonly referred to as "pi-p-y" and "pi-p-z") are thus applied to the bonding orbitals, and $\pi_{p_y}^*$ and $\pi_{p_z}^*$ ("pi-p-y star" and "pi-p-z star") to the antibonding orbitals. Except for their orientation, the π_{p_y} and π_{p_z} orbitals are identical and have the same energy; that is, they are **degenerate.** The $\pi_{p_y}^*$ and $\pi_{p_z}^*$ antibonding orbitals are also identical except for their orientation and are degenerate.

A total of six molecular orbitals results from the combinations of the six p atomic orbitals in two atoms. If the σ_p and σ_p^* orbitals are oriented along the x axis and the two sets of π_p and π_p^* orbitals are oriented along the y and z axes, the six molecular orbitals are identified as σ_{p_x} and $\sigma_{p_x}^*$, π_{p_y} and $\pi_{p_y}^*$, and π_{p_z} and $\pi_{p_z}^*$.

7.2 Molecular Orbital Energy Diagrams

The relative energy levels of the lower-energy atomic and molecular orbitals of a homonuclear diatomic molecule are typically as shown in Fig. 7.4. Each colored disk represents one atomic or molecular orbital that can hold either one or two electrons. Bonding and antibonding molecular orbitals are joined by dashed lines to the atomic orbitals that combine to form them.

The distribution of electrons in the molecular orbitals of a homonuclear diatomic molecule can be predicted by filling these molecular orbitals in the same way that atomic orbitals are filled, by the aufbau process (Section 5.12). The number of electrons in each orbital is indicated with a superscript. Thus a molecule containing seven electrons would have the molecular electron configuration $(\sigma_{1s})^2(\sigma_{1s}*)^2(\sigma_{2s})^2(\sigma_{2s}*)^1$.

In some cases the energy levels of the σ_p and the two π_p bonding orbitals are reversed; the π_{p_y} and π_{p_z} bonding orbitals are slightly lower in energy than the σ_{p_x} orbital. There is a logical reason for this. We have assumed in our discussion thus far that s orbitals interact only with s orbitals to form σ_s molecular orbitals and that p

degenerate means orbitals with exact same energy levels

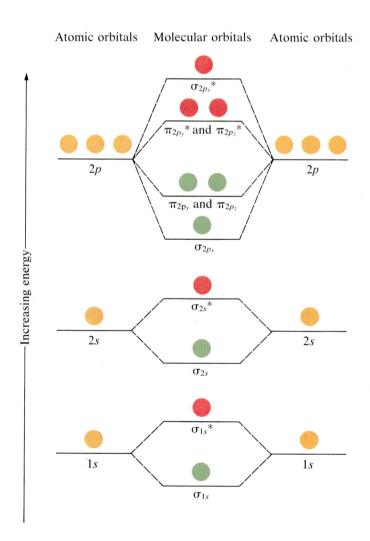

Atomic orbitals   Molecular orbitals   Atomic orbitals

Increasing energy →

$\sigma_{2p_x}*$

$\pi_{2p_y}*$ and $\pi_{2p_z}*$

2p   2p

π_{2p_y} and π_{2p_z}

σ_{2p_x}

$\sigma_{2s}*$

2s   2s

σ_{2s}

$\sigma_{1s}*$

1s   1s

σ_{1s}

Figure 7.4

Molecular orbital energy diagram for a diatomic molecule containing identical atoms. A colored disk represents an atomic or molecular orbital that can hold one or two electrons. Atomic orbitals are orange, bonding molecular orbitals are green, and antibonding molecular orbitals are red.

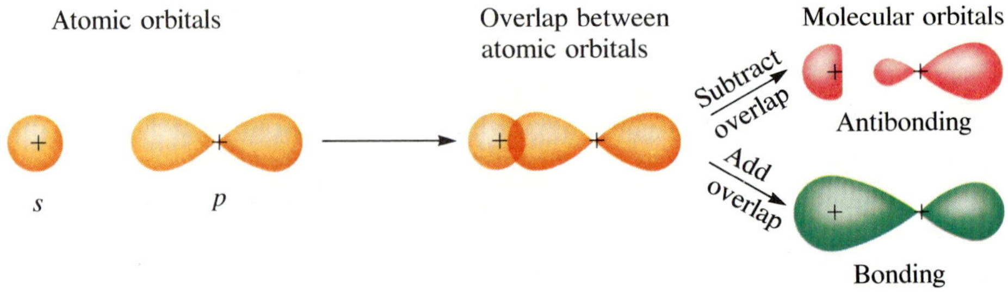

Atomic orbitals · Overlap between atomic orbitals · Molecular orbitals

Figure 7.5
A representation of the two molecular orbitals formed by the combination of an s atomic orbital with a p atomic orbital.

orbitals interact only with p orbitals to form σ_p or π_p molecular orbitals. The main interactions do indeed occur between orbitals that are identical in energy, but an s orbital can interact with a p orbital, as shown in Fig. 7.5., if the s and p orbitals are similar in energy. The mixing of s and p orbitals shifts the energies of the molecular orbitals and in some cases even changes the relative positions of their energy levels. Hence the π_{p_y} and π_{p_z} molecular orbitals, which normally have a higher energy than the σ_{p_x} orbital, sometimes have a slightly lower energy than the σ_{p_x} orbital. Such energy shifts are greatest when the energy difference between the s and p orbitals is small.

7.3 Bond Order

As electrons fill molecular orbitals such as those shown in Fig. 7.4 for a diatomic molecule, some electrons enter bonding molecular orbitals, and others may enter antibonding molecular orbitals. Thus some electrons contribute to the stability of the molecule, whereas others may destabilize the molecule. The net contribution of the electrons to the stability of a molecule can be identified by determining the net order of the bond that results from the filling of the molecular orbitals by electrons.

When using Lewis formulas to describe the distribution of electrons in molecules, the order of a bond (bond order) between two atoms can be defined as the number of bonding pairs of electrons between the atoms (see exercise 70, Chapter 6). Bond order is defined differently when the molecular-orbital description of the distribution of electrons is used, but the resulting bond order is the same. In the molecular orbital model the **bond order for a given bond is the net number of pairs of bonding electrons and is equal to one-half of the difference between the number of bonding electrons and the number of antibonding electrons.**

The order of a covalent bond is a guide to its strength; a bond between two given atoms becomes stronger as the bond order increases. If the distribution of electrons in the molecular orbitals between two atoms is such that the resulting bond would have a negative bond order or a bond order of zero, a stable bond will not be formed. The strength of the bond between two atoms increases as the bond order increases.

7.4 The Dihydrogen Molecule, H_2

A dihydrogen molecule (H_2) forms from two hydrogen atoms, each with one electron in a $1s$ atomic orbital. When the atomic orbitals of the two atoms combine, the electrons seek the molecular orbital of lowest energy, the σ_{1s} bonding orbital. Each molecular orbital can hold two electrons, so both electrons are in the σ_{1s} bonding orbital. The electronic configuration is $(\sigma_{1s})^2$. This can be represented by a molecular orbital energy diagram (Fig. 7.6) in which one electron in an orbital is indicated by an

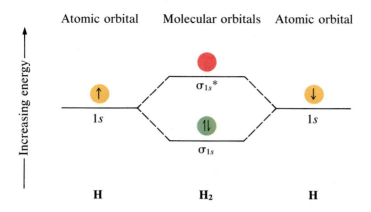

Atomic orbital   Molecular orbitals   Atomic orbital

Figure 7.6
Molecular orbital energy diagram for the dihydrogen molecule, H_2.

arrow, ①. Two electrons of opposite spin in an orbital are designated ①. A dihydrogen molecule, H_2, readily forms from two hydrogen atoms, because the energy of a H_2 molecule is lower than that of two H atoms. The σ_{1s} orbital that contains both electrons is lower in energy than either of the two $1s$ atomic orbitals.

The order of the bond in a dihydrogen molecule is equal to one-half of the difference between the number of bonding electrons and the number of antibonding electrons (Section 7.3). A dihydrogen molecule contains two bonding electrons and no antibonding electrons.

$$\text{Bond order in } H_2 = \frac{(2-0)}{2} = 1$$

Since the bond order for the hydrogen-hydrogen bond is equal to 1, the bond is a single bond.

7.5 Two Diatomic Helium Species: He_2^+ and He_2

A helium atom has two electrons, both of which are in its $1s$ orbital (Table 5.5). The helium ion (He^+) results from the loss of one of these electrons. When a helium atom and a helium ion combine to give a dihelium ion, He_2^+, two of the electrons occupy the σ_{1s} bonding orbital. The third electron must go into the molecular orbital that is next lowest in energy, the σ_{1s}^* antibonding orbital, so He_2^+ has the configuration $(\sigma_{1s})^2(\sigma_{1s}^*)^1$. The locations of the electrons are illustrated in the molecular orbital energy-level diagram of Fig. 7.7.

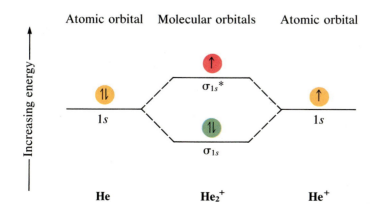

Atomic orbital   Molecular orbitals   Atomic orbital

Figure 7.7
Molecular orbital energy diagram for the He_2^+ ion.

The net energy of a dihelium ion is slightly lower than the sum of the energies of a helium atom and a helium ion, because there are two electrons in the lower-energy molecular orbital and only one in the higher-energy molecular orbital. The lower net energy for the dihelium ion means a helium atom and a helium ion are likely to combine to form a dihelium ion.

The bond order of the dihelium ion indicates that the bond in the ion is weak. The bond order is $\frac{1}{2}$ [$(2 - 1)/2 = \frac{1}{2}$].

Two helium atoms do not combine to form a dihelium molecule, He_2, with four electrons, because the two electrons in the lower-energy bonding orbital would be balanced by the two electrons in the higher-energy molecular orbital. [The configuration would be $(\sigma_{1s})^2(\sigma_{1s}*)^2$.] Hence the net energy change would be zero, so there is no driving force for helium atoms to form the diatomic molecule. In fact, helium exists as discrete atoms rather than as diatomic molecules.

The bond order in a hypothetical dihelium molecule would be zero [$(2 - 2)/2$]. A calculated bond order of zero indicates that no bond is formed between two atoms.

7.6 The Dilithium Molecule, Li_2

The combination of two lithium atoms to form a dilithium molecule, Li_2, is analogous to the formation of H_2, except that the atomic orbitals principally involved are the $2s$ orbitals. Each of the two lithium atoms, with an electronic configuration $1s^2 2s^1$, has one valence electron. Hence two valence electrons are available to go into the σ_{2s} bonding molecular orbital. (The lower-lying $1s$ electrons in the $1s$ orbital of each atom are not appreciably involved in the bonding.) Since both valence electrons occupy the lower-energy σ_{2s} bonding orbital (Fig. 7.8), Li_2 should be a stable molecule. The molecule is, in fact, present in appreciable concentration in lithium vapor at temperatures near the boiling point of the element.

In general, electrons in inner shells of atoms are not involved in forming molecular orbitals. Because of the relatively high effective nuclear charge experienced by the electrons in inner shells, the inner shells have small radii. Consequently, inner shells of adjacent atoms do not overlap. To indicate that inner shells do not form molecular orbitals, we include letters that stand for them when we write the electronic configuration of diatomic molecules. The configuration for Li_2 is KK $(\sigma_{2s})^2$, where the two K's indicate that the $1s$ orbitals (the K shells) on each atom are filled but do not enter into the bonding.

In calculating the bond order for Li_2, the result is the same if we use all of the electrons in the molecule [two σ_{1s} electrons, two $\sigma_{1s}*$ electrons, and two σ_{2s} electrons;

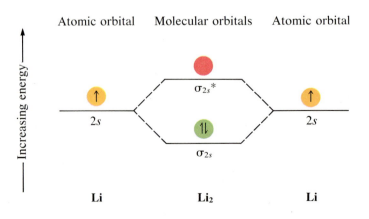

Figure 7.8
Molecular orbital energy diagram for Li_2. Lower-energy electrons in the $1s$ orbital of each atom are not shown.

bond order = $(6 - 4)/2 = 1$] or if we use only the valence electrons [two σ_{2s} electrons; bond order = $(2 - 0)/2 = 1$]. Either way the bond order is calculated to be 1, corresponding to a lithium-lithium single bond.

7.7 The Instability of the Diberyllium Molecule, Be_2

The diatomic molecule of beryllium, Be_2, unlike the diatomic lithium molecule, would not be expected to be particularly stable, inasmuch as two of the four valence electrons would go to the σ_{2s} bonding orbital and the other two to the σ_{2s}^{*} antibonding orbital (Fig. 7.9). The electron configuration would be $KK(\sigma_{2s})^2(\sigma_{2s}^{*})^2$. The net energy change, just as is the case with diatomic helium, is zero, indicating no tendency for beryllium atoms to combine to form the diatomic beryllium molecule. This is in accord with the experimental finding that no stable Be_2 molecule is known. The bond order of a hypothetical diberyllium molecule would be zero [$(4 - 4)/2 = 0$].

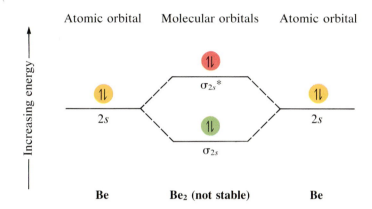

Figure 7.9

Molecular orbital energy diagram for Be_2 (not stable). Lower-energy electrons in the $1s$ orbital of each atom are not shown.

7.8 The Diboron Molecule, B_2

The element boron has the electronic structure $1s^2 2s^2 2p^1$. Hence the p orbitals make an important contribution to the bonding of boron (and to that of the remaining elements in the period). We have noted that combinations of p atomic orbitals give rise to both σ and π molecular orbitals and that the two π_p bonding orbitals are of equal energy (degenerate). The filling of degenerate molecular orbitals is analogous to the filling of degenerate atomic orbitals (Section 5.11). **Whenever two or more molecular orbitals have the same energy, electrons fill all of these orbitals singly with parallel spins before any pairing of electrons takes place within these orbitals.**

We have noted that sometimes the energy of the σ_p orbital is slightly lower than the energies of the two π_p orbitals, and that sometimes the reverse is true. Experimental magnetic data for the B_2 molecule tell us that the molecule contains two unpaired electrons. This indicates that *in this particular case* the two π_p orbitals are lower in energy than the σ_p orbital and are filled first, with one electron going singly into each. If the σ_p orbital were lower in energy than the two π_p orbitals, the two electrons would be expected to pair within that orbital. Figure 7.10 is the molecular orbital energy diagram for B_2 and shows the two unpaired electrons. The electron configuration of B_2 is $KK(\sigma_{2s})^2(\sigma_{2s}^{*})^2(\pi_{2p_y}, \pi_{2p_z})^2$. The four electrons in the σ_{2s} and σ_{2s}^{*} orbitals balance each other in terms of energy and do not make a significant contribution to the stability

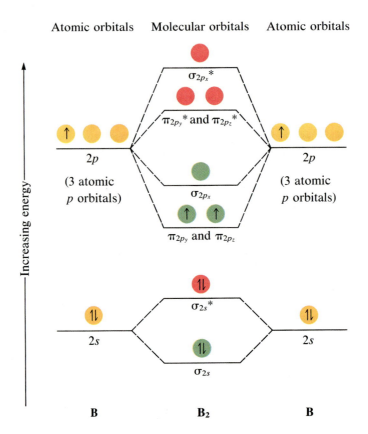

Figure 7.10

Molecular orbital energy diagram for B_2. Lower-energy electrons in the $1s$ orbital of each atom are not shown.

of the B_2 molecule. However, the two π_p electrons in bonding orbitals stabilize the diboron molecule.

Four electrons in the valence shell of a diboron molecule occupy bonding molecular orbitals, and two occupy antibonding molecular orbitals. Thus the bond order of the boron-boron bond is $(4 - 2)/2 = 1$, indicating a single bond.

7.9 The Dicarbon Molecule, C_2

The dicarbon molecule C_2, is similar to B_2. As in B_2, the π_{2p} orbitals of C_2 have a lower energy than that of the σ_{2p} orbital. The two additional electrons (one from each atom) fill the π_{2p_y} and π_{2p_z} molecular orbitals, giving the electronic structure $KK(\sigma_{2s})^2(\sigma_{2s}{}^*)^2(\pi_{2p_y},\pi_{2p_z})^4$. The calculated bond order for the carbon-carbon bond is $(6 - 2)/2 = 2$, indicating a double bond.

7.10 The Dinitrogen Molecule, N_2

The dinitrogen molecule, N_2, is well known and very stable. Experiment shows that the two $2s$ and three $2p$ electrons from each nitrogen atom are paired in the molecular orbitals of the N_2 molecule and that the π_{2p} orbitals are lower in energy than the σ_{2p} orbital (Fig. 7.11). The electron configuration is $KK(\sigma_{2s})^2(\sigma_{2s}{}^*)^2(\pi_{2p_y}, \pi_{2p_z})^4(\sigma_{2p_x})^2$.

The exceptional stability of the dinitrogen molecule is consistent with the assumption that the three bonding molecular orbitals resulting from the overlap of all $2p$

Atomic orbitals   Molecular orbitals   Atomic orbitals

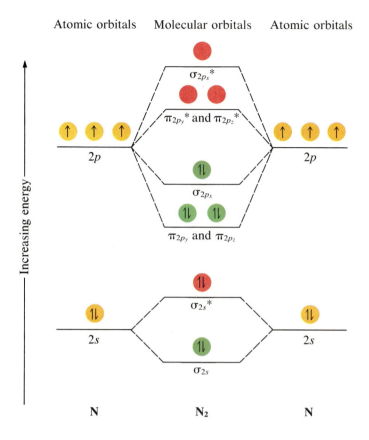

Increasing energy

$\sigma_{2p_x}^*$

$\pi_{2p_y}^*$ and $\pi_{2p_z}^*$

$2p$ $2p$

σ_{2p_x}

π_{2p_y} and π_{2p_z}

σ_{2s}^*

$2s$ $2s$

σ_{2s}

N N_2 N

Figure 7.11

Molecular orbital energy diagram for N_2. Lower-energy electrons in the $1s$ orbital of each atom are not shown.

atomic orbitals are fully occupied with six electrons. This leads to a bond order of $(8 - 2)/2 = 3$, corresponding to a triple bond.

7.11 The Dioxygen Molecule, O_2

The order of the σ_{2p} and π_{2p} orbitals in the dioxygen molecule differs from that for B_2, C_2, and N_2; the σ_{2p} orbital in oxygen is at a lower energy than the π_{2p} orbitals (Fig. 7.12).

The electronic configuration for O_2 [$KK(\sigma_{2s})^2(\sigma_{2s}^*)^2(\sigma_{2p_x})^2(\pi_{2p_y}, \pi_{2p_z})^4(\pi_{2p_y}^*, \pi_{2p_z}^*)^2$] is consistent with the fact that an oxygen molecule has two unpaired electrons. This proved to be difficult to explain on the basis of valence electron formulas, but molecular orbital theory explains it quite simply. In fact, the unpaired electrons of the oxygen molecule provide one of the strong pieces of support for molecular orbital theory.

The bond in the dioxygen molecule is weaker than the bond in the dinitrogen molecule; the bond energy of the O=O bond is 498 kilojoules per mole, whereas that of the N≡N bond is 946 kilojoules per mole (Section 6.9). The two electrons in the antibonding orbitals make the bond in the oxygen molecule weaker than the bond in the nitrogen molecule, in which no electrons are in antibonding orbitals.

A dioxygen molecule contains eight bonding electrons in its valence shell (in the σ_{2s}, σ_{2p}, π_{2p_y}, and π_{2p_z} molecular orbitals) and four antibonding electrons (in the σ_{2s}^*, $\pi_{2p_y}^*$, and $\pi_{2p_z}^*$ molecular orbitals). Thus the bond order in the dioxygen molecule is $(8 - 4)/2 = 2$, corresponding to a double bond. The double bond in O_2 is not as strong as the triple bond in N_2.

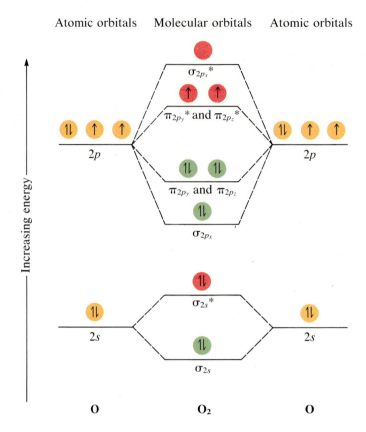

Figure 7.12

Molecular orbital energy diagram for O_2. Lower-energy electrons in the $1s$ orbital of each atom are not shown.

7.12 The Difluorine Molecule, F_2

The difluorine molecule has no unpaired electrons; six of the p electrons are in bonding orbitals and four in antibonding orbitals. The electronic structure is $KK(\sigma_{2s})^2(\sigma_{2s}^*)^2(\sigma_{2p_x})^2(\pi_{2p_y}, \pi_{2p_z})^4(\pi_{2p_y}^*, \pi_{2p_z}^*)^4$. Because F_2 has two more electrons in antibonding molecular orbitals than O_2, the difluorine molecule has a weaker bond than the dioxygen molecule. In fact, the F_2 bond is one of the weakest of the covalent bonds; the bond energy is only 160 kilojoules per mole.

The bond order of the difluorine molecule is $(8 - 6)/2 = 1$, corresponding to a single bond.

7.13 The Instability of the Dineon Molecule, Ne_2

The Ne_2 molecule has no appreciable tendency to form, because it would have the same number of electrons in antibonding orbitals as in bonding orbitals. As with helium, the atoms remain as discrete atoms. A hypothetical Ne_2 molecule would have the electron configuration $KK(\sigma_{2s})^2(\sigma_{2s}^*)^2(\sigma_{2p_x})^2(\pi_{2p_y}, \pi_{2p_z})^4(\pi_{2p_y}^*, \pi_{2p_z}^*)^4 (\sigma_{2p_x}^*)^2$ and a bond order of zero.

7.14 Diatomic Systems with Two Different Elements

Thus far we have considered only molecular orbitals in molecules and ions containing two identical atoms, but molecular orbital theory can be extended to **heteronuclear diatomic species,** that is, diatomic molecules and ions with bonds between atoms of two different elements.

Molecular orbitals of heteronuclear diatomic species are not fundamentally different from those of homonuclear diatomic molecules. Electrons still move over the whole molecule in molecular orbitals that can be approximated by combinations of atomic orbitals from each atom. However, in heteronuclear molecules the equivalent atomic orbitals for each contributing atom do not have the same energy. Orbitals for the more electronegative element are at a lower energy. Figure 7.13 shows a molecular orbital energy diagram for nitric oxide, NO. The energies of the atomic orbitals of oxygen are lower than those of nitrogen due to the greater charge of the oxygen nucleus. Consequently, some molecular orbitals of NO are closer in energy to the energy of the oxygen atomic orbitals than to that of the nitrogen atomic orbitals. Other molecular orbitals are closer in energy to the energy of the nitrogen atomic orbitals than to that of the oxygen atomic orbitals.

Unlike the molecular orbitals in a diatomic molecule composed of identical atoms, the molecular orbitals in a heteronuclear diatomic molecule are not symmetrical. The

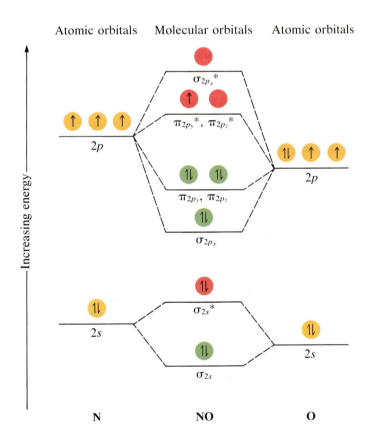

Figure 7.13

Molecular orbital energy diagram for NO. Lower-energy electrons in the 1s orbital of each atom are not shown.

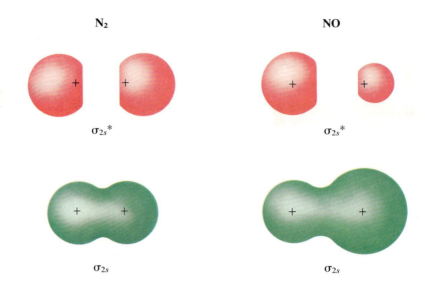

Figure 7.14

Sigma (σ) molecular orbitals found in N_2 and NO. The plus signs (+) indicate the locations of nuclei.

electron density is greater nearer the atom whose atomic orbital is closest to the energy of the molecular orbital. Figure 7.14 illustrates the distribution of electron density in the σ_{2s} and σ_{2s}^{*} orbitals of N_2 and NO. In the N_2 molecule symmetrical σ_{2s} and σ_{2s}^{*} orbitals are formed by combination of two $2s$ orbitals that have identical energies. In the NO molecule unsymmetrical σ_{2s} and σ_{2s}^{*} orbitals are formed by combination of the $2s$ orbital of oxygen with the higher-energy $2s$ orbital of nitrogen. The σ_{2s}^{*} orbital is closer in energy to the nitrogen $2s$ atomic orbital (Fig. 7.13); hence this antibonding molecular orbital has the greatest electron density near the nitrogen atom. The electron density of the bonding σ_{2s} orbital is greatest near the oxygen atom, because the energy of this molecular orbital is closer to that of the oxygen $2s$ atomic orbital. The pair of electrons in the σ_{2s} molecular orbital spends more time, on the average, closer to the oxygen atom than to the nitrogen atom. The electrons in the σ_{2s}^{*} molecular orbital spend more time closer to the nitrogen atom.

In spite of the asymmetry of the orbitals in heteronuclear diatomic molecules, they are filled in the same way as are orbitals of homonuclear diatomic molecules. Each molecular orbital still holds only two electrons, and the lowest-energy orbitals are filled first. The electron configuration of NO is $KK(\sigma_{2s})^2(\sigma_{2s}^{*})^2(\sigma_{2p_x})^2(\pi_{2p_y}, \pi_{2p_z})^4(\pi_{2p_y}^{*}, \pi_{2p_z}^{*})^1$. Since there are more electrons in bonding molecular orbitals with greater electron density near the oxygen atom than in antibonding orbitals with greater electron density near the nitrogen atom, the bond is polar (Section 6.6).

Molecular orbital energy diagrams like the one for NO can be used to describe the electron distribution in many other heteronuclear diatomic molecules and ions, such as CO, NO^+, CN^-, and NO^-.

To explain the bonding in a molecule like HF, it is necessary to introduce the idea that only atomic orbitals that are similar in energy can be combined to give molecular orbitals. If the energies of two atomic orbitals are very different, molecular orbitals do not form from them. In HF the $1s$ orbital of hydrogen has about the same energy as the $2p$ orbitals of fluorine (Fig. 7.15). Thus the molecular orbitals form from the hydrogen $1s$ orbital and the fluorine $2p_x$ orbital (Fig. 7.16). The energy difference between the hydrogen $1s$ orbital and the fluorine $2s$ orbital is too great for these atomic orbitals to combine.

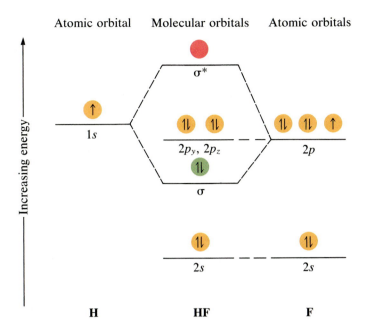

Atomic orbital   Molecular orbitals   Atomic orbitals

Figure 7.15

Molecular orbital energy diagram for HF. The filled 1s orbital of F is not shown.

When an *s* orbital combines with a *p* orbital, it does so in the head-to-tail fashion shown in Fig. 7.16. In HF this gives the bonding and antibonding orbitals shown in Fig. 7.15. The $2s$, $2p_y$, and $2p_z$ orbitals of the fluorine atom are shown at the same energies as those of the original fluorine atomic orbitals in Fig. 7.15. To a first approximation, the energies of the $2s$, $2p_y$, and $2p_z$ orbitals in the molecule do not differ from their energies in a free fluorine atom, since they do not combine with orbitals of the hydrogen atom. As may be seen in the molecular orbital energy diagram of HF (Fig. 7.15), one bonding and one antibonding orbital are formed. The eight valence electrons are located in the lowest-energy orbitals in the molecule—the $2s$, $2p_y$, and $2p_z$ atomic orbitals on fluorine and the σ bonding orbital. The molecule is stable because the two electrons in the σ orbital have a lower energy than they have in the isolated atoms.

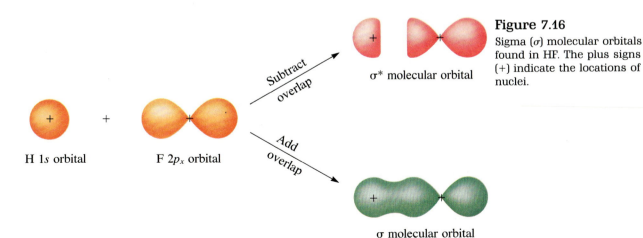

Figure 7.16

Sigma (σ) molecular orbitals found in HF. The plus signs (+) indicate the locations of nuclei.

σ^* molecular orbital

H 1s orbital   F $2p_x$ orbital

σ molecular orbital

For Review

Summary

Molecular orbital theory describes the behavior of an electron in a molecule by a wave function, which may be used to determine the energy of the electron and the region of space around the molecule where it is most likely to be found. The electron occupies an orbital that is called a **molecular orbital** since it may extend over all of the atoms in the molecule. An electron in a **bonding molecular orbital** stabilizes a molecule. An electron in an **antibonding molecular orbital** makes a molecule less stable. A σ_s molecular orbital is a bonding orbital created by the addition of the overlap of two s orbitals of adjacent atoms. A σ_p molecular orbital is a bonding orbital created from the addition of the end-to-end overlap of p orbitals of adjacent atoms. A π_p molecular orbital is a bonding orbital resulting from the addition of side-to-side overlap of p orbitals of adjacent atoms. The σ_s^*, σ_p^*, and π_p^* molecular orbitals are antibonding orbitals resulting from the subtraction of the overlap of atomic orbitals.

One order of increasing energies of the molecular orbitals in a **homonuclear diatomic molecule** is given by the electron configuration $(\sigma_{1s})(\sigma_{1s}^*)(\sigma_{2s})(\sigma_{2s}^*)(\sigma_{2p_x})$ $(\pi_{2p_y}, \pi_{2p_z})(\pi_{2p_y}^*, \pi_{2p_z}^*)(\sigma_{2p_x}^*)$. In some molecules the order of the σ_{2p} and π_{2p} levels is reversed. Electrons fill the orbitals of lowest energy first, with both π_{2p} orbitals (or π_{2p}^* orbitals) being occupied by one electron before either will accept two electrons. According to their molecular orbital electron configurations, H_2 and He_2^+ are stable, whereas He_2, which would contain the same number of bonding and antibonding electrons, is not. Similarly, Li_2, B_2, C_2, N_2, O_2, and F_2 are expected to be stable, whereas Be_2 and Ne_2 are not. Experimental evidence confirms these expectations. **Heteronuclear diatomic species** are similar to homonuclear molecules in the formation and filling of orbitals, but the molecular orbitals are not symmetrically distributed between the two atoms due to the differences in energy of the two types of atomic orbitals. Polar bonds result.

The **bond order** of a bond between two atoms is determined by the numbers of bonding and antibonding electrons it contains.

Key Terms and Concepts

antibonding orbital (7.1)
bond order (7.3)
bonding orbital (7.1)
degenerate orbitals (7.1)

heteronuclear diatomic molecule (7.14)
homonuclear diatomic molecule (7.1)
molecular orbital (7.1)
molecular orbital energies (7.2)

overlap (7.1)
pi (π) orbital (7.1)
sigma (σ) orbital (7.1)

Exercises

Molecular Orbitals

1. Why does an electron in a bonding molecular orbital formed from two atomic orbitals have a lower energy than in either of the individual atomic orbitals?

2. How do the following differ and how are they similar?
   (a) Molecular orbitals and atomic orbitals
   (b) Bonding orbitals and antibonding orbitals
   (c) σ orbitals and π orbitals

3. Formulate a rule, similar to Hund's rule (Section 5.11), for the filling of molecular orbitals and give specific examples of its application.

4. Describe the similarities and differences between σ orbitals formed from two s atomic orbitals, from two p atomic orbitals, and from an s and a p atomic orbital.

5. Draw diagrams showing the molecular orbitals that can be formed by combining two s orbitals, by combining two p orbitals, and by combining an s and a p orbital.

6. Describe the similarities and differences between the set of molecular orbitals formed by end-to-end overlap of p orbitals and the set of molecular orbitals formed by side-by-side overlap of p orbitals.

7. Consider the molecular orbitals represented by the following outlines [each plus sign (+) indicates the location of a nucleus]:

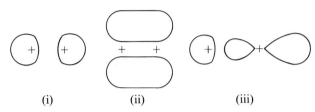

(i)	(ii)	(iii)

   (a) Indicate how each of these is formed either by addition or by subtraction of the overlap between s and/or p orbitals.

   (b) Which of these is a member of a degenerate pair of molecular orbitals?

Homonuclear Diatomic Molecules

8. Using molecular orbital energy diagrams and the electron occupancy in the molecular orbitals, compare the stability of Li_2, Be_2^+, and Be_2.

9. Consider the diatomic ions X_2^{2+}, where X is any one of the first ten elements of the Periodic Table except hydrogen. (a) Write the molecular orbital electron configurations of these ions. (b) Based on the molecular orbital electron configuration, which of these ions should not be stable?

10. The acetylide ion, C_2^{2-}, is a component of calcium acetylide. Draw the molecular orbital energy diagram of this ion and write the molecular orbital electron configuration.

11. Diatomic molecules of sulfur, S_2, are found in the vapor above molten sulfur. Draw a molecular orbital energy diagram for S_2, using only the orbitals in the valence shell of sulfur that are occupied in the free atom. What is the electron configuration of S_2?

12. Draw a molecular orbital energy diagram for the ion S_2^{2-}, using only the orbitals in the valence shell of sulfur that are occupied in the free atom. What is the electron configuration of S_2^{2-}?

Heteronuclear Diatomic Molecules

13. How does the molecular orbital energy diagram for a diatomic molecule involving atoms of two different elements differ from that for a diatomic molecule made up of two atoms of the same element?

14. Under what conditions does an s atomic orbital on one atom combine with a p atomic orbital on a second atom rather than with the s atomic orbital on the second atom?

15. What does the presence of a polar bond in a molecule imply about the shapes of its molecular orbitals?

16. Draw the molecular orbital energy diagrams for CO and NO^+. Show that CO and NO^+ are isoelectronic and have the same bond order. How does the molecular orbital electron configuration of these species compare with that for NO?

17. Draw the molecular orbital energy diagram of BrCl, using only those valence orbitals occupied in the free atoms.

18. Will the unpaired electron in NO spend more time near the nitrogen nucleus or near the oxygen nucleus, or will it spend equal amounts of time near both? Explain your answer.

19. Draw the molecular orbital energy diagram of HCl.

Bond Order

20. Determine the bond orders of the homonuclear diatomic molecules formed by the first ten elements of the Periodic Table.

21. Determine the bond orders of the homonuclear ions, X_2^{2+}, described in exercise 9.

22. Arrange the following in increasing order of bond energy: F_2, F_2^+, F_2^-.

23. The Lewis structure of the nitric oxide anion ($N\!=\!O^-$) indicates a bond order of 2 for this ion. Beginning with the molecular orbital energy diagram for NO, work out the bond order for NO^-, using molecular orbital theory.

24. What is the bond order of NO^+? Is the bond in NO^+ stronger or weaker than the one in NO?

25. Use the molecular orbital theory to determine the bond orders of O_2^+, O_2, O_2^-, and O_2^{2-}. Arrange these species in order of increasing bond energy.

26. Use the molecular orbital theory to determine the bond orders of N_2^{2+}, N_2, and N_2^{2-}. Arrange these species in order of increasing bond energy.

Additional Exercises

27. Like the ionization energy of an atom (Section 5.15), the ionization energy of a molecule is the energy necessary to remove the least tightly bound electron from the molecule. Arrange the molecules or ions in each of the following groups in increasing order of first ionization energy:
   (a) N_2, N_2^{2+}, N_2^-.
   (b) NO, NO^-, NO^+.

28. Predict whether each of the following atoms, ions, and molecules contains only paired electrons or contains some unpaired electrons.

(a) Li	(b) Be^+	(c) Be_2^{2+}
(d) N	(e) N_2	(f) O
(g) O_2	(h) CO	(i) NO
(j) NO^-	(k) H_2^+	(l) F_2^+

29. Consider the molecular orbitals represented by the following outlines (each plus sign indicates the location of a nucleus):

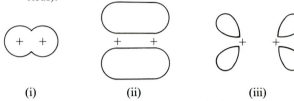

(i)   (ii)   (iii)

(a) What is the maximum number of electrons that can be placed in molecular orbital (i)? In orbital (ii)? In orbital (iii)?

(b) How many orbitals of type (i) are found in the valence shell of F_2? How many orbitals of type (iii)?

(c) What homonuclear diatomic molecule formed by an element in the third period of the Periodic Table has its two highest-energy electrons in orbitals of type (iii)?

(d) What homonuclear diatomic molecule formed by the elements in the third period of the Periodic Table has its four highest-energy electrons in orbitals like type (iii)?

30. Identify the homonuclear diatomic molecules or ions that have the following electron configurations:

(a) X_2^+: $KK(\sigma_{2s})^2(\sigma_{2s}^*)^1$

(b) X_2: $KK(\sigma_{2s})^2(\sigma_{2s}^*)^2(\pi_{2p_y}, \pi_{2p_z})^2$

(c) X_2^-: $KK(\sigma_{2s})^2(\sigma_{2s}^*)^2(\pi_{2p_y}, \pi_{2p_z})^1$

(d) X_2^-: $KK(\sigma_{2s})^2(\sigma_{2s}^*)^2(\sigma_{2p_x})^2(\pi_{2p_y}, \pi_{2p_z})^3$

31. The following outline indicates a bonding molecular orbital in the molecule AB. Which element, A or B, is more electronegative? Explain your answer.

Molecular Structure and Hybridization

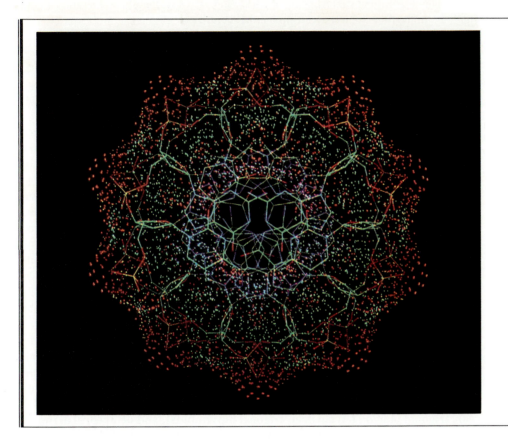

Computer-generated
model of DNA
showing tetrahedral,
trigonal planar, and
bent geometries.

The properties of molecules often reflect the spatial distribution of their constituent atoms. For example, we can digest starch from wheat and corn, but not the cellulose that forms the bodies of these plants. Starch and cellulose have the same chemical formula and are practically identical except for small differences in the three-dimensional arrangement of their atoms. This three-dimensional arrangement has a very important role in the chemical behavior of starch and cellulose, as well as in other molecules. In this chapter we will examine how to predict the three-dimensional arrangement of atoms in a molecule—the molecular structure of the molecule.

When one or more atoms are located around a central atom, the orbitals of that central atom behave differently than they do in a free atom. For example, the atomic orbitals on a free carbon atom are different from those on a carbon atom in a CO_2 molecule or a CH_4 molecule. Once we are able to predict the molecular structure about an atom, we can describe the changes in its atomic orbitals. The concept of hybridization is used to describe these changes, and this concept will also be discussed in this chapter.

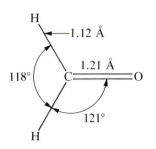

Figure 8.1

Bond distances and angles in the formaldehyde molecule, H_2CO (1 Å = 100 pm = 10^{-10} m).

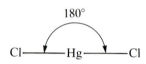

Figure 8.2

The linear structure of the $HgCl_2$ molecule. The bonds are on opposite sides of the Hg atom.

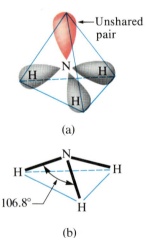

Figure 8.3

(a) Regions occupied by the unshared pair (shown in red) and bonds (black) in NH_3 and (b) the resulting trigonal pyramidal molecular structure.

Valence Shell Electron-Pair Repulsion Theory

8.1 Prediction of Molecular Structures

The three-dimensional, geometrical arrangement of the atoms in a molecule is called its **molecular structure,** or **molecular geometry.** Molecular structures are normally described in terms of bond distances and bond angles. A **bond distance** is the distance between the nuclei of two bonded atoms along a straight line joining the nuclei. Bond distances are measured in picometers (1 pm = 10^{-12} m) or in angstrom units (1 Å = 100 pm = 10^{-10} m). A **bond angle** is the angle between any two bonds that include a common atom. The bond distances and angles in the formaldehyde molecule, H_2CO, are illustrated in Fig. 8.1.

Approximate bond angles around a central atom in a molecule can be predicted from the number of bonds and unshared electron pairs on the central atom using the **valence shell electron-pair repulsion theory (VSEPR theory).** The electrons in the valence shell of the central atom of a molecule are present in regions of high electron density either as bonding pairs, located primarily between bonded atoms, or as unshared pairs occupying regions of space shaped rather like those occupied by the bonding pairs. The VSEPR theory is based on the idea that the electrostatic repulsion of these electrons is reduced to a minimum when the various regions of high electron density assume positions as far from each other as possible. For example, in a molecule such as $HgCl_2$ with only two bonds to the central atom, the bonds are as far apart as possible, and the electrostatic repulsion between these regions of high electron density are reduced to a minimum when they are on opposite sides of the central atom. The bond angle is 180° (Fig. 8.2). Another example is illustrated in Fig. 8.3, which shows the regions of space occupied by the unshared pair and bonding pairs in the ammonia molecule. The repulsions among the four regions of electron density in NH_3 are minimized if they are directed toward the corners of a tetrahedron, as shown in Fig. 8.3(a). This and other geometries that reduce the repulsions among regions of high electron density (bonds and/or unshared pairs) to a minimum are illustrated in Table 8.1. **Linear, trigonal planar, tetrahedral, trigonal bipyramidal, and octahedral arrangements of bonds and unshared electron pairs result for two, three, four, five, and six regions of high electron density, respectively.**

If the regions of high electron density are not identical, the bond angles may differ by several degrees from the ideal values given in Table 8.1. Nevertheless, the structures given are usually good approximations for the distribution of electron density around a central atom. In NH_3, for example, the central nitrogen atom is surrounded by three N—H single bonds and one unshared pair of electrons. Since the regions of high electron density are not alike (three are bonds and one is an unshared pair), the bond angle is 106.8°, as shown in Fig. 8.3(b), instead of 109.5°, as shown in Table 8.1 for symmetrical tetrahedral structures. In formaldehyde, H_2CO, the regions of high electron density consist of two single bonds and one double bond, and the bond angles differ from one another by about 3° (Fig. 8.1).

Even though the arrangement of unshared pairs and bonds in a molecule may correspond to one of the arrangements in Table 8.1, its molecular structure may look different. The presence of an unshared pair affects the structure of a molecule, but the unshared pair is invisible to the experimental techniques used to determine structure.

Table 8.1   Arrangement of unshared pairs and bonds as a result of electron-pair repulsions

Two regions of high electron density (bonds and/or unshared pairs)	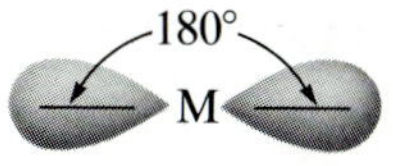	Linear. 180° angle.
Three regions of high electron density (bonds and/or unshared pairs)	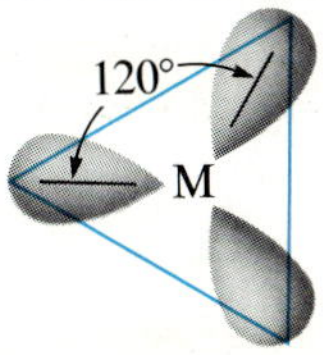	Trigonal planar. All angles 120°.
Four regions of high electron density (bonds and/or unshared pairs)	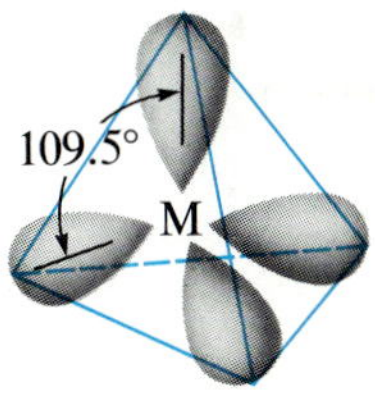	Tetrahedral. All angles 109.5°.
Five regions of high electron density (bonds and/or unshared pairs)	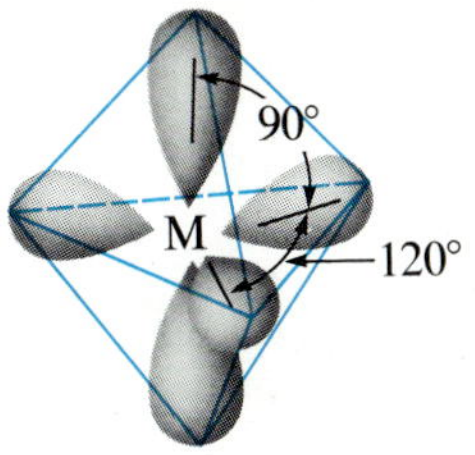	Trigonal bipyramidal. Angles of 90° or 120° (an attached atom may be equatorial, in the plane of the triangle, or axial, above or below the plane of the triangle). 180°
Six regions of high electron density (bonds and/or unshared pairs)	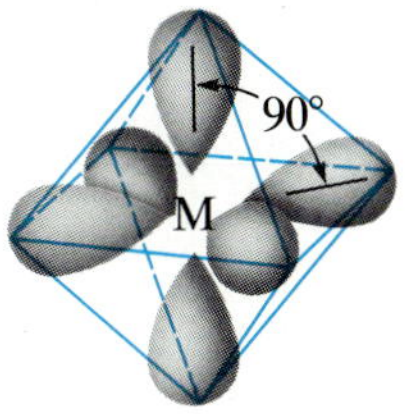	Octahedral. All angles 90°.

Consequently, the molecular structure is described as if the unshared pair were not there. The NH_3 molecule, for example, has a tetrahedral arrangement of regions of electron density, Fig. 8.3(a). However, one of these regions is an unshared pair and is not observed when the molecular structure is determined experimentally. The molecular structure (the arrangement of atoms only) is a trigonal pyramid, Fig. 8.3(b), with the nitrogen atom at the apex and three hydrogen atoms forming the base. Table 8.2 illustrates the structures that are observed for various combinations of unshared pairs and bonds.

For several of the examples in Table 8.2, a different arrangement of the locations of unshared pairs and bonds would give a different molecular structure. For example, the molecular structure of ClF_3 is T-shaped, as shown in Fig. 8.4(a). However, two other possible arrangements for the three bonds and two unshared pairs could be written: a trigonal planar arrangement, Fig. 8.4(b), and a trigonal pyramidal arrangement,

Table 8.2   Molecular structures based on the valence shell electron-pair repulsion theory

Regions of high electron density (bonds and unshared pairs)	Molecular structures and examples (chemical bonds are indicated in black, unshared pairs in red)			
Three: trigonal planar arrangement of bonds and/or unshared pairs	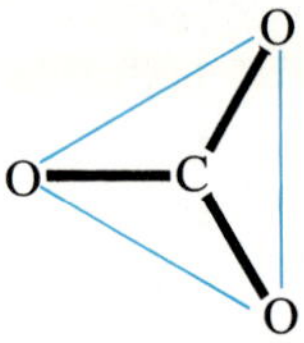 3 Bonds 0 Unshared pairs Trigonal planar CO_3^{2-}, BF_3, NO_3^-	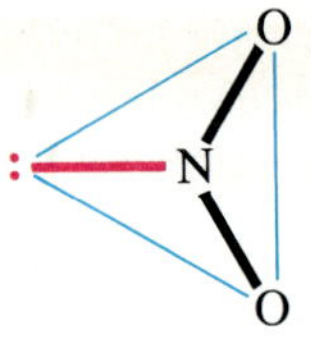 2 Bonds 1 Unshared pair Angular (120°) NO_2^-, ClNO		
Four: tetrahedral arrangement of bonds and/or unshared pairs	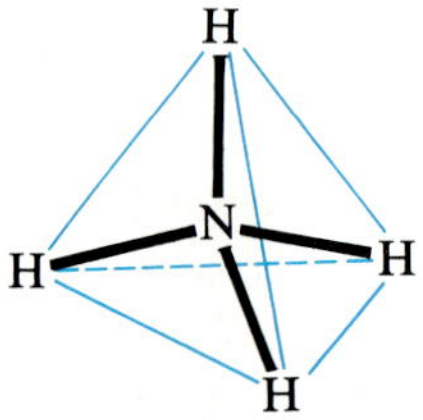 4 Bonds 0 Unshared pairs Tetrahedral NH_4^+, CH_4	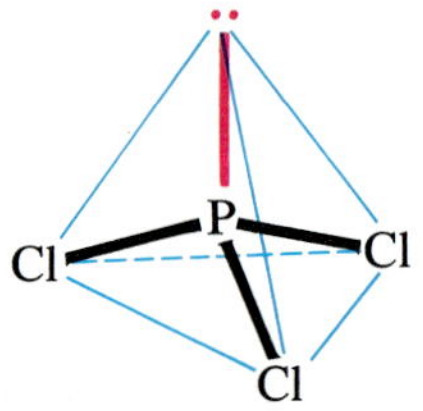 3 Bonds 1 Unshared pair Trigonal pyramidal H_3O^+, PCl_3, NH_3	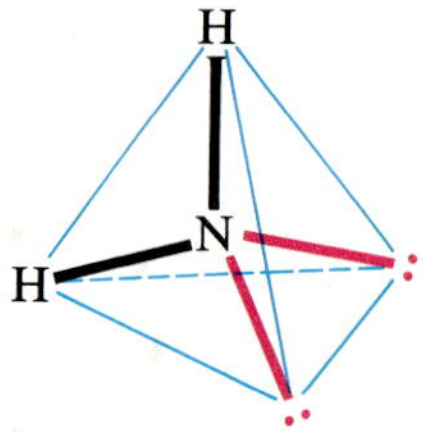 2 Bonds 2 Unshared pairs Angular (109.5°) NH_2^-, H_2O	
Five: trigonal bipyramidal arrangement of bonds and/or unshared pairs	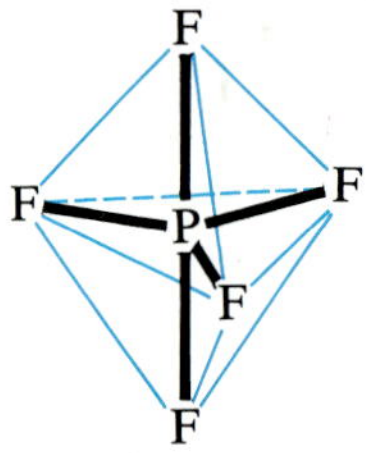 5 Bonds 0 Unshared pairs Trigonal bipyramidal PF_5, $SnCl_5^-$	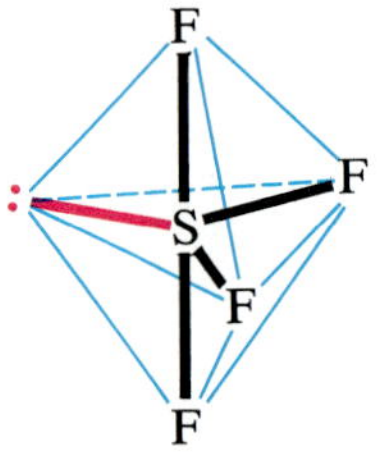 4 Bonds 1 Unshared pair Seesaw SF_4, ClF_4^+	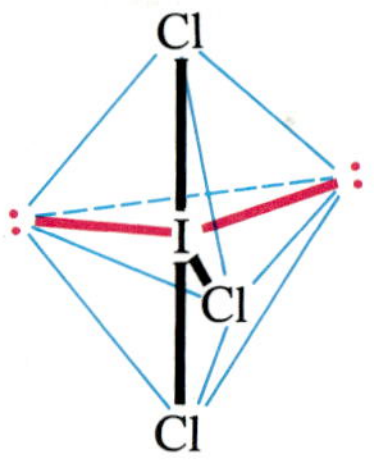 3 Bonds 2 Unshared pairs T-shaped ICl_3, ClF_3	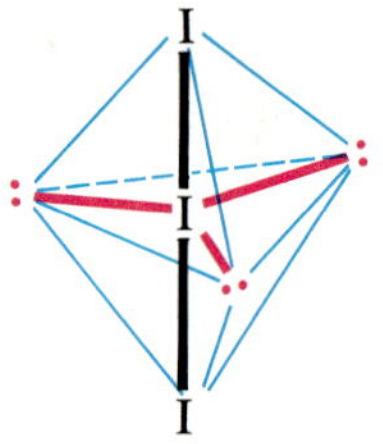 2 Bonds 3 Unshared pairs Linear I_3^-, ClF_2^-
Six: octahedral arrangement of bonds and/or unshared pairs	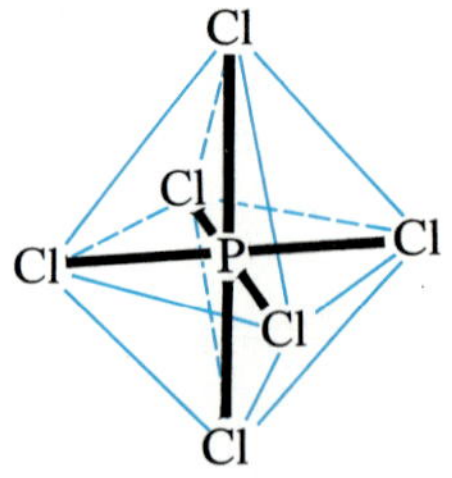 6 Bonds 0 Unshared pairs Octahedral PCl_6^-, SF_6, IF_6^+	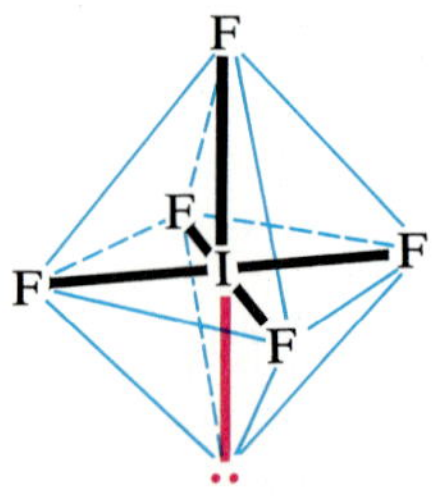 5 Bonds 1 Unshared pair Square pyramidal IF_5, XeF_5^+	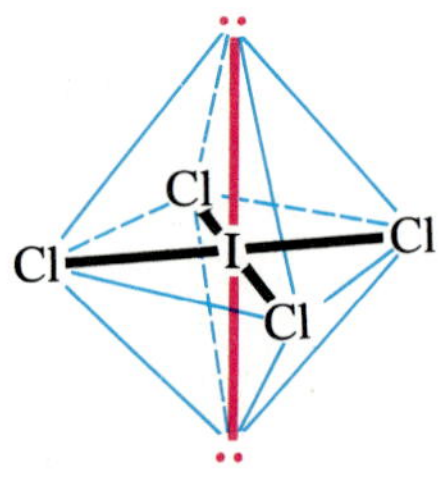 4 Bonds 2 Unshared pairs Square planar ICl_4^-, XeF_4	

Fig. 8.4(c). The stable structure is the one that puts the unshared pairs as far apart as possible and thus minimizes electron-pair repulsions. In trigonal bipyramidal arrangements (such as that of ClF_3), the positions that minimize repulsions are those in the triangular plane of the molecule [shown in Fig. 8.4(a)], because the electrons in these positions are 120° apart.

The amount of repulsive force varies with the kinds of electron pairs involved. The repulsive forces between two unshared pairs are the largest, between an unshared pair and a bond are intermediate, and between two bonds are smallest. An alternative way of thinking about these repulsions is that unshared pairs require more room than bonds. When several possible structures can be written, the one that minimizes as many of the strong repulsions as possible will be the most stable.

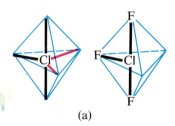

(a)

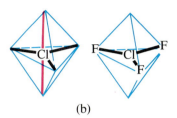

(b)

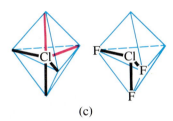

(c)

Figure 8.4

Possible arrangements of two unshared pairs (shown in red) and three bonds (black) in ClF_3, and also the resulting molecular structures showing only the positions of the atoms without the unshared pairs. (a) shows the stable *T*-shaped arrangement.

8.2 Rules for Predicting Molecular Structures

To use the VSEPR theory to predict molecular structures, we apply the following procedure:

1. Write the Lewis structure of the molecule as described in Section 6.7.
2. Count the regions of high electron density (unshared pairs and chemical bonds) around the central atom in the Lewis structure. A single, double, or triple bond counts as one region of high electron density. An unpaired electron counts as an unshared pair.
3. Identify the most stable arrangement of the regions of high electron density as linear, trigonal planar, tetrahedral, trigonal bipyramidal, or octahedral (Table 8.1).
4. If more than one arrangement of unshared pairs and chemical bonds is possible, choose the one that will minimize unshared-pair repulsions. In trigonal bipyramidal arrangements repulsion is minimized when every unshared pair is in the plane of the triangle. In an octahedral arrangement with two unshared pairs, repulsion is minimized when the unshared pairs are on opposite sides of the central atom.
5. Identify the molecular structure (the arrangement of atoms) from the locations of the atoms at the ends of the bonds (Table 8.2).

Predict the molecular structure of a gaseous $BeCl_2$ molecule.

EXAMPLE 8.1

The Lewis structure of $BeCl_2$ is

The central beryllium atom has no unshared pairs but is bonded to two atoms. There are therefore two regions of high electron density around it. Two regions of high electron density arrange themselves on opposite sides of the central atom with a bond angle of 180° (Table 8.1). $BeCl_2$ should be a linear molecule.

Predict the molecular structure of BCl_3.

EXAMPLE 8.2

The Lewis structure of BCl_3 is

$$: \overset{..}{\underset{..}{Cl}} :$$
$$|$$
$$: \overset{..}{\underset{..}{Cl}} - B - \overset{..}{\underset{..}{Cl}} :$$

BCl_3 contains three bonds, and there are no unshared pairs on boron. The arrangement of three regions of high electron density will be trigonal planar (Table 8.1). The bonds in BCl_3 should lie in a plane with 120° angles between them, a trigonal planar arrangement (Fig. 8.5).

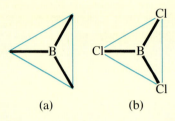

(a)   (b)

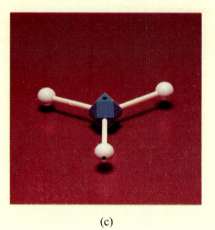

(c)

Figure 8.5
(a) Trigonal planar arrangement of three bonds in BCl_3, (b) the resulting trigonal planar molecular structure, and (c) a model of BCl_3.

EXAMPLE 8.3   Predict the molecular structure of NH_4^+.

The Lewis structure of NH_4^+ is

$$H-\overset{\displaystyle H}{\underset{\displaystyle H}{N}}-H^+$$

NH_4^+ contains four bonds from the nitrogen atom to hydrogen atoms. Four regions of high electron density arrange themselves so that they point at the corners of a tetrahedron with the central atom in the middle (Table 8.1). The hydrogen atoms located at the ends of the bonds are located at the corners of a tetrahedron. The molecular structure of NH_4^+ (Fig. 8.6) is tetrahedral.

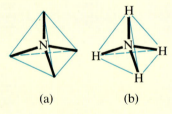

(a)   (b)

(c)

Figure 8.6
(a) Tetrahedral arrangement of four bonds in NH_4^+, (b) the resulting tetrahedral molecular structure, and (c) a model of the NH_4^+ ion.

EXAMPLE 8.4   Predict the molecular structure of H_2O.

The Lewis structure of H_2O,

$$
\begin{array}{c}
\text{H} \\
| \\
\text{:O—H} \\
\cdot\cdot
\end{array}
$$

indicates that there are four regions of high electron density around the oxygen atom—two unshared pairs and two chemical bonds. These four regions are arranged in a tetrahedral fashion [Fig. 8.7(a)], as indicated in Table 8.1. However, the arrangement of the atoms themselves (the molecular structure) in H_2O is angular with a bond angle of approximately 109.5° [Table 8.2 and Fig. 8.7(b)].

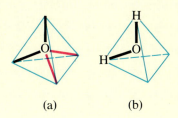

(a)   (b)

Figure 8.7

(a) Tetrahedral arrangement of two unshared pairs (shown in red) and two bonds (black) in H_2O and (b) the resulting bent molecular structure.

Predict the molecular structure of SF_4.

EXAMPLE 8.5

The Lewis structure of SF_4,

indicates five regions of high electron density about the sulfur atom—one unshared pair and four chemical bonds. These five regions are directed toward the corners of a trigonal bipyramid (Table 8.1). In order to minimize unshared pair-bond pair repulsions, the unshared pair occupies one of the locations in the plane of the triangle. The molecular structure (Fig. 8.8) is that of a seesaw (Table 8.2).

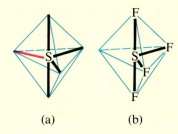

(a)   (b)

Figure 8.8

(a) Trigonal bipyramidal arrangement of one unshared pair (shown in red) and four bonds (black) in SF_4 and (b) the resulting seesaw-shaped molecular structure.

Predict the molecular structure of IF_4^-.

EXAMPLE 8.6

The Lewis structure of IF_4^-,

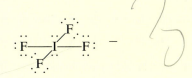

indicates six regions of high electron density around the iodine atom—two unshared pairs and four bonds. These six regions adopt an octahedral arrangement. The two possible arrangements of unshared pairs and bonds are illustrated in Figs. 8.9(a) and 8.9(b). To minimize repulsions the unshared pairs should be on opposite sides of the central atom, so the structure in Fig. 8.9(a) is the more stable of the two. The five atoms are all in the same plane and have a square planar configuration.

Figure 8.9

(a), (b) Possible octahedral arrangements of two unshared pairs (shown in red) and four bonds (black) in IF_4^- and (c) the stable square planar molecular structure.

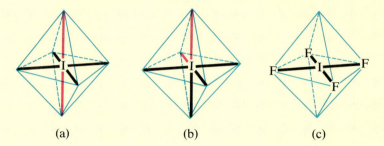

(a)   (b)   (c)

EXAMPLE 8.7

Predict the molecular structure of ClO_2F.

The least electronegative element, chlorine, is the central element, as shown in this Lewis structure:

$$
\begin{array}{c}
:\!\overset{\displaystyle ..}{O}\!: \\
| \\
:\!Cl\!-\!\overset{..}{\underset{..}{F}}\!: \\
| \\
:\!\overset{..}{\underset{..}{O}}\!:
\end{array}
$$

The four regions of high electron density are arranged tetrahedrally. The atoms are arranged to give a trigonal pyramidal structure, as shown in Fig. 8.10.

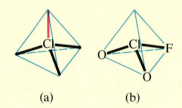

(a)   (b)

Figure 8.10

(a) Tetrahedral arrangement of one unshared pair (shown in red) and three bonds in ClO_2F and (b) the resulting trigonal pyramidal molecular structure.

Hybridization of Atomic Orbitals

Many of the molecules discussed in previous sections contain identical bonds joining a central atom with from two to six other atoms. These identical bonds form from various atomic orbitals on the central atom and can be described by hybridization, a concept introduced by Linus Pauling in 1935. Nonequivalent atomic orbitals can **hybridize** (combine) to give a set of equivalent hybrid orbitals whose orientation reflects the geometry of the molecule or ion. The theory uses atomic rather than molecular orbitals, but it is useful in explaining why the bonds in many molecules are equivalent and how the atomic orbitals on a central atom of a molecule interact with the orbitals on the other atoms.

8.3 Methane, CH$_4$, and Ethane, C$_2$H$_6$ (sp^3 Hybridization)

A molecule of methane, CH$_4$, consists of a carbon atom surrounded by four hydrogen atoms at the corners of a tetrahedron (see Fig. 8.12). However, an uncombined carbon atom has only two unpaired electrons in its ground state (lowest-energy state). The electron configuration of a carbon atom is $1s^2 2s^2 2p^2$, which we will write as $1s^2 2s^2 2p^1 2p^1 2p^0$ to emphasize that two of the $2p$ orbitals contain a single electron and the third is empty. With this configuration (two unpaired electrons) we might expect carbon to form only two covalent bonds, but it usually forms four, as in methane. How does the carbon atom get four equivalent electrons to form these four equivalent covalent bonds? The answer is by promoting (exciting) one of the electrons from the $2s$ orbital to the empty $2p$ orbital producing the electron configuration $1s^2 2s^1 2p^1 2p^1 2p^1$. The energy required for **promotion of electrons** is more than offset by the energy given off during bond formation. Therefore the energy state after electron promotion *and* bond formation is lower than it was before these steps; the decrease in energy corresponds to a more stable state.

The electron distribution in the ground state and after promotion can be shown as follows for the carbon atom. Each orbital is represented by a circle and each electron by an arrow; arrows pointing in opposite directions designate electrons of opposite spin.

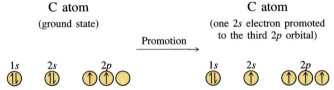

The three p orbitals of an uncombined carbon atom are oriented at right angles (Section 5.10); the s orbital is nondirectional because of its spherical shape. The four carbon-hydrogen bonds of methane, however, are equal in length and in strength and are pointed to the corners of a regular tetrahedron. When a carbon atom is bonded to four other atoms at the corners of a tetrahedron, the one $2s$ orbital and the three $2p$ orbitals of the carbon atom hybridize (combine) to form four equivalent hybrid orbitals, whose lobes point to the corners of the tetrahedron (Fig. 8.11). These are called sp^3 hybrid orbitals, to designate the hybridization of one s orbital and three p orbitals; sp^3 **hybridization** always results in a tetrahedral orientation of four orbitals.

When a methane molecule forms, each of the four hydrogen atoms shares its one electron (indicated by a colored arrow) with the electron in one of the four sp^3 hybrid orbitals.

CH$_4$ molecule
Tetrahedral structure

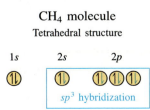

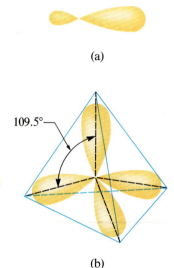

Figure 8.11
(a) An individual sp^3 hybrid orbital. (b) The four tetrahedral sp^3 hybrid orbitals of the carbon atom. For clarity, only the large lobe of each hybrid orbital is shown.

Thus the 1*s* orbital of each of the four hydrogen atoms overlaps with one of the four *sp*³ orbitals of the carbon atom to form an *sp*³-*s* sigma (σ) bond. (Notice the similarity between this and the overlap to produce molecular orbitals, Section 7.1.) This results in the formation of four very strong equivalent covalent bonds between one carbon atom and four hydrogen atoms to produce the methane molecule, CH₄ (Fig. 8.12).

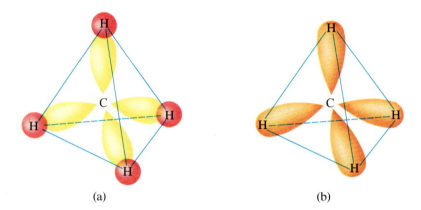

Figure 8.12

The methane molecule. (a) Diagram showing the overlap of the four tetrahedral *sp*³ hybrid orbitals (yellow) with four *s* orbitals (red) of four hydrogen atoms to produce the methane molecule.
(b) The overall outline of the four bonding orbitals (orange) in methane.

(a)

(b)

The structure of ethane, C₂H₆, is similar to that of methane in that each carbon in ethane has four neighboring atoms arranged at the corners of a tetrahedron—three H atoms and one C atom (Fig. 8.13). In ethane an *sp*³ orbital of one carbon atom overlaps end-to-end with an *sp*³ orbital of a second carbon atom to form a σ bond. Each of the other three *sp*³ hybrid orbitals of each carbon atom overlaps with an *s* orbital of a hydrogen atom to form additional σ bonds. The structure and overall outline of the bonding orbitals of ethane are shown in Fig. 8.13. Ethane is made up of two tetrahedra with one corner in common.

An *sp*³ hybrid orbital can also hold an unshared pair of electrons. For example, the nitrogen atom in ammonia [Fig. 8.3(a)] is surrounded by three bonding pairs of electrons and an unshared pair, all directed to the corners of a tetrahedron. Thus the nitrogen atom may be regarded as being *sp*³ hybridized with one hybrid orbital occupied by the unshared pair (represented by the pair of black arrows).

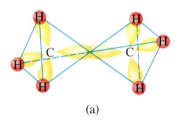

(a)

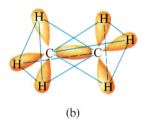

(b)

Figure 8.13

The ethane molecule. (a) The overlap diagram for the ethane molecule. (b) The overall outline of the seven bonding orbitals (orange) in ethane.

NH₃ molecule

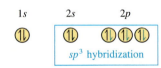

The molecular structure of water (Fig. 8.7) is consistent with a tetrahedral arrangement of two unshared pairs and two bonding pairs of electrons. Thus we say that the oxygen atom is *sp*³ hybridized, with two of the hybrid orbitals occupied by unshared pairs and two by bonding pairs.

Any atom with a tetrahedral arrangement of unshared pairs and bonding pairs may be regarded as being *sp*³ hybridized. But note that hybridization can occur *only* when the atomic orbitals involved have very similar energies. It is possible to hybridize 2*s* and 2*p* orbitals, or 3*s* and 3*p* orbitals, for example, but not 2*s* and 3*p* orbitals.

8.4 Beryllium Chloride, $BeCl_2$ (sp Hybridization)

The electron configuration of beryllium, Be, is $1s^2 2s^2$, and it appears that in its ground state the atom should not form covalent bonds at all. However, in an excited state an electron from the 2s orbital has been promoted to a 2p orbital so that the configuration is $1s^2 2s^1 2p^1$, which means that two unpaired electrons are available for forming covalent bonds with atoms that can share electrons.

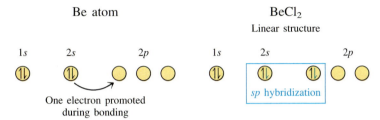

In the gaseous state $BeCl_2$ is a linear molecule, Cl—Be—Cl; all three atoms lie in a straight line. The middle atom in a linear set of three atoms hybridizes so that the valence s orbital and one of the valence p orbitals combine to give two equivalent sp hybrid orbitals with a linear (straight-line) geometry (Fig. 8.14). In $BeCl_2$ these sp orbitals overlap with p orbitals of the chlorine atoms to form σ bonds.

Other atoms that exhibit sp hybridization include the mercury atom in the linear $HgCl_2$ molecule (Fig. 8.2) and the zinc atom in $Zn(CH_3)_2$, which contains a linear C—Zn—C arrangement.

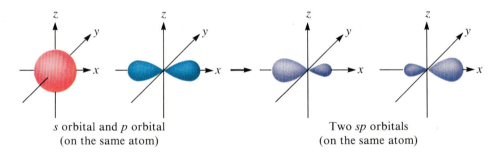

s orbital and p orbital (on the same atom)   Two sp orbitals (on the same atom)

Figure 8.14

Hybridization of an s orbital (red) and a p orbital of the same atom (blue) to produce two sp hybrid orbitals (purple).

8.5 Boron Trifluoride, BF_3 (sp^2 Hybridization)

An atom surrounded by a trigonal planar arrangement of unshared pairs and bonding pairs can form sp^2 hybrid orbitals directed at the corners of an equilateral triangle. One s orbital and two p orbitals combine to form these sp^2 hybrid orbitals.

The structure of BF_3 suggests sp^2 hybridization for boron in this compound. Experimental evidence shows that there are three equivalent B—F bonds in boron trifluoride. All four atoms of this molecule lie in the same plane, with the boron atom in the center and the three fluorine atoms at the corners of an equilateral triangle (a trigonal planar structure, Fig. 8.15).

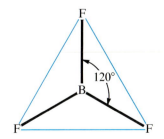

Figure 8.15

The trigonal planar structure of BF_3.

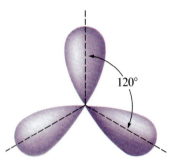

Figure 8.16
The shape and spatial orientation of trigonal planar, sp^2, hybrid orbitals.

The electron configuration of boron in the ground state is $1s^2 2s^2 2p^1$. During a chemical reaction with fluorine, a $2s$ electron of boron appears to be promoted to a $2p$ orbital, giving the configuration $1s^2 2s^1 2p^1 2p^1$. The $2s$ orbital and two of the $2p$ orbitals hybridize to form three sp^2 hybrid orbitals (see Fig. 8.16).

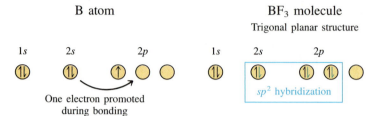

Other atoms that exhibit sp^2 hybridization include the boron atom in BCl_3 (Fig. 8.5), the nitrogen atom in NO_3^- and ClNO, the sulfur atom in SO_2, and the carbon atom in CH_3^+.

8.6 Other Types of Hybridization

An atom with either a trigonal bipyramidal or an octahedral arrangement of bonding electron pairs and unshared pairs cannot form hybrid orbitals using only its four s and p valence orbitals. To accommodate five pairs of electrons in a trigonal bipyramidal arrangement, an atom must hybridize five atomic orbitals—the s orbital, the three p orbitals, and one of the d orbitals in its valence shell (or shells)—giving five sp^3d hybrid orbitals. With an octahedral arrangement of six electron pairs, an atom hybridizes six atomic orbitals—the s orbital, the three p orbitals, and two of the d orbitals in its valence shell—giving six sp^3d^2 hybrid orbitals.

In a molecule of phosphorus(V) chloride, PCl_5 (Fig. 8.17), there are five P—Cl bonds (and thus five pairs of valence electrons about the phosphorus atom) directed toward the corners of a trigonal bipyramid. The electron configuration of an uncombined phosphorus atom is $1s^2 2s^2 2p^6 3s^2 3p^1 3p^1 3p^1 3d^0$. To form a set of sp^3d hybrid orbitals on the phosphorus atom in PCl_5, one of the s valence electrons must be promoted into a d orbital and then the singly occupied orbitals must hybridize to give the five sp^3d orbitals.

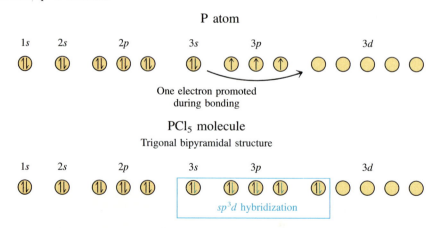

Other atoms that exhibit sp^3d hybridization include the sulfur atom in SF_4 (Fig. 8.8) and the chlorine atom in ClF_3 [Fig. 8.4(a)] and in ClF_4^+.

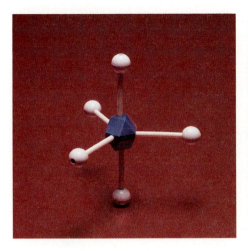

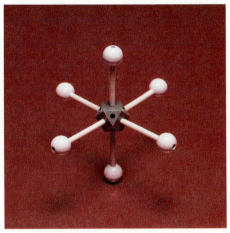

Figure 8.17 (left)
Structure of phosphorus (V) chloride, PCl_5; sp^3d hybridization.

Figure 8.18 (right)
The octahedral configuration of SF_6; sp^3d^2 hybridization.

The sulfur atom in sulfur(VI) fluoride, SF_6, exhibits sp^3d^2 hybridization. A molecule of sulfur(VI) fluoride has six fluorine atoms surrounding a single sulfur atom (Fig. 8.18). To bond six fluorine atoms, the sulfur atom must provide six bonding orbitals. The electron configuration of sulfur in the ground state is $1s^2 2s^2 2p^6 3s^2 3p^2 3p^1 3p^1 3d^0 3d^0$. Promotion of an s and a p valence electron to d orbitals gives the configuration $1s^2 2s^2 2p^6 3s^1 3p^1 3p^1 3p^1 3d^1 3d^1$, and hybridization of the singly occupied orbitals gives a set of six sp^3d^2 hybrid orbitals directed toward the corners of an octahedron.

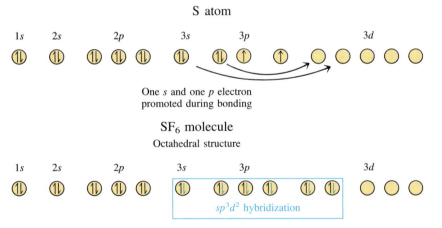

S atom

One s and one p electron promoted during bonding

SF_6 molecule
Octahedral structure

sp^3d^2 hybridization

Other atoms that exhibit sp^3d^2 hybridization include the phosphorus atom in PCl_6^-, the iodine atom in IF_6^+, IF_5, ICl_4^- (Table 8.2), and IF_4^- [Fig. 8.9(a)], and the xenon atom in XeF_4.

8.7 Prediction of Hybrid Orbitals

Hybridization is a model that describes how hybrid orbitals occur and result in placement of electrons in the regions between bonded atoms in a given geometry. In the preceding four sections we have described hybrid orbitals for the common distributions of bonding electrons and unshared pairs. These were assigned after the geometry

Table 8.3   Geometrical arrangements characteristic of hybrid orbitals

Hybrid orbitals		Geometrical arrangement
sp	Linear (180° angle)	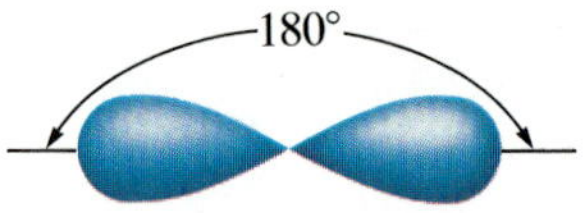
sp^2	Trigonal planar (120° angle)	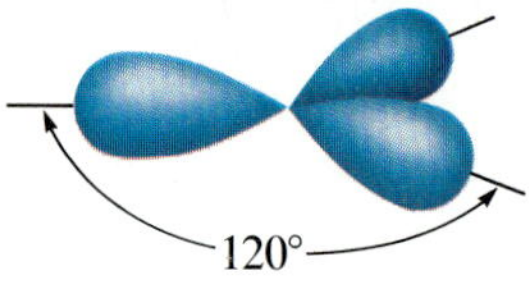
sp^3	Tetrahedral (109.5° angle)	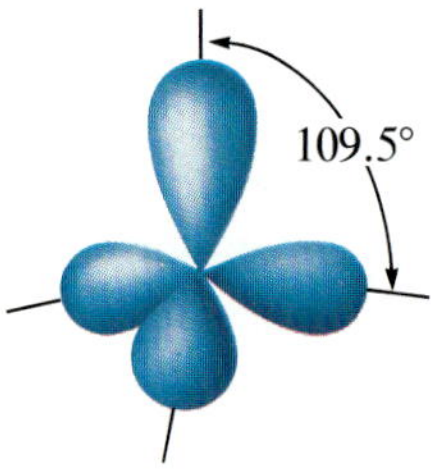
sp^3d	Trigonal bipyramidal (120° and 90° angles)	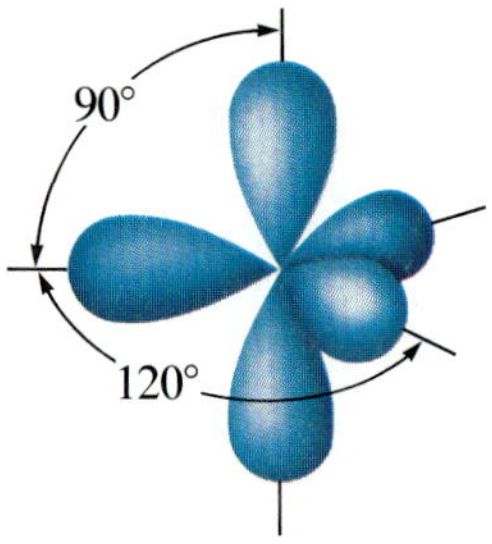
sp^3d^2	Octahedral (90° angle)	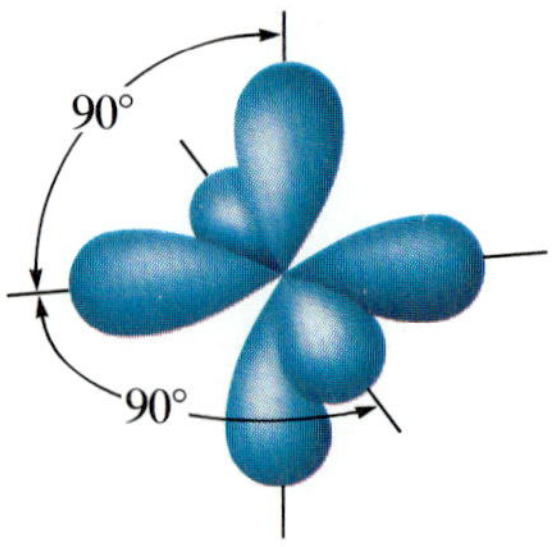

around the hybridized atom had been determined. Hybrid orbitals have little predictive value—we cannot say in advance, for example, that the oxygen atom in water must use sp^3 hybrid orbitals.

The geometrical arrangements characteristic of the various sets of hybrid orbitals are shown in Table 8.3. To find the hybridization of a central atom, we first determine the geometry of its regions of high electron density and then assign the set of hybridized orbitals from Table 8.3 that corresponds to this arrangement.

EXAMPLE 8.8

What is the hybridization of the nitrogen atom in the ammonium ion, NH_4^+?

As described in Example 8.3, the nitrogen atom in NH_4^+ has a tetrahedral arrangement of regions of high electron density. This corresponds to sp^3 hybridization of nitrogen (Table 8.3).

EXAMPLE 8.9

Urea, $NH_2C(O)NH_2$, is sometimes used as a source of nitrogen in fertilizers. What is the hybridization of the nitrogen and carbon atoms in urea?

The Lewis structure of urea is

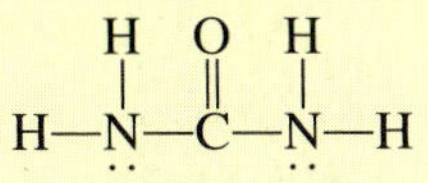

The nitrogen atoms are surrounded by four regions of high electron density, which will arrange themselves in a tetrahedral geometry (Table 8.1). The hybridization in a tetrahedral arrangement is sp^3 (Table 8.3). This is the hybridization of the nitrogen atoms in urea.

The carbon atom is surrounded by three regions of electron density, positioned in a trigonal planar arrangement (Table 8.1). The hybridization in a trigonal planar arrangement is sp^2 (Table 8.3), which is the hybridization of the carbon atom in urea.

8.8 Ethylene, C_2H_4, and Acetylene, C_2H_2 (π Bonding)

If we apply the VSEPR theory (Section 8.1) to the Lewis structure of ethylene, C_2H_4,

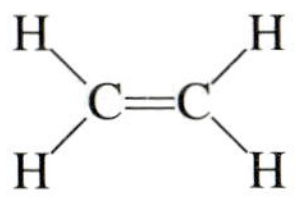

each carbon atom should be surrounded by one carbon atom and two hydrogen atoms, forming a trigonal planar array. Thus the σ bonds from each carbon atom should be formed using a set of sp^2 hybrid orbitals directed toward the apexes of a triangle. These orbitals form the C—H single bonds and one part of the C=C double bond, the σ bond (Fig. 8.19). The second part of the C=C bond, which gives it a bond order of 2, results from the $2p$ orbital that is not involved in the sp^2 hybridization.

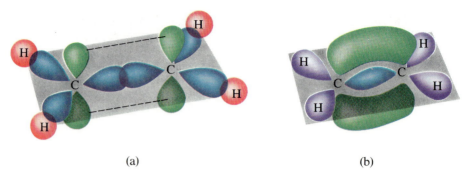

(a)   (b)

Figure 8.19

The ethylene molecule. (a) The overlap diagram of three sp^2 hybrid orbitals (blue) on carbon atoms and four s orbitals (red) from four hydrogen atoms. There are four C—H σ bonds and a carbon-carbon double bond involving one C—C σ bond and one C—C π bond. The dashed lines, each connecting two lobes, indicate the side-by-side overlap of the two unhybridized p orbitals (green). The sp^2 hybrid orbitals lie in a plane with the unhybridized p orbitals extending above and below the plane and perpendicular to it. (b) The overall outline of the bonding molecular orbitals in ethylene. The two portions of the π bonding orbital (shown in green), resulting from the side-by-side overlap of the unhybridized p orbitals, are above and below the plane. Carbon-hydrogen bonds are shown in purple.

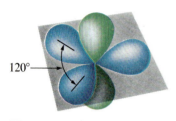

Figure 8.20

Diagram illustrating the three trigonal sp^2 hybrid orbitals of the carbon atom, which lie in the same plane, and the one unhybridized p orbital (shown in green), which is perpendicular to the plane.

Only two $2p$ orbitals hybridize with the $2s$ orbital, so one $2p$ orbital is left unhybridized. As shown in Fig. 8.20, the unhybridized p orbital (shown in green) is perpendicular to the plane of the sp^2 orbitals. Thus when two sp^2 hybridized carbon atoms come together, one sp^2 orbital on each of them can overlap to form a σ bond, while the unhybridized $2p$ orbitals can overlap in a side-by-side fashion. This overlap results in a pi (π) bond of the type discussed in Chapter 7. The overlap to form the π bond is not as efficient as the overlap to form the σ bond, and the π bond is therefore weaker than the σ bond. The two carbon atoms of ethylene are thus bound together by two kinds of bonds—one σ and one π—giving a double bond.

Note that in an ethylene molecule, the four hydrogen atoms and the two carbon atoms are all in the same plane. If the two planes of sp^2 hybrid orbitals are tilted, the π bond will be weakened since the p orbitals that form it cannot overlap effectively if they are not parallel. A planar configuration for the ethylene molecule is the most stable form.

As we have seen for $BeCl_2$, if just one $2p$ orbital hybridizes with a $2s$ orbital, two sp hybrid orbitals result. This arrangement leaves two $2p$ orbitals unhybridized (Fig. 8.21). When sp hybrid orbitals of two carbon atoms combine, they overlap end-to-end

Figure 8.21

Diagram of the two linear sp hybrid orbitals of the carbon atom, which lie in a straight line, and the two unhybridized p orbitals (shown in green).

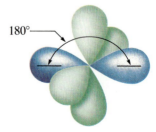

to form a σ bond (Fig. 8.22). The remaining *sp* orbital on each carbon may be used to bond with another atom such as hydrogen, forming a linear molecule such as acetylene, H—C≡C—H. In addition to this, as indicated in Fig. 8.22, the two sets of unhybridized *p* orbitals are positioned so that they overlap side-by-side and hence form two π bonds. The two carbon atoms of acetylene are thus bound together by one σ bond and two π bonds, giving a triple bond.

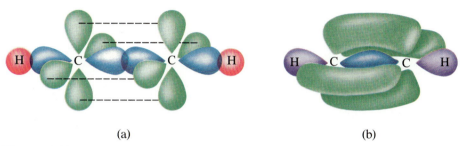

(a)   (b)

Figure 8.22

The acetylene molecule. (a) The overlap diagram of two *sp* hybrid carbon atoms and two *s* orbitals from two hydrogen atoms. There are two C—H σ bonds and a carbon-carbon triple bond involving one C—C σ bond and two C—C π bonds. The dashed lines, each connecting two lobes, indicate the side-by-side overlap of the four unhybridized *p* orbitals. The *sp* hybrid orbitals are shown in blue, and the unhybridized *p* orbitals are shown in green. (b) The overall outline of the bonding orbitals in acetylene. The π bonding orbitals (in green) are positioned with one above and below the line of the σ bonds and the other behind and in front of the line of the σ bonds.

For Review

Summary

The **molecular structure,** or three-dimensional arrangement of the atoms in a molecule or ion, is described in terms of **bond distances** and **bond angles.** Approximate bond angles can be predicted using the **valence shell electron-pair repulsion (VSEPR) theory.** According to this theory, regions of high electron density (either bonding pairs or unshared pairs) located about a central atom repel each other, and the most stable arrangement is reached when they are as far away from each other as possible. For two regions of high electron density, the most stable is a **linear structure;** for three, a **trigonal planar structure;** for four, a **tetrahedral structure;** for five, a **trigonal bipyramidal structure;** and for six, an **octahedral structure.** Since an unshared pair of electrons is undetectable by the techniques used to determine molecular structure, the expected arrangement of the regions of high electron density and the actual molecular structure may appear different. In ammonia three bonding pairs and one unshared pair are in a tetrahedral arrangement, but the nitrogen and hydrogen atoms form a trigonal pyramid. Other geometries are illustrated in Table 8.2.

The orbitals a central atom uses to form bonds with other atoms may be described as **hybrid orbitals,** combinations of some or all of its valence atomic orbitals. These hybrid orbitals may form **sigma (σ) bonds** directed toward other atoms of the molecule or may contain unshared pairs. The type of hybridization about a central atom may be determined from the geometry of the regions of high electron density about it. Two such regions imply sp hybridization; three, sp^2 hybridization; four, sp^3 hybridization; five, sp^3d hybridization; and six, sp^3d^2 hybridization.

Atomic orbitals that are not used in hybridization are available to form **pi (π) bonds.** The carbon atoms in ethylene, C_2H_4, are sp^2 hybridized, and the remaining p orbital on each is used to form a π bond, resulting in a double bond. The carbon atoms in acetylene, HCCH, are sp hybridized, and the two remaining p orbitals on each are used to form two π bonds, resulting in a triple bond.

Key Terms and Concepts

bond angle (8.1)
bond distance (8.1)
double bond (8.8)
hybridization (8.3)
linear structure (8.1)
molecular structure (8.1)

octahedral structure (8.1)
pi (π) bond (8.8)
promotion of electrons (8.3)
sigma (σ) bond (8.3, 8.8)
tetrahedral structure (8.1)
trigonal bipyramidal structure (8.1)

trigonal planar structure (8.1)
triple bond (8.8)
valence shell electron-pair repulsion (VSEPR) theory (8.1)

Exercises

Molecular Structure

1. How many regions of high electron density are required around an atom to form a linear arrangement of these regions about the atom? A trigonal planar arrangement? A tetrahedral arrangement? A trigonal bipyramidal arrangement? An octahedral arrangement? What are the angles between the regions in each of these arrangements?

2. Predict the structure of each of the following molecules:
   (a) BH_3 (b) BeH_2
   (c) PCl_5 (d) SeF_6
   (e) SiH_4

3. Predict the structure of each of the following ions:
   (a) XeF_6^{2+} (b) BF_4^-
   (c) CH_3^+ (d) SF_5^+
   (e) BH_2^+

4. Predict the structure of each of the following molecules:
   (a) PH_3 (b) BrF_5
   (c) SeF_4 (d) H_2O
   (e) IF_3 (f) XeF_4

5. Predict the structure of each of the following ions:
   (a) CH_3^- (b) IF_4^-
   (c) ClF_4^+ (d) NH_2^-
   (e) $SnCl_3^-$

6. Predict the structure of each of the following molecules:
   (a) CO_2 (b) IOF_5
   (c) F_2CO (d) SO_2F_2

7. Predict the structure of each of the following molecules or ions. In each case xenon or the least electronegative atom is the central atom.
   (a) NO_2^- (b) SOF_2 (c) SeO_2
   (d) FNO (e) $ClSO^+$ (f) ClO_3^-
   (g) XeO_2F_2

8. Predict, the geometry around the indicated atom or atoms:
   (a) The nitrogen atom in nitrous acid, HNO_2 (HONO).
   (b) The nitrogen atom and the two carbon atoms in glycine, the simplest amino acid. The skeleton structure of glycine is

   (c) The nitrogen atoms in hydrazine, H_2NNH_2.
   (d) The phosphorus atom in phosphoric acid, H_3PO_4 [$(HO)_3PO$].
   (e) The central oxygen atom in the ozone molecule, O_3.
   (f) The carbon atom in the freon molecule, CF_2Cl_2.
   (g) The carbon atoms in cyclohexane. The skeleton structure of cyclohexane is

(h) The carbon atom in the carbonate ion, CO_3^{2-}; the bicarbonate ion, HCO_3^- ($HOCO_2^-$); and carbonic acid, H_2CO_3 [$(HO)_2CO$].

9. Which of the following molecules, which contain only two carbon atoms and several hydrogen atoms, would contain the strongest carbon-carbon bond: (a) a molecule in which both C atoms have a tetrahedral molecular structure about them; (b) a molecule in which both C atoms have a trigonal planar structure about them; or (c) a molecule in which both C atoms have a linear structure about them?

Hybrid Orbitals

10. What is a hybrid orbital? Illustrate your answer with the hybrid orbitals that may be formed by carbon.

11. What are the angles between the hybrid orbitals in each of the following sets: sp, sp^2, sp^3, sp^3d, sp^3d^2?

12. Identify the hybridization of each carbon atom in each of the following molecules:
    (a) vinyl chloride, the compound used to make the plastic PVC (polyvinyl chloride), H_2C=$CHCl$
    (b) carbon tetrafluoride, CF_4
    (c) acetic acid, the compound that gives vinegar its acidic taste,

$$H-\overset{\overset{\displaystyle H}{|}}{\underset{\underset{\displaystyle H}{|}}{C}}-\overset{\overset{\displaystyle O}{\|}}{C}-OH$$

    (d) calcium cyanamide, $Ca^{2+}(NCN^{2-})$, used as a fertilizer and in preparation of the plastic Melmac

13. Identify the hybridization of the indicated atom in each of the following molecules:
    (a) S in SF_4 (b) Sb in $SbCl_5$ (c) B in BH_3
    (d) N in NO_2^+ (e) S in SO_4^{2-}

14. Identify the hybridization of the indicated atom in each of the following molecules or ions:
    (a) N in NH_2^- (b) O in H_3O^+ (c) S in SO_3^{2-}
    (d) P in PCl_6^- (e) I in IF_4^+ (f) Cl in ClF_3

15. Show the distribution of electrons in the valence orbitals of the central atom in each of the following molecules or ions (1) when it is the free atom in the ground state, (2) following electron promotion, and (3) following bond formation:
    (a) SiH_4 (b) ClF_3
    (c) AlH_3 (d) IF_5
    (e) IF_6^+ (consider this ion as being formed from I^+ and 6 F atoms)

16. Formaldehyde, H_2CO, contains a carbon-oxygen double bond. Describe the hybrid orbitals of the carbon atoms that form the σ bonds in this molecule and the atomic orbitals of oxygen and carbon that give the π bond. Note that it is not necessary to hybridize the oxygen atom.

17. Hydrogen cyanide, HCN, contains a carbon-nitrogen triple bond. Describe the hybrid orbitals of the carbon atom that form the σ bonds in this molecule and the atomic orbitals

of nitrogen and carbon that give the π bonds. Note that it is not necessary to hybridize the nitrogen atom.

18. Write Lewis structures for the molecules and ions listed below. Utilizing both the valence shell electron-pair repulsion theory and the concept of hybridization, and remembering that single bonds are σ bonds, double bonds are a σ plus a π bond, and triple bonds are a σ plus two π bonds, indicate the type of hybridization of each central atom in the listed molecules and ions.
    (a) BO_3^{3-} (b) SO_2
    (c) ClNO   (d) BF_4^-
    (e) SO_4^{2-} (f) PH_3
    (g) SeO_2 (h) N_2O_4 (contains an N—N bond)
    (i) $PO_2F_2^-$ (j) BrCN

19. Some forms of solid phosphorus(V) chloride contain PCl_4^+ and PCl_6^- ions. Upon heating, gaseous molecules of PCl_5 are formed. Describe the hybridization of the phosphorus atom in PCl_4^+, PCl_6^-, and PCl_5.

20. Which of the following carbon-carbon bonds would be the strongest: (a) a bond between two sp^3 hybridized carbon atoms, (b) a bond between two sp^2 hybridized carbon atoms, or (c) a bond between two sp hybridized carbon atoms?

21. The hybridization of phosphorus in a molecule containing 1 P atom and several F atoms is sp^3. What is the formula of the compound?

Additional Exercises

22. Predict the geometry about the indicated atoms in each of the following compounds and describe its hybridization:
    (a) The sulfur in sulfuric acid, H_2SO_4 [$(HO)_2SO_2$].
    (b) The nitrogen atom in pyridine, C_5H_5N. The skeleton structure of pyridine is

$$\begin{array}{c}
H \qquad\qquad H \\
\diagdown \qquad\quad \diagup \\
C-C \\
\diagup \qquad\qquad \diagdown \\
H-C \qquad\qquad N \\
\diagdown \qquad\qquad \diagup \\
C-C \\
\diagup \qquad\qquad \diagdown \\
H \qquad\qquad H
\end{array}$$

    (c) The boron atom in trimethylboron, $B(CH_3)_3$.
    (d) The sulfur atoms in disulfur decafluoride, S_2F_{10} (contains an S—S bond).
    (e) The oxygen atoms in hydrogen peroxide, HOOH.
    (f) The arsenic atom in arsenic trichloride difluoride, $AsCl_3F_2$.
    (g) The carbon and internal oxygen atoms in the industrial solvent methyl acetate, with the skeleton structure.

$$H-\overset{\overset{\displaystyle H}{|}}{\underset{\underset{\displaystyle H}{|}}{C}}-\overset{\overset{\displaystyle O}{\|}}{C}-O-\overset{\overset{\displaystyle H}{|}}{\underset{\underset{\displaystyle H}{|}}{C}}-H$$

23. XF_3 has a trigonal planar molecular structure. To which main group of the Periodic Table does X belong?

24. In the compound H—X=O, X exhibits sp^2 hybridization. To which main group of the Periodic Table does X belong?

25. If XCl_3 exhibits a dipole moment, X could be a member of either of two groups in the Periodic Table. Which ones?

26. Elemental sulfur consists of covalently bonded, puckered rings of eight sulfur atoms, S_8. The S—S—S angles in the ring are 107.9°. The production of sulfuric acid, $SO_2(OH)_2$, from sulfur involves oxidation of sulfur to SO_2, then to SO_3, and finally reaction with water to give sulfuric acid. Trace the changes in the hybridization of the sulfur atom during this sequence of reactions.

27. Although the location of the unshared pairs in its hybrid orbitals cannot be observed, the oxygen atom in water is believed to exhibit sp^3 hybridization. Explain.

28. One-half mol of a compound contains 0.5 mol of C atoms and 1.0 mol of S atoms. Identify the molecular structure of the compound and the hybridization of the carbon atom.

29. One mol of S_8 reacts exactly with 16 mol of F_2 to give a compound containing one sulfur and several fluorine atoms in each molecule. Describe the hybridization of the sulfur atom in the molecules.

30. In 1986, 3.7 million tons of a compound with a molecular weight of 32 and containing 37.5% C, 12.5% H, and 50.0% O were prepared in the United States for use as a solvent and to manufacture other chemicals. Write the Lewis structure of this compound, describe the geometry around the C and O atoms in the molecule, and identify the hybridization of the C and O atoms.

31. A compound with a molecular weight of 30 contains 40% C, 6.7% H, and 53.3% O. Determine the molecular formula of the compound, write its Lewis structure, and identify the hybridization of the C atom.

Chemical Reactions and the Periodic Table

9

The reaction of potassium with water.

Chemistry is the study of the properties of matter and of the reactions that convert matter from one form to another. In the preceding four chapters we have considered some electronic and structural properties of atoms, molecules, and ions. In this chapter we will examine their chemical properties. In particular, we will look at the kinds of chemical reactions that interconvert common elements and compounds, and we will see how the products of many common types of chemical reactions can be predicted.

Over 1.6×10^{11} kilograms (350,000,000,000 pounds) of inorganic elements and chemicals and 8.5×10^{10} kilograms (188,000,000,000 pounds) of organic chemicals were produced in the United States during 1986. Large amounts were also produced in other industrialized countries. The chemistry of several of these elements and compounds will be used to illustrate common and important types of chemical behavior.

9.1 Chemical Behavior and the Periodic Table

Much of the behavior of an element can be predicted from its position in the Periodic Table. As an example, consider the use that Mendeleev made of his periodic table in predicting the properties of unknown elements. Mendeleev's table included only 62 elements, the number known at that time. It also contained six places that he left vacant so that known elements of similar chemical properties would fall in the same group. Mendeleev predicted the properties of the elements that would fit these gaps from the known chemical behavior of their neighbors in the Periodic Table. The elements are scandium (Sc), gallium (Ga), germanium (Ge), technetium (Tc), rhenium (Re), and polonium (Po), and they all have properties similar to those predicted by Mendeleev.

A comparison of the properties predicted by Mendeleev for germanium, which he called eka-silicon, and those determined experimentally for the element after it was isolated is given in Table 9.1. You can see that the properties are intermediate between those of silicon and tin, the neighbors of germanium in Group IVA.

Table 9.1   Predicted properties for eka-silicon and observed properties of silicon, tin, and germanium

	Predicted for eka-silicon	Silicon	Germanium	Tin
Atomic weight	72	28	72.59	118
Specific gravity	5.5	2.3	5.3	7.3
Color	Gray metal	Gray nonmetal	Gray metal	White metal
Oxidation number with oxygen	4	4	4	4
Reaction with acid	Very slow reaction	No reaction	Slow reaction with conc. acid	Slow reaction
Reaction with base	Slow reaction	Slow reaction	Slow reaction with conc. base	Slow reaction
Formula of chloride	$EkCl_4$	$SiCl_4$	$GeCl_4$	$SnCl_4$
Specific gravity of chloride	1.9	1.5	1.88	2.2
Boiling point of chloride (°C)	below 100°	57.6°	83°	114°

You can apply the same principles Mendeleev used to predict the properties of elements to correlate and recall their behavior. The Periodic Table provides a powerful framework for organizing the chemical behavior of the elements.

9.2 Classification of Chemical Compounds

Before we can begin our examination of the properties of elements, we need to consider several of the types of compounds that they form.

1. SALTS. A **salt** is an ionic compound composed of positive ions (cations) and negative ions (anions). (See Section 6.1.) Simple salts are formed by the combination of a metal and a nonmetal—the metal forms the cation, and the nonmetal, the anion. In more complex salts the cation may be a polyatomic positive ion such as NH_4^+ or PCl_4^+, and the anion a polyatomic negative ion such as NO_3^- or SO_4^{2-}. Ionic compounds that contain hydroxide or oxide ions are called bases rather than salts. (See Part 2 of this section.)

Compounds that are composed of ions are held together in the solid state by ionic bonds, strong electrostatic attractions between oppositely charged ions (Section 6.1). When soluble salts dissolve in water, the ions separate (Fig. 9.1) and are free to move about independently. The process for the dissolution (dissolving) of the salt sodium chloride in water may be represented by the following equation, in which the abbreviation *aq* indicates that the ions are separated, surrounded by water molecules, and moving independently:

$$NaCl(s) \xrightarrow{H_2O(l)} Na^+(aq) + Cl^-(aq) \tag{1}$$

2. ACIDS AND BASES. A compound that donates a hydrogen ion (H^+), or proton, to another compound is called a **Brønsted acid,** or simply an **acid,** after Johannes Brønsted, a Danish chemist. A compound that accepts a hydrogen ion is called a **Brønsted base,** or simply a **base.** Other definitions have been developed that emphasize other aspects of the behavior of acids and bases (see Chapter 16, Section 16.15), but for now we will concentrate on this one. The following reactions illustrate the transfer of a hydrogen ion from a Brønsted acid to a Brønsted base:

$$\begin{array}{cc} \text{Acid} & \text{Base} \\ HCl(g) + NH_3(g) & \longrightarrow NH_4Cl(s) \end{array} \tag{2}$$

$$H_2SO_4(l) + NaOH(s) \longrightarrow NaHSO_4(s) + H_2O(l) \tag{3}$$

Salts are produced in these reactions: NH_4Cl contains the ions NH_4^+ and Cl^-, and $NaHSO_4$ contains Na^+ and HSO_4^-.

When dissolved in water, acids donate hydrogen ions to water molecules, forming **hydronium ions, H_3O^+,** plus whatever anion is produced when the acid loses a hydrogen ion (Fig. 9.2).

$$HCl(g) + H_2O(l) \longrightarrow H_3O^+(aq) + Cl^-(aq) \tag{4}$$

$$H_2SO_4(l) + H_2O(l) \longrightarrow H_3O^+(aq) + HSO_4^-(aq) \tag{5}$$

$$HSO_4^-(aq) + H_2O(l) \longrightarrow H_3O^+(aq) + SO_4^{2-}(aq) \tag{6}$$

Thus an acid forms hydronium ions when dissolved in water. For convenience the hydronium ion is sometimes written as H^+ or $H^+(aq)$, but remember that H^+ and $H^+(aq)$ are abbreviations; a hydrogen ion in water is always associated with at least one water molecule. Note that in forming a hydronium ion water accepts a hydrogen ion and therefore behaves as a base.

When a base is added to an aqueous solution of an acid, protons are donated from the hydronium ions to the base. For example, a solution of hydrogen chloride in water contains hydronium ions and chloride ions. If gaseous ammonia is bubbled through the

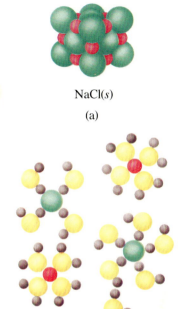

NaCl(s)

(a)

$Na^+(aq) + Cl^-(aq)$

(b)

Figure 9.1

(a) The distribution of ions in solid sodium chloride. Red spheres represent Na^+; green spheres represent Cl^-. Each ion is in contact with six ions of opposite charge. (b) The distribution of ions in an aqueous solution of sodium chloride. Each ion is free to move independently because it is surrounded by water molecules and separated from ions of opposite charge.

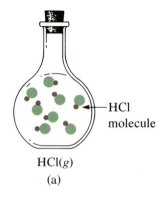

HCl molecule

HCl(g)

(a)

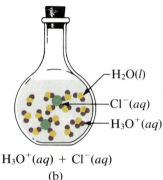

H₂O(l)

Cl⁻(aq)

H₃O⁺(aq)

$H_3O^+(aq) + Cl^-(aq)$

(b)

Figure 9.2

(a) Molecules of hydrogen chloride in the gas phase. Each chlorine atom (green sphere) is bonded to a hydrogen atom (brown sphere) by a covalent bond. (b) A solution of hydrogen chloride in water (hydrochloric acid). The hydrogen ions are bonded to water molecules by a coordinate covalent bond giving a solution of H_3O^+ and Cl^- ions.

solution, the ammonia molecules pick up protons from the hydronium ions and thereby form ammonium ions, NH_4^+, which dissolve in the solution.

$$NH_3(g) + H_3O^+(aq) + Cl^-(aq) \longrightarrow NH_4^+(aq) + Cl^-(aq) + H_2O(l) \qquad (7)$$

The product of this reaction is a solution of ammonium chloride, NH_4Cl, a salt.

An acid is classified as strong or weak, depending on the extent to which it reacts with water to form hydronium ions. A **strong acid** dissolves in water giving a 100% yield (or very nearly so) of hydronium ion and the anion of the acid. The strong acid HCl dissolves in water and gives a 100% yield of H_3O^+ and Cl^-.

$$HCl(g) + H_2O(l) \longrightarrow H_3O^+(aq) + Cl^-(aq) \qquad (8)$$

Sulfuric acid, H_2SO_4, dissolves in water to give a 100% yield of hydronium ion and the hydrogen sulfate ion, HSO_4^-.

The six common strong acids and the principal anions they form in water are listed in Table 9.2

Table 9.2 Common strong acids and their anions

Molecular formula	Name	Anion	Name of anion
HCl	Hydrogen chloride	Cl^-	Chloride
HBr	Hydrogen bromide	Br^-	Bromide
HI	Hydrogen iodide	I^-	Iodide
HNO_3	Nitric acid	NO_3^-	Nitrate
$HClO_4$	Perchloric acid	ClO_4^-	Perchlorate
H_2SO_4	Sulfuric acid	HSO_4^-	Hydrogen sulfate

A **weak acid** ionizes in dilute solution, giving a small percent yield of hydronium ion (ordinarily 10% or less). Thus a solution of a weak acid consists primarily of covalent molecules (or ions) of the original acid, with lesser amounts of hydronium ion and the anion of the acid. When acetic acid, CH_3CO_2H, a weak acid, is added to water, the majority of the acid remains as acetic acid molecules. Only about 1% of the acetic acid molecules in a 0.1 molar solution react with water to form hydronium ions and acetate ions, $CH_3CO_2^-$. When sodium hydrogen sulfate, $NaHSO_4$, is added to water, the weakly acidic hydrogen sulfate ions, HSO_4^-, react with water to give a small yield of hydronium ions and sulfate ions, but the majority do not react and remain as HSO_4^-.

Some common weak acids and the principal anions they form in water are listed in Table 9.3. A more extensive listing of weak acids is given in Appendix F.

Table 9.3 Common weak acids and their anions

Molecular formula	Name	Anion	Name of anion
CH_3CO_2H	Acetic acid	$CH_3CO_2^-$	Acetate
H_2CO_3	Carbonic acid	HCO_3^-	Hydrogen carbonate
HF	Hydrogen fluoride	F^-	Fluoride
H_2S	Hydrogen sulfide	HS^-	Hydrogen sulfide
HNO_2	Nitrous acid	NO_2^-	Nitrite
H_3PO_4	Phosphoric acid	$H_2PO_4^-$	Dihydrogen phosphate
HSO_4^-	Hydrogen sulfate ion	SO_4^{2-}	Sulfate

A **strong base** dissolves in water and gives a 100% yield (or very nearly so) of hydroxide ions and the cation of the base. Examples of strong bases are the following:

1. *Soluble metal hydroxides*. Metal hydroxides are ionic; they are composed of cations and hydroxide ions. If such hydroxides dissolve, they give a 100% yield of hydroxide ions in solution. The soluble metal hydroxides include the hydroxides of the metals of Group IA and of strontium and barium of Group IIA.

$$NaOH(s) \xrightarrow{H_2O} Na^+(aq) + OH^-(aq) \tag{9}$$

$$Ba(OH)_2(s) \xrightarrow{H_2O} Ba^{2+}(aq) + 2OH^-(aq) \tag{10}$$

2. *Soluble ionic metal oxides*. Soluble compounds that contain oxide ions, O^{2-}, are strong bases because the oxide ion reacts with water to give hydroxide ions:

$$O^{2-} + H_2O \longrightarrow 2OH^-$$

Potassium oxide, K_2O, for example, is a strong base; it contains the O^{2-} ion and dissolves in water giving hydroxide ion in 100% yield because of the reaction

$$K_2O(s) + H_2O(l) \longrightarrow 2K^+(aq) + 2OH^-(aq) \tag{11}$$

3. *Soluble compounds containing anions that can be formed by removing one or more hydrogen ions from the binary hydrogen compounds of the nonmetals of Groups IVA, VA, and VIA*. For example, the salt Li_3N, which forms from the reaction of Li with N_2, contains the N^{3-} ion, an anion that can also be formed by removing three hydrogen ions from ammonia, NH_3. This salt acts as a strong base when it reacts with water and gives a 100% yield of hydroxide ions.

$$Li_3N(s) + 3H_2O(l) \longrightarrow 3Li^+(aq) + 3OH^-(aq) + NH_3(aq) \tag{12}$$

It should also be noted that metal oxides, discussed in paragraph (2) above, can also fit in this category, because the oxide ion can be formed by removing two hydrogen ions from H_2O.

Some typical strong bases are listed in Table 9.4.

Table 9.4 Some typical strong bases and the products that result when they dissolve in water

Molecular formula	Name	Species produced upon dissolution
NaOH	Sodium hydroxide	$Na^+(aq) + OH^-(aq)$
$Ba(OH)_2$	Barium hydroxide	$Ba^{2+}(aq) + 2OH^-(aq)$
Li_2O	Lithium oxide	$Li^+(aq) + OH^-(aq)$
SrO	Strontium oxide	$Sr^{2+}(aq) + 2OH^-(aq)$
K_2S	Potassium sulfide	$2K^+(aq) + OH^-(aq) + HS^-(aq)$
$NaNH_2$	Sodium amide	$Na^+(aq) + OH^-(aq) + NH_3(aq)$
Ca_3N_2	Calcium nitride	$3Ca^{2+}(aq) + 6OH^-(aq) + 2NH_3(aq)$
$LiCH_3$	Methyl lithium	$Li^+(aq) + OH^-(aq) + CH_4(g)$

When an acid is added to a solution containing hydroxide ions, its molecules donate protons to the hydroxide ions. Adding nitric acid to a solution of barium hydroxide, for example, gives a solution of the salt barium nitrate, $Ba(NO_3)_2$, and water.

$$2HNO_3(l) + Ba^{2+}(aq) + 2OH^-(aq) \longrightarrow$$
$$Ba^{2+}(aq) + 2NO_3^-(aq) + 2H_2O(l) \tag{13}$$

A **weak base** is a base that gives only a low concentration of hydroxide ions in water. Hydroxides such as $Mg(OH)_2$, $Ca(OH)_2$, and $Al(OH)_3$ are weak bases, even though they are ionic, because they are not very soluble. Only small amounts of hydroxide ion enter the solution when they dissolve. Other weak bases are molecules or ions that may be quite soluble but give only a low yield (ordinarily 10% or less) of hydroxide ion when they react with water. Ammonia is a weak base of this kind; a solution of ammonia in water consists primarily of solvated ammonia molecules, $NH_3(aq)$. However, water acts as an acid with ammonia, and ammonia accepts protons from water to a very limited extent.

$$NH_3(aq) + H_2O(l) \longrightarrow NH_4^+(aq) + OH^-(aq) \qquad (14)$$

A one molar solution of NH_3 contains 99.6% $NH_3(aq)$ and only 0.4% $NH_4^+(aq)$. Ammonia is regarded as a weak base since it gives only a low yield (0.4%) of hydroxide ion in water. When an acid is added, protons are transferred both to the ammonia molecules and to the hydroxide ions. However, the principal reaction (99.6% of the reactions that occur) is the reaction of hydrogen ion with ammonia molecules producing ammonium ions, NH_4^+, and anions.

$$HNO_3(l) + NH_3(aq) \longrightarrow NH_4^+(aq) + NO_3^-(aq) \qquad (15)$$

The anion of a weak acid is also a base; it will accept a proton to reform the weak acid. For example, sodium acetate, the sodium salt of acetic acid, reacts with a solution of perchloric acid to form acetic acid and sodium perchlorate.

$$NaCH_3CO_2 + HClO_4 \longrightarrow CH_3CO_2H + NaClO_4 \qquad (16)$$

When the salt of a weak acid dissolves in water, the anions will accept protons from the water to a very limited extent. An acetate ion (for example, from sodium acetate) will react with water and give a small yield of acetic acid molecules and hydroxide ions.

$$CH_3CO_2^-(aq) + H_2O(l) \longrightarrow CH_3CO_2H(aq) + OH^-(aq) \qquad (17)$$

A 0.05 M solution of sodium gives only a 0.01% yield of acetic acid molecules and hydroxide ion.

Note that water can behave as either an acid [Equation (14)] or a base [Equations (4) through (6)], depending upon the nature of the substance dissolved in it. This dual behavior is characteristic of other compounds too.

3. ELECTROLYTES AND NONELECTROLYTES. **Electrolytes** are compounds that dissolve and give solutions that contain ions. Ionic compounds such as sodium hydroxide and potassium nitrate (which dissolve in water giving solutions of ions), or covalent compounds such as sulfuric acid and ammonia (which react with water to form ions) are electrolytes. Ionic compounds and those covalent compounds (such as strong acids) that give essentially a 100% yield of ions are called **strong electrolytes.** Compounds (such as weak acids and ammonia) that give a low percentage yield of ions in water, are called **weak electrolytes. Nonelectrolytes** are compounds that do not ionize when they dissolve in water. Only covalent compounds can be nonelectrolytes. Many compounds of carbon, such as methane, CH_4, benzene, C_6H_6, ethanol, C_2H_5OH, ether, $(C_2H_5)_2O$, and formaldehyde, CH_2O, are nonelectrolytes. A few inorganic compounds such as nitrous oxide, N_2O, phosphine, PH_3, and nitrogen(III) chloride, NCl_3, are nonelectrolytes.

9.3 Classification of Chemical Reactions

Just as it is convenient to classify the elements as metals and nonmetals, it is convenient to classify chemical reactions. Several common types of reactions are discussed below. However, note that the classifications overlap, and some reactions fall into more than one class.

1. ADDITION, OR COMBINATION, REACTIONS. An **addition reaction** occurs when two or more substances combine to give another substance.

$$S + O_2 \xrightarrow{\Delta} SO_2 \qquad (1)$$

$$Ca + Br_2 \longrightarrow CaBr_2 \qquad (2)$$

$$2K_2S + 3O_2 \xrightarrow{\Delta} 2K_2SO_3 \qquad (3)$$

$$CaO + SO_3 \longrightarrow CaSO_4 \qquad (4)$$

Remember that most compounds that contain both metals and nonmetals are ionic. Thus the products in Equations (2), (3), and (4) are salts.

2. DECOMPOSITION REACTIONS. A **decomposition reaction** occurs when one compound breaks down (decomposes) into two or more substances.

$$2HgO \xrightarrow{\Delta} 2Hg + O_2 \qquad (5)$$

$$CaCO_3 \xrightarrow{\Delta} CaO + CO_2 \qquad (6)$$

$$2Cu(NO_3)_2 \xrightarrow{\Delta} 2CuO + 4NO_2 + O_2 \qquad (7)$$

$$2NaHSO_4 \xrightarrow{\Delta} Na_2SO_4 + SO_3 + H_2O \qquad (8)$$

3. METATHETICAL REACTIONS. A **metathetical reaction** is a reaction in which two compounds exchange parts. We can expect a metathetical reaction to occur when an insoluble compound, a gas, a nonelectrolyte, or a weak electrolyte is formed as a product.

When solutions of the salts calcium chloride, $CaCl_2$, and silver nitrate, $AgNO_3$, are mixed, insoluble solid silver chloride, $AgCl$, forms and a solution of the salt calcium nitrate, $Ca(NO_3)_2$, remains.

$$Ca^{2+}(aq) + 2Cl^-(aq) + 2Ag^+(aq) + 2NO_3^-(aq) \longrightarrow$$
$$2AgCl(s) + Ca^{2+}(aq) + 2NO_3^-(aq) \quad (9)$$

Solid sodium chloride reacts with concentrated sulfuric acid to give gaseous hydrogen chloride and sodium hydrogen sulfate.

$$NaCl(s) + H_2SO_4(l) \longrightarrow HCl(g) + NaHSO_4(s) \qquad (10)$$

A solution of KNO_2 the potassium salt of the weak acid HNO_2 (nitrous acid) reacts with a solution of HCl to give a solution containing HNO_2 molecules and the salt KCl (potassium chloride).

$$\underbrace{K^+(aq) + NO_2^-(aq)}_{\text{Solution of } KNO_2} + \underbrace{H_3O^+(aq) + Cl^-(aq)}_{\text{Solution of HCl}} \longrightarrow$$

$$\underbrace{HNO_2(aq) + K^+(aq) + Cl^-(aq) + H_2O(l)}_{\text{Solution of } HNO_2 \text{ and KCl}} \quad (11)$$

4. OXIDATION-REDUCTION REACTIONS. Many of the reactions described above can also be classified as oxidation-reduction reactions, reactions involving changes in oxidation numbers. When an atom, either free or in a molecule or ion, loses electrons, it is **oxidized,** and its oxidation number (Section 6.10) increases. When an atom, either free or in a molecule or ion, gains electrons, it is **reduced,** and its oxidation number decreases.

Oxidation and reduction always occur simultaneously, for if one atom gains electrons and is reduced, a second atom must provide the electrons and thereby be oxidized. Reactions involving oxidation and reduction are referred to as **oxidation-reduction, or redox, reactions.** An example is the reaction between sodium and chlorine, in which sodium is oxidized and chlorine is reduced.

$$\overset{0}{2Na} + \overset{0}{Cl_2} \longrightarrow \overset{+1-1}{2NaCl} \tag{12}$$

As sodium is oxidized, its oxidation number increases from 0 to $+1$, as indicated by the small numbers in Equation (12). As chlorine is reduced, its oxidation number decreases from 0 to -1. Other examples of oxidation-reduction reactions are given below. In each case, the species that is oxidized is written first.

$$\overset{0}{Zn} + \overset{+1}{H_2SO_4} \longrightarrow \overset{+2}{ZnSO_4} + \overset{0}{H_2} \tag{13}$$

$$\overset{+2}{SnCl_2} + \overset{+4}{PbCl_4} \longrightarrow \overset{+4}{SnCl_4} + \overset{+2}{PbCl_2} \tag{14}$$

$$\overset{+2}{2CO} + \overset{+2}{2NO} \longrightarrow \overset{+4}{2CO_2} + \overset{0}{N_2} \tag{15}$$

The **reducing agent** in an oxidation-reduction reaction gives up electrons to another reactant and causes the other reactant to be reduced. In Equations (12) through (15) the first reactant is the reducing agent. Since a reducing agent loses electrons, it is oxidized. The **oxidizing agent** in a redox reaction gains electrons and causes the reducing agent to be oxidized. The oxidizing agent picks up electrons during a redox reaction, so it is reduced.

Although Equation (12) is an oxidation-reduction reaction, it can also be classified as an addition reaction. Other oxidation-reduction reactions that are also addition reactions include Equations (1), (2), and (3) in this section. Oxidation-reduction reactions may also be decomposition reactions, as illustrated in Equations (5) and (7) in this section. These reactions are examples of the overlap of classifications noted at the beginning of this section.

5. ACID-BASE REACTIONS. An **acid-base** reaction occurs when a hydrogen ion is transferred from a Brønsted acid to a Brønsted base. The reaction of H_2SO_4 with $Mg(OH)_2$ is an acid-base reaction.

$$H_2SO_4 + Mg(OH)_2 \longrightarrow MgSO_4 + 2H_2O$$

Hydrogen ions are transferred from H_2SO_4 to the hydroxide ions in $Mg(OH)_2$ in this acid-base reaction. Reactions (2) through (8) and (11) through (17) in Part 2 of Section 9.2 are acid-base reactions, as are Reactions (10) and (11) in Part 3 of this section.

In general, an acid-base reaction gives a high yield of product when either a strong acid or a weak acid is added to a strong base, when a strong acid is added to a weak base, or when a strong acid is added to a salt containing the anion of a weak acid [Reaction (11) in Part 3 of this section]. These reactions occur when the pure substances or their solutions are mixed.

6. REVERSIBLE REACTIONS. A **reversible reaction** is a reaction that can proceed in either direction. When acetic acid is added to water, the acetic acid reacts to form hydronium ions and acetate ions in low yield. As soon as any hydronium ions and acetate ions form, some of them react with each other to give acetic acid. This reaction is the reverse of that which occurs when acetic acid is added to water. Both reactions, forward and reverse, then proceed simultaneously. Equations describing reactions that can proceed in either direction are written with a double arrow ($\rightleftharpoons$).

$$CH_3CO_2H(aq) + H_2O(l) \rightleftharpoons H_3O^+(aq) + CH_3CO_2^-(aq) \qquad (16)$$

The oxidation-reduction reaction of nitrogen with hydrogen is also a reversible reaction.

$$N_2(g) + 3H_2(g) \rightleftharpoons 2NH_3(g) \qquad (17)$$

When a closed container of nitrogen and hydrogen at high pressure is heated at 300°C, ammonia is produced with a yield of about 50% once the reaction is complete (when the concentration of ammonia stops changing). If the container is then heated to 400°C, the reverse reaction occurs and some of the ammonia decomposes to nitrogen and hydrogen. The percentage of ammonia in the container is reduced to about 35% by the time the concentration of ammonia again stops changing. The extent to which each of the two opposite reactions takes place can be varied by changing the quantities of the reactants or by changing the temperature, pressure, or volume. The properties of reversible reactions will be considered more fully in Chapter 15. At this stage of your study, it is sufficient for you to recognize that a reversible reaction is one that can proceed in two opposing directions, and that the two opposing reactions proceed simultaneously.

9.4 Metals, Nonmetals, and Semi-Metals

The elements may be divided into three broad groups: metals, nonmetals, and semi-metals. A pure **metal** is generally a good conductor of heat and electricity. It shows a metallic luster and is malleable and ductile (that is, it can be bent or drawn into sheets or wires without breaking). A pure **nonmetal** is generally a poor conductor. It normally shows no metallic luster and is brittle and nonductile in the solid state. Some elements, called **semi-metals,** or **metalloids,** cannot be satisfactorily identified as being either metals or nonmetals, for they possess some of the properties of each. The semi-metal silicon, for example, exhibits a bright metallic luster, but it is not a good conductor and it is brittle. Fig. 9.3 shows the metal aluminum, the nonmetal sulfur, and the semi-metal silicon.

Figure 9.3

Aluminum, a metal (left), silicon, a semi-metal (right), and sulfur, a nonmetal (top).

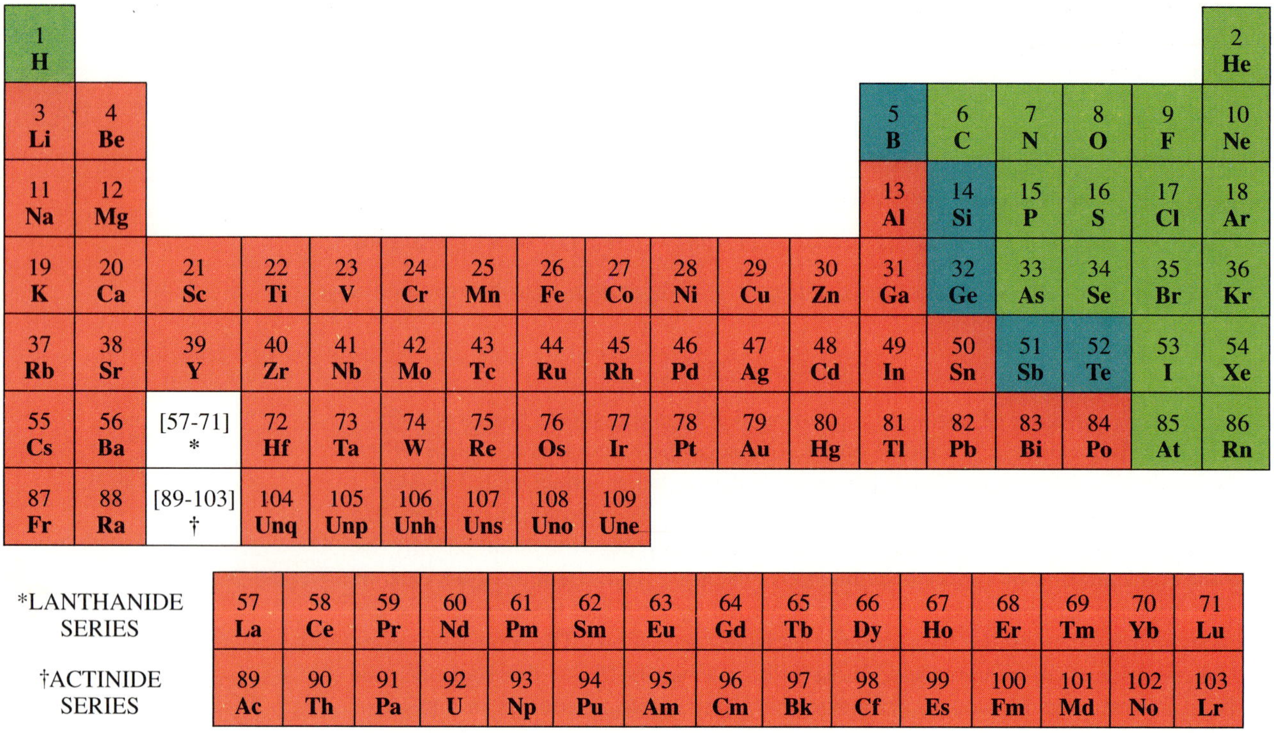

Figure 9.4

Metals, nonmetals, and semi-metals in the Periodic Table. The nonmetals are shaded in green, the semi-metals in blue, and the metals in red.

The distribution of metals, nonmetals, and semi-metals in the Periodic Table is shown in Fig. 9.4. The nonmetals lie to the right and above the heavy line running from between beryllium (Be) and boron (B) at the upper left to between polonium (Po) and astatine (At) at the lower right. The metals lie to the left and below this line. About three-fourths of the elements are metals. Semi-metals lie along the dividing line between metals and nonmetals.

If you are familiar with the general behavior patterns of metals and nonmetals, you can reasonably predict a great deal of the chemical behavior of an element simply from its position in the Periodic Table. The general types of chemical behavior characteristic of metals and nonmetals are outlined in Table 9.5. Semi-metals may exhibit either metallic or nonmetallic behavior, depending on the conditions under which they react.

As an example of metallic behavior, consider the behavior of sodium and its compounds. Sodium (Na) is clearly identifiable as a metal; it lies in Group IA at the left of the Periodic Table. It loses its one valence electron when it reduces nonmetals, giving ionic compounds containing the Na^+ ion and the anion formed by the nonmetal.

$$2Na(s) + H_2(g) \longrightarrow 2NaH(s) \qquad (Na^+ \text{ and } H^-)$$

$$2Na(s) + F_2(g) \longrightarrow 2NaF(s) \qquad (Na^+ \text{ and } F^-)$$

$$16Na(s) + S_8(s) \longrightarrow 8Na_2S(s) \qquad (Na^+ \text{ and } S^{2-})$$

$$12Na(s) + P_4(s) \longrightarrow 4Na_3P(s) \qquad (Na^+ \text{ and } P^{3-})$$

Sodium oxide, Na_2O, is a soluble ionic oxide that reacts with water giving sodium hydroxide, a base.

$$Na_2O(s) + H_2O(l) \longrightarrow 2Na^+(aq) + 2OH^-(aq)$$

Table 9.5 Chemical behavior of metals and nonmetals

Metals	Nonmetals
1. Reduce elemental nonmetals (except noble gases)	1. Oxidize elemental metals and often oxidize less electronegative nonmetals
2. Form oxides that may react with water to give hydroxides	2. Form oxides that may react with water to give acids
3. Form basic hydroxides	3. Form acidic hydroxides (oxyacids)
4. React with O_2, F_2, H_2, and other nonmetals, usually giving ionic compounds	4. React with O_2, F_2, H_2, and other nonmetals, giving covalent compounds
5. Form binary metal hydrides, which, if soluble, are strong bases	5. Form binary hydrides, which may be acidic
6. React with other metals, giving metallic compounds	6. React with metals, often giving ionic compounds
7. Exhibit lower electronegativity values	7. Exhibit higher electronegativity values
8. Have one to five electrons in outermost shell; usually not more than three	8. Usually have four to eight electrons in outermost shell
9. Readily form cations by loss of electrons	9. Readily form anions by accepting electrons to fill outermost shell (except noble gases)

The hydride of sodium, NaH, is an ionic compound containing H^- ions. The hydride ion, H^-, is a very strong base and accepts hydrogen ion from water to form OH^- and H_2.

$$NaH(s) + H_2O(l) \longrightarrow NaOH(aq) + H_2(g)$$

As an example of nonmetallic behavior, consider the behavior of chlorine and its compounds. Chlorine lies at the upper right in the Periodic Table in Group VIIA and thus may be readily identified as a nonmetal. Because chlorine is a nonmetal, we expect it to oxidize metals and to form ionic compounds that contain chloride ion. This ion forms when a chlorine atom picks up one electron and fills its valence shell. Ionic metal chlorides are formed when chlorine oxidizes metals lying to the left of it in the Periodic Table.

$$2Li(s) + Cl_2(g) \longrightarrow 2LiCl(s) \qquad (Li^+ \text{ and } Cl^-)$$
$$Ca(s) + Cl_2(g) \longrightarrow CaCl_2(s) \qquad (Ca^{2+} \text{ and } 2Cl^-)$$
$$Mn(s) + Cl_2(g) \longrightarrow MnCl_2(s) \qquad (Mn^{2+} \text{ and } 2Cl^-)$$

As a nonmetal, chlorine is also expected to oxidize less electronegative nonmetals (hydrogen and the nonmetals lying to the left and below chlorine in the Periodic Table) to form covalent molecules. This behavior is in fact observed; for example, the products in the following reactions are covalent:

$$H_2(g) + Cl_2(g) \longrightarrow 2HCl(g)$$
$$P_4(s) + 10Cl_2(g) \longrightarrow 4PCl_5(s)$$
$$Si + 2Cl_2(g) \longrightarrow SiCl_4(l)$$

As is expected for a compound of hydrogen and a nonmetal, HCl can act as an acid.

$$HCl(g) + H_2O(l) \longrightarrow H_3O^+(aq) + Cl^-(aq)$$
$$HCl(g) + NH_3(aq) \longrightarrow NH_4^+(aq) + Cl^-(aq)$$

The covalent oxide of chlorine Cl_2O_7 reacts with water to give the acid $HClO_4$.

$$Cl_2O_7 + H_2O \longrightarrow 2HClO_4$$

Written in ionic form, this reaction is as follows:

$$Cl_2O_7(l) + 3H_2O(l) \longrightarrow 2H_3O^+(aq) + 2ClO_4{}^-(aq)$$

Perchloric acid, $HClO_4$, contains three oxygen atoms and one hydroxide group (OH group) covalently bonded to the chlorine:

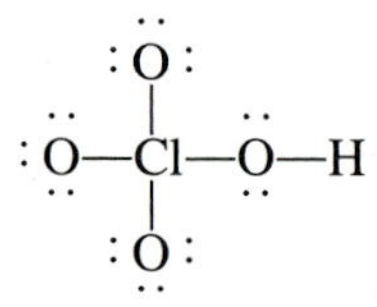

Thus $HClO_4$ may be regarded as a hydroxide of the nonmetal, chlorine. Such molecules are more commonly called **oxyacids.** Other common oxyacids that contain hydroxide groups include nitric acid, which can be expressed by either the formula HNO_3 or the formula $HONO_2$; sulfuric acid, H_2SO_4 or $(HO)_2SO_2$; and phosphoric acid, H_3PO_4 or $(HO)_3PO$.

9.5 Variation in Metallic and Nonmetallic Behavior of the Representative Elements

Elements that easily lose electrons exhibit metallic behavior. The elements with only one or two electrons in the valence shell and the heavier elements lying at the bottoms of the groups of the Periodic Table generally lose electrons most readily. Atoms of nonmetals fill their valence shells either by sharing electrons with other nonmetals or by using electrons transferred from metal atoms. These are the atoms that lie in the upper right-hand portion of the table (Section 6.2). Sodium, the active metal at the left end of the third period, enters into chemical combinations by the loss of its single valence electron. An atom of chlorine, a typical nonmetal at the right end of the third period, combines by adding one electron to its valence shell of seven electrons, either by gaining an electron from a metal or by sharing an electron with a nonmetal.

Proceeding across the third period from sodium to chlorine, we encounter five other elements. In general, going across a period, valence electrons are lost with increasing difficulty. Thus the chemical behavior becomes decreasingly metallic and increasingly nonmetallic as we go from left to right. The changeover from metallic to nonmetallic behavior is gradual, but can be said to occur between aluminum and silicon in this period. Both aluminum and silicon, however, may exhibit metallic or nonmetallic properties under the appropriate conditions. Such behavior is called **amphoteric** behavior.

To demonstrate this gradual changeover, let us examine the variation in the properties of the representative elements across the third period. Sodium, magnesium, and aluminum are shiny metals that conduct heat and electricity well. Silicon is a semimetal. It has a luster characteristic of a metal, but it is a semiconductor (a poor conductor of electricity). Phosphorus, sulfur, and chlorine are dull in appearance and are nonconducting elements. Sodium oxide, Na_2O, reacts with water, giving the strong base sodium hydroxide, $NaOH$. Magnesium oxide, MgO, reacts slowly with water, giving the less strongly basic magnesium hydroxide, $Mg(OH)_2$. Aluminum oxide, Al_2O_3, does not react with water, but the reaction of a solution of an aluminum salt such as aluminum nitrate, $Al(NO_3)_3$, with a solution of a base produces aluminum hydroxide, $Al(OH)_3$, an amphoteric compound. Aluminum hydroxide can act either as a very weak base (reacting with strong acids) or as a weak acid (reacting with strong bases). Silicon dioxide, SiO_2, will not react with water, but many covalent silicon compounds such as silicon tetrachloride, $SiCl_4$, will, giving gelations precipitates of silicic acid, $Si(OH)_4$, a very weak acid. The covalent oxides of phosphorus, sulfur, and chlorine in which these elements exhibit their highest oxidation numbers (P_4O_{10}, SO_3, and Cl_2O_7) react with water, giving phosphoric acid, H_3PO_4, sulfuric acid, H_2SO_4, and perchloric acid, $HClO_4$, respectively. Of these acids, phosphoric acid is the weakest and perchloric acid, the strongest. Thus, from left to right across any period of representative elements, the metallic character decreases and the nonmetallic character increases. *Hence the properties characteristic of metallic behavior become less pronounced and the properties characteristic of nonmetallic character become more pronounced from left to right across a period of representative elements.* For example, base strength, ionic character, and strength as a reducing agent decrease, whereas acid strength, covalent character, and strength as an oxidizing agent increase.

Because atoms at the top of a group lose electrons with more difficulty than those at the bottom of the group, *the metallic character of the elements decreases and the nonmetallic character increases as we go up a group of representative elements.* For example, bismuth, at the bottom of Group VA, is a metallic element; antimony is a semi-metal; and nitrogen, phosphorus, and arsenic, at the top of the group, are nonmetals (Fig. 9.5). Elemental fluorine, at the top of Group VIIA, is a stronger oxidizing agent than is elemental iodine, near the bottom of the group. Iodine has sufficient metallic character to possess a metallic luster.

Figure 9.5
The solid elements of Group VA. From left to right, phosphorus and arsenic (nonmetals), antimony (a semi-metal), and bismuth (a metal). Nitrogen, the lightest member of the group, is a colorless gas and is a nonmetal.

The behavior of an element also varies with its oxidation number. *The metallic behavior of an element decreases and its nonmetallic behavior increases as the positive oxidation number of the element in its compounds increases.* For instance, in perchloric acid, $HClO_4$ ($HOClO_3$), the oxidation number of chlorine is $+7$; in hypochlorous acid, $HOCl$, chlorine has an oxidation number of $+1$. Perchloric acid is a very strong acid (a characteristic of a compound of an element with pronounced nonmetallic behavior), whereas hypochlorous acid is a weak acid (less pronounced nonmetallic behavior). Thallium(III) chloride, $TlCl_3$, is a covalent compound (a characteristic of a nonmetal chloride) whereas thallium(I) chloride, $TlCl$, is ionic (a characteristic of a metal chloride). The chlorides of tin(IV) and lead(IV), $SnCl_4$ and $PbCl_4$, are liquids containing tetrahedral covalent molecules. The chlorides of tin(II) and lead(II), $SnCl_2$ and $PbCl_2$, are solids with much more ionic character in their bonds than the chlorides of tin(IV) and lead(IV). Thus, in addition to the position of an element in the Periodic Table, we must consider its oxidation number when predicting the behavior of its compounds.

9.6 Periodic Variation of Oxidation Numbers

As shown in Section 6.11, if we know the oxidation numbers of the elements in a compound, we can write the formula of the compound. Thus a knowledge of the oxidation numbers possible for an element can help in predicting its reaction products.

Some regularities in the oxidation numbers commonly observed for the representative elements are related to their position in the Periodic Table, as shown in Table 9.6. This array of numbers may look formidable at first, but there are many regularities that will simplify your recall of them. You should refer to Table 9.6 (p. 217) as you study the regularities outlined below.

1. The maximum positive oxidation number found in any group of representative elements is equal to the number of the group. Thus the maximum possible positive oxidation number increases from $+1$ for the alkali metals (Group IA) to $+7$ for all of the halogens (Group VIIA) except fluorine, which only has an oxidation number of -1 in its compounds. With the exceptions of thallium at the bottom of Group IIIA and mercury at the bottom of Group IIB, the maximum positive oxidation number is the only common oxidation number displayed by the metallic elements of Groups IA, IIA, IIB, and IIIA.

2. Metallic elements usually exhibit only positive oxidation numbers.

3. The most negative oxidation number of a group of representative elements is equal to the group number minus 8. Thus, for the elements in Group VA, the most negative oxidation number possible is $5 - 8 = -3$. This corresponds to the number of electrons that would have to be gained to provide an outer shell of eight electrons for the elements.

4. Negative oxidation numbers are commonly limited to nonmetals and semimetals and are observed only when these elements are combined with less electronegative elements. Since fluorine is the most electronegative element known, it never has a positive oxidation number.

5. Elements commonly exhibit positive oxidation numbers only when combined with more electronegative elements. Oxygen exhibits a positive oxidation number only in the few compounds it forms with the more electronegative element fluorine.

Table 9.6   Commonly observed oxidation numbers of the representative elements when in compounds

IA	IIA	IIB	IIIA	IVA	VA	VIA	VIIA	VIIIA
H +1 −1							**H** +1 −1	**He**
Li +1	**Be** +2		**B** +3	**C** +4 to −4	**N** +5 to −3	**O** −1 −2	**F** −1	**Ne**
Na +1	**Mg** +2		**Al** +3	**Si** +4	**P** +5 +3 −3	**S** +6 +4 −2	**Cl** +7 +5 +3 +1 −1	**Ar**
K +1	**Ca** +2	**Zn** +2	**Ga** +3	**Ge** +4	**As** +5 +3 −3	**Se** +6 +4 −2	**Br** +7 +5 +3 +1 −1	**Kr** +4 +2
Rb +1	**Sr** +2	**Cd** +2	**In** +3	**Sn** +4 +2	**Sb** +5 +3 −3	**Te** +6 +4 −2	**I** +7 +5 +3 +1 −1	**Xe** +8 +6 +4 +2
Cs +1	**Ba** +2	**Hg** +2 +1	**Tl** +3 +1	**Pb** +4 +2	**Bi** +5 +3	**Po** +2	**At** −1	**Rn**
Fr +1	**Ra** +2							

6. With the exceptions of boron, carbon, nitrogen, oxygen, and mercury, each representative element that exhibits multiple oxidation numbers in its compounds commonly has either all even or all odd oxidation numbers. Carbon commonly exhibits oxidation numbers from +4 to −4; nitrogen, from +5 to −3; and oxygen, −2 or −1. Mercury forms both a diatomic ion, Hg_2^{2+}, in which the oxidation number of each atom is +1, and a monatomic mercury(II) ion, Hg^{2+}, for which the oxidation number is +2.

Before we proceed, a word of explanation about the meaning of the phrase *common oxidation numbers* is necessary. The common oxidation numbers of an element are those observed in a majority of its compounds. In some cases, as for the elements in

Figure 9.6

When an iron nail is immersed in a solution of copper sulfate (top), copper (a less active metal than iron) is deposited as the iron reduces the Cu^{2+} ion to metallic copper and is itself oxidized to Fe^{2+} (bottom).

Groups IA and IIA, this majority may be all or nearly all of the known compounds of the element. Of all the thousands of known sodium compounds, for example, there is only one very reactive compound in which sodium has been shown to exhibit an oxidation number of -1. For other elements, the minority may be relatively large. Boron has an oxidation number of $+3$ in the majority of its compounds; however, in one series of compounds it has an oxidation number of $+2$, and in another it exhibits oxidation numbers that are nonintegers. Since most compounds discussed in this text contain elements having their common oxidation numbers, memorizing these numbers now will simplify studying the compounds later.

9.7 The Activity Series

Certain metals, such as sodium and potassium, react readily with cold water, displacing hydrogen and forming metal hydroxides. Other metals, such as magnesium and iron, react with water only when heated. Sodium and potassium react much more vigorously with acids than do magnesium and iron. Experimental observations like these are used to arrange the metals in order by their chemical activities and thereby to establish an **activity,** or **electromotive, series.** A brief form of the activity series containing only the common elements is given in Table 9.7. Elements exhibiting more metallic behavior are at the top of the series (Groups IA and IIA). In general, the less the metallic character of an element, the lower it appears in the series.

Potassium is the most reactive of the common metals and therefore heads the activity series. Each succeeding metal in the series is less reactive, and gold, the least reactive of all, is found at the bottom of the series. In theory, any metal in the series, in its elemental form, will reduce the ion, in water, of any metal below it in the series. For example, the copper ion in an aqueous solution of a copper(II) salt is reduced to copper metal by iron, which is above copper in the series (Fig. 9.6).

$$Fe(s) + Cu^{2+}(aq) \longrightarrow Cu(s) + Fe^{2+}(aq)$$

In a similar manner aqueous silver ions are reduced by metallic copper, and aqueous mercury ions are reduced by both copper and silver. Any metal above hydrogen in the series will liberate hydrogen from aqueous acids (solutions of acids in water); those from cadmium to the top will liberate hydrogen from hot water or steam; and those from sodium to the top will liberate hydrogen even from cold water. The metals below hydrogen do not displace hydrogen from water or from aqueous acids.

The reactivity of the metals toward oxygen decreases down the series, as do the heat of formation and the stability of the compounds formed. It is evident then that the activity series is very useful since it indicates the possibility of a reaction of a given metal with water, acids, salts of other metals, and oxygen. In addition, the series provides some indication of the stability of the compounds formed. It should be noted, however, that the order in which the elements are placed in the series depends somewhat on the conditions under which the activity is observed. A series determined from observations of the activities of the metals with respect to their ions in water will be slightly different from that reflecting the order of activity at high temperatures and in the absence of a solvent. The series shown in Table 9.7 is applicable to reactions in water.

The activity series of metals will be discussed in more detail in Chapter 21.

Table 9.7 Activity series of common metals

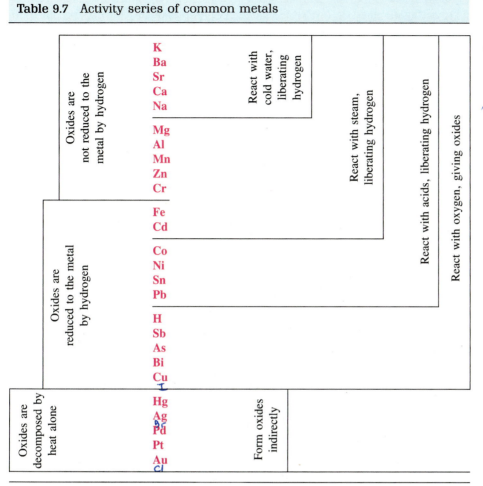

Metals	Oxides	Water	Steam	Acids	Oxygen
K, Ba, Sr, Ca, Na	Oxides are not reduced to the metal by hydrogen	React with cold water, liberating hydrogen	React with steam, liberating hydrogen	React with acids, liberating hydrogen	React with oxygen, giving oxides
Mg, Al, Mn, Zn, Cr					
Fe, Cd	Oxides are reduced to the metal by hydrogen				
Co, Ni, Sn, Pb					
H, Sb, As, Bi, Cu, I					
Hg, Ag, Br, Pd, Pt, Au, Cl	Oxides are decomposed by heat alone				Form oxides indirectly

top replace things underneath not vice versa

9.8 Prediction of Reaction Products

Many factors determine whether or not a chemical reaction will occur and what the products will be. These include the kinds of reactants and the conditions under which the reaction is run. Some chemists spend years developing a "feel" for the types of reactants and conditions that are likely to give a desired product. However, even they must resort to the ultimate test of their ideas; they try a proposed reaction in the laboratory to see if it works. The design and testing of reactions in order to prepare specific types of compounds or to improve the ease, yield, or simplicity with which known compounds are prepared constitute the field known as **chemical synthesis.**

In theory it should be possible to predict accurately the products of any chemical reaction using concepts of chemical bonding, kinetics, thermodynamics, solution behavior, and electrochemistry. In practice, however, most chemical systems are simply too complicated to be treated precisely by these tools. However, as we shall see in subsequent chapters, such tools do provide valuable insights into the behavior of chemical systems.

In this section we shall examine some of the general guidelines that can be helpful in answering the question ''What are the likely products, if any, that will result from the reaction of two or more substances?'' These guidelines are based on the metallic or nonmetallic behavior of the representative elements involved, the common oxidation numbers that they are likely to exhibit, and the similarities of the compounds involved to those types already discussed. We shall consider only those reactions that occur at room temperature in water or that occur when the pure substances are heated. Obviously, different conditions may lead to the formation of different products, but these guidelines can serve as a foundation for our considerations of chemical reactions in subsequent chapters. The following examples illustrate the approach.

EXAMPLE 9.1   As shown in the photograph, gaseous ammonia, NH_3, reacts with gaseous hydrogen iodide, HI. What is the likely product of this reaction?

From our discussion of acids and bases, we recognize that ammonia is a base and that hydrogen iodide is one of the six strong acids. Thus we can expect an acid-base reaction in which a hydrogen ion is transferred from HI, the acid, to NH_3, the base, forming the salt ammonium iodide, NH_4I.

$$NH_3 + HI \longrightarrow NH_4I$$

This reaction is analogous to the reaction of ammonia with HCl [HCl + $NH_3 \longrightarrow NH_4Cl$]. The similarity between the predicted reaction and known behavior supports our prediction.

When colorless gaseous ammonia and colorless gaseous hydrogen iodide combine, they react and form very finely divided solid ammonium iodide.

EXAMPLE 9.2   Predict the product of the reaction that occurs when elemental gallium (Ga) and sulfur (S_8) are warmed together.

First consider what we know about gallium and sulfur from their respective positions in the Periodic Table. Gallium is located in Group IIIA and thus should exhibit metallic behavior. As a metal, it can be expected to form compounds in which it exhibits a positive oxidation number. Since it is a member of Group IIIA, the expected oxidation number is +3. Sulfur is a member of Group VIA and thus should exhibit nonmetallic behavior. Its common oxidation numbers are $+6$ (its maximum positive oxidation number, equal to the group number), $+4$, and -2 (its most negative oxidation number, equal to the group number minus 8). Since sulfur is being combined with the less electronegative gallium, it is expected to exhibit the negative oxidation number of -2. As a metal, gallium can act as a reducing agent; as a nonmetal, sulfur can act as an oxidizing agent. Thus we can expect an oxidation-reduction reaction between gallium and sulfur, giving a product containing gallium with an oxidation number of $+3$ and sulfur with an oxidation number of -2. This leads us to formulate the product (Section 6.11) as Ga_2S_3.

The predicted chemical reaction therefore is as follows:

$$16Ga + 3S_8 \longrightarrow 8Ga_2S_3$$

Ga_2S_3 is, in fact, the product of the reaction of gallium with sulfur.

As a general approach to predicting the products of a reaction, examine the reactants to see if they are analogous to compounds or elements whose behavior you know. If, as in Example 9.1, the reactants are familiar, you can often readily predict their behavior. If analogies are not obvious, try the following method of analysis of the system.

1. Identify each element in the reactants as a metal or a nonmetal from its position in the Periodic Table and find its oxidation number (as described in Section 6.10), as well as the oxidation numbers it commonly exhibits in its compounds (Section 9.6).

2. Consider the possibility of an oxidation-reduction reaction (Section 9.3, Part 4). The following general guidelines are helpful:

(a) An elemental metal will reduce an elemental nonmetal.

$$Mg + Cl_2 \longrightarrow MgCl_2 \tag{1}$$

(b) A less electronegative metal (a more active metal) in its elemental form will reduce the ion of a more electronegative metal (Section 9.7).

$$Mg + SnCl_2 \longrightarrow MgCl_2 + Sn \tag{2}$$

(c) A more electronegative nonmetal in its elemental form will oxidize a less electronegative nonmetal.

$$S + O_2 \longrightarrow SO_2 \tag{3}$$

$$2SO_2 + O_2 \longrightarrow 2SO_3 \tag{4}$$

(d) The metals of Group IA and calcium, strontium, and barium of Group IIA are active enough to reduce the hydrogen in water or acids to hydrogen gas (Section 9.7).

$$2Na + 2H_2O \longrightarrow 2NaOH + H_2 \tag{5}$$

$$Ba + 2HCl \longrightarrow BaCl_2 + H_2 \tag{6}$$

The other representative metals, with the exception of lead (among the least metallic of the representative metals), will only reduce the hydrogen in acids.

$$4Al + 6H_2SO_4 \longrightarrow 2Al_2(SO_4)_3 + 6H_2 \tag{7}$$

3. Consider the possibility of an acid-base reaction (Section 9.3, Part 5). Remember that the oxides and hydroxides of nonmetals are generally acidic and that the oxides and hydroxides of metals are generally basic. The monatomic anions of the nonmetals of Groups IV, V, and VI are very strongly basic, and soluble compounds containing these ions accept protons from weak acids, even those as weak as water. Thus, we might find acid-base reactions involving familiar types of acids and bases:

$$2HI + Sr(OH)_2 \longrightarrow SrI_2 + 2H_2O \qquad (8)$$

involving acidic oxides and bases:

$$SO_3 + Ca(OH)_2 \longrightarrow CaSO_4 + H_2O \qquad (9)$$

involving acids and basic oxides:

$$6HCl + Al_2O_3 \longrightarrow 2AlCl_3 + 3H_2O \qquad (10)$$

or involving acids and other types of basic anions:

$$2HCl + Na_2S \longrightarrow 2NaCl + H_2S \qquad (11)$$

$$3H_2O + Li_3N \longrightarrow 3LiOH + NH_3 \qquad (12)$$

4. Consider the possibility of a metathetical reaction (Section 9.3, Part 3). Metathetical reactions between ionic compounds generally occur to form at least one product that is a solid, a gas, a weak electrolyte, or a nonelectrolyte.

5. Consider whether or not the products of the reaction may react with each other or with the solvent (if any).

Some specific illustrations of the application of these ideas are provided by the following examples.

EXAMPLE 9.3   Write the balanced chemical equation for the reaction of elemental calcium (Ca) with iodine (I_2).

Let us assume that we have no specific knowledge of the chemistry of this system, and analyze it using the steps discussed above. Calcium is a member of Group IIA and is thus an active metal. Iodine is a member of Group VIIA and is thus a nonmetal. An active metal reacts with a nonmetal in an oxidation-reduction reaction. Iodine will be reduced to -1, its only available negative oxidation number as a member of Group VIIA, while calcium will be oxidized to $+2$, the only oxidation number that a Group IIA element exhibits in compounds. With these oxidation numbers, the product must be CaI_2. The balanced equation is

$$Ca + I_2 \longrightarrow CaI_2$$

EXAMPLE 9.4   Write a balanced chemical equation for the reaction that occurs when tin(II) oxide, SnO, is added to a solution of perchloric acid, $HClO_4$.

$HClO_4$ is a strong acid, so we look for an acid-base reaction. Tin is a member of Group IVA. It is located near the bottom of the group and has a low oxidation number in SnO, so it should exhibit metallic properties. The oxides of metals are basic; hence we can expect an acid-base reaction between SnO and $HClO_4$ to produce a salt and water.

$$SnO + 2HClO_4 \longrightarrow Sn(ClO_4)_2 + H_2O$$

EXAMPLE 9.5   As shown in the photograph, phosphorus burns in a chlorine atmosphere. What are the two possible products of the reaction of phosphorus, P_4, with chlorine, Cl_2?

Both phosphorus (Group VA) and chlorine (Group VIIA) are nonmetals. Phosphorus has common oxidation numbers of $+5$, $+3$, and -3; chlorine, $+7$, $+5$, $+3$, $+1$, and -1. Chlorine lies to the right of phosphorus in the third period, so it is more electronegative than phosphorus. Thus we recognize that it can oxidize phosphorus. Because phosphorus is oxidized, it will exhibit a positive oxidation number—probably $+3$ or $+5$, since these are the common positive oxidation numbers exhibited by a member of Group VA. Chlorine will exhibit a negative oxidation number because it is combined with a less electronegative element. The only negative oxidation number for chlorine is -1. Thus P_4 and Cl_2 could react to give two compounds. In one phosphorus has an oxidation number of $+3$, and chlorine has an oxidation number of -1. In the other phosphorus has an oxidation number of $+5$, and chlorine has an oxidation number of -1. Thus the two compounds are PCl_3 and PCl_5, respectively.

Note that either phosphorus(III) chloride or phosphorus(V) chloride can in fact be formed, depending on the choice of reaction conditions. The reaction of 1 mol of P_4 with 6 mol of Cl_2 gives PCl_3. If the amount of Cl_2 is increased to 10 mol, then PCl_5 is produced. Hence an excess of Cl_2 favors the formation of PCl_5; conversely, an excess of P_4 favors the formation of PCl_3.

$$P_4 + 6Cl_2 \longrightarrow 4PCl_3$$
$$P_4 + 10Cl_2 \longrightarrow 4PCl_5$$

Phosphorus burns in an atmosphere of chlorine.

Predict the likely products of the reaction of sodium, Na, with sulfur trioxide, SO_3.   **EXAMPLE 9.6**

Sodium (Group IA) is an active metal. Sulfur and oxygen are both nonmetals located in Group VIA. Elemental sodium has an oxidation number of zero and can be oxidized only to $+1$. In SO_3 sulfur has an oxidation number of $+6$; oxygen, an oxidation number of -2.

In this system sodium can function only as a reducing agent. The only element that can be reduced is sulfur. Oxygen already exists with its minimum (most negative) oxidation number of -2. Thus we can expect the reaction of Na with SO_3 to give products in which the sulfur is reduced to an oxidation number of $+4$, 0, or -2. With an oxidation number of $+4$, sulfur would still be combined with oxygen, the only element in the system that exhibits a negative oxidation number. The reaction in this case would be

$$2Na + SO_3 \longrightarrow Na_2SO_3 \quad \text{(ratio of mol Na to mol } SO_3 = 2:1)$$

With additional sodium the reaction could give sulfur with an oxidation number of zero, that is, elemental sulfur, S_8.

$$48Na + 8SO_3 \longrightarrow 24Na_2O + S_8 \quad \text{(ratio of mol Na to mol } SO_3 = 6:1)$$

The sodium in this reaction must combine with oxygen to form sodium oxide, since sodium can only be oxidized to an oxidation number of $+1$. Finally, even more sodium could reduce the sulfur in SO_3 to an oxidation number of -2.

$$8Na + SO_3 \longrightarrow Na_2S + 3Na_2O \quad \text{(ratio of mol Na to mol } SO_3 = 8:1)$$

Which reaction actually occurs will depend on the relative amounts of sodium and sulfur trioxide reacting, but in each case sodium functions as a reducing agent and is itself oxidized.

Element or compound	1986 U.S. production (in billion-lb units)
H_2SO_4	73.6
N_2	48.6
O_2	33.0
CaO, $Ca(OH)_2$	30.3
NH_3	28.0
$NaOH$	22.0
Cl_2	21.0
H_3PO_4	18.4
Na_2CO_3	17.2
HNO_3	13.1
NH_4NO_3	11.1
CO_2	8.5
HCl	6.0
$(NH_4)_2SO_4$	4.2

Table 9.8 Important industrial inorganic chemicals and elements

Reprinted with permission from *Chemical and Engineering News*. Copyright 1987 American Chemical Society.

9.9 Chemical Properties of Some Important Industrial Chemicals

The compounds and elements listed in Table 9.8 constituted about 96% of the inorganic elements and compounds produced in the United States during 1986. The chemical properties of these compounds and elements clearly show the differences between the chemistry of metals and nonmetals. They also illustrate the types of compounds and reactions, as well as the oxidation numbers, described in the preceding sections. The behavior of organic materials will be described in Chapter 30.

1. SULFURIC ACID. With 14% of the industrial chemical production of the United States devoted to its manufacture, sulfuric acid (Fig. 9.7) ranks as the single most significant industrial chemical. Per capita use of sulfuric acid has been taken as one index of the technical development of a nation. Perhaps surprisingly, sulfuric acid and the sulfate ion rarely appear in finished materials. Sulfuric acid is used extensively as an acid (a source of hydrogen ions) because it is the cheapest strong acid. It is used to manufacture fertilizer, leather, tin plate, and other chemicals, to purify petroleum, and to make and dye fabrics.

Sulfuric acid is prepared from elemental sulfur or sulfides (Fig. 9.8). Sulfur is burned in air, and the nonmetal sulfur is oxidized by the more electronegative nonmetal oxygen. Covalent molecules of gaseous sulfur dioxide, with sulfur having one of its common oxidation numbers (+4), result.

$$S_8 + 8O_2 \longrightarrow 8SO_2$$

The sulfur is then oxidized to its highest oxidation number (+6) by oxygen giving sulfur trioxide in a reversible reaction.

$$2SO_2 + O_2 \overset{\triangle}{\rightleftharpoons} 2SO_3$$

Even though the yield from this reaction is highest at lower temperatures, sulfur trioxide forms slowly at these temperatures. At higher temperatures, it forms more rapidly, but the yield is lower. In order to get as high a yield as possible, the reaction is run at lower temperatures with vanadium(V) oxide, V_2O_5, as a catalyst. A **catalyst** is a substance that changes the speed of a chemical reaction without affecting the yield and without undergoing a permanent chemical change itself.

Since sulfur trioxide is an oxide of a nonmetal in which the nonmetal has a high oxidation number, we expect it to react with water to form an acid—in this case, sulfuric acid.

$$H_2O + SO_3 \longrightarrow H_2SO_4$$

This reaction does occur, for example, when SO_3 from polluted air dissolves in rain drops to give acid rain. However, it is more efficient industrially to combine sulfur trioxide with sulfuric acid to produce pyrosulfuric acid, $H_2S_2O_7$. This acid then reacts with water to form sulfuric acid.

$$SO_3 + H_2SO_4 \longrightarrow H_2S_2O_7$$
$$H_2S_2O_7 + H_2O \longrightarrow 2H_2SO_4$$

Oxygen will also oxidize sulfur with an oxidation number of −2 in compounds. Thus sulfuric acid is sometimes prepared from the sulfur dioxide produced by burning

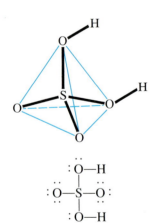

Figure 9.7
The molecular structure and Lewis structure of sulfuric acid.

Figure 9.8 (left)
A sulfuric acid plant.

Figure 9.9 (below)
The phosphate rock in this mine is converted to soluble phosphates by treatment with sulfuric acid in the plant shown.

hydrogen sulfide, an impurity separated from natural gas, or from the sulfur dioxide produced by roasting metal sulfides such as Cu_2S in air.

$$2H_2S + 3O_2 \xrightarrow{\triangle} 2SO_2 + 2H_2O$$
$$Cu_2S + 2O_2 \xrightarrow{\triangle} 2CuO + SO_2$$

Sulfuric acid is both a strong acid and an oxidizing agent. Its largest use is as a source of hydrogen ion. About 65% of the sulfuric acid produced in the United States is used to manufacture fertilizers (Fig. 9.9) by acid-base reactions with ammonia or with calcium phosphates [typically $Ca_5(PO_4)_3F$, found principally in North Carolina and Florida as phosphate rock]. The reaction with calcium phosphates converts these extremely insoluble compounds into somewhat soluble dihydrogen phosphates that can enter a plant's roots and provide it with necessary phosphorus.

$$2Ca_5(PO_4)_3F(s) + 7H_2SO_4(l) + 3H_2O(l) \longrightarrow$$
$$3Ca(H_2PO_4)_2 \cdot H_2O(s) + 2HF(g) + 7CaSO_4(s)$$

This is an acid-base reaction in which a strong acid reacts with the anions (PO_4^{3-} and F^-) of two weak acids. The solid mixture of the salts $Ca(H_2PO_4)_2 \cdot H_2O$ and $CaSO_4$ is used directly as the fertilizer. The hydrogen fluoride is recovered and used in the preparation of fluorocarbons and other fluorides. If the amount of sulfuric acid is increased, phosphoric acid is produced in a related acid-base reaction.

$$Ca_5(PO_4)_3F(s) + 5H_2SO_4(l) \longrightarrow 5CaSO_4(s) + 3H_3PO_4(l) + HF(g)$$

Both of these acid-base reactions are also metathetical reactions, which proceed because solid and gaseous products are formed.

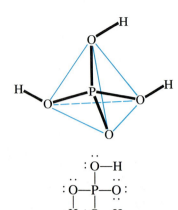

Figure 9.10

The molecular structure and Lewis structure of phosphoric acid.

Phosphoric acid (Fig. 9.10) is used primarily to make ammonium hydrogen phosphate and "triple superphosphate" fertilizers. The reactions are acid-base reactions.

$$2NH_3(g) + H_3PO_4(l) \longrightarrow (NH_4)_2HPO_4(s)$$

$$Ca_5(PO_4)_3F(s) + 7H_3PO_4(l) \longrightarrow 5Ca(H_2PO_4)_2(s) + HF(g)$$

The acid behavior of sulfuric acid is also evident in the acid-base reaction used to prepare the salt ammonium sulfate, another soluble solid fertilizer providing nitrogen.

$$2NH_3(g) + H_2SO_4(l) \longrightarrow (NH_4)_2SO_4(s)$$

In 1986 about 4.2 billion pounds of $(NH_4)_2SO_4$ were prepared in the United States.

Many metal oxides behave as bases and react with sulfuric acid to form ionic sulfates. For example, copper sulfate (used as a fungicide and in electroplating), aluminum sulfate (used in water treatment and in papermaking), and magnesium sulfate (Epsom salts) are prepared this way.

$$CuO(s) + H_2SO_4(aq) \longrightarrow Cu^{2+}(aq) + SO_4^{2-}(aq) + H_2O(l)$$

$$Al_2O_3(s) + 3H_2SO_4(aq) \longrightarrow 2Al^{3+}(aq) + 3SO_4^{2-}(aq) + 3H_2O(l)$$

$$MgO(s) + H_2SO_4(aq) \longrightarrow Mg^{2+}(aq) + SO_4^{2-}(aq) + H_2O(l)$$

Since these products are ionic, they exist in solution as ions. They can be recovered as solids by evaporating the water.

Sulfuric acid also undergoes an acid-base reaction with sodium chloride, giving hydrogen chloride and sodium hydrogen sulfate. The sodium hydrogen sulfate will then react with additional sodium chloride if the mixture is heated.

$$H_2SO_4(l) + NaCl(s) \longrightarrow NaHSO_4(s) + HCl(g)$$

$$NaHSO_4(s) + NaCl(s) \longrightarrow Na_2SO_4(s) + HCl(g)$$

Even though chloride ion is a very weak base, these metathetical reactions proceed because covalent gaseous HCl is formed and escapes from the reaction.

Sulfuric acid is an oxidizing agent, but it is not often used industrially as one. In fact, its oxidizing ability prevents it from being used in two potentially useful reactions: the preparation of hydrogen bromide and of hydrogen iodide by reaction with the respective metal halides. A strong acid should react with NaBr or NaI to form gaseous molecules of HBr or HI in a manner analogous to the reaction of H_2SO_4 with NaCl. Unfortunately, sulfuric acid will oxidize the bromide ion to bromine and iodide ion to iodine, and so it cannot be used.

2. AMMONIA. Ammonia (Fig. 9.11) is one of the few compounds that can be prepared readily from elemental nitrogen. It is prepared by the reversible reaction of a mixture of hydrogen and nitrogen at high pressure and temperature (400–600°C) in the presence of a catalyst containing iron. The catalyst speeds up the reaction but does not increase the yield.

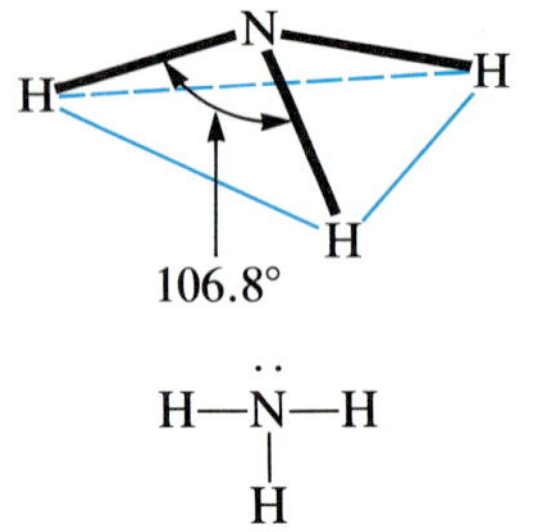

$$N_2(g) + 3H_2(g) \underset{\text{High pressure}}{\overset{\substack{400-600°C \\ \text{Fe catalyst}}}{\rightleftharpoons}} 2NH_3(g)$$

This reaction is simply the oxidation of a less electronegative nonmetal by a more electronegative nonmetal, giving a covalent compound.

Pure ammonia is a gas that is very soluble in water. The uses of ammonia can be attributed to two principal chemical properties of the molecule: (1) ammonia is a base,

Figure 9.11

The molecular structure and Lewis structure of ammonia.

and (2) the nitrogen atom in ammonia is reactive and can readily be oxidized to higher oxidation numbers, particularly by plants, in the production of other nitrogen-containing compounds. Thus ammonia and its derivatives form an important source of nitrogen fertilizers.

Approximately 15 billion pounds of ammonium salts are prepared as fertilizers each year in the United States (Fig. 9.12). The reactions are acid-base reactions that give salts containing the NH_4^+ ion. The principal reactions involve nitric acid, HNO_3, sulfuric acid, H_2SO_4, and phosphoric acid, H_3PO_4.

$$NH_3(g) + HNO_3(l) \longrightarrow NH_4NO_3(s)$$
$$2NH_3(g) + H_2SO_4(l) \longrightarrow (NH_4)_2SO_4(s)$$
$$NH_3(g) + H_3PO_4(l) \longrightarrow NH_4H_2PO_4(s)$$

The nitrogen in ammonia can be oxidized by oxygen. Ammonia burns in oxygen, forming nitrogen(II) oxide, NO, which can then react with additional oxygen to give nitrogen(IV) oxide, NO_2. The latter oxide reacts with water to give the oxyacid nitric acid. This reaction is more complex than most reactions of water with nonmetal oxides, because oxidation-reduction is also involved. The series of reactions is as follows:

$$4NH_3 + 5O_2 \longrightarrow 4NO + 6H_2O$$
$$2NO + O_2 \longrightarrow 2NO_2$$
$$3NO_2 + H_2O \longrightarrow 2HNO_3 + NO$$

Nitric acid (Fig. 9.13) is important industrially as a strong acid and oxidizing agent.

Figure 9.12
Ammonia is used as a fertilizer. Pure ammonia is a gas at room temperature and pressure; thus it is transported in a pressurized container. The ammonia is retained in the soil because it is very soluble in the water present and because it reacts with acids in the soil.

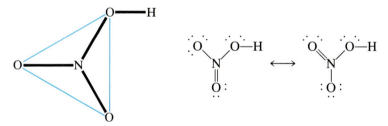

Figure 9.13
The molecular structure and Lewis structure (showing resonance forms) of nitric acid.

3. CALCIUM OXIDE AND CALCIUM HYDROXIDE. Calcium oxide and calcium hydroxide are inexpensive bases used extensively in chemical processing, although most of the useful products prepared from them do not contain calcium.

Calcium oxide, CaO, is made by heating calcium carbonate, $CaCO_3$, which is widely and inexpensively available as limestone.

$$CaCO_3(s) \xrightarrow{\Delta} CaO(s) + CO_2(g)$$

Although this decomposition reaction is reversible and CaO and CO_2 react to give $CaCO_3$, a 100% yield of CaO is obtained if the CO_2 is allowed to escape. Calcium hydroxide is prepared by the familiar acid-base reaction of a soluble metal oxide with water.

$$CaO(s) + H_2O(l) \longrightarrow Ca(OH)_2(s)$$

This reaction produces a lot of heat. In fact, farmers used to worry about the weather, for good reason, when they went to market in their wooden wagons to get calcium

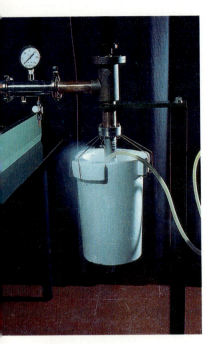

Figure 9.14
Liquid nitrogen is used in this trap to freeze out gases and thereby improve the vacuum in the line.

oxide (lime) for their fields. The reaction caused by a sudden rain on the CaO could make a wagon so hot that it would catch fire.

Since Ca^{2+} is the ion of an active metal and is therefore very difficult to reduce, and since oxygen with its oxidation number of -2 is difficult to oxidize, the oxidation-reduction chemistry of both CaO and $Ca(OH)_2$ is limited. These compounds are useful because they accept protons and thus neutralize acids.

4. OXYGEN AND NITROGEN. Both pure oxygen and pure nitrogen can be isolated from air. Nitrogen is surprisingly unreactive for so electronegative a nonmetal. This has been attributed to the very strong nitrogen-nitrogen triple bond in the N_2 molecule (Section 7.10). The principal uses of nitrogen gas are actually based on this nonreactivity. It is used as an inert atmosphere blanket in the food industry to prevent spoilage due to oxidation. Liquid nitrogen boils at 77 K (-196 °C). It is used to store biological materials such as blood and tissue samples at very low temperatures and as a coolant to produce low temperatures (Fig. 9.14).

Oxygen behaves as one would expect for a very electronegative nonmetal: it is a strong oxidizing agent. Its principal uses are in oxidations in the steel and petrochemical industries. During steel production oxygen is blown through hot liquid iron to oxidize impurities. Two common reactions are

$$C + O_2 \longrightarrow CO_2$$
$$Si + O_2 \longrightarrow SiO_2$$

Silicon dioxide, SiO_2, is less dense than liquid iron and floats to the surface where it is removed; CO_2 escapes as a gas. Note that the two Group IVA elements carbon and silicon are each oxidized to their maximum oxidation number. During these reactions oxygen is reduced to its most negative oxidation number (-2).

5. CHLORINE AND SODIUM HYDROXIDE Although they are very different chemically, chlorine and sodium hydroxide are paired here because they are prepared simultaneously in a very important electrochemical industrial process. The process utilizes sodium chloride, which occurs in large deposits in several parts of the country. Sodium chloride is composed of sodium ions and chloride ions, both of which are very resistant to chemical reduction or oxidation. (Recall that, although sulfuric acid does oxidize bromide and iodide ions, it does not oxidize chloride ions.) An electrochemical process, however, can be used to oxidize chloride ion to chlorine.

In the electrochemical production of chlorine, two electrodes are placed in a concentrated solution of sodium chloride (Fig. 9.15). When a direct current of electricity is passed through the solution, the chloride ions migrate to the positive electrode and are oxidized to gaseous chlorine by giving up an electron to the electrode.

$$2Cl^-(aq) \longrightarrow Cl_2(g) + 2e^- \text{ (at the positive electrode)}$$

The electrons are transferred through the outside electrical circuit to the negative electrode. Although the positive sodium ions migrate toward this negative electrode, metallic sodium is not produced because sodium ions are too difficult to reduce under the conditions used. (Recall that metallic sodium is active enough to react with water and hence, even if produced, would immediately react with water to produce sodium ions again.) Instead, water molecules, which are more easily reduced, pick up electrons from the electrode and are reduced, giving hydrogen gas and hydroxide ions.

$$2H_2O(l) + 2e^- \text{ (from the negative electrode)} \longrightarrow H_2(g) + 2OH^-(aq)$$

These changes convert the aqueous solution of NaCl into an aqueous solution of NaOH, gaseous Cl_2, and gaseous H_2.

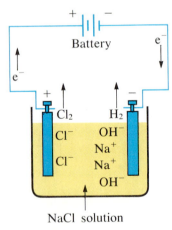

Figure 9.15
The electrolysis of an aqueous solution of NaCl. The net reaction produces hydrogen gas and hydroxide ions at the negative electrode and chlorine gas at the positive electrode, resulting in a solution containing sodium hydroxide and unreacted sodium chloride.

$$2Na^+(aq) + 2Cl^-(aq) + 2H_2O(l) \xrightarrow{\text{Electrolysis}}$$

Sodium chloride

$$2Na^+(aq) + 2OH^-(aq) + Cl_2(g) + H_2(g)$$

Sodium hydroxide

The nonmetal chlorine is more electronegative than any other element except fluorine, oxygen, and nitrogen (Table 6.2). Thus we would expect elemental chlorine to oxidize all of the other elements except for these three (and the noble gases, which are quite nonreactive). Its oxidizing property, in fact, is responsible for its principal use, as an oxidizing agent. For example, phosphorus(V) chloride, an important intermediate in the preparation of insecticides, is manufactured by oxidizing the less electronegative nonmetal phosphorus with chlorine.

$$P_4 + 10Cl_2 \longrightarrow 4PCl_5$$

A great deal of chlorine is also used to oxidize and destroy organic or biological materials in water purification and in bleaching.

Sodium hydroxide is a strong, very water-soluble (420 g/L) base used to make concentrated solutions of base.

$$NaOH(s) \xrightarrow{H_2O} Na^+(aq) + OH^-(aq)$$

It is used in acid-base reactions during chemical processing and in the preparation of soaps, rayon, and paper.

6. NITRIC ACID. Nitric acid (Fig. 9.13) is an important industrial chemical (13.1 billion pounds produced in the United States in 1986). It is now prepared primarily by oxidation of ammonia as discussed in Part 2 of this section. It was previously manufactured by the acid-base reaction of sulfuric acid with sodium nitrate, mined principally in Chile.

$$NaNO_3(s) + H_2SO_4(l) \longrightarrow NaHSO_4(s) + HNO_3(l)$$

The nitrate ion can be easily metabolized by most plants. Thus nitrates, particularly ammonium nitrate, NH_4NO_3, are good fertilizers. In addition, nitric acid is both a strong acid and a strong oxidizing agent and is therefore used in both acid-base and oxidation-reduction reactions to prepare soluble metal salts. Its acid-base reactions include those with hydroxides, oxides, and carbonates.

$$KOH + HNO_3 \longrightarrow KNO_3 + H_2O$$
$$CuO + 2HNO_3 \longrightarrow Cu(NO_3)_2 + H_2O$$
$$CaCO_3 + 2HNO_3 \longrightarrow Ca(NO_3)_2 + H_2CO_3$$
$$ \longrightarrow H_2O + CO_2$$

Carbonic acid, H_2CO_3, is not very stable and decomposes into gaseous carbon dioxide and water. This decomposition reaction is reversible.

Nitrogen exhibits a wide range of oxidation numbers in its compounds. Oxidation-reduction reactions of nitric acid generally involve reduction of the nitrogen (oxidation number of $+5$) to either an oxidation number of $+4$ with formation of NO_2 or an oxidation number of $+2$ with formation of NO. Reactions involving concentrated HNO_3 usually give predominantly NO_2. Dilute HNO_3 often gives NO.

$$Cu + 4HNO_3(conc) \longrightarrow Cu(NO_3)_2 + 2NO_2 + 2H_2O$$
$$3Ag + 4HNO_3(dilute) \longrightarrow 3AgNO_3 + NO + 2H_2O$$

Since gold is not oxidized by nitric acid, the latter reaction is used to separate silver from gold. Note that, whereas most acids react with active metals such as zinc with evolution of hydrogen, nitric acid reacts by reduction of the nitrate group (Fig. 9.16).

<big>For Review</big>

Summary

The chemical behavior of an element reflects its position in the Periodic Table. Those elements that lie to the left or toward the bottom of the table are **metals,** and all have similar characteristic behavior. **Nonmetals,** which lie in the upper right portion of the table, have characteristic nonmetallic behavior. The chemical behavior characteristic of metals becomes less pronounced from left to right across a period and from bottom to top in a group. The chemical behavior characteristic of nonmetals becomes more pronounced from left to right and from bottom to top. An element exhibits behavior increasingly more characteristic of a nonmetal as its oxidation number in its compounds increases. The elements that lie near the dividing line between metals and nonmetals, the **semi-metals,** cannot be identified clearly as metals or nonmetals, and they may exhibit metallic or nonmetallic behavior depending on the conditions under which they react.

Elemental metals generally react with elemental nonmetals, resulting in **oxidation** of the metal and **reduction** of the nonmetal. The oxidation numbers of the elements in the products can often be predicted from their positions in the Periodic Table. Most representative metals have a single common positive oxidation number in their compounds, and nonmetals have a single common negative oxidation number when they form monatomic anions. Generally, representative metals and nonmetals form ionic **salts,** which dissolve in water giving solutions of **electrolytes.** The hydroxides of metals act as **bases** unless the metals exhibit unusually high oxidation numbers. These hydroxides will accept protons from acids. Soluble ionic metal oxides are also bases because the oxide ion will pick up a hydrogen ion from water to form a hydroxide ion.

In addition to oxidizing metals a nonmetal will often oxidize less electronegative nonmetals, giving covalent compounds. Fluorine, chlorine, and oxygen are particularly good oxidizing agents. Oxidation of hydrogen by nonmetals gives binary hydrides, which can act as **acids,** although HCl, HBr, and HI are the only binary **strong acids.** The other nonmetal hydrides range from **weak acids** to essentially nonacidic compounds as the nonmetallic character of the element involved decreases. Water and ammonia will also act as a base. Some oxides of the nonmetals react with water to produce **oxyacids** (nonmetal hydroxides), of which $HClO_4$, H_2SO_4, and HNO_3 are common strong acids. Most other oxyacids are weak.

The reactivity of unfamiliar species can be predicted by comparison with the chemistry of analogous compounds whose behavior is known or by consideration of the metallic or nonmetallic character of the elements involved, using the analysis discussed in this chapter.

The inorganic chemicals used industrially illustrate the differences between metals and nonmetals. Sulfuric acid and nitric acid are oxyacids formed from nonmetal oxides. Hydrochloric acid is a nonmetal hydride. Calcium oxide is a metal oxide that is used as a base and that reacts with water to give calcium hydroxide, another base. The hydroxide of sodium is also a base. In the reactions described involving the metal ions Na^+ and Ca^{2+}, the oxidation numbers of the metal ions do not change. The oxidation numbers of the nonmetals, however, vary widely. The nitrogen atom, for example,

Figure 9.16
When nitric acid reacts with copper, the nitrate ion is reduced producing brown $NO_2(g)$ or colorless $NO(g)$. The copper is converted to copper nitrate.

varies from an oxidation number of -3 in NH_3 to $+5$ in HNO_3. The stability of the most electronegative nonmetals in their reduced states is illustrated by the difficulty of preparing elemental chlorine.

Key Terms and Concepts

acid (9.2)
acid-base reaction (9.3)
activity series (9.7)
addition reaction (9.3)
ammonia (9.9)
amphoteric behavior (9.5)
base (9.2)
Brønsted acid (9.2)
Brønsted base (9.2)
calcium hydroxide (9.9)
calcium oxide (9.9)
catalyst (9.9)
chemical synthesis (9.8)
chlorine (9.9)
combination reaction (9.3)

decomposition reaction (9.3)
electrolyte (9.2)
hydronium ion (9.2)
metal (9.4)
metalloid (9.4)
metathetical reaction (9.3)
nitric acid (9.9)
nitrogen (9.9)
nonelectrolyte (9.2)
nonmetal (9.4)
oxidation (9.3)
oxidation number (9.6)
oxidation-reduction reaction (9.3)
oxidizing agent (9.3)
oxyacid (9.4)

oxygen (9.9)
reducing agent (9.3)
reduction (9.3)
reversible reaction (9.3)
salt (9.2)
semi-metal (9.4)
sodium hydroxide (9.9)
strong acid (9.2)
strong base (9.2)
strong electrolyte (9.2)
sulfuric acid (9.9)
weak acid (9.2)
weak base (9.2)
weak electrolyte (9.2)

Exercises

Types of Compounds

1. In what ways does a salt differ from a metal hydroxide? From the hydroxide of a nonmetal (an oxyacid)? In what ways are salts and metal hydroxides similar?

2. Identify the cations and anions in the following compounds:
   (a) $LiCl$ (b) CaI_2 (c) $NaOH$
   (d) Ga_2O_3 (e) K_2SO_4 (f) NH_4Cl
   (g) $Ca(H_2PO_4)_2$ (h) $SrHPO_4$ (i) $[PCl_4][PCl_6]$

3. When the following compounds dissolve in water, do they give solutions of acids, bases, or salts?
   (a) KI (b) P_4O_{10} (c) H_3PO_4
   (d) HCl (e) NH_4Cl (f) $RbOH$
   (g) $NaClO_4$ (h) $HClO_2$ (i) CaO
   (j) SO_3 (k) NO_2 (l) Li_2S

4. Explain why sulfuric acid, H_2SO_4, which is a covalent molecule, dissolves in water and behaves as an electrolyte.

5. Indicate whether each of the following compounds and elements dissolves in water to give a solution of a strong electrolyte, a weak electrolyte, or a nonelectrolyte:
   (a) H_2S (b) $In(NO_3)_3$ (c) $Ba(OH)_2$
   (d) SO_3 (e) Li_2O (f) NH_4NO_3
   (g) Cl_2O_7 (h) CH_4 (i) CaO
   (j) KIO_4 (k) H_3PO_4 (l) NH_3
   (m) Xe

6. Sodium hydrogen carbonate, also called sodium bicarbonate or baking soda, $NaHCO_3$, is sometimes kept in chemical laboratories because it can neutralize either acid or base spills. Explain how it can act either as a base or as an acid.

7. Soil that is too acidic is often treated with slaked lime, $Ca(OH)_2$, or with calcium carbonate, $CaCO_3$. Explain why treatment with these materials neutralizes the acids in soil.

Types of Chemical Reactions

8. How do addition reactions that are also oxidation-reduction reactions differ from addition reactions that are acid-base reactions? Write balanced chemical equations illustrating each type.

9. Classify the following as acid-base reactions or oxidation-reduction reactions:
   (a) $Na_2S + 2HCl \longrightarrow 2NaCl + H_2S$
   (b) $2Na + 2HCl \longrightarrow 2NaCl + H_2$
   (c) $Mg + Cl_2 \longrightarrow MgCl_2$
   (d) $MgO + 2HCl \longrightarrow MgCl_2 + H_2O$
   (e) $K_3P + 2O_2 \longrightarrow K_3PO_4$
   (f) $3KOH + H_3PO_4 \longrightarrow K_3PO_4 + 3H_2O$

10. Identify the atoms that are oxidized and reduced, the change in oxidation number for each, and the oxidizing and reducing agents in each of the following equations:
    (a) $Mg + NiCl_2 \longrightarrow MgCl_2 + Ni$
    (b) $PCl_3 + Cl_2 \longrightarrow PCl_5$
    (c) $C_2H_4 + 3O_2 \longrightarrow 2CO_2 + 2H_2O$
    (d) $Zn + H_2SO_4 \longrightarrow ZnSO_4 + H_2$
    (e) $2K_2S_2O_3 + I_2 \longrightarrow K_2S_4O_6 + 2KI$
    (f) $3Cu + 8HNO_3 \longrightarrow 3Cu(NO_3)_2 + 2NO + 4H_2O$

Metals and Nonmetals

11. From the positions of their components in the Periodic Table, predict which member in each of the following pairs will:
    (a) conduct electricity: Cl_2 or Al
    (b) be ionic: KF or HF
    (c) react as a base: $P(OH)_3$ or $Al(OH)_3$
    (d) oxidize Si: O_2 or Al
    (e) reduce Sn^{2+} to Sn: Mg or Cl_2
    (f) oxidize S_8: Na or F_2
    (g) contain ions: KNO_2 or $ClNO_2$
    (h) react with water to give a solution of an acid: In_2O_3 or P_4O_6

12. From the positions of the elements in the Periodic Table, predict which member in each of the following pairs will:
    (a) reduce S_8: Cl_2 or Mg
    (b) neutralize H_2SO_4: KOH or BrOH
    (c) oxidize Si: Ca or Cl_2
    (d) oxidize SO_2: Mg or F_2
    (e) reduce SeO_2: Ca or Cl_2
    (f) reduce Cu^{2+}: Fe or H_2O

13. From the positions of the elements in the Periodic Table, predict which member of each of the following will be:
    (a) more basic: CsOH or BrOH
    (b) more acidic: $In(OH)_3$ or $B(OH)_3$
    (c) more acidic: H_2SO_3 or $HClO_3$
    (d) a stronger oxidizing agent: S_8 or Ge
    (e) more easily reduced: F_2 or I_2
    (f) an ionic compound: RbCl or BrCl
    (g) more easily oxidized: Ca or S

14. Assume that you have an unlabeled bottle containing a colorless gas. The gas reacts with oxygen and with water to give a solution of an acid. Could this gaseous compound be Na_2SO_4? N_2? SO_2? SO_3? Explain your answers.

15. A sample of an element that conducts electricity is broken into small pieces by cracking it with a hammer. These pieces react with chlorine to form a volatile liquid chloride. Could the element be aluminum? Silicon? Iodine? Explain your answers.

16. An element is oxidized by fluorine but not by chlorine. Could the element be sodium? Aluminum? Silicon? Sulfur? Oxygen? Explain your answers.

Common Oxidation Numbers

17. Explain why the maximum oxidation number of an atom of Group IIA is equal to the number of electrons in its valence shell.

18. Explain why the most negative oxidation number of an atom of Group VIA is equal to the number of electrons required to fill its valence shell.

19. Which of the following elements will exhibit a positive oxidation number when combined with phosphorus?
    (a) Mg   (b) F   (c) Al
    (d) O   (e) Na   (f) Cl
    (g) Fe

20. With which elements will sulfur form binary compounds in which it has a positive oxidation number?

21. Write the formula of a binary compound (a compound containing only two elements) for each of the following elements so that the element has the indicated oxidation number:
    (a) Ca, +2   (b) N, −3
    (c) N, +3   (d) S, −2
    (e) Cl, +1   (f) Br, +7
    (g) P, +3   (h) Pb, +2

22. From their positions in the Periodic Table and without reference to Table 9.6, predict the common oxidation numbers of the following elements in their compounds:
    (a) Al   (b) Cl
    (c) Ca   (d) Cs
    (e) As   (f) S
    (g) Ge   (h) H

23. Without reference to Table 9.6, identify those representative metals that may display two or more common oxidation numbers in their compounds.

The Activity Series

24. Under what chemical conditions may the activity series be relied on to predict chemical behavior?

25. One of the elements in the activity series (Table 9.7) will not produce hydrogen when it reacts with steam, but it will form hydrogen when it reacts with hydrochloric acid. Reaction of this element with a solution of $Ni(NO_3)_2$ produces Ni(s). What is this element?

26. Zinc and mercury are both members of Group IIB, but ZnS and HgS behave differently when roasted in air. Write the equations for the reactions of ZnS and HgS with O_2 when roasted.

27. With the aid of the activity (electromotive) series (Table 9.7), predict whether or not the following reactions will take place:
    (a) $Mg + Co^{2+} \longrightarrow Mg^{2+} + Co$
    (b) $Al_2O_3 + 3H_2 \longrightarrow 2Al + 3H_2O$
    (c) $2Au + Fe^{2+} \longrightarrow 2Au^+ + Fe$
    (d) $Ni + 2H^+ \longrightarrow Ni^{2+} + H_2$
    (e) $PtO_2 \xrightarrow{\Delta} Pt + O_2$
    (f) $K_2O + H_2 \longrightarrow 2K + H_2O$
    (g) $3Cd + 2Bi^{3+} \longrightarrow 2Bi + 3Cd^{2+}$
    (h) $3Pd + 2Au^{3+} \longrightarrow 2Au + 3Pd^{2+}$

28. Lithium and beryllium do not appear in the activity series given in Table 9.7. From the position of these elements in the Periodic Table, locate their approximate position in the activity series and write equations for their reactions, if any, with water, steam, oxygen, and acids. Will the oxides of Li and Be be reduced by hydrogen? Are the oxides reduced by heat alone?

29. What mass of copper can be recovered from a solution of copper(II) sulfate by addition of 2.11 kg of aluminum?
    Ans. 7.45 kg

Industrial Chemicals

30. The following reactions are all similar to those of the industrial chemicals described in Section 9.9. Complete and balance the equations for these reactions.

    (a) reaction of a weak base and a strong acid

    $$NH_3 + HClO_4 \longrightarrow$$

    (b) pickling of steel in hydrochloric acid

    $$Fe_2O_3 + HCl \longrightarrow$$

    (c) preparation of a soluble source of calcium and phosphorus used in animal feeds

    $$Ca_3(PO_4)_2 + H_3PO_4 \longrightarrow$$

    (d) formation of an air pollutant when burning coal that contains iron sulfide

    $$FeS + O_2 \longrightarrow$$

    (e) preparation of a soluble silver salt for silver plating

    $$Ag_2CO_3 + HNO_3 \longrightarrow$$

    (f) neutralization of a basic solution of nylon in order to precipitate the nylon

    $$Na^+(aq) + OH^-(aq) + H_2SO_4 \longrightarrow$$

    (g) hardening of plaster containing slaked lime

    $$Ca(OH)_2 + CO_2 \longrightarrow$$

    (h) removal of sulfur dioxide from the flue gas of power plants

    $$CaO + SO_2 \longrightarrow$$

    (i) neutralization of acid drainage from coal mines

    $$CaCO_3 + H_2SO_3 \longrightarrow$$

    (j) the reaction of baking powder that produces carbon dioxide gas and causes bread to rise

    $$NaHCO_3 + NaH_2PO_4 \longrightarrow$$

    (k) separation of silver from gold with concentrated nitric acid

    $$Ag + HNO_3 \longrightarrow$$

    (l) preparation of strontium hydroxide by electrolysis of a solution of strontium chloride

    $$SrCl_2(aq) + H_2O(l) \xrightarrow{\text{Electrolysis}}$$

    (m) preparation of pure phosphoric acid for use in soft drinks (two reactions)

    $$P_4 + O_2 \longrightarrow X$$
    $$X + H_2O \longrightarrow$$

Prediction of Reaction Products

31. Complete and balance the equations for the following oxidation-reduction reactions. In some cases there may be more than one correct answer depending on the amounts of reactant used, as illustrated in Examples 9.5 and 9.6.

    (a) $Mg + O_2 \longrightarrow$ 2MgO

    (b) $Al + 3Cl_2 \longrightarrow$ 2AlCl$_3$

    (c) $K + S_8 \longrightarrow$ 8K$_2$S

    (d) $S_8 + O_2 \longrightarrow$ 8SO

    (e) $P_4 + O_2 \longrightarrow$ P$_2$O$_5$

    (f) $Mg + N_2 \longrightarrow$ Mg$_3$N$_2$

    (g) $P_4O_6 + O_2 \longrightarrow$ P$_4$O$_8$

    (h) $Ca(s) + 2HBr(g) \longrightarrow$ CaBr$_2$ + H$_2$

    (i) $Cs(s) + H_2O(l) \longrightarrow$

    (j) $Li(s) + 2HI(aq) \longrightarrow$ 2LiI + H$_2$

    (k) $Sr(s) + CH_3CO_2H(aq) \longrightarrow$

    (l) $Al(s) + SnCl_4(g) \xrightarrow{\Delta}$

    (m) $Mg(s) + PbO(s) \xrightarrow{\Delta}$

    (n) $H_2Se + O_2 \xrightarrow{\Delta}$

    (o) $AlP + O_2 \xrightarrow{\Delta}$

    (p) $CaSO_3 + O_2 \xrightarrow{\Delta}$

32. When heated to 700–800°C, diamonds, which are pure carbon, are oxidized by atmospheric oxygen. (They burn!) Write the balanced equation for this reaction.

33. Military lasers use the very intense light produced when fluorine combines explosively with hydrogen. What is the balanced equation for this reaction?

34. Predict the formula for the product of the reaction of phosphorus with an excess of sulfur.

35. Small amounts of hydrogen gas and oxygen gas are often produced by electrolysis of water containing a little sulfuric acid as a catalyst. Write the balanced equation for this reaction.

36. Dow Chemical Company for many years prepared bromine from the sodium bromide, NaBr, in Michigan brine by treating the brine with chlorine gas. Write a balanced equation for the process.

37. Metallic copper is produced by roasting ore containing Cu_2S in air, followed by reduction of the product with excess carbon. The gas escaping from the reduction process is burned to provide part of the heat for the roasting process. Write the balanced equation for each of these three steps.

38. Complete and balance the equations for the following acid-base reactions. If the reactions are run in water as a solvent, write the reactants and products as solvated ions. In some cases there may be more than one correct answer, depending on the amounts of reactants used.

    (a) $HBr(g) + In_2O_3(s) \longrightarrow$

    (b) $Mg(OH)_2(s) + HClO_4(aq) \longrightarrow$

    (c) $Al(OH)_3(s) + HNO_3(aq) \longrightarrow$

    (d) $Na_2O(s) + H_2O(l) \longrightarrow$

(e) $Li_2O(s) + CH_3CO_2H(l) \longrightarrow$

(f) $SrO(s) + H_2SO_4(l) \longrightarrow$

(g) $SO_3(g) + H_2O(l) \longrightarrow$

(h) $NH_3(g) + HI(g) \longrightarrow$

(i) $NH_3(aq) + HBr(aq) \longrightarrow$

(j) $CaO(s) + SO_3(g) \longrightarrow$

(k) $Li_2O(s) + N_2O_5(l) \longrightarrow$

(l) $NH_3(g) + NaH_2PO_4(s) \longrightarrow$

(m) $KCl(s) + H_2SO_4(l) \longrightarrow$

39. The following salts contain anions of weak acids. Write a balanced equation for the reaction of a solution of each with an excess of the indicated acid.

(a) $Li[CH_3CO_2] + H_3PO_4 \longrightarrow$

(b) $SrF_2 + H_2SO_4 \longrightarrow$

(c) $NaCN + HCl \longrightarrow$

(d) $CaCO_3 + HCl \longrightarrow$

(e) $CaHPO_4 + HClO_4 \longrightarrow$

(f) $KHCO_3 + HCl \longrightarrow$

(g) $KNO_2 + HBr \longrightarrow$

(h) $Li_2SO_3 + HCl \longrightarrow$

40. Complete and balance the equations of the following reactions, each of which is used to remove hydrogen sulfide from natural gas:

(a) $Ca(OH)_2(s) + H_2S(g) \longrightarrow$

(b) $NaOH(s) + H_2S(g) \longrightarrow$

(c) $Na_2CO_3(aq) + H_2S(g) \longrightarrow$

(d) $K_3PO_4(aq) + H_2S(g) \longrightarrow$

41. The medical laboratory test for cyanide ion, CN^-, involves separation of the cyanide ion from a blood, urine, or tissue sample by the addition of sulfuric acid. The gaseous product is absorbed in a sodium hydroxide solution and then analyzed. Write balanced equations that describe the separation and absorption.

42. Calcium propionate is sometimes added to bread to retard spoilage. This compound can be prepared by the reaction of calcium carbonate, $CaCO_3$, with propionic acid, $C_2H_5CO_2H$, which has properties similar to those of acetic acid. Write the balanced equation for the formation of calcium propionate.

43. A teaspoon of baking soda, $NaHCO_3$, in a glass of water can be used to treat acid stomach. Stomach acid is hydrochloric acid. Write a balanced equation for the reaction that occurs. Milk of magnesia, $Mg(OH)_2$, is a somewhat milder antacid. Write a balanced equation describing the action of milk of magnesia. Other antacids contain aluminum hydroxide, $Al(OH)_3$. Write the balanced equation that describes its action.

44. Derivatives of fatty acids occur in plant and animal cells. A fatty acid is a carboxylic acid (containing a —CO_2H group), like acetic acid, in which a large hydrocarbon group replaces the CH_3 group of acetic acid. Write a balanced equation for each of the following reactions, which involve fatty acids:

(a) stearic acid, $C_{17}H_{35}CO_2H$, with Na_2CO_3

(b) palmitic acid, $C_{15}H_{31}CO_2H$, with NaOH

(c) myristic acid, $C_{13}H_{27}CO_2H$, with SrO

(d) sodium palmitate, $C_{15}H_{31}CO_2Na$, with $HClO_4$

(e) capric acid, $C_9H_{19}CO_2H$, with CaH_2

(f) myristic acid, $C_{13}H_{27}CO_2H$, with Mg

Additional Exercises

45. Complete and balance the following equations. If the reactions are run in water as a solvent, write the reactants and products as solvated ions. In some cases there may be more than one correct answer, depending on the amounts of reactants used.

(a) $Ca(OH)_2(s) + HCl(g) \longrightarrow$

(b) $Sr(OH)_2 + HNO_3 \xrightarrow{H_2O}$

(c) $Ca + S_8 \longrightarrow$

(d) $Cs + P_4 \longrightarrow$

(e) $MgH_2 + H_2O \longrightarrow$

(f) $In + SnSO_4 \xrightarrow{H_2O}$

(g) $Si + F_2 \longrightarrow$

(h) $K(s) + H_2O(l) \longrightarrow$

(i) $Ca(CH_3CO_2)_2 + H_2SO_4 \xrightarrow{H_2O}$

(j) $Cs_2O + SO_3 \longrightarrow$

(k) $H_2 + Br_2 \longrightarrow$

(l) $P + F_2 \longrightarrow$

(m) $NH_3(g) + H_2S(g) \longrightarrow$

(n) $Sb + S_8 \longrightarrow$

(o) $Al(s) + HClO_4(aq) \longrightarrow$

(p) $Cl_2 + I_2 \longrightarrow$

(q) $NaF + HNO_3 \xrightarrow{H_2O}$

46. Hydrogen sulfide, H_2S, is removed from natural gas by passing the raw gas through a solution of ethanolamine, $HOCH_2CH_2NH_2$, whose behavior is similar to that of NH_3 (which can be viewed as HNH_2) in its acid-base reactions. After the solution is saturated, the H_2S can be recovered by heating the solution to reverse the reaction. Part of the recovered H_2S is burned with air, and the product is allowed to react with the remaining H_2S to produce sulfur, S_8, which is easier to ship than H_2S. Write the reaction of ethanolamine with hydrogen sulfide, showing the Lewis structures of the reactants and products. Write the balanced equations for the reactions that lead to the formation of sulfur from hydrogen sulfide.

47. Write a balanced chemical equation for the reaction used to prepare each of the following compounds from the given starting material(s). Several steps and additional reactants may be required.

(a) H_2SO_4 from CuS

(b) $(NH_4)NO_3$ from N_2

(c) HBr from Br_2

(d) H_2S from Zn and S

(e) K_2CO_3 from KCl

(f) NaH from H_3PO_4

48. Write balanced chemical equations for the preparation of each of the following compounds (a) by a metathetical

reaction, (b) by an oxidation-reduction reaction, and (c) by an acid-base reaction: $CaCl_2$, K_2SO_4, $Mg(H_2PO_4)_2$. Write different reactions for (a), (b), and (c) in each case.

49. Calcium cyclamate $Ca(C_6H_{11}NHSO_3)_2$, an artificial sweetener now banned by the FDA, was purified industrially by converting it to the barium salt through reaction of the acid $C_6H_{11}NHSO_3H$ with barium carbonate, treatment with sulfuric acid (barium sulfate is very insoluble), and then neutralization with calcium hydroxide. Write the balanced equations for these reactions.

50. Calcium hydrogen sulfite, $Ca(HSO_3)_2$, is used in the paper industry in the ''cooking liquor'' that is important in converting cellulose into paper. $Ca(HSO_3)_2$ is prepared from sulfur, calcium carbonate, air, and water. Write balanced equations for the three steps in its preparation.

51. Indicate whether water acts as an acid, a base, an oxidizing agent, or a reducing agent in each of the following reactions. Write an equation for each reaction.
    (a) Na is added to water
    (b) $NaHSO_4$ is added to water
    (c) Na_2O is added to water
    (d) CoF_3 is added to water, producing a solution of CoF_2 and HF
    (e) NaF is added to water

52. What is the percent by mass of phosphorus in the compound formed by the reaction of phosphorus with magnesium? (Use four significant figures.)   *Ans. 45.93%*

53. What is the percent by mass of sulfur in the compound formed when 0.050 mol of Tl reacts with H_2SO_4 producing a thallium(I) compound and 0.025 mol of H_2?   *Ans. 6.351%*

54. What is the hydroxide ion concentration in a solution formed by dissolution of 1.00 g of calcium metal in enough water to form 250 mL of the solution?   *Ans. 0.200 M*

55. What is the perchloric acid concentration in a solution formed by dissolution of 2.35 g of Cl_2O_7 in enough water to form 1.00 L of solution?   *Ans. 2.57×10^{-2} M*

56. Describe the molecular structure of the nonmetal oxide that reacts with water to form a solution of H_2CO_3.

57. What is the molecular structure of the molecular compound that forms when Na_3P is added to water?

58. How does the hybridization of the sulfur atom change when SO_2 is dissolved in water?

59. Would you expect arsenic to behave like a semi-metal? Bismuth? Iodine? Give convincing reasons to support each of your answers.

The Gaseous State and the Kinetic-Molecular Theory

10

Hotter air is less dense than cooler air, thus a balloon floats.

Gases have played an important part in the development of chemistry. The identification of oxygen as a component of air by Priestley and by Lavoisier was crucial to the development of the atomic theory. The belief that water is an element changed when it was prepared by the reaction of the gases oxygen and hydrogen and was thereby shown to be a compound. The densities of gases played an important role in the determination of atomic weights. Conclusions regarding chemical stoichiometry and the molecular nature of matter followed from observations of the volumes of gases that combine in chemical reactions. The variations in the pressure and volume of a gas with temperature led to the discovery of the concept of absolute zero and the development of the kinetic-molecular theory of gas behavior. Studies of gases also led to the first quantitative models for the description of the behavior of matter.

In this chapter we will consider how the temperature, pressure, volume, and mass of a sample of gas are related. The equations that describe these relationships are the tools used to describe the quantitative behavior of gases. We will see how to use them to convert physical measurements to moles of gas present, to the molecular weight of a gas, or to the quantities of gases involved in chemical changes. The kinetic-molecular model of gases will be described, and the experimental behavior of gases will be compared with the predictions of this theoretical model. This will help us to determine if the assumptions used in the theory to describe the nature of molecules of gases lead to a useful model for the behavior of gases.

The Physical Behavior of Gases

10.1 Behavior of Matter in the Gaseous State

We are surrounded by an ocean of gas—the atmosphere—so many of the properties of gases are familiar to us from our daily activities. We know that squeezing a balloon decreases the volume of the gas inside. In fact, the volume of any gas, unlike that of a solid or a liquid, can be decreased greatly by increasing the pressure upon the gas. This property is known as **compressibility**. The gas in a cylinder, such as that shown in Fig. 10.1, can be compressed by increasing the weight on the piston. The gas confined in the cylinder decreases in volume until it exerts enough pressure to support the greater weight.

If we heat a gas in a closed container, such as a cola bottle or can, its pressure increases. Heating a gas in a closed cylinder like the one shown in Fig. 10.1 will increase the pressure inside the cylinder, unless the piston moves. Raising the piston increases the volume and allows the gas to expand. This expansion can keep the pressure constant. The ability of a gas to expand and fill a container of any size—the property called **expansibility**—is characteristic of all gases.

A pungent gas released into the corner of a room can, in time, be smelled all over the room even if the air is still. When a sample of gas is introduced into an evacuated container, it almost instantly fills the container uniformly. These are examples of the **diffusion** of gases. Diffusion results because the molecules of a gas are in constant high-speed motion. Two gases introduced into the same container mix by diffusion. Each gas is said to *permeate* the other. Thus we say that gases possess the properties of **diffusibility** (or **diffusion**) and **permeability**.

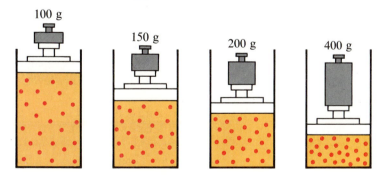

Figure 10.1

These figures illustrate how the gas molecules in a container move closer together as they are subjected to more and more pressure by placing heavier and heavier weights on a movable, gas-tight piston at constant temperature. The number of molecules represented is a minute fraction of the number present in such a volume.

10.2 Measurement of Gas Pressures

Gases exert pressure on surfaces. When you squeeze a filled balloon, it pushes back because the gas inside the balloon exerts pressure on the inside surface of the balloon. It is harder to push a dent into a balloon when it is filled with a gas at a

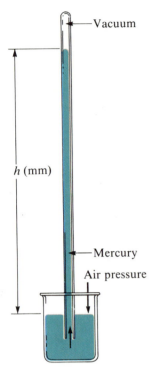

Figure 10.2

A mercury barometer. The height, h, of the mercury column is proportional to the pressure; thus the pressure can be given as the height of the column in millimeters, leading to the pressure unit mmHg.

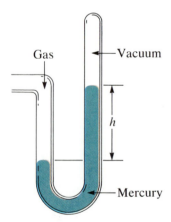

Figure 10.3

A simple two-armed mercury barometer. Such barometers are sometimes called manometers.

high pressure than when it is filled with a gas at a lower pressure. At the higher pressure the gas pushes back harder.

The pressure of a gas is measured by the force that it exerts upon a unit area of surface. In the United States pressure is often measured in units of pounds of force on a square inch of area (pounds per square inch, or psi). The pressure unit in the International System of Units (SI) is the **pascal (Pa),** which is the pressure of a force of one newton on an area of one square meter; 1 Pa = 1 newton/m^2. A **newton (N)** is the force that, when applied for 1 second, will give a 1-kilogram mass a speed of 1 meter per second. However, it may be more helpful to think of a newton as the force with which a 0.102-kilogram (3.6-ounce) object presses on a table. Thus a pascal would be the pressure exerted by 0.102 kilogram lying on a square with an edge length of 1 meter. One pascal is a very small pressure; 0.102 kilogram of water would cover a square meter to a depth of only 0.01 centimeter, and this film of water would not exert much pressure. (See exercise 6 to calculate this value.)

The units of a newton are kg m/s^2, or kg m s^{-2}. Since pressure is force per unit area, the pressure in pascals is expressed in terms of newtons per square meter. Thus,

$$Pa = N/m^2 = kg\ m\ s^{-2}/m^2 = kg\ s^{-2}/m = kg/s^2\ m$$

In many cases, units of **kilopascals (kPa)** are more convenient than units of pascals (1 kPa = 1000 Pa).

Pressure is also commonly measured in units of **atmospheres (atm).** One atmosphere of pressure is defined as the average pressure of the air at sea level at a latitude of 45°. One atmosphere is equal to 101.325 kPa.

The pressure exerted by the air (a mixture of gases) may be measured by a simple mercury **barometer** (Fig. 10.2). A barometer can be made by filling an 80-centimeter-long glass tube, closed at one end, with mercury and inverting it in a container of mercury. The mercury in the tube falls until the pressure exerted by the atmosphere upon the surface of the mercury in the container is just sufficient to support the weight of the mercury in the tube. Because the pressure of the atmosphere is proportional to the height of the mercury column (the vertical distance between the surface of the mercury in the tube and that in the open vessel), pressure is sometimes expressed in terms of **millimeters of mercury (mmHg).** A pressure of 1 mmHg is generally referred to as a **torr** after Evangelista Torricelli, who invented the barometer in 1643. A column of mercury exactly 1 millimeter high exerts a pressure of exactly 1 torr.

A second type of barometer, a **manometer,** has two arms, one closed and one open to the atmosphere or connected to a container filled with gas (Fig. 10.3). The pressure exerted by the atmosphere or by a gas in a container is proportional to the height of the column of mercury, the distance between the mercury levels in the two arms of the tube (h in the diagram).

One atmosphere of pressure (the average pressure of the air at sea level at a latitude of 45°) will support a column of mercury 760 millimeters in height. Thus 1 atmosphere of pressure is exactly 760 millimeters of mercury, or 760 torr. The pressure of the atmosphere varies with the distance above sea level and with climatic changes. At a higher elevation, the air is less dense, so the pressure is less. The atmospheric pressure at 20,000 feet is only one-half of that at sea level because about half of the atmosphere is below this elevation.

The conversions between these units are

$$1\ atmosphere = 760\ mmHg = 760\ torr = 101.325\ kPa$$

In this text we will use pressure units of atmospheres, torr, or pascals.

EXAMPLE 10.1

Weather reports in the United States often give barometric pressures in inches of mercury rather than atm or torr (29.92 in Hg = 760 torr). Convert a pressure of 29.2 in of mercury to torr, atm, and kPa.

EXAMPLE 10.1

This problem involves conversion of a measurement in one set of units to other sets of units. It is analogous to the conversions described in Section 1.9. The problem gives the relationship between inches of mercury and torr (760 torr/29.92 in Hg). First convert the pressure in inches of mercury to torr:

$$29.2 \text{ in Hg} \times \frac{760 \text{ torr}}{29.92 \text{ in Hg}} = 742 \text{ torr}$$

Now convert torr into atm and into kPa, using the conversion factors given in this section.

$$742 \text{ torr} \times \frac{1 \text{ atm}}{760 \text{ torr}} = 0.976 \text{ atm}$$

$$742 \text{ torr} \times \frac{101.325 \text{ kPa}}{760 \text{ torr}} = 98.9 \text{ kPa}$$

10.3 Relation of Volume to Pressure at Constant Temperature; Boyle's Law

Experiments show that the volume of a sample of gas is reduced to half when the pressure on the gas is doubled (Fig. 10.4), provided that the temperature does not change. Conversely, its volume doubles when the pressure is halved. Observations of this sort were summarized by the English physicist and chemist Robert Boyle in 1660 in a statement now known as **Boyle's law: The volume of a given mass of gas held at constant temperature is inversely proportional to the pressure under which it is measured,** or

$$V \propto \frac{1}{P}$$

where V is the volume of the gas, P is the pressure, and $\propto$ means "is proportional to."

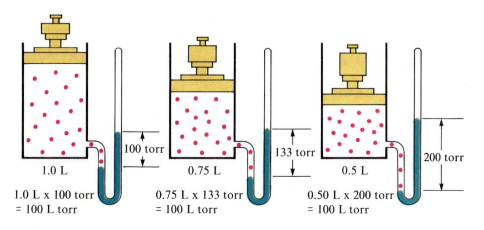

1.0 L

1.0 L x 100 torr
= 100 L torr

100 torr

0.75 L

0.75 L x 133 torr
= 100 L torr

133 torr

0.5 L

0.50 L x 200 torr
= 100 L torr

200 torr

Figure 10.4

An illustration of the variation of the pressure of a gas with volume (for a constant quantity of gas at constant temperature). Note how the product of the volume of the gas times its pressure is always equal to the same value.

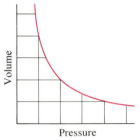

Figure 10.5

A graphical illustration of Boyle's law. Note how the volume decreases as the pressure increases and, conversely, how the volume increases as the pressure decreases. This graph illustrates an inversely proportional relationship.

The proportionality may be changed to an equality by including a constant, k, referred to as a *proportionality constant*.

$$V = \text{constant} \times \frac{1}{P}, \quad \text{or} \quad V = k \times \frac{1}{P}$$

Hence

$$PV = \text{constant}, \quad \text{or} \quad PV = k \qquad (1)$$

The value of the constant changes if the mass of gas or the temperature changes. It does not change if only the pressure or the volume of a particular mass of gas changes at constant temperature. The meaning of Equation (1) is that **the product of the pressure of a given mass of gas times its volume is always constant if the temperature does not change.** Thus as the pressure of gas increases at constant temperature, the volume must decrease in order for the product to remain constant. If the pressure decreases, the volume must increase at constant temperature. A graph showing this relationship is shown in Figure 10.5. The following examples illustrate the application of Boyle's law.

EXAMPLE 10.2

A balloon contains 14.0 L of air at a pressure of 760 torr. What will the volume of the air be when the balloon is taken to a depth of 10 ft in a swimming pool, where the pressure is 981 torr? The temperature of the air does not change.

This problem gives us the volume (14.0 L) of a gas at one pressure (760 torr) and asks for the volume of the gas at a second pressure (981 torr). From Boyle's law we know that $PV = $ constant. We can find the constant from the volume (14.0 L) at a pressure of 760 torr.

$$k = PV = 760 \text{ torr} \times 14.0 \text{ L} = 10{,}640 \text{ torr L}$$

Using this value of the constant, we can find the volume at 981 torr by rearranging the equation $PV = k$ to solve for V.

$$V = \frac{k}{P}$$

$$V = \frac{10{,}640 \text{ torr L}}{981 \text{ torr}} = 10.8 \text{ L}$$

There is a less formal way of solving this problem. From Boyle's law, Equation (1), we know that an increase in the pressure at constant temperature will result in a decrease in the volume. The new volume can be determined by multiplying the original volume by the ratio of the two pressures with the old pressure on top. In this example the ratio must have the smaller pressure in the numerator and the larger pressure in the denominator, so that when the original volume is multiplied by this ratio the calculation will show a change to a smaller volume in accord with our prediction.

$$14.0 \text{ L} \times \frac{760 \text{ torr}}{981 \text{ torr}} = 10.8 \text{ L}$$

Thus the volume is less than the original volume. Note that the unit torr appears in both the numerator and the denominator of the expression, thus it cancels out, leaving L as the unit of measurement in the answer.

A sample of Freon gas used in an air conditioner has a volume of 325 L and a pressure of 96.3 kPa at 20°C. What will the pressure of the gas be when its volume is 975 L at 20°C?

EXAMPLE 10.3

First find the constant in Boyle's law, $PV = k$, at the first pressure and volume.

$$k = 96.3 \text{ kPa} \times 325 \text{ L} = 3.130 \times 10^4 \text{ kPa L}$$

Then find the second pressure from the value of k and the second volume.

$$PV = k$$

$$P = \frac{k}{V}$$

$$P = \frac{3.130 \times 10^4 \text{ kPa } L}{975 \; L} = 32.1 \text{ kPa}$$

There is also a less formal way of solving this problem. At constant temperature, the pressure of a gas decreases as the volume of the gas increases. The new pressure will be determined by the ratio of the two volumes, with the old value on top. If the smaller volume is the numerator and the larger volume the denominator, multiplying the original pressure by this ratio will give the new (lower) pressure.

$$96.3 \text{ kPa} \times \frac{325 \; L}{975 \; L} = 32.1 \text{ kPa}$$

The new pressure must be one-third the initial pressure to permit the volume to increase threefold.

10.4 Relation of Volume to Temperature at Constant Pressure; Charles's Law

If a filled balloon is cooled so that the gas inside becomes cold, the balloon will contract (Fig. 10.6). This is an example of the effect of temperature on the volume of a confined gas. Studies of the effect of temperature on the volume of confined gases at constant pressure by the French physicist S. A. C. Charles in 1787 led to the

Figure 10.6
The effect of temperature on a gas. When a balloon filled at room temperature, $T = 25°C$, is placed in contact with liquid nitrogen, $T = -196°C$, the volume of the gas in the balloon decreases as it cools.

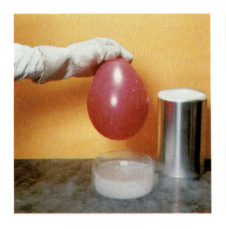

generalization known as **Charles's law: The volume of a given mass of gas is directly proportional to its temperature on the Kelvin scale when the pressure is held constant,** or

$$V \propto T$$

If we use a proportionality constant that depends on the mass of gas and its pressure, we get the equation

$$V = \text{constant} \times T, \quad \text{or} \quad V = k \times T$$

Hence

$$\frac{V}{T} = \text{constant}, \quad \text{or} \quad \frac{V}{T} = k \qquad (1)$$

This constant is different from that in Boyle's law. It does not vary unless the mass of the gas or the pressure changes. Equation (1) means that, **as the Kelvin temperature of a given mass of gas changes with no change in pressure, its volume also changes so that the ratio V/T remains the same.** A decrease in T results in a decrease in V, and an increase in T results in an increase in V. The relationship of volume and temperature is shown in Fig. 10.7. (See also Fig. 10.9.)

Charles's Law applies to the volume of a gas and *its Kelvin temperature*. In Section 1.14 we saw that the relationship between the Kelvin and Celsius temperature scales is $K = °C + 273.15$. (Recall that temperatures on the Kelvin scale are by convention reported without a degree sign.)

The following examples illustrate the application of Charles's law.

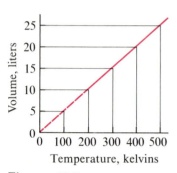

Figure 10.7

A graphical illustration of Charles's law. Note how the volume increases with increasing Kelvin temperature at constant pressure. This graph illustrates a directly proportional relationship.

EXAMPLE 10.4   A sample of carbon dioxide, CO_2, occupies 300 mL at 10°C and 750 torr. What volume will the gas have at 30°C and 750 torr? (Following the convention established in Section 1.10, assume all figures are significant.)

The relationship between volume and temperature requires the use of the Kelvin scale, so we must first convert the Celsius temperatures to the Kelvin scale:

$$10°C + 273.15 = 283 \text{ K}$$
$$30°C + 273.15 = 303 \text{ K}$$

Using the first Kelvin temperature (283 K) and the corresponding volume (300 mL), find the constant in the Charles's law equation.

$$\frac{V}{T} = k$$

$$k = \frac{300 \text{ mL}}{283 \text{ K}} = 1.06 \text{ mL/K}$$

Now solve for V at the second temperature.

$$V = T \times k$$
$$V = 303 \text{ K} \times 1.06 \text{ mL/K} = 321 \text{ mL}$$

This problem can also be solved in a less formal way. From Charles's law, we know that when the temperature of a gas is raised at constant pressure, the gas expands.

The final volume can be found by multiplying the initial volume by that ratio of the Kelvin temperatures that gives the larger volume (high temperature on top in the ratio).

$$300 \text{ mL} \times \frac{303 \text{ K}}{283 \text{ K}} = 321 \text{ mL}$$

EXAMPLE 10.5

Temperature is sometimes measured with a gas thermometer by measuring the change in the volume of the gas with temperature. A particular hydrogen gas thermometer has a volume of 150.0 cm^3 when immersed in a mixture of ice and water (0.00°C). When immersed in boiling liquid ammonia, the volume of the hydrogen, at the same pressure, is 131.7 cm^3. Find the temperature of boiling ammonia on the Kelvin and Celsius scales.

This problem asks us to find the temperature of a 131.7-cm^3 sample of hydrogen gas that has a volume of 150.0 cm^3 at a temperature of 0.00°C and the same pressure. Since the initial temperature of 0.00°C is 273.15 K, the constant for this sample of hydrogen in the Charles's law equation is

$$k = \frac{V}{T} = \frac{150.0 \text{ cm}^3}{273.15 \text{ K}} = 0.54915 \text{ cm}^3/\text{K}$$

The temperature at which the volume is 131.7 cm^3 can then be determined.

$$\frac{V}{T} = k$$

$$T = \frac{V}{k} = \frac{131.7 \text{ cm}^3}{0.54915 \text{ cm}^3/\text{K}} = 239.8 \text{ K}$$

Subtracting 273.15 from 239.8 K, we find that the temperature of the boiling ammonia on the Celsius scale is −33.3°C.

10.5 The Kelvin Temperature Scale

In Section 1.14 we discussed the Kelvin temperature scale and its relationship to the Celsius temperature scale. Now, with some knowledge of the effect of temperature on gases, we are ready to examine the basis for the Kelvin scale.

As we have already seen, experiments show that when a gas is heated, it must expand for its pressure to remain constant. Likewise, cooling any gas at constant pressure results in a decrease in its volume. When the pressure remains constant and 546 milliliters of gas at 0°C is warmed to 1°C, its volume increases by 2 milliliters to 548 milliliters; at 20°C its volume is 586 milliliters; at 273°C its volume is 1092 milliliters, twice its volume at 0°C; and so on (Fig. 10.8). This means that the volume of the gas increases by $\frac{1}{273}$ of its volume at 0° for each rise of 1° on the Celsius scale. The volume of a gas decreases in the same proportion when the temperature falls. If the temperature of 273 milliliters of a gas could be lowered from 0°C to −273°C, then the gas should have no volume at −273°C, because its volume should decrease at the rate of $\frac{1}{273}$ of its volume at 0°C for each fall of 1°C in temperature. Before the temperature of −273°C is reached, however, all gases become liquids or solids.

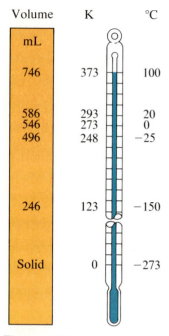

Volume	K	°C
mL		
746	373	100
586	293	20
546	273	0
496	248	−25
246	123	−150
Solid	0	−273

Figure 10.8

The volume of a gas at constant pressure is directly proportional to its Kelvin temperature.

Figure 10.9

Charles's law behavior for several gases, each at constant pressure. All gases extrapolate to a volume of 0 at −273.15°C. The dashed lines represent the extrapolated portion of each line for temperatures below the boiling point.

A plot of data obtained from several volume and temperature experiments is shown in Fig. 10.9, where the volumes of samples of several gases at constant pressure are plotted against the temperature in degrees Celsius. The straight lines reflect the direct proportionality between the volume and temperature of each sample, as described by Charles's law. The data stop at those temperatures at which the gases condense. If the straight lines are extended back (extrapolated) to the temperatures at which the gases would have a zero volume if they did not condense, the lines all intersect the temperature axis at −273.15°C. Regardless of the gas, these measurements yield −273.15°C. Below this temperature gases would theoretically have a negative volume. Since negative volumes are impossible, −273.15°C must be the lowest temperature possible. This temperature, called **absolute zero,** is taken as the zero point of the Kelvin temperature scale. The freezing point of water (0.00°C) is therefore 273.15 kelvin, and the boiling point (100.00°C) is 373.15 kelvin (Fig. 10.8).

To lower the temperature of a substance, we remove heat. Absolute zero (0 K) is the temperature reached when all possible heat has been removed from a substance. Obviously, a substance cannot be cooled any further after all possible heat has been removed from it.

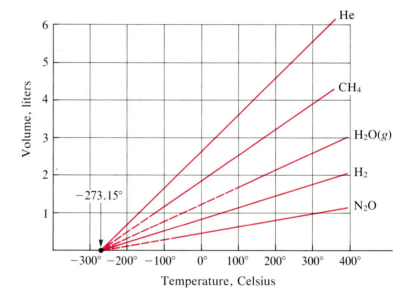

10.6 Reactions Involving Gases; Gay-Lussac's Law

Gases combine, or react, in definite and simple proportions by volume. For example, it has been determined by experiment that one volume of nitrogen will combine with three volumes of hydrogen to give two volumes of ammonia gas, provided that the volumes of the reactants and product are measured under the same conditions of temperature and pressure.

$$N_2(g) + 3H_2(g) \longrightarrow 2NH_3(g)$$

1 volume   3 volumes   2 volumes

The units, of course, must be the same: if the volume of nitrogen is measured in liters, the volumes of hydrogen and ammonia must also be measured in liters. Experimental observations on the volumes of combining gases were summarized by Joseph Louis Gay-Lussac (1788–1850) in **Gay-Lussac's law: The volumes of gases involved in a reaction, at constant temperature and pressure, can be expressed as a ratio of small whole numbers.** It has been shown that the small whole numbers in such a ratio are equal to the coefficients in the balanced equation that describes the reaction.

EXAMPLE 10.6

What volume of $O_2(g)$ measured at 25°C and 760 torr is required to react with 1.0 L of methane [$CH_4(g)$, the principal component of natural gas] measured under the same conditions of temperature and pressure?

The ratio of the volumes of CH_4 and O_2 will be equal to the ratio of their coefficients in the balanced equation for the reaction.

$$CH_4(g) + 2O_2(g) \longrightarrow CO_2(g) + 2H_2O(g)$$

1 volume   2 volumes   1 volume   2 volumes

From the equation we see that one volume of CH_4, in this case 1 L, will react with two volumes of O_2.

$$1.0 \text{ L CH}_4 \times \frac{2 \text{ L O}_2}{1 \text{ L CH}_4} = 2.0 \text{ L O}_2$$

A volume of 2.0 L of O_2 will be required to react with 1.0 L of CH_4.

It is important to remember that this law applies only to substances in the gaseous state measured at the same temperature and pressure. The volumes of any solids or liquids involved in the reactions are not considered.

EXAMPLE 10.7

What volume of $CO_2(g)$, measured at the same temperature and pressure, is formed by the reaction of 150 mL of $O_2(g)$ with carbon, a solid?

$$C(s) + O_2(g) \longrightarrow CO_2(g)$$

A solid   1 volume   1 volume

We can say nothing about the volume of carbon involved in this reaction, since it is a solid. However, the coefficients in the equation indicate that one volume of O_2 (in units of mL in this case) reacts to give one volume of CO_2.

$$150 \text{ mL O}_2 \times \frac{1 \text{ mL CO}_2}{1 \text{ mL O}_2} = 150 \text{ mL CO}_2$$

Thus 150 mL of O_2 produces 150 mL of CO_2 in this reaction.

EXAMPLE 10.8

In a recent year a volume of 683 billion cubic feet of gaseous ammonia, measured at 25°C and 1 atm, was manufactured in the United States. What volume of $H_2(g)$, measured under the same conditions, was required to prepare this amount of ammonia by reaction with N_2?

The ratio of the volumes of H_2 and N_2 will be equal to the ratio of the coefficients in the equation

$$N_2(g) + 3H_2(g) \longrightarrow 2NH_3(g)$$

1 volume   3 volumes   2 volumes

We see that two volumes of NH_3, in this case in units of billion ft^3, will be formed from three volumes of H_2. Thus

$$683 \text{ billion ft}^3 \text{ NH}_3 \times \frac{3 \text{ billion ft}^3 \text{ H}_2}{2 \text{ billion ft}^3 \text{ NH}_3} = 1.02 \times 10^3 \text{ billion ft}^3 \text{ H}_2$$

The manufacture of 683 billion ft^3 of NH_3 required 1020 billion ft^3 of H_2. (This is the volume of a cube with an edge length of approximately 1.9 mi.)

10.7 An Explanation of Gay-Lussac's Law; Avogadro's Law

Gay-Lussac's law concerning combining volumes of gases is explained by the molecular nature of gases if **equal volumes of all gases, measured under the same conditions of temperature and pressure, contain the same number of molecules.** The Italian physicist Amedeo Avogadro advanced this hypothesis in 1811 to account for the behavior of gases. His hypothesis, which has since been experimentally proven, is now accepted as fact and is known as **Avogadro's law.**

The explanation of Gay-Lussac's law by Avogadro's law may be illustrated by the reaction of the gases hydrogen and oxygen at 200°C to give gaseous water (water vapor), with the reactants and product at the same temperature and pressure. Gay-Lussac's Law indicates that two volumes of H_2 and one volume of O_2 will combine to give two volumes of H_2O.

$$2H_2(g) \; + \; O_2(g) \; \longrightarrow \; 2H_2O(g)$$
$$\text{2 volumes} \qquad \text{1 volume} \qquad\qquad \text{2 volumes}$$

According to Avogadro's law equal volumes of gaseous H_2, O_2, and H_2O contain the same number of molecules. Since two molecules of water are formed from two molecules of hydrogen and one molecule of oxygen, the volume of hydrogen required and the volume of gaseous water produced are both twice as great as the volume of oxygen required. This relationship is illustrated in Fig. 10.10.

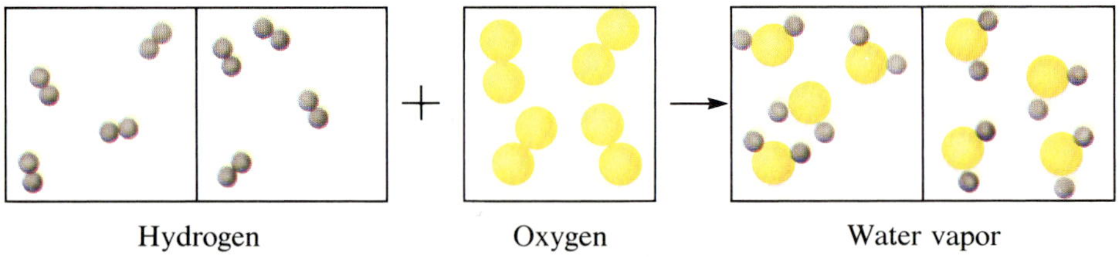

Figure 10.10
Two volumes of hydrogen combine with one volume of oxygen to yield two volumes of water vapor.

Hydrogen   Oxygen   Water vapor

10.8 The Ideal Gas Equation

Boyle's law, Charles's law, and Avogadro's law are special cases of one equation, the **ideal gas equation,** that relates pressure, volume, temperature, and number of moles of an **ideal gas** (a gas that follows these gas laws perfectly). Although it strictly applies only to an ideal gas, under normal conditions of temperature and pressure it also applies very well to real gases because real gases behave very nearly like ideal gases under these conditions. In this text we will approximate the behavior of all gases as ideal. Although we will not attempt to describe deviations from this behavior quantitatively, we shall consider them qualitatively (Section 10.18).

The ideal gas equation is

$$PV = nRT$$

where P is the pressure of a gas; V, its volume; n, the number of moles of the gas; T, its temperature on the Kelvin scale; and R, a constant called the **gas constant.** The following paragraphs show how the equation reduces to either Boyle's law, Charles's law, or Avogadro's law under the appropriate conditions.

Boyle's law (Section 10.3) states that for a given amount of a gas that exhibits ideal behavior the product of its volume, V, and its pressure, P, is a constant at constant temperature. Since n, R, and T do not change under the conditions for which Boyle's law holds, their product is a constant. Therefore the ideal gas equation, $PV = nRT$, becomes

$$PV = \text{constant}$$

which is the mathematical expression for Boyle's law.

Charles's law (Section 10.4) states that the volume, V, of a sample of an ideal gas divided by its temperature, T, on the Kelvin scale is equal to a constant at constant pressure. If we rearrange the ideal gas equation so that the terms that do not vary (n, R, and P) are on the right side and V and T are on the left, we obtain the equation expressing Charles's law.

$$\frac{V}{T} = \frac{nR}{P} = \text{constant}$$

If equal volumes of all gases, under the same conditions of temperature and pressure, contain the same number of molecules (Avogadro's law, Section 10.7), then they contain the same number of moles, n, of gas. This may be written in the form of an equation as follows:

$$V = \text{constant} \times n$$

The ideal gas equation may be rearranged to yield this expression of Avogadro's law by placing all the constant quantities (R, T, and P) together:

$$V = \frac{RT}{P} \times n$$

At constant P and T,

$$V = \text{constant} \times n$$

The numerical value of R in the ideal gas equation can be determined by substituting experimental values for P, V, n, and T in the equation. Exactly 1 mole of any gas that very closely approximates ideal behavior occupies a volume of 22.414 liters at 273.15 K and a pressure of exactly 1 atmosphere. Substituting these values in $PV = nRT$ gives

$$(1 \text{ atm})(22.414 \text{ L}) = (1 \text{ mol})(R)(273.15 \text{ K})$$

$$R = \frac{22.414 \text{ L} \times 1 \text{ atm}}{1 \text{ mol} \times 273.15 \text{ K}}$$

$$R = 0.08206 \text{ L atm/mol K}$$

The numerical value for R using other units for P, V, n, and T will be different. For example, if we take the pressure in kilopascals instead of atmospheres (1 atm = 101.325 kPa), we calculate another value of R as follows:

$$R = \frac{22.414 \text{ L} \times 101.325 \text{ kPa}}{1 \text{ mol} \times 273.15 \text{ K}}$$

$$R = 8.314 \text{ L kPa/mol K}$$

The units of pressure, volume, temperature, and amount of gas used in the ideal gas equation must always be the same as the units for R.

The following problems illustrate some of the ways that the ideal gas equation can be used to relate P, V, T, and n.

EXAMPLE 10.9

Calculate the volume occupied by 0.54 mol of N_2 at 15°C and 0.967 atm. Use the value of R with the units of L atm/mol K.

The ideal gas equation contains five terms (P, V, n, R, and T). If we know any four of these, we can rearrange the equation and find the fifth. In this problem we are given 0.54 mol, 15°C, and 0.967 atm and told to use 0.08206 L atm/mol K to find V. Thus we must rearrange $PV = nRT$ to solve for V.

$$V = \frac{nRT}{P}$$

Since the equation requires that the temperature be in kelvins, we convert the temperature from °C to K.

$$T = 15°C + 273.15 = 288 \text{ K}$$

Substitution gives

$$V = \frac{nRT}{P} = \frac{(0.54 \text{ mol})(0.08206 \text{ L atm/mol K})(288 \text{ K})}{0.967 \text{ atm}}$$

$$V = 13 \text{ L (to two significant figures, as justified by the data)}$$

Note that the units cancel to liters, an appropriate unit for volume. Cancellation of units is a very simple way of checking to see that the units for R are consistent with those for P, V, n, and T.

EXAMPLE 10.10

Methane, CH_4, can be used as fuel for an automobile; however, it is a gas at normal temperatures and pressures, which causes some problems with storage. One gallon of gasoline could be replaced by 655 g of CH_4. What is the volume of this much methane at 25°C and 745 torr? Use the value of R with the units of L atm/mol K.

As shown in Example 10.9 the volume of an ideal gas may be obtained from the equation

$$V = \frac{nRT}{P}$$

The units of R (liters, atmospheres, moles, and kelvins) require that the pressure be expressed in atmospheres rather than in torr as given in the problem, the amount of gas in moles rather than in grams, and the temperature in K. Once these conversions are completed, the values can be substituted into the equation and the volume calculated.

$$P = 745 \text{ torr} \times \frac{1 \text{ atm}}{760 \text{ torr}} = 0.980 \text{ atm}$$

$$n = 655 \text{ g CH}_4 \times \frac{1 \text{ mol}}{16.04 \text{ g CH}_4} = 40.8 \text{ mol}$$

$$T = 25°C + 273.15 = 298 \text{ K}$$

Now we substitute into the rearranged equation.

$$V = \frac{nRT}{P}$$

$$V = \frac{(40.8 \text{ mol})(0.08206 \text{ L atm/mol K})(298 \text{ K})}{0.980 \text{ atm}}$$

$$V = 1.02 \times 10^3 \text{ L}$$

Thus it would require 1020 L (269 gal) of gaseous methane at about 1 atm of pressure to replace 1 gal of gasoline. It requires a large container to hold enough methane to replace several gallons of gasoline (see the photograph).

During the Second World War drivers sometimes used fuel gas for their automobiles because gasoline was rationed. The bag on the top of this car was required to hold the large volume of gas necessary.

EXAMPLE 10.11

While resting, the average human male consumes 200 mL of O_2 per hour at 25°C and 1.0 atm for each kilogram of weight. How many moles of O_2 are consumed by a 70-kg man while resting for 1 h?

For the purpose of this problem, let us assume the oxygen consumption per kilogram per hour is good to two significant figures. The volume of O_2 consumed by a resting 70-kg male is then 14,000 mL/h, or 14 L/h.

To solve for moles, rearrange the ideal gas equation, $n = PV/RT$. Since the pressure and volume are given in atmospheres and liters, use $R = 0.08206$ L atm/mol K. The temperature is $273.15 + 25 = 298$ K. These values can now be used to find n.

$$n = \frac{PV}{RT} = \frac{1.0 \text{ atm} \times 14 \text{ L}}{0.08206 \text{ L atm/mol K} \times 298 \text{ K}} = 0.57 \text{ mol } O_2$$

EXAMPLE 10.12

What is the pressure in kilopascals in a 35.0-L balloon at 25°C filled with dried hydrogen gas produced by the reaction of 38.9 g of NaH with water?

To use the ideal gas equation, $PV = nRT$, for determining P, we must know V, n, R, and T ($P = nRT/V$). Values for V and T are given; R must be 8.314 L kPa/mol K in order for the pressure units to be kilopascals; and n can be calculated from the weight of NaH and the equation for the reaction.

$$NaH + H_2O \longrightarrow NaOH + H_2$$

$$38.9 \text{ g NaH} \times \frac{1 \text{ mol NaH}}{24.00 \text{ g NaH}} = 1.62 \text{ mol NaH}$$

$$1.62 \text{ mol NaH} \times \frac{1 \text{ mol } H_2}{1 \text{ mol NaH}} = 1.62 \text{ mol } H_2$$

so $\qquad n = 1.62$ mol

Now we convert T to the Kelvin scale.

$$T = 25°C + 273.15 = 298 \text{ K}$$

Using the ideal gas equation gives

$$P = \frac{nRT}{V} = \frac{1.62 \text{ mol} \times 8.314 \text{ L kPa/mol K} \times 298 \text{ K}}{35.0 \text{ L}} = 115 \text{ kPa}$$

The pressure in the balloon (115 kPa) is about 14% greater than atmospheric pressure (101 kPa).

EXAMPLE 10.13   What volume of hydrogen at 27°C and 723 torr may be prepared by the reaction of 8.88 g of gallium with an excess of hydrochloric acid?

$$2\text{Ga}(s) + 6\text{HCl}(aq) \longrightarrow 2\text{GaCl}_3(aq) + 3\text{H}_2(g)$$

In order to calculate the volume of hydrogen produced we need to know the number of moles, n, of hydrogen produced. This can be determined from the number of moles of gallium reacting and the chemical equation. The moles of gallium reacting can be determined from the mass of gallium used and its atomic weight. Thus this problem requires the following steps:

| Mass of Ga | → | Moles of Ga | → | Moles of H_2 | → | Volume of H_2 |

$$88.8 \text{ g Ga} \times \frac{1 \text{ mol Ga}}{69.72 \text{ g Ga}} = 0.127 \text{ mol Ga}$$

$$0.127 \text{ mol Ga} \times \frac{3 \text{ mol H}_2}{2 \text{ mol Ga}} = 0.191 \text{ mol H}_2$$

Now we use the ideal gas equation to get the volume of H_2, after converting the temperature to kelvins and the pressure to atmospheres.

$$V = \frac{nRT}{P}$$

$$V = \frac{(0.191 \text{ mol H}_2)(0.08206 \text{ L atm/mol K})(300 \text{ K})}{0.951 \text{ atm}}$$

$$V = 4.94 \text{ L H}_2$$

10.9 Standard Conditions of Temperature and Pressure

As the foregoing discussion shows, the volume of a given quantity of gas and the quantity (either moles or mass) of a given volume of gas vary with changes in pressure and temperature. In order to simplify comparisons of gases, chemists have adopted a set of **standard conditions of temperature and pressure (STP).** Accordingly, 273.15 K (0°C) and 1 atmosphere (760 torr or 101.325 kilopascals) are commonly used as standard conditions for reporting properties of gases.

Calculate the volume, in liters, occupied by 1.00 mol of an ideal gas at STP.   **EXAMPLE 10.14**

For units of R in L atm/mol K, T is exactly 273.15 K and P is exactly 1 atm. Thus

$$V = \frac{nRT}{P}$$

$$= \frac{(1.00 \text{ mol})(0.08206 \text{ L atm/mol K})(273.15 \text{ K})}{1.00 \text{ atm}}$$

$$= 22.4 \text{ L}$$

Note that the units cancel to liters, an appropriate unit for volume.

The volume of one mole of any gas at 0°C and 1 atmosphere pressure is 22.4 L and is referred to as the **molar volume** (see Fig. 10.11).

28.2 cm, or 11.1 in

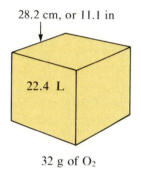

22.4 L

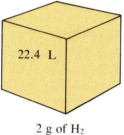

22.4 L

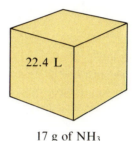

22.4 L

32 g of O_2 2 g of H_2 17 g of NH_3

Figure 10.11
A mole of any gas occupies a volume of approximately 22.4 L at 0°C and 1 atm of pressure. A cube with an edge length of 28.2 cm, or 11.1 in, contains 22.4 L.

The volume of a gas at one temperature and pressure can be converted to its volume at standard conditions or at any other temperature and pressure by using either the ideal gas equation or a combination of Boyle's and Charles's laws.

A sample of ammonia is found to occupy 250 mL under laboratory conditions of 27°C   **EXAMPLE 10.15**
and 740 torr. Correct the volume to standard conditions of 0°C and 760 torr. (All figures in the data are significant.)

   1. *Using the ideal gas equation.* Put all constant terms in the equation on the right side and the variable terms on the left. In this example the moles of ammonia and R do not vary, so

$$\frac{PV}{T} = nR = \text{constant}$$

Now we evaluate the constant term, nR, at the initial conditions (250 mL, 27 + 273.15 = 300 K, and 740 torr).

$$740 \text{ torr} \times \frac{250 \text{ mL}}{300 \text{ K}} = nR = 616.7 \text{ mL torr/K}$$

Now we calculate the volume at standard conditions (0°C = 273.15 K, 760 torr = 1 atm). Since the amount of ammonia does not change, nR remains constant:

$$\frac{PV}{T} = \text{constant} = nR$$

Rearranging the equation gives

$$V = nR \times \frac{T}{P}$$

$$V = 616.7 \text{ mL torr/K} \times \frac{273.15 \text{ K}}{760 \text{ torr}}$$

$$V = 222 \text{ mL}$$

Notice that the units cancel leaving mL, a correct unit for volume.

   2. *Using a combination of Boyle's and Charles's laws.* First convert the Celsius temperatures to the Kelvin scale: 27°C = 300 K, and 0°C = 273.15 K. A decrease in temperature from 300 K to 273.15 K alone will cause a decrease in the volume of ammonia. Therefore we multiply the original volume by a fraction made up of the two temperatures and having a value of less than 1 to correct for the temperature change:

$$250 \text{ mL} \times \frac{273.15 \text{ K}}{300 \text{ K}}$$

The increase in pressure from 740 torr to 760 torr alone will also decrease the volume, so we multiply by a fraction with the smaller pressure in the numerator. This factor may be included in the same expression with the temperature factor to obtain the corrected volume:

$$250 \text{ mL} \times \frac{273.15 \text{ K}}{300 \text{ K}} \times \frac{740 \text{ torr}}{760 \text{ torr}} = 222 \text{ mL}$$

10.10 Densities of Gases

   The **density** (mass per unit volume, Section 1.13) of a gas is the mass of one liter of the gas. For example, at STP the density of oxygen is 1.43 grams per liter, whereas that of hydrogen is 0.0899 grams per liter. Note that we must specify both the temperature and pressure of the gas when reporting a density, since the number of moles of gas (and thus the mass of the gas) in a liter changes with both temperature and pressure. Gas densities are commonly reported at STP.

   The density of a gas can be calculated by dividing the mass of a sample of the gas by its volume. Using the ideal gas equation, we can readily determine the mass of one liter of a gas and calculate its density from this mass.

EXAMPLE 10.16   Calculate the density of ethane, C_2H_6, at a pressure of 183.4 kPa and at a temperature of 25°C.

The density of C_2H_6 under these conditions is the mass of exactly 1 L of C_2H_6. The moles of C_2H_6 in exactly 1 L can be determined using the ideal gas equation and then can be converted to the mass of C_2H_6 in 1 L.

$$n = \frac{PV}{RT}$$

$$n = \frac{183.4 \text{ kPa} \times 1 \text{ L}}{8.314 \text{ L kPa/mol K} \times 298 \text{ K}} = 0.0740 \text{ mol}$$

$$\text{g } C_2H_6 = 0.0740 \text{ mol } C_2H_6 \times \frac{30.1 \text{ g } C_2H_6}{1 \text{ mol } C_2H_6} = 2.23 \text{ g } C_2H_6$$

$$\text{Density} = \frac{2.23 \text{ g}}{1 \text{ L}} = 2.23 \text{ g/L}$$

Since the pressure of C_2H_6 was expressed in kPa, the value of R with kPa in the units was used.

The density of a gas can also be calculated using a relationship derived from the ideal gas equation:

$$PV = nRT$$

The number of moles of a sample of gas, n, is equal to the mass, m, of the sample divided by the molecular weight, M, of the gas.

$$n = \frac{m}{M}$$

Thus

$$PV = \frac{m}{M}RT$$

Rearranging gives

$$PM = \frac{m}{V}RT$$

Because $\dfrac{m}{V} = d$ (where d = density)

$$PM = dRT$$

or

$$d = \frac{PM}{RT} \qquad (1)$$

Calculate the density of butane, C_4H_{10}, at a pressure of 117.4 kPa and a temperature of 125°C, using the derived relationship given in Equation (1) above.   **EXAMPLE 10.17**

$$d = \frac{PM}{RT}$$

Using $R = 8.314$ L kPa/mol K,

$$d = \frac{117.4 \text{ kPa} \times 58.12 \text{ g/mol}}{8.314 \text{ L kPa/mol K} \times 398 \text{ K}} = 2.06 \text{ g/L}$$

Equal volumes of different gases at the same temperature and pressure contain the same number of molecules. Consequently, the ratio of the masses of the gases contained in the same volume is equal to the ratio of their molecular weights, and the ratio of the densities of two gases is also equal to the ratio of their molecular weights, as demonstrated below.

As shown in Equation (1) in this section,

$$d = \frac{PM}{RT}$$

At a given temperature and pressure (R being a constant), P, T, and R are all constant and can be grouped into a single constant term k ($= P/RT$). Hence

$$d = kM$$

where M is the molecular weight of the gas. For two gases at the same temperature and pressure, and using the same value of R, the constant k will be the same. Thus

$$d_1 = kM_1$$

and

$$d_2 = kM_2$$

Taking the ratio of the two densities

$$\frac{d_1}{d_2} = \frac{kM_1}{kM_2}$$

$$\frac{d_1}{d_2} = \frac{M_1}{M_2}$$

Thus the density of a gas can be calculated by comparison to the density of a second gas of known density.

EXAMPLE 10.18   The density of oxygen at STP is 1.429 g/L. Calculate the density of carbon dioxide at STP.

The ratio of the densities of these gases at STP is the same as the ratio of their molecular weights.

$$\frac{d_{CO_2}}{d_{O_2}} = \frac{\text{mol. wt } CO_2}{\text{mol. wt } O_2}$$

$$d_{CO_2} = d_{O_2} \times \frac{M_{CO_2}}{M_{O_2}}$$

$$d_{CO_2} = 1.429 \text{ g/L} \times \frac{44.01}{32.00} = 1.965 \text{ g/L}$$

10.11 Determination of the Molecular Weights of Gases or Volatile Compounds

The molecular weight of a gas can be found experimentally by determining the mass of a given volume of the gas at a known temperature and pressure. The number of moles of gas is determined by using the ideal gas equation, and the molar mass and the molecular weight are determined from the number of moles of gas and its mass (Section 2.3).

Cyclopropane is a commonly used anesthetic. If a 2.00-L flask contains 3.11 g of cyclopropane gas at 684 torr and 23°C, what is the molecular weight of cyclopropane ($R = 0.08206$ L atm/mol K)?

EXAMPLE 10.19

To use the ideal gas equation, we must express temperature in kelvins, and the units of P and V must correspond to the units of R. Using 0.08206 L atm/mol K as the value of R requires that P be expressed in atmospheres.

$$P = 684 \text{ torr} \times \frac{1 \text{ atm}}{760 \text{ torr}} = 0.900 \text{ atm}$$

$$T = 23°C + 273.15 = 296 \text{ K}$$

Now we calculate the number of moles of cyclopropane using the ideal gas equation.

$$n = \frac{PV}{RT}$$

$$n = \frac{0.900 \text{ atm} \times 2.00 \text{ L}}{0.08206 \text{ L atm/mol K} \times 296 \text{ K}} = 0.0741 \text{ mol}$$

We use the mass of cyclopropane and the number of moles to calculate the molar mass.

$$\text{Molar mass of } C_3H_6 = \frac{m}{n} = \frac{3.11 \text{ g}}{0.0741 \text{ mol}} = 42.0 \text{ g/mol}$$

A molar mass of 42.0 g/mol corresponds to a molecular weight of 42.0 (Section 2.3).

The molecular weight could also be calculated using the relationship derived from the ideal gas equation shown in Equation (1) in Section 10.10.

$$PV = \frac{m}{M}RT \qquad \text{(where } m = \text{mass and } M = \text{molar mass)}$$

Rearranging to solve for M,

$$M = \frac{mRT}{PV}$$

$$= \frac{3.11 \text{ g} \times 0.08206 \text{ L atm/mol K} \times 296 \text{ K}}{0.900 \text{ atm} \times 2.00 \text{ L}}$$

$$= 42.0 \text{ g/mol}$$

which corresponds to a molecular weight of 42.0.

EXAMPLE 10.20

A sample of phosphorus vapor weighing 3.243×10^{-2} g at 550°C exerted a pressure of 31.89 kPa in a 56.0-mL bulb. What is the molecular weight and molecular formula of phosphorus vapor?

Since the pressure is given in kPa, we must take 8.314 L kPa/mol K as the value of R. The units of volume must be liters and the temperature, kelvins.

$$T = 550 + 273.15 = 823 \text{ K}$$

$$V = 56.0 \text{ mL} \times \frac{1 \text{ L}}{1000 \text{ mL}} = 0.0560 \text{ L}$$

$$n = \frac{PV}{RT}$$

$$n = \frac{31.89 \text{ kPa} \times 0.0560 \text{ L}}{8.314 \text{ L kPa/mol K} \times 823 \text{ K}} = 2.61 \times 10^{-4} \text{ mol}$$

Using the mass of phosphorus vapor,

$$\text{Molar mass} = \frac{3.243 \times 10^{-2} \text{ g}}{2.61 \times 10^{-4} \text{ mol}} = 124 \text{ g/mol}$$

Hence the molecular weight is 124.

Since the molecular weight of phosphorus vapor is 124 and the atomic weight of a single phosphorus atom is 31, there must be 124/31, or 4, phosphorus atoms in a molecule of phosphorus vapor. The molecular formula is therefore P_4. The P_4 molecule is a tetrahedron (Fig. 10.12).

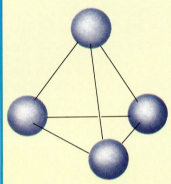

Figure 10.12

The molecular structure of gaseous phosphorus, P_4.

Alternatively, the molar mass can be calculated by using Equation (1) from Section 10.10.

$$M = \frac{mRT}{PV}$$

$$= \frac{(3.243 \times 10^{-2} \text{ g}) \times 8.314 \text{ L kPa/mol K} \times 823 \text{ K}}{31.89 \text{ kPa} \times 0.0560 \text{ L}}$$

$$= 124 \text{ g/mol}$$

10.12 The Pressure of a Mixture of Gases; Dalton's Law

If nitrogen gas is at a pressure of 100 torr in a 1-liter flask, and oxygen gas is at a pressure of 100 torr in a second 1-liter flask, transfer of the oxygen into the flask with the nitrogen gives a mixture with a total pressure of 200 torr (provided the temperature does not change). In the absence of chemical interaction between the components of a mixture of gases, the individual gases do not affect one another's pressures, and each gas exerts the same pressure before and after it is mixed. The pressure exerted by each gas is called the **partial pressure** of that gas, and the total pressure of the mixture of gases is the sum of the partial pressures of all the gases present. This is **Dalton's law of partial pressures,** which may be stated as follows: **The total pressure of a mixture of ideal gases is equal to the sum of the partial pressures of the component gases.**

$$P_T = P_A + P_B + P_C + \cdots$$

In this equation P_T is the total pressure of a mixture of gases; P_A is the pressure of gas A; P_B, the pressure of gas B; etc.

EXAMPLE 10.21

What is the total pressure in atmospheres in a 10.0-L vessel containing 2.50×10^{-3} mol of H_2, 1.00×10^{-3} mol of He, and 3.00×10^{-4} mol of Ne at 35°C?

The total pressure in the 10.0-L vessel is the sum of the partial pressures of the gases, since they do not react with each other.

$$P_T = P_{H_2} + P_{He} + P_{Ne}$$

The partial pressure of each gas can be determined from the ideal gas equation, using $P = nRT/V$.

$$P_{H_2} = \frac{(2.50 \times 10^{-3} \text{ mol})(0.08206 \text{ L atm/mol K})(308 \text{ K})}{10.0 \text{ L}} = 6.32 \times 10^{-3} \text{ atm}$$

$$P_{He} = \frac{(1.00 \times 10^{-3} \text{ mol})(0.08206 \text{ L atm/mol K})(308 \text{ K})}{10.0 \text{ L}} = 2.53 \times 10^{-3} \text{ atm}$$

$$P_{Ne} = \frac{(3.00 \times 10^{-4} \text{ mol})(0.08206 \text{ L atm/mol K})(308 \text{ K})}{10.0 \text{ L}} = 7.58 \times 10^{-4} \text{ atm}$$

$$P_T = (0.00632 + 0.00253 + 0.00076) \text{ atm} = 9.61 \times 10^{-3} \text{ atm}$$

The pressure of a sample of a gas that does not react with water can be measured by collecting the gas in a container over water and making its pressure equal to air pressure (which can be measured by a barometer). This is easily accomplished by adjusting the water level (Fig. 10.13) so that it is the same both inside and outside the container. The gas is then at the existing atmospheric pressure.

However, another factor must be considered when determining pressure by this method. There is always a little gaseous water (water vapor) above a sample of water. Thus when a gas is collected over water, it soon becomes saturated with water vapor. The total pressure of the mixture equals the sum of the partial pressures of the gas and the water vapor, according to Dalton's law. The pressure of the pure gas is therefore equal to the total pressure minus the pressure of the water vapor.

The pressure of the water vapor above a sample of water in a closed container like that in Fig. 10.13 depends on the temperature, as shown in Fig. 10.14 and Table 10.1.

Pressure of gas and water vapor is 760 torr

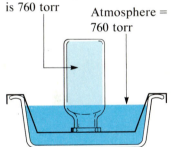

Atmosphere = 760 torr

Figure 10.13

Method by which the pressure on a confined mixture of gases can be made equal to the atmospheric pressure. The level of water inside and outside the vessel is made the same. In the diagram a typical atmospheric pressure of 760 torr is indicated. The actual pressure depends on temperature and altitude.

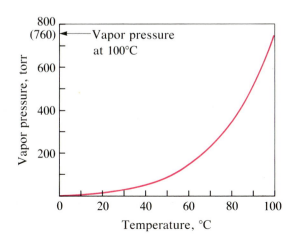

Figure 10.14

A graph illustrating change of water vapor pressure with change in temperature.

Table 10.1 Vapor pressure of ice and water at various temperatures

Temperature, °C	Pressure, torr	Temperature, °C	Pressure, torr	Temperature, °C	Pressure, torr
−10	1.95	12	10.5	27	26.7
−5	3.0	13	11.2	28	28.3
−2	3.9	14	12.0	29	30.0
−1	4.2	15	12.8	30	31.8
0	4.6	16	13.6	35	42.2
1	4.9	17	14.5	40	55.3
2	5.3	18	15.5	50	92.5
3	5.7	19	16.5	60	149.4
4	6.1	20	17.5	70	233.7
5	6.5	21	18.7	80	355.1
6	7.0	22	19.8	90	525.8
7	7.5	23	21.1	95	633.9
8	8.0	24	22.4	99	733.2
9	8.6	25	23.8	100.0	760.0
10	9.2	26	25.2	101.0	787.6
11	9.8				

At 23°C, the vapor pressure of water is 21.1 torr (Table 10.1). Thus, at 23°C the total pressure of 760 torr in the bottle in Fig. 10.13 would consist of 21 torr due to water vapor and 739 torr due to the other gases. At 29°C, water has a vapor pressure of 30 torr. At 29°C the bottle would contain water vapor at 30 torr, and the other gases at 730 torr.

EXAMPLE 10.22 If 0.200 L of argon is collected over water at a temperature of 26°C and a pressure of 750 torr in a system like that shown in Fig. 10.13, what is the partial pressure of argon?

According to Dalton's law, the total pressure in the bottle (750 torr) is the sum of the pressure of argon and the pressure of gaseous water.

$$P_T = P_{Ar} + P_{H_2O}$$

Rearranging this equation to solve for the pressure of argon gives

$$P_{Ar} = P_T - P_{H_2O}$$

The pressure of water vapor above a sample of liquid water at 26°C is 25.2 torr (Table 10.1). So,

$$P_{Ar} = 750 \text{ torr} - 25.2 \text{ torr} = 725 \text{ torr}$$

EXAMPLE 10.23 A 1.34-g sample of calcium carbide, CaC_2, was allowed to react with water. The acetylene, C_2H_2, produced by the reaction

$$CaC_2(s) + 2H_2O(l) \longrightarrow C_2H_2(g) + Ca(OH)_2(s)$$

was collected over water in a system like that shown in Fig. 10.15. If 471 mL of acetylene gas was collected at a temperature of 23°C and a pressure of 743 torr, what was the percent yield of the reaction?

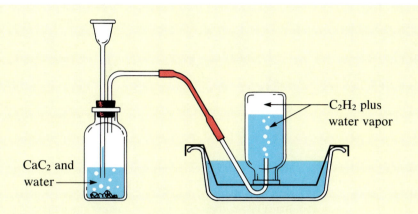

CaC₂ and water →

C₂H₂ plus water vapor

Figure 10.15
Acetylene from the reaction of CaC₂ with water is collected over water.

To determine the percent yield (Section 2.10), we need the actual yield and the theoretical yield. We determine the theoretical yield as follows:

$$1.34 \text{ g CaC}_2 \times \frac{1 \text{ mol CaC}_2}{64.1 \text{ g CaC}_2} = 2.09 \times 10^{-2} \text{ mol CaC}_2$$

$$2.09 \times 10^{-2} \text{ mol CaC}_2 \times \frac{1 \text{ mol C}_2\text{H}_2}{1 \text{ mol CaC}_2} = 2.09 \times 10^{-2} \text{ mol C}_2\text{H}_2$$

To determine the actual yield we need to calculate the moles of C_2H_2 produced. This is available from the ideal gas equation, $PV = nRT$, and the data in the problem. However, the pressure in the equation must be the pressure of $C_2H_2(g)$, not the pressure of a mixture of $H_2O(g)$ and $C_2H_2(g)$. So we use Dalton's law to get that partial pressure. The vapor pressure of water at 23°C is 21.1 torr (Table 10.1). Therefore,

$$P_{C_2H_2} = P_T - P_{H_2O}$$
$$P_{C_2H_2} = 743 \text{ torr} - 21.1 \text{ torr} = 722 \text{ torr (or 0.950 atm)}$$

Using the ideal gas equation with P in atmospheres, T in kelvins, and V in liters, we find n, the moles of C_2H_2 produced.

$$n_{C_2H_2} = \frac{PV}{RT}$$

$$n_{C_2H_2} = \frac{0.950 \text{ atm} \times 0.471 \text{ L}}{0.08206 \text{ L atm/mol K} \times 296 \text{ K}} = 1.84 \times 10^{-2} \text{ mol}$$

Now we calculate the percent yield.

$$\text{Percent yield} = \frac{\text{actual yield}}{\text{theoretical yield}} \times 100$$

$$\text{Percent yield} = \frac{1.84 \times 10^{-2} \text{ mol}}{2.09 \times 10^{-2} \text{ mol}} \times 100 = 88.0\%$$

10.13 Diffusion of Gases; Graham's Law

When a sample of gas is set free from one part of a closed container, it very quickly diffuses throughout the container (see Fig. 10.16).

If a mixture of gases is placed in a container with porous walls, diffusion of the gases through the walls occurs. The lighter gases diffuse through the small openings of

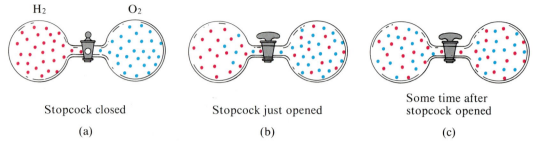

Stopcock closed   Stopcock just opened   Some time after stopcock opened

(a)   (b)   (c)

Figure 10.16

Two gases, H_2 and O_2, separated by a stopcock (a) intermingle when the stopcock is opened (b); they mix by diffusing together. The rapid motion of the molecules and the relatively large spaces between them explain why diffusion occurs. Note that H_2, the lighter of the two gases, diffuses faster than O_2. Thus the instant the stopcock is opened (b), more H_2 molecules move to the O_2 side than O_2 molecules move to the H_2 side. After some time has passed (c), the slower-moving O_2 molecules and the faster-moving H_2 molecules have both distributed themselves evenly on either side of the vessel.

the porous walls more rapidly than the heavier ones. In 1832 Thomas Graham studied the rates of diffusion of different gases and formulated **Graham's law: The rates of diffusion of gases are inversely proportional to the square roots of their densities (or their molecular weights).**

$$\frac{\text{Rate of diffusion of gas A}}{\text{Rate of diffusion of gas B}} = \frac{\sqrt{\text{density B}}}{\sqrt{\text{density A}}} = \frac{\sqrt{\text{mol. wt B}}}{\sqrt{\text{mol. wt A}}}$$

EXAMPLE 10.24   Calculate the ratio of the rate of diffusion of hydrogen to the rate of diffusion of oxygen.

Using densities. The density of hydrogen is 0.0899 g/L, and that of oxygen is 1.43 g/L.

$$\frac{\text{Rate of diffusion of hydrogen}}{\text{Rate of diffusion of oxygen}} = \frac{\sqrt{1.43 \text{ g/L}}}{\sqrt{0.0899 \text{ g/L}}} = \frac{1.20}{0.300} = \frac{4}{1}$$

Using molecular weights.

$$\frac{\text{Rate of diffusion of hydrogen}}{\text{Rate of diffusion of oxygen}} = \frac{\sqrt{32}}{\sqrt{2}} = \frac{\sqrt{16}}{\sqrt{1}} = \frac{4}{1}$$

Hydrogen diffuses four times as rapidly as oxygen.

Rates of diffusion depend on the speeds of the molecules—the faster the molecules move, the faster they diffuse. Molecules of smaller mass move with faster speeds than do molecules of larger mass; thus the former diffuse more rapidly. Differences in diffusion rates of gaseous substances are used in the separation of light isotopes from heavy ones. Diffusion through a porous barrier from a region of higher pressure to one of lower pressure is used for the large-scale separation of gaseous $^{235}_{92}UF_6$ from $^{238}_{92}UF_6$ at the atomic energy installation in Oak Ridge, Tennessee. It is said that for separation to be complete, a given volume of UF_6 must be diffused some two million times.

The Molecular Behavior of Gases

10.14 The Kinetic-Molecular Theory

Bernoulli in 1738, Poule in 1851, and Kronig in 1856 suggested that properties of gases such as compressibility, expansibility, and diffusibility could be explained by assuming that they consist of continuously moving molecules. During the latter half of the nineteenth century, Clausius, Maxwell, Boltzmann, and others developed this hypothesis into the detailed **kinetic-molecular theory** of gases. This theory was developed for an ideal gas using the following assumptions about the nature of the molecules of such a gas:

1. Gases are considered to be composed of molecules that are in continuous, completely random motion in all directions. The molecules move in straight lines and change direction only when they collide with other molecules or with the walls of a container.

2. The pressure of a gas in a container results from the bombardment of the walls of the container by the gas molecules.

3. At a given temperature, the pressure in a container does not change with time. Thus the collisions among molecules and between molecules and walls must be elastic; that is, the collisions involve no loss of energy due to friction.

4. At relatively low pressures the average distance between gas molecules is large compared to the size of the molecules.

5. At relatively low pressures the attractive forces between gas molecules, which depend on the separation of the molecules, are of no significance and can be ignored since the molecules are relatively widely separated.

6. Since the molecules are small compared to the distances between them, they may be considered to have no volume.

7. The average kinetic energy of the molecules (the energy due to their motion) is proportional to the temperature of the gas on the Kelvin scale and is the same for all gases at the same temperature.

As we will show in the next section, the assumptions of the kinetic-molecular theory explain the gas laws in a qualitative way.

10.15 Relation of the Behavior of Gases to the Kinetic-Molecular Theory

The gas laws may be explained qualitatively in light of the kinetic-molecular theory. The true test of the theory, however, is its ability to describe the behavior of a gas *quantitatively*. We shall show that the various gas laws can be derived from the assumptions of the theory; this has led chemists to believe that the assumptions of the theory represent the properties of gas molecules.

1. BOYLE'S LAW. The pressure exerted by a gas in a container is caused by the bombardment of the walls of the container by rapidly moving molecules of the gas. The pressure varies directly with the number of molecules hitting the walls per unit of time. Reducing the volume of a given mass of gas by half will double the number of molecules per unit of volume. The number of impacts per unit of time on the same area of wall surface will also be doubled. Doubling the number of impacts per unit of area

doubles the pressure. These results are in accordance with Boyle's law relating volume and pressure.

2. CHARLES'S LAW. At constant volume an increase in temperature increases the pressure of a gas. This increase in pressure reflects the increase in average speed and kinetic energy of the molecules as the temperature is raised. An increase in the average speed results in more frequent and harder impacts on the walls of the container—that is, in greater pressure.

In order for the pressure to remain constant with the increasing temperature, the volume must increase so that each molecule travels farther, on the average, before hitting a wall. The smaller number of molecules striking any wall at a given time exactly offsets the greater force with which each molecule hits, making it possible for the pressure to remain constant. This may be visualized by picturing a cylinder fitted with a piston. As the temperature increases, the molecules hit the piston harder. The piston must move up, thereby increasing the volume, if the pressure is to remain the same.

3. DALTON'S LAW. Because of the relatively large distance between the molecules of a gas, the molecules of one component of a mixture of gases will bombard the walls of the container with the same frequency in the presence of other kinds of molecules as in their absence. Thus the total pressure of a mixture of gases will be the sum of the partial pressures of the individual gases.

4. GRAHAM'S LAW. The facts that the molecules of a gas are in rapid motion and that the spaces between the molecules are relatively large explain the phenomenon of diffusion. Gas molecules can move past each other easily.

The rate of diffusion depends on the mass of the molecules of a gas since, at the same temperature, molecules of all gases have the same average kinetic energy. The kinetic energy, KE, of a moving body, in joules, is equal to $\frac{1}{2}mu^2$ where m is its mass in kilograms and u is its speed in meters per second. The average kinetic energy, KE_{avg}, of the molecules of a gas consisting of one type of molecule is equal to $\frac{1}{2}m(u^2)_{avg}$, where $(u^2)_{avg}$ is the average of the squares of the speeds of the molecules. Thus at a given temperature, lighter molecules, like hydrogen, must move at higher speeds than heavier molecules, like oxygen, since both have the same average kinetic energy (Fig. 10.17).

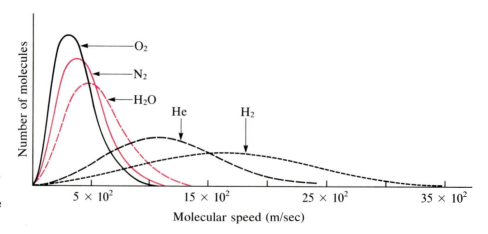

Figure 10.17

The distribution of the molecular speeds of different gases. The lighter gases move with faster average speeds.

For two different gases at the same temperature and pressure,

$$\text{Average kinetic energy for first gas} = \tfrac{1}{2}m_1(u_1{}^2)_{\text{avg}}$$

$$\text{Average kinetic energy for second gas} = \tfrac{1}{2}m_2(u_2{}^2)_{\text{avg}}$$

where m_1 and m_2 are the masses of individual molecules of the two gases and $(u_1{}^2)_{\text{avg}}$ and $(u_2{}^2)_{\text{avg}}$ are the averages of the squares of their speeds.

The kinetic energies of the two gases are equal according to the kinetic-molecular theory. Hence

$$\tfrac{1}{2}m_1(u_1{}^2)_{\text{avg}} = \tfrac{1}{2}m_2(u_2{}^2)_{\text{avg}}$$

Dividing both sides of the equation by $\tfrac{1}{2}$ gives

$$m_1(u_1{}^2)_{\text{avg}} = m_2(u_2{}^2)_{\text{avg}}$$

On rearranging, we have

$$\frac{(u_1{}^2)_{\text{avg}}}{(u_2{}^2)_{\text{avg}}} = \frac{m_2}{m_1}$$

Taking square roots gives

$$\frac{(u_1)_{\text{rms}}}{(u_2)_{\text{rms}}} = \frac{\sqrt{m_2}}{\sqrt{m_1}}$$

where $(u_1)_{\text{rms}}$ and $(u_2)_{\text{rms}}$ are the root-mean-square speeds of the respective molecules—that is, the square roots of the averages of the squares of the speeds:

$$(u_1)_{\text{rms}} = \sqrt{(u_1{}^2)_{\text{avg}}}, \qquad (u_2)_{\text{rms}} = \sqrt{(u_2{}^2)_{\text{avg}}}$$

The rate at which a gas will diffuse, R, is proportional to u_{rms}, the root-mean-square speed of its molecules. The molecular weights, M_1 and M_2, of two gases are proportional to the masses of the individual molecules. Thus

$$\frac{R_1}{R_2} = \frac{\sqrt{M_2}}{\sqrt{M_1}}$$

This equation states that the ratio of the rates of diffusion of two gases, held at the same temperature and pressure, equals the inverse ratio of the square roots of their molecular weights.

As shown in Section 10.10, the ratio of the densities (d_1 and d_2) of two gases is the same as the ratio of their molecular weights. Since $d_2/d_1 = M_2/M_1$,

$$\frac{R_1}{R_2} = \frac{\sqrt{d_2}}{\sqrt{d_1}}$$

Therefore Graham's law can equally well be expressed in terms of the square roots of the molecular weights and in terms of the square roots of the densities of the gases.

Hydrogen is the lightest gas and therefore diffuses the most rapidly. The root-mean-square speed of its molecules at room temperature is about 1 mile per second; that of oxygen molecules is about $\tfrac{1}{4}$ mile per second. Collisions between molecules make diffusion rates much lower than would be expected from such speeds. A molecule in hydrogen gas at standard conditions has about 11 billion collisions per second. Hydrogen molecules travel about 1.7×10^{-5} centimeter between collisions and, on the average, collide about 60,000 times in traveling between two points 1 centimeter apart.

10.16 The Distribution of Molecular Velocities

The individual molecules of a gas travel at different speeds. Because of the enormous number of collisions, the speeds of the molecules vary from practically zero to thousands of meters per second. The exchange of energy accompanying each collision can change the speed of a given molecule over a wide range. However, since a large number of molecules is involved, the distribution of the molecular speeds of the total number of molecules is constant.

The distribution of molecular speeds is shown in Fig. 10.18. The vertical axis represents the number of molecules, and the horizontal axis represents the molecular speed, u. The most probable speed is α, and the root-mean-square speed of the molecules is u_{rms}. The graph indicates that very few molecules move at very low or very high speeds. The number of molecules with intermediate speeds increases rapidly up to a maximum and then drops off rapidly.

At the higher temperature the whole curve is shifted to higher speeds. Thus at a higher temperature more molecules have higher speeds and fewer molecules have lower speeds. This is in accord with expectations based on kinetic-molecular theory.

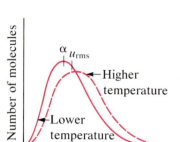

Figure 10.18

Distributions of molecular speed versus number of molecules of a gas at different temperatures.

10.17 Derivation of the Ideal Gas Equation from Kinetic-Molecular Theory

The mathematical expression of the ideal gas equation, $PV = nRT$, may be derived from the kinetic-molecular theory of gases. Consider **N** molecules, each having a mass m, confined within a cubical container, as shown in Fig. 10.19, with an edge length of l centimeters. Although the molecules are moving in all possible directions, we may assume that $\frac{1}{3}$ of the molecules ($\frac{1}{3}$**N**) are moving in the direction of the x axis, $\frac{1}{3}$ in the direction of the y axis, and $\frac{1}{3}$ in the direction of the z axis. This assumption is valid since the motion of the molecules is entirely random and no particular direction is preferred.

Initially, let us assume that all molecules are moving with the same speed, and consider the collisions of molecules with wall A (Fig. 10.19). A molecule, on the average, travels $2l$ centimeters between two consecutive collisions with wall A as it moves back and forth across the container. If the speed of the molecule is u centimeters per second, it will collide $u/2l$ times per second with wall A. Since the collisions are perfectly elastic, the molecule will rebound with a speed of $-u$, having lost no kinetic energy as a result of the collision. Because momentum is defined as the product of mass and speed, the average momentum before a collision is mu and after a collision is $-mu$. Thus the average change in momentum per molecule per collision is $2mu$. There are $u/2l$ collisions per second, making the average change in momentum per molecule per second

$$2mu \times \frac{u}{2l} = \frac{mu^2}{l}$$

The total change in momentum per second for the **N**/3 molecules that can collide with wall A is

$$\frac{\mathbf{N}}{3} \times \frac{mu^2}{l}$$

This is the average force on wall A, because force is equal to change of momentum per

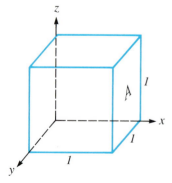

Figure 10.19

Drawing illustrating the container used in the derivation of the ideal gas equation from the kinetic-molecular theory.

second. Pressure is force per unit area. Since the area of A is l^2, the pressure P on A is

$$P = \frac{\text{force}}{\text{area}} = \frac{1}{l^2} \times \frac{N}{3} \times \frac{mu^2}{l} = \frac{N}{3} \times \frac{mu^2}{l^3}$$

However, l^3 is equal to the volume, V, of the cubical container; thus we have

$$P = \frac{N}{3} \times \frac{mu^2}{V}$$

or
$$PV = \tfrac{1}{3}Nmu^2$$

Actually, molecules move with different speeds, so we must use the average of the squares of their speeds. Thus

$$PV = \tfrac{1}{3}Nm(u^2)_{\text{avg}}$$

This is the fundamental equation of the kinetic-molecular theory of gases. It describes the behavior of an ideal gas exactly, and it is a very good approximation for the behavior of real gases at ordinary pressures.

The equation above may be written as follows:

$$PV = \tfrac{2}{3}N[\tfrac{1}{2}m(u^2)_{\text{avg}}]$$

The term in the set of brackets, $\tfrac{1}{2}m(u^2)_{\text{avg}}$, is the expression for the average kinetic energy of the gas, KE_{avg} (Section 10.15, Part 4). Thus

$$PV = \tfrac{2}{3}N(KE_{\text{avg}})$$

The average kinetic energy of the molecules, KE_{avg}, is directly proportional to the Kelvin temperature, T.
$$KE_{\text{avg}} = kT$$

The number of molecules, N, is proportional to the number of moles of molecules, n.
$$N = k'n$$

Substituting the expressions for KE_{avg} and N in the equation for PV gives

$$PV = \tfrac{2}{3}(k'n)(kT) = n(\tfrac{2}{3}k'k)T$$

The expression $\tfrac{2}{3}k'k$ is the constant (being made up entirely of constants) that we write as R and call the gas constant. Therefore,

$$PV = nRT$$

Thus, from the assumptions of the kinetic-molecular theory (Section 10.14), we can derive the ideal gas equation. Since this equation describes the behavior of gases very well, the assumptions used in its derivation are strongly supported.

10.18 Deviations from Ideal Gas Behavior

When a gas at ordinary temperature and pressure is compressed, the volume is reduced by crowding the molecules closer together. This reduction in volume is really a reduction in the amount of empty space between the molecules. At high pressures the molecules are crowded so close together that the actual volume of the molecules, the molecular volume, is a relatively large fraction of the total volume, including the empty space between molecules. Because the volume of the molecules themselves cannot be compressed, only a fraction of the entire volume is affected by a further increase in pressure. Thus at very high pressures the whole volume is not inversely proportional to the pressure, as predicted by Boyle's law. It should be noted, nevertheless, that even at moderately high pressures most of the volume is not occupied by molecules.

The molecules in a real gas at relatively low pressures have practically no attraction for one another because they are far apart. Thus they behave almost like molecules of ideal gases. However, as the molecules are crowded closer together at high pressures, the force of attraction between the molecules increases. This attraction has the same effect as does an increase in external pressure. Consequently, when external pressure is applied to a volume of gas, especially at low temperatures, there is a slightly greater decrease in volume than should be achieved by the pressure alone. This slightly greater decrease in volume caused by intermolecular attraction is more pronounced at low temperatures because the molecules move more slowly, their kinetic energy is smaller relative to the attractive forces, and they fly apart less easily after collisions with one another.

In 1879 van der Waals expressed the deviations of real gases from the ideal gas laws quantitatively in what is now known as the **van der Waals equation:**

$$\left(P + \frac{n^2a}{V^2}\right)(V - nb) = nRT$$

The constant a represents the attraction between molecules, and van der Waals assumed that this force varies inversely as the square of the total volume of the gas. (See Section 11.11 for a discussion of the nature of the van der Waals force.) Since this force augments the pressure and thus tends to make the volume smaller, it is added to the term P.

The term b in the equation represents the total volume of all of the molecules themselves and is subtracted from V, the total volume of the gas. When V is large, both nb and n^2a/V^2 are negligible, and the van der Waals equation reduces to the simple gas equation, $PV = nRT$.

At low pressures the correction for intermolecular attraction, a, is more important than the one for molecular volume, b. At high pressures and small volumes the correction for the volume of the molecules becomes important, because the molecules themselves are incompressible and constitute an appreciable fraction of the total volume. At some intermediate pressure the two corrections cancel one another, and the gas appears to follow the relationship given by $PV = nRT$ over a small range of pressures.

Strictly speaking, then, the ideal gas equation applies exactly only to gases whose molecules do not attract one another and do not occupy an appreciable part of the whole volume. Because no real gases have these properties, we speak of such hypothetical gases as *ideal* gases. Under ordinary conditions, however, the deviations from the gas laws are so slight that they may be disregarded.

For Review

Summary

A gas takes the shape and volume of its container. Gases are readily compressible and capable of infinite expansion. The pressure of a gas may be expressed (in terms of force per unit area) in the SI units of **pascals** or **kilopascals.** Other units used for pressure include **torr** and **atmospheres** (1 atm = 760 torr = 101.325 kPa).

The behavior of gases can be described by a number of laws based on experimental observations of their properties. For example, the volume of a given quantity of a gas is inversely proportional to the pressure of the gas, provided that the temperature does not change **(Boyle's law).** The volume of a given quantity of gas is directly proportional to its temperature on the Kelvin scale, provided that the pressure does not change **(Charles's law).** Gases react in simple proportions by volume **(Gay-Lussac's law),** since under the same conditions of temperature and pressure, equal volumes of all gases contain the same number of molecules **(Avogadro's law).** The rates of diffusion of gases are inversely proportional to the square roots of their densities or to the square roots of their molecular weights **(Graham's law).** In a mixture of gases the total pressure is equal to the sum of the partial pressures of the gases present. The **partial pressure** of a gas is the pressure that the gas would exert if it were the only gas present. The volumes and densities of gases are often reported under **standard conditions of temperature and pressure (STP):** 0°C and 1 atmosphere.

The equations describing Boyle's law, Charles's law, and Avogadro's law are special cases of the **ideal gas equation,** $PV = nRT$, where P is the pressure of the gas; V, its volume; n, the number of moles of the gas; and T, its Kelvin temperature. R is the **gas constant.** The value used for the gas constant when applying the equation depends on the units used for P and V.

The ideal gas equation has been derived for a gas that obeys the hypotheses of the **kinetic-molecular theory** of gases. This theory assumes that gases consist of molecules of negligible volume, which are widely separated with no attraction between them. The molecules move randomly and change direction only when they collide with one another or with the walls of the container. These collisions are elastic. The pressure of a gas results from the bombardment of the walls of the container by its molecules. The average kinetic energy of the molecules is proportional to the temperature of the gas and is the same for all gases at a given temperature.

All molecules of a gas are not moving at the same speed at the same instant of time. Some move relatively slowly, while others move relatively rapidly. The speeds are distributed over a wide range. The average of the squares of the molecular speeds, $(u^2)_{avg}$, is directly proportional to the temperature of the gas on the Kelvin scale and inversely proportional to the masses of the individual molecules. Heating a gas increases both the value of $(u^2)_{avg}$ and the number of molecules moving at higher speeds. As shown by Graham's law, heavier molecules diffuse more slowly than do lighter ones. Heavier molecules move more slowly, on the average.

The molecules in a real gas (in contrast to those in an ideal gas) possess a finite volume and attract each other slightly. At relatively low pressures the molecular volume can be neglected. At temperatures well above the temperature at which the gas liquefies, the attractions between molecules can be neglected. Under these conditions the ideal gas equation is a good approximation for the behavior of a real gas. However, at lower temperatures or higher pressures or both, corrections for molecular volume and molecular attractions are required. The **van der Waals equation** can be used to describe real gases under these conditions.

Key Terms and Concepts

absolute zero (10.5)	gas constant (10.8, 10.17)	partial pressure (10.12)
Avogadro's law (10.7)	Gay-Lussac's law (10.6, 10.7)	pressure (10.2)
barometer (10.2)	Graham's law (10.13, 10.15)	rate of diffusion (10.13, 10.15)
Boyle's law (10.3, 10.15)	kinetic energy (10.15)	real gas (10.18)
Charles's law (10.4, 10.15)	kinetic-molecular theory (10.14)	standard conditions, STP (10.9)
Dalton's law (10.12, 10.15)	ideal gas (10.8, 10.14)	temperature (10.4, 10.5)
density (10.10)	ideal gas equation (10.8, 10.17)	van der Waals equation (10.18)
diffusion (10.1, 10.13, 10.15)	molar volume (10.9)	volume (10.3, 10.4, 10.6)

Exercises

Pressure

1. A typical barometric pressure in Denver, Colorado, is 615 torr. What is this pressure in atmospheres and kilopascals? *Ans. 0.809 atm; 82.0 kPa*

2. European tire gauges are marked in units of kilopascals. What reading on such a gauge corresponds to 32 lb/in^2 (psi) (1 atm = 14.7 psi)? *Ans. 2.2 × 10^2 kPa*

3. A medical laboratory catalog describes the pressure in a cylinder of a gas as 14.82 MPa (1 MPa = 10^6 Pa). What is the pressure of this gas in atmospheres and torr? *Ans. 146.3 atm; 1.112 × 10^5 torr*

4. Arrange the following gases in increasing order by pressure: He at 375 torr, N$_2$O at 0.25 atm, CO$_2$ at 198 Pa. *Ans. CO$_2$ < N$_2$O < He*

5. A biochemist adds carbon dioxide, CO$_2$, to an evacuated bulb like the one shown in the figure at 19.8°C and stops when the difference in the heights of the mercury columns, h, is 4.75 cm. What is the pressure of CO$_2$ in the bulb in atmospheres, torr, and kilopascals? *Ans. 6.25 × 10^{-2} atm; 47.5 torr; 6.33 kPa*

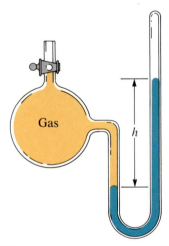

6. A pascal is equal to the pressure exerted by 0.102 kg of water lying on a square with an edge exactly 1 m long (Section 10.2). Calculate the thickness of this film of water, assuming that the density of water is 1 g/cm^3. *Ans. 0.01 cm*

The Gas Laws

7. Using Boyle's law, set up a mathematical equation relating the initial volume, V_1, and the final volume, V_2, of a given quantity of an ideal gas at constant temperature as the pressure changes from the initial pressure, P_1, to the final pressure, P_2.

8. Explain the meaning of the statement, "The pressure of a given sample of gas is inversely proportional to its volume if the temperature does not change."

9. The volume of a sample of carbon monoxide, CO, is 1.40 L at 2.25 atm and 467 K. What volume will it occupy at 4.50 atm and 467 K? *Ans. 0.700 L*

10. A sample of oxygen, O$_2$, occupies 38.9 L at 0°C and 917 torr. What volume will it occupy at STP? *Ans. 46.9 L*

11. The volume of a sample of ethane, C$_2$H$_6$, is 3.24 L at 477 torr and 27°C. What volume will it occupy at 27°C and 110.8 kPa? *Ans. 1.86 L*

12. A sample of nitrogen gas, N$_2$, occupies a volume of 1.94 L at a pressure of 98.74 kPa. What volume will it occupy at the same temperature if the pressure is 452 torr? *Ans. 3.18 L*

13. A typical scuba tank has a volume of 13.2 L. What volume of air, in liters, at 0.950 atm is required to fill such a scuba tank to a pressure of 153 atm, assuming no change in temperature? *Ans. 2.13 × 10^3 L*

14. A cylinder of oxygen for medical use contains 35.4 L of oxygen at a pressure of 150 atm. What is the volume of this oxygen, in liters, at 0.98 atm of pressure and the same temperature? *Ans. 5.4 × 10^3 L*

15. Using Charles's law, set up a mathematical equation relating the initial volume, V_1, and temperature, T_1, of a given quantity of an ideal gas at constant pressure to some other volume, V_2, and temperature, T_2.

16. Explain the meaning of the statement, "The temperature of a gas in kelvins is directly proportional to its volume if the pressure remains unchanged."

17. What is the volume of a sample of ethane, C$_2$H$_6$, at 467 K and 2.25 atm if it occupies 1.405 L at 300 K and 2.25 atm? *Ans. 2.19 L*

18. What is the volume of a sample of neon, Ne, at 3°C and 7.44 kPa if it occupies 13.3 L at 55°C and 7.44 kPa? *Ans. 11.2 L*

19. The gas in a 1.00-L bottle at 25°C can be put into a 1-qt (0.946-L) bottle at the same pressure if the temperature is reduced. What temperature is required? *Ans. 9°C.*

20. A 748-mL volume of hydrogen measured at the normal boiling point of nitrogen, −210.0°C, is warmed to the normal boiling point of water, 100°C. Calculate the new volume of the gas, assuming ideal behavior and no change in pressure. *Ans. 4.41 × 10^3 mL*

21. A gas occupies 275 mL at 100°C and 380 Pa. What final temperature is required to decrease the pressure to 305 Pa if the volume is held constant? *Ans. 26°C*

22. A gas occupies a volume of 46.25 mL at 726 torr and 125°C.
    (a) Calculate its volume at 705 torr and 108°C. *Ans. 45.6 mL*
    (b) Calculate its volume at STP. *Ans. 30.3 mL*

23. A gas occupies a volume of 12 L at 685.4 torr and 85.6°C. What would be its volume at 98.7 kPa and 64.8°C? *Ans. 10 L*

24. A gas at a pressure of 96 torr occupies 416 L at 56.2°C.

What would be the pressure if the temperature changed to 75.6°C and the volume changed to 442 L?

Ans. 96 torr   *(In this specific case, the changes in the volume and the temperature offset each other, and the pressure remains the same.)*

25. A high-altitude balloon is filled with 1.41×10^4 L of hydrogen at a temperature of 21°C and a pressure of 745 torr. What is the volume of the balloon at a height of 20 km, where the temperature is −48°C and the pressure is 63.1 torr?   *Ans. 1.27×10^5 L*

26. If the CO_2 in the bulb in exercise 5 is warmed from 19.8°C to 52.5°C, what will the height, *h*, of the mercury volume be? Ignore the change in volume due to the motion of the mercury; that is, assume that the volume remains constant.   *Ans. 5.28 cm*

27. A spray can is used until it is empty, except for the propellant gas, which has a pressure of 1344 torr at 23°C. If the can is thrown into a fire (*T* = 475°C), what will be the pressure (in atm) in the hot can?
   Ans. 3.40×10^3 torr

28. How many moles of carbon monoxide, CO, are contained in a 327.2-mL bulb at 48.1°C if the pressure is 149.3 kPa?   *Ans. 1.829×10^{-2} mol*

29. How many moles of chlorine gas, Cl_2, are contained in a 10.3-L tank at 21.2°C if the pressure is 63.3 atm?
   Ans. 27.0 mol

30. How many moles of oxygen gas, O_2, are contained in a cylinder of medical oxygen with a volume of 35.4 L, a pressure of 151 atm, and a temperature of 30.5°C?
   Ans. 2.14×10^2 mol

31. A small cylinder of neon used in chemistry lectures has a volume of 340 mL. How many moles of neon are contained in such a cylinder at a pressure of 151 atm and a temperature of 27°C?   *Ans. 2.09 mol*

32. What is the temperature of a 0.274-g sample of methane, CH_4, confined in a 300.0-mL bulb at a pressure of 198.7 kPa?   *Ans. 420 K or 147°C*

33. What is the volume of a bulb that contains 8.17 g of helium, He, at 13°C and a pressure of 8.73 atm?
   Ans. 5.49 L

34. Assume that 453.6 g (1 lb) of dry ice, $CO_2(s)$, is placed in an evacuated 3.785-L closed tank (a 1-gal tank). What is the pressure in the tank in atmospheres at a temperature of 33°C after all the $CO_2(s)$ has been converted to gas?
   Ans. 68.4 atm

35. How many grams of gas are present in each of the following cases?
   (a) 0.100 L of CO_2 at 307 torr and 26°C
     Ans. 7.25×10^{-2} g
   (b) 8.75 L of C_2H_4 at 378.3 kPa and 483 K
     Ans. 23.1 g
   (c) 73.3 mL of I_2 at 0.642 atm and 225°C
     Ans. 0.292 g

   (d) 4.3410 L of BF_3 at 1.220 atm and 788.0 K
     Ans. 5.553 g
   (e) 221 mL of Ar at 0.23 torr and −54°C
     Ans. 1.5×10^{-4} g

36. Calculate the volume in liters of each of the following quantities of gas at STP.
   (a) 6.72 g of NH_3 *Ans. 8.84 L*
   (b) 0.588 g of C_2H_4 *Ans. 0.470 L*
   (c) 1.47 kg of BF_3 *Ans. 486 L*
   (d) 0.72 mol of CO   *Ans. 16 L*
   (e) 13.5 mol of NO   *Ans. 302 L*
   (f) 0.027 mol of NOCl   *Ans. 0.60 L*

37. (a) What mass of laughing gas, N_2O, occupies a volume of 0.250 L at a temperature of 325 K and a pressure of 113.0 kPa?   *Ans. 0.460 g*
   (b) What is the density of N_2O under these conditions?
     Ans. 1.84 g/L

38. Calculate the density of chlorine gas, Cl_2, at STP, and at 35.0°C and 625 torr.   *Ans. 3.17 g/L; 2.31 g/L*

39. Calculate the density of Freon 12, CF_2Cl_2, at 30.0°C and 0.954 atm.   *Ans. 4.64 g/L*

40. Calculate the density of nitrogen, N_2, at a temperature of 373 K and a pressure of 108.3 kPa.   *Ans. 0.978 g/L*

41. A liter of methane gas, CH_4, contains more atoms of hydrogen than does a liter of pure hydrogen gas, H_2, under the same conditions. Using Avogadro's law as a starting point, explain why.

42. For 1 mol of H_2 showing ideal gas behavior, draw labeled graphs of:
   (a) the variation of *V* with *P* at *T* = 0°C.
   (b) the variation of *T* with *V* at *P* = 101.3 kPa.
   (c) the variation of *T* with *P* at *V* = 1.0 L.
   (d) the variation of the average velocity of the gas molecules with *T*

43. A cylinder of a standard gas for calibration of blood gas analyzers contains 3.5% CO_2, 10.0% O_2, and the remainder N_2 at a total pressure of 145 atm. What is the partial pressure of each component of this gas? (The percentages given indicate the percent of the total pressure due to each component.)
   Ans. CO_2, 5.1 atm; O_2, 14.5 atm; N_2, 125 atm

44. Most mixtures of hydrogen gas with oxygen gas are explosive. However, a mixture that contains less than 3.0% O_2 is not. If enough O_2 is added to a cylinder of H_2 at 33.2 atm to bring the total pressure to 33.5 atm, is the mixture explosive?
Ans. No   *(P_{O_2} = 0.3 atm, 0.9% of the total pressure)*

45. A 5.73-L flask at 25°C contains 0.0388 mol of N_2, 0.147 mol of CO, and 0.0803 mol of H_2. What is the pressure in the flask in atmospheres, torr, and kilopascals?   *Ans. 1.136 atm; 863.1 torr; 115.1 kPa*

46. A mixture of 0.200 g He, 1.00 g O_2 and 0.820 g Ne is contained in a closed container at STP. What is the vol-

ume of the container, assuming that the gases exhibit ideal behavior?   *Ans. 2.732 L*

47. A sample of carbon monoxide was collected over water at a total pressure of 756 torr and a temperature of 18°C. What is the pressure of the carbon monoxide? (See Table 10.1 for the vapor pressure of water.)   *Ans. 740 torr*

48. A sample of oxygen collected over water at a temperature of 29.0°C and a pressure of 714 torr has a volume of 0.560 L. What volume would the dried oxygen have under the same conditions of temperature and pressure?   *Ans. 0.536 L*

49. The volume of a sample of a gas collected over water at 30.0°C and 0.932 atm is 627 mL. What will the volume of the dried gas be at STP?   *Ans. 503 mL*

50. A 491-mL sample of gaseous nitric oxide, NO, was collected over mercury (which has a negligible vapor pressure) at 23°C and 734.0 torr. If the same sample of nitric oxide were collected over water at 19°C and 714.0 torr, what would its volume be?   *Ans. 510 mL*

51. Which is denser at the same temperature and pressure, dry air or air saturated with water vapor? Explain.
                                                      Ans. Dry air.

52. (a) What is the concentration of the atmosphere in molecules per milliliter at 25°C and 1.00 atm?
                  Ans. 2.46×10^{19} molecules/mL
    (b) At a height of 150 km (about 94 mi) the atmospheric pressure is about 4.0×10^{-9} atm with a temperature of 420 K. What is the concentration of the atmosphere in molecules per milliliter at 150 km?
                   Ans. 7.0×10^{10} molecules/mL

Combining Volumes, Molecular Weights, Stoichiometry

53. What volume of O_2 at STP is required to oxidize 20.0 L of CO at STP to CO_2? What volume of CO_2 is produced at STP?   *Ans. 10.0 L O_2; 20.0 L CO_2*

54. Methanol (sometimes called wood alcohol), CH_3OH, is produced industrially by the following reaction.

$$CO(g) + 2H_2(g) \xrightarrow[\text{300°C, 300 atm}]{\text{Copper catalyst}} CH_3OH(g)$$

Assuming that the gases behave as ideal gases, what is the ratio of the final volume to the total volume of the reactants?   *Ans. 1 to 3*

55. Calculate the volume of oxygen required to burn 5.00 L of butane gas, C_4H_{10}, to produce carbon dioxide and water, if the volumes of both the propane and the oxygen are measured under the same conditions.
                                                      Ans. 32.5 L

56. A 2.50-L sample of a brown gas at STP was decomposed to give 2.50 L of N_2 and 5.00 L of O_2 at STP. What is the colorless gas?   *Ans. N_2O_4*

57. An acetylene tank for an oxyacetylene welding torch will provide 9340 L of acetylene gas, C_2H_2, at STP. How many tanks of oxygen, each providing 7.00×10^3 L of

O_2 at STP, will be required to burn the acetylene, forming CO_2 and H_2O?   *Ans. 3.34 tanks*

58. How could you show by experiment that the molecular formula of cyclobutane is C_4H_8 and not CH_2?

59. What is the molecular weight of a gas if 125 mL of the gas at a pressure of 99.5 kPa at 22°C has a mass of 0.157 g?   *Ans. 31.0*

60. One of the noble gases of Group VIIIA was isolated in 1894 by allowing a large volume of air to react with magnesium so that only the noble gas, contaminated with traces of other noble gases, remained. A 32-mL sample of this gas weighs 0.054 g at 9°C and 748 torr. Determine the apparent molecular weight of this monatomic gas. (The traces of other gases are negligible.) Which noble gas is it?   *Ans. 40; Ar*

61. A sample of an oxide of nitrogen isolated from the exhaust of an automobile was found to weigh 0.571 g and to occupy 1.00 L at 356 torr and 27°C. Calculate the molecular weight of this oxide, and determine if it was N_2O, NO, NO_2, N_2O_4, or N_2O_5.
                                    Ans. Mol. wt = 30.0; NO

62. The density of a certain gaseous fluoride of phosphorus is 5.63 g/L at STP. Calculate the molecular weight of this fluoride, and determine its molecular formula.
                                            Ans. 126; PF_5

63. Cyclopropane is a gas containing only carbon and hydrogen and is often used as an anaesthetic for major surgery. If 250 mL of cyclopropane at 120°C and 0.72 atm reacts with O_2 to give 750 mL of CO_2 and 750 mL of $H_2O(g)$ at the same temperature and pressure, what is the molecular formula of cyclopropane?   *Ans. C_3H_6*

64. Joseph Priestley first prepared pure oxygen by heating mercuric oxide, HgO.

$$2HgO(s) \xrightarrow{\triangle} 2Hg(l) + O_2(g)$$

What volume of O_2 at 23°C and 0.975 atm is produced by the decomposition of 5.36 g of HgO?
                                                      Ans. 0.308 L

65. Cavendish prepared hydrogen in 1766 by passing steam through a red-hot rifle barrel.

$$4H_2O(g) + 3Fe(s) \xrightarrow{\triangle} Fe_3O_4(s) + 4H_2(g)$$

What volume of H_2 at a pressure of 745 torr and a temperature of 20°C can be prepared from the reaction of 15.0 g of H_2O?   *Ans. 20.4 L*

66. Hydrogen gas will reduce Fe_3O_4 in the reverse of the reaction shown in exercise 65.
    (a) What volume of H_2 at 200 atm and 19°C is required to reduce 1.00 metric ton (1000 kg = 1 metric ton) of Fe_3O_4 to Fe?   *Ans. 2.07×10^3 L*
    (b) If the reduction of 1.00 metric ton of Fe_3O_4 is run at 600°C and 10.0 atm, what volume of water vapor is produced?   *Ans. 1.24×10^5 L*

67. Gaseous hydrogen chloride, $HCl(g)$, is prepared com-

mercially by the reaction of NaCl with H_2SO_4 (Section 9.9, Part 1). What mass of NaCl is required to prepare enough HCl(g) to fill a 35.4-L cylinder to a pressure of 122 atm at 25°C?   *Ans. 10.3 kg*

68. If the oxygen consumed by a resting human male (Section 10.8, Example 10.11) is used to produce energy by the oxidation of glucose,

$$C_6H_{12}O_6 + 6O_2 \longrightarrow 6CO_2 + 6H_2O$$

what is the mass of glucose required per hour for a resting 70-kg male?   *Ans. 17 g*

69. Sulfur dioxide is an intermediate in the preparation of sulfuric acid (Section 9.9, Part 1). What volume of SO_2 at 443°C and 2.21 atm is produced by the combustion of 1.00 kg of sulfur?   *Ans. 829 L*

70. What volume of oxygen at 423.0 K and a pressure of 105.4 kPa will be produced by the decomposition of 192.7 g of BaO_2 to BaO and O_2?   *Ans. 18.99 L*

71. In a common freshman laboratory experiment, $KClO_3$ is decomposed by heating to give KCl and O_2. What mass of $KClO_3$ must be decomposed to give 238 mL of O_2 at a temperature of 28°C and a pressure of 752 torr?
                    Ans. 0.779 g

72. In a laboratory determination a 0.1009-g sample of a compound containing boron and chlorine gave 0.3544 g of silver chloride upon reaction with silver nitrate. A 0.06237-g sample of the compound exerted a pressure of 6.52 kPa at 27°C in a volume of 147 mL. What is the molecular formula of the compound?   *Ans. B_2Cl_4*

73. As 1.00 g of the radioactive element radium decays over 1 yr, it produces 1.16×10^{18} alpha particles (helium nuclei). Each alpha particle becomes an atom of helium gas. What is the pressure, in Pa, of the helium gas produced if it occupies a volume of 125 mL at a temperature of 25°C?   *Ans. 38.2 Pa*

Kinetic-Molecular Theory, Graham's Law

74. Use the postulates of the kinetic-molecular theory and explain why a gas will evenly fill any container of any shape.

75. Show how Boyle's law, Charles's law, and Dalton's law follow from the assumptions of the kinetic-molecular theory.

76. Describe what happens to the average kinetic energy of ideal gas molecules when the conditions are changed as follows:
    (a) The volume of the gas is increased by decreasing the pressure at constant temperature.
    (b) The volume of the gas is increased by increasing the temperature at constant pressure.
    (c) The average velocity of the molecules is decreased by a factor of 0.5.

77. Can the speed of a given molecule in a gas double at constant temperature? Explain your answer.

78. What is the ratio of the kinetic energy of a SO_2 molecule to that of an oxygen molecule in a mixture of two gases at the same temperature and pressure? What is the ratio of the root-mean-square speeds, u_{rms}, of the two gases?

$$Ans. \ \frac{u_{SO_2}}{u_{O_2}} = 0.7068$$

79. A 1-L sample of He initially at STP is heated to 546°C, and its volume is increased to 2 L.
    (a) What effect do these changes have on the number of collisions of the molecules of the gas per unit area of the container wall?
    (b) What is the effect on the average kinetic energy of the molecules?
    (c) What is the effect on the root-mean-square speed of the molecules?

80. The root-mean-square speed of H_2 molecules at 25°C is about 1.6 km/s. What is the root-mean-square speed of a N_2 molecule at 25°C?   *Ans. 0.43 km/s*

81. Show that the ratio of the rate of diffusion of gas 1 to the rate of diffusion of gas 2, R_1/R_2, is the same at 20°C and 200°C.

82. Heavy water, D_2O (mol. wt = 20.03), can be separated from ordinary water, H_2O (mol. wt = 18.01), as a result of the difference in the relative rates of diffusion of the molecules in the gas phase. Calculate the relative rates of diffusion of H_2O and D_2O.
                    Ans. H_2O diffuses 1.055 times faster than D_2O

83. Which of the following gases diffuse more slowly than oxygen: F_2, Ar, NO_2, CH_4, NO, Br_2, PH_3?
                    Ans. F_2, Ar, NO_2, Br_2, PH_3

84. Calculate the relative rate of diffusion of H_2 compared to D_2 (at. wt D = 2.0) and of O_2 compared to O_3.

$$Ans. \ \frac{r_{H_2}}{r_{D_2}} = 1.4; \ \frac{r_{O_2}}{r_{O_3}} = 1.22474$$

85. A gas of unknown identity diffuses at the rate of 16.0 mL/s in a diffusion apparatus in which a second gas, whose molecular weight is 28.0, diffuses at the rate of 24.2 mL/s. Calculate the molecular weight of the first gas.   *Ans. 64.1*

86. When two cotton plugs, one moistened with ammonia and the other with hydrobromic acid, are simultaneously inserted into opposite ends of a glass tube 87.0 cm long, a white ring of NH_4Br forms where gaseous NH_3 and gaseous HBr first come into contact.

$$NH_3(g) + HBr(g) \longrightarrow NH_4Br(s)$$

At what distance from the ammonia-moistened plug does this occur?   *Ans. 59.6 cm*

Nonideal Behavior of Gases

87. Graphs showing the behavior of several different gases are given on the following page. Which of these gases exhibit behavior significantly different from that expected for ideal gases?

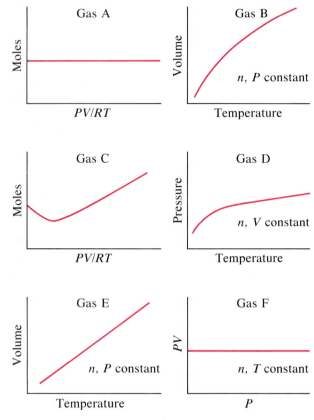

88. Describe the factors responsible for the deviation of the behavior of real gases from that of an ideal gas.

89. For which of the following sets of conditions will a real gas behave most like an ideal gas, and for which conditions would a real gas be expected to deviate from ideal behavior? Explain.
    (a) low volume, high pressure
    (b) low temperature, high pressure
    (c) low pressure, high temperature

90. For which of the following gases should the correction for the molecular volume be largest: CO, CO_2, H_2, He, NH_3, SF_6?   *Ans. SF_6*

Additional Exercises

91. A commercial mercury vapor analyzer can detect concentrations of gaseous Hg atoms (which are poisonous) in air as low as 2×10^{-6} mg/L of air. What is the partial pressure of gaseous mercury if the atmospheric pressure is 733 torr at 26°C?
    Ans. 2×10^{-10} atm or 2×10^{-7} torr

92. (a) What is the total volume of the $CO_2(g)$ and $H_2O(g)$ at 600°C and 0.888 atm produced by the combustion of 1.00 L of $C_2H_6(g)$ measured at STP?
    Ans. 18.0 L
    (b) What is the partial pressure of H_2O in the product gases?   *Ans. 0.533 atm*

93. Butene, an important petrochemical used in the production of synthetic rubber, is composed of carbon and hydrogen. A 0.0124-g sample of this compound produces 0.0390 g of CO_2 and 0.0159 g of H_2O when burned in an oxygen atmosphere. If the volume of a 0.125-g gaseous sample of the compound at 735 torr and 45°C is 60.2 mL, what is the molecular formula of the compound?   *Ans. C_4H_8*

94. Ethanol, C_2H_5OH, is produced industrially from ethylene, C_2H_4, by the following sequence of reactions:

$$3C_2H_4 + 2H_2SO_4 \longrightarrow$$
$$C_2H_5HSO_4 + (C_2H_5)_2SO_4$$

$$C_2H_5HSO_4 + (C_2H_5)_2SO_4 + 3H_2O \longrightarrow$$
$$3C_2H_5OH + 2H_2SO_4$$

What volume of ethylene at STP is required to produce 1.000 kg of ethanol if the overall yield of ethanol is 93.1%?   *Ans. 522 L*

95. Thin films of amorphous silicon for electronic applications are prepared by decomposing silane gas, SiH_4, on a hot surface at low pressures.

$$SiH_4(g) \xrightarrow{\triangle} Si(s) + 2H_2(g)$$

What volume of SiH_4 at 150 Pa and 825 K is required to produce a 20.0-cm-by-20.0-cm film that is 200 Å thick ($1 \text{ Å} = 10^{-8}$ cm)? The density of amorphous silicon is 1.9 g/cm³.   *Ans. 2.5 L*

96. One molecule of hemoglobin will combine with four molecules of oxygen. If 1.0 g of hemoglobin combines with 1.53 mL of oxygen at body temperature (37°C) and a pressure of 743 torr, what is the molecular weight of hemoglobin?   *Ans. 6.8×10^4*

97. Ethanol, C_2H_5OH, is often produced by the fermentation of sugars. For example, the preparation of ethanol from the sugar glucose is represented by the unbalanced equation

$$C_6H_{12}O_6(aq) \xrightarrow{\text{Yeast}} C_2H_5OH(aq) + CO_2(g)$$

What volume of CO_2 at STP is produced by the fermentation of 1.00 metric ton (1 metric ton = 1000 kg) of glucose if the reaction has a yield of 95.2%?
    Ans. 2.37×10^5 L

98. A sample prepared by pumping 15.0 L of air at STP and 2.0 L of CO at STP into a 3.0-L closed flask was heated until all the CO had been converted to CO_2. Assuming that air is 20% oxygen and 80% nitrogen by volume, calculate the partial pressures of N_2, O_2, and CO_2 in the flask at 0°C after the reaction has taken place.
    Ans. $P_{O_2} = 0.67$ atm; $P_{N_2} = 4.0$ atm; $P_{CO_2} = 0.67$ atm

99. One method of analysis of amino acids is the van Slyke method. The characteristic amino groups (—NH_2) in protein material are allowed to react with nitrous acid, HNO_2, to form N_2 gas. From the volume of the gas, the amount of amino acid can be determined. A 0.0604-g

sample of a biological material containing glycine, $CH_2(NH_2)COOH$, was analyzed by the van Slyke method, giving 3.70 mL of N_2 collected over water at a pressure of 735 torr and 29°C. What was the percentage of glycine in the sample?

$$CH_2(NH_2)COOH + HNO_2 \longrightarrow$$
$$CH_2(OH)COOH + H_2O + N_2$$

Ans. 17.2%

100. Natural gas often contains hydrogen sulfide, H_2S, a gas that is itself a pollutant and that produces another pollutant, sulfur dioxide, upon combustion. Hydrogen sulfide is removed from raw natural gas by the reaction

$$HOC_2H_4NH_2 + H_2S \longrightarrow (HOC_2H_4NH_3)HS$$

How many grams of ethanolamine, $HOC_2H_4NH_2$, is required to remove the H_2S from 1000 ft^3 (28,300 L) of natural gas at STP if the partial pressure of H_2S is 205 Pa?   *Ans. 156 g*

101. One step in the production of sulfuric acid (Section 9.9, Part 1) involves oxidation of $SO_2(g)$ to $SO_3(g)$ using air at 400°C with vanadium(V) oxide as a catalyst. Assuming that air is 21% oxygen by volume, what volume of air at 31°C and 0.912 atm is required to oxidize enough SO_2 to give 1.00 metric ton (1000 kg = 1 metric ton) of sulfuric acid, H_2SO_4?   *Ans. 6.6×10^5 L*

102. A sample of a compound of xenon and fluorine was confined in a bulb with a pressure of 24 torr. Hydrogen was added to the bulb until the pressure was 72 torr. Passage of an electric spark through the mixture produced Xe and HF. After the HF was removed by reaction with solid KOH, the final pressure of xenon and unreacted hydrogen in the bulb was 48 torr. What is the empirical formula of the xenon fluoride in the original sample? (Note: Xenon fluorides contain only one xenon atom per molecule.)   *Ans. XeF_2*

103. A 0.500-L bottle contains 250.0 mL of a 3.0% (by mass) solution of hydrogen peroxide. How much will the pressure (in atm) in the bottle increase if the H_2O_2 decomposes to $H_2O(l)$ and $O_2(g)$ at 24°C? Assume that the density of the solution is 1.00 g/cm^3, that the volume of the liquid does not change during the decomposition, and that the solubility of O_2 in water can be disregarded. Does this calculation explain why hydrogen peroxide solutions are sold in bottles with pressure relief valves? Explain.   *Ans. P increases by 11 atm*

104. The total pressure of a sample of hydrogen collected above water in a 425-mL bottle at 35°C is 763 torr. What is the volume of the sample when the temperature falls to 23°C, assuming that its pressure does not change?   *Ans. 397 mL*

Condensed Matter: Liquids and Solids

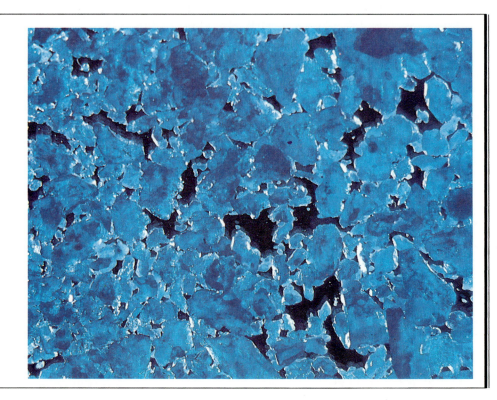

11

Crystalline copper
sulfate

The difference between the volume of a solid or liquid and the volume of the same amount of that substance when it is a gas is striking. A given mass of a liquid or solid substance occupies a much smaller volume than the same mass of that substance in the gas phase. For example, 44 grams of dry ice (one mole of solid CO_2) has a volume of 26 milliliters, about the size of a golf ball. If the same amount of dry ice is placed in an empty balloon at 25°C and one atmosphere and allowed to warm and form a gas, the balloon expands to a volume of about 25,000 milliliters, the volume of a beachball with a diameter of about 14 inches. This is an increase of almost 1000 times between the two volumes. The volume of 44 grams of liquid CO_2 is only 40 milliliters. Because of their small volumes relative to the gas phase, liquids and solids are often called **condensed phases.**

Liquids and solids are familiar forms of matter with many different properties. We know that liquids flow and assume the shape of any container into which they are poured. We also know that solids are rigid. Many liquids will evaporate; most solids will not. Liquids have a disordered structure; most solids have a regular internal structure. However, in spite of the apparent differences in their behavior, many of the properties of liquids and solids are similar. For example, both solid and liquid CO_2 can be used for refrigeration, because as either liquid or solid CO_2 is converted to a gas, heat is absorbed and the surroundings are cooled.

The similarities and differences in the behavior of liquids and gases can be explained in terms of the same kinetic-molecular theory we used to describe the properties of gases (Section 10.15). In this chapter we will explore the similarities and differences in the behavior of these two forms of condensed matter and examine how the kinetic-molecular theory can help us understand their properties. We will also examine the regularities in structure that result when atoms, molecules, or ions arrange themselves into crystalline materials.

11.1 The Formation of Liquids and Solids

We saw in Section 10.14 that the molecules in a gas are in constant and very rapid motion and, in general, the space between them is large compared to the sizes of the molecules themselves. As a consequence of their motion, gas molecules possess a kinetic energy that overcomes the forces of attraction between them. However, as a gas is cooled, the average speed and the kinetic energy of the molecules decreases. If the gas is cooled sufficiently, the intermolecular attractions overcome the tendency of the molecules to move apart, the gas condenses, and the molecules come into constant contact. This condensation produces either a liquid or a solid. You have probably seen liquid water form on a cold glass or beverage can as the water vapor in the air is cooled by the cold glass or can. The frost that forms in a freezer and on automobiles during some winter nights (Fig. 11.1) results when water vapor comes in contact with a very cold surface and is cooled sufficiently that it is converted directly to solid water without ever going through the liquid phase.

An increased pressure will bring the molecules of a gas closer together, so that the average distance between the molecules is decreased and the average attractions between them become strong relative to their kinetic energy. Gases can be liquefied by compressing them if the temperature is not too high. Carbon dioxide is a gas at room temperature and 1 atmosphere, but it generally is a liquid in CO_2 fire extinguishers

Figure 11.1

Frost forms when water vapor in the air is cooled and is converted directly to ice.

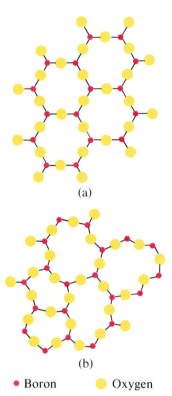

(a)

(b)

● Boron   ● Oxygen

Figure 11.2

(a) A two-dimensional illustration of the ordered arrangement of atoms in a crystal of boric oxide. (b) An illustration showing the disorder in amorphous boric oxide.

because the pressure is greater than about 65 atmospheres. At this pressure gaseous carbon dioxide condenses to a liquid at room temperature.

Although molecules in the liquid state are close together or in contact with each other, they still move about, though in a restricted manner. Because the molecules move, a liquid can change its shape, take the shape of a container, diffuse, and evaporate. An increase in the pressure on a liquid can only reduce the distance between the closely packed molecules slightly, so the volume of a liquid decreases very little with increased pressure. Liquids are relatively incompressible, compared to gases. The molecules in a liquid move past one another in a random fashion (diffuse), but because of the much shorter distance that molecules can move before they collide with other molecules, they diffuse much more slowly than do gases.

It is difficult to describe the structure of a liquid, because it changes from moment to moment as the molecules move. We can say that the molecules that make up a liquid are relatively closely packed, and it is likely that the arrangement of molecules is random (like that in amorphous solids, which will be described later in this section).

When the temperature of a liquid becomes sufficiently low, or the pressure on the liquid sufficiently high, the molecules of the liquid no longer have sufficient kinetic energy to move past each other, and a solid is formed. Solids are rigid. They cannot expand like gases or be poured like liquids, because the molecules cannot change position easily.

Most molecules in a solid do not move about, although they do vibrate. Diffusion takes place to a very limited extent in the solid state, and some crystalline compounds show a vapor pressure. This indicates that the molecules in a solid are not completely fixed. As the temperature of a solid is lowered, the motion of the molecules gradually decreases to a minimum at absolute zero ($-273°C$).

As most liquids are cooled, they eventually freeze; that is, their molecules assume ordered positions and form a crystalline solid. A **crystalline solid** is a homogeneous solid in which the atoms, ions, or molecules are arranged in a definite repeating pattern [Fig. 11.2(a)]. Although some solids, such as diamonds, sugar, and table salt, are composed of individual single crystals, most common crystalline solids are aggregates of many small crystals. Common examples of the latter are sandstone, chunks of ice, and metal objects.

Some liquids can be cooled well below their freezing temperatures before crystallization begins. A liquid that has been cooled below its freezing temperature yet remains in the liquid state is called a **supercooled liquid.** Such a liquid is not stable; it may crystallize spontaneously. Crystallization can often be induced by the introduction of a seed crystal of the substance; this provides an ordered structure to which additional molecules can become attached.

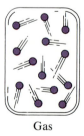

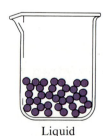

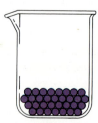

Gas   Liquid   Crystalline solid

Figure 11.3

The arrangement of molecules in a gas, a liquid, and a crystalline solid. The density of the gaseous molecules is exaggerated for the purpose of illustration.

Widely separated, disordered molecules in continuous motion.

Closely spaced, disordered molecules in continuous motion.

Ordered molecules in contact and in relatively fixed positions.

Liquid materials such as molten glass, melted butter, or molten asphalt, which contain large cumbersome molecules that cannot move readily into ordered positions, often show great tendencies to supercool. As the temperature is lowered, their large molecules move more and more slowly, and finally they stop in random positions before they can adopt an ordered arrangement. The resulting rigid solids are called **amorphous solids,** or **glasses** [Fig. 11.2(b)]. Amorphous solids lack an ordered internal structure. They do not melt at a definite temperature but gradually soften and become less viscous when heated.

A comparison of gases, liquids, and crystalline solids based on the kinetic-molecular model of the behavior of their molecules is presented in Fig. 11.3.

11.2 Evaporation of Liquids and Solids

The level of water present in an open vessel drops upon standing due to **evaporation,** the change of a liquid to a gas. Some liquids, such as ether, alcohol, and gasoline, evaporate more rapidly than water under the same conditions. Others evaporate more slowly. Motor oil and ethylene glycol (antifreeze) evaporate so slowly that they seem not to evaporate at all.

Some solids evaporate too. Chemists refer to this direct conversion from the solid phase to the gas phase as **sublimation.** Over a long period of time an ice cube in a freezer may become smaller as it sublimes. Snow sublimes at temperatures below its melting point. As a mothball sublimes, you can smell the molecules in the gas phase above it. When solid iodine is heated, the solid sublimes, and a beautiful purple vapor forms. The molecules of these solids pass directly from the solid into the gas phase.

Evaporation and sublimation may be explained in terms of the kinetic-molecular theory. In a liquid, just as in a gas, the molecules move randomly—some slowly, many at intermediate rates, and some very rapidly. A few rapidly moving molecules near the surface of the liquid may possess enough kinetic energy to overcome the attraction of their neighbors and escape—i.e., evaporate—into the gas above the liquid (Fig. 11.4) and become gas-phase molecules. If they are not confined, these gaseous molecules diffuse away, and more fast-moving molecules leave the liquid phase and appear in the gaseous phase above the liquid as evaporation continues.

Although the molecules in a solid do not move through the solid, they do possess energy that causes them to vibrate—some gently, many with intermediate energy, and some wildly. A few wildly vibrating molecules on the surface of the solid may possess enough energy to overcome the attraction of their neighbors and escape into the gas phase. In an open container, these gaseous molecules diffuse away, and more molecules leave the solid as sublimation continues.

If a solid or a liquid is in a closed container, molecules that evaporate cannot escape from the container. They strike the walls, rebound, and some eventually strike the surface of the solid or liquid and move back into it. This change from the vapor back to a condensed state is called **condensation.** When the rate of condensation becomes equal to the rate of evaporation, the vapor in the container is in **equilibrium** with the solid or liquid. At equilibrium, neither the amount of the condensed phase nor the amount of the vapor in the container changes, although some molecules from the vapor condense at the same time that an equal number of molecules from the solid or liquid evaporate.

$$\text{Liquid (or solid)} \underset{\text{Condensation}}{\overset{\text{Evaporation}}{\rightleftharpoons}} \text{Vapor}$$

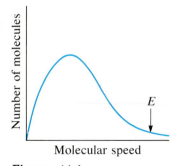

Figure 11.4

Distribution of the molecular speeds of molecules of a liquid. Molecules with speeds greater than E possess sufficient kinetic energy to escape through the surface of the liquid into the gas phase.

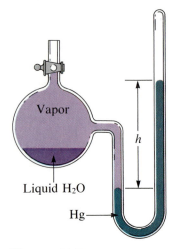

Figure 11.5

A closed vessel at 25°C containing only liquid water and water vapor. The vapor pressure is proportional to the difference in height, h, of the mercury in the two columns of the manometer.

Because both processes operate simultaneously and counterbalance each other, the equilibrium is called a **dynamic equilibrium.**

The pressure exerted by the vapor in equilibrium with a solid or a liquid at a given temperature is called the **vapor pressure** of the condensed phase. Neither the area of the surface of the solid or liquid in contact with the vapor nor the size of the vessel has any effect upon the vapor pressure. Vapor pressures are commonly measured by introducing a solid or a liquid into a closed container and measuring the increase in pressure due to the vapor in equilibrium with the condensed phase with a manometer (Section 10.2 and Fig. 11.5).

Figure 11.6 depicts the vapor pressures of four liquids with different vapor pressures. These differences are related to the attractive forces between the molecules in each liquid, which are least for ether and greatest for ethylene glycol, in these cases. Thus, the vapor pressure of a liquid is dependent upon the particular kind of molecule composing the liquid. The liquids that evaporate most rapidly, ethyl ether and ethyl alcohol, have the smaller intermolecular forces, as is evident from their higher vapor pressures. The very low vapor pressure of ethylene glycol (too low even to be shown on the graph at room temperature, about 23°C) is a reflection of the strong forces between its molecules that cause its very slow rate of evaporation.

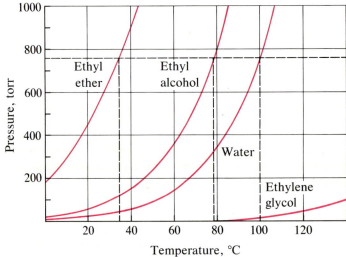

Figure 11.6

The vapor pressures of four common liquids at various temperatures. The intersection of a curve with the dashed line at 760 torr, referred to the temperature axis, indicates the normal boiling point of that substance. (The normal boiling point of ethyl ether is 34.6°C; of ethyl alcohol, 78.4°C; of water, 100°C; of ethylene glycol, 198°C, which is off the scale in this figure.)

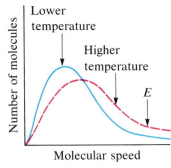

Figure 11.7

Distribution of molecular speeds in a liquid at two different temperatures. At the higher temperature, more molecules have the necessary speed, E, to escape from the liquid into the gas phase.

The vapor pressures of all liquids increase as their temperatures increase, because the rates of motion of their molecules increase with increasing temperature. At a higher temperature more molecules move rapidly enough to escape from the liquid, as shown in Fig. 11.7. The escape of more molecules per unit of time and the greater speed of each molecule that escapes both contribute to the higher vapor pressure.

As one might expect, the vapor pressures of solids behave analogously to those of liquids. Those solids in which the intermolecular forces are weak exhibit measurable vapor pressures at room temperature. The vapor pressures of solids increase with increasing temperature. For example, the vapor pressure of solid iodine is 0.2 torr at 20°C and 90 torr at 114°C, its melting point.

Figure 11.8
The crystals of iodine at the top of the test tube resulted from sublimation of the iodine in the bottom of the tube.

The combined process of a solid passing directly into the vapor state without melting and then recondensing into the solid state also is called **sublimation.** Many substances, such as iodine, may be purified by sublimation (Fig. 11.8) if the impurities have low vapor pressures.

11.3 Boiling of Liquids

As heat is added continuously to a liquid, its temperature and its vapor pressure increase until the vapor pressure equals the pressure of the gas above it; at this point the liquid begins to boil. Even as additional heat is added, the temperature of the boiling liquid remains constant as long as any liquid remains. Any heat that is added to a boiling liquid is used to convert a portion of the liquid to the gas phase rather than to raise the temperature of the liquid. When a liquid boils, bubbles of vapor form within it and then rise to the surface, where they burst and release the vapor.

The **normal boiling point** of a liquid is the temperature at which its vapor pressure equals one atmosphere (760 torr, Fig. 11.6). A liquid will boil at temperatures higher than its normal boiling point when the external pressure is greater than one atmosphere; alternatively, the boiling point is lowered by decreasing the pressure (Fig. 11.9). At

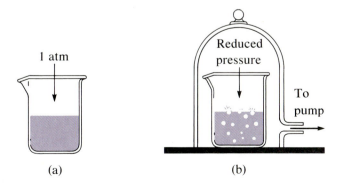

Figure 11.9
A liquid such as ethyl alcohol that (a) will not boil at room temperature under 1 atmosphere of pressure (b) may boil at that same temperature when the pressure is reduced by placing the liquid in a bell jar and pumping out the air.

high altitudes, where the atmospheric pressure is less than 760 torr, water boils at temperatures below its normal boiling point of 100°C. Food in boiling water cooks more slowly at high altitudes, because the temperature of boiling water is lower there than nearer sea level. The temperature of boiling water in a pressure cooker is higher than the normal boiling point because of the increased pressure in the cooker. Thus foods cook faster in a pressure cooker than in open vessels.

EXAMPLE 11.1   From the graph presented in Fig. 11.6, determine the boiling point of water contained in a bell jar such as that shown in Fig. 11.9(b) at a pressure of 510 torr. (This is a common atmospheric pressure in Leadville, Colorado, elevation 10,200 ft.)

The graph of the vapor pressure of water versus temperature indicates that the vapor pressure of water is 510 torr at about 90°C. Thus at about 90°C the vapor pressure of water will be equal to that of the atmosphere in the bell jar, and the water will boil.

11.4 Distillation

Dissolved materials in a liquid may make it unsuitable for a particular purpose. For example, water containing dissolved minerals should not be used in automobile batteries or steam irons, because it shortens their life. Water and other liquids may be separated from such impurities by a process known as **distillation.** When impure water is boiled in a distilling flask (Fig. 11.10), it vaporizes and passes into the condenser. The vapor is condensed to the liquid in the water-cooled condenser, and the liquid flows into the receiving vessel. Dissolved mineral matter, such as calcium sulfate or iron carbonate, is not volatile at the boiling point of water and stays in the flask. Distillation takes advantage of the facts that (1) addition of heat to a liquid speeds up its rate of evaporation (an endothermic change) and (2) cooling a vapor favors its condensation (an exothermic change). The distillation of two or more volatile substances from a mixture is described more fully in Section 12.19.

Figure 11.10
Laboratory distillation apparatus. When impure water is distilled with this apparatus, nonvolatile substances remain in the distilling flask (the flask that is heated). The water is vaporized, condensed in a water-cooled condenser, and finally collected as distillate in the receiving flask.

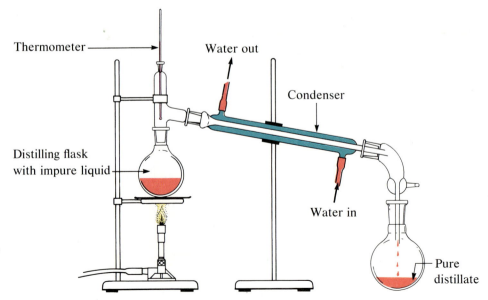

Thermometer

Water out

Condenser

Distilling flask with impure liquid

Water in

Pure distillate

11.5 Melting of Solids

As heat is added to a crystalline solid, its temperature and the average energy of its molecules increase. At some point in the addition of heat, this energy becomes large enough to overcome some of the intermolecular forces holding the molecules in their fixed positions, and the solid begins to melt. If heating is continued, all of the solid will melt, but the temperature of the mixture of solid and liquid will not increase as long as any solid remains. Any heat that is added to a melting solid is used to convert a portion of the solid to a liquid rather than to raise the temperature of the solid. If, however, the heating is stopped and no heat is allowed to escape, the solid and liquid phases remain, in equilibrium. The changes continue, but the rate of melting is just equal to the rate of freezing, and the quantities of solid and liquid remain constant. The temperature at which the solid and liquid phases of a given substance are in equilibrium is called the **melting point** of the solid, or the **freezing point** of the liquid. With the notable exception of water, most substances expand when they melt.

The melting point of a crystalline solid is determined by the strength of the attractive forces between the units present in the crystal. Molecules with weak attractive forces form crystals with low melting points (Section 11.11). Crystals consisting of molecules with stronger attractive forces melt at higher temperatures.

Crystalline solids, such as quartz crystals, have a sharp melting temperature because the forces holding their atoms, ions, or molecules together are all of the same strength. Thus the bonds in any one of these materials have the same strength, and all of its bonds require the same amount of energy to be broken. The gradual softening of amorphous materials, as opposed to the sharp melting of crystalline solids, results from the structural nonequivalence of the atoms. Some bonds are weaker than others, and when an amorphous material is heated, the weakest bonds break first. As the temperature is further increased, the stronger bonds are broken. Thus amorphous materials soften over a range of temperatures.

Quartz, SiO_2

11.6 Heat of Vaporization

As a liquid evaporates, energy must be supplied to overcome the intermolecular interactions that hold the molecules together in the liquid state. If the liquid is insulated from the surroundings, this energy comes from the liquid itself. Thus the internal energy of the remaining liquid is reduced, and it cools as evaporation proceeds.

Considering the process of evaporation is helpful in understanding why energy is lost from an insulated liquid as it evaporates. Evaporation occurs as molecules of high kinetic energy (faster-moving molecules) escape from the surface of the liquid (Section 11.2). The loss of such molecules leaves behind molecules with lower average kinetic energies and, consequently, results in a lower temperature ($T \propto KE_{avg}$, Section 10.17). The cooling effect due to evaporation of water is very evident to a swimmer emerging from the water. Some of the water on the swimmer's skin evaporates and leaves behind cooler water, which causes the skin to feel cold.

For the temperature of a liquid to remain constant as it evaporates, heat must be supplied to offset the cooling effect brought about by the escaping molecules. The energy that must be supplied to evaporate a given quantity of liquid at a constant specified temperature is known as the **heat of vaporization, ΔH_{vap}.** It requires 44.01 kilojoules to evaporate one mole of water at 25°C; thus the heat of vaporization of water is 44.01 kilojoules per mole ($\Delta H_{vap} = 44.01$ kJ/mol) at 25°C. That of ammonia is 23.31 kilojoules per mole at its boiling point, $-33°C$. At higher temperatures, less heat is required per mole of evaporated liquid. At its boiling point ($100°C$), the heat of vaporization of water is 40.67 kilojoules per mole.

$$H_2O(l) \longrightarrow H_2O(g) \qquad \Delta H_{vap} \text{ (at } 100°C) = 40.67 \text{ kJ/mol}$$

EXAMPLE 11.2

One mechanism for the removal of excess heat generated by the metabolic processes in the body is evaporation of the water in sweat. In a hot, dry climate as much as 1.5 L of water (1500 g) per day may be lost by one person through such evaporation. Although sweat is not pure water, we can get an *approximate* value of the amount of heat removed by evaporation by assuming that it is. Calculate the amount of heat required to evaporate this much water at $T = 46°C$ ($115°F$); $\Delta H_{vap} = 43.02$ kJ/mol at $46°C$.

To evaporate 1 mol of water at $46°C$ requires 43.02 kJ. So we must determine the number of moles of water in 1500 g and then determine the amount of heat necessary to vaporize this much water.

$$1500 \text{ g } H_2O \times \frac{1 \text{ mol } H_2O}{18 \text{ g } H_2O} = 83.3 \text{ mol } H_2O$$

$$83.3 \text{ mol } H_2O \times \frac{43.02 \text{ kJ}}{1 \text{ mol } H_2O} = 3.6 \times 10^3 \text{ kJ}$$

Thus 3600 kJ (or 860 kcal) of heat are removed by the evaporation of 1.5 L of water.

The quantity of heat evolved as a liquid condenses equals that absorbed during evaporation. At $100°C$

$$H_2O(l) \longrightarrow H_2O(g) \qquad \Delta H = \Delta H_{vap} = 40.67 \text{ kJ/mol}$$
$$H_2O(g) \longrightarrow H_2O(l) \qquad \Delta H = -\Delta H_{vap} = -40.67 \text{ kJ/mol}$$

A refrigerator cools by evaporating a liquid refrigerant. Heat leaves the refrigerator when the refrigerant (usually CCl_2F_2, a Freon) absorbs the energy needed to change it to a gas. In the gaseous state, the refrigerant is then circulated through a compressor outside the refrigerated compartment and again liquefied by combined cooling and compression. To be an effective refrigerant, a substance must be readily convertible from a gas to a liquid at the refrigerator's temperature and have a high heat of vaporization.

Evaporation is a physical process that can be described by the chemical equation

$$\text{Liquid} \longrightarrow \text{Gas}$$

The heat of vaporization of the liquid is equal to the enthalpy change of the process and can be determined using Hess's law (Section 4.5)

EXAMPLE 11.3

Using the data in Appendix I, determine the heat of vaporization of methanol, CH_3OH, under standard state conditions.

The heat of vaporization of methanol is the enthalpy change of the process described by the following equation

$$CH_3OH(l) \longrightarrow CH_3OH(g)$$

The enthalpy change of this process can be determined under standard state conditions from the standard molar heats of formation of liquid CH_3OH, -238.66 kJ/mol, and of gaseous CH_3OH, -200.66 kJ/mol, using Hess's law.

$$CH_3OH(l) \rightarrow C(s) + \frac{1}{2}O_2(g) + 2H_2(g) \quad \Delta H^\circ_{298} = 238.66 \text{ kJ}$$

$$C(s) + \frac{1}{2}O_2(g) + 2H_2(g) \rightarrow CH_3OH(g) \quad\quad \Delta H^\circ_{298} = -200.66 \text{ kJ}$$

$$CH_3OH(l) \rightarrow CH_3OH(g) \quad\quad \Delta H^\circ_{298} = 238.66 + (-200.66) \text{ kJ}$$
$$= 38.00 \text{ kJ}$$

The enthalpy change for the evaporation of 1 mol of $CH_3OH(l)$ is 38.00 kJ, so the heat of vaporization is 38.00 kJ/mol.

11.7 Heat of Fusion

When heat is added to a crystalline solid, the temperature rises until the melting point is reached. At this temperature any additional heat results in disrupting intermolecular forces so that the solid begins to melt. The temperature remains constant as additional heat is applied until all of the solid has melted. After that, the temperature rises again (Fig. 11.11). The quantity of heat needed to change a given quantity of a substance from the solid state to the liquid state at constant temperature is known as the **heat of fusion, ΔH_{fus}** of the substance. The heat of fusion of ice is 6.01 kilojoules per mole at 0°C.

$$H_2O(s) \longrightarrow H_2O(l) \quad \Delta H = \Delta H_{fus} = 6.01 \text{ kJ/mol}$$

The quantity of heat liberated during crystallization (freezing) equals that absorbed during fusion.

$$H_2O(l) \longrightarrow H_2O(s) \quad \Delta H = -\Delta H_{fus} = -6.01 \text{ kJ}$$

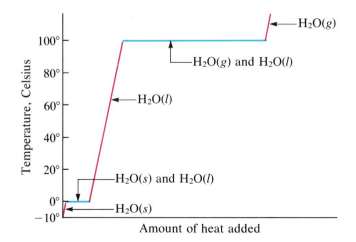

Figure 11.11

Heating curve for water. When ice is heated, the temperature increases until the melting point is reached (red line). The temperature then remains constant until all of the ice is melted (blue line), then it increases again (red line). When the water begins to boil, the temperature again remains constant (blue line) until all of the water has vaporized. The temperature then increases again (red line).

11.8 Critical Temperature and Pressure

At 25°C a sample of water sealed in an evacuated tube exists as liquid water and water vapor with a pressure of 23.8 torr (Table 10.1). There is a clear boundary between the two phases [Fig. 11.12(a)]. As the temperature is increased, the pressure of the vapor increases. At 100°C, for example, the pressure is 760 torr, but liquid water and water vapor are both present and the boundary between them is obvious [Fig. 11.12(b)]. However, at 374°C the boundary disappears [Fig. 11.12(c)], and all of the

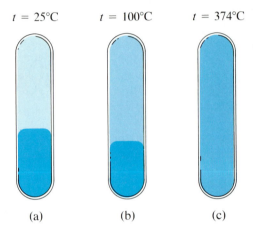

Figure 11.12

A sample of liquid water and water vapor (a) at 25°C, (b) at 100°C, and (c) at 374°C. At the critical temperature, 374°C, the boundary between liquid and vapor disappears.

water in the tube is physically identical. No distinction can be made between liquid and vapor. The temperature above which it is no longer possible to distinguish a liquid from its vapor is called the **critical temperature.** Above this temperature a gas cannot be liquefied, *no matter how much pressure is applied.* The pressure required to liquefy a gas at its critical temperature is called the **critical pressure.** The critical temperatures and critical pressures of some common substances are given in Table 11.1.

Table 11.1	Critical temperatures and pressures of some common substances	
	Critical temperature, K	Critical pressure, atm
Hydrogen	33.24	12.8
Nitrogen	126.0	33.5
Oxygen	154.3	49.7
Carbon dioxide	304.2	73.0
Ammonia	405.5	111.5
Sulfur dioxide	430.3	77.7
Water	647.1	217.7

EXAMPLE 11.4   If a carbon dioxide fire extinguisher is shaken on a cool day (temperature = 18°C), liquid CO_2 can be heard sloshing around in the cylinder. However, the same cylinder appears to contain no liquid on a hot summer day (temperature = 35°C). Explain these observations.

On the cool day the temperature of the CO_2 is below its critical temperature, which is 304 K, or 31°C (Table 11.1), so liquid CO_2 is present in the cylinder. On the hot day the temperature is greater than the critical temperature, so no liquid CO_2 can form.

Above the critical temperature of a substance, the average kinetic energy of its molecules is sufficient to overcome their mutually attractive forces. Therefore they will not cling together to form a liquid, no matter how great the pressure. If the temperature is decreased, the average kinetic energy of the molecules decreases. At the critical temperature the intermolecular forces are just large enough to liquefy the gas, provided that the pressure is equal to or greater than the critical pressure. The pressure helps the intermolecular forces bring the molecules close enough to one another to liquefy. Below the critical temperature the pressure required for liquefaction decreases with decreasing temperature until it reaches 1 atmosphere at the normal boiling temperature. Substances with strong intermolecular attractions, like water and ammonia, have high critical temperatures; substances with weak intermolecular attractions, like hydrogen and nitrogen, have low critical temperatures.

11.9 Phase Diagrams

If water at 80°C is confined in a syringe with a pressure of 760 torr on the plunger, only liquid water is found to be present (Fig. 11.13). As heat is added, the temperature of the water increases, but as long as the pressure is held at 760 torr, only liquid water will be present. When the temperature reaches 100°C, addition of more heat produces water vapor, provided the pressure remains at 760 torr [Fig. 11.13(b)]. With the further addition of heat, the temperature remains constant at 100°C until all of the liquid water changes to vapor. If the addition of heat is stopped and no heat is

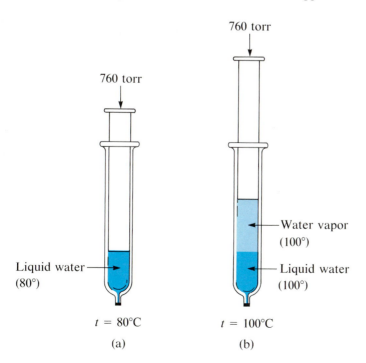

760 torr

760 torr

Water vapor
(100°)

Liquid water
(80°)

Liquid water
(100°)

$t = 80°C$

$t = 100°C$

(a)

(b)

Figure 11.13

Syringes containing water at 760 torr. (a) At 80°C and 760 torr, only liquid water is contained in the syringe. (b) If the water is warmed to 100°C and then a little additional heat is added, the syringe will contain both liquid water and water vapor at 100°C and 760 torr.

Realgar, AsS

allowed to escape while both liquid water and water vapor are present, the mixture remains at equilibrium.

When heat is removed from water vapor at 110°C and a constant pressure of 760 torr, the vapor cools until it reaches 100°C, at which point liquid water begins to form. If more heat is removed, the temperature does not change until all of the vapor condenses. Thus 100°C and 760 torr is found to be one point at which liquid water and water vapor are at equilibrium. At a lower pressure of 600 torr, the equilibrium between liquid and vapor is found at 94°C. At 380 torr, the equilibrium is found at 80°C.

Water in a syringe at 10°C and a pressure of 380 torr cools as heat is removed. When the temperature reaches 0.005°C, further loss of heat results in formation of ice with no change in temperature until all of the liquid is frozen. Ice and liquid water are in equilibrium at 0.005°C at a pressure of 380 torr. At 760 torr, the equilibrium between ice and water occurs at 0°C.

All of the information concerning the temperatures and pressures at which the various phases of a substance are in equilibrium can be summarized in its **phase diagram,** a diagram showing the pressures and temperatures at which gaseous, liquid, and solid phases of a substance can exist. Such a diagram is constructed by plotting points showing the pressures and temperatures at which two different phases of a substance are in equilibrium. A part of the phase diagram for water is given in Fig. 11.14.

The red line *BC* in the phase diagram of water (Fig. 11.14) is a plot of the vapor pressure of water versus temperature. It shows the combinations of temperature and pressure at which liquid water and water vapor are in equilibrium. This diagram shows that when the pressure is 760 torr the liquid and vapor are at equilibrium only when the temperature is 100°C, as we have seen in the experiment with the syringes described above. The figure also indicates that the equilibrium occurs at 94°C when the pressure is 600 torr and at 80°C when the pressure is 380 torr.

If the figure were large enough, line *BC* would be seen to terminate at a temperature of 374°C and a pressure of 165,500 torr (217.7 atmospheres), the critical temperature and pressure of water. Beyond the critical temperature, liquid and gaseous

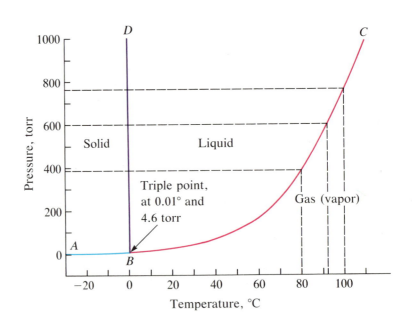

Figure 11.14
Phase diagram for water.

water cannot be in equilibrium because liquid water cannot exist above the critical temperature.

The blue line *AB* in Fig. 11.14 is a plot of the vapor pressure of ice versus temperature; at any combination of temperature and pressure on this line, ice and water vapor are in equilibrium. The purple line *BD* in Fig. 11.14 shows the temperature at which ice and liquid water are in equilibrium as the pressure changes.

For each value of pressure and temperature falling within the portion of the phase diagram (Fig. 11.14) labeled ''Solid'' (760 torr and −10°C, for example), water exists only in the solid form (ice). Similarly, for values of pressure and temperature within the portion labeled ''Liquid'' (760 torr and 80°C, for example), water exists only as a liquid. For any combination of pressure and temperature in the region labeled ''Gas'' (760 torr and 110°C, for example), water exists only in the gaseous state. At −5°C and 760 torr, the point on the diagram is in the region indicating that only solid water is stable. Moving across the diagram horizontally at a constant pressure of 760 torr, we see that as the temperature increases water becomes a liquid at 0°C, which is its *normal melting temperature*. At this temperature and pressure, both solid and liquid water can exist in equilibrium. Water is stable only as a liquid at a pressure of 760 torr for temperatures between 0°C and 100°C. It changes to a gas at 100°C, its *normal boiling temperature,* at which liquid water at 760 torr and water vapor at 760 torr can exist at equilibrium. At temperatures above 100°C at 760 torr, water is stable only as a gas.

The almost vertical orange line *BD* in Fig. 11.14 shows us that as the pressure increases, the melting temperature remains almost constant, actually decreasing very slightly. On the other hand, the diagram indicates clearly that the boiling temperature increases markedly with an increase in pressure. Representative vapor pressures for water are given in Table 10.1.

The lines that separate the various regions in Fig. 11.14 represent points at which an equilibrium exists between two phases. One point exists where all three phases are in equilibrium. This point, referred to as the **triple point,** occurs at point *B* in Fig. 11.14, where the three lines intersect. At the pressure and temperature of the triple point [4.6 torr and 0.01°C (273.16 K)], all three states are in equilibrium with each other.

Aquamarine, $Be_3Al_2Si_6O_{18}$

11.10 Surface Tension

The molecules within a liquid are attracted equally in all directions by neighboring molecules; the resultant force on any one molecule within the liquid is therefore zero. However, the molecules on the surface of a liquid are attracted only into the liquid and to either side (Fig. 11.15). This unbalanced molecular attraction tends to pull the surface molecules back into the liquid, and a condition of equilibrium is reached only when as many molecules as possible have been pulled into the liquid. At this point the minimum number of molecules possible are on the surface, and the surface area is reduced to a minimum. A small drop of liquid tends to assume a spherical shape because in a sphere the ratio of surface area to volume is at a minimum. **Surface tension** is defined as the force that causes the surface of a liquid to contract. The surface of a liquid acts as if it were a stretched membrane. A steel needle carefully placed on water will float. Some insects, even though they are denser than water, move on its surface because they are supported by the surface tension. One of the forces causing water to rise in capillary tubes (tubes with a very small bore) is the surface tension. Water is brought from the soil up through the roots and into the portion of a plant above the soil by this capillary action.

Surface

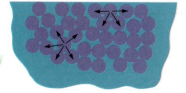

Figure 11.15
Attractive forces experienced by a molecule at the surface and by a molecule in the bulk of a liquid.

Attractive Forces Between Molecules

11.11 Intermolecular Forces

We saw in Section 11.1 that the condensation of a gas or the freezing of a liquid is due to the mutual forces of attraction between its constituent molecules. The strengths of these intermolecular attractive forces vary widely, although the *inter*molecular forces between small molecules are weak when compared to the *intra*molecular forces that bond atoms together within molecules. For example, to overcome the intermolecular forces in a mole of liquid HCl and convert it to gaseous HCl requires only about 17 kilojoules. However, to break a mole of the bonds between the hydrogen and chlorine atoms in hydrogen chloride requires about 430 kilojoules.

Intermolecular attractive forces, which are collectively called **van der Waals forces,** are electrical in nature and result from the attraction of charges of opposite sign. One type of van der Waals force results from the electrostatic attraction of the positive end of one polar molecule (Section 6.6) for the negative end of another. The BrCl molecule is a polar molecule because the more electronegative chlorine atom bears a partial negative charge and the less electronegative bromine atom bears a partial positive charge. The BrCl molecule has a dipole moment. The electrostatic attraction of the positive end of one BrCl molecule for the negative end of another constitutes an attractive force between two dipoles, a **dipole-dipole attraction** (Fig. 11.16), that causes BrCl molecules to attract each other.

An additional component of van der Waals forces, found with both nonpolar and polar molecules, is called the **dispersion force,** or **London force** (after Fritz London, who in 1930 first explained it). Because of the constant motion of its electrons, an atom that on average is nonpolar sometimes becomes polar, and while it is, it has a temporary dipole. A second atom, in turn, is distorted by the appearance of the dipole in the first atom. Its nucleus is attracted toward the negative end of the first atom (Fig. 11.17). Thus a negative end on the second atom is formed away from the negative end on the first atom, and the two rapidly fluctuating, temporary dipoles attract each other. If the two atoms are located in different molecules, the two molecules may attract each other. London forces between two atoms are weak compared to dipole-dipole attractions and are significant only when the atoms are very close together. Atoms must almost touch for London forces to be significant. These forces cause nonpolar substances such as the noble gases and the halogens to condense into liquids and to freeze into solids when the temperature is lowered sufficiently.

For molecules with similar shapes, the magnitude of London forces increases with the size of the molecule. For example, about 8 kilojoules will convert a mole of liquid CH_4 (mol. wt = 16) to a gas, but about 13 kilojoules is needed to convert a mole of liquid SiH_4 (mol. wt = 32) to a gas. This suggests that the larger and heavier SiH_4 molecule has stronger London forces. The size dependence is readily explained by London's theory. The valence electrons in a larger molecule are, on the average,

Figure 11.16

Two arrangements of polar molecules that allow an attractive interaction between the dipoles. These arrangements place the negative end of one molecule close to the positive end of another.

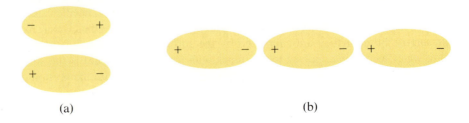

(a)   (b)

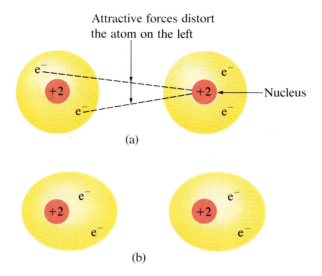

Attractive forces distort
the atom on the left

(a)

(b)

Figure 11.17

A representation of the formation of the temporary dipoles that give rise to London forces. The temporary dipole on the atom on the right in (a) produces a dipole in the atom on the left. The attraction of the two resulting dipoles (b) results in the attractive force.

farther from the nucleus than those in a smaller molecule. Thus they are less tightly held and can more easily form the temporary dipoles that produce the attraction. The increase in intermolecular attraction with increasing size is reflected in the rise in boiling point in the series of related substances shown in Table 11.2.

Table 11.2   Molecular weights and boiling points

Substance	He	Ne	Ar	Kr	Xe	Rn
Molecular weight	4.0	20.18	39.95	83.8	131.3	222
Boiling point, °C	−268.9	−245.9	−185.7	−152.9	−107.1	−61.8

Substance	H_2	F_2	Cl_2	Br_2	I_2
Molecular weight	2.016	38.0	70.91	159.8	253.8
Boiling point, °C	−252.7	−187	−34.6	58.78	184.4

The shapes of molecules also affect the magnitudes of the London forces between them. For example, two compounds with the same chemical formula, C_5H_{12}, boil at different temperatures. A sample of *n*-pentane [Fig. 11.18(a)] boils at 36°C, while one of neopentane [Fig. 11.18(b)] boils at 9.5°C. Boiling points increase as the forces of attraction between molecules increase. Thus the London forces in *n*-pentane are larger than those in neopentane. The neopentane molecule is shaped so that the valence electrons in the carbon-carbon bonds are well inside the molecule. *n*-Pentane is shaped so that the valence electrons in the carbon-carbon bonds are closer to the surface of the molecule. Thus any London forces that involve electrons in carbon-carbon bonds must act over a longer distance with neopentane, and this results in weaker total attractive forces.

In general, condensed phases composed of discrete molecules with no permanent dipole moments have low melting points, low boiling points, low heats of sublimation, and low heats of vaporization relative to polar molecules with equivalent molecular weights, because only the weak London forces must be overcome during vaporization. Examples include molecules of the noble gases and the halogens and other symmetrical molecules such as CH_4, SiH_4, CF_4, SiF_4, SF_6, and UF_6. Molecular substances such

(a)

(b)

Figure 11.18

The Lewis structures of
(a) *n*-pentane and (b) neopentane.

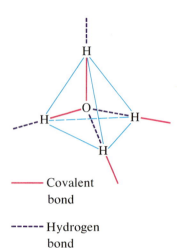

—— Covalent
bond

----- Hydrogen
bond

Figure 11.19

A tetrahedral arrangement of covalent bonds (solid red lines) and hydrogen bonds (broken purple lines) about an oxygen atom in water. Two hydrogen atoms participate in covalent bonds and two participate in hydrogen bonds to the oxygen. Each bond from a hydrogen atom and away from the tetrahedron leads to an oxygen of another water molecule.

as $CHCl_3$, NO, and PCl_3, which have permanent dipole moments, have higher melting points, boiling points, heats of sublimation, and heats of vaporization than nonpolar molecules with equivalent molecular weights.

11.12 Hydrogen Bonding

Nitrosyl fluoride (ONF, mol. wt 49) has a boiling point of −56°C. Water (H_2O, mol. wt 18) has a much higher boiling point of 100°C, even though it has a lower molecular weight. The difference in the boiling points of these two compounds cannot be attributed to London forces; both molecules have about the same shape, and ONF is the heavier and larger molecule. It cannot be attributed to differences in the dipole moments of the molecules; both molecules have the same dipole moment. The large difference in the boiling points of these two molecules is due to a special kind of dipole-dipole attraction called **hydrogen bonding.**

Hydrogen bonds can form when hydrogen is bonded to one of the more electronegative elements such as fluorine, oxygen, nitrogen, or chlorine. The large difference in electronegativity between hydrogen (2.1) and the second element (4.1 for fluorine, 3.5 for oxygen, 3.0 for nitrogen, 2.8 for chlorine) leads to a highly polar covalent bond in which the hydrogen bears a large fractional positive charge and the second element bears a large fractional negative charge. The electrostatic attraction between the partially positive hydrogen atom in one molecule and the partially negative atom of the more electronegative element in another molecule gives rise to the strong dipole-dipole attraction called a **hydrogen bond.** Hydrogen bonds are stronger than other dipole-dipole attractions and London forces. The strengths are about 5–10% of those of ordinary covalent bonds. Examples of hydrogen bonds include $HF\cdots HF$, $H_2O\cdots HOH$, $H_3N\cdots HNH_2$, $H_3N\cdots HOH$, and $H_2O\cdots HOCH_3$. Figure 11.19 illustrates the hydrogen bonds about a water molecule in liquid water.

Figure 11.20

Illustrations of (a) the boiling points and (b) the heats of vaporization of several binary hydrides of Groups IVA–VIIA. Each line connects compounds of the elements in a given group. H_2O, HF, and NH_3 have hydrogen bonding and exhibit abnormally high boiling points and heats of vaporization. CH_4 does not have hydrogen bonding. Plots for four of the noble gases are included to indicate typical behavior due to increasing London forces resulting from increasing mass.

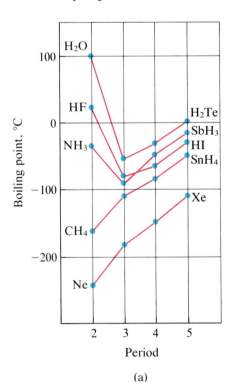

(a)

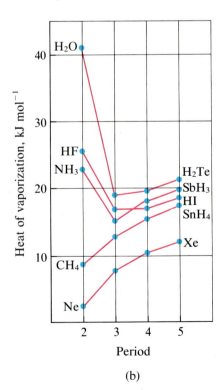

(b)

Because hydrogen bonds are relatively strong intermolecular forces, they can have a pronounced effect on the properties of condensed phases. In general, we expect boiling points and heats of vaporization to decrease with decreasing molecular or atomic weight (Section 11.11). The noble gases and the hydrides of Group IVA (CH_4 through SnH_4) exhibit this trend (Fig. 11.20). The trend is also observed for the hydrides of Groups VA, VIA, and VIIA, except for H_2O, HF, and NH_3, compounds that possess appreciable hydrogen bonding. Energy is required to break hydrogen bonds in order to bring a material to the gaseous (vapor) state. Hence these substances that possess hydrogen bonds have abnormally high boiling points and heats of vaporization. Methane (CH_4) and neon (Ne), however, do not have hydrogen bonding and so have the lower boiling points and heats of vaporization expected by extrapolation from the corresponding heavier members in their groups.

The Structures of Crystalline Solids

11.13 Types of Crystalline Solids

There are several different types of crystalline solids. Some solids, such as sodium chloride and potassium nitrate, are composed of positive and negative ions. Other solids, such as ice and sucrose (table sugar), are composed of neutral molecules. These examples illustrate two important types of solids: **ionic solids,** which are composed of ions, and **molecular solids,** composed of molecules. The structure of ice, a molecular solid, is shown in Fig. 11.21.

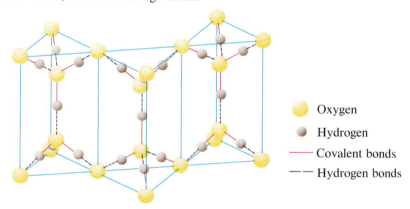

Oxygen
Hydrogen
— Covalent bonds
- - Hydrogen bonds

Figure 11.21

The arrangement of water molecules in ice, a molecular solid, showing each oxygen atom (yellow) with four hydrogen atoms (brown) as nearest neighbors, two attached by covalent bonds (red) and two by hydrogen bonds (purple). The oxygen atoms at the corners of a prism are also at the corners of adjoining prisms, as this is only a portion of a continuous array.

Crystals of some solids may be regarded as giant molecules. These include crystals of metals such as copper, aluminum and iron; crystals of diamond, silicon, and some other nonmetals; and crystals of some covalent compounds such as silicon dioxide, silicon carbide, and gallium arsenide. The atoms in all of these solids are held together by a three-dimensional network of covalent bonds; it is not possible to identify individual molecules in them. Solids composed of metal atoms are called **metallic solids,** while the other solids that can be regarded as giant molecules are called **covalent solids.** The structures of metallic solids are discussed in Section 11.15. Structures of some covalent solids are shown in Fig. 11.22.

The strengths of the attractive forces between the units present in different crystals vary widely, as indicated by the melting points of the crystals. Small, symmetrical molecules, such as H_2, N_2, O_2, and F_2, have weak attractive forces and form molecular crystals with very low melting points (below $-200°C$). Molecular crystals consisting of larger, nonpolar molecules have stronger attractive forces and melt at higher

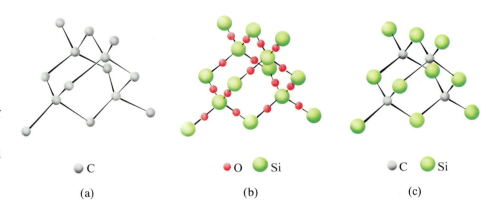

Figure 11.22
A covalent crystal contains a three-dimensional network of covalent bonds, as illustrated by the structures of (a) diamond, (b) silicon dioxide, and (c) silicon carbide. Lines between atoms indicate covalent bonds.

C

O   Si

C   Si

(a)   (b)   (c)

Figure 11.23
Ionic metal oxides are used to line this crucible. They do not melt when in contact with molten iron because they have very high melting temperatures.

temperatures, whereas molecular crystals composed of asymmetrical molecules with permanent dipole moments melt at still higher temperatures; examples are ice (mp 0°C) and sugar (mp 185°C). Diamond is a covalent crystal in which carbon atoms are held together by strong covalent bonds (Fig. 11.22); the melting point of diamond is very high (above 3500°C). The atoms in the crystals of most metals are strongly bonded together by modified covalent bonds (Section 25.3), and most metals have high melting points (often above 1000°C). The electrostatic forces of attraction between the ions in ionic solids can be quite strong (Fig. 11.23), so many ionic crystals also have high melting points (as high as 3000°C).

11.14 Crystal Defects

In a crystalline solid the atoms, ions, or molecules are arranged in a definite repeating pattern, but occasional defects may occur in the pattern. Several types of defects are known. It is common for some positions that should contain atoms or ions to be vacant [Fig. 11.24(a)]. Less commonly, some atoms or ions in a crystal may be located in positions, called **interstitial sites,** that are located between the regular positions for atoms [Fig. 11.24(b)]. Certain distortions occur in some impure crystals, as for example when the cations, anions, or molecules of the impurity are too large to fit into the regular positions without distorting the structure. Minute amounts of impurities are sometimes added to a crystal to cause imperfections in the structure so that the electrical conductivity (Section 25.3) or some other physical properties of the crystal will change. This has practical applications in the manufacture of semiconductors and printed circuits.

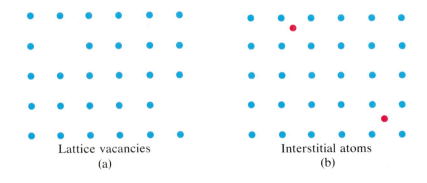

Figure 11.24
Representations of two types of crystal defects.

Lattice vacancies
(a)

Interstitial atoms
(b)

11.15 The Structures of Metals

A pure metal is a crystalline solid in which the metal atoms are packed closely together in a repeating array, or pattern. In most metals the atoms pack together as if they were spheres. If spheres of equal size are packed together as closely as possible in a plane, they arrange themselves as shown in Fig. 11.25(a), with each sphere in contact with six others. This arrangement, called **closest packing,** can extend indefinitely in a single layer. Crystals of many metals can be described as stacks of such closest packed layers.

Two types of stacking of closest packed layers are observed in simple metallic crystalline structures. In both types a second layer (B) is placed on the first layer (A) so that each sphere in the second layer is in contact with three spheres in the first layer, as shown in Fig. 11.25(b). A third layer can be positioned in one of two ways.

In one positioning each sphere in the third layer lies directly above a sphere in the first layer [Fig. 11.26(a)]. The third layer is also type A. The stacking continues with alternate type B and type A close packed layers (ABABAB···). This arrangement is called **hexagonal closest packing.** Metals that crystallize this way have a **hexagonal closest packed structure.** Examples include Be, Cd, Co, Li, Mg, Na, and Zn. (Those elements or compounds that crystallize with the same structure have **isomorphous structures.**)

In the second positioning the third layer is located so that its spheres are not directly above those in either layer A or layer B [Fig. 11.26(b)]. This layer is type C. The stacking continues with alternating layers of type A, type B, and type C (ABCABC···), an arrangement called **cubic closest packing.** Metals crystallizing

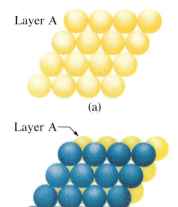

Layer A

(a)

Layer A

Layer B

(b)

Figure 11.25

(a) A portion of a layer of closest packed spheres in the same plane. Each sphere touches 6 others. (b) Spheres in two closest packed layers. Each sphere in layer B (the blue layer) touches 3 spheres in layer A (the yellow layer). In an actual crystal many more than two planes would exist. Each sphere contacts 6 spheres in its own layer, 3 spheres in the layer below, and 3 spheres in the layer above; each sphere touches a total of 12 other spheres.

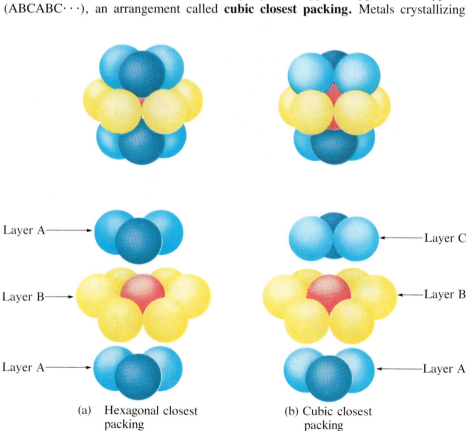

Layer A

Layer B

Layer A

(a) Hexagonal closest packing

Layer C

Layer B

Layer A

(b) Cubic closest packing

Figure 11.26

A portion of two types of crystal structures in which spheres are packed as compactly as possible. The lower diagrams show the structures expanded for clarification. Note that the first and third layers have identical orientations in (a). The first and third layers have different orientations in (b). In both structures, each sphere is surrounded by 12 others in an infinite extension of the structure and is said to have a coordination number of 12.

this way have a **cubic closest packed,** or **face-centered cubic, structure.** Examples include Ag, Al, Ca, Cu, Ni, Pb, and Pt.

In crystals of metals with either hexagonal closest packing or cubic closest packing, each atom touches 12 equidistant neighbors, 6 in its own plane and 3 in each adjacent plane. This gives each atom a coordination number of 12. The **coordination number** of an atom or ion is the number of neighbors nearest to it. About two-thirds of all metals crystallize in closest packed arrays with coordination numbers of 12.

Most of the remaining metals crystallize in a **body-centered cubic structure,** which contains planes of spheres that are not closest packed. Each sphere in a plane is surrounded by four nearest neighbors [Fig. 11.27(a)], rather than six, as in closest packed planes. The spheres in such a plane *do not touch.* The structure consists of repeating layers of these planes. The second layer is stacked on top of the first so that a sphere in the second layer *touches* four spheres in the first layer [Fig. 11.27(b)]. The spheres of the third layer are positioned directly above the spheres of the first layer (Fig. 11.28); those of the fourth, above the second; etc. Any atom in this structure touches four atoms in the layer above it and four atoms in the layer below it. An atom in a body-centered cubic structure thus has a coordination number of 8. Isomorphous metals with a body-centered cubic structure include Ba, Cr, Mo, W, and Fe at room temperature.

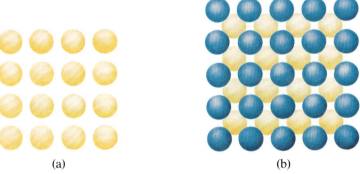

(a)   (b)

Figure 11.27
(a) A portion of a plane of spheres found in a body-centered cubic structure. Note that the spheres do not touch. (b) Spheres in two layers of a body-centered cubic structure. Each sphere in one layer touches four spheres in the adjacent layer but none in its own layer.

(a)   (b)

Figure 11.28
(a) A portion of a body-centered cubic structure, showing parts of three layers. (b) An expanded view of a body-centered cubic structure. In an infinite extension of this structure, each sphere touches four spheres in the layer above it and four spheres in the layer below and is said to have a coordination number of 8.

Polonium (Po) crystallizes in the **simple cubic structure,** which is rare for metals. It contains planes in which each sphere *touches* its four nearest neighbors [Fig. 11.29(a)]. Thus the structure is not closest packed. The planes are stacked directly above each other, so that an atom in the second layer touches only one atom in the first layer [Fig. 11.29(b)]. The coordination number of a polonium atom in a simple cubic array is 6; an atom touches four other atoms in its own layer, one atom in the layer above, and one atom in the layer below.

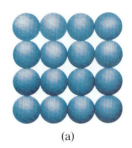

(a)

(b)

Figure 11.29

(a) A portion of a plane of spheres found in a simple cubic structure. Note that the spheres are in contact. (b) A portion of a simple cubic structure, showing two of the planes.

11.16 The Structures of Ionic Crystals

Ionic crystals consist of two or more different kinds of ions, usually having different sizes. The packing of these ions into a crystal structure is more complex than the packing of metal atoms that are the same size and kind.

Most monatomic ions behave as charged spheres; their attraction for ions of opposite charge is the same in every direction. Consequently, stable structures result (1) when ions of one charge are surrounded by as many ions as possible of the opposite charge and (2) when the cations and anions are in contact with each other. The structures are determined by two factors: the relative sizes of the ions and the relative numbers of positive and negative ions required to maintain the electrical neutrality of the crystal as a whole.

In simple ionic structures the anions, which are normally larger than the cations, are usually arranged in a closest packed array. The spaces remaining between the anions, called **holes,** or **interstices,** are occupied by the smaller cations. Figures 11.30 and 11.31 illustrate the two most common types of holes. The smaller of these is found between three spheres in one plane and one sphere in an adjacent plane [Fig. 11.30(a)]. The four spheres that bound this hole are arranged at the corners of a tetrahedron [Fig. 11.30(b)]; the hole is called a **tetrahedral hole.** The larger type of hole is found at the center of six spheres (three in one layer and three in an adjacent layer) located at the corners of an octahedron (Fig. 11.31). Such a hole is called an **octahedral hole.**

Depending on the relative sizes of the cations and anions, the cations of an ionic compound can occupy tetrahedral or octahedral holes. As will be discussed in Section 11.17, relatively small cations occupy tetrahedral holes, and larger cations occupy octahedral holes. If the cations are too large to fit into the octahedral holes, the packing of the anions may change to give a more open structure, such as a simple cubic array

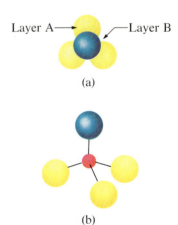

Layer A ——→   ←—— Layer B

(a)

(b)

Figure 11.30

(a) Spheres in two adjacent closest packed layers that form a tetrahedral hole. (b) A cation (smaller sphere) located in a tetrahedral hole surrounded by four anions (larger spheres) from a different perspective. The structure has been expanded to show the geometrical relationships.

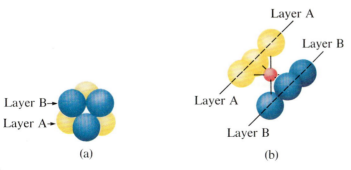

Layer B →
Layer A →

(a)

Layer A
Layer B
Layer A
Layer B

(b)

Figure 11.31

(a) Spheres in two adjacent closest packed layers that form an octahedral hole. (b) A cation (red sphere) located in an octahedral hole surrounded by six anions (larger spheres) from a different perspective. The structure has been expanded to show the geometrical relationships.

Figure 11.32
A cation in the cubic hole in a simple cubic array of anions.

(Fig. 11.29). The larger cations can then occupy the larger cubic holes made possible by the more open spacing (Fig. 11.32).

In either a hexagonal closest packed or a cubic closest packed array of anions, there are two tetrahedral holes for each anion in the array. The isomorphous compounds Li_2O, Na_2O, Li_2S, Na_2S, and Li_2Se, among others, crystallize with a cubic closest packed array of anions with the relatively small cations in tetrahedral holes. The ratio of tetrahedral holes to anions is 2 to 1; thus all of these holes must be filled by cations, since the cation-to-anion ratio is 2 to 1 in these compounds. A compound that crystallizes in a closest packed array of anions with cations in the tetrahedral holes can have a maximum cation-to-anion ratio of 2 to 1; all of the tetrahedral holes are filled at this ratio. Compounds with a ratio of less than 2 to 1 may also crystallize in a closest packed array of anions with cations in the tetrahedral holes, if the ionic sizes fit. In these compounds, however, some of the tetrahedral holes remain vacant.

EXAMPLE 11.5

Zinc sulfide crystallizes with zinc ions occupying $\frac{1}{2}$ of the tetrahedral holes in a closest packed array of sulfide ions. What is the formula of zinc sulfide?

Since there are 2 tetrahedral holes per anion (sulfide ion) and $\frac{1}{2}$ of these are occupied by zinc ions, there must be $\frac{1}{2} \times 2$, or 1, zinc ion per sulfide ion. Thus the formula is ZnS.

The ratio of octahedral holes to anions in either a hexagonal or a cubic closest packed structure is 1 to 1. Thus compounds with cations in octahedral holes in a closest packed array of anions can have a maximum cation-to-anion ratio of 1 to 1. In NiO, MnS, NaCl, and KH, for example, all of the octahedral holes are filled. Ratios of less than 1 to 1 are observed when some of the octahedral holes remain empty.

EXAMPLE 11.6

Aluminum oxide crystallizes with aluminum ions in $\frac{2}{3}$ of the octahedral holes in a closest packed array of oxide ions. What is the formula of aluminum oxide?

Since there is 1 octahedral hole per anion (oxide ion) and only $\frac{2}{3}$ of these holes are occupied, the ratio of aluminum to oxygen must be $\frac{2}{3}$ to 1, which would give $Al_{2/3}O$. The simplest whole-number ratio is 2 to 3, so the formula is Al_2O_3.

In a simple cubic array of anions (Fig. 11.32), there is one cubic hole that can be occupied by a cation for each anion in the array. In CsCl, and in other compounds with the same structure, all of the cubic holes are occupied. Half of the cubic holes are occupied in SrH_2, UO_2, $SrCl_2$, and CaF_2.

Different types of ionic compounds crystallize in the same structure because the relative sizes of the ions and the stoichiometry (the two principal features that determine structure) are similar.

11.17 The Radius Ratio Rule for Ionic Compounds

The structure of an ionic compound is largely the result of stoichiometry and of simple geometric and electrostatic relationships that depend on the relative sizes of the cation and anion. A relatively large cation can touch a large number of anions and so occupies a cubic or an octahedral hole, whereas a relatively small cation can touch only a few anions and so occupies a tetrahedral hole.

Consider a cation, M^+, with a coordination number of 6. As shown in Fig. 11.33(a), the M^+ ion touches four X^- ions in a plane. In addition, although they are not shown in the figure, there is an X^- ion above the M^+ ion and touching it and another below and touching it. Note that the M^+ ion is large enough to expand the array of X^- ions so that the X^- ions are not in contact with one another. As long as the expansion is not great enough to allow still another anion to touch the cation, this is a stable situation; the cation-anion contacts are maintained. Figure 11.33(b) illustrates what happens when the size of the M^+ ion is decreased somewhat. Here the X^- ions touch each other, as do the M^+ and X^- ions. If the size of M^+ is further decreased, it becomes impossible to get a structure with a coordination number of 6. The anions touch [Fig. 11.33(c)], but there is no contact between the M^+ and X^- ions—this is an unstable structure. In this case a more stable structure would be formed with only four anions about the cation. The limiting condition for the formation of a structure with a coordination number of 6 is illustrated in Fig. 11.33(b): the X^- ions touch one another, and the M^+ and X^- ions touch. This occurs when the sizes of the ions are such that the **radius ratio** (the radius of the positive ion, r^+, divided by the radius of the negative ion, r^-) is equal to 0.414.

There is a minimum radius ratio (r^+/r^-) for each coordination number. Below this value an ionic structure having that coordination number is generally not stable. The approximate limiting values for the radius ratios for ionic compounds are given in Table 11.3.

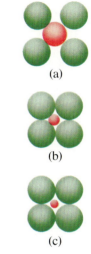

(a)

(b)

(c)

Figure 11.33
Packing of anions (green spheres) around cations of varying size (red spheres). Decreasing size of cations is illustrated successively in (a), (b), and (c).

Table 11.3 Limiting values for the radius ratio for ionic compounds (r^+ is radius of cation; r^- is radius of anion)

Coordination number	Type of hole occupied	Approximate limiting values of r^+/r^-
8	Cubic	Above 0.732
6	Octahedral	0.414 to 0.732
4	Tetrahedral	0.225 to 0.414

Predict the coordination number of Cs^+ ($r^+=1.69$ Å) and of Na^+ ($r^+=0.95$ Å) in CsCl and NaCl, respectively. The radius of a chloride ion is 1.81 Å.

EXAMPLE 11.7

For CsCl:

$$\frac{r^+}{r^-} = \frac{1.69\ \text{Å}}{1.81\ \text{Å}} = 0.934$$

The radius ratio is greater than 0.732 (Table 11.3), which indicates that a coordination number of 8 is likely for Cs^+ in CsCl.

For NaCl:

$$\frac{r^+}{r^-} = \frac{0.95\ \text{Å}}{1.81\ \text{Å}} = 0.52$$

This radius ratio is between 0.414 and 0.732 and so indicates that a coordination number of 6 is likely for Na^+ in NaCl (Table 11.3).

The radius ratio rule is only a guide to the type of structure that may form. It applies strictly only to ionic crystals and in some cases fails with them. In compounds in which the bonds are covalent, the rule may not hold. In spite of its limitations, however, the radius ratio rule is a useful guide for predicting many structures. It also underlines one of the most significant features responsible for the structures of ionic solids: the relative sizes of the cations and anions.

11.18 Crystal Systems

The atoms in a crystal are arranged in a definite repeating pattern, so any one point in the crystal matches many other points having identical environments. The collection of all of the points within the crystal that have identical environments is called a **space lattice.** A simple three-dimensional cubic space lattice is shown in Fig. 11.34. (The atoms around the points have been omitted so that you can see the lattice.)

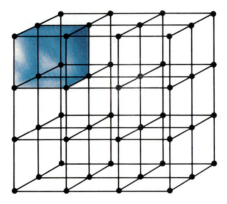

Figure 11.34

A portion of a simple cubic space lattice. One unit cell is shaded in blue.

Figure 11.35

A portion of the structure of NaCl, illustrating three ways to position a space lattice in the structure. Small red spheres represent Na^+; large green spheres, Cl^-. (a) Lattice points in the center of the Na^+ ions. (b) Lattice points in the center of the Cl^- ions. (c) Lattice points with a chloride ion to the left and a sodium ion to the right. Black lines connect points that define a unit cell in each lattice. The unit cell of NaCl is usually described as indicated in (a) or (b). Note that this structure has been expanded for clarity; the ions touch in the actual structure.

Figure 11.35 illustrates a portion of the structure of sodium chloride. This structure will be described more fully later in this section, but for now let us consider only the space lattice associated with it. There are an infinite number of ways to construct a space lattice in the sodium chloride structure. The points of one possible space lattice are located at the centers of the sodium ions, as shown in Fig. 11.35(a). Alternatively, points with identical environments could be located at the centers of chloride ions [Fig. 11.35(b)] or between sodium and chloride ions [Fig. 11.35(c)]. In each case the resulting space lattice is the same; only the locations of the points of the lattice differ.

That part of a space lattice that will generate the entire lattice if repeated in three dimensions is called a **unit cell.** The cube shaded in Fig. 11.34 is a unit cell for the simple cubic space lattice. The cubes outlined in Fig. 11.35 illustrate three ways in which unit cells may be selected for the sodium chloride structure. The structure of a crystal is specified by describing the size and shape of the unit cell and indicating the arrangement of its contents. The crystal can be built up by repeating the unit cell in three dimensions.

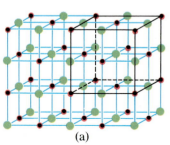

(a)

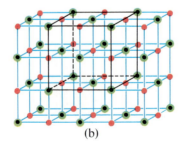

(b)

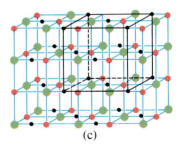

(c)

Thus far we have considered only unit cells shaped like a cube, but there are others. In general, a unit cell is a parallelpiped for which the size and shape are defined by the lengths of three axes (a, b, and c) and the angles (α, β, and γ) between them (Fig. 11.36). The axes are defined as being the lengths between points in the space lattice. Consequently, **unit cell axes join points with identical environments.** Unit cells must have one of the seven shapes indicated in Table 11.4.

Variations in the number and location of lattice points in these seven unit cells give rise to 14 types of space lattices, shown in Fig. 11.37. In some of these, points with

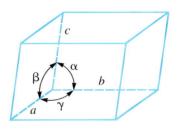

Figure 11.36
A unit cell.

Table 11.4 Unit cells of the seven crystal systems

System	Axes	Angles
Cubic	$a = b = c$	$\alpha = \beta = \gamma = 90°$
Tetragonal	$a = b \neq c$	$\alpha = \beta = \gamma = 90°$
Orthorhombic	$a \neq b \neq c$	$\alpha = \beta = \gamma = 90°$
Monoclinic	$a \neq b \neq c$	$\alpha = \gamma = 90°; \beta \neq 90°$
Triclinic	$a \neq b \neq c$	$\alpha \neq \beta \neq \gamma \neq 90°$
Hexagonal	$a = b \neq c$	$\alpha = \beta = 90°; \gamma = 120°$
Rhombohedral	$a = b = c$	$\alpha = \beta = \gamma \neq 90°$

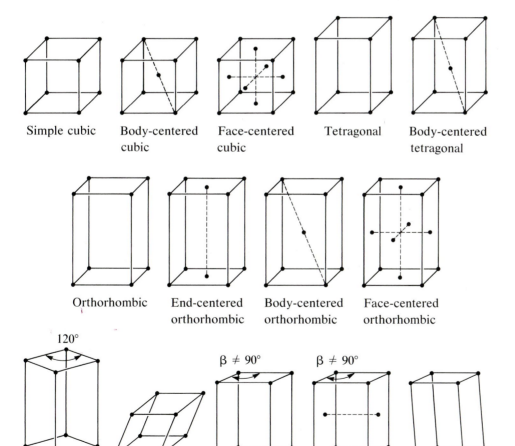

Simple cubic   Body-centered cubic   Face-centered cubic   Tetragonal   Body-centered tetragonal

Orthorhombic   End-centered orthorhombic   Body-centered orthorhombic   Face-centered orthorhombic

Hexagonal   Rhombohedral   Monoclinic   End-centered monoclinic   Triclinic

Figure 11.37
Unit cells of the 14 types of space lattices.

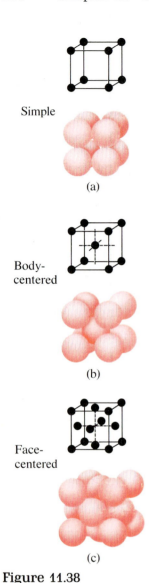

Simple

(a)

Body-centered

(b)

Face-centered

(c)

Figure 11.38
Cubic unit cells showing the locations of lattice points in the upper figures and metal atoms located on the lattice points, in the lower figures.

identical surroundings (lattice points) are found only at the corners of the unit cell. In others the points of the lattice occur both at the corners and in the centers of some or all of the faces of the unit cell. In still others the points of the lattice are found both at the corners and in the center of the unit cell.

Students sometimes have trouble reconciling the number of points or atoms given as the number per unit cell with the diagram of the unit cell. Some of the points or atoms shown in a diagram of a unit cell may be shared by other unit cells and therefore not lie completely within the unit cell shown. In order to determine the number of lattice points or of atoms in a unit cell, use the following rules:

1. A point or atom lying completely within a unit cell belongs to that unit cell only and is therefore counted as 1 when totaling the number of points or atoms in the unit cell.
2. A point or atom lying on a face of a unit cell is shared equally by two unit cells and is therefore counted as $\frac{1}{2}$ when totaling the number of points or atoms in a unit cell.
3. A point or atom lying on an edge is shared by four unit cells and is therefore counted as $\frac{1}{4}$.
4. A point or atom lying at a corner is shared by eight unit cells and is therefore counted as $\frac{1}{8}$.

Now let us look more closely at the contents of some cubic unit cells. The lattice points associated with the space lattice of each of the three cubic unit cells are indicated in Fig. 11.38. There is 1 lattice point ($8 \times \frac{1}{8}$) in the unit cell of the simple cubic lattice [Fig. 11.38(a)]. Since a unit cell containing one lattice point is called a **primitive cell,** the simple cubic lattice is sometimes called a **primitive cubic lattice.** The second cell [Fig. 11.38(b)] has 2 lattice points, 1 at the corners ($8 \times \frac{1}{8}$) and 1 in the center of the cube. This is called a **body-centered cubic cell.** Such a cell has points with identical surroundings at the corners and at its center. The third cell [Fig. 11.38(c)] has 4 lattice points (points with identical environments), one at the corners ($8 \times \frac{1}{8}$) and 3 from the 6 face centers ($6 \times \frac{1}{2}$). This is called a **face-centered cubic cell.**

The structure of polonium (Section 11.15) may be described as consisting of a simple cubic space lattice with an atom located at each of the lattice points [Fig. 11.38(a)]. The length of the unit cell edge is 3.36 Å. Since the polonium atoms touch along the edges of the cell, the nearest distance between the centers of polonium atoms is 3.36 Å, and the radius of each polonium atom is thus 1.68 Å. (Remember that the distance between two atoms is measured between their nuclei.)

A metal with a body-centered cubic structure consists of a space lattice composed of body-centered cubic unit cells [Fig. 11.38(b)]. One metal atom is located on each lattice point, so there are two identical metal atoms (atoms with identical environments) in the unit cell. The atoms touch along the diagonal of the cubic unit cell.

A metal with a cubic closest packed structure may be described as consisting of a space lattice made up of face-centered cubic unit cells [Fig. 11.38(c)]. One metal atom is located on each lattice point, so there are four equivalent metal atoms in each unit cell. The structure in Fig. 11.38(c) is the same as that in Fig. 11.26(b), but the perspective is different. Note that the atoms touch along the diagonals of the faces of the cell.

Ionic compounds can also crystallize with cubic unit cells; CsCl, NaCl, and one form of ZnS (zinc blende) crystallize with cubic space lattices. Another form of ZnS (wurtzite) crystallizes with a hexagonal space lattice. The assumption of two or more crystal structures by the same substance is called **polymorphism.**

The structure of CsCl is simple cubic. Chloride ions are located on the lattice points at the corners of the cell, and the cesium ion is located at the center of the cell

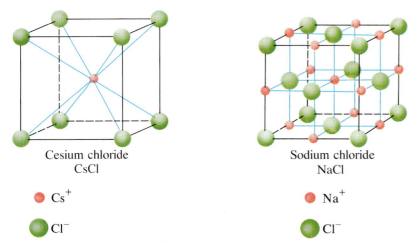

Cesium chloride
CsCl

● = Cs⁺

● = Cl⁻

Zinc blende
ZnS

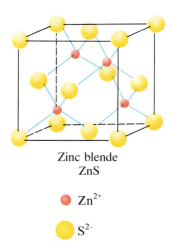

Sodium chloride
NaCl

● = Na⁺

● = Cl⁻

● = Zn²⁺

● = S²⁻

(Fig. 11.39). The cesium ion and the chloride ion touch along the diagonal of the cubic cell. There are one cesium ion and one chloride ion per unit cell, giving the 1-to-1 stoichiometry required by the formula for cesium chloride. There is no lattice point in the center of the cell because a cesium ion is not identical to a chloride ion.

Sodium chloride crystallizes with a face-centered cubic unit cell (Fig. 11.39). Chloride ions are located on the lattice points of a face-centered cubic unit cell. Sodium ions are located in the octahedral holes in the middle of the cell edges and in the center of the cell. The sodium and chloride ions touch each other. The unit cell contains four sodium ions and four chloride ions, giving a 1-to-1 stoichiometry as required by the formula, NaCl.

The cubic form of zinc sulfide, zinc blende, also crystallizes in a face-centered cubic unit cell (Fig. 11.39). This structure contains sulfide ions on the lattice points of a face-centered cubic lattice. (The arrangement of sulfide ions is identical to the arrangement of chloride ions in sodium chloride.) Zinc ions are located in alternate tetrahedral holes; that is, in one-half of the tetrahedral holes. There are four zinc ions and four sulfide ions in the unit cell, making the unit cell neutral in net charge.

A calcium fluoride unit cell is also a face-centered cubic unit cell (Fig. 11.40), but in this case the cations are located on the lattice points—equivalent calcium ions are located on the lattice points of a face-centered cubic lattice. All of the tetrahedral sites in the face-centered cubic array of calcium ions are occupied by fluoride ions. There are four calcium ions and eight fluoride ions in a unit cell, giving a calcium-to-fluorine ratio of 1-to-2, as required by the formula of calcium fluoride, CaF_2. Close examination of Fig. 11.40 will reveal the simple cubic array of fluoride ions with calcium ions in one-half of the cubic holes. The structure cannot be described in terms of a space lattice of points on the fluoride ions, since the fluoride ions do not all have identical environments. The orientation of the four calcium ions about the fluoride ions differs.

Figure 11.39

The unit cells of some ionic compounds of the general formula MX. The red spheres represent positive ions (cations), and the green or yellow spheres represent negative ions (anions). These structures have been expanded to show the geometrical relationships. In the crystal the cations and anions touch.

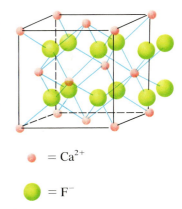

● = Ca²⁺

● = F⁻

Figure 11.40

The unit cell of CaF_2. The red spheres represent calcium ions, Ca^{2+}, and the yellow-green spheres, fluoride ions, F^-. Note the face-centered cubic array of Ca^{2+} and the simple cubic array of F^-. This structure has been expanded to show the geometrical relationships. In the crystal the cations and anions touch.

11.19 Calculation of Ionic Radii

If the edge length of a unit cell and the positions of the constituent ions are known, ionic radii for the ions in the crystal lattice can be calculated, making certain assumptions. The following examples illustrate the method and assumptions for cubic structures.

EXAMPLE 11.8   The edge length of the unit cell of LiCl (NaCl-like structure, face-centered cubic) is 5.14 Å. Assuming anion-anion contact, calculate the ionic radius for the chloride ion.

The NaCl structure (Fig. 11.39) contains a right triangle involving two chloride ions and one sodium ion. In the isomorphous LiCl structure, the lithium ion is so small that all ions in the structure touch, as in Fig. 11.33(b).

Because a, the distance between the center of a chloride ion and the center of a lithium ion, is $\frac{1}{2}$ of the edge length of the cubic unit cell,

$$a = \frac{5.14 \text{ Å}}{2}$$

Similarly, b is the distance between the center of a chloride ion and the center of a lithium ion and hence is also $\frac{1}{2}$ of the edge length of the cubic unit cell.

$$b = \frac{5.14 \text{ Å}}{2}$$

By the Pythagorean theorem, c, the distance between the centers of two chloride ions, can be calculated.

$$c^2 = a^2 + b^2$$
$$c^2 = \left(\frac{5.14}{2}\right)^2 + \left(\frac{5.14}{2}\right)^2 = 13.21$$
$$c = \sqrt{13.21} = 3.63 \text{ Å}$$

Since the anions are assumed to touch each other, c is twice the radius of one chloride ion. Hence the radius of the chloride ion is $\frac{1}{2}c$.

$$r_{\text{Cl}^-} = \tfrac{1}{2}c = \tfrac{1}{2} \times 3.63 \text{ Å} = 1.81 \text{ Å}$$

EXAMPLE 11.9   The edge length of the unit cell of KCl (NaCl structure, face-centered cubic) is 6.28 Å. Assuming anion-cation contact, calculate the ionic radius for the potassium ion.

Inspection of the structure in Fig. 11.39 shows that the distance between the center of a potassium ion and the center of a chloride ion is $\frac{1}{2}$ of the edge length of the cubic unit cell for KCl, or

$$\tfrac{1}{2} \times 6.28 \text{ Å} = 3.14 \text{ Å}$$

Assuming anion-cation contact, 3.14 Å is the sum of the ionic radii for K^+ and Cl^-.

$$r_{\text{K}^+} + r_{\text{Cl}^-} = 3.14 \text{ Å}$$

In Example 11.8, r_{Cl^-} was calculated as 1.81 Å. Therefore

$$r_{\text{K}^+} = 3.14 \text{ Å} - 1.81 \text{ Å} = 1.33 \text{ Å}$$

Note that the chloride ions do not touch in solid potassium chloride.

It is important to realize that values for ionic radii calculated from the edge lengths of unit cells depend on numerous assumptions, such as a perfect spherical shape for ions, which are approximations at best. Hence such calculated values are themselves approximate, and comparisons cannot be pushed too far. Nevertheless, this is one method that has proved useful for calculating ionic radii from experimental measurements such as X-ray crystallographic determinations.

11.20 The Lattice Energies of Ionic Crystals

The **lattice energy, U,** of an ionic compound may be defined as the energy required to separate the ions in a mole of the compound by infinite distances. For the ionic solid MX, the lattice energy is the enthalpy change of the *endothermic* process

$$MX(s) \longrightarrow M^{n+}(g) + X^{n-}(g) \qquad U = \Delta H^{\circ}_{298}$$

The same amount of energy is released when a mole of an ionic compound is formed by bringing positive and negative ions together from infinite distances. Lattice energies can be calculated from basic principles, or they can be measured experimentally.

The calculation of lattice energies is based primarily on the work of Born, Lande, and Mayer. They derived the following equation to express the lattice energy, U, of an ionic crystal:

$$U = \frac{C(Z^+ Z^-)}{R_0}$$

where C is a constant that depends on the type of crystal structure and the electronic structures of the ions; Z^+ and Z^- are the charges on the ions; and R_0 is the interionic distance (the sum of the radii of the positive and negative ions). For a given structure the principal factors determining the lattice energy are Z^+, Z^-, and R_0. The lattice energy increases rapidly with the charges on the ions. Keeping all other parameters constant, doubling the charge on both cation and anion quadruples the lattice energy. For example, the lattice energy of LiF (Li^+ and F^-; $Z^+ = 1$, $Z^- = -1$) is 1023 kilojoules per mole, while that of MgO (Mg^{2+} and O^{2-}; $Z^+ = 2$, $Z^- = -2$) is 3900 kilojoules per mole (R_0 is nearly the same for both compounds).

The lattice energy also increases rapidly with decreasing interionic distance in the crystal lattice, which may result from decreasing either the cation radius, the anion radius, or both. Some crystals that exhibit a large difference in interionic distances and lattice energies are lithium fluoride and rubidium chloride (table at right).

The large lattice energies of many ionic compounds result from strong electrostatic forces between the ions in the crystal. These strong interionic forces are also responsible for the relatively large heats of fusion, sublimation, and vaporization and for the high melting and boiling temperatures of these compounds, as compared to most molecular compounds. These forces must be overcome in order to vaporize a crystal to separated gaseous ions, for example. However, in the case of fusion, the energy requirement is less than the total lattice energy because of the interactions of neighboring ions in the liquid phase. The same strong forces are responsible for the fact that ionic crystals are generally hard, dense, rigid, and nonvolatile. While ionic compounds have such properties, some covalent substances, such as diamond and silicon carbide, also have them (Fig. 11.22). In such cases the very strong forces in the crystal are due to strong covalent bonds, which extend in three dimensions throughout the crystal, rather than to the weaker London forces usually found in molecular compounds.

	Interionic distance, R_0	Lattice energy, U
LiF	2.008 Å	1023 kJ/mol
RbCl	3.28 Å	680 kJ/mol

11.21 The Born-Haber Cycle

The lattice energies of only a few ionic crystals have been measured, because it is not possible to measure most lattice energies directly. However, a cyclic process can be used to calculate the lattice energy from other quantities. This **Born-Haber cycle** is a cycle that involves ΔH°_f, the enthalpy of formation of the compound (Section 4.4); I, the ionization energy of the metal (Section 5.15, Part 3); *E.A.,* the electron affinity of the nonmetal (Section 5.15, Part 4); ΔH°_s, the enthalpy of sublima-

Rhombic sulfur, S_8

tion of the metal; D, the bond dissociation energy of the nonmetal (Section 6.9); and U, the lattice energy of the compound. The Born-Haber cycle for sodium chloride analyzes the formation of $NaCl(s)$ from $Na(s)$ and $\frac{1}{2} Cl_2(g)$ as a step-by-step process that may be expressed diagrammatically as follows, with the overall change in color and the individual steps in black:

$$
\begin{array}{ccc}
Na(s) + \frac{1}{2}Cl_2(g) & \xrightarrow{\;\Delta H_f^\circ\;} & NaCl(s) \\
\downarrow \Delta H_s^\circ \quad \downarrow \frac{1}{2}D & & \nearrow \\
Na(g) \quad Cl(g) & \quad -U & \\
\downarrow I \quad \downarrow -E.A. & & \\
Na^+ \quad + \quad Cl^- & &
\end{array}
$$

This diagram indicates hypothetical steps in the formation of sodium chloride from 1 mole of sodium metal and $\frac{1}{2}$ mole of chlorine gas. First we assume that the sodium metal is vaporized and the bonds in the diatomic chlorine molecules are broken. Then the sodium atoms are ionized, and the electrons from them are transferred to the chlorine atoms to form chloride ions. The gaseous sodium ions and chloride ions thus formed come together to give solid sodium chloride. The enthalpy change in this step is the negative of the lattice energy, which is the amount of energy required to produce one mole of gaseous sodium ions and one mole of gaseous chloride ions from one mole of solid sodium chloride. The total energy evolved in this hypothetical preparation of sodium chloride is equal to the experimentally determined enthalpy of formation, ΔH_f° of the compound from its elements. Hess's law (Section 4.5) can be used to show the relationship between the enthalpies of the individual steps and the enthalpy of formation as follows.

1. Enthalpy of sublimation of $Na(s)$

    $Na(s) \longrightarrow Na(g)$ $\qquad\qquad\qquad\qquad \Delta H = \Delta H_s^\circ \quad = 109$ kJ

2. One-half of the bond energy of Cl_2

    $\frac{1}{2}Cl_2(g) \longrightarrow Cl(g)$ $\qquad\qquad\qquad \Delta H = \frac{1}{2}D \quad = 122$ kJ

3. Ionization energy of $Na(g)$

    $Na(g) \longrightarrow Na^+(g) + e^-$ $\qquad\qquad \Delta H = I \qquad = 496$ kJ

4. Negative of the electron affinity of Cl

    $Cl(g) + e^- \longrightarrow Cl^-(g)$ $\qquad\qquad \Delta H = -E.A. = -368$ kJ

5. Negative of the lattice energy of $NaCl(s)$

    $Na^+(g) + Cl^-(g) \longrightarrow NaCl(s)$ $\qquad \Delta H = -U \quad = \;?$

6. Enthalpy of formation of $NaCl(s)$, Appendix I: add Steps 1–5

    $Na(s) + \frac{1}{2}Cl_2(g) \longrightarrow NaCl(s)$ $\qquad \Delta H = \Delta H_f^\circ \quad = -411$ kJ

    $$\Delta H_f^\circ = \Delta H_s^\circ + \tfrac{1}{2}D + I + (-E.A.) + (-U)$$

The value of ΔH_f° is accurately known for many substances. If the other thermochemical values are available, we can solve for the lattice energy, U, by rearranging the equation for ΔH_f° as follows:

$$U = -\Delta H_f^\circ + \Delta H_s^\circ + \tfrac{1}{2} D + I - E.A.$$

For sodium chloride, using the above data, the lattice energy is

$$U = (411 + 109 + 122 + 496 - 368) \text{ kJ} = 770 \text{ kJ}$$

The Born-Haber cycle may be used to calculate any one of the quantities in the equation for lattice energy, provided all of the others are known. Usually, ΔH_f°, ΔH_s°, I, and D are known. The direct measurement of electron affinities is rather difficult, and accurate values of electron affinity have been determined only for the halogens. For halogen compounds, then, the Born-Haber cycle can be used to calculate lattice energies, which are found to agree well with those obtained by other means. Having verified that the cycle gives satisfactory calculated values, chemists use it to calculate electron affinities that cannot be measured directly using values of U calculated by the method of Born, Lande, and Mayer (Section 11.20).

11.22 X-Ray Diffraction

The size of the unit cell and the arrangement of atoms in a crystal can be determined from measurements of the **diffraction** of X rays by the crystal. X rays are electromagnetic radiation with wavelengths (Section 5.8) about as long as the distance between neighboring atoms in crystals (about 2 Å).

W. H. Bragg, an English physicist, showed that the diffraction of X rays by a crystal can be interpreted if the crystal is thought of as a reflection grating. The reflection of X rays from planes of lattice points in the crystal may be compared with the reflection of a beam of light from a stack of thin glass plates of equal thickness. It is known that light of a single wavelength is reflected from such a stack only at definite angles, which depend on the wavelength of the light and the thickness of the plates. The planes within a crystal correspond to the plates of glass.

There is a simple mathematical relation between the wavelength, λ, of the X rays, the distance between the planes in the crystal, and the angle of diffraction (reflection). When a beam of monochromatic X rays strikes two planes of a crystal at an angle θ, it is reflected (Fig. 11.41). The rays from the lower plane travel farther than those from the upper plane by a distance equal to the sum of the distances BC and CD. If this distance is equal to 1, 2, 3, or any other integral number of wavelengths (so $n\lambda = BC + CD$), the two beams emerging from the crystal will be in phase and will reinforce each other. From the right triangle ABC of Fig. 11.41,

$$\sin BAC = \frac{BC}{AC}$$

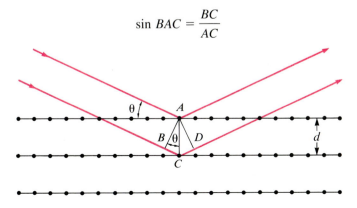

Figure 11.41

Diffraction of a monochromatic beam of X rays by two planes of a crystal.

Selenite, $CaSO_4 \cdot 2H_2O$

Rearranging gives

$$BC = AC \sin BAC$$

The distance AC is the distance between the crystal planes, which we will call d. Angle BAC is equal to θ. Then

$$BC = d \sin \theta$$

But the difference in the path lengths of the two beams is equal to $BC + CD$. Because $BC = CD$, the total difference in path length is $2BC$. X rays will be reflected, provided that, as indicated previously,

$$n\lambda = BC + CD = 2BC$$

Hence

$$n\lambda = 2d \sin \theta$$

which is the **Bragg equation.**

If the X rays strike the crystal at any angle other than θ, the extra distance $BC + CD$ will not be equal to an integral number of wavelengths, and there will be interference of the reflected rays. This will destroy the intensity of the reflected rays.

The reflection corresponding to $n = 1$ is called the first-order reflection; that corresponding to $n = 2$ is the second-order reflection; and so forth. Each successive order has a larger angle.

By rotating a crystal and varying the angle θ at which the X-ray beam strikes it, the values of θ at which diffraction occurs can be determined by looking for the maximum diffracted intensities. If the wavelength of the X rays is known, the spacing, d, of the planes within the crystal can be determined from these values. On the other hand, if the X rays strike planes for which d is known, their wavelength may be determined by measuring the angles.

For Review

Summary

The physical properties of condensed matter (liquids and solids) can be explained by the kinetic-molecular theory, which was also used with gases. In a liquid, the intermolecular attractive forces hold the molecules in contact, although they still have sufficient kinetic energy to move past each other (**diffuse**). Since the molecules are in contact, liquids are not very compressible. The speeds of the molecules in a liquid vary; some molecules have enough kinetic energy to escape from the liquid, resulting in **evaporation.** The pressure of a vapor in dynamic equilibrium with a liquid is called the **vapor pressure** of the liquid. The vapor pressure of a liquid increases with increasing temperature, and the **boiling point** of the liquid is reached when the vapor pressure equals the external pressure on the liquid. The **normal boiling point** of a liquid is the temperature at which its vapor pressure is equal to 1 atmosphere. The amount of energy necessary to evaporate 1 mole of a liquid at a constant temperature is called the **heat of vaporization** of the liquid.

When the temperature of a liquid becomes sufficiently low, or the pressure on the liquid sufficiently high, the molecules can no longer move past each other, and a rigid solid forms. Many solids are **crystalline;** that is, they are composed of a repeating pattern of atoms, molecules, or ions. Others are supercooled liquids, sometimes called **amorphous solids,** in which the disordered arrangement of molecules found in liquids is retained. Crystalline solids melt when enough energy is added so that the kinetic energy of their molecules just overcomes the intermolecular attractive forces holding the molecules in their fixed positions in the crystal. The temperature at which a solid and its liquid are in equilibrium is called the **melting point** of the solid or the **freezing point** of the liquid. Amorphous solids do not have a melting point but soften over a range of temperature. The **heat of fusion** of a crystalline solid is the amount of energy that must be added to convert 1 mole of a solid to a liquid at the same temperature. Some solids exhibit a measurable vapor pressure and will **sublime.**

The conditions under which a solid and a liquid, a liquid and a gas, or a solid and a gas are in equilibrium are described by a **phase diagram.** At the **triple point** of water (0.01°C and 4.6 torr), all three phases of water are in equilibrium.

Intermolecular attractive forces, collectively referred to as **van der Waals forces,** are responsible for the behavior of molecular liquids and solids and are electrostatic in nature. **Dipole-dipole attractions** result from the electrostatic attraction of the negative end of one dipolar molecule for the positive end of another. The temporary dipole that forms due to the motion of the electrons in an atom can induce a dipole in an adjacent atom and give rise to the **dispersion force,** or **London force.** London forces increase with increasing molecular size. **Hydrogen bonding** is a special type of dipole-dipole attraction that results when hydrogen is bonded to a very electronegative element such as F, O, N, or Cl. These various intermolecular forces are responsible for the critical temperature and critical pressure of a substance. The **critical temperature** of a substance is that temperature above which the substance cannot be liquefied no matter how much pressure is applied. The **critical pressure** is the pressure required to liquefy a gas at its critical temperature.

The structures of crystalline metals and simple ionic compounds can be described in terms of packing of spheres. Metal atoms can pack in **hexagonal closest packed structures, cubic closest packed structures, body-centered structures** and **simple cubic structures.** The anions in simple ionic structures commonly adopt one of these structures, with the cations occupying the spaces remaining between the anions. Small cations may occupy **tetrahedral holes** in a closest packed array of anions. Larger cations can occupy **octahedral holes.** Still larger cations can occupy **cubic holes** in a simple cubic array of anions. The **radius ratio rule** serves as a guide to the **coordination number** of the cation. The structure of a solid can be described by indicating the size and shape of a **unit cell** and the contents of the cell. The dimensions of a unit cell can be determined by a study of **diffraction** of X rays from a crystal.

The energy required to separate the ions in 1 mole of an ionic compound by an infinite distance is called the **lattice energy** of the compound. The lattice energy is proportional to the product of the charges on the ions and inversely proportional to the distance between their centers. Lattice energies may be calculated, or may be determined by using Hess's law in a **Born-Haber cycle.**

Key Terms and Concepts

amorphous solid (11.1)
body-centered cubic structure (11.15)
body-centered cubic unit cell (11.18)

boiling point (11.3)
Born-Haber cycle (11.21)
Bragg equation (11.22)
condensation (11.2)
coordination number (11.15)

covalent solid (11.13)
critical pressure (11.8)
critical temperature (11.8)
crystal defect (11.14)
crystalline solid (11.1)

cubic closest packed structure
   (11.15)
dipole-dipole attraction (11.11)
distillation (11.4)
dynamic equilibrium (11.2)
evaporation (11.2)
face-centered cubic structure
   (11.15)
face-centered cubic unit cell (11.18)
freezing point (11.5)
glass (11.1)
heat of fusion (11.7)
heat of vaporization (11.6)
hexagonal closest packed structure
   (11.15)

hydrogen bond (11.12)
intermolecular force (11.11)
intramolecular force (11.11)
interstitial site (11.14)
ionic radius (11.17, 11.19)
ionic solid (11.13)
isomorphous structures (11.15)
lattice energy (11.20, 11.21)
London force (11.11)
melting point (11.5)
metallic solid (11.13)
molecular solid (11.13)
normal boiling point (11.3)
octahedral hole (11.16)
phase diagram (11.9)

polymorphism (11.18)
radius ratio (11.17)
simple cubic structure (11.15)
space lattice (11.18)
sublimation (11.2)
supercooling (11.1)
surface tension (11.10)
tetrahedral hole (11.16)
triple point (11.9)
unit cell (11.18)
van der Waals force (11.11)
vapor pressure (11.2)
X-ray diffraction (11.22)

Exercises

Kinetic-Molecular Theory and the Condensed State

1. In what ways are liquids similar to solids? In what ways are liquids similar to gases?
2. The density of liquid SO_2 is 1.4 g/mL; the density of gaseous SO_2 at STP is 0.003 g/mL. Explain the difference in the densities of these two phases.
3. Describe how the motion of the molecules changes as a substance changes from a solid to a liquid and from a liquid to a gas.
4. The types of intermolecular forces in a substance are identical whether it is a solid, a liquid, or a gas. Why then does a substance change phase from a gas to a liquid or to a solid?
5. Explain why a liquid will assume the shape of any container into which it is poured, whereas a solid is rigid and retains its shape.
6. What types of liquid materials tend to supercool readily and form glasses?
7. Explain why a sample of amorphous boric oxide, (B_2O_3, Fig. 11.2) is considered as a form of a liquid even though it looks like a solid, is rigid, and will shatter if struck with a hammer.
8. The density of liquid chloroethane, C_2H_5Cl, is 0.903 g/mL at its boiling temperature of 13°C. What is the density of gaseous chloroethane in grams per milliliter at the same temperature and a pressure of 1.00 atm?

            Ans. 2.75×10^{-3} g/mL

Melting, Evaporation, and Boiling

9. What feature characterizes the equilibrium between a liquid and its vapor in a closed container as dynamic?
10. Why does spilled gasoline evaporate more rapidly on a hot day than on a cold day?
11. Explain why the vapor pressure of a liquid decreases as its temperature decreases.
12. How does boiling of a liquid differ from its evaporation?
13. How does the boiling point of a liquid differ from its normal boiling point?
14. What is the approximate boiling point of water in Kansas City when the atmospheric pressure is 700 torr? (See Fig. 11.6 for data.)
15. The liquid C_4H_{10} in a butane lighter has a boiling point of −1°C; the octane (C_8H_{18}) in gasoline has a boiling point of 125°C. Explain why the boiling point of $C_4H_{10}(l)$ is lower than that of $C_8H_{18}(l)$ even though the types of intermolecular forces are the same in both liquids.
16. How do we know that some solids such as ice and moth balls have vapor pressures sufficient to evaporate?
17. Why does iodine sublime more rapidly at 110°C than at 25°C?
18. Explain why ice, which is a crystalline solid, has a sharp melting temperature of 0°C whereas butter, which is a supercooled liquid (amorphous solid), softens over a range of temperature.
19. What is the relationship between the intermolecular forces in a solid and its melting temperature?
20. Chloroethane (bp = 13°C) has been used as a local anaesthetic. When the liquid is sprayed on the skin, it cools the skin enough to freeze and numb it. Explain the cooling effect of liquid chloroethane.

Heat of Fusion and Heat of Vaporization

21. A syringe like that shown in Fig. 11.13(a) at a temperature of 20°C is filled with liquid ether so that there is no space for any vapor. In one experiment the temperature is kept constant and the plunger is withdrawn somewhat, forming

a volume that can be occupied by vapor. What is the approximate pressure of the vapor produced? If the system were insulated in a second experiment so that no heat could enter or leave, would the temperature of the liquid increase, decrease, or remain constant as the plunger is withdrawn? Explain.

22. The heat of vaporization of water is larger than its heat of fusion. Explain why.

23. Explain why steam produces much more severe burns than does the same mass of boiling water.

24. How much heat in kilocalories is required to evaporate 1 mol of water at 25°C.   *Ans. 10.52 kcal*

25. How much heat in kilojoules is required to change 100 g of water at 100°C to steam at 100°C?   *Ans. 226 kJ*

26. How much heat is absorbed by a 30-g (1-oz) ice cube when it melts at 0°C?   *Ans. 10 kJ*

27. In hot, dry climates water is cooled by allowing some of it to evaporate slowly. How much water must evaporate to cool 1.0 L (1.0 kg) from 39°C to 21°C? Assume that the heat of vaporization of water is constant between 21° and 39° and is equal to the value at 25°C.   *Ans. 31 g*

28. The specific heat of iron is 0.443 J/g K. What weight of steam at 100°C must be converted to water at 100°C to raise the temperature of a 150-g iron block from 25°C to 100°C?   *Ans. 2.2 g*

29. How much water at 0°C could be converted to ice at 0°C by the cooling action of vaporizing 2.0 mol of liquid ammonia at its boiling point?   *Ans. 1.4×10^2 g*

30. Calculate the heat of vaporization under standard state conditions of each of the following, using the data in Appendix I, and identify which member of each pair has the stronger intermolecular forces:
    (a) $Al(s)$ and $I_2(s)$ *Ans. 326 kJ; 62.438 kJ; Al*
    (b) $CHCl_3(l)$ and $CCl_4(l)$
                         Ans. 31.4 kJ; 32.5 kJ; CCl_4
    (c) $CH_3OH(l)$ and $C_2H_5OH(l)$
                       Ans. 38.0 kJ; 42.6 kJ; C_2H_5OH
    (d) $C_2H_5Cl(l)$ and $C_2H_5OH(l)$
                       Ans. 24.3 kJ; 42.6 kJ; C_2H_5OH

Phase Diagrams, Critical Temperature and Pressure

31. Is it possible to liquefy oxygen at room temperature (about 20°C)? Is it possible to liquefy ammonia at room temperature? Explain your answers.

32. Draw a rough graph showing how the pressure, initially at 145 atm, inside a cylinder of nitrogen decreases as $N_2(g)$ is released from the cylinder at 25°C. Draw a similar curve for a cylinder of sulfur dioxide kept at 25°C, containing $SO_2(l)$ and $SO_2(g)$ at a pressure of 50 atm.

33. From the phase diagram for water (Fig. 11.14), determine the physical state of water at:
    (a) 800 torr and −10°C   (b) 800 torr and 25°C
    (c) 400 torr and 50°C   (d) 400 torr and 90°C

(e) 100 torr and 80°C   (f) 2 torr and −10°C
(g) 2 torr and 20°C

34. What phase changes can water undergo as the temperature changes if the pressure is held at 4 torr? If the pressure is held at 200 torr?

35. What phase changes can water undergo as the pressure changes if the temperature is held at 0.005°C? If the temperature is held at −20°C? At 20°C?

36. Explain what is meant by the triple point of a substance.

37. If one continues beyond point *C* on the line *BC* in Fig. 11.14, at what temperature and pressure does the line end?

38. Using the phase diagram for water (Fig. 11.14), determine the following:
    (a) At 400 torr what is the approximate temperature necessary to convert water from a solid to a liquid? From a liquid to a gas?
    (b) What is the approximate pressure at which water changes from a gas to a liquid at 20°C? At 66°C?

39. Will a sample of carbon dioxide in a cylinder at a pressure of 20 atm and a temperature of −30°C exist as a solid, liquid, or a gas? (A portion of the phase diagram of carbon dioxide is shown.)

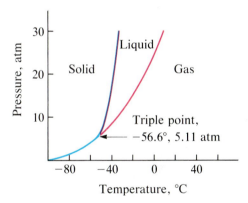

40. Dry ice, $CO_2(s)$, does not melt at atmospheric pressure. It sublimes at a temperature of −78°C. What is the lowest pressure at which $CO_2(s)$ will melt to give $CO_2(l)$? What is the approximate temperature at which this will occur? (See exercise 39 for a portion of the phase diagram for CO_2.)

41. What phase changes can carbon dioxide undergo as the temperature changes if the pressure is held at 20 atm? If the pressure is held at 1 atm? (See exercise 39 for a portion of the phase diagram for CO_2.)

42. What phase changes can carbon dioxide undergo as the pressure changes if the temperature is held at −50°C? If the temperature is held at −20°C? (See exercise 39 for a portion of the phase diagram for CO_2.)

Attractive Forces in Liquids and Solids

43. Why do the boiling points of the noble gases increase in the order He < Ne < Ar < Kr < Xe?

44. Describe the most important forces responsible for forming the solids of the following:
    (a) Xe   (b) CO
    (c) NH_3 (d) C_2H_5Cl
    (e) C_2H_5OH (f) KCl
    (g) C (diamond)   (h) Al
    (i) NH_4F (j) N_2
    (k) H_2Te (l) CH_4

45. Predict the member of each of the following pairs that has the stronger attractive forces. Verify your answers using the heats of vaporization calculated in exercise 30.
    (a) Al(s) or I_2(s)
    (b) $CHCl_3$(l) or CCl_4(l)
    (c) CH_3OH(l) or C_2H_5OH(l)
    (d) C_2H_5Cl(l) or C_2H_5OH(l)

46. Silane, SiH_4, phosphine, PH_3, and hydrogen sulfide, H_2S, melt at $-185°C$, $-133°C$, and $-85°C$, respectively. What does this suggest about the polar character and intermolecular attractions in the three compounds?

47. The heat of vaporization of CO_2(l) is 226 J/g. Would you expect the heat of vaporization of CS_2(l) to be 364 J/g, 226 J/g, or 119 J/g? Explain your answer.

48. The melting point of N_2(s) is $-210°C$. Would you expect the melting point of O_2(s) to be $-202°C$, $-210°C$, or $-218°C$? Explain your answer.

49. The melting point of H_2O(s) is $0°C$. Would you expect the melting point of H_2S(s) to be $-85°C$, $0°C$, or $185°C$? Explain your answer.

50. Which member of each of the following pairs will have the higher vapor pressure at the same temperature?
    (a) Cl_2(l) or Br_2(l)
    (b) HF(s) or Ar(s)
    (c) LiF(l) or MgO(l)

51. Explain why a hydrogen bond between two water molecules, HOH···OH_2, is stronger than a hydrogen bond between two ammonia molecules, H_2NH···NH_3.

52. The individual hydrogen bonds in liquid hydrogen fluoride, HF, are stronger than the individual hydrogen bonds in liquid water, yet the heat of vaporization of liquid hydrogen fluoride is less than that of water. Explain.

53. Arrange each of the following sets of compounds in order of increasing boiling temperature:
    (a) HF, H_2O, CH_4
    (b) F_2, Cl_2, Br_2, I_2
    (c) CH_4, C_2H_6, C_3H_8
    (d) H_2, HF, LiF

Properties of Crystalline Solids

54. Identify the type of crystalline solid (metallic, covalent, ionic, or molecular) formed by each of the following substances:
    (a) B_2O_3 (b) CO
    (c) NH_3 (d) C_2H_5OH
    (e) KCl   (f) C (diamond)

(g) Al   (h) NH_4F
(i) N_2 (j) Cu
(k) $BaSO_4$

55. Magnesium metal crystallizes in a hexagonal closest packed structure. What is the coordination number of a magnesium atom?

56. What is the coordination number of a molybdenum atom in the body-centered cubic structure of molybdenum?

57. Describe the crystal structure of copper, which crystallizes with four equivalent metal atoms in a cubic unit cell.

58. The free space in a metal may be found by subtracting the volume of the atoms in a unit cell from the volume of the cell. Calculate the percentage of free space in each of the three cubic lattices if all atoms in each are of equal size and touch their nearest neighbors. From the calculations, determine which of these structures represents the most efficient packing—that is, which packs with the least amount of unused space?   *Ans. Face-centered cubic*

59. Barium crystallizes in a body-centered cubic unit cell with an edge length of 5.025 Å.
    (a) What is the atomic radius of barium in this structure?
                              Ans. 2.176 Å
    (b) Calculate the density of barium.
                              Ans. 3.595 g/cm³

60. Platinum (atomic radius = 1.38 Å) crystallizes in a face-centered cubic unit cell with the nearest neighbors in contact.
    (a) Calculate the edge length of the unit cell.
                              Ans. 3.90 Å
    (b) Calculate the density of platinum.
                              Ans. 21.8 g/cm³

61. Aluminum crystallizes in a face-centered cubic unit cell with an aluminum atom on each lattice point with the edge length of the unit cell equal to 4.050 Å.
    (a) Calculate the atomic radius of aluminum.
                              Ans. 1.432 Å
    (b) Calculate the density of aluminum.
                              Ans. 2.698 g/cm³

62. What is the formula of a compound with a structure that may be described as a closest packed array of fluoride ions with cobalt atoms in $\frac{1}{2}$ of the octahedral holes? Explain your answer.   *Ans. CoF_2*

63. Thallous iodide crystallizes in a simple cubic array of iodide ions with thallium ions in all of the cubic holes. What is the formula of thallous iodide? Explain your answer.
                              Ans. TlI

64. Cadmium sulfide, sometimes used as a yellow pigment by artists, crystallizes with cadmium occupying $\frac{1}{2}$ of the tetrahedral holes in a closest packed array of sulfide ions. What is the formula of cadmium sulfide?   *Ans. CdS*

65. Although it is a covalent compound, silicon carbide can be described as a cubic closest packed array of silicon atoms, with carbon atoms in $\frac{1}{2}$ of the tetrahedral holes. What is the coordination number of carbon in silicon carbide? What is

the hybridization of the silicon and carbon atoms? Explain your answer. *Ans. 4; sp^3 for both*

66. What is the empirical formula of the magnetic oxide of cobalt, used on recording tapes, that crystallizes with cobalt atoms occupying $\frac{1}{8}$ of the tetrahedral holes and $\frac{1}{2}$ of the octahedral holes in a closest packed array of oxide ions? Explain your answer. *Ans. Co_3O_4*

67. A compound containing zinc, aluminum, and sulfur crystallizes with a closest packed array of sulfide ions. Zinc ions are found in $\frac{1}{8}$ of the tetrahedral holes and aluminum ions in $\frac{1}{2}$ of the octahedral holes. What is the empirical formula of the compound? Explain your answer. *Ans. $ZnAl_2S_4$*

68. Why are compounds that are isomorphous with ZnS generally not observed when the radius ratio for the compound is greater than 0.42?

69. Explain why the chemically similar alkali metal chlorides NaCl and CsCl have different structures, whereas chemically different NaCl and MnS have the same structure.

70. Each of the following compounds crystallizes in a structure matching that of either NaCl, CsCl, ZnS, or CaF₂. From the radius ratio, predict which structure is formed by each. Show your work.
    (a) ZnTe *Ans. ZnS str.*
    (b) BaF₂ *Ans. CaF₂ str.*
    (c) KBr *Ans. NaCl str.*
    (d) AlP *Ans. ZnS str.*
    (e) CaS *Ans. NaCl str.*
    (f) SrF₂ *Ans. CaF₂ str.*
    (g) NiO *Ans. NaCl str.*
    (h) CsBr *Ans. CsCl str.*
    (i) MgO *Ans. NaCl str.*
    (j) BeO *Ans. ZnS str.*

71. A cubic unit cell contains iodide ions at the corners and a thallium ion in the center. What is the formula of the compound? Explain your answer. *Ans. TlI*

72. A cubic unit cell contains manganese ions at the corners and fluoride ions at the center of each edge.
    (a) What is the empirical formula of this compound? Explain your answer. *Ans. MnF_3*
    (b) What is the coordination number of the Mn^{3+} ion? *Ans. 6*
    (c) Calculate the edge length of the unit cell if the radius of a Mn^{3+} ion is 0.65 Å. *Ans. 4.02 Å*
    (d) Calculate the density of the compound. *Ans. 2.86 g/cm³*

73. A cubic unit cell contains a cobalt ion in the center of the cell, potassium ions at the corners of the cell, and fluoride ions in the center of each face. What is the empirical formula of this compound? What is the oxidation number of the cobalt ion? Explain your answer. *Ans. $KCoF_3$; +2*

74. Thallium bromide crystallizes with the same structure as CsCl. The edge length of the unit cell of TlBr is 3.97 Å.
    (a) Calculate the ionic radius of Tl⁺. (The ionic radius of

Br⁻ may be found on the inside of the back cover.) *Ans. 1.49 Å*
    (b) Calculate the density of TlBr. *Ans. 7.54 g/cm³*

75. NaH crystallizes with the same crystal structure as NaCl. The edge length of the cubic unit cell of NaH is 4.880 Å.
    (a) Calculate the ionic radius of H⁻. (The ionic radius of Na⁺ may be found on the inside of the back cover.) *Ans. 1.49 Å*
    (b) Calculate the density of NaH. *Ans. 1.372 g/cm³*
    (c) Which contains a greater mass of hydrogen, 1.00 cm³ of liquid H₂ (density 0.070 g/cm³) or 1.00 cm³ of NaH? *Ans. Liquid H_2*

76. The unit cell edge length of CaF₂ is 5.46295 Å. The density of CaF₂ is 3.1805 g/cm³. From these data and the atomic weights of calcium and fluorine, calculate Avogadro's number. *Ans. 6.023×10^{23}*

77. The lattice energy of LiF is 1023 kJ/mol, and the Li-F distance is 2.008 Å. MgO crystallizes in the same structure as LiF but with a Mg—O distance of 2.05 Å. Which of the following values most closely approximates the lattice energy of MgO: 255, 890, 1023, 2046, or 4008 kJ/mol? Explain your choice. *Ans. 4008 kJ/mol*

78. The lattice energy of KF is 794 kJ/mol, and the interionic distance is 2.69 Å. The Na—F distance in NaF, which has the same structure as KF, is 2.31 Å. Is the lattice energy of NaF about 682, 794, 924, 1588, or 3175 kJ/mol? Explain your answer. *Ans. 924 kJ/mol*

79. What X-ray wavelength would give a second-order reflection ($n = 2$) with a θ angle of 20.40° from planes with a spacing of 4.00 Å? *Ans. 1.39 Å*

80. What is the spacing between crystal planes that diffract X rays with a wavelength of 1.541 Å at an angle, θ of 15.55° (first-order reflection)? *Ans. 2.874 Å*

81. Gold crystallizes in a face-centered cubic unit cell. The second-order reflection ($n = 2$) of X rays for the planes that make up the tops and bottoms of the unit cells is at $\theta = 22.20°$. The wavelength of the X rays is 1.54 Å. What is the density of metallic gold? *Ans. 19.3 g/cm³*

Additional Exercises

82. Most reactions in a chemical laboratory are run in the gaseous or liquid state. If molecules have to come in contact in order to react, explain why a mixture of solid reagents does not, in general, react very rapidly.

83. Explain why some molecules in a solid may have enough energy to sublime away from the solid even though the solid does not contain enough energy to melt.

84. How much energy is released when 250 g of steam at 135°C is converted to ice at −20°C? The specific heat of steam is 2.00 J/g K; of liquid water, 4.18 J/g K; of ice, 2.04 J/g K. *Ans. 780 kJ*

85. How much energy is required to convert 135.0 g of ice at −8.0°C to steam at 225°C? (See exercise 84 for additional information.) *Ans. 442.2 kJ*

86. During the fermentation step in the production of beer, 4.5×10^6 kcal of heat are evolved per 1000 gal of beer produced. How many liters of cooling water are required to maintain the optimum fermentation temperature of 58°F for 1000 gal of beer? How many gallons? The cooling water enters with a temperature of 5°C and is discharged with a temperature of 13°C.

Ans. 6×10^5 L; 2×10^5 gal

87. (a) A river is 20 ft wide and has an average depth of 5 ft and a current of 3 mi/h. A power plant dissipates 2.1×10^5 kJ of waste heat into the river every second. What is the temperature difference between the water upstream and downstream from the plant?

Ans. 4°C

(b) If the heat dissipated by the power plant is to be removed by evaporating water at 25°C, how many gallons of water per day would be needed?

Ans. 2.0×10^6 gal/day

88. By referring to Figure 11.6, determine the approximate boiling points of ethyl ether at 400 torr, of ethyl alcohol at 0.25 atm, and of water at 55 kPa. At what temperature will the vapor pressure of ethyl alcohol be equal to the vapor pressure of ethyl ether at 20°C?

89. Is work done on a liquid when it evaporates to give a gas or is work done by the liquid?

90. The melting points of NaCl, KCl, and RbCl decrease with increasing atomic number of the alkali metal. Suggest an explanation.

91. How much heat must be removed to condense 2.00 L of $HCN(g)$ at 25°C and 1.0 atm to a liquid under the same conditions?

Ans. 2.1 kJ

92. Using information about the structure of water molecules and their interaction with one another, discuss the fact that the heat of fusion of water is only 334 J/g but the heat of vaporization of water is 2258 J/g.

93. (a) To break each hydrogen bond in ice requires 3.5×10^{-20} J. The measured heat of fusion of ice is 6.01 kJ/mol. Essentially all of the energy involved in the heat of fusion goes to break hydrogen bonds. What percentage of the hydrogen bonds are broken when ice is converted to liquid water?

Ans. 14%

(b) How much additional heat would be required, per mole, to break the remaining hydrogen bonds?

Ans. 37 kJ

94. Which of the following elements reacts with sulfur to form a solid in which the sulfur atoms form a closest packed array with all of the octahedral holes occupied: Li, Na, Be, Ca, Al, O_2?

95. The carbon atoms in the unit cell of diamond occupy the same positions as both the zinc and sulfur atoms in cubic zinc sulfide. How many carbon atoms are found in the unit cell of diamond? The bonds between the carbon atoms are covalent. What is the hybridization of a carbon atom in diamond?

Ans. 8; sp^3

96. The density of diamond is 3.51 g/cm^3. Calculate the length of the unit cell edge of diamond. (The structure of diamond is described in exercise 95 and shown in Fig. 11.22.)

Ans. 3.57 Å

97. In terms of its internal structure, explain the high melting point and hardness of diamond. (The structure of diamond is described in exercise 95 and shown in Fig. 11.22.)

98. When an electron in an excited molybdenum atom falls from the L to the K shell, an X ray is emitted. These X rays are diffracted at an angle of 7.75° by planes with a separation of 2.64 Å. What is the difference in energy between the K and the L shell in molybdenum? (Assume a first-order reflection.)

Ans. 2.79×10^{-8} erg or 1.74×10^4 eV

99. What is the percent by mass of titanium in rutile, a mineral that contains titanium and oxygen, if the structure of rutile can be described as a closest packed array of oxygen atoms with titanium in $\frac{1}{2}$ of the octahedral holes? What is the oxidation number of titanium?

Ans. 59.95%; +4

Solutions; Colloids

Water and air inclusions fixed in amber.

S olutions are crucial to life and to many other processes that involve chemical reactions. When food is converted into a form that the body can use, the nutrients go into solution to pass through the walls of the digestive tract into the blood. There, they are carried throughout the body in solution.

The dissolution of substances from the air and the earth is important in the conversion of rocks to soil, in altering the fertility of the soil, and in changing the form of the earth's surface. Deposits of many minerals are the result of reactions that have taken place in solution, followed by the evaporation of the solvent.

Most chemical reactions take place in solution. In a gaseous or liquid solution, molecules and ions can move freely, come into contact with each other, and react. In solids, molecules and ions cannot move freely; chemical reactions between solids, if they occur at all, are generally very slow.

In this chapter we will consider what kind of solution may form when a solute and a solvent are mixed, what factors determine whether or not a solution will form, and the resulting properties of the solution. In addition, we will examine colloids—systems that resemble solutions but consist of dispersions of particles somewhat larger than ordinary molecules or ions.

12.1 The Nature of Solutions

When sugar is stirred with enough water, the sugar dissolves, and a solution of sugar in water is formed. The solution consists of the **solute** (the substance that dissolves—in this case, sugar) and the **solvent** (the substance in which a solute dissolves—in this case, water). The molecules of sugar are uniformly distributed among the molecules of water; that is, the solution is a **homogeneous mixture** of solute and solvent molecules. The molecules of sugar diffuse continuously through the water, and although sugar molecules are heavier than water molecules, the sugar does not settle out on standing.

When potassium dichromate, $K_2Cr_2O_7$, dissolves in water (Fig. 12.1), the potassium ions and dichromate ions from the crystalline solid become uniformly distributed throughout the water. This solution is a homogeneous mixture of water molecules, potassium ions, and dichromate ions. The solute particles (ions in this case) diffuse through the water just as molecular solutes do; they do not settle upon standing.

A solution with a small proportion of solute to solvent is **dilute;** one with a large proportion of solute to solvent is **concentrated** (Fig. 12.1). A solution is **unsaturated** when more solute will dissolve in it; it is **saturated** when no more solute will dissolve. If excess solute and a saturated solution are in contact, the dissolved solute is in equilibrium with the excess solute. The **solubility** of a solute is the quantity that will dissolve in a given amount of solvent to produce a saturated solution. Since solutions can have different concentrations, it is evident that the composition of a solution may vary within certain limits. Thus a solution is not a compound, because a compound always has the same composition. Figure 12.1 shows two solutions of potassium dichromate with different concentrations.

All solutions—whether they contain dissolved molecules or dissolved ions—exhibit some similar properties: (1) homogeneity, (2) absence of settling, (3) the mo-

Figure 12.1

Two solutions and a sample of solid potassium dichromate, $K_2Cr_2O_7$. The solution on the left was prepared by dissolving 1.0 g of $K_2Cr_2O_7$; the solution on the right, by dissolving 10.0 g of $K_2Cr_2O_7$. Note that both solutions are homogeneous; you can tell by the color that the orange $Cr_2O_7^{2-}$ ion is uniformly distributed. The more intense color on the right shows that this solution contains more of the $Cr_2O_7^{2-}$ ion than the solution on the left and thus is the more concentrated solution.

lecular or ionic state of subdivision of the components, and (4) a composition that can be varied continuously within limits.

When solid sugar is added to a saturated solution of sugar, it falls to the bottom, and no more seems to dissolve. Actually, molecules of sugar continue to leave the solid and dissolve, but at the same time, molecules of sugar in solution collide with the solid and take up positions on the crystal. Enough molecules return to the solid that the process of crystallization counterbalances that of dissolution, and a state of equilibrium exists.

If a saturated solution is prepared at an elevated temperature and undissolved solute is removed, the solution can sometimes be cooled without crystallization of solute. If the cool solution contains more solute than it would if the dissolved solute were in equilibrium with undissolved solute, it is a **supersaturated solution.** Such solutions are unstable, and the excess dissolved solute may crystallize spontaneously. Alternatively, agitation of the solution or the addition of a seed crystal of the solute may start crystallization of the excess solute. After crystallization is complete a saturated solution remains, in equilibrium with the crystals of solute. Some syrups are supersaturated solutions of sugar in water.

In chemistry we usually use liquid solvents. Water is used so often as a solvent that the word *solution* has come to imply a water solution to many people. However, almost any gas, liquid, or solid can act as a solvent for other gases, liquids, or solids. Many alloys are solid solutions of one metal dissolved in another: for example, nickel coins contain nickel dissolved in copper. Air is a gaseous solution, a homogeneous mixture of gases. Oxygen (a gas), alcohol (a liquid), and sugar (a solid) will each dissolve in water (a liquid) to form liquid solutions.

Although it is easy to identify the solute and solvent in most cases (as when one gram of sugar is dissolved in 100 milliliters of water), sometimes the identification is difficult. For example, it is not possible to distinguish solute from solvent in a solution of equal amounts of ethanol and water. In such cases, the choice is arbitrary.

12.2 Solutions of Gases in Liquids

The amount of gas that will dissolve in a liquid depends on the nature of the gas and the solvent. For example, at 0°C and 1 atmosphere, 0.049 milliliters of oxygen, 1.7 liters of carbon dioxide, 80 liters of sulfur dioxide, or 1180 liters of ammonia dissolve in 1 liter of water. At 20°C and 1 atmosphere, 0.200 liters of stibine, SbH_3, will dissolve in 1 liter of water, but 25 liters will dissolve in 1 liter of carbon disulfide.

The pressure of the gas and the temperature also affect the solubility of a gas. The solubility of a gas that does not react with the solvent increases as the pressure of the gas increases. You can see the effect of pressure upon solubility in bottled carbonated beverages. Pressure forces carbon dioxide into solution in the beverage, and the bottle is tightly capped to maintain this pressure. When you open the bottle, the pressure decreases and some of the gas escapes from the solution (Fig. 12.2). The escape of bubbles of a gas from a liquid is known as **effervescence.**

If 1 gram of a gas dissolves in 1 liter of water at 1 atmosphere of pressure, 5 grams will dissolve at 5 atmospheres. This direct proportionality is expressed quantitatively by **Henry's law: The quantity of a gas that dissolves in a definite volume of liquid is directly proportional to the pressure of the gas,** or

$$C_g = kP_g$$

Figure 12.2
When the pressure of the *gaseous* carbon dioxide above this beverage was reduced by opening the bottle, the *dissolved* carbon dioxide escaped from the solution and formed the bubbles shown.

where C_g is the solubility of the gas in the solution, P_g is the pressure of the gas over the solution, and k is a proportionality constant that depends on the identity of the gas and of the solvent.

EXAMPLE 12.1

The solubility of O_2 is 0.035 g or 1.1×10^{-3} mole per liter of water at 0°C and 0.50 atm. What is k in Henry's law for this solubility, if the solubility is expressed in moles per liter?

According to Henry's law, the solubility, C_g, of a gas (1.1×10^{-3} mol/L, in this case) is directly proportional to the pressure, P_g, of the gas; that is, the solubility is equal to a proportionality constant, k, times the pressure (0.50 atm, in this case).

$$C_g = kP_g$$

Since we know C_g and P_g, we need to rearrange this expression to solve for k.

$$k = \frac{C_g}{P_g}$$

$$k = \frac{(1.1 \times 10^{-3}\ \text{mol/L})}{0.50\ \text{atm}}$$

$$k = 2.2 \times 10^{-3}\ \text{mol/L atm}$$

The effect of pressure does not follow Henry's Law when a chemical reaction takes place between the gas and the solvent. Thus the solubility of ammonia in water does not increase as rapidly with increasing pressure as predicted by the law, because ammonia, being a base, reacts to some extent with water to form ammonium ions and hydroxide ions.

$$NH_3 + H_2O \rightleftharpoons NH_4^+ + OH^-$$

The solubility of a gas in a liquid decreases with an increase in temperature, provided that the gas does not react with the solvent (Fig. 12.3). For example, 48.9 milliliters of oxygen dissolve in one liter of water at 1 atmosphere and 0°C, but only 31.6 milliliters dissolve at 25°, 24.6 milliliters at 50°, and 23.0 milliliters at 100°. This relationship is not one of inverse proportion, however, and the solubility of a gas in a liquid at a given temperature must be determined experimentally. The decreasing solubility of gases such as oxygen with increasing temperature is a very important factor in thermal pollution of natural waters. A heat discharge that increases the temperature by a few degrees can reduce the solubility of oxygen in the water to a level too low for many forms of aquatic life to survive.

Gases can form supersaturated solutions. If a solution of a gas in a liquid is prepared either at low temperature or under pressure (or both), as the solution warms or as the gas pressure is reduced, the solution may become supersaturated. For example, a bottle of a carbonated beverage may not liberate the excess dissolved carbon dioxide when the carbon dioxide pressure above the solution is reduced by opening the bottle, but it does if it is shaken or stirred.

Most gases can be expelled from solvents by boiling their solutions in an open container. The gases oxygen, nitrogen, carbon dioxide, and sulfur dioxide, for example, can be removed from water by boiling the solution for a few minutes. The fact that a gas is expelled from a solution by boiling is not due just to the rise in temperature of

Figure 12.3

The small bubbles of air in this glass of water formed when the water warmed and the solubility of its dissolved air thereby decreased.

the solution. When the solution is boiled in an open container, part of the vapor of the solvent escapes, carrying with it some of the solute gas above the solution. This lowers the partial pressure of that gas above the solution, and (in accord with Henry's law) more gas escapes.

12.3 Solutions of Liquids in Liquids (Miscibility)

Two liquids that mix with each other in all proportions are said to be completely **miscible.** Ethyl alcohol, sulfuric acid, and ethylene glycol (antifreeze), for example, are completely miscible with water (Fig. 12.4). Liquids that mix with water in all proportions are usually polar substances (see Section 6.6) or substances that form hydrogen bonds (Section 11.12). For such liquids, the dipole-dipole attractions (or hydrogen bonding) of the solute molecules with the solvent molecules are at least as strong as those between molecules in the pure solute or in the pure solvent. Hence the two kinds of molecules mix easily.

Two liquids that do not mix are called **immiscible.** Two layers are formed when two immiscible liquids are poured into the same container. Gasoline, carbon disulfide, benzene, carbon tetrachloride, and many other nonpolar liquids are immiscible with water (Fig. 12.5). There is no effective attraction between the molecules of such nonpolar liquids and polar water molecules. The only strong attractions in such a mixture are between the water molecules, so they effectively squeeze out the molecules of the nonpolar liquid. Nonpolar liquids may be miscible with each other, however, because of the absence of an appreciable tendency of molecules either to attract other

Figure 12.4
Water and antifreeze are miscible; the yellow mixture of the two is homogeneous.

Figure 12.5
Water and oil are immiscible. Eventually the oil will float to the surface and two separate layers will form.

Figure 12.6

Bromine (a deep orange liquid) and water are partially miscible. The top layer in this mixture is a dilute (saturated) solution of bromine in water; the bottom layer is a dilute (saturated) solution of water in bromine.

molecules like themselves or molecules of another nonpolar liquid. The solubility of polar molecules in polar solvents and of nonpolar molecules in nonpolar solvents is an illustration of the old chemical axiom, "Like dissolves like."

Two liquids, such as ether and water or bromine and water, that are slightly soluble in each other are said to be partially miscible. Two partially miscible liquids usually form two layers when mixed. Each layer is a saturated solution of one liquid in the other (Fig. 12.6), and a dynamic equilibrium occurs between the two layers. When the partially miscible liquids bromine and water are in contact and at equilibrium, bromine molecules leave the bromine layer and enter the water layer at the same rate that bromine molecules leave the water layer and return to the bromine layer. Similarly, water molecules leave the water layer and enter the bromine layer at the same rate that water molecules leave the bromine layer and return to the water layer.

12.4 The Effect of Temperature on the Solubility of Solids in Liquids

The dependence of solubility upon temperature for a number of inorganic substances in water is shown by the solubility curves in Fig. 12.7. Generally, the solubility of a solid increases with increasing temperature, although there are exceptions. A sharp break in a solubility curve indicates the formation of a new compound, with a different solubility. For example, when $Na_2SO_4 \cdot 10H_2O$ (Glauber's salt) in equilibrium with a saturated solution is heated to 32.4°C, it forms the anhydrous salt, Na_2SO_4. The red curve in Fig. 12.7 shows the effect of increasing temperature upon the solubility of $Na_2SO_4 \cdot 10H_2O$ up to 32.4°C (it increases with temperature), and above this point the curve shows the effect on the solubility of Na_2SO_4 (it decreases with increasing temperature).

Figure 12.7

Graph showing the effect of temperature on the solubility of several inorganic substances. The break in the red curve occurs at the temperature (32.4°C) where solid $Na_2SO_4 \cdot 10 H_2O$ decomposes to Na_2SO_4 and water.

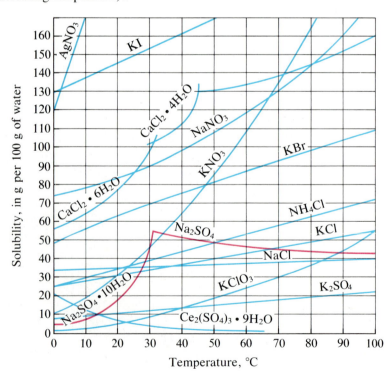

12.5 Rate of Solution

The amount of solute entering a solution per unit of time is called the **rate of solution** of the solute. This rate depends upon several factors.

1. *Solubility of the solute:* More soluble substances dissolve more rapidly than less soluble ones.

2. *State of subdivision of the solute:* A solid dissolves only at its surface, so the more finely divided the solute, the greater the surface area per unit mass and the more rapidly it dissolves.

3. *Mixing:* Dissolved molecules diffuse away from the solid solute relatively slowly, so the solution around the solid approaches saturation. Stirring or shaking the mixture brings unsaturated solution in contact with the solute and thus increases the rate of solution.

4. *Heating* often increases the solubility of a solid. In addition, if the solute and solvent react during the solution process, heating accelerates the reaction and hence increases the rate of solution.

The last three effects are familiar to those who use sugar in their tea or coffee. Finely divided sugar dissolves more rapidly than coarse sugar, stirring speeds up the dissolution of sugar, and sugar dissolves more rapidly in a hot beverage than in a cold one.

12.6 Electrolytes and Nonelectrolytes

Figure 12.8
Apparatus for demonstrating the conductivity of solutions. The solution on the left contains the nonelectrolyte sugar and has no conductivity. The solution in the middle contains a small quantity of NaCl, (100% ionized) in the water—a strong electrolyte. The solution on the right is a 5% acetic acid solution. The acetic acid is only partially ionized—a weak electrolyte.

Substances such as acids, bases, and salts that give solutions that conduct an electric current are called **electrolytes** (Section 9.2). Substances that conduct an electric current when molten are also called electrolytes. Other substances, such as sugar and alcohol, form solutions that do not conduct and are called **nonelectrolytes.**

Electrolytes and nonelectrolytes may be classified experimentally by setting up a simple conductivity apparatus, as shown in Fig. 12.8. A source of electric current is connected through a lamp to two electrodes in a beaker. When the beaker is filled with pure water, the lamp does not light because very little current flows. When a nonelectrolyte such as sugar, alcohol, or glycerin is added to the water, the lamp still does not

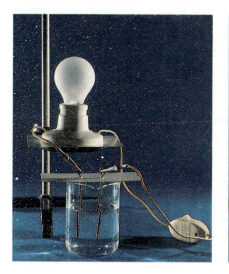
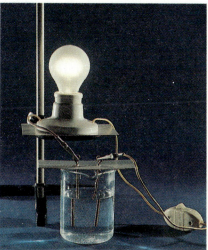
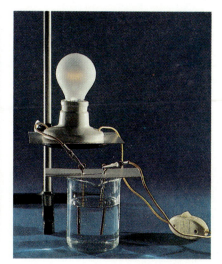

light. If, however, an electrolyte such as hydrochloric acid is dissolved in the water, the lamp glows brightly because the solution conducts an electric current.

With a 0.1 M solution of an acid such as hydrochloric, nitric, or sulfuric acid, of a base such as potassium, sodium, or barium hydroxide, or of most salts, the lamp in the circuit will glow brightly, showing that these solutions are good conductors of electricity. Substances whose aqueous solutions are good conductors of electricity are known as **strong electrolytes.** When a 0.1 M solution of acetic acid or ammonia is placed in the beaker, the lamp dims, showing that these solutions are poor conductors. Substances whose solutions are poor conductors of electricity are called **weak electrolytes.**

Svante Arrhenius, a Swedish chemist, first successfully explained electrolytic conduction. Although his theory has been modified by subsequent knowledge of atomic structure and chemical bonding, the current theory of electrolytes embodies most of the principal postulates of Arrhenius's theory.

According to the current theory, a solution of an electrolyte contains positive and negative ions that move independently. When a solution conducts an electric current, the positive ions (cations) move toward the negative electrode, while the negative ions (anions) move toward the positive electrode. The movement of ions toward the electrode of opposite charge accounts for electrolytic conduction. A solution of a nonelectrolyte contains *molecules* of the nonelectrolyte rather than ions and thus cannot conduct an electric current.

If a solution is a good conductor of electricity, the solute consists principally of ions, and if a solution is a poor conductor, the solute consists principally of molecules. Thus compounds that form solutions composed principally of ions are strong electrolytes. Compounds that exist principally as unreacted molecules are weak electrolytes. The partial ionization of acetic acid, a weak acid (and thus a weak electrolyte), is such that the acid in a 0.1 M solution is only about 1% ionized. This means that 99% of the acid is in the molecular form.

Pure water is an extremely poor conductor of electricity, indicating very slight ionization (actually only about 2 molecules out of every 100,000,000 ionize). Water ionizes when one molecule of water gives up a hydrogen ion to another molecule of water, yielding a hydronium ion and a hydroxide ion.

$$H_2O + H_2O \rightleftharpoons H_3O^+ + OH^-$$

To show how ions migrate during electrolytic conduction, partially fill a U-tube (Fig. 12.9) with a colorless solution of potassium nitrate, acidified with a few drops of sulfuric acid. Then carefully introduce a solution of copper(II) permanganate into the bottom of the U-tube without mixing the two solutions. When a current is passed

Figure 12.9

When an electric current flows in the system shown, blue Cu^{2+} ions and purple MnO_4^- ions travel toward different electrodes, visual evidence that the current is carried by the ions.

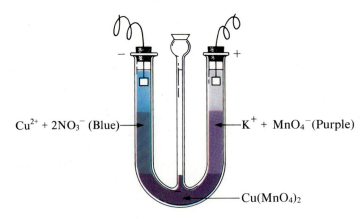

$Cu^{2+} + 2NO_3^-$ (Blue) →

← $K^+ + MnO_4^-$ (Purple)

← $Cu(MnO_4)_2$

through the U-tube, blue hydrated copper(II) ions, $Cu^{2+}(aq)$, move through the colorless potassium nitrate solution toward the negative electrode, and the purple permanganate ions, MnO_4^-, toward the positive electrode. The separation of colors shows that the current is carried by ions and that ions of opposite charge move independently in the solution.

12.7 The Solubilities of Common Metal Compounds

Knowledge of the solubilities of metallic compounds is very useful to students and chemists. Memorizing solubilities of individual compounds is unnecessary; it is simpler to learn the following general rules. (Remember that these rules are for simple compounds of the more common metals; there are exceptions for less common metals and complex compounds.)

1. Most nitrates and acetates are soluble in water; silver acetate, chromium(II) acetate, and mercury(I) acetate are slightly soluble; bismuth acetate reacts with water and forms insoluble bismuth oxyacetate, $BiO(CH_3CO_2)$.

2. All chlorides are soluble except those of mercury(I), silver, lead(II), and copper(I); lead(II) chloride is soluble in hot water.

3. All sulfates, except those of strontium, barium, and lead(II) are soluble; calcium sulfate and silver sulfate are slightly soluble.

4. Carbonates, phosphates, borates, arsenates, and arsenites are insoluble, except those of the ammonium ion and the alkali metals.

5. The hydroxides of the alkali metals and of barium and strontium are soluble, and other hydroxides are insoluble; calcium hydroxide is slightly soluble.

6. Most sulfides are insoluble. However, the sulfides of the alkali metals are soluble, but they react with water to give solutions of the hydroxide and hydrogen sulfide ion, HS^-; the sulfides of the alkaline earth metals and of aluminum also react to give OH^- and HS^- (or H_2S if the metal hydroxide is insoluble).

12.8 Solid Solutions

If a mixture consisting of equal amounts of lithium chloride and sodium chloride is melted, mixed well, and allowed to cool, the resulting crystalline solid contains an array of chloride ions, with a random distribution of lithium ions and sodium ions in holes in the array (Fig. 12.10). The crystal is a **solid solution** of LiCl and NaCl. It is homogeneous, just like a liquid solution. Its composition can be varied (from pure LiCl to pure NaCl), and neither NaCl nor LiCl separate out on standing.

Solid solutions of ionic compounds result when ions of one type randomly replace other ions of about the same size in a crystal. Ruby, for example, is a solid solution containing about 1% Cr_2O_3 in Al_2O_3. Application of fluoride to tooth enamel, which is mostly hydroxyapatite, forms a decay-resistant solid solution of hydroxyapatite, $Ca_5(PO_4)_3OH$, and fluoroapatite, $Ca_5(PO_4)_3F$.

Some ionic substances appear to be **nonstoichiometric compounds;** that is, their chemical formulas deviate from ideal ratios or are variable. In other respects, however, these substances resemble compounds; they are not heterogeneous mixtures but are instead homogeneous throughout. These so-called nonstoichiometric compounds are, in fact, solid solutions of two or more compounds. A sample of ruby, for example, with the formula $Cr_{0.02}Al_{1.98}O_3$ is a solid solution of Cr_2O_3 in Al_2O_3 with a 1 to 99

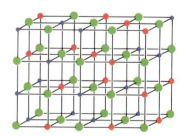

Figure 12.10

A portion of the structure of a solid solution of LiCl and NaCl. The chloride ions (large green spheres) form a face-centered cubic array, with the octahedral holes occupied by a random distribution of lithium ions (smaller purple spheres) and sodium ions (larger red spheres).

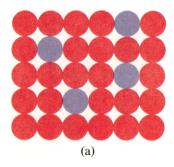

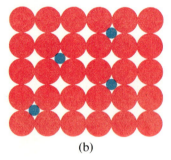

(a)

(b)

Figure 12.11
Two-dimensional representations of alloys. (a) A substitutional solid solution in which solute atoms (purple spheres) replace atoms of the solvent crystal (red spheres). (b) An interstitial solid solution in which small solute atoms (blue spheres) occupy holes in the lattice of the solvent crystal (red spheres).

ratio of Cr to Al. Some nonstoichiometric compounds are solid solutions containing one ion in two different oxidation states. For example, $TiO_{1.8}$ contains both Ti^{3+} and Ti^{4+} ions and may be considered to be a solid solution of Ti_2O_3 and TiO_2.

Some **alloys** are solid solutions composed of two or more metals. In such an alloy atoms of one of the component metals take up positions in the crystal lattice of the other. The solute atoms may randomly replace some of the atoms of the solvent crystal to form a class of solid solutions called **substitutional solid solutions** [Fig. 12.11(a)]. For example, chromium dissolves in nickel to form a solid solution in which the chromium atoms replace nickel atoms in the face-centered cubic structure of nickel. The solubility may be limited (zinc and copper; chromium and nickel) or practically infinite (nickel and copper).

Small atoms (hydrogen, carbon, boron, and nitrogen) may occupy the holes in the lattice of a metal, forming another class of solid solutions called **interstitial solid solutions** [Fig. 12.11(b)]. A solid solution of carbon in iron (austenite) is an example; the iron atoms are on the lattice points of a face-centered cubic lattice, and the carbon atoms occupy the interstitial positions.

Not all alloys are solid solutions. Some are *heterogeneous* mixtures in which the component metals are mutually insoluble and the solid alloy is composed of an intimate mixture of crystals of each metal. For example, tin and lead (in plumber's solder) are insoluble in each other in the solid state. Other alloys, such as Cu_5Zn_8 and Ag_3Al, are actually compounds that form with only one specific stoichiometric composition. In general, the formulas of such intermetallic compounds are not those that might be predicted on the basis of the usual valence rules.

The Process of Dissolution

12.9 Dissolution of Nonelectrolytes

When ethanol is added to water, a solution forms without the input of outside energy. The solution forms by a **spontaneous process.** Chemists recognize two factors involved in such a spontaneous process: (1) changes in energy, and (2) changes in the amount of disorder of the components of the solution. In the process of dissolution the change in disorder results from the mixing of the solute and solvent. The energy changes occur because the intermolecular attractions change from solute-solute and solvent-solvent attractions to solute-solvent attractions. First, let us consider formation of a solution in which energy changes are insignificant, so we can concentrate on how changes in disorder contribute to the formation of solutions.

When the strengths of the intermolecular forces of attraction between solute and solvent molecules (or ions) are the same as those between the molecules in the separate components, a solution is formed with no accompanying energy change. Such a solution is called an **ideal solution.** An ideal solution obeys Raoult's Law exactly (Section 12.17). Solutions of ideal gases (or of gases such as helium and argon, which closely approach ideal behavior) contain molecules with no significant intermolecular attractions, and thus these solutions behave as ideal solutions.

If a stopcock is opened between 500-milliliter bulbs containing helium and argon [Fig. 12.12(a)], the gases will spontaneously diffuse together and form a solution [Fig. 12.12(b)]. This solution forms because the disorder of the helium and argon molecules increases when they mix. They occupy a volume twice as large as that which each occupied before mixing, and the molecules are randomly distributed among one another.

Other solutions that closely approximate ideal behavior include solutions of pairs of chemically similar substances, such as the liquids methanol, CH_3OH, and ethanol, C_2H_5OH, or the liquids chlorobenzene, C_6H_5Cl, and bromobenzene, C_6H_5Br. If the samples of helium and argon in Fig. 12.12(a) were replaced with methanol and ethanol (or with chlorobenzene and bromobenzene), when the stopcock was opened, the molecules of the liquids would diffuse together spontaneously (although at a much slower rate than the gases), giving a solution with a disorder greater than that of the pure liquids. These examples show that **processes in which the disorder of the system increases tend to occur spontaneously.** Moving molecules will become randomly distributed among one another unless something holds them back.

Intermolecular forces of attraction can keep molecules from mixing. These forces are small and negligible in gases, so gases are mutually soluble in all proportions. However, the molecules in liquids and solids are relatively close together, and their intermolecular attractions are stronger and much more important. Sometimes, solute molecules attract one another strongly but attract solvent molecules weakly. The solute molecules thus remain in contact with one another and do not dissolve, even though formation of a solution would increase their disorder. If the solvent molecules attract one another strongly but do not attract the solute molecules, the solvent molecules will not separate to let the solute dissolve. A solution forms only when the attractions between solute and solvent molecules are about equal to (or are greater than) the combination of the attractions between solute molecules and the attractions between solvent molecules.

To see why gasoline and water don't mix, let us consider what happens when octane, a typical hydrocarbon in gasoline, is added to water. Water molecules are held in contact by hydrogen bonds (Section 11.12); octane molecules are held in contact by London forces (Section 11.11). When the two liquids are brought together, the forces of attraction between the octane molecules and the water molecules are not strong enough to overcome the hydrogen bonds between the water molecules. A great deal of stirring can put some molecules of water among the octane molecules. This stirring breaks hydrogen bonds between water molecules and destroys some of the London forces between the octane molecules. Relatively weak London forces then form between octane and water molecules. When the stirring stops, however, diffusion of the water molecules brings them back in contact with one another, and their relatively strong hydrogen bonds hold them clustered together. Eventually, water and octane separate into two layers.

On the other hand, if we add methanol to water, a solution forms readily. Methanol molecules hydrogen-bond about equally well to other methanol molecules and to water molecules. Water-water and water-methanol hydrogen bonds are of about equal strengths. Thus there is no stronger hydrogen bonding to cause the water molecules or the methanol molecules to cluster together, and a solution forms because of the increase in disorder.

Now let us consider the second factor involved in a spontaneous solution process, changes in energy. If the solute-solvent attractions are stronger than the solute-solute and solvent-solvent attractions, then heat is released during dissolution as the stronger attractions form (the dissolution process is exothermic). If the solute-solvent attractions are weaker than the solute-solute and solvent-solvent attractions, heat is absorbed (the dissolution process is endothermic). Although sulfuric acid is not a nonelectrolyte, its dissolution provides an excellent example. When one mole of sulfuric acid is dissolved in nine moles of water, the resulting acid-water interactions are stronger than the combination of the acid-acid attractive forces in pure sulfuric acid and the water-water attractive forces in pure water. The solution becomes very hot (Fig. 12.13) because

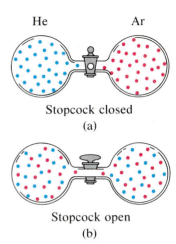

He   Ar

Stopcock closed
(a)

Stopcock open
(b)

Figure 12.12

The spontaneous mixing of helium and argon to give a solution. When the stopcock between samples of the two pure gases (a) is opened, they mix by diffusion to give a solution (b) in which the disorder of the molecules of the two gases is increased.

Figure 12.13
When concentrated sulfuric acid dissolves in water, the solution gets hot, as the production of steam indicates.

63.2 kilojoules of heat is produced. The loss of energy, as heat, when sulfuric acid dissolves in water indicates that the solution contains less energy than did the separate components before mixing. The **heat of solution** is the amount of heat absorbed or evolved when a solute dissolves in a specified amount of solvent.

   Processes in which the energy content of the system decreases tend to occur spontaneously. Thus both a loss of energy and an increase in disorder favor a spontaneous process such as the formation of a solution. For a solution to form, however, it is not necessary for both the energy change and the change in disorder to favor a spontaneous process. When ammonium nitrate is added to water, the ammonium nitrate dissolves even though the solution cools as heat is absorbed from the water. In this case the solute-solute attractions and solvent-solvent attractions combined are larger than the solute-solvent attractions. A solution forms, in spite of the endothermic nature of the process, because the increase in disorder is large enough that it more than compensates for the increase in the energy content. (Incidentally, ammonium nitrate is used to make instant ''ice'' packs for treatment of athletic injuries. A thin-walled plastic bag of NH_4NO_3 is sealed inside a larger bag filled with water. When the smaller bag is broken, a cold solution of NH_4NO_3 forms.)

12.10 Dissolution of Ionic Compounds

   When ionic compounds dissolve in water, the associated ions in the solid separate because water reduces the strong electrostatic forces between them. Let us consider the dissolution of potassium chloride in water. The hydrogen (positive) end of a polar water molecule is attracted to a negative chloride ion at the surface of the solid, and the oxygen (negative) end is attracted to a positive potassium ion. The water molecules surround individual K^+ and Cl^- ions at the surface of the crystal and penetrate between them, reducing the strong interionic forces that bind them together and letting them move off into solution as hydrated ions (Fig. 12.14). Several water molecules associate with each ion in solution as a result of the electrostatic attraction be-

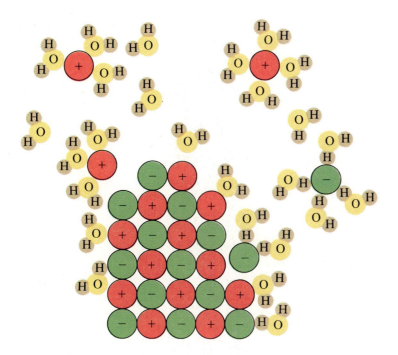

Figure 12.14
The dissolution of potassium chloride in water and the hydration of its ions. Water molecules in front of and behind the ions are not shown. The red positive spheres represent potassium ions; the green negative spheres, chloride ions.

tween the charged ion and the dipole of the water. Such an attraction is called an **ion-dipole attraction.** The increase in the distance between oppositely charged ions due to the layer of water molecules around each ion reduces the electrostatic attraction between them. In addition, the layers of water act as insulators, which further reduces the electrostatic attraction. This permits the independent motion of each hydrated ion in a dilute solution, resulting in an increase in the disorder of the system as the ions change from their ordered positions in the crystal to a much more disordered and mobile state in solution. This increased disorder is responsible for the dissolution of many ionic compounds, including potassium chloride, that dissolve with absorption of heat. In other cases the electrostatic attractions between the ions in a crystal are so large that an increase in disorder could not compensate for the energy required to separate the ions, and the crystal is insoluble.

Ionic compounds dissolve in polar solvents only when the polar solvent molecules can solvate and insulate the ions. In general, the larger the dielectric constant of a solvent, the greater its ability to insulate solvated ions and to reduce the electrostatic attractions between them, and the greater the solubility of an ionic compound in the solvent. This phenomenon is strikingly illustrated by the data of Table 12.1. Ionic

← what is dielectric constant

Table 12.1 The solubility of sodium chloride and the dielectric constant of the solvent

Solvent	Solubility of NaCl (grams per 100 g of solvent, 25°C)	Dielectric constant of the solvent
Water, H_2O	36.12	80.0
Methyl alcohol, CH_3OH	1.3	33.1
Carbon tetrachloride, CCl_4	0.00	2.2

substances in general do not dissolve appreciably in nonpolar solvents, such as benzene or carbon tetrachloride, because the nonpolar solvent molecules are not strongly attracted to ions and because nonpolar solvents have low dielectric constants.

12.11 Dissolution of Molecular Electrolytes

Gaseous hydrogen chloride consists of covalent HCl molecules and contains no ions; pure gaseous hydrogen chloride does not conduct an electric current. A solution of hydrogen chloride in a nonpolar solvent such as benzene is also a nonelectrolyte and so does not contain ions. However, a solution of hydrogen chloride in water is a strong electrolyte. The water molecules play an important part in causing ionization; hydrogen chloride reacts with water to form hydronium ions, H_3O^+, and chloride ions, Cl^-. As shown by the Lewis structures, a hydrogen ion (proton) shifts from a hydrogen chloride molecule to a lone pair of electrons on the water molecule.

$$H\!:\!\overset{..}{\underset{\overset{|}{H}}{O}}\!: \; + \; H\!:\!\overset{..}{\underset{..}{Cl}}\!: \; \longrightarrow \; H\!:\!\overset{..}{\underset{\overset{|}{H}}{O}}\!:\!H^+ \; + \; :\!\overset{..}{\underset{..}{Cl}}\!:^-$$

The hydronium ions and chloride ions conduct the current in the solution. All common strong acids react with water when they dissolve and are therefore strong electrolytes.

Many other compounds dissolve in water as hydrated molecules. In most cases these molecules are weak electrolytes, and a small fraction of the molecules undergo ionization, causing the solution to conduct electricity weakly. For example, cyanic acid (HOCN, a weak acid) dissolves in water principally as hydrated molecules. A few of these dissolved molecules ionize under ordinary conditions.

$$H\!:\!\overset{..}{\underset{\overset{|}{H}}{O}}\!: \; + \; H\!:\!\overset{..}{\underset{..}{O}}\!:\!C\!:\;:\;:\!\overset{..}{N}\!: \; \rightleftharpoons \; H\!:\!\overset{..}{\underset{\overset{|}{H}}{O}}\!:\!H^+ \; + \; :\!\overset{..}{\underset{..}{O}}\!:\!C\!:\;:\;:\!\overset{..}{N}\!:^-$$

Because the cyanate ion, OCN^-, binds a hydrogen ion more strongly than does a water molecule, the reaction gives only a 5.6% yield of ions in a 0.1 M solution of HOCN at 25°C. Other weak acids such as acetic acid, CH_3CO_2H, nitrous acid, HNO_2, and hydrogen cyanide, HCN, are also soluble in water, and only a small fraction of their hydrated molecules ionize at any one time.

Weak bases also dissolve to give solutions of hydrated molecules, some of which undergo ionization. For example, a solution of ammonia in water consists primarily of hydrated ammonia molecules, $NH_3(aq)$, with small amounts of ammonium ions, $NH_4^+(aq)$, and hydroxide ions, $OH^-(aq)$, which result from the reaction of ammonia with water.

$$H\!:\!\overset{\overset{\textstyle H}{\textstyle |}}{\underset{\overset{|}{H}}{N}}\!: \; + \; H\!:\!\overset{..}{\underset{..}{O}}\!: \; \rightleftharpoons \; H\!:\!\overset{\overset{\textstyle H}{\textstyle |}}{\underset{\overset{|}{H}}{N}}\!:\!H^+ \; + \; :\!\overset{..}{\underset{..}{O}}\!:\!H^-$$

The halides and cyanides of mercury and cadmium are also weak electrolytes. When these compounds dissolve they give solutions of molecules. A small fraction of the metal-halide or metal-cyanide bonds break, and the few resulting hydrated metal ions and halide or cyanide ions give the low conductivities of the solutions.

Expressing Concentration

12.12 Percent Composition

To express the concentration of a solute, we can state the mass of solute in a given mass of solvent, for example, 1 gram of NaCl in 100 grams of water, or we can give its composition as **percent by mass** (Section 3.4). A 10% NaCl solution by mass may contain 10 grams of NaCl in 100 grams of solution (10 grams of NaCl and 90 grams of water), or it may contain any other ratio of NaCl to solution for which the mass of NaCl is 10% of the total mass, for example, 20 milligrams of NaCl in 200 milligrams of solution, 1.5 grams of NaCl in 15 grams of solution, or 7.4 kilograms of NaCl in 74 kilograms of solution.

$$\% \text{ solute} = \frac{\text{mass of solute}}{\text{mass of solution}} \times 100$$

EXAMPLE 12.2

A bottle of a certain ceramic tile cleanser, which is essentially a solution of hydrogen chloride, contains 130 g of HCl and 750 g of water. What is the percent by mass of HCl in this cleanser?

The percent by mass of the solute is

$$\% \text{ solute} = \frac{\text{mass solute}}{\text{mass solution}} \times 100$$

$$= \frac{130 \text{ g}}{130 \text{ g} + 750 \text{ g}} \times 100 = 14.8\%$$

$$\frac{130}{130 + 750} \times 100 =$$

When using percent composition by mass in calculations, it is convenient to use grams of solute per 100 grams of solution (g solute/100 g solution), since the number of grams of solute in 100 grams of solution is equal to the percent by mass of solute. To calculate the mass of solute in a given *volume* of solution, given the percent composition by mass, you must know the specific gravity (or density) of the solution, the mass of 1 milliliter of the solution.

EXAMPLE 12.3

Concentrated hydrochloric acid, a saturated solution of hydrogen chloride, HCl, in water, is often used in the general chemistry laboratory. It has a specific gravity of 1.19 and contains 37.2% HCl by mass. What mass of HCl is contained in exactly 1 L of this concentrated acid?

This problem requires the following steps.

| Volume of solution | $\longrightarrow$ | Mass of solution | $\longrightarrow$ | Mass of HCl |

From the specific gravity we know that the mass of exactly 1 mL of concentrated hydrochloric acid is 1.19 g. Since 1 L equals 1000 mL,

$$\frac{1.19 \text{ g solution}}{1 \text{ mL}} \times 1000 \text{ mL} = 1190 \text{ g solution}$$

The solution contains 37.2% HCl by mass (37.2 g HCl/100.0 g solution).

$$1190 \text{ g solution} \times \frac{37.2 \text{ g HCl}}{100.0 \text{ g solution}} = 443 \text{ g HCl}$$

EXAMPLE 12.4   A student needs 125 g of HCl to prepare a metal chloride. What volume of concentrated hydrochloric acid with a density of 1.19 g/mL and containing 37.2% HCl by mass contains 125 g of HCl?

This problem requires the following steps.

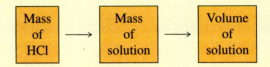

Since the solution contains 37.2% HCl, there are 37.2 g of HCl per 100.0 g of solution giving the conversion factor 37.2 g HCl/100.0 g solution or 100.0 g solution/37.2 g HCl.

$$125 \text{ g HCl} \times \frac{100.0 \text{ g solution}}{37.2 \text{ g HCl}} = 336 \text{ g solution}$$

The mass of 1 mL of solution is 1.19 g (1 mL/1.19 g).

$$336 \text{ g solution} \times \frac{1 \text{ mL solution}}{1.19 \text{ g solution}} = 282 \text{ mL solution}$$

Thus 282 mL of the concentrated hydrochloric acid contains 125 g of HCl.

12.13 Molarity

We have already considered the concept of molarity. In Section 3.1 the **molarity, M,** of a solution was defined as the number of moles of solute in exactly 1 liter of solution. Molarity may be calculated by dividing the moles of solute in a solution by the volume of the solution.

$$\text{Molarity} = \frac{\text{moles of solute}}{\text{liters of solution}}$$

Because 1 mole of any substance contains the same number of molecules as 1 mole of any other substance, equal volumes of 1 M solutions contain the same number of molecules of solute. The use of molarity to express the concentration of a solution makes it easy to select a desired number of moles, molecules, or ions of the solute by measuring out the appropriate volume of solution. For example, if 1 mole of sodium hydroxide is needed for a given reaction, we can use 40 grams of solid sodium hydroxide (1 mole), or 1 liter of a 1 M solution of the base, or 2 liters of a 0.5 M solution.

Figure 12.15
In one method of preparing a 1.00 M solution of cobalt(II) sulfate, 155 g of $CoSO_4$ (1 mol) is added to a flask that is calibrated to hold 1.000 L, and enough water is added to make 1.000 L of solution. The second photograph shows the 1.00 M solution after the solution has been shaken to assure uniform mixing.

To prepare a solution of known molarity, measure out the required amount of the solute, in moles, and add enough solvent to give the desired volume of solution. To prepare 1 liter of a 1.00 M solution, you could dissolve 1.00 mole of pure cobalt sulfate (155 g of $CoSO_4$) in enough water to form 1 liter of solution (Fig. 12.15).

Examples of the use of molar concentrations in stoichiometry calculations were presented in Chapter 3. However, here is one more as a reminder.

Concentrated sulfuric acid is a solution with a density of 1.84 g/mL and containing 98.3% H_2SO_4 by mass. What is the molarity of this acid?

EXAMPLE 12.5

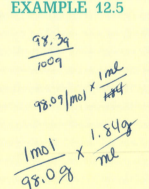

Calculating the moles of H_2SO_4 in 1.00 L of concentrated sulfuric acid requires the following steps.

$$\boxed{\begin{array}{c}\text{Volume of}\\\text{solution}\\\text{(1.00 L)}\end{array}} \longrightarrow \boxed{\begin{array}{c}\text{Mass}\\\text{of}\\\text{solution}\end{array}} \longrightarrow \boxed{\begin{array}{c}\text{Mass}\\\text{of}\\H_2SO_4\end{array}} \longrightarrow \boxed{\begin{array}{c}\text{Moles}\\\text{of}\\H_2SO_4\end{array}}$$

The mass of 1 mL of solution is 1.84 g; the mass of 1.00 L is given by the equation

$$1000 \text{ mL solution} \times \frac{1.84 \text{ g solution}}{1.00 \text{ mL solution}} = 1.84 \times 10^3 \text{ g solution}$$

There are 98.3 g of H_2SO_4 per 100 g of solution, since the solution is 98.3% H_2SO_4 by mass, so the mass of H_2SO_4 may be calculated.

$$1.84 \times 10^3 \text{ g solution} \times \frac{98.3 \text{ g } H_2SO_4}{100 \text{ g solution}} = 1.81 \times 10^3 \text{ g } H_2SO_4$$

Now we calculate the number of moles of H_2SO_4 in 1.00 L to obtain the molarity.

$$1.81 \times 10^3 \text{ g } H_2SO_4 \times \frac{1 \text{ mol } H_2SO_4}{98.0 \text{ g } H_2SO_4} = 18.5 \text{ mol } H_2SO_4$$

$$\frac{18.5 \text{ mol } H_2SO_4}{1.00 \text{ L}} = 18.5 \text{ M}$$

The calculations can be carried out in one step as follows:

$$\left(\frac{1.84 \text{ g solution}}{1 \text{ mL solution}} \times 1000 \text{ mL solution} \times \frac{98.3 \text{ g } H_2SO_4}{100 \text{ g solution}} \times \frac{1 \text{ mol } H_2SO_4}{98.0 \text{ g } H_2SO_4} \right) / 1.00 \text{ L} = 18.5 \text{ M}$$

A concentration unit called *normality* is sometimes used to describe the concentrations of acids and bases or of oxidizing and reducing agents. It is related to molarity but differs from it in one important way: the normality of a solute is dependent on the reaction that the solute undergoes. Normality will be described in the chapters dealing with acid-base reactions (Section 16.14) and with electrochemical reactions (Section 21.5).

12.14 Molality

The **molality, m,** of a solution is the number of moles of solute in exactly 1 kilogram of solvent. Molality may be calculated by dividing the moles of solute in a solution by the mass of the solvent in kilograms.

$$\text{Molality} = \frac{\text{moles of solute}}{\text{kilograms of solvent}}$$

Note that *kilograms of solvent* rather than liters of solution are specified. This is the difference between molality and molarity. Molality is a useful method for expressing concentration because it does not change its value as the temperature changes.

EXAMPLE 12.6

What is the molality of a solution that contains 0.850 g of ammonia, NH_3, in 125 g of water?

After the mass of NH_3 has been converted to moles of NH_3, the molality may be determined by dividing by the kilograms of water, the solvent.

$$0.850 \text{ g } NH_3 \times \frac{1 \text{ mol } NH_3}{17.0 \text{ g } NH_3} = 5.00 \times 10^{-2} \text{ mol } NH_3$$

$$\text{Molality} = \frac{\text{moles of solute}}{\text{kilograms of solvent}} = \frac{\text{mol } NH_3}{\text{kg } H_2O}$$

$$= \frac{5.00 \times 10^{-2} \text{ mol } NH_3}{0.125 \text{ kg } H_2O} = 0.400 \text{ m}$$

EXAMPLE 12.7

Calculate the molality of an aqueous solution of sodium chloride if 0.250 kg of the solution contains 40.0 g of NaCl.

The mass of solvent is the difference between the mass of the solution, 0.250 kg, and the mass of NaCl in the solution, 0.0400 kg: 0.250 kg − 0.0400 kg = 0.210 kg. The molality is determined by converting the mass of NaCl to moles of NaCl and dividing by the mass of the solvent, water, in kilograms.

$$40.0 \text{ g NaCl} \times \frac{1 \text{ mol NaCl}}{58.5 \text{ g NaCl}} = 0.684 \text{ mol NaCl}$$

$$\text{Molality of NaCl} = \frac{0.684 \text{ mol NaCl}}{0.210 \text{ kg H}_2\text{O}} = 3.26 \text{ m}$$

12.15 Mole Fraction

The **mole fraction, X,** of each component in a solution is the number of moles of the component divided by the total number of moles of all components present. The mole fraction of substance A in a solution (or other mixture) of substances A, B, C, etc., is expressed as follows:

$$\text{Mole fraction of A} = X_A = \frac{\text{moles A}}{\text{moles A} + \text{moles B} + \text{moles C} + \cdots}$$

The sum of the mole fractions of all components of a system always equals 1. Like molality, the mole fraction of a component of a solution does not change with temperature.

Calculate the mole fraction of each component in a solution containing 42.0 g CH_3OH, 35 g C_2H_5OH, and 50.0 g C_3H_7OH.   **EXAMPLE 12.8**

The number of moles of each component is calculated first.

$$42.0 \text{ g CH}_3\text{OH} \times \frac{1 \text{ mol CH}_3\text{OH}}{32.0 \text{ g CH}_3\text{OH}} = 1.31 \text{ mol CH}_3\text{OH}$$

$$35 \text{ g C}_2\text{H}_5\text{OH} \times \frac{1 \text{ mol C}_2\text{H}_5\text{OH}}{46.0 \text{ g C}_2\text{H}_5\text{OH}} = 0.76 \text{ mol C}_2\text{H}_5\text{OH}$$

$$50.0 \text{ g C}_3\text{H}_7\text{OH} \times \frac{1 \text{ mol C}_3\text{H}_7\text{OH}}{60.0 \text{ g C}_3\text{H}_7\text{OH}} = 0.833 \text{ mol C}_3\text{H}_7\text{OH}$$

The mole fractions are

$$X_{\text{CH}_3\text{OH}} = \frac{1.31}{1.31 + 0.76 + 0.833} = \frac{1.31}{2.90} = 0.452$$

$$X_{\text{C}_2\text{H}_5\text{OH}} = \frac{0.76}{1.31 + 0.76 + 0.833} = \frac{0.76}{2.90} = 0.26$$

$$X_{\text{C}_3\text{H}_7\text{OH}} = \frac{0.833}{2.90} = 0.287$$

Note that the sum of the mole fractions, 0.452 + 0.26 + 0.287, is 1.00.

EXAMPLE 12.9   Calculate the mole fraction of solute and solvent for a 3.0 m solution of sodium chloride.

A 3.0 m solution of sodium chloride contains 3.0 mol of NaCl dissolved in exactly 1 kg, or 1000 g, of water. Once we know the number of moles of water, we can calculate the mole fractions.

$$1000 \text{ g H}_2\text{O} \times \frac{1 \text{ mol H}_2\text{O}}{18.0 \text{ g H}_2\text{O}} = 55.6 \text{ mol H}_2\text{O}$$

$$X_{\text{NaCl}} = \frac{3.0}{3.0 + 55.6} = 0.051$$

$$X_{\text{H}_2\text{O}} = \frac{55.6}{3.0 + 55.6} = 0.949$$

Note that the sum of the two mole fractions, 0.051 + 0.949, is 1.000.

12.16 Applications of Concentration Calculations

When a solution is diluted, the volume is increased by adding more solvent. Although the concentration is decreased, the total amount of solute is constant.

EXAMPLE 12.10   If 0.750 L of a 5.00 M solution of copper nitrate, $Cu(NO_3)_2$, is diluted to a volume of 1.80 L by adding water (see photograph), what is the molarity of the resulting diluted solution?

Simple addition of water to a 0.750-L sample of a 5.00 M solution of $Cu(NO_3)_2$ (left) produces 1.80 L of a diluted (2.08 M) solution (right).

Since the number of moles of copper nitrate does not change on dilution, the problem can be solved by the following steps.

$$\boxed{\begin{array}{c}\text{Volume of}\\\text{concentrated}\\\text{solution}\end{array}} \longrightarrow \boxed{\begin{array}{c}\text{Moles of}\\\text{Cu(NO}_3)_2\end{array}} \longrightarrow \boxed{\begin{array}{c}\text{Molarity of}\\\text{diluted}\\\text{solution}\end{array}}$$

$$0.750 \text{ L solution} \times \frac{5.00 \text{ mol Cu(NO}_3)_2}{1.00 \text{ L solution}} = 3.75 \text{ mol Cu(NO}_3)_2$$

$$\frac{3.75 \text{ mol Cu(NO}_3)_2}{1.80 \text{ L}} = 2.08 \text{ M}$$

The solution was diluted from 5.00 M to 2.08 M.

How many milliliters of water will be required to dilute 11 mL of a 0.45 M acid solution to a concentration of 0.12 M?   **EXAMPLE 12.11**

Again the number of moles of solute does not change. The following steps are necessary to solve this problem.

$$\boxed{\begin{array}{c}\text{Volume of}\\\text{concentrated}\\\text{solution}\end{array}} \longrightarrow \boxed{\begin{array}{c}\text{Moles of}\\\text{acid}\end{array}} \longrightarrow \boxed{\begin{array}{c}\text{Volume of}\\\text{diluted}\\\text{solution}\end{array}} \longrightarrow \boxed{\begin{array}{c}\text{Volume of}\\\text{water}\\\text{added}\end{array}}$$

We convert the volume to liters and solve for the moles of acid present.

$$11 \text{ mL} \times \frac{1 \text{ L}}{1000 \text{ mL}} = 1.1 \times 10^{-2} \text{ L}$$

$$1.1 \times 10^{-2} \text{ L solution} \times \frac{0.45 \text{ mol acid}}{1.00 \text{ L solution}} = 4.95 \times 10^{-3} \text{ mol acid}$$

Rearrangement of the expression for molarity (molarity = moles/liters) gives the following expression.

$$\text{Liters} = \frac{\text{moles}}{\text{molarity}}$$

$$\text{Liters of dilute solution} = 4.95 \times 10^{-3} \text{ mol acid} \times \frac{1.00 \text{ L solution}}{0.12 \text{ mol acid}}$$

$$= 4.1 \times 10^{-2} \text{ L solution}$$

We convert to milliliters.

$$4.1 \times 10^{-2} \text{ L} \times \frac{1000 \text{ mL}}{1.00 \text{ L}} = 41 \text{ mL solution}$$

The final volume of the solution minus the original volume is equal to the volume of water added in the dilution.

$$41 \text{ mL} - 11 \text{ mL} = 30 \text{ mL}$$

EXAMPLE 12.12   A sulfuric acid solution containing 571.6 g of H_2SO_4 per liter of solution at 20°C has a density of 1.3294 g/mL. Calculate (a) the molarity, (b) the molality, (c) the percent by mass of H_2SO_4, and (d) the mole fractions for the solution.

(a)
$$571.6 \text{ g } H_2SO_4 \times \frac{1 \text{ mol } H_2SO_4}{98.08 \text{ g } H_2SO_4} = 5.828 \text{ mol } H_2SO_4$$

$$\text{Molarity} = \frac{5.828 \text{ mol } H_2SO_4}{1.000 \text{ L}} = 5.828 \text{ M}$$

(b) Since we know the number of moles of H_2SO_4 in 1 L of solution, to calculate the molality we need to find the mass of water in 1 L of solution. The following steps are required.

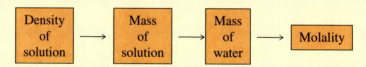

The mass of 1 L of solution is given by rearranging the expression for density (density = mass/volume).

$$\text{Density} \times \text{volume} = \text{mass}$$

$$\frac{1.3294 \text{ g}}{1 \text{ mL}} \times 1000 \text{ mL} = 1329.4 \text{ g}$$

Thus 1 L of solution weighs 1329.4 g and contains 571.6 g of H_2SO_4. The mass of water is therefore

$$1329.4 \text{ g} - 571.6 \text{ g} = 757.8 \text{ g (or 0.7578 kg)}$$

$$\text{Molality} = \frac{5.828 \text{ mol } H_2SO_4}{0.7578 \text{ kg } H_2O} = 7.691 \text{ m}$$

(c) The solution contains 571.6 g of H_2SO_4 in 1329.4 g of solution.

$$\text{Percent by mass} = \frac{571.6 \text{ g } H_2SO_4}{1329.4 \text{ g solution}} \times 100 = 43.00\% \ H_2SO_4 \text{ (by mass)}$$

(d) The number of moles of water present in 1 L of the solution is given by

$$757.8 \text{ g } H_2O \times \frac{1 \text{ mol}}{18.02 \text{ g } H_2O} = 42.05 \text{ mol } H_2O$$

$$X_{H_2SO_4} = \frac{\text{mol } H_2SO_4}{\text{mol } H_2SO_4 + \text{mol } H_2O}$$

$$= \frac{5.828}{5.828 + 42.05} = \frac{5.828}{47.88} = 0.1217$$

$$X_{H_2O} = \frac{42.05}{47.88} = 0.8782$$

Note that the sum of the mole fractions, 0.1217 + 0.8782, is 0.9999 (or 1.0000 within the customary uncertainty of 1 in the last significant figure).

Colligative Properties of Solutions

12.17 Lowering the Vapor Pressure of the Solvent

The freezing temperature of a solution, the vapor pressure of the solvent above a solution, and the boiling temperature of a solution change as the concentration of solute particles changes. Interestingly, however, the changes are independent of the nature (kind, size, and charge) of the solute particles, provided that the solution approximates ideal behavior (Section 12.9). The changes depend only on the *concentration* of solute particles. For example, solutions of 1 mole of solid sugar ($C_{12}H_{22}O_{11}$) dissolved in 1 kilogram of water, 1 mole of liquid ethylene glycol [$C_2H_4(OH)_2$, antifreeze] in 1 kilogram of water, and 1 mole of gaseous nitrous oxide (N_2O) in 1 kilogram of water each begin to freeze at $-1.86°C$. A solution containing 0.5 mole of ammonium ions (NH_4^+) and 0.5 mole of chloride ions (Cl^-), resulting from the dissolution of ammonium chloride in 1 kilogram of water, also freezes at $-1.86°C$; this solution contains a total of 1 mole of dissolved particles. The lowering of the freezing temperature of a solvent due to the presence of a solute and the other properties of solutions that depend only upon the concentration of solute species are called **colligative properties.**

When a nonvolatile substance (or one with such a low vapor pressure that we can disregard it) is dissolved, the vapor pressure of the solvent is lowered. Sugar (and most other solids) are nonvolatile; thus the vapor pressure of an aqueous sugar solution at 20°C is less than that of pure water at 20°C.

The vapor pressure of a liquid is determined by the frequency of escape of molecules from its surface. In a sugar solution both sugar and water molecules are found at the surface. Consequently, the number of water molecules at the surface is less than in pure water; thus the frequency of escape of water molecules, and hence the vapor pressure, is lower than in pure water.

For an ideal solution the decrease in vapor pressure is proportional to the ratio of the number of solute molecules to the total number of solute and solvent molecules. The greater the number of solute molecules, the lower the vapor pressure of the solution. These considerations are summed up in **Raoult's law: The vapor pressure of the solvent in an ideal solution, P_{solv}, is equal to the mole fraction of the solvent, X_{solv}, times the vapor pressure of the pure solvent, P^0_{solv}.**

$$P_{solv} = X_{solv}P^0_{solv} \qquad (1)$$

As was noted in Section 12.9, solutions of chemically similar pairs of substances often closely approximate ideal behavior. Dilute solutions also tend to exhibit ideal behavior. The decrease in vapor pressure, ΔP, of an ideal solution, compared to that of the pure solvent, is equal to the mole fraction of solute, X_{solute}, times the vapor pressure of pure solvent, P^0_{solv}.

$$\Delta P = X_{solute}P^0_{solv} \qquad (2)$$

Calculate the vapor pressure of an ideal solution of 92.1 g of glycerin, $C_3H_8O_3$, in 184 g of ethanol, C_2H_5OH, at 40°C. The vapor pressure of pure ethanol at 40°C is 135.3 torr. Glycerin is essentially nonvolatile at this temperature.

EXAMPLE 12.13

46.1 g/m   4 mol
                1 mol

We can find the vapor pressure of ethanol from Equation (1), $P_{solv} = X_{solv}P^0_{solv}$, but first we need to determine its mole fraction.

$$92.1 \text{ g C}_3\text{H}_8\text{O}_3 \times \frac{1 \text{ mol C}_3\text{H}_8\text{O}_3}{92.1 \text{ g C}_3\text{H}_8\text{O}_3} = 1.00 \text{ mol C}_3\text{H}_8\text{O}_3$$

$$184 \text{ g C}_2\text{H}_5\text{OH} \times \frac{1 \text{ mol C}_2\text{H}_5\text{OH}}{46.0 \text{ g C}_2\text{H}_5\text{OH}} = 4.00 \text{ mol C}_2\text{H}_5\text{OH}$$

$$X_{\text{C}_2\text{H}_5\text{OH}} = \frac{4.00}{1.00 + 4.00} = 0.800$$

Now we can calculate the vapor pressure of ethanol.

$$P_{\text{C}_2\text{H}_5\text{OH}} = X_{\text{C}_2\text{H}_5\text{OH}}P^0_{\text{C}_2\text{H}_5\text{OH}} = 0.800 \times 135.5 \text{ torr} = 108 \text{ torr}$$

The vapor pressure of the solvent (ethanol) has been lowered from 135.3 torr to 108 torr by adding the glycerin. A change of 27.1 torr can be calculated from Equation (2).

$$\Delta P = X_{solute}P^0_{solv} = 0.200 \times 135.3 \text{ torr} = 27.1 \text{ torr}$$

12.18 Elevation of the Boiling Point of the Solvent

A liquid is at its boiling point when its vapor pressure equals the external pressure on its surface (Section 11.3). Adding a solute lowers the vapor pressure of a liquid, so a higher temperature is needed to increase the vapor pressure to the point where the solution boils. The elevation of the boiling point of the solvent in a solution does not depend on the kind of solute, provided that it is a nonvolatile substance that does not dissociate into ions. The elevation of the boiling point is the same when solutions of the same concentration are considered. For example, 1 mole of sucrose, $C_{12}H_{22}O_{11}$, and 1 mole of glucose, $C_6H_{12}O_6$, each dissolved in 1 kilogram of water, have the same molal concentration and form solutions that have the same boiling point, 100.512°C at 760 torr, an increase of 0.512° above the boiling point of pure water. One mole of any nonelectrolyte dissolved in 1 kilogram of water raises the boiling point by 0.512°C. The difference between the boiling point of a dilute solution of a nonelectrolyte and the boiling point of the pure solvent is directly proportional to the *molal* concentration of the solute. Thus 0.200 mole of a nonelectrolyte dissolved in 1 kilogram of water (a 0.200 m solution) will increase the boiling point by

$$0.200 \times 0.512°C = 0.102°C$$

The change in boiling point of a dilute solution, ΔT, from that of the pure solvent is given by the expression

$$\Delta T = K_b \text{ m} \qquad (1)$$

where m is the molal concentration of the solute in the solvent and K_b is the increase in boiling point for a 1 m solution. K_b is called the **molal boiling-point elevation constant.** The values of K_b for several solvents are listed in Table 12.2; you can see that the value of K_b varies for each solvent.

Table 12.2 Boiling points, freezing points, and molal boiling-point
elevation and freezing-point depression constants for several
solvents

Solvent	Boiling point, °C (760 torr)	K_b(°C/m)	Freezing point, °C	K_f(°C/m)
Water	100.0	0.512	0	1.86
Acetic acid	118.1	3.07	16.6	3.9
Benzene	80.1	2.53	5.48	5.12
Chloroform	61.26	3.63	−63.5	4.68
Nitrobenzene	210.9	5.24	5.67	8.1

It should be emphasized that the extent to which the vapor pressure of a solvent is lowered and the boiling point is elevated depends on the number of solute particles present in a given amount of solvent, not on the mass or size of the particles. A mole of sodium chloride forms 2 moles of ions in solution and causes nearly twice as great a rise in boiling point as does 1 mole of nonelectrolyte. One mole of sugar contains 6.022×10^{23} particles (as molecules), whereas 1 mole of sodium chloride contains $2 \times 6.022 \times 10^{23}$ particles (as ions). Calcium chloride, $CaCl_2$, which consists of three ions, causes nearly three times as great a rise in boiling point as does sugar. Section 12.25 explains why the elevation is not exactly twice (for NaCl) or exactly three times (for $CaCl_2$) that of the boiling-point elevation for a nonelectrolyte.

How much does the boiling point of water change when 1.00 g of glycerin, $C_3H_8O_3$, is dissolved in 47.8 g of water? **EXAMPLE 12.14**

Since the change in boiling point is proportional to the molal concentration of glycerin, we first calculate its molal concentration.

$$1.00 \text{ g } C_3H_8O_3 \times \frac{1 \text{ mol } C_3H_8O_3}{92.1 \text{ g } C_3H_8O_3} = 1.09 \times 10^{-2} \text{ mol } C_3H_8O_3$$

$$\frac{1.09 \times 10^{-2} \text{ mol } C_3H_8O_3}{0.0478 \text{ kg } H_2O} = 0.228 \text{ m}$$

The change in boiling point is equal to the molal concentration of the glycerin multiplied by K_b (0.512°C/m for water, Table 12.2).

$$\Delta T = K_b \text{ m} = 0.512°C/m \times 0.228 \text{ m} = 0.117°C$$

What is the boiling point of a solution of 92.1 g of iodine, I_2, in 800.0 g of chloroform, $CHCl_3$, assuming that the iodine is nonvolatile? **EXAMPLE 12.15**

First we calculate the molality of iodine in the solution since the change in boiling point of a dilute solution is proportional to m.

3.63 °C/m

$$92.1 \text{ g } I_2 \times \frac{1 \text{ mol } I_2}{253.8 \text{ g } I_2} = 0.363 \text{ mol } I_2$$

$$\frac{0.363 \text{ mol } I_2}{0.8000 \text{ kg } CHCl_3} = 0.454 \text{ m}$$

The value of the molal boiling-point elevation constant for chloroform (Table 12.2) is 3.63°C/m. Thus, for a 0.4536 m solution,

$$\Delta T = K_b\ m = 3.63°C/m \times 0.454\ m = 1.65°C$$

Since the boiling point of chloroform, 61.26°C, is raised by 1.65°C, the boiling point of the solution will be

$$61.26°C + 1.65°C = 62.91°C$$

12.19 Fractional Distillation

In many cases pure solvent may be recovered from a solution by distillation (Section 11.4). If the solute is nonvolatile, an apparatus like the one in Fig. 11.10 may be used. Distillation makes use of the facts that heating a liquid speeds up its rate of evaporation and increases its vapor pressure and that cooling a vapor favors condensation.

The boiling point of a solution is the temperature at which the total vapor pressure of the mixture is equal to the atmospheric pressure. The total vapor pressure of a solution composed of two volatile substances depends on the concentration and the vapor pressure of each substance and will be

1. between the vapor pressures of the pure components,
2. less than the vapor pressure of either pure component, or
3. greater than that of either pure component.

Usually the total vapor pressure, and thus the boiling point, of a mixture of two liquids lies between those of the two components.

When a solution of type 1 is distilled, the vapor produced at the boiling point is richer in the lower-boiling component (the component with the higher vapor pressure) than was the original mixture. When this vapor is condensed, the resulting distillate contains more of the lower-boiling component than did the original mixture. This means that the composition of the boiling mixture constantly changes, the amount of the lower-boiling component of the mixture constantly decreases, and the boiling point rises as distillation continues. The vapor (and the distillate) contains more and more of the higher-boiling component and less and less of the lower-boiling component as distillation proceeds. Changing the receiver at intervals yields successive fractions, each one increasingly richer in the less volatile (higher-boiling) component. If this process of **fractional distillation** is repeated several times, relatively pure samples of the two liquids may be obtained.

Fractionating columns have been devised that achieve separation of liquids that would require a great number of simple fractional distillations of the type just described. Crude oil, a complex mixture of hydrocarbons, is separated into its components by fractional distillation on an enormous scale (Fig. 12.16).

The vapor pressure of a solution of nitric acid and water is less than that of either component (a solution of type 2). When a dilute solution is distilled, the first fraction of distillate consists mostly of water, because water has a higher vapor pressure than

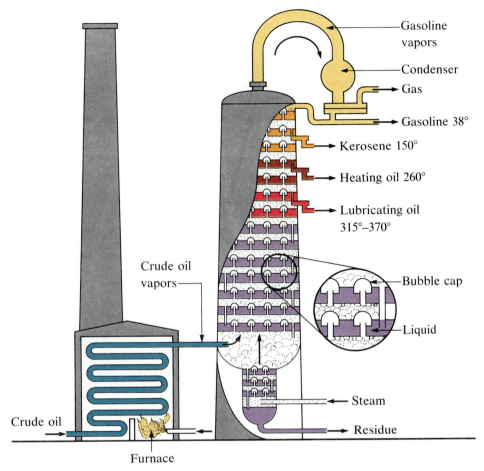

Gasoline vapors

Condenser

Gas

Gasoline 38°

Kerosene 150°

Heating oil 260°

Lubricating oil 315°–370°

Bubble cap

Liquid

Crude oil vapors

Steam

Residue

Crude oil

Furnace

Figure 12.16

Fractional distillation of crude oil. Oil heated to about 425°C in the furnace vaporizes when it enters the tower at its base. The vapors rise through bubble caps in a series of trays in the tower. As the vapors gradually cool, fractions of higher, then of lower, boiling points condense to liquids and are drawn off. The fraction of highest boiling point is drawn off at the bottom as a residue. It is heavy fuel oil. In modern refineries these fractions, which still consist of mixtures of hydrocarbons, are further processed.

nitric acid under distillation conditions. As distillation is continued, the solution remaining in the distilling flask becomes richer in nitric acid. When a concentration of 68% HNO_3 by mass is reached, the solution boils at a constant temperature of 120.5°C (at 760 torr). At this temperature, the solution and the vapor have the same composition, and the solution distills without any further change in composition.

If a nitric acid solution more concentrated than 68% is distilled, the vapor first formed contains a large amount of HNO_3. The solution that remains in the distilling flask contains a greater percentage of water than at first, and the concentration of the nitric acid in the distilling flask decreases as the distillation is continued. Finally, a concentration of 68% HNO_3 is reached, and the solution again boils at the constant temperature of 120.5°C (at 760 torr). The 68% solution of nitric acid is referred to as a **constant boiling solution.**

Solutions of both type 2 and type 3 produce constant boiling solutions when distilled. A constant boiling solution forms a vapor with the same composition as the solution; thus it distills without a change in concentration. Constant boiling solutions are also called **azeotropic mixtures.** Other common and important substances that form azeotropic mixtures with water are HCl (20.24%, 110°C at 760 torr) and H_2SO_4 (98.3%, 338°C at 760 torr).

12.20 Depression of the Freezing Point of the Solvent

Solutions freeze at lower temperatures than pure liquids. We use solutions of antifreezes like ethylene glycol in automobile radiators because they freeze at lower temperatures than pure water. Sea water, with its large salt content, freezes at a lower temperature than fresh water.

The freezing point of a solution or of a pure liquid is the temperature at which the solid and liquid are in equilibrium and at which both have the same vapor pressure. For example, pure water and ice have the same vapor pressure at 0°C. When aqueous solutions freeze, the solid that separates is almost always pure ice. Thus the vapor pressure of the ice is not affected by the presence of the solute. However, the water in the solution has a lower vapor pressure than ice at 0°C. If the temperature falls, the vapor pressure of the ice decreases more rapidly than does that of the water in the solution, and at some temperature below 0°C the ice and the water have the same vapor pressure. This is the temperature at which the solution and ice again are in equilibrium; it is the freezing point of the solution.

If ice and an aqueous solution are placed in contact and kept at 0°C (or at some temperature above the freezing point of the solution), the ice melts. This property permits the use of sodium chloride and calcium chloride to melt ice on streets and highways (Fig. 12.17).

One mole of a nonelectrolyte, such as sucrose, glycerin, ethylene glycol, or alcohol, when dissolved in 1 kilogram of water, gives a solution that freezes at −1.86°C. In general, the difference, ΔT, between the freezing point of a pure solvent and the freezing point of a solution of a nonelectrolyte dissolved in that solvent is directly proportional to the *molal* concentration of the solute.

$$\Delta T = K_f\, m$$

The constant K_f, the **molal freezing-point depression constant,** is the change in freezing point for a 1 m solution and varies for each solvent. Values of K_f for several solvents are listed in Table 12.2.

Figure 12.17

Salt, calcium chloride, or a mixture of the two is used to melt ice.

EXAMPLE 12.16

What is the freezing point of the solution of 92.1 g of I_2 in 800.0 g of $CHCl_3$ described in Example 12.15?

The molal concentration of I_2 was shown to be 0.454 m. The value of the molal freezing-point depression constant, K_f, in Table 12.2 is 4.68°C/m. Thus for a 0.454 m solution

$$\Delta T = K_f \, m = 4.68°C/m \times 0.454 \, m = 2.12°C$$

The freezing point of the solution will be 2.12° lower than that of pure $CHCl_3$, or

$$\text{Freezing point of solution} = -63.5°C - 2.12°C = -65.6°C$$

A mole of sodium chloride in 1 kilogram of water will show nearly twice the freezing-point depression produced by a molecular compound. Each individual ion produces the same effect on the freezing point as a single molecule does. In Section 12.25 we shall consider why the lowering produced by sodium chloride is not exactly twice that produced by a similar amount of a nonelectrolyte.

EXAMPLE 12.17

Assume that the ions in calcium chloride, $CaCl_2$, each have the same effect on the freezing point of water as a nonelectrolyte molecule. Calculate the difference in freezing points of water and of a solution of 0.724 g of $CaCl_2$ in 175 g of water.

The difference in freezing points is the freezing point depression and is proportional to the molal concentration of dissolved species, in this case, Ca^{2+} and Cl^- ions. One mol of $CaCl_2$ contains 3 mol of ions (1 mol of Ca^{2+} ions and 2 mol of Cl^- ions), so the molality of ions in the solution is three times greater than the molality of $CaCl_2$. The following chain of calculations is one way to determine the difference in freezing points.

Mass of $CaCl_2$	→	Moles of $CaCl_2$	→	Moles of ions	→	Molality of solution	→	Freezing point depression

$$0.724 \text{ g } CaCl_2 \times \frac{1 \text{ mol } CaCl_2}{111.0 \text{ g } CaCl_2} = 6.52 \times 10^{-3} \text{ mol } CaCl_2$$

$$6.52 \times 10^{-3} \text{ mol } CaCl_2 \times \frac{3 \text{ mol ions}}{1 \text{ mol } CaCl_2} = 1.96 \times 10^{-2} \text{ mol ions}$$

$$\text{Molality of ions} = \frac{1.96 \times 10^{-2} \text{ mol ions}}{0.175 \text{ kg } H_2O} = 0.112 \text{ m}$$

$$\Delta T = K_f \, m = 1.86°C/m \times 0.112 \, m = 0.208°C$$

12.21 Phase Diagram for an Aqueous Solution of a Nonelectrolyte

Figure 12.18 contains a phase diagram (in red) for an aqueous solution of a nonelectrolyte, such as sucrose, $C_{12}H_{22}O_{11}$. The phase diagram for pure water (Section 11.9) is included as a blue line for comparison. You can see that the freezing point for the solution is lower than that for pure water (the red line, separating solid and liquid

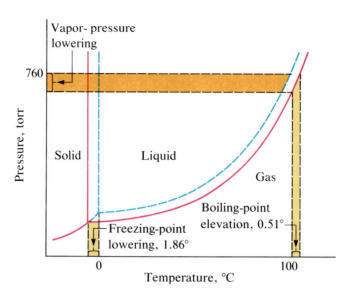

Figure 12.18
Phase diagram for a 1 m aqueous solution of a non-electrolyte (red solid lines) compared to that for pure water (blue dashed lines). The diagram is not to scale in order to show more clearly the differences between the solution and pure water.

states, is displaced to the left of the blue one). Correspondingly, the higher boiling point for the solution is shown by the displacement of the solid line separating the liquid and gas states to the right of the dashed one. The decrease in vapor pressure of the solution, at any given temperature, is indicated by the vertical distance (shown in orange) between the dashed line and the solid line. On the diagram the freezing-point depression and the boiling-point elevation are the horizontal distances (shown in yellow) between the broken line and the solid line near 0°C and near 100°C, respectively, at a pressure of 760 torr.

12.22 Osmosis and Osmotic Pressure of Solutions

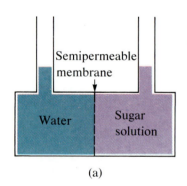

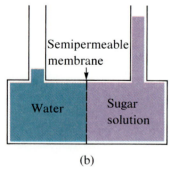

Figure 12.19
Apparatus for demonstrating osmosis. The levels are equal at the start (a), but at equilibrium (b) the level of the sugar solution is higher due to the net transfer of water molecules into it.

When a solution and its pure solvent are separated by a semipermeable membrane (one through which the solvent can pass but not the solute), the pure solvent will diffuse through the membrane and dilute the solution. This process is known as **osmosis.** Actually, the solvent diffuses through the membrane in both directions simultaneously; however, its rate of diffusion is greater from the pure solvent to the solution than in the opposite direction, so there is an increase in the number of solvent molecules in the solution.

When a solution and a pure solvent are separated by a semipermeable membrane as shown in Fig. 12.19, the volume of the solution increases because the net movement of solvent molecules is from the pure solvent to the solution. The liquid in the tube above the solution slowly rises. If the membrane is strong enough to withstand the pressure and if the tube is long enough, the liquid will rise until its hydrostatic pressure (due to the weight of the column of solution in the tube) is great enough to prevent the further osmosis of solvent molecules into the solution. The pressure required to stop the osmosis from a pure solvent into a solution is called the **osmotic pressure, π,** of the solution. The osmotic pressure of a dilute solution can be calculated from the expression

$$\pi = MRT$$

where M is the molar concentration of the solute, R is the gas constant, and T is the temperature of the solution on the Kelvin scale.

osmosis
solvent → solution

EXAMPLE 12.18

What is the osmotic pressure in atmospheres of a 0.30 M solution of glucose in water that is used for intravenous infusion at body temperature, 37°C?

.3 × .0821 · 310

The osmotic pressure, π, in atmospheres can be found using the formula $\pi = MRT$ with T on the Kelvin scale (310 K) and with a value of R that includes the unit of atm (0.08206 L atm/mol K).

7.6 atm

$$\pi = MRT$$
$$\pi = 0.30 \;\cancel{\text{mol/L}} \times 0.08206 \;\cancel{\text{L}} \; \text{atm}/\cancel{\text{mol K}} \times 310 \;\cancel{\text{K}} = 7.6 \; \text{atm}$$

If a solution is placed in an apparatus like the one in Fig. 12.20, applying pressure greater than the osmotic pressure of the solution reverses the osmosis and increases the volume of the pure solvent. This technique of **reverse osmosis** is used for desalting sea water. Plants using this principle produce water for the city of Key West, and other parts of the world.

The effects of osmosis are particularly evident in biological systems, since cells are surrounded by semipermeable membranes. Carrots and celery that have become limp due to loss of water to the atmosphere can be made crisp again by placing them in water. Water moves into the carrot or celery cells by osmosis. A cucumber placed in a concentrated salt solution loses water by osmosis and becomes a pickle. Osmosis can also affect animal cells. Solute concentrations are particularly important when solutions are injected into the body. Solutes in body cell fluids and blood serum give these solutions an osmotic pressure of approximately 7.7 atmospheres. Solutions injected into the body must have the same osmotic pressure as blood serum; that is, they should be **isotonic** with blood serum. If a less concentrated solution, a **hypotonic solution,** is injected in sufficient quantity to dilute the blood serum, water from the diluted serum will pass into the blood cells by osmosis, causing the cells to expand and rupture. If a more concentrated solution, a **hypertonic solution,** is injected, the cells will lose water to the more concentrated solution, shrivel, and possibly die.

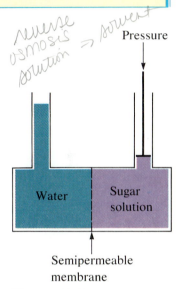

reverse osmosis solution → solvent

Figure 12.20

Application of a pressure greater than the osmotic pressure will reverse the osmosis.

12.23 Determination of Molecular Weights of Substances in Solution

The effect of a solute on the freezing point, boiling point, vapor pressure, or osmotic pressure of a solution can be measured. Changes in these properties are directly proportional to the concentration of solute present, so the molecular weight of the solute can be determined from the change.

EXAMPLE 12.19

A solution of 35.7 g of an organic nonelectrolyte in 220.0 g of chloroform has a boiling point of 64.5°C. What is the molecular weight of this organic compound?

This problem requires the following steps.

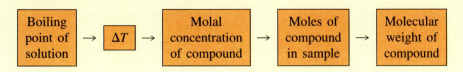

From the boiling point of pure chloroform (61.26°C, Table 12.2), we can calculate ΔT, the increase in boiling temperature.

$$\Delta T = 64.5°C - 61.26°C = 3.2°C$$

From K_b for chloroform (3.63°C/m, Table 12.2), we can calculate the molal concentration of the electrolyte (mol solute/mass solvent) by rearranging the equation $\Delta T = K_b m$.

$$m = \frac{\Delta T}{K_b} = \frac{3.2°C}{3.63°C/m} = 0.88 \ m$$

The number of moles of solute in 0.2200 kg (220.0 g) of solvent is then calculated as follows:

$$\text{Moles of solute} = \frac{0.88 \text{ mol solute}}{1.00 \text{ kg solvent}} \times 0.2200 \text{ kg solvent}$$

$$= 0.19 \text{ mol solute}$$

We can then calculate the molar mass.

$$\text{Molar mass} = \frac{35.7 \text{ g}}{0.19 \text{ mol}} = 180 \text{ g/mol, or } 1.8 \times 10^2 \text{ g/mol}$$

A molar mass of 180 g/mol corresponds to a molecular weight of 180. (Note that only two significant figures are justified, despite the fact that all data are expressed to at least three significant figures. Why is this?)

EXAMPLE 12.20   If 4.00 g of a certain nonelectrolyte is dissolved in 55.0 g of benzene, the resulting solution freezes at 2.32°C. Calculate the molecular weight of the nonelectrolyte.

The steps for this problem are

Freezing point of solution	→	ΔT	→	Molal concentration of compound	→	Moles of compound in sample	→	Molecular weight of compound

According to Table 12.2, K_f for benzene is 5.12°C/m and its freezing point is 5.48°C. Since the freezing point of the solution is 2.32°C, the 4.00 g of solute has lowered the freezing point from 5.48°C to 2.32°C. Thus $\Delta T = 3.16°C$.

The molality of the solution is

$$m = \frac{\Delta T}{K_f} = \frac{3.16°C}{5.12°C/m} = 0.617 \ m$$

The number of moles of solute in 0.0550 kg (55.0 g) of solvent is

$$\text{Moles of solute} = \frac{0.617 \text{ mol}}{1.000 \text{ kg solvent}} \times 0.0550 \text{ kg solvent}$$

$$= 0.0339 \text{ mol}$$

The molar mass can then be calculated:

$$\text{Molar mass} = \frac{4.00 \text{ g}}{0.0339 \text{ mol}} = 118 \text{ g/mol}$$

$$\text{Mol. wt} = 118$$

One liter of an aqueous solution containing 20.0 g of hemoglobin has an osmotic pressure of 5.9 torr at 22°C. What is the molecular weight of hemoglobin?

EXAMPLE 12.21

The steps for this problem are

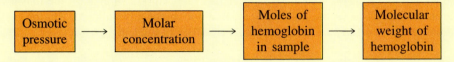

From $\pi = MRT$, we can relate the osmotic pressure to the molar concentration provided the pressure is expressed in atmospheres to match the units in R (0.08206 L atm/mol K) and the temperature is expressed on the Kelvin scale ($T = 273 + 22 = 295$ K).

$$\pi = 5.9 \text{ torr} \times \frac{1 \text{ atm}}{760 \text{ torr}} = 7.8 \times 10^{-3} \text{ atm}$$

Rearrangement of $\pi = MRT$ gives $M = \pi/RT$.

$$M = \frac{7.8 \times 10^{-3} \text{ atm}}{(0.08206 \text{ L atm/mol K})(295 \text{ K})} = 3.2 \times 10^{-4} \text{ mol/L} = 3.2 \times 10^{-4} \text{ M}$$

Since the volume of the solution containing the 20.0 g of hemoglobin is 1 L, 20.0 g of hemoglobin is equal to 3.2×10^{-4} mol.

$$\text{Molar mass} = \frac{20.0 \text{ g}}{3.2 \times 10^{-4} \text{ mol}} = 62,000 \text{ g/mol}$$

$$\text{Mol. wt} = 62,000$$

12.24 The Effect of Electrolytes on Colligative Properties

The effect of nonelectrolytes on the colligative properties of a solution is dependent only on the number, not on the kind, of particles dissolved. For example, 1 mole of any nonelectrolyte dissolved in 1 kilogram of water produces the same lowering of the freezing point as does 1 mole of any other nonelectrolyte because 1 mole of any nonelectrolyte contains 6.022×10^{23} molecules. However, the lowering of the freezing point produced by 1 mole of a strong electrolyte is much greater than that from 1 mole of a nonelectrolyte, because the electrolyte ionizes when it dissolves. When 1 mole of sodium chloride is dissolved in 1 kilogram of water, the solution freezes at −3.37°C. The freezing point is lowered 1.81 times as much as for a nonelectrolyte. This illustrates the fact that 1 mole of an electrolyte produces more than 6.022×10^{23} solute particles. Almost all acids, bases, and salts behave this way when dissolved.

12.25 Ion Activities

If the ions of sodium chloride were completely separated in aqueous solution, it would lower the freezing point and raise the boiling point of the solvent twice as much as would an equal molal concentration of a nonelectrolyte, since it gives 2 moles of ions per mole of compound. However, 1 mole of sodium chloride lowers the freezing point of water only 1.81 times as much. A similar discrepancy occurs for the boiling-point elevation. Apparently, sodium chloride (and other strong electrolytes) are not completely dissociated in solution.

Peter J. W. Debye and Erich Hückel proposed a theory to explain the apparent incomplete ionization of strong electrolytes. They suggested that although interionic attraction in an aqueous solution is very greatly reduced by hydration of the ions and the insulating action of the polar solvent, it is not completely nullified. The residual attractions prevent the ions from behaving as totally independent particles (Fig. 12.21).

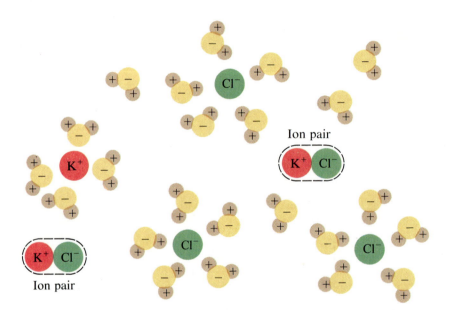

Figure 12.21

Diagrammatic representation of the various species thought to be present in a solution of potassium chloride.

In some cases a positive and negative ion may actually touch, giving a solvated unit called an **ion pair.** Thus the **activity,** or the effective concentration, of any particular kind of ion is less than that indicated by the actual concentration. This is evident in the colligative properties of the solution and in the amount of electric current conducted by it. Ions become more and more widely separated the more dilute the solution, and the residual interionic attractions become less and less. Thus in extremely dilute solutions the effective concentrations of the ions (their activities) are essentially equal to the actual concentrations.

Colloid Chemistry

Solutions are dispersions of discrete molecules or ions of a solute. We shall now consider dispersions of particles that are significantly larger than simple molecules.

12.26 Colloidal Matter

Although the mixture obtained when powdered starch is heated with water is not homogeneous, the particles of insoluble starch do not settle out but remain in suspension indefinitely. Such a system is called a **colloidal dispersion;** the finely divided starch is called the **dispersed phase,** and the water is called the **dispersion medium.**

The term **colloid**—from the Greek words *kolla,* meaning glue, and *eidos,* meaning like—was first used in 1861 by Thomas Graham to classify substances such as starch and gelatin. We now know that colloidal properties are not limited to this class of substances; any substance can exist in colloidal form. Many colloidal particles are aggregates of hundreds, or even thousands, of molecules, but others, such as proteins, viruses, and polymer molecules, consist of a single large molecule. Viruses are giant molecules with molecular weights ranging from several hundred thousands to billions. Tobacco mosaic virus has a molecular weight of about 40,000,000. Proteins and synthetic polymer molecules may have weights ranging from a few thousands to many millions.

Colloids may be dispersed in a gas, a liquid, or a solid, and the dispersed phase may be a gas, a liquid, or a solid. However, a gas dispersed in another gas is not a colloidal system, because the particles are of molecular dimensions. A classification of colloidal systems is given in Table 12.3.

Table 12.3 Colloidal systems

Dispersed phase	Dispersion medium	Examples	Common name
Solid	Gas	Smoke, dust	Solid aerosol
Solid	Liquid	Starch suspension, some inks, paints, milk of magnesia	Sol
Solid	Solid	Colored gems, some alloys	Solid sol
Liquid	Gas	Clouds, fogs, mists, sprays	Liquid aerosol
Liquid	Liquid	Milk, mayonnaise, butter	Emulsion
Liquid	Solid	Jellies, gels, opal (SiO_2 and H_2O), pearl ($CaCO_3$ and H_2O)	Solid emulsion
Gas	Liquid	Foams, whipped cream, beaten egg whites	Foam
Gas	Solid	Pumice, floating soaps	

12.27 Preparation of Colloidal Systems

A colloidal system is prepared by producing particles of colloidal dimensions and distributing these particles through the dispersion medium. Particles of colloidal size are formed by two methods:

1. **dispersion methods,** that is, the subdivision of larger particles. For example, paint pigments are produced by dispersing large particles by grinding in special mills.
2. **condensation methods,** that is, growth from smaller units, such as molecules or ions. For example, clouds form when water molecules condense and form very small droplets.

A few solid substances, when brought into contact with water, disperse spontaneously and form colloidal systems. Gelatin, glue, and starch behave in this manner, and are said to undergo **peptization.** The particles are already of colloidal size; the water simply disperses them. Some atomizers produce colloidal dispersions. Powdered milk with particles of colloidal size is produced by dehydrating milk spray.

An **emulsion** may be prepared by shaking together two immiscible liquids. Agitation breaks one liquid into droplets of colloidal size, which then disperse throughout the other liquid. The droplets of the dispersed phase, however, tend to coalesce, forming large drops, and separation of the liquids into two layers follows. Therefore, emulsions are usually stabilized by substances called **emulsifying agents.** For example, a little soap will stabilize an emulsion of kerosene in water. Milk is an emulsion of butterfat in water with casein as the emulsifying agent. Mayonnaise is an emulsion of oil in vinegar with egg yolk as the emulsifying agent. Oil spills in the ocean are particularly difficult to clean up because the oil and the water form an emulsion.

Condensation methods form colloidal particles by aggregation of smaller particles. If the particles grow beyond the colloidal size range, precipitates form, and no colloidal system results.

Condensation methods often employ chemical reactions. A dark red colloidal suspension of iron(III) hydroxide may be prepared by mixing a concentrated solution of iron(III) chloride with hot water.

$$Fe^{3+} + 3Cl^- + 6H_2O \longrightarrow Fe(OH)_3 + 3H_3O^+ + 3Cl^-$$

A colloidal suspension of arsenic(III) sulfide is produced by the reaction of hydrogen sulfide with arsenic(III) oxide dissolved in water.

$$As_2O_3 + 3H_2S \longrightarrow As_2S_3 + 3H_2O$$

A colloidal gold sol results from the reduction of a very dilute solution of gold chloride by such reducing agents as formaldehyde, tin(II) chloride, or iron(II) sulfate.

$$Au^{3+} + 3e^- \longrightarrow Au$$

Some gold sols prepared by Faraday in 1857 are still perfectly clear.

12.28 The Cleansing Action of Soaps and Detergents

Soaps are made by boiling either fats or oils with a strong base, such as sodium hydroxide. Pioneers made soap by boiling fat with a strongly basic solution made by leaching potassium carbonate, K_2CO_3, from wood ashes with hot water. When animal fat is treated with sodium hydroxide, glycerol and sodium salts of fatty acids such as palmitic, oleic, and stearic acid are formed. The sodium salt of stearic acid, sodium stearate, has the formula $C_{17}H_{35}CO_2Na$ and contains a nonpolar hydrocarbon chain, the $-C_{17}H_{35}$ unit, and an ionic carboxylate group, the $-CO_2^-$ unit.

Detergents (soap substitutes) also contain nonpolar hydrocarbon chains, such as $-C_{12}H_{25}$, and an ionic group, such as a sulfate, $-OSO_3^-$, or a sulfonate, $-SO_3^-$. Soaps form insoluble calcium and magnesium compounds in hard water; detergents form water-soluble products—a definite advantage for detergents.

The cleansing action of soaps and detergents can be explained in terms of the structures of the molecules involved.

Hydrocarbon chain   Ionic end

CH_3 — CH_2 — CH_2 — CH_2 — CH_2 — CH_2 — CH_2 — CH_2 — CH_2 — CH_2 — CH_2 — CH_2 — CH_2 — CH_2 — CH_2 — CH_2 — CH_2 — $CO_2^- Na^+$

Sodium stearate (soap)

CH_3 — CH_2 — CH_2 — CH_2 — CH_2 — CH_2 — CH_2 — CH_2 — CH_2 — CH_2 — CH_2 — $OSO_3^- Na^+$

Sodium lauryl sulfate (detergent)

The hydrocarbon end of a soap or detergent molecule is attracted by dirt, oil, or grease particles, and the ionic end is attracted by water (Fig. 12.22). The result is an orientation of the molecules at the interface between the dirt particles and the water in such a way that the dirt particles become suspended as colloidal particles and are readily washed away.

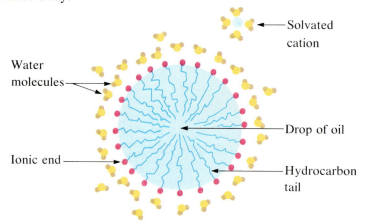

Solvated cation

Water molecules

Drop of oil

Ionic end

Hydrocarbon tail

Figure 12.22

Diagrammatic cross section of an emulsified drop of oil in water with soap or detergent as the emulsifier. The negative ions of the emulsifier are oriented at the interface between the oil particle and water. The positive ions are solvated and move independently through the water. The nonpolar hydrocarbon end of the ion is in oil, and the ionic end ($—CO_2^-$ for soap, $—SO_3^-$ for detergent) is in water.

12.29 Brownian Movement

When a colloidal system against a black background in a darkened space is viewed with a microscope at right angles to an intense beam of light, the colloidal particles look like tiny bright flashes of light in irregular rapid, dancing motion (Fig. 12.23). This motion is called **Brownian movement** after the botanist Robert Brown, who first observed it in 1828. He could not explain it, but we now know that the colloidal particles are so small that bombardment by molecules of the dispersion medium makes them move irregularly. Any particle in suspension is bombarded on all sides by the moving molecules of the dispersion medium. Particles larger than colloidal size do not move because bombardment on one side is likely to be counterbalanced by an equal bombardment on the opposite side. For particles of colloidal size, however, the probability of equal and simultaneous bombardments on opposite sides is slight, so they move irregularly. This movement explains why dispersed particles in colloidal systems do not settle, even though they are denser than the dispersion medium. The kinetic-molecular theory received one of its earliest confirmations as a result of studies of Brownian movement.

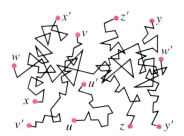

Figure 12.23

A diagram showing how six particles in suspension might move due to unequal bombardment by molecules of the dispersion medium.

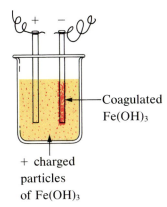

Figure 12.24

Colloidal iron(III) hydroxide is coagulated in an electrolytic cell.

12.30 Electrical Properties of Colloidal Particles

Dispersed colloidal particles are usually electrically charged. A particle of an iron(III) hydroxide sol, for example, does not contain enough hydroxide ions to compensate exactly for the positive charges on the iron(III) ions. Thus each individual colloidal particle bears a positive charge, and the colloidal dispersion consists of charged colloidal particles and hydrated ions of opposite charge, which keep the dispersion electrically neutral. In an electrolytic cell (Fig. 12.24), the dispersed particles move to the negative electrode, the cathode. Since opposite charges attract each other, this is good evidence that the iron(III) hydroxide particles are positively charged. At the cathode the colloidal particles lose their charge and coagulate as a precipitate. All colloidal particles in any one system have charges of the same sign. This helps to keep them dispersed, since like charges repel each other. Most metal hydroxide colloids have positive charges, while most metal sulfides and the metals themselves form negatively charged disperions.

The carbon and dust particles in smoke are often colloidally dispersed and electrically charged. Frederick Cottrell, an American chemist, developed a process to remove these particles. The charged particles are attracted to highly charged electrodes, where they are neutralized and deposited as dust (Fig. 12.25). The process is also important in the recovery of valuable products from the smoke of smelters, furnaces, and kilns.

12.31 Gels

Under certain conditions, the dispersed phase in a colloidal system coagulates so that the whole mass, including the liquid, sets to an extremely viscous body known as a **gel.** For example, a hot aqueous ''solution'' of gelatin sets to a gel on cooling. Because the formation of a gel is accompanied by the taking up of water or some other solvent, the gel is said to be hydrated or solvated. Apparently the fibers of the dispersed substance form a complex three-dimensional network, the interstices being filled with the liquid medium or a dilute solution of the dispersed phase.

Pectin, a carbohydrate from fruit juices, is a gel-forming substance important in jelly making. Silica gel, a colloidal dispersion of hydrated silicon dioxide, is formed when dilute hydrochloric acid is added to a dilute solution of sodium silicate. Canned Heat is a gel made by mixing alcohol and a saturated aqueous solution of calcium acetate. The wall of a living cell is colloidal in character, and within the cell is a gel. In fact, all living tissue is colloidal, and life processes depend on the chemistry of colloids.

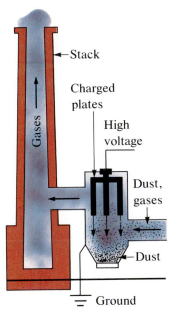

Figure 12.25

In a Cottrell precipitator, positively and negatively charged particles in smoke are precipitated as they pass over the electrically charged plates.

For Review

Summary

A **solution** forms when one substance, a **solute,** dissolves in a second substance, a **solvent,** giving a homogeneous mixture of atoms, molecules, or ions. The solute may be a solid, a liquid, or a gas. The solvent is usually a liquid, although solutions of gases in other gases as well as solutions of gases, liquids, and solids in solids are possible. Substances that dissolve and give solutions that contain ions and conduct electricity are

called **electrolytes.** Electrolytes may be molecular compounds that react with the solvent to give ions, or they may be ionic compounds. **Nonelectrolytes** are substances that dissolve to give solutions of molecules. These solutions do not conduct electricity.

The extent to which a gas dissolves in a liquid solvent is proportional to the pressure of the gas, provided that the gas does not react with the solvent. Generally, the solubility of a gas decreases with increasing temperature. Polar liquids tend to be soluble **(miscible)** in water, whereas nonpolar liquids tend to be insoluble **(immiscible)** in water. Generally, the solubility of solids in water increases with increasing temperature, but there are exceptions.

Solutions that obey **Raoult's law** exactly, **ideal solutions,** form when the average strength of the intermolecular forces of attraction between the solute and solvent is equal to the average strength of the forces of attraction between pure solute molecules and between pure solvent molecules. Ideal solutions form by a **spontaneous process** because the disorder of the molecules of solute and solvent increases when the solution forms. An increase in the average strength of the solute-solvent intermolecular attractive forces, compared to those in the pure solute and solvent, also favors spontaneous formation of a solution. The relative strengths of the solute-solute and solvent-solvent attractions compared to the solute-solvent attractions can be measured by the **heat of solution** of the solute in the solvent. In many cases, the increase in disorder compensates for the decrease in the average strength of the solvent-solute attractions compared to those of the pure components of a solution.

The relative amounts of solute and solvent in a solution can be described quantitatively as the **concentration** of the solution. Units of concentration include **percent composition, molar concentration, molal concentration,** and **mole fraction.**

Properties of a solution that depend only on the concentration of solute particles are called **colligative properties.** They include changes in the vapor pressure, boiling point, and freezing point of the solvent in the solution. The magnitudes of these properties depend only on the total concentration of solute particles in solution, not on the type of particles. The total concentration of solute particles in a solution also affects its **osmotic pressure.** This is the pressure that must be applied to the solution in order to prevent diffusion of molecules of pure solvent through a semipermeable membrane into the solution.

A **colloid** is a suspension of small, insoluble particles, which are usually larger than molecules but small enough not to settle from the dispersion medium. Examples of colloids include smoke, clouds, paints, mayonnaise, and foams.

Key Terms and Concepts

alloy (12.8)
azeotropic mixture (12.19)
boiling-point elevation (12.18)
Brownian movement (12.29)
colligative properties (12.17)
colloid (12.26)
constant boiling solution (12.19)
electrolyte (12.6)
emulsion (12.27)
fractional distillation (12.19)
freezing-point depression (12.20)
gel (12.31)

Henry's law (12.2)
ideal solution (12.9)
ion activity (12.24)
ion-dipole attraction (12.10)
miscibility (12.3)
molality (12.14)
molarity (12.13)
mole fraction (12.15)
nonelectrolyte (12.6)
nonstoichiometric compound
    (12.8)
osmosis (12.22)

osmotic pressure (12.22)
percent composition (12.12)
Raoult's law (12.17)
saturated solution (12.1)
solid solution (12.8)
solubility (12.1)
spontaneous process (12.9)
strong electrolyte (12.6)
supersaturated solution (12.1)
unsaturated solution (12.1)
vapor-pressure lowering (12.17)
weak electrolyte (12.6)

Exercises

The Nature of Solutions

1. How do solutions differ from compounds? From ordinary mixtures?

2. Explain how solutions of (a) water vapor in air, (b) sugar, $C_{12}H_{22}O_{11}$, in water, and (c) hydrogen chloride in water exhibit the principal characteristics of solutions.

3. Why are the majority of chemical reactions most readily carried out in solution?

4. Define the following terms when they are used in connection with solutions:
   - (a) solute
   - (b) solvent
   - (c) weak electrolyte
   - (d) strong electrolyte
   - (e) nonelectrolyte
   - (f) saturated
   - (g) supersaturated
   - (h) unsaturated

5. Describe the effect on the solubility of a gas of (a) the pressure of the gas, (b) the temperature of the solution, and (c) the reaction of the gas with the solvent.

6. At 0°C, 3.36 g of CO_2 at 1.00 atm will dissolve in exactly 1 L of water. (a) How many grams will dissolve at the same temperature if the pressure is 4.00 atm? (b) What is the value of k in Henry's law for CO_2 under these conditions?

   Ans. 13.4 g; 3.36 g/atm or 0.0763 mol/atm

7. In order to prepare supersaturated solutions of most solids, saturated solutions are cooled. Supersaturated solutions of gases are prepared by warming saturated solutions. Explain the reasons for the difference in the two procedures.

8. Why is a solution of perchloric acid, $HClO_4$, in water a good conductor of an electric current when both pure perchloric acid and pure water are not?

9. Explain why ionic compounds that are fused (melted) are good conductors but solid ionic compounds are not.

10. Using the general rules given in Section 12.7, answer the following questions:
    - (a) What common cations might be present in a solution if a white precipitate forms when Na_2SO_4 is added to the solution?
    - (b) The addition of a solid metal sulfide to water produces H_2S and an insoluble solid. What common metal sulfides would exhibit this behavior?
    - (c) What common metal cations are not present in a solution if addition of HCl gives no precipitate?
    - (d) What common metal cations might be present in a solution if no precipitate forms when a solution of Na_2CO_3 is added to it?
    - (e) A solution of $AgNO_3$ was added to a second solution, and no precipitate formed. What common anions might be present in the second solution?

11. Explain why HCl is a nonelectrolyte when dissolved in carbon tetrachloride and an electrolyte when dissolved in water.

12. Suppose you are presented with a clear solution of ammonium nitrate, NH_4NO_3. How could you determine whether the solution is unsaturated, saturated, or supersaturated?

The Process of Dissolution

13. Explain why the ions Na^+ and Cl^- are strongly solvated in water but not in benzene.

14. Why are nonpolar compounds such as benzene and carbon tetrachloride generally poor solvents for ionic compounds?

15. Which intermolecular attractions (Sections 11.11, 11.12, 12.9, 12.10) should be most important in each of the following solutions:
    - (a) NH_3 in water
    - (b) methanol, CH_3OH, in ethanol, C_2H_5OH
    - (c) the strong acid, sulfuric acid, H_2SO_4, in water
    - (d) HCl in benzene, C_6H_6
    - (e) carbon tetrachloride, CCl_4, in Freon, CF_2Cl_2
    - (f) $NaNO_3$ in water
    - (g) the weak acid, acetic acid, CH_3CO_2H, in water

16. Which of the following spontaneous processes occurs with an increase in the disorder of the system?
    - (a) evaporation of methanol, CH_3OH
    - (b) precipitation of $PbSO_4$ from solution
    - (c) condensation of iodine vapor to solid iodine
    - (d) mixing of $CH_4(g)$ and $C_2H_6(g)$
    - (e) dissolution of NH_4NO_3 in water

17. Compare the processes that occur when ethanol, hydrogen chloride, ammonia, and sodium hydroxide dissolve in water. Write equations and prepare sketches showing the form in which each of these compounds is present in its respective solution.

18. Solid water (ice) is soluble in ethanol, C_2H_5OH. Compare the process that occurs when ice dissolves in ethanol with the one that occurs when sodium chloride dissolves in water. (See Section 11.13 for a description of the structure of ice.)

19. Explain, in terms of the intermolecular attractions between solute and solvent, why methanol, CH_3OH, is miscible with water in all proportions, but only 0.11 g of butanol, $CH_3CH_2CH_2CH_2OH$, will dissolve in 100 g of water.

20. Suggest an explanation for the observations that ethanol, C_2H_5OH, is completely miscible with water and that ethanethiol, C_2H_5SH, is soluble only to the extent of 1.5 g/100 mL of water.

Units of Concentration

21. Distinguish between 1 M and 1 m solutions.

22. Is the following statement true or false? In all cases, the molality of a solution has a value that is larger than that of the molarity of the same solution. Explain your answer.

23. There are about 10 g of calcium, as Ca^{2+}, in 1.0 L of milk. What is the molarity of Ca^{2+} in milk?

Ans. 0.25 M

24. Calculate the number of moles and the mass of the solute in each of the following solutions:
    (a) 0.100 L of a 5.04×10^{-3} M solution of cholesterol, $C_{27}H_{46}O$, in serum, the average concentration of cholesterol in human serum

    Ans. 5.04×10^{-4} mol; 0.195 g

    (b) 0.500 L of 0.500 M NH_3, the concentration of NH_3 in household ammonia   *Ans. 0.250 mol; 4.26 g*

    (c) 393.6 mL of 9.087 M ethylene glycol, $C_2H_4(OH)_2$, an antifreeze solution   *Ans. 3.577 mol; 222.0 g*

    (d) 1.00 L of 9.74 M isopropanol, C_3H_7OH, the concentration of isopropanol in rubbing alcohol

    Ans. 9.74 mol; 585 g

    (e) 325 mL of 1.8×10^{-6} M $FeSO_4$, the minimum concentration of iron sulfate detectable by taste in drinking water   *Ans. 5.8×10^{-7} mol; 8.9×10^{-5} g*

25. Calculate the molarity of each of the following solutions:
    (a) 1.82 kg of H_2SO_4 per liter of concentrated sulfuric acid   *Ans. 18.6 M*

    (b) 1.9×10^{-4} g of NaCN per 100 mL, the minimum lethal concentration of sodium cyanide in blood serum   *Ans. 3.9×10^{-5} M*

    (c) 27 g of glucose, $C_6H_{12}O_6$, in 500 mL of solution used for intravenous injection   *Ans. 0.30 M*

    (d) 2.20 kg of formaldehyde, H_2CO, in 5.50 L of a solution used to "fix" tissue samples   *Ans. 13.3 M*

    (e) 0.029 g of I_2 in 0.100 L solution, the solubility of I_2 in water at 20°C   *Ans. 1.1×10^{-3} M*

26. Calculate the molality of each of the following solutions:
    (a) 69 g of sodium hydrogen carbonate (baking soda), $NaHCO_3$, in 1.00 kg of water, a saturated solution at 0°C   *Ans. 0.82 m*

    (b) 583 g of H_2SO_4 in 1.50 kg of water, the acid solution used in an automobile battery   *Ans. 3.96 m*

    (c) 120 g of NH_4NO_3 in 250 g of water, a mixture used to make an instant "ice pack"   *Ans. 6.00 m*

    (d) 0.86 g of NaCl in 100 g of water, a solution of sodium chloride for intravenous injection

    Ans. 0.15 m

    (e) 46.85 g of codeine, $C_{18}H_{21}NO_3$, in 125.5 g of ethanol, C_2H_5OH *Ans. 1.247 m*

    (f) 25 g of iodine, I_2, in 125 g of ethanol, C_2H_5OH

    Ans. 0.79 m

27. Calculate the mole fractions of solute and solvent in each of the solutions in exercise 26.

    Ans. (a) $NaHCO_3$, 0.015; H_2O, 0.985
    (b) H_2SO_4, 0.0666; H_2O, 0.9334
    (c) NH_4NO_3, 0.0974; H_2O, 0.9026
    (d) NaCl, 0.0026; H_2O, 0.9974
    (e) $C_{18}H_{21}NO_3$, 0.05433; C_2H_5OH, 0.94567
    (f) I_2, 0.035; C_2H_5OH, 0.965

28. What mass of concentrated sulfuric acid (a 95% solution of H_2SO_4 by mass) is needed to prepare 500 g of a 10.0% solution of H_2SO_4 by mass?   *Ans. 53 g*

29. What mass of sodium hydroxide (50.2% by mass) is needed to prepare 200.0 g of a 20.0% solution of NaOH by mass?   *Ans. 79.7 g*

30. What mass of 3.00% KOH solution by mass contains 5.1 g of KOH?   *Ans. 1.7×10^2 g*

31. What mass of HCl is contained in 45.0 mL of an HCl solution with a density of 1.19 g/mL and containing 37.21% HCl by mass?   *Ans. 19.9 g*

32. What mass of solid NaOH (97.0% NaOH by mass) is required to prepare 1.00 L of a 10.0% solution of NaOH by mass? The density of the 10.0% solution is 1.109 g/cm^3.   *Ans. 114 g*

33. The hardness of water (hardness count) is usually expressed as parts per million (by mass of $CaCO_3$), which is equivalent to milligrams of $CaCO_3$ per liter of water. What is the molar concentration of Ca^{2+} ions in a water sample with a hardness count of 175?

    Ans. 1.75×10^{-3} M

34. The Safe Drinking Water Act sets the maximum permissible amount of cadmium in drinking water at 0.01 mg/L. What is the maximum permissible molar concentration of cadmium in drinking water?

    Ans. 9×10^{-8} M

35. The concentration of glucose, $C_6H_{12}O_6$, in normal spinal fluid is 75 mg/100 g. What is the molal concentration?   *Ans. 4.2×10^{-3} m*

36. What volume of a 0.20 M Li_2SO_4 solution contains 5.7 g of Li_2SO_4?   *Ans. 0.26 L*

37. A 1.577 M solution of $AgNO_3$ has a density of 1.220 g/cm^3. What is the molality of the solution?

    Ans. 1.656 m

38. Calculate what volume of a sulfuric acid solution (specific gravity = 1.070, and containing 10.00% H_2SO_4 by mass) contains 18.50 g of pure H_2SO_4 at a temperature of 25°C. The density of water at 25°C is 0.99709 g/mL.   *Ans. 173.4 mL*

39. A 15.0% solution of K_2CrO_4 by mass has a density of 1.129 g/cm^3. Calculate the molarity of the solution.

    Ans. 0.872 M

40. Equal volumes of 0.050 M H_2SO_4 and 0.400 M KOH are mixed. Calculate the molarity of each ion present in the final solution.
    Ans. K^+, 0.200 M; SO_4^{2-}, 0.025 M; OH^-, 0.150 M

41. What volume of 0.333 M HNO_3 is required to react completely with 1.25 g of sodium hydrogen carbonate?

    $$NaHCO_3 + HNO_3 \longrightarrow NaNO_3 + CO_2 + H_2O$$
    Ans. 44.7 mL

42. Calculate the volume of 0.0500 M NaBr necessary to precipitate the silver contained in 12.0 mL of 0.0500 M $AgNO_3$. $Ag^+ + Br^- \longrightarrow AgBr(s)$

    Ans. 12.0 mL

43. What volume of a 0.33 M solution of hydrochloric acid, HCl, would be required to neutralize completely 1.00 L of 0.215 M barium hydroxide, Ba(OH)$_2$?

Ans. 1.3 L

44. To 10.0 mL of a 0.100 M K$_2$Cr$_2$O$_7$ solution is added 10.0 mL of a 0.100 M Pb(NO$_3$)$_2$ solution. What mass of PbCr$_2$O$_7$ forms? *Ans. 0.423 g*

45. A gaseous solution was found to contain 25% H$_2$, 20% CO, and 55% CO$_2$ by mass. What is the mole fraction of each component?

Ans. H$_2$, 0.86; CO, 0.050; CO$_2$, 0.087

46. A sample of lead glass is prepared by melting together 20.0 g of silica, SiO$_2$, and 80.0 g of lead(II) oxide, PbO. What is the mole fraction of SiO$_2$ and of PbO in the glass? *Ans. SiO$_2$, 0.482; PbO, 0.518*

47. Calculate the mole fractions of methanol, CH$_3$OH, ethanol, C$_2$H$_5$OH, and water in a solution that is 50% methanol, 30% ethanol, and 20% water by mass.

Ans. CH$_3$OH, 0.47; C$_2$H$_5$OH, 0.20; H$_2$O, 0.33

48. Concentrated hydrochloric acid is 37.0% HCl by mass and has a specific gravity of 1.19 at 25°C. The density of water is 0.99707 g/mL. Calculate (a) the molarity of the solution, (b) the molality of the solution, and (c) the mole fraction of HCl and of H$_2$O.

Ans. (a) 12.0 M; (b) 16.1 m; (c) HCl, 0.225; H$_2$O, 0.775

49. Calculate (a) the percent composition and (b) the molality of an aqueous solution of sucrose, C$_{12}$H$_{22}$O$_{11}$, if the mole fraction of sucrose is 0.0677.

Ans. 58.0% C$_{12}$H$_{22}$O$_{11}$; 4.03 m

50. What volume of 10.00 M nitric acid, a solution of HNO$_3$, is required to prepare 1.000 L of 0.050 M nitric acid? *Ans. 5.0 mL*

51. What volume of a sulfuric acid solution that is 96.0% by mass with a specific gravity of 1.84 at 25°C is required to prepare 8.00 L of a 1.50 M solution of sulfuric acid at 25°C? The density of water at 25°C is 0.99707 g/mL.

Ans. 668 mL

52. What volume of 0.200 M nitric acid, a solution of HNO$_3$, can be prepared from 250 mL of 14.5 M HNO$_3$?

Ans. 18.1 L

53. (a) What volume of 0.75 M HBr is required to prepare 1.0 L of 0.33 M HBr? *Ans. 0.44 L*

    (b) What volume of 3.50% KOH by mass (density 1.012 g/mL) can be prepared from 0.150 L of 30.0% KOH by mass (density 1.288 g/mL)?

Ans. 1.64 L

Colligative Properties

54. A solution of ethanol (a nonelectrolyte) and a solution of sodium chloride (an electrolyte) both freeze at −0.5°C. What other physical properties of the two solutions are identical?

55. Which will evaporate faster under the same conditions, 50 mL of distilled water or 50 mL of sea water?

56. The triple point of air-free water (Section 11.9) is defined as 273.15 K. Why is it important that the water be air-free?

57. Why will a mole of hydrogen chloride depress the freezing point of 1 kg of water by almost twice as much as a mole of ethanol?

58. (a) Sodium chloride, NaCl, calcium chloride, CaCl$_2$, and the nonelectrolyte urea, NH$_2$CONH$_2$, are frequently used to melt ice on streets and highways. Rank the three from high to low in terms of effectiveness per mole and per pound in lowering the freezing point. Explain your ranking.

    (b) Would these substances be as effective in melting ice on a day when the maximum temperature is −20°F? Explain why or why not.

59. Explain what is meant by osmotic pressure. How is the osmotic pressure of a solution related to its concentration?

60. The cell walls of red and white blood cells are semipermeable membranes. The concentration of solute particles in the blood is about 0.6 m. What will happen to blood cells that are placed in pure water? in a 1 m sodium chloride solution?

61. A 1 m solution of HCl in benzene has a freezing point of 0.4°C. Is HCl an electrolyte in benzene? Explain.

62. Arrange the following solutions in decreasing order by their freezing points: 0.1 m Na$_3$PO$_4$, 0.1 m C$_2$H$_5$OH, 0.01 m CO$_2$, 0.15 m NaCl, and 0.2 m CaCl$_2$.

63. Some mammals, including humans, excrete hypertonic urine (Section 12.22) in order to conserve water. In humans, some parts of the ducts of the kidney (which contain the urine) pass through a fluid that contains a much more concentrated salt solution than that normally found in the body. Explain how this could help conserve water in the body.

64. A solution of 5.00 g of an organic compound in 25.00 g of carbon tetrachloride (bp 76.8°C; K_b = 5.02°C/m) boils at 81.5°C at 1 atm. What is the molecular weight of the compound? *Ans. 2.1 × 10^2*

65. A solution contains 15.00 g of urea, CO(NH$_2$)$_2$, per 0.200 kg of water. If the vapor pressure of water at 25°C is 23.7 torr, what is the vapor pressure of the solution at 25°C? *Ans. 23.2 torr*

66. A 12.0 g sample of a nonelectrolyte is dissolved in 80.0 g of water. The solution freezes at −1.94°C. Calculate the molecular weight of the substance. *Ans. 144*

67. A sample of an organic compound (a nonelectrolyte) weighing 1.350 g lowered the freezing point of 10.0 g of benzene by 3.66°C. Calculate the molecular weight of the organic compound. *Ans. 189*

68. Calculate the boiling point elevation of 0.100 kg of water containing 0.020 mol of NaCl, 0.010 mol of Na$_2$SO$_4$, and 0.040 mol of MgCl$_2$, assuming complete dissociation of these electrolytes. *Ans. 0.973°C*

69. What is the approximate freezing point of a 0.37 m aque-

ous solution of sodium bromide? Assume complete dissociation of this electrolyte.   *Ans. −1.4°C*

70. If you had 150 g of glycerine, $C_3H_8O_3$, what mass of a 3.8 m aqueous solution could you prepare? What is the freezing point of this solution?   *Ans. 580 g; −7.1°C*

71. If 26.4 g of the nonelectrolyte dibromobenzene, $C_6H_4Br_2$, is dissolved in 0.250 kg of benzene, what is (a) the freezing point of the solution and (b) the boiling point of the solution at 1 atm?
    Ans. (a) 3.19°C; (b) 81.2°C

72. A sample of phosphorus weighing 0.101 g was dissolved in 17.8 g of carbon disulfide, CS_2 ($K_b = 2.34°C/m$). If the boiling-point elevation was 0.107°C, what is the formula of a phosphorus molecule in carbon disulfide?
    Ans. P_4

73. What is the boiling point, at 1 atm, of a solution containing 115.0 g of sucrose, $C_{12}H_{22}O_{11}$, in 350.0 g of water?
    Ans. 100.5°C

74. Lysozyme is an enzyme that cleaves cell walls. A 0.100-L sample of a solution of lysozyme that contains 0.0750 g of the enzyme exhibits an osmotic pressure of 1.32×10^{-3} atm at 25°C. What is the molecular weight of lysozyme?   *Ans. 1.39×10^4*

75. The osmotic pressure of solution containing 7.0 g of insulin per liter is 23 torr at 25°C. What is the molecular weight of insulin?   *Ans. 5.6×10^3*

76. The osmotic pressure of human blood is 7.6 atm at 37°C. What mass of glucose, $C_6H_{12}O_6$, is required to make 1.00 L of aqueous solution for intravenous feeding if the solution must have the same osmotic pressure as blood at body temperature, 37°C?   *Ans. 54 g*

Colloids

77. What is the difference between dispersion methods and condensation methods for preparing colloidal systems?

78. How do colloidal dispersions differ from true solutions with regard to dispersed particle size and homogeneity?

79. Identify the dispersed phase and the dispersion medium in each of the following colloidal systems: starch dispersion, smoke, fog, pearl, whipped cream, floating soap, jelly, milk, and opal.

80. Explain the cleansing action of soap.

81. Discuss the similarities and differences of soaps and detergents.

82. How can it be demonstrated that colloidal particles are electrically charged?

83. What is the structure of a gel?

84. Explain the phenomenon of Brownian movement. What is its relationship to the stability of the dispersed particles of a colloidal system?

Additional Exercises

85. For which of the following gases is solubility in water not directly proportional to the pressure of the gas: HCl, H_2, SO_2, NH_3, CH_4, N_2? Explain.

86. Why will ice melt when placed in an aqueous solution that is kept at 0°C?

87. What is a constant boiling solution? Describe two ways to prepare a constant boiling solution of hydrochloric acid.

88. The salt lithium nitrate, $LiNO_3$, does not depress the freezing point of water twice as much a nonelectrolyte. Explain.

89. The approximate radius of a water molecule, assuming a spherical shape, is 1.40 Å ($1 \text{ Å} = 10^{-8}$ cm). Assume that water molecules cluster around each metal ion in a solution so that the water molecules essentially touch both the metal ion and each other. On this basis, and assuming that 4, 6, 8, and 12 are the only possible coordination numbers, what is the maximum number of water molecules that can hydrate each of the following ions?
    (a) Mg^{2+} (radius 0.65Å)   *Ans. 6*
    (b) Al^{3+} (0.50 Å)   *Ans. 4*
    (c) Rb^+ (1.48 Å)   *Ans. 12*
    (d) Sr^{2+} (1.13 Å)   *Ans. 8*

90. Hydrogen gas dissolves in the metal palladium, with hydrogen atoms going into the holes between metal atoms. Determine the molarity, molality, and percent by mass of hydrogen atoms in a solution (density = 10.8 g/cm^3) of 0.89 g of hydrogen atoms in 215 g of palladium metal.
    Ans. 44 M; 4.1 m; 0.41%

91. How many liters of HCl(g) at 25°C and 1.26 atm are required to prepare 2.50 L of a 1.50 M solution of HCl?
    Ans. 72.8 L

92. A 1.80-g sample of an acid, H_2X, required 14.00 mL of KOH solution for neutralization of all the hydrogen ions. Exactly 14.2 mL of this same KOH solution was found to neutralize 10.0 mL of 0.750 M H_2SO_4. Calculate the molecular weight of H_2X.   *Ans. 244*

93. A 0.300-L volume of gaseous ammonia measured at 28.0°C and 754 torr was absorbed in 0.100 L of water. How many milliliters of 0.200 M hydrochloric acid are required in the neutralization of this aqueous ammonia? What is the molarity of the aqueous ammonia solution? (Assume no change in volume when the gaseous ammonia is added to the water.)   *Ans. 60.2 mL; 0.120 M*

94. Calculate the percent by mass and the molality in terms of Na_2SO_4 for a solution prepared by dissolving 11.5 g of $Na_2SO_4 \cdot 10H_2O$ in 0.100 kg of water. Remember to consider the water released from the hydrate.
    Ans. 4.55%; 0.335 m

95. Calculate the volume of the concentrated acid and the volume of water required to produce 0.525 L of 0.105 M nitric acid by diluting 10.00 M nitric acid. Assume that the volumes of nitric acid and water are additive.
    Ans. 5.51 mL conc. acid; 519 mL H_2O

96. What is the molarity of H_3PO_4 in a solution that is prepared by dissolving 35.08 g P_4O_{10} in sufficient water to make 0.7500 L of solution?   *Ans. 0.6590 M*

97. A solution of sodium carbonate having a volume of 0.250 L was prepared from 2.032 g of $Na_2CO_3 \cdot 10H_2O$.

Calculate the molarity of this solution.

Ans. 0.0284 M

98. The sulfate in 50.0 mL of dilute sulfuric acid was precipitated using an excess of barium chloride. The mass of $BaSO_4$ formed was 0.682 g. Calculate the molarity of the sulfuric acid solution.   *Ans. 0.0584 M*

99. A sample of $HgCl_2$ weighing 9.41 g is dissolved in 32.75 g of ethanol, C_2H_5OH ($K_b = 1.20°C/m$). The boiling-point elevation of the solution is 1.27°C. Is $HgCl_2$ an electrolyte in ethanol? Show your calculations.

Ans. No

100. A salt is known to be an alkali metal fluoride. A quick approximate freezing-point determination indicates that 4 g of the salt dissolved in 100 g of water produces a solution that freezes at about −1.4°C. What is the formula of the salt? Show your calculations.   *Ans. RbF*

101. A solution of 0.045 g of an unknown organic compound in 0.550 g of camphor melts at 158.4°C. The melting point of pure camphor is 178.75°C. K_f for camphor is 37.7°C/m. The solute contains 93.46% C and 6.54% H by mass. What is the molecular formula of the solute? Show your calculations.   *Ans. $C_{12}H_{10}$*

102. How many liters of $N_2O_3(g)$, measured at 30.0°C and 755 torr, are required to prepare 2.50 L of a 0.15 M solution of HNO_2?

$$H_2O + N_2O_3 \longrightarrow 2\ HNO_2$$

Ans. 4.7 L

103. The sugar fructose contains 40.0% C, 6.7% H, and 53.3% O by mass. A solution of 11.7 g of fructose in

325 g of ethanol has a boiling point of 78.59°C. The boiling point of ethanol is 78.35°C, and K_b for ethanol is 1.20°C/m. What is the molecular formula of fructose?

Ans. $C_6H_{12}O_6$

104. The vapor pressure of methanol, CH_3OH, is 94 torr at 20°C. The vapor pressure of ethanol, C_2H_5OH, is 44 torr at the same temperature. Calculate the mole fraction of methanol and of ethanol in a solution of 50.0 g of methanol and 50.0 g of ethanol. Ethanol and methanol form a solution that behaves like an ideal solution. Calculate the vapor pressure of methanol and of ethanol above the solution at 20°C. Calculate the mole fraction of methanol and of ethanol in the vapor above the solution.

Ans. In solution: CH_3OH, 0.589; C_2H_5OH, 0.411;
P_{CH_3OH}, 55 torr; $P_{C_2H_5OH}$, 18 torr
In vapor: CH_3OH, 0.75; C_2H_5OH, 0.25

105. Calculate the molarity, the molality, and the percent by mass of the following solutions of commercially available acids at 25°C. The density of water at 25°C is 0.99707 g/mL.

(a) a solution of 452 g of HCl in 1.0 L (specific gravity 1.19)

(b) a solution of 994 g of HNO_3 in 1.0 L (specific gravity 1.42)

(c) a solution of 1.75 kg of H_2SO_4 in 1.0 L of solution (specific gravity 1.84)

Ans. (a) HCl, 12 M, 17 m, 38%; (b) HNO_3, 16 M, 37 m, 70%; (c) H_2SO_4, 18 M, 2.1×10^2 m, 96%

The Active Metals

13

Aluminum ingots at a factory in Quebec.

Twelve of the metals found in Groups IA, IIA, and IIIA of the Periodic Table exhibit remarkably similar chemical behavior. They all react with acids, forming hydrogen gas and salts; with water they form hydrogen gas and hydroxides. They react with most nonmetals. All of these metals are easily oxidized, and the resulting ions are reduced with great difficulty. Each exhibits only a single oxidation number in its compounds, and they all form many simple ionic compounds. The twelve active metals are lithium, sodium, potassium, rubidium, cesium, and francium from Group IA; magnesium, calcium, strontium, barium, and radium from Group IIA; and aluminum from Group IIIA (Fig. 13.1).

Several of these active metals are among the most common of the elements. They do not occur in nature as free elements because of their extreme reactivity (most react with water and air). However, their compounds are widely distributed. Large mineral deposits of relatively pure compounds of sodium, potassium, magnesium, calcium, and aluminum are found in many parts of the world. Chlorides of sodium, magnesium, calcium, and potassium are the most abundant compounds in sea water. Potassium compounds are found in all plants. Both sodium and potassium are essential to animal life. Magnesium and calcium are found in silicate rocks, dolomite, and limestone. The

357

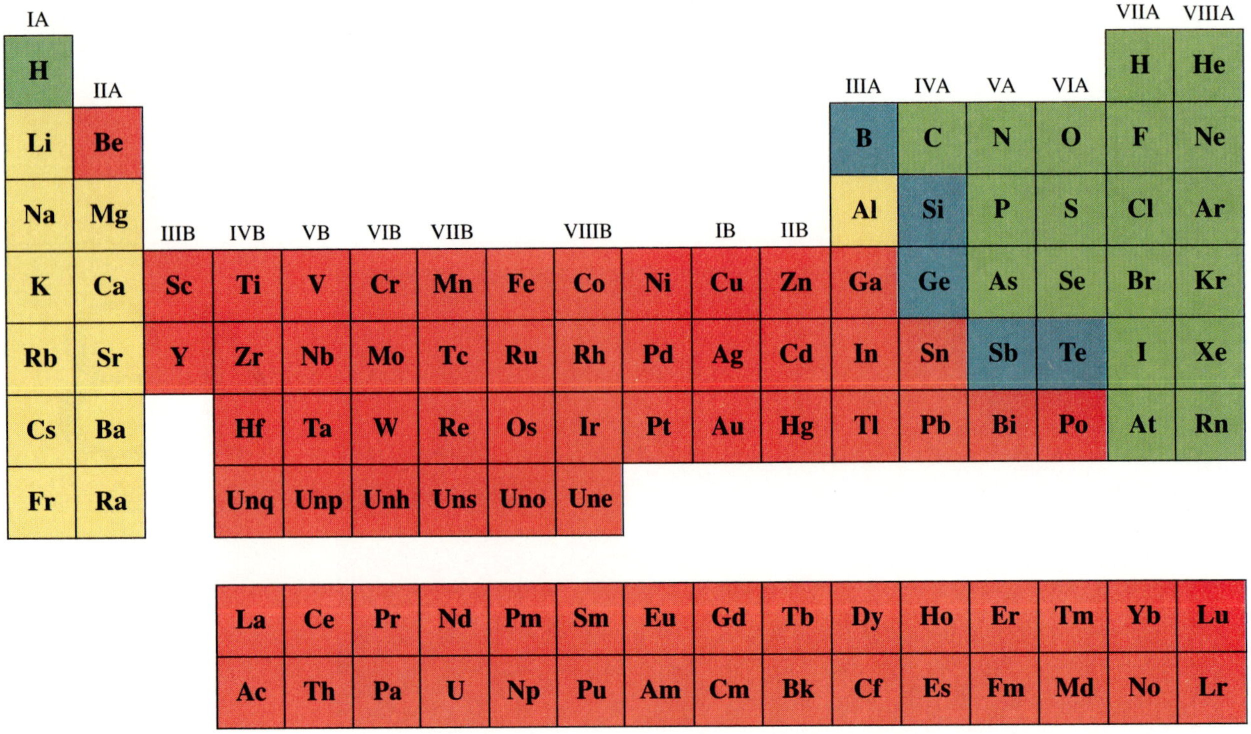

Figure 13.1

The location of the active metals (shown in yellow) in the Periodic Table. Nonmetals are shown in green; semimetals, in blue; and the remaining metals, in red.

chlorophyll in green plants is a magnesium-containing compound. Calcium is a component of bones, teeth, and nerve cells. Aluminum is present in most silicate rocks. Many compounds of these elements are major items of trade, commerce, and the chemical industry, and we use them routinely in our day-to-day existence.

In this chapter we will examine the kinds of compounds the active metals form and the chemical behavior of these compounds. We will begin with a survey of the behavior of these elements in relation to their positions in the Periodic Table. We will see how these very reactive elements can be isolated and consider their more important uses. Then the formation and uses of some of their more important compounds will be described. Finally, we will use the principles developed in preceding chapters to explain the similarities and differences in the behavior of these ionic compounds.

The Elemental Active Metals

13.1 Periodic Relationships among Groups IA, IIA, and IIIA

Although not all of the elements in Groups IIA and IIIA are considered to be active elements, it is helpful to consider the periodic relationships of all the elements of Groups IA, IIA, and IIIA in order to fix the behavior of the active metals from the perspective of the Periodic Table.

1. GROUP IA, THE ALKALI METALS. The alkali metals, lithium, sodium, potassium, rubidium, cesium, and francium, constitute Group IA of the Periodic Table. The

heaviest of these, francium, is a highly radioactive element that occurs in nature in very small quantities. It is not well characterized. Although hydrogen is often shown in Group IA (and in Group VIIA) in Periodic Tables, it is a nonmetal and is therefore not considered in this chapter.

The alkali metals show a closer relationship in their properties than do any other family of elements in the Periodic Table. The single electron in the outermost (valence) shell of each is not tightly bound, so these metals have the largest atomic radii (Table 13.1 on p. 360) in their respective periods. The valence electron is easily lost; thus, these metals readily form stable positive ions with a charge of +1. The corresponding difficulty with which these ions are reduced accounts for the difficulty of isolating the elements. No samples were isolated in a pure state until Sir Humphry Davy, in 1807, produced sodium and potassium by the electrolytic reduction of their hydroxides. Of the members of the group that have been extensively studied, the valence electron is least tightly bound in cesium, and this is the most reactive. However, francium, with a larger radius than cesium, should be still more reactive.

The alkali metals all react vigorously with water (Fig. 13.2), with the violence of the reaction increasing toward the bottom of the group. Gaseous hydrogen and solutions of the corresponding metal hydroxide are formed. For example, the reaction of sodium with water is

$$2Na + H_2O \longrightarrow 2NaOH + H_2$$

These metals react directly with oxygen, sulfur, hydrogen, phosphorus, and the halogens, giving ionic compounds containing the +1 metal ions. Most of the salts of the alkali metals are soluble in water, but Li_2CO_3, Li_3PO_4, and LiF are relatively insoluble. In this respect, lithium resembles magnesium, the second member of Group IIA.

2. GROUP IIA, THE ALKALINE EARTH METALS.

The alkaline earth metals, beryllium, magnesium, calcium, strontium, barium, and radium, constitute Group IIA of the Periodic Table. The heaviest of these, radium, is a radioactive element that occurs in nature in only small quantities.

Each of the two electrons in the outermost (valence) shell of each of the alkaline earth metals is more tightly bound than the single electron in the preceding alkali metal in the same period. Thus, each alkaline earth atom is smaller than the preceding alkali metal atom (Section 5.15 and Table 13.1) and not as reactive (Fig. 13.3). The alkaline earth metals are nevertheless very reactive elements, with the reactivity increasing with increasing size. Both valence electrons of the Group IIA elements are involved in chemical reactions, and these elements readily form compounds in which they exhibit the group oxidation number of +2. Like the alkali metals, the alkaline earth elements cannot be readily prepared by chemical techniques.

The lightest alkaline earth metal, beryllium, forms compounds in which it exhibits predominantly covalent character. Compounds of magnesium often show some covalent character, but the magnesium in them generally can be regarded as a +2 ion. Calcium, strontium, and barium form ionic compounds. The gradation from covalent to ionic character and the increase in reactivity from beryllium to barium and radium are due to the increasing atomic radius and the consequent ease of loss of the valence electrons in these atoms.

Beryllium does not react with water, even when heated, and combines with oxygen only when heated above 600°C. Magnesium reacts slowly with hot water or steam and with oxygen upon warming. The reactivity of magnesium metal at room temperature is often masked because the surface of the metal reacts with air and becomes

Figure 13.2
A sample of lithium on water. Lithium floats because it is less dense than water.

Figure 13.3
A sample of calcium in water. The reaction is not as vigorous as that of sodium or potassium with water.

Table 13.1 Properties of the elements of groups IA-IIIA with the active metals in the highlighted block

	Group IA	Group IIA	Group IIIA
	Lithium	Beryllium	Boron
Common oxidation number	+1	+2	+3
Atomic radius (Å)	1.52	1.11	0.88
Ionic radius (Å)	0.60	0.31	0.20
Density (g/cm^3)	0.534	1.85	2.34
Melting point, (°C)	180	1278	2300
Boiling point, (°C)	1326	2970	—
	Sodium	Magnesium	Aluminum
Common oxidation number	+1	+2	+3
Atomic radius (Å)	1.86	1.60	1.43
Ionic radius (Å)	0.95	0.65	0.50
Density (g/cm^3)	0.968	1.74	2.70
Melting point, (°C)	98	651	660
Boiling point, (°C)	889	1107	2467
	Potassium	Calcium	Gallium
Common oxidation number	+1	+2	+3
Atomic radius (Å)	2.31	1.97	1.22
Ionic radius (Å)	1.33	0.99	0.62
Density (g/cm^3)	0.856	1.55	5.91
Melting point, (°C)	63.4	850	29.8
Boiling point, (°C)	757	1490	2403
	Rubidium	Strontium	Indium
Common oxidation number	+1	+2	+3
Atomic radius (Å)	2.44	2.15	1.62
Ionic radius (Å)	1.48	1.13	0.81
Density (g/cm^3)	1.532	2.54	7.31
Melting point, (°C)	38.8	770	156.6
Boiling point, (°C)	679	1384	2000
	Cesium	Barium	Thallium
Common oxidation number	+1	+2	+1,+3
Atomic radius (Å)	2.62	2.17	1.71
Ionic radius (Å)	1.69	1.35	0.95 (+3)
Density (g/cm^3)	1.87	3.5	11.85
Melting point, (°C)	28.7	704	303.5
Boiling point, (°C)	690	1638	1457
	Francium	Radium	
Common oxidation number	+1	+2	
Atomic radius (Å)	2.7	2.20	
Ionic radius (Å)		1.52	
Density (g/cm^3)		about 5	
Melting point, (°C)		700	
Boiling point, (°C)		below 1737	

covered with a tightly adhering layer of magnesium oxycarbonate that protects the metal and prevents additional reaction. Calcium, strontium, and barium react with water and oxygen (Fig. 13.4) at room temperature, with barium showing the most vigorous reaction. The products of the reactions with water are hydrogen and the metal

Figure 13.4
These calcium turnings appear dull because the metal surface is covered with calcium oxide from the reaction of the metal with the oxygen of the air.

Figure 13.5
Aluminum oxide will not adhere to aluminum foil where it has been treated to produce a very thin film of mercury on its surface. Thus such aluminum foil is not protected by the customary oxide film and "rusts" in air, producing white aluminum oxide. Note the white Al_2O_3 that is forming on the surface and falling off.

hydroxide or, when magnesium reacts with steam, magnesium oxide. These metals react directly with acids, sulfur, phosphorus, the halogens, and, with the exception of beryllium, with hydrogen. Unlike the salts of the alkali metals, many of the common salts of the alkaline earth metals are insoluble in water. The solubility of the hydroxides increases, but the solubility of the carbonates and sulfates decreases with increasing atomic number.

3. GROUP IIIA. Group IIIA contains the elements boron, aluminum, gallium, indium, and thallium. The increase in metallic character moving down a column of the Periodic Table, which is only just apparent in Group IIA, is clearly illustrated by the elements of Group IIIA. The lightest element, boron, is semiconducting, and its compounds, which are covalent, display distinctively nonmetallic properties. For example, boron oxide, B_2O_3, is acidic and its hydroxide, $B(OH)_3$, is an acid (boric acid). The remaining elements of the group are metals, but the oxides and hydroxides change character. The oxides and hydroxides of aluminum and gallium exhibit both acidic and basic behavior, whereas those of indium and thallium exhibit only basic behavior.

Two of the three electrons in the valence shell of each of the Group IIIA elements are located in the outermost s orbital, and the third is in the outermost p orbital. Aluminum uses all of its valence electrons when it reacts, giving compounds in which it has an oxidation number of $+3$. Although many of these compounds are covalent, others, such as AlF_3 and Al_2O_3, are ionic. Aqueous solutions of aluminum salts contain the cation $Al^{3+}(aq)$. Gallium, indium, and thallium also form ionic compounds containing the M^{3+} ions. A few compounds with a $+1$ oxidation number are known for gallium and indium, and both the $+1$ and $+3$ oxidation numbers are commonly observed with thallium. In aqueous solution, the $Tl^+(aq)$ ion is the stable simple thallium ion.

The metals of Group IIIA (Al, Ga, In, Tl) are all reactive. However, like magnesium, metallic aluminum and, to a lesser extent, metallic gallium and indium do not appear to be as reactive as they actually are. These elements are coated by a hard, thin, tough film of the metal oxide that forms when they are exposed to air and protects them from chemical attack. When the film is broken, these elements react with water and oxygen (Fig. 13.5) giving compounds with oxidation numbers of $+3$ for the metals. Thallium also reacts with water and oxygen but gives thallium(I) derivatives. The metals of Group IIIA all react directly with nonmetals such as sulfur, phosphorus, and the halogens.

Unlike the elements of Groups IA and IIA, in which reactivity increases down the group, gallium, indium, and thallium are progressively less reactive. They are also less reactive than aluminum. They exhibit two oxidation numbers in their compounds, which is distinctly different from the behavior of the active metals. These differences are reflections of differences in the electronic structures of the elements.

Aluminum and the metals of Groups IA and IIA form ions which have noble gas electron configurations (Table 13.2). The neutral atoms have their valence electrons outside of a filled inner shell that contains only 8 electrons. The +3 ions of gallium, indium, and thallium have pseudo noble gas electron configurations (Section 6.2). Their neutral atoms have their valence electrons outside of a filled inner shell that contains 18 electrons. Gallium, indium and thallium are members of the group of elements that follow the transition metals in their respective periods. The metals of this set are called the **post-transition metals;** we will describe their chemistry when we consider the behavior of all of the post-transition metals in Chapter 28.

Table 13.2 Electron configurations with valence electrons in the neutral atoms shown in color

Al:	$1s^2 2s^2 2p^6 3s^2 3p^1$
Al^{3+}:	$1s^2 2s^2 2p^6 \equiv$ [Ne]
Ga:	$1s^2 2s^2 2p^6 3s^2 3p^6 3d^{10} 4s^2 4p^1$
Ga^{3+}:	$1s^2 2s^2 2p^6 3s^2 3p^6 3d^{10} \equiv$ [Ar]$3d^{10}$
In:	$1s^2 2s^2 2p^6 3s^2 3p^6 3d^{10} 4s^2 4p^6 4d^{10} 5s^2 5p^1$
In^{3+}:	$1s^2 2s^2 2p^6 3s^2 3p^6 3d^{10} 4s^2 4p^6 4d^{10} \equiv$ [Kr]$4d^{10}$
Tl:	$1s^2 2s^2 2p^6 3s^2 3p^6 3d^{10} 4s^2 4p^6 4d^{10} 4f^{14} 5s^2 5p^6 5d^{10} 6s^2 6p^1$
Tl^{3+}:	$1s^2 2s^2 2p^6 3s^2 3p^6 3d^{10} 4s^2 4p^6 4d^{10} 4f^{14} 5s^2 5p^6 5d^{10} \equiv$ [Xe]$4f^{14} 5d^{10}$

13.2 Occurrence and Isolation of the Active Metals

The active metals are not found as the free elements in nature because of their extreme reactivity. Compounds of many of these elements are common, although compounds of others are rare. For example, there is enough sodium chloride dissolved in the oceans to cover the entire continent of North America with solid salt 800 meters (2500 feet) deep. On the other hand, there is only about 15 grams of francium in the top kilometer of the earth's entire crust. Several of the active metals are among the most abundant of the elements (Section 1.5, Table 1.1), and their compounds occur extensively. These include simple compounds such as chlorides, oxides or hydroxides, carbonates, sulfates, and nitrates, as well as more complex silicates. The natural abundances of these metals and the more common minerals that contain them are listed in Table 13.3; sources from which the elements are isolated are in boldface type.

Although compounds containing active metals are abundant, the metals are difficult to isolate. The active metals are easily oxidized, so it is difficult to reduce their ions back to neutral atoms. Consequently, the largest quantities of the pure metals are prepared by electrolysis, in which the input of electrical energy forces reduction to occur. Electrolysis is often used to carry out oxidation-reduction reactions of species that are oxidized or reduced with difficulty.

Table 13.3 Natural abundances of the active metals in the earth's crust and their sources

Element	Abundance (% by mass)	Minerals and sources
Lithium	0.0018	**Lepidolite,** spodumene ($LiAlSi_2O_6$)
Sodium	2.63	**Rock salt** (NaCl), trona [$Na_2CO_3 \cdot NaHCO_3 \cdot (H_2O)_2$], Chile saltpeter ($NaNO_3$)
Potassium	2.40	**Sylvite** (KCl), carnallite [$KMgCl_3(H_2O)_6$]
Rubidium	0.0078	**Lepidolite**
Cesium	0.0003	**Pollucite** ($Cs_4Al_4Si_9O_{26}H_2O$)
Francium	trace	
Magnesium	1.93	Magnesite ($MgCO_3$), dolomite [$(Mg,Ca)CO_3$], epsomite [$MgSO_4(H_2O)_7$], silicates, **sea water, brines**
Calcium	3.39	Dolomite [$(Mg,Ca)CO_3$]; aragonite, marble, and **limestone** ($CaCO_3$); silicates
Strontium	0.02	**Celestite** ($SrSO_4$)
Barium	0.04	**Barite** ($BaSO_4$)
Radium	1×10^{-10}	**Uranium-containing ores**
Aluminum	7.50	**Bauxite** [$AlO(OH)$, $Al(OH)_3$], various silicates and clays, corrundum (Al_2O_3)

1. THE PREPARATION OF LITHIUM AND SODIUM. The most important method for the production of both lithium and sodium is the electrolysis of the molten chloride. For sodium the reaction involved in the process is

$$2NaCl(l) \xrightarrow[600°C]{\text{Electrolysis}} 2Na(l) + Cl_2(g)$$

The electrolysis is carried out in a cell (Fig. 13.6) that contains molten sodium chloride (melting point 801°C), to which calcium chloride has been added to lower the melting point to 600°C. When a direct current is passed through the cell, of which one type is known as a Downs cell, the sodium ions migrate to the negatively charged cathode, pick up electrons, and are reduced to sodium metal. Chloride ions migrate to the

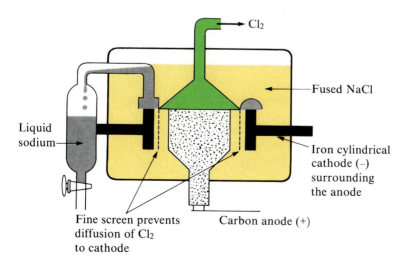

Figure 13.6

A section of a Downs cell for the production of sodium and chlorine.

positively charged anode, give up electrons, and are oxidized to chlorine gas. The overall change is obtained by adding the following reactions, in which e^- represents an electron:

At the cathode:	$2Na^+ + 2e^- \longrightarrow 2Na(l)$
At the anode:	$2Cl^- \longrightarrow Cl_2(g) + 2e^-$

Overall change:	$2Na^+ + 2Cl^- \longrightarrow 2Na(l) + Cl_2(g)$

Separation of the molten sodium and chlorine prevents recombination. The liquid sodium floats to the surface of the melt and flows into the collector. The gaseous chlorine is collected in tanks at high pressure. Chlorine, which will be discussed in Chapter 22, is as valuable a product as the sodium.

Metallic lithium is also prepared by electrolysis, using a molten mixture of lithium chloride and potassium chloride at about 450°C.

2. THE PREPARATION OF POTASSIUM, RUBIDIUM AND CESIUM. Potassium, rubidium, and cesium are produced on a relatively small scale, because they offer few advantages over the less expensive sodium. Although potassium could be produced by the electrolytic reduction of molten potassium chloride, the process is technologically difficult. Most potassium is produced by the reaction of sodium metal with molten potassium chloride.

$$Na(l) + KCl(l) \longrightarrow K(g) + NaCl(l)$$

The sodium is added to the bottom of a column of molten potassium chloride, and potassium vapor escapes from the top. This vapor is condensed and cast into sticks (Fig. 13.7). Note that the reduction of potassium ions by sodium metal is not predicted from the activity series (Section 9.7). Even though potassium is above sodium in the series, this reaction proceeds because potassium is more volatile than sodium and distills out of the reaction mixture. As in a metathetical reaction, the production of a gas as a product drives the reaction.

A similar process using calcium metal and rubidium chloride or cesium chloride is used to produce metallic rubidium or cesium, respectively.

3. THE PREPARATION OF MAGNESIUM, CALCIUM, STRONTIUM, AND BARIUM. Magnesium metal is prepared by the electrolysis of magnesium chloride, which may be obtained from several different sources, including sea water (which contains about 0.54% magnesium chloride). Nearly 6 million tons of magnesium as the chloride is contained in each cubic mile of seawater, but almost 800 tons of sea water must be processed to obtain 1 ton of magnesium.

The preparation of magnesium begins with the production of calcium hydroxide by addition of calcium oxide, CaO, to water.

$$CaO + H_2O \longrightarrow Ca(OH)_2$$

The calcium oxide is readily available by roasting limestone or oyster shells from shallow waters along the coast. Both limestone and oyster shells are calcium carbonate, $CaCO_3$, which decomposes upon heating to form the oxide.

$$CaCO_3(s) \longrightarrow CaO(s) + CO_2(g)$$

Addition of calcium hydroxide, $Ca(OH)_2$, to sea water separates magnesium as insoluble magnesium hydroxide, $Mg(OH)_2$.

$$Mg^{2+}(aq) + [2Cl^-(aq)] + Ca(OH)_2(s) \longrightarrow Mg(OH)_2(s) + Ca^{2+}(aq) + [2Cl^-(aq)]$$

Figure 13.7

Potassium for laboratory use is supplied as $\frac{1}{2}$-pound sticks stored under kerosene or in sealed containers to prevent contact with air and water.

After filtration, the magnesium hydroxide is treated with hydrochloric acid to produce a solution of magnesium chloride.

$$Mg(OH)_2(s) + 2H^+(aq) + 2Cl^-(aq) \longrightarrow Mg^{2+}(aq) + 2Cl^-(aq) + 2H_2O(l)$$

This solution is evaporated to recover $Mg(H_2O)_6Cl_2$, more commonly written as $MgCl_2 \cdot 6H_2O$. This crystallized salt is then partially dehydrated by heating. A melt of the partially dehydrated magnesium chloride with the composition $MgCl_2 \cdot 1.5H_2O$ [$Mg_2(H_2O)_3Cl_4$] is electrolyzed. Magnesium ions are reduced to molten magnesium metal at the cathode, and chloride ions are oxidized to chlorine gas at the anode.

$$MgCl_2 \cdot 1.5H_2O(l) \xrightarrow[700°C]{\text{Electrolysis}} Mg(l) + Cl_2(g) + 1.5H_2O(g)$$

A typical commercial electrolysis cell contains 10 tons of molten magnesium and molten salts. Each cell operates at about 6 volts with a current of 80,000 to 100,000 amps and produces an 80% yield of magnesium based on the number of electrons passed through the cell. The chlorine produced is used to manufacture more hydrochloric acid for the process.

Electrolysis of their molten chlorides is also the most important method for producing calcium, strontium, and barium. However, barium is also produced by the **King process,** the reduction of a mixture of barium oxide and barium peroxide with aluminum. At temperatures of 950–1100°C and at low pressures (10^{-3}–10^{-4} torr) barium distills out of the mixture and is collected on a cold surface.

4. THE PREPARATION OF ALUMINUM. Aluminum is prepared by a process invented in 1886 by Charles M. Hall, who began work on the problem while a student at Oberlin College. The process was discovered independently a month or two later by Paul L. T. Héroult in France.

The first step in the production of aluminum from the mineral bauxite involves purification of the mineral. The reaction of bauxite, AlO(OH), with hot sodium hydroxide forms soluble sodium aluminate, while clay and other impurities remain undissolved.

$$AlO(OH)(s) + NaOH(aq) + H_2O(l) \longrightarrow Na[Al(OH)_4](aq)$$

After the impurities are removed by filtration, aluminum hydroxide is reprecipitated by adding acid to the aluminate.

$$Na[Al(OH)_4](aq) + H^+(aq) \longrightarrow Al(OH)_3(s) + Na^+(aq) + H_2O(l)$$

The precipitated aluminum hydroxide is removed by filtration and heated, forming the oxide, Al_2O_3, which is then dissolved in a molten mixture of cryolite, Na_3AlF_6, and calcium fluoride, CaF_2. This solution is electrolyzed in a cell like that shown in Figure 13.8. Aluminum ions are reduced to the metal at the cathode, and oxygen, carbon monoxide, and carbon dioxide are liberated at the anode.

Aluminum prepared by electrolysis is about 99% pure; the impurities are small amounts of copper, iron, silicon, and aluminum oxide. If aluminum of exceptionally high purity (99.9%) is needed, it may be obtained by the electrolytic **Hoopes process** (Fig. 13.9), which employs a molten bath consisting of three layers. The bottom layer is a molten alloy of copper and impure aluminum, the top layer is pure molten aluminum, and the middle layer (the electrolyte) consists of a molten mixture of the fluorides of barium, aluminum, and sodium with nearly enough aluminum oxide to saturate it. The densities of the layers are such that their separation is maintained during electrolysis. The bottom layer serves as the anode and the top layer serves as the cathode. As

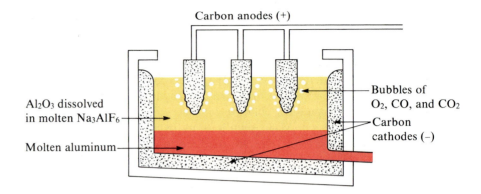

Carbon anodes (+)

Al₂O₃ dissolved in molten Na₃AlF₆

Al_2O_3 dissolved in molten Na_3AlF_6

Bubbles of O_2, CO, and CO_2

Carbon cathodes (−)

Molten aluminum

Figure 13.8
A cell for the production of aluminum.

Figure 13.9
The Hoopes electrolytic cell for the purification of aluminum: (a) is the bottom portion of the cell in contact with the anode layer; (b) is an alloy of impure molten aluminum and copper, which acts as the anode and at which the oxidation takes place; (c) is an electrolyte consisting of a mixture of molten fluorides; (d) is the molten layer of pure aluminum at the cathodes; (e) is a funnel arrangement by which impure aluminum may be added to the bottom of the cell.

electrolysis proceeds, the aluminum in the bottom layer is oxidized and passes into solution in the electrolyte as Al^{3+}, leaving the impurities behind; they are not oxidized under these conditions. The aluminum ion is reduced at the cathode. During electrolysis, purified aluminum is drawn off from the upper layer, and the impure metal is added to the lower layer.

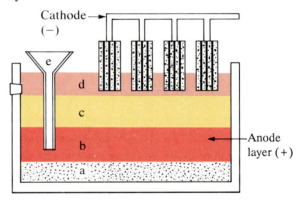

Cathode (−)

e

d

c

b

a

Anode layer (+)

13.3 Uses of the Active Metals

The alkali metals are not suitable for structural uses because of their reactivity and softness. They react with air and can be cut with a knife (Fig. 13.10). Their major utility stems from their reactivity and low melting points, although lithium is used to a limited extent in some alloys. Lithium-lead alloys are used in bearings, and lithium-magnesium and lithium-aluminum alloys are used in aviation and aerospace applications because of their lightness.

The alkali metals are silvery-white in color and are excellent conductors of heat and electricity. The metals are generally stored under kerosene or in sealed containers to prevent reaction with air or moisture (Fig. 13.7). The heat evolved when sodium and potassium react with water may cause the hydrogen produced or the metal to ignite. Rubidium and cesium ignite in contact with water. Never touch an alkali metal with your fingers; the heat of the reaction of the metal with the moisture of your skin could cause ignition of the metal.

Sodium is used as a reducing agent in the production of other metals (such as potassium, titanium, zirconium, and the heavier alkali metals) from their chlorides or oxides. Lithium and sodium are used as reducing agents in the manufacture of certain organic compounds including dyes, drugs, and perfumes. Sodium and its compounds

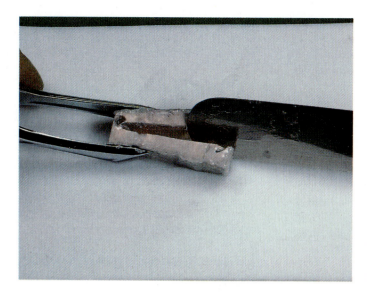

Figure 13.10
Sodium is so soft that it can be cut with a knife.

Figure 13.11
Heating sodium or sodium salts causes emission of a bright yellow light.

impart a yellow color to a flame (Fig. 13.11). The yellow light penetrates fog well, so sodium is used in street lights. The synthetic rubber industry consumes large amounts of sodium, and the metal is used to prepare compounds such as sodium peroxide and sodium oxide that cannot be made from sodium chloride. Potassium has no major uses for which sodium cannot be substituted, so the uses of potassium are limited.

Magnesium is a silvery-white metal that is malleable and ductile at high temperatures. Although very reactive, it does not undergo extensive reaction with air or water at room temperature, due to the protective oxycarbonate film that forms on its surface (Section 13.1, Part 2). Magnesium is the lightest of the widely used structural metals; most of the magnesium produced is used in making lightweight alloys, the most important of which are those with aluminum and zinc.

The potent reducing power of hot magnesium is utilized in preparing many metals and nonmetals, such as silicon and boron, from their oxides. Indeed, the affinity of magnesium for oxygen is so great that burning magnesium will react with carbon dioxide, reducing the carbon of the oxide to elemental carbon.

$$2Mg + CO_2 \longrightarrow 2MgO + C$$

(Thus a CO_2 fire extinguisher cannot be used to put out a magnesium fire.) The brilliant white light emitted by burning magnesium makes it useful in flashbulbs, flares, and fireworks.

Calcium, strontium, and barium are all silvery-white metals that are crystalline, malleable, and ductile. Calcium is harder than lead, strontium is as about as hard as lead, and barium is quite soft. Calcium is used as a dehydrating agent for certain organic solvents, as a reducing agent in the production of certain metals, as a scavenger to remove gases in molten metals in metallurgy, as a hardening agent for lead used for covering cables and making storage battery grids and bearings, in steel-making when alloyed with silicon, and for many other purposes.

Elemental strontium is not abundant and has no commercial uses. Barium is used as a degasing agent in the manufacture of vacuum tubes, and alloys of barium and nickel are used in vacuum tubes and spark plugs because of their high thermionic electron emission. It is interesting that Mg^{2+} and Ca^{2+} are not poisonous but Be^{2+} and Ba^{2+} are very much so.

Figure 13.12

The thermite reaction, the reaction of aluminum with iron(III) oxide producing a temperature of about 3000°C.

Freshly cut aluminum has a silvery appearance, but the surface soon becomes oxidized and assumes a duller luster. The tenacious oxide coating that forms protects the metal from further corrosion (Section 13.1, Part 3). The metal is very light, possesses high tensile strength, and, weight for weight, is twice as good an electrical conductor as copper.

The facts that aluminum is an excellent conductor of heat and is lightweight and corrosion-resistant account for its use in the manufacture of cooking utensils. The most important uses of aluminum are in the aircraft and other transportation industries, which depend on the lightness, toughness, and tensile strength of the metal. About half of the aluminum produced in this country is converted to alloys for special uses. Aluminum is also used in the manufacture of electrical transmission wire, as a paint pigment, and, in the form of foil, as a wrapping material.

Aluminum is one of the best reflectors of heat and light, including ultraviolet light. For this reason it is used as an insulating material and as a mirror in reflecting telescopes.

When powdered aluminum and iron(III) oxide are mixed and ignited by means of a magnesium fuse, a vigorous and highly exothermic reaction occurs.

$$2Al(s) + Fe_2O_3(s) \longrightarrow 2Fe(s) + Al_2O_3(s) \qquad \Delta H° = -851.4 \text{ kJ}$$

This is known as the **thermite reaction** and is applied in the **Goldschmidt process.** The temperature of the reaction mixture rises to about 3000°C, so the iron and aluminum oxide become liquid. The process is frequently used in welding large pieces of iron or steel (Fig. 13.12). The thermite reaction is used in the reduction of metallic oxides that are not readily reduced by carbon, such as MoO_3 and WO_3, or that do not give pure metals when reduced by carbon, such as MnO_2 and Cr_2O_3. The reaction is also of particular value for the production of carbon-free alloys, such as ferrotitanium.

Compounds of the Active Metals

With few exceptions, the simple compounds of the active metals are ionic. They contain cations of the metals and anions composed of nonmetals. The anions may be monatomic anions such as Cl^-, O^{2-}, and S^{2-}, or polyatomic anions such as OH^-, CO_3^{2-}, NO_3^-, and SO_4^{2-}. Much of the chemical behavior of these compounds reflects the chemical behavior of the anion.

13.4 Compounds of the Active Metals with Oxygen: Oxides, Peroxides, and Superoxides

Compounds of the active metals with oxygen fall into three categories: (1) **oxides,** containing oxide ions, O^{2-}; (2) **peroxides,** containing peroxide ions, O_2^{2-}, with single oxygen-oxygen covalent bonds; and (3) **superoxides,** containing superoxide ions, O_2^-, with oxygen-oxygen covalent bonds having bond orders of $1\frac{1}{2}$. The metals of Group IA form oxides, M_2O; peroxides, M_2O_2; and superoxides, MO_2. The metals of Group IIA form oxides, MO, and peroxides, MO_2. Aluminum forms two oxides, both with the formula Al_2O_3 but having different structures.

1. PREPARATION OF THE OXIDES, PEROXIDES, AND SUPEROXIDES. Oxides of most metals are produced by heating the corresponding hydroxides, nitrates,

or carbonates. However, most alkali metal salts containing these anions are extremely stable and do not decompose to the oxides at elevated temperatures. Burning the metals in air is not a satisfactory method for preparation of alkali metal oxides either. When burned in air, lithium forms lithium oxide (contaminated with some Li_2O_2), sodium forms sodium peroxide (contaminated with some Na_2O), and potassium, rubidium, and cesium form superoxides.

All of the alkali metal oxides can be prepared by the reaction of the nitrates or hydroxides with the elemental metal.

$$2MNO_3 + 10M \longrightarrow 6M_2O + N_2 \qquad (M = Li, Na, K, Rb, Cs)$$
$$2MOH + 2M \longrightarrow 2M_2O + H_2 \qquad (M = Li, Na, K, Rb, Cs)$$

Pure Li_2O and Na_2O can be prepared by the reaction of the metal in a closed container with a carefully controlled amount of oxygen.

$$4M + O_2 \longrightarrow 2M_2O \qquad (M = Li, Na)$$

Sodium peroxide is formed when Na or Na_2O is heated in pure oxygen.

$$2Na + O_2 \longrightarrow Na_2O_2$$
$$2Na_2O + O_2 \longrightarrow 2Na_2O_2$$

The peroxides of potassium, rubidium, and cesium can be prepared by heating the metal or its oxide in a carefully controlled amount of oxygen. With an excess of oxygen, KO_2, RbO_2, and CsO_2 form. Using potassium as an example,

$$2K + O_2 \longrightarrow K_2O_2 \qquad (2 \text{ mol K per mol } O_2)$$
$$K + O_2 \longrightarrow KO_2 \qquad (1 \text{ mol K per mol } O_2)$$

Note that the stability of the peroxides and superoxides of the alkali metals increases as the size of the cation increases.

The oxides of the alkaline earths are usually prepared by heating the respective carbonates, MCO_3 (Fig. 13.13). The hydroxides, $M(OH)_2$, also decompose at elevated temperatures and form oxides.

$$MCO_3 \longrightarrow MO + CO_2(g) \qquad (M = Mg, Ca, Sr, Ba)$$
$$M(OH)_2 \longrightarrow MO + H_2O(g) \qquad (M = Mg, Ca, Sr, Ba)$$

The stability of the carbonates increases as the size of the cation increases, as does that of the peroxides and the superoxides of the alkali metals. The temperature required to give a carbon dioxide pressure of 1 atmosphere is about 540°C for $MgCO_3$, 900°C for $CaCO_3$, 1290°C for $SrCO_3$, and 1360°C for $BaCO_3$.

Pure samples of the alkaline earth oxides cannot be prepared by burning the metals in air. Magnesium burns vigorously, but the MgO produced is contaminated with magnesium nitride, Mg_3N_2, produced by the reaction of magnesium with the nitrogen of the air. Calcium gives CaO contaminated with calcium peroxide, CaO_2, and calcium nitride, Ca_3N_2. When strontium and barium burn, they form the peroxides SrO_2 and BaO_2, respectively.

If magnesium oxide, sometimes called **magnesia,** is prepared from $MgCO_3$ at a temperature between 600°C and 800°C, the product is a light fluffy powder that retains a small percentage of magnesium carbonate. It reacts slowly with water, forming impure magnesium hydroxide, $Mg(OH)_2$, and dissolves readily in acids. On the other hand, when $MgCO_3$ is heated above 1400°C, the product contains no magnesium carbonate. This form of magnesium oxide is much more dense than the light form of the oxide and does not react with water.

Figure 13.13

For centuries lime (CaO) has been produced by heating limestone ($CaCO_3$). This more modern lime kiln is in Baton Rouge, Louisiana.

Aluminum also forms both a more reactive oxide and a less reactive oxide. Heating aluminum hydroxide, $Al(OH)_3$, or aluminum oxyhydroxide, $AlO(OH)$, below 450°C drives off water and produces **gamma-alumina** (γ-alumina), a form of aluminum oxide, Al_2O_3, which reacts with water and dissolves in acids. Above 1000°C the reaction of aluminum with oxygen or loss of water from $Al(OH)_3$ or $AlO(OH)$ produces **alpha-alumina** (α-alumina), a very hard form of Al_2O_3 that does not react with water and is not attacked by acids. Alpha-alumina occurs in nature as the mineral **corundum.**

2. REACTIONS AND USES OF THE OXIDES, PEROXIDES, AND SUPEROXIDES. The oxides of the active metals all contain the oxide ion, a very powerful hydrogen ion acceptor. With the exception of the very insoluble α-alumina and the dense form of magnesium oxide, these oxides react with water to form hydroxides and with acids to form salts.

$$M_2O + H_2O \longrightarrow 2MOH \qquad (M = Li,\ Na,\ K,\ Rb,\ Cs)$$
$$MO + H_2O \longrightarrow M(OH)_2 \qquad (M = Mg,\ Ca,\ Sr,\ Ba)$$
$$M_2O + 2HCl \longrightarrow 2MCl + H_2O \qquad (M = Li,\ Na,\ K,\ Rb,\ Cs)$$
$$MO + 2HNO_3 \longrightarrow M(NO_3)_2 + H_2O \qquad (M = Mg,\ Ca,\ Sr,\ Ba)$$
$$\gamma\text{-}Al_2O_3 + 3H_2SO_4 \longrightarrow Al_2(SO_4)_3 + 3H_2O$$

The oxides of the active metals also react with many nonmetal oxides and form oxyanions. Examples are shown in Table 13.4.

Table 13.4   Formation of common oxyanions

Name	Formula	Parent nonmetal oxide	Reaction
Borate	BO_3^{3-}	B_2O_3	$B_2O_3 + 3O^{2-} \longrightarrow 2BO_3^{3-}$
Carbonate	CO_3^{2-}	CO_2	$CO_2 + O^{2-} \longrightarrow CO_3^{2-}$
Nitrate	NO_3^-	N_2O_5	$N_2O_5 + O^{2-} \longrightarrow 2NO_3^-$
Phosphate	PO_4^{3-}	P_4O_{10}	$P_4O_{10} + 6O^{2-} \longrightarrow 4PO_4^{3-}$
Sulfite	SO_3^{2-}	SO_2	$SO_2 + O^{2-} \longrightarrow SO_3^{2-}$
Sulfate	SO_4^{2-}	SO_3	$SO_3 + O^{2-} \longrightarrow SO_4^{2-}$
Perchlorate	ClO_4^-	Cl_2O_7	$Cl_2O_7 + O^{2-} \longrightarrow 2ClO_4^-$

When a metal oxide reacts with a nonmetal oxide, a coordinate covalent bond (Section 6.4) is formed. The oxide ion acts as an electron pair donor and the nonmetal oxide acts as an electron pair acceptor. The reaction of sodium oxide with sulfur dioxide is shown as an example.

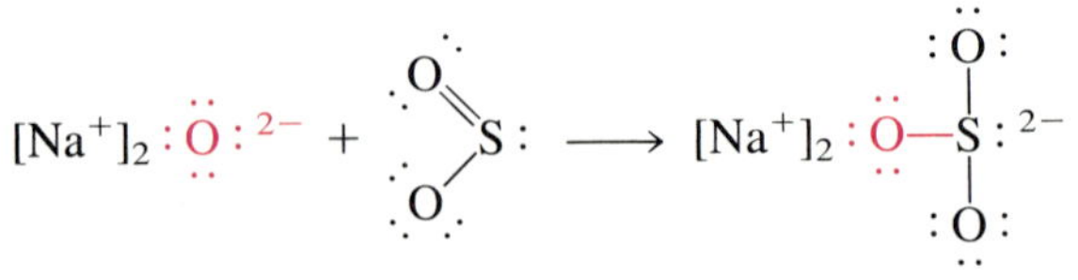

The oxides of the alkali metals are used in the laboratory as sources of the metal ions and of the oxide ion, but they have little industrial utility. On the other hand, magnesium oxide, calcium oxide, and aluminum oxide do. The dense form of magnesium oxide is used widely in making fire brick, crucibles, furnace linings, and thermal insulation, which are all applications that require chemical and thermal stability. Dense magnesium oxide is not very reactive and melts at 2800°C. Calcium oxide, sometimes called **quicklime** or **lime** in the industrial market, is very reactive, and its principal uses reflect its reactivity. The reaction of calcium oxide with water is vigorous and exothermic; consequently, it is used as a drying agent in applications such as the preparation of anhydrous alcohol and the drying of ammonia. Calcium oxide is used in the isolation of magnesium from sea water (Section 13.2, Part 3). Pure calcium oxide emits an intense white light when heated to a high temperature. Blocks of calcium oxide heated by gas flames were used as stage lights in theaters before electricity was available. This is the source of the phrase ''in the limelight.''

Alpha-alumina occurs in nature as the mineral **corundum,** a very hard substance that is used as an abrasive for grinding and polishing. Several precious stones consist of aluminum oxide with certain impurities that impart color: ruby (red, by chromium compounds); sapphire (blue, by compounds of cobalt, chromium, or titanium); oriental amethyst (violet, by manganese compounds); and oriental topaz (yellow, by iron). Artificial rubies and sapphires are now manufactured by melting aluminum oxide (mp = 2050°C) with small amounts of oxides to produce the desired colors and then cooling the melt in such a way as to produce large crystals. These gems are indistinguishable from natural stones, except for microscopic, rounded air bubbles in the synthetic ones and flattened bubbles in the natural stones. The synthetic gems are used not only as jewelry but also as bearings (''jewels'') in some watches and other instruments and as dies through which wires are drawn. Ruby lasers use synthetic ruby crystals. Very finely divided aluminum oxide, called **activated alumina,** is used as a dehydrating agent and a catalyst.

The peroxide ion is also a powerful hydrogen ion acceptor; the peroxides of Groups IA and IIA are strong bases. Solutions of these peroxides are basic due to the reaction of the peroxide ion with water, which functions as a weak acid in this case.

$$O_2^{2-} + H_2O \longrightarrow O_2H^- + OH^-$$
$$O_2H^- + H_2O \rightleftharpoons H_2O_2 + OH^-$$

At one time hydrogen peroxide was prepared commercially by the acid-base reaction of barium peroxide with sulfuric acid. In addition to the basicity of the O_2^{2-} ion, the insolubility of barium sulfate helped force the reaction to completion.

$$BaO_2(aq) + H_2SO_4(aq) \longrightarrow BaSO_4(s) + H_2O_2(aq)$$

Peroxides also are strong oxidizing agents. Sodium peroxide is used as an oxidizing and bleaching agent. It bleaches by oxidizing colored compounds to colorless compounds.

Superoxides are also strong oxidizing agents and are used to provide emergency supplies of oxygen. For example, potassium superoxide is used in gas masks for mine safety work. Moisture from the breath reacts with the superoxide to give oxygen for the wearer of the mask to breathe.

$$4KO_2 + 2H_2O \longrightarrow 4KOH + 3O_2$$

When the breath is exhaled through the mask, carbon dioxide is absorbed by the potassium hydroxide.

$$CO_2 + KOH \longrightarrow KHCO_3$$

The wearer of the mask needs no air from outside the mask.

13.5 Hydroxides of the Active Metals

All of the active metals form ionic **hydroxides.** These compounds contain the metal ion combined with hydroxide ions with the metal ion located in the holes in an extended array of hydroxide ions.

1. PREPARATION OF HYDROXIDES. With the exception of magnesium and aluminum hydroxides, the hydroxides of the active metals can be prepared in the laboratory by reaction of the respective metals with water, a reaction that produces hydrogen and the hydroxide (Figs. 13.2 and 13.3). However, these reactions are violent and can be dangerous, so the soluble hydroxides are generally prepared from the reaction of the respective oxide with water.

$$M_2O + H_2O \longrightarrow 2MOH \qquad (M = Li, Na, K, Rb, Cs)$$
$$MO + H_2O \longrightarrow M(OH)_2 \qquad (M = Ca, Sr, Ba)$$

These reactions are very exothermic; laboratory equipment in which they are run can get very hot.

The insoluble hydroxides of magnesium and aluminum can be prepared by addition of a solution of a soluble hydroxide, such as sodium hydroxide, to a solution of a magnesium or an aluminum salt such as a chloride, nitrate, or sulfate (Fig. 13.14).

$$Mg(NO_3)_2(aq) + 2NaOH(aq) \longrightarrow Mg(OH)_2(s) + 2NaNO_3(aq)$$
$$Al_2(SO_4)_3(aq) + 6NaOH(aq) \longrightarrow 2Al(OH)_3(s) + 3Na_2SO_4(aq)$$

Figure 13.14
Aluminum hydroxide, Al(OH)₃, forms as a white gelatinous precipitate when a solution of NaOH is added to a solution of an aluminum salt, such as Al(NO₃)₃.

An excess of hydroxide solution must be avoided in the preparation of aluminum hydroxide, or it will redissolve with the formation of a complex ion. This ion is commonly written as $[Al(OH)_4]^-$, but it also contains water molecules bonded to the aluminum ion by coordinate covalent bonds and is better written as $[Al(H_2O)_2(OH)_4]^-$.

$$Al(OH)_3(s) + OH^-(aq) + 2H_2O(l) \longrightarrow [Al(H_2O)_2(OH)_4]^-(aq)$$

2. REACTIONS AND USES OF THE HYDROXIDES. The active metal hydroxides exhibit reactions typical of compounds that contain the strongly basic hydroxide ion. They react with acids, and insoluble hydroxides dissolve, forming salts. They react with many of the oxides of the nonmetals to form salts of oxyacids similar to those described in Table 13.4.

The reaction of a metal oxide or a metal hydroxide with an acid is common behavior. However, a small number of metal oxides and hydroxides (Al_2O_3 and $Al(OH)_3$, for example) will also dissolve in a basic solution. When aluminum hydroxide dissolves in an acid, the hydroxide ions combine with the hydrogen ions of the acid to form water and the aluminum salt of the acid. In such a reaction aluminum hydroxide behaves like a base.

$$Al(OH)_3 + 3H^+ \longrightarrow Al^{3+} + 3H_2O$$

When aluminum hydroxide dissolves in base, hydroxide ions and water molecules bind to the aluminum ions, forming the complex ion $[Al(H_2O)_2(OH)_4]^-$ (see Part 1 of this Section). Because aluminum hydroxide removes free hydroxide ions from the solution, it is regarded as an acid in this case. Thus, aluminum hydroxide exhibits both acidic and basic behavior, depending on the nature of the other components in the solution. Compounds that can behave either as an acid or a base are referred to as **amphoteric compounds.** Such substances will be discussed in greater detail in Chapter 16.

Because they are very soluble, the alkali metal hydroxides form the most concentrated solutions of all of the hydroxides. **Sodium hydroxide,** NaOH, is frequently called **caustic soda** in commerce, and its solutions are sometimes referred to as **lye** or **soda lye.** The large quantities of sodium hydroxide used industrially are prepared from sodium chloride, because it is a less expensive starting material than the oxide. Sodium hydroxide was the seventh-ranked chemical in production in the United States in 1986 (over 11 million tons), and this production was almost entirely by electrolysis of solutions of sodium chloride (Fig. 13.15).

$$[2Na^+(aq)] + 2Cl^-(aq) + 2H_2O(l) \xrightarrow{\text{Electrolysis}}$$
$$[2Na^+(aq)] + 2OH^-(aq) + H_2(g) + Cl_2(g)$$

During electrolysis the sodium ions are not reduced. Instead, water, which is more easily reduced than sodium ions, is reduced to hydrogen gas and hydroxide ions at the cathode. The hydroxide ions replace the chloride ions that are oxidized to chlorine gas at the anode, and sodium hydroxide accumulates in the solution. After electrolysis the solution still contains some sodium chloride, but this can be removed relatively easily because it is less soluble than sodium hydroxide in the solution. Upon evaporation of some of the water from the solution, the sodium chloride crystallizes before any solid sodium hydroxide forms and can be removed by filtration. Such a process is called **fractional crystallization.**

Sodium hydroxide is an ionic compound that melts and boils without decomposition. It is very soluble in water, giving off a great deal of heat and forming very basic solutions; 40 grams of sodium hydroxide will dissolve in only 60 grams of water. Sodium hydroxide is used in the production of other sodium compounds and to neutralize acidic solutions during the production of other chemicals such as petrochemicals and polymers.

Potassium hydroxide is manufactured by an electrolytic process similar to that used for sodium hydroxide. It, too, is very soluble in water and forms very alkaline solutions. Potassium hydroxide is more expensive than sodium hydroxide because potassium chloride is more expensive than sodium chloride. Inasmuch as sodium hydroxide can almost always replace potassium hydroxide as a base, the use of potassium hydroxide is limited to the production of other potassium-containing compounds.

The hydroxides of the active metals of Group IIA show a smooth gradation in properties and a steady increase in solubility as the atomic weight of the metal increases. Magnesium hydroxide, $Mg(OH)_2$, is almost insoluble. An aqueous suspension

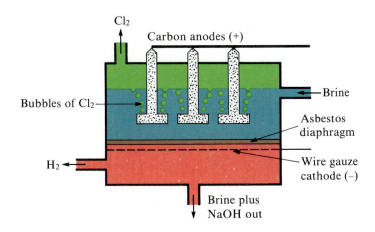

Figure 13.15

A diaphragm cell for the production of sodium hydroxide. Brine, a solution of NaCl, enters the cell at the top and flows through the diaphragm. $Cl_2(g)$ is produced at the anode ($2Cl^- \longrightarrow Cl_2 + 2e^-$); $H_2(g)$ and $OH^-(aq)$ at the cathode ($2H_2O + 2e^- \longrightarrow H_2 + 2OH^-$).

Figure 13.16
Slaked lime, Ca(OH)$_2$, is used to treat acid soil.

of magnesium hydroxide is used as the antacid **milk of magnesia.** Calcium hydroxide, Ca(OH)$_2$, and strontium hydroxide, Sr(OH)$_2$, exhibit limited solubility, whereas barium hydroxide, Ba(OH)$_2$, is sufficiently soluble to form solutions approaching those of the alkali metal hydroxides in hydroxide ion concentration strength.

Calcium hydroxide, Ca(OH)$_2$, known commercially as **slaked lime** or **hydrated lime,** is prepared from calcium oxide by reaction with water (slaking). Because of its activity and cheapness, calcium hydroxide is used more extensively in commercial applications than any other base (Fig. 13.16). Calcium hydroxide has a limited solubility (1.7 grams dissolves in 1 liter of water), but when suspended in water with substances that will react with hydroxide ions, it furnishes a ready supply of hydroxide ions because the solid continues to dissolve as hydroxide ions are removed from solution by reaction. A saturated solution of calcium hydroxide is called **limewater;** a suspension, **milk of lime.**

During 1986 over 15 million tons of lime were produced in the United States, making it fifth in industrial production, after sulfuric acid, nitrogen, oxygen, and ethylene. The manufacture of at least 150 industrial chemicals requires the use of either calcium hydroxide or calcium oxide. Calcium hydroxide is also used in the construction trades. Mortar is a mixture of slaked lime, sand, and water. It dries and hardens (sets) when exposed to air because it reacts with carbon dioxide to form calcium carbonate.

$$Ca(OH)_2 + CO_2 \longrightarrow CaCO_3 + H_2O$$

The crystals of calcium carbonate bind the particles of sand together.

Aluminum hydroxide is widely used to fix dyes to fabrics. Cloth soaked in a hot solution of aluminum acetate becomes impregnated with aluminum hydroxide. If aluminum hydroxide is precipitated from a solution containing a dye, the precipitate is colored. The aluminum hydroxide adsorbs the dye and holds it fast to the cloth. When used in this way, aluminum hydroxide is called a **mordant.**

Aluminum hydroxide is also used in the purification of water, because its gelatinous character enables it to carry down any suspended material in the water, including most of the bacteria. The aluminum hydroxide for water purification is ordinarily produced by the reaction of aluminum sulfate with lime.

$$2Al^{3+} + 3SO_4^{2-} + 3Ca^{2+} + 6OH^- \longrightarrow 2Al(OH)_3(s) + 3CaSO_4(s)$$

Lime, Ca(OH)$_2$, is dumped on a lake to neutralize the effects of acid rain.

13.6 Carbonates and Hydrogen Carbonates of the Active Metals

The active metals of Groups IA and IIA form ionic **carbonates**—compounds that contain the carbonate ion, CO_3^{2-}. The reaction of solutions of carbonates with carbon dioxide yields **hydrogen carbonates**—compounds containing the HCO_3^- ion.

1. PREPARATION OF THE CARBONATES AND HYDROGEN CARBONATES. Most solid oxides and hydroxides of the alkali metals and of the alkaline earth metals react with carbon dioxide to form carbonates.

$$M_2O + CO_2 \longrightarrow M_2CO_3 \qquad (M = Li, Na, K, Rb, Cs)$$
$$2MOH + CO_2 \longrightarrow M_2CO_3 + H_2O \qquad (M = Li, Na, K, Rb, Cs)$$
$$MO + CO_2 \longrightarrow MCO_3 \qquad (M = Ca, Sr, Ba)$$
$$M(OH)_2 + CO_2 \longrightarrow MCO_3 + H_2O \qquad (M = Mg, Ca, Sr, Ba)$$

Solid magnesium oxide does not react with carbon dioxide. Aluminum carbonate is unknown.

Alkali metal hydrogen carbonates, $MHCO_3$, form in solution when solutions of the hydroxides are saturated with carbon dioxide; the formation of sodium hydrogen carbonate is shown by the equation

$$[Na^+(aq)] + OH^-(aq) + CO_2(g) \longrightarrow [Na^+(aq)] + HCO_3^-(aq)$$

The use of less carbon dioxide or the addition of an equivalent of the hydroxide to a solution of the hydrogen carbonate gives a solution of the carbonate ion; again, the sodium salts are used as examples.

$$[2Na^+(aq)] + 2OH^-(aq) + CO_2(g) \longrightarrow [2Na^+(aq)] + CO_3^{2-}(aq) + H_2O(l)$$
$$[Na^+(aq)] + HCO_3^-(aq) + [Na^+(aq)] + OH^-(aq) \longrightarrow$$
$$[2Na^+(aq)] + CO_3^{2-}(aq) + H_2O(l)$$

Lithium carbonate is not very soluble and precipitates from solution more readily than the other alkali metal carbonates.

Alkaline earth carbonates are not soluble; they precipitate when solutions of soluble alkali metal carbonates are added to solutions of soluble alkaline earth metal salts. The following equation uses barium chloride and sodium carbonate as an example.

$$Ba^{2+}(aq) + [2Cl^-(aq)] + [2Na^+(aq)] + CO_3^{2-}(aq) \longrightarrow$$
$$BaCO_3(s) + [2Na^+(aq)] + [2Cl^-(aq)]$$

Although insoluble in pure water, the alkaline earth carbonates dissolve readily in water that contains dissolved carbon dioxide because of the formation of hydrogen carbonate salts. The formation of calcium hydrogen carbonate is typical.

$$CaCO_3(s) + CO_2(aq) + H_2O(l) \longrightarrow Ca^{2+}(aq) + 2HCO_3^-(aq)$$

Hydrogen carbonates of the alkaline earth metals are stable only in solution; evaporation of the solution produces the carbonates.

Sodium carbonate is prepared industrially in the United States by extraction from the mineral **trona**, $Na_2CO_3 \cdot NaHCO_3 \cdot 2H_2O$. Following recrystallization from water to remove clay and other impurities, trona is roasted to produce Na_2CO_3.

$$2(Na_2CO_3 \cdot NaHCO_3 \cdot 2H_2O) \xrightarrow{\Delta} 3Na_2CO_3 + 5H_2O(g) + CO_2(g)$$

Sodium carbonate is also manufactured by the **Solvay process.** This process is based on the reaction of ammonium hydrogen carbonate with a saturated solution of sodium chloride. Sodium hydrogen carbonate precipitates, since it is only slightly soluble in the reaction medium.

$$Na^+ + [Cl^-] + [NH_4^+] + HCO_3^- \longrightarrow NaHCO_3(s) + [NH_4^+] + [Cl^-]$$

The basic raw materials used in the process are limestone and common salt. The carbon dioxide is generated by heating limestone.

$$CaCO_3 \xrightarrow{\triangle} CaO + CO_2 \tag{1}$$

The ammonia is obtained by treating ammonium chloride, formed as a byproduct of the process, with calcium hydroxide.

$$2NH_4^+ + [2Cl^-] + [Ca^{2+}] + 2OH^- \longrightarrow$$
$$2NH_3(g) + 2H_2O + [Ca^{2+}] + [2Cl^-] \tag{2}$$

Calcium hydroxide results from slaking the lime produced during generation of carbon dioxide [Equation (1)].

$$CaO + H_2O \longrightarrow Ca^{2+} + 2OH^- \tag{3}$$

Aqueous ammonia reacts with the carbon dioxide, yielding ammonium hydrogen carbonate.

$$NH_3 + CO_2 + H_2O \longrightarrow NH_4^+ + HCO_3^- \tag{4}$$

Sodium hydrogen carbonate is also a valued product of the Solvay process; after being freed of ammonium chloride by recrystallization, it is made commercially available. The major portion of the sodium hydrogen carbonate, however, is converted to sodium carbonate by heating.

$$2NaHCO_3 \xrightarrow{\triangle} Na_2CO_3 + H_2O + CO_2 \tag{5}$$

The carbon dioxide from this reaction is used to produce more sodium hydrogen carbonate. The only by-product of the entire process not reused in the process is calcium chloride.

When a solution of sodium carbonate is evaporated below 35.2°C, **sodium carbonate 10-hydrate, Na$_2$CO$_3$·10H$_2$O,** known as **washing soda,** crystallizes out; above this temperature the monohydrate Na$_2$CO$_3$·H$_2$O crystallizes out. Heating either hydrate produces the anhydrous compound.

2. REACTIONS AND USES OF THE CARBONATES AND HYDROGEN CARBONATES. Carbonates are moderately strong bases. Aqueous solutions are basic because the carbonate ion accepts a hydrogen ion from water, as shown by the equation

$$CO_3^{2-} + H_2O \rightleftharpoons HCO_3^- + OH^-$$

They react with acids forming salts of the metal, gaseous carbon dioxide, and water (Fig. 13.17); potassium carbonate and nitric acid illustrate the reaction.

$$K_2CO_3(s) + 2HNO_3(aq) \longrightarrow 2KNO_3(aq) + CO_2(g) + H_2O(l)$$

Hydrogen carbonates are also bases, but much weaker bases than the carbonates.

Sodium carbonate, commonly called **soda ash** or **soda** in the industrial market, is an important chemical. In 1986 it was eleventh among industrial chemicals in the United States with a production of 8.6 million tons. It is used extensively in the

Figure 13.17

The reaction of an acid and a carbonate salt produces gaseous carbon dioxide.

manufacture of glass, other chemicals, soap, paper and pulp, cleansers, and water softeners and in the refining of petroleum.

Sodium hydrogen carbonate, $NaHCO_3$, is commonly called **bicarbonate of soda** or **baking soda.** Baking powders contain a mixture of baking soda and an acid such as potassium hydrogen tartrate, $KHC_4H_4O_6$ (cream of tartar), or calcium dihydrogen phosphate, $Ca(H_2PO_4)_2$. As long as the powder is dry no reaction occurs, but as soon as water is added the acid reacts with the hydrogen carbonate anion and carbon dioxide is formed.

$$HC_4H_4O_6^- + HCO_3^- \longrightarrow C_4H_4O_6^{2-} + CO_2 + H_2O$$

$$H_2PO_4^- + HCO_3^- \longrightarrow HPO_4^{2-} + CO_2 + H_2O$$

The carbon dioxide is trapped in the dough and expands when warmed, producing the texture characteristic of biscuits and other quickbreads.

The lactic acid of sour milk and buttermilk or the acetic acid in vinegar will also furnish hydrogen ion and thus serve the same purpose as the acidic component of baking powders. Thus it is possible to make dough rise using sour milk, buttermilk, or vinegar with ordinary sodium hydrogen carbonate (baking soda).

It is rarely necessary to manufacture **calcium carbonate** because it is widely distributed as limestone, a mineral second in abundance only to silicate rocks in the earth's crust. Coral reefs are composed of calcium carbonate deposited by marine animals. Pearls are concentric layers of calcium carbonate deposited around a foreign particle, such as a grain of sand, that has entered the shell of an oyster. Magnesium carbonate, $MgCO_3$, is found in nature as the mineral **magnesite.**

Limestone is the most widely used naturally occurring building stone used in the United States; most of it is quarried in southern Indiana. Most limestone is composed of particles of **calcite,** the low-temperature form of calcium carbonate that results when precipitation occurs below 30°C. Above 30°C calcium carbonate crystallizes in the high-temperature form known as **aragonite.** A single crystal of calcite possesses the property of birefringence, or double refraction; when a beam of light enters the crystal, it is broken into two beams (Fig. 13.18).

Figure 13.18
Calcite, or Iceland spar, has the unusual property of birefringence, or double refraction. Notice that the calcite produces two images of the word CALCITE, while the glass produces only one image.

When water containing carbon dioxide comes into contact with limestone rocks, they dissolve because **calcium hydrogen carbonate** is formed. This reaction is responsible for the formation of caves in beds of limestone rock. If water containing calcium hydrogen carbonate finds its way into a cave where the $Ca(HCO_3)_2$ can liberate carbon dioxide, calcium carbonate may again be deposited.

$$Ca^{2+}(aq) + 2HCO_3^-(aq) \longrightarrow CaCO_3(s) + CO_2(g) + H_2O(l)$$

This deposition gives the beautiful limestone formations that are found in many caves (Fig. 13.19).

Figure 13.19
Stalactites and stalagmites are cave formations of calcium carbonate. These are in the Luray Caverns in Virginia.

Calcium hydrogen carbonate is readily converted to calcium carbonate when its solutions are heated. This reaction is often responsible for the formation of scale in kettles and boilers and is the basis for one method of removing carbonate hardness from water (Section 29.11).

13.7 Halides of the Active Metals

The binary compounds of the active metals with the halogens are called **halides.** With the exception of $AlCl_3$, $AlBr_3$, and AlI_3, the active metal halides are ionic compounds.

1. PREPARATION OF THE HALIDES. The halides of the active metals form readily when the metals react directly with elemental halogens or with solutions of HF, HCl, HBr, or HI in water. However, with the partial exceptions of magnesium and aluminum, these reactions can be extremely violent and dangerous. Less vigorous reactions are generally used to prepare the halides.

Solutions of the halides of the active metals are prepared in the laboratory by addition of hydrofluoric acid, HF(*aq*), hydrochloric acid, HCl(*aq*), hydrobromic acid, HBr(*aq*), or hydriodic acid, HI(*aq*), to carbonates, hydroxides, oxides, or other compounds containing basic anions. Sample reactions are

$$Na_2CO_3(s) + 2H^+(aq) + 2F^-(aq) \longrightarrow 2Na^+(aq) + 2F^-(aq) + H_2O(l) + CO_2(g)$$

$$Mg(OH)_2(s) + 2H^+(aq) + 2Br^-(aq) \longrightarrow Mg^{2+}(aq) + 2Br^-(aq) + 2H_2O(l)$$

$$\gamma\text{-}Al_2O_3(s) + 6H^+(aq) + 6I^-(aq) \longrightarrow 2Al^{3+}(aq) + 6I^-(aq) + 3H_2O(l)$$

Evaporation of the water gives anhydrous halides of sodium, potassium, rubidium, and cesium. The other active metals form hydrated halides such as $LiCl \cdot 3H_2O$, $CaCl_2 \cdot 6H_2O$, $BaCl_2 \cdot 8H_2O$, and $AlCl_3 \cdot 6H_2O$, where the center dots in the formulas indicate that the water is present in the solids as **water of hydration;** that is, water that is present as identifiable molecules. These molecules are held in the solid by ion-dipole attractions (Section 12.10) between the metal ions and the polar water molecules and by hydrogen bonds (Section 11.12) between the hydrogen of the water molecule and the negatively charged halide anions. With the exception of magnesium and aluminum halides, the hydrated halides may be converted to anhydrous halides by heating.

Anhydrous halides of magnesium and aluminum cannot be prepared simply by heating the hydrated halides. The hydrated halides decompose upon heating, giving water vapor, the hydrogen halide, and the metal oxide; two typical examples are

$$MgCl_2 \cdot 6H_2O(s) \xrightarrow{\text{heat}} MgO(s) + 2HCl(g) + 5H_2O(g)$$

$$2(AlBr_3 \cdot 6H_2O)(s) \xrightarrow{\text{heat}} Al_2O_3(s) + 6HBr(g) + 9H_2O(g)$$

The oxides are formed in these cases because they are stabilized by large lattice energies (Section 11.20) resulting from the small metal ions with charges of +2 and +3. Anhydrous halides of magnesium and aluminum are prepared by heating the metal or the hydrated halide in a stream of the gaseous hydrogen halide, which prevents formation of the oxides.

$$Mg + 2HBr(g) \xrightarrow{\text{Heat}} MgBr_2 + H_2$$

$$AlCl_3 \cdot 6H_2O \xrightarrow[\text{heat}]{HCl(g)} AlCl_3 + 6H_2O(g)$$

Anhydrous aluminum chloride, bromide, and iodide are aluminum compounds that exhibit some covalent character by virtue of the relatively high charge (+3) and small size of the aluminum ion. Aluminum chloride is a white crystalline solid that sublimes at 180°C and is soluble in organic solvents that do not solvate metal ions (Section 12.10). Measurement of the density of aluminum chloride vapor shows that it is dimeric in the gas phase, with a formula of Al_2Cl_6. In these molecules each aluminum atom is tetrahedrally surrounded by four chlorine atoms, with two of the chlorine atoms (shown in color) bridging the two aluminum atoms.

Aluminum
chloride

Each aluminum atom possesses an octet of electrons. X-ray studies show that there are no discrete Al_2Cl_6 molecules in solid $AlCl_3$. The structure consists of a closest packed array of chloride ions (Section 11.16), with aluminum ions in the octahedral holes. Thus the aluminum changes its coordination number from 6 to 4 when the solid melts or sublimes.

2. REACTIONS AND USES OF THE HALIDES. **Sodium chloride** or **rock salt,** NaCl, the chloride used most extensively in the chemical industry, is one of the most abundant minerals (Fig. 13.20). Sea water contains 2.7% sodium chloride, and the waters of the Dead Sea contain 27%. There are vast deposits of rock salt in Germany, and in the United States in New York, Michigan, West Virginia, and California; a very extensive bed (about 400–500 feet thick) underlies parts of Oklahoma, Texas, and Kansas. Most of the salt consumed comes from salt beds rather than from sea water. The salt is removed from these underground beds by mining or by forcing water down into the deposits to form saturated brines, which are then pumped to the surface. Natural sodium chloride contains other soluble salts that make it unsuitable for many uses, so it is purified by dissolving it in water, concentrating the solution by evaporation (often under reduced pressure), and allowing the crystals to form again. Some of the impurities are more soluble than sodium chloride and remain in solution when the

Figure 13.20
Salt is mined from beds up to 500 feet thick. This mine is near Cleveland, Ohio.

salt crystallizes. Some impurities that are less soluble than salt do not crystallize out because they are present in small amounts. Other less soluble impurities, such as calcium sulfate, would crystallize with the salt but are removed by taking advantage of their decreasing solubility in sodium chloride solution as the temperature is raised above 80°C. Thus, if the brine is filtered hot, the resulting solution will not contain enough of these impurities to be saturated when cooled. Hence they cannot precipitate with sodium chloride, and sodium chloride crystallizes relatively free of the impurities.

Sodium chloride is the usual source of sodium or chlorine in almost all other compounds containing these elements. Its most extensive use is in the manufacture of chlorine, hydrochloric acid, sodium hydroxide, and sodium carbonate. Sodium chloride is an essential constituent of the foods of animals. It not only makes food more palatable but is also the source of chlorine for the production of hydrochloric acid, a constituent of gastric juice. Sodium chloride is also a constituent of the blood and is essential for the life processes of the human body. However, excessive consumption of salt should be avoided, especially by those with potential heart problems or high blood pressure.

Calcium chloride, $CaCl_2$, is obtained from brines and can be prepared commercially and in the laboratory by treating calcium oxide, hydroxide, or carbonate with hydrochloric acid. The salt crystallizes from water as the hexahydrate $CaCl_2 \cdot 6H_2O$, which is converted by heating to the monohydrate $CaCl_2 \cdot H_2O$. Complete dehydration occurs when the monohydrate is heated to a higher temperature. The anhydrous salt and the monohydrate are used extensively as drying agents for gases and liquids. Calcium chloride 6-hydrate is very soluble in water and with ice makes an excellent freezing mixture, giving temperatures as low as −55°C. Solutions of calcium chloride are quite commonly used as cooling brines in refrigeration plants. Because it is a very deliquescent salt, $CaCl_2 \cdot 6H_2O$ is used to keep dust down on highways and in coal mines. Calcium chloride is frequently used to remove snow and ice from roads and walks.

Strontium chloride 6-hydrate, $SrCl_2 \cdot 6H_2O$, is used, as are other salts of strontium, to produce a red color in fireworks and flares (Fig. 13.21).

Figure 13.21
The red colors in these fireworks result when strontium salts are heated.

Barium chloride, BaCl$_2$, is prepared by the high-temperature reduction of barium sulfate with carbon to form barium sulfide, which is then treated with hydrochloric acid.

$$BaSO_4 + 4C \xrightarrow{\Delta} BaS(s) + 4CO(g)$$

$$BaS + 2H^+(aq) + [2Cl^-(aq)] \longrightarrow Ba^{2+}(aq) + [2Cl^-(aq)] + H_2S(g)$$

When solutions of barium chloride are evaporated, the dihydrate BaCl$_2 \cdot$ 2H$_2$O separates. When a solution of the barium ion is required, this salt is usually employed. Like all other soluble barium salts, the chloride is very poisonous.

The aluminum halides are commonly used as catalysts. The antiperspirant "aluminum chlorohydrate," Al$_2$(OH)$_5$Cl $\cdot$ 2H$_2$O, is an astringent; it acts by constricting the openings of sweat glands to reduce the amount of perspiration that escapes. Other aluminum compounds are used as astringents to control bleeding from cuts and scrapes.

13.8 Hydrides of the Active Metals

With the exception of aluminum, the active metals react with hydrogen at slightly elevated temperatures giving ionic metal hydrides.

$$2M + H_2 \longrightarrow 2MH \qquad (M = Li, Na, K, Rb, Cs)$$

$$M + H_2 \longrightarrow MH_2 \qquad (M = Mg, Ca, Sr, Ba)$$

Aluminum hydride can be prepared by the metathetical reaction of anhydrous aluminum trichloride with lithium aluminum hydride, LiAlH$_4$, in solution in diethyl ether; aluminum hydride precipitates from the solution.

$$3LiAlH_4 + AlCl_3 \xrightarrow{Ether} 4AlH_3 + 3LiCl$$

The binary hydrides of the metals of Groups IA and IIA are ionic compounds containing the hydride anion, H$^-$. They crystallize with structures similar to those of the halides, with the hydride ion in the location occupied by the halide ion. For example, sodium hydride, NaH, and potassium hydride, KH, crystallize with the sodium chloride structure (Fig. 13.22). Aluminum hydride is one of the covalent compounds of aluminum.

The hydrides are readily decomposed by water with the liberation of hydrogen. Hydride ions are strongly basic and pick up hydrogen ions from water,

$$H^- + H_2O \longrightarrow H_2(g) + OH^-(aq)$$

leaving hydroxide ions in solution. Hydrides are also strong reducing agents, because hydrogen is not particularly stable, having an oxidation number of -1.

13.9 Other Salts of the Active Metals

Thousands of salts of the active metals have been prepared. Generally these are formed from the metals or from the common compounds that we have just discussed by acid-base, oxidation-reduction, or metathetical reactions.

1. PREPARATION OF SALTS. Examples of reactions that can be used to form a variety of additional salts of the active metals are described below. In many cases these reactions are very similar to those used to produce the various compounds described in the preceding sections.

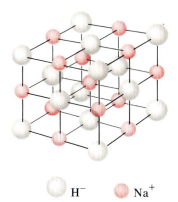

H$^-$ Na$^+$

Figure 13.22

The crystal structure of NaH. This is the same structure as that of NaCl (Fig. 11.39, Section 11.18). The smaller spheres (red) represent sodium ions; the larger ones (brown) hydride ions. The upper structure is an "expanded" view that shows the geometry more clearly.

1. *Reactions of acids with basic metal compounds.* Oxides, hydroxides, and carbonates of the active metals react with solutions of most acids to form solutions of salts, which can be isolated by evaporation of the solvent. Like the halides, the solid compounds may contain water of hydration. The reaction of magnesium hydroxide with a perchloric acid solution is an example.

$$Mg(OH)_2 + 2HClO_4 + 4H_2O \longrightarrow Mg(ClO_4)_2 \cdot 6H_2O$$

If the acid is a liquid, the reaction can be carried out without a solvent. Calcium propionate, a preservative used in baked goods, can be prepared by the reaction of calcium carbonate and pure propionic acid, a liquid.

$$CaCO_3 + 2C_2H_5CO_2H \longrightarrow Ca(C_2H_5CO_2)_2 + CO_2 + H_2O$$

The product contains the propionate anion, $C_2H_5CO_2^-$.

2. *Oxidation-reduction reactions.* Because the active metals are generally very reactive, oxidation-reduction reactions utilizing the metals are difficult to control and are generally restricted to preparation of compounds of magnesium or aluminum. The other metals are sometimes used when it is necessary to prepare a salt of a particularly unreactive substance or to prepare a compound that decomposes in water. Oxidation-reduction reactions involving the anions in compounds of the active metals are common.

Magnesium sulfide and aluminum sulfide decompose in water, but they can be prepared by the reaction of the metal with sulfur or by heating the metal in a stream of hydrogen sulfide.

$$Mg + H_2S(g) \longrightarrow MgS + H_2(g)$$
$$2Al + 3H_2S(g) \longrightarrow Al_2S_3 + 3H_2(g)$$

Some salts are prepared by oxidation or reduction of the anion in a second salt; for example, sodium sulfide is prepared by reduction of sodium sulfate with carbon.

$$Na_2SO_4 + 2C \xrightarrow{900°C} Na_2S + 2CO_2$$

3. *Metathetical reactions.* Insoluble salts can often be prepared by mixing solutions of two soluble compounds. Barium sulfate is prepared in this way from the soluble salts barium chloride and sodium sulfate.

$$Ba^{2+}(aq) + [2Cl^-(aq)] + [2Na^+(aq)] + SO_4^{2-}(aq) \longrightarrow$$
$$BaSO_4(s) + [2Na^+(aq)] + [2Cl^-(aq)]$$

Calcium dihydrogen phosphate monohydrate, $Ca(H_2PO_4)_2 \cdot H_2O$, used in baking powders, can be prepared by mixing solutions of calcium nitrate and potassium dihydrogen phosphate.

$$Ca^{2+}(aq) + [2NO_3^-(aq)] + [2K^+(aq)] + 2H_2PO_4^-(aq) + H_2O \longrightarrow$$
$$Ca(H_2PO_4)_2 \cdot H_2O(s) + [2K^+(aq)] + [2NO_3^-(aq)]$$

Other metathetical reactions occur because a volatile product is formed. Both sodium sulfate, Na_2SO_4, and sodium hydrogen sulfate, $NaHSO_4$, can be prepared from the reaction of sodium chloride and sulfuric acid, because hydrogen chloride is a volatile product.

$$NaCl + H_2SO_4 \longrightarrow NaHSO_4 + HCl(g)$$
$$2NaCl + H_2SO_4 \longrightarrow Na_2SO_4 + 2HCl(g)$$

2. REACTIONS AND USES OF SALTS. Simple salts of the active metals find a variety of industrial applications and some of them occur in nature as minerals. Salts of the alkali metals are common components of underground brines.

Potassium nitrate is formed in nature by the decay of organic matter and is prepared commercially from sodium nitrate and potassium chloride.

$$Na^+(aq) + [NO_3^-(aq)] + [K^+(aq)] + Cl^-(aq) \longrightarrow$$
$$NaCl(s) + [K^+(aq)] + [NO_3^-(aq)]$$

The sodium nitrate and potassium chloride are dissolved in hot water, and the solution is evaporated by boiling. Of the four compounds possible in a solution containing Na^+, NO_3^-, K^+, and Cl^-, sodium chloride is the least soluble in hot water, so it is the first to crystallize when the concentration of the solution is increased by evaporation. On the other hand, potassium nitrate is very soluble in hot water, making it possible to separate the two salts by filtration. Sodium chloride has about the same solubility in both cold and hot water, so very little more of it crystallizes out when the filtrate is cooled. However, potassium nitrate is only slightly soluble in cold water, and thus it crystallizes as the filtrate is cooled. Potassium nitrate is a valuable fertilizer because it furnishes both potassium and nitrogen in forms that are readily utilized by growing plants.

Extensive deposits of **sodium sulfate** occur in Canada, North Dakota, and the southwestern section of the United States, but most sodium sulfate is obtained as a by-product of the manufacture of hydrochloric acid from salt and sulfuric acid. The sodium sulfate thus formed is commonly called **salt cake** and is used in the manufacture of glass, paper, rayon, coal-tar dyes, and soap. Sodium sulfate ranked 47th in production among all chemicals in the United States in 1986, with a total production of 775,000 tons. **Potassium sulfate, K_2SO_4,** is also obtained from natural deposits and is used as a source of potassium in fertilizers.

Magnesium sulfate is found as the minerals kieserite, $MgSO_4 \cdot H_2O$, and epsomite, $MgSO_4 \cdot 7H_2O$. The heptahydrate in pure form is familiar as **Epsom salts.** It is used in medicine as a purgative, particularly in veterinary practice, in weighting cotton and silk, for polishing, and in insulation.

Anhydrous calcium sulfate occurs in nature as the mineral anhydrite, $CaSO_4$, and as the dihydrate, gypsum, $CaSO_4 \cdot 2H_2O$. The latter is found in enormous deposits. Even though it is low in solubility, calcium sulfate is largely responsible for the noncarbonate hardness of ground waters (Section 29.10).

When gypsum is heated, it loses water, forming the hemihydrate $(CaSO_4)_2 \cdot H_2O$.

$$2(CaSO_4 \cdot 2H_2O) \longrightarrow (CaSO_4)_2 \cdot H_2O + 3H_2O$$

The hemihydrate is known as **plaster of paris.** When it is ground to a fine powder and mixed with water, it sets by forming small interlocking crystals of gypsum. The setting results in an increase in volume, so the plaster fits tightly any mold into which it is poured. Plaster of paris is used extensively as a component of plaster for the interiors of buildings and in making statuary, stucco, wallboard, and casts of various kinds.

Gypsum is also used in making portland cement (Section 25.19), blackboard crayon (erroneously called chalk), plate glass, terra cotta, pottery, and orthopedic and dental plasters. Anhydrous calcium sulfate is sold under the name Drierite as a drying agent for gases and organic liquids.

Strontium sulfate occurs in nature in the mineral celestite, which is sometimes found in caves in the form of beautiful crystals whose faint tinge of blue suggested the name of the mineral.

The mineral barite, which is principally composed of **barium sulfate, BaSO$_4$,** is the usual source of barium in the production of other barium compounds. Since the sulfate is insoluble in acids, it is first converted to the sulfide by a high-temperature reduction with carbon. Barium sulfide is then dissolved in acids such as hydrochloric or nitric to produce the desired salts.

A mixture of BaSO$_4$ and ZnS is known as **lithopone** and is used as a white paint pigment. It is made by the reaction of aqueous solutions of barium sulfide and zinc sulfate.

$$Ba^{2+} + S^{2-} + Zn^{2+} + SO_4^{2-} \longrightarrow BaSO_4(s) + ZnS(s)$$

Barium sulfate is used in taking X-ray photographs of the intestinal tract (Fig. 13.23) because it is opaque to X rays. Even though Ba^{2+} is poisonous, BaSO$_4$ is so slightly soluble that it is nonpoisonous.

Aluminum sulfate 18-hydrate, Al$_2$(SO$_4$)$_3$ · 18H$_2$O, is prepared by treating bauxite or clay with sulfuric acid.

$$\underset{\text{Clay}}{Al_2Si_2O_5(OH)_4} + 6H^+ + [3SO_4^{2-}] \longrightarrow 2H_2SiO_3(s) + 2Al^{3+} + [3SO_4^{2-}] + 3H_2O$$

The metasilicic acid (H$_2$SiO$_3$) is removed by filtration. The sulfate is the cheapest soluble salt of aluminum, so it is used as a source of aluminum hydroxide for the purification of water, the dyeing and waterproofing of fabrics, and the sizing of paper.

When solutions of potassium sulfate and aluminum sulfate are mixed and concentrated by evaporation, crystals of a salt called **potassium alum, KAl(SO$_4$)$_2$ · 12H$_2$O,** are formed. It is the commonest of a large class of salts known as **alums** and is frequently referred to simply as alum. The unipositive cation in an alum may be an alkali metal ion, a silver ion, or the ammonium ion; the tripositive cation may be an aluminum, chromium, iron, manganese, or other ion. The selenate ion, SeO$_4^{2-}$, may replace the sulfate ion. Thus we have ammonium alum, NH$_4$Al(SO$_4$)$_2$ · 12H$_2$O, sodium iron(III) alum, NaFe(SO$_4$)$_2$ · 12H$_2$O, and potassium chromium alum, KCr(SO$_4$)$_2$ · 12H$_2$O. The alums are all isomorphous; that is, they have the same crystal structure.

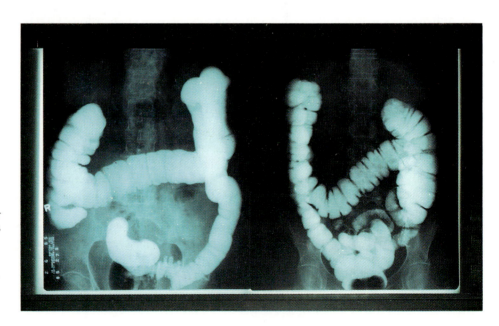

Figure 13.23

The contrast of the large intestine in this X-ray photograph has been enhanced by BaSO$_4$, which absorbs X rays to a much greater extent than body tissues or bone. The left part of the photograph shows diverticulitis of the colon; the right part shows a normal colon.

13.10 Ionic Sizes and the Chemistry of the Active Metals

For the most part, the active metals form compounds in which the metals are present as ions. These elements do not exhibit multiple oxidation states, and they are neither oxidized nor reduced during the vast majority of the reactions involving their compounds. Thus, any differences in the chemical behavior of these compounds must be attributable either to the anion present or to the size and charge of the metal ion. Within a group, any differences between compounds with the same anion must be due to differences in ionic sizes of the anions. As examples of the effect of ionic size, we will consider how the behavior of lithium differs from that of the other alkali metals, the solubilities of the active metal sulfates, and the thermal stabilities of the alkaline earth carbonates.

1. THE BEHAVIOR OF LITHIUM. Some distinctive differences in the behavior of lithium compounds, when compared to compounds of the other alkali metals, are related to the relatively small size of the lithium ion. The radius of Li^+ is 0.60 Å; the radii of Na^+, K^+, Rb^+, and Cs^+ ions are 0.95, 1.33, 1.48, and 1.69 Å, respectively. A table of atomic and ionic radii may be found inside the back cover of this book.

The stability of a solid ionic compound is related to its lattice energy (Section 11.20); compounds with larger lattice energies are more stable than those with smaller lattice energies. As discussed in Section 11.20, the lattice energy, U, of a solid depends on (1) the charge on the ions, Z^+ and Z^-, and (2) the distance, R_0, between the positive ion (the cation) and the negative ion (the anion).

$$U = \frac{C(Z^+Z^-)}{R_0}$$

Thus, lattice energies increase with increasing charge on the ions or with decreasing distance between the negative and positive ions (as the cations and/or anions become smaller).

Lithium hydride is stable to about 900°C, whereas sodium hydride decomposes at about 350°C. Lithium hydroxide and lithium carbonate decompose upon heating and form the oxide; the hydroxides and carbonates of the other alkali metals simply melt. Lithium fluoride and lithium carbonate are not soluble; the other fluorides and carbonates are. All of these differences reflect the smaller size of the Li^+ ion and the correspondingly larger lattice energies of lithium compounds, especially those with small anions, compared to the other alkali metals.

The solubilities and thermal stabilities of lithium compounds resemble those of magnesium compounds—behaviors that are reflected in the so-called diagonal relationships described in Section 13.11.

2. SOLUBILITIES OF GROUP IIA SULFATES. The solubilities of the sulfates of the active metals of Group IIA decrease regularly with increasing atomic number of the metal (Table 13.5). Magnesium sulfate, $MgSO_4$, is soluble; barium sulfate, $BaSO_4$, is insoluble. These differences in solubility can be interpreted as resulting from the change in the size of the metal ions in these compounds.

For an ionic compound to dissolve in water the crystal lattice must be broken down into its component ions. This requires energy, which is provided during hydration of the ions with the formation of ion-dipole bonds between the ions and water (Section 12.10). If the energy given up as the hydrated ions form is greater than the lattice

Table 13.5 Molar solubilities of alkaline earth sulfates

$MgSO_4$	2.80 mol/L
$CaSO_4$	1.4×10^{-3} mol/L
$SrSO_4$	7.6×10^{-5} mol/L
$BaSO_4$	1.1×10^{-6} mol/L

energy, the compound should be soluble. If it is less than the lattice energy, the compound should be relatively insoluble. As we proceed in sequence from $MgSO_4$ to $BaSO_4$, the size of the metal ion increases. In turn, the strength of the ion-dipole attractions between the positive ions and water molecules decreases because these attractions, like ion-ion attractions (Section 11.20), decrease as the distance increases. Thus progressively less energy is given up as the larger ions form ion-dipole attractions. On the other hand, the lattice energies of the sulfates do not change much upon going from magnesium to barium because the sulfate ion is large relative to the metal ions. In those compounds where the anion is large relative to the size of the cation, changes in the size of the cation have little effect on the lattice energy because the size of the anion dominates the cation-anion distance. With smaller anions, the cation plays a more significant role. Thus as we proceed from $MgSO_4$ to $BaSO_4$, the energy required to break up the lattice into ions does not change dramatically, but the energy available to break it up (that is, the energy released due to the formation of ion-dipole attractions) decreases. The solubility of these compounds reflects these changes in energy.

3. DECOMPOSITION OF THE GROUP IIA CARBONATES. As noted in Section 13.4, magnesium carbonate decomposes to the metal oxide and carbon dioxide at about 540°C, while barium carbonate decomposes at about 1360°C. The intermediate carbonates decompose at intermediate temperatures.

The increase in decomposition temperature with increasing atomic weight results from the increasing ionic size of the metal ions in these compounds. As in the series of sulfates, and for the same reason, the lattice energies of the Group IIA carbonates do not vary much with the size of the metal ion, because the anion (carbonate ion in this case) is a relatively large ion. However, the lattice energies of the oxides, which result from the decomposition, are much more sensitive to the changing size of the metal ion. This is because the oxide ion is smaller than the carbonate ion and is, in fact, about the same size as the cations in these systems. All of these decompositions are endothermic and require heating to proceed. However, the formation of magnesium oxide as a product, with its higher lattice energy, releases more energy to help compensate for the energy necessary to decompose the carbonate than does formation of barium oxide, with its smaller lattice energy. Thus the decomposition of the carbonates of the lighter metals is less endothermic and requires less heat from outside the system. Therefore, the decomposition reactions proceed at lower temperatures for the carbonates of the lighter alkaline earth metals than for those of the heavier ones.

13.11 Diagonal Relationships

In some respects the properties of the first elements of Groups IA and IIA resemble those of the second elements in the adjacent groups more than those of the heavier members of their own group. This behavior results in the **diagonal relationships** indicated by the arrows.

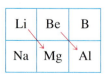

The diagonal relationship in the properties of these elements results from similarities in their ionic radii and their similar high charge densities. The charge density is simply the charge of an ion divided by its ionic radius. The radius of Li^+, 0.60 Å, is closer to that of Mg^{2+}, 0.65 Å, than to that of Na^+, 0.95 Å. Both Be^{2+} and Al^{3+} are small ions, with radii of 0.31 and 0.50 Å, respectively. The charge density of Be^{2+}, 6.5, is closer to that of Al^{3+}, 6.0, than to that of Mg^{2+}, 3.1.

A comparison of the properties of lithium compounds with compounds of the other alkali metals and of magnesium illustrate the relationship. Lithium is the only alkali metal to form an oxide when it burns in an excess of oxygen, and it forms a nitride, Li_3N, with nitrogen. The other alkali metals do not behave this way. Magnesium burns in oxygen forming MgO and reacts with N_2 forming the nitride Mg_3N_2. The fluorides, carbonates, and phosphates of both lithium and magnesium are insoluble, whereas those of the other alkali metals are soluble.

Beryllium resembles aluminum more than the other alkaline earth elements in the amphoteric behavior of its oxide and hydroxide. Like aluminum, and unlike the other alkaline earth elements, it forms covalent halides and hydrides.

For Review

Summary

The **active metals** are the **alkali metals** lithium, sodium, potassium, rubidium, cesium, and francium from Group IA; the **alkaline earth metals** magnesium, calcium, strontium, barium, and radium from Group IIA; and **aluminum** from Group IIIA. These metals exhibit similar chemical behavior. They all react with acids forming hydrogen gas and salts; with water they form hydrogen gas and basic hydroxides, although the reactivity of magnesium and aluminum is masked by inert coatings. Their oxides and hydroxides are basic. They react with most nonmetals. All are easily oxidized, and the resulting ions are reduced with great difficulty. Each of these metals exhibits a single oxidation number in ionic compounds. Several of the active metals are among the most common of the elements.

The atoms of the active metals are relatively large, and the metals have low densities. Lithium, sodium, and potassium will float on water as they react vigorously. The lightness of magnesium and aluminum makes them valuable structural materials.

The alkali metals show a closer relationship of properties than do any other group of elements. The single electron in the outermost shell of each is loosely bound and easily lost. These metals readily form stable positive ions with a charge of +1 in ionic compounds that are usually soluble.

Each of the two electrons in the outermost shell of each of the alkaline earth metals is more tightly bound than the single electron in the preceding alkali metal, so each alkaline earth atom is smaller than, and not as reactive as, the preceding alkali metal atom. These elements easily form compounds in which the metals exhibit an oxidation number of +2. Compounds of magnesium often show some covalent character, but the magnesium in them can generally be regarded as a +2 ion. Like the alkali metals, the alkaline earth elements are difficult to prepare by chemical methods; the pure elements are isolated by electrochemical techniques.

Aluminum forms both ionic and covalent compounds, in which it has an oxidation number of +3. Its reactive amphoteric oxide and hydroxide react with both acids and bases. Aluminum is also manufactured by an electrochemical process.

The differences in the stoichiometries of compounds of the various active metals reflect their different oxidation numbers. Differences in physical properties reflect ionic charge and the relative sizes of their ions. Both play an important role in determining the lattice energies of these compounds and the strengths of the ion-dipole attractions that are important in determining their solubility.

Key Terms and Concepts

acid-base reactions (13.4)
amphoteric compounds (13.5)
carbonates and hydrogen carbonates (13.6)
diagonal relationship (13.11)
electrolysis reactions (13.2)
electron configurations (13.1)
fractional crystallization (13.5)
halides (13.7)

hydrides (13.8)
hydroxides (13.5)
ionic size effects (13.10)
metathetical reactions (13.9)
nitrates (13.9)
occurrence (13.2)
oxidation-reduction reactions (13.9)

oxides (13.4)
periodic relationships (13.1)
peroxides (13.4)
preparation of the pure metals (13.2)
sulfates (13.9)
superoxides (13.4)

Exercises

The Alkali Metals

1. Why does the reactivity of the alkali metals increase from Li to Cs?
2. Is the reaction of sodium with water more or less vigorous than that of lithium? Than that of cesium?
3. Why should alkali metals never be handled with the fingers?
4. Why is calcium chloride added to the electrolyte in the electrolysis of molten sodium chloride?
5. By analogy with its reaction with water, suggest a chemical formula for the product of the reaction between sodium metal and $H_2S(g)$.
6. Why must the chlorine and sodium resulting from the electrolysis of sodium chloride be kept separate during the production of sodium metal?
7. Outline the chemistry of the Solvay process for the production of sodium carbonate.
8. Consult a handbook for solubilities and suggest a reason why is it that potassium hydrogen carbonate cannot be manufactured by the Solvay process.
9. What is the principal reaction involved when baking powder acts as a leavening agent?
10. What evidence is available to show that hydrogen is present as the negative hydride ion in sodium hydride?
11. Cite evidence that hydride ions are strongly basic, and that peroxide ions are strongly basic.
12. Explain why the stoichiometries of binary lithium compounds are similar to those of sodium, whereas their solubilities are more similar to those of the binary compounds of magnesium.
13. What is the difference that makes LiCl soluble in water whereas LiF is insoluble?

14. How could a solution of an acid be used to distinguish between NaOH and Na_2CO_3?
15. What does the existence of large surface deposits of Chile saltpeter indicate about the climate of Chile in the regions where these deposits are found?
16. A 25.00-mL sample of KOH solution is exactly neutralized with 35.27 mL of 0.1062 M HCl. What is the concentration of KOH?   *Ans. 0.1498 M*
17. How much anhydrous sodium carbonate contains the same number of moles of sodium as in 100 g of $Na_2CO_3 \cdot 10H_2O$?   *Ans. 37.0 g*
18. The usual saline solution, a solution of sodium chloride in water used for washing blood cells, contains 0.85 g of NaCl in 0.100 L of solution. What is the sodium ion concentration in this saline solution?   *Ans. 0.15 M*
19. Give balanced equations for the reactions occurring at the electrodes and the overall reaction for the electrolysis of lithium chloride.

The Alkaline Earth Metals

20. How do the metals of Group IIA of the Periodic Table differ from those of Group IA in atomic structure and general properties?
21. What property or properties of beryllium or its compounds keeps it from being classified as an active metal?
22. Account for the differences in size between the alkaline earth metal atoms and their respective dipositive ions. (See Table 13.1.)
23. Write balanced equations for the steps in the extraction of magnesium from sea water.
24. Magnesium is an active metal; it is burned in the form of ribbons and filaments to provide flashes of brilliant light.

Why is it possible to use magnesium in construction and even for the fabrication of cooking grills?

25. Why cannot a magnesium fire be extinguished by either water or carbon dioxide? Suggest a method of extinguishing such a fire.

26. How is metallic calcium produced commercially?

27. Identify each of the following: quicklime, slaked lime, milk of lime.

28. Explain the development of limestone caves and the limestone formations in these caves.

29. Write a balanced chemical equation describing the dehydrating action of calcium metal.

30. Give the chemistry of the preparation and setting of plaster of paris.

31. What is the essential reaction involved in the setting of mortar?

32. Show how you could prepare CaH_2 from limestone and H_2 in three steps. (Other reagents are required.)

33. Based solely on the lattice energy of the product, should you expect Mg or Ba to be more reactive with oxygen?

34. What soluble sodium compound could be added to two solutions, one of $MgCl_2$ and one of $BaCl_2$, to distinguish between the two?

35. Crushed limestone is used in the treatment of acidic soils. Write balanced equations for the reactions involved.

36. On the basis of atomic structure, explain why barium is more reactive than calcium.

37. Write balanced equations for the reaction of barium with each of the following: oxygen, hydrogen chloride, hydrogen, sulfur, and water.

38. The barium ion is poisonous. Why, then, is it safe to take barium sulfate internally for making photographs of the intestinal tract?

39. When $MgNH_4PO_4$ is heated to 1000°C, it is converted to $Mg_2P_2O_7$. A 1.203-g sample containing magnesium yielded 0.5275 g of $Mg_2P_2O_7$ after precipitation of $MgNH_4PO_4$ and heating. What percent by mass of magnesium was present in the original sample? *Ans. 9.577%*

40. Write balanced equations for the production of barium (a) by the King process and (b) by electrolytic reduction of $BaCl_2$.

Group IIIA

41. Describe the production of metallic aluminum by electrolytic reduction.

42. Write the equation for the reaction of aluminum with the H_2O in a solution of NaOH.

43. Illustrate the amphoteric nature of aluminum hydroxide by suitable equations.

44. Why can aluminum, which is an active metal, be used so successfully as a structural metal?

45. Why is it impossible to prepare anhydrous aluminum chloride by heating the hexahydrate to drive off the water?

46. What is the Goldschmidt process?

47. Write balanced chemical equations for the following reactions:
    (a) Gaseous hydrogen fluoride is bubbled through a suspension of bauxite in molten sodium fluoride.
    (b) Metallic aluminum is burned in air.
    (c) Aluminum is heated in an atmosphere of chlorine.
    (d) Aluminum sulfide is added to water.
    (e) Aluminum hydroxide is added to a solution of nitric acid.

48. How does B_2O_3 differ from Al_2O_3?

49. What properties of thallium or its compounds keep it from being classified as an active metal?

Additional Exercises

50. Write balanced chemical equations for the following reactions.
    (a) Lithium oxide is added to water.
    (b) Lithium oxide is added to a solution of HBr.
    (c) Rubidium carbonate is added to an excess of a solution of HF.
    (d) Cesium superoxide is added to water.
    (e) An electric current is passed through a sample of molten $CaBr_2$.
    (f) Gamma-aluminum oxide is added to a solution of $HClO_4$.
    (g) SO_2 is passed over a sample of barium oxide.
    (h) $HCl(g)$ is passed over barium hydride.
    (i) Aluminum is heated with barium oxide in a vacuum.
    (j) A solution of barium hydroxide is added to a solution of magnesium sulfate.
    (k) Calcium and calcium hydroxide are heated together.
    (l) Sodium and sodium peroxide are heated together in an inert atmosphere.
    (m) A solution of sodium carbonate is added to a solution of magnesium nitrate.
    (n) A solution of sodium carbonate is added to a solution of aluminum nitrate.
    (o) Sodium peroxide is warmed (very carefully) with sodium hydride.
    (p) Ti metal is produced from the reaction of $TiCl_4$ with sodium.

51. (a) A current of 1000 A flowing for 96.5 s contains 1 mol of electrons. How long will it take to produce 100 kg of sodium metal when a current of 50,000 A is passed through a Down's cell like that shown in Figure 13.6 if the yield of sodium is 100% of the theoretical yield? *Ans. 2.33 h*
    (b) What volume of chlorine at 25°C and 1.00 atm is produced? *Ans. 5.32 × 10⁴ L*

52. A current of 1000 A flowing for 96.5 s contains 1 mol of electrons. What mass of magnesium is produced when 100,000 A is passed through a $MgCl_2$ melt for 1.00 h if the yield of magnesium is 85% of the theoretical yield? *Ans. 39 kg*

53. Peroxides, like oxides, are basic. What volume of 0.250 M HCl solution is required to neutralize a solution containing 5.00 g of BaO_2?   *Ans. 236 mL*

54. (a) Write the electron configuration for the isoelectronic ions Na^+, Mg^{2+}, and Al^{3+}.
    (b) Write the electron configurations for the ions K^+, Ca^{2+}, and Ga^{3+} and indicate how the configuration of Ga^{3+} differs from that of the other two ions.

55. Determine the oxidation number of oxygen in aluminum oxide, barium peroxide, and cesium superoxide.

56. Which of the following reactions (some of which are hypothetical) would be the least endothermic?
    (a) $2LiO + O_2 \longrightarrow 2LiO_2$
    (b) $Na_2O_2 + O_2 \longrightarrow 2NaO_2$
    (c) $K_2O_2 + O_2 \longrightarrow 2KO_2$
    (d) $Cs_2O_2 + O_2 \longrightarrow 2CsO_2$

57. What volume of oxygen at 25°C and 1.00 atm is required to prepare 1.00 kg of Na_2O? To prepare 1.00 kg of Na_2O_2?   *Ans. Na_2O, 197 L; Na_2O_2, 314 L*

58. Under typical conditions of temperature and pressure an average resting male requires 0.20 L of O_2 per hour per kilogram of body mass. What mass of KO_2 is required to provide a 1.00-hour supply of oxygen at 0°C and 1.00 atm to an average 100-kg resting male?   *Ans. 85 g*

59. Calculate the density of vapor above a sample of aluminum trichloride at 171°C. The vapor pressure of $AlCl_3$ at 171°C is 400 torr.   *Ans. 3.85 g/L*

60. The alkaline earth oxides are all isomorphous with NaCl. Is the radius ratio rule violated for any of these oxides?

61. Alpha-alumina crystallizes with an approximately closest packed array of oxide ions, with aluminum ions in the octahedral holes. What fraction of the octahedral holes are occupied?

62. Identify the colligative property that is an important reason for the use of $CaCl_2$ in the NaCl melt used in the production of sodium metal.

Chemical Kinetics

14

A lizard warms itself
to speed up its
metabolism.

I f we plan to run a chemical reaction, we should ask whether or not it will produce
the desired products in useful quantities. We could answer this question through
the use of equilibrium and thermodynamic calculations, as discussed in the next
chapter and in Chapters 17–19, or we could use qualitative considerations, such as the
fact that the reaction of a strong acid with a strong base gives a salt or that an active
metal (a reducing agent) generally reacts with a nonmetal (an oxidizing agent). We
should also ask whether or not the reaction will proceed rapidly enough to be useful. A
reaction that takes 50 years to produce a product is almost as useless as one that will
never produce a product.

This chapter considers the latter question—that is, the rate at which a chemical
reaction yields products **(chemical kinetics).** We examine the factors influencing the
rates of chemical reactions, the mechanisms by which reactions proceed, and the
quantitative techniques used to determine and to describe the rates at which reactions
occur. Chapter 15 deals with chemical equilibrium and describes how it is related to
kinetics.

The photograph on the preceding page illustrates the importance of temperature on the rates at which reactions occur. A warm lizard runs faster than a cold one because the chemical reactions that move its muscles occur more rapidly at higher temperatures than at lower ones.

14.1 The Rate of Reaction

The rate at which a chemical process gives products is called its **rate of reaction.** Rates of reaction, like all rates, involve a division by some unit of time. For example, the rate of production of a well is measured in gallons *per minute*. The rate of use of coal by an electric generating plant is measured in tons *per hour*. **The rate of a chemical reaction is measured by the decrease in concentration of a reactant or the increase in concentration of a product during a given unit of time.** For example, pure hydrogen peroxide, dissolved in water, slowly decomposes according to the following equation:

$$2H_2O_2(aq) \longrightarrow 2H_2O(l) + O_2(g)$$

The change in concentration with time at 40°C of a solution that is initially 1 M in hydrogen peroxide is shown in Fig. 14.1 and in Table 14.1. From these data, which must be determined experimentally, we can determine the rate, or speed, at which the hydrogen peroxide decomposes during successive time intervals.

$$\text{Rate} = \frac{\text{change in concentration of reactant}}{\text{time interval}}$$

$$= -\frac{[H_2O_2]_{t_2} - [H_2O_2]_{t_1}}{t_2 - t_1} = -\frac{\Delta[H_2O_2]}{\Delta t}$$

Figure 14.1

The decomposition of H_2O_2 $(2H_2O_2 \longrightarrow 2H_2O + O_2)$ at 40°C. The intensity of the color represents the concentration of H_2O_2 at the indicated time after the reaction begins. Note that H_2O_2 is actually colorless; the color indicated here is symbolic of the decreasing concentration of H_2O_2 with time.

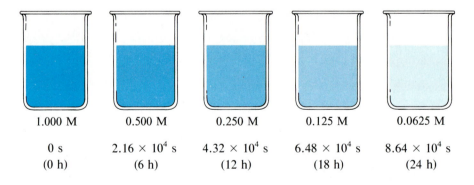

1.000 M	0.500 M	0.250 M	0.125 M	0.0625 M
0 s	2.16×10^4 s	4.32×10^4 s	6.48×10^4 s	8.64×10^4 s
(0 h)	(6 h)	(12 h)	(18 h)	(24 h)

Table 14.1   The variation in the rate of decomposition of H_2O_2 at 40°C

Time, s	$[H_2O_2]$, mol L^{-1}	$\Delta[H_2O_2]$, mol L^{-1}	Δt, s	Rate, mol L^{-1} s^{-1}
0	1.000			
		−0.500	2.16×10^4	2.31×10^{-5}
2.16×10^4	0.500			
		−0.250	2.16×10^4	1.16×10^{-5}
4.32×10^4	0.250			
		−0.125	2.16×10^4	0.579×10^{-5}
6.48×10^4	0.125			
		−0.062	2.16×10^4	0.29×10^{-5}
8.64×10^4	0.0625			

In these equations brackets are used to signify molar concentrations, and the symbol delta (Δ), to signify "the change in." Thus $[H_2O_2]_{t_1}$ represents the molar concentration of hydrogen peroxide at some time t_1; $[H_2O_2]_{t_2}$, the molar concentration of hydrogen peroxide at a later time t_2; and $\Delta[H_2O_2]$, the change in molar concentration of hydrogen peroxide during the time interval Δt (i.e., $t_2 - t_1$). The minus sign preceding the whole expression indicates that the concentration of the reactant is decreasing during the course of the reaction.

EXAMPLE 14.1

Calculate the average rate of decomposition of a 1.000 M solution of hydrogen peroxide at 40°C for the period of the reaction from 0 to 21,600 s, using the data in Table 14.1.

In Table 14.1 we find that for the period of the reaction from 0 to 21,600 s,

$$[H_2O_2]_{t_1} = 1.000 \text{ mol L}^{-1} \qquad (t_1 = 0 \text{ s})$$

and

$$[H_2O_2]_{t_2} = 0.500 \text{ mol L}^{-1} \qquad (t_2 = 21,600 \text{ s})$$

Thus

$$\Delta t = t_2 - t_1 = 21,600 \text{ s} - 0 \text{ s} = 21,600 \text{ s}$$
$$\Delta[H_2O_2] = [H_2O_2]_{t_2} - [H_2O_2]_{t_1}$$
$$= 0.500 \text{ mol L}^{-1} - 1.000 \text{ mol L}^{-1}$$
$$= -0.500 \text{ mol L}^{-1}$$

$$\text{Rate} = -\left(\frac{\Delta[H_2O_2]}{\Delta t}\right) = -\left(\frac{-0.500 \text{ mol L}^{-1}}{21,600 \text{ s}}\right)$$
$$= 2.31 \times 10^{-5} \text{ mol L}^{-1} \text{ s}^{-1}$$

The unit mol/L (moles per liter, or molarity) can be written as mol L^{-1}, since $1/X = X^{-1}$, according to the mathematical rules for exponents. The unit s^{-1} (per second) is similarly equivalent to 1/s.

Since the rate of a chemical reaction changes with time, the reaction rates presented in Table 14.1 are average rates for the time intervals indicated. The rate of decomposition of hydrogen peroxide is faster at the beginning of the decomposition ($t = 0$ seconds) than it is at 21,600 seconds. The rate of 2.31×10^{-5} mol L^{-1} s^{-1} is the average rate of decomposition between 0 seconds and 21,600 seconds.

EXAMPLE 14.2

Calculate the rate of decomposition of hydrogen peroxide at 40°C for the time interval from 0 seconds to 86,400 seconds, using the data in Table 14.1.

$$\text{Rate} = -\left(\frac{0.0625 \text{ mol L}^{-1} - 1.000 \text{ mol L}^{-1}}{86,400 \text{ s} - 0 \text{ s}}\right)$$

$$= -\left(\frac{-0.9375 \text{ mol L}^{-1}}{86,400 \text{ s}}\right) = 1.09 \times 10^{-5} \text{ mol L}^{-1} \text{ s}^{-1}$$

This value is the average of the rates for the four time intervals in Table 14.1.

A graph of the concentration of hydrogen peroxide versus time shows the rate of the reaction at any instant of time. This rate is given by the slope of a straight line tangent to the curve at that time. Such a graph is shown in Fig. 14.2, with a tangent drawn at $t = 40,000$ seconds. The slope of this line (-8.90×10^{-6} mol $L^{-1}s^{-1}$) is $\Delta[H_2O_2]/\Delta t$. Thus the rate of decomposition at 40,000 seconds is

$$\text{Rate} = 8.90 \times 10^{-6} \text{ mol } L^{-1} \text{ s}^{-1}$$

The initial rate of the reaction is equal numerically (but opposite in sign) to the slope of the tangent at time zero.

Figure 14.2

A graph of concentration versus time for the 1.000 M solution of H_2O_2 described in Fig. 14.1. The rate at any instant of time is equal to the slope of a line tangent to this curve at that time. A tangent (the purple line) is shown for $t = 40,000$ s.

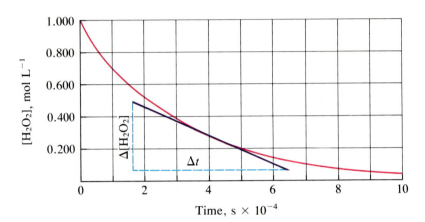

The rates of chemical reactions vary greatly. The initial rate of decomposition of 1 M hydrogen peroxide at 40°C is 3.2×10^{-5} mol L^{-1} s^{-1}, and the reaction takes many hours to go to completion. The initial rate of formation of water when 1 M hydrochloric acid and 1 M sodium hydroxide react at 25°C is 1.4×10^{11} mol L^{-1} s^{-1}. This acid-base reaction occurs as rapidly as the two solutions can be mixed—almost instantaneously. The initial rate of reaction of hydrochloric acid with sodium hydroxide is 10^{15} times faster than that for the decomposition of hydrogen peroxide.

The rates at which reactants disappear during reactions are rarely constant. As the reactants are consumed, rates usually decrease until they reach zero. In an irreversible reaction, such as the decomposition of hydrogen peroxide, the rate of the reaction reaches zero when all of the reactant has been consumed. The decomposition of hydrogen peroxide is effectively irreversible because the oxygen produced by the reaction escapes from the open beaker. In a reversible reaction, a reaction in which the products react to reform the reactants, the rate becomes zero when equilibrium is established.

The rate of a reversible reaction that becomes zero at equilibrium is the *net* rate, taking into account the forward and reverse reactions. At equilibrium, the forward and reverse reactions both continue to operate but at equal rates, so that the *reactants* are being used up in the forward reaction at the same rate that they are being produced by the reverse reaction. Similarly, the *products* of the reaction are being produced by the forward reaction at the same rate that they are being used up by the reverse reaction.

The reaction of nitrogen with hydrogen to form ammonia in a closed system is an example of a reversible reaction (Section 9.3, Part 6). In order to establish the rate of a reversible chemical reaction, the reaction is studied at the beginning, before equilibrium is reached.

14.2 Rate of Reaction and the Nature of the Reactants

The rate of a reaction is strongly influenced by the nature of the reacting substances. The reaction of an acid with a base, a process involving the simple combination of two ions of opposite charge,

$$H_3O^+ + OH^- \longrightarrow 2H_2O$$

is much faster than the decomposition of hydrogen peroxide, a process involving the rearrangement of molecules,

$$2H_2O_2 \longrightarrow 2H_2O + O_2$$

Even similar reactions have different rates under the same conditions if different reactants are involved. The active metals calcium and sodium both react with water to form hydrogen gas and the metal hydroxide. But calcium reacts at a moderate rate, whereas sodium reacts so rapidly that its reaction is almost explosive.

14.3 Rate of Reaction and the State of Subdivision of the Reactants

Except for those reactions involving substances in the gaseous state or in solution, reactions occur at the boundary, or interface, between two phases. The rate of a reaction between two phases depends to a great extent on the area of surface contact between them. Reactions involving solids occur on their surfaces. Finely divided solids, because of the greater surface area available, react more rapidly than do large pieces of the same substances. For example, large pieces of wood smolder slowly, smaller pieces burn rapidly, and grain dust (which, like wood, is composed of cellulose) may burn at an explosive rate (Fig. 14.3).

Figure 14.3
Grain dust (very finely divided grain) can burn at an explosive rate. If the dust is suspended in a confined space, such as a grain elevator, an explosion can result. A grain dust explosion in this storage elevator destroyed several silos.

14.4 Rate of Reaction and Temperature

Chemical reactions are accelerated by increases in temperature. The initial rate of decomposition of a 1 M hydrogen peroxide solution is 3.2×10^{-5} mol L^{-1} s^{-1} at 40°C and is slightly more than twice as fast, 7.2×10^{-5} mol L^{-1} s^{-1}, at 50°C. The oxidation of either iron or coal is very slow at ordinary temperatures but proceeds rapidly at high temperatures. Foods cook faster at higher temperatures than at lower ones. We heat reactions in the laboratory to increase their rates. In many cases the rate of a reaction in a homogeneous system is *approximately* doubled by an increase in temperature of only 10°C (Fig. 14.4). This rule, however, is a rough approximation and applies only to reactions that last longer than a second or two.

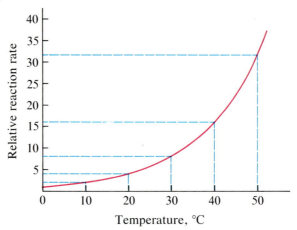

Figure 14.4

The effect of temperature on reaction rate, shown graphically. The rate of reaction is often approximately doubled for each rise in temperature of 10°C.

14.5 Rate of Reaction and Concentration; Rate Equations

At a fixed temperature the rates of most reactions depend on the concentrations of the reactants. The rates at which limestone buildings deteriorate due to reaction with the air pollutant sulfur dioxide depend on the amount of sulfur dioxide in the air, for example. In a highly polluted atmosphere where the concentration is high, they crumble more rapidly than they would in less polluted air. The rate of decomposition of hydrogen peroxide (Table 14.1) decreases with decreasing concentration. The rate at any instant of time has been found to be directly proportional to the concentration of hydrogen peroxide at that time.

$$\text{Rate} = k[H_2O_2] \qquad (1)$$

The proportionality constant, k, is called the **rate constant.** The brackets signify the molar concentration of the reactant. The rate constant is independent of the concentration of the reactant but does vary with temperature. At 40°C, k is 3.2×10^{-5} s^{-1}; at 50°C, 7.2×10^{-5} s^{-1}. The rate for the reaction of hydrochloric acid with sodium hydroxide is described by a similar expression:

$$\text{Rate} = k[H_3O^+][OH^-] \qquad (2)$$

The value of k in this equation, 1.4×10^{11} L mol^{-1} s^{-1}, differs from that in Equation (1) since the rate constant is dependent on the nature of the reacting substances.

Equations (1) and (2) are examples of **rate equations,** or **rate laws.** These equations give the relationship between reaction rate and concentrations of reactants.

In general, a rate equation has the form

$$\text{Rate} = k[A]^m[B]^n[C]^p \ldots$$

in which [A], [B], and [C] represent molar concentrations of reactants (or sometimes products or other substances), k is the rate constant for the particular reaction, and the exponents m, n, and p are usually positive integers (although fractions and negative numbers sometimes appear). *Both k and the exponents m, n, and p must be determined experimentally from the variation of the rate of a reaction as the concentrations of the reactants are varied.*

EXAMPLE 14.3

One of the reactions occurring in the ozone layer in the upper atmosphere is the combination of nitric oxide, NO, with ozone, O_3.

$$NO + O_3 \longrightarrow NO_2 + O_2$$

This reaction has been studied in the laboratory, and the following rate data obtained at 25°C.

[NO], mol L^{-1}	[O_3], mol L^{-1}	$\dfrac{\Delta[NO_2]}{\Delta t}$, mol L^{-1} s^{-1}
1.00×10^{-6}	3.00×10^{-6}	0.660×10^{-4}
1.00×10^{-6}	6.00×10^{-6}	1.32×10^{-4}
1.00×10^{-6}	9.00×10^{-6}	1.98×10^{-4}
2.00×10^{-6}	9.00×10^{-6}	3.96×10^{-4}
3.00×10^{-6}	9.00×10^{-6}	5.94×10^{-4}

Determine the rate equation for the reaction at 25°C.

The rate equation will have the form

$$\text{Rate} = k[NO]^m[O_3]^n$$

The values of m and n cannot be determined from the chemical equation, but must be found from the experimental data. In the first three lines in the data table, [NO] is held constant and [O_3] is allowed to vary. The reaction rate changes in direct proportion to the change in [O_3]. When [O_3] doubles, the rate doubles; when [O_3] increases by a factor of 3, the rate increases by a factor of 3. Thus the rate is directly proportional to [O_3], or

$$\text{Rate} = k'[O_3]$$

Hence n in the rate equation must be equal to 1. Writing [O_3] is equivalent to writing [O_3]^1.

In the last three lines in the data table, [NO] varies as [O_3] is held constant. When [NO] doubles the rate doubles, and when [NO] triples, the rate also triples. Thus the rate is also directly proportional to [NO], and m in the rate equation is also equal to 1.

The rate equation is thus

$$\text{Rate} = k[NO][O_3]$$

The value of k can be determined from one set of concentrations and the corresponding rate.

$$k = \frac{\text{Rate}}{[NO][O_3]}$$

$$= \frac{0.660 \times 10^{-4} \text{ mol L}^{-1} \text{ s}^{-1}}{(1.00 \times 10^{-6} \text{ mol L}^{-1})(3.00 \times 10^{-6} \text{ mol L}^{-1})}$$

$$= 2.20 \times 10^{7} \text{ L mol}^{-1} \text{ s}^{-1}$$

EXAMPLE 14.4   Acetaldehyde decomposes when heated to yield methane and carbon monoxide according to the equation

$$CH_3CHO(g) \longrightarrow CH_4(g) + CO(g)$$

Determine the rate equation and the rate constant for the reaction from the following experimental data:

$[CH_3CHO]$, mol L^{-1}	$\dfrac{\Delta[CH_3CHO]}{\Delta t}$, mol L^{-1} s^{-1}
1.75×10^{-3}	2.06×10^{-11}
3.50×10^{-3}	8.24×10^{-11}
7.00×10^{-3}	3.30×10^{-10}

The rate equation will have the form

$$\text{Rate} = k[CH_3CHO]^n$$

From the experimental data we can see that when $[CH_3CHO]$ doubles, the rate increases by a factor of 4; when $[CH_3CHO]$ increases by a factor of 4, the rate increases by a factor of 16. Thus the rate is proportional to the square of the concentration, or

$$\text{Rate} = k[CH_3CHO]^2$$

The value of k can be determined from one concentration and the corresponding rate.

$$k = \frac{\text{Rate}}{[CH_3CHO]^2}$$

$$= \frac{3.30 \times 10^{-10} \text{ mol L}^{-1}\text{s}^{-1}}{(7.0 \times 10^{-3} \text{ mol L}^{-1})^2}$$

$$= 4.71 \times 10^{-8} \text{ L mol}^{-1} \text{ s}^{-1}$$

It is not unusual to find that a rate equation does not reflect the overall stoichiometry of a chemical reaction. In some cases one or more of the reactants may not even appear in the rate equation (see Example 14.5).

14.6 Order of a Reaction

The order of a reaction is determined from the exponents found in its rate equation. The order can be given as the order with respect to a specific reactant or as the overall order of the reaction. **The order of a reaction with respect to one of the**

reactants is equal to the power to which the concentration of that reactant is raised in the rate equation.

Consider a reaction for which the rate equation is

$$\text{Rate} = k[A]^m[B]^n$$

If the exponent m is 1, the reaction is said to be **first order** with respect to A. If m is 2, the reaction is said to be **second order** with respect to A. If n is 2, the reaction is said to be second order with respect to B. If a reaction is **zero order** with respect to a reactant, the rate of the reaction does not change as the concentration of that reactant changes. If this reaction were zero order with respect to B, n would be 0. Note that $[B]^0$ is equal to 1, no matter what the value of $[B]$.

The overall order of a reaction is the sum of the orders with respect to each reactant. If m is 1 and n is 1, the reaction is first order in A and first order in B. The overall order of the reaction is given by the sum of m and n; therefore the overall order of this particular reaction is second order.

Experiment shows that the reaction of nitrogen dioxide with carbon monoxide   **EXAMPLE 14.5**

$$NO_2 + CO \longrightarrow NO + CO_2$$

is second order in NO_2 and zero order in CO at 200°C. What is the rate equation for the reaction?

The rate equation will have the form

$$\text{Rate} = k[NO_2]^m[CO]^n$$

$R = K[NO_2]^2$

The reaction is second order in NO_2; thus $m = 2$. The reaction is zero order in CO; thus $n = 0$. The rate equation is

$$\text{Rate} = k[NO_2]^2[CO]^0 = k[NO_2]^2$$

Remember that a number raised to the zero power is equal to 1, so $[CO]^0 = 1$.

What are the orders with respect to each reactant and the overall order of the reaction   **EXAMPLE 14.6**

$$NO + O_3 \longrightarrow NO_2 + O_2$$

in Example 14.3, Section 14.5?

The rate equation for the reaction was found to be

$$\text{Rate} = k[NO][O_3]$$

Thus the reaction is first order with respect to both NO and O_3. The overall reaction is second order since the sum of the exponents is 2.

The rate equation for the reaction   **EXAMPLE 14.7**

$$H_2(g) + 2NO(g) \longrightarrow N_2O(g) + H_2O(g)$$

is $$\text{Rate} = k[NO]^2[H_2]$$

What are the orders with respect to each reactant and the overall order of the reaction?

The reaction is second order with respect to NO and first order with respect to H_2. The overall reaction is third order, since the sum of the exponents $(2 + 1)$ is 3.

14.7 Half-Life of a Reaction

The half-life of a reaction, $t_{1/2}$, is the time required for half of the original concentration of the limiting reactant to be consumed. In each succeeding half-life, half of the remaining concentration of the reactant will be used up. The decomposition of hydrogen peroxide illustrated in Fig. 14.1 displays the concentration after each of several successive half-lives. During the first half-life (from 0 seconds to 2.16×10^4 seconds), the concentration decreases from 1.000 M to 0.500 M. During the second half-life (from 2.16×10^4 seconds to 4.32×10^4 seconds), it decreases from 0.500 M to 0.250 M; during the third half-life, from 0.250 M to 0.125 M. The concentration decreases by half during each successive period of 2.16×10^4 seconds. The decomposition of hydrogen peroxide is a first-order reaction, and, as shown below, the half-life of a first-order reaction is independent of the concentration of the reactant. However, half-lives of higher-order reactions are not independent of the concentrations of the reactants.

1. FIRST-ORDER REACTIONS. An equation relating the rate constant k to the initial concentration $[A_0]$ and the concentration $[A]$ present after any given time t can be derived for a first-order reaction. The derivation requires more advanced mathematics than we have been using so we will not present it here, but the resulting equation is

$$\log \frac{[A_0]}{[A]} = \frac{kt}{2.303} \tag{1}$$

Note that the logarithm in this equation is to the base 10. The common notation is *log* for a logarithm to the base *10* and *ln* for a logarithm to the base *e*. The following example demonstrates the use of this equation.

EXAMPLE 14.8   The rate constant for the first-order decomposition of cyclobutane, C_4H_8 (Section 14.11), at 500°C is 9.2×10^{-3} s^{-1}. How long will it take for 90.0% of a sample of 0.100 M C_4H_8 to decompose—that is, for the concentration of C_4H_8 to decrease to 0.0100 M?

The initial concentration of C_4H_8, $[A_0]$, is 0.100 mol L^{-1}; the concentration at time t, $[A]$ is 0.0100 mol L^{-1}; and k is 9.2×10^{-3} s^{-1}.

$$\log \frac{[A_0]}{[A]} = \frac{kt}{2.303}$$

$$t = \log \frac{[A_0]}{[A]} \times \frac{2.303}{k}$$

$$t = \log \left(\frac{0.100 \text{ mol L}^{-1}}{0.0100 \text{ mol L}^{-1}} \right) \times \frac{2.303}{9.2 \times 10^{-3} \text{ s}^{-1}} = 2.5 \times 10^2 \text{ s}$$

The equation for determining the half-life for a first-order reaction can be derived from Equation (1) as follows.

$$\log \frac{[A_0]}{[A]} = \frac{kt}{2.303}$$

$$t = \log \frac{[A_0]}{[A]} \times \frac{2.303}{k}$$

If the time t is the half-life, $t_{1/2}$, the concentration of the limiting reactant at the end of this time, [A], is equal to $\frac{1}{2}$ of the initial concentration. Hence, at time $t_{1/2}$, [A] = $\frac{1}{2}[A_0]$.

Therefore

$$t_{1/2} = \log \frac{[A_0]}{\frac{1}{2}[A_0]} \times \frac{2.303}{k}$$

$$= \log 2 \times \frac{2.303}{k}$$

$$= 0.301 \times \frac{2.303}{k} = \frac{0.693}{k}$$

Thus

$$t_{1/2} = \frac{0.693}{k} \qquad (2)$$

You can see in Equation (2) that the half-life of a first-order reaction is inversely proportional to the rate constant k. Hence, a fast reaction with a large k has a short half-life; a slow reaction with a smaller k has a longer half-life. Also note that the half-life for a first-order reaction is independent of the concentrations of reactants.

Calculate the rate constant for the first-order decomposition of hydrogen peroxide in water at 40°C using the data from Fig. 14.1

EXAMPLE 14.9

The half-life for the decomposition of H_2O_2 is 2.16×10^4 s.

$$t_{1/2} = \frac{0.693}{k}$$

$$k = \frac{0.693}{t_{1/2}} = \frac{0.693}{2.16 \times 10^4 \text{ s}} = 3.21 \times 10^{-5} \text{ s}^{-1}$$

---

2. SECOND-ORDER REACTIONS. Let us consider two types of second-order reactions: (1) those that are second order with respect to one reactant and (2) those that are first order with respect to each of two reactants *and* in which the initial concentration of each reactant is the same. In such cases the equation relating the rate constant k to the initial concentration $[A_0]$ of any of the reactants and to the concentration [A] present after any given time t is

$$\frac{1}{[A]} - \frac{1}{[A_0]} = kt \qquad (3)$$

The reaction of an organic ester (a compound with the general formula RCOOR', where R and R' are alkyl groups; see Chapter 30) with a strong base is second order with a rate constant equal to 4.50 L mol^{-1} min^{-1}. If the initial concentrations of ester and base are both 0.0200 M, what is the concentration remaining after 10.0 min?

EXAMPLE 14.10

$$\frac{1}{[A]} - \frac{1}{[A_0]} = kt$$

$$[A_0] = 0.0200 \text{ mol L}^{-1}$$

$$k = 4.50 \text{ L mol}^{-1} \text{ min}^{-1}$$

$$t = 10.0 \text{ min}$$

Therefore, $\dfrac{1}{[A]} - \dfrac{1}{0.0200\ \text{mol L}^{-1}} = 4.50\ \text{L mol}^{-1}\ \text{min}^{-1} \times 10.0\ \text{min}$

or $\dfrac{1}{[A]} = 4.50\ \text{L mol}^{-1}\ \text{min}^{-1} \times 10.0\ \text{min} + \dfrac{1}{0.0200\ \text{mol L}^{-1}}$

$= 45.0\ \text{L mol}^{-1} + \dfrac{1}{0.0200}\ \text{mol}^{-1}\ \text{L}$

$= 45.0\ \text{L mol}^{-1} + 50.0\ \text{L mol}^{-1}$

$= 95.0\ \text{L mol}^{-1}$

Then $[A] = \dfrac{1}{95.0\ \text{L mol}^{-1}} = 0.0105\ \text{mol L}^{-1}$

Hence 0.0105 mol of both the ester and the strong base remains per liter at the end of 10.0 min, compared to the 0.0200 mol per liter of each originally present.

The equation used for calculating the half-life of a second-order reaction in which the initial concentration of each reactant is the same can be derived from Equation (3) as follows.

$$\frac{1}{[A]} - \frac{1}{[A_0]} = kt$$

If $t = t_{1/2}$, then

$$[A] = \tfrac{1}{2}[A_0]$$

Therefore,

$$\frac{1}{\tfrac{1}{2}[A_0]} - \frac{1}{[A_0]} = kt_{1/2}$$

$$\frac{1 - \tfrac{1}{2}}{\tfrac{1}{2}[A_0]} = kt_{1/2}$$

$$\frac{\tfrac{1}{2}}{\tfrac{1}{2}[A_0]} = kt_{1/2}$$

$$\frac{1}{[A_0]} = kt_{1/2}$$

$$t_{1/2} = \frac{1}{k[A_0]} \qquad (4)$$

For a second-order reaction $t_{1/2}$ depends on the concentration, and the half-life increases as the reaction proceeds. The half-life is independent of the concentration for a first-order reaction. Consequently, the use of the half-life concept is more complex for second-order reactions. For example, the rate constant of a second-order reaction cannot be calculated directly from the half-life, or vice versa, unless the initial concentration is known; this is not true for first-order reactions.

EXAMPLE 14.11   Calculate the half-life for the second-order reaction described in Example 14.10 ($k = 4.50\ \text{L mol}^{-1}\ \text{min}^{-1}$) if the initial concentrations of both reactants are 0.0200 M.

The half-life of this second-order reaction can be calculated from the equation

$$t_{1/2} = \frac{1}{k[A_0]}$$

Thus

$$t_{1/2} = \frac{1}{k[A_0]} = \frac{1}{(4.50 \text{ L mol}^{-1} \text{ min}^{-1})(0.0200 \text{ mol L}^{-1})} = 11.1 \text{ min}$$

The calculations in Example 14.10 showed that slightly more than half of each of the reactants (0.0105 mol L^{-1}) remains after 10.0 min. The half-life, therefore, must be a little bit longer than 10.0 min. This estimate provides us with a useful order of magnitude to check the calculated answer. In accord with our estimate the half-life is a little longer (11.1 min) than 10.0 min when the initial concentrations are 0.0200 M.

EXAMPLE 14.12

What is the half-life for the reaction in Examples 14.10 and 14.11 if the initial concentrations of both reactants are 0.0300 M?

$$t_{1/2} = \frac{1}{k[A_0]} = \frac{1}{(4.50 \text{ L mol}^{-1} \text{ min}^{-1})(0.0300 \text{ mol L}^{-1})} = 7.41 \text{ min}$$

Example 14.11 and this one clearly show the dependence of half-life on initial concentrations for a second-order reaction. The increase in the initial concentrations from 0.0200 M to 0.0300 M shortens the half-life for this particular reaction from 11.1 min to 7.41 min.

14.8 Collision Theory of the Reaction Rate

Chemists believe that before atoms, molecules, or ions can react, they must collide with one another. In a few reactions every collision between reactants leads to products, and the rates of such reactions are determined solely by how rapidly the reactants can diffuse together. These reactions are called **diffusion-controlled reactions.** Diffusion-controlled reactions are very fast; hence they have large rate constants. For a typical diffusion-controlled second-order gas-phase reaction at 25°C, such as the reaction of an oxygen atom with a nitrogen molecule

$$O + N_2 \longrightarrow NO + N$$

the rate constant falls in the range 10^{10}–10^{12} L mol^{-1} s^{-1}. The diffusion-controlled reaction between hydronium ions and hydroxide ions in water at 25°C

$$H_3O^+ + OH^- \longrightarrow 2H_2O$$

has a rate constant of 1.4×10^{11} L mol^{-1} s^{-1}. At these rates over 95% of the reactants would be consumed in 10^{-11} second.

Most reactions, however, occur at a much slower rate, because only a very small fraction of collisions in these slower reactions give products. In the majority of collisions the reactants simply bounce away unchanged. For a collision to lead to a reaction, the following must occur:

1. The reacting species usually must have a particular orientation during the collision so that the atoms that are bonded together in the product come in contact.

2. The collision must occur with enough energy for the valence shells of the reacting species to penetrate into each other so that the electrons can rearrange and form new bonds (and new chemical species).

The gas-phase reaction of nitrogen dioxide with carbon monoxide

$$NO_2 + CO \longrightarrow NO + CO_2$$

above 225°C illustrates the factors necessary for effective collision. During the course of the reaction, an oxygen atom is transferred from an NO_2 molecule to a CO molecule. There are many orientations of the NO_2 and CO molecules during collision that will not place an oxygen atom of the NO_2 molecule close to the carbon atom of the CO molecule. Three of these are indicated in Fig. 14.5 (a, b, and c). These collisions will not be effective in producing a chemical reaction. Only a collision in which an oxygen atom strikes the carbon atom [Fig. 14.5(d)] can produce a reaction.

(a)   (b)   (c)   (d)

Figure 14.5

Some possible collisions between NO_2 and CO molecules. Only in (d) are the molecules correctly oriented for transfer of an oxygen atom from NO_2 to CO to give NO and CO_2.

Even if the orientation is correct, a collision may still not lead to reaction. As the oxygen atom of an NO_2 molecule approaches the carbon atom of a CO molecule, the electrons in the two molecules begin to repel each other. Unless the molecules possess a kinetic energy greater than a certain minimum value, the two molecules will bounce away from each other before they can get close enough to react. If the molecules are moving fast enough that their kinetic energy exceeds this minimum, then the repulsion between their electrons is not strong enough to keep them apart, and the molecules can get close enough for a C—O bond to begin to form as the N—O bond begins to break. Using dots to represent partially formed or broken bonds, we can write the resulting species as follows:

$$NO_2 + CO \longrightarrow \left[O{=}N{\cdots}O{\cdots}C{\equiv}O \right]$$

The species $O{=}N{\cdots}O{\cdots}C{\equiv}O$, which contains the partially formed $C{\cdots}O$ and partially broken $N{\cdots}O$ bonds, is called the **activated complex,** or **transition state.** An activated complex is a combination of reacting molecules that is intermediate between reactants and products and in which some bonds have weakened and new bonds have begun to form. Ordinarily, an activated complex cannot be isolated. It breaks down to give either reactants or products, depending on the conditions under which the reaction takes place.

$$NO_2 + CO \longrightarrow \left[O{=}N{\cdots}O{\cdots}C{\equiv}O \right] \begin{array}{c} \nearrow NO_2 + CO \\ \text{or} \\ \searrow NO + CO_2 \end{array}$$

The collision theory shows why reaction rates decrease as concentrations decrease. With decreased concentration of any of the reacting substances, the chances for collisions between molecules are decreased because fewer molecules are present per unit of volume. Fewer collisions mean a slower reaction rate.

14.9 Activation Energy and the Arrhenius Equation

The minimum energy necessary to form an activated complex during a collision between reactants is called the **activation energy, E_a** (Fig. 14.6). In a slow reaction the activation energy is much larger than the average energy content of the molecules, and the energies of a large fraction of the molecules in a system are close to the average value. Hence, only a few of the molecules, the fast-moving ones, have relatively high energies. The collisions between the fast-moving molecules are most apt to result in reactions. In very fast reactions the fraction of molecules possessing the necessary activation energy is large, and most collisions between molecules result in reaction.

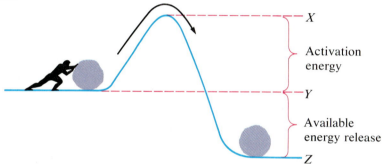

Figure 14.6

Illustration of activation energy in relation to available energy release. The boulder can release the energy created by falling the distance from height Y to height Z. However, energy must be put into the system (the activation energy) to lift it over the barrier before it can fall to Z. The activation energy is released as the boulder falls through the distance X to Y, and the additional energy is then released as it falls from Y to Z. The net energy released is that provided by the fall from Y to Z.

The energy relationships for the general reaction of a molecule of A with a molecule of B to form molecules of C and D are shown in Fig. 14.7.

$$A + B \rightleftharpoons C + D$$

The figure shows that after the activation energy, E_a, is exceeded, and as C and D begin to form, the system loses energy until its total energy is lower than that of the initial mixture. The forward reaction (that between molecules A and B) therefore tends to take place readily if sufficient energy is available in any one collision to exceed the

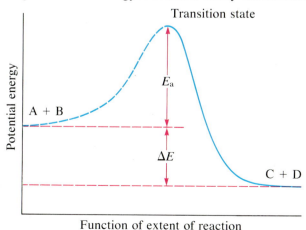

Figure 14.7

Potential energy relationships for the reaction A + B $\rightleftharpoons$ C + D. The energy represented by the broken portion of the curve is that for the system with a molecule of A and a molecule of B present; the energy represented by the solid portion of the curve is that for the system with a molecule of C and a molecule of D present. The activation energy for the forward reaction is represented by E_a, the activation energy for the reverse reaction by $(E_a + \Delta E)$. The species present at the peak maximum corresponds to the transition state.

activation energy, E_a. In Fig. 14.7, ΔE represents the difference in energy between the two molecules of reactants (A and B) and the two molecules of products (C and D). The sum of E_a and ΔE represents the activation energy for the reverse reaction,

$$C + D \longrightarrow A + B$$

For a given reaction, the rate constant is related to the activation energy by a relationship known as the **Arrhenius equation.**

$$k = A \times 10^{-E_a/2.303RT}$$

R is a constant with the value 8.314 J mol^{-1} K^{-1}, T is temperature on the Kelvin scale, E_a is the activation energy in joules per mole, and A is a constant called the **frequency factor,** which is related to the frequency of collisions and the orientation of the reacting molecules. A indicates how many collisions have the correct orientation to lead to products. The remainder of the equation, $10^{-E_a/2.303RT}$, gives the fraction of the collisions in which the energy of the reacting species is greater than E_a, the activation energy for the reaction.

The Arrhenius equation describes quantitatively much of what we have discussed about reaction rates. Reactions with similar frequency factors but different activation energies, E_a, will have different rate constants because the rate constants are proportional to $10^{-E_a/2.303RT}$. For two reactions at the same temperature, the reaction with the higher activation energy will have the lower rate constant and the slower rate. The larger value of E_a results in a smaller value for $10^{-E_a/2.303RT}$, reflecting the smaller fraction of molecules with sufficient energy to react. Alternatively, the reaction with the smaller E_a will have a larger fraction of molecules with the necessary energy to react (Fig. 14.8). This will be reflected as a larger value of $10^{-E_a/2.303RT}$, a larger rate constant, and a faster rate for the reaction. An increase in temperature has the same effect as a decrease in activation energy. A larger fraction of molecules has the necessary energy to react (Fig. 14.9), as indicated by an increase in the value of $10^{-E_a/2.303RT}$. The rate constant is also directly proportional to the frequency factor, A. Hence a change in conditions or reactants that increases the fraction of collisions in which the orientation of the molecules is right for reaction results in an increase in A and, consequently, an increase in k.

In order to determine E_a for a reaction, we must measure k at different temperatures and evaluate E_a from the Arrhenius equation. The Arrhenius equation may be rewritten as follows:

$$\log k = \log A - \frac{E_a}{2.303\ RT}$$

Figure 14.8

As the activation energy of a reaction decreases, the number of molecules with at least this much energy increases, as shown by the yellow shaded areas.

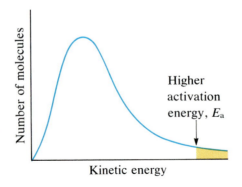

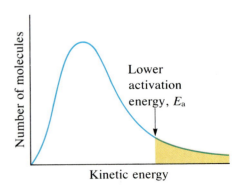

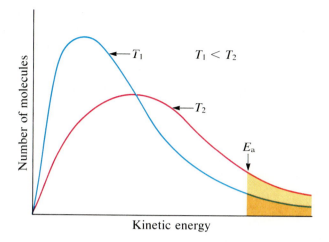

Number of molecules

T_1

$T_1 < T_2$

T_2

E_a

Kinetic energy

Figure 14.9

At a higher temperature, T_2, more molecules have an energy greater than E_a, as shown by the yellow shaded area.

A plot of log k against $1/T$ gives a straight line whose slope is $-E_a/2.303\ R$, from which E_a may be determined.

The variation of the rate constant with temperature for the decomposition of HI(g) to $H_2(g)$ and $I_2(g)$ is given in the table. What is the activation energy for the reaction?   **EXAMPLE** 14.14

T, K	$1/T$, K^{-1}	k, L mol^{-1} s^{-1}	log k
555	1.80×10^{-3}	3.52×10^{-7}	-6.453
575	1.74×10^{-3}	1.22×10^{-6}	-5.913
645	1.55×10^{-3}	8.59×10^{-5}	-4.066
700	1.43×10^{-3}	1.16×10^{-3}	-2.936
781	1.28×10^{-3}	3.95×10^{-2}	-1.403

A graph of log k versus $1/T$ is given in Fig. 14.10. In order to determine the slope of the line, we need two values of log k, which are determined from the line at two values

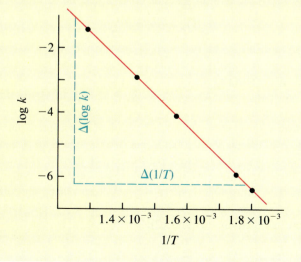

log k

$\Delta(\log k)$

-2

-4

-6

$\Delta(1/T)$

1.4×10^{-3} 1.6×10^{-3} 1.8×10^{-3}

$1/T$

Figure 14.10

A graph of the linear relationship between log k and $1/T$ for the reaction $2HI \longrightarrow H_2 + I_2$ according to the Arrhenius equation.

of $1/T$ (one near each end of the line is preferable). For example, the value of log k determined from the line when $1/T = 1.25 \times 10^{-3}$ is -1.126; the value when $1/T = 1.78 \times 10^{-3}$ is -6.273. The slope of this line is given by the following expression:

$$\text{Slope} = \frac{\Delta(\log k)}{\Delta(1/T)}$$

$$= \frac{(-6.273) - (-1.126)}{(1.78 \times 10^{-3} \text{ K}^{-1}) - (1.25 \times 10^{-3} \text{ K}^{-1})}$$

$$= \frac{-5.147}{0.53 \times 10^{-3} \text{ K}^{-1}} = -9.7 \times 10^3 \text{ K}$$

$$\text{Slope} = -\frac{E_a}{2.303 \, R}$$

Thus

$$-E_a = \text{slope} \times 2.303 \, R = -9.7 \times 10^3 \, \text{K} \times 2.303 \times 8.314 \text{ J mol}^{-1} \text{ K}^{-1}$$

$$E_a = 1.9 \times 10^5 \text{ J mol}^{-1} = 1.9 \times 10^2 \text{ kJ mol}^{-1}$$

14.10 Elementary Reactions

A balanced equation for a chemical reaction indicates what is reacting and what is produced, but it says nothing about how the reaction actually takes place. The process, or pathway, by which a reaction occurs is called the **reaction mechanism,** or the **reaction path.**

Reactions often occur in steps. The decomposition of ozone, for example, is believed to follow a mechanism with two steps.

$$O_3 \longrightarrow O_2 + O \qquad (1)$$

$$O + O_3 \longrightarrow 2O_2 \qquad (2)$$

The two steps add up to the overall reaction for the decomposition,

$$2O_3 \longrightarrow 3O_2$$

The oxygen atom produced in the first step is used in the second and thus does not appear as a final product. Species that are produced in one step and consumed in another are called **intermediates.**

Each of the steps in a reaction mechanism is called an **elementary reaction.** Elementary reactions occur exactly as they are written and cannot be broken down into simpler steps. An overall reaction can often be broken down into a number of steps. Thus, although the overall reaction indicates that two molecules of ozone react to give three molecules of oxygen, the reaction path does not involve the collision of two ozone molecules. Instead, a molecule of ozone decomposes to an oxygen molecule and an intermediate oxygen atom, and then the oxygen atom reacts with a second ozone molecule to give two oxygen molecules. These two elementary reactions occur exactly as they are written in Equations (1) and (2).

14.11 Unimolecular Reactions

An elementary reaction is **unimolecular** if the rearrangement of a *single* molecule (or ion) produces one or more molecules of product. A unimolecular reaction may be one of several elementary reactions in a complex mechanism, or it may be the only reaction in a mechanism. Equation (1) in Section 14.10,

$$O_3 \longrightarrow O_2 + O$$

illustrates a unimolecular elementary reaction occurring in a two-step reaction mechanism. The gas-phase decompositions of dinitrogen pentaoxide, N_2O_5, and cyclobutane, C_4H_8, each occur with one-step unimolecular mechanisms.

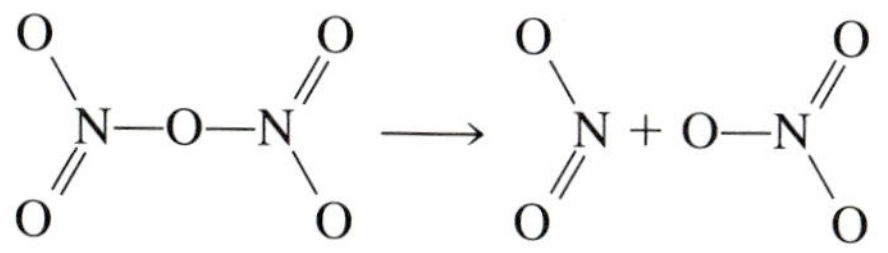

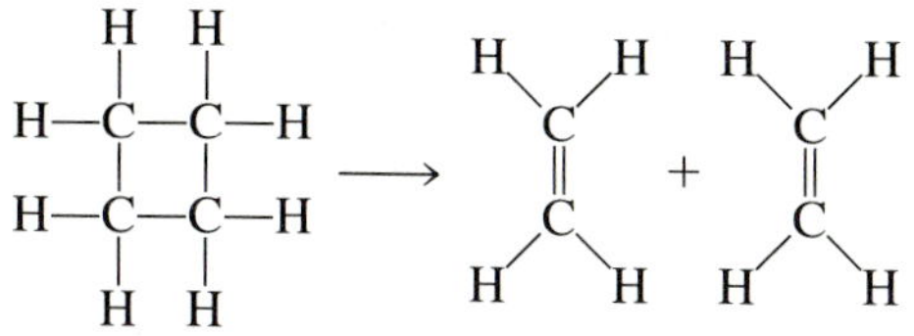

All that is required for each of these three reactions to occur is the separation of the single reactant molecule into two parts. The latter two reactions also show that an overall reaction may be an elementary reaction.

Chemical bonds do not simply fall apart during chemical reactions. The decomposition of C_4H_8, for example, requires the input of 261 kilojoules of energy per mole to distort the molecules into activated complexes that can decompose into products. (Thus the activation energy for this reaction is 261 kilojoules per mole of C_4H_8 reacting.)

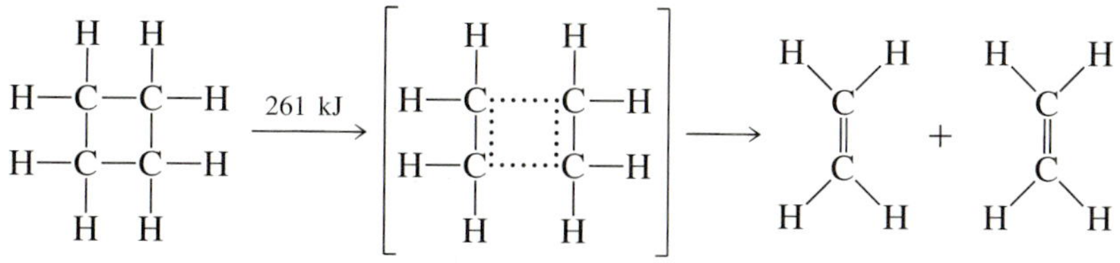

In a sample of C_4H_8, very few molecules contain enough energy to form activated complexes. Those that do, react by the process indicated above. Others collide with rapidly moving molecules and pick up additional energy. (The kinetic energy of the rapidly moving molecules is converted into activation energy in the other molecules.) If this energy is sufficient to form the activated complex, then the activated molecules too can undergo reaction. In effect, a particularly energetic collision will knock the reacting molecules into the geometry of the activated complex. However, only a small fraction of gas molecules travel at sufficiently high speeds with large enough kinetic energies to accomplish this (Fig. 14.8). Hence at any one time only a few molecules pick up enough energy from collisions to climb the activation energy barrier and react.

Doubling the concentration of C_4H_8 molecules in a sample gives twice as many molecules per liter. Although the fraction of molecules with enough energy to react will be the same, the total number of such molecules will be twice as great. Consequently, the change in the amount of C_4H_8 per liter and thus the reaction rate $(-\Delta[C_4H_8]/\Delta t)$ will be twice as great. The reaction rate is directly proportional to the concentration of C_4H_8.

$$\text{Rate} = k[C_4H_8]$$

This relationship applies to any *unimolecular elementary reaction;* the reaction rate is directly proportional to the concentration of the reactant, and the reaction exhibits first-order behavior. Hence the proportionality constant is the rate constant for the particular unimolecular reaction.

14.12 Bimolecular Reactions

The collision *and combination* of two reactants to give an activated complex in an elementary reaction is called a **bimolecular reaction.** Equation (2) in Section 14.10,

$$O + O_3 \longrightarrow 2O_2$$

is an example of a bimolecular elementary reaction that occurs in a two-step reaction mechanism. The reaction of nitrogen dioxide with carbon monoxide (Section 14.8) and the decomposition of two hydrogen iodide molecules to give hydrogen, H_2, and iodine, I_2 (Fig. 14.11), are examples of reactions whose mechanisms consist of a single bimolecular elementary reaction.

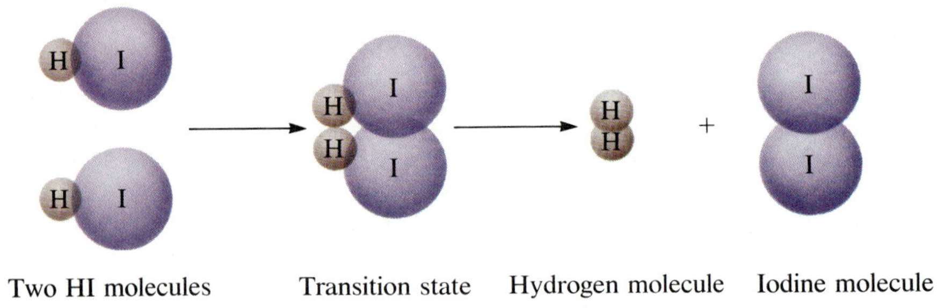

Figure 14.11

Probable mechanism for the dissociation of two HI molecules to produce one molecule of H_2 and molecule of I_2.

Two HI molecules   Transition state   Hydrogen molecule   Iodine molecule

For the bimolecular elementary reaction

$$A + B \longrightarrow \text{products}$$

the rate equation is first order in A and first order in B.

$$\text{Rate} = k[A][B]$$

Doubling the concentration of either A or B doubles the total number of molecular collisions and the number of effective collisions between molecules A and B. Thus if the concentration of A is doubled (and that of B is left the same), the rate of reaction is doubled. Doubling the concentrations of both A and B quadruples the number of collisions and the rate of reaction (Fig. 14.12). If the initial concentrations of both A and B are tripled, then the reaction proceeds nine times as fast. For the bimolecular elementary reaction

$$A + A \longrightarrow \text{products}$$

the rate equation is second order in A.

$$\text{Rate} = k[A][A] = k[A]^2$$

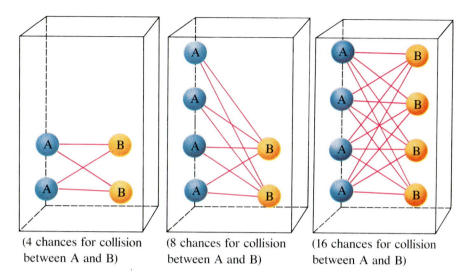

(4 chances for collision between A and B)

(8 chances for collision between A and B)

(16 chances for collision between A and B)

Figure 14.12

A schematic representation of the effect of concentration on the number of possible collisions and hence the rate of reaction.

14.13 Termolecular Reactions

An elementary **termolecular reaction** involves the simultaneous collision of any combination of three atoms, molecules, or ions. Termolecular elementary reactions are uncommon because the probability of three particles colliding simultaneously is less than a thousandth of that of two particles colliding. There are, however, a few established termolecular elementary reactions. The reaction of nitric oxide with oxygen,

$$2NO + O_2 \longrightarrow 2NO_2$$

the reaction of nitric oxide with chlorine,

$$2NO + Cl_2 \longrightarrow 2NOCl$$

and the reaction of hydrogen with iodine

$$H_2 + I_2 \longrightarrow 2HI$$

all involve termolecular steps.

We might expect the reaction of hydrogen with iodine to involve a bimolecular reaction of H_2 with I_2. In fact, however, two iodine atoms, produced by the dissociation of an iodine molecule, $(I_2 \rightarrow 2I)$, collide with a hydrogen molecule to give a transition state, which then splits to give two hydrogen iodide molecules (Fig. 14.13).

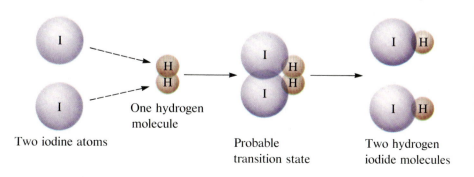

Two iodine atoms

One hydrogen molecule

Probable transition state

Two hydrogen iodide molecules

Figure 14.13

The termolecular step in the mechanism for the reaction of H_2 and I_2 to produce two HI molecules. The first step (not shown) involves the dissociation of an iodine molecule into two iodine atoms $(I_2 \longrightarrow 2I)$. In the termolecular step (shown here) two iodine atoms and one hydrogen molecule combine to produce two HI molecules.

For the termolecular elementary reaction

$$A + B + C \longrightarrow \text{products}$$

the rate equation is first order in A, B, and C.

$$\text{Rate} = k[A][B][C]$$

For the termolecular elementary reaction

$$2A + B \longrightarrow \text{products}$$

the rate equation is second order in A and first order in B.

$$\text{Rate} = k[A]^2[B]$$

14.14 Reaction Mechanisms

The stepwise sequence of elementary reactions that converts reactants into products is called the reaction mechanism, or reaction path. The decomposition of C_4H_8 ($C_4H_8 \rightarrow 2C_2H_4$), for example, has a one-step mechanism (Section 14.11); the decomposition of ozone ($2O_3 \rightarrow 3O_2$), a two-step mechanism (Section 14.10). Since elementary reactions involving three or more reactants are rare, it is reasonable to expect that complex reactions take place in several steps. For a reaction such as the following one to take place in one step,

$$2MnO_4^- + 10Cl^- + 16H_3O^+ \longrightarrow 2Mn^{2+} + 5Cl_2 + 24H_2O$$

2 permanganate ions (MnO_4^-), 10 chloride ions, and 16 hydronium ions would have to collide simultaneously in an effective collision. This, of course, is very unlikely.

Even an apparently simple reaction such as the formation of ethylene, C_2H_4, from ethane, C_2H_6, may proceed by a complex mechanism. Ethylene is one of the most important reagents used by the chemical industry for manufacture of polyethylene, polyester fiber, synthetic rubber, and some detergents. Ethylene may be prepared by the pyrolysis (decomposition by heating) of ethane, a component of natural gas, at temperatures of 500°–800°C.

$$C_2H_6 \longrightarrow C_2H_4 + H_2$$

The actual mechanism is much more complex than suggested by this equation and involves the following elementary reactions:

1. *Initiation*. The reaction begins with a unimolecular reaction in which a C_2H_6 molecule splits at its weakest point, the C—C bond.

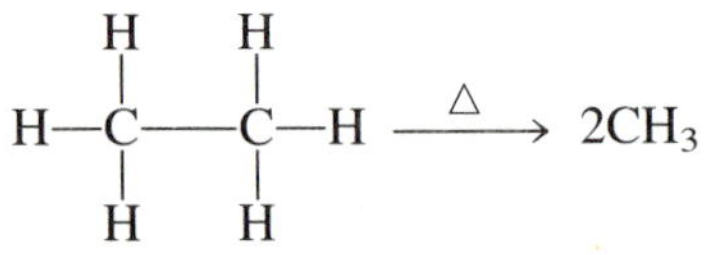

$$\underset{\substack{| \\ H}}{\overset{\substack{H \\ |}}{H-C}}\!\!-\!\!\underset{\substack{| \\ H}}{\overset{\substack{H \\ |}}{C}}-H \xrightarrow{\;\triangle\;} 2CH_3 \qquad (1)$$

The CH_3 group abstracts a hydrogen atom from another C_2H_6 molecule in a bimolecular reaction.

$$CH_3 + C_2H_6 \longrightarrow CH_4 + C_2H_5 \qquad (2)$$

2. *Propagation*. The C_2H_5 group undergoes a unimolecular dissociation.

$$C_2H_5 \longrightarrow C_2H_4 + H \qquad (3)$$

The hydrogen atom produced in Equation (3) reacts with another C_2H_6 molecule in a bimolecular reaction.

$$H + C_2H_6 \longrightarrow C_2H_5 + H_2 \qquad (4)$$

The C_2H_5 group produced in Equation (4) reacts according to Equation (3), producing another hydrogen atom to undergo reaction as in Equation (4) and produce another C_2H_5. The reactions given by Equations (3) and (4) produce C_2H_4 and H_2, which are the principal products of the pyrolysis. Normally, each initiation [Equations (1) and (2)] is followed by about a hundred cycles of propagation [Equations (3) and (4)] before termination.

3. *Termination*. The chain of propagation reactions may be stopped by one of the following bimolecular reactions:

$$2C_2H_5 \longrightarrow C_4H_{10}$$
$$2C_2H_5 \longrightarrow C_2H_4 + C_2H_6$$

The C_2H_5 produced in Equation (4) can react according to Equation (3) to produce another hydrogen atom for Equation (4). This reinitiates the series of elementary reactions represented by Equations (3) and (4), and they repeat themselves over and over. Such a mechanism, involving repeating reactions, is known as a **chain mechanism.**

Some of the elementary reactions in a reaction path are relatively slow. The slowest reaction step determines the maximum rate, since a reaction can proceed no faster than its slowest step. The slowest step, therefore, is the **rate-determining step** of the reaction.

We can write the rate equation for each elementary reaction in a reaction mechanism (Sections 14.11–14.13), *but we cannot ordinarily write a correct rate equation or establish the reaction order for an overall reaction that involves several steps simply by inspection of the overall balanced equation.* We must determine the overall rate equation from experimental data. The reaction of nitrogen dioxide and carbon monoxide (Example 14.5 and Section 14.8) is an excellent example.

$$NO_2 + CO \longrightarrow CO_2 + NO$$

For temperatures above 225°C the reaction proceeds by a single one-step bimolecular elementary reaction (Section 14.12). For temperatures above 225°C, therefore, the reaction is first order with respect to NO_2 and first order with respect to CO. The rate equation is

$$\text{Rate} = k[NO_2][CO]$$

The overall order of the reaction is second order.

At temperatures below 225°C the reaction proceeds by two elementary reactions, the first of which is slow and is therefore the rate-determining step.

$$NO_2 + NO_2 \longrightarrow NO_3 + NO \qquad \text{(slow)}$$
$$NO_3 + CO \longrightarrow NO_2 + CO_2 \qquad \text{(fast)}$$

(A summation of the two equations represents the net overall reaction.) At lower temperatures the rate equation, based on the first step as the rate-determining step, is

$$\text{Rate} = k[NO_2]^2$$

Note that the rate equation for the reaction at lower temperatures does not even include the concentration of CO. At temperatures below 225°C, therefore, the reaction is second order with respect to NO_2 and zero order with respect to CO; the overall order of the reaction is still second order.

In general, when the rate-determining step is the first step, it provides the rate equation for the overall mechanism. However, when the rate-determining step occurs later in the mechanism, the rate equation may be complex. The oxidation of iodide ion by hydrogen peroxide illustrates this point.

$$H_2O_2 + 3I^- + 2H_3O^+ \longrightarrow 4H_2O + I_3^-$$

In a solution with a high concentration of acid, one reaction pathway has the following rate equation:

$$\text{Rate} = k[H_2O_2][I^-][H_3O^+]$$

This rate equation is consistent with several mechanisms, three of which follow:

Mechanism A

$H_2O_2 + H_3O^+ + I^- \longrightarrow 2H_2O + HOI$	(slow)
$HOI + H_3O^+ + I^- \longrightarrow 2H_2O + I_2$	(fast)
$I_2 + I^- \longrightarrow I_3^-$	(fast)

Mechanism B

$H_3O^+ + I^- \longrightarrow HI + H_2O$	(fast)
$H_2O_2 + HI \longrightarrow H_2O + HOI$	(slow)
$HOI + H_3O^+ + I^- \longrightarrow 2H_2O + I_2$	(fast)
$I_2 + I^- \longrightarrow I_3^-$	(fast)

Mechanism C

$H_3O^+ + H_2O_2 \longrightarrow H_3O_2^+ + H_2O$	(fast)
$H_3O_2^+ + I^- \longrightarrow H_2O + HOI$	(slow)
$HOI + H_3O^+ + I^- \longrightarrow 2H_2O + I_2$	(fast)
$I_2 + I^- \longrightarrow I_3^-$	(fast)

In mechanism A the slow step is the first elementary reaction, so the rate equation for the overall reaction is equal to that for this step. In mechanisms B and C the rate-determining step is the second elementary reaction, and the overall rate equation is not simply the rate equation for this step. To derive the rate equations for mechanisms B and C requires some familiarity with equilibrium constants, which will be introduced in the next chapter. However, since mechanisms A, B, and C all have the same overall rate equation, it is not possible to distinguish among them only on this basis. Additional experimental information is required to distinguish which reaction pathway actually leads to product. Moreover, the rate equation provides no information about what happens after the rate-determining step. The subsequent steps must be worked out from other chemical knowledge or from other measurements.

The determination of the mechanism of a reaction is important in selecting conditions that provide a good yield of the desired product. Knowing a reaction's mechanism sometimes helps a chemist to prepare a previously unknown compound. Compared to the number of known chemical reactions, rather few reaction mechanisms have been completely characterized. The study of reaction mechanisms, or the **kinetics of reaction,** is a very active research area.

14.15 Catalysts

The rate of many reactions can be accelerated by **catalysts,** substances that are not themsleves used up by the reaction. Such substances may be divided into two general classes: **homogeneous catalysts** and **heterogeneous catalysts.**

1. HOMOGENEOUS CATALYSTS. A homogeneous catalyst is present in the same phase as the reactants. It interacts with a reactant forming an intermediate substance, which then decomposes or reacts with another reactant to regenerate the original catalyst and give product.

The ozone in the stratosphere (the upper atmosphere) that protects the earth from ultraviolet radiation is formed by the following mechanism when ultraviolet light (hv) interacts with oxygen molecules.

$$O_2(g) \xrightarrow{hv} 2O(g) \tag{1}$$
$$O(g) + O_2(g) \longrightarrow O_3(g) \tag{2}$$

As shown in Section 14.10, ozone decomposes by the following mechanism.

$$O_3(g) \longrightarrow O_2(g) + O(g) \tag{3}$$
$$O(g) + O_3(g) \longrightarrow 2O_2(g) \tag{4}$$

The rate of the decomposition of ozone is influenced by the presence of nitric oxide, NO. Nitric oxide catalyzes the decomposition of ozone by the following mechanism:

$$NO(g) + O_3(g) \longrightarrow NO_2(g) + O_2(g) \tag{5}$$
$$O_3(g) \longrightarrow O_2(g) + O(g) \tag{6}$$
$$NO_2(g) + O(g) \longrightarrow NO(g) + O_2(g) \tag{7}$$

The overall chemical change for Equations (5) through (7) is the same as for Equations (3) and (4):

$$2O_3 \longrightarrow 3O_2$$

Since the nitric oxide is not permanently used up in these reactions, it is a catalyst. The rate of the decomposition of ozone is greater in the presence of nitric oxide because of its catalytic activity. In the presence of a constant concentration of nitric oxide, the rate at which ozone forms in the upper atmosphere according to Equations (1) and (2) is equal to the rate at which it decomposes according to Equations (3) and (4) and Equations (5) through (7). A state of equilibrium exists, and the total concentration of ozone does not change significantly over long periods. However, if additional nitric oxide were introduced into the stratosphere by atmospheric nuclear explosions or from the exhaust of high-flying supersonic aircraft, the rate of decomposition would increase and the total amount of ozone present would decrease. (Certain compounds that contain chlorine also catalyze the decomposition of ozone.)

Many biological reactions are catalyzed by **enzymes,** which are complex substances produced by living organisms. A pure solution of sugar in water will not react with oxygen, even if it stands for years. However, the enzyme zymase, produced by yeast cells, catalyzes the reaction between sugar and oxygen to form alcohol in a matter of hours. A part of the body's energy is produced by the oxidation of sugar in a multistep process that is catalyzed by enzymes. Many other reactions in living organisms are also catalyzed by enzymes. There may be as many as 30,000 different enzymes in the human body, each of which is a protein constructed to be effective as a catalyst for a specific chemical reaction useful to the body. The subject of enzymes and their kinetic activity will be examined in greater detail in Chapter 31.

2. HETEROGENEOUS CATALYSTS. Heterogeneous catalysts act by furnishing a surface at which a reaction can occur. Gas-phase and liquid-phase reactions catalyzed by heterogeneous catalysts typically occur on the surface of the catalyst rather than within the gas or liquid phase. For this reason heterogeneous catalysts are sometimes called *contact catalysts*.

Heterogeneous catalysis has at least four steps: (1) adsorption of the reactant onto the surface of the catalyst, (2) activation of the adsorbed reactant, (3) reaction of the adsorbed reactant, and (4) diffusion of the product from the surface into the gas or liquid phase (desorption). Any one of these steps may be slow and thus may be the rate-determining step. In general, however, the overall rate of the reaction is faster than the rate would be if the reactants were in the gas or liquid phase. The steps that are believed to occur in the reaction of compounds containing a carbon-carbon double bond with hydrogen on a nickel catalyst are illustrated in Fig. 14.14. This is the catalyst used in the hydrogenation of polyunsaturated fats and oils (which contain several carbon-carbon double bonds) to produce saturated fats and oils (which contain only carbon-carbon single bonds).

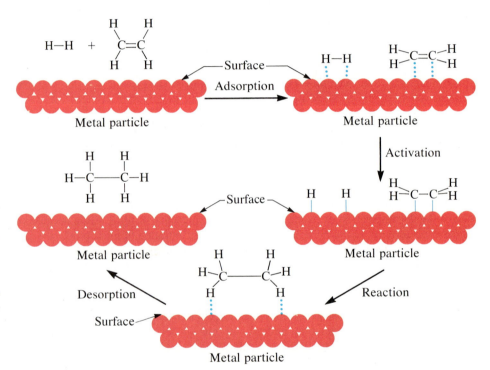

Figure 14.14

Steps in the catalysis by nickel of the reaction $C_2H_4 + H_2 \longrightarrow C_2H_6$. Both molecules are adsorbed by weak attractive forces. Activation occurs when the bonding electrons in the molecules rearrange to form bonds to metal atoms. Following the reaction of the activated atoms, the weakly adsorbed C_2H_6 molecule escapes from the surface.

Other significant industrial processes involving the use of contact catalysts include the preparation of sulfuric acid (Section 9.9, Part 1), the preparation of ammonia (Section 9.9, Part 2), the oxidation of ammonia to nitric acid (Section 23.13), and the synthesis of methanol, CH_3OH (Section 30.9).

Heterogeneous catalysts are also used in the catalytic converters found on most gasoline-powered automobiles (Fig. 14.15). The exhaust gases are mixed with air and passed over a mixture of metal oxides. The catalyst promotes oxidation of carbon monoxide and unburned hydrocarbons to relatively harmless carbon dioxide and water. The newer catalysts also promote the decomposition of nitrogen oxides to nitrogen and oxygen. However, these catalysts also catalyze the oxidation of SO_2 to SO_3, which reacts with water vapor and forms a sulfuric acid mist. Since sulfur-containing fuels produce SO_2, these fuels must be avoided.

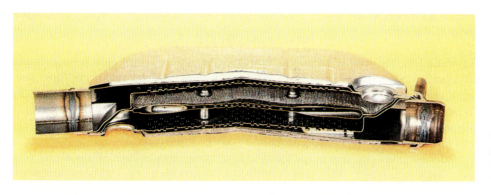

Figure 14.15
A catalytic converter used to decrease pollutants in the exhaust of automobiles.

Both homogeneous and heterogeneous catalysts function by providing a reaction path with a lower activation energy than would be found in the absence of the catalyst (Fig. 14.16). This lower activation energy results in an increase in rate (Section 14.9). Note that a catalyst decreases the activation energy for both the forward and the reverse reactions and hence *accelerates both the forward and the reverse reactions*.

Some substances, called **inhibitors,** decrease the rate of a chemical reaction. In many cases these are substances that react with and "poison" some catalyst in the system and thereby prevent its action. Many biological poisons are inhibitors that reduce the catalytic activity of an organism's enzymes. Catalytic converters are poisoned by lead, so lead-free fuels must be used in automobiles equipped with such converters.

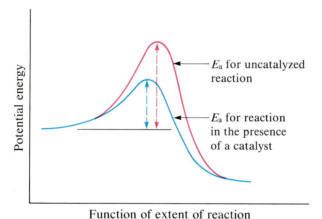

E_a for uncatalyzed reaction

E_a for reaction in the presence of a catalyst

Potential energy

Function of extent of reaction

Figure 14.16
Potential energy diagram showing the effect of a catalyst on the activation energy of a reaction.

For Review

Summary

The **rate** at which a chemical reaction proceeds can be defined as either the decrease in concentration of a reactant or the increase in concentration of a product per unit of time. In general, the rate of a given reaction increases as the temperature or the concentration of a reactant increases. The rate of a reaction can also be increased by a **catalyst,** a substance that is not permanently changed by the reaction. Reactions involving two phases proceed more rapidly, the more finely divided the condensed phase. The rate of a given reaction can be described by an experimentally determined **rate equation** of the form

$$\text{Rate} = k[A]^m[B]^n[C]^p \ldots$$

in which [A], [B], and [C] represent molar concentrations of reactants (or sometimes products or other substances); m, n, and p are usually, but not always, positive integers; and k is the rate constant. The exponents, m, n, p, etc., describe the **order of the reaction** with respect to each specific reactant. The **overall order of the reaction** is the sum of the exponents. The **half-life** of a reaction is the time required for half of the original reactant to be consumed.

Before atoms, molecules, or ions can react, they must collide. When every collision leads to reaction, the rate is controlled by how rapidly the reactants diffuse together. These **diffusion-controlled reactions** are very fast. Most reactions occur more slowly because the reacting species must be oriented correctly when they collide and because they must possess a certain minimum energy, the **activation energy,** in order to form an activated complex, or transition state. The rate constant for a reaction is related to these effects by the **Arrhenius equation:**

$$k = A \times 10^{-E_a/2.303RT}$$

The collection of individual steps, or **elementary reactions,** by which reactants are converted into products during the course of an overall reaction is called the **reaction mechanism,** or **reaction path.** The overall rate of a reaction is determined by the rate of the slowest step, the **rate-determining step.** Although it is not possible to write a rate equation for an overall reaction without data from experimental observations, once the elementary reactions have been determined, rate equations can be written by inspection for each elementary reaction. **Unimolecular elementary reactions** have first-order rate equations; **bimolecular elementary reactions,** second-order rate equations; and **termolecular elementary reactions** (which are uncommon), third-order rate equations.

Key Terms and Concepts

activated complex (14.8)
activation energy (14.9)
Arrhenius equation (14.9)
bimolecular reaction (14.12)
catalysis (14.15)
chain mechanism (14.14)
collision theory of reaction (14.8)
diffusion-controlled reactions (14.8)

elementary reaction (14.10)
frequency factor (14.9)
half-life of a reaction (14.7)
heterogeneous catalysts (14.15)
homogeneous catalysts (14.15)
order of a reaction (14.6)
rate constant (14.5)
rate-determining step (14.14)

rate equations (14.5)
rate of reaction (14.1–14.5)
reaction mechanism (14.10, 14.14)
termolecular reaction (14.13)
transition state (14.8)
unimolecular reaction (14.11)

Exercises

Reaction Rates and Rate Equations

1. State the factors that determine the rate of reaction and explain how each is responsible for changing the rate.
2. Doubling the concentration of a reactant increases the rate of a reaction four times. What is the order of the reaction with respect to that reactant?

    Ans. Second order

3. Explain the difference between the rate of a reaction and its rate constant.

4. Nitrosyl chloride, NOCl, decomposes to NO and Cl_2.

$$2NOCl(g) \longrightarrow 2NO(g) + Cl_2(g)$$

Determine the rate equation and the rate constant for this reaction from the following data:

[NOCl], M	0.10	0.20	0.30
Rate, mol L^{-1} s^{-1}	8.0×10^{-10}	3.20×10^{-9}	7.2×10^{-9}

Ans. Rate = $k[NOCl]^2$; $k = 8.0 \times 10^{-8}$ L mol^{-1} s^{-1}

Rate$k[NOCl]^2$

5. The following data have been determined for the reaction

$$I^- + OCl^- \longrightarrow IO^- + Cl^-$$

[I$^-$], M	0.20	0.40	0.60
[OCl$^-$], M	0.050	0.050	0.010
Rate, mol L^{-1} s^{-1}	6.10×10^{-4}	1.22×10^{-3}	3.66×10^{-4}

Determine the rate equation and rate constant for this reaction.

Ans. Rate = k[I$^-$][OCl$^-$];
k = 6.1 × 10^{-2} L mol^{-1} s^{-1}

6. The rate constant for the decomposition at 45°C of nitrogen(V) oxide, N_2O_5, dissolved in chloroform, $CHCl_3$, is 6.2×10^{-4} min^{-1}.

$$2N_2O_5 \longrightarrow 4NO_2 + O_2$$

The decomposition is first order in N_2O_5.
(a) What is the rate of the reaction when [N_2O_5] = 0.25 M?

Ans. 1.6 × 10^{-4} mol L^{-1} min^{-1}

(b) What is the concentration of N_2O_5 remaining at the end of 3.5 h if the initial concentration of N_2O_5 was 0.25 M?

Ans. 0.22 M

7. Hydrogen reacts with nitrogen(II) oxide to form nitrogen(I) oxide, laughing gas, according to the equation

$$H_2(g) + 2NO(g) \longrightarrow N_2O(g) + H_2O$$

Determine the rate equation and the rate constant for the reaction from the following data:

[NO], M	0.40	0.80	0.80
[H$_2$], M	0.35	0.35	0.70
Rate, mol L^{-1} s^{-1}	5.040×10^{-3}	2.016×10^{-2}	4.032×10^{-2}

Ans. Rate = k[NO]^2[H$_2$]; k = 9.0 × 10^{-2} L^2 mol^{-2} s^{-1}

8. Nitrogen(II) oxide reacts with chlorine according to the equation

$$2NO(g) + Cl_2(g) \longrightarrow 2NOCl(g)$$

The following initial rates of reaction have been observed for certain reactant concentrations:

NO, mol L^{-1}	Cl$_2$, mol L^{-1}	Rate, mol L^{-1} h^{-1}
0.50	0.50	1.14
1.00	0.50	4.56
1.00	1.00	9.12

What is the rate equation describing the rate dependence on the concentrations of NO and Cl_2?

Ans. Rate = k[NO]^2[Cl]; k = 9.12 L^2 mol^{-2} h^{-1}

9. Most of the 13.1 billion pounds of HNO_3 produced in the United States during 1986 was prepared by the following sequence of reactions, each run in a separate reaction vessel.

$$4NH_3(g) + 5O_2(g) \longrightarrow 4NO(g) + 6H_2O(g) \quad (1) \text{ fast}$$
$$2NO(g) + O_2(g) \longrightarrow 2NO_2(g) \quad (2) \text{ slow}$$
$$3NO_2(g) + H_2O(l) \longrightarrow 2HNO_3(aq) + NO(g) \quad (3) \text{ fast}$$

The first reaction is run by burning ammonia in air over a platinum catalyst. This reaction is fast. The reaction in Equation (3) is also fast. The second reaction limits the rate at which nitric acid can be prepared from ammonia. If Equation (2) is second order in NO and first order in O_2, what is the rate of formation of NO_2 when the oxygen concentration is 0.45 M and the nitric oxide concentration is 0.70 M? The rate constant for the reaction is 5.8×10^{-6} L^2 mol^{-2} s^{-1}.

Ans. 1.3 × 10^{-6} mol L^{-1} s^{-1}

10. One of the reactions involved in the formation of photochemical smogs is

$$O_3(g) + NO(g) \longrightarrow O_2(g) + NO_2(g)$$

The rate constant for this reaction is 1.2×10^7 L mol^{-1} s^{-1}. The reaction is first order in O_3 and first order in NO. Calculate the rate of formation of NO_2 in air in which the O_3 concentration is 5×10^{-8} M and the NO concentration 8×10^{-8} M.

Ans. 5 × 10^{-8} mol L^{-1} s^{-1}

11. Determine the rate constant for the decomposition of H_2O_2 shown in Fig. 14.1 from the data given in the figure.

Ans. 3.21 × 10^{-5} s^{-1}

12. A liter of a 1 M solution of H_2O_2 slowly decomposes into H_2O and O_2. If 0.50 mol of H_2O_2 decomposes during the first 6 h of the reaction, explain why only 0.25 mol decomposes during the next 6-h period.

13. The decomposition of SO_2Cl_2 to SO_2 and Cl_2 is a first-order reaction with $k = 2.2 \times 10^{-5}$ s^{-1} at 320°C.
(a) Determine the half-life of the reaction.

Ans. t$_{1/2}$ = 3.2 × 10^4 s

(b) At 320°C, how much $SO_2Cl_2(g)$ would remain in a 1.00-L flask 2.5 h after the introduction of 0.0215 mol of SO_2Cl_2? Assume that the rate of the reverse reaction is so slow that it can be ignored.

Ans. 1.8 × 10^{-2} mol

14. Radioactive materials decay by a first-order process. The very dangerous isotope strontium-90 decays with a half-life of 28 years.

$$^{90}_{38}Sr \longrightarrow {}^{90}_{39}Y + e^-$$

What is the rate constant for this decay?

Ans. 2.5 × 10^{-2} yr^{-1}

15. The half-life for the radioactive decay of ^{14}C (first-order decay reaction) is 5730 yr. What is the rate constant in units of min^{-1} and h^{-1}?

Ans. 2.30 × 10^{-10} min^{-1}; 1.38 × 10^{-8} h^{-1}

16. For the reaction $A \rightarrow B + C$, the following data were obtained at 30.0°C:

[A], M	0.170	0.340	0.680
Rate, $mol\ L^{-1}\ h^{-1}$	0.0500	0.100	0.200

    (a) What is the order of the reaction with respect to [A], and what is the rate equation?

    (b) Calculate k for the reaction.   *Ans. 0.294 h^{-1}*

17. The reaction of compound A to give compounds C and D was found to be second order in A. The rate constant for the reaction was determined to be 2.42 $L\ mol^{-1}\ s^{-1}$. If the initial concentration is 0.0500 mol/L, what is the value of $t_{1/2}$?   *Ans. 8.26 s*

18. The half-life of a reaction of compound A to give compounds D and E is 8.50 min when the initial concentration of A is 0.150 mol/L. How long will it take for the concentration to drop to 0.0300 mol/L if the reaction is (a) first order with respect to A? (b) Second order with respect to A? In your calculations, give careful attention to the units for each quantity, showing how the units cancel to provide the proper units in each answer.

    Ans. (a) 19.8 min; (b) 34.0 min

Collision Theory of Reaction Rates

19. Chemical reactions occur when reactants collide. For what reasons may a collision fail to produce a chemical reaction?

20. If every collision between reactants leads to a reaction, what determines the rate at which the reaction will occur?

21. What is the activation energy of a reaction, and how is this energy related to the activated complex of the reaction?

22. Account for the relationship between the rate of a reaction and its activation energy.

23. If the rate of a reaction doubles for every 10°C rise in temperature, how much faster would the reaction proceed at 45°C than at 25°C? At 95°C than at 25°C?

    Ans. 4 times faster; 128 times faster

24. What is the effect on the rate of many reactions by an increase in temperature of 10°C? Explain this effect in terms of the collision theory of reaction rate.

25. In an experiment, a sample of $NaClO_3$ was 90% decomposed in 48 min. Approximately how long would this decomposition have taken if the sample had been heated 20° higher?   *Ans. 12 min*

26. The rate constant for the decomposition of acetaldehyde, CH_3CHO, to methane, CH_4, and carbon monoxide, CO, in the gas phase is 1.1×10^{-2} $L\ mol^{-1}\ s^{-1}$ at 703 K and 4.95 $L\ mol^{-1}\ s^{-1}$ at 865 K. Determine the activation energy for this decomposition.   *Ans. 191 kJ*

27. The rate constant at 325°C for the reaction $C_4H_8 \rightarrow 2C_2H_4$ (Section 14.11) is 6.1×10^{-8} s^{-1}, and the activation energy is 261 kJ per mole of C_4H_8. Determine the frequency factor for the reaction.   *Ans. 3.8×10^{15} s^{-1}*

Elementary Reactions, Reaction Mechanisms, Catalysts

28. Define the following:
    (a) unimolecular reaction
    (b) bimolecular reaction
    (c) elementary reaction
    (d) overall reaction

29. Which of the following equations, as written, could describe elementary reactions?
    (a) $Cl_2 + CO \longrightarrow Cl_2CO$;
        rate = $k[Cl_2]^{3/2}[CO]$
    (b) $PCl_3 + Cl_2 \longrightarrow PCl_5$;
        rate = $k[PCl_3][Cl_2]$
    (c) $2NO + 2H_2 \longrightarrow N_2 + 2H_2O$;
        rate = $k[NO][H_2]$
    (d) $2NO + O_2 \longrightarrow 2NO_2$;
        rate = $k[NO]^2[O_2]$
    (e) $NO + O_3 \longrightarrow NO_2 + O_2$;
        rate = $k[NO][O_3]$

    Ans. (b), (d), and (e)

30. In general, can we predict the effect of doubling the concentration of A on the rate of the overall reaction $A + B \rightarrow C$? Can we predict the effect if the reaction is known to be an elementary reaction?

31. Why are more elementary reactions involving three or more reactants very uncommon?

32. What is the rate equation for the elementary termolecular reaction $A + 2B \rightarrow$ products? For $3A \rightarrow$ products?

33. Describe how a homogeneous catalyst and a heterogeneous catalyst function.

34. Account for the increase in reaction rate brought about by a catalyst.

Additional Exercises

35. (a) It has been suggested that chlorine atoms resulting from decomposition of chlorofluoromethanes, such as CCl_2F_2, catalyze the decomposition of ozone in the ozone layer of the earth's atmosphere. One simplified mechanism for the decomposition is

$$O_3 \xrightarrow{\text{Sunlight}} O_2 + O \quad \text{(slow)}$$
$$O_3 + Cl \longrightarrow O_2 + ClO \quad \text{(fast)}$$
$$ClO + O \longrightarrow Cl + O_2 \quad \text{(fast)}$$

    Explain why chlorine atoms are catalysts in the gas-phase transformation

$$2O_3 \rightleftharpoons 3O_2$$

    (b) Nitric oxide is also involved in the decomposition of ozone by the mechanism

$$O_3 \xrightarrow{\text{Sunlight}} O_2 + O$$
$$O_3 + NO \longrightarrow NO_2 + O_2$$
$$NO_2 + O \longrightarrow NO + O_2$$

Is NO a catalyst for the decomposition? Explain your answer.

36. Write the rate equation for each of the elementary reactions given in both parts of exercise 35.

37. An elevated level of the enzyme alkaline phosphatase (ALP) in the serum is an indication of possible liver or bone disorder. The level of serum ALP is so low that it is very difficult to measure directly. However, ALP catalyzes a number of reactions, and its relative concentration can be determined by measuring the rate of one of these reactions under controlled conditions. One such reaction is the conversion of p-nitrophenyl phosphate (PNPP) to p-nitrophenoxide ion (PNP) and phosphate ion. Control of temperature during the test is very important; the rate of the reaction increases 1.47 times if the temperature changes from 30°C to 37°C. What is the activation energy for the ALP-catalyzed conversion of PNPP to PNP and phosphate? *Ans. 43 kJ mol^{-1}*

38. Some bacteria are resistant to the antibiotic penicillin because they produce penicillinase, an enzyme with a molecular weight of 30,000, which converts penicillin into inactive molecules. Although the kinetics of enzyme-catalyzed reactions can be complex, at low concentrations this reaction can be described by a rate equation that is first order in the catalyst (penicillinase) and that also involves the concentration of penicillin. From the following data for a 1.0-L solution containing 0.15 μg (0.15 × 10^{-6} g) of penicillinase, determine the order of the reaction with respect to penicillin and the value of the rate constant.

[Penicillin], M	Rate, mol L^{-1} min^{-1}
2.0×10^{-6}	1.0×10^{-10}
3.0×10^{-6}	1.5×10^{-10}
4.0×10^{-6}	2.0×10^{-10}

Ans. Rate = k[penicillinase][penicillin];
k = 1.0 × 10^7 L mol^{-1} min^{-1}

An Introduction to Chemical Equilibrium

15

Disruption of the oxygen-ozone equilibrium created a hole in the ozone layer above the Antarctic.

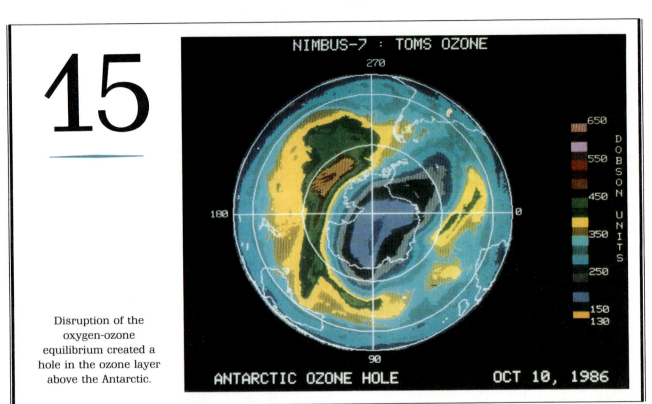

NIMBUS-7 : TOMS OZONE

ANTARCTIC OZONE HOLE   OCT 10, 1986

DOBSON UNITS

650
550
450
350
250
150
130

I n the previous chapter we learned how to describe how fast a chemical reaction proceeds quantitatively, using a rate law. We also learned how to predict the effect that changing the concentration has on the rate. However, this information says nothing about the percent conversion to products. In order to discuss the percent conversion, we will introduce the concept of chemical equilibrium in this chapter.

The opening photograph for this chapter was taken by a satellite above the southern hemisphere. The colors represent different concentrations of ozone and show that the ozone concentration is significantly depleted above the Antarctic. In the ozone layer, ozone is continually forming and decomposing simultaneously in such a way that its concentration normally is constant. Some effect above the Antarctic has upset this equilibrium, and consequently the ozone concentration has been reduced.

15.1 The State of Equilibrium

Equilibrium is reached in a system when the conversion of reactants into products and the conversion of products back into reactants occur simultaneously at the same rate. Reactants are used at equilibrium to form products at the same rate that they are replaced by the reverse reaction of the products to form reactants. Hence there is no *net* change, over a period of time, in the concentrations of reactants and products present in a system at equilibrium. For example, we have defined the vapor pressure of a liquid (Section 11.2) as the pressure exerted by its vapor in equilibrium with the liquid. The conversions in this case are evaporation and condensation, and at equilibrium there is no change in the quantity of liquid or of vapor over a period of time.

An equilibrium can be likened in some ways to two jugglers throwing Indian clubs back and forth simultaneously between them (Fig. 15.1). To avoid either juggler accumulating more clubs than can be handled at one time, the person on the left must throw clubs to the person on the right at the same rate as the one on the right throws clubs back. Thus, at equilibrium, the jugglers both receive clubs at the same rate that they throw clubs. The number of clubs each juggler has at a given time remains constant. Such an analogy cannot be pressed too far but may be helpful in understanding the basic equilibrium concept.

The point at which equilibrium occurs in a reversible reaction varies with the conditions under which the reaction takes place. Knowing the effect of a change in conditions on an equilibrium enables chemists to select conditions and to control the relative amounts of substances present at equilibrium. For example, it was known for many years that nitrogen and hydrogen react to form ammonia.

$$N_2 + 3H_2 \rightleftharpoons 2NH_3$$

However, ammonia was manufactured in useful quantities by this reaction only after the factors that influence its equilibrium were understood (Section 23.10). The advance calculation of yields of ammonia under various conditions of temperature and pressure, based upon equilibrium concepts, has been extremely important in the huge fertilizer industry, in the industrial preparation of nitric acid, and in other applications that use ammonia. In fact, Fritz Haber, a German chemist, received the 1918 Nobel prize in chemistry for his pioneering work using equilibrium concepts to develop the synthesis of ammonia on a commercial scale.

Figure 15.1

Indian club jugglers provide an illustration of equilibrium. Each person throws clubs to the other person at the same rate that clubs are received from the other person. Since clubs are being thrown continuously in both directions, the number of clubs moving in each direction is constant, and the number of clubs each juggler has at a given time remains constant.

15.2 Law of Mass Action

Whenever the products of a chemical reaction can react to reform the reactants, the two reactions occur simultaneously and make up a reversible reaction (Section 9.3, Part 6). A reversible reaction does not go to completion as long as the products do not escape, the reactants are not completely converted into products, and a state of equilibrium is attained. Most chemical reactions that occur in a closed system are reversible and do not go to completion.

Let us consider the oxidation of carbon monoxide by nitrogen dioxide in a closed flask at 300°C.

$$NO_2 + CO \rightleftharpoons NO + CO_2$$

At the beginning of the reaction the flask contains only NO_2 and CO. Above 225°C the rate of the forward reaction (Section 14.14) is given by the expression

$$\text{Rate}_1 = k_1[NO_2][CO]$$

At first no molecules of NO and CO_2 are present, so there can be no reverse reaction to reform NO_2 and CO. The rate of the reverse reaction, Rate_2, is zero (see Fig. 15.2). However, as soon as some of the products (NO and CO_2) are formed, they begin to react. Initially the rate is relatively slow because their concentrations are low. However, as the forward reaction between NO_2 and CO proceeds, the concentrations of NO and CO_2 increase and the rate of the reverse reaction, Rate_2, increases. The rate of the reverse reaction above 225°C is given by

$$\text{Rate}_2 = k_2[NO][CO_2]$$

Meanwhile, the concentrations of NO_2 and CO decrease, so the rate of the forward reaction, Rate_1, falls off. Consequently, the two reaction rates approach each other and finally become equal. When Rate_2 equals Rate_1, a condition of **dynamic equilibrium** is established. The opposing reactions are in full operation but at the same rate. Thus at equilibrium we may write

$$\text{Rate}_2 = \text{Rate}_1$$
$$k_2[NO][CO_2] = k_1[NO_2][CO]$$

or, by rearranging,

$$\frac{k_1}{k_2} = \frac{[NO][CO_2]}{[NO_2][CO]} \qquad \frac{K_1}{K_2} = \frac{expression_2}{expression_1}$$

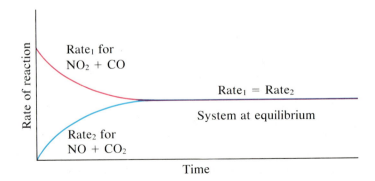

Figure 15.2

Rates of reaction of forward and reverse reactions for $NO_2(g) + CO(g) \rightleftharpoons NO(g) + CO_2(g)$, assuming only NO_2 and CO are present initially.

Rate of reaction

Rate_1 for $NO_2 + CO$

$\text{Rate}_1 = \text{Rate}_2$

System at equilibrium

Rate_2 for $NO + CO_2$

Time

Because k_1 and k_2 are constants, the ratio k_1/k_2 is also a constant, and the expression may be written

$$K = \frac{[NO][CO_2]}{[NO_2][CO]}$$

This is called the **equilibrium constant expression,** and K is called the **equilibrium constant** for the reaction. Just as the rate constants k_1 and k_2 are specific for each reaction at a definite temperature, K is a constant specific to the system in equilibrium at a given temperature. The values for the molar concentrations used to evaluate K from the above expression must always be the concentrations present after the reaction has reached equilibrium.

It is important to remember that while the rates of reaction, Rate$_1$ and Rate$_2$, are equal at equilibrium, the molar concentrations of the reactants and products in the equilibrium mixture are usually not equal. However, the individual concentration of each reactant and product remains constant *at equilibrium* because the rate at which any one reactant is being used up in one reaction is equal to the rate at which it is being formed by the opposite reaction.

A general equation for a chemical reaction may be written

$$mA + nB + \cdots \rightleftharpoons xC + yD + \cdots$$

When this reaction has reached equilibrium, the ratio

$$\frac{[C]^x[D]^y \ldots}{[A]^m[B]^n \ldots}$$

called the **reaction quotient,** is constant; that is,

$$\frac{[C]^x[D]^y \ldots}{[A]^m[B]^n \ldots} = K$$

The reaction quotient, for equilibrium conditions, becomes the equilibrium constant expression, with K being the equilibrium constant. A mathematical expression of the law of chemical equilibrium, or the **law of mass action, is: When a reversible reaction has attained equilibrium at a given temperature, the reaction quotient (the product of the molar concentrations of the substances to the right of the arrow divided by the product of the molar concentrations of the substances to the left, with each concentration raised to a power equal to the number of moles of that substance appearing in the equation) is a constant.**

The law of mass action is more precisely expressed in terms of the activities of the reactants and products (Section 12.25) rather than their concentrations. However, as we have seen, the activity of a dilute solute is closely approximated by its molar concentration, so concentrations are commonly used in equilibrium constants for dissolved species. The activity of a gas is approximated by its pressure (in atmospheres), so pressures are commonly used in equilibrium constants involving gases. Also, since the molar concentration of a gas is directly proportional to its pressure, molar concentrations are sometimes used in equilibrium constants involving gases. The activity of a pure solid, a pure liquid, or a solvent is 1.

The mathematical expression of the law of mass action means that, regardless of how we might change individual concentrations and temporarily upset an equilibrium, the composition of the system will always adjust itself to a new condition of equilibrium for which the reaction quotient, $[C]^x[D]^y \ldots /[A]^m[B]^n \ldots$, will again have the value K, provided that the temperature does not change. When a mixture of A, B, C, and D is prepared in such proportions that the reaction quotient is not equal to K, then the system is not in equilibrium, and its composition will change until equilibrium is established. If the reaction quotient is less than K, the rate of reaction of A with B will be greater than that of C with D, so that A and B will be used up faster than they are formed until the reaction quotient becomes equal to K. Conversely, if the reaction quotient is greater than K, the rate of reaction of C with D will be greater than that of A with B until the reaction quotient becomes equal to K.

The value of an equilibrium constant is a measure of the completeness of a reversible reaction. A large value for K indicates that equilibrium is attained only after the reactants A and B have been largely converted into the products C and D. When K is very small—much less than 1—equilibrium is attained when only small amounts of A and B have been converted to C and D.

EXAMPLE 15.1   Write the equilibrium constant expressions using concentrations for the reactions $3O_2 \rightleftharpoons 2O_3$ and $3H_2 + N_2 \rightleftharpoons 2NH_3$.

$$3O_2 \rightleftharpoons 2O_3 \qquad K = \frac{[O_3]^2}{[O_2]^3}$$

$$3H_2 + N_2 \rightleftharpoons 2NH_3 \qquad K = \frac{[NH_3]^2}{[H_2]^3[N_2]}$$

The use of square brackets indicates that molar concentrations (units of mol L^{-1}) are used in these equations.

Using the law of mass action we can determine the concentrations of reactants and products at equilibrium.

EXAMPLE 15.2   At 2000°C the equilibrium constant for the reaction

$$N_2(g) + O_2(g) \rightleftharpoons 2NO(g)$$

is 4.1×10^{-4} (no units; the units cancel). What is the concentration of $NO(g)$ in a mixture of $NO(g)$, $N_2(g)$, and $O_2(g)$ at equilibrium with $[N_2] = 0.036$ mol L^{-1} and $[O_2] = 0.0089$ mol L^{-1}?

At equilibrium the reaction quotient $[NO]^2/[N_2][O_2]$ is equal to the equilibrium constant. Thus we have

$$K = \frac{[NO]^2}{[N_2][O_2]}$$

Since we know K, $[N_2]$, and $[O_2]$, we can solve for $[NO]^2$ by rearranging the equation

$$[NO]^2 = K[N_2][O_2]$$
$$[NO] = \sqrt{K[N_2][O_2]}$$
$$= \sqrt{4.1 \times 10^{-4} \times 0.036 \text{ mol } L^{-1} \times 0.0089 \text{ mol } L^{-1}}$$
$$= \sqrt{1.31 \times 10^{-7} \text{ (mol } L^{-1})^2}$$
$$= 3.6 \times 10^{-4} \text{ mol } L^{-1}$$

Thus the concentration of $NO(g)$ is 3.6×10^{-4} mol L^{-1} at equilibrium under these conditions.

The equilibrium constant for the reaction of nitrogen and hydrogen to produce ammonia at a certain temperature is 6.00×10^{-2} mol^{-2} L^2. Calculate the equilibrium concentration of ammonia, if the equilibrium concentrations of nitrogen and hydrogen are 4.26 M and 2.09 M, respectively.

EXAMPLE 15.3

The balanced equation and the equilibrium constant expression are:

$$N_2(g) + 3H_2(g) \rightleftharpoons 2NH_3(g)$$

$$K = \frac{[NH_3]^2}{[N_2][H_2]^3} = 6.00 \times 10^{-2} \text{ mol}^{-2} \text{ L}^2$$

Substituting the known equilibrium concentrations of N_2 and H_2 into the equilibrium constant expression and solving for NH_3,

$$\frac{[NH_3]^2}{(4.26 \text{ mol L}^{-1})(2.09 \text{ mol L}^{-1})^3} = 6.00 \times 10^{-2} \text{ mol}^{-2} \text{ L}^2$$

$$[NH_3]^2 = (6.00 \times 10^{-2} \text{ mol}^{-2} \text{ L}^2)(4.26 \text{ mol L}^{-1})(2.09 \text{ mol L}^{-1})^3$$

$$[NH_3]^2 = 2.33 \text{ (mol L}^{-1})^2$$

$$[NH_3] = \sqrt{2.33 \text{ (mol) L}^{-1})^2} = 1.53 \text{ mol L}^{-1}$$

Units for equilibrium constants differ, depending upon the reaction and the corresponding equilibrium constant expression. In some cases the units cancel out. Several examples of units for equilibrium constants are:

Reaction	Equilibrium Equation	Units
$3O_2 \rightleftharpoons 2O_3$	$K = \dfrac{[O_3]^2}{[O_2]^3}$	$\dfrac{(\text{mol L}^{-1})^2}{(\text{mol L}^{-1})^3} = \dfrac{1}{\text{mol L}^{-1}} = \text{mol}^{-1} \text{ L}$
$N_2O_4 \rightleftharpoons 2NO_2$	$K = \dfrac{[NO_2]^2}{[N_2O_4]}$	$\dfrac{(\text{mol L}^{-1})^2}{\text{mol L}^{-1}} = \text{mol L}^{-1}$
$N_2 + 3H_2 \rightleftharpoons 2NH_3$	$K = \dfrac{[NH_3]^2}{[N_2][H_2]^3}$	$\dfrac{(\text{mol L}^{-1})^2}{(\text{mol L}^{-1})(\text{mol L}^{-1})^3}$ $= \dfrac{1}{(\text{mol L}^{-1})^2} = \text{mol}^{-2} \text{ L}^2$
$N_2 + O_2 \rightleftharpoons 2NO$	$K = \dfrac{[NO]^2}{[N_2][O_2]}$	$\dfrac{(\text{mol L}^{-1})^2}{(\text{mol L}^{-1})(\text{mol L}^{-1})} = \text{(no units)}$

It is common practice to omit the units of K; we will do so in most examples.

The equilibrium constant K for the reaction

EXAMPLE 15.4

$$PCl_5(g) \rightleftharpoons PCl_3(g) + Cl_2(g)$$

is 0.0211 mol L^{-1} at a certain temperature. What are the equilibrium concentrations of PCl_5, PCl_3, and Cl_2 starting with an initial concentration of PCl_5 of 1.00 M?

$$K = \frac{[PCl_3][Cl_2]}{[PCl_5]} = 0.0211$$

In this problem, we are not given *any* of the equilibrium concentrations. We do know that in establishing equilibrium, some of the initial quantity of PCl_5 will decompose to provide some PCl_3 and Cl_2 and that the concentrations of PCl_3 and Cl_2 at equilibrium will depend upon the amount of PCl_5 that decomposes.

Let us assign the letter x to represent the number of moles of PCl_5 that decompose per liter in establishing the equilibrium state. The *initial* concentration of PCl_5 is known to be 1.00 mol L^{-1}. Thus, the concentration of PCl_5 at equilibrium, $[PCl_5]$, must be $(1.00 - x)$ mol L^{-1}.

The equation tells us that the decomposition of 1 mole of PCl_5 produces 1 mole of PCl_3 and 1 mole of Cl_2. Hence the decomposition of x moles of PCl_5 will produce x moles of PCl_3 and also x moles of Cl_2. Thus, the equilibrium concentrations of PCl_3 and Cl_2, $[PCl_3]$ and $[Cl_2]$ respectively, are each x mol L^{-1}.

Summarizing, the *equilibrium* concentrations of the three species in terms of x are:

$$[PCl_5] = (1.00 - x) \text{ mol } L^{-1}$$
$$[PCl_3] = x \text{ mol } L^{-1}$$
$$[Cl_2] = x \text{ mol } L^{-1}$$

Substituting in the equilibrium constant expression:

$$K = \frac{[PCl_3][Cl_2]}{[PCl_5]} = 0.0211$$

$$\frac{(x)(x)}{1.00 - x} = 0.0211$$

$$x^2 = 0.0211(1.00 - x)$$

$$x^2 + 0.0211x - 0.0211 = 0$$

This is a quadratic equation, which can be solved in either of two ways: (1) by use of the quadratic formula (see Appendix A), or (2) by use of the method of successive approximations.

1. BY USE OF THE QUADRATIC FORMULA. For $ax^2 + bx + c = 0$,

$$x = \frac{-b \pm \sqrt{b^2 - 4ac}}{2a}$$

In this problem, $a = 1$, $b = 0.0211$, and $c = -0.0211$. Substituting in the quadratic formula:

$$x = \frac{-0.0211 \pm \sqrt{(0.0211)^2 - 4(1)(-0.0211)}}{2(1)}$$

$$= \frac{-0.0211 \pm \sqrt{(4.45 \times 10^{-4}) + (8.44 \times 10^{-2})}}{2}$$

$$= \frac{-0.0211 \pm 0.2913}{2}$$

Hence

$$x = \frac{-0.0211 + 0.2913}{2} \quad \text{or} \quad \frac{-0.0211 - 0.2913}{2}$$

$$= 0.135 \text{ mol/L} \quad \text{or} \quad -0.156 \text{ mol/L}$$

Quadratic equations have two solutions, one correct and one incorrect. The correct solution can ordinarily be determined easily, because the incorrect solution is often not physically possible. In the present case, for example, the second solution indicates a negative concentration, which has no physical meaning (an "extraneous root") and cannot be correct.

Thus

$$x = 0.135 \text{ M}$$

The equilibrium concentrations, then, are:

$$[PCl_5] = 1.00 - x = 1.00 - 0.135 = 0.86 \text{ M}$$
$$[PCl_3] = x = 0.135 \text{ M}$$
$$[Cl_2] = x = 0.135 \text{ M}$$

2. BY USE OF THE METHOD OF SUCCESSIVE APPROXIMATIONS.

$$\frac{(x)(x)}{1.00 - x} = 0.0211$$

First approximation. On a trial basis, we will neglect the x term in the denominator and approximate the value of $(1.00 - x)$ to be 1.00. (Such a simplifying assumption is valid for most purposes, in this kind of calculation, if x is less than 5% of the number from which it is to be subtracted, or added in some cases.)

$$K = \frac{x^2}{1.00} = 0.0211$$

$$x^2 = 0.0211$$

$$x = 0.145 \text{ M}$$

The value of x obtained from the first approximation is not less than 5% of 1.00 (it is 14.5% of it). This indicates that x cannot be neglected in the expression $(1.00 - x)$ for a final answer. Had x been less than 5% of 1.00, we could have stopped after the first approximation for most purposes.

However, the value of $(1.00 - x)$ is closer to $(1.00 - 0.145) = 0.855$ (using the value of x obtained in the first approximation) than it is to the assumed value of 1.00 in the first approximation, so a better estimate of the true value of $(1.00 - x)$ can be obtained using the value of x just calculated.

Second approximation.

$$1.00 - x = 1.00 - 0.145 = 0.855$$

$$K = \frac{x^2}{0.855} = 0.0211$$

$$x^2 = 0.0180$$

$$x = 0.134 \text{ M}$$

Since the values of x obtained from the first and second approximations differ by over 5% from each other, a third approximation is necessary.

Third approximation. A still better estimate for $(1.00 - x)$ using the value of x from the second approximation is

$$1.00 - x = 1.00 - 0.134 = 0.866$$

$$K = \frac{x^2}{0.866} = 0.0211$$

$$x^2 = 0.0183$$

$$x = 0.135 \text{ M}$$

Note that this third value of x differs from that obtained in the second approximation by less than 5% (and is in fact the same as the value calculated by means of the quadratic formula). Hence we can safely conclude that $[PCl_3] = 0.135$ M, $[Cl_2] = 0.135$ M, and $[PCl_5] = 1.00 - 0.135 = 0.86$ M.

The method of successive approximations is a useful, rapid method of calculations in many circumstances. However, it should be realized that it either is not feasible or would not provide a saving of time in some cases. Each problem must be evaluated individually as to the most advantageous method of calculation.

EXAMPLE 15.5   Acetic acid, CH_3CO_2H, reacts with ethanol, C_2H_5OH, to form water and ethyl acetate, $CH_3CO_2C_2H_5$, the solvent responsible for the odor in some nail polish removers.

$$CH_3CO_2H + C_2H_5OH \rightleftharpoons CH_3CO_2C_2H_5 + H_2O$$

The equilibrium constant for this reaction (when run in dioxane as a solvent) is 4.0. What mass of ethyl acetate is formed by the reaction of 0.10 mol of CH_3CO_2H and 0.15 mol of C_2H_5OH in enough dioxane to make 1.0 L of solution?

From the law of mass action, we know that at equilibrium

$$\frac{[CH_3CO_2C_2H_5][H_2O]}{[CH_3CO_2H][C_2H_5OH]} = 4.0$$

We need to solve this equation for the concentration of $CH_3CO_2C_2H_5$ at equilibrium.

When we are not given equilibrium concentrations, it is sometimes helpful to set up a table of initial concentrations (the concentrations of reactants before any reaction occurs) and equilibrium concentrations (the concentrations of reactants and products after equilibrium has been established). For this problem we have the following table. Entries in color are given in the problem statement. Entries in black are calculated as described below.

	$[CH_3CO_2H]$	$[C_2H_5OH]$	$[CH_3CO_2C_2H_5]$	$[H_2O]$
Initial concentrations, M	0.10	0.15	0	0
Equilibrium concentrations, M	$0.10 - x$	$0.15 - x$	x	x

The problem tells us the initial concentrations of CH_3CO_2H and C_2H_5OH, the concentrations before any reaction occurs. The concentration of the products, $CH_3CO_2C_2H_5$ and H_2O, is zero before the reaction starts.

Now let us consider the situation after equilibrium has been reached. We do not know the concentration of $CH_3CO_2C_2H_5$ at equilibrium, so we will set this concentration equal to x. From the chemical equation we can see that one mole of H_2O is formed for each mole of $CH_3CO_2C_2H_5$ produced, so the concentration of H_2O at equilibrium can also be set equal to x. From the equation we can see that one mole of CH_3CO_2H (or of C_2H_5OH) reacts for each mole of $CH_3CO_2C_2H_5$ that forms. Thus if x mol of $CH_3CO_2C_2H_5$ appears as product, x mol of CH_3CO_2H is consumed, and $(0.10 - x)$ mol/L remains. Also, $(0.15 - x)$ mol/L of C_2H_5OH remains at equilibrium.

If we substitute the equilibrium concentrations into the expression of the law of mass action, we have

$$K = \frac{[CH_3CO_2C_2H_5][H_2O]}{[CH_3CO_2H][C_2H_5OH]}$$

$$4.0 = \frac{(x)(x)}{(0.10 - x)(0.15 - x)}$$

At this stage, we can make a quick check to see if x is less than 5% of either 0.10 or 0.15, from which it is to be subtracted. This is done by calculating the value of x making the initial simplifying assumptions that $(0.10 - x)$ is equal to 0.10 and that $(0.15 - x)$ is equal to 0.15, a relatively easy calculation.

$$4.0 = \frac{(x)(x)}{(0.10)(0.15)}$$

$$x^2 = (4.0)(0.10)(0.15) = 0.060$$

$$x = \sqrt{0.060} = 0.24$$

The calculated value, based on the simplifying assumption, is seen thus to be actually larger than either 0.10 or 0.15, from which it is to be subtracted. Thus it is clear that for this case the simplifying assumption is not valid and that we must make the calculation without the simplifying assumption.

The original equation

$$4.0 = \frac{(x)(x)}{(0.10 - x)(0.15 - x)}$$

contains only one unknown and can be solved for x, the concentration of $CH_3CO_2C_2H_5$. Let us expand the equation, and solve the resulting quadratic using the technique shown in Appendix A.

$$4.0 = \frac{x^2}{x^2 - 0.25x + 0.015}$$

$$4.0(x^2 - 0.25x + 0.015) = x^2$$

$$4.0x^2 - x + 0.060 = x^2$$

$$3x^2 - x + 0.060 = 0$$

Now we use the quadratic formula.

$$x = \frac{-(-1) \pm \sqrt{(-1)^2 - 4(3)(0.060)}}{2(3)}$$

$$= \frac{1 \pm \sqrt{1 - 0.72}}{6}$$

$$= 0.25 \text{ and } 0.078$$

Only one of these solutions, 0.078, is physically reasonable. If $x = 0.25$, then $[CH_3CO_2C_2H_5] = 0.25$ M, but the maximum concentration can be only 0.10 M since 0.10 mol of CH_3CO_2H, the amount present before reaction, can only form 0.10 mol of product. The answer 0.25, therefore, has no physical meaning and is an "extraneous root," and

$$[CH_3CO_2C_2H_5] = 0.078 \text{ M (at equilibrium)}$$

$$\text{Mass } CH_3CO_2C_2H_5 = \frac{0.078 \text{ mol}}{1 \text{ L}} \times 1.0 \text{ L} \times \frac{88.1 \text{ g } CH_3CO_2C_2H_5}{1 \text{ mol}}$$

$$= 6.9 \text{ g}$$

It might be noted that this is an example of a quadratic equation for which the method of successive approximations, described in Example 15.4, is not feasible.

15.3 Determination of Equilibrium Constants

Although the values of equilibrium constants can be determined from the forward and reverse rate constants for a reaction, they are more commonly determined by measuring, *at equilibrium,* the concentrations of reactants and products of a reaction.

EXAMPLE 15.6

An equimolar mixture of hydrogen and iodine was heated at 400°C until no further change in the concentration of H_2, I_2, or HI was observed. At this point it was assumed that equilibrium had been reached, and it was found by analysis that $[H_2] = 0.221$ M, $[I_2] = 0.221$ M, and $[HI] = 1.563$ M. Calculate the equilibrium constant for the reaction at 400°.

$$H_2(g) + I_2(g) \rightleftharpoons 2HI(g)$$

The reaction quotient for this reaction is

$$\frac{[HI]^2}{[H_2][I_2]}$$

At equilibrium this ratio is a constant equal to the equilibrium constant.

$$K = \frac{[HI]^2}{[H_2][I_2]}$$

Substitution of the concentrations determined at equilibrium gives

$$K = \frac{[HI]^2}{[H_2][I_2]} = \frac{(1.563 \text{ mol L}^{-1})^2}{(0.221 \text{ mol L}^{-1})(0.221 \text{ mol L}^{-1})} = 50.0 \qquad \text{(units cancel out)}$$

If we start with pure HI at any molar concentration, or with any mixture of H_2 and I_2, or with any mixture of H_2, I_2, and HI, and hold the temperature of the system at 400°C until equilibrium is established, the molar concentrations of the three substances will have changed so that the reaction quotient $[HI]^2/[H_2][I_2]$ is equal to 50.0.

The following example illustrates that it is not always necessary to measure the concentration of each species present at equilibrium. The chemical equation can be used to relate concentrations.

EXAMPLE 15.7

Iodine molecules react reversibly with iodide ions [Fig. 15.4(c) in Section 15.4] to produce triiodide ions.

$$I_2(aq) + I^-(aq) \rightleftharpoons I_3^-(aq)$$

If a liter of solution prepared from 1.000×10^{-3} mol of I_2 and 1.000×10^3 mol of the strong electrolyte KI contains 6.61×10^{-4} mol of iodine at equilibrium, what is the equilibrium constant for the reaction?

From the law of mass action, we know that the equilibrium constant is given by the equation

$$K = \frac{[I_3^-]}{[I_2][I^-]}$$

In order to determine the value of K, we need to know the values of $[I_3^-]$, $[I_2]$, and $[I^-]$ at equilibrium.

We can set up the following table of initial concentrations and equilibrium concentrations. Entries in color are those given in the problem.

	$[I_2]$	$[I^-]$	$[I_3^-]$
Initial concentrations, M	1.000×10^{-3}	1.000×10^{-3}	0
Equilibrium concentrations, M	6.61×10^{-4}	6.61×10^{-4}	3.39×10^{-4}

First consider the table entries for initial concentrations. The data tell us that the solution was prepared from 1.000×10^{-3} mol of I_2 per liter and 1.000×10^{-3} mol of KI per liter. Thus, before the reaction begins, the concentration of I_2 is 1.000×10^{-3} mol/1 L or 1.000×10^{-3} M. Since KI is a strong electrolyte, the concentration of I^- is equal to the concentration of KI (1 mol I^-/1 mol KI), or 1.000×10^{-3} M. The initial concentration of I_3^- is zero, since no I_3^- has yet been formed.

At equilibrium the concentration of I_2 is given as 6.61×10^{-4} M. Before reaction (when the solution was first prepared) the concentration of I_2, $[I_2]_i$, was 1.000×10^{-3} M; however, it decreased to 6.61×10^{-4} M at equilibrium. The change in concentration of I_2 was

$$[I_2]_i - [I_2] = 1.000 \times 10^{-3} \text{ M} - (6.61 \times 10^{-4} \text{ M}) = 3.39 \times 10^{-4} \text{ M}$$

Thus 3.39×10^{-4} mol of I_2 per liter of solution was consumed as the reaction proceeded to equilibrium. From the chemical equation we know that one mole of I_3^- is formed for each mole of I_2 that reacts. Thus the concentration of I_3^- at equilibrium, $[I_3^-]$, must be

$$\frac{3.39 \times 10^{-4}\ \cancel{mol\ I_2}}{1\ L} \times \frac{1\ mol\ I_3^-}{1\ \cancel{mol\ I_2}} = \frac{3.39 \times 10^{-4}\ mol\ I_3^-}{1\ L} = 3.39 \times 10^{-4}\ M$$

Before reaction the concentration of I^-, $[I^-]_i$, was also 1.000×10^{-3} M. The concentration of I^- at equilibrium, $[I^-]$, is that amount that did not react to form I_3^-. Since the chemical equation tells us that one mole of I^- reacts for each mole of I_3^- formed, the amount of I^- that reacts is equal to the amount of I_3^- formed, or 3.39×10^{-4} M. Therefore

$$[I^-] = [I^-]_i - [I_3^-] = 1.000 \times 10^{-3}\ M - (3.39 \times 10^{-4}\ M) = 6.61 \times 10^{-4}\ M$$

Now we simply substitute these equilibrium concentrations into the expression for the equilibrium constant and evaluate it.

$$K = \frac{[I_3^-]}{[I_2][I^-]} = \frac{(3.39 \times 10^{-4}\ \cancel{mol\ L^{-1}})}{(6.61 \times 10^{-4}\ \cancel{mol\ L^{-1}})(6.61 \times 10^{-4}\ mol\ L^{-1})} = 776\ mol^{-1}\ L$$

15.4 Effect of Change of Concentration on Equilibrium

A chemical system may be shifted out of equilibrium by increasing the rate of the forward or the reverse reaction. When the system

$$A + B \rightleftharpoons C + D$$

is in equilibrium and an additional quantity of A is added, the rate of the forward reaction increases because the concentration of the reacting molecules increases [Fig. 15.3(a)]. The rate of the forward reaction becomes faster than that of the reverse reaction, so the system is out of equilibrium. However, as the concentrations of C and D increase, the rate of the reverse reaction increases, while the decrease in the concentrations of A and B causes the rate of the forward reaction to decrease. The rates of the two reactions become equal again, and a second state of equilibrium is reached. Although the concentrations of A, B, C, and D have changed (as have, also, the rates of the forward and the reverse reactions), the reaction quotient [C][D]/[A][B] is again equal to its original value of K.

The equilibrium is said to have been shifted to the right, since in the new state of equilibrium the products C and D are present in greater concentrations than they were before the addition of A. Also, B is present in smaller concentration and A is present in greater concentration than they were at the initial equilibrium, although the new equilibrium concentration of A is less than it was just after the additional A was added. Increasing the concentration of B would also shift the equilibrium to the right. Increasing the concentration of either C or D, or both, would shift the equilibrium to the left by increasing the rate of the *reverse* reaction and thereby increasing the concentrations of A and B.

A chemical system may also be shifted out of equilibrium by decreasing the rate of the forward or the reverse reaction [Fig. 15.3(b)]. The rate of the forward reaction can

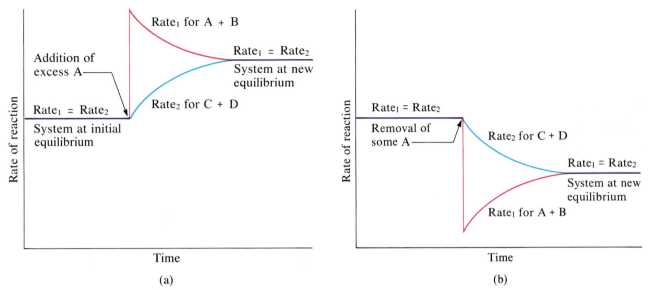

Figure 15.3
(a) The effect of adding some reactant A on the rates of the forward and reverse reactions for A + B ⇌ C + D when the system is initially at equilibrium. (b) The effect of removing some reactant A on the rates of the forward and reverse reactions for A + B ⇌ C + D when the system is initially at equilibrium.

be decreased and equilibrium shifted to the left by removing either A or B, or both. The forward reaction then proceeds more slowly than the reaction to the left until the reaction rates again become equal and a new equilibrium is attained. Removal of either C or D, or both, decreases the rate of the *reverse* reaction. The reverse reaction temporarily proceeds more slowly than the forward reaction until the reaction rates again become equal and a new equilibrium is reached; the equilibrium has been shifted to the right.

As a specific example of shifting a chemical system out of equilibrium by reducing the rate of the forward reaction, let us consider the reaction of 1.000×10^{-3} mole of I^- with 1.000×10^{-3} mole of I_2 in a one-liter aqueous solution to produce I_3^-.

$$I^-(aq) + I_2(aq) \rightleftharpoons I_3^-(aq)$$

As shown in Example 15.7, the concentrations at equilibrium were $[I_2] = 6.61 \times 10^{-4}$ mol L^{-1}, $[I^-] = 6.61 \times 10^{-4}$ mol L^{-1}, and $[I_3^-] = 3.39 \times 10^{-4}$ mol L^{-1}. I_2 is brown in aqueous solution, I^- is colorless, and I_3^- is pale yellow (Fig. 15.4 on page 436). The color at equilibrium results from the mixture of the three species. Inasmuch as the concentration of I_2 is a little more than 10 times the I_3^- concentration, the color of the equilibrium solution [Fig. 15.4(c)] is brown but a little lighter than that of a solution of I_2 alone in aqueous solution [Fig. 15.4(a)].

The reaction can be reversed by adding some carbon tetrachloride (CCl_4), which forms an immiscible layer with water but dissolves I_2 [Fig. 15.4(b and d)]. Most of the I_2 is extracted from the aqueous layer into the immiscible CCl_4 layer, thus decreasing the I_2 concentration in the aqueous layer and resulting in a fainter brown color in the aqueous layer [Fig. 15.4(d)]. The reduced I_2 concentration causes a decrease in the rate of the forward reaction, which then temporarily proceeds at a slower rate than does the reverse reaction but gradually increases in rate as I^- and I_2 are formed faster than they react. Simultaneously, the reverse reaction temporarily proceeds at a faster rate than does the forward reaction but gradually decreases in rate as I_3^- reacts faster than it is formed. The rates quickly become equal, and a new equilibrium is thus established. The equilibrium has been shifted to the left by removal of some of reactant I_2. The violet color of the CCl_4 layer indicates that an appreciable amount of the I_2 is now there.

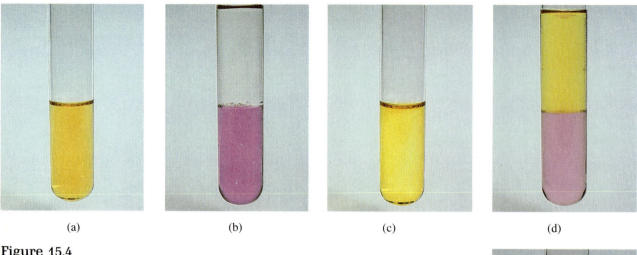

(a)   (b)   (c)   (d)

Figure 15.4
(a) 1.0×10^{-3} M I_2 in water solution. (b) The solution in (a) after adding colorless CCl_4; the CCl_4 extracts I_2 from the aqueous (top) layer to the CCl_4 (bottom) layer, giving a fainter brown I_2 color in the aqueous layer and the violet I_2 color in the CCl_4 layer. (c) 1.0×10^{-3} M I_2 in H_2O solution, to which 1.0×10^{-3} M KI has been added, producing an equilibrium mixture of I^-, I_2, and I_3^- [$I^-(aq) + I_2(aq) \rightleftharpoons I_3^-(aq)$]. The paler color, compared to (a), indicates a decrease in I_2 concentration (brown) and the formation of I_3^- (pale yellow). (d) The equilibrium solution of I_2, I^-, and I_3^- from (c) after CCl_4 is added. Note the fading of the color, as a result of the decrease in $[I_2]$ in the aqueous layer by extraction to the CCl_4 layer and the decrease in $[I_3^-]$ resulting from the subsequent shift of the equilibrium to the left. (e) The solution from (d) after addition of an excess of I^-, which reacts with some of the I_2 in the CCl_4 layer, shifting the equilibrium in the aqueous layer to the right. The corresponding fainter violet color of I_2 in the CCl_4 layer results because I_2 is extracted into the aqueous layer to replace that used to form I_3^- so that the ratio $[I_3^-/([I^-][I_2])$ remains constant.

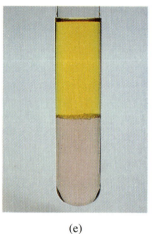

(e)

If an excess of I^- were then added to the mixture [Fig. 15.4(e)], the equilibrium would shift to the right as a result of the extra I^- pulling much of the I_2 back out of the CCl_4 layer by reacting with it, thereby decreasing the violet color in the CCl_4 layer and increasing the depth of brown color in the aqueous layer.

Figure 15.5 shows another system in which a shift in equilibrium is caused by a decrease in the concentration of a reactant. Thiocyanate ions, SCN^-, react with iron(III), Fe^{3+}, to form a soluble red ion, $[Fe(NCS)]^{2+}$.

$$Fe^{3+}(aq) + SCN^-(aq) \rightleftharpoons [Fe(NCS)]^{2+}(aq) \tag{1}$$

Figure 15.5(a) shows an aqueous solution of 0.1 M Fe^{3+}. Figure 15.5(b) shows the same 0.1 M Fe^{3+} solution as in (a) after some SCN^- is added; the red color of $[Fe(NCS)]^{2+}$ is readily apparent. Figure 15.5(c) shows the same solution as in (b) after addition of some silver nitrate, $AgNO_3$, which introduces silver ion, Ag^+, into the solution. Silver ion reacts with thiocyanate ion to form white insoluble silver thiocyanate, AgNCS.

$$Ag^+(aq) + SCN^-(aq) \rightleftharpoons AgNCS(s) \tag{2}$$

| (a) | (b) | (c) |

Figure 15.5

(a) Test tube contains 0.1 M Fe^{3+}.

(b) Thiocyanate ion has been added to solution in (a), forming the red $[Fe(NCS)]^{2+}$ ion.

$$Fe^{3+}(aq) + SCN^-(aq) \rightleftharpoons [Fe(NCS)]^{2+}(aq)$$

(c) Silver nitrate has been added to the solution in (b), precipitating some of the SCN^- as the white $AgNCS(s)$ seen in the bottom of the tube $[Ag^+(aq) + SCN^-(aq) \rightleftharpoons AgNCS(s)]$. The decrease in SCN^- concentration shifts the equilibrium between Fe^{3+}, SCN^-, and $[Fe(NCS)]^{2+}$ to the left, decreasing the concentration (and color) of the red $[Fe(NCS)]^{2+}$.

With the formation of the precipitate of AgNCS, the concentration of SCN^- in the solution is reduced, shifting the equilibrium of reaction (1) to the left and thereby decreasing the concentration of the red $[Fe(NCS)]^{2+}$. The fainter red color in the solution after addition of $AgNO_3$ is in complete accord with equilibrium theory, which calls for a decrease in $[Fe(NCS)]^{2+}$.

The effect of a change in concentration on a system in equilibrium is illustrated further by the equilibrium $H_2 + I_2 \rightleftharpoons 2HI$, for which K was found to be 50.0 at 400°C with $[H_2] = [I_2] = 0.221$ M and $[HI] = 1.563$ M (Example 15.6). If H_2 is introduced into the system quickly so that its concentration doubles before it begins to react, the rate of the reaction of H_2 with I_2 to form HI increases. When equilibrium is again reached, $[H_2] = 0.374$ M, $[I_2] = 0.153$ M, and $[HI] = 1.692$ M. If these new values are substituted in the expression for the equilibrium constant for this system, we have

$$\frac{[HI]^2}{[H_2][I_2]} = \frac{(1.692)^2}{(0.374)(0.153)} = 50.0 = K$$

Hence, by doubling the concentration of H_2, we have caused the formation of more HI, used up about one-third of the I_2 present at the first equilibrium, and used up some, but not all, of the excess H_2 added.

The effect of a change in concentration on a system in equilibrium is an important application of **Le Châtelier's principle: If a stress (such as a change in concentration, pressure, or temperature) is applied to a system in equilibrium, the equilibrium shifts in a way that tends to undo the effect of the stress.**

15.5 Effect of Change in Pressure on Equilibrium

Changes in pressure measurably affect systems in equilibrium only when gases are involved, and then only when the chemical reaction produces a change in the total number of gaseous molecules in the system. As the pressure on a gaseous system increases, the gases are compressed, the total number of molecules per unit of volume increases, and their molar concentrations increase. The stress in this case is an increase in the number of molecules per unit of volume. A chemical reaction that reduces the total number of molecules per unit of volume relieves the stress and will, in accord with Le Châtelier's principle, be favored.

Consider the effect of an increase in pressure on the system in which one molecule of nitrogen and three molecules of hydrogen interact to form two molecules of ammonia.

$$N_2(g) + 3H_2(g) \rightleftharpoons 2NH_3(g)$$

The formation of ammonia decreases the total number of molecules in the system by 50%, thus reducing the total pressure exerted by the system. Experiment shows that an increase in pressure does drive the reaction to the right. On the other hand, a decrease in the pressure on the system favors decomposition of ammonia into hydrogen and nitrogen. These observations are fully in accord with Le Châtelier's principle. (Table 23.3 shows equilibrium concentrations of ammonia at several temperatures and pressures.)

Let us now consider the reaction in which a molecule of nitrogen interacts with a molecule of oxygen with the formation of two molecules of nitric oxide.

$$N_2(g) + O_2(g) \rightleftharpoons 2NO(g)$$

Because there is no change in the total number of molecules in the system during reaction, a change in pressure does not favor either formation or decomposition of gaseous nitric oxide.

Whenever gases are involved in a system in equilibrium, the pressure of each gas can be substituted for its concentration in the expression for the equilibrium constant, because the concentration of a gas at constant temperature varies directly with the pressure. This relationship can be derived from the ideal gas equation

$$PV = nRT$$

The molar concentration of the gas, designated as C, is the number of moles of the gas, n, per liter. Substituting 1 L for V, and rearranging the equation,

$$P(1 \text{ L}) = nRT$$

$$P = \frac{n}{1 \text{ L}} RT$$

$$P = CRT$$

R is a constant. At a constant temperature, T is also constant. Hence the quantity RT is constant and can be designated by k.

$$P = kC \text{ (at constant temperature)}$$

Hence the pressure of a gas at constant temperature is shown to be directly proportional to the concentration of the gas. Thus for the system

$$N_2(g) + 3H_2(g) \rightleftharpoons 2NH_3(g)$$

(handwritten margin note:) —as pressure increases on a gas, so does molar volume (M) —as pressure decreases off a gass, so does the molar volume (M)

we may write

$$\frac{P_{NH_3}^2}{P_{N_2}P_{H_2}^3} = K_p$$

(We will use the symbol K_p to designate an equilibrium constant that utilizes pressures of gases.) If partial pressures of gases are substituted for their concentrations, the equilibrium constant, K_p, is still a constant, but its numerical value and units may change.

The units for K_p for the reaction to produce ammonia, if the three partial pressures are expressed in atmospheres, would be

$$K_p = \frac{atm^2}{atm \times atm^3} = \frac{1}{atm^2} = atm^{-2}$$

Although equilibrium constants have units (unless they happen to cancel out as they do for the reaction $N_2 + O_2 \rightleftharpoons 2NO$), their values are often given without units.

EXAMPLE 15.8

A vessel contains gaseous carbon monoxide, carbon dioxide, hydrogen, and water in equilibrium at 980°C. The pressures of these gases are CO, 0.150 atm; CO_2, 0.200 atm; H_2, 0.0900 atm; H_2O, 0.200 atm. Hydrogen is pumped into the vessel, and the equilibrium pressure of CO changes to 0.230 atm. Calculate the partial pressures of the other substances at the new equilibrium, assuming no change in temperature.

$$\underset{0.200}{CO_2(g)} + \underset{0.0900}{H_2(g)} \rightleftharpoons \underset{\substack{0.230 \\ 0.150}}{CO(g)} + \underset{0.200}{H_2O(g)}$$

Before we can calculate the pressures after the addition of hydrogen, we need the value of the equilibrium constant. This can be determined from the initial equilibrium pressures of the reactants and products.

$$K_p = \frac{P_{CO}P_{H_2O}}{P_{CO_2}P_{H_2}} = \frac{(0.150 \text{ atm})(0.200 \text{ atm})}{(0.200 \text{ atm})(0.0900 \text{ atm})} = 1.67$$

(Units cancel out, and the equilibrium constant in this case has no units.)

Now we set up a table showing what we know about the initial pressures in the system (the pressures *after* addition of H_2 *but before* the additional H_2 began to react) and about the *pressures* at the new equilibrium.

	P_{CO_2}	P_{H_2}	P_{CO}	P_{H_2O}
Initial pressures, atm	0.200	?	0.150	0.200
Equilibrium pressures, atm	?	?	0.230	?

At the new equilibrium P_{CO} is 0.230 atm. Therefore P_{CO} has increased by 0.080 atm (from 0.150 atm to 0.230 atm). The balanced equation tells us that for each increase of 1 mol of CO, the concentration of H_2O must also increase by 1 mol; hence the pressure increase for H_2O from its initial pressure of 0.200 atm will also be 0.080 atm. Thus P_{H_2O} at the new equilibrium is

$$0.200 \text{ atm} + 0.080 \text{ atm} = 0.280 \text{ atm}$$

The balanced equation also tells us that for every mole of CO produced, one mole of CO_2 must be consumed, so the pressure of CO_2 must decrease from its initial pressure of 0.200 atm by the same amount as the increase of pressure for CO (0.080 atm). Thus P_{CO_2} at the new equilibrium is

$$0.200 \text{ atm} - 0.080 \text{ atm} = 0.120 \text{ atm}$$

Now we have all of the pressures at the new equilibrium except that of H_2, which is our unknown and can be designated as x. (Note that it is not necessary to calculate the initial pressure of H_2, because this value is not required in the final calculation.)

	P_{CO_2}	P_{H_2}	P_{CO}	P_{H_2O}
Initial pressures, atm	0.200	—	0.150	0.200
Equilibrium pressures, atm	0.120	x	0.230	0.280

Now we can calculate the partial pressure of hydrogen at the new equilibrium by substituting P_{CO}, P_{H_2O}, and P_{CO_2} at the new equilibrium into the expression for K_p and solving for P_{H_2}:

$$K_p = \frac{P_{CO}P_{H_2O}}{P_{CO_2}P_{H_2}} = \frac{(0.230 \text{ atm})(0.280 \text{ atm})}{(0.120 \text{ atm})(x)} = 1.67$$

$$x = \frac{(0.230 \text{ atm})(0.280 \text{ atm})}{(0.120 \text{ atm})(1.67)} = 0.321 \text{ atm}$$

Therefore the partial pressures for CO, H_2O, CO_2, and H_2 at the new equilibrium are 0.230 atm, 0.280 atm, 0.120 atm, and 0.321 atm, respectively.

EXAMPLE 15.9   In a 3.0-L vessel the following partial pressures are measured at equilibrium: N_2, 0.380 atm; H_2, 0.400 atm; NH_3, 2.000 atm. Hydrogen is removed from the vessel until the partial pressure of nitrogen, at a new equilibrium, is equal to 0.450 atm. Calculate the partial pressures of the other substances under the new conditions.

$$N_2(g) + 3H_2(g) \rightleftharpoons 2NH_3(g)$$

$$K_p = \frac{P_{NH_3}^2}{P_{N_2}P_{H_2}^3}$$

$$K_p = \frac{(2.000 \text{ atm})^2}{(0.380 \text{ atm})(0.400 \text{ atm})^3} = 1.645 \times 10^2 \text{ atm}^{-2}$$

Since the amount of H_2 removed from the system is unknown, we let P_{H_2} at the new equilibrium be x.

The removal of H_2 shifts the equilibrium to the left, producing additional N_2 and therefore increasing the pressure of N_2. At the new equilibrium $P_{N_2} = 0.450$ atm. Hence the pressure of N_2 has increased by 0.070 atm. The balanced equation tells us that 2 mol of NH_3 must be used to produce 1 mol of N_2. The pressure of NH_3 must decrease in proportion to the decrease in its molar concentration. The pressure of NH_3 must therefore decrease by twice the amount that the pressure of N_2 increases; thus the pressure of NH_3 must decrease by 0.140 atm. At the new equilibrium

$$P_{NH_3} = 2.000 - 0.140 = 1.860 \text{ atm}$$

Now we can set up a table containing the initial pressures (*after* H_2 was removed, *but before* the reaction begins to shift to restore the equilibrium) and the new equilibrium pressures. Values given in the problem statement are shown in color.

	P_{N_2}	P_{H_2}	P_{NH_3}
Initial pressures, atm	0.380	—	2.000
Equilibrium pressures, atm	0.450	x	1.860

To calculate P_{H_2} at the new equilibrium, we substitute P_{N_2} and P_{NH_3} in the expression for K_p and solve for x.

$$K_p = \frac{P_{NH_3}^2}{P_{N_2}P_{H_2}^3} = \frac{(1.860 \text{ atm})^2}{(0.450 \text{ atm})(x^3)} = 1.645 \times 10^2 \text{ atm}^{-2}$$

$$x^3 = \frac{(1.860 \text{ atm})^2}{(0.450 \text{ atm})(1.645 \times 10^2 \text{ atm}^{-2})}$$

$$= 4.674 \times 10^{-2} \text{ atm}^3 = 46.74 \times 10^{-3} \text{ atm}^3$$

$$x = \sqrt[3]{46.74 \times 10^{-3} \text{ atm}^3} = 0.360 \text{ atm}$$

At the new equilibrium the partial pressures of NH_3 and H_2 are 1.860 atm and 0.360 atm, respectively.

15.6 Effect of Change in Temperature on Equilibrium

All chemical changes involve either the evolution or the absorption of energy. In every system in equilibrium an endothermic and an exothermic reaction are taking place simultaneously. The endothermic reaction ($\Delta H > 0$) is favored by an increase in temperature, that is, an increase in energy, and the exothermic reaction ($\Delta H < 0$) is favored by a decrease in temperature. Changing the temperature changes the value of the equilibrium constant.

The effect of temperature changes on systems in equilibrium is summarized by **van't Hoff's law: When the temperature of a system in equilibrium is raised, the equilibrium is displaced in such a way that heat is absorbed.** This generalization is a special case of Le Châtelier's principle.

In the reaction between gaseous hydrogen and gaseous iodine, heat is evolved.

$$H_2(g) + I_2(g) \longrightarrow 2HI(g) \qquad \Delta H = -9.4 \text{ kJ} \quad \text{(exothermic)}$$

Recall that ΔH refers to the forward reaction, as written if it went to completion (100% yield of HI). ΔH for the complete reverse reaction would have the same numerical value but be of opposite sign.

$$2HI(g) \longrightarrow H_2(g) + I_2(g) \qquad \Delta H = +9.4 \text{ kJ} \quad \text{(endothermic)}$$

The two equations could also be written

$$H_2(g) + I_2(g) \longrightarrow 2HI(g) + 9.4 \text{ kJ}$$

$$2HI(g) + 9.4 \text{ kJ} \longrightarrow H_2(g) + I_2(g)$$

At equilibrium lowering the temperature of this system favors formation of hydrogen iodide; raising it favors decomposition. Raising the temperature decreases the value of the equilibrium constant, since the concentration of HI at the new equilibrium decreases while the molar concentration of H_2 and I_2 increase. The value of the equilibrium constant

$$K = \frac{[HI]^2}{[H_2][I_2]}$$

decreases from 67.5 at 357°C to 50.0 at 400°C.

The equation for the formation of ammonia from hydrogen and nitrogen is

$$N_2 + 3H_2 \rightleftharpoons 2NH_3 \qquad \Delta H = -92.2 \text{ kJ}$$

The reaction is exothermic, and the equilibrium can be shifted to the right to favor the formation of more ammonia by lowering the temperature (see Table 23.3). However, equilibrium is reached more slowly because of the large decrease of reaction rate with decreasing temperature. In the commercial production of ammonia, it is not feasible to use temperatures much lower than 500°C. At lower temperatures, even in the presence of a catalyst, the reaction proceeds too slowly to be practical.

The effect of temperature on the equilibrium between NO_2 and N_2O_4 is indicated in Figure 15.6.

$$2NO_2 \rightleftharpoons N_2O_4 \qquad \Delta H = -57.20 \text{ kJ}$$

The negative ΔH value tells us that the reaction is exothermic and could be written

$$2NO_2(g) \rightleftharpoons N_2O_4(g) + 57.20 \text{ kJ}$$

The heat term can be treated like a product in this exothermic reaction (and like a reactant in an endothermic reaction) using Le Châtelier's principle.

At higher temperatures, the gas mixture has a deep brown color, indicative of a preponderance of brown NO_2 molecules. If, however, we put a stress on the system by cooling the mixture (withdrawing heat), the equilibrium shifts to the right to supply some of the heat lost by cooling. The concentration of colorless N_2O_4 increases, and the concentration of brown NO_2 decreases. The decreased concentration of NO_2 causes the brown color to fade.

Figure 15.6

Photographs showing the effect of temperature on the equilibrium between NO_2 and N_2O_4.

$2NO_2(g) \rightleftharpoons N_2O_4(g)$
$\Delta H = -57.20$ kJ (exothermic)
At the lower temperature (right photograph) the brown color of the NO_2 is much fainter than at the higher temperature (left photograph), due to a shift of the equilibrium from NO_2 to colorless N_2O_4 as heat is removed.

15.7 Effect of a Catalyst on Equilibrium

Iron powder is used as a catalyst in the production of ammonia from nitrogen and hydrogen to increase the rate of reaction of these two elements.

$$N_2 + 3H_2 \xrightarrow[\text{Fe}]{\triangle} 2NH_3$$

However, this same catalyst serves equally well to increase the rate of the reverse reaction, that is, the decomposition of ammonia into its constituent elements.

$$2NH_3 \xrightarrow[\text{Fe}]{\triangle} N_2 + 3H_2$$

Thus the net effect of iron in the reversible reaction

$$N_2 + 3H_2 \rightleftharpoons 2NH_3$$

is to cause equilibrium to be reached more rapidly. A *catalyst has no effect on the value of an equilibrium constant or on equilibrium concentrations*. It merely increases the rate of both the forward and the reverse reactions to the same extent, so that equilibrium is reached more rapidly.

15.8 Homogeneous and Heterogeneous Equilibria

A **homogeneous equilibrium** is an equilibrium within a single phase, such as a mixture of gases or a solution. Most of the equilibria that have been considered in this chapter are homogeneous equilibria involving reversible changes in only one phase, the gas or the liquid phase.

A **heterogeneous equilibrium** is an equilibrium between two or more different phases. Liquid water in equilibrium with ice, liquid water in equilibrium with water vapor, and a solid in contact with its saturated solution are examples of heterogeneous equilibria. Each of these equilibria involves some kind of boundary surface between two phases.

The equilibrium between calcium carbonate, calcium oxide, and carbon dioxide is a heterogeneous equilibrium.

$$CaCO_3(s) \rightleftharpoons CaO(s) + CO_2(g)$$

The expression for the equilibrium constant for this reversible reaction is

$$K = \frac{[CaO][CO_2]}{[CaCO_3]}$$

The concentration of a solid substance remains constant as long as the temperature does not change.

Students sometimes have trouble accepting the fact that the concentration of a pure solid or liquid remains constant. This can be visualized intuitively by realizing that as the quantity of a pure solid or liquid increases or decreases, the volume of that pure solid or liquid also increases or decreases in direct proportion. For example, 48 grams of pure solid CaO (0.86 mole) occupies twice the volume that 24 grams (or 0.43 mole) occupies. Hence the number of moles per unit volume (the concentration) remains the same.

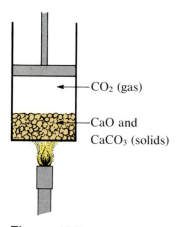

Figure 15.7

The thermal decomposition of calcium carbonate in a closed system is an example of a heterogeneous equilibrium.

Since the concentrations of solids involved in an equilibrium are constant, they can be included as a part of the equilibrium constant and need not appear in the reaction quotient obtained by the law of mass action. We may write

$$K \times \frac{[CaCO_3]}{[CaO]} = [CO_2]$$

The left-hand side of this equation contains only constant terms. These can be combined into a single constant, K'.

$$K' = [CO_2]$$

You can assume that all equilibrium constants for heterogeneous equilibria contain such constant terms.

The expression for the equilibrium constant for the preceding reaction in terms of pressure is

$$K_p = P_{CO_2}$$

This equation means that at a given temperature there is only one pressure at which gaseous carbon dioxide can be in equilibrium with solid calcium carbonate and calcium oxide. A sample of $CaCO_3$ placed in a cylinder with a movable piston (Fig. 15.7) and heated at 900°C will decompose into CaO and CO_2. At equilibrium the pressure of carbon dioxide in the cylinder will equal K_p (actually 1.04 atm), if the piston is held stationary.

$$K_p = P_{CO_2} = 1.04 \text{ atm} \qquad \text{(at 900°C)}$$

At this pressure, since equilibrium has been reached, $CaCO_3$ decomposes into CaO and CO_2 at the same rate that CaO and CO_2 react to produce $CaCO_3$. If we increase the pressure by pushing the piston down and compressing the gas, more CO_2 will combine with CaO to form $CaCO_3$, and the pressure will again drop to 1.04 atm. If, on the other hand, we decrease the pressure by raising the piston, just enough $CaCO_3$ will decompose to bring the pressure exerted by the CO_2 back to its equilibrium value of 1.04 atm.

In the commercial production of quicklime, CaO, from limestone, the carbon dioxide is continuously removed as fast as it is formed by means of a stream of air; this causes the reaction to proceed essentially only to the right. Since the pressure of carbon dioxide never reaches the equilibrium pressure, equilibrium is never established.

Another example of a heterogeneous equilibrium is the equilibrium of bromine vapor with liquid bromine (Fig. 15.8).

15.9 The Distribution Law and Extraction

A good example of heterogeneous equilibria is the distribution of a solute in two immiscible solvents that are in contact but separated by a phase boundary. As noted in Section 15.4, one such solute is iodine, which is soluble in both water and carbon tetrachloride. If an aqueous solution of iodine is shaken vigorously with a sufficient quantity of carbon tetrachloride, the greater part of the iodine will leave the water and go into the carbon tetrachloride. After shaking is stopped, the carbon tetrachloride, colored violet by the iodine, settles to the bottom of the container. The equilibrium between the iodine in the water and in the carbon tetrachloride may be represented by

$$I_2 \text{ (in } H_2O \text{ phase)} \rightleftharpoons I_2 \text{ (in } CCl_4 \text{ phase)}$$

Figure 15.8

A heterogeneous equilibrium of bromine vapor in equilibrium with liquid bromine.

Applying the law of mass action to the equilibrium system, we have

$$K = \frac{[I_2(CCl_4)]}{[I_2(H_2O)]}$$

The equilibrium constant for systems of this type is called the **distribution ratio,** and it has the value of 90 for this particular system at room temperature. In other words, the molar concentration of iodine in the carbon tetrachloride is 90 times that in the aqueous layer when equilibrium is established. Thus most of the iodine in a water solution can be removed by a single treatment with carbon tetrachloride. This process is known as **extraction.** If the aqueous layer is extracted a second time with a fresh portion of carbon tetrachloride, the concentration of iodine in the water solution is reduced to a negligibly small value.

For Review

Summary

A reaction is said to have reached **dynamic equilibrium** when, at some point during the course of a reaction, the rate of formation of products becomes equal to the rate at which the products reform reactants. At equilibrium there is no net change in the concentrations of reactants or products. For any reaction

$$mA + nB + \cdots \rightleftharpoons xC + yD + \cdots$$

at equilibrium the **reaction quotient** is equal to a constant K, the **equilibrium constant** for the reaction.

$$K = \frac{[C]^x[D]^y \cdots}{[A]^m[B]^n \cdots}$$

At equilibrium the values of [A], [B], [C], [D], etc., may vary, but the reaction quotient will always be equal to K. Addition or removal of a reactant or product may shift a reaction out of equilibrium, but the reaction will proceed so that the concentrations change and equilibrium is reestablished to produce the same value of K, provided the temperature remains the same.

A change in temperature changes the value of the equilibrium constant. A change in pressure measurably affects a system in equilibrium only when gases are involved. At a given temperature, a change in pressure does not change the value of the equilibrium constant or on the equilibrium concentrations.

A catalyst increases the rate of the forward and reverse reactions equally. Hence the net effect of a catalyst in a reversible reaction is to cause the reaction to come to equilibrium more rapidly, but the catalyst has no effect on the value of the equilibrium constant or on the equilibrium concentrations.

A **homogeneous equilibrium** is an equilibrium within a single phase. A **heterogeneous equilibrium** is an equilibrium between two or more phases and involves a boundary surface between any two phases.

The effect of a change in conditions is described by **Le Châtelier's principle:** If a stress such as a change in temperature, pressure, or concentration is applied to a system at equilibrium, the equilibrium shifts in a way that relieves the effects of the stress.

Key Terms and Concepts

concentration effects on
    equilibrium (15.4)
distribution ratio (15.9)
equilibrium (15.1)
equilibrium constant (15.2)
extraction (15.9)

heterogeneous equilibrium (15.8)
homogeneous equilibrium (15.8)
law of mass action (15.2)
Le Châtelier's principle (15.4)
pressure effects on equilibrium
    (15.5)

reaction quotient (15.2)
temperature effects on equilibrium
    (15.6)
van't Hoff's law (15.6)

Exercises

Equilibrium Constants

1. Using the law of mass action, derive the mathematical expression of the equilibrium constant for the following reversible reactions:
   - (a) $N_2(g) + O_2(g) \rightleftharpoons 2NO(g)$
   - (b) $NH_4Cl(s) \rightleftharpoons NH_3(g) + HCl(g)$
   - (c) $2H_2(g) + O_2(g) \rightleftharpoons 2H_2O(l)$
   - (d) $S_8(g) \rightleftharpoons 8S(g)$
   - (e) $N_2(g) + 3H_2(g) \rightleftharpoons 2NH_3(g)$
   - (f) $4NH_3(g) + 5O_2(g) \rightleftharpoons 4NO(g) + 6H_2O(g)$
   - (g) $2Pb(NO_3)_2(s) \rightleftharpoons 2PbO(s) + 4NO_2(g) + O_2(g)$
   - (h) $BaSO_3(s) \rightleftharpoons BaO(s) + SO_2(g)$
   - (i) $CO_2(g) + H_2(g) \rightleftharpoons CO(g) + H_2O(g)$
   - (j) $CH_4(g) + Cl_2(g) \rightleftharpoons CH_3Cl(g) + HCl(g)$
   - (k) $N_2O_4(g) \rightleftharpoons 2NO_2(g)$
   - (l) $2SO_2(g) + O_2(g) \rightleftharpoons 2SO_3(g)$

2. Nitrogen reacts with hydrogen to give ammonia according to the equation $N_2 + 3H_2 \rightleftharpoons 2NH_3$. An equilibrium mixture of the above substances at 400°C was found to contain 0.45 mol of nitrogen, 0.63 mol of hydrogen, and 0.24 mol of ammonia per liter. Calculate the equilibrium constant for the system. *Ans. 0.51*

3. A sample of PCl_5 is put into a 1.00-L vessel and heated. At equilibrium, the vessel contains 0.40 mol of $PCl_3(g)$ and 0.40 mol of $Cl_2(g)$. The value of the equilibrium constant for the decomposition of PCl_5 to PCl_3 and Cl_2 at this temperature is 0.50 mol L^{-1}. Calculate the concentration of PCl_5 at equilibrium. *Ans. 0.32 M*

4. The equilibrium constant for the gaseous reaction $H_2 + I_2 \rightleftharpoons 2HI$ is 50.2 at 448°C. Calculate the number of grams of HI that are in equilibrium with 1.25 mol of H_2 and 63.5 g of iodine at this temperature. *Ans. 507 g*

5. A sample of $NH_3(g)$ was formed from $H_2(g)$ and $N_2(g)$ at 500°C. If the equilibrium mixture was found to contain 1.35 mol H_2 per liter, 1.15 mol N_2 per liter, and 4.12×10^{-1} mol NH_3 per liter, what is the value of the equilibrium constant for the formation of NH_3? *Ans. 6.00×10^{-2} (M^{-2}, or $mol^{-2} L^2$)*

6. Most of the 73.6 billion lb of sulfuric acid produced in the United States during 1986 resulted from the reaction sequence

$$S_8(g) + 8O_2(g) \longrightarrow 8SO_2(g) \tag{1}$$

$$2SO_2(g) + O_2(g) \xrightarrow{V_2O_5} 2SO_3(g) \tag{2}$$

$$SO_3(g) + H_2O(l) \longrightarrow H_2SO_4(l) \tag{3}$$

V_2O_5 is required in Equation (2) because the oxidation of SO_2 to SO_3 is slow in the absence of a catalyst. Under a specific set of conditions at 500°C, equilibrium pressures of SO_2, O_2, and SO_3 in the reaction of Equation (2) have been determined to be 0.342 atm, 0.173 atm, and 0.988 atm, respectively. What is K_p for this reaction? *Ans. 48.2 atm^{-1}*

7. The vapor pressure of water is 0.0728 atm at 40°C. What is the equilibrium constant for the transformation $H_2O(l) \rightleftharpoons H_2O(g)$? *Ans. $K_p = 0.0728$ atm*

8. In general, the equilibrium constant for a reaction in the gas phase has different units and a different numerical value when pressures rather than concentrations are used to evaluate it. Show that such is not the case for the decomposition of HI into H_2 and I_2; show that the numerical value of the equilibrium constant for this particular reaction is independent of the units in which concentration is expressed.

9. (a) Calculate the equilibrium concentration of NO_2 in a solution prepared by dissolving 0.20 mol of N_2O_4 in 400 mL of chloroform. For the reaction $N_2O_4(g) \rightleftharpoons 2NO_2(g)$, in chloroform, $K = 1.07 \times 10^{-5}$. *Ans. 2.3×10^{-3} M*
   (b) What will be the percent decomposition of the original N_2O_4? *Ans. 0.23%*

10. Analysis of the gases in a 4.0-L sealed reaction vessel containing NH_3, N_2, and H_2 at equilibrium at 400°C established the presence of 4.8 mol N_2 and 0.64 mol H_2. At 400°C, $K = 0.50$ mol^2 L^{-2}.

$$2NH_3(g) \rightleftharpoons N_2(g) + 3H_2(g)$$

Calculate the equilibrium molar concentration of NH_3. *Ans. 9.9×10^{-2} M*

11. The rate of the reaction $H_2(g) + I_2(g) \rightarrow 2HI(g)$ at 25°C is given by

$$\text{Rate} = 1.7 \times 10^{-18} \, [H_2][I_2]$$

The rate of decomposition of gaseous HI to $H_2(g)$ and $I_2(g)$ at 25°C is given by

$$\text{Rate} = 2.4 \times 10^{-21} \, [HI]^2$$

What is the equilibrium constant for the formation of gaseous HI from the gaseous elements at 25°C?

Ans. 7.1 × 10²

12. If 60.0 g each of acetic acid (CH_3CO_2H) and ethanol (C_2H_5OH) are allowed to react in a 1.00-L sealed container until equilibrium is established, how many moles of the ester ($CH_3CO_2C_2H_5$) and water are formed, and how many moles of ethanol and acid remain? ($K = 4.00$.)

$$CH_3CO_2H + C_2H_5OH \rightleftharpoons CH_3CO_2C_2H_5 + H_2O$$

Ans. 0.252 mol of acid and 0.55 mol of ethanol;
0.747 mol each of ester and water

13. Acetic acid (CH_3CO_2H) and ethanol (C_2H_5OH) react in dioxane solvent to produce ethyl acetate ($CH_3CO_2C_2H_5$) and water.

$$CH_3CO_2H + C_2H_5OH \rightleftharpoons CH_3CO_2C_2H_5 + H_2O$$

The equilibrium constant is 4.0. If 1.00 mol of acetic acid and 2.00 mol of ethanol are mixed in enough dioxane to make 1.00 L of solution, how many moles of each reactant and product will be present in the solution when equilibrium is established?

Ans. CH₃CO₂H, 0.16 mol; C₂H₅OH, 1.16 mol;
CH₃CO₂C₂H₅, 0.84 mol; H₂O, 0.84 mol

14. (a) Write the mathematical expression of the law of chemical equilibrium for the reversible reaction $2NO_2 \rightleftharpoons 2NO + O_2$.

(b) At 1 atm and 25°C, NO_2 with an initial concentration of 1.00 M is $3.3 \times 10^{-3}\%$ decomposed into NO and O_2. Calculate the equilibrium constant.

Ans. K = 1.8 × 10⁻¹⁴ mol L⁻¹

15. The numerical value of K is 6.00×10^{-2} at 25°C, if concentrations are expressed in moles per liter, for the reaction

$$N_2(g) + 3H_2(g) \rightleftharpoons 2NH_3(g)$$

(a) What are the units for K?

(b) What is the value for K, including units, for the reverse reaction?

$$2NH_3(g) \rightleftharpoons N_2(g) + 3H_2(g)$$

(c) Will the numerical value of K change if the reaction is written as follows?

$$NH_3(g) \rightleftharpoons \frac{1}{2}N_2(g) + \frac{3}{2}H_2(g)$$

What are the units?

Effect of Changes in Concentration, Pressure, or Temperature on Equilibrium

16. State Le Châtelier's principle as it applies to chemical equilibria.

17. For each of the following reactions between gases at equilibrium, determine the effect on the equilibrium concentrations of the products when the temperature is decreased and when the external pressure on the system is decreased.

(a) $2H_2O(g) \rightleftharpoons 2H_2(g) + O_2(g) \qquad \Delta H = 484$ kJ
(b) $N_2(g) + O_2(g) \rightleftharpoons 2NO(g) \qquad \Delta H = 181$ kJ
(c) $N_2(g) + 3H_2(g) \rightleftharpoons 2NH_3(g) \qquad \Delta H = -92.2$ kJ
(d) $2O_3(g) \rightleftharpoons 3O_2(g) \qquad \Delta H = -285$ kJ
(e) $H_2(g) + F_2(g) \rightleftharpoons 2HF(g) \qquad \Delta H = 541$ kJ

18. Suggest four ways in which the equilibrium concentration of ammonia can be increased for the reaction in exercise 17(c).

19. For the reaction in exercise 17(b), the equilibrium constant at 2000°C is 6.2×10^{-4}. Consider each of the following situations, and decide whether or not a reaction will occur and, if so, in which direction it will predominantly proceed.

(a) A 5.0-L flask contains 0.26 mol of N_2, 0.0062 mol of O_2, and 0.0010 mol of NO at 2000°C.

(b) A 2.5-L flask contains 0.26 mol of N_2, 0.0062 mol of O_2, and 0.0010 mol of NO at 2500°C.

20. Under what conditions do changes in pressure affect systems in equilibrium?

21. How will an increase in temperature affect each of the following equilibria? An increase in pressure?

(a) $N_2(g) + 3H_2(g) \rightleftharpoons 2NH_3(g) \qquad \Delta H = -92.2$ kJ
(b) $H_2O(l) \rightleftharpoons H_2O(g) \qquad \Delta H = 41$ kJ
(c) $N_2(g) + O_2(g) \rightleftharpoons 2NO(g) \qquad \Delta H = 181$ kJ
(d) $3O_2(g) \rightleftharpoons 2O_3(g) \qquad \Delta H = 285$ kJ
(e) $CaCO_3(s) \rightleftharpoons CaO(s) + CO_2(g) \qquad \Delta H = 176$ kJ

22. (a) For the system

$$H_2(g) + CO_2(g) \rightleftharpoons H_2O(g) + CO(g)$$

the equilibrium constant is 1.6 at 990°C. Calculate the number of moles of each component in the equilibrium mixture that results from placing 1.00 mol of H_2 and 1.00 mol of CO_2 in a sealed 5.00-L reactor at 990°C.

Ans. 0.44 mol H₂; 0.44 mol CO₂; 0.56 mol H₂O;
0.56 mol CO

(b) Calculate the number of moles of each component at a second equilibrium if 1.00 mol of additional H_2 is added to the equilibrium mixture obtained in part (a).

Ans. 1.27 mol H₂; 0.27 mol CO₂; 0.73 mol H₂O;
0.73 mol CO

(c) Compare the answers in parts (a) and (b). Are they consistent with equilibrium theory and Le Châtelier's principle? Explain your conclusions.

23. (a) For the system described in exercise 22 (but starting with a new set of components), calculate the number of moles of each component in the equilibrium mixture that results from initially placing 2.00 mol of H_2, 1.00 mol of CO_2, 1.00 mol of H_2O, and 0.75 mol of CO in the sealed 5.00-L reactor at 990°C.

   Ans. 1.61 mol H_2; 0.61 mol CO_2; 1.39 mol H_2O; 1.14 mol CO

   (b) Calculate the number of moles of each component in a final equilibrium mixture if an additional 1.00 mol of H_2 and 1.00 mol of CO_2 are added to the equilibrium mixture produced in part (a).

   Ans. 2.03 mol H_2; 1.03 mol CO_2; 1.97 mol H_2O; 1.72 mol CO

   (c) Compare the answers in parts (a) and (b). Are they consistent with equilibrium theory and Le Châtelier's principle?

24. An equilibrium mixture of N_2, H_2, and NH_3 in a 1.00-L vessel is found to contain 0.300 mol of N_2, 0.400 mol of H_2, and 0.100 mol of NH_3. How many moles of H_2 must be introduced into the vessel in order to double the equilibrium concentration of NH_3 if the temperature remains unchanged?   *Ans. 0.425 mol*

25. At 25°C and atmospheric pressure, the partial pressures in an equilibrium mixture of N_2O_4 and NO_2 are: $P_{N_2O_4} = 0.70$ atm; $P_{NO_2} = 0.30$ atm. Calculate the partial pressures of these two gases when they are in equilibrium at 9.0 atm and 25°C.

   Ans. $P_{N_2O_4} = 8.0$ atm; $P_{NO_2} = 1.0$ atm

26. A 1.00-L vessel at 400°C contains the following equilibrium concentrations: N_2, 1.00 M; H_2, 0.50 M; and NH_3, 0.50 M. How many moles of hydrogen must be removed from the vessel in order to increase the concentration of nitrogen to 1.2 M?

   Ans. 0.94 (Note that more hydrogen is removed than was originally present as elemental H_2. Is this possible?)

27. 1.0 mol of PCl_5 is placed in a 4.0-L container. Reaction takes place according to the equation $PCl_5(g) \rightleftharpoons PCl_3(g) + Cl_2(g)$. At equilibrium 0.25 mol of Cl_2 is present. Calculate the value of the equilibrium constant for this reaction under the conditions of the experiment.

   Ans. 0.021 mol L^{-1}

28. The equilibrium constant for the reaction

   $$CO + H_2O \rightleftharpoons CO_2 + H_2$$

   is 5.0 at a given temperature.

   (a) Upon analysis, an equilibrium mixture of the substances present was found to contain 0.20 mol of CO, 0.30 mol of water vapor, and 0.90 mol of H_2 in a liter. How many moles of CO_2 were there in the equilibrium mixture?   *Ans. 0.33 mol*

   (b) Maintaining the same temperature, additional H_2 was added to the system, and some water vapor was removed by drying. A new equilibrium mixture was

thereby established containing 0.40 mol of CO, 0.30 mol of water vapor, and 1.2 mol of H_2 in a liter. How many moles of CO_2 were in the new equilibrium mixture? Compare this with the quantity in part (a) and discuss whether the second value is reasonable. Explain how it is possible for the water vapor concentration to be the same in the two equilibrium solutions even though some vapor was removed before the second equilibrium was established.   *Ans. 0.50 mol*

29. A 5.00-L vessel contains CO, CO_2, H_2, and H_2O in equilibrium at 980°C. The equilibrium partial pressures are: CO, 300 torr; CO_2, 300 torr; H_2, 90 torr; H_2O, 150 torr. Carbon monoxide is then pumped into the vessel until the equilibrium partial pressure of hydrogen is equal to 130 torr. Calculate the partial pressures of the other substances at the new equilibrium assuming that no temperature change occurs.

   Ans. CO_2, 340 torr; H_2O, 110 torr; CO, 670 torr

30. In a 3.0-L vessel, the following equilibrium partial pressures are measured: N_2, 190 torr; H_2, 317 torr; NH_3, 1000 torr. Hydrogen is removed from the vessel until the partial pressure of nitrogen, at equilibrium, is equal to 250 torr. Calculate the partial pressures of the other substances under the new conditions.

   Ans. P_{NH_3}, 880 torr; P_{H_2}, 266 torr

31. (a) Using the law of mass action, write the expression for the equilibrium constant for the reversible reaction

   $$N_2 + O_2 \rightleftharpoons 2NO \qquad \Delta H = 181 \text{ kJ}$$

   (b) What will happen to the concentration of NO at equilibrium if (1) more O_2 is added? (2) N_2 is removed? (3) the pressure on the system is increased? (4) the temperature of the system is increased?

   (c) This reaction produces nitrogen oxide pollutants during the operation of an internal combustion engine. NO forms in the cylinders during combustion and is swept out in the exhaust. The reaction for the decomposition of NO ($2NO \rightarrow N_2 + O_2$) has an equilibrium constant of $K_p = 3.0 \times 10^{31}$ at 25°C (the temperature of a warm day). At equilibrium what would be the atmospheric NO pressure ($P_{O_2} = 0.2$ atm, $P_{N_2} = 0.8$ atm)?   *Ans. 7×10^{-17} atm*

   (d) The actual concentration of NO along an expressway during rush-hour traffic is higher than this value. Suggest an explanation.

Heterogeneous Equilibria

32. A sample of ammonium chloride was heated in a closed container.

   $$NH_4Cl(s) \rightleftharpoons NH_3(g) + HCl(g)$$

   At equilibrium the pressure of $NH_3(g)$ was found to be 1.75 atm. What is the equilibrium constant for the decomposition at this temperature?   *Ans. $K_p = 3.06$ atm^2*

33. Sodium sulfate 10-hydrate, $Na_2SO_4 \cdot 10H_2O$, dehydrates according to the equation

$$Na_2SO_4 \cdot 10H_2O(s) \rightleftharpoons Na_2SO_4(s) + 10H_2O(g)$$

with $K = 4.08 \times 10^{-25}$ at 25°C. What is the pressure of water vapor in equilibrium with a sample of $Na_2SO_4 \cdot 10H_2O$?   *Ans. 3.64×10^{-3} atm*

34. Under what conditions will the reversible decomposition

$$CaCO_3(s) \rightleftharpoons CaO(s) + CO_2(g)$$

proceed to completion in a closed container so that no $CaCO_3$ remains?

35. What is the minimum weight of $CaCO_3$ required to establish equilibrium at a certain temperature in a 6.50-L container if the equilibrium constant is 0.050 mol L^{-1} for the decomposition reaction of $CaCO_3$ at that temperature? Justify the units given in the problem for the equilibrium constant.

$$CaCO_3(s) \rightleftharpoons CaO(s) + CO_2(g)$$

Ans. 33 g

36. 1.0 mol of solid ammonium chloride was placed in a cylinder, with a piston serving as the top surface, and heated. Partial decomposition occurred:

$$NH_4Cl(s) \rightleftharpoons NH_3(g) + HCl(g)$$

When the NH_3 partial pressure at equilibrium equals 1.6 atm, what is the equilibrium constant? What happens if the volume of the cylinder is decreased 50% by pressing down the piston? What happens if the piston is locked into position and a hole is punched in the cylinder?

Ans. $K = 2.6$ atm^2

Additional Exercises

37. The binding of oxygen by hemoglobin (Hb), giving oxyhemoglobin (HbO$_2$), is partially regulated by the concentration of H^+ and CO_2 in the blood. Although the equilibrium is complicated, it can be summarized as

$$HbO_2 + H^+ + CO_2 \rightleftharpoons CO_2-Hb-H^+ + O_2$$

(a) Write the equilibrium constant expression for this reaction.

(b) Explain why the production of lactic acid and CO_2 in a muscle during exertion stimulates release of O_2 from the oxyhemoglobin in the blood passing through the muscle.

38. For the reaction $A \longrightarrow B + C$ the following data were obtained at 30°C:

Experiment	[A] (mol L^{-1})	Rate (mol L^{-1} h^{-1})
1	0.170	0.0500
2	0.340	0.100
3	0.680	0.200

(a) What is the rate equation and order of the reaction?
Ans. First order
(b) Calculate k for the reaction   *Ans. 0.294 h^{-1}*
(c) The equilibrium constant for the reaction is 0.500. What is the rate constant for the reverse reaction?
Ans. 0.588 h^{-1}

39. The hydrolysis of the sugar sucrose to the sugars glucose and fructose

$$C_{12}H_{22}O_{11} + H_2O \longrightarrow C_6H_{12}O_6 + C_6H_{12}O_6$$

follows a first-order rate equation for the disappearance of sucrose.

$$Rate = k[C_{12}H_{22}O_{11}]$$

(The products of the reaction, glucose and fructose, have the same molecular formulas but differ in the arrangement of the atoms in their molecules.)

(a) In neutral solution, $K = 2.1 \times 10^{-11}$ s^{-1} at 27°C and 8.5×10^{-11} s^{-1} at 37°C. Determine the activation energy, the frequency factor, and the rate constant for this equation at 47°C.
Ans. $E_a = 108$ kJ; $A = 1.3 \times 10^8$ s^{-1}; $K = 3.1 \times 10^{-10}$ s^{-1} (An alternative and equally acceptable method of working the problem produces the answers $E_a = 109$ kJ; $A = 2.0 \times 10^8$ s^{-1}; $K = 3.2 \times 10^{-10}$ s^{-1}.)

(b) The equilibrium constant for the reaction is 1.36×10^5 at 27°C. What are the concentrations of glucose, fructose, and sucrose after a 0.150 M aqueous solution of sucrose has reached equilibrium? Remember that the activity of a solvent (the effective concentration) is 1.
Ans. [Sucrose] = 1.65×10^{-7} M; [glucose] = [fructose] = 0.150 M

(c) How long will the reaction of a 0.150 M solution of sucrose require to reach equilibrium at 27°C in the absence of a catalyst? Since the concentration of sucrose at equilibrium is so low, assume that the reaction is irreversible.   *Ans. 6.5×10^{11} s (21,000 years!)*

40. At high temperatures, the reaction of hydrogen and oxygen to form water is reversible and achieves equilibrium. Under such conditions, a 1.0-L container holding pure hydrogen at a pressure of 0.80 atm and another 1.0-L container holding pure oxygen at a pressure of 20.0 atm were connected and opened to each other so that the gases could mix and come to equilibrium with water vapor. Then the partial pressure of oxygen was measured and found to be *essentially* 10.0 atm. (Assume that the quantity of oxygen that reacts is negligible compared to the quantity originally present.) The equilibrium constant for the reaction at the temperature of the experiment has a value of 2.50 atm^{-1}. (a) Write the equilibrium constant expression for the reaction. (b) What are the equilibrium partial pressures of hydrogen and water? (c) Check whether the assumption made regarding the quantity of oxygen that reacts compared to

the quantity originally present is justifiable. Could a similar assumption be justified for hydrogen? Explain.

Ans. 0.07 atm H_2; 0.33 atm H_2O

41. The reaction between hydrogen, $H_2(g)$, and sulfur, $S_8(g)$, to produce H_2S is exothermic at 25°C. How should the pressure and temperature be adjusted in order to improve the equilibrium yield of H_2S, assuming that all reactants and products are in the gaseous state? How will these conditions affect the rate of attainment of equilibrium?

42. One possible mechanism for the reaction of H_2O_2 with I^- in acid solution (Section 14.14) is

$$H_3O^+ + I^- \longrightarrow HI + H_2O \qquad \text{(fast)}$$
$$H_2O_2 + HI \longrightarrow H_2O + HOI \qquad \text{(slow)}$$
$$HOI + H_3O^+ + I^- \longrightarrow 2H_2O + I_2 \qquad \text{(fast)}$$
$$I_2 + I^- \longrightarrow I_3^- \qquad \text{(fast)}$$

(a) Write the rate equation for the slow elementary reaction.   *Ans. Rate $= k[H_2O_2][HI]$*

(b) Write the equilibrium constant expression for the first elementary reaction.

$$\text{Ans. } K = \frac{[HI]}{[H_3O^+][I^-]}$$

(c) Solve the equilibrium constant expression from part (b) for [HI], and substitute this concentration into the rate equation from part (a) to obtain the overall rate equation for this mechanism.

Ans. Rate $= kK[H_3O^+][I^-][H_2O_2] =$
$k'[H_3O^+][I^-][H_2O_2]$

43. Using the expression for the equilibrium constant for the fast reaction of H_3O^+ with H_2O_2 and the rate equation for the slow elementary reaction of $H_3O_2^+$ with I^-, derive the overall rate equation based on mechanism C in Section 14.14 for the reaction of H_2O_2 with I^- in acid solution.

44. The equilibrium constant for the reaction

$$SbCl_5(g) \rightleftharpoons SbCl_3(g) + Cl_2(g)$$

at 448°C is 2.50×10^{-2}. What are the equilibrium concentrations of $SbCl_5$, $SbCl_3$, and Cl_2 if the initial concentration of $SbCl_5$ is 0.750 M? Use the quadratic formula to obtain your answers, and then use the method of successive approximations as an alternative method.

Ans. $SbCl_5$, 0.625 M; $SbCl_3$, 0.125 M; Cl_2, 0.125 M

Acids and Bases

16

Cave formations result from acid-base reactions involving $CaCO_3$.

A cids and bases have been defined in a number of ways. When Boyle first characterized them in 1680, he noted that acids dissolve many substances, change the color of certain natural dyes (for example, litmus from blue to red), and lose these characteristic properties after coming in contact with alkalies (bases). In the eighteenth century it was recognized that acids have a sour taste, react with limestone with the liberation of a gaseous substance (CO_2), and interact with alkalies to form neutral substances. In 1787 Lavoisier proposed that acids are binary compounds of oxygen and considered oxygen to be responsible for their acidic properties. The hypothesis that oxygen is an essential component of acids was disproved by Davy in 1811 when he showed that hydrochloric acid contains no oxygen. Davy contributed greatly to the development of the acid-base concept by concluding that hydrogen, rather than oxygen, is the essential constituent of acids. In 1814 Gay-Lussac concluded that acids are substances that can neutralize alkalies and that these two classes of substances can be defined only in terms of each other. Davy and Gay-Lussac provided the foundation for our concepts of acids and bases in aqueous solutions.

The significance of hydrogen was reemphasized in 1884 when Arrhenius defined an acid as a compound that dissolves in water to yield hydrogen ions, and a base as a compound that dissolves in water to yield hydroxide ions. In 1923 the close relationship between acids and bases was pointed out by both Johannes Brønsted, a Danish chemist, and Thomas Lowry, an English chemist. Their models for acid-base behavior define acids as proton donors and bases as proton acceptors. An even broader view of the relationship between acids and bases was developed by the American chemist G. N. Lewis. The Lewis theory defines acids as electron-pair acceptors and bases as electron-pair donors.

In this chapter we shall first consider the description of acid-base behavior presented by Brønsted and Lowry and then the more specific way in which the Brønsted-Lowry concept applies to aqueous solutions. Later in the chapter, we will proceed to the more generalized Lewis concept.

The Brønsted-Lowry Concept of Acids and Bases

16.1 The Protonic Concept of Acids and Bases

A compound that donates a proton (a hydrogen ion, H^+) to another compound is called a Brønsted acid, and a compound that accepts a proton is called a Brønsted base (Section 9.2). Stated simply, an acid is a proton donor, and a base is a proton acceptor.

An **acid-base reaction** is the transfer of a proton from a proton donor (acid) to a proton acceptor (base). A proton may be donated by a molecular acid, such as HCl, H_2SO_4, H_3PO_4, CH_3CO_2H, H_2S, or H_2O, or it may be donated by either an anion or a cation. Anions that can exhibit acidic behavior include HSO_4^-, $H_2PO_4^-$, HPO_4^{2-}, and HS^-. Cations that can exhibit acidic behavior include H_3O^+, NH_4^+, $[Cu(H_2O)_4]^{2+}$, and $[Fe(H_2O)_6]^{3+}$. A proton may be accepted by a molecular base, such as H_2O, NH_3, or CH_3NH_2, or by an anion, such as OH^-, HS^-, S^{2-}, HCO_3^-, CO_3^{2-}, HSO_4^-, SO_4^{2-}, HPO_4^{2-}, Cl^-, F^-, NO_3^-, or PO_4^{3-}, or by a cation, such as $[Fe(H_2O)_5OH]^{2+}$ or $[Cu(H_2O)_3OH]^+$.

The most familiar bases are ionic compounds such as NaOH and $Ca(OH)_2$, which contain the hydroxide ion, OH^-. The hydroxide ion accepts protons from acids, forming water.

$$H_3O^+ + OH^- \longrightarrow 2H_2O$$

In our discussions of acid-base properties, we will sometimes identify specific ions as being acids or bases. However, it must be reemphasized that no solid or solution can contain only cations or only anions. For every ion in a solid or solution there must be an ion of the opposite charge, a **counter ion,** so that the solid or solution is electrically neutral. If the counter ions do not affect the acid-base properties of the system, we may leave them out of equations, but they will be present in the actual solution.

When an acid loses a proton, it forms what is called the **conjugate base** of the acid. The conjugate base is a base because it can pick up a proton (and thereby reform the acid).

Acid	Proton	Conjugate base
HCl	$\longrightarrow$ H^+	$+$ Cl^-
H_2SO_4	$\longrightarrow$ H^+	$+$ HSO_4^-
H_2O	$\longrightarrow$ H^+	$+$ OH^-
HSO_4^-	$\longrightarrow$ H^+	$+$ SO_4^{2-}
NH_4^+	$\longrightarrow$ H^+	$+$ NH_3
$[Fe(H_2O)_6]^{3+}$	$\longrightarrow$ H^+	$+$ $[Fe(H_2O)_5(OH)]^{2+}$

When a base accepts a proton, it forms what is called the **conjugate acid** of the base. The conjugate acid is an acid because it can give up a proton (and thus reform the base).

Base	Proton	Conjugate acid
H_2O	$+$ H^+	$\longrightarrow$ H_3O^+
NH_3	$+$ H^+	$\longrightarrow$ NH_4^+
OH^-	$+$ H^+	$\longrightarrow$ H_2O
S^{2-}	$+$ H^+	$\longrightarrow$ HS^-
CO_3^{2-}	$+$ H^+	$\longrightarrow$ HCO_3^-
F^-	$+$ H^+	$\longrightarrow$ HF
$[Fe(H_2O)_5OH]^{2+}$	$+$ H^+	$\longrightarrow$ $[Fe(H_2O)_6]^{3+}$

In order for an acid to act as a proton donor, a base (proton acceptor) must be present to receive the proton. An acid does not form its conjugate base unless a second base is present to accept the proton. When the second base accepts the proton, it forms its conjugate acid, the second acid. When hydrogen chloride, HCl, reacts with anhydrous ammonia, NH_3, forming ammonium ions, NH_4^+, and chloride ions, Cl^-, hydrogen chloride (acid$_1$) gives up a proton forming chloride ion, its conjugate base (base$_1$). Ammonia acts as the proton acceptor and therefore is a base (base$_2$). The proton combines with ammonia to give its conjugate acid, the ammonium ion (acid$_2$). The equations for this and several other acid-base reactions are as follows:

Acid$_1$	Base$_2$	Acid$_2$	Base$_1$
HCl	$+$ NH_3	$\rightleftharpoons$ NH_4^+	$+$ Cl^-
HNO_3	$+$ F^-	$\rightleftharpoons$ HF	$+$ NO_3^-
HSO_4^-	$+$ CO_3^{2-}	$\rightleftharpoons$ HCO_3^-	$+$ SO_4^{2-}
NH_4^+	$+$ S^{2-}	$\rightleftharpoons$ HS^-	$+$ NH_3

In all of these acid-base reactions, the forward reaction is the transfer of a proton from acid$_1$ to base$_2$. The reverse reaction is the transfer of a proton from acid$_2$ to base$_1$. In effect, base$_1$ and base$_2$ are in competition for the proton, and the species that are present in the greatest concentrations will depend on which base competes most effectively for the proton. In all of the examples given in the table above, base$_2$ is the stronger base. As a result, at equilibrium the system will consist primarily of a mixture of acid$_2$ and base$_1$.

When a Brønsted acid stronger than water dissolves in water, protons are transferred from the acid molecules to water molecules, giving hydronium ions, H_3O^+. For example, when hydrogen chloride dissolves in water, a proton is transferred from the hydrogen chloride to a water molecule. Water is thus the proton acceptor and acts as a base.

$$HCl + H_2O \rightleftharpoons H_3O^+ + Cl^-$$

acid$_1$ base$_2$ acid$_2$ base$_1$

The hydronium ion is the conjugate acid of water. Hence the properties that are common to Brønsted acids in aqueous solution are those of the hydronium ion. The hydronium ion is the species actually formed, but the symbol H^+ is often used instead of H_3O^+ for the sake of convenience. It is important to remember, however, that H^+ is an abbreviation and that hydrogen ion is always hydrated in water solution. In a strict sense even the formula H_3O^+ is an abbreviation. For example, good evidence has been accumulated for the existence of the species $H_9O_4^+$, corresponding to $[H_3O(H_2O)_3]^+$.

When a base such as ammonia dissolves in water, water functions as a proton donor and hence as an acid.

$$H_2O + NH_3 \rightleftharpoons NH_4^+ + OH^-$$

acid$_1$ base$_2$ acid$_2$ base$_1$

The hydroxide ion is the conjugate base of water. Note that the hydroxide ion is a stronger base than ammonia, so this reaction gives only a small amount of NH_4^+ and OH^-; the majority of ammonia molecules do not react.

Under the appropriate conditions, then, water can function either as an acid or as a base, depending on the nature of the solute. The ability of water to act as either an acid or a base is also illustrated by the fact that a water molecule is capable of providing a proton to another water molecule to form a hydronium ion and a hydroxide ion.

$$H_2O + H_2O \rightleftharpoons H_3O^+ + OH^-$$

acid$_1$ base$_2$ acid$_2$ base$_1$

This type of reaction, in which a substance ionizes by means of one molecule of the substance spontaneously reacting with another molecule of the same substance, is referred to as **autoionization,** or **self-ionization.** Pure water is only slightly ionized through autoionization, as evidenced by the small value of the equilibrium constant, which is 1.0×10^{-14} at 25°C. Only about two out of every 10^9 molecules in a sample of pure water is ionized at 25°C.

16.2 Amphiprotic Species

Certain molecules and ions may exhibit either acidic or basic behavior under the appropriate conditions. A species that, like water, may either gain or lose a proton is said to be **amphiprotic.** Water may lose a proton to a base, such as NH_3, or gain a proton from an acid, such as HCl, and so is classified as amphiprotic.

The proton-containing anions given in Section 16.1 are also amphiprotic, which can be readily seen from the following equations involving HS^- and HCO_3^-.

Acid$_1$	Base$_2$		Acid$_2$	Base$_1$
HS^-	$+ OH^-$	$\rightleftharpoons$	H_2O	$+ S^{2-}$
HBr	$+ HS^-$	$\rightleftharpoons$	H_2S	$+ Br^-$
HCO_3^-	$+ CN^-$	$\rightleftharpoons$	HCN	$+ CO_3^{2-}$
H_3O^+	$+ HCO_3^-$	$\rightleftharpoons$	H_2CO_3	$+ H_2O$

The hydroxides of certain metals, especially those near the boundary between metals and nonmetals in the Periodic Table, are amphiprotic and so react either as acids or bases.

$Acid_1$	$Base_2$		$Acid_2$	$Base_1$
$[Al(H_2O)_3(OH)_3]\ +$	OH^-	$\rightleftharpoons$	H_2O	$+\ [Al(H_2O)_2(OH)_4]^-$
H_3O^+	$+\ [Al(H_2O)_3(OH)_3]$	$\rightleftharpoons$	$[Al(H_2O)_4(OH)_2]^+\ +$	H_2O

In the first reaction one of the water molecules of the hydrated aluminum hydroxide loses a proton to the hydroxide ion. In the second reaction the aluminum hydroxide accepts a proton from the hydronium ion. In both cases the insoluble hydrated aluminum hydroxide dissolves.

16.3 The Strengths of Brønsted Acids and Bases

The fundamental acid-base relationship for the reaction of any Brønsted acid with water is

$$HA + H_2O \rightleftharpoons H_3O^+ + A^-$$

where A^- is the conjugate base of the acid HA and the hydronium ion is the conjugate acid of water. Water is the base that reacts with the acid HA. **Strong acids** react with water and ionize in dilute solution, giving a 100% yield (or very nearly a 100% yield) of hydronium ion and the conjugate base of the acid. **Weak acids** ionize to a slight extent in aqueous solution, giving small yields of hydronium ion (ordinarily 10% or less). Thus the strength of an acid can be measured by the extent of its tendency to form hydronium ions in aqueous solution by donation of protons to water molecules (Fig. 16.1).

The extent to which protons will be donated to water molecules depends on the strength of the conjugate base, A^-, of the acid. If A^- is a strong base and accepts protons readily, there will be relatively little A^- and H_3O^+ in solution, and the acid, HA, is a weak one. If A^- is a weak base and does not readily accept protons, the solution will contain primarily A^- and H_3O^+, and the acid is a strong one. Strong acids form weak conjugate bases, and weak acids form relatively strong conjugate bases. The first six acids in Table 16.1 (p. 456) are the common strong acids, which are completely dissociated in aqueous solution. The conjugate bases of these acids are weaker bases than water. When one of these acids dissolves in water, the proton is transferred to water, the stronger base.

Relative acid strength

Relative conjugate base strength

Figure 16.1

A representation of the relative strengths of conjugate acid-base pairs in aqueous solution.

Table 16.1   The relative strengths of conjugate acid-base pairs

Acid				Base	
Perchloric acid	$HClO_4$			ClO_4^-	Perchlorate ion
Sulfuric acid	H_2SO_4			HSO_4^-	Hydrogen sulfate ion
Hydrogen iodide	HI	Stronger acids than		I^-	Iodide ion
Hydrogen bromide	HBr	H_3O^+; form H_3O^+ in		Br^-	Bromide ion
Hydrogen chloride	HCl	100% yield in H_2O.		Cl^-	Chloride ion
Nitric acid	HNO_3			NO_3^-	Nitrate ion
Hydronium ion	H_3O^+			H_2O	Water
Hydrogen sulfate ion	HSO_4^-			SO_4^{2-}	Sulfate ion
Phosphoric acid	H_3PO_4			$H_2PO_4^-$	Dihydrogen phosphate ion
Hydrogen fluoride	HF			F^-	Fluoride ion
Nitrous acid	HNO_2			NO_2^-	Nitrite ion
Acetic acid	CH_3CO_2H			$CH_3CO_2^-$	Acetate ion
Carbonic acid	H_2CO_3			HCO_3^-	Hydrogen carbonate ion
Hydrogen sulfide	H_2S			HS^-	Hydrogen sulfide ion
Ammonium ion	NH_4^+			NH_3	Ammonia
Hydrogen cyanide	HCN			CN^-	Cyanide ion
Hydrogen carbonate ion	HCO_3^-			CO_3^{2-}	Carbonate ion
Hydrogen sulfide ion	HS^-			S^{2-}	Sulfide ion
Water	H_2O			OH^-	Hydroxide ion
Ethanol	C_2H_5OH			$C_2H_5O^-$	Ethoxide ion
Ammonia	NH_3	Stronger bases than		NH_2^-	Amide ion
Hydrogen	H_2	OH^-; form OH^- in		H^-	Hydride ion
Methane	CH_4	100% yield in H_2O.		CH_3^-	Methide ion

Increasing acid strength (left margin, upward arrow)

Increasing base strength (right margin, downward arrow)

Those acids that lie between the hydronium ion and water in Table 16.1 form conjugate bases that can compete with water for possession of a proton. Both hydronium ions and nonionized acid molecules are present in equilibrium in a solution of one of these acids. One way of measuring their relative strengths is by measuring their percent ionizations in aqueous solutions of equal concentration, with the stronger acid forming the higher yield of hydronium ion. From the data given below, the order of acid strength for three of these acids is $HSO_4^- > HNO_2 > CH_3CO_2H$.

$$HSO_4^- + H_2O \rightleftharpoons H_3O^+ + SO_4^{2-} \qquad \text{(29\% in 0.10 M solution)}$$
$$HNO_2 + H_2O \rightleftharpoons H_3O^+ + NO_2^- \qquad \text{(6.5\% in 0.10 M solution)}$$
$$CH_3CO_2H + H_2O \rightleftharpoons H_3O^+ + CH_3CO_2^- \qquad \text{(1.3\% in 0.10 M solution)}$$

Other ways of measuring the strengths of Brønsted acids are discussed in Section 16.5.

Compounds that lie below water in the column of acids in Table 16.1 exhibit no observable acidic behavior when dissolved in water. Their conjugate bases are stronger than the hydroxide ion, and if any conjugate base were formed, it would react with water to reform the acid.

The fundamental reaction of a Brønsted base with water is given by the equation

$$H_2O + B \rightleftharpoons HB^+ + OH^-$$

where HB^+ is the conjugate acid of the base B, and the hydroxide ion is the conjugate base of water. Water is the acid that reacts with the base. **Strong bases** react with water giving a 100% yield (or very nearly a 100% yield) of hydroxide ion and the conjugate acid of the base. **Weak bases** react to a small extent with water giving small yields of hydroxide ion (ordinarily 10% or less). Soluble ionic hydroxides are strong bases since they dissolve in water and give a 100% yield of hydroxide ion.

The extent to which a base will form hydroxide ions in aqueous solution depends on the strength of the base relative to that of the hydroxide ion, as shown in the column of bases in Table 16.1. A strong base, such as one of those lying below hydroxide ion in the table, will accept a proton from water, forming the conjugate acid and hydroxide ion in 100% yield. Those bases lying between water and hydroxide ion in Table 16.1 will accept protons from water, but the result will be a mixture of the hydroxide ion and the base. Since stronger bases give higher concentrations of hydroxide ion, the relative strengths of these bases can be measured by the yield of hydroxide ion that each produces in aqueous solution. From the percent ionization data given below, the order of base strength for three of these bases is $NH_3 > CH_3CO_2^- > NO_2^-$.

$$NH_3 + H_2O \rightleftharpoons NH_4^+ + OH^- \qquad \text{(1.3\% in 0.10 M solution)}$$
$$CH_3CO_2^- + H_2O \rightleftharpoons CH_3CO_2H + OH^- \qquad \text{(0.0075\% in 0.10 M solution)}$$
$$NO_2^- + H_2O \rightleftharpoons HNO_2 + OH^- \qquad \text{(0.0015\% in 0.10 M solution)}$$

Other methods of measuring the strengths of Brønsted bases are discussed in Section 16.5.

Bases that are weaker bases than water (above water in the column of bases in Table 16.1) show no observable basic behavior in dilute aqueous solution.

According to the Brønsted-Lowry model, then, the strongest acids have the weakest conjugate bases, and the strongest bases have the weakest conjugate acids. Nitrous acid, HNO_2, a stronger acid than acetic acid, CH_3CO_2H, forms the nitrite ion, NO_2^-, as its conjugate base. The nitrite ion is a weaker base than the acetate ion, $CH_3CO_2^-$, the conjugate base of acetic acid. Table 16.1 illustrates the relationships between conjugate acid-base pairs. An acid will donate a proton to the conjugate base of any acid lying below it in the table. Thus, nitrous acid will react with acetate ion to form nitrite ion and acetic acid.

$$HNO_2 + CH_3CO_2^- \rightleftharpoons NO_2^- + CH_3CO_2H$$

16.4 Acid-Base Neutralization

An **acid-base neutralization** reaction occurs when stoichiometrically equivalent quantities of an acid and a base are mixed. This statement does not mean, however, that the resulting solution is always chemically neutral; it may contain either excess hydronium ions or excess hydroxide ions. The nature of the particular acid and base involved determines whether the resulting aqueous solution is acidic or basic. The following equations illustrate some reactions that take place when equivalent quantities of acids and bases are brought together in aqueous solution:

1. *A strong acid plus a strong base gives a neutral solution.*

$$H_3O^+(aq) + [Cl^-(aq)] + [Na^+(aq)] + OH^-(aq) \longrightarrow$$
$$2H_2O(l) + [Na^+(aq)] + [Cl^-(aq)]$$

Both acid and base are completely ionized. The reaction goes to completion, and water molecules, sodium ions, and chloride ions are the only products. The solution is neutral.

The sodium and chloride ions undergo no chemical change and appear as the crystalline salt sodium chloride if the solution is evaporated. Since the only change that occurs in the solution is the reaction of hydronium and hydroxide ions, the neutralization may be represented simply by the net equation

$$H_3O^+ + OH^- \longrightarrow 2H_2O$$

This equation may be used to represent the reaction of an aqueous solution of any strong acid with an aqueous solution of any strong base.

2. *A strong acid plus a weak base gives a weakly acidic solution.*

$$H_3O^+(aq) + [Cl^-(aq)] + NH_3(aq) \rightleftharpoons NH_4^+(aq) + [Cl^-(aq)] + H_2O(l)$$

Because ammonia is a moderately weak base, its conjugate acid (the ammonium ion) is a weak acid. A small fraction of the ammonium ions give up protons to water molecules, giving an acidic solution.

3. *A weak acid plus a strong base gives a weakly basic solution.*

$$CH_3CO_2H(aq) + [Na^+(aq)] + OH^-(aq) \rightleftharpoons$$
$$H_2O(l) + [Na^+(aq)] + CH_3CO_2^-(aq)$$

Because acetic acid is a moderately weak acid, its conjugate base, the acetate ion, is a moderately weak base. A small fraction of the acetate ions pick up protons from water, producing hydroxide ions and giving a basic solution.

4. *A weak acid plus a weak base.*

$$CH_3CO_2H(aq) + NH_3(aq) \rightleftharpoons NH_4^+(aq) + CH_3CO_2^-(aq)$$

This is the most complex of the four types of reactions and will be discussed in detail in Chapter 17. However, it is appropriate to note here that if the weak acid and the weak base have the same strengths, the solution will be neutral. Ammonia, for example, takes up about the same amount of protons to form ammonium ion as acetic acid gives up, and, although the reaction does not go to completion, the solution is neutral. If the weak acid and the weak base are of *unequal* strengths, the solution will be either acidic or basic, depending on the relative strengths of the two substances.

16.5 Measures of the Strengths of Brønsted Acids and Bases in Aqueous Solution

1. PERCENT IONIZATION. Examples of the use of percent ionization to determine the relative strengths of acids and bases were discussed in Section 16.3.

2. EQUILIBRIUM CONSTANT. As we have seen, Brønsted acids dissociate partially (weak acids) or completely (strong acids) in water to give hydrogen ion and the corresponding anion. For those acids that dissociate partially, an equilibrium constant can be written according to the law of mass action (Section 15.2). The following shows the equations (indicating hydrogen ion in its simplified form, H^+—see Section 16.1) and the equilibrium constant expressions for acetic acid and nitrous acid, respectively. The brackets in the equilibrium constant expressions indicate molar concentrations.

$$CH_3CO_2H \rightleftharpoons H^+ + CH_3CO_2^-$$
$$K = \frac{[H^+][CH_3CO_2^-]}{[CH_3CO_2H]} = 1.8 \times 10^{-5} \text{ mol L}^{-1}$$

$$HNO_2 \rightleftharpoons H^+ + NO_2^-$$

$$K = \frac{[H^+][NO_2^-]}{[HNO_2]} = 4.5 \times 10^{-4} \text{ mol L}^{-1}$$

(Values of equilibrium constants for weak acids are provided in Appendix F.)

The more a weak acid dissociates to form H^+ and the corresponding anion, the larger will be the concentrations of the ions, and the smaller will be the concentration of the undissociated acid at equilibrium. Inasmuch as the concentrations of the ions are in the numerator of the equilibrium constant expression and the concentration of the undissociated acid is in the denominator, the equilibrium constant of a stronger acid (more dissociated) will be larger than the equilibrium constant of a weaker acid (less dissociated).

Hence, since nitrous acid has a larger equilibrium constant than does acetic acid, nitrous acid is the stronger acid, in agreement with the order derived from percent ionizations in Section 16.3. [Note that the value of the equilibrium constant for strong acids (100% dissociated in water, 0% concentration of undissociated acid) is infinitely large and hence equal to infinity, ∞.]

Similarly, comparisons of strengths of weak Brønsted bases can be made from the equilibrium constant for the following reaction:

$$B + H_2O \longrightarrow BH^+ + OH^- \qquad K = \frac{[BH^+][OH^-]}{[B]}$$

Stronger bases have larger equilibrium constants.

3. CONCENTRATION OF HYDROGEN ION AND HYDROXIDE ION. If two acids are present in different solutions, but *in equal concentrations,* the stronger acid produces the greater quantity of hydrogen ion. Thus, the concentration of hydrogen ion is a measure of the relative strengths of the two acids. Similarly, a solution of a stronger base will contain a larger concentration of hydroxide ion than a solution of a weaker base if both solutions are prepared from the same molar amount of the respective bases.

4. pH AND pOH. As just noted, the concentration of hydrogen ion in a solution is a measure of its acidity or basicity. This concentration is often expressed in terms of the **pH of the solution, the negative logarithm of the hydrogen ion concentration.**

$$pH = -\log [H^+]$$

The pH value is the negative power to which 10 must be raised to give the hydrogen ion concentration.

$$[H^+] = 10^{-pH} \text{ M}$$

A neutral solution, one in which $[H^+] = [OH^-]$, has a pH of 7 at 25°C. The pH of an acidic solution ($[H^+] > [OH^-]$) is less than 7; the pH of a basic solution ($[OH^-] > [H^+]$) is greater than 7 (Fig. 16.2 on page 460). The pH of a solution can be calculated from its hydrogen ion concentration or measured with a pH meter (Fig. 17.1).

The concentration of hydroxide ion in a solution can be expressed in terms of the **pOH of the solution, the negative logarithm of the hydroxide ion concentration.**

$$pOH = -\log [OH^-]$$

A neutral solution has a pOH of 7 at 25°C, the pOH of an acidic solution is greater than 7, and the pOH of a basic solution is less than 7.

The relation of pH and pOH will be discussed in greater detail in Chapter 17.

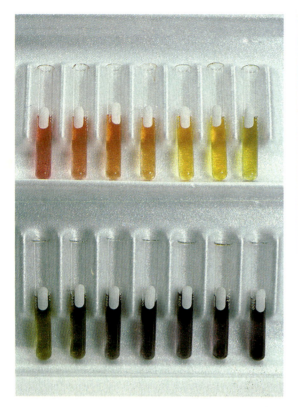

HYDROCHLORIC ACID   ACETIC ACID   AMMONIUM CHLORIDE   WATER   ANILINE   AMMONIUM HYDROXIDE   SODIUM HYDROXIDE

Figure 16.2

Universal indicator assumes a different color in solutions of different pH. Thus it can be added to a solution to determine the pH of the solution. The solutions in the 14 test tubes (left) show the gradations in color for pH 1 to 14, increasing in steps of one pH unit each. (above) 0.1 M solutions of the progressively weaker acids HCl (pH = 1), CH_3CO_2H (pH = 3), and NH_4Cl (pH = 5); pure water (a neutral substance, pH = 7); and 0.1 M solutions of the progressively stronger bases $C_6H_5NH_2$ (pH = 9), NH_3 (pH = 11), and NaOH (pH = 13).

16.6 The Relative Strengths of Strong Acids and Bases

The strongest acids, such as HCl, HBr, and HI, appear to have about the same strength in water. The water molecule is such a strong base compared to the conjugate bases Cl^-, Br^-, and I^- that ionization of the strong acids HCl, HBr, and HI is nearly complete in aqueous solutions. Hence in water these acids are all strong and appear to have equal strengths. In solvents less strongly basic than water, HCl, HBr, and HI differ markedly in their tendency to give up a proton to the solvent. In ethanol, a weaker base than water, the extent of ionization increases in the order HCl < HBr < HI, and it is evident that HI is the strongest of these acids. Water tends to equalize any differences in strength among strong acids; this is known as the **leveling effect** of water.

Water also exerts a leveling effect on the strengths of very strong bases. For example, the oxide ion, O^{2-}, and the amide ion, NH_2^-, are such strong bases that they react completely with water.

$$O^{2-} + H_2O \longrightarrow OH^- + OH^-$$

$$NH_2^- + H_2O \longrightarrow NH_3 + OH^-$$

Thus O^{2-} and NH_2^- appear to have the same base strength in water.

In binary compounds of hydrogen with nonmetals, the acid strength (the tendency to lose a proton) increases as the H—A bond strength decreases down a group in the Periodic Table. For Group VIIA the order of increasing acidity is HF < HCl < HBr < HI, in the absence of any leveling effect due to the solvent. Likewise, for Group VIA the order of increasing acid strength is $H_2O < H_2S < H_2Se < H_2Te$.

Across a row in the Periodic Table, the acid strength of binary hydrogen compounds increases with increasing electronegativity of the nonmetal atom as the polarity of the H—A bond increases. Thus the order of increasing acidity (for removal of one proton) across the second row is $CH_4 < NH_3 < H_2O < HF$, and across the third row is $SiH_4 < PH_3 < H_2S < HCl$.

Compounds containing oxygen and one or more hydroxide groups can be acidic, basic, or amphoteric depending on the position in the Periodic Table (and thus the electronegativity) of E, the element bonded to the oxygen and hydroxide group. Such compounds have the general formula $O_nE(OH)_m$ and include sulfuric acid, $O_2S(OH)_2$, sulfurous acid, $OS(OH)_2$, nitric acid, O_2NOH, perchloric acid, O_3ClOH, and aluminum hydroxide, $Al(OH)_3$. If the central atom, E, has a low electronegativity, its attraction for electrons is low, little tendency exists for it to form a strong covalent bond with the oxygen atom, and the bond between the element and oxygen is weaker than that between oxygen and hydrogen. Hence bond a is ionic, hydroxide ions are released to the solution, and the material behaves as a base. Large atomic size, small nuclear charge, and low oxidation number operate to lower the electronegativity of E and are characteristic of the more metallic elements.

$$-\overset{|}{\underset{|}{E}}-O-H$$

Bond b

Bond a

If, on the other hand, the element E has a relatively high electronegativity, it strongly attracts the electrons it shares with the oxygen atom, giving rise to a relatively strong bond between E and oxygen. The oxygen-hydrogen bond is thereby weakened, since electrons are displaced toward E. Bond b is polar and readily releases hydrogen ions to the solution, so the material behaves as an acid. Small atomic size, large nuclear charge, and high oxidation number operate to increase the electronegativity of E and are characteristic of the more nonmetallic elements. As the electronegativity of E increases, the O—H bond becomes weaker, and the acid strength increases.

Increasing the oxidation number of E also increases the acidity of an oxyacid, since this increases the attraction of E for the electrons it shares with oxygen and thereby weakens the O—H bond. Sulfuric acid, $O_2\overset{+6}{S}(OH)_2$, is more acidic than sulfurous acid, $O\overset{+4}{S}(OH)_2$; likewise nitric acid, $O_2\overset{+5}{N}OH$, is more acidic than nitrous acid, $O\overset{+3}{N}OH$. In each of these pairs the oxidation number of the central atom (indicated by the small superior number) is larger for the stronger acid.

The hydroxides of elements with intermediate electronegativities and relatively high oxidation numbers (for example, elements near the diagonal line separating the metals from the nonmetals in the Periodic Table) are usually **amphoteric.** This means that the hydroxides act as acids toward strong bases and as bases toward strong acids. The amphoterism of aluminum hydroxide is reflected in its solubility in both strong acids and strong bases. In strong bases the relatively insoluble hydrated aluminum hydroxide, $[Al(H_2O)_3(OH)_3]$, is converted to the soluble ion, $[Al(H_2O)_2(OH)_4]^-$, by reaction with hydroxide ion (Section 16.2). In strong acids it is converted in the first step to the soluble ion $[Al(H_2O)_4(OH)_2]^+$ by reaction with hydronium ion and in two additional steps to the soluble ion $[Al(H_2O)_6]^{3+}$. The net equation is

$$[Al(H_2O)_3(OH)_3] + 3H_3O^+ \rightleftharpoons [Al(H_2O)_6]^{3+} + 3H_2O$$

The different direction in the trend of acid strengths for the binary hydrogen halides arises from the fact that the halogen is attached directly to a hydrogen in the binary acid (instead of to an oxygen, which is in turn attached to a hydrogen, as in the oxyacids). In either kind of compound, as we go from a larger halogen to a smaller (i.e., from I to Br to Cl), a progressive shrinking of the electron shells takes place, including a pulling closer of the adjacent atom, whether the adjacent atom be hydrogen (as in the hydrogen halides) or oxygen (as in the oxyacids). As the adjacent atom is pulled closer, it is brought into a region of higher electron density. If you picture the valence electrons as being inside a smaller volume for the smaller halogen, you can visualize the higher electron density. If the adjacent atom being pulled closer is a hydrogen atom (as is the case with the hydrogen halides), it is held more tightly by the higher electron density and is pulled loose with greater difficulty. Conversely, the hydrogen atom is in a region of lower electron density when attached to a larger halogen, so it is attracted less tightly and is pulled away more easily.

16.7 Properties of Brønsted Acids in Aqueous Solution

Since aqueous solutions of acids contain higher concentrations of hydronium ions than pure water, these solutions all exhibit the following properties, which are due to the presence of the hydronium ion:

1. They have a sour taste.

2. They change the color of certain indicators. For example, they change litmus from blue to red and phenolphthalein from red to colorless.

3. They have a pH which is less than 7 (Fig. 16.2).

4. They react with metals above hydrogen in the activity series (Section 9.7) and liberate hydrogen; for example,

$$2H_3O^+(aq) + Zn(s) \longrightarrow Zn^{2+} + H_2(g) + 2H_2O(l)$$

5. They react with many basic metal oxides and hydroxides forming salts and water.

$$2H_3O^+ + FeO \longrightarrow Fe^{2+} + 3H_2O$$

$$2H_3O^+ + Fe(OH)_2 \longrightarrow Fe^{2+} + 4H_2O$$

6. They react with the salts of either weaker or more volatile acids, such as carbonates or sulfides, to give a new salt and a new acid.

$$2H_3O^+ + CaCO_3 \longrightarrow H_2CO_3 + Ca^{2+} + 2H_2O$$
$$\longrightarrow H_2O + CO_2(g)$$

$$2H_3O^+ + FeS \longrightarrow H_2S(g) + Fe^{2+} + 2H_2O$$

16.8 Preparation of Brønsted Acids

Brønsted acids may be prepared by any of the following methods:

1. *By the direct union of the constituent elements.* Since binary compounds of hydrogen with the more electronegative nonmetals are acids (Section 16.6), the direct reaction of hydrogen with such nonmetals as F_2, Cl_2, Br_2, and S_8 yields an acid. For example,

$$H_2 + Br_2 \longrightarrow 2HBr$$
$$8H_2 + S_8 \overset{\Delta}{\rightleftharpoons} 8H_2S$$

2. *By the reaction of water with an oxide of a nonmetal.* Most oxides of nonmetals are acidic (Section 9.4). The action of water on such a nonmetal oxide forms an acid.

$$SO_3 + H_2O \longrightarrow H_2SO_4$$
$$P_4O_{10} + 6H_2O \longrightarrow 4H_3PO_4$$
$$CO_2 + H_2O \rightleftharpoons H_2CO_3$$

3. *By the metathetical reaction (Section 9.3):*

a. *of a salt of a volatile acid with a nonvolatile or slightly volatile acid.*

$$NaF(s) + H_2SO_4(l) \longrightarrow NaHSO_4(s) + HF(g)$$

b. *of a salt with an acid to produce a second acid and an insoluble precipitate.*

$$Ca_3(PO_4)_2(s) + 3H_2SO_4(l) \longrightarrow 2H_3PO_4(l) + 3CaSO_4(s)$$

c. *of a salt of a weak acid with a strong acid.*

$$[Na^+(aq)] + CH_3CO_2^-(aq) + H_3O^+(aq) + [Cl^-(aq)] \longrightarrow$$
$$CH_3CO_2H(aq) + [Na^+(aq)] + [Cl^-(aq)] + H_2O$$

4. *By reaction of water with certain nonmetal halides containing polar bonds.*

$$PBr_3 + 3H_2O \longrightarrow H_3PO_3 + 3HBr$$
$$PCl_5 + 4H_2O \longrightarrow H_3PO_4 + 5HCl$$
$$SiI_4 + 4H_2O \longrightarrow Si(OH)_4 + 4HI$$

16.9 Polyprotic Acids

Acids can be classified by the number of protons per molecule that can be given up in a reaction. Acids such as HCl, HNO_3, and HCN that contain one ionizable hydrogen atom in each molecule are called **monoprotic acids.** Their reactions with water are as follows:

$$HCl + H_2O \longrightarrow H_3O^+ + Cl^-$$
$$HNO_3 + H_2O \longrightarrow H_3O^+ + NO_3^-$$
$$HCN + H_2O \rightleftharpoons H_3O^+ + CN^-$$

Acetic acid, CH_3CO_2H, is monoprotic because only one of the four hydrogen atoms in each molecule is given up as a proton in reactions with bases.

$$\text{H}-\overset{\displaystyle H}{\underset{\displaystyle H}{\overset{\displaystyle |}{\underset{\displaystyle |}{C}}}}-CO_2H + H_2O \rightleftharpoons H_3O^+ + \text{H}-\overset{\displaystyle H}{\underset{\displaystyle H}{\overset{\displaystyle |}{\underset{\displaystyle |}{C}}}}-CO_2^-$$

Diprotic acids contain two ionizable hydrogen atoms per molecule; ionization of such acids occurs in two stages. The primary ionization always takes place to a greater extent than the secondary ionization. For example, carbonic acid ionizes as follows:

$$H_2CO_3 + H_2O \rightleftharpoons H_3O^+ + HCO_3^- \qquad \text{(primary ionization)}$$
$$HCO_3^- + H_2O \rightleftharpoons H_3O^+ + CO_3^{2-} \qquad \text{(secondary ionization)}$$

Triprotic acids, such as phosphoric acid, ionize in three steps:

$$H_3PO_4 + H_2O \rightleftharpoons H_3O^+ + H_2PO_4^- \qquad \text{(primary ionization)}$$
$$H_2PO_4^- + H_2O \rightleftharpoons H_3O^+ + HPO_4^{2-} \qquad \text{(secondary ionization)}$$
$$HPO_4^{2-} + H_2O \rightleftharpoons H_3O^+ + PO_4^{3-} \qquad \text{(tertiary ionization)}$$

16.10 Brønsted Bases in Aqueous Solution

When a Brønsted base stronger than water dissolves in water, a solution containing a higher concentration of hydroxide ions than that found in pure water is formed. Strong bases give a 100% yield of hydroxide ion, whereas weak bases give only a small yield of hydroxide ion.

The following properties are common to Brønsted bases in aqueous solution and are due to the presence of the hydroxide ion:

1. They have a bitter taste.
2. They change the colors of certain indicators. For example, they change litmus from red to blue, and phenolphthalein from colorless to red.
3. They have a pH which is greater than 7 (Fig. 16.2) and a pOH less than 7.
4. They neutralize aqueous acids, forming water and a solution of a salt.

$$[M^+] + OH^- + H_3O^+ + [A^-] \longrightarrow [M^+] + [A^-] + 2H_2O$$

Some weak Brønsted bases also form enough hydroxide ion to have these properties.

16.11 Preparation of Hydroxide Bases

Bases containing hydroxide ion (hydroxide bases) may be prepared by the following methods:

1. *By the reaction of active metals with water*. The active metals that lie above magnesium in the activity series (Section 9.7) react directly with a stoichiometric amount of water to give strong bases.

$$2K + 2H_2O \longrightarrow 2KOH + H_2$$
$$Ca + 2H_2O \longrightarrow Ca(OH)_2 + H_2$$

If an excess of water is present, aqueous solutions of the bases are formed.

2. *By the reaction of oxides of active metals with water*. The oxides of active metals react with water to give bases (Sections 9.4 and 13.4).

$$Li_2O + H_2O \longrightarrow 2LiOH$$
$$SrO + H_2O \longrightarrow Sr(OH)_2$$

If an excess of water is present, aqueous solutions of the bases may form.

3. *By the metathetical reaction of a salt with a base to give a solution of a second base and an insoluble precipitate*.

$$[Ca^{2+}] + SO_4^{2-} + Ba^{2+} + [2OH^-] \longrightarrow BaSO_4(s) + [Ca^{2+}] + [2OH^-]$$

4. *By electrolysis of certain salt solutions.* (See Section 13.5)

$$[2Na^+] + 2Cl^- + 2H_2O \xrightarrow{\text{Electrolysis}} [2Na^+] + 2OH^- + H_2(g) + Cl_2(g)$$

16.12 Salts

The products of a neutralization reaction in aqueous solution are water and a salt. As described in Section 9.2, a salt is an ionic compound. Among the cations most frequently found in salts are simple metal ions such as Na^+, K^+, Ca^{2+}, and Ba^{2+}. Hydrated metal ions such as $[Al(H_2O)_6]^{3+}$ and more complex ions such as NH_4^+ (ammonium ion) and $(CH_3)_3NH^+$ (trimethylammonium ion) are also found. The simple anions of salts include F^-, Cl^-, Br^-, and I^-. Many salts contain polyatomic negative ions such as NO_3^-, SO_4^{2-}, PO_4^{3-}, and CN^-.

Salts can be formed by other reactions as well. For example, Brønsted acid-base reactions, metathetical reactions, and the direct reactions between metals and non-metals can all produce salts.

A salt that contains neither a replaceable hydrogen nor a hydroxide group is sometimes called a **normal salt;** examples are NaCl, K_2SO_4, and $Ca_3(PO_4)_2$. When only part of the acidic hydrogens of a polyprotic acid are replaced by a cation, the compound is known as a **hydrogen salt;** examples are sodium hydrogen carbonate, $NaHCO_3$ (often called sodium bicarbonate, or baking soda), and sodium dihydrogen phosphate, NaH_2PO_4.

When only part of the hydroxide groups of an ionic metal hydroxide are replaced by another anion, the compound is known as a **hydroxysalt;** examples are barium hydroxychloride, Ba(OH)Cl, and bismuth dihydroxychloride, $Bi(OH)_2Cl$. Some hydroxysalts readily lose the elements of one or more molecules of water and form **oxysalts.** For example, bismuth dihydroxychloride breaks down to give bismuth oxy-chloride, BiOCl.

$$Bi(OH)_2Cl \longrightarrow BiOCl + H_2O$$

16.13 Quantitative Reactions of Acids and Bases

Using chemical reactions to determine the concentration of a solution (to **standardize** a solution) or to determine the amount of a substance present in a sample using a **standard solution** (a solution of known concentration) makes up the branch of quantitative analysis called **volumetric analysis.** Since the end point of a titration (Section 3.2) involving a neutralization reaction can be detected readily by use of an indicator, solutions of acids and bases are often used in volumetric analyses.

Titration of a 40.00-mL sample of a solution of H_3PO_4 requires 35.00 mL of 0.1500 M KOH to reach the end point. Determine the molar concentration of H_3PO_4 if the reaction is

EXAMPLE 16.1

$$2KOH + H_3PO_4 \longrightarrow K_2HPO_4 + 2H_2O$$

35 ml   40 ml
.1500 M

The following steps are required for this problem.

$$\boxed{\text{Volume of KOH}} \longrightarrow \boxed{\text{Moles of KOH}} \longrightarrow \boxed{\text{Moles of } H_3PO_4} \longrightarrow \boxed{\text{Molarity of } H_3PO_4}$$

$$0.03500 \text{ L KOH} \times \frac{0.1500 \text{ mol KOH}}{1 \text{ L KOH}} = 5.250 \times 10^{-3} \text{ mol KOH}$$

$$5.250 \times 10^{-3} \text{ mol KOH} \times \frac{1 \text{ mol } H_3PO_4}{2 \text{ mol KOH}} = 2.625 \times 10^{-3} \text{ mol } H_3PO_4$$

$$\frac{2.625 \times 10^{-3} \text{ mol } H_3PO_4}{0.04000 \text{ L}} = 0.06562 \text{ M}$$

EXAMPLE 16.2   You may have used crystalline potassium hydrogen phthalate, $KHC_8H_4O_4$, as a standard acid in the laboratory because it is easy to purify and weigh. If you titrate 1.5024 g of this acid with 37.28 mL of NaOH solution, what is the concentration of the NaOH solution?

$$KHC_8H_4O_4 + NaOH \longrightarrow KNaC_8H_4O_4 + H_2O$$

You can determine concentration of NaOH from the number of moles of NaOH in 37.28 mL (0.03728 L) of the solution. To determine the concentration requires the following steps.

$$\boxed{\text{Mass of } KHC_8H_4O_4} \longrightarrow \boxed{\text{Moles of } KHC_8H_4O_4} \longrightarrow \boxed{\text{Moles of NaOH}} \longrightarrow \boxed{\text{Molarity of NaOH}}$$

$$1.5024 \text{ g } KHC_8H_4O_4 \times \frac{1 \text{ mol } KHC_8H_4O_4}{204.223 \text{ g } KHC_8H_4O_4}$$
$$= 7.3567 \times 10^{-3} \text{ mol } KHC_8H_4O_4$$

$$7.3567 \times 10^{-3} \text{ mol } KHC_8H_4O_4 \times \frac{1 \text{ mol NaOH}}{1 \text{ mol } KHC_8H_4O_4}$$
$$= 7.3567 \times 10^{-3} \text{ mol NaOH}$$

$$\frac{7.3567 \times 10^{-3} \text{ mol NaOH}}{0.03728 \text{ L}} = 0.1973 \text{ M}$$

We can solve titration problems using mole relationships, as described here and in Section 3.2. An alternate method, which uses equivalents, is described in the next section.

16.14 Equivalents of Acids and Bases

An **equivalent of a Brønsted acid** is the mass of the acid in grams that will provide 1 mole of protons (H^+) in a reaction. An **equivalent of a base** is the mass of the base in grams that will provide 1 mole of hydroxide ions in a reaction or will react with 1 mole of protons. An equivalent is sometimes called a **gram-equivalent weight.** In an acid-base reaction 1 equivalent of an acid will react with 1 equivalent of a base.

In an acid-base reaction protons (H^+) are transferred from the acid to the base, and the number of protons transferred is used in the calculation of the mass of an equivalent of the acid and the base. When hydrochloric acid reacts with sodium hydroxide, a proton is transferred to the hydroxide ion.

$$HCl + NaOH \longrightarrow NaCl + H_2O$$

When 1 mole of hydrochloric acid reacts with 1 mole of sodium hydroxide, 1 mole of protons (6.022×10^{23} protons) is transferred to 1 mole of hydroxide ions (6.022×10^{23} hydroxide ions). It follows, then, that 1 equivalent (equiv) of hydrochloric acid is the same as 1 mole (36.5 g) of HCl (1 equiv HCl/1 mol HCl), and 1 equivalent of sodium hydroxide is the same as 1 mole (40.0 g) of NaOH (1 equiv NaOH/1 mol NaOH). However, 1 mole of sulfuric acid can react with 2 moles of sodium hydroxide, transferring 2 moles of protons from the acid to the base.

$$H_2SO_4 + 2NaOH \longrightarrow Na_2SO_4 + 2H_2O$$

Hence 1 equivalent of sulfuric acid is equal to $\frac{1}{2}$ mole of the acid (2 equiv H_2SO_4/1 mol H_2SO_4).

Do not forget that the number of protons *actually transferred* in an acid-base reaction determines the mass of an equivalent of the acid or the base. For example, phosphoric acid may react with potassium hydroxide in any one of three ways:

$$H_3PO_4 + KOH \longrightarrow KH_2PO_4 + H_2O \qquad (1)$$
$$H_3PO_4 + 2KOH \longrightarrow K_2HPO_4 + 2H_2O \qquad (2)$$
$$H_3PO_4 + 3KOH \longrightarrow K_3PO_4 + 3H_2O \qquad (3)$$

The numbers of moles of protons transferred, per mole of H_3PO_4, in the three reactions are 1, 2, and 3 moles, respectively. Thus an equivalent of H_3PO_4 in the three cases is equal to 1 mole, $\frac{1}{2}$ mole, and $\frac{1}{3}$ mole, respectively. *It must be emphasized that the mass of an equivalent of an acid or a base must be deduced from the reaction, not merely from the formula of the substance.*

Calculate the mass of an equivalent of H_3PO_4 in Equations (1), (2), and (3).   **EXAMPLE 16.3**

The molecular weight of H_3PO_4 is 98.00. In Equation (1), 1 mol of H_3PO_4 provides 1 mol of H^+ (1 equiv H_3PO_4/1 mol H_3PO_4).

$$\frac{98.00 \text{ g } H_3PO_4}{1 \text{ mol } H_3PO_4} \times \frac{1 \text{ mol } H_3PO_4}{1 \text{ equiv } H_3PO_4} = \frac{98.00 \text{ g } H_3PO_4}{1 \text{ equiv } H_3PO_4}$$

Thus the mass of an equivalent of H_3PO_4 in Equation (1) is 98.00 g.

In Equation (2), 1 mol of H_3PO_4 provides 2 mol of H^+ (2 equiv H_3PO_4/1 mol H_3PO_4).

$$\frac{98.00 \text{ g } H_3PO_4}{1 \text{ mol } H_3PO_4} \times \frac{1 \text{ mol } H_3PO_4}{2 \text{ equiv } H_3PO_4} = \frac{49.00 \text{ g } H_3PO_4}{1 \text{ equiv } H_3PO_4}$$

The mass of an equivalent of H_3PO_4 in Equation (2), therefore, is 49.00 g.

In Equation (3), 1 mol of H_3PO_4 provides 3 mol of H^+ (3 equiv H_3PO_4/1 mol H_3PO_4).

$$\frac{98.00 \text{ g } H_3PO_4}{1 \text{ mol } H_3PO_4} \times \frac{1 \text{ mol } H_3PO_4}{3 \text{ equiv } H_3PO_4} = \frac{32.67 \text{ g } H_3PO_4}{1 \text{ equiv } H_3PO_4}$$

Hence the mass of an equivalent of H_3PO_4 in Equation (3) is 32.67 g.

The **normality, N,** of a solution is the number of equivalents of solute per liter of solution. (Compare this with the definition of molarity in Section 3.1.)

$$\text{Normality} = \frac{\text{equivalents of solute}}{\text{liters of solution}} = \frac{\text{equiv}}{\text{L}}$$

Normality may also be determined from the molarity of the solution, provided that the reaction is known.

$$\text{Normality} = \frac{\text{moles of solute}}{\text{liters of solution}} \times \frac{\text{equivalents of solute}}{\text{moles of solute}} = \frac{\text{equiv}}{\text{L}}$$

A solution that contains 1 equivalent of solute in 1 liter of solution is a 1 N (one normal) solution. Provided that both hydrogens react, a 1 N solution of sulfuric acid contains 1 equivalent, or 98 g/2 = 49 g, of H_2SO_4 per liter; a 2 N solution of H_2SO_4 contains 2 equivalents, or 98 g, per liter; and a 0.01 N solution contains 0.49 g per liter. A 1 N solution of sulfuric acid, if both hydrogens react, is identical to a 0.5 M solution of sulfuric acid; each contains 49 g of solute per liter of solution.

It should be evident that a 1 M solution of hydrochloric acid is also 1 N, since 1 mole of HCl is equal to 1 equivalent of HCl. However, a 1 M solution of sulfuric acid (98 g/L) is 2 N, since 1 mole of H_2SO_4 is equal to 2 equivalents of H_2SO_4, if both hydrogens react.

EXAMPLE 16.4   Calculate the normality of a solution containing 3.65 g of HCl per 0.500 L of solution.

The mass of an equivalent of HCl is 36.5 g.

$$3.65 \text{ g HCl} \times \frac{1 \text{ equiv HCl}}{36.5 \text{ g HCl}} = 0.100 \text{ equiv HCl}$$

$$\frac{0.100 \text{ equiv}}{0.500 \text{ L}} = 0.200 \text{ N HCl}$$

EXAMPLE 16.5   What is the normality of a solution of $Ba(OH)_2$ that contains 0.01713 g of $Ba(OH)_2$ per 0.1000 L of solution?

$$Ba(OH)_2 + 2HCl \longrightarrow BaCl_2 + 2H_2O$$

Since both of the hydroxide ions react, 1 mol of $Ba(OH)_2$ provides 2 mol of OH^-. Hence there is 1 mol of $Ba(OH)_2$ per 2 equiv of $Ba(OH)_2$. This problem requires the following steps.

Mass of $Ba(OH)_2$	→	Moles of $Ba(OH)_2$	→	Equivalents of $Ba(OH)_2$	→	Normality of solution

$$0.01713 \text{ g Ba(OH)}_2 \times \frac{1 \text{ mol Ba(OH)}_2}{171.34 \text{ g Ba(OH)}_2} = 1.000 \times 10^{-4} \text{ mol Ba(OH)}_2$$

$$1.000 \times 10^{-4} \text{ mol Ba(OH)}_2 \times \frac{2 \text{ equiv Ba(OH)}_2}{1 \text{ mol Ba(OH)}_2} = 2.000 \times 10^{-4} \text{ equiv Ba(OH)}_2$$

$$\frac{2.000 \times 10^{-4} \text{ equiv Ba(OH)}_2}{0.1000 \text{ L}} = 2.000 \times 10^{-3} \text{ N}$$

How many grams of H_2SO_4 are contained in 1.2 L of 0.50 N H_2SO_4 solution that reacts according to this equation?

EXAMPLE 16.6

$$H_2SO_4 + Ba(OH)_2 \longrightarrow BaSO_4 + 2H_2O$$

This problem may be solved by the following steps.

| Volume of solution | → | Equivalents of H_2SO_4 | → | Moles of H_2SO_4 | → | Mass of H_2SO_4 |

A 0.50 N H_2SO_4 solution contains 0.50 equiv/L.

$$1.2 \; L \times \frac{0.50 \text{ equiv } H_2SO_4}{1.0 \; L} = 0.60 \text{ equiv } H_2SO_4$$

Since each mole of H_2SO_4 provides two protons in the acid-base reaction, 1 mol of H_2SO_4 is equal to 2 equiv of H_2SO_4.

$$0.60 \text{ equiv } H_2SO_4 \times \frac{1 \text{ mol } H_2SO_4}{2 \text{ equiv } H_2SO_4} = 0.30 \text{ mol } H_2SO_4$$

$$0.30 \text{ mol } H_2SO_4 \times \frac{98.1 \text{ g } H_2SO_4}{1 \text{ mol } H_2SO_4} = 29 \text{ g } H_2SO_4$$

Calculate the normality of a solution containing 2.35 g of H_3PO_4 per 0.600 L of solution, for a reaction in which two of the hydrogens react.

EXAMPLE 16.7

The problem may be solved by the following steps.

| Mass of H_3PO_4 | → | Moles of H_3PO_4 | → | Equivalents of H_3PO_4 | → | Normality of H_3PO_4 |

(In the following solution, Steps 1 and 2 are combined in one overall calculation.) The molecular weight of H_3PO_4 is 98.00.

$$2.35 \text{ g } H_3PO_4 \times \frac{1 \text{ mol } H_3PO_4}{98.00 \text{ g } H_3PO_4} \times \frac{2 \text{ equiv } H_3PO_4}{1 \text{ mol } H_3PO_4} = 4.80 \times 10^{-2} \text{ equiv } H_3PO_4$$

$$\frac{4.80 \times 10^{-2} \text{ equiv}}{0.600 \text{ L}} = 8.00 \times 10^{-2} \text{ N } H_3PO_4$$

The Lewis Concept of Acids and Bases

16.15 Definitions and Examples

In 1923, G. N. Lewis proposed a generalized model of acid-base behavior in which acids and bases are not restricted to proton donors and proton acceptors. **According to the Lewis concept, an acid is any species (molecule or ion) that can accept a pair of electrons, and a base is any species (molecule or ion) that can donate a pair of electrons.** An acid-base reaction occurs when a base donates a pair of electrons to an acid. An acid-base adduct, a compound that contains a coordinate covalent bond between the Lewis acid and the Lewis base, is formed. The following equations show the general application of the Lewis model.

The boron atom in boron trifluoride, BF_3, has only six electrons in its valence shell. Consequently, BF_3 is a very good **Lewis acid** and reacts with many Lewis bases; fluoride ion is the **Lewis base** in this reaction:

$$\left[\ddot{:}\ddot{F}\ddot{:}\right]^- + \begin{matrix} :\ddot{F}: \\ B:\ddot{F}: \\ :\ddot{F}: \end{matrix} \longrightarrow \left[\begin{matrix} :\ddot{F}: \\ :\ddot{F}:B:\ddot{F}: \\ :\ddot{F}: \end{matrix}\right]^-$$

| Base | Acid | Acid-base adduct |

In the following reaction two ammonia molecules, Lewis bases, each donate a pair of electrons to a silver ion, the Lewis acid.

$$2H:\overset{H}{\underset{H}{\ddot{N}}}: + Ag^+ \longrightarrow \left[H:\overset{H}{\underset{H}{\ddot{N}}}:Ag:\overset{H}{\underset{H}{\ddot{N}}}:H\right]^+$$

| Base | Acid | Acid-base adduct |

$$2NH_3 + Ag^+ \longrightarrow [Ag(NH_3)_2]^+$$

An analogous reaction, in which four ammonia molecules serve as Lewis bases, each donating a pair of electrons to a copper ion that serves as the Lewis acid, is shown in Fig. 16.3.

$$4NH_3 + Cu^{2+} \longrightarrow [Cu(NH_3)_4]^{2+}$$

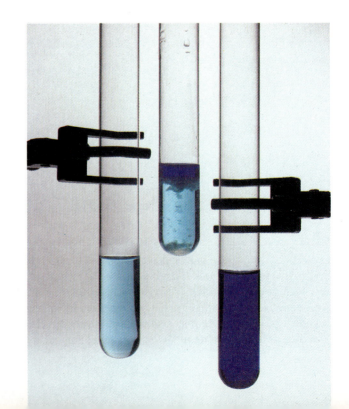

Figure 16.3

A photograph showing the reaction of $NH_3(aq)$ in small quantity and in excess with copper ion in a 0.1 M solution of copper sulfate. The first test tube contains 0.1 M copper sulfate. The second test tube shows the precipitation of blue copper hydroxide by reaction of copper ion with hydroxide ion when a small quantity of $NH_3(aq)$ is added.

$$Cu^{2+}(aq) + 2NH_3(aq) + 2H_2O(l) \rightleftharpoons Cu(OH)_2(s) + 2NH_4^+$$

The third test tube shows the formation of the dark blue solution of $[Cu(NH_3)_4]^{2+}$ by the reaction of copper ion (a Lewis acid) with NH_3 (a Lewis base) when an excess of $NH_3(aq)$ is added. The precipitate of copper hydroxide is dissolved by a shift of the equilibrium of copper hydroxide and its ions, due to removal of copper ions through their reaction with ammonia to form $[Cu(NH_3)_4]^{2+}$

$$Cu(OH)_2(s) \rightleftharpoons Cu^{2+}(aq) + 2OH^-(aq)$$

$$Cu^{2+}(aq) + 4NH_3(aq) \rightleftharpoons [Cu(NH_3)_4]^{2+}(aq)$$

Nonmetal oxides act as Lewis acids and react with oxide ions, the Lewis base, to form oxyanions.

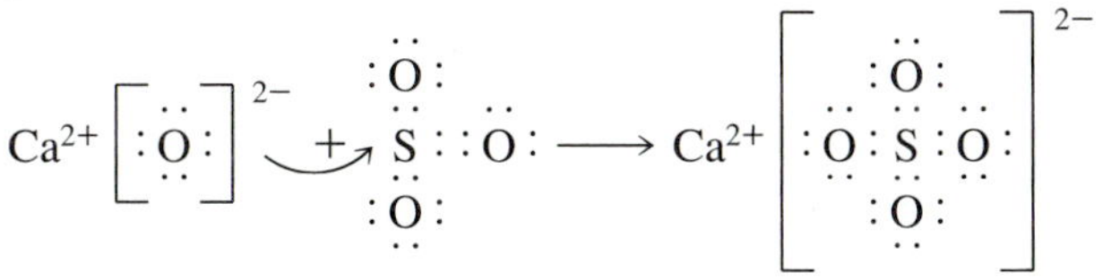

$$Ca^{2+} \left[:\overset{..}{\underset{..}{O}}: \right]^{2-} + \overset{:\overset{..}{O}:}{\underset{:\overset{..}{O}:}{S}}::\overset{..}{O}: \longrightarrow Ca^{2+} \left[\overset{:\overset{..}{O}:}{\underset{:\overset{..}{O}:}{:\overset{..}{O}:S:\overset{..}{O}:}} \right]^{2-}$$

Base   Acid   Acid-base adduct

Many Lewis acid-base reactions are displacement reactions in which one Lewis base displaces another Lewis base from an acid-base adduct or in which one Lewis acid displaces another Lewis acid.

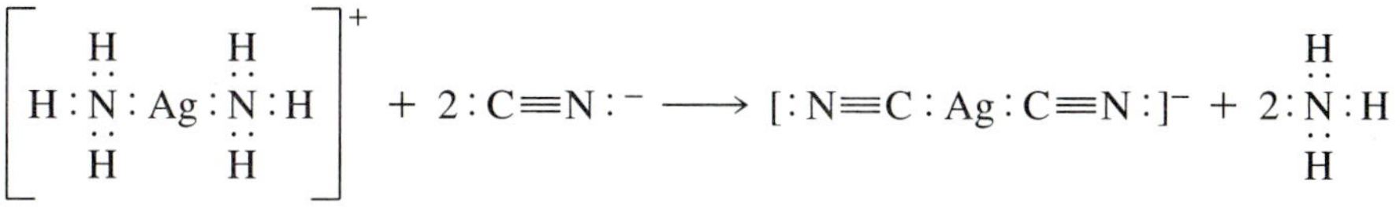

$$\left[\begin{matrix} H & & H \\ H:\overset{..}{N}:Ag:\overset{..}{N}:H \\ H & & H \end{matrix} \right]^{+} + 2:C\equiv N:^{-} \longrightarrow [:N\equiv C:Ag:C\equiv N:]^{-} + 2:\overset{H}{\underset{H}{N}}:H$$

Acid-base adduct   Base   New adduct   New base

$$Ca^{2+} \left[\overset{:\overset{..}{O}:}{\underset{:\overset{..}{O}:}{C}}::\overset{..}{O}: \right]^{2-} + SO_3 \longrightarrow Ca^{2+} \left[\overset{:\overset{..}{O}:}{\underset{:\overset{..}{O}:}{:\overset{..}{O}:S:\overset{..}{O}:}} \right]^{2-} + CO_2$$

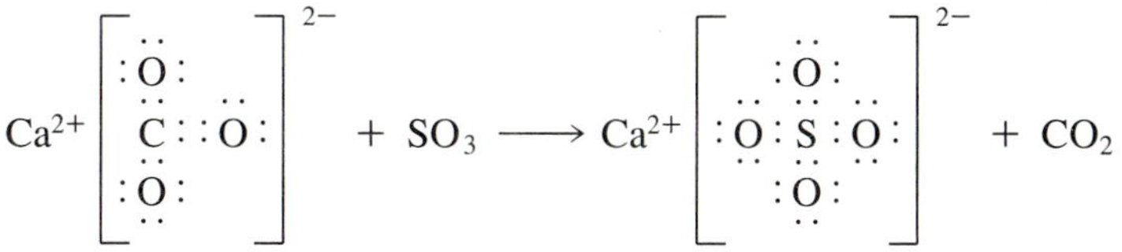

Acid-base adduct   Acid   New adduct   New acid

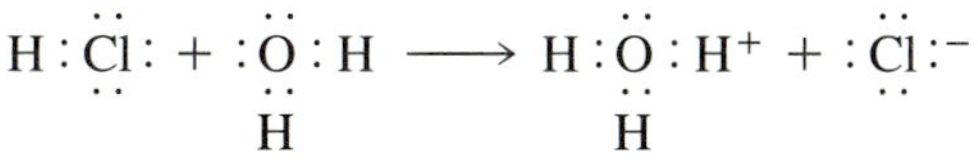

$$H:\overset{..}{\underset{..}{Cl}}: + :\overset{..}{\underset{H}{O}}:H \longrightarrow H:\overset{..}{\underset{H}{O}}:H^{+} + :\overset{..}{\underset{..}{Cl}}:^{-}$$

Acid-base adduct   Base   New adduct   New base

The last displacement reaction shows how the reaction of a Brønsted acid with a base fits into the Lewis model. A Brønsted acid such as HCl is an acid-base adduct according to the Lewis concept, and proton transfer occurs because a more stable acid-base adduct is formed. Thus, although the basis for defining acids and bases in the two theories is quite different, the theories overlap considerably.

For Review

Summary

A compound that can donate a proton (a hydrogen ion) to another compound is called a **Brønsted acid.** The compound that accepts the proton is called a **Brønsted base.** The species remaining after a Brønsted acid has lost a proton is the **conjugate base** of the acid. When a Brønsted base gains a proton, its **conjugate acid** forms. Thus an **acid-base reaction** occurs when a proton is transferred from a reactant acid to a reactant base, with formation of the conjugate base of the reactant acid and the conjugate acid of the reactant base. **Amphiprotic species** can act as both a proton donor and a proton acceptor. Water is the most important amphiprotic species. It can form both the hydronium ion, H_3O^+, and the hydroxide ion, OH^-.

Strong acids form weak conjugate bases, and weak acids form strong conjugate bases. Thus **strong acids** are completely ionized in aqueous solution because their conjugate bases are weaker bases than water. **Weak acids** are only partially ionized because their conjugate bases are strong enough to compete successfully with water for possession of protons. **Strong bases** react with water to give a 100% yield of hydroxide ion. **Weak bases** give only small yields of hydroxide ion. The strengths of the binary acids increase from left to right across a period of the Periodic Table ($CH_4 <$ $NH_3 < H_2O < HF$), and they increase down a group of the Table ($HF < HCl <$ $HBr < HI$). The strengths of oxyacids containing the same central element increase as the oxidation number of the element increases ($H_2SO_3 < H_2SO_4$). The strengths of oxyacids also increases as the electronegativity of the central element increases [$H_2SeO_4 < H_2SO_4$; $Al(OH)_3 < Si(OH)_4 < H_3PO_4 < H_2SO_4 < HClO_4$].

The characteristic properties of aqueous solutions of Brønsted acids are due to the presence of hydronium ions; those of aqueous solutions of Brønsted bases are due to the presence of hydroxide ions. The **neutralization** that occurs when aqueous solutions of acids and bases are mixed results from the reaction of the hydronium and hydroxide ions to form water. **Salts** are also formed in neutralization reactions.

The concentration of an acid or a base in solution may be described in terms of the number of moles per liter (**molarity**) or of the number of equivalents per liter (**normality**). An **equivalent,** or gram-equivalent weight, of an acid is the mass of the acid in grams that will provide 1 mole of protons in a reaction. An equivalent of a base is the mass of the base in grams that will provide 1 mole of hydroxide ions or will accept 1 mole of protons.

The strengths of Brønsted acids and bases in aqueous solutions can be studied in various ways, including measurements of percent ionization, comparison of equilibrium constants, measurements of hydrogen ion and hydroxide ion concentrations, and measurement of pH and pOH. The **pH** is the negative logarithm of the hydrogen ion concentration; the **pOH** is the negative logarithm of the hydroxide ion concentration.

A **Lewis acid** is a molecule or ion that can accept a pair of electrons. A **Lewis base** is the molecule or ion that donates the pair of electrons. Thus a Lewis acid-base reaction results in the formation of a coordinate covalent bond.

Key Terms and Concepts

amphiprotic (16.2)	hydroxysalt (16.12)	pOH (16.5)
amphoteric (16.6)	Lewis acid (16.15)	polyprotic acid (16.9)
Brønsted acid (16.1)	Lewis base (16.15)	salt (16.12)
Brønsted base (16.1)	monoprotic acid (16.9)	strong acid (16.3)
conjugate acid (16.1)	neutralization (16.4)	strong base (16.3)
conjugate base (16.1)	normal salt (16.12)	triprotic acid (16.9)
diprotic acid (16.9)	normality (16.14)	weak acid (16.3)
equivalent (16.14)	oxysalt (16.12)	weak base (16.3)
hydrogen salt (16.12)	pH (16.5)	

Exercises

Brønsted Acids and Bases

1. According to the Brønsted-Lowry concept, what is an acid? A base?
2. Show by suitable equations that each of the following species can act as a Brønsted acid:

   (a) H_3O^+ (b) HCl   (c) H_2O
   (d) CH_3CO_2H (e) NH_4^+ (f) HSO_4^-

3. Show by suitable equations that each of the following species can act as a Brønsted base:

   (a) H_2O (b) OH^- (c) NH_3
   (d) CN^- (e) S^{2-} (f) $H_2PO_4^-$

4. Write equations for the reaction of each of the following with water:
   (a) HCl   (b) HNO_3 (c) NH_3
   (d) NH_2^- (e) $HClO_4$ (f) F^-
   (g) NH_4^+
   What is the role played by water in each of these acid-base reactions?

5. (a) Identify the strong Brønsted acids and strong Brønsted bases in the list of important industrial inorganic compounds in Table 9.8.
   (b) List those compounds in Table 9.8 that can behave as Brønsted acids with strengths lying between those of H_3O^+ and H_2O.
   (c) List those compounds in Table 9.8 that can behave as Brønsted bases with strengths lying between those of H_2O and OH^-.

6. What is the conjugate acid formed when each of the following reacts as a base?
   (a) OH^- H_2O (b) F^- HF (c) H_2O H_3O^+
   (d) HSO_4^- H_2SO_4 (e) HCO_3^- H_2CO_3 (f) NH_3 NH_4^+
   (g) NH_2^- NH_3 (h) H^- H_2 (i) N^{3-} HN^{2-}
   Is each a strong or weak acid?

7. What is the conjugate base formed when each of the following reacts as an acid?
   (a) OH^- O^{2-} (b) H_2O H_3O^- (c) HCO_3^- CO_3^{2-}
   (d) HBr Br^- (e) HSO_4^- SO_4^{2-} (f) NH_3 NH_2^-
   (g) HS^- S^{2-} (h) PH_3 PH_2^- (i) H_2O_2 HO_2^-

8. Identify and label the Brønsted acid, its conjugate base, the Brønsted base, and its conjugate acid in each of the following equations:
   (a) $HNO_3 + H_2O \longrightarrow H_3O^+ + NO_3^-$
   (b) $CN^- + H_2O \longrightarrow HCN + OH^-$
   (c) $H_2SO_4 + Cl^- \longrightarrow HCl + HSO_4^-$
   (d) $HSO_4^- + OH^- \longrightarrow SO_4^{2-} + H_2O$
   (e) $O^{2-} + H_2O \longrightarrow 2OH^-$
   (f) $[Cu(H_2O)_3(OH)]^+ + [Al(H_2O)_6]^{3+} \longrightarrow$
   $[Cu(H_2O)_4]^{2+} + [Al(H_2O)_5(OH)]^{2+}$
   (g) $H_2S + NH_2^- \longrightarrow HS^- + NH_3$

9. What is the conjugate base of each of the following?
   (a) NH_4^+ NH_3 (b) HCl Cl^- (c) HNO_3 NO_3^-
   (d) $HClO_4$ ClO_4^- (e) CH_3CO_2H $CH_3CO_2^-$ (f) HCN CN^-
   Is each a strong or weak base?

10. Hydrogen chloride can be prepared by the acid-base reaction of pure sulfuric acid with solid sodium chloride (Section 9.9). Write the balanced chemical equation for the reaction, and identify the conjugate acid-base pairs in this reaction.

11. Gastric juice, the digestive fluid produced in the stomach, contains hydrochloric acid, HCl. Milk of magnesia, a suspension of solid $Mg(OH)_2$ in an aqueous medium, is sometimes used to neutralize excess stomach acid. Write a complete balanced equation for the neutralization reaction, and identify the conjugate acid-base pairs.

12. Nitric acid reacts with insoluble copper(II) oxide to form soluble copper(II) nitrate, $Cu(NO_3)_2$. Write the balanced chemical equation for the reaction of an aqueous solution of HNO_3 with CuO, and indicate the conjugate acid-base pairs.

13. What are amphiprotic species? Illustrate with suitable equations.

14. State which of the following species are amphiprotic and write chemical equations illustrating the amphiprotic character of the species:
    (a) H_2O (b) $H_2PO_4^-$ (c) S^{2-}
    (d) CH_4 (e) HSO_4^- (f) H_2CO_3

15. Explain what is an amphoteric substance. Illustrate with equations.

Strengths of Acids and Bases

16. Predict which acid in each of the following pairs is the stronger:
    (a) H_2O or HF   (b) NH_3 or H_2O
    (c) PH_3 or HI   (d) NH_3 or H_2S
    Explain your reasoning for each.

17. Predict which acid in each of the following pairs is the stronger:
    (a) HSO_4^- or $HSeO_4^-$ (b) $B(OH)_3$ or $Al(OH)_3$
    (c) HSO_3^- or HSO_4^- (d) NH_3 or PH_3
    (e) H_2O or H_2Te
    Explain your reasoning for each.

18. Rank the compounds in each of the following groups in order of increasing acidity, and explain the order you assign:
    (a) HCl, HBr, HI   HCl, HBr, HI
    (b) HOCl, HOBr, HOI   HOI, HOBr, HOCl
    (c) $NaHSO_3$, $NaHSeO_3$, $NaHSO_4$
    (d) $Mg(OH)_2$, $Si(OH)_4$, $ClO_3(OH)$
    (e) HF, H_2O, NH_3, CH_4
    (f) HOCl, HOClO, $HOClO_2$, $HOClO_3$

19. Rank the bases in each of the following groups in order of increasing base strength. Explain.
    (a) BrO_2^-, ClO_2^-, IO_2^-
    (b) H_2O, OH^-, H^-, Cl^-
    (c) NH_2^-, HS^-, HTe^-, PH_2^-
    (d) ClO_4^-, ClO^-, ClO_2^-, ClO_3^-

20. What is meant by the leveling effect of water on strong acids and strong bases?

21. Some porcelain cleansers contain sodium hydrogen sulfate, $NaHSO_4$. Is a solution of $NaHSO_4$ acidic, neutral, or basic? These cleansers remove the deposits due to hard water (primarily calcium carbonate, $CaCO_3$) very effectively. Write a balanced chemical equation for one reaction between $NaHSO_4$ and $CaCO_3$.

22. What is the relationship between the extent of ionization of an acid in aqueous solution and the strength of the acid?

23. How can the relative strengths of strong acids be measured?

24. Both HF and HCN ionize in water to a limited extent. Which of the conjugate bases, F^- or CN^-, is the stronger base? See Table 16.1.

25. Soaps are sodium and potassium salts of a family of acids called fatty acids, which are isolated from animal fats. These acids, which are related to acetic acid, CH_3CO_2H, all contain the carboxyl group, $—CO_2H$, and have about the same strength as acetic acid. Examples include palmitic acid, $C_{15}H_{31}CO_2H$, and stearic acid, $C_{17}H_{35}CO_2H$.
    (a) Write a balanced chemical equation indicating the formation of sodium palmitate, $C_{15}H_{31}CO_2Na$, from palmitic acid and sodium carbonate, and a corresponding equation for the formation of sodium stearate.
    (b) Is a soap solution acidic, basic, or neutral?

Solutions of Brønsted Acids and Bases

26. Write the equation for the essential reaction between aqueous solutions of strong acids and of strong bases.
27. Write equations to illustrate three typical and characteristic reactions of aqueous acids.
28. Containers of $NaHCO_3$, sodium hydrogen carbonate (sometimes called sodium bicarbonate), are often kept in chemical laboratories for use in neutralizing any acids or bases spilled. Write balanced chemical equations for the neutralization by $NaHCO_3$ of (a) a solution of HCl and (b) a solution of KOH.
29. Write equations to show the stepwise ionization of the following polyprotic acids: H_2S, H_2CO_3, and H_3PO_4.
30. In aqueous solution arsenic acid, H_3AsO_4, ionizes in three steps. Write an equation for each step, and label the conjugate acid-base pairs. Which arsenic-containing species are amphiprotic?
31. Contrast the action of water on oxides of metals with that of water on oxides of nonmetals.
32. Describe the chemical reaction, using equations where appropriate, that you would expect from the combination of a solution of $HClO_4$ with each of the following:
    (a) a potassium hydroxide solution
    (b) ammonia gas
    (c) aluminum
    (d) solid aluminum sulfide, Al_2S_3
    (e) solid sodium oxide
    (f) a blue-colored litmus solution
    (g) calcium carbonate
33. Give four methods of preparing hydroxide bases and of preparing Brønsted acids.
34. Define acid-base *neutralization*. Does your definition imply that the resulting solution is always neutral? Explain.
35. Define the term *salt*, and give examples of normal salts, hydrogen salts, hydroxysalts, and oxysalts.

pH and pOH

36. Define pH both in words and mathematically.
37. Calculate the pH and pOH of each of the following solutions. (Each is a strong acid or a strong base.)

(a) 0.100 M HBr   *Ans. pH = 1.000; pOH = 13.00*
(b) 1.45 M NaOH   *Ans. pH = 14.16; pOH = −0.161*
(c) 0.0071 M $Ba(OH)_2$ *Ans. pH = 12.15; pOH = 1.85*
(d) 0.03 M HNO_3 *Ans. pH = 1.5; pOH = 12.5*

Titration; Normality

38. Calculate the mass of an equivalent of each of the reactants in the following equations to one decimal place:
    (a) $H_2SO_4 + 2LiOH \longrightarrow Li_2SO_4 + 2H_2O$
        Ans. LiOH, 23.9 g; H_2SO_4, 49.0 g
    (b) $KOH + KHCO_3 \longrightarrow K_2CO_3 + H_2O$
        Ans. KOH, 56.1 g; $KHCO_3$, 100.1 g
    (c) $HBr + NH_3 \longrightarrow NH_4Br$
        Ans. HBr, 80.9 g; NH_3, 17.0 g
    (d) $Mg + 2HCl \longrightarrow MgCl_2 + H_2$
        Ans. Mg, 12.2 g; HCl, 36.5 g
    (e) $H_2S + HgCl_2 \longrightarrow HgS + 2HCl$
        Ans. H_2S, 17.0 g; $HgCl_2$, 135.7 g
    (f) $KOH + H_2SO_3 \longrightarrow KHSO_3 + H_2O$
        Ans. KOH, 56.1 g; H_2SO_3, 82.1 g
    (g) $Al(OH)_3 + 3HNO_3 \longrightarrow Al(NO_3)_3 + 3H_2O$
        Ans. $Al(OH)_3$, 26.0 g; HNO_3, 63.0 g
39. Calculate the normality of each of the following solutions:
    (a) 4.0 equivalents of HCl in 2.0 liters of solution.
        Ans. 2.0 N
    (b) 2.5×10^{-3} equivalent of HBr in 50 mL of solution.
        Ans. 0.050 N
    (c) 0.30 equivalents of H_2SO_4 in 400 mL of solution.
        Ans. 0.75 N
    (d) 150×10^{-3} equivalents of NaOH in 100 mL of solution.   *Ans. 1.50 N*
40. (a) Calculate the molarity of a solution of hydrochloric acid that contains 3.65 g of HCl in 0.50 L of solution.   *Ans. 0.20 M*
    (b) Calculate the normality of the solution in (a).
        Ans. 0.20 N
41. (a) How many grams of H_2SO_4 are contained in 1.2 L of 0.50 M solution?   *Ans. 59 g*
    (b) How many grams of H_2SO_4 are contained in 1.2 L of 0.50 N solution assuming both hydrogen ions react?
        Ans. 29 g
42. (a) What is the molarity of a solution of barium hydroxide, 92.62 mL of which contains 14.621 mg of $Ba(OH)_2$?   *Ans. 9.213×10^{-4} M*
    (b) What is the normality of the solution in (a) if all OH^- ions react?   *Ans. 1.843×10^{-3} N*
43. Titration of 15.60 mL of a 0.432 N HI solution with NH_3 requires 30.60 mL of the NH_3 solution to reach the end point. What is the normality of the NH_3 solution?
        Ans. 0.220 N
44. A 0.366 N solution of KOH is titrated with H_2SO_4, producing K_2SO_4. If a 32.00-mL sample of the KOH solution is used, what volume of 0.366 M H_2SO_4 is required to reach the end point? What volume of 0.366 N H_2SO_4?

(Assume both protons react.)

Ans. 16.0 mL of 0.366 M H_2SO_4;
32.0 mL of 0.366 N H_2SO_4

45. A sample of solid calcium hydroxide, $Ca(OH)_2$, is allowed to stand in water until a saturated solution is formed. A titration of 75.00 mL of this solution with 5.00×10^{-2} N HCl requires 36.6 mL of the acid to reach the end point.

$$Ca(OH)_2 + 2HCl \longrightarrow CaCl_2 + 2H_2O$$

What is the normality of a saturated $Ca(OH)_2$ solution? The molarity? What is the solubility of $Ca(OH)_2$ in grams per liter of solution?

Ans. 2.44×10^{-2} N; 1.22×10^{-2} M; 0.904 g/L

46. (a) A 0.1824-g sample of an acid is titrated with 0.3090 N NaOH, and 9.256 mL is required to reach the end point. What is the mass of an equivalent of the acid?

Ans. 63.77 g

(b) Which of the following acids was used: HCl, H_3PO_4, H_2SO_4, HNO_3, or CH_3CO_2H? Explain your answer.

Ans. HNO_3

47. What volume of 0.75 N phosphoric acid, H_3PO_4, would be required to neutralize completely 0.500 L of 0.70 M potassium hydroxide?

Ans. 0.47 L

48. What is the normality of a H_2SO_4 solution if 40.2 mL of the solution is required to titrate 6.50 g of $NaHCO_3$?

$$H_2SO_4 + 2NaHCO_3 \longrightarrow Na_2SO_4 + 2H_2O + 2CO_2$$

Ans. 1.92 N

49. Titration of 0.4500 g of an acid requires 282.00 mL of 6.00×10^{-3} N NaOH to reach the end point. What is the mass of an equivalent of the acid?

Ans. 266 g

50. What is the normality of a 40.0% aqueous solution of sulfuric acid, for which the specific gravity is 1.305?

Ans. 10.6 N

Lewis Acids and Bases

51. According to the Lewis concept, what is an acid? A base?

52. Write the Lewis structures of the reactants and product of each of the following equations, and identify the Lewis acid and the Lewis base in each:

(a) $CO_2 + OH^- \longrightarrow HCO_3^-$
(b) $B(OH)_3 + OH^- \longrightarrow B(OH)_4^-$
(c) $I^- + I_2 \longrightarrow I_3^-$
(d) $AlCl_3 + Cl^- \longrightarrow AlCl_4^-$
(e) $O^{2-} + SO_3 \longrightarrow SO_4^{2-}$

53. The dissolution of solid $Al(NO_3)_3$ in water is accompanied by both a Lewis and a Brønsted acid-base reaction. Write balanced chemical equations for the reactions.

54. (a) Write a balanced chemical equation for the Lewis acid-base reaction between $[Al(OH)_4]^-$ and CO_2, which causes the precipitation of $Al(OH)_3$ when CO_2 is added to an aqueous solution of $[Al(OH)_4]^-$.

(b) Write the corresponding equation using the more accurate hydrated forms for the ion, $[Al(H_2O)_2(OH)_4]^-$, and for the precipitate, $[Al(H_2O)_3(OH)_3]$.

55. The reaction of SO_3 with H_2SO_4 to give pyrosulfuric acid, $H_2S_2O_7$ (Section 9.9), is a Lewis acid-base reaction that results in the formation of S—O—S bonds. Write the chemical equation for this reaction showing the Lewis structures of the reactants and product, and identify the Lewis acid and Lewis base.

56. Calcium oxide is prepared by the decomposition of calcium carbonate at elevated temperatures (Section 9.9). This is the reverse of the Lewis acid-base reaction between CaO and CO_2. Write the chemical equation for the reaction of CaO with CO_2 showing the Lewis structures of the reactants and product, and identify the Lewis acid and Lewis base.

57. Each of the species given below may be considered to be a Lewis acid-base adduct of a proton with a Lewis base. For each of the following pairs, indicate which adduct is the stronger acid:

(a) HCl or HCN   (b) HBr or H_3O^+
(c) H_3O^+ or NH_4^+ (d) H_2O or NH_3
(e) HSO_4^- or HCO_3^-

Additional Exercises

58. The concentration of hydrochloric acid secreted by the stomach after a meal is about 1.2×10^{-3} M. What is the pH of stomach acid?

Ans. 2.92

59. In dilute aqueous solution HF acts as a weak acid. However, pure liquid HF (b.p. = 19.5°C) is a strong acid. In liquid HF, HNO_3 acts like a base and accepts protons. The acidity of liquid HF can be increased by adding one of several inorganic fluorides that are Lewis acids and accept F^- ion (for example, BF_3 or SbF_5). Write balanced chemical equations for the reaction of pure HNO_3 with pure HF and of pure HF with BF_3. Write the Lewis structures of the reactants and products, and identify the conjugate acid-base pairs.

60. In papermaking the use of wood pulp as a source of cellulose and of alum-rosin sizing (which keeps ink from soaking into the pages and blurring) produces a paper that deteriorates in 25–50 years. This is quite satisfactory for most purposes, but it is not suitable for the needs of libraries and archives. The deterioration is due to formation of acids. Alum-rosin slowly produces sulfuric acid under humid conditions. Also, wood pulp contains lignin, which forms carboxylic acids as it ages. These acids are similar to acetic acid in that they contain the acidic —CO_2H group. In order to stop this acidification, books are sometimes soaked in magnesium hydrogen carbonate solution; treated with cyclohexylamine, $C_6H_{11}NH_2$, a base like ammonia but with one hydrogen atom replaced by a —C_6H_{11} group; and treated with gaseous diethyl zinc, $(C_2H_5)_2Zn$, a covalent molecule containing Zn—C bonds. Diethyl zinc is a source of $C_2H_5^-$, which is very much like CH_3^- in its properties. Write balanced equations for these reactions. Use the formula RCO_2H for the carboxylic acids formed

from lignin; the exact nature of R is not important in these reactions.

61. Using Lewis structures, write balanced equations for the following reactions:
   (a) $HF + NH_3 \longrightarrow$
   (b) $H_3O^+ + CN^- \longrightarrow$
   (c) $Na_2O + SO_3 \longrightarrow$
   (d) $Li_3N + H_2O \longrightarrow$
   (e) $NH_3 + CH_3^- \longrightarrow$

62. Amino acids are biologically important molecules that are the building blocks of proteins. The simplest amino acid is glycine, $H_2NCH_2CO_2H$. The common feature of amino acids is that they contain the groups indicated in color: amino, $—NH_2$, and carboxyl, $—CO_2H$. An amino acid can function as either an acid or a base. For glycine the acid strength of the carboxyl group is about the same as that of acetic acid, CH_3CO_2H, and the base strength of the amino group is slightly greater than that of ammonia, NH_3.
   (a) Write the Lewis structures of the ions that form when glycine is dissolved in 1 M HCl and in 1 M KOH.
   (b) Write the Lewis structure of glycine when this amino acid is dissolved in water. (*Hint:* Consider the relative base strengths of the $—NH_2$ and $—CO_2^-$ groups.)

63. Predict the products of the following reactions, and write balanced equations for each. There may be more than one reasonable equation depending on the stoichiometry assumed for the reactants.
   (a) $Fe_2O_3(s) + H_2SO_4(l) \longrightarrow$
   (b) $Li_2O(s) + SiO_2(s) \xrightarrow{\triangle}$
   (c) $HCN(g) + Na \longrightarrow$
   (d) $NaHCO_3(s) + NaNH_2(s) \longrightarrow$
   (e) $Li_3N(s) + NH_3(l) \longrightarrow$
   (f) $BaO(s) + Cl_2O_7(l) \longrightarrow$
   (g) $NaF(s) + H_3PO_4(l) \longrightarrow$
   (h) $MgH_2(s) + H_2S(g) \longrightarrow$
   (i) $NaCH_3(s) + NH_3(g) \longrightarrow$
   (j) $KHCO_3(s) + KHS(s) \longrightarrow$
   (k) $H_2SO_4(l) + NaCH_3CO_2(s) \longrightarrow$

64. How many milliliters of a 0.1500 M solution of KOH will be required to titrate 40.00 mL of a 0.0656 M solution of H_3PO_4?

$$H_3PO_4 + 2KOH \longrightarrow K_2HPO_4 + 2H_2O$$
Ans. 35.0 mL

65. A 46.28-mL sample of a sulfuric acid solution is required to titrate 50.00 mL of a 0.1050 M LiOH solution. What is the molar concentration of the sulfuric acid solution?

$$H_2SO_4 + 2LiOH \longrightarrow Li_2SO_4 + 2H_2O$$
Ans. 0.05672 M

66. A 0.3420-g sample of potassium acid phthalate, $KHC_8H_4O_4$, reacts with 35.73 mL of a NaOH solution in a titration. What is the molar concentration of the NaOH?

$$KHC_8H_4O_4 + NaOH \longrightarrow KNaC_8H_4O_4 + H_2O$$
Ans. 0.04687 M

67. Trichloroacetic acid, CCl_3CO_2H, is amphiprotic.

$$CCl_3CO_2H + B \longrightarrow CCl_3CO_2^- + BH$$
$$HA + CCl_3CO_2H \longrightarrow CCl_3CO_2H_2^+ + A^-$$

Write equations for the reaction of pure $CCl_3CO_2H(l)$ with $H_2O(l)$, $HClO_4(l)$, $HBr(g)$, $NH_3(g)$, and $CH_3CO_2H(l)$. Trichloroacetic acid has an acid strength between those of H_3O^+ and HSO_4^-.

68. The reaction of WCl_6 with Al at about 400°C gives black crystals of a compound containing only tungsten and chlorine. A sample of this compound, when reduced with hydrogen, gives 0.2232 g of tungsten metal and hydrogen chloride, which is absorbed in water. Titration of the hydrochloric acid thus produced requires 46.2 mL of 0.1051 M NaOH to reach the end point. What is the empirical formula of the black tungsten chloride?
Ans. WCl_4

69. Write equations for three methods using acids of preparing the insoluble salt $BaSO_4$.

70. The reaction of 0.871 g of sodium with an excess of liquid ammonia containing a trace of $FeCl_3$ as a catalyst produced 0.473 L of pure H_2 measured at 25°C and 745 torr. What is the equation for the reaction of sodium with liquid ammonia? Show your calculations.
Ans. $2Na + 2NH_3 \longrightarrow 2NaNH_2 + H_2$

71. What mass of magnesium when treated with an excess of dilute sulfuric acid will produce the same volume of hydrogen gas as produced by 30.0 g of aluminum under the same conditions?
Ans. 40.5 g

Ionic Equilibria of Weak Electrolytes

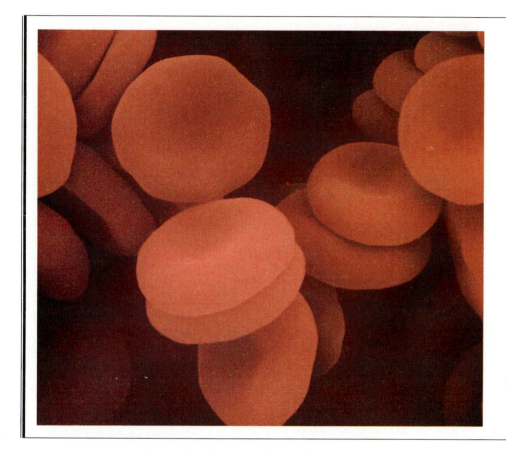

17

Red blood cells are immersed in a complex weakly basic buffer system.

Water is probably the most important solvent on our planet. Our lives depend on reactions that occur in aqueous solution in our cells. Many aqueous solutions contain ions that result from complete ionization of strong electrolytes or from partial ionization of weak electrolytes. The concentrations of these ions are very important in determining the rates, mechanisms, and yields of reactions in such solutions. In this chapter we will only consider the solvent water. However, many of the concepts applied to water can also be applied to other polar solvents.

The concentrations of ions in a solution of a strong electrolyte can be calculated from the concentration of the solute, since it is 100% ionized in solution. However, the concentrations of ions in a solution of a weak electrolyte cannot be determined so easily. Both ions and nonionized molecules of the weak electrolyte are present, and the concentrations of each of these species must be calculated from the concentration of the weak electrolyte and the extent to which it ionizes. These calculations involve the concepts of equilibria introduced in Chapter 15.

Weak acids and weak bases are probably the most significant weak electrolytes. They are important in many purely chemical processes and in those of biological interest. For example, amino acids are both weak acids and weak bases. In this chapter we will consider some ways of expressing concentrations of hydronium ion and hydroxide ion in solutions of weak acids and weak bases and will then examine equilibria involving these weak electrolytes.

17.1 The Ionization of Water

Water is an extremely weak electrolyte that undergoes autoionization (or self-ionization) (Section 16.1), with the formation of very small, *but equal,* amounts of hydronium and hydroxide ions.

$$H_2O + H_2O \rightleftharpoons H_3O^+ + OH^-$$

This slight ionization plays an important role in acid-base reactions in water.

Since this reaction is reversible, we can apply the law of mass action (Section 15.2) and write

$$\frac{[H_3O^+][OH^-]}{[H_2O]^2} = K$$

The brackets stand for "molar concentration of"; for example, $[H_3O^+]$ means molar concentration of hydronium ion. For dilute solutions, the molar concentration of pure water may be considered to be constant. (With a density of 1.00 grams per milliliter, each liter of water contains 1000 grams of water, or $1000 \; g \times \dfrac{1 \; mol}{18.0 \; g} = 55.6 \; mol$, so $[H_2O] = 55.6 \; mol/L = 55.6 \; M$.) Thus we can write

$$[H_3O^+][OH^-] = K[H_2O]^2 = K_w$$

where K_w is the **ion product for water.** At 25°C, it has the value 1.0×10^{-14} $(mol/L)^2$. This is an important equilibrium constant; you should memorize it. The degree of ionization of water and the resulting concentrations of hydronium ion and hydroxide ion increase as the temperature increases. At 100°C, K_w is about 1×10^{-12} $(mol/L)^2$, 100 times larger than at 25°C.

Pure water ionizes to provide equal numbers of hydronium and hydroxide ions (and thus is neutral). Therefore, in pure water, $[H_3O^+] = [OH^-]$, and at 25°C,

$$[H_3O^+]^2 = [OH^-]^2 = 1.0 \times 10^{-14} \; (mol/L)^2$$
$$[H_3O^+] = [OH^-] = \sqrt{1.0 \times 10^{-14} \; (mol/L)^2} = 1.0 \times 10^{-7} \; mol/L$$

All aqueous solutions contain both hydronium ions and hydroxide ions. If an acid is added to pure water at 25°C, $[H_3O^+]$ becomes larger than 1.0×10^{-7} M, and $[OH^-]$ becomes less than 1.0×10^{-7} M, but not 0. If a base is added to pure water at 25°C, $[OH^-]$ becomes greater than 1.0×10^{-7} M, and $[H_3O^+]$ becomes less than 1.0×10^{-7} M, but not 0. The product of $[H_3O^+]$ and $[OH^-]$ at any specific temperature is always a constant, so the value of either concentration can come close to 0 but never equal 0.

One of the problems students have in chemistry is that chemists sometimes use different names or symbols to represent the same thing. For example, most American chemists call element 104 rutherfordium; Russian chemists call it kurchatovium; radiochemists generally call it element-104 or simply 104. Some chemists use the recommended name, unnilquadium (Section 5.13).

There is a similar lack of consistency in acid-base chemistry. Although almost all chemists recognize that a hydrogen ion is present in water as the hydronium ion, H_3O^+, many of them write it as H^+ or $H^+(aq)$ and talk about solutions of hydrogen ions or protons (which are, after all, hydrogen ions). In most of the literature dealing with equilibria, H^+ or $H^+(aq)$ is used instead of H_3O^+. The self-ionization of water is often written

$$H_2O \rightleftharpoons H^+ + OH^-$$

for which the equilibrium equation is

$$[H^+][OH^-] = K_w = 1.0 \times 10^{-14} \text{ (at 25°C)}$$

You can see this is the equation for the ion product of water, except that $[H^+]$ replaces $[H_3O^+]$. In fact, all equilibrium expressions for reactions in water can be written using $[H^+]$ to replace $[H_3O^+]$.

There is nothing wrong with using the term hydrogen ion rather than hydronium ion, provided you remember that a hydrogen ion is bound to at least one molecule of water in aqueous solutions. (However, some instructors will want you to use hydronium ion to show that you remember.) From this point, we will usually talk about hydrogen ions and use the symbol H^+ in dealing with equilibrium problems. If you want to use hydronium ions in place of hydrogen ions, you can replace H^+ with H_3O^+ in equilibria and chemical equations and add one water molecule for each hydronium ion on the other side of the chemical equations.

17.2 pH and pOH

In Section 16.5, Part 4, we defined the **pH** of a solution as the negative logarithm of the hydrogen ion concentration and **pOH** as the negative logarithm of the hydroxide ion concentration.

$$pH = -\log [H^+] \qquad \text{or} \qquad pH = \log \frac{1}{[H^+]}$$

$$pOH = -\log [OH^-] \qquad \text{or} \qquad pOH = \log \frac{1}{[OH^-]}$$

From these definitions, it follows that the pH value is the negative power to which 10 must be raised to give the hydrogen ion concentration, and the pOH value is the negative power to which 10 must be raised to give the hydroxide ion concentration.

$$[H^+] = 10^{-pH} \text{ M}$$
$$[OH^-] = 10^{-pOH} \text{ M}$$

Similarly, we may define pK_w as the negative logarithm of the ion product for water (K_w); that is, $pK_w = -\log K_w$. Now we can write the ion product equation for water in terms of pH, pOH, and pK_w.

$$[H^+][OH^-] = K_w$$
$$(-\log [H^+]) + (-\log [OH^-]) = -\log K_w$$
$$pH + pOH = pK_w$$

Since K_w has the value 1.0×10^{-14} at 25°C,

$$pK_w = -\log (1.0 \times 10^{-14}) = -(-14.00) = 14.00$$

it follows that

$$pH + pOH = 14.00$$
$$pH = 14.00 - pOH$$
$$pOH = 14.00 - pH$$

Several examples of hydrogen ion and hydroxide ion concentrations and the corresponding pH and pOH values are given in Table 17.1.

A neutral solution, one in which $[H^+] = [OH^-]$, has a pH of 7.00 at 25°C. The pH of an acidic solution ($[H^+] > [OH^-]$) is less than 7; the pH of a basic solution ($[OH^-] > [H^+]$) is greater than 7. The pH of a solution can be calculated from its hydrogen ion concentration or measured with a **pH meter** (Figs. 17.1 and 17.2) or with pH paper (Fig. 17.3).

Table 17.1 shows the relationships between $[H^+]$, $[OH^-]$, pH, and pOH and the values for some common substances at 25°C. Note that pH + pOH = 14. The lower the pH (and the higher the pOH), the higher is the acidity.

Table 17.1 Relationships of $[H^+]$, $[OH^-]$, pH and pOH

$[H^+]$	$[OH^-]$	pH	pOH	Sample solution
10^1	10^{-15}	-1	15	Strongly acidic
10^0 or 1	10^{-14}	0	14	← 1 M HCl
10^{-1}	10^{-13}	1	13	
10^{-2}	10^{-12}	2	12	← Lime juice 1 M CH_3CO_2H
10^{-3}	10^{-11}	3	11	← Stomach acid
10^{-4}	10^{-10}	4	10	← Wine
10^{-5}	10^{-9}	5	9	← Coffee
10^{-6}	10^{-8}	6	8	
10^{-7}	10^{-7}	7	7	← Pure water Neutral
10^{-8}	10^{-6}	8	6	← Blood
10^{-9}	10^{-5}	9	5	
10^{-10}	10^{-4}	10	4	
10^{-11}	10^{-3}	11	3	← Milk of magnesia
10^{-12}	10^{-2}	12	2	← Household ammonia, NH_3
10^{-13}	10^{-1}	13	1	
10^{-14}	10^0 or 1	14	0	← 1 M NaOH Strongly basic

Figure 17.1
A pH meter. The pH of a solution is determined electronically from the difference in electrical potentials between the two electrodes when they are dipped into the solution and is displayed on the scale for convenient reading.

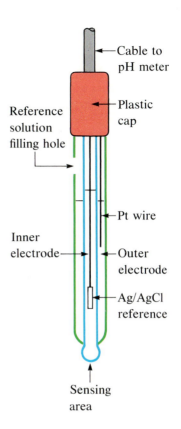

- Cable to pH meter
- Plastic cap

Reference solution filling hole

Inner electrode

- Pt wire
- Outer electrode
- Ag/AgCl reference

Sensing area

Figure 17.2
Some pairs of pH electrodes are combined into a single unit for convenience.

Figure 17.3
pH paper contains a mixture of indicators that give it different colors in solutions of differing pH.

EXAMPLE 17.1   The concentration of the hydrochloric acid secreted by the stomach after a meal is about 1.2×10^{-3} M. What is the pH of stomach acid?

Because hydrochloric acid is a strong acid, the hydrogen ion concentration of stomach acid is the same as its molar concentration. Thus we have

$$\text{pH} = -\log [H^+] = -\log (1.2 \times 10^{-3})$$
$$= -(-2.92) = 2.92$$

(The use of logarithms is explained in Appendix A. On a calculator take the logarithm of 1.2×10^{-3}; the pH is equal to its negative.)

The answer 2.92 represents *two* significant figures rather than three. The number to the left of the decimal in a logarithm is the *characteristic,* which merely establishes the decimal place in the number that the logarithm represents. The only significant figures in a logarithm are those to the right of its decimal. The number of significant figures in a logarithm should be equal to the number of significant figures in the number from which the logarithm is obtained. There are two significant figures in the hydrogen ion concentration (1.2×10^{-3}) and two in the pH.

EXAMPLE 17.2   Calculate the pH of a 0.10 M solution of nitric acid (a strong acid) and of 0.10 M solutions of acetic acid (1.3% ionized) and nitrous acid (6.5% ionized).

0.1 M nitric acid. Nitric acid is completely ionized in dilute solutions, so the concentration of the hydrogen ion is 0.10 M, or 1.0×10^{-1} M, in 0.10 M solution. Substituting, we have

$$\text{pH} = -\log [H^+] = -\log (1.0 \times 10^{-1}) = -(-1.00) = 1.00$$

(On a calculator, simply take the logarithm of 1.0×10^{-1}; the pH is equal to the negative of this logarithm.)

0.1 M acetic acid. The percent ionization for 0.10 M acetic acid solution is 1.3% (Section 16.3). Hence the number of moles of hydrogen ion in exactly 1 liter of a solution containing 0.10 mol of acetic acid is $(0.013)(0.10) = 0.0013$ mol. The concentration of H^+ is 0.0013 mol/1 L or 1.3×10^{-3} M. Thus

$$[H^+] = 1.3 \times 10^{-3} \text{ M}$$
$$\text{pH} = -\log [H^+] = -\log (1.3 \times 10^{-3}) = -(-2.89) = 2.89$$

Note that the logarithm 2.89 contains only *two* significant figures. The 2 is the characteristic and is not a significant figure.

0.1 M nitrous acid. The percent ionization for 0.10 M nitrous acid is 6.5% (Section 16.3). Hence the number of moles of hydrogen ion in exactly 1 liter of solution is $(0.065)(0.10) = 0.0065$ mol, and $[H^+] = 6.5 \times 10^{-3}$ M. Thus

$$\text{pH} = -\log [H^+] = -\log (6.5 \times 10^{-3}) = -(-2.19) = 2.19$$

Note that the strengths of the acids decrease as the pH values increase (given the same initial concentration). Nitric acid (pH = 1.00) is the strongest of the three acids, nitrous acid (pH = 2.19) is weaker, and acetic acid (pH = 2.89) is still weaker, in agreement with Section 16.3.

Water in equilibrium with air contains $4.4 \times 10^{-5}\%$ carbon dioxide. The resulting carbonic acid, H_2CO_3, gives the solution a hydrogen ion concentration of 2.0×10^{-6} M, about 20 times larger than that of pure water. Calculate the pH of the solution.

EXAMPLE 17.3

$$pH = -\log [H^+] = -\log (2.0 \times 10^{-6})$$
$$= -(-5.70) = 5.70$$

(On a calculator take the logarithm of 2.0×10^{-6}; the pH is equal to its negative.)

Thus we see that water in contact with air is acidic, rather than neutral, due to dissolved carbon dioxide.

The following examples illustrate the conversion of pH values to hydrogen ion concentrations.

Calculate the hydrogen ion concentration of a solution whose pH is 9.0.

EXAMPLE 17.4

$$pH = -\log [H^+] = 9.0$$
$$\log [H^+] = -9.0$$
$$[H^+] = 10^{-9.0} \quad \text{or} \quad [H^+] = \text{antilog of } -9.0$$
$$[H^+] = 1 \times 10^{-9} \text{ M}$$

(On a calculator take the antilog, or the "inverse" log, of -9.0, or calculate $10^{-9.0}$.)

Calculate the hydrogen ion concentration of blood, the pH of which is 7.3 (slightly alkaline).

EXAMPLE 17.5

$$pH = -\log [H^+] = 7.3$$
$$\log [H^+] = -7.3$$
$$[H^+] = 10^{-7.3} \quad \text{or} \quad [H^+] = \text{antilog of } -7.3$$
$$[H^+] = 5 \times 10^{-8} \text{ M}$$

(On a calculator simply take the antilog, or the "inverse" log, of -7.3, or calculate $10^{-7.3}$.)

What are the pOH and pH of a 0.0125 M solution of potassium hydroxide, KOH?

EXAMPLE 17.6

Potassium hydroxide is completely ionized in solution (Section 9.2, Part 2), so $[OH^-] = 0.0125$ M.

$$pOH = -\log [OH^-] = -\log 0.0125$$
$$= -(-1.903) = 1.903$$

The pH can be found from the pOH.

$$pH + pOH = 14.00$$
$$pH = 14.00 - pOH = 14.00 - 1.903 = 12.10$$

EXAMPLE 17.7   Calculate the pOH and the pH of a 0.10 M solution of acetate ion ($CH_3CO_2^-$) at 25°C.

$$CH_3CO_2^- + H_2O \rightleftharpoons CH_3CO_2H + OH^-$$

The percent yield of OH^- in this reaction is 0.0075%.

The percent reaction of acetate ion with water for a 0.10 M solution of acetate ion is 0.0075% at 25.0°C (Section 16.3). Hence the number of moles of acetic acid and of hydroxide ion in exactly 1 L of solution are $(0.000075)(0.10) = 7.5 \times 10^{-6}$ and the concentrations are 7.5×10^{-6} M. Thus

$$pOH = -\log (7.5 \times 10^{-6}) = -(-5.12) = 5.12$$
$$pH = 14.00 - pOH = 14.00 - 5.12 = 8.88$$

The calculated values of pOH and pH (pOH < 7; pH > 7) show that the acetate is a weak base in water solution, as expected.

Ionic Equilibria of Weak Acids and Weak Bases

17.3 Ion Concentrations in Solutions of Strong Electrolytes

Strong electrolytes are completely ionized in aqueous solution, so the ion concentrations may be found directly from the molar concentration. For example, in a 0.01 M solution of hydrochloric acid ($HCl \rightarrow H^+ + Cl^-$) the HCl is 100% ionized, with each molecule of HCl providing one hydrogen ion and one chloride ion. Hence $[H^+]$ and $[Cl^-]$ are each equal to 0.01 M. From the chemical equation we can find the conversions 1 mol H^+/1 mol HCl and 1 mol Cl^-/1 mol HCl. Thus a HCl concentration of 0.01 M (0.01 mol HCl/1 L) can be converted to $[H^+]$ and $[Cl^-]$ as follows:

$$[H^+] = \frac{0.01 \text{ mol HCl}}{1 \text{ L}} \times \frac{1 \text{ mol H}^+}{1 \text{ mol HCl}} = \frac{0.01 \text{ mol H}^+}{1 \text{ L}} = 0.01 \text{ M}$$

$$[Cl^-] = \frac{0.01 \text{ mol HCl}}{1 \text{ L}} \times \frac{1 \text{ mol Cl}^-}{1 \text{ mol HCl}} = \frac{0.01 \text{ mol Cl}^-}{1 \text{ L}} = 0.01 \text{ M}$$

Because 1 mole of the strong electrolyte potassium sulfate, K_2SO_4, contains 2 moles of potassium ions and 1 mole of sulfate ions, $[K^+]$ and $[SO_4^{2-}]$ in a 0.03 M solution of K_2SO_4 are given by

$$[K^+] = \frac{0.03 \text{ mol K}_2SO_4}{1 \text{ L}} \times \frac{2 \text{ mol K}^+}{1 \text{ mol K}_2SO_4} = 0.06 \text{ M}$$

$$[SO_4^{2-}] = \frac{0.03 \text{ mol K}_2SO_4}{1 \text{ L}} \times \frac{1 \text{ mol SO}_4^{2-}}{1 \text{ mol K}_2SO_4} = 0.03 \text{ M}$$

17.4 The Ionization of Weak Acids

We can tell by measuring the pH of an aqueous solution of a weak acid such as acetic acid, CH_3CO_2H, that only a fraction of the molecules are ionized into hydrogen cations and acetate anions (Fig. 17.4). The remaining acid is present as the nonion-

Figure 17.4
pH paper indicates that a 0.1 M solution of HCl (beaker on left) has a pH of 1 [also see Figs. 17.3 and 16.2]. A 0.1 M solution of CH_3CO_2H (beaker on right) is less acidic and has a pH of 3 because the weak acid CH_3CO_2H is only partially ionized.

ized, or molecular, form. Alternatively, we can calculate the concentrations of nonionized acetic acid, of hydrogen cations, and of acetate anions from the equilibrium constant for the ionization of acetic acid. For acetic acid the ionization reaction is

$$CH_3CO_2H(aq) \rightleftharpoons H^+(aq) + CH_3CO_2^-(aq)$$

Since equilibrium is attained almost immediately when an acid dissolves in water, we can use the law of mass action and write the equation for K_a, **the acid dissociation constant,** or **acid ionization constant.**

$$K_a = \frac{[H^+][CH_3CO_2^-]}{[CH_3CO_2H]} \tag{1}$$

Before we use this equilibrium expression, let's show that it is the same as the one using hydronium ion. The chemical equation using hydronium ion is written

$$CH_3CO_2H(aq) + H_2O(l) \rightleftharpoons H_3O^+(aq) + CH_3CO_2^-(aq)$$

For this equation the law of mass action gives

$$\frac{[H_3O^+][CH_3CO_2^-]}{[CH_3CO_2H][H_2O]} = K$$

For dilute solutions the number of moles of water consumed in the formation of hydronium ions is negligible compared to the number of moles of water present, and the molar concentration of water remains practically constant, as shown in the following Example.

Calculate the percent of the moles of water present that are consumed in the formation of hydronium ions in the ionization reaction for 0.10 M acetic acid (1.3% ionized).

EXAMPLE 17.8

$$CH_3CO_2H(aq) + H_2O \rightleftharpoons H_3O^+(aq) + CH_3CO_2^-(aq)$$

In Section 16.3 we noted that 1.3% of the acetic acid molecules are ionized in a 0.10 M aqueous solution of the acid. Hence the number of moles of acid that react with water is

$$(0.013)(0.10 \text{ mol/L}) = 0.0013 \text{ mol } CH_3CO_2H/L \text{ of solution}$$

The number of moles of water that react is

$$0.0013 \text{ mol } CH_3CO_2H \times \frac{1 \text{ mol } H_2O}{1 \text{ mol } CH_3CO_2H} = 0.0013 \text{ mol } H_2O/L$$

A liter of water contains 55.6 mol of H_2O (Section 17.1). Hence only 0.0013 mole of water out of the 55.6 moles present per liter of solution reacts, or only 0.0023% of the water molecules, which is a negligible quantity compared to the total present.

Because the concentration of water does not change we can write

$$\frac{[H_3O^+][CH_3CO_2^-]}{[CH_3CO_2H]} = [H_2O]K$$

$$\frac{[H_3O^+][CH_3CO_2^-]}{[CH_3CO_2H]} = K_a$$

Except for the replacement of $[H^+]$ with $[H_3O^+]$, this last equation is identical to Equation (1).

Concentrations of ions and molecules in the equation for K_a are generally expressed in moles per liter. The data in Table 17.2 show that K_a for acetic acid remains practically constant over a considerable range of concentrations. Note that the extent to which acetic acid ionizes increases with decreasing concentration. For example, 0.100 M acetic acid is 1.34% ionized (98.66% is in the molecular form), while 0.0100 M acetic acid is 4.15% ionized (95.85% is in the molecular form). The reason for this is that in a more dilute solution at equilibrium the ions are farther apart and have less tendency to combine to form undissociated acetic acid molecules.

Table 17.2 Ionization constants for acetic acid at different concentrations

Molarity	% Ionized	$[H^+]$ and $[CH_3CO_2^-]$	$[CH_3CO_2H]$	K_a at 25°C
0.100	1.34	0.00134	0.09866	1.82×10^{-5}
0.0800	1.50	0.00120	0.07880	1.83×10^{-5}
0.0300	2.45	0.000735	0.02927	1.85×10^{-5}
0.0100	4.15	0.000415	0.009585	1.80×10^{-5}

The ionization constants of a number of weak acids are given in Table 17.3; a more complete list is given in Appendix F. The acids in the table are listed in order of decreasing strength, as indicated by the decreasing size of the ionization constant (compare with Table 16.1 in Section 16.3). As the fraction of acid existing in the ionized form decreases, the ionization constant decreases. Thus HF with its K_a of 7.2×10^{-4} is a much stronger acid than HCN, which has a K_a of 4×10^{-10}.

The following examples exemplify the types of problems that are based on the partial ionization of weak acids.

Table 17.3 Ionization constants of some weak acids

Ionization reaction	K_a at 25°C
$HF \rightleftharpoons H^+ + F^-$	7.2×10^{-4}
$HNO_2 \rightleftharpoons H^+ + NO_2^-$	4.5×10^{-4}
$HNCO \rightleftharpoons H^+ + NCO^-$	3.46×10^{-4}
$HCO_2H \rightleftharpoons H^+ + HCO_2^-$	1.8×10^{-4}
$CH_3CO_2H \rightleftharpoons H^+ + CH_3CO_2^-$	1.8×10^{-5}
$HClO \rightleftharpoons H^+ + ClO^-$	3.5×10^{-8}
$HBrO \rightleftharpoons H^+ + BrO^-$	2×10^{-9}
$HCN \rightleftharpoons H^+ + CN^-$	4×10^{-10}

EXAMPLE 17.9

In a 0.0800 M solution acetic acid is 1.50% ionized. Calculate $[H^+]$, $[CH_3CO_2^-]$, and $[CH_3CO_2H]$ in the solution.

First, we write the equation for the ionization of acetic acid.

$$CH_3CO_2H \rightleftharpoons H^+ + CH_3CO_2^-$$

The equation shows that, for each mole of CH_3CO_2H that ionizes, one mole of H^+ and one mole of $CH_3CO_2^-$ are formed. Since 1.50% of 0.0800 mol/L of acid ionizes,

$$[H^+] = [CH_3CO_2^-] = 0.0800 \text{ M} \times 0.0150 = 0.00120 \text{ M}$$

and

$$[CH_3CO_2H] = 0.0800 \text{ M} - 0.00120 \text{ M} = 0.0788 \text{ M}$$

EXAMPLE 17.10

Using the concentrations found in Example 17.9, calculate the ionization constant of acetic acid.

First, we write the expression for the ionization constant of acetic acid and then substitute in it the values found above to solve for K_a.

$$K_a = \frac{[H^+][CH_3CO_2^-]}{[CH_3CO_2H]} = \frac{0.00120 \times 0.00120}{0.0788} = 1.83 \times 10^{-5}$$

EXAMPLE 17.11

The pH of a 0.0495 M solution of nitrous acid, HNO_2, is 2.34. What is K_a for nitrous acid?

The equation for the ionization of nitrous acid is

$$HNO_2 \rightleftharpoons H^+ + NO_2^-$$

for which

$$K_a = \frac{[H^+][NO_2^-]}{[HNO_2]}$$

Thus we need to know the concentrations of H^+, NO_2^-, and HNO_2 at equilibrium in order to calculate K_a.

In many equilibrium problems we are given concentrations of reactants (and sometimes of products) before equilibrium is established and are asked to determine concentrations at equilibrium. It is often helpful to set up a table such as the one below showing the initial concentrations (the concentrations before equilibrium is established) and the concentrations at equilibrium. The concentration given in the problem is shown in color; the entries given in black are calculated as described following the table.

	$[HNO_2]$	$[H^+]$	$[NO_2^-]$
Initial concentrations, M	0.0495	~0	0
Final concentrations, M	0.0449	0.0046	0.0046

We are given the pH (from which we can get $[H^+]$ at equilibrium) and the total concentration of HNO_2 (0.0495 M), which may be considered to be that before any ionizes. Thus before equilibrium is established, $[HNO_2]_i$ = 0.0495 M. The subscript letter i is used to indicate an initial concentration, the concentration before any reaction occurs. As shown in Table 17.1, $[H^+]$ in pure water (the concentration of H^+ before the reaction begins) is not zero. However, in most cases involving the dissolution of a weak acid in water, the initial concentration of H^+ is sufficiently small that it may be neglected, so we take $[H^+]_i$ = ~0. (The symbol ~ means "approximately," so ~0 means "approximately zero.") The concentration of NO_2^- is zero before any HNO_2 reacts with water.

Since pH = $-\log [H^+]$, we can calculate $[H^+]$ at equilibrium.

$$pH = -\log [H^+] = 2.34$$
$$\log [H^+] = -2.34$$
$$[H^+] = 10^{-2.34} = 0.0046 \text{ M}$$

From the chemical equation we see that one mole of NO_2^- forms for each mole of H^+ formed. Thus $[NO_2^-]$ also equals 0.0046 M.

For each mole of H^+ that forms, one mole of HNO_2 ionizes. Thus the equilibrium concentration of HNO_2 is equal to the initial concentration (0.0495 M) minus the amount that ionizes (0.0046 M).

$$[HNO_2] = 0.0495 \text{ M} - 0.0046 \text{ M} = 0.0449 \text{ M}$$

At equilibrium

$$K_a = \frac{[H^+][NO_2^-]}{[HNO_2]}$$
$$= \frac{0.0046 \times 0.0046}{0.0449}$$
$$= 4.7 \times 10^{-4}$$

EXAMPLE 17.12   Taking K_a for acetic acid to be 1.8×10^{-5}, calculate the percent ionization in a 0.100 M solution of the acid.

The equation for the ionization of acetic acid is

$$CH_3CO_2H \rightleftharpoons H^+ + CH_3CO_2^-$$

Here is a table of initial concentrations (concentrations before the reaction occurs) and equilibrium concentrations, with the one given in the problem in color.

	$[CH_3CO_2H]$	$[H^+]$	$[CH_3CO_2^-]$
Initial concentrations, M	0.100	~0	0
Equilibrium concentrations, M	$0.100 - x$	x	x

If we let x be the number of moles of H^+ formed by the ionization of the acid in 1 L of solution, the concentration of H^+ at equilibrium, $[H^+]$, is equal to x mol/1 L or x M. Since the ionization of acetic acid yields one mole of $CH_3CO_2^-$ for each mole of H^+ produced, the concentration of acetate ion at equilibrium, $[CH_3CO_2^-]$, is also equal to x M. The equilibrium concentration of nonionized acetic acid is equal to the initial concentration of CH_3CO_2H minus the amount that ionizes to give H^+ and $CH_3CO_2^-$.

$$[CH_3CO_2H] = [CH_3CO_2H]_i - x = 0.100 - x$$

At equilibrium

$$\frac{[H^+][CH_3CO_2^-]}{[CH_3CO_2H]} = K_a = 1.8 \times 10^{-5}$$

Substitution of equilibrium concentrations yields

$$\frac{(x)(x)}{0.100 - x} = \frac{x^2}{0.100 - x} = 1.8 \times 10^{-5}$$

The small value of K_a indicates that the ratio $x^2/(0.100 - x)$ is small, so x will be small relative to 0.100, from which it is to be subtracted. Thus $(0.100 - x)$ is virtually equal to 0.100, and the preceding equation may be simplified for an approximate solution as follows:

$$\frac{x^2}{0.100} = 1.8 \times 10^{-5}$$

$$x^2 = 1.8 \times 10^{-6}$$

$$x = \sqrt{1.8 \times 10^{-6}} = 1.3 \times 10^{-3} \text{ mol L}^{-1}$$

Thus the concentration of H^+ and of $CH_3CO_2^-$ at equilibrium is 1.3×10^{-3} M.

We were justified in neglecting x in the expression $(0.100 - x)$, since $0.100 - 0.0013 = 0.0987$ which is almost 0.100. When x is less than about 5% of the total concentration of the acid, use this approximation method; if x is greater than 5%, then solve by the method of successive approximations (Section 15.2) or use the complete quadratic equation (Section 15.2).

To calculate the percent ionization, divide the concentration of the acid in the ionized form, which is equal to $[H^+]$, by the total concentration of the acid, and then multiply by 100. Thus we have

$$\frac{\text{Ionized } CH_3CO_2H}{\text{Total } CH_3CO_2H} \times 100 = \frac{[H^+]}{[CH_3CO_2H]_i} \times 100$$

$$= \frac{1.3 \times 10^{-3}}{0.100} \times 100 = 1.3\% \text{ ionized}$$

The values calculated in this example do not exactly match the values given in Table 17.2 because only two significant figures were used here.

17.5 The Ionization of Weak Bases

The most common weak base is aqueous ammonia (Fig. 17.5). When gaseous ammonia is dissolved in water, the solution becomes basic because of the reaction

$$NH_3(aq) + H_2O(l) \rightleftharpoons NH_4^+(aq) + OH^-(aq)$$

Application of the law of mass action to this system gives the expression

$$\frac{[NH_4^+][OH^-]}{[NH_3][H_2O]} = K$$

Only a small fraction of the water is consumed in the reaction, because we are working with a dilute solution. Thus the concentration of water is practically constant, and we may write

$$\frac{[NH_4^+][OH^-]}{[NH_3]} = K[H_2O] = K_b = 1.8 \times 10^{-5}$$

Aqueous ammonia is about as strong a base as acetic acid is an acid; the ionization constants for the two substances are the same (to two significant figures).

Equilibria involving the ionization of weak bases may be treated mathematically in the same manner as those of weak acids. The ionization constants of several weak bases are given in Table 17.4 and in Appendix G.

Figure 17.5

pH paper indicates that a 0.1 M solution of NH_3 is weakly basic and has a pH of 11 because the weak base NH_3 only partially reacts with water. A 0.1 M solution of NaOH has a pH of 13 because NaOH is a strong base.

Base	Ionization	K_b at 25°C
$(CH_3)_2NH$ Dimethylamine	$+ H_2O \rightleftharpoons (CH_3)_2NH_2^+ + OH^-$	7.4×10^{-4}
CH_3NH_2 Methylamine	$+ H_2O \rightleftharpoons CH_3NH_3^+ + OH^-$	4.4×10^{-4}
$(CH_3)_3N$ Trimethylamine	$+ H_2O \rightleftharpoons (CH_3)_3NH^+ + OH^-$	7.4×10^{-5}
NH_3 Ammonia	$+ H_2O \rightleftharpoons NH_4^+ + OH^-$	1.8×10^{-5}
$C_6H_5NH_2$ Aniline	$+ H_2O \rightleftharpoons C_6H_5NH_3^+ + OH^-$	4.6×10^{-10}

Table 17.4   Ionization constants of some weak bases

EXAMPLE 17.13   The sedative Veronal, $C_8H_{12}N_2O_3$ (Fig. 17.6), is a weak base with an ionization constant, K_b, of 1.1×10^{-8}. What is $[OH^-]$ in a 0.010 M aqueous solution of Veronal?

Using Vr as the abbreviation for Veronal, the equation for the reaction with water is

$$Vr + H_2O \rightleftharpoons VrH^+ + OH^-$$

Before any Veronal reacts with water, [Vr] is 0.010 M. As shown in Table 17.1, the

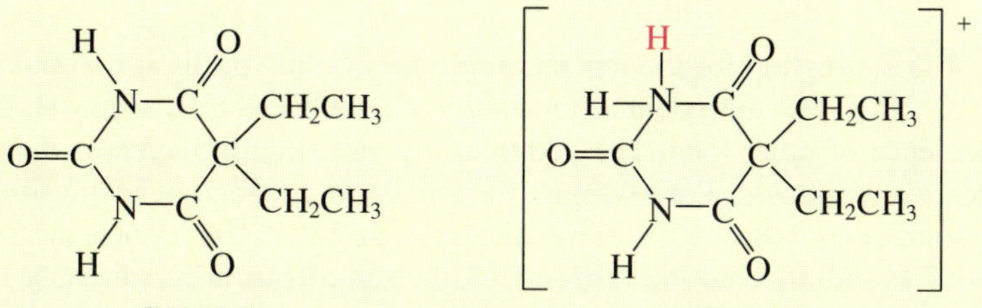

Figure 17.6

The structure of the base Veronal, $C_8H_{12}N_2O_3$, and of the cation formed when a proton is added to it.

concentration of OH^- in pure water ($[OH^-]$ before the reaction) is not zero. However, in most cases $[OH^-]_i$ can be taken to be approximately zero (~ 0). The concentration of VrH^+ is zero before any Veronal reacts with water.

The concentrations of VrH^+ and OH^- at equilibrium are equal, since the reaction of 1 mol of Vr produces 1 mol of VrH^+ and 1 mol of OH^-. Thus $[VrH^+] = [OH^-] = x$, and the concentration of Vr at equilibrium is $0.010 - x$, where x is the number of moles of Veronal that accept a proton per liter of solution.

The following table shows initial concentrations (concentrations before the reaction occurs) and equilibrium concentrations, with the one given in the problem shown in color.

	[Vr]	[VrH$^+$]	[OH$^-$]
Initial concentrations, M	0.010	0	~0
Equilibrium concentrations, M	$0.010 - x$	x	x

At equilibrium

$$\frac{[VrH^+][OH^-]}{[Vr]} = K_b = 1.1 \times 10^{-8}$$

or

$$\frac{(x)(x)}{0.010 - x} = \frac{x^2}{0.010 - x} = 1.1 \times 10^{-8}$$

If we assume that x is small relative to 0.010 (a reasonable assumption in view of the small K_b), then

$$x^2 = 0.010 \times 1.1 \times 10^{-8}$$
$$x = \sqrt{1.1 \times 10^{-10}} = 1.0 \times 10^{-5}$$

Thus, the concentration of OH^- is 1.0×10^{-5} M.

To check our assumption about the magnitude of x relative to 0.010, we need to check the percent ionization of Veronal, or the value of $[VrH^+]/[Vr]_i$ multiplied by 100. If the percent ionization is less than 5%, the approximation is justified.

$$\frac{[VrH^+]}{[Vr]_i} \times 100 = \frac{1.0 \times 10^{-5}}{0.010} \times 100 = 0.10\%$$

The approximation is justified.

17.6 The Salt Effect

The expression for the ionization constant holds for dilute solutions of weak electrolytes. However, the extent of ionization of a weak electrolyte may be influenced by the presence of other ions. The addition of a salt slightly increases the degree of ionization of a weak acid. For example, the ionization constant of acetic acid in dilute aqueous solution is 1.8×10^{-5}. This increases to 2.2×10^{-5} when the solution is made 0.1 M in sodium chloride. This effect, the **salt effect,** occurs because the activities of the ions (Section 12.25) of a weak electrolyte decrease due to interionic attractions for the ions of the strong electrolyte. An ionization constant is not strictly constant but varies slightly with the ionic strength of the solution. For most purposes, and for ours, the expression for the ionization constant may be used without regard to the slight errors that may arise due to the salt effect.

17.7 The Common Ion Effect

The acidity of an aqueous solution of acetic acid decreases when the strong electrolyte sodium acetate, $NaCH_3CO_2$, is added. This can be explained by Le Châtelier's principle (Section 15.4); the addition of acetate ions causes the equilibrium to shift to the left, increasing the concentration of CH_3CO_2H and reducing the concentration of H^+.

$$CH_3CO_2H \rightleftharpoons H^+ + CH_3CO_2^-$$

Since sodium acetate and acetic acid have the acetate ion in common, this influence on the equilibrium is known as the **common ion effect.**

The common ion effect is important in the adjustment of the hydrogen ion concentration for many of the precipitations and separations in a qualitative analysis scheme, and it also plays an essential role in blood and other biological fluids.

The extent to which the concentration of the hydrogen ion is decreased may be calculated from the expression for the ionization constant.

EXAMPLE 17.14   Calculate $[H^+]$ in a 0.10 M solution of CH_3CO_2H that is 0.50 M with respect to $NaCH_3CO_2$.

	$[CH_3CO_2H]$	$[H^+]$	$[CH_3CO_2^-]$
Initial concentrations, M	0.10	~0	0.50
Equilibrium concentrations, M	$0.10 - x$	x	$0.50 + x$

Before reaction $[CH_3CO_2H]_i$ is 0.10 M, $[H^+]_i$ is approximately zero, and $[CH_3CO_2^-]_i$ is 0.50 M. Sodium acetate is an ionic compound, and thus is completely ionized in solution. A 0.50 M solution of $NaCH_3CO_2$ is 0.50 M in Na^+ and 0.50 M in $CH_3CO_2^-$.

At equilibrium the unknown concentration of H^+ may be taken as x. Since one mole of $CH_3CO_2^-$ is formed for each mole of H^+ formed, the concentration of $CH_3CO_2^-$ at equilibrium is equal to $0.50 + x$, the sum of the amount of $CH_3CO_2^-$ present before the reaction and the amount of $CH_3CO_2^-$ produced as equilibrium is reached. The concentration of nonionized acetic acid at equilibrium is $0.10 - x$. Substituting in the expression for the ionization constant for acetic acid, we have

$$K_a = \frac{[H^+][CH_3CO_2^-]}{[CH_3CO_2H]} = \frac{x(0.50 + x)}{0.10 - x} = 1.8 \times 10^{-5}$$

Even in the absence of $NaCH_3CO_2$, the concentration of the acetate ion derived from the ionization of 0.10 M acetic acid would be small (0.0013 mol L^{-1}) relative to that from $NaCH_3CO_2$ (0.50 mol L^{-1}). Since the degree of ionization is even smaller in the presence of sodium acetate, the concentration of acetate ion may be taken as equal to the concentration of the sodium acetate; that is, x is negligible compared to 0.50 in the term $0.50 + x$. Likewise, the concentration of the nonionized acetic acid is very nearly equal to 0.10 M, and x in the term $0.10 - x$ may be dropped. Thus an approximate solution to the problem is

$$\frac{[H^+][CH_3CO_2^-]}{[CH_3CO_2H]} = \frac{x(0.50)}{0.10} = 1.8 \times 10^{-5}$$

$$x = \frac{0.10}{0.50} \times 1.8 \times 10^{-5} = 3.6 \times 10^{-6} \text{ M} = [H^+]$$

Table 17.2 shows that the concentration of the hydrogen ion in 0.100 M CH_3CO_2H is 0.00134 M. This is reduced to 0.0000036 M when the solution is also 0.50 M in sodium acetate. The simplifying assumption that x is negligible compared to either 0.50 or 0.10 is justified.

A 0.10 M solution of aqueous ammonia, also containing ammonium chloride, has a hydroxide ion concentration of 2.8×10^{-6} M. What is the concentration of ammonium ion in the solution?

EXAMPLE 17.15

	[NH$_3$]	[NH$_4^+$]	[OH$^-$]
Initial concentrations, M	0.10	—	—
Equilibrium concentrations, M	$0.10 - (2.8 \times 10^{-6})$	x	2.8×10^{-6}

Since we are calculating the equilibrium concentration of NH_4^+, we set $[NH_4^+] = x$. In order to determine the other concentrations in the table, we first write the equation for the ionization of ammonia.

$$NH_3 + H_2O \rightleftharpoons NH_4^+ + OH^-$$

This equation indicates that for each mole of OH^- formed, one mole of NH_3 is converted to NH_4^+, and the concentration of NH_3 decreases by one mole. Since we are given the equilibrium concentration of OH^-,

$$[OH^-] = 2.8 \times 10^{-6} \text{ M}$$

the initial concentration of NH_3 must have decreased by that amount to produce 2.8×10^{-6} mol of OH^- per liter. Thus

$$[NH_3] = [NH_3]_i - (2.8 \times 10^{-6}) = 0.10 - (2.8 \times 10^{-6})$$

We now have all the concentrations necessary to use the equilibrium constant expression and need not find $[NH_4^+]_i$ or $[OH^-]_i$.

Since 2.8×10^{-6} is small relative to 0.10, we can approximate $[NH_3]$ as 0.10, and

$$K_b = \frac{[NH_4^+][OH^-]}{[NH_3]} = \frac{x(2.8 \times 10^{-6})}{0.10} = 1.8 \times 10^{-5}$$

Solving for x:

$$x = \frac{(0.10)(1.8 \times 10^{-5})}{2.8 \times 10^{-6}} = 0.64$$

Thus we find the concentration of the ammonium ion is 0.64 M at equilibrium. Of this amount, 2.8×10^{-6} mol is formed by the ionization of ammonia, and the remainder comes from the added ammonium chloride.

Addition of a base to a weak acid produces a salt of the acid; addition of an acid to a weak base produces a salt of the base. Hence addition of a base to a solution of a weak acid can create a common ion effect if the amount of base is not sufficient to neutralize all of the weak acid. Thus it will produce a solution that contains both the weak acid and its salt. Similarly, addition of an acid to a solution of a weak base can create a common ion effect. Example 17.16 illustrates the creation of a common ion effect by the addition of a base to a solution of a weak acid.

EXAMPLE 17.16   Exactly 10 mL of 0.100 M sodium hydroxide solution is added to 25 mL of 0.100 M acetic acid solution. Calculate the hydrogen ion concentration of the resulting 35 mL of solution.

Sodium hydroxide, a strong base, reacts with acetic acid to form sodium acetate. A volume of 10 mL (0.010 L) of 0.100 M sodium hydroxide contains

$$0.010 \text{ L} \times 0.100 \text{ mol/L} = 1.0 \times 10^{-3} \text{ mol NaOH}$$

A volume of 25 mL (0.025 L) of 0.100 M acetic acid contains

$$0.025 \text{ L} \times 0.100 \text{ mol/L} = 2.5 \times 10^{-3} \text{ mol CH}_3\text{CO}_2\text{H}$$

The 1.0×10^{-3} mol of NaOH neutralizes 1.0×10^{-3} mol of CH_3CO_2H according to the equation

$$\text{NaOH}(aq) + \text{CH}_3\text{CO}_2\text{H}(aq) \longrightarrow \text{Na}^+(aq) + \text{CH}_3\text{CO}_2^-(aq) + \text{H}_2\text{O}(l)$$

producing 1.0×10^{-3} mol of $NaCH_3CO_2$ and leaving 1.5×10^{-3} mol of unreacted CH_3CO_2H. Thus a solution containing sodium acetate and acetic acid results.

The problem is a common ion problem like that in Example 17.14. In order to find the equilibrium concentration of H^+, we need to calculate the initial concentrations of 1.5×10^{-3} mol of CH_3CO_2H and 1.0×10^{-3} mol of $NaCH_3CO_2$ dissolved in the total 35 mL of solution.

$$[\text{CH}_3\text{CO}_2\text{H}]_i = \frac{1.5 \times 10^{-3} \text{ mol}}{0.035 \text{ L}} = 4.29 \times 10^{-2} \text{ M}$$

$$[\text{CH}_3\text{CO}_2^-]_i = [\text{NaCH}_3\text{CO}_2] = \frac{1.0 \times 10^{-3} \text{ mol}}{0.035 \text{ L}} = 2.86 \times 10^{-2} \text{ M}$$

	$[CH_3CO_2H]$	$[H^+]$	$[CH_3CO_2^-]$
Initial concentrations, M	4.29×10^{-2}	~ 0	2.86×10^{-2}
Equilibrium concentrations, M	$(4.29 \times 10^{-2}) - x$	x	$(2.86 \times 10^{-2}) + x$

$$\frac{[H^+][CH_3CO_2^-]}{[CH_3CO_2H]} = \frac{x[(2.86 \times 10^{-2}) + x]}{(4.29 \times 10^{-2}) - x} = 1.8 \times 10^{-5}$$

Assuming that x is small relative to either 2.86×10^{-2} or 4.29×10^{-2},

$$\frac{x(2.86 \times 10^{-2})}{4.29 \times 10^{-2}} = 1.8 \times 10^{-5}$$

$$x = \frac{4.29 \times 10^{-2}}{2.86 \times 10^{-2}} \times 1.8 \times 10^{-5}$$

$$= 2.7 \times 10^{-5}$$

Thus $[H^+] = 2.7 \times 10^{-5}$ M, a value low enough to justify our approximation based on neglecting x.

17.8 Buffer Solutions

Mixtures of weak acids and their salts or mixtures of weak bases and their salts are called **buffer solutions**, or **buffers**. They resist a change in pH when small amounts of acid or base are added (Fig. 17.7). An example of a buffer that consists of a weak acid and its salt is a solution of acetic acid and sodium acetate. An example of a buffer that consists of a weak base and its salt is a solution of ammonia and ammonium chloride.

1. HOW BUFFERS WORK. An acetic acid–sodium acetate mixture acts as a buffer to keep the hydrogen ion concentration (and hence the pH) almost constant because, as sodium hydroxide is added, the hydroxide ions react with the few hydrogen ions present, and then more of the acetic acid ionizes, restoring the hydrogen ion concentration to near its original value. If a small amount of hydrochloric acid is added to the

Figure 17.7

(left) The color of the indicator in the solutions shows that the buffered solution on the left and the unbuffered solution on the right have the same pH (pH 8). They are basic. (right) After the addition of 1 mL of a 0.01 M HCl solution, the buffered solution has not detectibly changed its pH but the unbuffered solution has become acidic.

buffer solution, most of the hydrogen ions from the hydrochloric acid combine with acetate ions, forming acetic acid molecules.

$$H^+ + CH_3CO_2^- \longrightarrow CH_3CO_2H$$

Thus there is very little increase in the concentration of the hydrogen ion, and the pH remains practically unchanged.

An ammonia–ammonium chloride buffer mixture keeps the hydrogen ion concentration almost constant, because as hydroxide ions are added, ammonium ions in the buffer mixture react with the hydroxide ions to form ammonia molecules and water, thus reducing the hydroxide ion concentration to almost its original value and leaving the hydrogen ion concentration almost constant.

$$NH_4^+ + OH^- \longrightarrow NH_3 + H_2O$$

If hydrogen ions are added, ammonia molecules in the buffer mixture react with the hydrogen ions to form ammonium ions, thereby reducing the hydrogen ion concentration to almost its original value.

$$NH_3 + H^+ \longrightarrow NH_4^+$$

EXAMPLE 17.17

A buffer consisting of equal concentrations of acetic acid and sodium acetate is acidic, as the indicator in this solution shows.

(a) Calculate the pH of a buffer that is a mixture containing 0.10 M acetic acid and 0.10 M sodium acetate. (See the photograph.)

Determination of the pH of the buffer solution requires the calculation of $[H^+]$ in a typical common ion equilibrium, as illustrated in Example 17.14.

	$[CH_3CO_2H]$	$[H^+]$	$[CH_3CO_2^-]$
Initial concentrations, M	0.10	~0	0.10
Equilibrium concentrations, M	$0.10 - x$	x	$0.10 + x$

(where x is the number of moles of CH_3CO_2H that dissociate per liter in establishing equilibrium).

Assuming that x is negligible compared to 0.10,

$$K_a = \frac{[H^+][CH_3CO_2^-]}{[CH_3CO_2H]} = \frac{x(0.10)}{0.10} = 1.8 \times 10^{-5}$$

$$[H^+] = x = \frac{0.10}{0.10} \times 1.8 \times 10^{-5} = 1.8 \times 10^{-5}$$

$$pH = -\log [H^+] = -\log (1.8 \times 10^{-5}) = 4.74$$

(b) Calculate the change in pH when 1.0 mL of 0.10 M NaOH is added to 100 mL of the buffer in (a).

On addition of 1.0 mL (0.0010 L) of 0.10 M sodium hydroxide to 100 mL (0.100 L) of this buffer solution, an equivalent amount of acetic acid is neutralized by the sodium hydroxide, and sodium acetate is formed.

$$CH_3CO_2H + OH^- \longrightarrow H_2O + CH_3CO_2^-$$

Now we will calculate the new concentrations of acetic acid and sodium acetate and then use them to calculate the concentration of hydrogen ion.

Before reaction, 0.100 L of the buffer solution contains

$$0.100 \, L \times \frac{0.10 \text{ mol } CH_3CO_2H}{1 \, L} = 1.0 \times 10^{-2} \text{ mol } CH_3CO_2H$$

and 1.0×10^{-2} mol of $NaCH_3CO_2$. Also, 1.0 mL (0.0010 L) of 0.10 M sodium hydroxide contains

$$0.0010 \, L \times \frac{0.10 \text{ mol NaOH}}{1 \, L} = 1.0 \times 10^{-4} \text{ mol NaOH}$$

The 1.0×10^{-4} mol of NaOH neutralizes 1.0×10^{-4} mol of CH_3CO_2H, leaving

$$(1.0 \times 10^{-2}) - (1.0 \times 10^{-4}) = 0.99 \times 10^{-2} \text{ mol of } CH_3CO_2H$$

and producing 1.0×10^{-4} mol of $NaCH_3CO_2$. This makes a total of

$$(1.0 \times 10^{-2}) + (1.0 \times 10^{-4}) = 1.01 \times 10^{-2} \text{ mol of } NaCH_3CO_2$$

After reaction, 0.99×10^{-2} mol of CH_3CO_2H and 1.01×10^{-2} mol of $NaCH_3CO_2$ are contained in 101 mL of solution, so the concentrations are 9.9×10^{-3} mol/0.101 L = 0.098 M CH_3CO_2H and 1.01×10^{-2} mol/0.101 L = 0.100 M $NaCH_3CO_2$. Now we calculate the pH of the solution, which is 0.098 M in CH_3CO_2H and 0.100 M in $NaCH_3CO_2$.

	$[CH_3CO_2H]$	$[H^+]$	$[CH_3CO_2^-]$
Initial concentrations, M	0.098	~0	0.100
Equilibrium concentrations, M	$0.098 - x$	x	$0.100 + x$

The buffer is still acidic with the same pH after addition of 1 mL of 0.1 M NaOH.

(where x is the number of moles of CH_3CO_2H that dissociate per liter in establishing a new equilibrium).

Assuming that x is small relative to 0.100 or 0.098,

$$K_a = \frac{[H^+][CH_3CO_2^-]}{[CH_3CO_2H]} = \frac{x(0.100)}{0.098} = 1.8 \times 10^{-5}$$

$$[H^+] = x = \frac{0.098}{0.100} \times 1.8 \times 10^{-5} = 1.76 \times 10^{-5}$$

$$pH = -\log [H^+] = -\log (1.76 \times 10^{-5}) = 4.75$$

Thus the addition of the base barely changes the pH of the solution.

(c) Calculate the change in pH when 1.0 mL of 0.10 M HCl is added to 100 mL of the buffer in (a).

On addition of 1.0 mL (0.0010 L) of 0.10 M HCl to 100 mL (0.100 L) of the buffer solution, an equivalent amount of acetate ion, $CH_3CO_2^-$, reacts with the H^+ to form undissociated acetic acid, CH_3CO_2H.

$$H^+ + CH_3CO_2^- \longrightarrow CH_3CO_2H$$

Thus we must calculate the new concentrations of acetic acid and acetate ion and then use them to calculate the new equilibrium concentration of hydrogen ion.

In (b) we calculated that the original buffer solution contains 1.0×10^{-2} mol CH_3CO_2H and 1.0×10^{-2} mol $NaCH_3CO_2$. The $NaCH_3CO_2$, being 100% ionized in water solution, would supply 1.0×10^{-2} mol $CH_3CO_2^-$ to the solution.

One mL (0.0010 L) of 0.10 M HCl contains

$$0.0010 \cancel{L} \times \frac{0.10 \text{ mol HCl}}{1 \cancel{L}} = 1.0 \times 10^{-4} \text{ mol HCl} = 1.0 \times 10^{-4} \text{ mol H}^+$$

The 1.0×10^{-4} mol of H^+ would react with 1.0×10^{-4} mol of $CH_3CO_2^-$, leaving

$$(1.0 \times 10^{-2}) - (1.0 \times 10^{-4}) = 0.99 \times 10^{-2} \text{ mol of } CH_3CO_2^-$$

and producing 1.0×10^{-4} mol of CH_3CO_2H. This makes a total of

$$(1.0 \times 10^{-2}) + (1.0 \times 10^{-4}) = 1.01 \times 10^{-2} \text{ mol of } CH_3CO_2H$$

Thus, after reaction, 0.99×10^{-2} mol of $CH_3CO_2^-$ and 1.01×10^{-2} mol of CH_3CO_2H are contained in 101 mL of solution. Hence the concentrations are 9.9×10^{-3} mol/0.101 L = 0.098 M $CH_3CO_2^-$ and 1.01×10^{-2} mol/0.101 L = 0.100 M CH_3CO_2H.

	$[CH_3CO_2H]$	$[H^+]$	$[CH_3CO_2^-]$
Initial concentrations, M	0.100	~0	0.098
Equilibrium concentrations, M	$0.100 - x$	x	$0.098 + x$

(where x is the number of moles of CH_3CO_2H that dissociate per liter in establishing a new equilibrium).

Assuming that x is small relative to 0.100 or 0.098,

$$K_a = \frac{[H^+][CH_3CO_2^-]}{[CH_3CO_2H]} = \frac{x(0.098)}{0.100} = 1.8 \times 10^{-5}$$

$$[H^+] = x = \frac{0.100}{0.098} \times 1.8 \times 10^{-5} = 1.8 \times 10^{-5}$$

$$pH = -\log [H^+] = -\log (1.8 \times 10^{-5}) = 4.74$$

The pH of the buffer is unchanged after addition of 1 mL of 0.1 M HCl.

Hence the addition of the acid did not change the pH of the original solution (to two significant figures).

(d) Calculate the pH of an unbuffered solution containing 1.8×10^{-5} M HCl.

HCl is 100% ionized. Hence in a 1.8×10^{-5} M HCl solution, $[H^+] = 1.8 \times 10^{-5}$ M. Thus

$$pH = -\log (1.8 \times 10^{-5}) = 4.74$$

The initial pH of the unbuffered 1.8×10^{-5} M HCl solution is the same as that of the buffer solution described in (a).

(e) Calculate the change in pH of the unbuffered 1.8×10^{-5} M HCl solution of (d): (1) after adding 1.0 mL of 0.10 M NaOH to 100 mL of the solution, and (2) after adding 1.0 mL of 0.10 M HCl to 100 mL of the unbuffered solution.

Compare the change in pH that occurs in each case with that which occurs for the buffered solution with the same initial pH [calculated in (b) and (c)].

Table 17.5   pH values, pOH values, and ratios of the weak acid to its salt, with increasing increments of a strong base to 100 mL of a buffer solution of 0.10 M CH_3CO_2H and 0.10 M $NaCH_3CO_2$

Volume of 0.10 M NaOH added, mL	pH	pOH	Ratio $\dfrac{[CH_3CO_2H]}{[CH_3CO_2^-]}$	
0	4.74	9.26	1.00	
1.0	4.75	9.25	0.98	
2.0	4.76	9.24	0.96	
5.0	4.79	9.21	0.90	
10.0	4.83	9.17	0.82	
20.0	4.92	9.08	0.67	
30.0	5.01	8.99	0.54	
40.0	5.11	8.89	0.43	
50.0	5.22	8.78	0.33	
60.0	5.35	8.65	0.25	
70.0	5.50	8.50	0.18	$[CH_3CO_2H]$
80.0	5.70	8.30	0.11	is 11% of
90.0	6.02	7.98	0.050	$[CH_3CO_2^-]$
95.0	6.34	7.66	0.030	
96.0	6.43	7.57	0.020	
97.0	6.56	7.44	0.015	
98.0	6.74	7.26	0.010	
99.0	7.04	6.96	0.005	
100.0	8.72	5.28	0	
101.0	10.70	3.30	0	
102.0	11.00	3.00	0	

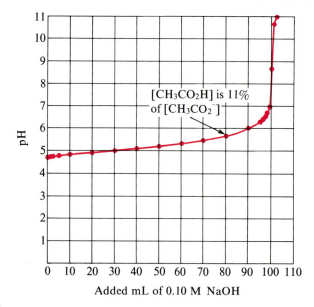

Figure 17.9

An illustration of buffering action. The graph shows change of pH as an increasing volume of 0.10 M NaOH solution is added to 100 mL of a buffer solution in which initially $[CH_3CO_2H] = 0.10$ M and $[CH_3CO_2^-] = 0.10$ M.

bringing $[CH_3CO_2H]$ to 0 and $[CH_3CO^{-2}]$ to its maximum possible value (ratio $[CH_3CO_2H]/[CH_3CO_2^-] = 0$). At that point, the solution has lost all of its buffering capacity for added NaOH; further additions of NaOH simply provide equivalent quantities of free hydroxide ion to the solution.

3. Weak acids and their salts are better as buffers for pH < 7; weak bases and their salts are better as buffers for pH > 7.

4. For a weak acid, HA, and its salt, the ratio $\dfrac{[H^+]}{K_a}$ should be as close to 1 as possible for effective buffer action. Similarly, for a weak base, B, and its salt, the ratio $\dfrac{[OH^-]}{K_b}$ should be as close to 1 as possible for effective buffer action. This concept can be developed as follows.

For HA $\rightleftharpoons$ H$^+$ + A$^-$ (where HA is a weak acid and A$^-$ is the anion of the weak acid and thus also the anion of the salt of the weak acid),

$$K_a = \frac{[H^+][A^-]}{[HA]}$$

$$\frac{[H^+]}{K_a} = \frac{[HA]}{[A^-]}$$

Since an ideal buffer has $[HA] = [A^-]$ $\left(\text{and hence the ratio } \dfrac{[HA]}{[A^-]} = 1\right)$, $\dfrac{[H^+]}{K_a}$ should also equal 1 for the most effective weak acid–salt buffer action.

For B + H$_2$O $\rightleftharpoons$ BH$^+$ + OH$^-$ (where B is a weak base and BH$^+$ is the cation of the salt of the weak base),

$$K_b = \frac{[BH^+][OH^-]}{[B]}$$

$$\frac{[OH^-]}{K_b} = \frac{[B]}{[BH^+]}$$

An ideal buffer should have the ratio $\dfrac{[B]}{[BH^+]} = 1$. Therefore, $\dfrac{[OH^-]}{K_b}$ should also equal 1 for the most effective weak base–salt buffer action. (See exercises 99 and 100 at the end of the chapter for additional relationships that are useful in selecting suitable buffer mixtures. This is a good time, while the subject is fresh in your mind, to work through those two exercises.)

Buffers play a very important role in chemical processes; they keep hydrogen ion and hydroxide ion concentrations approximately constant over a range of changes in condition, such as the addition of acid, base, or water. Common buffer pairs besides CH_3CO_2H and $CH_3CO_2^-$ or NH_3 and NH_4^+ include H_2CO_3 and HCO_3^-, and $H_2PO_4^-$ and HPO_4^{2-}.

Blood is an important example of a buffered solution, with the principal acid and ion responsible for the buffering action being carbonic acid, H_2CO_3, and the hydrogen carbonate ion, HCO_3^-. When an excess of hydrogen ion enters the blood stream, it is removed primarily by the reaction

$$H^+ + HCO_3^- \longrightarrow H_2CO_3$$

When an excess of the hydroxide ion is present, it is removed by the reaction

$$OH^- + H_2CO_3 \longrightarrow H_2O + HCO_3^-$$

The pH of human blood thus remains very near 7.35, slightly alkaline. Variations are usually less than 0.1 of a pH unit. A change of 0.4 of a pH unit is likely to be fatal.

17.9 The Ionization of Weak Diprotic Acids

Polyprotic acids, those from which more than one proton may be removed, ionize in water in successive steps (Section 16.9). An example of a **diprotic acid** is hydrogen sulfide. In aqueous solution the first step in the ionization of hydrogen sulfide yields hydrogen ions and hydrogen sulfide ions.

$$H_2S(aq) \rightleftharpoons H^+(aq) + HS^-(aq)$$

The expression for the ionization constant for the first ionization of H_2S is

$$\frac{[H^+][HS^-]}{[H_2S]} = K_{H_2S} = 1.0 \times 10^{-7}$$

The hydrogen sulfide ion in turn ionizes and forms hydrogen ions and sulfide ions.

$$HS^-(aq) \rightleftharpoons H^+(aq) + S^{2-}(aq)$$

The expression for the ionization constant for the hydrogen sulfide ion is

$$\frac{[H^+][S^{2-}]}{[HS^-]} = K_{HS^-} = 1.0 \times 10^{-19}$$

Note that K_{HS^-} is smaller than K_{H_2S} by a factor of 10^{12}. This means that very little of the HS^- formed by the ionization of H_2S ionizes to give H^+ and S^{2-}. Thus the concentrations of H^+ and HS^- are practically equal in a pure aqueous solution of H_2S.

When the first ionization constant of a polyprotic acid is much larger than the second, it is convenient and appropriate to treat the first ionization step separately and to calculate concentrations of species resulting from it before calculating concentrations of species resulting from subsequent ionization steps.

EXAMPLE 17.18

The concentration of H_2S in a saturated aqueous solution of the gas at room temperature is approximately 0.10 M. Calculate $[H^+]$, $[HS^-]$, and $[S^{2-}]$ of the solution.

We have the following table for the first ionization step

$$H_2S \rightleftharpoons H^+ + HS^-$$

	$[H_2S]$	$[H^+]$	$[HS^-]$
Initial concentrations, M	0.10	~0	0
Equilibrium concentrations, M	$0.10 - x$	x	x

If we let x equal $[H^+]$, then $[HS^-]$ will also be equal to x, and $[H_2S]$ will be equal to $0.10 - x$. Since x will be quite small compared to 0.10, it may be dropped from the term $0.10 - x$. Solving for x, we have

$$K_{H_2S} = \frac{[H^+][HS^-]}{[H_2S]} = \frac{x^2}{0.10} = 1.0 \times 10^{-7}$$

$$x^2 = 1.0 \times 10^{-8}$$

$$x = 1.0 \times 10^{-4} \text{ M} = [H^+] = [HS^-]$$

Now we treat the second step in the ionization of H_2S

$$HS^- \rightleftharpoons H^+ + S^{2-}$$

and let $[S^{2-}] = y$. Since, in any solution, all possible equilibria must be satisfied simultaneously, $[H^+]$ and $[HS^-]$ must be the same in the equilibria involved in the ionizations of both HS^- and H_2S.

	$[HS^-]$	$[H^+]$	$[S^{2-}]$
Initial concentrations, M	1.0×10^{-4}	1.0×10^{-4}	0
Equilibrium concentrations, M	$(1.0 \times 10^{-4}) - y$	$(1.0 \times 10^{-4}) + y$	y

Substituting in the expression for the ionization constant of HS^-, we have

$$K_{HS^-} = \frac{[H^+][S^{2-}]}{[HS^-]} = \frac{[(1.0 \times 10^{-4}) + y]\,y}{(1.0 \times 10^{-4}) - y} = 1.0 \times 10^{19}$$

Assuming that y is small relative to 1.0×10^{-4} and thus may be neglected,

$$[S^{2-}] = y = \frac{1.0 \times 10^{-4}}{1.0 \times 10^{-4}} \times 1.0 \times 10^{-19} = 1.0 \times 10^{-19}$$

Since y is in fact much smaller than 1.0×10^{-4}, the approximation that y can be neglected is justifiable.

(Note that in a pure aqueous solution of H_2S $[S^{2-}]$ is equal to K_{HS^-}. In fact, for any weak diprotic acid the concentration of the divalent anion is numerically equal to the secondary ionization constant.)

By multiplying the expressions for the ionization constants K_{H_2S} and K_{HS^-}, we obtain

$$\frac{[H^+][HS^-]}{[H_2S]} \times \frac{[H^+][S^{2-}]}{[HS^-]} = K_{H_2S} \times K_{HS^-}$$

$$\frac{[H^+]^2[S^{2-}]}{[H_2S]} = K_{H_2S}\,K_{HS^-} = 1.0 \times 10^{-26}$$

This relates $[H^+]$, $[S^{2-}]$, and $[H_2S]$ at equilibrium for the reaction

$$H_2S(aq) \rightleftharpoons 2H^+(aq) + S^{2-}(aq)$$

If two of the concentrations are known, the third can be calculated.

EXAMPLE 17.19 Calculate the concentration of sulfide ion at equilibrium in a saturated solution of hydrogen sulfide to which hydrochloric acid has been added to make the hydrogen ion concentration of the solution 0.10 M at equilibrium. (A saturated solution of H_2S is 0.10 M in hydrogen sulfide.)

To calculate the sulfide ion concentration, we can use the equilibrium

$$H_2S \rightleftharpoons 2H^+ + S^{2-}$$

	[H₂S]	[H⁺]	[S²⁻]
Equilibrium concentrations, M	0.10	0.10	x

The value of $[H^+]$ is 0.10 M at equilibrium since enough hydrochloric acid has been added to make it 0.10 M. For the reaction at equilibrium

$$H_2S \rightleftharpoons 2H^+ + S^{2-}$$

$$\frac{[H^+]^2[S^{2-}]}{[H_2S]} = \frac{(0.10)^2 x}{0.10} = 1.0 \times 10^{-26}$$

$$[S^{2-}] = x = \frac{(1.0 \times 10^{-26})(0.10)}{(0.10)^2} = 1.0 \times 10^{-25} \text{ M}$$

(We could use the same equation to calculate the hydrogen ion concentration necessary to produce a given sulfide ion concentration.)

Since a saturated solution of H_2S is 0.10 M, the general relationship, at equilibrium *for a saturated H_2S solution only,* becomes

$$\frac{[H^+]^2[S^{2-}]}{0.10} = 1.0 \times 10^{-26}$$

$$[H^+]^2[S^{2-}] = 1.0 \times 10^{-27} \qquad \text{(saturated } H_2S \text{ solutions only)}$$

A word of caution is necessary here. Combining the ionization steps for a diprotic acid and multiplying the ionization constants together is a convenient way to calculate the concentration of one species at equilibrium, if all of the other *equilibrium concentrations* for the reactants involved in the reaction are known. However, the use of a combined equation for calculating equilibrium concentrations from *initial concentrations* is very tricky, is likely to lead to unacceptably large errors, and will not be considered here. Such calculations should be made using each ionization step separately.

It is of interest to note that for many years the value of K_2 for hydrogen sulfide has been thought to be on the order of 1×10^{-13} to 1×10^{-14}. The exact value is exceedingly difficult to determine experimentally but is now known to be much smaller than formerly believed. Recent experimental evidence indicates that it is within the range 1×10^{-17} to 1×10^{-19}, with 1.0×10^{-19} being a reasonable value. We will use 1.0×10^{-19} in this text.

This is illustrative of an important point that should always be kept in mind. Many scientific matters are not known exactly. Research goes on continually to refine or correct our current beliefs and to extend scientific knowledge to new areas.

17.10 The Ionization of Weak Triprotic Acids

A typical example of a **triprotic acid** is phosphoric acid, which ionizes in water in three steps.

STEP 1	$H_3PO_4 \rightleftharpoons H^+ + H_2PO_4^-$
STEP 2	$H_2PO_4^- \rightleftharpoons H^+ + HPO_4^{2-}$
STEP 3	$HPO_4^{2-} \rightleftharpoons H^+ + PO_4^{3-}$

Each step in the ionization has a corresponding ionization constant. The expressions for the ionization constants of the three steps are

$$\frac{[H^+][H_2PO_4^-]}{[H_3PO_4]} = K_{H_3PO_4} = 7.5 \times 10^{-3}$$

$$\frac{[H^+][HPO_4^{2-}]}{[H_2PO_4^-]} = K_{H_2PO_4^-} = 6.3 \times 10^{-8}$$

$$\frac{[H^+][PO_4^{3-}]}{[HPO_4^{2-}]} = K_{HPO_4^{2-}} = 3.6 \times 10^{-13}$$

In each successive step the degree of ionization is less. This is a general characteristic of polyprotic oxyacids; successive ionization constants often differ by a factor of about 10^5 to 10^6. (See Section 17.9 for the binary acid H_2S.)

This set of three dissociation reactions appears to make calculations of equilibrium concentrations very complicated. However, because the successive ionization constants differ by a factor of 10^4 to 10^6, the calculations can be broken down into a series of steps similar to those for monoprotic acids.

EXAMPLE 17.20

Calculate the concentrations of all species—H_3PO_4, $H_2PO_4^-$, HPO_4^{2-}, PO_4^{3-}, H^+, and OH^-—present at equilibrium in a solution containing a total phosphoric acid concentration of 0.100 M.

Let x be the concentrations of H^+ and $H_2PO_4^-$ produced by the first step in this equilibrium.

$$H_3PO_4 \rightleftharpoons H^+ + H_2PO_4^-$$

The concentration of H_3PO_4 will be $0.100 - x$. A small amount of the $H_2PO_4^-$ produced in the first step will disappear by dissociation into H^+ and HPO_4^{2-} in the second step, and a small amount of additional H^+ will be produced in both the second and third steps. However, because K_2 and K_3 are very small compared to K_1, it is logical to assume that the decrease in $[H_2PO_4^-]$ in the second step and the increase in $[H^+]$ in the second and third steps are negligible compared to the quantities of each ion (x) produced in the first step.

Hence at equilibrium:

	$[H_3PO_4]$	$[H^+]$	$[H_2PO_4^-]$
Initial concentrations, M	0.100	~0	0
Equilibrium concentrations, M	$0.100 - x$	x	x

$$K_1 = \frac{[H^+][H_2PO_4^-]}{[H_3PO_4]} = 7.5 \times 10^{-3}$$

$$\frac{(x)(x)}{0.100 - x} = \frac{x^2}{0.100 - x} = 7.5 \times 10^{-3}$$

If we neglect x in the term $0.100 - x$, we have

$$\frac{x^2}{0.100} = 7.5 \times 10^{-3}$$

$$x^2 = 7.5 \times 10^{-4}$$

$$x = 2.7 \times 10^{-2}$$

This value of x corresponds to $(2.7 \times 10^{-2}/0.100) \times 100$, or 27% of the value of the initial concentration of H_3PO_4. Thus it is much too large to neglect. A relatively large value of K_a, such as 7.5×10^{-3}, should also serve to alert us that x probably should not be neglected. Hence an accurate value of x can be calculated by means of the quadratic formula (see Appendix A).

$$x^2 = (7.5 \times 10^{-4}) - (7.5 \times 10^{-3})x$$

$$x^2 + (7.5 \times 10^{-3})x - (7.5 \times 10^{-4}) = 0$$

$$x = \frac{-(7.5 \times 10^{-3}) \pm \sqrt{(7.5 \times 10^{-3})^2 - 4(-7.5 \times 10^{-4})}}{2(1)}$$

$$= \frac{(-7.5 \times 10^{-3}) \pm (5.5 \times 10^{-2})}{2}$$

Hence

$$x = \frac{-7.5 \times 10^{-3} + 5.5 \times 10^{-2}}{2} = 2.4 \times 10^{-2} \text{ M}$$

or

$$\frac{-7.5 \times 10^{-3} - 5.5 \times 10^{-2}}{2} = -3.1 \times 10^{-2} \text{ M}$$

(It is not possible to have a negative concentration, so this is an extraneous root.) Thus

$$x = 2.4 \times 10^{-2} \text{ M}$$

Alternatively, the equation can be solved by successive approximations (see Section 15.2, Example 15.4).

First approximation. On a trial basis, we will neglect the x term in the denominator:

$$K_1 = \frac{x^2}{0.100} = 7.5 \times 10^{-3}$$

$$x^2 = 7.5 \times 10^{-4}$$

$$x = 2.7 \times 10^{-2} \text{ M}$$

Second approximation. The value of x obtained from the first approximation is not negligible compared to 0.100 M. A better estimate of the equilibrium concentration of phosphoric acid, using the approximate value of x just obtained, is

$$0.100 - x = 0.100 - 0.027 = 0.073 \text{ M}$$

$$K_1 = \frac{x^2}{0.073} = 7.5 \times 10^{-3}$$

$$x^2 = (7.5 \times 10^{-3})(7.3 \times 10^{-2}) = 5.5 \times 10^{-4}$$

$$x = 2.3 \times 10^{-2} \text{ M}$$

Since the values of x obtained from the first and second approximations differ by 15%, a third approximation is necessary.

Third approximation. A still better estimate of the equilibrium concentration of phosphoric acid, using the value of x obtained from the second approximation, is

$$0.100 - x = 0.100 - 0.023 = 0.077 \text{ M}$$

$$K_1 = \frac{x^2}{0.077} = 7.5 \times 10^{-3}$$

$$x^2 = (7.5 \times 10^{-3})(0.077) = 5.8 \times 10^{-4}$$

$$x = 2.4 \times 10^{-2} \text{ M}$$

(Note that this value of x is the same as that calculated using the quadratic formula.)

The values of x from the last two approximations agree to within approximately 4%. Hence we can safely conclude that

$$[H^+] = x = 0.024 \text{ M}$$

$$[H_2PO_4^-] = x = 0.024 \text{ M}$$

$$[H_3PO_4] = 0.100 - x = 0.100 - 0.024 = 0.076 \text{ M}$$

$$[OH^-] = \frac{K_w}{[H^+]} = \frac{1.0 \times 10^{-14}}{2.4 \times 10^{-2}} = 4.2 \times 10^{-13} \text{ M}$$

To calculate the concentration of HPO_4^{2-}, we must use the second step of the acid dissociation.

$$H_2PO_4^- \rightleftharpoons H^+ + HPO_4^{2-}$$

Let y be the concentration of HPO_4^- produced in the second step.

	$[H_2PO_4^-]$	$[H^+]$	$[HPO_4^{2-}]$
Initial concentrations, M	0.024	0.024	0
Equilibrium concentrations, M	0.024 − y	0.024 + y	y

$$K_2 = \frac{(0.024 + y)\,y}{0.024 - y} = 6.3 \times 10^{-8}$$

Since K_2 is so small, we can neglect the additive and subtractive y terms. Then the preceding equation becomes

$$\frac{0.024\,y}{0.024} = y = 6.3 \times 10^{-8} \text{ M} = [HPO_4^{2-}]$$

Note that this provides a check for our earlier assumption that the increase in $[H^+]$ and the decrease in $[H_2PO_4^-]$ (now known to be 6.3×10^{-8} M) are negligible compared to the quantity of each ion produced in the first step (0.024 mol).

Now we consider the third dissociation step

$$HPO_4^{2-} \rightleftharpoons H^+ + PO_4^{3-}$$

and determine the concentration of PO_4^{3-}. Let z be the number of moles PO_4^{3-} formed by this step.

	$[HPO_4{}^{2-}]$	$[H^+]$	$[PO_4{}^{3-}]$
Initial concentrations, M	6.3×10^{-8}	0.024	0
Equilibrium concentrations, M	$(6.3 \times 10^{-8}) - z$	$0.024 + z$	z

$$K_3 = \frac{(0.024 + z)(z)}{(6.3 \times 10^{-8}) - z} = 3.6 \times 10^{-13}$$

Since K_3 is very small compared to K_2, the extent of the third dissociation is insignificant, and we can neglect the additive and subtractive z terms in the preceding equation. Thus we can easily solve for z:

$$\frac{0.024\,z}{6.3 \times 10^{-8}} = 3.6 \times 10^{-13}$$

$$z = [PO_4{}^{3-}]$$

$$= \frac{(3.6 \times 10^{-13})(6.3 \times 10^{-8})}{2.4 \times 10^{-2}} = 9.5 \times 10^{-19}\ M$$

The quantity of H^+ added to the solution and the decrease in $[HPO_4{}^{2-}]$, z in the third step, are indeed negligible compared to the amounts previously existing in the solution.

Summarizing the equilibrium concentrations requested in the problem:

$$[H_3PO_4] = 0.076\ M;\ [H_2PO_4{}^-] = 0.024\ M;\ [HPO_4{}^{2-}] = 6.3 \times 10^{-8}$$

$$[PO_4{}^{3-}] = 9.5 \times 10^{-19}\ M;\ [H^+] = 0.024\ M;\ [OH^-] = 4.2 \times 10^{-13}$$

17.11 Reactions of Salts with Water

The conjugate base of a weak acid reacts with water to increase the concentration of hydroxide ion (Section 16.3). A salt of a weak acid and a strong base, such as $NaCH_3CO_2$, contains the conjugate base ($CH_3CO_2{}^-$) of a weak acid and so forms aqueous solutions that are basic.

$$CH_3CO_2{}^- + H_2O \rightleftharpoons CH_3CO_2H + OH^-$$

The conjugate acid of a weak base reacts with water to increase the concentration of hydrogen ion. A salt of a weak base and a strong acid, such as NH_4Cl, contains the conjugate acid ($NH_4{}^+$) of a weak base and so forms aqueous solutions that are acidic.

$$NH_4{}^+ + H_2O \rightleftharpoons NH_3 + H_3O^+$$

Since both the conjugate base of a strong acid and the conjugate acid of a strong base are very weak, a salt such as NaCl or KNO_3 that is the product of a strong acid and a strong base forms aqueous solutions that are neutral. A salt of a weak acid and a weak base can react with water to form neutral, basic, or acidic solutions, depending on the relative strengths of the conjugate acid and conjugate base present in the salt.

17.12 Reaction of a Salt of a Strong Base and a Weak Acid with Water

A salt formed from a strong base and a weak acid contains the conjugate base of the weak acid. For instance, sodium acetate, $NaCH_3CO_2$, is the salt of the strong base sodium hydroxide and the weak acid acetic acid. Such salts dissolve in water to give basic solutions (Fig. 17.10), because the sodium ion (the ion of an alkali metal) has no appreciable reaction with water (see Section 17.15), whereas the acetate ion (the conjugate base of a weak acid) does react with water to produce hydroxide ions.

$$Na^+ + H_2O \longrightarrow \text{(no appreciable reaction)}$$

$$CH_3CO_2^- + H_2O \rightleftharpoons CH_3CO_2H + OH^-$$

Applying the law of mass action to the reaction, we have

$$\frac{[CH_3CO_2H][OH^-]}{[CH_3CO_2^-][H_2O]} = K$$

Because the number of moles of water involved in the reaction is negligible compared with the total number of moles of water present in dilute solution, we may assume that the concentration of water remains unchanged. Hence the preceding equation can be written

$$\frac{[CH_3CO_2H][OH^-]}{[CH_3CO_2^-]} = K[H_2O] = K_b$$

This equilibrium constant is simply the ionization constant for the base $CH_3CO_2^-$ (Section 17.5). When the value for K_b is small, the amount of reaction is slight; when it is large, the amount of reaction is extensive.

The value of the equilibrium constant for the base $CH_3CO_2^-$ may be calculated from the values for the ionization constants of water and the conjugate acid of the anion (acetic acid). For every aqueous solution

$$[H^+][OH^-] = K_w \qquad \text{or} \qquad [OH^-] = \frac{K_w}{[H^+]} \qquad \text{(See Section 17.1)}$$

Figure 17.10

Addition of an indicator to a colorless solution of $NaCH_3CO_2$ shows that the solution is slightly basic.

Substituting $K_w/[H^+]$ for $[OH^-]$ in the expression for the ionization constant, we obtain

$$\frac{[CH_3CO_2H][OH^-]}{[CH_3CO_2^-]} = \frac{[CH_3CO_2H]K_w}{[CH_3CO_2^-][H^+]} = K_b$$

But the expression

$$\frac{[CH_3CO_2H]}{[CH_3CO_2^-][H^+]}$$

is the reciprocal of K_a, or $1/K_a$, for acetic acid. Therefore

$$K_b = \frac{K_w}{K_a} = \frac{1.0 \times 10^{-14}}{1.8 \times 10^{-5}} = 5.6 \times 10^{-10}$$

For any salt of a strong base and a weak acid, the ionization constant of the conjugate base of the weak acid is given by the expression

$$K_b = \frac{K_w}{K_a}$$

where K_a is the ionization constant of the weak acid.

It is sometimes useful to rewrite the expression as

$$K_a \times K_b = K_w$$

for any conjugate acid-base pair. This shows clearly that for a relatively strong acid (large K_a), the conjugate base must be relatively weak (small K_b); conversely, for a relatively weak acid (small K_a), the conjugate base will be relatively strong (large K_b).

Calculate the hydroxide ion concentration, the percent reaction, and the pH of a 0.050 M solution of sodium acetate.   **EXAMPLE 17.21**

The equation for the reaction of sodium acetate with water is

$$CH_3CO_2^- + H_2O \rightleftharpoons CH_3CO_2H + OH^-$$

and the expression for the equilibrium constant is

$$\frac{[CH_3CO_2H][OH^-]}{[CH_3CO_2^-]} = K_b = \frac{K_w}{K_a} = 5.6 \times 10^{-10}$$

	$[CH_3CO_2^-]$	$[CH_3CO_2H]$	$[OH^-]$
Initial concentrations, M	0.050	0	~ 0
Equilibrium concentrations, M	$0.050 - x$	x	x

In this table, the information given in the problem is indicated in color.

If we let x equal the concentration of acetic acid formed by the reaction of the acetate ion with water, then the concentration of hydroxide ion will also be equal to x, and the concentration of acetate ion at equilibrium will be $0.050 - x$. Substituting in the equilibrium constant expression, we have

$$K_b = \frac{x^2}{0.050 - x} = 5.6 \times 10^{-10}$$

Since x is very small relative to 0.050, we may drop it from the denominator. Then

$$x^2 = 0.050 \times 5.6 \times 10^{-10} = 28 \times 10^{-12}$$

$$x = 5.3 \times 10^{-6} = [CH_3CO_2H] = [OH^-]$$

Thus the concentration of hydroxide ion is 5.3×10^{-6} M, and the percent reaction is equal to

$$\frac{[CH_3CO_2H]}{[CH_3CO_2^-]} \times 100 = \frac{5.3 \times 10^{-6}}{0.050} \times 100 = 0.011\%$$

We can calculate the pH of the solution by first finding the hydronium ion concentration:

$$[H_3O^+] = \frac{K_w}{[OH^-]} = \frac{1.0 \times 10^{-14}}{5.3 \times 10^{-6}} = 1.9 \times 10^{-9}$$

and then using that value.

$$pH = -\log [H_3O^+] = -\log (1.9 \times 10^{-9}) = 8.72$$

This pH corresponds to a basic solution, as expected.

EXAMPLE 17.22   Calculate the equilibrium constant for the reaction of the sulfide ion with water and the sulfide ion concentration in a 0.0010 M solution of sodium sulfide (see the photograph).

The sulfide ion reacts with water according to the equation

$$S^{2-} + H_2O \rightleftharpoons HS^- + OH^-$$

The hydrogen sulfide ion thus formed also undergoes reaction with water, but the extent of this is so slight compared with that of the sulfide ion that it can be neglected. The expression for the equilibrium constant of the sulfide ion is

$$\frac{[HS^-][OH^-]}{[S^{2-}]} = K_b = \frac{K_w}{K_a \text{ (for } HS^-)} = \frac{1.0 \times 10^{-14}}{1.0 \times 10^{-19}} = 1.0 \times 10^5$$

The indicator shows that a 0.001 M solution of Na_2S is very basic.

The large value for the equilibrium constant indicates that essentially all of the sulfide ion reacts with water. Thus $[HS^-]$ will be close to 0.0010 M. If we let x equal $[S^{2-}]$ at equilibrium, then the number of moles of S^{2-} per liter that react to form HS^- and OH^- is $(0.0010 - x)$, and the concentrations of HS^- and OH^- thus formed are each $(0.0010 - x)$ M. With our choice of x, its value is very small compared to 0.0010 and can thus be neglected. This illustrates how by planning ahead, a judicious choice of x can sometimes simplify the resulting calculation.

	$[S^{2-}]$	$[HS^-]$	$[OH^-]$
Initial concentrations, M	0.0010	0	~0
Equilibrium concentrations, M	x	$0.0010 - x$	$0.0010 - x$

Substituting these values in the expression for K_b, we have

$$K_b = \frac{[HS^-][OH^-]}{[S^{2-}]} = \frac{(0.0010 - x)(0.0010 - x)}{x} = 1.0 \times 10^5$$

Since the reaction is nearly complete, x is small compared to 0.0010 and probably can be neglected in the numerator. The expression then becomes

$$\frac{(0.0010)^2}{x} = 1.0 \times 10^5$$

$$x = \frac{(0.0010)^2}{1.0 \times 10^5} = 1.0 \times 10^{-11} = [S^{2-}]$$

The value of x is sufficiently small that our approximation was justified. The sulfide ion concentration is reduced from 0.0010 M to 1.0×10^{-11} M by the reaction with water.

17.13 Reaction of a Salt of a Weak Base and a Strong Acid with Water

Ammonium chloride, NH_4Cl, is a salt of weak base and a strong acid. The ammonium ion, the conjugate acid of the weak base ammonia, is acidic and reacts with water to produce ammonia and hydronium ions. Note that the chloride ion, being the conjugate base of a strong acid (HCl), has no significant tendency to attract a proton from water and hence has no appreciable effect on the H^+ and OH^- concentrations in the solution.

$$NH_4^+ + H_2O \rightleftharpoons NH_3 + H_3O^+$$
$$Cl^- + H_2O \longrightarrow \text{(no appreciable reaction)}$$

From the law of mass action, we obtain

$$\frac{[NH_3][H_3O^+]}{[NH_4^+]} = [H_2O]K = K_a$$

The equilibrium constant is simply the ionization constant for the acid NH_4^+.

The value of K_a for NH_4^+ may be determined from K_w and the ionization constant, K_b, for NH_3. The expression for the acidity constant, based on the equation

$$NH_4^+ + H_2O \rightleftharpoons NH_3 + H_3O^+$$

is written

$$\frac{[NH_3][H_3O^+]}{[NH_4^+]} = K_a$$

Substituting $K_w/[OH^-]$ for $[H_3O^+]$ in the above equation, we obtain

$$\frac{[NH_3]K_w}{[NH_4^+][OH^-]} = K_a$$

The expression $\dfrac{[NH_3]}{[NH_4^+][OH^-]}$ is equal to the reciprocal of K_b for NH_3 and equals

$\dfrac{1}{K_b \text{ (for } NH_3)}$, hence

$$K_a = \frac{K_w}{K_b} = \frac{1.0 \times 10^{-14}}{1.8 \times 10^{-5}} = 5.6 \times 10^{-10}$$

The equilibrium constant for the ionization of the conjugate acid of any weak base may be calculated from the expression

$$K_a = \frac{K_w}{K_b}$$

where K_b is the ionization constant of the weak base.

17.14 Reaction of a Salt of a Weak Base and a Weak Acid with Water

When we dissolve ammonium acetate, $NH_4CH_3CO_2$, in water, we have a solution of ammonium ion (an acid) and acetate ion (a base), both of which undergo reaction with water.

$$NH_4^+ + H_2O \rightleftharpoons NH_3 + H_3O^+$$

$$CH_3CO_2^- + H_2O \rightleftharpoons CH_3CO_2H + OH^-$$

The extent to which these reactions take place is about the same because the reactions of ammonium ion and acetate ion with water have nearly equal equilibrium constants. Therefore the products are present in nearly equal concentrations at equilibrium. As H^+ and OH^- are produced by these reactions, they react to form water because the product of their concentrations cannot exceed 1.0×10^{-14}. Thus the solution remains neutral as the reactions proceed.

The net reaction may be seen as being the sum of three equations:

$$NH_4^+ + H_2O \rightleftharpoons NH_3 + H_3O^+$$

$$CH_3CO_2^- + H_2O \rightleftharpoons CH_3CO_2H + OH^-$$

$$\underline{H_3O^+ + OH^- \rightleftharpoons 2H_2O}$$

$$NH_4^+ + CH_3CO_2^- \rightleftharpoons CH_3CO_2H + NH_3$$

Applying the law of mass action to the net reaction, we obtain

$$\frac{[CH_3CO_2H][NH_3]}{[NH_4^+][CH_3CO_2^-]} = K$$

To express the equilibrium constant in terms of the ion product for water and the ionization constants for aqueous ammonia and acetic acid, we multiply the numerator and denominator of the above equation by $[H_3O^+][OH^-]$.

$$\frac{[NH_3]}{[NH_4^+][OH^-]} \times \frac{[CH_3CO_2H]}{[H_3O^+][CH_3CO_2^-]} \times \frac{[H_3O^+][OH^-]}{1} = K$$

Thus it may readily be seen that

$$K = \frac{K_w}{K_b \text{ (for NH}_3\text{)} \times K_a \text{ (for CH}_3\text{CO}_2\text{H)}}$$

$$= \frac{1.0 \times 10^{-14}}{(1.8 \times 10^{-5})(1.8 \times 10^{-5})} = 3.1 \times 10^{-5}$$

When the ions of a salt of a weak base and a weak acid do not have the same equilibrium constants, the aqueous solution of the salt is either basic or acidic, depending on which ion has the larger equilibrium constant. For example, ammonium cyanide, NH_4CN, reacts with water to give a basic solution, since K_b for CN^- is larger than K_a for NH_4^+. The reaction of the cyanide ion with water,

$$CN^- + H_2O \rightleftharpoons HCN + OH^-$$

is more extensive than that of the ammonium ion,

$$NH_4^+ + H_2O \rightleftharpoons NH_3 + H_3O^+$$

and an excess of hydroxide ions accumulates in the solution.

17.15 The Ionization of Hydrated Metal Ions

A large number of metal ions behave as acids in solution. For example, the aluminum(III) ion of aluminum nitrate reacts with water to give a hydrated aluminum ion, $[Al(H_2O)_6]^{3+}$.

$$Al(NO_3)_3(s) + 6H_2O(l) \longrightarrow [Al(H_2O)_6]^{3+}(aq) + 3NO_3^-(aq)$$

The ion $[Al(H_2O)_6]^{3+}$ is an acid and donates hydrogen ions to water (Fig. 17.11).

$$[Al(H_2O)_6]^{3+} + H_2O \rightleftharpoons [Al(OH)(H_2O)_5]^{2+} + H_3O^+$$

The simplified version of the equation for the reaction is usually used in calculations of the concentrations present at equilibrium.

$$[Al(H_2O)_6]^{3+} \rightleftharpoons [Al(OH)(H_2O)_5]^{2+} + H^+$$

Just like a polyprotic acid, the hydrated aluminum ion ionizes in stages, as shown by

$$[Al(H_2O)_6]^{3+} \rightleftharpoons [Al(OH)(H_2O)_5]^{2+} + H^+$$
$$[Al(OH)(H_2O)_5]^{2+} \rightleftharpoons [Al(OH)_2(H_2O)_4]^{+} + H^+$$
$$[Al(OH)_2(H_2O)_4]^{+} \rightleftharpoons [Al(OH)_3(H_2O)_3] + H^+$$

The ionization of a cation carrying more than one charge is not extensive beyond the first stage. Additional examples of the first stage in the ionization of hydrated metal ions are

$$[Fe(H_2O)_6]^{3+} \rightleftharpoons [Fe(OH)(H_2O)_5]^{2+} + H^+$$
$$[Cu(H_2O)_4]^{2+} \rightleftharpoons [Cu(OH)(H_2O)_3]^{+} + H^+$$
$$[Zn(H_2O)_6]^{2+} \rightleftharpoons [Zn(OH)(H_2O)_5]^{+} + H^+$$

Figure 17.11

The pH paper shows that a 0.1 M solution of aluminum nitrate has a pH of 3.

Ions known to be hydrated in solution are often indicated in abbreviated form without showing the hydration. For example, Al^{3+} is often written instead of $[Al(H_2O)_6]^{3+}$. However, when water participates in a reaction, the hydration becomes important and we use formulas that show the extent of hydration.

EXAMPLE 17.23   Let us calculate the pH of a 0.10 M solution of aluminum chloride that dissolves to give the hydrated aluminum ion, $[Al(H_2O)_6]^{3+}$, in solution. The ionization constant, K_a, is 1.4×10^{-5} for the reaction

$$[Al(H_2O)_6]^{3+} \rightleftharpoons [Al(OH)(H_2O)_5]^{2+} + H^+$$

The expression for the ionization constant is

$$\frac{[Al(OH)(H_2O)_5{}^{2+}][H^+]}{[Al(H_2O)_6{}^{3+}]} = K_a = 1.4 \times 10^{-5}$$

Let x equal the equilibrium concentration of hydrogen ion, which is equal to the concentration of $[Al(OH)(H_2O)_5]^{2+}$. Then $0.10 - x$ will equal the equilibrium concentration of $[Al(H_2O)_6]^{3+}$.

	$[Al(H_2O)_6{}^{3+}]$	$[H^+]$	$[Al(OH)(H_2O)_5{}^{2+}]$
Initial concentrations, M	0.10	~0	0
Equilibrium concentrations, M	$0.10 - x$	x	x

Substituting in the expression for the ionization constant, we have

$$K_a = \frac{[Al(OH)(H_2O)_5{}^{2+}][H^+]}{[Al(H_2O)_6{}^{3+}]} = \frac{x^2}{0.10 - x} = 1.4 \times 10^{-5}$$

Since K_a is small, x will also be small, and we may drop x from the term $0.10 - x$ in an approximate solution of the problem. Thus we obtain

$$\frac{x^2}{0.10} = 1.4 \times 10^{-5}$$

$$x = 1.2 \times 10^{-3} = [H^+]$$

$$pH = -\log(1.2 \times 10^{-3}) = 2.92 \quad \text{(an acid solution)}$$

The constants for the different stages of ionization are not known for many metal ions, so we cannot calculate the extent of their ionization. However, if the hydroxide of a metal ion is insoluble in water, the hydrated metal ion will ionize extensively to give a very acidic solution. In fact, practically all metal ions other than those of the alkali metals ionize to give acidic solutions. The ionization constant increases as the charge of the metal ion increases or as the size of the metal ion decreases.

17.16 Acid-Base Indicators

Certain organic substances change color in dilute solution when the hydrogen ion concentration reaches a particular value. For example, phenolphthalein is a colorless substance in any aqueous solution with a hydrogen ion concentration greater than 5.0×10^{-9} M (pH less than 8.3). In solutions with a hydrogen ion concentration less than 5.0×10^{-9} M (pH greater than 8.3), phenolphthalein is red or pink. Substances like phenolphthalein, which can be used to determine the pH of a solution, are called **acid-base indicators.** Acid-base indicators are either weak organic acids, HIn, or weak organic bases, InOH, where the letters In stand for a complex organic group.

The equilibrium in a solution of the acid-base indicator **methyl orange,** a weak acid, can be represented by the equation

$$HIn \rightleftharpoons H^+ + In^-$$

Red   Yellow

The anion of methyl orange is yellow, and the nonionized form is red. If acid is added to the solution, the increase in the hydrogen ion concentration shifts the equilibrium toward the red form in accordance with the law of mass action.

$$K_a = \frac{[H^+][In^-]}{[HIn]}$$

The indicator's color is the visible result of the ratio of the concentrations of the two species In^- and HIn. For methyl orange,

$$\frac{[In^-]}{[HIn]} = \frac{[\text{substance with yellow color}]}{[\text{substance with red color}]} = \frac{K_a}{[H^+]}$$

When $[H^+]$ has the same numerical value as K_a, the ratio of $[In^-]$ to $[HIn]$ is equal to 1, meaning that 50% of the indicator is present in the red acid form and 50% in the yellow ionic form, and the solution appears orange in color. When the hydrogen ion concentration increases to a pH of 3.1, about 90% of the indicator is present in the red form and 10% in the yellow form, and the solution turns red. No change in color is visible for any further increase in the hydrogen ion concentration.

Addition of a base to the system reduces the hydrogen ion concentration and shifts the equilibrium toward the yellow form. At a pH of 4.4 about 90% of the indicator is in the yellow ionic form, and a further decrease in the hydrogen ion concentration does not produce a visible color change. The pH range between 3.1 (red) and 4.4 (yellow) is the **color-change interval** of methyl orange; the pronounced color change takes place between these pH values.

The large number of known acid-base indicators cover a wide range of pH values and can be used to determine the approximate pH of an unknown solution by a process of elimination. Table 17.6 lists some of these indicators, their colors, and their color-

Table 17.6   Some acid-base indicators

Indicator	Color in the more acid range	pH range	Color in the more basic range
Methyl violet	Yellow	0–2	Violet
Thymol blue	Pink	1.2–2.8	Yellow
Bromophenol blue	Yellow	3.0–4.7	Violet
Methyl orange	Pink	3.1–4.4	Yellow
Bromocresol green	Yellow	4.0–5.6	Blue
Bromocresol purple	Yellow	5.2–6.8	Purple
Litmus	Red	4.7–8.2	Blue
Phenolphthalein	Colorless	8.3–10.0	Pink
Thymolphthalein	Colorless	9.3–10.5	Blue
Alizarin yellow G	Colorless	10.1–12.1	Yellow
Trinitrobenzene	Colorless	12.0–14.3	Orange

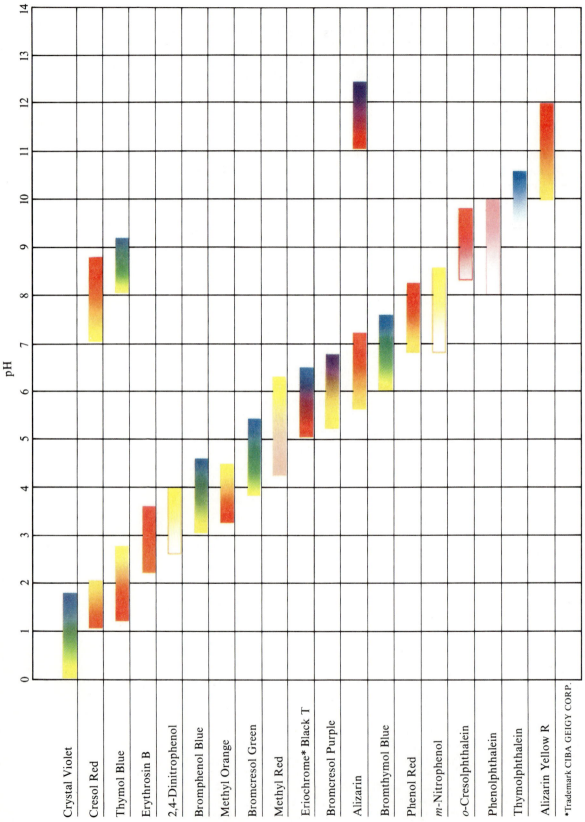

Figure 17.12

Ranges of color changes for several acid-base indicators.

The pH ranges shown are approximate. Specific transition ranges depend on the indicator solvent chosen.

*Trademark CIBA GEIGY CORP.

Crystal Violet

Cresol Red

Thymol Blue

Erythrosin B

2,4-Dinitrophenol

Bromphenol Blue

Methyl Orange

Bromcresol Green

Methyl Red

Eriochrome* Black T

Bromcresol Purple

Alizarin

Bromthymol Blue

Phenol Red

m-Nitrophenol

o-Cresolphthalein

Phenolphthalein

Thymolphthalein

Alizarin Yellow R

pH

change intervals (Also see Fig. 17.12). The use of these indicators for specific purposes will be discussed in Section 17.17.

17.17 Titration Curves

Titration curves, plots of the pH versus the volume of acid or base added in a titration (Section 3.2), show the point where equivalent quantities of acid and base are present (the equivalence point), which aids in the choice of a proper indicator.

The simplest acid-base reactions are those of a strong acid with a strong base. Let us consider the titration of a 25.0-milliliter sample of 0.100 M hydrochloric acid with 0.100 M sodium hydroxide. The values of the pH measured after successive additions of small amounts of NaOH are listed in Table 17.7, column (a), and shown in Fig. 17.13. The pH increases very slowly at first, increases very rapidly in the middle portion of the curve, and then increases very slowly again. The point of inflection (the midpoint of the vertical part of the curve) is the **equivalence point** for the titration. It indicates when equivalent quantities of acid and base are present. For the titration of a strong acid and a strong base, the equivalence point occurs at a pH of 7 [Table 17.7, column (a), and Fig. 17.13]. A solution with a pH of 7 is neutral; the hydrogen ion concentration is equal to hydroxide ion concentration.

Table 17.7   pH values (a) in the titration of a strong acid with a strong base and (b) in the titration of a weak acid with a strong base

Volume of 0.100 M NaOH added, mL	Moles of NaOH added	pH Values	
		(a) Titration of 25.00 mL of 0.100 M HCl	(b) Titration of 25.00 mL of 0.100 M CH_3CO_2H
0.0	0.0	1.00	2.87
5.0	0.00050	1.18	4.14
10.0	0.00100	1.37	4.57
15.0	0.00150	1.60	4.92
20.0	0.00200	1.95	5.35
22.0	0.00220	2.20	5.61
24.0	0.00240	2.69	6.13
24.5	0.00245	3.00	6.44
24.9	0.00249	3.70	7.14
25.0	0.00250	7.00	8.72
25.1	0.00251	10.30	10.30
25.5	0.00255	11.00	11.00
26.0	0.00260	11.29	11.29
28.0	0.00280	11.75	11.75
30.0	0.00300	11.96	11.96
35.0	0.00350	12.22	12.22
40.0	0.00400	12.36	12.36
45.0	0.00450	12.46	12.46
50.0	0.00500	12.52	12.52

(a) Titration of 25.00 mL of 0.100 M HCl (0.00250 mol of HCl) with 0.100 M NaOH
(b) Titration of 25.00 mL of 0.100 M CH_3CO_2H (0.00250 mol of CH_3CO_2H) with 0.100 M NaOH

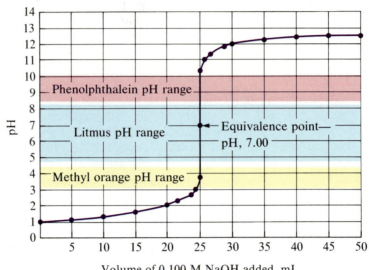

Figure 17.13

Titration curve for the titration of 25.00 mL of 0.100 M HCl (strong acid) with 0.100 M NaOH (strong base). The pH ranges for the color change of phenolphthalein, litmus, and methyl orange are indicated by the shaded areas.

The titration of a weak acid with a strong base (or of a weak base with a strong acid) is somewhat more complicated than that just discussed, but follows the same basic principles. Let us consider the titration of 25.0 milliliters of 0.100 M acetic acid with 0.100 M sodium hydroxide and compare the titration curve with that of the strong acid (Fig. 17.13). Table 17.7, column (b), gives the pH values during the titration; Fig. 17.14 shows the titration curve.

The similarities in the two titration curves are readily apparent, but there are also important differences. The titration curve for the weak acid begins at a higher pH value (less acidic) and maintains higher pH values up to the equivalence point. This is because acetic acid is a weak acid and releases only a small quantity of hydrogen ions. The pH at the equivalence point is also higher (8.72, rather than 7.00) because the solution contains sodium acetate. The acetate ion is a weak base and raises the pH as described in Section 17.12.

$$CH_3CO_2^- + H_2O \rightleftharpoons CH_3CO_2H + OH^-$$

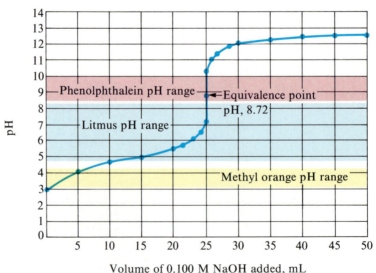

Figure 17.14

Titration curve for the titration of 25.00 mL of 0.100 M CH_3CO_2H (weak acid) with 0.100 M NaOH (strong base). The pH ranges for the color change of phenolphthalein, litmus, and methyl orange are indicated by the shaded areas.

After the equivalence point the two curves are identical (Table 17.7 and Figs. 17.13 and 17.14) because the pH is dependent on the excess of hydroxide ion in both cases. After an acid is neutralized, whether it is a strong acid (like HCl) or a weak acid (like CH_3CO_2H), it has no further significant effect on the concentration of the hydroxide ion.

Titration curves help us pick an indicator that will provide a sharp color change at the equivalence point. The pH ranges for the color changes of three indicators from Table 17.6 are indicated in Figs. 17.13 and 17.14.

Phenolphthalein has a sharp color change over a small added volume of NaOH and hence is suitable for titrations that are either strong acid–strong base or weak acid–strong base. The equivalence point is located in the steeply rising portion of the titration curve, where the curve passes through the color-change interval of the indicator.

Litmus is a suitable indicator for the HCl titration because the titration curve shows that it changes color within 0.10 milliliter of the equivalence point. However, litmus would be a poor choice for the CH_3CO_2H titration because the pH is within the color change interval of litmus when only about 12 milliliters of NaOH have been added and does not leave the range until 25 milliliters have been added. The end of the color change would be at about the equivalence point (a good feature), but the color change would be very gradual, taking place during the addition of 13 milliliters of NaOH, making it almost useless as an indicator of the equivalence point.

Methyl orange could be used for the HCl titration (Fig. 17.13), although its use would have two disadvantages: (1) it completes its color change slightly before the equivalence point is reached (but very close to it, so this is not too serious), and (2) it changes color, as the plot shows, during the addition of nearly 0.5 milliliter of NaOH, which is not as sharp a color change as that of litmus or phenolphthalein. The titration curve shows that methyl orange would be completely useless as an indicator for the CH_3CO_2H titration. Its color change begins after about 1 milliliter of NaOH has been added and ends when about 8 milliliters have been added. The color change is completed long before the equivalence point (which occurs when 25.0 milliliters of NaOH have been added) is reached and hence provides no indication of the equivalence point.

The pH range in which an equivalence point will be observed can also be determined by calculating a pH curve. For example, the hydrogen ion concentration of a 0.100 M solution of acetic acid can be calculated as described in Example 17.12 and then converted to pH. The determination of the hydrogen ion concentration in a solution resulting from the addition of a known amount of sodium hydroxide to 0.100 M acetic acid has been described in Example 17.16. At the equivalence point equimolar amounts of sodium hydroxide and acetic acid have been mixed. The problem therefore becomes the calculation of the pH of a solution of sodium acetate, a calculation described in Section 17.12. Beyond the equivalence point the pH is controlled by the excess sodium hydroxide present and may be determined from the concentration of hydroxide ion.

For Review

Summary

Water is an extremely weak electrolyte that undergoes **autoionization,** or **self-ionization,**

$$H_2O + H_2O \rightleftharpoons H_3O^+ + OH^- \quad \text{or} \quad H_2O \rightleftharpoons H^+ + OH^-$$

Thus pure water contains equal concentrations of hydrogen ion and hydroxide ion. The **ion product for water, K_w,** is the equilibrium constant for the self-ionization reaction; at 25°C.

$$K_w = [H^+][OH^-] = 1.0 \times 10^{-14}$$

All aqueous solutions contain both H^+ and OH^-, with the product $[H^+][OH^-]$ equal to 1.0×10^{-14} at 25°C.

The concentration of H^+ in a solution may be expressed as the pH of the solution; **pH = −log [H$^+$].** The concentration of OH^- may be expressed as the pOH of the solution: **pOH = −log [OH$^-$].** In pure water pH = 7.00 and pOH = 7.00. In any aqueous solution at 25°C, pH + pOH = 14.00.

Although strong electrolytes ionize completely in solution, a solution of a weak electrolyte contains a mixture of molecular and ionic species. The concentration of the nonionized molecules of a weak acid (HA), of hydrogen ion, and of the conjugate base (A^-) of the weak acid, can be determined from K_a, the **ionization constant of the weak acid.** K_a is the equilibrium constant for the reaction

$$HA \rightleftharpoons H^+ + A^-; \qquad K_a = \frac{[H^+][A^-]}{[HA]}$$

The **ionization constant for a weak base** (B) is the equilibrium constant for the reaction

$$B + H_2O \rightleftharpoons BH^+ + OH^-; \qquad K_b = \frac{[BH^+][OH^-]}{[B]}$$

The extent of the ionization of a weak acid in solution can be reduced by adding a compound containing the conjugate base of the weak acid **(the common ion effect)** or by adding another acid. Likewise, the extent of the ionization of a weak base can be reduced by adding the conjugate acid of the weak base or by adding hydroxide. A solution containing a mixture of an acid and its conjugate base, or of a base and its conjugate acid, is called a **buffer solution.** Unlike solutions of acids, bases, or most salts, the hydrogen ion concentration of a buffer solution does not change greatly when a small amount of acid or base is added to it.

The conjugate base of a weak acid reacts with water to give a basic solution. The conjugate acid of a weak base gives an acid solution. The equilibrium constant for the reaction of the conjugate base of a weak acid with water can be calculated from K_w and K_a for the weak acid

$$K_b = \frac{K_w}{K_a}$$

For the conjugate acid of a weak base,

$$K_a = \frac{K_w}{K_b}$$

where K_b is the ionization constant of the weak base.

The pH of a solution may be determined either by using a **pH meter** or with **acid-base indicators.** Acid-base indicators are also used to determine the equivalence point in acid-base titrations. The correct indicator for the titration of an acid and a base is selected by considering the **titration curve** for the acid-base pair.

Key Terms and Concepts

acid-base indicator (17.16)
autoionization (17.1)
buffer solution (17.8)
buffer capacity (17.8)
common ion effect (17.7)
equivalence point (17.17)
ionization constant (17.4)
ionization of hydrated metal ions
(17.15)

ionization of water (17.1)
ionization of weak acids (17.4,
17.9, 17.10)
ionization of weak bases (17.5)
ion product for water (17.1)
pH (17.2)
pK_w (17.2)

pOH (17.2)
polyprotic acid (17.9, 17.10)
reactions of salts with water
(17.11–17.15)
salt effect (17.6)
self-ionization (17.1)
titration curve (17.17)

Exercises

Values of K not given in the exercises may be found in Appendixes F and G.

Ionic Equilibria of Acids and Bases; pH and pOH

1. Calculate the pH and pOH of each of the following acidic solutions:
    (a) 0.0092 M HOCl ($[H^+] = 1.8 \times 10^{-5}$ M)
    Ans. pH = 4.74; pOH = 9.26
    (b) 0.0810 M HCN ($[H^+] = 6 \times 10^{-6}$ M)
    Ans. pH = 5.2; pOH = 8.8
    (c) 0.0992 M $HC_2O_4^-$ (percent ionization is 2.5%)
    Ans. pH = 2.60; pOH = 11.40
    (d) 0.138 M CH_3CO_2H (percent ionization is 1.2%)
    Ans. pH = 2.78; pOH = 11.22
2. Calculate the pH and pOH of each of the following basic solutions:
    (a) 0.0784 M $C_6H_5NH_2$ ($[OH^-] = 6.0 \times 10^{-6}$ M)
    Ans. pH = 8.78; pOH = 5.22
    (b) 0.1098 M $(CH_3)_3N$ ($[OH^-] = 2.8 \times 10^{-3}$ M)
    Ans. pH = 11.45; pOH = 2.55
    (c) 0.222 M CN^- (percent of ions that produce OH^- is 1.1%)

    $$CN^- + H_2O \rightleftharpoons HCN + OH^-$$
    Ans. pH = 11.39; pOH = 2.61
    (d) 0.300 M N_2H_4 (percent ionization is 0.3%)

    $$N_2H_4 + H_2O \rightleftharpoons N_2H_5^+ + OH^-$$
    Ans. pH = 11.0; pOH = 3.0
    (e) 1.13 M BrO^- (percent BrO^- that forms OH^- is 0.2%)

    $$BrO^- + H_2O \rightleftharpoons HOBr + OH^-$$
    Ans. pH = 11.3; pOH = 2.7
3. Calculate the $[H^+]$ and the percent ionization of each of the following solutions:

(a) 0.19 M HNO_2 (pH = 2.04)
Ans. 9.1×10^{-3} M; 4.8%
(b) 1.4×10^{-2} M HCNO (pH = 2.69)
Ans. 2.0×10^{-3} M; 14%
(c) 0.0407 M HNO_3 *Ans. 0.0407 M; 100%*
(d) 0.407 M $HC_2O_4^-$ (pH = 2.29)
Ans. 5.1×10^{-3} M; 1.3%
(e) 0.0407 M $HC_2O_4^-$ (pH = 2.79)
Ans. 1.6×10^{-3} M; 3.9%

4. Calculate the pH and pOH of pure water at 25°C. ($[H^+] = [OH^-] = 1.00 \times 10^{-7}$ M)
Ans. pH = 7.00; pOH = 7.00
5. Sodium hydrogen sulfate, $NaHSO_4$, is used as a solid acid in some porcelain cleansers because it reacts with calcium carbonate precipitates. The percent ionization of the HSO_4^- ion in a 0.50 M solution of $NaHSO_4$ is 14%. Calculate the pH and pOH.
Ans. pH = 1.15; pOH = 12.85
6. What ionic and molecular species are present in an aqueous solution of hydrogen fluoride, HF? A solution of sulfuric acid? A solution of SO_2 in water (sulfurous acid)?
7. Equilibrium calculations are not necessary when dealing with ionic concentrations in solutions of strong electrolytes such as NaOH and HCl. Why not?
8. Compare the extent of ionization of strong electrolytes and weak electrolytes in water.
9. Calculate the concentration of each of the ions in the following solutions of strong electrolytes:
    (a) 1.642 M HNO_3
    Ans. $[H^+] = 1.642$ M; $[NO_3^-] = 1.642$ M
    (b) 0.172 M $NiCl_2$
    Ans. $[Ni^{2+}] = 0.172$ M; $[Cl^-] = 0.344$ M
    (c) 0.107 M $[Al(H_2O)_6]_2(SO_4)_3$
    Ans. $[Al(H_2O)_6^{3+}] = 0.214$ M; $[SO_4^{2-}] = 0.321$ M
10. From the equilibrium concentrations given, calculate K for each of the following weak acids or weak bases:
(a) HCN: $[H^+] = 1.6 \times 10^{-5}$ M;
$[CN^-] = 1.6 \times 10^{-5}$ M; $[HCN] = 0.6$ M
Ans. 4×10^{-10}

(b) NH_3: $[NH_4^+] = 1.6 \times 10^{-3}$ M;
$[OH^-] = 1.6 \times 10^{-3}$ M; $[NH_3] = 0.14$ M
Ans. 1.8×10^{-5}

(c) HNO_2: $[H^+] = 0.018$ M; $[NO_2^-] = 0.0362$ M;
$[HNO_2] = 1.45$ M   *Ans. 4.5×10^{-4}*

(d) OCN^-: $[OH^-] = 4.00 \times 10^{-5}$ M;
$[HOCN] = 2.63 \times 10^{-7}$ M; $[OCN^-] = 0.364$ M
Ans. 2.89×10^{-11}

11. From the following data, calculate the value of the equilibrium constant and the pOH:

(a) 0.0088 M HClO: 0.20% ionization
Ans. $K = 3.5 \times 10^{-8}$; pOH = 9.25

(b) CH_3CO_2H: 1.1% ionization; $[H^+] = 1.6 \times 10^{-3}$ M
Ans. $K = 1.8 \times 10^{-5}$; pOH = 11.20

(c) 0.08 M HCN: pH = 5.24
Ans. $K = 4 \times 10^{-10}$; pOH = 8.8

(d) 0.0992 M $HC_2O_4^-$: $[OH^-] = 3.97 \times 10^{-12}$ M
Ans. $K = 6.40 \times 10^{-5}$; pOH = 11.401

12. Calculate the hydroxide ion concentration, the percent ionization of the weak base, the pH, and the pOH in each of the following solutions:

(a) 0.222 M CN^- ($K_b = 2.5 \times 10^{-5}$) $CN^- \overset{H_2O}{\rightleftharpoons} HCN + OH^-$
Ans. 2.4×10^{-3} M; 1.1%; 11.38; 2.62

(b) 1.13 M BrO^- ($K_b = 5 \times 10^{-6}$) $BrO^- \rightleftharpoons HBrO + OH^-$
Ans. 2×10^{-3} M; 0.2%; 11.3; 2.7

(c) 0.300 M N_2H_4 ($K_b = 3 \times 10^{-6}$ for

$$N_2H_4 + H_2O \rightleftharpoons N_2H_5^+ + OH^-)$$

Ans. 9×10^{-4} M; 0.3%; 11.0; 3.0

(d) 0.1098 M $(CH_3)_3N$ ($K_b = 7.4 \times 10^{-5}$)
Ans. 2.8×10^{-3} M; 2.6%; 11.45; 2.55

(e) 0.0784 M $C_6H_5NH_2$ ($K_b = 4.6 \times 10^{-10}$)
Ans. 6.0×10^{-6} M; 7.7×10^{-3}%; 8.78; 5.22

13. Calculate the percent ionization of each of the following solutions. Calculate (1) using the simplifying assumption that the amount of dissociation is negligible compared to the concentration of acid originally present and (2) without using simplifying assumption and with use of successive approximations or the quadratic formula. Compare the values.

(a) 0.020 M HCO_2H ($K_a = 1.8 \times 10^{-4}$)
Ans. (1) 9.5%; (2) 9.0%

(b) 1.0×10^{-4} M H_2O_2 ($K_a = 2.4 \times 10^{-12}$)
Ans. (1) and (2) 0.015%

(c) 1.0 M hydrazoic acid, HN_3 ($K_a = 1 \times 10^{-4}$ for

$$HN_3 \rightleftharpoons H^+ + N_3^-)$$

Ans. (1) and (2) 1%

(d) 0.10 M HF   ($K_a = 7.2 \times 10^{-4}$) $= CH-C \equiv$
Ans. (1) 8.5%; (2) 8.1%

14. Propionic acid, $C_2H_5CO_2H$ ($K_a = 1.34 \times 10^{-5}$), is used in the manufacture of calcium propionate, a food preservative. What is the hydrogen ion concentration in a 0.685 M solution of $C_2H_5CO_2H$? *Ans. 3.03×10^{-3} M*

15. Formic acid, HCO_2H, is the irritant that causes the body's reaction to an ant's sting. What is the concentration of hydrogen ion and the percent ionization in a 0.575 M solution of formic acid?
Ans. 1.0×10^{-2} M; 1.7%

16. Calculate the hydrogen ion concentration and percent ionization of the weak acid in each of the following solutions. Note that the ionization constants may be such that the change in electrolyte concentration cannot be neglected, and the quadratic formula or successive approximations may be required.

(a) 0.100 M HF   ($K_a = 7.2 \times 10^{-4}$)
Ans. 8.1×10^{-3} M; 8.1%

(b) 0.0184 M HCNO   ($K_a = 3.46 \times 10^{-4}$)
Ans. 2.36×10^{-3} M; 12.8%

(c) 0.10 M HSO_3NH_2 ($K_a = 1.0 \times 10^{-1}$)
Ans. 0.062 M; 62%

(d) 0.0655 M $[Fe(H_2O)_6]^{3+}$ ($K_a = 4.0 \times 10^{-3}$ for

$$[Fe(H_2O)_6]^{3+} \rightleftharpoons [Fe(OH)(H_2O)_5]^{2+} + H^+)$$

Ans. 1.4×10^{-2} M; 22%

(e) 0.02173 M CH_2ClCO_2H ($K_a = 1.4 \times 10^{-3}$)
Ans. 4.9×10^{-3} M; 22%

17. Calculate the hydroxide ion concentration and the percent ionization of the weak base in each of the following solutions. Note that the ionization constants are sufficiently large that it may be necessary to use the quadratic formula or successive approximations.

(a) 4.113×10^{-3} M CH_3NH_2 ($K_b = 4.4 \times 10^{-4}$)
Ans. 1.1×10^{-3} M; 28%

(b) 0.11 M $(CH_3)_2NH$ ($K_b = 7.4 \times 10^{-4}$)
Ans. 8.7×10^{-3} M; 7.9%

18. Calculate $[OH^-]$ and pH for each of the following:

(a) 0.23 M cyclohexylamine, $C_6H_{11}NH_2$ ($K_b = 4.6 \times 10^{-4}$)   *Ans. $[OH^-] = 0.010$ M; pH = 12.01*

(b) 0.00253 M $C_6H_5O^-$ ($K_b = 7.81 \times 10^{-5}$)
Ans. $[OH^-] = 4.07 \times 10^{-4}$ M; pH = 10.61

19. How would the extent of ionization of acetic acid in a solution be changed by the addition of potassium acetate? By the addition of hydrogen chloride?

20. Calculate the hydrogen ion concentration, the percent ionization of the HSO_4^- ion, the pH, and the pOH of a 0.65 M solution of $NaHSO_4$. (Note that K_a, which you can look up in Appendix F, is relatively large compared to that of most weak acids, indicating that the acid is more highly dissociated than most other weak acids—several times the dissociation of acetic acid, for example. Be sure to check whether the commonly used simplifying assumption is valid.)
Ans. 8.2×10^{-2} M; 13%; 1.08; 12.92

21. What is the fluoride ion concentration in 0.675 L of a solution that initially contains 0.1400 g of HF? (See note in exercise 20.)   *Ans. 2.4×10^{-3} M*

22. Calculate the percent ionization of a 0.50 M solution of

H_3PO_4. (Consider the second and third ionization steps to be negligible compared to the first.)   *Ans. 12%*

23. A solution that contains 20.0 g of lactic acid ($CH_3CHOHCO_2H$) in 1.00 L of solution has a hydrogen ion concentration of 5.43×10^{-3} M. What is the value of the ionization constant?   *Ans. 1.36×10^{-4}*

24. What is the concentration of acetic acid and the pH in a solution that is 0.25% ionized?   *Ans. 2.9 M; 2.14*

25. The artificial sweetener sodium saccharide (often referred to as sodium saccharin on soft drink cans) contains the basic ion $C_7H_4NSO_3^-$ ($K_b = 4.8 \times 10^{-3}$). What is the hydroxide ion concentration in a liter of water sweetened with 0.100 g of $Na(C_7H_4NSO_3)$?

    Ans. 4.5×10^{-4} M

26. What is the hydroxide ion concentration in a dilute solution of household ammonia that contains 0.3124 mol of NH_3 per liter?   *Ans. 2.4×10^{-3} M*

27. A 0.010 M solution of HF is 23% ionized. Calculate $[H^+]$, $[F^-]$, $[HF]$, and the ionization constant for HF.
    Ans. $[H^+] = [F^-] = 2.3 \times 10^{-3}$ M;
    $[HF] = 7.7 \times 10^{-3}$ M; $K_a = 6.9 \times 10^{-4}$

28. Calculate the ionization constants for each of the following solutes from the percent ionization and the concentration of the solute:

    (a) 1.0×10^{-4} M H_2O_2, 1.5×10^{-2}% ionized
    Ans. 2.2×10^{-12}

    (b) 0.300 M CH_2ClCO_2H, 6.60% ionized
    Ans. 1.40×10^{-3}

    (c) 0.10 M HF, 8.1% ionized   *Ans. 7.1×10^{-4}*

    (d) 0.050 M HClO, 8.4×10^{-2}% ionized
    Ans. 3.5×10^{-8}

    (e) 1.0 M NH_3, 0.42% ionized   *Ans. 1.8×10^{-5}*

    (f) 0.010 M HNO_2, 19% ionized   *Ans. 4.5×10^{-4}*

29. Describe the self-ionization of water.

30. Calculate the pH and pOH of pure water at 100°C ($K_w = 1 \times 10^{-12}$ at 100°C).   *Ans. pH = pOH = 6.0*

31. The ionization constant for water (K_w) is 9.614×10^{-14} at 60°C. Calculate $[H^+]$, $[OH^-]$, pH, and pOH for pure water at 60°C. *Ans. $[H^+] = [OH^-] = 3.101 \times 10^{-7}$ M;*
    pH = pOH = 6.5085

32. Calculate the pH of a solution resulting from mixing 0.10 L of 0.10 M NaOH with 0.40 L of 0.025 M HF.   *Ans. 7.72*

33. Calculate the pH of a solution prepared by mixing 10.0 mL of 0.10 M LiOH and 20.0 mL of a 0.500 M benzoic acid solution. K_a for benzoic acid is 6.7×10^{-5}.   *Ans. 3.22*

34. Calculate the pH of 0.500 L of a 0.120 M solution of HClO to which 0.010 mol of $Ca(OH)_2$ has been added.   *Ans. 7.15*

35. Calculate the pH of a solution made by mixing equal volumes of 0.30 M NH_3 and 0.030 M HNO_3.
    Ans. 10.21

36. Calculate the pH of 0.500 L of a 0.0880 M solution of

$NaHCO_3$ to which has been added 1.10×10^{-2} mol of HCl.   *Ans. 6.84*

Buffers

37. What constitutes a buffer solution?

38. Of what practical value are buffer solutions?

39. Explain why the pH does not change significantly when a small amount of an acid or a base is added to a solution containing equal amounts of H_3PO_4 and NaH_2PO_4.

40. Calculate the pH of each of the following buffer solutions, containing equal volumes of two solutes in the concentrations indicated:

    (a) 0.60 M NH_3; 0.60 M NH_4Cl *Ans. 9.26*
    (b) 0.30 M CH_3CO_2H, 0.30 M $NaCH_3CO_2$ *Ans. 4.74*
    (c) 0.25 M CH_3CO_2H, 0.10 M $NaCH_3CO_2$ *Ans. 4.35*
    (d) 0.20 M H_2S; 0.30 M NaHS   *Ans. 7.18*
    (e) 0.25 M H_3PO_4; 0.15 M NaH_2PO_4 *Ans. 1.99*

41. A buffer solution is prepared from 5.0 g of NH_4NO_3 and 0.100 L of 0.10 M NH_3. What is the pH of the buffer?   *Ans. 8.46*

42. Given a 0.1 M solution of ammonia that is also 0.1 M in ammonium chloride and a 1.8×10^{-5} M solution of NaOH, calculate (a) the initial pH in each solution, (b) the pH when a liter of each of the original solutions is treated with 0.03 mol of solid NaOH, and (c) the pH when a liter of each of the original solutions is treated with 0.03 mol of HCl. (Assume no volume change with the addition of the NaOH or the HCl.) Is the buffer solution effective in serving its intended function of holding $[H^+]$ relatively constant compared to the unbuffered one? *Ans. (a) 9.3 for each solution; (b) 9.5; 12.5; (c) 9.0; 1.5*

43. In venous blood the following equilibrium is set up by dissolved carbon dioxide:

$$H_2CO_3 \rightleftharpoons H^+ + HCO_3^-$$

If the pH of the blood is 7.4, what is the ratio of $[HCO_3^-]$ to $[H_2CO_3]$?   *Ans. 11 to 1*

44. How many moles of NH_4Cl must be added to 1.0 L of a 1.0 M NH_3 solution to prepare a buffer solution with a pH of 9.00? Of 9.50?   *Ans. 1.8 mol; 0.57 mol*

45. How many moles of sodium acetate must be added to 1.0 L of a 1.0 M acetic acid solution to prepare a buffer solution with a pH of 5.08? Of 4.20?
    Ans. 2.2 mol; 0.29 mol

46. A buffer solution is made up of equal volumes of 0.100 M acetic acid and 0.500 M sodium acetate. (a) What is the pH of this solution? (b) What is the pH that results from adding 1.00 mL of 0.100 M HCl to 0.200 L of the buffer solution? (Use 1.80×10^{-5} for the ionization constant of acetic acid.)   *Ans. 5.444; 5.438*

47. Calculate the change in pH when 5.0 mL of 0.050 M NaOH is added to 75 mL of a buffer solution made up of 0.10 M HClO and 0.10 M NaClO.
    Ans. pH changes from 7.46 to 7.48, or 0.02 pH units

Ionic Equilibria of Polyprotic Acids and Bases

48. Explain how successive ionization constants for polyprotic acids differ.

49. Evaluate the ionization constants $K_{H_2CO_3}$ and $K_{HCO_3^-}$ as well as the constant for the equilibrium

$$H_2CO_3 \rightleftharpoons 2H^+ + CO_3^{2-}$$

from the following concentrations found at equilibrium in a solution of H_2CO_3 and $NaHCO_3$:

$$[H^+] = 8.0 \times 10^{-6} \text{ M}$$
$$[HCO_3^-] = 1.6 \times 10^{-3} \text{ M}$$
$$[CO_3^{2-}] = 1.4 \times 10^{-8} \text{ M}$$
$$[H_2CO_3] = 3.0 \times 10^{-2} \text{ M}$$

Ans. $K_{H_2CO_3} = 4.3 \times 10^{-7}$; $K_{HCO_3^-} = 7.0 \times 10^{-11}$; $K_1 \cdot K_2 = 3.0 \times 10^{-17}$

50. Calculate the concentration of each species present in a 0.050 M solution of H_2S.
Ans. $[H^+] = 7.1 \times 10^{-5}$ M; $[OH^-] = 1.4 \times 10^{-10}$ M; $[H_2S] = 0.050$ M; $[HS^-] = 7.1 \times 10^{-5}$ M; $[S^{2-}] = 1.0 \times 10^{-19}$ M

51. Calculate the concentration of each species present in a 0.0250 M solution of Na_3PO_4.

$$PO_4^{3-} + H_2O \rightleftharpoons HPO_4^{2-} + OH^-$$
$$(K_b = 2.8 \times 10^{-2})$$
$$HPO_4^{2-} + H_2O \rightleftharpoons H_2PO_4^- + OH^-$$
$$(K_b = 1.6 \times 10^{-7})$$
$$H_2PO_4^- + H_2O \rightleftharpoons H_3PO_4 + OH^-$$
$$(K_b = 1.3 \times 10^{-12})$$

Ans. $[H^+] = 6.3 \times 10^{-13}$ M; $[OH^-] = 1.6 \times 10^{-2}$ M; $[PO_4^{3-}] = 9.1 \times 10^{-3}$ M; $[HPO_4^{2-}] = 1.6 \times 10^{-2}$ M; $[H_2PO_4^-] = 1.6 \times 10^{-7}$ M; $[H_3PO_4] = 1.3 \times 10^{-17}$ M; $[Na^+] = 0.0750$ M

52. Nicotine, $C_{10}H_{14}N_2$, is a base that will accept two protons ($K_1 = 7 \times 10^{-7}$, $K_2 = 1.4 \times 10^{-11}$). What is the concentration of each species present in a 0.050 M solution of nicotine?
Ans. $[H+] = 5 \times 10^{-11}$ M; $[OH^-] = 2 \times 10^{-4}$ M; $[C_{10}H_{14}N_2] = 0.050$ M; $[C_{10}H_{14}N_2H^+] = 2 \times 10^{-4}$ M; $[C_{10}H_{14}N_2H_2^{2+}] = 1.4 \times 10^{-11}$ M

53. Calculate the concentration of each species present in a 0.010 M solution of phthalic acid, $C_6H_4(CO_2H)_2$.

$$C_6H_4(CO_2H)_2 \rightleftharpoons C_6H_4(CO_2H)(CO_2^-) + H^+$$
$$(K_a = 1.1 \times 10^{-3})$$
$$C_6H_4(CO_2H)(CO_2^-) \rightleftharpoons C_6H_4(CO_2^-)_2 + H^+$$
$$(K_a = 3.9 \times 10^{-6})$$

Ans. $[H^+] = 2.8 \times 10^{-3}$ M; $[OH^-] = 3.6 \times 10^{-12}$ M; $[C_6H_4(CO_2H)_2] = 7 \times 10^{-3}$ M; $[C_6H_4(CO_2H)(CO_2^-)] = 2.8 \times 10^{-3}$ M; $[C_6H_4(CO_2^-)_2] = 3.9 \times 10^{-6}$ M

Titration Curves and Indicators

54. Calculate the pH after the addition of 20.0, 24.9, 25.0, 25.1, and 30.0 mL of NaOH in the titration of HCl described in Table 17.7 in Section 17.17.

55. Calculate the pH after the addition of 20.0, 24.9, 25.0, 25.1, 25.5, and 30.0 mL of NaOH in the titration of CH_3CO_2H described in Table 17.7 in Section 17.17.

56. Determine a theoretical titration curve for the titration of 20.0 mL of 0.100 M $HClO_4$, a strong acid, with 0.200 M KOH.

57. Determine a theoretical titration curve for the titration of 25.0 mL of 0.150 M NH_3 with 0.150 M HCl.

58. Why does an acid-base indicator change color over a range of pH values rather than at a specific pH?

59. Which of the indicators in Table 17.6 in Section 17.16 would be appropriate for the titration described in exercise 56? In exercise 57?

60. Using the data in Table 17.6, arrange the following indicators (which are weak acids) in increasing order of acid strength: methyl orange, litmus, thymol blue.
Ans. litmus < methyl orange < thymol blue

61. A 0.5000-g sample of an impure monobasic amine is titrated to the equivalence point with 75.00 mL of 0.100 M HCl. The pH after 40.00 mL of acid is added is 10.65. The molecular weight of the pure base is known to be 59.1. (a) Calculate the ionization constant of this base and (b) its percent purity. (c) Which indicator from the following list would be most suitable for this titration? State your reasoning.

Orange IV (red–yellow; pH 1.2–2.6)
Methyl orange (red–yellow; pH 3.1–4.4)
Methyl red (red–yellow; pH 4.8–6.0)
Bromthymol blue (yellow–blue; pH 6.0–7.6)
Phenolphthalein (colorless–pink; pH 8.3–10.0)
Thymolphthalein (colorless–blue; pH 9.3–10.5)
Ans. (a) 5.1×10^{-4}; (b) 88.7%; (c) methyl red

62. The hydrogen ion concentration at the half-equivalence point (the point at which half of the amount of base necessary to reach the equivalence point has been added) in the titration of a weak acid with a strong base is equal to K_a for the weak acid. Explain.

63. The indicator dinitrophenol is an acid with a K_a of 1.1×10^{-4}. In a 1.0×10^{-4} M solution, it is colorless in acid and yellow in base. Calculate the pH range over which it goes from 10% ionized (colorless) to 90% ionized (yellow). *Ans. 3.00–4.91*

Reaction of Salts with Water

64. Why does the salt of a strong acid and a strong base give an approximately neutral solution?

65. Why does the salt of a weak acid and a strong base give a basic solution?

66. Why does the salt of a strong acid and a weak base give an acidic solution?

67. Does the salt of a weak acid and a weak base give an acidic solution, a basic solution, or a neutral solution? Explain your answer.

68. Neutralization is defined in Section 16.4 as the reaction of an acid with a stoichiometric amount of a base. Why do some neutralization reactions give nonneutral solutions?

69. Explain why a reaction of a salt with water is an acid-base reaction.

70. Arrange the conjugate bases of the acids in Table 17.3 in order of increasing base strength.

71. Arrange the conjugate acids of the bases in Table 17.4 in order of decreasing acid strength.

72. (a) Write the equation for the reaction that makes potassium cyanide, KCN, behave as a base in water.
    (b) What is the K_b for this reaction?
    (c) What is the pH of a 0.255 M solution of KCN?
    Ans. (a) $CN^- + H_2O \rightleftharpoons HCN + OH^-$;
    (b) $K_b = 2 \times 10^{-5}$; (c) 11.4

73. Calculate the pH of each of the following solutions:
    (a) 0.1050 M NH_4NO_3 *Ans. 5.12*
    (b) 0.4735 M NaCN   *Ans. 11.5*
    (c) 0.333 M   [$(CH_3)_2NH_2]_2SO_4$ [Note   that $(CH_3)_2NH_2^+$ is the conjugate acid of the weak base $(CH_3)_2NH$, just as NH_4^+ is the conjugate acid of the weak base NH_3.]   *Ans. 5.52*
    (d) 0.0270 M $Ca(N_3)_2$ *Ans. 8.4*
    (e) 0.100 M NH_4F *Ans. 6.20*

74. Calculate the ionization constant (K_a or K_b) for each of the following acids or bases.
    (a) NO_2^- *Ans. 2.2×10^{-11}*
    (b) $HC_2O_4^-$ (as a base)   *Ans. 1.7×10^{-13}*
    (c) AsO_4^{3-} *Ans. 1×10^{-1}*
    (d) F^- *Ans. 1.4×10^{-11}*
    (e) $(CH_3)_2NH_2^+$ *Ans. 1.4×10^{-11}*
    (f) S^{2-} *Ans. 1.0×10^{-5}*
    (g) NH_4^+ *Ans. 5.6×10^{-10}*

75. Calculate the pH of a solution made by mixing equal volumes of 0.040 M aqueous aniline, $C_6H_5NH_2$, and 0.040 M nitric acid. (K_b for $C_6H_5NH_2 = 4.6 \times 10^{-10}$.)   *Ans. 3.18*

76. In a 0.0010 M solution of KCN, the KCN reacts with water to the extent of 14.0%. Calculate the value of the ionization constant of HCN.   *Ans. 4.4×10^{-10}*

77. What concentration of calcium acetate, $Ca(CH_3CO_2)_2$, will give a solution in which the percent reaction of $Ca(CH_3CO_2)_2$ with water is 0.10%?
    Ans. 2.8×10^{-4} M

78. Calculate the pH of a 0.470 M solution of lithium carbonate, Li_2CO_3.   *Ans. 11.9*

79. Sodium nitrite, $NaNO_2$, is used in meat processing. What are [H^+] and the pH of a 0.225 M $NaNO_2$ solution? Is the solution acidic or basic?
    Ans. 4.5×10^{-9} M; 8.35; basic

80. The ion HTe^- is an amphiprotic species; it can act as either an acid or a base. What is K_a for the acid reaction of HTe^- with H_2O? What is K_b for the reaction in which HTe^- functions as a base in water?
    Ans. 1×10^{-5}; 4.3×10^{-12}

81. Novocaine, $C_{13}H_{21}O_2N_2Cl$, is the salt of the base procaine and hydrochloric acid. The ionization constant for procaine is 7×10^{-6}. What is the pH of a 2.0% solution by mass of novocaine, assuming that the density of the solution is 1.0 g/cm^3?   *Ans. 5.0*

Additional Exercises

82. By calculation, check the values given in Section 16.3 for the percent ionization of 0.10 M solutions of the weak acids HSO_4^-, HNO_2, and CH_3CO_2H.

83. By calculation, check the values given in Section 16.3 for the percent ionization of 0.10 M solutions of the weak bases NH_3, $CH_3CO_2^-$, and NO_2^-.

84. Using the K_a values in Appendix F and in Section 17.15, place [$Al(H_2O)_6]^{3+}$ in the correct location in Table 16.1 in Section 16.3, the table of relative strengths of conjugate acid-base pairs.

85. How many moles of hydrogen chloride must be added to 1.50 L of a 0.450 M solution of CH_3NH_2 to give a pH of 10.95?   *Ans. 0.22 mol*

86. Sulfuric acid is a strong acid. The extent of the ionization, $H_2SO_4 \rightleftharpoons H^+ + HSO_4^-$, is effectively 100%. The hydrogen sulfate ion, HSO_4^-, is a weaker acid with $K_a = 1.2 \times 10^{-2}$.
    (a) What is the concentration of sulfate ion, SO_4^{2-}, in a 0.102 M solution of $NaHSO_4$?   *Ans. 0.029 M*
    (b) What is the concentration of sulfate ion in a 0.102 M solution of H_2SO_4?   *Ans. 9.9×10^{-3} M*

87. Sulfuric acid ionizes in two steps:

$$H_2SO_4 \rightleftharpoons H^+ + HSO_4^-$$
$$HSO_4^- \rightleftharpoons H^+ + SO_4^{2-}$$

Since sulfuric acid is a strong acid, the first step is effectively 100% complete ($K_{H_2SO_4}$ is large). The ionization constant for the second step, $K_{HSO_4^-}$, is 1.2×10^{-2}. Check the assumption that the first step goes completely to the right, based on the approximation that for a polyprotic acid, successive ionization constants often differ by a factor of 1×10^6 and calculate (a) $K_{H_2SO_4}$, (b) the concentration of H_2SO_4, (c) the percent ionization of H_2SO_4 into H^+ and HSO_4^- in a 0.100 M solution of H_2SO_4.

Ans. (a) 1×10^4; (b) 1×10^{-6} M; (c) 99.999%
(100% to two significant figures)

88. The pH of a sample of acid rain (Section 29.4) is 4.07.

Assuming that the acidity is due to the presence of HNO_2 in the rain, what is the total concentration of HNO_2?
Ans. 1.6×10^{-5} M

89. Lime juice is among the most acidic of fruit juices, with a pH of 1.92. If the acidity is due to citric acid, which we can abbreviate as H_3Cit, what is the ratio of each of the following to $[Cit^{3-}]$: $[H_3Cit]$; $[H_2Cit^-]$; $[HCit^{2-}]$?

$$H_3Cit \rightleftharpoons H^+ + H_2Cit^- \quad K_a = 8.4 \times 10^{-4}$$
$$H_2Cit^- \rightleftharpoons H^+ + HCit^{2-} \quad K_a = 1.8 \times 10^{-5}$$
$$HCit^{2-} \rightleftharpoons H^+ + Cit^{3-} \quad K_a = 4.0 \times 10^{-6}$$

Ans. 2.9×10^7; 2.0×10^6; 3.0×10^3 to 1

90. In many detergents phosphates have been replaced with silicates as water conditioners. If 125 g of a detergent that contains 8.0% Na_2SiO_3 by weight is used in 4.0 L of water, what are the pH and the hydroxide ion concentration in the wash water?

$$SiO_3^{2-} + H_2O \rightleftharpoons SiO_3H^- + OH^- \quad K_b = 1.6 \times 10^{-3}$$
$$SiO_3H^- + H_2O \rightleftharpoons SiO_3H_2 + OH^- \quad K_b = 3.1 \times 10^{-5}$$

Ans. pH = 11.70; $[OH^-] = 5.0 \times 10^{-3}$ M

91. The ionization constant for water, K_w, is 5.474×10^{-14} at 50°C. At 50°C, K_b for NH_3 is 1.892×10^{-5}. What is the ionization constant for NH_4^+ at 50°C?
Ans. 2.893×10^{-9}

92. What is the pH of 1.000 L of a solution of 100.0 g of glutamic acid ($C_5H_9NO_4$, a diprotic acid; $K_1 = 8.5 \times 10^{-5}$, $K_2 = 3.39 \times 10^{-10}$) to which has been added 20.0 g of NaOH during the preparation of monosodium glutamate, the flavoring agent?
Ans. 4.51

93. The solutions of ammonia used for cleaning windows usually contain about 10% NH_3 by mass. If a solution that is exactly 10% by mass has a density of 0.99 g/cm³, what are the pH and the hydroxide ion concentration?
Ans. 12.01; 1.0×10^{-2} M

94. Saccharin, $C_7H_4NSO_3H$, is a weak acid ($K_a = 2.1 \times 10^{-12}$). If 0.250 L of a soft drink with a pH of 5.48 was prepared from 2.00×10^{-3} g of sodium saccharide (often referred to as sodium saccharin on soft drink cans), $Na(C_7H_4NSO_3)$, what are the final concentrations of saccharin and sodium saccharide in the solution?
Ans. $[C_7H_4NSO_3H] = 3.9 \times 10^{-5}$ M; $[Na(C_7H_4NSO_3)] = 2.5 \times 10^{-11}$ M

95. A typical urine sample contains 2.3% by mass of the base urea, $CO(NH_2)_2$

$$CO(NH_2)_2 + H_2O \rightleftharpoons CO(NH_2)(NH_3)^+ + OH^-$$
$$K_b = 1.5 \times 10^{-14}$$

If the density of the urine sample is 1.06 g/cm³ and the pH of the sample is 6.35, calculate the concentration of $CO(NH_2)_2$ and of $CO(NH_2)(NH_3)^+$ in the sample.
Ans. $[CO(NH_2)_2] = 0.41$ M; $[CO(NH_2)(NH_3)^+] = 2.7 \times 10^{-7}$ M

96. If abominable snowmen were to exist in the world in the same proportion to humans as do hydrogen ions to water molecules in pure water, how many abominable snowmen would there be in the world? Assume a world population of 4 billion people.
Ans. 7

97. Show that the rate of reaction of acetic acid with sodium hydroxide is given by the following expression:

$$\text{Rate} = \frac{kK_a[CH_3CO_2H][OH^-]}{[CH_3CO_2^-]}$$

98. Calculate $[H^+]$, $[HCO_3^-]$, $[H_2CO_3]$, $[CO_3^{2-}]$, pH, pOH, pK_{a_1}, and pK_{a_2} for a 0.075 M H_2CO_3 solution that is also 0.050 M in $NaHCO_3$.
Ans. $[H^+] = 6.4 \times 10^{-7}$ M; $[HCO_3^-] = 0.050$ M; $[CO_3^{2-}] = 5 \times 10^{-6}$ M; $[H_2CO_3] = 0.075$ M; pH = 6.19; pOH = 7.81; $pK_{a_1} = 6.37$; $pK_{a_2} = 10.2$

99. It is sometimes said that one should try to select a buffer whose weak acid has a pK_a close to the desired pH, and, similarly, that one should try to select a buffer whose weak base has a pK_b close to the desired pOH. Are these suggestions logical? Explain your answer.

100. A useful relationship for buffer solutions of a weak acid and its salt (general equilibrium, $HA \rightleftharpoons H^+ + A^-$) is one called the Henderson-Hasselbalch equation:

$$pH = pK_a + \log \frac{[A^-]}{[HA]}$$

(a) Derive the equation mathematically, beginning with the expression for K_a.
(b) Use the equation to support the following statements: (1) pH should equal pK_a for an ideal buffer, where $[A^-]$ and $[HA]$ are equal; and (2) $[H^+]$ should equal K_a for an ideal buffer.

The Solubility Product Principle

18

Barium chromate is essentially insoluble.

I n chemical reactions that involve the formation of solids (precipitation reactions) we often work with slightly soluble substances that are completely ionized in solution. Equilibrium systems consisting of slightly soluble ionic substances in contact with their saturated aqueous solutions are particularly important in analytical chemistry. In this chapter we shall be concerned with the theory and application of the precipitation and dissolution of slightly soluble ionic substances.

The Formation of Precipitates

18.1 The Solubility Product

A slightly soluble electrolyte dissolves until a saturated solution, a solution of its ions in equilibrium with undissolved solute, is formed. The use of radioactive tracers shows that at equilibrium the solid (the undissolved solute) continues to dissolve, the process of crystallization occurs, and the opposing processes have equal rates. The equilibrium of a solution of silver chloride with solid silver chloride can be expressed by

$$AgCl(s) \underset{\text{Precipitation}}{\overset{\text{Dissolution}}{\rightleftharpoons}} Ag^+(aq) + Cl^-(aq) \qquad \text{(saturated solution)}$$

Applying the law of mass action to this system, we obtain

$$\frac{[Ag^+][Cl^-]}{[AgCl]} = K$$

Because the concentration of the solid silver chloride remains constant, we can write

$$[Ag^+][Cl^-] = K\,[AgCl] = K_{sp}$$

K_{sp} **is called the solubility product,** or sometimes the **solubility product constant.** The product of the concentrations of the dissolved ions must be equal to the solubility product when a saturated solution is in equilibrium with undissolved solute.

Calcium phosphate dissolves according to the equation

$$Ca_3(PO_4)_2(s) \rightleftharpoons 3Ca^{2+}(aq) + 2PO_4^{3-}(aq)$$

The expression for its solubility product is

$$[Ca^{2+}]^3[PO_4^{3-}]^2 = K_{sp}$$

For a salt like magnesium ammonium phosphate, which dissolves according to the equation

$$MgNH_4PO_4(s) \rightleftharpoons Mg^{2+}(aq) + NH_4^+(aq) + PO_4^{3-}(aq)$$

we have

$$[Mg^{2+}][NH_4^+][PO_4^{3-}] = K_{sp}$$

Solubility products have units, as do all equilibrium constants (Section 15.2), but we will follow the common practice and omit them.

The solubility product is constant only for solutions of *slightly* soluble electrolytes. It is not constant for moderately or highly soluble salts, such as NaCl and KNO₃. Furthermore, high concentrations of ions not involved in the equilibrium increase the solubility of a slightly soluble substance—the salt effect (Section 17.6). The salt effect varies with the concentrations and with the charges of the additional ions. Hence the solubility product is constant *only* for saturated solutions with extremely small ionic concentrations. However, the solubility product concept is very useful for calculating approximate concentrations of slightly soluble electrolytes.

18.2 Calculation of Solubility Products from Solubilities

The solubility product of a slightly soluble electrolyte can be calculated from its solubility in moles per liter. When concentrations are given in other units, such as grams per liter, they must be converted to moles per liter.

EXAMPLE 18.1   The solubility of lead chromate, $PbCrO_4$, the artist's pigment chrome yellow, is 4.3×10^{-5} g L⁻¹. Calculate the solubility product for $PbCrO_4$.

We need the solubility of $PbCrO_4$ in moles per liter. From this solubility we can find the equilibrium concentrations and, from them, K_{sp}. We can use the formula weight of $PbCrO_4$, 323.2, to find the molar solubility.

$$\frac{4.3 \times 10^{-5} \text{ g } PbCrO_4}{1 \text{ L}} \times \frac{1 \text{ mol } PbCrO_4}{323.2 \text{ g } PbCrO_4}$$

$$= \frac{1.33 \times 10^{-7} \text{ mol } PbCrO_4}{1 \text{ L}} = 1.33 \times 10^{-7} \text{ M}$$

$$PbCrO_4(s) \rightleftharpoons Pb^{2+}(aq) + CrO_4^{2-}(aq)$$

Since 1 mol of $PbCrO_4$ gives 1 mol of $Pb^{2+}(aq)$ and 1 mol of $CrO_4^{2-}(aq)$, both $[Pb^{2+}]$ and $[CrO_4^{2-}]$ are equal to the molar solubility of $PbCrO_4$.

A table of concentrations before the reaction (before any $PbCrO_4$ dissolves) and at equilibrium contains the following entries:

	$[Pb^{2+}]$	$[CrO_4^{2-}]$
Initial concentrations, M	0	0
Equilibrium concentrations, M	1.33×10^{-7}	1.33×10^{-7}

We substitute the equilibrium concentrations of Pb^{2+} and CrO_4^{2-} into the expression for K_{sp}.

$$[Pb^{2+}][CrO_4^{2-}] = K_{sp}$$

$$(1.33 \times 10^{-7})(1.33 \times 10^{-7}) = 1.8 \times 10^{-14} = K_{sp}$$

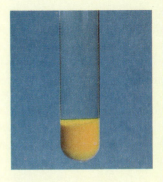

Yellow lead chromate, $PbCrO_4$, is not appreciably soluble in water.

In this example it was not really necessary to include the initial concentrations of the ions in the table, since they were both zero. However, it is good practice to do so, since initial concentrations will be helpful later in this chapter.

Calculate the solubility product, K_{sp}, of silver chromate, Ag_2CrO_4, whose solubility is 1.3×10^{-4} mol L^{-1}.

EXAMPLE 18.2

From the equation for the dissolution of silver chromate

$$Ag_2CrO_4(s) \rightleftharpoons 2Ag^+(aq) + CrO_4^{2-}(aq)$$

we note that K_{sp} is given by

$$[Ag^+]^2[CrO_4^{2-}] = K_{sp}$$

From the equation for the dissolution of Ag_2CrO_4, we can see that 1 mol of $CrO_4^{2-}(aq)$ is formed for each mole of Ag_2CrO_4 that dissolves (1 mol CrO_4^{2-}/1 mol Ag_2CrO_4), and 2 mol of $Ag^+(aq)$ are formed (2 mol Ag^+/1 mol Ag_2CrO_4).

$$\frac{1.3 \times 10^{-4} \text{ mol } Ag_2CrO_4}{1 \text{ L}} \times \frac{2 \text{ mol } Ag^+}{1 \text{ mol } Ag_2CrO_4} = \frac{2.6 \times 10^{-4} \text{ mol } Ag^+}{1 \text{ L}}$$

$$\frac{1.3 \times 10^{-4} \text{ mol } Ag_2CrO_4}{1 \text{ L}} \times \frac{1 \text{ mol } CrO_4^{2-}}{1 \text{ mol } Ag_2CrO_4} = \frac{1.3 \times 10^{-4} \text{ mol } CrO_4^{2-}}{1 \text{ L}}$$

Thus $[Ag^+] = 2.6 \times 10^{-4}$ M and $[CrO_4^{2-}] = 1.3 \times 10^{-4}$ M.

Silver chromate, Ag_2CrO_4 is not appreciably soluble in water.

Now we set up a table of concentrations before the reaction (before any Ag_2CrO_4 dissolves) and at equilibrium, containing the following entries:

	$[Ag^+]$	$[CrO_4^{2-}]$
Initial concentrations, M	0	0
Equilibrium concentrations, M	2.6×10^{-4}	1.3×10^{-4}

Substituting the equilibrium concentrations of Ag^+ and CrO_4^{2-} in the expression for the solubility product gives

$$[Ag^+]^2[CrO_4^{2-}] = K_{sp}$$
$$(2.6 \times 10^{-4})^2(1.3 \times 10^{-4}) = 8.8 \times 10^{-12} = K_{sp}$$

Note that $[Ag^+]$ is squared in this expression.

A table of solubility products of some slightly soluble electrolytes at 25°C is given in Appendix D.

18.3 Calculation of Molar Solubilities from Solubility Products

The solubility of a slightly soluble electrolyte can be readily calculated from its solubility product.

EXAMPLE 18.3

The solubility product for silver chloride is 1.8×10^{-10} (Appendix D). Calculate the molar solubility of silver chloride.

Silver chloride forms a saturated solution according to the equation

$$AgCl(s) \rightleftharpoons Ag^+(aq) + Cl^-(aq)$$

Let x be the solubility of AgCl in moles per liter. The molar concentrations of both Ag^+ and Cl^- will also be x since the salt is completely dissociated. Thus we have this table:

	$[Ag^+]$	$[Cl^-]$
Initial concentrations, M	0	0
Equilibrium concentrations, M	x	x

Substituting the equilibrium concentrations for Ag^+ and Cl^- in the solubility product expression for AgCl gives

$$[Ag^+][Cl^-] = K_{sp} = 1.8 \times 10^{-10}$$
$$x^2 = 1.8 \times 10^{-10}$$
$$x = \sqrt{1.8 \times 10^{-10}} = 1.3 \times 10^{-5}$$

The molar solubility of silver chloride is 1.3×10^{-5} M.

A precipitate of insoluble silver chloride, AgCl, forms when solutions of $AgNO_3$ and AgCl are mixed.

EXAMPLE 18.4

Although most mercury compounds are very poisonous, sixteenth-century physicians used calomel, Hg_2Cl_2, as a medication. Their patients (usually) did not die of mercury poisoning because calomel is quite insoluble. Calculate the molar solubility of Hg_2Cl_2 ($K_{sp} = 1.1 \times 10^{-18}$; see Appendix D).

The equation for the dissolution of Hg_2Cl_2 is

$$Hg_2Cl_2(s) \rightleftharpoons Hg_2^{2+}(aq) + 2Cl^-(aq)$$

If the solubility of Hg_2Cl_2 is x mol L^{-1}, then

$$[Hg_2^{2+}] = \frac{x \text{ mol } Hg_2Cl_2}{1 \text{ L}} \times \frac{1 \text{ mol } Hg_2^{2+}}{1 \text{ mol } Hg_2Cl_2} = \frac{x \text{ mol}}{1 \text{ L}}$$

$$[Cl^-] = \frac{x \text{ mol } Hg_2Cl_2}{1 \text{ L}} \times \frac{2 \text{ mol } Cl^-}{1 \text{ mol } Hg_2Cl_2} = \frac{2x \text{ mol}}{1 \text{ L}}$$

Thus x mol of Hg_2Cl_2 gives $2x$ mol of Cl^-.

	$[Hg_2^{2+}]$	$[Cl^-]$
Initial concentrations, M	0	0
Equilibrium concentrations, M	x	$2x$

Mercury (I) chloride is essentially insoluble in water.

We substitute the equilibrium concentrations of Hg_2^{2+} and Cl^- in the expression for K_{sp} and calculate the value of x.

$$[Hg_2^{2+}][Cl^-]^2 = K_{sp} = 1.1 \times 10^{-18}$$
$$(x)(2x)^2 = 4x^3 = 1.1 \times 10^{-18}$$
$$x = 6.5 \times 10^{-7}$$

The molar solubility of Hg_2Cl_2 is 6.5×10^{-7} M; $[Hg_2^{2+}] = x = 6.5 \times 10^{-7}$ M, and $[Cl^-] = 2x = 1.3 \times 10^{-6}$ M.

18.4 The Precipitation of Slightly Soluble Electrolytes

A slightly soluble electrolyte will precipitate if its ions are present in solution in such quantities that the product of their concentrations, raised to the appropriate powers, *exceeds the solubility product of the electrolyte.* We can show that silver chloride will precipitate if we mix equal volumes of a 2×10^{-4} M solution of silver nitrate and a 2×10^{-4} M solution of sodium chloride. The volume doubles when we mix equal volumes, so each concentration is reduced to half its initial value. Consequently, immediately upon mixing, $[Ag^+]$ and $[Cl^-]$ will both be equal to $\frac{1}{2}$ (2×10^{-4} M) = 1×10^{-4} M. The product $[Ag^+][Cl^-]$ will, *momentarily,* be greater than K_{sp} for AgCl.

$$K_{sp} = 1.8 \times 10^{-10}$$
$$[Ag^+][Cl^-] = (1 \times 10^{-4})(1 \times 10^{-4}) = 1 \times 10^{-8} > K_{sp}$$

The system will not be at equilibrium, and silver chloride will precipitate until the product $[Ag^+][Cl^-]$ equals the solubility product; that is, until

$$[Ag^+][Cl^-] = K_{sp}$$

The ions Ag^+ and Cl^- are present in equal concentrations in a saturated solution of AgCl in pure water in contact with solid AgCl, and the product $[Ag^+][Cl^-]$ just equals K_{sp}. If we add NaCl to such a solution, the added chloride ion will cause the product of $[Ag^+]$ and $[Cl^-]$ to be larger than the solubility product, and AgCl will form until the ion product again equals K_{sp}. At the new equilibrium $[Ag^+]$ will be less and $[Cl^-]$ greater than they were in the solution of AgCl in pure water. The greater we make $[Cl^-]$, the smaller $[Ag^+]$ will become. However, we can never reduce $[Ag^+]$ to zero, because the product $[Ag^+][Cl^-]$ will always be equal to the solubility product.

18.5 Calculation of the Concentration of an Ion Necessary to Form a Precipitate

When the concentration of one of the ions of a slightly soluble electrolyte and the value for the solubility product of the electrolyte are known, we can calculate the concentration that the other ion must exceed before precipitation will occur.

EXAMPLE 18.5

If a solution contains 0.001 mol of CrO_4^{2-} per liter, what concentration of Ag^+ ion must be exceeded (by adding solid $AgNO_3$ to the solution) before Ag_2CrO_4 will begin to precipitate, neglecting any increase in volume due to the addition of solid silver nitrate? ($K_{sp} = 9 \times 10^{-12}$; see Appendix D.)

The equilibrium involved in the problem is

$$Ag_2CrO_4(s) \rightleftharpoons 2Ag^+(aq) + CrO_4^{2-}(aq)$$

There is no equilibrium between Ag_2CrO_4, Ag^+, and CrO_4^{2-} until the first trace of solid Ag_2CrO_4 appears. Equilibrium is established only when enough $AgNO_3$ has been added to form the first trace of $Ag_2CrO_4(s)$, and only then can the relationship between $[Ag^+]$ and $[CrO_4^{2-}]$ be determined from the solubility product expression. Let x equal $[Ag^+]$ when equilibrium is established.

	$[Ag^+]$	$[CrO_4^{2-}]$
Initial concentrations, M	0	0.001
Equilibrium concentrations, M	x	0.001

Since enough $AgNO_3$ to form only a trace of solid Ag_2CrO_4 has been added, $[CrO_4^{2-}]$ has not changed significantly. Substitution in the solubility product expression gives

$$[Ag^+]^2[CrO_4^-] = K_{sp} = 9 \times 10^{-12}$$
$$(x^2)(0.001) = 9 \times 10^{-12}$$
$$x^2 = \frac{9 \times 10^{-12}}{0.001} = 9 \times 10^{-9}$$
$$x = 9.5 \times 10^{-5} \quad \text{or} \quad 1 \times 10^{-4} \text{ M (rounded)}$$

A concentration of Ag^+ greater than 1×10^{-4} M is necessary to cause precipitation of Ag_2CrO_4 under these conditions.

18.6 Calculation of the Concentration of an Ion in Solution after Precipitation

It is often useful to know the concentration of an ion remaining in solution after precipitation. This, too, can be determined using the solubility product.

EXAMPLE 18.6

Clothing washed in water with a manganese concentration exceeding 0.1 mg/L (1.8×10^{-6} M) may be stained by the manganese. A laundry wishes to add a base to the water to precipitate the manganese as the hydroxide $Mn(OH)_2$ ($K_{sp} = 4.5 \times 10^{-14}$). At what pH will $[Mn^{2+}]$ be equal to 1.8×10^{-6} M? The dissolution of $Mn(OH)_2$ is described by the equation

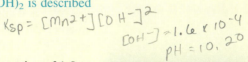

$$Mn(OH)_2(s) \rightleftharpoons Mn^{2+}(aq) + 2OH^-(aq)$$

Let x be the concentration of OH^- in equilibrium with a Mn^{2+} concentration of 1.8×10^{-6} M.

	$[Mn^{2+}]$	$[OH^-]$
Equilibrium concentrations, M	1.8×10^{-6}	x

Substitution into the expression for K_{sp} gives

$$[Mn^{2+}][OH^-]^2 = K_{sp} = 4.5 \times 10^{-14}$$
$$(1.8 \times 10^{-6})(x)^2 = 4.5 \times 10^{-14}$$
$$x^2 = \frac{4.5 \times 10^{-14}}{1.8 \times 10^{-6}} = 2.5 \times 10^{-8}$$
$$x = 1.6 \times 10^{-4} \text{ M}$$

Thus

$$[OH^-] = 1.6 \times 10^{-4} \text{ M}$$
$$pOH = -\log [OH^-] = -\log (1.6 \times 10^{-4}) = 3.80$$
$$pH = 14.00 - pOH = 10.20$$

Manganese hydroxide, $Mn(OH)_2$, precipitates when a solution of a base is added to a solution of a manganese (II) salt.

If the laundry adds a base to their water until the pH rises to 10.20, the manganese ion will be reduced to a concentration of 1.8×10^{-6} M; at or below that concentration, it will not stain.

18.7 Supersaturation of Slightly Soluble Electrolytes

When the product of the concentrations of the ions raised to their proper powers exceeds the solubility product, precipitation *usually* occurs. However, a supersaturated solution sometimes forms, and then precipitation does not occur immediately. For example, if magnesium is being precipitated as magnesium ammonium phosphate from a solution with a low concentration of magnesium ion, the solution may stand for several hours before a visible precipitate appears. Furthermore, even though precipitation may begin immediately, it may not be complete. The first crystals to precipitate are often very small and more soluble than are the larger ones that form as the solution stands in contact with the precipitate (see Section 18.8). To ensure maximum precipitation you should let a solution stand before filtering or centrifuging it.

18.8 Solubility and Crystal Size

The solubility of fine crystals of barium sulfate has been found to be about twice as great as that of coarse crystals. In fact, small crystals of any substance are more soluble than large ones because there are proportionately more ions inside the large crystal. Ions in the interior of a crystal are bound more tightly than are those at the faces and edges. The fraction of ions occupying surface positions is greater in a small crystal than in a large crystal, making the tendency for ions to enter a solution greater for the small crystal. Because small crystals are more soluble than large ones, the smallest crystals gradually dissolve while larger ones grow still larger. True equilibrium is not reached until large, perfect crystals are formed. Solubility products are determined for solutions in contact with relatively large crystals.

Very small crystals pass through filters and resist centrifugation; therefore precipitates are often digested (warmed in the solution) to increase the size of the crystals. Heat increases the rate at which large crystals grow from the smaller ones. Simply allowing a precipitate to stand in solution may also yield larger crystals.

18.9 Calculations Involving Both Ionization Constants and Solubility Products

In dealing with precipitation we often work with systems that involve more than one equilibrium. The following examples illustrate how to treat multiple equilibria.

EXAMPLE 18.7   Calculate the hydrogen ion concentration required to prevent the precipitation of ZnS in a solution that is 0.050 M in $ZnCl_2$ and saturated with H_2S (0.10 M in H_2S).

Zinc sulfide is insoluble in neutral solution (left), but dissolves in acidic solution (right). The pH paper shows that the pH of the solution is less than 1.

We need to consider two reactions in this system.

$$ZnS(s) \rightleftharpoons Zn^{2+}(aq) + S^{2-}(aq) \tag{1}$$

$$H_2S(aq) \rightleftharpoons 2H^+(aq) + S^{2-}(aq) \tag{2}$$

ZnS will not precipitate if the product $[Zn^{2+}][S^{2-}]$ is less than K_{sp} for ZnS (1×10^{-27}). Let us find the conditions under which the first trace of ZnS will appear.

	$[Zn^{2+}]$	$[S^{2-}]$
Equilibrium concentrations, M	0.050	x

$$[Zn^{2+}][S^{2-}] = K_{sp} = 1 \times 10^{-27}$$

$$(0.050)(x) = 1 \times 10^{-27}$$

$$x = \frac{1 \times 10^{-27}}{0.050} = 2 \times 10^{-26} \text{ M}$$

Hence, when $[Zn^{2+}]$ is 0.050 M, ZnS will begin to appear unless $[S^{2-}]$ is less than 2×10^{-26} M.

The concentration of S^{2-} can be adjusted by changing the concentration of H^+ (see Equation 2 above and Section 17.9).

$$\frac{[H^+]^2[S^{2-}]}{[H_2S]} = K_a = 1.0 \times 10^{-26}$$

Now we let $[H^+]$ be equal to y and calculate $[H^+]$ for a solution in which $[S^{2-}]$ is 2×10^{-26} M.

	$[H^+]$	$[S^{2-}]$	$[H_2S]$
Equilibrium concentrations, M	y	2×10^{-26}	0.10

$$\frac{[H^+]^2[S^{2-}]}{[H_2S]} = K_a = 1.0 \times 10^{-26}$$

$$\frac{(y)^2(2 \times 10^{-26})}{0.10} = 1.0 \times 10^{-26}$$

$$y^2 = \frac{0.10}{2 \times 10^{-26}} \times 1.0 \times 10^{-26} = 5.0 \times 10^{-2}$$

$$y = 2.2 \times 10^{-1} = 0.2 \text{ M}$$

Thus $[S^{2-}]$ is 2×10^{-26} M when $[H^+]$ is 0.2 M. If $[H^+]$ is greater than 0.2 M, $[S^{2-}]$ will be less than 2×10^{-26} M, and precipitation of ZnS will not occur.

As we mentioned in Section 17.9, it is now known that K_2, for the second step in the dissociation of hydrogen sulfide ($HS^- \rightleftharpoons H^+ + S^{2-}$), is smaller ($1.0 \times 10^{-19}$) than previously thought, by a factor of about 10^6.

Calculations using the new value for K_2 show that the number of sulfide ions per milliliter in a saturated hydrogen sulfide solution (0.10 M H_2S) is only 60 ions per milliliter. In a saturated H_2S solution that is acidified to 0.30 M hydrogen ion, commonly used in qualitative analysis procedures in the precipitation of metal sulfides, the number of sulfide ions is reduced to only 6×10^{-6} sulfide ions per milliliter. Hence sulfide ions appear to have a practically insignificant effect in precipitating a metal sulfide in a hydrogen sulfide solution. It is more logical to consider that either the hydrogen sulfide ion (HS^-) or hydrogen sulfide (H_2S) is the major precipitating agent. In acid solution,

$$M^{2+}(aq) + HS^-(aq) \rightleftharpoons MS(s) + H^+(aq) \tag{3}$$

or

$$M^{2+}(aq) + H_2S(aq) \rightleftharpoons MS(s) + 2H^+(aq) \tag{4}$$

It should be pointed out, nevertheless, that calculations using K_{sp} values do not depend upon the reaction mechanism assumed for obtaining the final product. Using S^{2-}, HS^-, or H_2S as the precipitating agent yields the same answer provided that the reactions chosen yield the final product or products and that the equilibrium constants used are consistent with the reactions selected. We usually use the traditional calculation method, considering S^{2-} as the precipitating agent, but it is instructive to work an example to illustrate the postulation of HS^- or H_2S as the precipitating agent. Example 18.8 shows the calculation for Example 18.7 considering H_2S as the precipitating agent.

EXAMPLE 18.8

Calculate the hydrogen ion concentration required to prevent the precipitation of ZnS in a 0.050 M $ZnCl_2$ solution saturated with H_2S (as in Example 18.7), considering H_2S as the precipitating agent.

We need to consider the following reaction, which is the reverse equation for reaction (4) above, applied to ZnS:

$$ZnS(s) + 2H^+(aq) \rightleftharpoons Zn^{2+}(aq) + H_2S(aq)$$

We can calculate a revised K_{sp}, for this reaction, which we may call K_{sp_a}, by the following procedure:

$$K_{sp_a} = \frac{[Zn^{2+}][H_2S]}{[H^+]^2}$$

Multiplying by $\dfrac{[S^{2-}]}{[S^{2-}]}$, which is equal to 1, does not change the value and provides a useful result.

$$K_{sp_a} = \frac{[Zn^{2+}][H_2S]}{[H^+]^2} \times \frac{[S^{2-}]}{[S^{2-}]} = [Zn^{2+}][S^{2-}] \times \frac{[H_2S]}{[H^+]^2[S^{2-}]}$$

$[Zn^{2+}][S^{2-}]$ is K_{sp} for ZnS (1×10^{-27}).

$\dfrac{[H_2S]}{[H^+]^2[S^{2-}]}$ is the reciprocal of the combined K_a for H_2S, or $\dfrac{1}{K_a}$.

Therefore, K_{sp_a} becomes

$$K_{sp_a} = K_{sp} \times \frac{1}{K_a} = \frac{K_{sp}}{K_a} = \frac{1 \times 10^{-27}}{1.0 \times 10^{-26}} = 1 \times 10^{-1}$$

Thus,

$$K_{sp_a} = \frac{[Zn^{2+}][H_2S]}{[H^+]^2} = 0.1$$

Now that we have the value of the equilibrium constant for the reaction, we are ready to work the problem.

At equilibrium

$$[Zn^{2+}] = 0.050 \text{ M}$$

$$[H_2S] = 0.10 \text{ M}$$

$$[H^+] = x \text{ M}$$

Substituting in K_{sp_a},

$$K_{sp_a} = \frac{[Zn^{2+}][H_2S]}{[H^+]^2} = 0.1$$

$$\frac{(0.050)(0.10)}{x^2} = 0.1$$

$$x^2 = \frac{(0.050)(0.10)}{0.1}$$

$$x^2 = 0.050$$

$$x = 2.2 \times 10^{-1}$$

Thus 0.2 M H^+ is required to prevent precipitation of H_2S.

Note that the answer for Example 18.8 is the same answer we calculated in Example 18.7, where we considered S^{2-} as the precipitating agent. This confirms that although one mechanism may be the more logical of several choices, the end result of the calculation does not depend upon the mechanism selected. We need only be sure that the reactions chosen lead to the correct final product or products and that the equilibrium constants used are consistent with the selected reactions.

EXAMPLE 18.9

Calculate the concentration of ammonium ion, supplied by NH_4Cl, required to prevent the precipitation of $Mg(OH)_2$ in a liter of solution containing 0.10 mol of NH_3 and 0.10 mol of Mg^{2+}.

Two equilibria are involved in this system.

$$Mg(OH)_2(s) \rightleftharpoons Mg^{2+}(aq) + 2OH^-(aq)$$

$$NH_3(aq) + H_2O(l) \rightleftharpoons NH_4^+(aq) + OH^-(aq)$$

If $Mg(OH)_2$ is not to form, then the product $[Mg^{2+}][OH^-]^2$ must be less than K_{sp} for $Mg(OH)_2$. The $[OH^-]$ can be reduced by the addition of NH_4^+, which shifts the second reaction to the left and thus reduces $[OH^-]$.

First we determine the hydroxide ion concentration at which the first trace of $Mg(OH)_2$ will form when $[Mg^{2+}]$ is 0.10 M. This is the hydroxide concentration for which $[Mg^{2+}][OH^-]^2$ equals K_{sp}.

	$[Mg^{2+}]$	$[OH^-]$
Equilibrium concentrations, M	0.10	x

$$[Mg^{2+}][OH^-]^2 = K_{sp} = 1.5 \times 10^{-11}$$

$$(0.10)(x)^2 = 1.5 \times 10^{-11}$$

$$x^2 = \frac{1.5 \times 10^{-11}}{0.10} = 1.5 \times 10^{-10}$$

$$x = 1.2 \times 10^{-5} \text{ M}$$

Magnesium hydroxide precipitates in a 0.1 M solution of NH_3 but redissolves when NH_4Cl is added.

Thus, $Mg(OH)_2$ will not begin to precipitate in a 0.10 M Mg^{2+} solution if $[OH^-]$ is less than 1.2×10^{-5} M.

Now we must find what $[NH_4^+]$ is necessary to reduce $[OH^-]$ to 1.2×10^{-5} M.

	$[NH_3]$	$[NH_4^+]$	$[OH^-]$
Equilibrium concentrations, M	0.10	y	1.2×10^{-5}

$$\frac{[NH_4^+][OH^-]}{[NH_3]} = K_b = 1.8 \times 10^{-5}$$

$$\frac{(y)(1.2 \times 10^{-5})}{0.10} = 1.8 \times 10^{-5}$$

$$y = \frac{0.10}{1.2 \times 10^{-5}} \times 1.8 \times 10^{-5} = 0.15 \text{ M}$$

If $[NH_4^+]$ equals 0.15 M, $[OH^-]$ will be 1.2×10^{-5} M. Any $[NH_4^+]$ greater than 0.15 M will reduce $[OH^-]$ below 1.2×10^{-5} M and prevent formation of $Mg(OH)_2$.

18.10 Fractional Precipitation

When two anions form slightly soluble compounds with the same cation or when two cations form slightly soluble compounds with the same anion, the less soluble compound will precipitate first as the precipitating agent (the precipitant) is added to a solution containing both ions. Almost all of the less soluble compound will precipitate before any of the more soluble one does; then, however, any remaining less soluble compound will precipitate along with the more soluble one (**coprecipitate**) as more precipitant is added.

For example, a solution containing both sodium iodide and sodium chloride is treated with silver nitrate. Silver iodide, being less soluble than silver chloride, is precipitated first. Only after most of the iodide is precipitated will the chloride begin to precipitate.

EXAMPLE 18.10   A solution contains 0.010 mol of KI and 0.10 mol of KCl per liter. $AgNO_3$ is gradually added to this solution. Which will be precipitated first, AgI or AgCl?

We must find $[Ag^+]$ at which AgI will begin to precipitate and $[Ag^+]$ at which AgCl will begin to precipitate. The salt that forms at the lower $[Ag^+]$ will precipitate first.

For AgI. Let x be equal to $[Ag^+]$ at which precipitation begins.

$$[Ag^+][I^-] = K_{sp} = 1.5 \times 10^{-16}$$

$$(x)(0.010) = 1.5 \times 10^{-16}$$

$$x = \frac{1.5 \times 10^{-16}}{0.010} = 1.5 \times 10^{-14} \text{ M}$$

A precipitate of silver iodide, AgI (left) and of silver chloride, AgCl (right).

Thus AgI will begin to precipitate when $[Ag^+]$ is 1.5×10^{-14} M.

For AgCl. Let y be equal to $[Ag^+]$ at which precipitation begins.

$$[Ag^+][Cl^-] = K_{sp} = 1.8 \times 10^{-10}$$
$$(y)(0.10) = 1.8 \times 10^{-10}$$
$$y = \frac{1.8 \times 10^{-10}}{0.10} = 1.8 \times 10^{-9} \text{ M}$$

Thus $[Ag^+]$ must be 1.8×10^{-9} M for AgCl to precipitate.

AgI begins to precipitate at a lower silver ion concentration than that at which AgCl begins to precipitate, so AgI will begin to precipitate first.

(a) What is the concentration of I^- in the solution described in Example 18.10 when AgCl begins to precipitate, and what fraction of the original I^- remains in solution at this point?

EXAMPLE 18.11

When $[Ag^+]$ is 1.8×10^{-9} M, AgCl will begin to precipitate. Let x be equal to $[I^-]$ when $[Ag^+]$ is 1.8×10^{-9} M.

$$[Ag^+][I^-] = K_{sp} = 1.5 \times 10^{-16}$$
$$(1.8 \times 10^{-9})(x) = 1.5 \times 10^{-16}$$
$$x = \frac{1.5 \times 10^{-16}}{1.8 \times 10^{-9}} = 8.3 \times 10^{-8} \text{ M}$$

$[I^-]$ will be 8.3×10^{-8} M when AgCl begins to precipitate.

The fraction of I^- remaining at this point may be determined as follows:

$$\frac{[I^-] \text{ when precipitation of AgCl begins}}{[I^-] \text{ originally present}} = \frac{8.3 \times 10^{-8}}{0.010} = 8.3 \times 10^{-6}$$

Only about 8 parts in a million of the original I^- remain in solution when AgCl begins to precipitate.

(b) What will the concentration of I^- in the solution be after half of the Cl^- initially present is precipitated as AgCl?

When half of the Cl^- initially present has been precipitated, $[Cl^-]$ will have been reduced to 0.050 M.

$$[Ag^+] = \frac{K_{AgCl}}{[Cl^-]} = \frac{1.8 \times 10^{-10}}{0.050} = 3.6 \times 10^{-9} \text{ M}$$

$$[I^-] = \frac{K_{AgI}}{[Ag^+]} = \frac{1.5 \times 10^{-16}}{3.6 \times 10^{-9}} = 0.417 \times 10^{-7} = 4.2 \times 10^{-8} \text{ M}$$

Hence half of the very small amount of I^- remaining in solution when AgCl begins to precipitate (8.3×10^{-8} M) is coprecipitated as half of the Cl^- is precipitated.

The Dissolution of Precipitates

When the product of the initial molar concentrations of the ions raised to their appropriate powers is less than ($<$) the solubility product, a solid either dissolves completely or dissolves until the ion concentration product is equal to the solubility product. For example, calcium carbonate dissolves in water as long as

$$[Ca^{2+}][CO_3^{2-}] < K_{sp}$$

For a precipitate to dissolve completely, it is necessary to make the concentration of at least one of its ions low enough that the ion concentration product is less than the solubility product. Ions can be removed (1) by the formation of a weak electrolyte, (2) by the conversion of an ion to another species, or (3) by the formation of a complex ion.

18.11 Dissolution of a Precipitate by the Formation of a Weak Electrolyte

Slightly soluble solids derived from weak acids are often soluble in strong acids. For example, $CaCO_3$, FeS, and $Ca_3(PO_4)_2$ dissolve in HCl because their anions react to form weak acids (H_2CO_3, H_2S, and $H_2PO_4^-$).

When hydrochloric acid is added to solid calcium carbonate, the hydrogen ion from the acid combines with the carbonate ion and forms the hydrogen carbonate ion, a weak acid.

$$H^+(aq) + CO_3^{2-}(aq) \rightleftharpoons HCO_3^-(aq)$$

Additional hydrogen ion reacts with the hydrogen carbonate ion according to the equation

$$H^+(aq) + HCO_3^-(aq) \rightleftharpoons H_2CO_3(aq)$$

Finally, the solution becomes saturated with the slightly ionized and unstable carbonic acid, and carbon dioxide gas is evolved.

$$H_2CO_3(aq) \rightleftharpoons H_2O(l) + CO_2(g)$$

These reactions reduce the carbonate ion concentration to such a low level that the product of the calcium ion and carbonate ion concentrations is less than the solubility product of calcium carbonate.

$$[Ca^{2+}][CO_3^{2-}] < K_{sp}$$

Consequently, the calcium carbonate dissolves. Even acetic acid provides enough hydrogen ion to dissolve calcium carbonate because, although weak, acetic acid is still a stronger acid than either the hydrogen carbonate ion or carbonic acid. (Check the K_a values for acetic acid, the hydrogen carbonate ion, and carbonic acid in Appendix F to verify this statement.)

Lead sulfate dissolves in solutions of ammonium acetate because the formation of slightly ionized (but soluble) lead acetate reduces the product of the lead ion and sulfate ion concentrations below the value of the solubility product for lead sulfate.

$$PbSO_4(s) \rightleftharpoons Pb^{2+}(aq) + SO_4^{2-}(aq)$$
$$Pb^{2+}(aq) + 2CH_3CO_2^-(aq) \rightleftharpoons Pb(CH_3CO_2)_2(aq)$$
$$[Pb^{2+}][SO_4^{2-}] < K_{sp}$$

The formation of water causes most metal hydroxides, such as $Al(OH)_3$, $Mg(OH)_2$, and $Fe(OH)_3$, to dissolve in acid solutions.

$$Al(OH)_3(s) \rightleftharpoons Al^{3+}(aq) + 3OH^-(aq)$$
$$3H^+(aq) + 3OH^-(aq) \rightleftharpoons 3H_2O(l)$$
$$\textit{Net: } Al(OH)_3(s) + 3H^+(aq) \longrightarrow Al^{3+}(aq) + 3H_2O(l)$$
$$[Al^{3+}][OH^-]^3 < K_{sp}$$
$$Mg(OH)_2(s) \rightleftharpoons Mg^{2+}(aq) + 2OH^-(aq)$$
$$2NH_4^+(aq) + 2OH^-(aq) \rightleftharpoons 2NH_3(aq) + 2H_2O(l)$$
$$\textit{Net: } Mg(OH)_2(s) + 2NH_4^+(aq) \longrightarrow Mg^{2+}(aq) + 2H_2O(l) + 2NH_3(aq)$$
$$[Mg^{2+}][OH^-]^2 < K_{sp}$$

18.12 Dissolution of a Precipitate by the Conversion of an Ion to Another Species

The solubility products for some metal sulfides are sufficiently large that the hydrogen ion from a strong acid lowers the sulfide ion concentration (by forming the weak electrolyte hydrogen sulfide) enough to dissolve the sulfide. For example, iron(II) sulfide is readily dissolved by hydrochloric acid according to the equation

$$FeS(s) \rightleftharpoons Fe^{2+}(aq) + S^{2-}(aq)$$
$$2H^+(aq) + S^{2-}(aq) \rightleftharpoons H_2S(g)$$
$$\textit{Net: } FeS(s) + 2H^+(aq) \rightleftharpoons Fe^{2+}(aq) + H_2S(g)$$
$$[Fe^{2+}][S^{2-}] < K_{sp}$$

However, the solubility products for other metal sulfides, such as lead sulfide, are so small that even very high concentrations of hydrogen ion are not sufficient to lower the sulfide ion concentration enough to dissolve them. To dissolve these sulfides, the sulfide ion concentration must be decreased by oxidizing the sulfide to elemental sulfur.

$$3S^{2-}(aq) + 2NO_3^-(aq) + 8H^+(aq) \longrightarrow 3S(s) + 2NO(g) + 4H_2O(l)$$

Lead sulfide dissolves in nitric acid, then, because the sulfide ion is oxidized to sulfur, making the product of the lead ion and sulfide ion concentrations less than the solubility product:

$$[Pb^{2+}][S^{2-}] < K_{sp}$$

18.13 Dissolution of a Precipitate by the Formation of a Complex Ion

Many slightly soluble electrolytes dissolve as a result of the formation of complex ions. Several examples that are important to qualitative analysis follow.

$$AgCl(s) + 2NH_3(aq) \rightleftharpoons Ag(NH_3)_2^+(aq) + Cl^-(aq)$$
$$CuCN(s) + CN^-(aq) \rightleftharpoons Cu(CN)_2^-(aq)$$
$$Zn(OH)_2(s) + 2OH^-(aq) \rightleftharpoons Zn(OH)_4^{2-}(aq)$$
$$Sn(OH)_2(s) + OH^-(aq) \rightleftharpoons Sn(OH)_3^-(aq)$$
$$Al(OH)_3(s) + OH^-(aq) \rightleftharpoons Al(OH)_4^-(aq)$$
$$As_2S_3(s) + 3S^{2-}(aq) \rightleftharpoons 2AsS_3^{3-}(aq)$$
$$HgS(s) + S^{2-}(aq) \rightleftharpoons HgS_2^{2-}(aq)$$
$$Sb_2S_3(s) + 3S^{2-}(aq) \rightleftharpoons 2SbS_3^{3-}(aq)$$

The stability of a complex ion is described by an equilibrium constant for the formation of the complex ion from its components in solution. The larger the equilibrium constant, the more stable the complex. For example, the complex ion $Cu(CN)_2^-$ forms by the reaction

$$Cu^{2+} + 2CN^- \rightleftharpoons Cu(CN)_2^-$$

The equilibrium constant for this reaction is the **formation constant, K_f.**

$$K_f = \frac{[Cu(CN)_2^-]}{[Cu^{2+}][CN^-]^2} = 1 \times 10^{16}$$

This constant is sometimes referred to as the **stability constant** or **association constant.** Appendix E contains a table of formation constants. The larger the formation constant, the more stable the complex.

Alternatively, the stability of a complex ion can be described by its **dissociation constant, K_d,** the equilibrium constant for the decomposition of the complex ion into its components in solution. For $Cu(CN)_2^-$ the dissociation is

$$Cu(CN)_2^- \rightleftharpoons Cu^{2+} + 2CN^-$$

and

$$K_d = \frac{[Cu^{2+}][CN^-]^2}{[Cu(CN)_2^-]} = 1 \times 10^{-16}$$

It should be apparent that

$$K_d = \frac{1}{K_f}$$

The smaller the dissociation constant, the more stable the complex.

As an example of dissolution by complex ion formation, let us consider the dissolution of silver chloride in aqueous ammonia. Silver chloride dissolves in water giving a small concentration of Ag^+ ($[Ag^+] = 1.3 \times 10^{-5}$ M; see Example 18.3). However, if NH_3 is present in the water, the complex diamminesilver ion, $Ag(NH_3)_2^+$, can form according to the equation

$$Ag^+(aq) + 2NH_3(aq) \rightleftharpoons Ag(NH_3)_2^+(aq)$$

The formation constant, K_f, for $Ag(NH_3)_2^+$ is 1.6×10^7.

$$\frac{[Ag(NH_3)_2^+]}{[Ag^+][NH_3]^2} = K_f = 1.6 \times 10^7$$

The large value for this formation constant indicates that most of the free silver ions produced by the dissolution of AgCl combine with NH_3 to form $Ag(NH_3)_2^+$. As a consequence, the concentration of silver ions, $[Ag^+]$, is reduced, and the product $[Ag^+][Cl^-]$ becomes less than the solubility product of silver chloride.

$$[Ag^+][Cl^-] < K_{sp}$$

More silver chloride then dissolves. If the concentration of ammonia is great enough, all of the silver chloride will dissolve.

18.14 Calculations Involving Complex Ions

The following examples illustrate calculations of equilibrium concentrations in reactions involving complex ions.

Calculate the concentration of silver ions in a solution that is 0.10 M with respect to $Ag(NH_3)_2^+$.

EXAMPLE 18.12

The formation of the complex ion may be represented by the equation

$$Ag^+(aq) + 2NH_3(aq) \rightleftharpoons Ag(NH_3)_2^+(aq)$$

The reverse reaction shows the dissociation of $Ag(NH_3)_2^+$. Each mole of $Ag(NH_3)_2^+$ that dissociates gives 1 mol of Ag^+ and 2 mol of NH_3. Let x = the number of moles of $Ag(NH_3)_2^+$ that dissociates per liter of solution. When x mol of $Ag(NH_3)_2^+$ dissociates, x mol of Ag^+ and $2x$ mol of NH_3 will be produced. Hence at equilibrium

$$[Ag^+] = x, \qquad [NH_3] = 2x, \qquad \text{and} \qquad [Ag(NH_3)_2^+] = 0.10 - x$$

	$[Ag^+]$	$[NH_3]$	$[Ag(NH_3)_2^+]$
Initial concentrations, M	0	0	0.10
Equilibrium concentrations, M	x	$2x$	$0.10-x$

$$\frac{[Ag(NH_3)_2{}^+]}{[Ag^+][NH_3]^2} = K_f = 1.6 \times 10^7$$

$$\frac{0.10 - x}{(x)(2x)^2} = 1.6 \times 10^7$$

The formation constant of $Ag(NH_3)_2{}^+$ is large, so we can assume that $[Ag(NH_3)_2{}^+]$ is close to 0.10 M and that x is negligible in the term $0.10 - x$. Therefore

$$\frac{0.10}{(x)(2x)^2} = K_f = 1.6 \times 10^7$$

$$0.10 = (4x^3)(1.6 \times 10^7)$$

$$x^3 = \frac{0.10}{(4)(1.6 \times 10^7)} = 1.6 \times 10^{-9}$$

$$x = 1.2 \times 10^{-3} \text{ M} = [Ag^+]$$

$$2x = 2.4 \times 10^{-3} \text{ M} = [NH_3]$$

(Since only 1.2% of the $Ag(NH_3)_2{}^+$ dissociates to Ag^+ and NH_3, neglecting x in the term $0.10 - x$ is justified.)

EXAMPLE 18.13 Unexposed silver halides are removed from photographic film when they react with a solution of $Na_2S_2O_3$ (hypo) and form the complex ion $Ag(S_2O_3)_2{}^{3-}$ ($K_f = 4.7 \times 10^{13}$). What mass of $Na_2S_2O_3$ is required to prepare 1 L of a solution that will dissolve 1.00 g of silver bromide, AgBr, by the formation of $Ag(S_2O_3)_2{}^{3-}$?

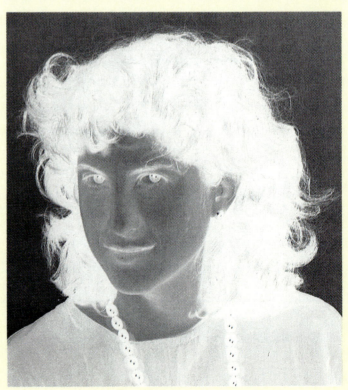

An exposed and developed photographic film. The transparent or light grey parts are the portions of the film that were not exposed to light or were lightly exposed. At these portions of the film the silver halide is dissolved during the developing process by the sodium thiosulfate (hypo) through formation of the soluble $Ag(S_2O_3)_2{}^{3-}$ complex ion. The darker grey or black parts are the portions of the film that were exposed to light, reducing the metal halide to metallic silver (grey or black depending on degree of exposure).

Two equilibria are involved when AgBr dissolves in a solution containing the thiosulfate ion, $S_2O_3^{2-}$

$$AgBr(s) \rightleftharpoons Ag^+(aq) + Br^-(aq)$$

$$Ag^+(aq) + 2S_2O_3^{2-}(aq) \rightleftharpoons Ag(S_2O_3)_2^{3-}(aq)$$

In order for the AgBr(s) to dissolve completely, the maximum $[Ag^+]$ must be low enough that the product $[Ag^+][Br^-]$ is less than K_{sp} for AgBr(s).

Let us first calculate the concentration of Br^- produced by the dissolution of 1.00 g of AgBr and then determine the maximum possible concentration of Ag^+ that will not result in the reprecipitation of AgBr.

$$1.00 \text{ g AgBr} \times \frac{1 \text{ mol AgBr}}{187.77 \text{ g AgBr}} = 5.33 \times 10^{-3} \text{ mol AgBr}$$

$$5.33 \times 10^{-3} \text{ mol AgBr} \times \frac{1 \text{ mol Br}^-}{1 \text{ mol AgBr}} = 5.33 \times 10^{-3} \text{ mol Br}^-$$

$$\frac{5.33 \times 10^{-3} \text{ mol Br}^-}{1.000 \text{ L}} = 5.33 \times 10^{-3} \text{ M} = [Br^-]$$

Now find the maximum $[Ag^+]$ without reprecipitation of AgBr taking place. At this point $[Br^-]$ is 5.33×10^{-3} M, and $[Ag^+]$ is set equal to x.

$$[Ag^+][Br^-] = K_{sp} = 3.3 \times 10^{-13}$$

$$(x)(5.33 \times 10^{-3}) = 3.3 \times 10^{-13}$$

$$x = \frac{3.3 \times 10^{-13}}{5.33 \times 10^{-3}} = 6.19 \times 10^{-11} \text{ M}$$

The maximum $[Ag^+]$ is 6.19×10^{-11} M. Below this concentration, when $[Br^-]$ is 5.33×10^{-3} M, AgBr will not form because

$$[Ag^+][Br^-] < K_{sp}$$

Now we determine $[S_2O_3^{2-}]$ when $[Ag^+]$ is 6.19×10^{-11} M. Since 5.33×10^{-3} mol of AgBr dissolves,

$$(5.33 \times 10^{-3}) - (6.19 \times 10^{-11}) = 5.33 \times 10^{-3} \text{ mol Ag}(S_2O_3)_2^{3-}$$

must form. Thus $[Ag(S_2O_3)_2^{3-}]$ is equal to 5.33×10^{-3} M. To form 5.33×10^{-3} mol of $Ag(S_2O_3)_2^{3-}$ would require $2 \times (5.33 \times 10^{-3})$ mol of $S_2O_3^{2-}$. However, the total number of moles of $S_2O_3^{2-}$ that must be added to the solution is equal to the number of moles that react in forming the complex *plus* the number of moles of free $S_2O_3^{2-}$ in solution at equilibrium. At equilibrium $[S_2O_3^{2-}]$ is set equal to y, $[Ag^+]$ is 6.19×10^{-11} M, and $[Ag(S_2O_3)_2^{3-}]$ is 5.33×10^{-3} M.

$$\frac{[Ag(S_2O_3)_2^{3-}]}{[Ag^+][S_2O_3^{2-}]^2} = K_f = 4.7 \times 10^{13}$$

$$\frac{5.33 \times 10^{-3}}{(6.19 \times 10^{-11})(y)^2} = 4.7 \times 10^{13}$$

$$y^2 = \frac{5.33 \times 10^{-3}}{6.19 \times 10^{-11}} \times \frac{1}{4.7 \times 10^{13}} = 1.83 \times 10^{-6}$$

$$y = 1.35 \times 10^{-3} \text{ M} = [S_2O_3^{2-}]$$

If $[S_2O_3{}^{2-}]$ is increased above 1.35×10^{-3} M, $[Ag^+]$ will decrease below 6.19×10^{-11} M, and no solid AgBr will remain. The total amount of $S_2O_3{}^{2-}$ that must be added to the solution by adding $Na_2S_2O_3$ is

$$2 \times (5.33 \times 10^{-3} \text{ mol}) + 1.35 \times 10^{-3} \text{ mol}$$
$$= 1.20 \times 10^{-2} \text{ mol } S_2O_3{}^{2-} = 1.20 \times 10^{-2} \text{ mol } Na_2S_2O_3$$

$$1.20 \times 10^{-2} \text{ mol } Na_2S_2O_3 \times \frac{158.1 \text{ g } Na_2S_2O_3}{1 \text{ mol } Na_2S_2O_3} = 1.9 \text{ g } Na_2S_2O_3$$

Thus 1 L of a solution prepared from 1.9 g of $Na_2S_2O_3$ will dissolve 1.0 g of AgBr.

18.15 Effect of the Reaction of Salts with Water on the Dissolution of Precipitates

Many salts react with water when they dissolve in aqueous solution, some with the formation of precipitates, the evolution of gases, or both. For example, hydrated aluminum hydroxide precipitates and hydrogen sulfide is produced when aluminum sulfide, Al_2S_3, dissolves in water. The aluminum ions and the sulfide ions combine with water.

$$Al_2S_3(aq) \rightleftharpoons 2Al^{3+}(aq) + 3S^{2-}(aq)$$
$$2Al^{3+}(aq) + 12H_2O(l) \longrightarrow 2[Al(H_2O)_6]^{3+}(aq)$$
$$3S^{2-}(aq) + 6H_2O(l) \rightleftharpoons 3H_2S(aq) + 6OH^-(aq)$$
$$2[Al(H_2O)_6]^{3+}(aq) + 6OH^-(aq) \rightleftharpoons 2[Al(H_2O)_3(OH)_3](s) + 6H_2O(l)$$

The net reaction is given by

$$Al_2S_3(s) + 12H_2O(l) \rightleftharpoons 2[Al(H_2O)_3(OH)_3](s) + 3H_2S(aq)$$

If the initial amount of aluminum sulfide is high enough, the solubility of hydrogen sulfide may be exceeded and it will evolve as a gas. When a relatively insoluble sulfide such as lead sulfide, PbS, dissolves in water, an appreciable fraction of the sulfide ion reacts with water to form the hydrogen sulfide ion and, in some cases, hydrogen sulfide.

$$PbS(s) \rightleftharpoons Pb^{2+}(aq) + S^{2-}(aq)$$
$$S^{2-}(aq) + H_2O(l) \rightleftharpoons HS^-(aq) + OH^-(aq)$$
$$HS^-(aq) + H_2O(l) \rightleftharpoons H_2S(aq) + OH^-(aq)$$

Under these conditions, the lead ion concentration is not the same as that of the sulfide ion but equals the sum of the concentrations of sulfide ion, hydrogen sulfide ion, and hydrogen sulfide.

$$[Pb^{2+}] = [S^{2-}] + [HS^-] + [H_2S]$$

This means that in calculating the solubility of a relatively insoluble sulfide from the solubility product, we must consider the reaction of the basic sulfide ion with water.

Calculate the solubility of lead sulfide ($K_{sp} = 6.5 \times 10^{-34}$) in water, neglecting the reaction of the sulfide ion with water.

EXAMPLE 18.14

Let x equal the molar solubility of PbS in water. Then x will be equal to $[Pb^{2+}]$ at equilibrium and, if we assume that the reaction of the sulfide ion with water is negligible, will also be equal to $[S^{2-}]$. Substituting in the expression for the solubility product, we obtain

$$[Pb^{2+}][S^{2-}] = K_{sp}$$
$$(x)(x) = x^2 = 6.5 \times 10^{-34}$$
$$x = 2.5 \times 10^{-17} \text{ mol L}^{-1}$$

If the reaction of the sulfide ion with water is neglected, we find the solubility of lead sulfide in water to be 2.5×10^{-17} mol L^{-1}.

Now solve the problem of Example 18.14 correctly, taking into consideration the reaction of the sulfide ion with water.

EXAMPLE 18.15

We shall consider three steps in the dissolution of PbS in water.

$$PbS(s) \rightleftharpoons Pb^{2+} + S^{2-} \qquad K_{sp} = 6.5 \times 10^{-34}$$

$$S^{2-} + H_2O \rightleftharpoons HS^- + OH^- \qquad K_{b_1} = \frac{[HS^-][OH^-]}{[S^{2-}]} = \frac{K_w}{K_a \text{ (for HS}^-)}$$

$$= \frac{1.0 \times 10^{-14}}{1.0 \times 10^{-19}} = 1.0 \times 10^{5}$$

$$HS^- + H_2O \rightleftharpoons H_2S + OH^- \qquad K_{b_2} = \frac{[H_2S][OH^-]}{[HS^-]} = \frac{K_w}{K_a \text{ (for H}_2\text{S)}}$$

$$= \frac{1.0 \times 10^{-14}}{1.0 \times 10^{-7}} = 1.0 \times 10^{-7}$$

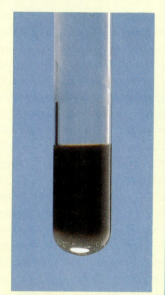

A precipitate of lead sulfide, PbS.

Since K_{b_2} is quite small compared to K_{b_1}, we might be tempted to consider only the first reaction with water and neglect the second. This is often a valid way to proceed. In Example 17.22 we found that essentially 100% of the sulfide ion in a 0.0010 M solution of sodium sulfide reacts with water and produces 1.0×10^{-3} mol of OH$^-$. Since this quantity of OH$^-$ is large compared to the 1.0×10^{-7} mol furnished by the water, the total concentration of OH$^-$ in solution was found to be 1.0×10^{-3} mol L^{-1}. We can calculate the ratio of $[H_2S]$ to $[HS^-]$, from K_{b_2}, to establish the relative amount of H_2S produced in the second step of the reaction with water.

$$\frac{[H_2S][OH^-]}{[HS^-]} = \frac{[H_2S](1.0 \times 10^{-3})}{[HS^-]} = K_{b_2} = 1.0 \times 10^{-7}$$

$$\frac{[H_2S]}{[HS^-]} = 1.0 \times 10^{-4}$$

We find for that example that $[H_2S]$ is only 0.0001 times $[HS^-]$, indicating that the second step in the reaction of S^{2-} with water was negligible compared to the first step and can be neglected *in that calculation*.

However, in this example, we have a much smaller amount of sulfide ion. Our rough calculation for the solubility of lead sulfide, neglecting the reaction of sulfide ion with water, in Example 18.14 indicated only 2.5×10^{-17} mol of lead sulfide per liter. The hydroxide ion concentration due to the reaction of the sulfide ion with water

$$S^{2-} + H_2O \rightleftharpoons HS^- + OH^-$$

is on the order of 2.5×10^{-17} M. However, this hydroxide ion concentration is so small relative to that provided by the water (1.0×10^{-7} M) that $[OH^-]$ of the solution actually is 1.0×10^{-7} M. Even some increase in the solubility of PbS and the additional OH^- produced in the second step (the reaction of HS^- with H_2O) would not be likely to add sufficient OH^- to be significant compared with the amount furnished by the water. We can now calculate the ratio of $[H_2S]$ to $[HS^-]$ from K_{b_2}, for a solution in which $[OH^-]$ is 1×10^{-7} M (a neutral solution).

$$\frac{[H_2S][OH^-]}{[HS^-]} = \frac{[H_2S](1.0 \times 10^{-7})}{[HS^-]} = K_{b_2} = 1.0 \times 10^{-7}$$

Therefore

$$\frac{[H_2S]}{[HS^-]} = 1$$

Hence, for our neutral solution, $[H_2S]$ is equal to $[HS^-]$, indicating that the second step in the reaction of sulfide ion with water (that is, the reaction of the hydrogen sulfide ion with water) *is* important in this calculation and cannot be neglected. For $[HS^-]$ to equal $[H_2S]$, half of the HS^- ions produced in the first step must react with water to produce H_2S in the second step.

Now, consider the three equilibria listed earlier for this system. As PbS dissolves, equal quantities of Pb^{2+} and S^{2-} are produced initially. The high value of K_{b_1} (1.0×10^5) indicates that essentially all of the S^{2-} that is produced reacts with water and is hence converted to HS^- in the first step of the reaction with water. At this point and not yet taking the second step into account, $[Pb^{2+}]$ is equal to $[HS^-]$. However, we have established that half of the HS^- that is produced reacts with water in the second stage, thereby reducing the concentration of HS^- to half of its former value and thus to half of the concentration of Pb^{2+}. Hence at final equilibrium

$$[HS^-] = \tfrac{1}{2}[Pb^{2+}]$$

Using 1.0×10^{-7} M for $[OH^-]$, we can calculate the ratio of $[HS^-]$ to $[S^{2-}]$ by substituting in the expression for K_{b_1} and then solving for $[S^{2-}]$.

$$\frac{[HS^-][OH^-]}{[S^{2-}]} = \frac{[HS^-](1.0 \times 10^{-7})}{[S^{2-}]} = K_{b_1} = 1.0 \times 10^5$$

$$\frac{[HS^-]}{[S^{2-}]} = 1.0 \times 10^{12}, \quad \text{or} \quad [S^{2-}] = \frac{[HS^-]}{1.0 \times 10^{12}}$$

Substituting the above value for $[S^{2-}]$ in the expression for the solubility product for lead sulfide, we have

$$[Pb^{2+}][S^{2-}] = [Pb^{2+}] \frac{[HS^-]}{1.0 \times 10^{12}} = K_{sp} = 6.5 \times 10^{-34}$$

Because [HS$^-$] is equal to $\frac{1}{2}$ [Pb^{2+}], we can write

$$[Pb^{2+}] \left(\frac{\frac{1}{2}[Pb^{2+}]}{1.0 \times 10^{12}} \right) = 6.5 \times 10^{-34}$$

$$\frac{\frac{1}{2}[Pb^{2+}]^2}{1.0 \times 10^{12}} = 6.5 \times 10^{-34}$$

$$[Pb^{2+}]^2 = 1.3 \times 10^{-21}$$

$$[Pb^{2+}] = 3.6 \times 10^{-11} \text{ M} = \text{molar solubility of PbS}$$

We have found, then, that the solubility of lead sulfide in water is 3.6×10^{-11} mol L^{-1} when the reaction of the sulfide ion with water is considered. This value is over a million times greater than that calculated (2.5×10^{-17} mol L^{-1}) by neglecting the reaction of sulfide ion with water. If we were to take into account the fact that the lead ion reacts slightly with water to give Pb(OH)$^+$ and Pb(OH)$_2$, the solubility of lead sulfide in water would be found to be even greater.

Finally, note that the reaction of the sulfide ion with water would produce 3.6×10^{-11} mol of OH$^-$ per liter in the first step and half of that amount, or 1.8×10^{-11} mol of OH$^-$ per liter, in the second step, for a total of 5.4×10^{-11} mol of OH$^-$ per liter. This amount is indeed negligible compared to the 1.0×10^{-7} mol of OH$^-$ per liter furnished by the water. Hence the assumption that [OH$^-$] is equal to 1.0×10^{-7} M for this particular case is entirely justified.

The insoluble carbonates behave similarly to the sulfides. When a relatively insoluble carbonate such as barium carbonate, BaCO$_3$, is dissolved in water, its solubility is increased due to reaction of the basic carbonate ion with water.

$$CO_3^{2-} + H_2O \rightleftharpoons HCO_3^- + OH^-$$
$$HCO_3^- + H_2O \rightleftharpoons H_2CO_3 + OH^-$$

The concentration of the barium ion is equal to the sum of the concentrations of the carbonate ion, CO$_3^{2-}$, the hydrogen carbonate ion, HCO$_3^-$, and carbonic acid, H$_2$CO$_3$.

For Review

Summary

An equilibrium involving the precipitation or dissolution of a slightly soluble electrolyte is described by the **solubility product, K_{sp}**, the equilibrium constant for the equilibrium between the solid and its ions in solution. The solubility product of a compound can be calculated from its solubility, and the solubility of the compound can be calculated from its solubility product.

A slightly soluble electrolyte will begin to precipitate when the product of the concentrations of its ions raised to their appropriate powers exceeds the solubility product of the electrolyte. Precipitation will continue until this ion concentration product equals the solubility product. Consequently, for a solution containing one of the ions of a slightly soluble electrolyte, we can calculate how much of the other ion of the electrolyte must be added either to cause precipitation of the electrolyte to begin or to reduce the concentration of the first ion to any desired value.

Many slightly soluble electrolytes contain anions of weak acids. These electrolytes often dissolve in acidic solutions because reaction of the anion with water to give the weak acid reduces the concentration of the anion to the point where the ion concentration product is less than the solubility product. The ion concentration product can also be reduced by the conversion of an ion to another species or by the formation of a complex ion. The stability of a complex ion is described by its **formation constant, K_f.**

Reaction of the ions of a precipitate with water can have a significant effect on its solubility in water.

Key Terms and Concepts

complex ion (18.13, 18.14)
coprecipitation (18.10)
dissociation constant (18.13)
dissolution of precipitates (18.11, 18.12, 18.13)

formation constant (18.13)
fractional precipitation (18.10)
molar solubility (18.3)

precipitation (18.4)
solubility product (18.1)
supersaturated solution (18.7)

Exercises

Solubility Products

1. Write the expression for the solubility product of each of the following slightly soluble electrolytes:
   (a) $AgBr$ (b) Ag_2S
   (c) $PbCl_2$ (d) $MgNH_4AsO_4$
   (e) $Ba_3(PO_4)_2$

2. Under what circumstances, if any, will a sample of solid $AgCl$ completely dissolve in pure water?

3. How do the concentrations of Ag^+ and CrO_4^{2-} in a liter of water above 1.0 g of solid Ag_2CrO_4 change when 100 g of solid Ag_2CrO_4 is added to the system? Explain.

4. Refer to Appendix D for solubility products for calcium salts. Determine which of the calcium salts listed is most soluble in moles per liter. Determine which is most soluble in grams per liter.

5. How do the concentrations of Pb^{2+} and S^{2-} change when K_2S is added to a saturated solution of PbS?

6. Calculate the solubility product of each of the following from the solubility given:
   (a) $CaCO_3$, 6.9×10^{-3} g L^{-1} *Ans. 4.8×10^{-9}*
   (b) $AgBr$, 5.7×10^{-7} mol L^{-1} *Ans. 3.2×10^{-13}*
   (c) PbF_2, 2.1×10^{-3} mol L^{-1} *Ans. 3.7×10^{-8}*
   (d) Ag_2CrO_4, 4.3×10^{-2} g L^{-1} *Ans. 8.7×10^{-12}*
   (e) Ag_2SO_4, 4.47 g L^{-1} *Ans. 1.18×10^{-5}*

7. A saturated solution of a slightly soluble electrolyte in contact with some of the solid electrolyte is said to be a system in equilibrium. Explain. Why is such a system called a heterogeneous equilibrium?

8. Solid silver bromide is in equilibrium with a saturated solution of its ions, Ag^+ and Br^-. How, if at all, will this equilibrium be affected if (a) more solid silver bromide is added? (b) Silver nitrate is added? (c) Sodium bromide is added? (d) The temperature is raised? (The solubility increases with temperature.)

9. Which of the following compounds will precipitate from a solution initially containing the indicated concentrations of ions (see Appendix D for K_{sp} values)?
   (a) $CaCO_3$: $[Ca^{2+}] = 0.003$ M,
                   $[CO_3^{2-}] = 0.003$ M
                              Ans. Precipitates
   (b) $CaHPO_4$: $[Ca^{2+}] = 0.01$ M,
                     $[HPO_4^{2-}] = 2 \times 10^{-6}$ M
                              Ans. Does not precipitate
   (c) Ag_2S: $[Ag^+] = 1 \times 10^{-10}$ M,
                $[S^{2-}] = 1 \times 10^{-13}$ M
                              Ans. Precipitates
   (d) $Co(OH)_2$: $[Co^{2+}] = 0.01$ M,
                   $[OH^-] = 1 \times 10^{-7}$ M
                              Ans. Does not precipitate
   (e) $Pb_3(PO_4)_2$: $[Pb^{2+}] = 0.01$ M,
                         $[PO_4^{3-}] = 1 \times 10^{-13}$ M
                              Ans. Precipitates

10. The *Handbook of Chemistry and Physics* gives solubilities for the compounds listed below in grams per 100 mL of water. Since these compounds are only slightly soluble, assume that the volume does not change on dissolution, and calculate the solubility product for each.
    (a) $TlCl$, 0.29 g/100 mL   *Ans. 1.5×10^{-4}*
    (b) $Ce(IO_3)_4$, 1.5×10^{-2} g/100 mL   *Ans. 4.6×10^{-17}*
    (c) $Gd_2(SO_4)_3$, 3.98 g/100 mL   *Ans. 1.36×10^{-4}*
    (d) InF_3, 4.0×10^{-2} g/100 mL   *Ans. 7.9×10^{-10}*

11. Calculate the molar solubility of each of the following minerals from its K_{sp}:
    (a) alabandite, MnS: $K_{sp} = 4.3 \times 10^{-22}$
    Ans. 2.1×10^{-11} M
    (b) anglesite, PbSO$_4$: $K_{sp} = 1.8 \times 10^{-8}$
    Ans. 1.3×10^{-4} M
    (c) brucite, Mg(OH)$_2$: $K_{sp} = 1.5 \times 10^{-11}$
    Ans. 1.6×10^{-4} M
    (d) fluorite, CaF$_2$: K_{sp}: $= 3.9 \times 10^{-11}$
    Ans. 2.1×10^{-4} M

12. Calculate the concentrations of ions in a saturated solution of each of the following (see Appendix D for solubility products):
    (a) AgI   *Ans. $[Ag^+] = [I^-] = 1.2 \times 10^{-8}$ M*
    (b) Ag$_2$SO$_4$ *Ans. $[Ag^+] = 2.86 \times 10^{-2}$ M; $[SO_4^{2-}] = 1.43 \times 10^{-2}$ M*
    (c) Mn(OH)$_2$ *Ans. $[Mn^{2+}] = 2.2 \times 10^{-5}$ M; $[OH^-] = 4.5 \times 10^{-5}$ M*
    (d) Sr(OH)$_2 \cdot 8$H$_2$O   *Ans. $[Sr^{2+}] = 4.3 \times 10^{-2}$ M; $[OH^-] = 8.6 \times 10^{-2}$ M*

13. Calculate the concentration of Tl$^+$ when TlCl ($K_{sp} = 1.9 \times 10^{-4}$) just begins to precipitate from a solution that is 0.0250 M in Cl$^-$.   *Ans. 7.6×10^{-3} M*

14. (a) Calculate $[Ag^+]$ in a saturated aqueous solution of AgBr ($K_{sp} = 3.3 \times 10^{-13}$).   *Ans. 5.7×10^{-7} M*
    (b) What will $[Ag^+]$ be when enough KBr has been added to make $[Br^-] = 0.050$ M?   *Ans. 6.6×10^{-12} M*
    (c) What will $[Br^-]$ be when enough AgNO$_3$ has been added to make $[Ag^+] = 0.020$ M?   *Ans. 1.6×10^{-11} M*

15. Iron concentrations greater than 5.4×10^{-6} M in water used for laundry purposes can cause staining. What $[OH^-]$ is required to reduce $[Fe^{2+}]$ to this level by precipitation of Fe(OH)$_2$?   *Ans. 3.8×10^{-5} M*

16. Public Health Service standards for drinking water set a maximum of 250 mg/L of SO$_4^{2-}$, or $[SO_4^{2-}] = 2.60 \times 10^{-3}$ M, because it is a laxative. Does natural water that is saturated with CaSO$_4$ (''gyp'' water) as a result of passing through soil containing gypsum, CaSO$_4 \cdot 2$H$_2$O, meet these standards? What is $[SO_4^{2-}]$ in such water?   *Ans. No; $[SO_4^{2-}] = 4.9 \times 10^{-3}$ M*

17. The first step in the preparation of magnesium metal is the precipitation of Mg(OH)$_2$ from sea water by the addition of Ca(OH)$_2$. The concentration of Mg^{2+}(aq) in sea water is 5.37×10^{-2} M. Using the solubility product for Mg(OH)$_2$, calculate the pH at which $[Mg^{2+}]$ is reduced to 5.37×10^{-5} M by the addition of Ca(OH)$_2$. *Ans. 10.72*

18. Most barium compounds are very poisonous; however, barium sulfate is often administered internally as an aid in the X-ray examination of the lower intestinal tract. This use of BaSO$_4$ is possible because of its insolubility. Calculate the molar solubility of BaSO$_4$ ($K_{sp} = 1.08 \times 10^{-10}$) and the mass of barium present in 1.00 L of water saturated with BaSO$_4$. *Ans. 1.04×10^{-5} M; 1.43×10^{-3} g*

19. The solubility product of CaSO$_4 \cdot 2$H$_2$O is 2.4×10^{-5}. What mass of this salt will dissolve in 1.0 L of 0.010 M K$_2$SO$_4$?   *Ans. 0.3 g*

20. In one experiment a precipitate of BaSO$_4$ ($K_{sp} = 1.08 \times 10^{-10}$) was washed with 0.100 L of distilled water; in another experiment a precipitate of BaSO$_4$ was washed with 0.100 L of 0.010 M H$_2$SO$_4$. Calculate the quantity of BaSO$_4$ that dissolved in each experiment, assuming that the wash liquid became saturated with BaSO$_4$.   *Ans. 1.04×10^{-6} mol; 1.08×10^{-9} mol*

21. Calculate the concentration of F$^-$ required to begin precipitation of CaF$_2$ in a solution that is 0.010 M in Ca^{2+}.   *Ans. 6.2×10^{-5} M*

22. Calculate the concentration of Sr^{2+} when SrF$_2$ ($K_{sp} = 3.7 \times 10^{-12}$) starts to precipitate from a solution that is 0.0025 M in F$^-$.   *Ans. 5.9×10^{-7} M*

23. A volume of 0.800 L of a 2×10^{-4} M Ba(NO$_3$)$_2$ solution is added to 0.200 L of 5×10^{-4} M Li$_2$SO$_4$. Will BaSO$_4$ precipitate? Explain your answer.   *Ans. Yes*

24. Calculate the maximum concentration of lead(II) ion in a solution of lead(II) sulfate in which the concentration of sulfate ion is 0.0045 M.   *Ans. 4.0×10^{-6} M*

Solubility Products and Ionization Constants

25. A solution of 0.060 M MnBr$_2$ is saturated with H$_2$S ($[H_2S] = 0.10$ M). What is the minimum pH at which MnS ($K_{sp} = 4.3 \times 10^{-22}$) will precipitate?   *Ans. 3.4*

26. Calculate the molar solubility of CaF$_2$ in a 0.100 M solution of HNO$_3$.   *Ans. 3.4×10^{-3} M*

27. Using the solubility product, calculate the molar solubility of AgCN (a) in pure water and (b) in a buffer solution with a pH of 3.00.
    Ans. (a) 1.1×10^{-8} M; (b) 2×10^{-5} M

28. Which of the following compounds, when dissolved in a 0.01 M solution of HClO$_4$, will have a solubility greater than in pure water: CuCl, CaCO$_3$, MnS, PbBr$_2$, CaF$_2$? Explain your answers.   *Ans. CaCO$_3$, MnS, CaF$_2$*

29. For each of the following solutions, calculate the $[H^+]$, and the corresponding pH, needed to prevent the precipitation of the metal sulfide when the solution is saturated with hydrogen sulfide:
    (a) 0.125 M Mn(NO$_3$)$_2$
    Ans. $[H^+] = 5.4 \times 10^{-4}$ M; pH = 3.27
    (b) 0.125 M AgNO$_3$
    Ans. $[H^+] = 1.4 \times 10^{14}$ M; pH = -14.15
    (c) 0.125 M TlNO$_3$ *Ans. $[H^+] = 4.1$ M; pH = -0.61*
    (d) 0.125 M Bi(NO$_3$)$_3$
    Ans. $[H^+] = 17$ M; pH = -1.23
    Note that $[H^+]$ calculated for (b) and (d) are impossible to reach in aqueous solution.

30. To a 0.10 M solution of Pb(NO$_3$)$_2$ is added enough HF(g) to make $[HF] = 0.10$ M. (a) Will PbF$_2$ precipitate from

this solution? (b) What is the minimum pH at which PbF_2 will precipitate?   *Ans. (a) Yes; (b) 0.93*

31. (a) What are the concentrations of Ca^{2+} and CO_3^{2-} in a saturated solution of $CaCO_3$ ($K_{sp} = 4.8 \times 10^{-9}$)?
    Ans. $[Ca^{2+}] = [CO_3^{2-}] = 6.9 \times 10^{-5}$ M

    (b) What are the concentrations of Ca^{2+} and CO_3^{2-} in a buffer solution with a pH of 4.55 in contact with an excess of $CaCO_3$?
    Ans. $[Ca^{2+}] = 0.35$ M; $[CO_3^{2-}] = 1.4 \times 10^{-8}$ M

32. Calculate the molar solubility of $Sn(OH)_2$ (a) in pure water (pH = 7) and (b) in a buffer solution containing equal concentrations of NH_3 and NH_4^+.
    Ans. (a) 5×10^{-12} M; (b) 2×10^{-16} M

33. Calculate the concentration of Cd^{2+} resulting from the dissolution of $CdCO_3$ in a solution that is 0.250 M in CH_3CO_2H, 0.375 M in $NaCH_3CO_2$, and 0.010 M in H_2CO_3.   *Ans. 1×10^{-5} M*

34. A volume of 50 mL of 1.8 M NH_3 is mixed with an equal volume of a solution containing 0.95 g of $MgCl_2$. What mass of NH_4Cl must be added to the resulting solution to prevent the precipitation of $Mg(OH)_2$?   *Ans. 7.1 g*

Separation of Ions

35. A solution is 0.15 M in both Pb^{2+} and Ag^+. If Cl^- is added to this solution, what is $[Ag^+]$ when $PbCl_2$ begins to precipitate?   *Ans. 1.7×10^{-8} M*

36. A solution that is 0.10 M in both Pb^{2+} and Fe^{2+} and 0.30 M in HCl is saturated with H_2S ($[H_2S] = 0.10$ M). What concentrations of Pb^{2+} and Fe^{2+} remain in the solution?
    Ans. $[Pb^{2+}] = 5.8 \times 10^{-8}$ M; $[Fe^{2+}] = 0.10$ M

37. (a) With what volume of water must a precipitate containing $NiCO_3$ be washed to dissolve 0.100 g of this compound? Assume that the wash water becomes saturated with $NiCO_3$ ($K_{sp} = 1.36 \times 10^{-7}$).   *Ans. 2.28 L*

    (b) If the $NiCO_3$ were a contaminant in a sample of $CoCO_3$ ($K_{sp} = 1.0 \times 10^{-12}$), what mass of $CoCO_3$ would have been lost? Keep in mind that both $NiCO_3$ and $CoCO_3$ dissolve in the same solution.
    Ans. 7.3×10^{-7} g

38. A solution is 0.010 M in both Cu^{2+} and Cd^{2+}. What percentage of Cd^{2+} remains in the solution when 99.9% of the Cu^{2+} has been precipitated as CuS by addition of sulfide?   *Ans. 100%*

39. What reagent might be used to separate the ions in each of the following mixtures, which are 0.1 M with respect to each ion? In some cases it may be necessary to control the pH. (*Hint:* Consider the K_{sp} values in Appendix D).
    (a) Hg_2^{2+} and Cu^{2+} (b) SO_4^{2-} and Cl^-
    (c) Hg^{2+} and Co^{2+} (d) Zn^{2+} and Sr^{2+}
    (e) Ba^{2+} and Mg^{2+} (f) CO_3^{2-} and OH^-

40. The maximum allowable chloride ion concentration in drinking water is 0.25 g/L (7.1×10^{-3} M). A commercial kit for the analysis of chloride ion in water contains

K_2CrO_4, which is dissolved in a water sample as an indicator, and a standard solution of $AgNO_3$, which is used as a titrant. As the $AgNO_3$ solution is added to the water sample drop by drop, insoluble white AgCl is formed. After "all" of the chloride ion has been precipitated, the next drop of $AgNO_3$ solution reacts with the K_2CrO_4 to form an orange-colored precipitate of Ag_2CrO_4, which indicates the end of the titration. What percentage of the initial chloride ion content remains unprecipitated when the first trace of Ag_2CrO_4 forms in a solution for which initial $[Cl^-] = 7.1 \times 10^{-3}$ M and $[CrO_4^{2-}] = 1.0 \times 10^{-4}$ M? Assume that the titration does not change the volume of the solution.   *Ans. 0.008%*

Formation Constants of Complex Ions

41. Calculate the concentration of Ni^{2+} in a 1.0 M solution of $[Ni(NH_3)_6](NO_3)_2$.   *Ans. 0.014 M*

42. Calculate the concentration of Zn^{2+} in a 0.30 M solution of $[Zn(CN)_4]^{2-}$.   *Ans. 4×10^{-5} M*

43. Calculate the silver ion concentration, $[Ag^+]$, of a solution prepared by dissolving 1.00 g of $AgNO_3$ and 10.0 g of KCN in sufficient water to make 1.00 L of solution.
    Ans. 3×10^{-21} M

44. What are the concentrations of Ag^+, CN^-, and $Ag(CN)_2^-$ in a saturated solution of AgCN?
    Ans. $[Ag^+] = 1 \times 10^{-6}$ M;
    $[CN^-] = 1 \times 10^{-10}$ M;
    $[Ag(CN)_2^-] = 1 \times 10^{-6}$ M

45. Sometimes equilibria for complex ions are described in terms of dissociation constants, K_d. For the complex ion AlF_6^{3-} the dissociation reaction is

$$AlF_6^{3-} \rightleftharpoons Al^{3+} + 6F^-$$

and

$$K_d = \frac{[Al^{3+}][F^-]^6}{[AlF_6^{3-}]} = 2 \times 10^{-24}$$

    (a) Calculate the value of the formation constant, K_f, for AlF_6^{3-}.   *Ans. 5×10^{23}*

    (b) Using the value of the formation constant for the complex ion $Co(NH_3)_6^{2+}$, calculate the dissociation constant.   *Ans. 1.2×10^{-5}*

46. The equilibrium constant for the reaction

$$Hg^{2+}(aq) + 2Cl^-(aq) \rightleftharpoons HgCl_2(aq)$$

is 1.6×10^{13}. Is $HgCl_2$ a strong electrolyte or a weak electrolyte? What are the concentrations of Hg^{2+} and Cl^- in a 0.015 M solution of $HgCl_2$?
    Ans. Weak electrolyte; $[Hg^{2+}] = 6.2 \times 10^{-6}$ M;
    $[Cl^-] = 1.2 \times 10^{-5}$ M

47. Calculate the cadmium ion concentration, $[Cd^{2+}]$, in a solution prepared by mixing 0.100 L of 0.0100 M $Cd(NO_3)_2$ with 0.150 L of 0.100 M $NH_3(aq)$. *Ans. 2.5×10^{-4} M*

Dissolution of Precipitates

48. Both AgCl and AgI dissolve in NH_3.
    (a) What mass of AgI will dissolve in 1.0 L of 1.0 M NH_3?   *Ans. 1.2×10^{-2} g*
    (b) What mass of AgCl will dissolve in 1.0 L of 1.0 M NH_3?   *Ans. 6.9 g*

49. Calculate the minimum concentration of ammonia needed in 1.0 L of solution to dissolve 3.0×10^{-3} mol of silver bromide.   *Ans. 1.3 M*

50. Calculate the solubility of CuS in a 0.25 M solution of $NH_3(aq)$. For the reaction

    $$Cu^{2+}(aq) + 4NH_3(aq) \rightleftharpoons Cu(NH_3)_4^{2+}$$

    K_f is 1.2×10^{12}.   *Ans. 1.8×10^{-16} M*

51. Calculate the minimum number of moles of cyanide ion that must be added to 100 mL of solution to dissolve 2×10^{-2} mol of silver cyanide, AgCN.
                                 Ans. 2×10^{-2} mol

52. (a) Which of the following slightly soluble compounds will have a solubility greater than that calculated from its solubility product due to reaction of the anion with water? $PbCO_3$, Tl_2S, CuI, $CoSO_3$, $KClO_4$, $PbCl_2$
                                 Ans. $CoSO_3$, $PbCO_3$, Tl_2S
    (b) For which compound in part (a) will reaction of the anion with water be most extensive?   *Ans. Tl_2S*

53. (a) Calculate the maximum concentration of sulfide ion that can exist in a solution that is 0.0020 M in $Cd(NO_3)_2$.   *Ans. 1.4×10^{-32} M*
    (b) Calculate the maximum amount of sulfide ion that can exist in a solution that is 0.0020 M in $AgNO_3$.
                                 Ans. 1.9×10^{-52} M
    (c) Calculate the maximum amount of sulfide ion that can exist in a solution that is 0.0020 M in $Bi(NO_3)_3$.
                                 Ans. 5.7×10^{-29} M

54. (a) Calculate the solubility of CdS in water, not taking into account the reaction of the sulfide ion with water.   *Ans. 5.3×10^{-18} mol L^{-1}*
    (b) Calculate the solubility of CdS in water, taking into account the reaction of the sulfide ion with water.
                         Ans. 7.5×10^{-12} mol L^{-1}
    (c) Calculate the ratio of solubility of CdS, taking into account the reaction of the sulfide ion with water, to the solubility not taking into account the reaction of the sulfide ion with water. Determine the effect the reaction of sulfide ion with water has on the solubility of CdS.
    Ans. 1.4×10^6. (Hence the solubility calculated taking into account the reaction of the sulfide ion with water is 1.4 million times greater than that calculated not taking into account the reaction of the sulfide ion with water.)

55. Calculate the volume of 1.50 M CH_3CO_2H required to dissolve a precipitate composed of 350 mg each of $CaCO_3$, $SrCO_3$, and $BaCO_3$.   *Ans. 10.2 mL*

56. A roll of 35-mm black and white film contains about 0.27 g of unexposed AgBr before developing. What mass of $Na_2S_2O_3 \cdot 5H_2O$ (hypo) in 1.0 L of developer is required to dissolve the AgBr as $Ag(S_2O_3)_2^{3-}$ ($K_f = 4.7 \times 10^{13}$)?   *Ans. 0.80 g*

57. Calculate the solubility of $Al(OH)_3$ in water ($K_{sp} = 1.9 \times 10^{-33}$). Do not neglect the reaction of aluminum ions with water (Sections 17.15 and 18.15).   *Ans. 2.7×10^{-10} M*

Additional Exercises

58. Calculate $[HgCl_4^{2-}]$ in a solution prepared by adding 8.0×10^{-3} mol of NaCl to 0.100 L of a 0.040 M $HgCl_2$ solution.   *Ans. 0.040 M*

59. What mass of NaCN must be added to 1 L of 0.010 M $Mg(NO_3)_2$ in order to produce the first trace of $Mg(OH)_2$?   *Ans. 5×10^{-3} g*

60. Even though $Ca(OH)_2$ is an inexpensive base, its limited solubility restricts its use. What is the pH of a saturated solution of $Ca(OH)_2$?   *Ans. 12.40*

61. About 50% of urinary calculi (kidney stones) consist of calcium phosphate, $Ca_3(PO_4)_2$. The normal midrange calcium content excreted in the urine is 0.10 g of Ca^{2+} per day. The normal midrange amount of urine passed may be taken as 1.4 L per day. What is the maximum concentration of phosphate ion possible in urine before a calculus begins to form?   *Ans. 4×10^{-9} M*

62. In a titration of cyanide ion, 28.72 mL of 0.0100 M $AgNO_3$ is added before precipitation begins. (The reaction of Ag^+ with CN^- goes to completion, producing the $[Ag(CN)_2]^-$ complex. Precipitation of solid AgCN takes place when excess Ag^+ is added to the solution, above the amount needed to complete the formation of $[Ag(CN)_2]^-$.) How many grams of NaCN were in the original sample?   *Ans. 0.0281 g*

63. The calcium ions in human blood serum are necessary for coagulation. In order to prevent coagulation when a blood sample is drawn for laboratory tests, an anticoagulant is added to the sample. Potassium oxalate, $K_2C_2O_4$, can be used as an anticoagulant because it removes the calcium as a precipitate of $CaC_2O_4 \cdot H_2O$. In order to prevent coagulation, it is necessary to remove all but 1.0% of the Ca^{2+} in serum. If normal blood serum with a buffered pH of 7.40 contains 9.5 mg of Ca^{2+} per 100 mL of serum, what mass of $K_2C_2O_4$ is required to prevent the coagulation of a 10-mL blood sample that is 55% serum by volume? [All volumes are accurate to two significant figures. Note that the volume of fluid (serum) in a 10-mL blood sample is 5.5 mL. Assume that the K_{sp} value for CaC_2O_4 in serum is the same as in water.]   *Ans. 2.2×10^{-3} g*

64. A 0.010-mol sample of solid AgCN is rendered soluble in 1 L of solution by adding just enough excess cyanide ion to form $[Ag(CN)_2]^-$. When all of the solid silver cyanide has just dissolved, the concentration of free cyanide ion is 1.125×10^{-7} M. Neglecting reaction of the cyanide ion

with water, determine the concentration of free, uncomplexed silver ion in the solution. If more cyanide ion is added (without changing the volume) until the equilibrium concentration of free cyanide ion is 1.0×10^{-6} M, what will be the equilibrium concentration of free silver ion?

Ans. 8×10^{-9} M; 1×10^{-10} M

65. The pH of a normal urine sample is 6.30, and the total phosphate concentration ($[PO_4^{3-}] + [HPO_4^{2-}] + [H_2PO_4^-] + [H_3PO_4]$) is 0.020 M. What is the minimum concentration of Ca^{2+} necessary to induce calculus formation? (See exercise 61 for additional information.)

Ans. 3×10^{-3} M

66. Magnesium metal (a component of alloys used in aircraft and a reducing agent used in the production of uranium, titanium, and other active metals) is isolated from sea water by the following sequence of reactions:

$$Mg^{2+}(aq) + Ca(OH)_2(aq) \longrightarrow Mg(OH)_2(s) + Ca^{2+}(aq)$$

$$Mg(OH)_2(s) + 2HCl \longrightarrow MgCl_2(s) + 2H_2O$$

$$MgCl_2(l) \xrightarrow{\text{Electrolysis}} Mg(s) + Cl_2(g)$$

Sea water has a density of 1.026 g/cm³ and contains 1272 parts per million of magnesium as $Mg^{2+}(aq)$ by mass. What mass, in kilograms, of $Ca(OH)_2$ is required to precipitate 99.9% of the magnesium in 1.00×10^3 L of sea water?

Ans. 3.97 kg

67. In the most commonly used qualitative analysis procedure, several metal ions are separated on the basis of relative solubilities of metal sulfides. Group II metal ions (Hg^{2+}, Pb^{2+}, Bi^{3+}, Cu^{2+}, Cd^{2+}, As^{3+}, Sb^{3+} and Sn^{2+}) precipitate as sulfides in a saturated solution of hydrogen sulfide (0.1 M H_2S) that is 0.30 M in H^+, whereas Group III metal ions (Co^{2+}, Ni^{2+}, Mn^{2+}, Fe^{2+}, Al^{3+}, Cr^{3+} and Zn^{2+}) do not.

Calculate for each of the above ions (for which K_{sp} values of the metal sulfides are given in Appendix D) whether the ion should precipitate as the sulfide under the above conditions, if the metal ion concentration is 0.050 M. Compare your calculated results with the experimental results described above. (Note that, of the Group II and Group III metals for which K_{sp} values are given in the appendix, Bi^{3+} is a +3 ion; the others are +2 ions.)

Ans.

Group II	
HgS	Yes, should precipitate
PbS	Yes
CuS	Yes
CdS	Yes
Bi_2S_3	Yes
SnS	Yes

Group III	
CoS(α)	No
CoS(β)	Yes
NiS(α)	No
NiS(β)	Yes
MnS	No
FeS	No
ZnS	No

Chemical Thermodynamics

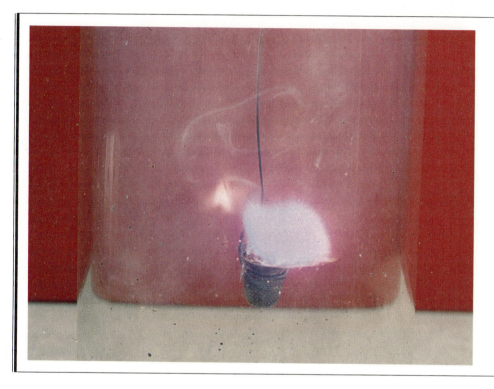

The combustion of sulfur in pure oxygen is an exothermic reaction.

W hen we consider a potential chemical reaction, such as the oxidation of glucose to carbon dioxide and water in the body, we often have several questions. Will the substances react when put together? If they do react, what are the final concentrations of reactants and products when equilibrium is established? If a reaction occurs, how much heat, if any, will be produced? How fast will the reaction go? What is its mechanism? The last two questions can be answered from a consideration of the kinetics of the reaction (Chapter 14). The first three are the topic of this chapter, in which we will consider energy changes and changes in randomness or disorder that accompany chemical processes. We will see that these changes can be determined quantitatively and that their values can be used to predict whether or not a chemical reaction will occur. Finally, we will see that the equilibrium constant for a reaction can be calculated from these values.

In several of the preceding chapters we have pointed out chemical processes that are accompanied by changes in the internal energy of the system. For example, removing the electron from a hydrogen atom, breaking a carbon-hydrogen bond, evaporating water, melting ice, and forming calcium oxide and carbon dioxide from calcium carbonate all proceed with an increase in internal energy. These changes are endothermic (Fig. 19.1), and because energy must be added to the system to get them to go, the

Figure 19.1

The decomposition of $CaCO_3$ to CaO and CO_2 is an endothermic change; heating is required.

Figure 19.2

Carbon burns in an exothermic reaction; heat is evolved.

internal energy of the system increases. On the other hand, burning of carbon (coal), digestion of food, combustion of hydrogen and oxygen, condensation of water vapor to a solid or a liquid, reaction of hydrochloric acid and sodium hydroxide, and reaction of sodium with water all proceed spontaneously, with the evolution of heat (Fig. 19.2) and a decrease in internal energy. When we discussed the formation of solutions, we noted that processes that increase the disorder of the system also tend to be spontaneous. These familiar concepts of changes in internal energy or in the amount of disorder can be used in analyzing the behavior of chemical systems. They will prove helpful in predicting whether or not a reaction will proceed, the amount of reactants and products present at equilibrium, and the energy changes associated with a reaction.

Internal Energy and Enthalpy

19.1 Spontaneous Processes, Energy, and Disorder

Chemical thermodynamics is the study of the energy transformations and transfers that accompany chemical and physical changes. The part of the universe that undergoes a change (the part we have under study) is called the **system;** the rest of the universe is the **surroundings.** In our study of chemical thermodynamics we will examine the energy transformations that occur within a system and any transfer of energy that may occur between the system and the surroundings.

Two fundamental laws of nature are particularly important in thermodynamics.

1. Systems tend to attain a state of minimum potential energy. For example, let us suppose that we have a system that consists of a box containing an assembled jigsaw puzzle. If we drop the box, it falls to the floor. During this change in the box's position, part of its potential energy (due to its height above the floor) is converted to kinetic energy. When the box hits the floor, the kinetic energy is converted to heat. The loss of heat indicates that the system (the box and the puzzle) has gone to a lower energy state. It certainly has a lower potential energy, since it cannot fall as far as it could before the change in its position.

2. Systems tend toward a state of maximum disorder. The assembled jigsaw puzzle in the box would almost certainly be partly disassembled (more disordered) after the fall. The latter fact may not seem very important or fundamental, but no one would try to assemble a jigsaw puzzle by dropping the separated pieces on the floor, whereas the reverse process is readily accomplished by dropping the assembled puzzle. A system tends to become less orderly because there are so many more ways to be disorderly than to be orderly. The probability of a system becoming more disordered or more random is greater than that of its becoming more ordered.

A physical or a chemical change that occurs spontaneously is accompanied by one, or both, of these two shifts—a shift toward a minimum energy and/or a shift to a more disordered state.

19.2 Internal Energy, Heat, and Work

The **internal energy of a system, E,** is simply the total of all of the possible kinds of energy present in that portion of the universe that we choose to study—the total of kinetic energies, ionization energies of the electrons, bond energies, lattice energies, etc., present in the system. In Chapter 4 we described how the internal energy of a system varies with the loss or gain of heat by the system and with the work done by or on the system as it undergoes a chemical or physical change. In this section we will consider how to calculate the energy change associated with one type of work, called expansion work, and review how the combination of work and loss or gain of heat changes the internal energy of a system.

We cannot measure the value of E for a system; we can only determine the change in internal energy, ΔE, that accompanies a change in the system. In Section 6.9 we saw that for a hydrogen-hydrogen bond, the bond energy is 436 kilojoules per mole of bonds. Thus a system consisting of one mole of hydrogen molecules can be converted into two moles of hydrogen atoms with the input of 436 kilojoules of heat or other energy. During this process the internal energy changes (increases) by 436 kilojoules. Although we cannot measure its initial internal energy, E_1, when the system exists as a mole of hydrogen molecules or its final energy, E_2, when it exists as two moles of hydrogen atoms, we can measure the difference in internal energy between the two states. The internal energy of the two moles of hydrogen atoms is 436 kJ greater than the internal energy of one mole of hydrogen molecules because 436 kJ of energy has been added to the system. The difference between E_2 and E_1 is equal to the change in internal energy.

$$\Delta E = E_2 - E_1$$

The value of a ΔE is positive when energy is transferred from the surroundings to the system. When energy is transferred from the system to the surroundings, ΔE is negative.

Energy is transferred into or out of a system in one, or both, of two ways: by heat transfer and/or by work. In Section 4.2 we discussed how to measure the amount of heat, q, transferred. We saw that when heat flows into the system from the surroundings, the value of q is positive ($q > 0$). When heat flows from the system to the surroundings, the value of q is negative ($q < 0$). For example, if a system consisting of 18 grams of steam at 100°C condensed to liquid water at 100°C and did no work, the internal energy of the system would decrease, because 40.7 kilojoules of heat would flow from the system: $q = -40.7$ kJ.

In Section 4.3 we saw that if a system does work on the surroundings, the internal energy of the system decreases by the amount of work it does; if work is done on the system by the surroundings, the internal energy of the system increases by the amount of work done on it. The symbol for the amount of work done by a system is w. When energy is transferred from the system to the surroundings as work, work is done on the surroundings and the value of w is positive ($w > 0$). When energy is transferred to the system from the surroundings as work, the surroundings do work on the system and the value of w is negative ($w < 0$). In this chapter we will limit our consideration of work to that of expansion against a constant pressure. This kind of work is called **expansion work.**

Gases expanding against a restraining pressure are capable of doing work, as in a steam engine or an automobile engine (Fig. 19.3). A system with an initial volume V_1 that goes at a constant pressure to a larger volume V_2 does expansion work on the surroundings. If the restraining pressure, P, remains constant, the amount of work done is equal to $P(V_2 - V_1)$. The work term $P(V_2 - V_1)$ is an energy term with energy units. Pressure is defined as force divided by area and therefore has units of force/(length)^2; volume has units of (length)^3. Thus the units for the product are

$$P(V_2 - V_1) = \frac{\text{force}}{\text{length}^2} \times \text{length}^3 = \text{force} \times \text{length}$$

Units of force × length are units of work, or energy. If the pressure is expressed in pascals (newtons/meter^2) and the volume change in cubic meters (m^3), the resulting

Figure 19.3

The expansion of gaseous water in the cylinder of this steam engine results in expansion work that is converted into kinetic energy.

product will have units of pascal meters^3, which is the same as newton meters, or joules (Section 4.1).

A system can contract instead of expand. When a system contracts, the volume decreases ($V_2 - V_1 < 0$), and work is done on the system by the surroundings. A change in volume is often referred to in thermodynamics as an *expansion;* an increase in volume is a positive expansion and a decrease in volume, a negative expansion.

EXAMPLE 19.1

Determine the amount of work done when a system consisting of 1.0 cm^3 of liquid water at 0°C freezes under a constant pressure of 1.0 atm (101,325 Pa) and forms 1.1 cm^3 of ice (see the figure).

As the water freezes it expands against a restraining pressure of 1 atm and pushes back the atmosphere. Thus the system does work on its surroundings. The amount of work done on the surroundings, w, can be calculated from the expression $w = P(V_2 - V_1)$. The original volume of the system, V_1, was given as 1.0 cm^3; the final volume, V_2, as 1.1 cm^3. These volumes must be converted to cubic meters (m^3) and the pressure expressed in pascals in order to get units of joules from the equation.

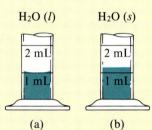

When 1.0 cm^3 of liquid water freezes, it expands to a volume of 1.1 cm^3.

$$V_1 = 1.0 \text{ cm}^3 \times \frac{1 \text{ m}^3}{1 \times 10^6 \text{ cm}^3} = 1.0 \times 10^{-6} \text{ m}^3$$

$$V_2 = 1.1 \text{ cm}^3 \times \frac{1 \text{ m}^3}{1 \times 10^6 \text{ cm}^3} = 1.1 \times 10^{-6} \text{ m}^3$$

$$w = P(V_2 - V_1)$$
$$= 101,325 \text{ Pa} \times (1.1 \times 10^{-6} \text{ m}^3 - 1.0 \times 10^{-6} \text{ m}^3)$$
$$= 101,325 \text{ Pa} \times 1 \times 10^{-7} \text{ m}^3 = 1.0 \times 10^{-2} \text{ Pa M}^3$$
$$= 1 \times 10^{-2} \text{ J}$$

The positive value of w indicates that work is done by the system. The magnitude of w indicates that very little work is done by this expansion. As will be demonstrated in Examples 19.2 and 19.3, other changes may involve a great deal more work.

19.3 The First Law of Thermodynamics

The **first law of thermodynamics** is actually the law of conservation of energy (Section 1.3): **The total amount of energy in the universe is constant.** The first law is often considered in a rather special form. If an amount of heat, q, is added to a system with an internal energy E_1, and the system then does some work, w, on the surroundings, the system ends up with a new internal energy, E_2. The law of conservation of energy requires that the final internal energy of the system be equal to its initial internal energy plus the energy added as heat from the surroundings minus the energy lost as work done on the surroundings.

$$E_2 = E_1 + q - w$$

or

$$E_2 - E_1 = q - w$$

$E_2 - E_1$ is the change in the internal energy of the system, or ΔE.

$$\Delta E = E_2 - E_1 = q - w$$

In other words, the change in the internal energy of a given system equals the heat transferred to the system from the surroundings minus the work transferred from the system to the surroundings. The value of ΔE may be either positive or negative depending on the relative values of q and w.

If ΔE is the energy change for the system and ΔE_{sur} is the energy change for the surroundings

$$\Delta E + \Delta E_{sur} = 0$$

This emphasizes the fact expressed by the first law that the total amount of energy in the universe is constant.

EXAMPLE 19.2

(a) If 600 J of heat is added to a system in energy state E_1 and the system does 450 J of work on the surroundings, what is the energy change of the system?

To calculate ΔE we use the expression relating ΔE, q, and w.

$$\Delta E = q - w$$

Since heat is added to the system, q is positive ($q > 0$), so q is +600 J. The value of w is also positive ($w > 0$), since work is done on the surroundings; w is +450 J.

$$\Delta E = q - w = (+600 \text{ J}) - (+450 \text{ J}) = +150 \text{ J}$$

The internal energy of the system increases by 150 J.

(b) What is the energy change of the surroundings?

$$\Delta E + \Delta E_{sur} = 0$$
$$150 \text{ J} + \Delta E_{sur} = 0$$
$$\Delta E_{sur} = -150 \text{ J}$$

The energy of the surroundings decreases by 150 J.

(c) What is the internal energy of the system in the new energy state, E_2?

Since we cannot determine the internal energy of the system, all we can say is that its final internal energy, E_2, is 150 J greater than its initial internal energy, E_1.

$$\Delta E = E_2 - E_1 = 150 \text{ J}$$
$$E_2 = E_1 + 150 \text{ J}$$

EXAMPLE 19.3

A system consisting of 18.02 g (one mole) of $H_2O(g)$ at 100°C and 1 atm with a volume of 30.12 L (0.03012 m³) condenses at a constant pressure of 1 atm to $H_2O(l)$ at 100°C and 1 atm with a density of 0.9584 g/cm³. There is 40,668 J of heat released to the surroundings. Calculate ΔE for this process.

Since

$$\Delta E = q - w$$

we need two values to determine ΔE: the amount of heat transferred to or from the system and the amount of work done on or by the system. The problem tells us that 40,668 J of heat is transferred from the water to the surroundings, so q is $-40,668$ J. The sign is negative since heat leaves the system.

The amount of work (in joules) involved in this change can be calculated from the expression

$$w = P(V_2 - V_1)$$

if P is expressed in pascals (1 atm = 101,325 Pa; Section 10.2) and V_2 and V_1, in cubic meters.

The value of V_1 is given as 0.03012 m³. The value of V_2 may be calculated from the mass of water and its density, as follows:

$$V_2 = 18.02 \text{ g} \times \frac{1 \text{ cm}^3}{0.9584 \text{ g}} = 18.80 \text{ cm}^3$$

$$= 18.80 \text{ cm}^3 \times \left(\frac{1 \text{ m}}{100 \text{ cm}}\right)^3 = 1.880 \times 10^{-5} \text{ m}^3$$

Now the value of w can be determined.

$$
\begin{aligned}
w &= P(V_2 - V_1) \\
&= 101,325 \text{ Pa} \times (1.880 \times 10^{-5} \text{ m}^3 - 3012 \times 10^{-5} \text{ m}^3) \\
&= -3050 \text{ Pa m}^3 = -3050 \text{ J}
\end{aligned}
$$

As indicated in Section 19.2, units of Pa m³ are equivalent to joules. Note that the value of w is negative, indicating that work is done on the system as its volume decreases from 0.03012 m³ to 0.0000188 m³ at a constant pressure of 101,325 Pa (1 atm). Now that we know both q (−40,668 J) and w (−3050 J), we can calculate ΔE.

$$
\begin{aligned}
\Delta E &= q - w \\
&= (-40,668 \text{ J}) - (-3050 \text{ J}) \\
&= -37,618 \text{ J} = -37.618 \text{ kJ}
\end{aligned}
$$

The internal energy of the system decreases by 37,618 J ($\Delta E < 0$), as the system gives up 40,668 J to the surroundings ($q < 0$), and the surroundings do 3050 J of work on the system ($w < 0$).

19.4 State Functions

A mole of water that condenses from a gas to a liquid at 100°C and 1 atmosphere (as in Example 19.3) undergoes a change in internal energy, ΔE, that is simply the difference between the internal energies of 1 mole of liquid H_2O at 100°C and of 1 mole of gaseous H_2O at 100°C. This difference does not depend upon how we convert the steam to liquid water. We could convert the gas directly to liquid at 100°C and 1 atmosphere

$$H_2O(g) \longrightarrow H_2O(l)$$

or we could use a two-step process. We could first convert the $H_2O(g)$ to 1 mole of $H_2(g)$ and $\frac{1}{2}$ mole of $O_2(g)$ and then reconvert the $H_2(g)$ and $O_2(g)$ to 1 mole of $H_2O(l)$ at 100°C and 1 atmosphere.

$$H_2O(g) \longrightarrow H_2(g) + \tfrac{1}{2}O_2(g) \longrightarrow H_2O(l)$$

Either way, ΔE for the process would be −37.618 kilojoules.

When a property of a system (its density or its internal energy, for example) does not depend on how the system gets to the state that exhibits that property, the property is said to be a **state function.** How a system goes from an initial state to a final state is irrelevant to the value of a state function of the system in its final state. The change in the value of a state function equals its value in the final state minus its value in the initial state. For example, the change in internal energy that accompanies the condensation of a mole of water (-37.618 kJ) is a change in a state function—the internal energy of one mole of water.

A second example of a state function is the change in the potential energy of a tennis ball as it is moved from a second-floor window ledge to a third-floor window ledge. (The potential energy is a measure of how hard the ball would hit the ground if it fell.) The change in potential energy is the same whether the ball is taken directly from the second to the third floor or it is carried to the roof and then back to the third floor. The potential energy of the tennis ball is a state function, since a difference in the path by which it is moved from the second to the third floor will make no difference in how hard the ball will hit the ground if it falls from the third-floor window.

In contrast to state functions, there are other functions whose values depend on the paths followed. The distinction between the two types of functions can be clarified by the expression

$$\Delta E = E_2 - E_1 = q - w$$

The internal energy, E, is a state function; ΔE of a reaction is the same regardless of how that reaction is carried out. On the other hand, the values of q and w are not constant; they vary with the process used. As an example of how an energy change can be independent of the path used to cause it, consider the tennis ball again. More work is required to carry the tennis ball from the second floor to the roof and back to the third floor of a ten-story building than to carry it directly from the second floor to the third floor, but the change in its potential energy is the same for either path. Similarly, the change in energy, ΔE, associated with the transformation of $H_2O(g)$ to $H_2O(l)$ at a given temperature and pressure is always the same and is a state function. However, q and w depend on the way in which the transformation is carried out, and thus they are not state functions. Quantities that are state functions are designated by capital letters; those that are not, by lower-case letters.

19.5 Enthalpy Changes

The amount of heat, q, absorbed or released by a system undergoing a chemical or physical change is usually not a state function; it varies with the way the process occurs. However, if the change occurs in such a way that the only work done is due to a change in volume of the system at constant pressure, q is a state function. The amount of heat, q, exchanged when the only work done by the system is expansion work at constant pressure is called the **enthalpy change, ΔH,** of the system ($\Delta H = q$). Most chemical reactions occur under the essentially constant pressure of the atmosphere. Consequently the enthalpy change of these, or any other changes that occur at constant pressure, can be determined by measuring the amount of heat transferred during the process (Fig. 19.4).

The change in internal energy for such a reaction can be calculated from the equation

$$\Delta E = q - w = \Delta H - w$$

Since the reaction occurs at constant pressure and the only work done results from the change in volume of the system, the work term in this case is equal to $P(V_2 - V_1)$, or $P\Delta V$.

At constant pressure *and* constant volume,

$$P\Delta V = 0 \qquad \text{and} \qquad \Delta H = \Delta E$$

Figure 19.4

The heat released when 2 moles of Al and 1 mole of Fe_2O_3 at 25°C react and form 1 mole of Al_2O_3 and 2 moles of Fe at 25°C is the enthalpy change of the reaction $2Al + Fe_2O_3 \rightarrow Al_2O_3 + 2Fe$. The heat is released in two stages. This photograph shows that heat is released as the reaction occurs; additional heat is released as the molten iron and hot Al_2O_3 cool back down to 25°C.

Recall that if ΔH for a chemical reaction conducted at constant pressure (state 1 = reactants, state 2 = products) is negative, the system evolves heat to the surroundings, and the reaction is exothermic. If ΔH for a chemical reaction conducted at constant pressure is positive, heat is absorbed by the system from the surroundings, and the reaction is endothermic (Section 4.1).

At this point we need to digress briefly and discuss the notation used to indicate enthalpy changes for processes that occur at different temperatures with different pressures and concentrations of reactants and products. The use of 298.15 K (25°C) and 1 atmosphere as standard state conditions was noted in Section 4.4. A process that occurs under standard state conditions and that converts a system's reactants, at unit activities, to products, at unit activities, has an enthalpy change that is labeled as ΔH°_{298}. The subscript 298 is used to specify that the temperature is 298.15 K. A solid, a pure liquid, or a solvent has an activity of one (unit activity). The activity of a gas is approximated by its pressure (in atmospheres); thus we will consider a gas with a pressure of 1 atmosphere to have unit activity. The activity of a dissolved solute is approximated by its concentration, and we will consider a 1 molar solution to have unit activity.

The enthalpy change of a process that occurs at some temperature other than 298.15 K but that involves reactants and products that all have unit activities is given the symbol ΔH°. The symbol ΔH is used to indicate an enthalpy change with no restrictions on the temperature or the activities of the components. The various symbols used to indicate enthalpy changes are summarized in Table 19.1.

The difference between the enthalpy change for a process at 298.15 K and at some other temperature is generally small and is often neglected. It is common to make the approximation that ΔH°_{298} equals ΔH° for the same process.

The enthalpy changes that we have seen in preceding chapters include heat of combustion and standard molar enthalpy of formation (Section 4.4), heat of vaporization (Section 11.6), and heat of fusion (Section 11.7). In addition, enthalpy changes are used to define ionization energy and electron affinity (Section 5.15), covalent bond energy (Section 6.9), and lattice energy (Section 11.21).

We have seen how to calculate the enthalpy change for a reaction that can be carried out in a series of steps using Hess's law (Section 4.5). The enthalpy change of a process that can be written as the sum of several steps is equal to the sum of the enthalpy changes of each of the steps.

Table 19.1   Symbols for various thermodynamic parameters

Enthalpy change	Free energy change	Entropy change	Conditions
ΔH°_{298}	ΔG°_{298}	ΔS°_{298}	Standard state conditions, unit activities ($T = 298.15$ K, pure solids, pure liquids, $P = 1$ atm, 1 M concentrations)
ΔH°	ΔG°	ΔS°	Unit activities (Pure solids, pure liquids, $P = 1$ atm, 1 M concentrations)
ΔH	ΔG	ΔS	No restrictions

EXAMPLE 19.4   Calculate the enthalpy change for the following reaction at 25°C and 1 atm

$$S(s) + 1\tfrac{1}{2}O_2(g) \longrightarrow SO_3(g)$$

from the enthalpy of formation of $SO_2(g)$ (-296.8 kJ mol^{-1}) and the enthalpy change for the following reaction

$$SO_2(g) + \tfrac{1}{2}O_2(g) \longrightarrow SO_3(g) \qquad \Delta H^{\circ}_{298} = -98.9 \text{ kJ}$$

The enthalpy change for the process that converts a system at 25°C consisting of 1 mole of $S(s)$ and $1\tfrac{1}{2}$ mole of $O_2(g)$ (at 1 atmosphere pressure) to 1 mole of $SO_3(g)$ (at 1 atmosphere pressure), ΔH°_{298}, is the sum of the enthalpy changes of two reactions: the reaction that converts $S(s)$ to $SO_2(g)$, and the reaction that converts $SO_2(g)$ to $SO_3(g)$. The enthalpy change of the first reaction is simply the standard molar enthalpy of formation of $SO_2(g)$. Thus

$$S(s) + O_2(g) \longrightarrow SO_2(g) \qquad \Delta H^{\circ}_{298} = \Delta H^{\circ}_{f_{SO_2(g)}} = -296.8 \text{ kJ}$$
$$SO_2(g) + \tfrac{1}{2}O_2(g) \longrightarrow SO_3(g) \qquad\qquad\qquad \Delta H^{\circ}_{298} = -98.9 \text{ kJ}$$
$$\overline{}$$
$$S(s) + 1\tfrac{1}{2}O_2(g) \longrightarrow SO_3(g) \qquad\qquad\qquad \Delta H^{\circ}_{298} = -395.7 \text{ kJ}$$

The value of ΔH°_{298} for the bottom reaction is the sum of the ΔH°_{298} values for the two reactions that add up to it.

We have also seen how Hess's law can be used to determine the standard state enthalpy change of any reaction if the standard molar enthalpies of formation of the reactants and products are available (Section 4.5). The steps that we used were the decompositions of the reactants into their component elements, for which the enthalpy changes are proportional to the negative of the enthalpies of formation of the reactants, followed by recombinations of the elements to give the products, for which the enthalpy changes are proportional to the enthalpies of formation of the products.

EXAMPLE 19.5   Using the standard molar enthalpies of formation of the compounds involved, calculate the standard state enthalpy change, ΔH°_{298}, for the reaction

$$Na_2O(s) + CO_2(g) \longrightarrow Na_2CO_3(s)$$

The standard state enthalpy change can be calculated from the following sum:

$$Na_2O(s) \longrightarrow 2Na(s) + \tfrac{1}{2}O_2(g) \qquad \Delta H_1 = -\Delta H^{\circ}_{f_{Na_2O(s)}}$$
$$CO_2(g) \longrightarrow C(s) + O_2(g) \qquad \Delta H_2 = -\Delta H^{\circ}_{f_{CO_2(g)}}$$
$$2Na(s) + C(s) + \tfrac{3}{2}O_2(g) \longrightarrow Na_2CO_3(s) \qquad \Delta H_3 = \Delta H^{\circ}_{f_{Na_2CO3(s)}}$$
$$\overline{}$$
$$Na_2O(s) + CO_2(g) \longrightarrow Na_2CO_3(s) \qquad H^{\circ}_{298} = \Delta H_1 + \Delta H_2 + \Delta H_3$$

$$\Delta H^{\circ}_{298} = \Delta H_1 + \Delta H_2 + \Delta H_3 = -\Delta H^{\circ}_{f_{Na_2O(s)}} - \Delta H^{\circ}_{f_{CO_2(g)}} + \Delta H^{\circ}_{f_{Na_2CO_3(s)}}$$

Using the values of the standard molar enthalpies of formation, the value of ΔH°_{298} is found to be

$$\Delta H^{\circ}_{298} = -(-415.9 \text{ kJ}) - (-393.51 \text{ kJ}) + (-1130.8 \text{ kJ})$$
$$= -321.4 \text{ kJ}$$

The equation used to calculate ΔH in Example 19.5, $\Delta H^{\circ}_{298} = -\Delta H^{\circ}_{f_{Na_2O(s)}} - \Delta H^{\circ}_{f_{CO_2(g)}} + \Delta H^{\circ}_{f_{Na_2CO_3(s)}}$, is a specific example of a useful version of Hess's law for that type of calculation: *The enthalpy change for a chemical reaction run under standard state conditions is equal to the sum of the standard molar enthalpies of formation of all the products, each multiplied by the number of moles of the product in the balanced chemical equation, minus the corresponding sum for the reactants.*

$$\Delta H^{\circ}_{298} = \sum \Delta H^{\circ}_{f_{products}} - \sum \Delta H^{\circ}_{f_{reactants}}$$

For the reaction

$$mA + nB \longrightarrow xC + yD$$

$$\Delta H^{\circ}_{298} = [x \times \Delta H^{\circ}_{f_C} + y \times \Delta H^{\circ}_{f_D}] - [m \times \Delta H^{\circ}_{f_A} + n \times \Delta H^{\circ}_{f_B}]$$

When using this version of Hess's law, it is important to remember that the standard molar enthalpy of formation of an element in its most stable state (gas, liquid, or solid) is zero.

EXAMPLE 19.6

Calculate ΔH° at 298.15 K for the reaction

$$2Ag_2S(s) + 2H_2O(l) \longrightarrow 4Ag(s) + 2H_2S(g) + O_2(g)$$

The standard molar enthalpies of formation, ΔH°_f, of the compounds involved may be found in Table 19.2 on page 570.

$$\Delta H^{\circ}_{298} = 4\,\Delta H^{\circ}_{f_{Ag(s)}} + 2\,\Delta H^{\circ}_{f_{H_2S(g)}} + \Delta H^{\circ}_{f_{O_2(g)}} - 2\,\Delta H^{\circ}_{f_{Ag_2S(s)}} - 2\,\Delta H^{\circ}_{f_{H_2O(l)}}$$

The standard molar enthalpy of formation of an element in its most stable state is zero. Thus

$$\Delta H^{\circ}_{298} = 4\ \text{mol Ag}(s) \times (0\ \text{kJ mol}^{-1}) + 2\ \text{mol H}_2\text{S}(g) \times (-20.6\ \text{kJ mol}^{-1})$$
$$+\ 1\ \text{mol O}_2(g) \times (0\ \text{kJ mol}^{-1}) - 2\ \text{mol Ag}_2\text{S}(s) \times (-32.6\ \text{kJ mol}^{-1})$$
$$-\ 2\ \text{mol H}_2\text{O}(l) \times (-285.8\ \text{kJ mol}^{-1})$$
$$= [-41.2 - (-65.2) - (-571.6)]\ \text{kJ} = 595.6\ \text{kJ}$$

Hence the reaction is endothermic.

$$2Ag_2S(s) + 2H_2O(l) \longrightarrow 4Ag(s) + 2H_2S(g) + O_2(g) \qquad \Delta H^{\circ}_{298} = 595.6\ \text{kJ}$$

EXAMPLE 19.7

Calculate ΔH° at 298.15 K for the reaction

$$2Na(s) + 2H_2O(l) \longrightarrow 2NaOH(s) + H_2(g)$$

The standard molar enthalpies of formation for the reactants and products involved are as follows: $H_2O(l)$, -285.8 kJ mol^{-1}; $NaOH(s)$, -426.8 kJ mol^{-1}; Na and H_2, 0 kJ mol^{-1}.

$$\Delta H^{\circ}_{298} = 2\,\Delta H^{\circ}_{f_{NaOH(s)}} + \Delta H^{\circ}_{f_{H_2(g)}} - 2\,\Delta H^{\circ}_{f_{Na(s)}} - 2\,\Delta H^{\circ}_{f_{H_2O(l)}}$$
$$\Delta H^{\circ}_{298} = 2\ \text{mol NaOH}(s) \times (-426.8\ \text{kJ mol}^{-1}) + 1\ \text{mol H}_2(g) \times (0\ \text{kJ mol}^{-1})$$
$$-\ 2\ \text{mol Na}(s) \times (0\ \text{kJ mol}^{-1}) - 2\ \text{mol H}_2\text{O}(l) \times (-285.8\ \text{kJ mol}^{-1})$$
$$= [-853.5 + 571.6]\ \text{kJ} = -281.9\ \text{kJ}$$

Hence the reaction evolves heat.

Entropy and Free Energy

19.6 The Spontaneity of Chemical and Physical Changes

A **spontaneous change,** in a thermodynamic sense, is a change in a system that proceeds without any outside influence on the system (Fig. 19.5). For example, when we add solid potassium chloride to water (Section 12.10), the solid dissolves without any outside influence—the dissolution is spontaneous. Liquid water freezes to ice spontaneously at $-1°C$; when a solution of an acid is added to a solution of a base, the hydrogen ions and hydroxide ions combine spontaneously; and carbon combines spontaneously with oxygen to give carbon monoxide or carbon dioxide. Since the definition of a spontaneous change does not include time, changes may be spontaneous even though they proceed very slowly. For those who own diamonds (which are essentially pure carbon) it is fortunate that the spontaneous reaction of carbon with oxygen is very slow at room temperature.

Those exothermic changes for which ΔH is large (a large amount of heat given off) are frequently spontaneous. Enough heat to melt and ignite the metal may be produced by the spontaneous reaction of sodium with water; the reaction is exothermic (Example 19.7).

$$2Na(s) + 2H_2O(l) \longrightarrow 2NaOH(s) + H_2(g) \qquad \Delta H^\circ_{298} = -281.9 \text{ kJ}$$

Because of the large negative value of ΔH, we would expect this reaction to occur spontaneously at 25°C and 1 atmosphere, and it does. The reaction in Example 19.6

$$2Ag_2S(s) + 2H_2O(l) \longrightarrow 4Ag(s) + 2H_2S(g) + O_2(g) \qquad \Delta H^\circ_{298} = 595.6 \text{ kJ}$$

is endothermic, and we should not expect it to occur spontaneously. However, the reverse reaction ($\Delta H^\circ_{298} = -595.6$ kJ) should be expected to occur spontaneously, and it does. Silver exposed to hydrogen sulfide and air tarnishes rapidly.

Predictions based solely on the value of ΔH° are not always valid, particularly if ΔH° has a small value. For example, a reaction that proceeds spontaneously at one temperature and pressure may not proceed spontaneously at another temperature and pressure. At 25°C and 1 atmosphere of pressure, the reaction

$$2Ag(s) + \tfrac{1}{2}O_2(g) \longrightarrow Ag_2O(s)$$

proceeds spontaneously; ΔH°_{298} is -31 kilojoules. However, above about 200°C this reaction is not spontaneous, even though ΔH changes very little with the change in temperature. A crystalline solid melts spontaneously above its melting temperature even though melting is an endothermic change ($\Delta H > 0$). Below its melting temperature it does not melt spontaneously, yet the ΔH value changes very little. Whether chemical reactions and physical changes proceed spontaneously or not can depend on temperature and pressure. Yet temperature and pressure changes often have virtually no effect on the value of ΔH for a reaction.

It is evident, then, that another factor in addition to ΔH must be considered when determining whether or not a given reaction will proceed spontaneously. This factor is the entropy change that occurs as the change in the system takes place.

Figure 19.5

When a solution of Na_2S is added to a solution of $Mn(NO_3)_2$, pink solid MnS and a solution of $NaNO_3$ form spontaneously.

19.7 Entropy and Entropy Changes

Consider two changes that are spontaneous but not exothermic: the melting of ice at room temperature.

$$H_2O(s) \longrightarrow H_2O(l) \qquad \Delta H° = 6.0 \text{ kJ}$$

and the decomposition of calcium carbonate at high temperature,

$$CaCO_3(s) \longrightarrow CaO(s) + CO_2(g) \qquad \Delta H° = 178 \text{ kJ}$$

The disorder of both of these systems increases when these reactions occur. Water molecules are held in fixed positions in a regular, repeating array in an ice crystal (Fig. 11.21). When the ice melts, the water molecules are free to move through the liquid as well as to change their orientations. The molecules in the liquid are much more randomly distributed than those in the solid. Thus the amount of order is higher in the solid than in the liquid. As calcium carbonate decomposes, the system changes from an ordered array of calcium ions and carbonate ions in solid calcium carbonate to an ordered array of calcium and oxide ions in solid calcium oxide plus a disordered collection of carbon dioxide molecules in the gas phase. The arrangement of the carbon dioxide molecules in the gas phase is even more random (the disorder is greater) than that of an equal number of water molecules in the liquid phase.

The randomness, or the amount of disorder, of a system can be determined quantitatively and is referred to as the entropy, S, of the system. The greater the randomness, or disorder, in a system, the higher is its entropy. Entropy is one of the most important of scientific concepts; as we shall see, an entropy increase corresponding to an increase in disorder, is the major driving force in many chemical and physical processes.

Every substance has an entropy as one of its characteristic properties, just as it has color, hardness, volume, melting point, density, and enthalpy. Like E, the internal energy of a system, the entropy, S, of a system is a state function. The entropy of a system is equal to the sum of the entropies of its components. The **entropy change, ΔS,** for a chemical change is equal to the sum of the entropies of the products of that change minus the sum of the entropies of the reactants. For the reaction $mA + nB \rightarrow xC + yD$, the entropy change is

$$\Delta S = \Sigma S_{products} - \Sigma S_{reactants}$$
$$= [(x \times S_C) + (y \times S_D)] - [(m \times S_A) + (n \times S_B)]$$

where S_A is the entropy of A, S_B is the entropy of B, etc.

A positive value for ΔS ($\Delta S > 0$) indicates an increase in randomness, or disorder, during the change; a negative value for ΔS ($\Delta S < 0$), a decrease in randomness, or an increase in order. The change to a more random and less ordered system when ice melts or when calcium carbonate decomposes corresponds to an increase in entropy ($\Delta S > 0$). Similarly, as liquid molecules go to the still more disordered and random state characteristic of gases, entropy increases. The amount of change in the entropy value is a measure of the increase in the randomness, or disorder, of a system.

The notation used to indicate entropy changes for processes that occur at different temperatures with different pressures and concentrations of reactants and products is analogous to that used for enthalpy changes (Table 19.1). The entropy change for a process that occurs under standard state conditions and that converts a system's components, at unit activities, to products, at unit activities, is labeled $\Delta S°_{298}$. Remember that a solid, a pure liquid, or a solvent has an activity of 1 (unit activity), the activity of a gas is approximated by its pressure in atmospheres (we consider a gas with a pressure of 1 atmosphere to have unit activity), and the activity of a dissolved solute is approximated by its concentration (we consider a 1 molar solution to have unit activity).

The entropy change of a process that occurs at some temperature other than 298.15 K but that involves reactants and products that all have unit activities is given the symbol $\Delta S°$. The symbol ΔS is used to indicate an entropy change with no restrictions on the temperature or the activities of the components. The difference between the entropy *change* for a process at 298.15 K and at some other temperature is generally small and is often neglected. It is common to make the approximation that $\Delta S°_{298}$ equals $\Delta S°$ for the same process.

It is possible to measure the absolute entropy of a substance using the technique described in Section 19.13. Absolute standard molar entropy values for several substances are listed in Table 19.2 and in Appendix I. Each of these $S°_{298}$ values represents the actual entropy content of 1 mole of a substance. The superscript symbol in $S°_{298}$ indicates that the value is for a substance in a standard state (298.15 K, 1 atm, unit activities). Note that solids (most ordered) tend to have lower entropies than liquids (less ordered) and liquids tend to have lower entropies than gases (least ordered). In the

Table 19.2 Standard molar enthalpies of formation, standard molar free energies of formation, and absolute standard molar entropies (298.15 K, 1 atm). (See Appendix I for additional values.)

Substance	$\Delta H°_f,$ kJ mol^{-1}	$\Delta G°_f,$ kJ mol^{-1}	$S°_{298},$ J mol^{-1} K^{-1}
Carbon			
C(s) (graphite)	0	0	5.740
C(g)	716.68	671.29	157.99
CO(g)	−110.5	−137.2	197.56
CO$_2$(g)	−393.5	−394.4	213.6
CH$_4$(g)	−74.81	−50.75	186.15
Chlorine			
Cl$_2$(g)	0	0	222.96
Cl(g)	121.7	105.7	165.09
Copper			
Cu(s)	0	0	33.15
CuS(s)	−53.1	−53.6	66.5
Hydrogen			
H$_2$(g)	0	0	130.57
H(g)	218.0	203.3	114.60
H$_2$O(g)	−241.8	−228.6	188.71
H$_2$O(l)	−285.8	−237.2	69.91
HCl(g)	−92.31	−95.30	186.80
H$_2$S(g)	−20.6	−33.6	205.7
Oxygen			
O$_2$(g)	0	0	205.03
Silver			
Ag$_2$O(s)	−31.0	−11.2	121
Ag$_2$S(s)	−32.6	−40.7	144.0

same physical state, substances with simple molecules tend to have lower entropies than those of substances with more complicated molecules. Since the latter have larger numbers of atoms that can move about, they can exhibit greater randomness or disorder; thus they have higher entropies. Hard substances, such as diamond, tend to be more ordered and have lower entropies than do softer materials, such as graphite or sodium.

It is important to remember that the entropy of a single substance varies with temperature. In subsequent sections we will assume that the entropy *change* for a process, ΔS, is independent of temperature, but we cannot make the same assumption about the entropy, S, of a single substance.

The following examples show how absolute standard molar entropies can be used to calculate entropy changes.

EXAMPLE 19.8

Determine the entropy change for the process $H_2O(l) \longrightarrow H_2O(g)$ when both $H_2O(l)$ and $H_2O(g)$ are in their standard states at 25°C.

The entropy change for this reaction is the sum of the entropies of the products, only $H_2O(g)$ in this case, minus the sum of the entropies of the reactants, $H_2O(l)$ in this case. Entropy values are listed in Table 19.2.

$$\Delta S^\circ_{298} = S^\circ_{H_2O(g)} - S^\circ_{H_2O(l)}$$
$$= 1 \text{ mol } H_2O(g) \times (188.71 \text{ J mol}^{-1} \text{K}^{-1}) - 1 \text{ mol } H_2O(l)$$
$$\times (69.91 \text{ J mol}^{-1} \text{K}^{-1})$$
$$= 118.80 \text{ J K}^{-1}$$

The value for ΔS°_{298} is positive, indicating greater disorder in gaseous H_2O than in liquid H_2O. This is in accord with the fact that the molecules are in much more rapid and random motion in the gaseous state than they are in the liquid state.

EXAMPLE 19.9

Calculate the entropy change for the following reaction when the reactants and products are in their standard states at 25°C:

$$2H_2(g) + O_2(g) \longrightarrow 2H_2O(l)$$

Note that absolute standard molar entropies of the elements are not zero at 25°C. Thus

$$\Delta S^\circ_{298} = 2S^\circ_{H_2O(l)} - 2S^\circ_{H_2(g)} - S^\circ_{O_2(g)}$$
$$= 2 \text{ mol } H_2O(l) \times (69.91 \text{ J mol}^{-1} \text{K}^{-1}) - 2 \text{ mol } H_2(g)$$
$$\times (130.57 \text{ J mol}^{-1} \text{K}^{-1}) - 1 \text{ mol } O_2(g) \times (205.03 \text{ J mol}^{-1} \text{K}^{-1})$$
$$= -326.35 \text{ J K}^{-1}$$

As this reaction proceeds, the entropy *of the system* decreases. We will see in a subsequent section that the entropy of the surroundings increases.

19.8 Free Energy Changes

The effects of randomness and enthalpy on a chemical reaction are such that, when possible, reactions proceed spontaneously toward a state of minimum energy (ΔH negative) and maximum disorder (ΔS positive) (Section 19.1). The enthalpy change and the entropy change of a reaction can be combined to give the change in another state function, the **free energy, G,** of the system. The relationship between the change in free energy, ΔG, the enthalpy change, and the entropy change for a reaction is given by the expression

$$\Delta G = \Delta H - T\Delta S$$

where T is the temperature of the reaction on the Kelvin scale. The free energy change gives an unambiguous prediction of the spontaneity of a chemical reaction run at constant temperature and pressure, because it combines the effects of both ΔH and ΔS. **Reactions for which the value of ΔG is negative ($\Delta G < 0$) are spontaneous.**

The notation used to indicate free energy changes for processes that occur at different temperatures with different pressures and concentrations of reactants and products is similar to that used for enthalpy changes and entropy changes (Table 19.1). The free energy change for a process that occurs under standard state conditions and that converts a system's reactants, at unit activities, to products, at unit activities, is labeled ΔG°_{298}. The free energy change of a process that occurs at some temperature other than 298.15 K but that involves reactants and products that all have unit activities is given the symbol ΔG°. The symbol ΔG is used to indicate a free energy change, with no restrictions on the temperature or the activities of the components.

In the free energy equation, $\Delta G = \Delta H - T\Delta S$, if $T\Delta S$ is small compared to ΔH, then ΔG and ΔH have nearly the same value and each predicts the spontaneity of a given reaction. However, when the value of $T\Delta S$ is significant compared to that of ΔH, only the ΔG value can be used to predict spontaneity in a reliable way. The ΔH term of the free energy equation represents the energy, in the form of heat, involved in a reaction at constant pressure; hence it corresponds to a difference in energy between the initial and final states in a process. The ΔS term of the free energy equation (and hence the $T\Delta S$ term), on the other hand, corresponds to the difference in the amount of order of the atoms in the products and reactants.

When the initial and final states have the same enthalpy (ΔH is zero) and ΔS is not zero, the entropy change, indicated by the $T\Delta S$ term, controls whether or not the reaction will occur. When the amounts of order of the initial and final states are the same (ΔS, and hence $T\Delta S$, is zero) and ΔH is not zero, the enthalpy change, indicated by the ΔH term, determines whether or not the reaction will occur. If ΔH and ΔS both differ from zero, a combination of the two determines whether or not the reaction will occur.

The free energy equation $\Delta G = \Delta H - T\Delta S$ tells us that a reaction that is exothermic (negative ΔH) and produces more numerous molecules (more disorder, positive ΔS) will proceed spontaneously (both ΔH and ΔS contribute toward a more negative value for ΔG). A reaction that is endothermic (positive ΔH) and produces a more ordered system (negative ΔS) will not proceed spontaneously (both ΔH and ΔS contribute toward a more positive value for ΔG). If a reaction is exothermic (negative ΔH) but produces a more ordered system (negative ΔS) or if a reaction is endothermic (positive ΔH) but produces a less ordered system (positive ΔS), then ΔH and ΔS are working in opposite directions, and the relative sizes of ΔH and $T\Delta S$ will determine the prediction.

Note that T will always be positive because it is measured on the Kelvin scale; therefore ΔS (which can be either positive or negative) determines the sign of the $T\Delta S$ term. Also, as T increases, the value of $T\Delta S$ gets larger, so ΔS plays an increasingly important role in the value of ΔG.

Again, it should be emphasized that predicting whether or not a given reaction will proceed spontaneously says *nothing whatever about the rate*. A reaction with a negative ΔG value will proceed spontaneously, but it may do so at a very rapid rate or at an incredibly slow rate.

19.9 Determination of Free Energy Changes

Free energy is a state function. Hence values of ΔG°_{298}, the free energy change for a reaction run under standard state conditions, can be calculated from standard molar free energies of formation, ΔG°_{f}, in the same way that ΔH°_{298} values are calculated from standard molar enthalpies of formation. The free energy change of a chemical process equals the sum of the free energies of the products minus the sum of the free energies of the reactants. For the reaction

$$m\text{A} + n\text{B} \longrightarrow x\text{C} + y\text{D}$$

the free energy change is

$$\Delta G^{\circ}_{298} = \Sigma\, \Delta G^{\circ}_{\text{products}} - \Sigma\, \Delta G^{\circ}_{\text{reactants}}$$
$$= [(x \times \Delta G^{\circ}_{f_C}) + (y \times \Delta G^{\circ}_{f_D})] - [(m \times \Delta G^{\circ}_{f_A}) + (n \times \Delta G^{\circ}_{f_B})]$$

Standard molar free energy of formation values for several substances are given in Table 19.2 and in Appendix I for the standard state condition of 25°C and 1 atmosphere. Note that the standard molar free energy of formation of any free element in its most stable state is zero.

EXAMPLE 19.10

Calculate the standard free energy change, ΔG°_{298}, for the reaction

$$2\text{Ag}_2\text{S}(s) + 2\text{H}_2\text{O}(l) \longrightarrow 4\text{Ag}(s) + 2\text{H}_2\text{S}(g) + \text{O}_2(g)$$

Standard molar free energies of formation of the compounds involved may be found in Table 19.2 or Appendix I; for $\text{Ag}_2\text{S}(s)$ ΔG°_{f} is -40.7 kJ mol^{-1}; for $\text{H}_2\text{O}(l)$, -237.2 kJ mol^{-1}; for $\text{H}_2\text{S}(g)$, -33.6 kJ mol^{-1}; and for $\text{Ag}(s)$ and $\text{O}_2(g)$, 0 kJ mol^{-1}. Thus

$$\Delta G^{\circ}_{298} = 4\, \Delta G^{\circ}_{f_{\text{Ag}(s)}} + 2\, \Delta G^{\circ}_{f_{\text{H}_2\text{S}(g)}} + \Delta G^{\circ}_{f_{\text{O}_2(g)}} - 2\, \Delta G^{\circ}_{f_{\text{Ag}_2\text{S}(s)}} - 2\, \Delta G^{\circ}_{f_{\text{H}_2\text{O}(l)}}$$

$$= 4 \text{ mol Ag}(s) \times (0 \text{ kJ mol}^{-1}) + 2 \text{ mol H}_2\text{S}(g) \times (-33.6 \text{ kJ mol}^{-1})$$
$$+ 1 \text{ mol O}_2(g) \times (0 \text{ kJ mol}^{-1}) - 2 \text{ mol Ag}_2\text{S}(s) \times (-40.7 \text{ kJ mol}^{-1})$$
$$- 2 \text{ mol H}_2\text{O}(l) \times (-237.2 \text{ kJ mol}^{-1})$$

$$= 488.6 \text{ kJ}$$

The positive value of ΔG°_{298} indicates that the reaction should not occur spontaneously as written, but that the reverse reaction ($\Delta G^{\circ}_{298} = -488.6$ kJ) should at 25°C and 1 atm.

EXAMPLE 19.11

The reaction of calcium oxide with the pollutant, sulfur trioxide,

$$\text{CaO}(s) + \text{SO}_3(g) \longrightarrow \text{CaSO}_4(s)$$

has been proposed as one way of removing SO_3 from the smoke resulting from burning high-sulfur coal. Using the following ΔH°_{f} and S°_{298} values, calculate the standard free energy change for the reaction under standard state conditions.

$\text{CaO}(s)$: $\Delta H^{\circ}_{f} = -635.5$ kJ mol^{-1}; $S^{\circ}_{298} = 40$ J mol^{-1} K^{-1}

$\text{SO}_3(g)$: $\Delta H^{\circ}_{f} = -395.7$ kJ mol^{-1}; $S^{\circ}_{298} = 256.6$ J mol^{-1} K^{-1}

$\text{CaSO}_4(s)$: $\Delta H^{\circ}_{f} = -1432.7$ kJ mol^{-1}; $S^{\circ}_{298} = 107$ J mol^{-1} K^{-1}

This calculation can be approached in two ways.

1. ΔH°_{298} and ΔS°_{298} for the reaction can be calculated from the data, as shown in Sections 19.5 and 19.7, and then ΔG°_{298} can be calculated from the free energy equation.

$$\Delta G^\circ_{298} = \Delta H^\circ_{298} - T\Delta S^\circ_{298}$$

2. Using S°_{298} and ΔH°_f for each compound, ΔG°_f for each compound can be determined, and ΔG°_{298} for the reaction can be calculated from these data using the technique illustrated in Example 19.10.

We will use the first approach.

$$
\begin{aligned}
\Delta H^\circ_{298} &= \Delta H^\circ_{f_{CaSO_4(s)}} - \Delta H^\circ_{f_{CaO(s)}} - \Delta H^\circ_{f_{SO_3(g)}} \\
&= 1\ \text{mol CaSO}_4(s) \times (-1432.7\ \text{kJ mol}^{-1}) - 1\ \text{mol CaO}(s) \\
&\quad \times (-635.5\ \text{kJ mol}^{-1}) - 1\ \text{mol SO}_3(g) \times (-395.7\ \text{kJ mol}^{-1}) \\
&= -401.5\ \text{kJ} = -401{,}500\ \text{J} \\
\Delta S^\circ_{298} &= S^\circ_{CaSO_4(s)} - S^\circ_{CaO(s)} - S^\circ_{SO_3(g)} \\
&= 1\ \text{mol CaSO}_4(s) \times (107\ \text{J mol}^{-1}\ \text{K}^{-1}) - 1\ \text{mol CaO}(s) \\
&\quad \times (40\ \text{J mol}^{-1}\ \text{K}^{-1}) - 1\ \text{mol SO}_3(g) \times (256.6\ \text{J mol}^{-1}\ \text{K}^{-1}) \\
&= -189.6\ \text{J K}^{-1} \\
\Delta G^\circ_{298} &= \Delta H^\circ_{298} - T\ \Delta S^\circ_{298} \\
&= -401{,}500\ \text{J} - [298.15\ \text{K} \times (-189.6\ \text{J K}^{-1})] \\
&= -401{,}500\ \text{J} + 56{,}529\ \text{J} \\
&= -344{,}971\ \text{J} = -345.0\ \text{kJ}
\end{aligned}
$$

Note that the units of ΔH were initially in *kilojoules* but were converted to joules because the units of ΔS were in *joules* per kelvin. We must use the same energy units for ΔH and ΔS when calculating ΔG.

19.10 The Second Law of Thermodynamics

If a process occurs at a constant temperature and pressure in a closed system, a system in which no heat can get in or out, ΔH must be zero. If the process is spontaneous, the free energy change, ΔG, is negative, and since ΔH is zero, the entropy change, ΔS, must be positive. Thus the entropy increases in a closed system when a spontaneous process occurs. The universe is a closed system; the universe includes everything; nothing can get in or out. **The second law of thermodynamics states that any spontaneous change that occurs in the universe must be accompanied by an increase in the entropy of the universe.**

It is difficult to apply the second law in its pure form. However, a useful modification can be developed from the free energy equation presented in Section 19.8.

$$\Delta G = \Delta H - T\Delta S \qquad \text{(all values pertaining to the system)}$$

Dividing by T gives

$$\frac{\Delta G}{T} = \frac{\Delta H}{T} - \Delta S \qquad \text{(for the system)}$$

It can be shown that

$$\Delta S_{sur} = -\frac{\Delta H_{sys}}{T} \qquad \text{(at constant temperature and pressure)}$$

where ΔS_{sur} is the change in entropy for the surroundings, and ΔH_{sys} is the change in enthalpy for the system.

Thus

$$\frac{\Delta G_{sys}}{T} = -\Delta S_{sur} - \Delta S_{sys}$$

or

$$-\frac{\Delta G_{sys}}{T} = \Delta S_{sur} + \Delta S_{sys}$$

$$\Delta S_{univ} = \Delta S_{sur} + \Delta S_{sys}$$

Hence,

$$-\frac{\Delta G_{sys}}{T} = \Delta S_{univ}$$

This equation is valid only at constant temperature and pressure, because the relationships from which it is derived are valid only at constant temperature and pressure.

This equation indicates that if ΔG_{sys} is negative, corresponding to a spontaneous change, the entropy of the universe must increase, that is, ΔS_{univ} must be positive. Thus it can be seen that **for a spontaneous change to take place at constant temperature and pressure, the entropy of the universe must increase.**

19.11 Free Energy Changes and Nonstandard States

Many chemical reactions occur when the reactants and products are not in their standard states (Fig. 19.6). The free energy change, ΔG, of such a reaction can be determined from the free energy change, $\Delta G°$, of the same reaction when both the reactants and the products are at unit activities (pure solids or liquids, $P = 1$ atm, concentrations = 1 M) by using the equation

$$\Delta G = \Delta G° + 2.303RT \log Q$$

In this equation T is the temperature on the Kelvin scale, R is a constant (8.314 J K^{-1}), and Q is the reaction quotient of the chemical reaction. For the chemical reaction

$$mA + nB + \cdots \longrightarrow xC + yD + \cdots$$

the reaction quotient is

$$Q = \frac{[C]^x[D]^y \cdots}{[A]^m[B]^n \cdots}$$

Thus the reaction quotient has the same form as the equilibrium constant, K, for the reaction. However, Q has no fixed value; its magnitude is determined by the concentrations of reactants and products at whatever stage in the reaction we choose to evaluate Q. When A and B are first mixed, no products are present, and Q is equal to zero. The value of Q increases as the reaction proceeds. Only when the reaction has reached equilibrium is Q equal to K. As with the equilibrium constant (Section 15.2), strictly speaking, the activities of the reactants and products should be used to evaluate Q. However, we will continue to use pressure (in atmospheres) for gases, concentrations

Figure 19.6
The free energy change for the reaction of methane with oxygen in this flame [$CH_4(g) + 2O_2(g) \rightarrow CO_2(g) + 2H_2O(g)$] is not equal to $\Delta G°_{298}$ for the reaction. Neither is it equal to $\Delta G°$. The reactants (CH_4 and O_2) and the products (CO_2 and H_2O) are far from unit activities (1 atm) and the temperature is not equal to 25°C.

(in moles per liter) for dissolved species, and unity for solvents and for pure solids and liquids as good approximations of the activities of these species.

EXAMPLE 19.12

The reaction of calcium oxide with sulfur trioxide (Example 19.11) rarely occurs under standard state conditions. Calculate ΔG for this reaction at 25°C when the pressure of SO_3 is 0.15 atm.

To determine ΔG under these nonstandard state conditions at 25°C (298 K) we use the equation

$$\Delta G = \Delta G^\circ_{298} + 2.303RT \log Q$$

For the reaction $CaO(s) + SO_3(g) \rightarrow CaSO_4(s)$

$$Q = \frac{[CaSO_4]}{[CaO]P_{SO_3}}$$

CaO and $CaSO_4$ are solids, so their concentrations (activities) are 1. The activity of the gas SO_3 is taken as its pressure in atmospheres, 0.15.

$$Q = \frac{1}{(1) \times (0.15)} = 6.67$$

In Example 19.11 ΔG°_{298} was shown to be -345.0 kJ. Thus

$$\Delta G = \Delta G^\circ_{298} + 2.303RT \log Q$$

$$\Delta G = -345.0 \text{ kJ} + (2.303)(8.314 \text{ J K}^{-1})\left(\frac{1 \text{ kJ}}{10^3 \text{ J}}\right)(298.15 \text{ K})(\log 6.67)$$

$$= -345.0 \text{ kJ} + 4.7 \text{ kJ}$$

$$= -340.3 \text{ kJ}$$

The change of conditions caused a change in ΔG from -345.0 kJ to -340.3 kJ, but the reaction is still spontaneous under these conditions. Note that R had units of J K^{-1}, which had to be converted to kJ K^{-1}.

EXAMPLE 19.13

The partial pressure of carbon dioxide in the atmosphere is 0.0033 atm. Calculate ΔG for the conversion of limestone ($CaCO_3(s)$) to quicklime ($CaO(s)$) in air at 1000°C.

$$CaCO_3(s) \longrightarrow CaO(s) + CO_2(g)$$

To determine ΔG we must consider a process that converts the initial state (1 mol of solid $CaCO_3$) to the final state (1 mol of solid CaO and 1 mol of gaseous CO_2 with a pressure of 0.0033 atm) under nonstandard conditions. First we need to calculate ΔG° for a CO_2 pressure of 1 atm at 1000°C, assuming that ΔH° and ΔS° for the reaction are equal to ΔH°_{298} and ΔS°_{298}, respectively. ΔH°_{298} and ΔS°_{298} can be calculated from the values in Appendix I.

$$\Delta H^\circ_{298} = \left(1 \text{ mol CaO(s)} \times \frac{-635.5 \text{ kJ}}{1 \text{ mol}}\right) + \left(1 \text{ mol CO}_2(g) \times \frac{-393.51 \text{ kJ}}{1 \text{ mol}}\right)$$

$$- \left(1 \text{ mol CaCO}_3(s) \times \frac{-1206.9 \text{ kJ}}{1 \text{ mol}}\right) = 177.9 \text{ kJ}$$

$$\Delta S^{\circ}_{298} = \left(1 \text{ mol } CaO(s) \times \frac{40 \text{ J}}{1 \text{ mol K}}\right) + \left(1 \text{ mol } CO_2(g) \times \frac{213.6 \text{ J}}{1 \text{ mol K}}\right)$$

$$- \left(1 \text{ mol } CaCO_3(s) \times \frac{92.9 \text{ J}}{1 \text{ mol K}}\right) = 161 \text{ J K}^{-1}$$

Now, assuming $\Delta H^{\circ} = \Delta H^{\circ}_{298}$ and $\Delta S^{\circ} = \Delta S^{\circ}_{298}$,

$$\Delta G^{\circ} = \Delta H^{\circ} - T \Delta S^{\circ} = 177.9 \text{ kJ} - 1273 \text{ K} \times 0.161 \text{ kJ K}^{-1} = -27 \text{ kJ}$$

(This value of ΔG° indicates that the decomposition of limestone is spontaneous at 1000°C.)

Now we can calculate ΔG.

$$\Delta G = \Delta G^{\circ} + 2.303RT \log Q$$

$$Q = \frac{P_{CO_2}[CaO]}{[CaCO_3]} = \frac{0.0033 \times 1}{1} = 0.0033$$

$$G = -27 \text{ kJ} + (2.303)(8.314 \text{ J K}^{-1})\left(\frac{1 \text{ kJ}}{1000 \text{ J}}\right)(1273 \text{ K})(\log 0.0033)$$

$$= -27 + (-60.5) = -87 \text{ kJ}$$

Changing the pressure of $CO_2(g)$ reduces the free energy change from -27 kJ (at a pressure of 1 atm) to -87 kJ (at a pressure of 0.0033 atm) and makes the reaction even more spontaneous at 1000°C.

19.12 The Relationship between Free Energy Changes and Equilibrium Constants

If ΔG for any reaction (A + B → C + D, for example) is negative, the reaction will occur spontaneously as written, and the quantities of products will increase. If ΔG is positive, the reverse reaction will occur spontaneously; that is, the species on the right will react to give increased quantities of the species on the left. What is the situation at equilibrium, which, by definition, is the state when the quantities of reactants and products do not change? At equilibrium, ΔG can be neither positive nor negative; thus it can only be zero. At equilibrium, we therefore have

$$\Delta G = 0 = \Delta G^{\circ} + 2.303RT \log Q$$

In this equation the value determined for ΔG° is for unit activities of the reactants and products at temperature T. (See Example 19.14.) Since the reaction is at equilibrium, the value of the reaction quotient must be equal to that of the equilibrium constant for the reaction (Q is equal to K), so we have

$$\Delta G = 0 = \Delta G^{\circ} + 2.303RT \log K$$

$$\Delta G^{\circ} = -2.303RT \log K \tag{1}$$

This derivation shows us that the value of the standard state free energy change, ΔG°, for a reaction can be used to determine the equilibrium constant for that reaction and that values of ΔG° can be determined experimentally from equilibrium constants (Fig. 19.7).

Figure 19.7
The value of the equilibrium constant ($K_p = 0.282$ atm) for the phase change $Br_2(l) \rightarrow Br_2(g)$ at 25°C can be used to determine ΔG°_{298} (3.14 kJ) for the change.

EXAMPLE 19.14   Calculate the standard state free energy change for the ionization of acetic acid at 0°C.

$$CH_3CO_2H(aq) \longrightarrow H^+(aq) + CH_3CO_2^-(aq)$$

The equilibrium constant for the reaction has been found to be 1.657×10^{-5}.

At equilibrium, $\Delta G° = -2.303RT \log K$, where

$$K = \frac{[H^+][CH_3CO_2^-]}{[CH_3CO_2H]} = 1.657 \times 10^{-5}$$

Thus $\quad \Delta G° = -(2.303)(8.314 \text{ J K}^{-1})(273.15 \text{ K})[\log (1.657 \times 10^{-5})]$

$\qquad\qquad = -(2.303)(8.314 \text{ J K}^{-1})(273.15 \text{ K})(4.7807) = 2.500 \times 10^4 \text{ J}$

$\qquad\qquad = 25.00 \text{ kJ}$

Therefore the free energy change for the *complete* transformation at 0°C of 1 L of 1 M acetic acid in water into 1 mol of hydrogen ion and 1 mol of acetate ion, both at 1 M concentrations in water, would be 25.00 kJ. Since $\Delta G°$ is positive, the reaction is not spontaneous. However, the reverse reaction,

$$H^+(aq) + CH_3CO_2^-(aq) \longrightarrow CH_3CO_2H(aq)$$

is spontaneous ($\Delta G° = -25.00 \text{ kJ mol}^{-1}$).

EXAMPLE 19.15   Using data in Appendix I, calculate the value of K at 298.15 K for the reaction

$$H_2(g) + \tfrac{1}{2}O_2(g) \longrightarrow H_2O(g)$$

The equilibrium constant for this reaction can be evaluated using the equation

$$\Delta G°_{298} = -2.303RT \log K$$

Since this reaction involves formation of 1 mol of $H_2O(g)$ from the elements, $\Delta G°_{298}$ for the reaction is equal to the standard molar free energy of formation of water vapor, -228.59 kJ (Appendix I) or $-228,590$ J. Therefore

$$-228,590 \text{ J} = (-2.303)(8.314 \text{ J K}^{-1})(298.15 \text{ K})(\log K)$$

$$\log K = \frac{-228,590 \text{ J}}{(-2.303)(8.314 \text{ J K}^{-1})(298.15 \text{ K})} = 40.04$$

$$K = 1.1 \times 10^{40}$$

Note that when reactants are gases, the equilibrium constant calculated from Equation (1) involves the pressures of the gases in atmospheres. Thus

$$K = \frac{P_{H_2O(g)}}{P_{H_2(g)} \times P_{O_2(g)}^{1/2}} = 1.1 \times 10^{40}$$

Now we see one reason why calculation of the value of $\Delta G°$ is useful. It allows us to predict the value of the equilibrium constant for a reaction and thus to determine whether or not a reaction will give significant amounts of products.

If we combine Equation (1) with the equation relating $\Delta G°$, $\Delta H°$, and $\Delta S°$, we have

$$\Delta G° = \Delta H° - T\Delta S° = -2.303RT \log K \qquad (2)$$

As the temperature varies, $\Delta H°$ and $\Delta S°$ change only slightly, so if we assume that they are indeed independent of temperature, we can obtain a simple equation relating the equilibrium constant and temperature.

Consider an equilibrium reaction at two different temperatures, T_1 and T_2, with equilibrium constants K_{T_1} and K_{T_2}.

$$\Delta H° - T_1 \, \Delta S° = -2.303RT_1 \log K_{T_1} \tag{3}$$

and
$$\Delta H° - T_2 \, \Delta S° = -2.303RT_2 \log K_{T_2} \tag{4}$$

Equations (3) and (4) may be rearranged by dividing Equation (3) by T_1 and Equation (4) by T_2. This gives

$$\frac{\Delta H°}{T_1} - \Delta S° = -2.303R \log K_{T_1} \tag{5}$$

and
$$\frac{\Delta H°}{T_2} - \Delta S° = -2.303R \log K_{T_2} \tag{6}$$

Subtracting Equation (6) from Equation (5) gives

$$\frac{\Delta H°}{T_1} - \frac{\Delta H°}{T_2} - \Delta S° - (-\Delta S°) = -2.303R \log K_{T_1} - (-2.303R \log K_{T_2})$$

or
$$\Delta H° \left(\frac{1}{T_1} - \frac{1}{T_2} \right) = 2.303R \log \frac{K_{T_2}}{K_{T_1}}$$

or, by multiplying the left side by T_1T_2/T_1T_2 and dividing both sides by $2.303R$,

$$\frac{\Delta H°(T_2 - T_1)}{2.303RT_1T_2} = \log \frac{K_{T_2}}{K_{T_1}} \tag{7}$$

Equations (1) and (7) prove to be quite useful. If we know $\Delta G°_{298}$ (at 298 K), we can determine the equilibrium constant, K, at 298 K using Equation (1). Then, if we know $\Delta H°_{298}$, we can obtain K for any other temperature by means of Equation (7) (within the limitation that both $\Delta H°$ and $\Delta S°$ are independent of T). The value of $\Delta G°$ at a temperature other than 298 K can be calculated if $\Delta G°_{298}$ and $\Delta H°_{298}$ are known, because the value of $\Delta S°_{298}$ can be determined using the equation

$$\Delta G°_{298} = \Delta H°_{298} - T\Delta S°_{298} \qquad (T = 298 \text{ K})$$

Then, since both $\Delta H°$ and $\Delta S°$ are essentially independent of temperature, $\Delta G°$ at other temperatures can be calculated.

(a) For the reaction

$$CuS(s) + H_2(g) \longrightarrow Cu(s) + H_2S(g)$$

EXAMPLE 19.16

calculate $\Delta G°$ and $\Delta H°$ at 298.15 K and 1 atm.

$$\Delta G°_{298} = \Delta G°_{f_{Cu(s)}} + \Delta G°_{f_{H_2S(g)}} - \Delta G°_{f_{CuS(s)}} - \Delta G°_{f_{H_2(g)}}$$

$$\Delta G°_{298} = 1 \text{ mol } Cu(s) \times (0 \text{ kJ mol}^{-1}) + 1 \text{ mol } H_2S(g) \times (-33.6 \text{ kJ mol}^{-1})$$
$$- 1 \text{ mol } CuS(s) \times (-53.6 \text{ kJ mol}^{-1}) - 1 \text{ mol } H_2(g) \times (0 \text{ kJ mol}^{-1})$$

$$\Delta G°_{298} = 20.0 \text{ kJ} \qquad \text{(reaction is not spontaneous)}$$

$$\Delta H°_{298} = \Delta H°_{f_{Cu(s)}} + \Delta H°_{f_{H_2S(g)}} - \Delta H°_{f_{CuS(s)}} - \Delta H°_{f_{H_2(g)}}$$

$$\Delta H°_{298} = 1 \text{ mol } Cu(s) \times (0 \text{ kJ mol}^{-1}) + 1 \text{ mol } H_2S(g) \times (-20.6 \text{ kJ mol}^{-1})$$
$$- 1 \text{ mol } CuS(s) \times (-53.1 \text{ kJ mol}^{-1}) - 1 \text{ mol } H_2(g) \times (0 \text{ kJ mol}^{-1})$$

$$\Delta H°_{298} = 32.5 \text{ kJ} \qquad \text{(reaction is endothermic)}$$

(b) Calculate the value for the equilibrium constant, K, at 298.15 K and 1 atm.

$$K = \frac{P_{H_2S}}{P_{H_2}} \qquad \text{(where } P \text{ is the partial pressure of a gas in atm)}$$

$$\Delta G^\circ_{298} = -2.303RT \log K \qquad (R = 8.314 \text{ J K}^{-1})$$

Then
$$\log K = \frac{\Delta G^\circ}{-2.303RT}$$

$$= \frac{20.0 \times 10^3 \text{ J}}{-(2.303)(8.314 \text{ J K}^{-1})(298.15 \text{ K})}$$

$$= -3.5038$$

$$K = 3.13 \times 10^{-4} \qquad \text{(at 298.15 K)}$$

Note that at 298.15 K the equilibrium constant has a value less than 1, indicating that the elements in the reaction are present in larger quantities as reactants than as products and that the equilibrium therefore lies far to the left.

(c) Estimate the value for K at 798 K and 1 atm.

$$\frac{\Delta H^\circ(T_2 - T_1)}{2.303RT_1T_2} = \log \frac{K_{T_2}}{K_{T_1}}$$

Then
$$\frac{32,500 \text{ J }(798 \text{ K} - 298 \text{ K})}{(2.303)(8.314 \text{ J K}^{-1})(798 \text{ K})(298 \text{ K})} = \log \frac{K_{798}}{K_{298}}$$

$$3.5689 = \log \frac{K_{798}}{K_{298}}$$

$$\log K_{798} - \log K_{298} = 3.5689$$

$$\log K_{298} = -3.5038 \qquad \text{[calculated in part (b)]}$$

Therefore
$$\log K_{798} = 3.5689 + (-3.5038) = 0.0651$$

$$K = 1.16 \qquad \text{(at 798 K)}$$

Note that the equilibrium constant at 798 K is greater than 1, indicating that the elements are present in greater quantities as products than as reactants and that the equilibrium has been displaced to the right.

(d) Calculate ΔS° at 298.15 K and 1 atm.

Although we could calculate ΔS° from absolute standard molar entropy values (Section 19.7), let us use a different method since we already have values for ΔG and ΔH.

$$\Delta G^\circ_{298} = \Delta H^\circ_{298} - T\Delta S^\circ_{298}$$

$$\Delta S^\circ_{298} = \frac{\Delta H^\circ_{298} - \Delta G^\circ_{298}}{T} = \frac{32,500 \text{ J} - 20,000 \text{ J}}{298.15 \text{ K}}$$

$$= 41.9 \text{ J K}^{-1}$$

Note that this value for ΔS°_{298} is positive and favors a spontaneous reaction, whereas the positive value of ΔH°_{298} is unfavorable for reaction. This is, therefore, a case where ΔS° is sufficiently large that ΔH° is not a valid indicator of spontaneity, and ΔG° must be considered. In particular, whether or not the reaction is spontaneous is expected to depend on the temperature.

We found in part (a) that $\Delta G^\circ_{298} = +20.0$ kJ indicating that at 25°C (298 K) the reaction is not spontaneous. Let us now calculate ΔG° at a higher temperature assuming that ΔH°_{298} and ΔS°_{298} do not change significantly as the temperature increases.

(e) Estimate ΔG° at 798 K and 1 atm.

$$\Delta G^\circ = \Delta H^\circ - T\Delta S^\circ$$
$$= 32{,}500 - (798 \text{ K})(41.9 \text{ J K}^{-1})$$
$$= -936 \text{ J} = -0.936 \text{ kJ}$$

At the higher temperature ΔG° is negative, showing that at 798 K the reaction *is* spontaneous. ΔH°, being relatively independent of temperature, still has a value of about 32.5 kJ at 798 K and hence does not indicate the change to spontaneity.

(f) Estimate the temperature at which ΔG° is equal to zero, assuming that ΔH° and ΔS° do not change significantly as the temperature increases.

$$\Delta G^\circ = \Delta H^\circ - T\Delta S^\circ \qquad [\Delta S^\circ = 41.9 \text{ J K}^{-1}, \text{ as calculated in part (d)}]$$
$$0 = 32{,}500 \text{ J} - T(41.9 \text{ J K}^{-1})$$
$$T = \frac{32{,}500 \text{ J}}{41.9 \text{ J K}^{-1}} = 776 \text{ K}$$

Hence $\Delta G^\circ = 0$ at 776 K. At temperatures below 776 K the values of ΔG° are positive; at temperatures above 776 K they are negative. Therefore, above 776 K the reaction is spontaneous; below 776 K the reaction as written is not spontaneous—indeed, it is spontaneous in the opposite direction.

(a) Determine the temperature at which liquid water and gaseous water are in equilibrium with each other at 1 atm.

EXAMPLE 19.17

Since the two states are in equilibrium, the free energy change, ΔG°, in going from the liquid to the gas is zero. Assuming again that ΔH° and ΔS° are independent of temperature, for the reaction $H_2O(l) \rightarrow H_2O(g)$,

$$\Delta H^\circ_{298} = \Delta H^\circ_{f_{H_2O(g)}} - \Delta H^\circ_{f_{H_2O(l)}} = 1 \text{ mol } H_2O(g) \times (-241.8 \text{ kJ mol}^{-1})$$
$$- 1 \text{ mol } H_2O(l) \times (-285.8 \text{ kJ mol}^{-1})$$
$$= 44.0 \text{ kJ} = 44{,}000 \text{ J}$$

$$\Delta S^\circ_{298} = S^\circ_{H_2O(g)} - S^\circ_{H_2O(l)}$$
$$= 1 \text{ mol } H_2O(g) \times (188.71 \text{ J mol}^{-1} \text{ K}^{-1})$$
$$- 1 \text{ mol } H_2O(l) \times (69.91 \text{ J mol}^{-1} \text{ K}^{-1})$$
$$= 118.80 \text{ J K}^{-1}$$

$$\Delta G^\circ = \Delta H^\circ - T\Delta S^\circ = 0 \qquad (\Delta G^\circ = 0 \text{ at equilibrium})$$

Since we assume ΔH° and ΔS° to be independent of temperature, $\Delta H^\circ = \Delta H^\circ_{298}$ and $\Delta S^\circ = \Delta S^\circ_{298}$. Thus

$$T = \frac{\Delta H^\circ}{\Delta S^\circ} = \frac{44{,}000 \text{ J}}{118.80 \text{ J K}^{-1}}$$
$$= 370 \text{ K} = 97°C$$

The correct answer, of course, is 373 K (100°C), the boiling point of water at 1 atm, but the calculation of the value 370 K was based on the assumptions that $\Delta H°$ and $\Delta S°$ are independent of temperature. This is a good place to reemphasize that these assumptions are only approximately correct. They are sufficiently close to be highly useful, but for really exact calculations values of $\Delta H°$ and $\Delta S°$ at the temperature in question must be used.

(b) Recalculate the temperature at which liquid water and gaseous water are in equilibrium with each other at 1 atm, this time using the values of $\Delta H°$ and $\Delta S°$ at the temperature 97°C.

$$\Delta H° \text{ at } 97°C = 40,720 \text{ J mol}^{-1}; \Delta S° \text{ at } 97°C = 109.1 \text{ J mol}^{-1} \text{ K}^{-1}$$

Note that the values of $\Delta H°_{298}$ (44,000 J) and of $\Delta S°_{298}$ (118.80 J K^{-1}) are close enough, as stated earlier, to the values at 97°C for an approximate calculation but are not sufficiently close for an exact calculation.

Again using

$$\Delta G° = \Delta H° - T\Delta S° = 0 \qquad (\Delta G° = 0 \text{ at equilibrium})$$

$$T = \frac{\Delta H°_{370}}{\Delta S°_{370}} = \frac{40,720 \text{ J}}{109.1 \text{ J K}^{-1}} = 373.2 \text{ K}$$

or $\qquad\qquad 373.2 - 273.2 = 100.0°C$

In this part of the example we obtained the experimental value of 100°C. If the calculation is repeated with the values of $\Delta H°$ and $\Delta S°$ at 100°C (40,656 J and 108.95 J K^{-1}), a value of $T = 100.0°C$ is again obtained, indicating that the values for $\Delta H°$ and $\Delta S°$ at 97°C are close enough to give the correct temperature of 100.0°C (to four significant figures).

At temperatures below 373 K the value of $T\Delta S°$ for the vaporization of liquid water becomes smaller, and hence a smaller term is subtracted from $\Delta H°$ to give $\Delta G°(\Delta G° = \Delta H° - T\Delta S°)$. Thus $\Delta G°$ is positive (it equals zero at the equilibrium state at 373 K), indicating that *at temperatures below 373 K (100°C) at 1 atmosphere of pressure the spontaneous change is from gaseous H_2O to liquid H_2O*. At temperatures higher than 373 K, $T\Delta S°$ becomes larger, and hence a larger term is subtracted from $\Delta H°$. Thus $\Delta G°$ is negative, indicating that *at temperatures above 373 K (100°C) at 1 atmosphere of pressure the spontaneous change is from liquid H_2O to gaseous H_2O*.

Here the heat effect ($\Delta H°$) and the entropy effect ($\Delta S°$) work at cross purposes, since $\Delta S°$ is positive (increasing disorder, indicating favorable conditions for the change), and $\Delta H°$ is positive (endothermic, indicating unfavorable conditions for the change). Neither $\Delta H°$ nor $\Delta S°$ alone is sufficient to predict spontaneity. $\Delta G°$ (which combines $\Delta H°$, $\Delta S°$, and T) is the quantity on which to base such predictions. As shown above, $\Delta G°$ is positive at temperatures below 373 K (100°C) and negative at temperatures above 373 K.

19.13 The Third Law of Thermodynamics

In the preceding sections we have considered the changes in state functions that accompany chemical or physical changes; we have not tried to obtain values for the functions in any particular state. In fact, it is not possible to measure values for E,

H, and G. However, it is possible to measure the **absolute entropy** of a pure substance at any given temperature. The reason for this is found in the **third law of thermodynamics: The entropy of any pure, perfect crystalline element or compound at absolute zero (0 K) is equal to zero.** At absolute zero all molecular motion is at a minimum, and at that temperature a pure crystalline substance has no disorder (its entropy is zero). All molecular motion is also at a minimum in an impure substance at absolute zero, but the impurity can be distributed in different ways, giving rise to disorder (and a nonzero value for entropy).

If we take a pure crystalline substance at absolute zero (with an entropy of zero) and measure the entropy change as its temperature is increased to any temperature T, then we have measured the absolute entropy of the substance at the temperature T. Hence we can find *absolute entropies* of pure substances; in contrast, for free energy and enthalpy, we can find only differences between two values, not the absolute values. Absolute entropies allow us to compare the relative amounts of disorder present in different pure substances, and they can be used to determine entropy changes (Section 19.7). Caution must be exercised when using absolute entropies, however, since the absolute entropy of a pure elemental substance at standard state conditions will *not* be equal to zero. Table 19.2 and Appendix I contain absolute standard molar entropies, S°_{298}, of some common substances at standard state conditions.

For Review

Summary

Chemical thermodynamics is the study of the energy transformations and transfers that accompany chemical and physical changes. From such a study we can determine if a change will occur spontaneously and how far it will proceed—the equilibrium position of the change. The spontaneity of a chemical change is determined by the changes in potential energy and in disorder that accompany the change.

The **first law of thermodynamics** states that in any change that occurs in nature the total energy of the universe remains constant. Thus by measuring the energy lost or gained by the system we can determine the energy change within it. Energy can be lost from a system as the system does **work, w,** on the surroundings ($w > 0$) or as the surroundings do work on the system ($w < 0$). The amount of work due to expansion is given by the expression $P(V_2 - V_1)$, when the expansion is carried out at constant pressure. Energy can also be transferred as **heat, q.** In an **endothermic process** heat is transferred into the system from the surroundings ($q > 0$). In an **exothermic process** heat is transferred from the system to the surroundings ($q < 0$).

Although the amounts of heat and work accompanying a change may vary depending on how the change is carried out, the change in internal energy of the system is independent of how the change is accomplished. A property such as internal energy that does not depend on how the system gets from one state to another is called a **state function.** If a chemical change is carried out at constant pressure such that the only work done is expansion work, q is a state function and is called the **enthalpy change, ΔH,** of the reaction. The enthalpy change accompanying the formation of 1 mole of a substance from the elements in their most stable states, all at 298.15 K and 1 atm (a **standard state**), is called the **standard molar enthalpy of formation of the substance, ΔH°_f.** Using **Hess's law** the enthalpy change of any reaction can be described as the sum of the standard molar enthalpies of all of the products minus those of all of the reactants.

A **spontaneous change,** a change in a system that will proceed without any outside influence, is favored when the change is exothermic or when the change leads to an increase in the randomness, or disorder, of the system. The randomness of a system can be determined quantitatively and is called the **entropy, S,** of the system. S is a state function, and values of S°_{298} for many substances in a standard state have been measured and tabulated. The entropy change, ΔS°, for many reactions can be calculated as the sum of the **absolute standard molar entropies** of all products minus the sum of those of all reactants. A positive value of ΔS° ($\Delta S^{\circ} > 0$) indicates that the disorder of the system has increased, whereas a negative value of ΔS° ($\Delta S^{\circ} < 0$) indicates that the disorder of the system has decreased. The **second law of thermodynamics** states that any spontaneous change that occurs in the universe must be accompanied by an increase in the entropy of the universe.

The **free energy change, ΔG,** of a reaction is the difference between the enthalpy change of the reaction and the product of the temperature and entropy change of the reaction.

$$\Delta G = \Delta H - T\Delta S$$

For a reaction at constant temperature and pressure, a negative value of ΔG ($\Delta G < 0$) indicates that the reaction is spontaneous. Values of ΔG°_{298} can be calculated from **standard molar free energies of formation, ΔG°_f,** or from the values of ΔH° and ΔS° for the reaction.

The free energy change of a reaction that involves reactants or products at concentrations other than 1 M or pressures other than 1 atm, ΔG, can be determined using the equation

$$\Delta G = \Delta G^{\circ} + 2.303RT \log Q$$

where R is equal to 8.314 $J\ K^{-1}$ and Q is the reaction quotient. When a reaction has reached equilibrium, Q is equal to K, the equilibrium constant for the reaction, and ΔG is equal to zero. Thus at equilibrium

$$\Delta G^{\circ} = -2.303RT \log K$$

The **third law of thermodynamics** states that the entropy of a pure crystalline substance at 0 K is equal to zero. Thus a measurement of the entropy change of such a substance as it is heated from 0 K to a higher temperature gives the absolute entropy of the substance at the higher temperature. The entropy change for a chemical reaction can be determined by subtracting the sum of the absolute molar entropies of all of the reactants from the sum of those of the products.

$$\Delta S^{\circ} = \Sigma\ S^{\circ}_{products} - \Sigma\ S^{\circ}_{reactants}$$

Key Terms and Concepts

absolute entropy (19.13)
chemical thermodynamics (19.1)
endothermic process (19.5)
enthalpy change (19.5)
entropy (19.7)
entropy change (19.7)
equilibrium constant (19.12)
exothermic process (19.5)
expansion work (19.2)

first law of thermodynamics
   (19.3)
free energy change (19.9)
heat (19.2)
Hess's law (19.5)
internal energy (19.2)
reaction quotient (19.11)
second law of thermodynamics
   (19.10)

spontaneous change (19.6)
standard molar entropy (19.7)
state function (19.4)
surroundings (19.1)
system (19.1)
third law of thermodynamics
   (19.13)
work (19.2)

Exercises

Heat, Work, and Internal Energy

1. State the first law of thermodynamics in words and by using an equation.

2. Calculate the missing value of ΔE, q, or w for a system, given the following data:
   - (a) $q = 570$ J; $w = -300$ J   Ans. $\Delta E = 870$ J
   - (b) $\Delta E = -7500$ J; $w = 4500$ J   Ans. $q = -3000$ J
   - (c) $\Delta E = -250$ J; $q = 300$ J   Ans. $w = 550$ J
   - (d) The system absorbs 2.000 kJ of heat and does 1425 J of work on the surroundings.   Ans. $\Delta E = 575$ J

3. In which of the following changes at constant pressure is work done by the surroundings on the system? By the system on the surroundings? Is essentially no work done? What is the value of w in each case: $w > 0$, $w < 0$, or $w = 0$ or almost 0?

Initial State	Final State
(a) $H_2O(s)$	$H_2O(l)$
(b) $H_2O(g)$	$H_2O(s)$
(c) $2Na(s) + 2H_2O(l)$	$2NaOH(s) + H_2(g)$
(d) $H_2(g) + Cl_2(g)$	$2HCl(g)$
(e) $3H_2(g) + N_2(g)$	$2NH_3(g)$
(f) $CaCO_3(s)$	$CaO(s) + CO_2(g)$
(g) $NO_2(g) + CO(g)$	$NO(g) + CO_2(g)$

   $w = P(V_2 - V_1)$

4. Calculate the work involved in compressing a system consisting of exactly 2 mol of H_2O as it changes from a gas at 373 K (volume = 61.2 L) to a liquid at 373 K (volume = 37.8 mL) under a constant pressure of 1.00 atm. Does this work add to or decrease the internal energy of the system?   Ans. $w = -6.20$ kJ, increasing E

5. What is the change in internal energy of a gas that absorbs 225 J of heat and expands from 10.0 L to 25.0 L against a constant pressure of 0.75 atm?   Ans. -915 J

6. What work is done when 1.00 mol of solid $CaCO_3$ (volume 34.2 mL) decomposes at 25°C and a pressure of exactly 1.00 atm to give solid CaO (volume 16.9 mL) and $CO_2(g)$?   Ans. 2.48 kJ

7. Assume that the only change in volume is due to the production of hydrogen and calculate w, the work done, when 2.00 mol of Zn dissolves in hydrochloric acid giving H_2 at 35°C and 1.00 atm.

   $$Zn + 2HCl \longrightarrow ZnCl_2 + H_2(g)$$
   2   4   2   2   Ans. 5.12 kJ

8. Assume that the gases exhibit ideal behavior and calculate w, the work done, when 17.0 g of $CH_4(g)$ is burned at 775°C and 2.00 atm according to the following reaction:

   $$CH_4(g) + 2O_2(g) \longrightarrow CO_2(g) + 2H_2O(g)$$
   Ans. 0

Enthalpy Changes; Hess's Law

9. What is the difference between ΔE and ΔH for a system undergoing a change at constant pressure?

10. The enthalpy, H, of a system has been referred to as the heat content of the system. Why might this follow from the relationship between ΔH and q?

11. (a) Using the data in Appendix I, calculate the standard enthalpy change, ΔH°_{298}, for each of the following reactions:

    (1) $2Al(s) + 3F_2(g) \longrightarrow 2AlF_3(s)$
    Ans. -3008 kJ

    (2) $N_2(g) + O_2(g) \longrightarrow 2NO(g)$
    Ans. 180.5 kJ

    (3) $CaO(s) + H_2O(l) \longrightarrow Ca(OH)_2(s)$
    Ans. -65.3 kJ

    (4) $Fe_2O_3(s) + 3CO(g) \longrightarrow 2Fe(s) + 3CO_2(g)$
    Ans. -24.8 kJ

    (5) $2LiOH(s) + CO_2(g) \longrightarrow Li_2CO_3(s) + H_2O(l)$
    Ans. -133.5 kJ

    (6) $CaSO_4 \cdot 2H_2O(s) \longrightarrow CaSO_4(s) + 2H_2O(g)$
    Ans. 104.8 kJ

    (7) $CH_4(g) + N_2(g) \longrightarrow HCN(g) + NH_3(g)$
    Ans. 164 kJ

    (8) $CS_2(g) + 3Cl_2(g) \longrightarrow CCl_4(g) + S_2Cl_2(g)$
    Ans. -238 kJ

    (b) Which of these reactions are exothermic?
    Ans. 1,3,4,5,8

12. The decomposition of hydrogen peroxide, H_2O_2, has been used to provide thrust in the control jets of various space vehicles. How much heat is produced by the decomposition of 0.500 kg of H_2O_2 under standard conditions?

    $$2H_2O_2(l) \longrightarrow 2H_2O(g) + O_2(g)$$
    Ans. 794 kJ

13. How many kilojoules of heat will be liberated when 27.89 g of manganese is burned to form $Mn_3O_4(s)$ at standard state conditions?   Ans. 234.9 kJ

14. The standard molar heat of formation of $OsO_4(s)$ is -391 kJ mol^{-1}, and the heat of sublimation of $OsO_4(s)$ under standard state conditions is 56.4 kJ mol^{-1}. What is ΔH°_{298} for the process $OsO_4(g) \longrightarrow Os(s) + 2O_2(g)$?   Ans. 335 kJ

15. The oxidation of the sugar glucose, $C_6H_{12}O_6$, is described by the following equation:

    $$C_6H_{12}O_6(s) + 6O_2(g) \longrightarrow 6CO_2(g) + 6H_2O(l)$$
    $$\Delta H^\circ_{298} = -2816 \text{ kJ}$$

    Metabolism of glucose gives the same products, although the glucose reacts with oxygen in a series of steps in the body. (a) How much heat in kilojoules is produced by the metabolism of 1.0 g of glucose? (b) How many nutritional Calories (1 cal = 4.184 J; 1 nutritional Cal = 1000 cal) are produced by the metabolism of 1.0 g glucose?
    Ans. 16 kJ; 3.7 Cal

16. The white pigment TiO_2 is prepared by the hydrolysis of titanium tetrachloride, $TiCl_4$, in the gas phase.

    $$TiCl_4(g) + 2H_2O(g) \longrightarrow TiO_2(s) + 4HCl(g)$$

How much heat is evolved in the production of 1.00 kg of $TiO_2(s)$ under standard state conditions of 25°C and 1 atm? *Ans. 840 kJ*

17. A sample of $WO_2(s)$ with a mass of 0.745 g was "burned" in oxygen at constant pressure, giving $WO_3(s)$ and 1.143 kJ of heat. The enthalpy of formation of $WO_3(s)$ under these conditions is -842.91 kJ mol^{-1}. Calculate the enthalpy of formation of $WO_2(s)$.
Ans. −512 kJ mol^{-1}

Entropy Changes and Absolute Entropy

18. What is the connection between entropy and disorder?
19. What is the absolute standard molar entropy of a pure substance?
20. Arrange the following systems, each of which consists of 1 mol of substance, in order of increasing entropy: $H_2O(g)$ at 100°C, $N_2(s)$ at −215°C, $C_2H_5OH(g)$ at 100°C, $H_2O(l)$ at 25°C, $H_2O(s)$ at −215°C, $C_2H_5OH(s)$ at 0 K.
21. State the second law of thermodynamics in terms of entropy changes for the universe. State the second law of thermodynamics in terms of free energy changes.
22. State the third law of thermodynamics.
23. Does the entropy of each of the following systems increase, decrease, or not change in going from the initial to the final state? If the entropy does change, give the sign of ΔS and explain your answers.

Initial State	*Final State*
(a) $NaCl(s)$ at 298 K	$NaCl(s)$ at 0 K *decrease*
(b) $H_2O(s)$ at 273 K and 1 atm *increase*	$H_2O(l)$ at 273 K and 1 atm
(c) 1 mol Si and 1 mol O_2 *decrease*	1 mol SiO_2
(d) 1 mol $CaCO_3$ *decrease*	1 mol CaO and 1 mol CO_2
(e) egg in shell *increase*	scrambled egg
(f) chicken feed and baby chick *decrease*	full grown chicken

24. (a) Using the data in Appendix I, calculate ΔS°_{298}, the standard entropy change for each of the following reactions:
   (1) $2Al(s) + 3F_2(g) \longrightarrow 2AlF_3(s)$
   Ans. −531.8 J K^{-1}
   (2) $N_2(g) + O_2(g) \longrightarrow 2NO(g)$
   Ans. 24.8 J K^{-1}
   (3) $CaO(s) + H_2O(l) \longrightarrow Ca(OH)_2(s)$
   Ans. −34 J K^{-1}
   (4) $Fe_2O_3(s) + 3CO(g) \longrightarrow 2Fe(s) + 3CO_2(g)$
   Ans. 15.3 J K^{-1}
   (5) $2LiOH(s) + CO_2(g) \longrightarrow Li_2CO_3(s) + H_2O(l)$
   Ans. −153.7 J K^{-1}
   (6) $CaSO_4 \cdot 2H_2O(s) \longrightarrow CaSO_4(s) + 2H_2O(g)$
   Ans. 290 J K^{-1}
   (7) $CH_4(g) + N_2(g) \longrightarrow HCN(g) + NH_3(g)$
   Ans. 16.4 J K^{-1}
   (8) $CS_2(g) + 3Cl_2(g) \longrightarrow CCl_4(g) + S_2Cl_2(g)$
   Ans. −265.5 J K^{-1}

(b) For which of these reactions are the entropy changes favorable for the reaction to proceed spontaneously?
Ans. 2,4,6,7

25. What is the entropy change for condensation of 59.7 g of chloroform, $CHCl_3(g) \rightarrow CHCl_3(l)$, under standard state conditions? *Ans. −47 J K^{-1}*

26. What is ΔS°_{298} for the following reaction?
$$N_2(g) + 3H_2(g) \longrightarrow 2NH_3(g)$$
Ans. −198.6 J K^{-1}

27. What is ΔS°_{298} for the formation of ozone, $O_3(g)$, from oxygen under standard state conditions?
Ans. −68.7 J K^{-1}

Free Energy Changes

28. Why is the emphasis in thermodynamics on the free energy change as a system changes rather than on the values of the free energies of the initial and final state?
29. What is the difference between ΔH and ΔG for a system undergoing a change at constant temperature and pressure?
30. What is a spontaneous reaction?
31. Explain why the free energy change of a reaction varies with temperature.
32. As ammonium nitrate dissolves spontaneously in water at constant pressure, heat is absorbed and the solution gets cold. What is the sign of ΔH for this process? Is it possible to identify the sign of ΔS for this process? Why?
33. As sulfuric acid dissolves spontaneously in water at constant pressure, heat is produced and the solution gets hot. What is the sign of ΔH for this process? Is it possible to identify the sign of ΔS for this process? Why?
34. The reaction $3O_2(g) \rightarrow 2O_3(g)$ is endothermic and proceeds with a decrease in the entropy of the system. Is this likely to be a spontaneous reaction? Explain your reasoning.
35. Under what conditions is ΔG equal to ΔG° for the reaction $2H_2(g) + O_2(g) \rightarrow 2H_2O(l)$?
36. Using the data in Table 19.2, show that H, S, and G are state functions by calculating ΔH°_{298}, ΔS°_{298}, and ΔG°_{298} for the formation of $HCl(g)$ from $H_2(g)$ and $Cl(g)$ by two pathways, both at standard state conditions.
   Path 1: $H_2(g) + Cl_2(g) \longrightarrow 2HCl(g)$
   Path 2: $H_2(g) \longrightarrow 2H(g)$
   $Cl_2(g) \longrightarrow 2Cl(g)$
   $2H(g) + 2Cl(g) \longrightarrow 2HCl(g)$
37. (a) Using the data in Appendix I, calculate the standard free energy changes for the following reactions:
   (1) $2Al(s) + 3F_2(g) \longrightarrow 2AlF_3(s)$
   Ans. −2850 kJ
   (2) $N_2(g) + O_2(g) \longrightarrow 2NO(g)$
   Ans. 173.14 kJ
   (3) $CaO(s) + H_2O(l) \longrightarrow Ca(OH)_2(s)$
   Ans. −55.4 kJ
   (4) $Fe_2O_3(s) + 3CO(g) \longrightarrow 2Fe(s) + 3CO_2(g)$
   Ans. −29.4 kJ

(5) $2LiOH(s) + CO_2(g) \longrightarrow Li_2CO_3(s) + H_2O(l)$
    Ans. -87.4 kJ

(6) $CaSO_4 \cdot 2H_2O(s) \longrightarrow CaSO_4(s) + 2H_2O(g)$
    Ans. 18.2 kJ

(7) $CH_4(g) + N_2(g) \longrightarrow HCN(g) + NH_3(g)$
    Ans. 159 kJ

(8) $CS_2(g) + 3Cl_2(g) \longrightarrow CCl_4(g) + S_2Cl_2(g)$
    Ans. -159.0 kJ

(b) Which of those reactions are spontaneous?
    Ans. 1,3,4,5,8

(c) For which of these reactions does the value of $\Delta H°$ favor spontaneity (exercise 11)? The value of $\Delta S°$ (exercise 24)?

38. The standard molar enthalpies of formation of $NO(g)$, $NO_2(g)$, and $N_2O_3(g)$ are 90.25, 33.2, and 83.72 kJ mol^{-1}, respectively. Their standard entropies are 210.65, 239.9, and 312.2 J mol^{-1} K^{-1}, respectively.

(a) Use this data to calculate the standard state free energy change for the following reaction at 25°C.

$$NO(g) + NO_2(g) \longrightarrow N_2O_3(g)$$
    Ans. 1.6 kJ

(b) Repeat the calculation for 0°C and 100°C, assuming that the enthalpy and entropy changes do not change with temperature. Is the reaction spontaneous at 0.00°C? at 100.0°C?
    Ans. -1.9 kJ, 11.9 kJ; spontaneous at 0°C, but not at 100°C

39. For a certain process at 300 K, $\Delta G = -17.0$ kJ and $\Delta H = 6.9$ kJ. Find the entropy change for this process at 300 K.
    Ans. 79.7 J K^{-1}

40. Hydrogen chloride, $HCl(g)$, and ammonia, $NH_3(g)$, escape from bottles of their solutions and react to form the white glaze often seen on glass in chemistry laboratories.

$$HCl(g) + NH_3(g) \longrightarrow NH_4Cl(s)$$

(a) Calculate the free energy change, $\Delta G°$, for this reaction.
    Ans. -89.7 kJ

(b) At what temperature will $\Delta G°$ for the reaction be equal to zero?   *Ans. 618.6 K (345.4°C)*

Free Energy Changes and Equilibrium Constants

41. Explain why equilibrium constants change with temperature.

42. No matter what their bond energy, all compounds will decompose if heated to a sufficiently high temperature. Why will the reaction $AB \rightarrow A + B$, where A and B represent atoms, eventually become spontaneous with $K > 1$ as the temperature of the system is increased?

43. Calculate the equilibrium constant for the decomposition of solid $CaSO_4 \cdot 2H_2O$ to solid $CaSO_4$ and water vapor at 25°C. At 300°C.   *Ans. 6.5×10^{-4}; 3.9×10^5*

44. Calculate the equilibrium constant at 25°C for the decomposition of solid NH_4Cl to $HCl(g)$ and $NH_3(g)$. $\Delta G°$ for the reaction is 89.7 kJ (exercise 40).   *Ans. 2×10^{-16}*

45. Calculate the equilibrium constant for the decomposition of solid NH_4Br to $HBr(g)$ and $NH_3(g)$ at 25°C.
    Ans. 4×10^{-19}

46. Consider the reaction

$$2ICl(g) \longrightarrow I_2(g) + Cl_2(g)$$

(a) For this reaction $\Delta H°_{298} = 26.9$ kJ and $\Delta S°_{298} = -11.3$ J K^{-1}. Calculate $\Delta G°_{298}$ for the reaction.
    Ans. 30.3 kJ

(b) Calculate the equilibrium constant for this reaction at 25°C.   *Ans. 4.9×10^{-6}*

47. If the entropy of vaporization of H_2O is equal to 109 J mol^{-1} K^{-1} and the enthalpy of vaporization is 40.62 kJ mol^{-1}, calculate the normal boiling temperature of water in °C.   *Ans. 100°C*

48. Will the conversion of graphite to diamond become spontaneous at any temperature?   *Ans. No; $\Delta H > 0$, $\Delta S < 0$*

49. The standard molar enthalpy of formation of $BaCO_3(s)$ is -1219 kJ mol^{-1} and the standard free energy of formation is -1139 kJ mol^{-1}. For the decomposition of $BaCO_3(s)$ into $BaO(s)$ and $CO_2(g)$ at 1 atm,

(a) Estimate the temperature at which the equilibrium pressure of CO_2 would be 1 atm.
    Ans. 1560 K (1290°C)

(b) Calculate the equilibrium vapor pressure of $CO_2(g)$ above $BaCO_3(s)$ in a closed container at 298 K.
    Ans. 1×10^{-38} atm

50. If the enthalpy of vaporization of CH_2Cl_2 is 29.0 kJ mol^{-1} at 25.0°C and the entropy of vaporization is 92.5 J mol^{-1} K^{-1}, calculate the normal boiling temperature of CH_2Cl_2, the temperature at which the equilibrium pressure of the vapor is 1 atm.   *Ans. 314 K (41°C)*

51. (a) What is the vapor pressure of methyl alcohol, $CH_3OH(l)$, at 25°C?   *Ans. 0.17 atm*

(b) What is the boiling point of methyl alcohol under an external pressure of 0.450 atm?   *Ans. 318 K (45°C)*

52. The equilibrium constant, K_p, for the reaction $N_2O_4(g) \rightarrow 2NO_2(g)$ is 0.142 at 298 K. What is $\Delta G°_{298}$ for the reaction?   *Ans. 4.84 kJ*

53. Calculate $\Delta G°_{298}$ for the following reaction from the equilibrium constant and temperature given.

$$I_2(aq) + I^-(aq) \longrightarrow I_3^-(aq) \qquad T = 25°C, K = 771$$
    Ans. -16.5 kJ

54. (a) Calculate $\Delta G°$ for each of the following reactions from the equilibrium constant at the temperature given.

(1) $N_2(g) + O_2(g) \longrightarrow 2NO(g)$ $\qquad T = 2000°C$, $K_p = 4.1 \times 10^{-4}$

(2) $H_2(g) + I_2(g) \longrightarrow 2HI(g)$ $\qquad T = 400°C$, $K_p = 50.0$

(3) $CO_2(g) + H_2(g) \longrightarrow CO(g) + H_2O(g)$ $\qquad T = 980°C$, $K_p = 1.67$

(4) $CaCO_3(s) \longrightarrow CaO(s) + CO_2(g)$ $\qquad T = 900°C$, $K_p = 1.04$
    Ans. 147; -21.9; -5.343; -0.3826 kJ

(b) Assume that $\Delta H°$ does not vary with temperature and calculate $\Delta S°$ for these reactions.

Ans. 14.7; 111; 37.1; 152 J K^{-1}

55. Calculate $\Delta G°_{298}$ for the reaction of 1 mol of $H^+(aq)$ with 1 mol of $OH^-(aq)$, using the equilibrium constant for the self-ionization of water at 298 K.

$$H_2O \longrightarrow H^+(aq) + OH^-(aq), \qquad K_w = 1.00 \times 10^{-14}$$

Ans. 79.9 kJ

56. At what temperature does the decomposition of dinitrogen trioxide become spontaneous?

$$N_2O_3(g) \longrightarrow NO(g) + NO_2(g)$$

Ans. 287 K (14°C)

57. Hydrogen sulfide, a pollutant found in some natural gas, is removed by the reaction

$$2H_2S(g) + SO_2(g) \longrightarrow 3S(s) + 2H_2O(g)$$

What is the equilibrium constant for this reaction at 25°C? Is the reaction endothermic or exothermic?

Ans. 5 × 10^{15}; exothermic

Additional Exercises

58. In 1774 oxygen was prepared by Joseph Priestley by heating red mercury(II) oxide with sunlight focused through a lens. How much heat is required to decompose enough $HgO(s)$ to form 200 mL of $O_2(g)$ at 25°C and 1.00 atm?

Ans. 1.49 kJ

59. The structure of solid NaCl and a solution of NaCl are illustrated in Figure 9.1. (a) What is the sign of ΔS for the reaction $NaCl(s) \rightarrow NaCl(aq)$? (b) Sodium chloride spontaneously dissolves in water. What is the sign of ΔG for this reaction? (c) The heat of solution of $NaCl(s)$ is 3.88 kJ mol^{-1}. Is this consistent with a spontaneous reaction? (d) What is the driving force for the reaction?

60. For the vaporization of bromine liquid to bromine gas, $Br_2(l) \rightarrow Br_2(g)$, at 25°C,

(a) Calculate the change in enthalpy and the change in entropy at standard state conditions.

Ans. 30.91 kJ; 93.12 J K^{-1}

(b) Discuss the relative disorder in bromine liquid compared to bromine gas. State what you can about the spontaneity of the vaporization.

(c) Estimate the value of $\Delta G°_{298}$ for the vaporization of bromine from the values of ΔH and ΔS determined in (a).

Ans. 3.146 kJ

(d) State what you can about the spontaneity of the process from the value you obtained for ΔG in part (c).

(e) Estimate the temperature at which liquid Br_2 and gaseous Br_2 with a pressure of 1 atm are in equilibrium

(assume that $\Delta H°$ and $\Delta S°$ are independent of temperature).

Ans. 331.9 K (58.8°C)

(f) State in which direction the process would be spontaneous at 298 K and at 398 K, using the temperature value obtained in part (e).

(g) Compare $\Delta H°$, $\Delta S°$, and $\Delta G°$ in terms of their usefulness in predicting the spontaneity of the vaporization of bromine.

61. Carbon dioxide decomposes into CO and O_2 at elevated temperatures. What is the equilibrium partial pressure of oxygen in a sample at 1000°C for which the initial pressure of CO_2 was 1.15 atm?

Ans. 1.3 × 10^{-5} atm

62. Carbon tetrachloride, an important industrial solvent, is prepared by the chlorination of methane, $CH_4(g) + 4Cl_2(g) \longrightarrow CCl_4(g) + 4HCl(g)$, at 850 K.

(a) What is the equilibrium constant for the reaction at 850 K?

(b) Will the reaction vessel need to be heated or cooled to keep the temperature of the reaction constant?

Ans. $K_p = 2.05 \times 10^{23}$; cooled, $\Delta H° = -397.3$ kJ

63. Acetic acid, CH_3CO_2H, can form a dimer, $(CH_3CO_2H)_2$, in the gas phase.

$$2CH_3CO_2H(g) \longrightarrow (CH_3CO_2H)_2(g)$$

The dimer is held together by two hydrogen bonds,

with a total strength of 66.5 kJ per mole of dimer. At 25°C the equilibrium constant for the dimerization is 1.3×10^3 (pressure in atm). What is ΔS for the reaction?

Ans. −163 J K^{-1}

64. At 1000 K the equilibrium constant for the decomposition of bromine molecules, $Br_2(g) \rightleftharpoons 2Br(g)$, is 2.8×10^4 (pressure in atm). What is $\Delta G°$ for the reaction? Assume that the bond energy of Br_2 does not change between 298 K and 1000 K and calculate the approximate value of $\Delta S°$ for the reaction at 1000 K.

Ans. $\Delta G° = -85.2$ kJ; $\Delta S° = 278$ J K^{-1}

65. Nitric acid, NHO_3, can be prepared by the following sequence of reactions:

$$4NH_3(g) + 5O_2(g) \longrightarrow 4NO(g) + 6H_2O(g)$$
$$2NO(g) + O_2(g) \longrightarrow 2NO_2(g)$$
$$3NO_2(g) + H_2O(l) \longrightarrow 2HNO_3(l) + NO(g)$$

How much heat is evolved when 1 mol of $NH_3(g)$ is converted to $HNO_3(l)$? Assume that all reactants and products are in their standard states at 25°C and 1 atm.

Ans. 307.3 kJ

A Survey of the Nonmetals

20

Enormous quantities of
sulfur are produced
each year.

The nonmetals are the elements located in the upper right-hand portion of the Periodic Table. At room temperature and pressure they exist as gases, one liquid, and some of the softest and hardest of solids. The nonmetals exhibit a rich variety of chemical behavior. They include the most reactive and the most nonreactive of the elements and form many different ionic and covalent compounds, many of which are acids and bases.

In this chapter we present an overview of the chemical behavior of the nonmetals so that you can get a feel for their overall behavior before you deal with one or more of the specific elements. The chemical behavior of selected nonmetals is presented in more detail in subsequent chapters.

20.1 The Nonmetals

As noted in Chapter 9, there are no distinct points at which changes from metallic to nonmetallic behavior occur in periods or groups in the Periodic Table. Consequently, one can quibble about whether some elements should be regarded as nonmetals or as semi-metals. We will consider hydrogen and those elements with electronegativities equal to or larger than that of hydrogen as nonmetals. The remaining elements that exhibit borderline nonmetallic character will be considered in our discussion of the semi-metals (Chapter 25). The noble gases will be included among the nonmetals. Even though their electronegativities are not known, they are not metals, and the limited number of compounds they form are covalent and exhibit behavior characteristic of nonmetals. For our purposes the nonmetals are hydrogen (Group IA or VIIA); carbon (Group IVA); nitrogen, phosphorus, and arsenic (Group VA); oxygen, sulfur, and selenium (Group VIA); fluorine, chlorine, bromine, iodine, and astatine (Group VIIA); and helium, neon, argon, krypton, xenon, and radon (Group VIIIA). (See Fig. 20.1.)

The chemistry of carbon is extremely varied; it is described in detail in two subsequent chapters. Chapter 30 introduces organic chemistry, the chemistry of compounds that contain carbon-carbon bonds. Chapter 31 discusses biochemistry, the chemistry of carbon compounds found in living systems. In this chapter we will concentrate on the inorganic chemistry of carbon.

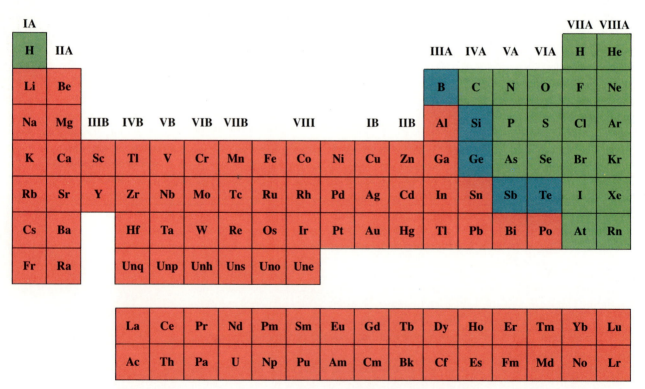

Figure 20.1

The location of the nonmetals (shown in green) in the Periodic Table. Semi-metals are shown in blue, metals in red.

20.2 The General Behavior of the Nonmetals

Although the nonmetals exhibit widely differing kinds of chemical behavior, there are similarities and trends in their behavior and in the behavior of their compounds that can help simplify your classification and recall of this chemistry.

Nonmetals do not form monatomic positive ions; their ionization energies are too large (Section 5.15). Thus the only monatomic ions formed by nonmetals are negative ions such as the iodide ion, I^-, or nitride ion, N^{3-}, and these are formed only with active metals, transition metals, or post-transition metals with relatively low oxidation numbers. Nonmetals form covalent bonds with other nonmetals and with some transition metals or post-transition metals with oxidation numbers higher than +3.

The common oxidation numbers that the nonmetals exhibit in their compounds have been discussed in Section 9.6. These are reproduced in Table 20.1. Remember that an element exhibits a positive oxidation number when it is combined with a more electronegative element and a negative oxidation number with a less electronegative element.

The variety of oxidation numbers displayed by most of the nonmetals means that many of their chemical reactions involve changes in these oxidation numbers in oxidation-reduction reactions (Section 9.3). A number of general aspects of the oxidation-reduction chemistry of the nonmetals simplify recall:

1. Elemental nonmetals oxidize most elemental metals (Fig. 20.2). The oxidation number of the metal becomes positive as it is oxidized and that of the nonmetal becomes negative.

2. With the exception of nitrogen and carbon, which are poor oxidizing agents, a more electronegative nonmetal will oxidize a less electronegative nonmetal or the anion of the nonmetal (Fig. 20.3). Fluorine and oxygen are the strongest oxidizing agents within their respective groups; either will oxidize another element lying below it in the group. Within a period the strongest oxidizing agent is found in Group VIIA, and a nonmetal will often oxidize another element lying to its left in the same period.

3. The stronger a nonmetal is as an oxidizing agent, the more difficult it is to remove electrons from the anion formed by the nonmetal. Thus the most stable negative ions are formed by elements at the top of their group or in Group VIIA of their period.

Figure 20.2
Hot iron powder is vigorously oxidized by the oxygen of air, forming iron oxides.

Table 20.1 Common oxidation numbers of the nonmetals in compounds, by group

IA	IVA	VA	VIA	VIIA	VIIIA
H	C	N	O	F	
+1	+4	+5	−1	−1	
−1	to	to	−2		
	−4	−3			
		P-As	S-Se	Cl-I	Xe
		+5	+6	+7	+8
		+3	+4	+5	+6
		−3	−2	+3	+4
				+1	+2
				−1	

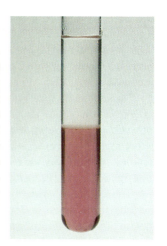

Figure 20.3

When it burns in air, sulfur is oxidized by the more electronegative element oxygen (left). Chlorine oxidizes I$^-$ to I$_2$, as indicated by the purple color of I$_2$ extracted into the lower layer (right).

Most of the oxides of the nonmetals are acidic. They react with water to form oxyacids, and many react with oxide ions or other bases in Lewis acid-base reactions. The notable exceptions are carbon monoxide, CO, nitrous oxide, N$_2$O, nitric oxide, NO, and water, H$_2$O, itself. The behavior of the acidic oxides may be generalized as follows:

1. Those oxides such as SO$_2$ and N$_2$O$_5$, in which the nonmetal exhibits one of its common oxidation numbers, react with water with no change in oxidation number. The product is an oxyacid. For example,

$$SO_2 + H_2O \longrightarrow H_2SO_3$$
$$N_2O_5 + H_2O \longrightarrow 2HNO_3$$

2. Those oxides such as NO$_2$ and ClO$_2$, in which the nonmetal does not exhibit one of its common oxidation numbers, react with water. In this reaction the nonmetal is both oxidized and reduced. For example,

$$3NO_2 + H_2O \longrightarrow 2HNO_3 + NO$$
$$6ClO_2 + 3H_2O \longrightarrow 5HClO_3 + HCl$$

Reactions in which the same element is both oxidized and reduced are called **disproportionation reactions.**

3. The acid strength of an oxyacid increases as the electronegativity of the central atom in the acid increases and as the oxidation number of the central atom increases (Section 16.6).

The binary hydrides of the nonmetals also exhibit acidic behavior, although only HCl, HBr, and HI are strong acids. The acid strength of the nonmetal hydrides increases from left to right across a period and down a group (Section 16.6).

The chemical behavior of carbon, nitrogen, oxygen, and fluorine, the lightest members of their respective groups in the periodic table, differs significantly from that of the remaining members of their respective groups in the following ways:

1. The valence electron shell of each of these atoms consists of 2s and 2p orbitals. These atoms are limited to a maximum of eight electrons in their valence shell (Section 6.4).

2. Carbon, nitrogen, and oxygen form stronger π bonds than the heavier members of their groups. As a consequence, molecules containing these elements may possess double or triple bonds, whereas those containing only the heavier nonmetals possess single bonds (see Section 20.3).

3. Nitrogen, oxygen, and fluorine are resistant to oxidation. Oxygen and fluorine are much stronger oxidizing agents than the heavier members of their groups, and these two elements cannot be oxidized to the maximum positive oxidation number possible for their respective groups.

4. Nitrogen, oxygen, and fluorine are the three elements with the largest electronegativities and are much more electronegative than the other elements in their respective groups.

5. Fluorine does not form compounds in which it exhibits a positive oxidation number. Oxygen exhibits a positive oxidation number only when combined with fluorine; nitrogen, only when combined with oxygen and/or fluorine.

20.3 Structures of the Elemental Nonmetals

The structures of the elemental nonmetals differ dramatically from those of metals. Metals crystallize in closely packed arrays with no obvious molecules or covalent bonds (Section 11.15). On the other hand, the presence of covalent bonds is obvious in the structures of the nonmetals, and many nonmetals consist of individual molecules.

The noble gases are found as monatomic gases. Hydrogen, nitrogen, oxygen, and the halogens exist as the diatomic molecules H_2, N_2, O_2, F_2, Cl_2, Br_2, I_2, and At_2, respectively. Elemental oxygen also occurs as the unstable allotrope **ozone, O_3,** a triatomic molecule (Fig. 20.4). (**Allotropes** are two forms of the same element with different structures in the same state.) All of these molecules are gases at room temperature except bromine, which is a liquid, and iodine and astatine, which are solids. The remaining nonmetals are solids with extensive covalent bonding.

Elemental carbon exists in two allotropic forms, **diamond** and **graphite** (Fig. 20.5). Charcoal, coke, and carbon black are microcrystalline, or amorphous, forms of carbon with the graphite structure.

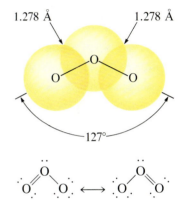

Figure 20.4

The molecular structure of ozone. Its electronic structure is described by the resonance forms indicated.

Figure 20.5

Diamond is colorless; graphite is black. A specimen of diamond ore is shown on the left.

Figure 20.6

The crystal structure of diamond. Each sphere represents a carbon atom, with the darker spheres indicating the atoms at the corners and the centers of the faces of a cubic unit cell, and the red spheres, the atoms in tetrahedral sites. The covalent bonds are wedge-shaped to help show the perspective. (b) The crystal structure of graphite.

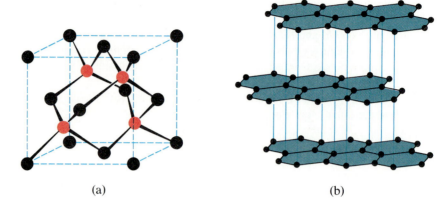

(a)   (b)

The atoms in a crystal of diamond [Fig. 20.6(a)] are bonded into a single giant molecule by single bonds. Each atom is covalently bonded to four others at the corners of a tetrahedron (sp^3 hybridization). The bonds are very strong, extending throughout the crystal in three dimensions, making the crystal very hard and giving it a high melting point (probably the highest of all elements). Diamond does not conduct electricity because it has no mobile electrons; they are all used in bond formation.

Graphite, also called **plumbago,** or **black lead,** is distinctly different from diamond. It is a soft, grayish black solid that crystallizes in hexagonal plates that have a metallic luster and conduct electricity. Graphite is the more stable of the two crystalline forms of carbon.

The comparative softness and electrical conductivity of graphite are related to its structure [Fig. 20.6(b)]. A crystal of graphite consists of layers of carbon atoms, each atom having three atoms as near neighbors in a trigonal planar arrangement. An atom forms three σ bonds, one to each of its nearest neighbors in the layer, by formation of sp^2 hybrid orbitals. The unhybridized p orbital on each carbon atom projects above and below the layer (Fig. 20.7) and can overlap with the unhybridized p orbital on the adjacent carbon atoms. There is one electron for each unhybridized orbital, and these π electrons form π bonds to adjacent carbon atoms.

Many resonance formulas are necessary to describe the electronic structure of a graphite layer; two are shown in Fig. 20.8. The σ and π bonds bind the atoms tightly together within the layers. However, layers can be separated easily because only weak van der Waals forces hold them together. These weak bonds give graphite the flaky, soft character that makes it useful as the "lead" in lead pencils. The loosely held electrons of the mobile double bonds are responsible for the electrical conductivity and the black color of graphite.

Figure 20.7

The orientation of unhybridized p orbitals of the carbon atoms in graphite. Each p orbital is perpendicular to the plane of carbon atoms.

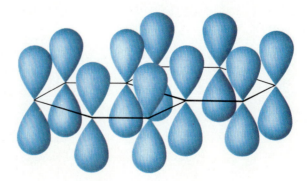

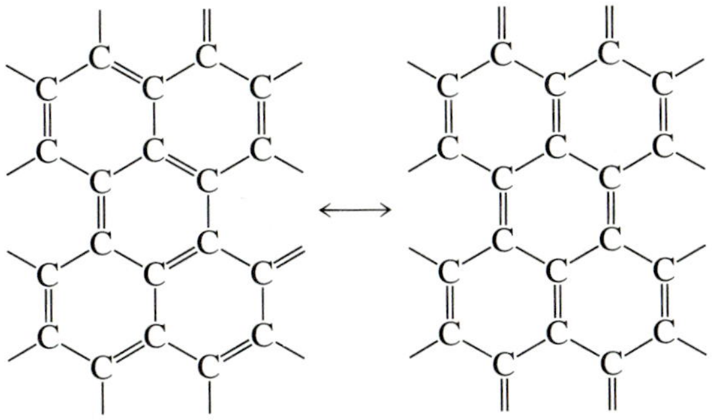

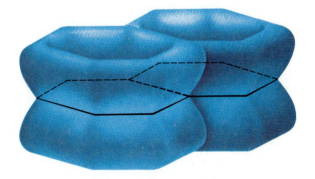

Figure 20.8

Two of the resonance forms of graphite necessary to describe its electronic structure, which is a resonance hybrid.

The π bonding in graphite can also be described by molecular orbital theory (Chapter 7). The unhybridized p orbitals on each carbon atom overlap to give molecular orbitals that extend over the entire layer, so an electron in one of these orbitals can move over the entire layer. Since the electron can move, graphite conducts electricity. A portion of one molecular orbital is shown in Fig. 20.9.

Figure 20.9

A portion of one molecular orbital in graphite, which contains delocalized π electrons.

Phosphorus exists in many allotropic forms, although we will restrict our attention to only three of these: **white phosphorus, red phosphorus,** and **black phosphorus.** Phosphorus forms three single bonds in each of these forms.

White phosphorus is a white, translucent, waxlike solid. It is very soluble in carbon disulfide, less so in ether, chloroform, and other organic solvents, and very nearly insoluble in water. It melts at 44.2°C and boils at 280°C. Either as a solid or in solution, white phosphorus exists as P_4 molecules. The four phosphorus atoms are arranged at the corners of a regular tetrahedron, and each atom is covalently bonded to the three other atoms of the molecule by single covalent bonds [Fig. 20.10(a)]. Phosphorus vapor just above the boiling point contains P_4 molecules. Above 800°C, P_2 molecules form in the vapor; these are believed to have the electronic structure $:P\equiv P:$, which is similar to that of the N_2 molecule.

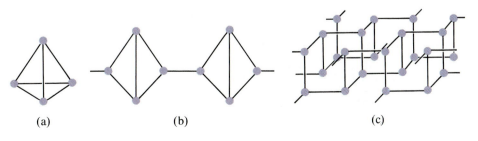

(a)   (b)   (c)

Figure 20.10

(a) The P_4 molecule found in white phosphorus. (b) The chain structure of red phosphorus. (c) The sheet structure of black phosphorus.

Red phosphorus is made by heating white phosphorus between 230°C and 300°C, with air excluded. It is insoluble in the solvents that dissolve white phosphorus. The structure of red phosphorus is not definitely known, but it is believed to consist of chains of P_4 tetrahedra formed from P_4 molecules by the rupture of one P—P bond and a subsequent linking of the ruptured tetrahedra by single bonds [Fig. 20.10(b)]. However, the P—P bond is relatively weak, so the bonds between P_4 units break when red phosphorus is heated and P_4 molecules sublime from the solid.

The presence of chains of P_4 tetrahedra renders red phosphorus much less soluble and less reactive than white phosphorus. White phosphorus can inflame spontaneously upon exposure to air. Red phosphorus does not ignite in air unless it is heated to about 250°C. However, the products of the reactions of red phosphorus are the same as those of the white form.

Black phosphorus is made by heating white phosphorus under high pressure. The structure of black phosphorus [Fig. 20.10(c)] contains puckered sheets of phosphorus atoms with single bonds to each of three other atoms. As in red phosphorus, this interconnected structure gives black phosphorus a higher melting point, higher boiling point, and lower reactivity than white phosphorus.

Arsenic also crystallizes with a structure that contains sheets of arsenic atoms with single bonds to three other atoms within the sheet. This structure is similar to that of black phosphorus.

Sulfur exists in several allotropic forms. Naturally occurring sulfur is a yellow solid, called **rhombic sulfur** because of the shape of its crystals [Fig. 20.11(a)]. This form is soluble in carbon disulfide. Crystals of rhombic sulfur melt at 113°C and form a straw-colored liquid. When this liquid cools and crystallizes, long needles of **monoclinic sulfur** are formed [Fig. 20.11(b)]. This is the stable form above 96°C. It melts at 119°C and is soluble in carbon disulfide. At room temperature it gradually changes to the rhombic form.

Rhombic sulfur, monoclinic sulfur, the straw-colored liquid, and solutions of rhombic sulfur in carbon disulfide all contain S_8 molecules in which the atoms form

Figure 20.11
(below) Crystals of rhombic sulfur.
(right) Crystals of monoclinic sulfur.

     (a)   (b)

Figure 20.12

(a) An S_8 molecule and (b) a chain of sulfur atoms.

eight-membered, puckered rings [Fig. 20.12(a)]. Each atom of sulfur is linked to each of its two neighbors in the ring by single covalent bonds.

The straw-colored liquid form of sulfur is quite mobile; its viscosity is low because S_8 molecules are essentially spherical and offer relatively little resistance as they move past each other. As the temperature rises, S—S bonds in the rings break and chains of sulfur atoms result [Fig. 20.12(b)]. These chains combine end to end, forming still longer chains that become entangled with one another, so the viscosity of the liquid increases. The liquid gradually darkens in color and becomes so viscous that finally (at about 230°C) it will not pour easily. The unpaired electrons at the ends of the chains of sulfur atoms are responsible for the dark red color. When the liquid is cooled rapidly, a rubberlike amorphous mass, called **plastic sulfur,** results. This form is insoluble in carbon disulfide. After standing for several days at room temperature, plastic sulfur, like all of the other allotropes, changes to the rhombic form.

Sulfur boils at 445°C and forms a vapor consisting of S_2, S_6, and S_8 molecules; at about 1000°C the vapor density corresponds to the formula S_2.

The structural behavior of selenium is similar to that of sulfur. Selenium crystallizes in a form that contains Se_8 rings. This allotrope is unstable and transforms to a stable form with chains of Se atoms.

The structural behavior of the nonmetals illustrates the following two important features of their chemistry.

1. Combined atoms of the nonmetals are usually found with eight electrons in their valence shells. As individual atoms combine to give samples of the elements, they form enough covalent bonds to supplement the electrons already present in the atom. For example, in both diamond and graphite, carbon forms four covalent bonds and thus supplements its four valence electrons. The members of Group VA have five valence electrons and require only three additional electrons to fill their valence shell. These elements form three covalent bonds in their free state; triple bonds in the N_2 and P_2 molecules and three single bonds to three different atoms in arsenic and in each of the allotropes of phosphorus containing the P_4 unit. The elements of Group VIA require only two additional electrons. Oxygen forms a double bond in the O_2 molecule, and sulfur, selenium, and tellurium form two single bonds in various rings and chains. The halogens form molecules containing one single bond per atom in order to provide the one electron required to supplement the seven electrons in the valence shells of the individual atoms. An atom of a noble gas contains eight electrons in its valence shell, and noble gases do not form covalent bonds to other noble gas atoms.

2. As noted in Section 20.2, only carbon, nitrogen, and oxygen atoms form strong multiple bonds. Multiple bonds are observed in the structures of these elements. Double bonds occur in the resonance forms of graphite (Fig. 20.8); N_2 contains a triple bond; and O_2 contains a double bond. Other elements form multiple bonds, but these are not as strong as the corresponding bonds in N_2 and O_2. Phosphorus forms a weak triple bond in the gaseous P_2 molecule and sulfur, a weak double bond in the gaseous S_2 molecule. However, these are obtained only when energy is added by heating the more stable singly bonded structures.

Chemical Behavior of the Nonmetals

20.4 Hydrogen

An uncombined hydrogen atom consists of one proton and one valence electron in the $1s$ orbital. The assignment of hydrogen to a periodic group is not clear-cut. It can be considered a member of Group IA because it has only one valence electron, but it can also be considered a member of Group VIIA because it is one electron short of having a filled valence shell. A hydrogen atom can fill its valence shell by picking up one electron and forming a hydride ion, H^-, or by sharing one electron and forming a single covalent bond.

At ordinary temperatures hydrogen is relatively inactive chemically, but when heated it enters into many chemical reactions. It reacts directly with the halogens, oxygen, sulfur, and nitrogen, producing covalent hydrides. These hydrides (as well as hydrides of elements that do not react *directly* with hydrogen) can also be formed by the reaction of an anion of the respective nonmetal with an acid. With the exception of the halide ions, all monatomic negative ions such as the hydride ion, H^-, nitride ion, N^{3-}, and phosphide ion, P^{3-}, are very strong bases and pick up hydrogen ions even from very weak acids such as water.

When heated, hydrogen reacts with the elements of Group IA and with Ca, Sr, and Ba (the more active elements in Group IIA) and forms crystalline ionic hydrides that contain hydrogen as the anion H^-. Hydrogen is absorbed (dissolved) by many transition metals, lanthanide metals, and actinide metals.

Hydrogen gas, H_2, is a reducing agent (Fig. 20.13). It reduces the heated oxides of many metals, with the formation of the metal and water (Section 9.7). For example, when hydrogen is passed over heated CuO, copper and water are formed.

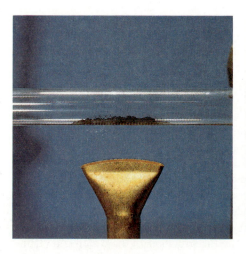

Figure 20.13

(left) A sample of CuO. (right) Copper produced from the CuO by heating in an atmosphere of H_2.

20.5 Carbon

Carbon is the first member of Group IVA; the other members are silicon, germanium, tin, and lead. Carbon is predominantly nonmetallic in character. Silicon and germanium are semi-metals (see Chapter 25); tin and lead are post-transition metals (see Chapter 28). Although their properties vary widely, gaseous atoms of each of these elements have four valence electrons with a valence shell electron configuration of ns^2np^2.

With few exceptions, carbon is covalent in its compounds. There are no d orbitals in the valence shell of carbon, so it can contain a maximum of eight valence electrons. Generally, all four of its valence electrons are used in bonding, with either sp, sp^2, or sp^3 hybridization at the carbon atom. Thus carbon can have a coordination number of 2, 3, or 4, with a linear, trigonal planar, or tetrahedral geometry, respectively.

$$H-C\equiv C-H \qquad :\ddot{O}=C=\ddot{O}: \qquad \overset{H}{\underset{H}{\diagdown}}C=\ddot{O}: \qquad H-\overset{\overset{\displaystyle H}{|}}{\underset{\underset{\displaystyle H}{|}}{C}}-H$$

Elemental carbon is not very reactive. It will burn and form carbon monoxide, CO, if there is not enough oxygen to oxidize it fully, or form carbon dioxide, CO_2, in an excess of oxygen. Carbon is also oxidized by sulfur at high temperatures, forming carbon disulfide, CS_2, and by fluorine forming carbon tetrafluoride, CF_4.

20.6 Nitrogen, Phosphorus, and Arsenic

Nitrogen is the first member of Group VA, which also includes phosphorus, arsenic, antimony, and bismuth. There is a regular gradation in the family, from the nonmetallic behavior of nitrogen, phosphorus, and arsenic to the semi-metallic behavior of antimony and the metallic behavior of bismuth. As in the case of fluorine, oxygen, and carbon, some of the properties of nitrogen are anomalous, and the expected trends are followed by the other members of the family.

The valence shell electron configuration of the gaseous atoms of Group VA elements is ns^2np^3, with the three p electrons distributed in p_x, p_y, and p_z orbitals. There are no d orbitals in the valence shell of nitrogen, so it contains a maximum of eight electrons. Nitrogen forms sp, sp^2, and sp^3 hybrid orbitals in covalent compounds. It is one of the most electronegative elements; only fluorine and oxygen have higher electronegativities.

Nitrogen, phosphorus, and arsenic fill their valence shells in several ways. Each can gain three electrons from the active metals of Groups IA and IIA, forming ionic compounds that contain the very strongly basic nitride ion, N^{3-}, phosphide ion, P^{3-}, and arsenide ion, As^{3-}, respectively. Transition metals and the less active metals also form nitrides, phosphides, and arsenides, but most of these are not ionic. The octet can be completed by the formation of three single bonds, as in NH_3 and PCl_3. Nitrogen readily forms multiple bonds—for example, in molecular nitrogen, N_2, cyanide ion, CN^-, and nitrous acid, HNO_2.

$$:N\equiv N: \qquad :C\equiv N:^- \qquad H-\ddot{O}-N\overset{\displaystyle \cdot\cdot}{\underset{\displaystyle \cdot\cdot}{\diagup}}\overset{\displaystyle O}{}$$

Aside from the stoichiometries of some of their simpler compounds (NH_3 and PH_3 or NF_3 and AsF_3, for example), marked differences exist between nitrogen and the heavier nonmetals of Group VA. Nitrogen is quite inert except at elevated temperatures, whereas phosphorus and arsenic are reactive even at low temperatures. The union of nitrogen with oxygen is an endothermic reaction (Fig. 20.14); with excess oxygen the product is nitrogen(IV) oxide, NO_2. Phosphorus and arsenic burn readily in either an excess of air or in oxygen, forming phosphorus(V) oxide, P_4O_{10}, and

Figure 20.14

Lightning provides the energy necessary for the reaction of nitrogen with oxygen in the atmosphere. About 40 million tons of N_2 are converted to NO by lightning each year.

arsenic(V) oxide, As_2O_5, respectively. Phosphorus and arsenic combine exothermically with the halogens; nitrogen does not react with chlorine, bromine, or iodine. When heated, phosphorus and arsenic combine with sulfur and many of the metals; nitrogen does not react with sulfur nor with most metals.

Compounds of nitrogen in all of the possible oxidation states from -3 to $+5$ are known. The most commonly observed oxidation states of phosphorus and arsenic are -3 with metals or less electronegative nonmetals and $+3$ and $+5$ with more electronegative nonmetals. Much of the chemistry of the elements of Group VA involves oxidation-reduction reactions, which convert these elements from one oxidation state to another.

20.7 Oxygen, Sulfur, and Selenium

Oxygen is the first member of Group VIA, which also includes the nonmetals sulfur and selenium, the semi-metal tellurium, and the metal polonium. Gaseous atoms of each element in Group VIA have six valence electrons, with valence shell electron configuration ns^2np^4. These atoms can fill their valence shells by picking up two electrons and forming an anion, X^{2-}, with two negative charges, or by sharing electrons in covalent bonds.

Oxygen is more electronegative than any element but fluorine, so it usually has an oxidation state of -2 in its compounds. The other, less electronegative elements of the group have both negative and positive oxidation numbers—most commonly -2, $+4$, and $+6$.

Oxygen exhibits pronounced nonmetallic behavior. It forms ionic compounds with metals and covalent compounds with nonmetals. Oxygen forms the oxide ion, O^{2-}, in many compounds with metals; it can also form two single covalent bonds as in water, H_2O, a double bond as in formaldehyde, $H_2C{=}O$, or a triple bond in carbon monoxide, $:C{\equiv}O:$.

Oxygen forms compounds with all of the elements except He, Ne, Ar, and Kr, and it generally has an oxidation number of -2 in these compounds. In compounds that contain O—O single bonds, oxygen is present with a -1 oxidation number. Elemental oxygen is a strong oxidizing agent and will eventually oxidize most organic substances—fortunately, at a slow rate. A fast oxidation rate would make life on earth impossible; all organic matter, including animal life, would burn up.

Both elemental sulfur and selenium are reactive elements, although generally less so than oxygen. Only the noble gases and the elements iodine, nitrogen, tellurium, gold, platinum, and palladium do not combine directly with elemental sulfur. Selenium is only a little less reactive than sulfur, but selenium compounds usually decompose more readily than the corresponding compounds of sulfur (or oxygen).

Sulfur and selenium also exhibit distinctly nonmetallic behavior. They oxidize metals (Fig. 20.15), giving metal sulfides and selenides, respectively. The simplest metal compounds contain the sulfide ion, S^{2-}, or selenide ion, Se^{2-}, respectively. Adding excess sulfur yields compounds containing the disulfide ion, S_2^{2-}, or polysulfide ions, for example, S_4^{2-} and S_5^{2-} (Section 22.12).

$$2Na + S \longrightarrow Na_2S$$
$$Na_2S + S \longrightarrow Na_2S_2$$
$$Na_2S + 4S \longrightarrow Na_2S_5$$

The diselenide ion, Se_2^{2-}, may form when metals react with an excess of selenium.

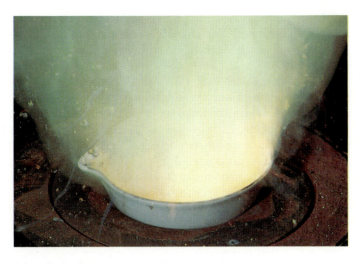

Both sulfur and selenium form compounds with nonmetals in which sulfur and selenium have two covalent bonds.

The oxidation number of sulfur or selenium in these compounds is either $+2$ or -2, depending upon whether the nonmetal is more electronegative (Cl) or less electronegative (C, H) than sulfur or selenium (Section 6.5). Sulfur and selenium also form compounds containing four, five, or six bonding electron pairs with more electronegative nonmetals (Section 6.4). In compounds such as SO_2, SeO_2, H_2SO_4, H_2SeO_4, $SOCl_2$, SeF_4, and SF_6, they exhibit an oxidation state of $+4$ or $+6$.

Sulfur and selenium are oxidized by nonmetals, such as oxygen and the halogens, that are more electronegative than they are. The reduced stability of the highest oxidation number of the heavier members of a group is illustrated by the reactions of sulfur and selenium with oxygen. When sulfur burns in air, it forms sulfur dioxide and a little sulfur trioxide. More sulfur trioxide will form when sulfur or sulfur dioxide is heated with oxygen in the presence of a catalyst. The heavier element selenium forms selenium dioxide when heated in air or oxygen, but selenium trioxide cannot be made by heating selenium or selenium dioxide with oxygen because it decomposes above 180°C.

20.8 Fluorine, Chlorine, Bromine, and Iodine

Fluorine, chlorine, bromine, iodine, and astatine constitute Group VIIA of the Periodic Table (the halogens). Astatine is a very radioactive element that decays rapidly; it will not be considered in this chapter.

The valence shell of a gaseous halogen atom has the electron configuration ns^2np^5. Thus a halogen atom forms only one single covalent bond with a less electronegative nonmetal, although it can form several bonds to atoms of the more electronegative nonmetals. In compounds with metals, a halogen atom generally accepts one electron to form a stable univalent negative ion. Like other nonmetals, the halogens are oxidizing agents. Fluorine is the strongest oxidizing agent of all of the elements. Halogens become progressively weaker oxidizing agents as their nonmetallic character decreases down the group.

All of the halogens except fluorine form some compounds in which they exhibit a positive oxidation number. With this one exception, the chemical properties of the halogens differ in degree rather than in kind. The high ionization potentials make the formation of positive halogen ions highly unlikely, except possibly for iodine. On the other hand, positive oxidation numbers of $+1$, $+3$, $+5$, and $+7$, and sometimes other oxidation numbers, are observed in covalent compounds of the halogens with oxygen in oxides, oxyacids, and salts of these oxyacids, or with more electronegative halogens in compounds such as ICl_3 or BrF_5. Since fluorine is the most electronegative element, it does not form compounds in which it has a positive oxidation number.

The halogens will oxidize a variety of metals, nonmetals, and ions, giving ionic or covalent products in which the halogen has an oxidation number of -1.

$$2Na + I_2 \longrightarrow 2NaI$$
$$2Al + 3Br_2 \longrightarrow 2AlBr_3$$
$$Co + Cl_2 \longrightarrow CoCl_2$$
$$P_4 + 10F_2 \longrightarrow 4PF_5$$

Like oxygen, fluorine oxidizes almost all of the other elements; it also oxidizes most compounds. Chlorine is a strong oxidizing agent, and bromine and iodine are good oxidizing agents.

An elemental halogen will oxidize less electronegative halogens (poorer oxidizing agents) and halide ions that are formed from less electronegative halogens. For example, fluorine will oxidize chlorine, bromine, and iodine (or their anions), whereas chlorine will only oxidize bromine and iodine (or their anions).

20.9 Helium, Neon, Argon, Krypton, Xenon, and Radon

These elements of Group VIIIA are essentially unreactive; only xenon has been shown to exhibit any extensive chemistry. Helium, neon, and argon exhibit no reactivity. Krypton reacts with fluorine, forming unstable KrF_2. Xenon is oxidized by fluorine, forming a series of stable but very reactive fluorides: XeF_2, XeF_4, and XeF_6, depending on the stoichiometry of the reaction. Xenon exhibits oxidation numbers of $+2$, $+4$, and $+6$, respectively, in these compounds. The disproportionation of xenon when XeF_6 reacts with hydroxide ion, forming XeO_6^{4-} and Xe, produces compounds containing xenon with an oxidation number of $+8$. All of the compounds of xenon are strong oxidizing agents and form xenon gas when they undergo oxidation-reduction reactions.

20.10 Compounds of the Nonmetals

The nonmetals form binary ionic compounds with the active metals and with most other metals with oxidation numbers of $+1$, $+2$, and $+3$. Nonmetals form covalent binary compounds with other nonmetals and semi-metals. Most metals with oxidation numbers of $+4$ or higher form covalent compounds, or compounds with an extensive component of covalent behavior, with the nonmetals.

1. HYDRIDES. Hydrogen compounds of the metals of Group IA and of Ca, Sr, and Ba are ionic compounds that contain hydrogen as the anion H^-. The hydride ion is a strong reducing agent and a strong base, so these hydrides react vigorously with water and other acids, forming hydrogen gas (Fig. 20.16).

$$H^- + H_2O \longrightarrow H_2 + OH^-$$

Hydrogen dissolves in many transition metals, lanthanide metals, and actinide metals, producing materials called *metallic hydrides* because they exhibit many metallic properties. For example, they are electrical conductors. These hydrides do not contain hydride ions; instead, hydrogen atoms are located in the spaces between the metal atoms, and the compounds are generally nonreactive.

Most hydrides of the nonmetals exhibit acidic behavior. The compounds HCl, HBr, and HI are strong acids; HF, H_2O, and H_2S are weak acids. Ammonia and phosphine are very weak acids, too weak to give up a proton to water; they usually function as bases. With the exception of acetylene, H—$C\equiv C$—H, and similar compounds containing a —$C\equiv C$—H grouping, the binary hydrides of carbon exhibit no acidic behavior in water. Acetylene is a very weak acid.

Chlorine, bromide, and iodide ions are very weak bases, and fluoride ion is a moderately weak base. All other monatomic negative ions, such as the hydride ion, H^-, nitride ion, N^{3-}, and sulfide ion, S^{2-}, are very strong bases and pick up hydrogen ions from acids (even weak acids such as water), forming the hydrides.

2. OXIDES Oxygen forms binary compounds with all of the elements except He, Ne, Ar, and Kr. Metals and nonmetals can exhibit their highest oxidation numbers in compounds with oxygen (and fluorine; see Part 3 of this section). The oxides of the alkali metals, of the alkaline earth metals, and of the transition metals and post-transition metals in their lower oxidation states ($+1$, $+2$, $+3$) are ionic in their behavior; they contain the oxide ion, O^{2-}. The oxides of the metals with high charges and small sizes or with high oxidation numbers are covalent in character. Nonmetal oxides are covalent.

The oxide ion is strongly basic, and all soluble and most insoluble *ionic* oxides react with acids, giving water and a salt of the metal. Ionic metal oxides that are soluble react with water to form metal hydroxides (Section 13.4). The soluble ionic oxides include most of the oxides of the active metals. Ionic transition metal and post-transition metal oxides are generally not soluble in water.

Most oxides of the nonmetals are acidic. Some nonmetal oxides, such as CO and N_2O, do not react with water, but most do, forming oxyacids.

Covalent oxides of the transition metals and post-transition metals with higher oxidation numbers also exhibit acidic behavior. These react with water and form acids or neutralize bases and form salts.

$$CrO_3(s) + H_2O \longrightarrow H_2CrO_4$$
$$SnO_2 + Ca(OH)_2 \longrightarrow CaSnO_3 + H_2O$$

The ease with which elemental oxygen picks up electrons is mirrored by the difficulty of removing electrons from oxygen in most oxides. Of the elements, only the very reactive fluorine molecule will oxidize oxides to form oxygen.

3. HALIDES. Halogens form compounds, usually called halides, with almost all of the other elements, including other halogens. Metals and nonmetals with their highest oxidation numbers are usually found as fluorides (or oxides). For example, sulfur exhibits an oxidation number of $+6$ in sulfur(VI) fluoride, SF_6, but the highest oxidation number for sulfur in a binary chloride is $+4$, found in sulfur(IV) chloride, SCl_4.

Figure 20.16
Calcium hydride reacts rapidly with water, producing $H_2(g)$ and $Ca(OH)_2$.

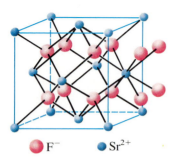

F^- Sr^{2+}

SrF$_2$, a solid

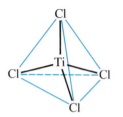

TiCl$_4$, a liquid

Figure 20.17
The structures of SrF$_2$, an ionic solid, and TiCl$_4$, a covalent molecule.

The highest bromide is sulfur(I) bromide, S_2Br_2, and sulfur iodides are not known. The halides become progressively better reducing agents as they become heavier, reflecting the increasing ease of removing electrons from larger atoms or ions. The iodide ion is a better reducing agent than the bromide ion, which in turn is a better reducing agent than the chloride ion. Thus the heavier a halide ion, the more likely it is to reduce a metal or nonmetal with a high oxidation number, and the less stable are halides of the heavier halogens with other elements with higher oxidation numbers.

The variation in properties among the halides ranges from those of ionic species to those of covalent molecules (Fig. 20.17). The halides of the alkali metals, of the alkaline earth metals, and of the transition metals in their lower oxidation states ($+1$, $+2$, $+3$) are ionic in their behavior. These ionic compounds tend to have high melting points (above 400°C) and high electrical conductivities in the molten state. Those that are soluble in water form hydrated metal ions and halide ions.

The halides of the nonmetals as well as those of the metals with high charges and small sizes and/or high oxidation states are covalent in character. In the covalent metal halides the attraction of the metal for the valence electrons of the halide ion produces the covalent behavior. These halides are characterized by low melting points, by volatility, by solubility in nonpolar solvents, and by lack of electrical conductivity in the molten state.

Except for the halides of carbon and nitrogen, most covalent halides react with water to form a covalent oxide or hydroxide and the corresponding hydrohalic acid, as indicated by the reactions of PI_3 and WCl_6 with water.

$$PI_3 + 3H_2O \longrightarrow P(OH)_3 + 3HI$$
$$WCl_6 + 3H_2O \longrightarrow WO_3 + 6HCl$$

The formula of $P(OH)_3$ is usually written as H_3PO_3, because the compound is an acid, the weak acid called phosphorous acid.

As might be expected, some metal halides are intermediate in character between the ionic and covalent ones. For example, iron(III) chloride, $FeCl_3$, is volatile and has a low melting point. It also reacts with water, giving an acid solution; however, it is a good conductor of electricity in the molten state.

For Review

Summary

The noble gases and elements with electronegativities greater than or equal to that of hydrogen are considered nonmetals. These elements are hydrogen (Group IA or VIIA); carbon (Group IVA); nitrogen and phosphorus (Group VA); oxygen, sulfur, and selenium (Group VIA); fluorine, chlorine, bromine, iodine, and astatine (Group VIIA); and helium, neon, argon, krypton, xenon, and radon (Group VIIIA). Compounds of the nonmetals with active metals, or with most other metals with oxidation numbers of $+1$, $+2$, and $+3$, are ionic and contain the nonmetal as a negative ion. Nonmetals form covalent compounds with other nonmetals and with metals with large charges and small sizes or high oxidation numbers. With the exception of the hydrides of C, N, and P, the hydrides of the nonmetals generally are acidic, as are the oxides. The nonmetals become progressively better oxidizing agents as their electronegativity increases (except for nitrogen, which is an anomalously weak oxidizing agent).

Hydrogen has the chemical properties of a nonmetal with a relatively low electro-negativity. It forms ionic hydrides with active metals, covalent compounds in which it has an oxidation number of -1 with less electronegative elements, and covalent compounds in which it has an oxidation number of $+1$ with more electronegative non-metals. It will react with the halogens, oxygen, sulfur, and nitrogen. Hydrogen will reduce the oxides of many metals, forming the metal and water.

Carbon (Group IVA) exhibits oxidation numbers ranging from -4 to $+4$. It is covalently bonded in the vast majority of its compounds. The only nonmetals that oxidize carbon are oxygen, sulfur, and fluorine; carbon is a relatively nonreactive nonmetal.

Nitrogen (Group VA) exhibits all oxidation numbers from -3 to $+5$. Phosphorus and arsenic commonly exhibit an oxidation number of -3 with active metals and $+3$ and $+5$ with more electronegative nonmetals. Phosphorus and arsenic are oxidized by halogens and by oxygen. Elemental nitrogen is not very reactive.

The nonmetals of Group VIA (O, S, Se) form compounds with almost all of the elements. Oxygen is a strong oxidizing agent, and sulfur and selenium are good oxidizing agents. Except in a few compounds with fluorine, oxygen exhibits negative oxidation numbers in all of its compounds; the most common is -2, in oxides. The -1 oxidation number is found in compounds that contain O—O bonds (peroxides). Sulfur and selenium exhibit oxidation numbers of -2, $+2$, $+4$, and $+6$.

The halogens are members of Group VIIA. The heavier halogens form compounds in which they have oxidation numbers of -1, $+1$, $+3$, $+5$, and $+7$, although they sometimes exhibit other oxidation numbers. Fluorine is a very potent oxidizing agent and forms compounds in which it always exhibits an oxidation number of -1. The oxidizing ability of the elemental halogens and the resistance of halide anions to oxidation decreases as halogen size increases.

The noble gas elements exhibit a very limited chemistry. Only xenon will react readily with fluorine, forming the xenon fluorides XeF_2, XeF_4, and XeF_6. Xenon oxides may be prepared by the reaction of water with the fluorides.

Key Terms and Concepts

acid-base behavior (20.2)	halide (20.10)	peroxide (20.7)
allotrope (20.3)	hydride (20.10)	polysulfide (20.7)
disproportionation (20.2)	oxide (20.10)	sulfide (20.7)
disulfide (20.7)		

Exercises

1. Both hydrogen and oxygen are colorless, odorless, and tasteless gases. How may we distinguish between them using chemical properties?
2. Both H_2 and O_2 are diatomic molecules. How do the chemical bonds in the molecules differ, and how are they similar?
3. Describe five chemical properties of sulfur, or of compounds of sulfur, that characterize it as a nonmetal.
4. Compare and contrast the chemical properties of elemental phosphorus and nitrogen.
5. With what elements can chlorine form binary compounds in which it exhibits a positive oxidation number?

6. Which of the following elements will exhibit a positive oxidation number when combined with sulfur? With oxygen?
   (a) Al   (b) Br   (c) Ca   (d) Cl
   (e) F   (f) N   (g) Si
7. With which elements will phosphorus form compounds in which it has a positive oxidation number?
8. Why does nitrogen form a maximum of four single covalent bonds?
9. From the positions of the elements in the Periodic Table, predict which of the following pairs will:
   (a) reduce S, Cl_2, or C

(b) neutralize H_2SO_4, Li_2O, or ClO_2
(c) oxidize S, P_4, or Cl_2
(d) oxidize P_4, Al, or Cl_2
(e) reduce Se, Zn, or O_2
(f) neutralize $NH_3(aq)$, Li_2O, or SO_2

10. Which anion in each of the following groups is most difficult to oxidize?
    (a) Cl^-, F^-, I^-
    (b) O^{2-}, S^{2-}, Se^{2-}
    (c) As^{3-}, N^{3-}, P^{3-}
    (d) F^-, N^{3-}, O^{2-}

11. Write a complete and balanced equation for the reaction of each of the following with water.
    (a) CO_2
    (b) N_2O_3
    (c) N_2O_5
    (d) P_4O_6
    (e) P_4O_{10}
    (f) SO_2
    (g) SO_3
    (h) Cl_2O_7
    (i) I_2O_5

12. Arrange the members of each of the following groups in order of increasing acid strength.
    (a) CH_4, NH_3, H_2O, HF
    (b) H_2O, H_2S, H_2Se
    (c) NH_3, PH_3, AsH_3
    (d) HI, H_2S, NH_3

13. Write a complete and balanced equation for the reaction of each of the following. In some cases there may be more than one correct answer, depending on the amounts of reactants used.
    (a) $NaF + H_2SO_4$
    (b) $Li_3N + H_2O$
    (c) $Mg_3P_2 + HCl$
    (d) $FeS + HCl$
    (e) $NiO + H_2SO_4$
    (f) $HC{\equiv}C^- + H_2O$
    (g) $Na_2O + H_2O$
    (h) $CaO + HF$

Hydrogen

14. Why does a hydrogen atom form only one single bond in compounds?

15. Why does hydrogen not exhibit an oxidation number of -1 when bonded to nonmetals?

16. Write a balanced equation for the reaction of an excess of hydrogen with each of the following: PtO, Na, Fe_2O_3, Cl_2, O_3.

17. Which of the binary hydrides of the nonmetals are strong acids? Which is the weakest acid?

18. The reaction of NaH with H_2O may be characterized as an acid-base reaction. Identify the acid and the base of the reactants. The reaction is also an oxidation-reduction reaction. Identify the oxidizing agent, the reducing agent, and the changes in oxidation number that occur in the reaction.

19. What mass of hydrogen would result from the reaction of 27.2 g of BaH_2 with water? *Ans. 0.787 g*

20. What mass of LiH is required to provide enough hydrogen by reaction with water to fill a balloon at 0°C and 1.0 atm pressure with a volume of 3.5 L? *Ans. 1.2 g*

Carbon

21. Why does carbon form a maximum of four single covalent bonds?

22. Describe the crystal structures of graphite and diamond and relate the physical properties of each to their structures.

23. (a) Compare and contrast the bonding in diamond and graphite.
    (b) Compare and contrast the bonding in graphite with that in ethylene and acetylene (Section 8.8).

24. Write the Lewis structure, including resonance forms where necessary, for each of the following molecules or ions: CO_2; CO_3^{2-}; CCl_4. Identify the hybridization of the carbon atom in each.

25. Write the chemical equation for the reaction of CO_2 with water. With an aqueous solution of NaOH.

26. What is ΔG°_{298} for the following reaction?

$$C(s,diamond) \longrightarrow C(s,\ graphite)$$

Is the reaction spontaneous? *Ans. -2.900 kJ; yes*

27. How could the heat of combustion of graphite and diamond be used to show that the conversion

$$C(s,\ graphite) \longrightarrow C(s,diamond)$$

is endothermic?

Nitrogen, Phosphorus, and Arsenic

28. Explain why nitrogen occurs as diatomic molecules, whereas the isoelectronic phosphorus and arsenic atoms form solids with sheets or chains of atoms.

29. Although PF_5 and AsF_5 are stable, nitrogen does not form NF_5 molecules. Explain this difference between members of the same group.

30. Write Lewis structures for the following:
    (a) N_2 and P_2
    (b) NH_3 and PH_3
    (c) NF_3, PF_3, and AsF_3
    (d) PF_5 and AsF_5
    (e) NH_4^+ and PH_4^+
    (f) NO_3^- [the stable oxyanion of nitrogen(V)] and PO_4^{3-} [the stable oxyanion of phosphorus(V)]

31. What are the products of the reaction between aluminum nitride and water?

32. Contrast the properties of white, red, and black phosphorus.

33. The bond energy of a phosphorus-phosphorus single bond is 215 kJ per mole of bonds and of a phosphorus-phosphorus triple bond, 490 kJ per mole of bonds. What is the approximate value of ΔH° for the following reaction?

$$P_4(g) \longrightarrow 2P_2(g)$$

Ans. 310 kJ

34. Complete and balance the following equations (*xs* indicates that the reactant is present in excess):
    (a) $P_4 + xsS \longrightarrow$
    (b) $P_4 + Li \longrightarrow$
    (c) $P_4 + xsO_2 \longrightarrow$
    (d) $P_4 + Mg \longrightarrow$
    (e) $P_4 + xsF_2 \longrightarrow$

35. Is an As-As triple bond stronger or weaker than three As-As single bonds. Explain your answer.

Oxygen, Sulfur, Selenium

36. Write the Lewis structures of the two allotropes of oxygen.
37. Why does oxygen form diatomic molecules, whereas sulfur forms eight membered rings?
38. What are the allotropic forms of solid sulfur and how may they be produced?
39. What are the molecular structures of sulfur found in the liquid, solid, and gaseous states?
40. Describe the oxygen-oxygen bonding in the peroxide ion in terms of the Lewis structure of O_2^{2-}.
41. Write Lewis structures for the following:
    (a) O_2 and S_2
    (b) H_2O, H_2S, and H_2Se
    (c) OF_2, SF_2, and SeF_2
    (d) O_2^{2-}, S_2^{2-}, and Se_2^{2-}
    (e) O_3 and SO_2
    (f) OH^-, SH^-, and SeH^-
42. There is an extensive chemistry of sulfur and selenium in which these elements exhibit positive oxidation numbers, but oxygen has a very limited chemistry with positive oxidation numbers. Explain.
43. Write a balanced equation for the reaction that occurs when each of the following is burned in air.
    (a) C   (b) Mg   (c) Se
    (d) Ge   (e) Al   (f) As
    (g) S (two reactions)
44. What mass and volume of $O_2(g)$ (density = 1.429 g/L) are required to prepare 41.3 g of MgO?
    Ans. 16.4 g; 11.5 L
45. Elements generally exhibit their highest oxidation numbers in compounds containing oxygen or fluorine. What is the oxidation number of osmium in 0.3789 g of an osmium prepared by the reaction of 0.2827 g of Os with O_2? Write the equation for the reaction that gives this compound.
    Ans. +8; $Os + 2O_2 \longrightarrow OsO_4$
46. Predict the products of and write balanced equations for the reactions of each of the following with oxygen, sulfur, and selenium. (There may be more than one correct answer, depending on your choice of stoichiometries.)
    (a) Li   (b) Ga   (c) P
    (d) H_2 (e) Si   (f) F_2
    (g) C
47. Write chemical equations describing three chemical reactions in which elemental sulfur acts as an oxidizing agent. Three in which it acts as a reducing agent.
48. Determine the oxidation number of sulfur in each of the following species:
    (a) H_2S (b) SCl_2 (c) SO_2
    (d) SF_6 (e) SO_3 (f) Na_2S
    (g) $CaSO_3$
49. Why are solutions of oxides or of sulfides basic? Write equations.
50. How does the hybridization of the sulfur atom change when S_8 reacts with oxygen to form SO_2? With F_2 to form SF_6? With Cl_2 to form SCl_4?

Halogens

51. Why does fluorine not form allotropes like oxygen does?
52. (a) Which of the halogen atoms has the lowest first ionization energy?
    (b) Which of the halide ions is the strongest base?
53. Write Lewis structures for the following.
    (a) HF, HCl, HBr, and HI
    (b) ClF, Cl_2, ClBr, and ClI
    (c) PF_3, PCl_3, PBr_3, and PI_3
    (d) CF_4, CCl_4, CBr_4, and CI_4
    (e) KF, KCl, KBr, and KI
    (f) HOCl, HOBr, and HOI
    (g) HOClO, HOBrO, and HOIO
    (h) ClF_3, BrF_3, and IF_3
54. Write balanced chemical equations that describe the reactions of chlorine with each of the following:
    (a) lithium
    (b) magnesium
    (c) hydrogen
    (d) an excess of phosphorus
    (e) phosphorus (with an excess of Cl_2)
    (f) iodine (with a 1:1 and with a 1:3 $I_2:Cl_2$ ratio)
    (g) sulfur (with an excess of Cl_2)
55. Using the reactions of the halogens with sulfur, show that the oxidizing abilities of the halogens decrease as their size increases.
56. Write chemical equations describing the changes that occur in each of the following cases.
    (a) Sulfur is burned in a fluorine atmosphere.
    (b) Potassium bromide is treated with chlorine.
    (c) Phosphorus is heated with an excess of bromine.
    (d) Hydrogen and fluorine are mixed.
    (e) Aluminum is heated with iodine.
57. Write a balanced chemical equation describing the reaction that occurs in each of the following cases.
    (a) Calcium is added to hydrobromic acid.
    (b) Potassium hydroxide is added to hydrofluoric acid.
    (c) Ammonia is bubbled through hydrofluoric acid.
    (d) Silver oxide is added to hydrobromic acid.
    (e) Chlorine is bubbled through hydrobromic acid.
58. The reaction of titanium metal with HF(g) at 250°C for 2 days yielded a titanium fluoride that contained 45.67% titanium. Write the chemical equation that describes the reaction.   *Ans. $2Ti + 6HF \longrightarrow 2TiF_3 + 3H_2$*
59. What mass of PCl_3 can be prepared by the reaction of chlorine with 13.55 g of phosphorus?
    Ans. 60.08 g
60. From the data in Appendix I, calculate the free energy change for the reaction of hydrogen with each of the halogens. Determine which of these reactions is not spontaneous at 25°C.   *Ans. $H_2 + I_2 \longrightarrow 2HI$*
61. Calculate the bond energies of the F—F, Cl—Cl, Br—Br, and I—I bonds from the data in Appendix I, and arrange these molecules in increasing order of their X—X bond energy.   *Ans. $I_2 < F_2 < Br_2 < Cl_2$*

Noble Gases

62. Why do the noble gas elements exist as monatomic molecules in the free state?

63. Write the Lewis structures for and describe the molecular structures of XeF_2 and XeF_4.

64. A mixture of xenon and fluorine was heated at 400°C for 1 h. A sample of the white solid that formed reacted with hydrogen to give 54.0 mL of xenon at STP and hydrogen fluoride, which was collected in water, giving a solution of hydrofluoric acid. The hydrofluoric acid solution was titrated, and 45.62 mL of 0.2115 M sodium hydroxide was required to reach the equivalence point. Determine the empirical formula for the white solid and write balanced chemical equations for the reactions involving xenon.

$$Ans.\ Xe + 2F_2 \longrightarrow XeF_4$$
$$XeF_4 + 2H_2 \longrightarrow Xe + 4HF$$

65. A 0.492-g sample of the XeF_4 described in exercise 64 exerted a pressure of 47.8 torr at 127°C in a bulb with a volume of 1238 mL. Determine the molecular formula of xenon tetrafluoride from these data.   *Ans. XeF_4*

66. Basic solutions of Na_4XeO_6 are powerful oxidants. What mass of $Mn(NO_3)_2 \cdot 6H_2O$ will react with 125.0 mL of a 0.1717 M basic solution of Na_4XeO_6 if the products include Xe and a solution of sodium permanganate?   *Ans. 9.857 g*

Additional Exercises

67. Which of the following compounds will react with water to give an acid solution? Which will give a basic solution? Write a balanced equation for the reaction of each with water.
    (a) NaH   (b) H_2S (c) HCl
    (d) SO_3 (e) CaO   (f) Na_3P
    (g) PCl_3 (h) N_2O_5 (i) SF_4

68. (a) Which is a stronger acid, H_2S or H_2Se?
    (b) Which is a stronger base, OH^- or SH^-?
    (c) Which is a stronger base, SH^- or S^-?
    Explain each answer.

69. Air contains 20.99% oxygen by volume. What volume of air in cubic meters at 27.0°C and 0.9868 atm is required to oxidize 1.00 metric ton (1000 kg) of sulfur to sulfur dioxide, assuming that all of the oxygen reacts with sulfur?   *Ans. $3.71 \times 10^3\ m^3$*

70. The bond length in the O_2 molecule is 1.209 Å, while that in the O_3 molecule is 1.278 Å. Why does ozone have a longer bond?

71. Iodine reacts with liquid chlorine at −40°C to give an orange compound containing 54.3% iodine and 45.7% chlorine. Write the balanced equation for the reaction.   *Ans. $I_2 + 3Cl_2 \longrightarrow 2ICl_3$*

72. The combustion of 9.180 g of sodium to sodium oxide is exothermic; 100.74 kJ of heat is evolved. What is the heat of formation of sodium oxide?   *Ans. $-504.6\ kJ\ mol^{-1}$*

73. The heat of formation of $CS_2(l)$ is +21.4 kcal mol^{-1}. How much heat is required for the reaction of 10.00 g of S_8 with carbon according to the following equation?

$$4C(s) + S_8(s) \longrightarrow 4CS_2(l)$$

Is the reaction exothermic or endothermic?   *Ans. 3.34 kcal; endothermic*

74. When phosphorus burns to give $P_4O_{10}(s)$, an unusually large amount of energy is involved. Determine ΔH°_{298} for the reaction of exactly 1 mol of P_4, using the data in Appendix I.   *Ans. −2984 kJ*
    Determine ΔG°_{298} for the reaction.   *Ans. −2698 kJ*

75. At 422 K the partial pressures of gaseous PCl_5, PCl_3, and Cl_2 in a closed system were 0.453 atm, 6.08×10^{-2} atm, and 6.08×10^{-2} atm, respectively.
    Determine K_P for the oxidation

$$PCl_3(g) + Cl_2(g) \longrightarrow PCl_5(g)$$
    Ans. 1.23×10^2

76. Thallium(I) chloride is one of the few insoluble chlorides. What is the molar solubility of TlCl, whose solubility product is 1.8×10^{-4}?   *Ans. $1.3 \times 10^{-2}\ M$*

Electrochemistry and Oxidation-Reduction

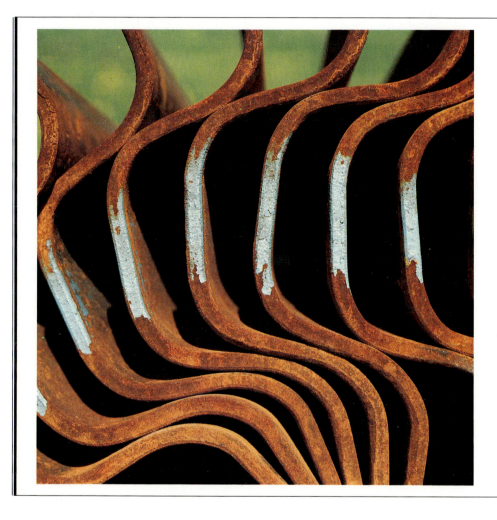

21

Corrosion of iron is an electrochemical process.

The most apparent applications of electrochemistry in our daily lives are the batteries we use to power our portable radios, cassette players, toys, and power tools and to start our cars. These batteries are voltaic cells, which produce electrical energy by chemical reactions. In other applications, which may not be so obvious, electrical energy is used to bring about chemical changes. The aluminum in your soft drink cans, the chlorine used to purify swimming pools, and many other commercial products are manufactured in electrolytic cells, in which electrical energy is used to bring about a chemical change. When you recharge a battery, you are using electrical energy to accomplish a chemical change that returns the battery to the state where it can again deliver electricity.

Electrochemistry deals with chemical changes produced by an electric current and with the production of electricity by chemical reactions. Energy changes in chemical reactions (Chemical Thermodynamics, Chapter 19) are studied electrochemically when possible, because the quantity of electrical energy produced or consumed during electrochemical changes can be measured very accurately. The free energy change of a reaction (ΔG) and its equilibrium constant can be determined by electrochemical measurements. Furthermore, an understanding of the reactions that take place at the electrodes of electrochemical cells clarifies the process of oxidation and reduction and chemical reactivity.

Many electrochemical processes are important in science and industry. Electrical energy is used in the manufacture of hydrogen, oxygen, ozone, hydrogen peroxide, chlorine, sodium hydroxide, and oxygen compounds of the halogens (Chapter 22). Electrical energy is also used in production of many other chemicals, electrorefining of metals (Chapter 13), electroplating of metals and alloys, and production of metal articles by electrodeposition.

21.1 The Conduction of Electricity

The movement of charged particles (either electrons, positive ions, or negative ions) through a conductor produces a current of electricity. **Metallic conduction** occurs when electrons move through a metal with no changes in the metal and with no movement of the metal atoms. **Electrolytic, or ionic, conduction** occurs when ions move through a molten substance, a solution, or, occasionally, a solid.

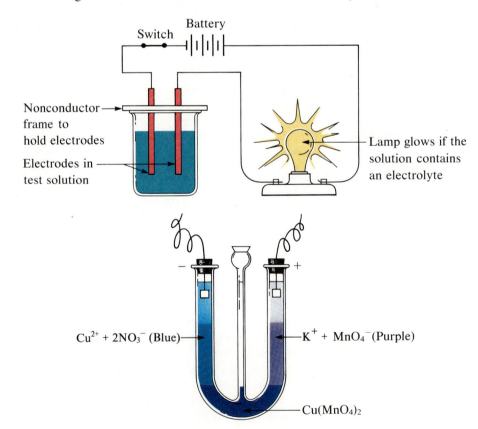

Figure 21.1

An electrolytic cell used to determine the conductivity of a solution.

Figure 21.2

When an electric current flows through a solution of $Cu(MnO_4)_2$ and KNO_3, the positive Cu^{2+} ions (blue) can be seen to move towards the negatively charged cathode, and the negative MnO_4^- ions (purple) towards the positively charged anode.

Substances that conduct an electric current by the movement of ions are called **electrolytes** (Section 12.6). When two electrodes that are good conductors dip into an electrolyte, electrons can be forced onto one of the electrodes and withdrawn from the other by a battery or a generator (Fig. 21.1). The electrode onto which the electrons are forced becomes negatively charged, and the electrode from which electrons are withdrawn becomes positively charged. The positive ions (cations) in the electrolyte are attracted by the negatively charged cathode and move toward it, and the negative ions (anions) are attracted by the positively charged anode and move toward it (Fig. 21.2). The movement of the ions results in the passage of a current.

Electrolytic Cells

21.2 The Electrolysis of Molten Sodium Chloride

In Section 13.2 we saw that lithium, sodium, magnesium, calcium, and aluminum are extracted from their compounds by **electrolysis,** the input of electrical energy as a direct current of electricity, which forces a nonspontaneous reaction to occur. As an example of this type of reaction, let us examine the electrolysis of molten sodium chloride a little more carefully than we did in that section.

The Downs cell used for the commercial electrolysis of molten sodium chloride (Fig. 13.6) is complicated because of the need to prevent the sodium produced from reacting with air and from reacting with the chlorine produced. A simpler representation of the cell is given in Figure 21.3. The essential elements of this cell are a container of molten sodium chloride, two inert electrodes, and a separator that permits diffusion of ions from one side of the cell to the other but prevents the sodium produced at one electrode from reacting with the chlorine produced at the other.

Molten sodium chloride contains equal numbers of sodium ions and chloride ions, which move about with considerable freedom. When an electrochemical cell containing molten NaCl is connected to a battery or some other source of direct current, the positive sodium ions are attracted to the cathode, where they combine with electrons to form sodium atoms (metallic sodium). The process at the cathode, called a **half-reaction,** is

$$Na^+ + e^- \longrightarrow Na$$

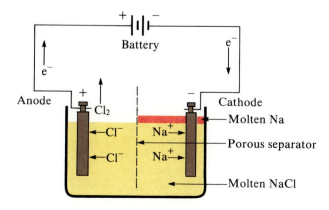

Figure 21.3

The electrolysis of molten sodium chloride. The reduction of sodium ions at the cathode produces sodium metal, and the oxidation of chloride ions at the anode produces chlorine gas. The separator prevents the reaction of the sodium and chlorine thus produced.

The oxidation number of sodium decreases from $+1$ to 0 in this reduction reaction, sometimes called cathodic reduction because the **cathode** is the electrode at which reduction occurs. The negative chloride ions are attracted to the anode, where they give up one electron each and become chlorine atoms, which then combine to form molecules of chlorine gas. The anode half-reaction is

$$2Cl^- \longrightarrow Cl_2 + 2e^-$$

The oxidation number of chlorine increases from -1 to 0 in this oxidation reaction, sometimes called anodic oxidation because the **anode** is the electrode at which oxidation occurs.

The net reaction for the electrolytic production of sodium and chlorine from molten sodium chloride is the sum of the two electrode half-reactions. The number of electrons produced by the oxidation is equal to the number of electrons consumed in the reduction, so 2 Na$^+$ ions must be reduced for each Cl$_2$ molecule that forms. Thus the cathode reaction is multiplied by 2 so that the complete equation will be balanced.

$$
\begin{aligned}
\textit{At the cathode:} \quad & 2Na^+ + 2e^- \longrightarrow 2Na \\
\textit{At the anode:} \quad & 2Cl^- \longrightarrow Cl_2 + 2e^- \\
\hline
\textit{Overall reaction:} \quad & 2Na^+ + 2Cl^- \longrightarrow 2Na + Cl_2 \\
\textit{or} \quad & 2NaCl(l) \longrightarrow 2Na + Cl_2
\end{aligned}
$$

The electrolysis of molten LiCl, MgCl$_2$, and CaCl$_2$ (Section 13.2) is very similar to that of NaCl. The metal is produced by reduction of the metal ion at the cathode, and chlorine is produced by oxidation of the chloride ion at the anode.

21.3 The Electrolysis of Aqueous Solutions

When an electrolyte is dissolved in water and an electrical current is passed through it in an electrolytic cell, several reactions are possible at the electrodes. The ions of the electrolyte or the water itself can be oxidized or reduced. The half-reactions that occur depend on the ease of oxidation or reduction of the ions relative to that of water.

1. ELECTROLYSIS OF HYDROCHLORIC ACID. When an aqueous solution of fairly concentrated hydrochloric acid is electrolyzed, the positive hydrogen ions are attracted to the cathode, accept electrons from it, and are reduced to hydrogen gas.

$$2H^+ + 2e^- \longrightarrow H_2$$

The negatively charged chloride ions are attracted to the anode. Two oxidations are possible at the anode, the oxidation of chloride ion or the oxidation of water:

$$2Cl^- \longrightarrow Cl_2 + 2e^- \tag{1}$$

$$2H_2O \longrightarrow O_2 + 4H^+ + 4e^- \tag{2}$$

Chloride ion and water are oxidized with almost equal ease, so the concentration of the chloride ion plays a significant role in determining the product. With a high chloride ion concentration, Reaction (1) occurs, and chlorine is formed at the anode. If the chloride concentration is low, Reaction (2) occurs as well, and oxygen is formed in addition to the chlorine. In a very dilute solution of hydrochloric acid, very little chlorine is formed, and the primary product is oxygen.

In a concentrated solution of hydrochloric acid, chlorine is produced at the anode, and the overall cell reaction may be obtained by adding the electrode reactions.

$$\begin{array}{ll}
\text{\textit{At the cathode:}} & 2H^+ + 2e^- \longrightarrow H_2 \\
\text{\textit{At the anode:}} & \underline{\quad\quad 2Cl^- \longrightarrow Cl_2 + 2e^-} \\
& 2H^+ + 2Cl^- \longrightarrow H_2 + Cl_2
\end{array}$$

2. ELECTROLYSIS OF A SOLUTION OF SODIUM CHLORIDE. Hydrogen gas, chlorine, and an aqueous solution of sodium hydroxide are produced when a concentrated solution of sodium chloride is electrolyzed during the production of sodium hydroxide (Section 13.5 and Fig. 21.4). Chlorine is formed at the anode, and hydrogen gas and hydroxide ion are formed at the cathode. In order to account for these products we need to consider all of the possible electrode reactions.

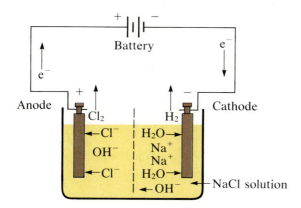

Figure 21.4

The electrolysis of an aqueous solution of sodium chloride. The reduction of water at the cathode produces hydrogen gas and hydroxide ions, and the oxidation of chloride ions at the anode produces chlorine gas. A solution of sodium hydroxide remains.

There are two species that might be reduced at the cathode: the sodium ion and water.

$$Na^+ + e^- \longrightarrow Na \tag{3}$$

$$2H_2O + 2e^- \longrightarrow H_2 + 2OH^- \tag{4}$$

Water is more easily reduced than sodium ion, so hydroxide ion and hydrogen gas form according to Equation (4). The hydroxide ions migrate toward the anode.

As we have just seen, both chloride ion and water can be oxidized at the anode.

$$2Cl^- \longrightarrow Cl_2 + 2e^- \tag{5}$$

$$2H_2O \longrightarrow O_2 + 4H^+ + 4e^- \tag{6}$$

In order to optimize the production of chlorine, a concentrated sodium chloride solution is used. The net reaction for the electrolysis of concentrated aqueous sodium chloride can be obtained by adding the electrode reactions.

$$\begin{array}{ll}
\text{\textit{At the anode:}} & 2H_2O + 2e^- \longrightarrow H_2 + 2OH^- \\
\text{\textit{At the cathode:}} & \underline{\quad\quad 2Cl^- \longrightarrow Cl_2 + 2e^-} \\
& 2H_2O + 2Cl^- \longrightarrow H_2 + Cl_2 + 2OH^-
\end{array}$$

Hydroxide ions are formed during the electrolysis as chloride ions are removed from solution. Because the sodium ions remain unchanged, sodium hydroxide accumulates as electrolysis proceeds.

3. ELECTROLYSIS OF A SOLUTION OF SULFURIC ACID. Sulfuric acid ionizes in water in two steps:

$$H_2SO_4 \longrightarrow H^+ + HSO_4^- \quad \text{(essentially complete)}$$

$$HSO_4^- \rightleftharpoons H^+ + SO_4^{2-}$$

Figure 21.5
The electrolysis of a dilute
solution of H_2SO_4.

As in a solution of hydrochloric acid, the only reduction possible at the cathode is

$$2H^+ + 2e^- \longrightarrow H_2$$

Two oxidations are possible at the anode: oxidation of the sulfate ion to the peroxydisulfate ion (Section 22.16) or oxidation of water.

$$2SO_4^{2-} \longrightarrow S_2O_8^{2-} + 2e^-$$
$$2H_2O \longrightarrow O_2 + 4H^+ + 4e^-$$

Oxygen and hydrogen ions form because water is more easily oxidized than sulfate ion.

The net reaction is the sum of the anode and cathode reactions. (The cathode reaction is multiplied by 2 to balance the number of electrons in the oxidation and reduction.)

$$\begin{array}{ll} \textit{At the cathode:} & 4H^+ + 4e^- \longrightarrow 2H_2 \\ \textit{At the anode:} & \underline{2H_2O \longrightarrow O_2 + 4H^+ + 4e^-} \\ & 2H_2O \longrightarrow 2H_2 + O_2 \end{array}$$

The electrolysis of an aqueous solution of sulfuric acid produces hydrogen and oxygen in a 2:1 ratio (Fig. 21.5). Although hydrogen ion from the sulfuric acid is consumed by cathodic reduction, it is regenerated at the anode at the same rate at which it disappears at the cathode. Water is consumed, and the sulfuric acid becomes more concentrated as electrolysis proceeds.

21.4 Electrolytic Refining of Metals

The electrodes used in the electrolytic cells considered thus far are electrically conducting, unreactive materials such as graphite or platinum. However, when a more reactive metal is used as the anode of an electrolytic cell, the anode reaction may involve oxidation of the metal.

When a strip of metallic copper is used as the anode in the electrolysis of a solution of copper(II) sulfate (Fig. 21.6), the copper dissolves as the following anodic reaction occurs:

$$Cu \longrightarrow Cu^{2+} + 2e^-$$

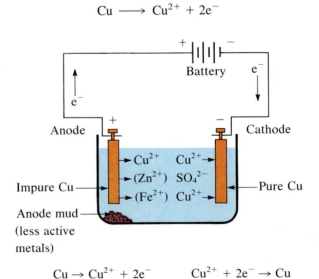

Figure 21.6
The electrolytic refining of
copper.

Figure 21.7

In the electrochemical purification of copper, ultrapure sheets of copper (the cathodes) are placed between sheets of impure copper (the anodes) in a tank containing a solution of $CuSO_4$. During the electrolysis, pure copper is deposited on the cathodes.

The oxidation of copper is easier than the oxidation of water, so the copper anode dissolves as copper(II) ions. The copper(II) ions formed at the anode migrate to the cathode, where they are more readily reduced than the water.

$$Cu^{2+} + 2e^- \longrightarrow Cu$$

Thus pure copper plates out on the cathode.

Crude copper is refined electrochemically (Fig. 21.7) to improve its electrical conductivity. If impurities are not removed, they can reduce copper's conductivity by over 10%, and impure copper wires can get dangerously hot when they conduct electricity. When impure copper is used as an anode, impurities such as gold and silver, which do not oxidize as easily as copper, do not dissolve (are not oxidized) and fall to the bottom of the cell, forming a mud from which they are readily recovered. Impurities such as zinc and iron, which oxidize more easily than copper, go into the solution as ions. If the electrical potential between the electrodes is carefully regulated, these ions are not reduced at the cathode; only the more easily reduced copper is deposited. The refined copper is deposited either upon a thin sheet of pure copper, which serves as a cathode, or upon some other metal from which the deposit may be stripped.

A similar electrochemical process is used to refine aluminum by the Hoopes process (Section 13.2).

21.5 Faraday's Law of Electrolysis

In 1832–1833 Michael Faraday performed experiments demonstrating that the amount of a substance undergoing a chemical change at each electrode during electrolysis is directly proportional to the quantity of electricity that passes through the electrolytic cell (Fig. 21.8). This observation has come to be known as **Faraday's law of electrolysis.**

The quantity of electricity can be expressed as the number of electrons, or the number of moles of electrons, passing through a cell. The quantity of a substance

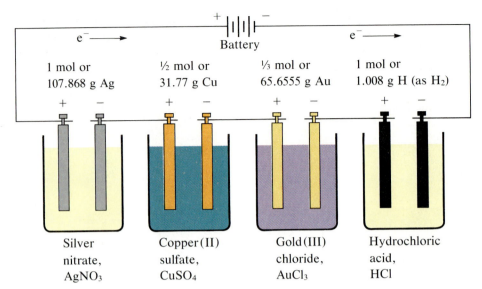

Figure 21.8
Amounts of various elements discharged at the cathode by 1 faraday of electricity (96,485 C, or 1 mol of electrons).

undergoing chemical change is related to the number of electrons that are involved in its half-reaction and can be expressed in terms of the moles of substance or in terms of equivalents of the substance. An electrochemical **equivalent** is the mass of a substance, in grams, that combines with or releases 1 mole of electrons. Thus, one equivalent of an oxidizing agent combines with 1 mole of electrons; one equivalent of a reducing agent releases 1 mole of electrons. The number of equivalents of solute in a liter of solution (equivalents per liter) is called the **normality** of the solution. A solution that contains 1 equivalent of a substance per liter is called a 1 N solution. One liter of a 1 N solution will combine with or release 1 mole of electrons.

Experimental results show that 1 electron reduces 1 silver ion, whereas 2 electrons are required to reduce 1 copper(II) ion.

$$Ag^+ + e^- \longrightarrow Ag$$
$$Cu^{2+} + 2e^- \longrightarrow Cu$$

Therefore 6.022×10^{23} electrons reduce 6.022×10^{23} silver ions (1 mole) to silver atoms. The mass of silver produced is 107.9 grams; thus the mass of 1 equivalent of silver is the same as the mass of 1 mole. Because 2 electrons are required to reduce 1 Cu^{2+} ion to atomic copper, 6.022×10^{23} electrons would reduce only $\frac{1}{2}$ of a mole of copper. Thus the mass of 1 equivalent of copper is $\frac{1}{2}$ the mass of 1 mole.

A **faraday, F,** is the charge on 1 mole of electrons. Thus when 1 faraday is passed through an electrochemical cell, 1 mole of electrons passes through the cell, 1 equivalent of a substance is reduced at the cathode, and 1 equivalent of a substance is oxidized at the anode. Other units of electricity commonly encountered are the coulomb and the ampere. The relationships between these units are as follows:

1 faraday = 6.022×10^{23} electrons = 1 mole of electrons
= 96,485 coulombs (C)

1 coulomb = quantity of electricity involved when a current of 1 ampere (A) flows for 1 second
= 1 A s

1 ampere = 1 coulomb per second = $\dfrac{1\ C}{1\ s}$

Calculations involving passage of an electric current (measured in amperes) through a cell may be handled like any other stoichiometry problem: convert the current in amperes to coulombs, then to faradays, and, finally, to moles of electrons. Use the moles of electrons as you would moles of any other reactant.

Calculate the mass of copper produced by the reduction of Cu^{2+} ion at the cathode during the passage of 1.600 ampere of current through a solution of copper(II) sulfate for 1.000 hour.   **EXAMPLE 21.1**

Copper(II) ion is reduced at the cathode according to the equation

$$Cu^{2+} + 2e^- \longrightarrow Cu$$

If we know the number of moles of copper produced by the current, we can determine the mass of copper produced. The moles of copper can be calculated if we know the moles of electrons that pass through the cell. The moles of electrons can be determined from the amperage and time of the current flow. The chain of calculation is

| Amps and time | → | Coulombs | → | Faradays, moles of electrons | → | Moles of copper | → | Mass of copper |

$$1.600 \text{ A} \times 1.000 \text{ h} \times \frac{60 \text{ min}}{1 \text{ h}} \times \frac{60 \text{ s}}{1 \text{ min}} = 5760 \text{ A s} = 5760 \text{ C}$$

$$5760 \text{ C} \times \frac{1 \text{ faraday}}{96,485 \text{ C}} = 5.970 \times 10^{-2} \text{ faraday}$$

$$5.970 \times 10^{-2} \text{ faraday} \times \frac{1 \text{ mol e}^-}{1 \text{ faraday}} = 5.970 \times 10^{-2} \text{ mol e}^-$$

The above half-reaction at the cathode is used to relate moles of electrons to the mass of copper produced.

$$5.970 \times 10^{-2} \text{ mol e}^- \times \frac{1 \text{ mol Cu}}{2 \text{ mol e}^-} = 2.985 \times 10^{-2} \text{ mol Cu}$$

$$2.985 \times 10^{-2} \text{ mol Cu} \times \frac{63.55 \text{ g Cu}}{1 \text{ mol Cu}} = 1.897 \text{ g Cu}$$

Very large currents are used in many industrial electrolytic cells. How much time is required to produce exactly 1 metric ton (1000 kg) of magnesium by passage of a current of 150,000 A through molten $MgCl_2$? Assume a 100% yield of magnesium based on the current.   **EXAMPLE 21.2**

We need to determine the time required to produce 1000 kg of magnesium. If we know the number of coulombs, C, passed through the melt, we can calculate the time, in seconds, from the relationship C = A × s, because we know the number of amperes, A. The number of coulombs is available from the number of faradays (moles of electrons) passed through the cell. The number of moles of electrons can be determined from the number of moles of Mg produced; 2 mol of electrons are used per mole of Mg produced, $Mg^{2+} + 2 e^- \rightarrow Mg$. The moles of Mg can be determined from the mass of Mg to be produced. The chain of calculations is

$$\boxed{\begin{array}{c}\text{Mass} \\ \text{of} \\ \text{Mg}\end{array}} \longrightarrow \boxed{\begin{array}{c}\text{Moles} \\ \text{of} \\ \text{Mg}\end{array}} \longrightarrow \boxed{\begin{array}{c}\text{Moles} \\ \text{of} \\ \text{electrons}\end{array}} \longrightarrow \boxed{\text{Coulombs}} \longrightarrow \boxed{\text{Time}}$$

$$1000 \text{ kg Mg} \times \frac{1000 \text{ g}}{1 \text{ kg}} \times \frac{1 \text{ mol Mg}}{24.305 \text{ g Mg}} = 4.114 \times 10^4 \text{ mol Mg}$$

$$4.114 \times 10^4 \text{ mol Mg} \times \frac{2 \text{ mol e}^-}{1 \text{ mol Mg}} = 8.229 \times 10^4 \text{ mol e}^-$$

$$8.229 \times 10^4 \text{ mol e}^- \times \frac{96,485 \text{ C}}{1 \text{ mol e}^-} = 7.940 \times 10^9 \text{ C}$$

We can rearrange the equation $C = A \times s$ and solve for s ($s = C/A$), because we know the current in amps and the number of coulombs passed through the melt.

$$s = \frac{C}{A} = \frac{7.940 \times 10^9 \text{ C}}{150,000 \text{ C s}^{-1}} = 5.293 \times 10^4 \text{ s}$$

It requires 5.293×10^4 s, or 14.70 h, to produce one metric ton of magnesium with a current of 150,000 A.

Electrode Potentials

21.6 Electrodes and Electrode Potentials

In our discussion of electrolysis and electrochemical cells we mentioned electrodes, the conductors that deliver electricity into a cell without necessarily entering into the cell reaction. Chemists also use the term **electrode** to refer to a system in which a conductor is in contact with a mixture of oxidized and reduced forms of some chemical species. One such electrode consists of a strip of metal placed in a solution containing ions of the metal. The metal strip is both the conductor and the reduced species; the metal ions are the oxidized species. Such an electrode is also called a **half-cell.**

Let us consider the behavior of an electrode, or half-cell, consisting of a zinc strip placed in a solution of a zinc compound with a 1 molar zinc ion concentration. When the strip is first placed in the solution, a few of the atoms leave the metal and pass into solution as ions. The valence electrons lost in forming the ions remain on the metal. The change may be expressed by

$$Zn \longrightarrow Zn^{2+}(aq) + 2e^-(\text{on the strip})$$

The negative charge generated on the zinc strip attracts positive zinc ions, and the electrons on the strip can reduce them back to the metal:

$$Zn^{2+} + 2e^- \longrightarrow Zn$$

Thus the change is reversible, and an equilibrium is rapidly established.

$$Zn^{2+} + 2e^- \rightleftharpoons Zn$$

At equilibrium the zinc strip is negatively charged, because it contains slightly more electrons than zinc atoms. The solution is positively charged; it contains slightly more zinc ions than negative ions.

At equilibrium, the difference in charge between a metal strip and a solution of its metal ions depends on (a) the ease with which the metal forms ions, (b) the ease with which the metal ions can be reduced to the metal, and (c) the concentration of the metal ions. More active metals (more easily oxidized metals) have greater tendencies to form ions and are more negatively charged. The zinc strip in a zinc half-cell is more negative than the copper strip in a copper half-cell because zinc is more easily oxidized than copper. The difference in charge between the metal strip and the solution also varies with the concentration of the metal ion. The charge on a zinc strip changes as the concentration of zinc ions in solution changes. As the zinc ion concentration increases, the number of electrons produced must decrease because of the equilibrium involving these species.

$$Zn^{2+} + 2e^- \rightleftharpoons Zn$$

Thus the difference in charge between the metal and the solution decreases. A decrease in the concentration of zinc ions increases the number of electrons, and the charge difference increases. Since the zinc electrode is a solid, its concentration is constant, and the size of the electrode does not influence the magnitude of the difference.

When a metal strip is placed in a 1 molar solution of ions of the metal, electrons spontaneously accumulate on the strip. To move them back into the solution would require the expenditure of energy. The amount of energy required varies with the difference between the negative charge on the strip and the positive charge in the solution; that is, it varies with the difference in electrical potential between the strip and the solution. This difference is called the **electrode potential** of the electrode. The larger the electrode potential, the greater the energy involved in moving an electron from the strip to the solution.

The unit used to measure a difference in electrical potential is the **volt.** In order to move 1/96,485 mole of electrons (1 coulomb of charge) from a lower potential to a higher potential, 1 joule of energy is required when the potential difference is 1 volt; $1 J = 1 C \times 1 V$. The higher the potential difference, the more work is required to move electrons; the lower the potential, the less work is required. The differences in potential associated with the operation of electrochemical cells are expressed in volts.

There is no satisfactory method for measuring the difference in electrical potential between a metal strip and a solution of its ions. We can only measure the difference between the electrode potentials of two electrodes. However, if we could measure the difference between an unknown electrode and a reference electrode with an electrode potential of zero volts, the measured difference would be the electrode potential of the unknown electrode. The electrode that is used as a reference electrode, assigned an electrode potential of exactly 0 volts, is the **standard hydrogen electrode** (Fig. 21.9). The potential of all other electrodes is reported relative to the standard hydrogen electrode. (This is similar to arbitrarily establishing sea level as zero elevation and reporting all elevations in terms of how much higher or lower than this level they are.)

The standard hydrogen electrode is a **gas electrode.** Gas electrodes are half-cells with a gas as either the oxidized or the reduced species. Such an electrode is set up by bubbling the gas around an inert conductor that carries electrons but does not enter the

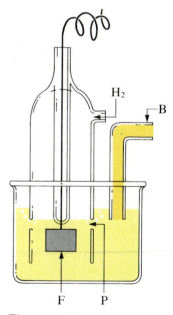

Figure 21.9

Diagram of a simple standard hydrogen electrode. F indicates the platinized platinum foil; P, the port for escape of hydrogen bubbles; B, part of the salt bridge.

electrode reaction. A hydrogen electrode is constructed so that hydrogen gas is bubbled around a platinum foil or wire covered with very finely divided platinum and immersed in a solution of hydrogen ions. Some molecules of hydrogen give up electrons to the inert platinum electrode and form hydrogen ions. At the same time, hydrogen ions are reduced to hydrogen molecules by acquiring electrons from the electrode. Thus a difference in potential between the electrode and the solution is established in the same manner as for zinc.

$$2H^+ + 2e^- \rightleftharpoons H_2$$

The platinum acts as an inert conductor and as a catalyst that quickly brings the system to equilibrium.

The potential of a gas electrode changes with changing pressure of the gas. A standard hydrogen electrode is prepared by bubbling hydrogen gas at a temperature of 25°C and a pressure of 1 atmosphere around platinum immersed in a solution containing a 1 M concentration of hydrogen ions. *By international agreement a value of zero is assigned as the electrode potential of the standard hydrogen electrode.*

To be precise we should use activities rather than concentrations of ions and pressures of gases. However, the approximations that the activity of an ionic species is equal to its concentration and that the activity of a gas is equal to its pressure will be satisfactory for our purposes (Section 19.5).

21.7 Measurement of Electrode Potentials

To measure the potential of a metal electrode relative to the standard hydrogen electrode, we need an electrochemical cell consisting of a metal strip in contact with a solution of its ions as one half of the cell and a standard hydrogen electrode as the other half. A cell for measuring the potential of a zinc electrode is shown in Figure 21.10. In this cell the zinc strip is in contact with a 1 molar solution of zinc ion. The two halves of the cell are connected by a **salt bridge** consisting of a concentrated solution of an electrolyte such as potassium chloride. This bridge allows K^+ and Cl^- ions to migrate from cell to cell and thus allows a current to flow, but the two solutions cannot mix. The zinc strip in the Zn^{2+} solution and the platinum strip of the hydrogen electrode are connected to a voltmeter, which shows the **cell potential** (the potential difference, in volts, between the two electrodes).

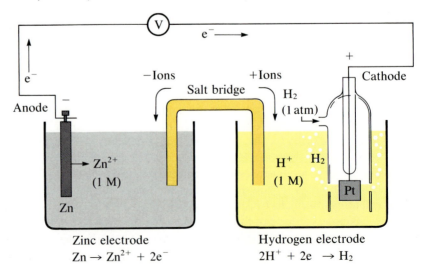

Figure 21.10

Measurement of the electrode potential of zinc, using the standard hydrogen electrode as the reference.

Zinc electrode
$Zn \rightarrow Zn^{2+} + 2e^-$

Hydrogen electrode
$2H^+ + 2e^- \rightarrow H_2$

Before the circuit is closed, that is, before the zinc and platinum strips are connected by a wire, we have the following electrode half-reactions, each at equilibrium.

$$2H^+ + 2e^- \rightleftharpoons H_2$$
$$Zn^{2+} + 2e^- \rightleftharpoons Zn$$

Because zinc has a greater tendency to form ions than hydrogen has, the zinc strip acquires a higher electron density (from the electrons released as the zinc ions are formed) than does the platinum of the hydrogen electrode. Therefore, when the electrical circuit is closed, the greater electron density, or electron pressure, at the zinc electrode causes electrons to flow from the zinc electrode to the hydrogen electrode. The electron density increases at the hydrogen electrode, and its equilibrium is shifted to the left, causing hydrogen ions to be reduced to hydrogen gas. The following reaction occurs at the hydrogen electrode as current flows:

$$2H^+ + 2e^- \longrightarrow H_2$$

At the same time, the electron density at the zinc electrode decreases because electrons are flowing to the hydrogen electrode. The equilibrium at the zinc electrode is shifted to the right, and more zinc is oxidized to zinc ions. The reaction that occurs at the zinc electrode as current flows is

$$Zn \longrightarrow Zn^{2+} + 2e^-$$

The overall reaction that takes place when the zinc-hydrogen cell is operating is an oxidation-reduction reaction.

Anodic oxidation:	$Zn \longrightarrow Zn^{2+} + 2e^-$	
Cathodic reduction:	$2H^+ + 2e^- \longrightarrow H_2$	
Cell reaction:	$Zn + 2H^+ \longrightarrow Zn^{2+} + H_2(g)$	

This same reaction will take place when zinc metal is immersed in a solution of hydrochloric acid (Fig. 21.11). Electrons are transferred directly from zinc to hydrogen ions. However, in the zinc-hydrogen cell, the reaction takes place without contact of the reactants. The electrons involved in the reaction are transferred from the zinc to the hydrogen ions through a wire. This is important from a practical point of view, because the current can be used to do work in an electric motor, to light an electric lamp, or to produce some other form of energy.

The voltmeter, V, shows the cell potential, the difference in potential between the half-cells. This potential is also called the **electromotive force (emf)** of the cell, which in this case is 0.76 volt. This 0.76-volt potential causes electrons to move from the zinc electrode to the hydrogen electrode. Since the potential of the hydrogen electrode is assumed to be zero volt, we assign the entire electromotive force of the cell to the zinc electrode. We say that the potential of the zinc electrode for the oxidation of zinc metal to zinc ion,

$$Zn \longrightarrow Zn^{2+} + 2e^-$$

is 0.76 volt above that of the hydrogen electrode and is positive (+0.76 V). For the reverse reaction, the reduction of zinc ion to metallic zinc,

$$Zn^{2+} + 2e^- \longrightarrow Zn$$

the electrode potential has the same absolute value but is of opposite sign (−0.76 V).

When copper metal in contact with a 1 M solution of copper(II) ions is the second electrode in a cell containing a standard hydrogen electrode, we find that hydrogen has a greater tendency to form positive ions than copper has. The formation of hydrogen

Figure 21.11

A zinc strip dissolves in hydrochloric acid, producing hydrogen gas and a solution of Zn^{2+} (zinc chloride).

ions from hydrogen leaves more electrons on the platinum than are left on the copper when the latter forms positive ions. Consequently, electrons flow from the hydrogen electrode to the copper electrode, where they reduce copper(II) ions.

$$
\begin{aligned}
\textit{Anodic oxidation:} \qquad\qquad\qquad H_2 &\rightleftharpoons 2H^+ + 2e^- \\
\textit{Cathodic reduction:} \qquad Cu^{2+} + 2e^- &\rightleftharpoons Cu \\
\hline
\textit{Cell reaction:} \qquad H_2 + Cu^{2+} &\rightleftharpoons Cu + 2H^+
\end{aligned}
$$

The emf of the cell as measured by a voltmeter is 0.337 volt. Since copper ion is reduced to the free element more easily than is hydrogen ion, the potential of the copper electrode for the reduction

$$ Cu^{2+} + 2e^- \longrightarrow Cu $$

has a positive value (+0.337 V) with respect to the hydrogen electrode.

By international consent, the values of electrode potentials are given for the reduction process. **Standard electrode potentials, or standard reduction potentials, $E°$,** are potentials measured with respect to a standard hydrogen electrode at 25°C with 1 M concentration (unit activity) of each ion and 1 atmosphere of pressure (unit activity) of each gas involved. The standard reduction potential, $E°$, for zinc is −0.76 volt. The standard reduction potential for any metal electrode having an electron density greater than the hydrogen electrode has a negative sign. An electrode that has less electron density than the hydrogen electrode (copper, for example) is positive with reference to the hydrogen electrode and hence has a positive potential. The standard electrode potential for copper is +0.337 volt.

Compared with the standard hydrogen electrode potential of 0.000 volt, the standard reduction potential of sodium is −2.71 volt, that of zinc is −0.76 volt, that of copper is +0.337 volt, and that of silver is +0.7991 volt. As the standard reduction potential for a series of half-cells becomes more negative, it becomes progressively more difficult to carry out the reduction. Thus Na^+ is more difficult to reduce than Zn^{2+}, which is more difficult to reduce than Cu^{2+}, which is more difficult to reduce than Ag^+. Also, the more negative the standard reduction potential, the more positive is the potential for the reverse reaction, the oxidation

$$ M \longrightarrow M^{n+} + ne^- $$

and the more easily oxidized is the metal.

Any cell that generates an electric current by an oxidation-reduction reaction is called a **voltaic cell.** The cell composed of zinc and hydrogen electrodes is a voltaic cell that can be diagrammed as follows:

$$ \xrightarrow{\quad e^- \quad} $$

$$ Zn \,|\, Zn^{2+}(1\ M) \quad \| \quad H^+(1\ M) \,|\, H_2(1\ atm) \,|\, Pt $$

This diagram indicates zinc metal in contact with a 1 molar solution of zinc ions. The solution of zinc is connected by a salt bridge, represented by $\|$, to a 1 molar solution of hydrogen ions in a hydrogen electrode with a gaseous hydrogen pressure of 1 atmosphere. A single vertical line, $|$, is used to separate two different phases; two different species in the same phase would be separated by a semicolon. The anode (the electrode at which oxidation occurs) is written on the left in such a diagram.

The purpose of the salt bridge in a cell is to establish a path for electrolytic conduction between the solutions of the two electrodes. An excess of zinc ions accumulates in the half-cell where zinc goes into solution. Similarly, an excess of anions accumulates in the half-cell where hydrogen ions are reduced. Migration of ions through the salt bridge (negative ions into the zinc half-cell solution and positive ions into the hydrogen half-cell solution) compensates for this imbalance and maintains neutral solutions. Without the salt bridge no current would flow in the external circuit, and the cell reactions would not take place. The ions that migrate through the salt bridge to maintain neutrality do not need to be the ions participating in the electrode reactions. In the zinc-hydrogen cell, they are in fact largely chloride ions and potassium ions supplied by the salt bridge.

21.8 The Activity, or Electromotive, Series

The positions of the metals in the activity series (Section 9.7) can be determined from their standard reduction potentials. When the standard reduction potentials of a number of metals are arranged in order from the most negative to the most positive, we have the activity series, shown in Table 21.1. This table also gives the reduction potentials of a few nonmetals. (Additional reduction potentials are given in Table 21.2 and in Appendix H.)

Table 21.1 Standard reduction potentials

	Oxidized form	Reduced form	Standard reduction potential, $E°$, V
Potassium	$K^+ + e^-$	$\rightleftharpoons$ K	−2.925
Calcium	$Ca^{2+} + 2e^-$	$\rightleftharpoons$ Ca	−2.87
Sodium	$Na^+ + e^-$	$\rightleftharpoons$ Na	−2.714
Magnesium	$Mg^{2+} + 2e^-$	$\rightleftharpoons$ Mg	−2.37
Aluminum	$Al^{3+} + 3e^-$	$\rightleftharpoons$ Al	−1.66
Manganese	$Mn^{2+} + 2e^-$	$\rightleftharpoons$ Mn	−1.18
Zinc	$Zn^{2+} + 2e^-$	$\rightleftharpoons$ Zn	−0.763
Chromium	$Cr^{3+} + 3e^-$	$\rightleftharpoons$ Cr	−0.74
Iron	$Fe^{2+} + 2e^-$	$\rightleftharpoons$ Fe	−0.440
Cadmium	$Cd^{2+} + 2e^-$	$\rightleftharpoons$ Cd	−0.40
Cobalt	$Co^{2+} + 2e^-$	$\rightleftharpoons$ Co	−0.277
Nickel	$Ni^{2+} + 2e^-$	$\rightleftharpoons$ Ni	−0.250
Tin	$Sn^{2+} + 2e^-$	$\rightleftharpoons$ Sn	−0.136
Lead	$Pb^{2+} + 2e^-$	$\rightleftharpoons$ Pb	−0.126
Hydrogen	$2H^+ + 2e^-$	$\rightleftharpoons$ H_2	0.00
Copper	$Cu^{2+} + 2e^-$	$\rightleftharpoons$ Cu	+0.337
Iodine	$I_2 + 2e^-$	$\rightleftharpoons$ $2I^-$	+0.5355
Mercury	$Hg_2^{2+} + 2e^-$	$\rightleftharpoons$ 2Hg	+0.789
Silver	$Ag^+ + e^-$	$\rightleftharpoons$ Ag	+0.7991
Bromine	$Br_2(l) + 2e^-$	$\rightleftharpoons$ $2Br^-$	+1.0652
Platinum	$Pt^{2+} + 2e^-$	$\rightleftharpoons$ Pt	+1.2
Oxygen	$O_2 + 4H^+ + 4e^-$	$\rightleftharpoons$ $2H_2O$	+1.23
Chlorine	$Cl_2 + 2e^-$	$\rightleftharpoons$ $2Cl^-$	+1.3595
Gold	$Au^+ + e^-$	$\rightleftharpoons$ Au	+1.68
Fluorine	$F_2 + 2e^-$	$\rightleftharpoons$ $2F^-$	+2.87

Figure 21.12 (left)
When a zinc strip is immersed in a solution containing Cu^{2+}, copper metal and Zn^{2+} form.

Figure 21.13 (right)
Iodide ion reduces bromine in a solution of bromine in water, producing iodine and colorless bromide ion. The iodine has been extracted into an organic layer, giving the characteristic purple color.

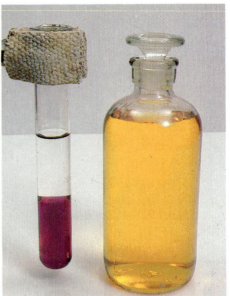

The activity series correlates many chemical properties of the elements; some of the more important ones are given in the following list.

1. The metals with large negative reduction potentials at the top of the series are good reducing agents in the free state. They are the metals most easily oxidized to their ions by the removal of electrons (active metals).

2. The elements with large positive reduction potentials at the bottom of the series are good oxidizing agents when in the oxidized form, that is, when the metals are in the form of ions and the nonmetals are in the elemental state.

3. The reduced form of any element will reduce the oxidized form of any element below it. For example, metallic zinc will reduce copper(II) ions (Fig. 21.12) according to the equation

$$Zn + Cu^{2+} \longrightarrow Cu + Zn^{2+}$$

Most metals will reduce the halogens that are found at the bottom of the series, and iodide ion (the reduced form of iodine) will reduce elemental bromine (Fig. 21.13), forming iodine and bromide ion (the reduced form of bromine).

21.9 Calculation of Cell Potentials

The net cell reaction of a voltaic cell made up of a standard zinc electrode and a standard copper electrode (Fig. 21.14), is obtained by adding the two half-reactions together.

Anode half-reaction:	$Zn \rightleftharpoons Zn^{2+} + 2e^-$
Cathode half-reaction:	$Cu^{2+} + 2e^- \rightleftharpoons Cu$
Net cell reaction:	$Zn + Cu^{2+} \rightleftharpoons Zn^{2+} + Cu$

The emf of the cell can be calculated by adding the standard potential of the copper electrode to that of the zinc electrode, taking into account the appropriate sign on each potential, based on whether the half-reaction taking place is oxidation or reduction. The standard reduction potential for zinc is −0.76 volt. This is the potential for the half-reaction

$$Zn^{2+} + 2e^- \longrightarrow Zn$$

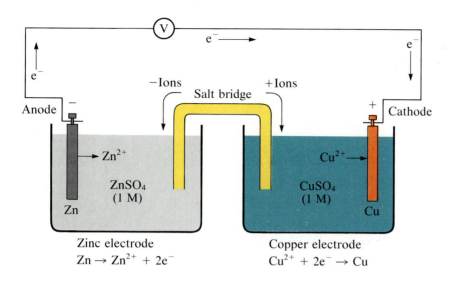

Figure 21.14

A zinc-copper voltaic cell.

Zinc electrode
$Zn \rightarrow Zn^{2+} + 2e^-$

Copper electrode
$Cu^{2+} + 2e^- \rightarrow Cu$

The oxidation of Zn to Zn^{2+}, the reverse of that half-reaction, has a potential of $+0.76$ volt. The standard reduction potential for copper is $+0.337$ volt. Since Cu^{2+} is being reduced to Cu, the potential for the copper half-reaction is $+0.337$ volt. The emf for the cell is

$$(+0.76 \text{ V}) + (+0.337 \text{ V}) = 1.10 \text{ V}$$

A positive emf value, such as that calculated for the zinc-copper cell, indicates that the net cell reaction will proceed spontaneously to the right and that the cell will deliver an electric current. A negative emf value for a cell indicates that the net cell reaction will not proceed spontaneously to the right but will proceed spontaneously to the left.

EXAMPLE 21.3

Calculate the emf of a cell composed of copper and silver electrodes in 1 M solutions of their respective ions.

Because copper has a lower $E°$ than silver (Cu is higher in the electromotive series than Ag), copper metal will reduce silver ion.

$$Cu + 2Ag^+ \rightleftharpoons Cu^{2+} + 2Ag$$

To obtain this net cell reaction, we add the half-reactions. Note that the silver half-reaction must be multiplied by 2 to have the same number of electrons involved in both the copper and silver half-reactions; this does not change the $E°$ value of the half cell or the cell potential. To obtain the cell potential, we add the half cell potentials:

$$
\begin{array}{ll}
Cu \rightleftharpoons Cu^{2+} + 2e^- & E° = -0.337 \text{ V} \\
2Ag^+ + 2e^- \rightleftharpoons 2Ag & E° = +0.799 \text{ V} \\
\hline
Cu + 2Ag^+ \rightleftharpoons Cu^{2+} + 2Ag & E° = (-0.337 \text{ V}) + (+0.779 \text{ V}) \\
& \quad = +0.442 \text{ V}
\end{array}
$$

The emf of the cell is 0.442 V. Since it is positive, the reaction proceeds spontaneously to the right as written, and the cell delivers current. Note that the potential of the copper half-reaction is equal in magnitude but of opposite sign to that listed in Table 21.1 for the reduction half-reaction

$$Cu^{2+} + 2e^- \longrightarrow Cu$$

EXAMPLE 21.4   Calculate the emf of a cell made up of standard bromine and chlorine electrodes.

The half-reactions and standard potentials (Table 21.1) are

$$
\begin{array}{ll}
Cl_2 + 2e^- \rightleftharpoons 2Cl^- & E° = +1.3595 \text{ V} \\
\underline{2Br^- \rightleftharpoons Br_2 + 2e^-} & \underline{E° = -1.0652 \text{ V}} \\
2Br^- + Cl_2 \rightleftharpoons Br_2 + 2Cl^- & E° = +0.2943 \text{ V}
\end{array}
$$

The emf of the cell is thus found to be 0.2943 V, meaning that the net cell reaction proceeds spontaneously to the right as written (Cl_2 oxidizing Br^-). If the half-reactions are reversed and added as follows:

$$
\begin{array}{ll}
2Cl^- \rightleftharpoons Cl_2 + 2e^- & E° = -1.3595 \text{ V} \\
\underline{Br_2 + 2e^- \rightleftharpoons 2Br^-} & \underline{E° = +1.0652 \text{ V}} \\
2Cl^- + Br_2 \rightleftharpoons Cl_2 + 2Br^- & E° = -0.2943 \text{ V}
\end{array}
$$

the negative emf value indicates that the reaction proceeds spontaneously to the left, rather than to the right. Chlorine oxidizes bromide ion to free bromine, but bromine will not oxidize chloride ion to free chlorine (Section 20.8).

Standard reduction potentials (such as those in Table 21.1) refer to standard conditions with concentrations of 1 M in aqueous solution, pressures of 1 atm, and a temperature of 25°C. If the conditions change, the potentials change, and some members may even change places in the table. The effect of the concentration on potential is illustrated by the reduction potentials for the hydrogen half-reaction at two different concentrations.

$$
\begin{array}{ll}
2H^+(1 \text{ M}) + 2e^- \rightleftharpoons H_2(1 \text{ atm}) & E° = 0.00 \text{ V} \\
2H^+(10^{-7} \text{ M}) + 2e^- \rightleftharpoons H_2(1 \text{ atm}) & E° = -0.41 \text{ V}
\end{array}
$$

The values given in Table 21.1 do not apply to nonaqueous solutions or to molten salts, although similar electromotive series can be constructed for them. The reduction potentials are different from those in water, and the members of the series fall in a different order because solvation energies vary from solvent to solvent.

21.10 Electrode Potentials for Other Half-Reactions

In addition to reactions between elements, there are many other oxidation-reduction reactions that can take place in half-cells. Table 21.2 and Appendix H contain standard reduction potentials for a number of half-reactions in which the conductor in the half-cell is an inactive substance such as carbon, which undergoes no change itself and merely serves as a carrier of electrons.

The standard reduction potential of the Pt, Sn^{4+}, Sn^{2+} electrode (Table 21.2) is that of an inert platinum strip immersed in a solution that is 1 M in both Sn^{4+} and Sn^{2+} ions. Similarly, the standard reduction potential of a Pt, Fe^{3+}, Fe^{2+} electrode is that when Fe^{3+} and Fe^{2+} are both 1 M. The emf of a cell composed of these two electrodes

$$
Pt \mid Sn^{2+}(1 \text{ M}); Sn^{4+}(1 \text{ M}) \xrightarrow{e^-} \parallel Fe^{3+}(1 \text{ M}); Fe^{2+}(1 \text{ M}) \mid Pt
$$

Table 21.2 Standard reduction potentials for some selected half-reactions

	Electrode reaction		Standard reduction potential, V
Electrode	Oxidized form	Reduced form	
Fe, Fe(OH)$_2$, OH$^-$	Fe(OH)$_2$ + 2e$^-$ $\rightleftharpoons$ Fe + 2OH$^-$		-0.877
Pb, PbSO$_4$, SO$_4^{2-}$	PbSO$_4$ + 2e$^-$ $\rightleftharpoons$ Pb + SO$_4^{2-}$		-0.356
Pt, Sn^{4+}, Sn^{2+}	Sn^{4+} + 2e$^-$ $\rightleftharpoons$ Sn^{2+}		$+0.15$
Ag, AgCl, Cl$^-$	AgCl + e$^-$ $\rightleftharpoons$ Ag + Cl$^-$		$+0.222$
Hg, Hg$_2$Cl$_2$, Cl$^-$	Hg$_2$Cl$_2$ + 2e$^-$ $\rightleftharpoons$ 2Hg + 2Cl$^-$		$+0.27$
NiO$_2$, Ni(OH)$_2$, OH$^-$	NiO$_2$ + 2H$_2$O + 2e$^-$ $\rightleftharpoons$ Ni(OH)$_2$ + 2OH$^-$		$+0.49$
Pt, Fe^{3+}, Fe^{2+}	Fe^{3+} + e$^-$ $\rightleftharpoons$ Fe^{2+}		$+0.771$
Pt, Cr$_2$O$_7^{2-}$, H$^+$, Cr^{3+}	Cr$_2$O$_7^{2-}$ + 14H$^+$ + 6e$^-$ $\rightleftharpoons$ 2Cr^{3+} + 7H$_2$O		$+1.33$
Pt, MnO$_4^-$, H$^+$, Mn^{2+}	MnO$_4^-$ + 8H$^+$ + 5e$^-$ $\rightleftharpoons$ Mn^{2+} + 4H$_2$O		$+1.51$
PbO$_2$, PbSO$_4$, H$_2$SO$_4$	PbO$_2$ + SO$_4^{2-}$ + 4H$^+$ + 2e$^-$ $\rightleftharpoons$ PbSO$_4$ + 2H$_2$O		$+1.685$

can be calculated as follows:

$$
\begin{array}{ll}
\text{Sn}^{2+} \rightleftharpoons \text{Sn}^{4+} + 2e^- & E° = -0.15 \text{ V} \\
\underline{2\text{Fe}^{3+} + 2e^- \rightleftharpoons 2\text{Fe}^{2+}} & \underline{E° = +0.771 \text{ V}} \\
\text{Sn}^{2+} + 2\text{Fe}^{3+} \rightleftharpoons \text{Sn}^{4+} + 2\text{Fe}^{2+} & E° = +0.62 \text{ V}
\end{array}
$$

The emf of the cell is 0.62 V when standard electrodes are used. This means that the reaction will proceed spontaneously to the right as written and will deliver a current. It also means that when solutions containing tin(II) and iron(III) are mixed, Sn^{2+} will reduce Fe^{3+}.

21.11 The Nernst Equation

Standard reduction potentials (Section 21.8), $E°$, are tabulated for the standard state condition with 1 M solutions of ions, 1 atmosphere of pressure for gases, and a temperature of 25°C. However, as pointed out in Section 21.9, the reduction potential of a half-reaction changes when the concentrations change. The **Nernst equation** can be used to calculate emf values at other concentrations. The Nernst equation for the emf of a half-cell at 25°C is

$$
E = E° - \frac{0.05915}{n} \log Q
$$

where E = the electrode potential at the new concentration

$E°$ = the *standard* electrode potential

n = the number of moles of electrons stated in the half-reaction

Q = the reaction quotient (Section 19.11) for the half-reaction, an expression that takes the same form as the equilibrium constant for the half-reaction, disregarding the electrons.

For example, a typical half-reaction and the corresponding Nernst equation is

$$Sn^{4+} + 2e^- \rightleftharpoons Sn^{2+}$$

$$E = E° - \frac{0.05915}{2} \log \frac{[Sn^{2+}]}{[Sn^{4+}]}$$

or, with the value of the standard reduction potential, $E°$, from Table 21.2 inserted,

$$E = 0.15 - \frac{0.05915}{2} \log \frac{[Sn^{2+}]}{[Sn^{4+}]}$$

where the brackets, as usual, mean *molar concentration of,* and log signifies the common logarithm.

EXAMPLE 21.5   Calculate E for the half-reaction

$$Sn^{4+} + 2e^- \rightleftharpoons Sn^{2+}$$

at 25°C if the concentration of Sn^{2+} is 0.40 M and the concentration of Sn^{4+} is 0.10 M.

$$E = 0.15 - \frac{0.05915}{2} \log \frac{[Sn^{2+}]}{[Sn^{4+}]}$$

$$= 0.15 - \frac{0.05915}{2} \log \frac{0.40}{0.10}$$

$$= 0.15 - \frac{0.05915}{2} (0.6021)$$

$$= 0.13 \text{ V}$$

In Section 21.9 we noted that the standard reduction potential, $E°$, for the reduction of $2H^+$ to H_2 at standard state conditions is 0.00 volts, but the emf value, E, for the same reduction at 10^{-7} M is -0.41 volts. The following example shows the calculation of this value.

EXAMPLE 21.6   Calculate the emf value for the reduction of H^+ at 1.0×10^{-7} M to H_2 at 1.0 atm.

$$2H^+(1.0 \times 10^{-7} \text{ M}) + 2e^- \rightleftharpoons H_2(1.0 \text{ atm})$$

$$E = E° - \frac{0.05915}{n} \log \frac{P_{H_2}}{[H^+]^2}$$

$$= 0.00 - \frac{0.05915}{2} \log \frac{1.0}{(1.0 \times 10^{-7})^2}$$

$$= 0.00 - \frac{0.05915}{2} \log \frac{1.0}{1.0 \times 10^{-14}}$$

$$= 0.00 - \frac{0.05915}{2} (14.00)$$

$$= -0.41 \text{ V}$$

Hence for the half-reaction

$$2H^+(1.0 \times 10^{-7} \text{ M}) + 2e^- \rightleftharpoons H_2(1.0 \text{ atm}) \qquad E = -0.41 \text{ V}$$

21.12 Relationship of the Free Energy Change, the Cell Potential, and the Equilibrium Constant

Electrochemical measurements are very helpful to chemists and other scientists, because they can be used to gather data that can be used to calculate thermodynamic parameters and equilibrium constants for a wide variety of chemical changes. The standard free energy change of an electrochemical reaction, ΔG°_{298} (Section 19.8), is related to the cell potential at 25°C, E°, and the equilibrium constant at 25°C, K (Section 19.12), by the following two equations:

$$\Delta G^{\circ}_{298} = -nFE^{\circ} \quad \text{and} \quad \Delta G^{\circ}_{298} = -2.303\,RT \log K$$

where $n =$ the number of moles of electrons exchanged in the reaction,
$\quad F =$ a constant known as the Faraday constant, whose value is
$\qquad$ 96.485 kJ V^{-1}, or 96,485 coulombs,
$\quad E^{\circ} =$ the standard cell potential at 25°C.
$\quad R =$ the gas constant (8.314 J K^{-1}),
$\quad T =$ the Kelvin temperature, 298.15 K
$\quad K =$ the equilibrium constant at 25°C

and log indicates a logarithm to base 10.

Combining the two expressions for ΔG°_{298}, we have

$$-nFE^{\circ} = -2.303\,RT \log K$$

Then

$$E^{\circ} = \frac{2.303\,RT}{nF} \log K$$

At 298.15 K (25°C) with substitution for R and F, the equation becomes

$$E^{\circ} = \frac{0.05915}{n} \log K$$

Thus the standard free energy change and the equilibrium constant of a reaction can be determined from the standard reduction potentials of two half-cell reactions, which add up to give the reaction.

Under nonstandard state conditions (when the temperature is not equal to 25° and/or when all concentrations and pressures do not approximate unit activity) the value of ΔG for a reaction may be related to the cell potential, E, of a cell with the nonstandard temperature and/or concentrations.

$$\Delta G = -nFE$$

EXAMPLE 21.7

Calculate the standard free energy change at 25°C for the reaction

$$Cd + Pb^{2} \rightleftharpoons Cd^{2+} + Pb$$

The half-reactions and calculation of the emf of the cell are

$$
\begin{array}{lll}
Cd \rightleftharpoons Cd^{2+} + 2e^{-} & E^{\circ} = +0.40 \text{ V} \\
\underline{Pb^{2+} + 2e^{-} \rightleftharpoons Pb} & \underline{E^{\circ} = -0.13 \text{ V}} \\
Cd + Pb^{2+} \rightleftharpoons Cd^{2+} + Pb & E^{\circ} = +0.27 \text{ V}
\end{array}
$$

Hence the standard cell potential is 0.27 V.

$$\Delta G^{\circ}_{298} = -nFE^{\circ} = -2(96.485 \text{ kJ V}^{-1})(0.27 \text{ V}) = -52 \text{ kJ}$$

The negative value for ΔG°_{298} indicates that the net cell reaction proceeds spontaneously to the right (Section 19.8). A positive E° means a negative ΔG°_{298}, so a positive E° also indicates a spontaneous forward reaction.

EXAMPLE 21.8   Calculate the equilibrium constant at 25°C for the reaction in Example 21.7.

$$E^{\circ} = \frac{0.05915}{n} \log K$$

$$0.27 = \frac{0.05915}{2} \log K$$

$$\log K = 0.27 \times \frac{2}{0.05915} = 9.13$$

$$K = 10^{9.13} = 1.3 \times 10^9$$

At equilibrium the molar concentration of cadmium ion in the solution is more than a billion times that of lead ion!

EXAMPLE 21.9   Calculate the free energy change and the equilibrium constant for the following reaction at 25°C:

$$Zn + Cu^{2+}(0.20 \text{ M}) \rightleftharpoons Zn^{2+}(0.0050 \text{ M}) + Cu$$

Under standard state conditions

$$
\begin{array}{llr}
Zn \rightleftharpoons Zn^{2+} + 2e^- & & E^{\circ} = +0.76 \text{ V} \\
Cu^{2+} + 2e^- \rightleftharpoons Cu & & E^{\circ} = +0.34 \text{ V} \\
\hline
Zn + Cu^{2+} \rightleftharpoons Zn^{2+} + Cu & & E^{\circ} = +1.10 \text{ V}
\end{array}
$$

Hence for the cell reaction *under standard state conditions* (Cu^{2+} and Zn^{2+} at concentrations of 1 M or, more accurately, at unit activity), E° is +1.10 V.

The Nernst equation lets us calculate the potential, E, when $[Zn^{2+}]$ and $[Cu^{2+}]$ are 0.0050 M and 0.20 M, respectively.

Zinc half-reaction:

$$E = E^{\circ} - \frac{0.05915}{n} \log \frac{[Zn^{2+}]}{[Zn^{\circ}]} = 0.76 - \frac{0.05915}{2} \log \frac{0.0050}{1}$$

$$= 0.76 - \frac{0.05915}{2}(-2.30) = 0.76 + 0.07 = +0.83 \text{ V}$$

Copper half-reaction:

$$E = E^{\circ} - \frac{0.05915}{n} \log \frac{[Cu^{\circ}]}{[Cu^{2+}]} = +0.34 - \frac{0.05915}{2} \log \frac{1}{0.20}$$

$$= +0.34 - \frac{0.05915}{2}(0.699) = +0.34 - 0.02 = +0.32 \text{ V}$$

Thus, for the cell reaction under these conditions,

$$E = (+0.83) + (+0.32) = +1.15 \text{ V}$$

(compared to the $E°$ value of $+1.10$ V for standard state conditions).

$$\Delta G = -nFE = -2(96.485 \text{ kJ V}^{-1})(1.15 \text{ V}) = -222 \text{ kJ}$$

The negative value of ΔG indicates that the reaction proceeds spontaneously to the right.

The value of K at 25°C must be calculated from $E°$.

$$E° = \frac{0.05915}{n} \log K$$

$$1.10 = \frac{0.05915}{2} \log K$$

$$\log K = 1.10 \times \frac{2}{0.05915} = 37.19$$

$$K = 10^{37.19} = 1.6 \times 10^{37}$$

Voltaic Cells

21.13 Primary Voltaic Cells

A voltaic cell is an electrochemical cell in which a chemical reaction is used to produce electrical energy—that is, a battery. **Primary cells,** such as most flashlight batteries, cannot be recharged, because the electrodes and electrolytes cannot be restored to their original states by an external electrical potential. The Daniell cell and dry cells are examples of primary cells.

1. THE DANIELL CELL. Daniell cells (Fig. 21.15) are not commonly used, but their simple design makes them a useful starting point for our discussion. The cell consists of a copper cathode (C) with crystals of copper(II) sulfate in a saturated solution of copper(II) sulfate and a zinc anode (A) suspended near the top of the cell in a dilute solution of zinc sulfate. The zinc sulfate solution floats on the denser solution of copper(II) sulfate. When the two metals are connected by a wire, electrons flow from the more easily oxidized zinc to the less easily oxidized copper.

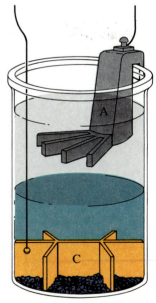

Figure 21.15
The Daniell, or gravity, cell. (C) sheet copper surrounded by saturated copper(II) sulfate solution and crystals of copper sulfate; (A) zinc plate surrounded by zinc sulfate solution.

At the anode:	$Zn \longrightarrow Zn^{2+} + 2e^-$
At the cathode:	$Cu^{2+} + 2e^- \longrightarrow Cu$
Cell reaction:	$Cu^{2+} + Zn \longrightarrow Zn^{2+} + Cu$

As the zinc ion concentration increases a (slightly) lower voltage is produced.

2. DRY CELLS. The familiar flashlight battery, or Leclanche cell, (Fig. 21.16) is one form of dry cell. It consists of a zinc container, which serves as the anode; a carbon (graphite) cathode; and a moist mixture of ammonium chloride, manganese dioxide, zinc chloride, and an inert filler such as sawdust. This mixture is separated from the zinc anode by a porous paper liner. When the cell delivers a current, the zinc anode is oxidized, forming zinc ions and electrons, which are involved in the reduction of

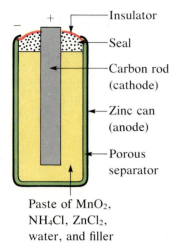

Paste of MnO_2, NH_4Cl, $ZnCl_2$, water, and filler

Figure 21.16

Cross section of a flashlight battery, a dry cell.

manganese (IV) oxide at the cathode. This reduction is not completely understood, but one possible reaction involves the reduction of the manganese from an oxidation state of $+4$ in MnO_2 to $+3$ in $MnO(OH)$

$$e^- + NH_4^+ + MnO_2 \longrightarrow NH_3 + MnO(OH)$$

The ammonia formed unites with some of the zinc ions, forming the complex ion $[Zn(NH_3)_4]^{2+}$. This reaction helps to hold down the concentration of zinc ions, thereby keeping the potential of the zinc electrode more nearly constant. It also prevents the accumulation of an insulating layer of ammonia molecules on the cathode, a condition called **polarization,** which would stop the action of the cell.

A second type of dry cell, the alkaline cell, also uses zinc and manganese oxide, but the electrolyte contains potassium hydroxide instead of ammonium chloride. Many of the small button-shaped batteries used in cameras, calculators, and watches are alkaline cells. Figure 21.17 illustrates a cutaway view of such a cell.

The small Rubin-Mallory cell is used in hearing aids. It utilizes a zinc container as the anode and a mercury-mercury oxide electrode with a carbon rod as the cathode. The electrolyte is sodium or potassium hydroxide. The cell produces a potential of 1.35 volt. The half-reactions and net cell reaction are

Anode:	$Zn + 2OH^- \longrightarrow Zn(OH)_2 + 2e^-$
Cathode:	$HgO + H_2O + 2e^- \longrightarrow Hg + 2OH^-$
Net cell reaction:	$Zn + HgO + H_2O \longrightarrow Zn(OH)_2 + Hg$

Figure 21.17

Cross section of a small alkaline cell used in watches and calculators.

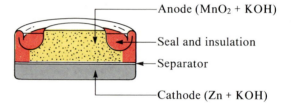

Anode (MnO_2 + KOH)

Seal and insulation

Separator

Cathode (Zn + KOH)

21.14 Secondary Voltaic Cells

Secondary cells are voltaic cells that can be regenerated by passing a current of electricity through the cell in the reverse direction of that of discharge. This recharges the cell. The lead storage battery and the nickel-cadmium cell are examples of secondary cells.

1. THE LEAD STORAGE BATTERY. The battery in an automobile is a lead storage battery (a lead-acid battery). Its electrodes are lead alloy plates in the form of grids. The openings of one set of grids are filled with lead(IV) oxide and the openings of the other with spongy lead metal. Dilute sulfuric acid serves as the electrolyte. When the battery (Fig. 21.18) discharges (delivers a current), the spongy lead is oxidized to lead ions, which combine with sulfate ions of the electrolyte and coat the lead electrode with insoluble lead sulfate. The net lead electrode reaction when the battery discharges is

$$Pb + SO_4^{2-} \longrightarrow PbSO_4(s) + 2e^-$$

During discharge the lead electrode is the anode, since it is the electrode at which oxidation occurs.

Electrons from the lead electrode flow through an external circuit and enter the lead(IV) oxide electrode. The lead(IV) oxide is reduced to lead(II) ions, and water is formed. Again lead ions combine with sulfate ions of the sulfuric acid electrolyte, and this plate also becomes coated with lead sulfate. The net reaction at the positive lead(IV) oxide electrode when the battery is discharging is

$$PbO_2 + 4H^+ + SO_4^{2-} + 2e^- \longrightarrow PbSO_4 + 2H_2O$$

The positively charged lead(IV) oxide electrode is the cathode, since it is the electrode at which reduction occurs.

The lead storage battery can be recharged by passing electrons in the reverse direction through each cell by applying an external potential. The cell becomes an electrolytic cell during recharging. The half-reactions are just the reverse of those that occur when the cell is operating as a voltaic cell producing a current. Electrons forced into the lead electrode reduce lead ions from the lead sulfate so this electrode becomes the cathode during recharging. Electrons are withdrawn from the lead(IV) oxide electrode, making it the anode, and lead ions from the lead sulfate are oxidized.

The charge and discharge at the two plates may be summarized as follows:

$$Pb + SO_4^{2-} \underset{\text{Charge}}{\overset{\text{Discharge}}{\rightleftharpoons}} PbSO_4(s) + 2e^- \qquad \text{(at the lead plate)}$$

$$PbO_2 + 4H^+ + SO_4^{2-} + 2e^- \underset{\text{Charge}}{\overset{\text{Discharge}}{\rightleftharpoons}} PbSO_4(s) + 2H_2O$$

(at the lead dioxide plate)

During recharging the sulfate ion of the electrolyte is regenerated, and lead sulfate is converted back to spongy lead at the lead electrode. Hydrogen ions and sulfate ions of the electrolyte are regenerated, and lead sulfate is converted back to lead(IV) oxide at the lead(IV) oxide electrode.

The net cell reaction of the lead storage battery is obtained by adding the two electrode reactions together and is:

$$Pb + PbO_2 + 4H^+ + 2SO_4^{2-} \underset{\text{Charge}}{\overset{\text{Discharge}}{\rightleftharpoons}} 2PbSO_4(s) + 2H_2O$$

This equation indicates that the amount of sulfuric acid decreases as the cell discharges and increases as the cell is charged. Sulfuric acid is much denser than water, so a lead storage battery may be tested by determining the specific gravity of the electrolyte. As can be calculated from the potentials in Table 21.2, a single standard lead cell has a potential of

$$(0.36) + (1.69) = 2.05 \text{ V}$$

The potential falls off slowly as the cell is used. A 12-volt automobile battery contains six lead storage cells.

2. THE NICKEL-CADMIUM CELL. The rechargeable cells used in calculators and battery-operated tools are based on nickel and cadmium electrodes. Cadmium metal serves as the anode, while nickel(IV) oxide is reduced to nickel(II) hydroxide at the cathode. The electrolyte is a hydroxide solution. When the cell delivers a current, cadmium is oxidized at the anode.

$$\textit{Anode:} \quad Cd + 2OH^- \longrightarrow Cd(OH)_2 + 2e^-$$

NiO_2 is reduced at the cathode.

$$\textit{Cathode:} \quad NiO_2 + 2H_2O + 2e^- \longrightarrow Ni(OH)_2(s) + 2OH^-$$

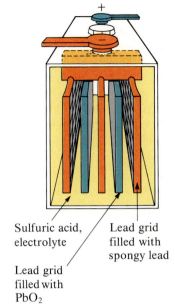

Sulfuric acid, electrolyte

Lead grid filled with spongy lead

Lead grid filled with PbO_2

Figure 21.18

One cell of a lead storage battery. A 12-volt automobile battery contains 6 of these cells.

These reactions are reversed during charging. The net reaction is

$$Cd + NiO_2 + 2H_2O \underset{\text{Charge}}{\overset{\text{Discharge}}{\rightleftharpoons}} Cd(OH)_2 + Ni(OH)_2$$

Note that the anode is positive and the cathode negative in an electrolytic cell (see Fig. 21.3, for example), whereas the anode is negative and the cathode positive in a voltaic cell (Fig. 21.14). *For all types of cells, however, the electrode where oxidation occurs is the anode and the electrode where reduction occurs is the cathode.* Figure 21.19 shows diagrammatically the relationship between a voltaic cell and an electrolytic cell. The sign convention is such that when a voltaic cell is supplying the current for an electrolytic cell, the negative electrode of one cell is connected to the negative electrode of the other cell, and likewise the positive electrodes of the two cells are connected.

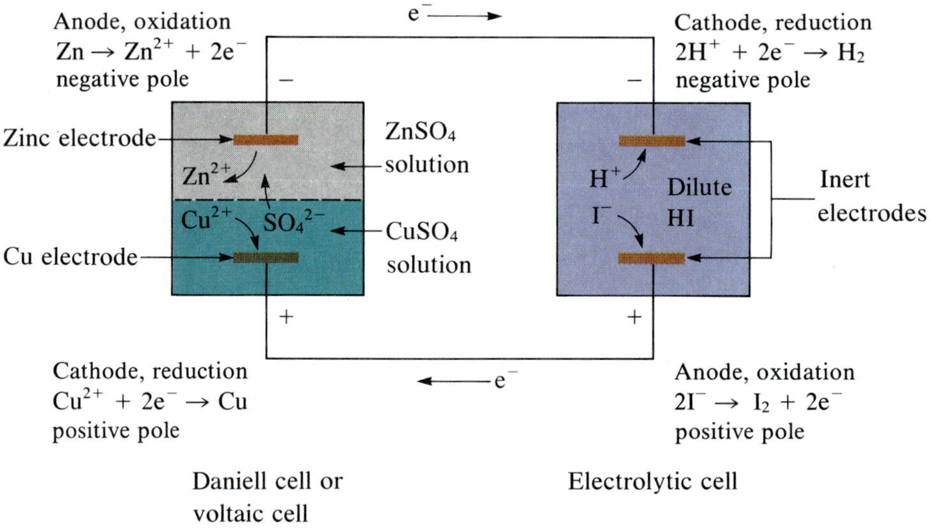

Figure 21.19

Illustration of the relationship between a voltaic cell and an electrolytic cell. The voltaic cell (*left*) provides the electric current to run the electrolytic cell (*right*). Note that the signs for the electrodes are opposite for the two types of cell, resulting in electrodes of like sign being connected. Note also that regardless of whether the cell is voltaic or electrolytic, oxidation occurs at the anode and reduction at the cathode.

Anode, oxidation
$Zn \rightarrow Zn^{2+} + 2e^-$
negative pole

Zinc electrode — ZnSO$_4$ solution

Zn^{2+}
Cu^{2+} SO_4^{2-}

Cu electrode — CuSO$_4$ solution

Cathode, reduction
$Cu^{2+} + 2e^- \rightarrow Cu$
positive pole

Daniell cell or voltaic cell

e^-

Cathode, reduction
$2H^+ + 2e^- \rightarrow H_2$
negative pole

H^+ Dilute
I^- HI

Inert electrodes

Anode, oxidation
$2I^- \rightarrow I_2 + 2e^-$
positive pole

Electrolytic cell

21.15 Fuel Cells

Fuel cells are voltaic cells in which electrode materials, usually in the form of gases, are supplied continuously to an electrochemical cell and consumed to produce electricity.

A typical fuel cell, of the type currently used in the space shuttle, is based on the reaction of hydrogen (the fuel) and oxygen (the oxidizer) to form water. Hydrogen gas is diffused through the anode, a porous electrode with a catalyst such as finely divided platinum or palladium on its surface. Oxygen is diffused through the cathode, a porous electrode impregnated with cobalt oxide, platinum, or silver as catalyst. The two electrodes are separated by a concentrated solution of sodium hydroxide or potassium hydroxide (Fig. 21.20) as an electrolyte.

Hydrogen diffuses through the anode and is absorbed on the surface in the form of hydrogen atoms, which react with hydroxide ions of the electrolyte to form water.

$$H_2 \xrightarrow{\text{Catalyst}} 2H$$
$$2H + 2OH^- \longrightarrow 2H_2O + 2e^-$$

Net anode reaction: $H_2 + 2OH^- \longrightarrow 2H_2O + 2e^-$

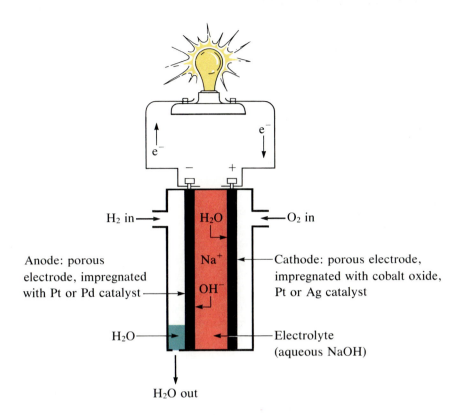

Figure 21.20
Diagram of a hydrogen-oxygen fuel cell. It would take several such cells to light an ordinary 110-V bulb.

The electrons produced at the anode flow through the external circuit to the cathode. The oxygen adsorbed on the cathode's surface is reduced to hydroxide ions.

Cathode reaction: $O_2 + 2H_2O + 4e^- \longrightarrow 4OH^-$

The hydroxide ions replace those that react at the anode. As in all voltaic cells, the electrical output of the cell results from the flow of electrons through the external circuit from anode to cathode.

The overall cell reaction is the combination of hydrogen and oxygen to produce water.

Anodic oxidation:	$2H_2 + 4OH^- \longrightarrow 4H_2O + 4e^-$
Cathodic reduction:	$O_2 + 2H_2O + 4e^- \longrightarrow 4OH^-$
Cell reaction:	$2H_2 + O_2 \longrightarrow 2H_2O$

A great deal of effort is being expended in investigating other fuels, such as methane and other hydrocarbons, and other electrode systems.

The efficiency of the fuel cell is better than that of conventional generating equipment, because electric current is produced directly from the reaction of fuel and oxidizer without the inherently wasteful intermediate conversion of chemical energy to heat. This conversion can waste 60–80% of the available chemical energy. Fuel cells can, at the present stage of development, increase conversion efficiencies from about 30% in conventional power plants to over 40%. It is estimated that sufficient power for 20,000 people can be produced in a unit that is 18 feet high and covers less than $\frac{1}{2}$ acre of ground. In addition, a fuel cell produces no pollution. The chemical products of such a cell would be carbon dioxide and water vapor. However, power needs for the process, such as the energy required to produce the fuel and to pump the cell components into the cell, must be taken into account in assessing the overall efficiency of fuel cells for producing power commercially.

21.16 Corrosion

Many metals, particularly iron, undergo corrosion when exposed to air and water (Fig. 21.21). Losses caused by corrosion of metals total billions of dollars annually in the United States.

It has been shown that iron will not rust in dry air or in water that is free from dissolved oxygen. Both air and water are involved in the corrosion process. The presence of an electrolyte in the water accelerates corrosion, particularly when the solution is acidic. Heated portions of a metal corrode more rapidly than unheated ones. Finally, iron in contact with a less active metal, such as tin, lead, or copper, corrodes more rapidly than iron that is either alone or in contact with a more active metal, such as zinc or magnesium.

Corrosion appears to be an electrochemical process (Fig. 21.22). When iron is in contact with a drop of water, the iron tends to oxidize.

$$\textit{Anodic oxidation:} \qquad Fe \longrightarrow Fe^{2+} + 2e^-$$

The electrons from the oxidation pass through the iron to the edge of the drop, where they reduce oxygen from the air to hydroxide ion.

$$\textit{Cathodic reduction:} \qquad O_2 + 2H_2O + 4e^- \longrightarrow 4OH^-$$

The iron(II) ions and hydroxide ions diffuse together and combine, forming insoluble iron(II) hydroxide.

$$Fe^{2+} + 2OH^- \longrightarrow Fe(OH)_2(s)$$

This precipitate is rapidly oxidized by oxygen to rust, an iron(III) compound with the approximate composition $Fe_2O_3 \cdot H_2O$.

Many methods and devices have been employed to prevent or retard corrosion. Iron can be protected against corrosion by coating it with an organic material such as paint, lacquer, grease, or asphalt; with another metal such as zinc, copper, nickel, chromium, or tin; with a ceramic enamel, like that used on sinks, bathtubs, stoves, refrigerators, and washers; or with an adherent oxide, which may be formed by exposing iron to superheated steam, thereby giving it an adherent coating of Fe_3O_4. Some alloys of iron are corrosion-resistant. Typical examples are stainless steel (Fe, Cr, and Ni) and duriron (Fe and Si).

Another method of preventing the corrosion of iron or steel involves an application of electrochemistry called cathodic protection. For example, corrosion of iron or steel water tanks can be retarded by suspending several stainless steel anodes in the tank; the tank serves as cathode. A small current is passed continuously through the system; the natural salts in the water make it conducting. A slight cathodic evolution of a protective coat of hydrogen takes place on the wall of the tank.

Figure 21.21

Iron rusts when exposed to air and water.

Figure 21.22

The electrochemical corrosion of iron. The iron in the anodic region dissolves, forming Fe^{2+} ions and causing the pit to form. The electrons travel to the cathodic region where they react with O_2 with the formation of OH^- ions. The combination of Fe^{2+} and OH^-, followed by oxidation, produces rust.

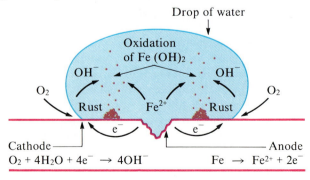

Cathodic protection is also used on iron and steel, such as underground pipeline, that is in contact with soil. If the iron is connected by a wire to a more active metal, such as zinc, aluminum, or magnesium, the iron becomes a cathode at which oxygen is reduced, rather than an anode where iron is oxidized. The difference in activity of the two metals causes a current to flow between them, producing corrosion of the more active metal and furnishing protection to the iron. With magnesium as the more active metal,

At the anode:	$2Mg \longrightarrow 2Mg^{2+} + 4e^-$
At the cathode:	$O_2 + 2H_2O + 4e^- \longrightarrow 4OH^-$
Net reaction:	$2Mg + O_2 + 2H_2O \longrightarrow 2Mg^{2+} + 4OH^-$

In this series of reactions no iron is oxidized. However, the active metal is slowly consumed and must be replaced periodically, but this is less expensive than replacing a pipeline.

Some of the more active metals such as aluminum and magnesium, which might be expected to corrode rapidly, are protected by a tightly adhering oxide coat of the metal oxide that forms when the metal is exposed to air (Section 13.1, Parts 2 and 3). The metal is made passive by this coating.

Oxidation-Reduction Reactions

Oxidation-reduction reactions (redox reactions) involve oxidation and reduction of the reactants (Section 9.3, Part 4). The chemical changes occurring in electrochemical cells involve oxidation of one species at the anode and reduction of another species at the cathode; thus the net reactions of most electrochemical cells are oxidation-reduction reactions. Those net reactions with a positive cell potential proceed spontaneously either in a cell or when the reactants are mixed together. For example, metallic zinc transfers electrons directly to a copper(II) ion and reduces it to metallic copper leaving zinc(II) ions when a piece of zinc metal is placed in a solution of copper nitrate, or it transfers electrons through an external wire when a zinc half-cell and a copper half-cell are combined in an electrochemical cell, as described in Section 21.9.

21.17 Spontaneity, Free Energy Change, and Equilibrium Constants of Redox Reactions

Since a spontaneous oxidation-reduction reaction will occur either in an electrochemical cell or on direct mixing of the reactants, standard reduction potentials and the Nernst equation can be used to predict the spontaneity of an oxidation-reduction reaction and to calculate its free energy change and its equilibrium constant (Section 21.12). The oxidation-reduction reaction, however, must be written as the sum of two half-reactions with known standard reduction potentials.

Chloride ion can be oxidized to chlorine in an acid solution by permanganate ion.   **EXAMPLE 21.10**

$$2MnO_4^- + 10Cl^- + 16H^+ \longrightarrow 2Mn^{2+} + 5Cl_2 + 8H_2O$$

Show that this reaction is spontaneous at standard state conditions and calculate ΔG°_{298} and K for it.

In order to solve this problem, we need to write the reaction as the sum of two half-reactions, which are found in Appendix H:

$$Cl_2 + 2e^- \longrightarrow 2Cl^- \qquad\qquad E° = 1.3595\ V$$
$$MnO_4^- + 8H^+ + 5e^- \longrightarrow Mn^{2+} + 4H_2O \qquad E° = 1.51\ V$$

The oxidation-reduction reaction and its emf can be written as the following sum:

$$5(2Cl^- \longrightarrow Cl_2 + 2e^-) \qquad\qquad E° = -1.3595\ V$$
$$\underline{2(MnO_4^- + 8H^+ + 5e^- \longrightarrow Mn^{2+} + 4H_2O) \qquad E° = +1.51\ V}$$
$$2MnO_4^- + 16H^+ + 10Cl^- \longrightarrow 2Mn^{2+} + 5Cl_2 + 8H_2O \qquad E° = +0.15\ V$$

Since the potential for a cell involving this net reaction is positive, the reaction is spontaneous.

$\Delta G_{298}°$ and K can be determined as described in Section 21.12

$$\Delta G_{298}° = -nFE°$$
$$= -10 \times (96.487\ kJ\ V^{-1}) \times (0.15\ V)$$
$$= -1.4 \times 10^2\ kJ$$

$$E° = \frac{0.05915}{n} \log K$$

$$\log K = E° \times \frac{n}{0.05915} = 0.15 \times \frac{10}{0.05915}$$

$$= 25.359$$
$$K = 10^{25.359} = 2.3 \times 10^{25}$$

21.18 Balancing Redox Equations by the Half-Reaction Method

Equations for oxidation-reduction reactions can become quite complicated and difficult to balance by trial and error. The half-reaction method for balancing redox equations provides a systematic approach. In this method the overall reaction is broken down into two half-reactions, one representing the oxidation step and the other the reduction step. Each half-reaction is balanced separately, the number of electrons in each half-reaction is made the same, then the two half-reactions are added to give the balanced equation.

EXAMPLE 21.11   Iron(II) is oxidized to iron(III) by chlorine. Use the half-reaction method to balance the equation for the reaction

$$Fe^{2+} + Cl_2 \longrightarrow Fe^{3+} + Cl^-$$

The oxidation half-reaction is oxidation of Fe^{2+} to Fe^{3+}.

$$Fe^{2+} \longrightarrow Fe^{3+}$$

To balance this half-reaction in terms of ion charges and electrons, we add 1 electron to the right side of the equation.

$$Fe^{2+} \longrightarrow Fe^{3+} + e^-$$
$$+2 \quad = \quad +3 \quad -1$$

The half-reaction for the reduction of Cl_2 to Cl^- has 2 electrons on the left side of the equation.

$$Cl_2 + 2e^- \longrightarrow 2Cl^-$$
$$-2 \quad = \quad -2$$

In the oxidation half-reaction 1 electron is lost, while in the reduction half-reaction 2 electrons are gained. The oxidation half-reaction is multiplied by 2 to balance the electrons.

$$2Fe^{2+} \longrightarrow 2Fe^{3+} + 2e^-$$

When the balanced half-reactions are added together, the two electrons on either side of the equation cancel one another, and the balanced equation is obtained.

$$2Fe^{2+} \longrightarrow 2Fe^{3+} + 2e^-$$
$$Cl_2 + 2e^- \longrightarrow 2Cl^-$$
$$\overline{2Fe^{2+} + Cl_2 \longrightarrow 2Fe^{3+} + 2Cl^-}$$

EXAMPLE 21.12

The dichromate ion will oxidize ion(II) to iron(III) in acid solution. Balance the equation

$$Cr_2O_7{}^{2-} + H^+ + Fe^{2+} \longrightarrow Cr^{3+} + Fe^{3+} + H_2O$$

by the half-reaction method.

The oxidation half-reaction is balanced as shown in Example 21.11.

$$Fe^{2+} \longrightarrow Fe^{3+} + e^-$$

The reduction half-reaction involves the change

$$Cr_2O_7{}^{2-} \longrightarrow 2Cr^{3+}$$

In acid solution, excess oxide combines with hydrogen ion to give water; thus this half-reaction also involves H^+ as a reactant and H_2O as a product. By inspection we can see that the 7 atoms of oxygen from the $Cr_2O_7{}^{2-}$ ion require 14 H^+ ions to produce 7 molecules of water.

$$Cr_2O_7{}^{2-} + 14H^+ \longrightarrow 2Cr^{3+} + 7H_2O$$

To balance this reduction half-reaction in terms of ion charges and electrons, we add 6 electrons on the left side of the equation.

$$Cr_2O_7{}^{2-} + 14H^+ + 6e^- \longrightarrow 2Cr^{3+} + 7H_2O$$
$$-2 \quad +14 \quad -6 \quad = \quad +6$$

Now we multiply the oxidation half-reaction by 6 so that both half-reactions involve the same number of electrons.

$$6Fe^{2+} \longrightarrow 6Fe^{3+} + 6e^-$$

On addition of the balanced half-reactions, the 6 electrons on either side of the equation cancel one another, and we have

$$6Fe^{2+} \longrightarrow 6Fe^{3+} + 6e^-$$
$$\underline{Cr_2O_7^{2-} + 14H^+ + 6e^- \longrightarrow 2Cr^{3+} + 7H_2O}$$
$$Cr_2O_7^{2-} + 14H^+ + 6Fe^{2+} \longrightarrow 2Cr^{3+} + 6Fe^{3+} + 7H_2O$$

EXAMPLE 21.13   In acidic solution hydrogen peroxide oxidizes Fe^{2+} to Fe^{3+}. Use the half-reaction method to balance the equation.

$$H_2O_2 + H^+ + Fe^{2+} \longrightarrow H_2O + Fe^{3+}$$

The oxidation half-reaction is

$$Fe^{2+} \longrightarrow Fe^{3+} + e^-$$

The reduction half-reaction involves the reduction of H_2O_2 to H_2O in acid solution.

$$H_2O_2 \longrightarrow ?H_2O$$

By inspection we can see there is sufficient oxygen in 1 molecule of H_2O_2 to form 2 molecules of water. This requires 2 hydrogen ions.

$$H_2O_2 + 2H^+ \longrightarrow 2H_2O$$

Then 2 electrons are needed to balance the positive charges.

$$H_2O_2 + 2H^+ + 2e^- \longrightarrow 2H_2O$$

So that both half-reactions involve the same number of electrons, we multiply the oxidation half-reaction by 2.

$$2Fe^{2+} \longrightarrow 2Fe^{3+} + 2e^-$$

Now we add the two half-reactions.

$$2Fe^{2+} \longrightarrow 2Fe^{3+} + 2e^-$$
$$\underline{H_2O_2 + 2H^+ + 2e^- \longrightarrow 2H_2O}$$
$$H_2O_2 + 2H^+ + 2Fe^{2+} \longrightarrow 2H_2O + 2Fe^{3+}$$

When using the half-reaction method in basic solution, you must remember that in *a basic solution* excess oxide combines with water to give hydroxide ion,

$$O^{2-} + H_2O \longrightarrow 2OH^-$$

and excess hydrogen combines with hydroxide ion to give water,

$$H^+ + OH^- \longrightarrow H_2O$$

EXAMPLE 21.14   Hydrogen peroxide reduces MnO_4^- to MnO_2 in basic solution. Write a balanced equation for the reaction.

Initially we have

$$H_2O_2 + MnO_4^- \longrightarrow MnO_2$$

and we know that the reaction is in basic solution. The reduction half-reaction involves the change

$$MnO_4^- \longrightarrow MnO_2$$

Since it is in basic solution, the excess oxide combines with water to give hydroxide ion. Thus H_2O is needed as a reactant and OH^- is a product.

$$MnO_4^- + 2H_2O \longrightarrow MnO_2 + 4OH^-$$

Adding electrons to balance the charge gives us the following half-reaction:

$$MnO_4^- + 2H_2O + 3e^- \longrightarrow MnO_2 + 4OH^-$$

The oxidation half-reaction involves oxidation of hydrogen peroxide. Since oxygen already has an oxidation number of -1 in H_2O_2, it can only be oxidized to O_2, with an oxidation number of zero. The oxidation half-reaction must involve the change

$$H_2O_2 \longrightarrow O_2$$

In basic solution excess hydrogen combines with hydroxide ion to give water.

$$H_2O_2 + 2OH^- \longrightarrow O_2 + 2H_2O$$

Addition of electrons to balance the charge gives us

$$H_2O_2 + 2OH^- \longrightarrow O_2 + 2H_2O + 2e^-$$

After multiplying to get the same number of electrons in both half-reactions, we add them to give the overall equation.

$$2MnO_4^- + 4H_2O + 6e^- \longrightarrow 2MnO_2 + 8OH^-$$
$$3H_2O_2 + 6OH^- \longrightarrow 3O_2 + 6H_2O + 6e^-$$
$$\overline{2MnO_4^- + 4H_2O + 3H_2O_2 + 6OH^- \longrightarrow 2MnO_2 + 8OH^- + 3O_2 + 6H_2O}$$

Cancellation of identical species on the left and right sides of this equation gives

$$2MnO_4^- + 3H_2O_2 \longrightarrow 2MnO_2 + 3O_2 + 2OH^- + 2H_2O$$

21.19 Some Half-Reactions

When redox equations are balanced by the half-reaction method, the correct half-reactions must be used. The half-reactions in Table 21.3 on page 642 should be studied carefully, for they are frequently encountered in the balancing of redox equations. The products listed are the *major products* under the conditions given but often are not the sole products.

21.20 Balancing Redox Equations by the Change in Oxidation Number Method

Oxidation and reduction always occur together and to the same extent; the total increase in oxidation number due to oxidation must equal the total decrease in oxidation number due to reduction. This makes it possible to use changes in oxidation number in balancing redox equations.

Table 21.3 Important half-reactions

$X_2 + 2e^- \longrightarrow 2X^-$	The halogens F_2, Cl_2, Br_2, and I_2 as oxidizing agents
$O_2 + 4H^+ + 4e^- \longrightarrow 2H_2O$	Oxygen as an oxidizing agent in acid solution
$2H^+ + 2e^- \longrightarrow H_2$	The hydrogen ion as an oxidizing agent
$H_2 \longrightarrow 2H^+ + 2e^-$	Molecular hydrogen as a reducing agent in aqueous solution
$M \longrightarrow M^{n+} + ne^-$	Metals as reducing agents
$M^{n+} + ne^- \longrightarrow M$	Metal ions as oxidizing agents
$M^{n+} \longrightarrow M^{n+x} + xe^-$	Oxidation of a metal ion to a higher oxidation number; for example, $$Cr^{2+} \longrightarrow Cr^{3+} + e^-$$ $$Fe^{2+} \longrightarrow Fe^{3+} + e^-$$ $$Sn^{2+} \longrightarrow Sn^{4+} + 2e^-$$
$M^{n+} + xe^- \longrightarrow M^{n-x}$	Reduction of a metal ion to a lower oxidation number; for example, $$Fe^{3+} + e^- \longrightarrow Fe^{2+}$$
$H_2O_2 + 2H^+ + 2e^- \longrightarrow 2H_2O$	Hydrogen peroxide as an oxidizing agent in acid solution
$H_2O_2 \longrightarrow O_2 + 2H^+ + 2e^-$	Hydrogen peroxide as a reducing agent in acid solution
$MnO_4^- + 8H^+ + 5e^- \longrightarrow Mn^{2+} + 4H_2O$	Permanganate ion as an oxidizing agent in acid solution
$MnO_4^- + 2H_2O + 3e^- \longrightarrow MnO_2 + 4OH^-$	Permanganate ion as an oxidizing agent in neutral or basic solution
$Cr_2O_7^{2-} + 14H^+ + 6e^- \longrightarrow 2Cr^{3+} + 7H_2O$	Dichromate ion as an oxidizing agent in acid solution
$NO_3^- + 2H^+ + e^- \longrightarrow NO_2 + H_2O$	Concentrated nitric acid as an oxidizing agent toward less active metals, such as Cu, Ag, and Pb, and certain anions, such as Br^-
$NO_3^- + 4H^+ + 3e^- \longrightarrow NO + 2H_2O$	Dilute nitric acid as an oxidizing agent toward less active metals and certain nonmetals, such as Br^-
$2NO_3^- + 10H^+ + 8e^- \longrightarrow N_2O + 5H_2O$	Dilute nitric acid as an oxidizing agent toward moderately active metals, such as Zn and Fe
$H_2SO_4 + 2H^+ + 2e^- \longrightarrow SO_2 + 2H_2O$	Concentrated sulfuric acid as an oxidizing agent toward HBr, C, and Cu
$H_2SO_4 + 6H^+ + 6e^- \longrightarrow S + 4H_2O$	Concentrated sulfuric acid as an oxidizing agent toward H_2S
$H_2SO_4 + 8H^+ + 8e^- \longrightarrow H_2S + 4H_2O$	Concentrated sulfuric acid as an oxidizing agent toward HI
$NO_2^- + 2H^+ + e^- \longrightarrow NO + H_2O$	Nitrous acid as an oxidizing agent
$NO_2^- + H_2O \longrightarrow NO_3^- + 2H^+ + 2e^-$	Nitrous acid as a reducing agent
$HClO + H^+ + 2e^- \longrightarrow Cl^- + H_2O$	Hypochlorous acid as an oxidizing agent

EXAMPLE 21.15

Use the changes in oxidation number to balance the equation for the reaction of antimony and chlorine to form antimony(III) chloride.

First we write the reactants and products of the reaction and then indicate the oxidation number of each atom in the reactants and products (Section 6.10. Remember that elements in the free state are always assigned an oxidation number of zero.)

$$\overset{0}{Sb} + \overset{0}{Cl_2} \longrightarrow \overset{+3\;-1}{SbCl_3}$$

We indicate the changes in oxidation number that occur during the reaction. An ascending arrow (↑) denotes an increase in oxidation number, and a descending arrow (↓) denotes a decrease in oxidation number.

$$\overset{0}{Sb} \longrightarrow \overset{+3}{Sb} \;\uparrow 3 \qquad \text{(total increase of 3 in oxidation number per formula unit, Sb)}$$

$$\overset{0}{Cl_2} \longrightarrow \overset{-1}{2Cl} \;\downarrow 2 \qquad \text{(total decrease of 2 in oxidation number per formula unit, } Cl_2\text{)}$$

To balance the equation we must select the proper numbers of antimony and chlorine atoms so that the total increase in oxidation number will equal the total decrease. If we use 2 atoms of antimony and 3 molecules (6 atoms) of chlorine, then there will be both an increase and a decrease of 6 units of oxidation number.

$$\overset{0}{Sb} \longrightarrow \overset{+3}{Sb} \;\uparrow \qquad 3 \times 2 = 6$$

$$\overset{0}{Cl_2} \longrightarrow \overset{-1}{2Cl} \;\downarrow \qquad 2 \times 3 = 6$$

The balanced equation, then, is

$$2Sb + 3Cl_2 \longrightarrow 2SbCl_3$$

EXAMPLE 21.16

The chloride ion is oxidized by the permanganate ion in acid solution to give the manganese(II) ion, molecular chlorine, and water. Balance the equation

$$MnO_4^- + Cl^- + H^+ \longrightarrow Mn^{2+} + Cl_2 + H_2O$$

using the changes in oxidation number.

After we assign oxidation numbers, it becomes evident that manganese is reduced from +7 to +2 and chlorine is oxidized from −1 to 0.

$$\overset{+7}{MnO_4^-} + Cl^- + H^+ \longrightarrow Mn^{2+} + \overset{0}{Cl_2} + H_2O$$

$$2Cl^- \longrightarrow \overset{0}{Cl_2} \;\uparrow \qquad 2 \times 5 = 10$$

$$\overset{+7}{Mn} \longrightarrow Mn^{2+} \;\downarrow \qquad 5 \times 2 = 10$$

We can see that 10 chloride ions increase by 10 units of oxidation number in going to

5 chlorine molecules (10 chlorine atoms), and 2 manganese atoms decrease by 10 units of oxidation number.

$$2MnO_4^- + 10Cl^- + ?H^+ \longrightarrow 2Mn^{2+} + 5Cl_2 + ?H_2O$$

Now we balance the ion charges on both sides of the equation. On the right side of the equation the only charged particles are the 2 manganese ions, with a total ionic charge of $+4$. The sum of the charges on the left must also be equal to $+4$. The sum of the charges on the 2 permanganate ions and the 10 chloride ions is -12, so 16 hydrogen ions are necessary to give a sum of $+4$.

$$-2 - 10 + 16 = +4$$

The 16 hydrogen ions produce 8 molecules of water.

$$2MnO_4^- + 10Cl^- + 16H^+ \longrightarrow 2Mn^{2+} + 5Cl_2 + 8H_2O$$

To check the balancing, note that there are 8 oxygen atoms on each side of the arrow.

EXAMPLE 21.17   Very active reducing agents such as zinc can reduce dilute nitric acid to ammonium ion. Using the changes in oxidation number, balance the equation for this reduction with zinc.

First we write the reactants and products of the reaction and assign oxidation numbers to the atoms that undergo a change in oxidation number.

$$\overset{0}{Zn} + \overset{+5}{NO_3^-} + H^+ \longrightarrow Zn^{2+} + \overset{-3}{NH_4} + H_2O$$

Then we balance the equation with regard to the atoms that change in oxidation number.

$$\overset{0}{Zn} \longrightarrow Zn^{2+} \uparrow \quad 2 \times 4 = 8$$

$$\overset{+5}{N} \longrightarrow \overset{-3}{N} \quad \downarrow \quad 8 \times 1 = 8$$

$$4Zn + NO_3^- + ?H^+ \longrightarrow 4Zn^{2+} + NH_4^+ + ?H_2O$$

Next, we balance the ion charges to find the number of hydrogen ions required.

$$NO_3^- + ?H^+ \longrightarrow 4Zn^{2+} + NH_4^+$$

$$4Zn + NO_3^- + 10H^+ \longrightarrow 4Zn^{2+} + NH_4^+ + ?H_2O$$
$$\quad\quad\quad -1 \quad\quad +10 \quad = \quad +8 \quad\quad +1$$

Of the 10 hydrogen atoms on the left, 4 are found in the ammonium ion on the right. Thus the other 6 hydrogen atoms form 3 molecules of water on the right.

$$4Zn + NO_3^- + 10H^+ \longrightarrow 4Zn^{2+} + NH_4^+ + 3H_2O$$

To check the balancing of the equation, we note that there are 3 oxygen atoms on each side of the arrow. We also note that the net charge on each side of the arrow is $9+$. Charges and atoms thus balance respectively on each side.

For Review

Summary

In an **electrolytic cell,** electrical energy is used to produce a chemical change; in a **voltaic cell,** chemical changes are used to produce electrical energy. In either type of cell the electrode at which oxidation occurs is the **anode,** and the electrode at which reduction occurs, the **cathode.**

Typical changes carried out in electrochemical cells include electrolysis of molten sodium chloride, producing sodium and chlorine; of an aqueous solution of hydrogen chloride, producing hydrogen and chlorine; of an aqueous solution of sodium chloride, producing hydrogen, chlorine, and a solution of sodium hydroxide; and of a solution of sulfuric acid, producing hydrogen and oxygen. The extent of the chemical changes in these processes can be related to the amount of electricity that passes through the cell. A **faraday,** which corresponds to passage of 1 mole of electrons, reduces 1 **equivalent** of a substance at the cathode or is produced by the oxidation of 1 equivalent of a substance at the anode. The amount of electrical charge possessed by a mole of electrons is 96,485 coulombs; 1 **coulomb** is the quantity of electricity involved when a current of 1 ampere flows for 1 second.

The potential that causes electrons from a voltaic cell to move from one electrode to the other through an external circuit is called the **electromotive force,** or **emf,** of the cell. The emf of a cell can be regarded as the sum of the electrode potential at the anode plus the electrode potential at the cathode. A positive emf value indicates that the cell reaction will occur spontaneously as written. **Standard electrode potentials,** or **standard reduction potentials,** are tabulated for a variety of reduction (cathode) **half-reactions** with ion concentrations of 1 M, gas pressures of 1 atmosphere, and a temperature of 25°C. The potential of an oxidation half-cell (the anode half-reaction) is the same magnitude as and opposite in sign to the reduction potential. The more positive the potential associated with a half-reaction, the greater is the tendency of that reaction to occur as written. Electrode potentials vary with the concentrations of the reactants and products involved in the half-reaction and can be calculated from standard electrode potentials using the **Nernst equation:**

$$E = E° - \frac{0.05915}{n} \log Q$$

where n is the number of moles of electrons involved in the cell reaction.

The standard free energy change, $\Delta G°_{298}$, for a cell reaction can be calculated from the emf of a cell based on the reaction. At standard conditions

$$\Delta G°_{298} = -nFE°$$

($F = 96.485$ kJ V^{-1}, or 96,485 coulombs.) The equilibrium constant for the reaction is related to the standard free energy change and thus to $E°$.

$$E° = \frac{0.05915}{n} \log K$$

Examples of voltaic cells include the Daniell cell and the familiar flashlight battery, a dry cell. These are **primary cells,** which cannot be recharged. The lead storage battery is an example of a **secondary cell,** which is rechargeable.

Oxidation-reduction (redox) reactions can be considered to consist of two half-reactions. The standard free energy change and equilibrium constant of an oxidation-reduction reaction can be calculated from the standard electrode potentials of the two half-reactions. After the half-reactions are balanced, they can be added to obtain the overall balanced equation for the reaction. Redox equations can also be balanced by considering the changes in oxidation number that accompany the reaction.

Key Terms and Concepts

activity series (21.8)
anode (21.2)
cathode (21.2)
cathodic protection (21.16)
cell potential (21.7)
change in oxidation number
    method for balancing redox
    equations (21.20)
coulomb (21.5)
electrode (21.6)
electrode potential (21.6)
electrolysis (21.2)
electrolytes (21.1)
electrolytic cell (21.2)

electrolytic conduction (21.1)
electromotive force, emf (21.7)
equilibrium constants (21.12)
equivalent (21.5)
faraday, F (21.5)
Faraday's law (21.5)
free energy change (21.12)
fuel cell (21.15)
gas electrode (21.6)
half-cell (21.6)
half-reaction (21.2)
half-reaction method for balancing
    redox equations (21.18)
metallic conduction (21.1)

Nernst equation (21.11)
normality (21.5)
primary cell (21.13)
salt bridge (21.7)
secondary cell (21.14)
spontaneity of redox reactions
    (21.17)
standard electrode potential (21.7)
standard hydrogen electrode
    (21.6)
volt (21.6)
voltaic cell (21.7)

Exercises

Can you explain #7,#8

1. How does conduction of electricity in a metal differ from conduction of electricity in a solution of an electrolyte?
2. How does a voltaic cell differ from an electrolytic cell?
3. Define an anode and a cathode.
4. Complete and balance the following half-reactions. (In each case indicate whether oxidation or reduction occurs.)
    (a) $Cd \longrightarrow Cd^{2+}$
    (b) $I_2 \longrightarrow I^-$
    (c) $AgCl \longrightarrow Ag + Cl^-$
    (d) $H_2O \longrightarrow H_2$ (in base)
    (e) $H_2O_2 \longrightarrow O_2$ (in acid)
    (f) $ClO_3^- \longrightarrow ClO_4^-$ (in acid)
    (g) $MnO_4^- \longrightarrow MnO_2$ (in base)
    (h) $Cl^- \longrightarrow ClO_3^-$ (in acid)
    (i) $[Co(NH_3)_6]^{3+} \longrightarrow Co + NH_3$

Electrolytic Cells and Faraday's Law of Electrolysis

5. Describe the electrolytic purification of aluminum, the Hoopes process, using half-reactions to describe the changes at the anode and cathode.
6. In this and previous chapters we have described commercial electrochemical preparations of hydrogen, oxygen, chlorine, magnesium, and sodium hydroxide. Write equations for the net cell reaction and the individual electrode half-reactions describing these commercial processes.

do we need to know this

7. Using a diagram of the type

$$\xrightarrow{e^-}$$

Anode | anode soln ‖ cathode soln | cathode

diagram the electrolytic cell for each of the following cell reactions:
    (a) $2NaBr \longrightarrow 2Na + Br_2$ (using Fe and inert carbon as electrodes)
    (b) $2NaCl(aq) + 2H_2O \longrightarrow$
                $2NaOH(aq) + H_2(g) + Cl_2(g)$
      (using inert electrodes of titanium, Ti)
    (c) $Cu + 2Ag^+ \longrightarrow Cu^{2+} + 2Ag$
    (d) $Br_2 + 2I^- \longrightarrow I_2 + 2Br^-$ (using platinum electrodes)
8. Write the anode half-reaction, the cathode half-reaction, and the cell reaction of the following electrolytic cells:
    (a) $C \,|\, NaCl(l) \,|\, Cl_2 \,\|\, NaCl(l) \,|\, Na(l) \,|\, Fe$ $\xrightarrow{e^-}$
    (b) $Pt \,|\, Cl_2 \,|\, Cl^- \,\|\, H^+ \,|\, H_2 \,|\, Pt$ $\xrightarrow{e^-}$
    (c) $Fe \,|\, Fe^{2+}; Fe^{3+} \,\|\, MnO_4^-; OH^- \,|\, MnO_2 \,|\, Pt$ $\xrightarrow{e^-}$

(d) $Pb \mid PbSO_4 \mid H_2SO_4(aq) \parallel$

$$H_2SO_4(aq) \mid PbSO_4 \mid PbO_2 \mid Pb$$

9. Tarnished silverware is coated with Ag_2S. The tarnish can be removed by placing the silverware in an aluminum pan and covering it with a solution of an inert electrolyte such as NaCl. Explain the electrochemical basis for this procedure.

10. Soon after an iron rod is placed in a copper nitrate solution, Fe^{2+} ions are observed in the solution and copper metal has been deposited on the rod. Can the iron rod be considered to be an electrode? If so, is it an anode or a cathode?

11. State Faraday's law. What is a faraday of electricity?

12. Calculate the value of the Faraday constant, F, from the charge on a single electron, 1.6021×10^{-19} C.

Ans. 9.648×10^4 C

13. How many moles of electrons are involved in the following electrochemical changes?
    (a) 0.800 mol of I^- is converted to I_2 *Ans. 0.800 mol*
    (b) 118 g of Fe^{2+} is converted to Fe   *Ans. 4.23 mol*
    (c) 27.6 g of SO_3^- is converted to SO_3^{2-} *Ans. 0.689 mol*
    (d) 1.174 g of MnO_4^- is converted to MnO_2
    2.961×10^{-2} mol
    (e) 1.0 L of H_2 at STP is converted to H_2O in acid solution   *Ans. 8.9×10^{-2} mol*
    (f) The Cu^{2+} in 100 mL of 0.250 M $CuSO_4$ solution is converted to Cu 5.00×10^{-2} mol
    (g) The Mn^{2+} in 12.58 mL of 0.1145 M $Mn(NO_3)_2$ is converted to MnO_4^- *Ans. 7.202×10^{-3} mol*

14. How many faradays of electricity are involved in the electrochemical changes described in exercise 13?
    Ans. (a) 0.800 F; (b) 4.23 F; (c) 0.689 F; (d) 2.961 × 10^{-2} F; (e) 8.9 × 10^{-2} F; (f) 5.00 × 10^{-2} F; (g) 7.202 × 10^{-3} F

15. How many coulombs of electricity are involved in the electrochemical changes described in exercise 13?
    Ans. (a) 7.72×10^4 C; (b) 4.08×10^5 C; (c) 6.65×10^4 C; (d) 2.857×10^3 C; (e) 8.6×10^3 C; (f) 4.82×10^3 C; (g) 694.9 C

16. Aluminum is manufactured by the electrolysis of a molten mixture of Al_2O_3 and Na_3AlF_6. How many moles of electrons are required to convert 1.0 mol Al^{3+} to Al? How many faradays? How many coulombs?
    Ans. 3.0 mol; 3.0 F; 2.9×10^5 C

17. How many moles of electrons are required to prepare 1.000 metric ton (1000 kg) of sodium hydroxide by electrolysis of an aqueous solution of sodium chloride as described in Section 21.3? How many faradays? How many coulombs? How long will the electrolysis take if a current of 100,000 A is used? Assume that the efficiency of the electrochemical cell is 100%; that is, every electron involved results in the production of a chlorine atom. (In an actual commercial cell the efficiency is about 65%.)
    Ans. 2.500×10^4 mol; 2.500×10^4 F; 2.412×10^9 C; 2.412×10^4 s (6.701 h)

18. Ammonium perchlorate, NH_4ClO_4, used in the solid fuel in the booster rockets on the space shuttle, is prepared from sodium perchlorate, $NaClO_4$, which is produced commercially by the electrolysis of a hot, stirred solution of sodium chloride.

$$NaCl + 4H_2O \longrightarrow NaClO_4 + 4H_2$$

How many moles of electrons are required to produce 1.00 kg of sodium perchlorate? How many faradays? How many coulombs?   *Ans. 65.3 mol; 65.3 F; 6.30×10^6 C*

19. Electrolysis of a sulfuric acid solution with a certain amount of current produces 0.3718 g of oxygen. What mass of silver would be produced by the same amount of current? What mass of copper?
    Ans. 5.013 g Ag; 1.477 g Cu

20. How many grams of tin will be deposited from a solution of tin(II) nitrate by 3.40 F of electricity?   *Ans. 202 g*

21. An experiment is conducted using the apparatus depicted in Figure 21.8. How many grams of gold would be plated out of solution by the current required to plate out 4.97 g of silver? How many moles of hydrogen gas simultaneously are released from the hydrochloric acid solution? How many moles of oxygen gas are freed at each anode in the copper sulfate and silver nitrate solutions?
    Ans. 3.03 g Au; 2.30×10^{-2} mol H_2; 1.15×10^{-2} mol O_2

22. How many moles of electrons flow through a lamp that draws a current of 2.0 A for 1.0 h?
    Ans. 7.5×10^{-2} mol

23. How many grams of cobalt will be deposited from a solution of cobalt(II) chloride electrolyzed with a current of 10.0 A for 30.0 min?   *Ans. 5.50 g*

24. How many faradays of electricity would be required to reduce 21.0 g of $CdCl_2$ to metallic cadmium? How long would this take (in minutes) with a current of 7.5 A?
    Ans. 0.229 F; 49 min

25. Chromium metal can be plated electrochemically from an acidic aqueous solution of CrO_4^{2-}. (a) What is the half-reaction for the process? (b) What mass of chromium, in grams, will be deposited by a current of 250 A passing for 20.0 min? (c) How long will it take to deposit 1.0 g of chromium using a current of 10.0 A?
    Ans. (a) $CrO_4^{2-} + 8H^+ + 6e^- \longrightarrow Cr + 4H_2O$; (b) 26.9 g; (c) 1100 s

26. A single commercial electrolytic cell for the production of chlorine draws a current of 150,000 A. Assume that the efficiency of the cell is 100% and calculate the mass of chlorine in kilograms produced by such a cell in 1.00 h.   *Ans. 198 kg*

27. Which metals in Table 21.1 could be purified by an electrolysis similar to that used to purify copper?

28. Why are different products obtained when molten $ZnCl_2$ is electrolyzed than when a solution of $ZnCl_2$ is electrolyzed? Why do a concentrated solution of $CuCl_2$ and molten $CuCl_2$ give the same products on electrolysis?

Standard Reduction Potentials and Electromotive Force

29. Diagram voltaic cells having the following net reactions:
    (a) $2Li + Cl_2 \longrightarrow 2LiCl$
    (b) $Mn + 2Ag^+ \longrightarrow Mn^{2+} + Ag$
    (c) $Sn^{4+} + H_2 \longrightarrow Sn^{2+} + 2H^+$
    (d) $Cr_2O_7^{2-} + 14H^+ + 6I^- \longrightarrow 3I_2 + 2Cr^{3+} + 7H_2O$
    (e) $2H_2 + O_2 \longrightarrow 2H_2O$

30. Define *standard reduction potential*.

31. Why is the potential for the standard hydrogen electrode listed at 0.00?

32. Which is the better oxidizing agent in each of the following pairs at standard conditions?
    (a) Al^{3+} or Cu^{2+}
    (b) Br_2 or Mn^{2+}
    (c) MnO_4^- or $Cr_2O_7^{2-}$ (in acid solution)
    (d) $AgCl$ or Hg_2Cl_2
    (e) MnO_4^- in acid or MnO_4^- in base

33. Which is the better reducing agent in each of the following pairs at standard state conditions?
    (a) F^- or Cl^-
    (b) H_2 or Mn^{2+}
    (c) Sn^{2+} or Co^{2+}
    (d) Fe^{2+} or MnO_2
    (e) H_2O_2 in acid or H_2O_2 in base

34. Calculate the emf of each cell at standard conditions based on the following reactions:
    (a) $Co + Cl_2 \longrightarrow Co^{2+} + 2Cl^-$ *Ans. 1.636 V*
    (b) $Cd + Sn^{2+} \longrightarrow Cd^{2+} + Sn$ *Ans. 0.26 V*
    (c) $Pt^{2+} + 2Cl^- \longrightarrow Pt + Cl_2$ *Ans. −0.2 V*
    (d) $MnO_4^- + 8H^+ + 5Ag \longrightarrow Mn^{2+} + 4H_2O + 5Ag^+$ *Ans. 0.71 V*
    (e) $Mn + 2Hg_2Cl_2 \longrightarrow Mn^{2+} + 2Cl^- + 2Hg$ *Ans. 1.45 V*
    (f) $Zn + NiO_2 + 2H_2O \longrightarrow Zn(OH)_2 + Ni(OH)_2$ *Ans. 1.74 V*

35. Determine the standard emf for each of the following cells:

    (a) $\xrightarrow{e^-}$
    $Co \mid Co^{2+}(1\ M) \parallel Cr^{3+}(1\ M) \mid Cr$
    Ans. −0.46 V

    (b) $\xrightarrow{e^-}$
    $Ni \mid Ni^{2+}(1\ M) \parallel Br^-(1\ M) \mid Br_2(l) \mid Pt$
    Ans. +1.315 V

    (c) $\xrightarrow{e^-}$
    $Pb \mid PbSO_4(s) \mid SO_4^{2-}(1\ M) \parallel$
    $H^+(1\ M) \mid H_2(1\ atm) \mid Pt$
    Ans. +0.356 V

    (d) $\xrightarrow{e^-}$
    $Pt \mid Mn^{2+}(1\ M);\ MnO_4^-(1\ M);\ H^+(1\ M) \parallel$
    $Fe^{3+}(1\ M);\ Fe^{2+}(1\ M) \mid Pt$
    Ans. −0.74 V

36. Write the cell reaction for a voltaic cell based on each of the following pairs of half-reactions, and calculate the emf of the cell under standard conditions.
    (a) $La^{3+} + 3e^- \longrightarrow La$
    $Ni^{2+} + 2e^- \longrightarrow Ni$ *Ans. 2.27 V*
    (b) $S + 2e^- \longrightarrow S^{2-}$
    $I_2 + 2e^- \longrightarrow 2I^-$ *Ans. 1.02 V*
    (c) $ZnS + 2e^- \longrightarrow Zn + S^{2-}$
    $CdS + 2e^- \longrightarrow Cd + S^{2-}$ *Ans. +0.23 V*
    (d) $Co(OH)_3 + e^- \longrightarrow Co(OH)_2 + OH^-$
    $Cr(OH)_4^- + 3e^- \longrightarrow Cr + 4OH^-$ *Ans. +1.4 V*
    (e) $HClO_2 + 2H^+ + 2e^- \longrightarrow HClO + H_2O$
    $ClO_3^- + 3H^+ + 2e^- \longrightarrow HClO_2 + H_2O$ *Ans. +0.43 V*

37. Rechargeable nickel-cadmium cells are used in calculators and other battery-powered devices. The Telstar communication satellite also uses these cells. The cell reaction is

    $$NiO_2 + Cd + 2H_2O \longrightarrow Ni(OH)_2 + Cd(OH)_2$$

    Calculate the emf of such a cell, using the following half-cell potentials:

    $$NiO_2 + 2H_2O + 2e^- \longrightarrow Ni(OH)_2 + 2OH^-$$
    $$E° = +0.49\ V$$

    $$Cd(OH)_2 + 2e^- \longrightarrow Cd + 2OH^-$$
    $$E° = -0.81\ V$$
    Ans. 1.30 V

38. (a) Calculate the emf at standard state conditions of a single lead storage cell.   *Ans. 2.041 V*
    (b) A lead storage battery, such as the one used in an automobile, contains six lead storage cells. What maximum voltage would be expected of such a battery?   *Ans. 12.246 V*

39. Under standard conditions the potential of the cell diagrammed below is +0.74 V. What is the metal M? Show your calculations.

    $$M \mid M^{n+} \parallel Cu^{2+} \mid Cu$$
    Ans. Cd

40. What is the cell with the highest emf that could be constructed from the metals magnesium, nickel, cobalt, and cadmium?   *Ans. $Mg|Mg^{2+} \parallel Ni^{2+}|Ni$*

The Nernst Equation

41. What is the utility of the Nernst equation in electrochemistry?

42. Why does the potential of a single electrode change as the concentrations of the species involved in the half-reaction change?

43. Which is the better oxidizing agent in each of the following pairs? (a) MnO_2 in acidic solution or MnO_2 in basic solution
    (b) $HClO$ in acidic solution or $HClO$ in neutral solution
    (c) HO_2^- in neutral solution or HO_2^- in basic solution

44. Calculate the emf for each of the following half-reactions:
    (a) $Sn^{2+}(0.0100\ M) + 2e^- \longrightarrow Sn$ *Ans. −0.195 V*

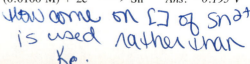

(b) $Hg \longrightarrow Hg^{2+}(0.2500\ M) + 2e^-$ *Ans. −0.836 V*

(c) $O_2(0.0010\ atm) + 4H^+(0.100\ M) + 4e^- \longrightarrow$
$$2H_2O(l)$$ *Ans. 1.13 V*

(d) $Cr_2O_7{}^{2-}(0.150\ M) + 14H^+(0.100\ M) + 6e^- \longrightarrow$
$$2Cr^{3+}(0.000100\ M) + 7H_2O(l)$$ *Ans. 1.26 V*

(e) $Mn^{2+}(0.0125\ M) + 4H_2O(l) \longrightarrow$
$$MnO_4{}^-(0.0125\ M) + 8H^+(0.100\ M) + 5e^-$$ *Ans. −1.42 V*

(f) $Sn^{4+}(0.00010\ M) + 2e^- \longrightarrow Sn^{2+}(4.0\ M)$ *Ans. 0.01 V*

45. Hypochlorous acid, HOCl, is a stronger oxidizing agent in acidic solution than in neutral solution. Calculate the potential for the reduction of HOCl to Cl^- in a solution with a pH of 7.00 in which [HOCl] and [Cl^-] both equal 1.00 M.   *Ans. +1.28 V*

46. The standard reduction potential of oxygen in acidic solution is 1.23 V ($O_2 + 4H^+ + 4e^- \longrightarrow 2H_2O$). Calculate the standard reduction potential of oxygen in basic solution and compare your results with the value in Appendix H. (*Hint:* What is [H^+] when [OH^-] is 1 M? *Ans. 0.40 V*

47. The Nernst equation for the net reaction of an electrochemical cell is

$$E = E° - \frac{0.05915}{n} \log Q$$

where Q is the reaction quotient for the reaction and n is the total number of moles of electrons transferred in the net reaction. Show that the Nernst equation for the net reaction

$$2Au + 3Cl_2 \longrightarrow 2Au^{3+} + 6Cl^-$$

is identical to that derived by addition of the Nernst equations for the two half-reactions

$$Au \longrightarrow Au^{3+} + 3e^- \qquad E° = -1.50\ V$$
$$Cl_2 + 2e^- \longrightarrow 2Cl^- \qquad E° = +1.3595\ V$$

48. Calculate the voltage produced by each of the following cells:

(a) $Zn \mid Zn^{2+}(0.0100\ M) \parallel Cu^{2+}(1.00\ M) \mid Cu$ *Ans. +1.16 V*

(b) $Al \mid Al^{3+}(0.250\ M) \parallel Co^{2+}(0.0500\ M) \mid Co$ *Ans. +1.36 V*

(c) $Pt \mid Br_2(l) \mid Br^-(0.450\ M) \parallel$
$$Cl^-(0.0500\ M) \mid Cl_2(g)(0.900\ atm) \mid Pt$$ *Ans. +0.3494 V*

(d) $Pt \mid H_2(g)(0.790\ atm) \mid H^+(0.100\ M) \parallel$
$$Cl^-(0.0500\ M) \mid Cl_2(g)(0.100\ atm) \mid Pt$$ *Ans. 1.4630 V*

(e) $Cu \mid Cu^{2+}(2.0\ M) \parallel Cu^{2+}(0.010\ M) \mid Cu$ *Ans. −0.068 V*

49. What is the theoretical potential required to electrolyze a 0.0250 M solution of nickel(II) chloride, producing metallic nickel and chlorine at 0.300 atm of pressure?   *Ans. 1.718 V*

50. A standard zinc electrode is combined with a hydrogen

electrode with H_2 at 1 atm. If the emf of the cell is 0.46 V, what is the pH of the electrolyte in the hydrogen electrode?   *Ans. 5.07*

Free Energy Changes and Equilibrium Constants

51. For a cell based on each of the following reactions run at standard state conditions, calculate the emf of the cell, the standard free energy change of the reaction, and the equilibrium constant of the reaction:

(a) $Mn(s) + Cd^{2+}(aq) \longrightarrow Mn^{2+}(aq) + Cd(s)$
    Ans. +0.78 V; −150 kJ; 1 × 10²⁶ (log K = 26)

(b) $2Al(s) + 3Co^{2+}(aq) \longrightarrow 2Al^{3+}(aq) + 3Co(s)$
    Ans. +1.38 V; −799 kJ; 1 × 10¹⁴⁰ (log K = 140)

(c) $2Br^-(aq) + I_2(s) \longrightarrow Br_2(l) + 2I^-(aq)$
    Ans. −0.5297 V; 102.2 kJ; 1.2 × 10⁻¹⁸

(d) $Cr_2O_7{}^{2-}(aq) + 3Fe(s) + 14H^+(aq) \longrightarrow$
$$2Cr^{3+}(aq) + 3Fe^{2+}(aq) + 7H_2O(l)$$ *Ans. 1.77 V; −1020 kJ; 1 × 10¹⁸⁰*

(e) $2Mn^{2+}(aq) + 8H_2O(l) + 5Fe^{2+}(aq) \longrightarrow$
$$2MnO_4{}^-(aq) + 16H^+(aq) + 5Fe(s)$$ *Ans. −1.95 V; 1880 kJ; 1 × 10⁻³³⁰*

52. Calculate the standard free energy change and equilibrium constant for the reaction

$$2Br^- + Cl_2 \longrightarrow 2Cl^- + Br_2$$ *Ans. 56.79 kJ; 8.93 × 10⁹*

53. Using the standard reduction potentials for the half-reactions in the hydrogen-oxygen fuel cell, calculate the free energy change and the equilibrium constant for the combustion of hydrogen:

$$2H_2 + O_2 \longrightarrow 2H_2O$$ *Ans. −475 kJ; 1.5 × 10⁸³*

54. Copper(I) salts disproportionate in water to form copper(II) salts and copper metal:

$$2Cu^+ \longrightarrow Cu^{2+} + Cu$$

What concentration of Cu^+ remains at equilibrium in 1.00 L of a solution prepared from 1.00 mol of Cu_2SO_4?   *Ans. 7.8 × 10⁻⁴ M*

55. Use the emf values of the following half-cells and show that hydrogen peroxide, H_2O_2, is unstable with respect to decomposition into oxygen and water:

$$H_2O_2 + 2H^+ + 2e^- \longrightarrow 2H_2O \qquad E° = +1.77\ V$$
$$O_2 + 2H^+ + 2e^- \longrightarrow H_2O_2 \qquad E° = +0.68\ V$$

56. Should each of the following compounds be stable in a 1 M aqueous solution? (*Hint:* Check the possibility of oxidation or reduction of the anion by the cation.)

(a) $Sn(MnO_4)_2$ (in 1 M H^+)   *Ans. No*

(b) FeI_3 *Ans. No*

(c) $Fe[HgBr_4]$ *Ans. Yes*

(d) $[Co(NH_3)_6](ClO)_2$ *Ans. No*

(e) $Na_2[Cd(CN)_4]$ *Ans. Yes*

Oxidation-Reduction Reactions

57 Balance the following redox equations.
(a) $H_2SO_4 + HBr \longrightarrow SO_2 + Br_2 + H_2O$
(b) $Zn + H^+ + NO_3^- \longrightarrow Zn^{2+} + N_2 + H_2O$
(c) $MnO_4^- + S^{2-} + H_2O \longrightarrow MnO_2 + S + OH^-$
(d) $NO_3^- + I_2 + H^+ \longrightarrow IO_3^- + NO_2 + H_2O$
(e) $Cu + H^+ + NO_3^- \longrightarrow Cu^{2+} + NO_2 + H_2O$
(f) $Zn + H^+ + NO_3^- \longrightarrow Zn^{2+} + N_2O + H_2O$
(g) $Cu + H^+ + NO_3^- \longrightarrow Cu^{2+} + NO + H_2O$
(h) $MnO_4^- + H_2S + H^+ \longrightarrow Mn^{2+} + S + H_2O$
(i) $H_2SO_4 + HI \longrightarrow H_2S + I_2 + H_2O$
(j) $MnO_4^- + NO_2^- + H_2O \longrightarrow MnO_2 + NO_3^- + OH^-$
(k) $OH^- + Cl_2 \longrightarrow ClO_3^- + Cl^- + H_2O$
(l) $H_2O_2 + MnO_4^- + H^+ \longrightarrow Mn^{2+} + H_2O + O_2$
(m) $MnO_4^{2-} + H_2O \longrightarrow MnO_4^- + OH^- + MnO_2$
(n) $Br_2 + SO_2 + H_2O \longrightarrow H^+ + Br^- + SO_4^{2-}$

58. Balance the following redox equations.
(a) $MnO_4^{2-} + Cl_2 \longrightarrow MnO_4^- + Cl^-$
(b) $OH^- + NO_2 \longrightarrow NO_3^- + NO_2^- + H_2O$
(c) $HBrO \longrightarrow H^+ + Br^- + O_2$
(d) $Br_2 + CO_3^{2-} \longrightarrow Br^- + BrO_3^- + CO_2$
(e) $CuS + H^+ + NO_3^- \longrightarrow Cu^{2+} + S + NO + H_2O$
(f) $NH_3 + O_2 \longrightarrow NO + H_2O$
(g) $C + HNO_3 \longrightarrow NO_2 + H_2O + CO_2$
(h) $ZnS + O_2 \longrightarrow ZnO + SO_2$
(i) $NO_3^- + Zn + OH^- + H_2O \longrightarrow NH_3 + Zn(OH)_4^{2-}$
(j) $HClO_3 \longrightarrow HClO_4 + ClO_2 + H_2O$
(k) $H_2S + H_2O_2 \longrightarrow S + H_2O$
(l) $ClO_3^- + H_2O + I_2 \longrightarrow IO_3^- + Cl^- + H^+$
(m) $Al + H^+ \longrightarrow Al^{3+} + H_2$
(n) $Cr_2O_7^{2-} + HNO_2 + H^+ \longrightarrow Cr^{3+} + NO_3^- + H_2O$

59. Complete and balance the following equations. When the reaction occurs in acidic solution, H^+ and/or H_2O may be added on either side of the equation, as necessary, to balance the equation properly; when the reaction occurs in basic solution, OH^- and/or H_2O may be added, as necessary, on either side of the equation. No indication of the acidity of the solution is given if neither H^+ nor OH^- is involved as a reactant or product.
(a) $Ag + NO_3^- \longrightarrow Ag^+ + NO$ (acidic solution)
(b) $C_2H_4 + MnO_4^- \longrightarrow Mn^{2+} + CO_2$ (acidic solution)
(c) $H_2S + I_2 \longrightarrow S + I^-$ (acidic solution)
(d) $Br_2 \longrightarrow BrO_3^- + Br^-$ (basic solution)
(e) $Ag^+ + AsH_3 \longrightarrow Ag + H_3AsO_3$ (acidic solution)
(f) $PbO_2 + Cl^- \longrightarrow Pb^{2+} + Cl_2$ (acidic solution)
(g) $CN^- + MnO_4^- \longrightarrow CNO^- + MnO_2$ (basic solution)
(h) $HgS + Cl^- + NO_3^- \longrightarrow [HgCl_4]^{2-} + NO_2 + S$ (acidic solution)
(i) $H_2O_2 + ClO_2 \longrightarrow ClO_2^- + O_2$ (basic solution)
(j) $UF_6^- + H_2O_2 \longrightarrow UO_2^{2+} + HF$ (acidic solution)

60. Complete and balance the following equations (see instructions for exercise 59).
(a) $Fe^{2+} + MnO_4^- \longrightarrow Fe^{3+} + Mn^{2+}$ (acidic solution)

(b) $Cr_2O_7^{2-} + I^- \longrightarrow Cr^{3+} + I_2$ (acidic solution)
(c) $Hg_2Cl_2 + NH_3 \longrightarrow Hg + HgNH_2Cl + NH_4^+ + Cl^-$
(d) $Fe^{3+} + I^- \longrightarrow Fe^{2+} + I_2$
(e) $MnO_2 + Cl^- \longrightarrow Mn^{2+} + Cl_2$ (acidic solution)
(f) $CN^- + [Fe(CN)_6]^{3-} \longrightarrow CNO^- + [Fe(CN)_6]^{4-}$ (basic solution)
(g) $Fe^{2+} + Cr_2O_7^{2-} \longrightarrow Fe^{3+} + Cr^{3+}$ (acidic solution)
(h) $I_2 + H_3AsO_3 \longrightarrow I^- + H_3AsO_4$ (acidic solution)
(i) $Sn^{2+} + HgCl_2 + Cl^- \longrightarrow SnCl_6^{2-} + Hg_2Cl_2$
(j) $CrO_4^{2-} + HSnO_2^- \longrightarrow HSnO_3^- + CrO_2^-$ (basic solution)
(k) $CH_2O + [Ag(NH_3)_2]^+ \longrightarrow Ag + HCO_2^- + NH_3$ (basic solution)

61. Complete and balance the following equation (see instructions for exercise 59). This is a challenging equation.

$Pb(N_3)_2 + Cr(MnO_4)_2 \longrightarrow$
$\quad Cr_2O_3 + MnO_2 + Pb_3O_4 + NO$ (basic solution)

Additional Exercises

62. When gold is plated electrochemically from a basic solution of $[Au(CN)_4]^-$, O_2 forms at one electrode and Au is deposited at the other. Write the half-reactions occurring at each electrode and the net reaction for the electrochemical cell. (The cyanide ion, CN^-, is not oxidized or reduced under these conditions.)

63. A lead storage battery is used to electrolyze a solution of hydrogen chloride. Sketch the two cells, label the cathode and anode in each cell, give the sign of each electrode, indicate the direction of flow of electrons through the system, show the movements of ions in the cells, and write the half-reactions occurring at each electrode.

64. A current of 15.0 A flowed for 25.0 min through water containing a small quantity of sodium hydroxide. How many liters of gas were formed at the anode at 27.0°C and 1.03 atm pressure? *Ans. 1.39 L*

65. A lead storage battery has 9.00 kg of lead and 9.00 kg of PbO_2, plus excess H_2SO_4. Theoretically, how long could this cell deliver a current of 50.0 A, without recharging, if it were possible to operate it so that the reaction goes to completion? *Ans. 40.3 h*

66. A total of 69,500 coulombs of electricity was required to reduce 37.7 g of M^{3+} to the metal. What is M? *Ans. Gd*

67. A current of 20.0 A is applied for 0.50 h to 1.0 L of a solution containing 1.00 mol of HCl. Calculate the pH of the solution after this time. *Ans. 0.20*

68. A current of 1.0 A is applied for 0.50 h to 50 mL of a 1.0 M solution of NaCl. Calculate the pH of the solution at the end of this time. *Ans. 13.57*

69. Describe briefly how you could determine the solubility product of AgCl ($K_{sp} \approx 1.8 \times 10^{-10}$), using an electrochemical measurement.

70. When chlorine dissolves in water, it disproportionates, producing chloride ion and hypochlorous acid. At what

hydrogen ion concentration does the potential (emf) for the disproportionation of chlorine change from a negative value to a positive value, assuming 1.00 atm of pressure and concentrations of 1.00 M for all species except hydrogen ion? (The standard electrode potential for the reduction of chlorine to chloride ion is 1.36 V, and for hypochlorous acid to chlorine, 1.63 V). Could chlorine be produced from hypochlorite and chloride ions in solution, through the reverse of the disproportionation reaction, by acidifying the solution with strong acid? Explain.

Ans. 2.7 × 10⁻⁵ M; yes.

71. The standard reduction potentials for the reactions

$$Ag^+ + e^- \longrightarrow Ag$$

and $$AgCl + e^- \longrightarrow Ag + Cl^-$$

are +0.7991 V and +0.222 V, respectively. From these data and the Nernst equation, calculate a value for the solubility product (K_{sp}) for AgCl. Compare your answer with the value given in Appendix D.   *Ans. 1.7 × 10⁻¹⁰*

72. Calculate the standard reduction potential for the reaction $H_2O + e^- \longrightarrow \frac{1}{2}H_2 + OH^-$ using the Nernst equation and the fact that the standard reduction potential for the reaction $H^+ + e^- \longrightarrow \frac{1}{2}H_2$ is, by definition, equal to 0.00 V.   *Ans. −0.83 V*

73. The standard reduction potentials for the reactions

$$Ag^+ + e^- \longrightarrow Ag$$

and $$[Ag(NH_3)_2]^+ + e^- \longrightarrow Ag + 2NH_3$$

are +0.7991 V and +0.373 V, respectively. From these values and the Nernst equation, determine K_f for the $[Ag(NH_3)_2]^+$ ion. Compare your answer with the value given in Appendix E.   *Ans. 1.6 × 10⁷*

The Nonmetals, Part 1: Hydrogen, Oxygen, Sulfur, and the Halogens

22

Ninety percent of the atoms in the universe are hydrogen atoms.

I n Chapter 20 we discussed the general behavior of the nonmetals as a group. In this chapter and the next we will examine the chemical behavior of some of these elements in more detail. This chapter treats hydrogen, oxygen, sulfur, and the halogens. The next chapter will treat carbon, nitrogen, phosphorus, and the noble gases.

Hydrogen

Early in the sixteenth century the Swiss-German physician Paracelsus noted that a flammable gas was formed by the reaction of sulfuric acid with iron. However, it was not until 1766 that Cavendish, an Englishman, recognized this gas as a distinct substance and prepared it by the action of various acids on certain metals and by the novel method of passing steam through a red-hot gun barrel. Lavoisier, a French chemist, named the gas *hydrogen,* meaning ''water producer,'' because water was formed when the gas burned in air.

22.1 Occurrence and Preparation of Hydrogen

Hydrogen is the most abundant element in the universe. The sun and other stars appear to be composed largely of hydrogen, as do the gases found in interstellar space. It is estimated that 90% of the atoms in the universe are hydrogen atoms. However, hydrogen is only the ninth most abundant element in the earth's crust, and only negligible quantities are found in the uncombined state on the earth.

Hydrogen comprises nearly 11% of the weight of water, its most abundant compound. It is an important part of the tissues of all plants and animals, petroleum, many minerals, cellulose and starch, sugar, fats, oils, alcohols, acids and bases, and thousands of other substances. Hydrogen is a component of more compounds than any other element.

At ordinary temperatures, hydrogen is a colorless, odorless, and tasteless gas consisting of diatomic molecules, H_2. It is the lightest known substance at STP. If cooled and compressed, hydrogen changes to a liquid that boils at $-253°C$ (20 K) and freezes at $-259°C$ (14 K).

Hydrogen is more expensive to isolate than elements like oxygen and nitrogen that are found in the air in an uncombined state. Hydrogen must be obtained from compounds by breaking chemical bonds, and this requires much more energy than simply condensing an element to separate it from other substances found in air. The most common methods of preparing hydrogen follow.

1. FROM HYDROCARBONS. Hydrogen is produced commercially in large quantities from the hydrocarbons in oil and natural gas. **Hydrocarbons** are compounds which contain only carbon and hydrogen. When a mixture of methane (CH_4, the principal component of natural gas) and steam is heated to a high temperature in the presence of catalysts, a gaseous mixture of carbon monoxide, carbon dioxide, and hydrogen is produced.

$$CH_4 + H_2O \xrightarrow{\text{Catalyst}} CO + 3H_2$$

$$CO + H_2O \xrightarrow{\text{Catalyst}} CO_2 + H_2$$

These are typical reactions of hydrocarbons, and other hydrocarbons may be substituted for methane.

Hydrocarbons can decompose and rearrange at high temperatures in the presence of a catalyst. These reactions, called **cracking reactions,** are used in the refining of petroleum and may produce hydrogen as a by-product. One example, using Lewis structures to illustrate the rearrangements, is

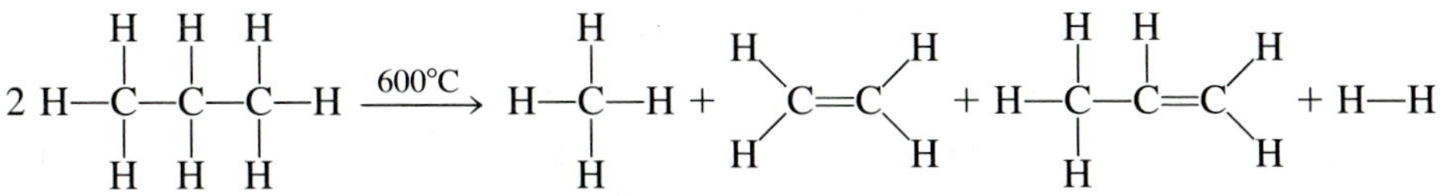

2. ELECTROLYSIS. Hydrogen is liberated when a direct current of electricity is passed through water containing a small amount of an electrolyte such as H_2SO_4,

NaOH, or Na_2SO_4 (Fig. 22.1). Bubbles of hydrogen are formed at the cathode, and oxygen is evolved at the anode. The net reaction can be summarized by the equation

$$2H_2O(l) + \text{electrical energy} \longrightarrow 2H_2(g) + O_2(g)$$

Because electricity is so expensive, electrolytic hydrogen is produced in large quantities only when very pure gas is needed. Hydrogen is also a product of the electrolytic production of sodium hydroxide and chlorine (Sections 13.5 and 21.3).

$$[2Na^+(aq)] + 2Cl^-(aq) + 2H_2O + \text{electrical energy} \longrightarrow$$
$$[2Na^+(aq)] + 2OH^-(aq) + Cl_2(g) + H_2(g)$$

3. REACTION OF METALS WITH ACIDS OR WATER. Iron reduces the hydrogen ion in acids, producing hydrogen gas and an iron salt (Fig. 22.2).

$$Fe(s) + 2H^+(aq) + [2Cl^-(aq)] \longrightarrow Fe^{2+}(aq) + [2Cl^-(aq)] + H_2(g)$$

Sodium rapidly reduces the hydrogen in water at room temperature, producing hydrogen gas and a solution of sodium hydroxide.

$$2Na + 2H_2O \longrightarrow 2Na^+(aq) + 2OH^-(aq) + H_2(g)$$

This reaction is too vigorous to use for the laboratory production of hydrogen. Sometimes it gets hot enough to ignite the metal as well as the hydrogen as it is produced.

Except for their low cost and ready availability, there is nothing special about iron and sodium that causes them, rather than other metals, to be used in the laboratory preparation of hydrogen. Other active metals with about the same electronegativity as sodium (1.0) will displace hydrogen from water, and metals with about the same electronegativity as iron (1.6) will react with acids.

Iron and other less active metals will displace hydrogen from water at elevated temperatures. When steam is passed over magnesium, for example, hydrogen is produced rapidly.

$$Mg + H_2O \longrightarrow H_2 + MgO$$

Figure 22.1 (left)
The electrolysis of water produces hydrogen and oxygen.

Figure 22.2 (right)
The reaction of iron with an acid produces hydrogen. This example shows the reaction of iron with hydrochloric acid.

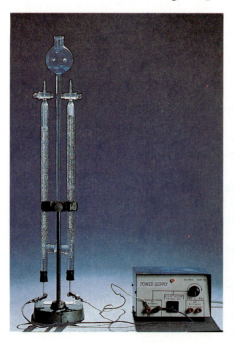

4. REACTION OF CARBON WITH WATER. White-hot carbon (1500–1600°C) will react with steam in an endothermic reaction, producing a mixture of carbon monoxide and hydrogen called **water gas.**

$$C(s) + H_2O(g) \longrightarrow CO(g) + H_2(g)$$

Because both carbon monoxide and hydrogen will burn in oxygen or in air, water gas is a valuable industrial fuel. When water gas is mixed with steam and passed over a catalyst such as iron oxide or thorium oxide at a fairly high temperature (500°C), carbon dioxide and additional hydrogen are produced in the **water gas shift reaction.**

$$CO(g) + H_2O(g) \longrightarrow CO_2(g) + H_2(g)$$

5. REACTION OF CERTAIN ELEMENTS WITH STRONG BASES. Aluminum and many post-transition metals and semi-metals such as zinc and silicon react with concentrated solution of strong bases, with the liberation of hydrogen (Fig. 22.3).

$$2Al(s) + 2OH^-(aq) + 6H_2O(l) \longrightarrow 2Al(OH)_4^-(aq) + 3H_2(g)$$
$$Zn(s) + 2OH^-(aq) + 2H_2O(l) \longrightarrow Zn(OH)_4^{2-}(aq) + H_2(g)$$

6. REACTION OF IONIC METAL HYDRIDES WITH WATER. Hydrogen can also be produced by the reaction of hydrides of the active metals, which contain the very strongly basic H^- anion, with water (Fig. 22.4).

$$CaH_2(s) + 2H_2O(l) \longrightarrow Ca^{2+}(aq) + 2OH^-(aq) + 2H_2(g)$$

Metal hydrides are expensive but convenient sources of hydrogen, especially where space and weight are important factors, as they are in the inflation of life jackets, life rafts, and military balloons.

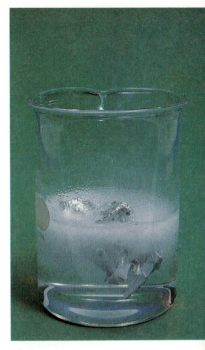

Figure 22.3
Hydrogen is produced when aluminum metal dissolves in a basic solution.

22.2 Chemical Properties of Hydrogen

At ordinary temperatures hydrogen is relatively inactive chemically, but when heated it enters into many chemical reactions.

1. REACTION WITH ELEMENTS. When heated, hydrogen reacts with the metals of Group IA and with Ca, Sr, and Ba (the more active elements in Group IIA). The compounds formed are crystalline **ionic hydrides** that contain the hydride anion, a strong reducing agent and a strong base, which reacts vigorously with water and other acids to form hydrogen gas.

Hydrogen is absorbed (dissolved) by many transition metals, lanthanide metals, and actinide metals. The materials thus produced are **metallic hydrides.** Most of these hydrides are solid solutions; they have no fixed atomic composition. In palladium hydride, for example, the H-to-Pd ratio can vary from 0.4 to 0.7. Because of this solubility, hydrogen readily passes through the heated walls of a palladium container.

The reactions of hydrogen with the nonmetals generally produce **acidic hydrides.** The reactions become more exothermic and vigorous as the electronegativity of the nonmetal increases. Hydrogen reacts with nitrogen and sulfur only when heated, but it reacts explosively with fluorine (producing gaseous HF) and, under some conditions, with chlorine (producing gaseous HCl). A mixture of hydrogen and oxygen will explode if ignited (Fig. 22.5). Because of the explosive nature of the reaction, caution should be used in handling hydrogen (or any other combustible gas) in order to avoid the formation of an explosive mixture in a confined space. Most hydrides of the non-

Figure 22.4
Hydrogen is produced when an ionic hydride reacts with water. The example shows the reaction of calcium hydride with water.

Figure 22.5
Explosion of hydrogen on the airship *Hindenburg* while landing at Lakehurst, New Jersey, May 6, 1937. Following the ignition of hydrogen in one compartment, ignition of the other compartments took place within seconds, resulting in the complete destruction of the giant airliner.

metals are acidic (Section 16.6). However, ammonia and phosphine (PH_3) are very weak acids and can also function as bases. The reactions of hydrogen with the elements are summarized in Table 22.1.

As we saw in Chapter 9, the formulas of the binary hydrides can generally be determined from the position of the combining element in the Periodic Table, the relative electronegativities of the two elements, and the resulting oxidation number of the hydrogen. Thus the reaction of the Group IA element lithium with hydrogen gives LiH (oxidation number of Li $= +1$, oxidation number of H with a less electronegative

Table 22.1   Chemical reactions of hydrogen with the elements

General equation	Comments
$\frac{n}{2} H_2 + M \longrightarrow MH_n$	Ionic hydrides with Group IA and Ca, Sr, Ba; metallic hydrides with transition metals
$H_2 + C \longrightarrow$ (no reaction)	
$3H_2 + N_2 \longrightarrow 2NH_3$	Requires high pressure and temperature; low yield
$2H_2 + O_2 \longrightarrow 2H_2O$	Exothermic and potentially explosive
$H_2 + S \longrightarrow H_2S$	Requires heating; low yield
$H_2 + X_2 \longrightarrow 2HX$	X = F, Cl, Br, I; explosive with F_2; low yield with I_2

element = -1); the reaction of nitrogen, a Group VA element, gives NH_3 (oxidation number of N = -3, oxidation number of H with a more electronegative element = $+1$).

2. REACTION WITH COMPOUNDS. Hydrogen reduces the heated oxides of many metals, with the formation of the metal and water (Fig. 20.13). For example, when hydrogen is passed over heated CuO, copper and water are formed.

$$H_2 + CuO \longrightarrow Cu + H_2O$$

Hydrogen may also reduce some metal oxides to lower oxides.

$$H_2 + MnO_2 \longrightarrow MnO + H_2O$$

22.3 Isotopes of Hydrogen; Heavy Water

Hydrogen is composed of three isotopes: (1) ordinary hydrogen, ^1_1H; (2) **deuterium**, ^2_1H; and (3) **tritium**, ^3_1H. In a naturally occurring sample of hydrogen, there is one atom of deuterium for every 7000 ^1_1H atoms and one atom of tritium for every 10^{18} ^1_1H atoms. The chemical properties of the different isotopes are very similar, since they have identical electronic structures. Some physical properties are different because of their differing atomic masses. Elemental deuterium and tritium have lower vapor pressures than does ordinary hydrogen. Consequently, when liquid hydrogen evaporates, the heavier isotopes are somewhat concentrated in the last portions to evaporate. Deuterium is isolated by the electrolysis of heavy water, D_2O. Tritium is made by a nuclear reaction (Section 24.13).

Water composed of deuterium and oxygen, D_2O, is called **heavy water,** or deuterium oxide. Heavy water is produced by an exchange process followed by electrolysis. In the process liquid water and gaseous hydrogen sulfide are mixed, and the deuterium atoms exchange between sulfur and oxygen. At low temperatures deuterium preferentially binds to oxygen as D_2O, and at high temperatures, with sulfur as D_2S.

$$H_2S + D_2O \underset{\text{Cold}}{\overset{\text{Hot}}{\rightleftharpoons}} D_2S + H_2O$$

Hydrogen sulfide, bubbled through water at an elevated temperature, is enriched with respect to D_2S through exchange. After the enriched hydrogen sulfide is drawn off, the reverse exchange at lower temperatures produces water that contains a larger proportion of D_2O than before. Many repetitions of the process yield water with a D_2O concentration of about 15%, which is then enriched to 99.8% D_2O by electrolysis. During the electrolysis, molecules containing deuterium migrate to a cathode more slowly than light water molecules, so the heavier water molecules tend to remain behind while the light water is decomposed. Several hundred tons of heavy water are now produced per year.

Heavy water resembles ordinary water in appearance and chemical behavior but differs from it slightly in its physical properties. For example, the density of D_2O is 10% higher than that of H_2O, and the melting and boiling points are 3.79°C and 1.4°C higher, respectively. Reactions in D_2O are somewhat slower than in H_2O. Heavy water is one of the moderators used in nuclear reactors to reduce the speed of neutrons so that the fission process can take place (Chapter 24). It is a convenient source of deuterium for preparation of compounds in which all of the hydrogen atoms have been replaced by deuterium. For example, NaOD can be prepared by the reaction of Na or Na_2O with D_2O.

It is possible to detect the presence of as little as 1 part of deuterium in 100,000 parts of hydrogen in a sample. For this reason, heavy water and deuterium serve as valuable tracers in the study of both chemical and physiological changes. For example, by replacing ordinary hydrogen with deuterium in molecules of food, the processes of digestion and metabolism in the body can be studied.

22.4 Uses of Hydrogen

Two-thirds of the world's hydrogen production is used to manufacture ammonia, which is used primarily as a fertilizer and in the manufacture of nitric acid. It is used extensively in the process of **hydrogenation,** in which vegetable oils are changed from liquids to solids. Crisco is an example of a hydrogenated oil. The change results from the addition of H_2 to carbon-carbon double bonds to give single bonds. Ethane, C_2H_4, is hydrogenated in a similar reaction.

$$
\begin{array}{c}
H \\
\diagdown \\
C=C \\
\diagup \quad \diagdown \\
H \qquad H
\end{array}
+ \; H{-}H \longrightarrow
\begin{array}{c}
H \; H \\
| \; | \\
H{-}C{-}C{-}H \\
| \; | \\
H \; H
\end{array}
$$

Methyl alcohol, an important industrial solvent and raw material, is produced synthetically by the catalyzed reaction of hydrogen with carbon monoxide:

$$2H_2 + CO \xrightarrow{\text{Catalyst}} CH_3OH$$

Water gas, a mixture of hydrogen with carbon monoxide, is an important industrial fuel. Hydrogen is used in the reduction of certain oxides to obtain the free metal.

Hydrogen will burn without explosion under controlled conditions. The very high heat of combustion of hydrogen with pure oxygen makes it possible to achieve temperatures up to 2800°C with an oxygen-hydrogen torch. The hot flame of this torch can be used in ''cutting'' thick sheets of many metals.

Oxygen

The discovery of oxygen marked the beginning of modern chemistry. Credit for the discovery of oxygen is usually given to Joseph Priestley, an English clergyman and scientist, who prepared oxygen in 1774 by focusing the sun's rays on mercury(II) oxide by means of a lens (Fig. 22.6). He tested the gaseous product with a burning candle and noted that the candle burned more brightly than in ordinary air. Shortly thereafter, Antoine-Laurent Lavoisier recognized that the products of the combustion of metals weighed more than the pure metals because oxygen combined with the metals during combustion.

22.5 Occurrence and Preparation of Oxygen

Oxygen is the most abundant element in the earth's crust, and it forms compounds with almost all of the other elements. It is essential to combustion and to respiration in most plants and animals. Oxygen occurs as O_2 molecules and, to a limited extent, as O_3 molecules in air and forms about 23% of the mass of the air. About 89% of water by mass consists of combined oxygen. About 50% of the mass and

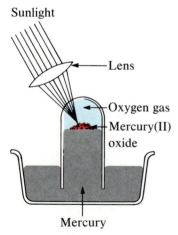

Figure 22.6
Apparatus used by Priestley to prepare oxygen. When HgO is heated, it decomposes, producing gaseous oxygen and mercury.

90% of the volume of the earth's crust is oxygen combined with other elements, principally silicon. In combination with carbon, hydrogen, and nitrogen, oxygen is a large part of the weight of the bodies of plants and animals.

In 1986, 16.5 million tons of oxygen were produced in the United States, making it third in production for all chemicals. Approximately 97% of the oxygen produced commercially comes from air and 3% from the electrolysis of water (Fig. 22.1), because air and water are abundant, cheap, and easy to process. Methods for the preparation of oxygen follow.

1. FRACTIONAL EVAPORATION OF LIQUID AIR. Oxygen is separated from the air in commercial quantities by cooling and compressing air until it liquefies and then distilling off the lower-boiling nitrogen and some other elements (see Section 29.2).

2. ELECTROLYSIS OF WATER. The electrolysis of water to produce hydrogen and water is described in Section 22.1.

3. DECOMPOSITION OF METAL OXIDES AND PEROXIDES. Small amounts of oxygen are sometimes produced by decomposition reactions. The oxides of metals located near the bottom of the activity series (Section 9.7) are not thermally stable and lose oxygen when heated. For example, when red mercury(II) oxide, HgO, is heated, metallic mercury and oxygen are formed.

$$2HgO \longrightarrow 2Hg + O_2$$

This is the preparation of oxygen used by Priestley and Lavoisier (Fig. 22.6).

In some metal oxides in which the metal has a high oxidation number, the metal is a strong enough oxidizing agent to oxidize oxide ions to oxygen gas when heated. These compounds undergo thermal decomposition reactions in which part of the oxygen is liberated. The remainder is combined in a metal oxide in which the metal has a lower, more stable oxidation number. Examples include the decomposition of lead(IV) oxide to lead(II) oxide and chromium(VI) oxide to chromium(III) oxide.

$$2PbO_2 \longrightarrow 2PbO + O_2$$
$$4CrO_3 \longrightarrow 2Cr_2O_3 + 3O_2$$

Ionic peroxides decompose to the corresponding oxides, with evolution of oxygen. For example,

$$2Ba^{2+}(:\overset{..}{O}:\overset{..}{O}:^{2-}) \longrightarrow 2Ba^{2+}(:\overset{..}{O}:^{2-}) + O_2$$

4. HEATING CERTAIN SALTS THAT CONTAIN OXYGEN. The nitrate salts of certain metals yield oxygen when heated.

$$2NaNO_3 \longrightarrow 2NaNO_2 + O_2$$
$$2Cu(NO_3)_2 \longrightarrow 2CuO + 4NO_2 + O_2$$

If a mixture of manganese(IV) oxide and potassium chlorate is heated, the latter decomposes quite rapidly at about 270°C, producing oxygen and potassium chloride. The manganese(IV) oxide is a catalyst for the decomposition.

$$2KClO_3 \xrightarrow{MnO_2} 2KCl + 3O_2$$

The preparation of oxygen by heating potassium chlorate can be dangerous when done carelessly. Explosions can occur when combustible materials such as carbon, sulfur, rubber, or a glowing wood splint come in contact with hot potassium chlorate. The danger involved in this experiment cannot be overemphasized. Proceed with caution when preparing oxygen by this method.

Figure 22.7

The paramagnetic oxygen molecules in liquid oxygen are attracted to a magnet and will remain suspended between the poles of the magnet until the liquid evaporates.

22.6 Properties of Oxygen

Oxygen is a colorless, odorless, and tasteless gas at ordinary temperatures. It is pale blue in the liquid state and boils at $-183°C$. Solid oxygen, also pale blue, melts at $-218.4°C$. Oxygen is slightly more dense than air. Although it is only slightly soluble in water (49 mL of gas dissolve in one liter at STP), its solubility is very important to aquatic life.

The oxygen molecule is **paramagnetic;** that is, it is attracted by a magnetic field (Fig. 22.7). This paramagnetism is an indication that the diatomic molecule contains unpaired electrons. Since an oxygen atom has six valence electrons, we might expect two pairs of electrons to be shared between the atoms in a diatomic oxygen molecule, with all electrons paired. However, careful measurement of the magnitude of the paramagnetism indicates two unpaired electrons per molecule. Their presence is consistent with the molecular orbital description of the bonding in O_2 (Section 7.11), in which the unpaired electrons are distributed singly in the two $\pi_p{}^*$ antibonding molecular orbitals.

Oxygen is a very reactive substance that reacts with most other elements and with many compounds.

1. REACTION WITH ELEMENTS. Oxygen reacts directly at room temperatures or at elevated temperatures with all other elements (Fig. 22.8) except the noble gases, the halogens, and a few second- and third-row transition metals of low reactivity [those below copper in the electromotive series, whose oxides decompose upon heating (Section 9.7)]. The more active metals form peroxides or superoxides. Less active metals and the nonmetals give oxides. Oxides of the halogens, of some of the noble gases, and of the metals at the bottom of the activity series (Section 9.7) can be prepared, but not by the direct action of the elements with oxygen. The preparations and the chemical behavior of the oxides are described in the sections dealing with the oxides of the other elements. The reactions of oxygen with other elements are summarized in Table 22.2.

As we saw in Chapter 9, the formulas of the binary oxides, peroxides, and superoxides can generally be determined from the position of the combining element in the Periodic Table and the oxidation number of the oxygen. Thus the reaction the Group IA element lithium with oxygen at ordinary pressures gives Li_2O (oxidation number of Li = $+1$, oxidation number of O = -2) and gallium, a Group IIIA element, gives Ga_2O_3 (oxidation number of Ga = $+3$, oxidation number of O = -2).

2. REACTION WITH COMPOUNDS. Elemental oxygen will also react with some compounds. If a compound is composed of elements that will combine with oxygen when free, the compound may be expected to react with oxygen to form oxides of the constituent elements, provided that at least one of the atoms in the compound does not exhibit its maximum oxidation number. For example, hydrogen sulfide, H_2S, contains sulfur with an oxidation number of -2. Since the sulfur does not exhibit its maximum oxidation number, and since free sulfur will react with oxygen, we would expect H_2S would react with oxygen. It does, giving water and sulfur dioxide. Oxides containing an element with a lower oxidation number than is usually formed when the element combines with an excess of oxygen, such as CO and P_4O_6, will also react with additional oxygen. These and other examples of compounds that react with oxygen and the equations for the reactions may be found in Table 22.2.

Elemental oxygen is a strong oxidizing agent and will eventually oxidize most organic substances, though (fortunately) at a slow rate. A fast rate would make life on earth impossible; all organic matter, including animal life, would burn up.

Figure 22.8

Magnesium combines with oxygen in an exothermic reaction.

Table 22.2 Chemical properties of elemental oxygen

General equation	Comments
Reactions with elements	
$mM + \dfrac{m}{2}O_2 \longrightarrow M_nO_m$	Oxygen reacts with most metals, M, except those at the bottom of the activity series (Section 9.7)
$2Na + O_2 \longrightarrow Na_2O_2$	A peroxide forms with Na (Section 13.4)
$M + O_2 \longrightarrow MO_2$	Peroxides form with M = Ca, Sr, Ba (Section 13.4)
$M + O_2 \longrightarrow MO_2$	M = K, Rb, Cs form superoxides (Section 13.4)
$2H_2 + O_2 \longrightarrow 2H_2O$	Potentially explosive reaction
$2C + O_2 \longrightarrow 2CO$	With a stoichiometric amount of O_2
$E + O_2 \longrightarrow EO_2$	With heating in excess O_2; E = lighter members of Group IVA: C, Si, Ge, Sn
$2Pb + O_2 \longrightarrow 2PbO$	Pb is the heaviest member of Group IVA
$N_2 + O_2 \longrightarrow 2NO$	High temperature required; low yield of product
$P_4 + 3O_2 \longrightarrow P_4O_6$	50% yield, requires a stoichiometric amount of O_2
$E_4 + 5O_2 \longrightarrow E_4O_{10}$	In excess O_2; E = lighter members of Group VA: P, As, Sb
$4Bi + 3O_2 \longrightarrow 2Bi_2O_3$	With heating in excess O_2; Bi is the heaviest member of Group VA
$E + O_2 \longrightarrow EO_2$	E = heavier members of Group VIA: S, Se, Te
Reactions with compounds	
$2CO + O_2 \longrightarrow 2CO_2$	Requires ignition
$2NO + O_2 \longrightarrow 2NO_2$	Spontaneous reaction
$P_4O_6 + 2O_2 \longrightarrow P_4O_{10}$	
$2SO_2 + O_2 \longrightarrow 2SO_3$	Requires heat and Pt catalyst
$C_nH_m + \left(n + \dfrac{m}{4}\right)O_2 \longrightarrow nCO_2 + \dfrac{m}{2}H_2O$	This is the general reaction for the burning of a hydrocarbon
(For example, $CH_4 + 2O_2 \longrightarrow CO_2 + 2H_2O$)	
$4NH_3 + 5O_2 \longrightarrow 4NO + 6H_2O$	Ammonia burns in O_2
$2H_2S + 3O_2 \longrightarrow 2SO_2 + 2H_2O$	
$CS_2 + 3O_2 \longrightarrow CO_2 + 2SO_2$	
$2MS + 3O_2 \longrightarrow 2MO + 3SO_2$	Most metal sulfides form metal oxides and SO_2 when heated in air

The ease with which elemental oxygen picks up electrons is mirrored by the difficulty of removing electrons from oxygen in most oxides. Of the elements, only the very reactive fluorine molecule will oxidize oxides to form oxygen.

22.7 Ozone

When dry oxygen is passed between the two electrically charged plates (Fig. 22.9), a decrease in the volume of the gas occurs, and **ozone** (O_3, Fig. 22.10), a

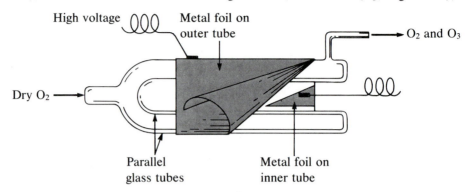

Figure 22.9

Laboratory apparatus for preparing ozone.

substance possessing a distinctive odor, is formed. Ozone can also be prepared by the electrolysis of very cold, concentrated sulfuric acid.

The formation of ozone from oxygen is an endothermic reaction, in which the energy is furnished in the form of an electrical discharge, heat, or ultraviolet light.

$$3O_2 \xrightarrow[\text{discharge}]{\text{Electric}} 2O_3 \qquad \Delta H° = 286 \text{ kJ}$$

The sharp odor associated with sparking electrical equipment is due, in part, to ozone.

Ozone is formed in the upper atmosphere by the action of ultraviolet light from the sun upon the oxygen there. This ozone acts as a protective barrier to harmful ultraviolet light from the sun. There is concern that residual chlorofluorocarbon propellants (Freons) from aerosol cans and other sources may react and catalyze the decomposition of ozone in the upper atmosphere, decreasing its ozone concentration. This would lessen the ozone layer's effectiveness as a protective barrier to ultraviolet light. If this happens, it could have a drastic effect on life on the earth.

Ozone produced in the lower atmosphere reacts with engine exhaust and contributes to smog (Fig. 22.11 and Section 29.4). A great deal of the deterioration of automobile tires is due to contact with ozone in the air.

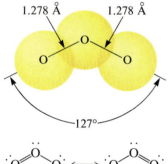

Figure 22.10

The bent O_3 molecule and the resonance structures necessary to describe its bonding.

Figure 22.11

A mixture of ozone, nitrogen oxides, and unburned hydrocarbons forms smog when exposed to sunlight.

Ozone is pale blue as a gas and deep blue as a liquid. It has a sharp, irritating odor, produces headaches, and is poisonous. Energy is absorbed when ozone is formed from oxygen, so ozone is more active chemically than oxygen and decomposes readily into oxygen by the exothermic reaction

$$2O_3 \longrightarrow 3O_2 \qquad \Delta H° = -286 \text{ kJ}$$

As a consequence, it is less stable and more reactive than oxygen.

Ozone is a powerful oxidizing agent and forms oxides with many elements at temperatures where O_2 will not react. It can be readily detected in a gas by passing the gas through a solution of an iodide containing some starch.

$$2I^-(aq) + O_3(g) + H_2O(l) \longrightarrow 2OH^-(aq) + I_2 + O_2(g)$$

The O_3 oxidizes I^- to I_2, elemental iodine, which in combination with excess I^- imparts a blue color to the starch.

22.8 The Importance of Oxygen to Life

The energy required for maintenance of normal body functions in human beings and other animals is derived from the slow oxidation of chemical compounds in the body. Oxygen is the source of the oxidizing agents in these reactions. Oxygen passes from the lungs into the blood, where it combines with hemoglobin, producing oxyhemoglobin. In this form, oxygen is carried by the blood to the various tissues of the body, where it is released and consumed in reactions with oxidizable compounds. Products are mainly carbon dioxide and water. The blood carries the carbon dioxide through the veins to the lungs, where it gives up the carbon dioxide and collects another supply of oxygen. The digestion and assimilation of food regenerates the materials consumed by oxidation in the body, with the same amount of energy being liberated as if the food had been burned outside the body.

The oxygen in the atmosphere is continually replenished through the action of algae and higher plants by a process called **photosynthesis** (Fig. 22.12). The products of photosynthesis may vary, but the process can be generalized as the conversion of carbon dioxide and water to glucose (a sugar) and oxygen by chlorophyll, using the energy of light.

$$6CO_2 + 6H_2O \xrightarrow[\text{Light}]{\text{Chlorophyll}} C_6H_{12}O_6 + 6O_2$$

$$\text{Carbon} \qquad \text{Water} \qquad\qquad\qquad \text{Glucose} \qquad \text{Oxygen}$$
$$\text{dioxide}$$

Thus the oxygen that is converted to carbon dioxide and water by the metabolic processes in plants and animals is returned to the atmosphere by these photosynthetic reactions.

22.9 Uses of Oxygen

Oxygen is essential in combustion processes such as the burning of fuels for the production of heat; oxygen is also required for the decay of organic matter. Plants and animals use the oxygen from the air in respiration. Oxygen-enriched air is used in medical practice when a patient receives an inadequate supply of oxygen due to shock, pneumonia, or some other illness. Health services use about 13% of the oxygen produced commercially.

Figure 22.12
The bubbles of oxygen are produced by photosynthesis in the algae.

Approximately 30% of all oxygen produced commercially is used to remove carbon from iron during steel production (Section 27.22). Large quantities of pure oxygen are also consumed in metal fabrication and in the cutting and welding of metals with oxyhydrogen and oxyacetylene torches. The chemical industry uses oxygen for oxidation of many substances.

Liquid oxygen is used as an oxidizing agent in the space shuttle and other rocket engines (Fig. 22.13). It is also used to provide gaseous oxygen for life support in space.

The uses of ozone depend on its readiness to react with other substances. As a bleaching agent for oils, waxes, fabrics, and starch, it reacts and oxidizes the colored compounds in these substances to colorless compounds. It is sometimes used instead of chlorine to purify water.

Sulfur

Sulfur has been known from very early times, because it occurs free in nature as a solid. In fact, sulfur (or brimstone, as it was sometimes called) and carbon were the only nonmetallic elements known to the ancients. Sulfur, along with sulfuric acid, is a major component of the atmosphere of the planet Venus, and the surface (Fig. 22.14) of Jupiter's moon Io is believed to be covered with sulfur.

22.10 Occurrence and Preparation of Sulfur

The principal deposits of sulfur in the United States are in Texas and Louisiana. There are also extensive deposits in Mexico and in the volcanic regions of Italy and Japan. Sulfides of iron, zinc, lead, and copper and sulfates of sodium, calcium, barium, and magnesium are widespread and abundant as minerals. Hydrogen sulfide is a common component of natural gas and occurs in many volcanic gases. Sulfur compounds are also found in coal. Sulfur is a constituent of some proteins and therefore exists in the combined state in animal and vegetable matter.

Figure 22.13

A mixture of aluminum powder and ammonium perchlorate is the principal ingredient in the boosters used to launch the space shuttle. The large clouds of smoke contain aluminum oxide and aluminum chloride. The large central fuel tank contains liquid hydrogen and liquid oxygen, which are burned in the engines at the rear of the space craft, giving a pale blue flame and invisible water vapor.

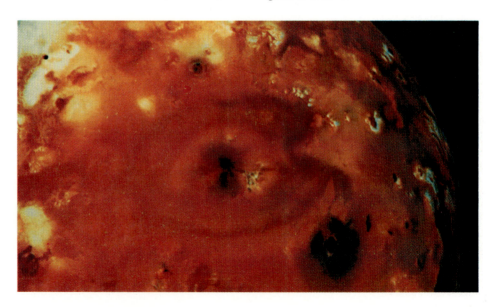

Figure 22.14

The surface of Io, a moon of Jupiter, is believed to be covered with sulfur.

Free sulfur is mined by the **Frasch process** (Fig. 22.15) from enormous underground deposits in Texas and Louisiana. Superheated water (170°C and 10 atmospheres pressure) is forced down the outermost of three concentric pipes to the underground deposit. When the hot water melts the sulfur, compressed air is forced down the innermost pipe. The liquid sulfur mixed with air forms a foam that flows up through the outlet pipe. The emulsified sulfur is conveyed to large settling vats, where it solidifies on cooling (Fig. 22.16). Sulfur produced by this method is 99.5 to 99.9% pure and requires no purification for most uses.

Sulfur is also obtained, in quantities exceeding that from the Frasch process, from hydrogen sulfide recovered during the purification of "sour" natural gas (Section 22.12).

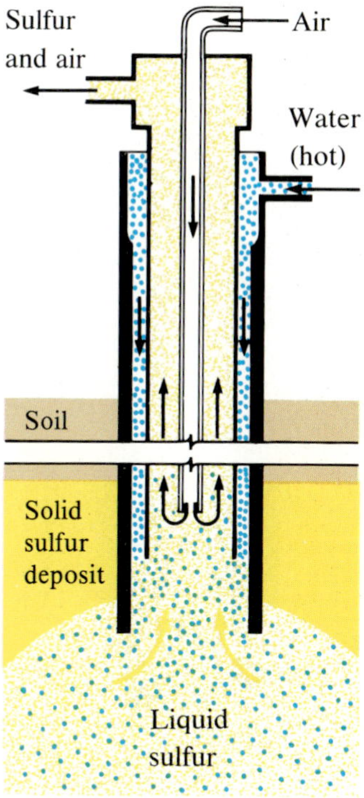

22.11 Properties of Sulfur

Sulfur exists in several allotropic forms (Section 20.3). The stable form at room temperature contains eight-membered rings and should be written as S_8 to be perfectly correct. However, chemists commonly use the symbol S to simplify the coefficients in chemical equations; we will follow this practice in the rest of this book.

Like oxygen, which is also a member of Group VIA, sulfur exhibits distinctly nonmetallic behavior. It oxidizes metals, giving a variety of binary sulfides in which sulfur exhibits a negative oxidation number. Elemental sulfur oxidizes less electronegative nonmetals and is oxidized by more electronegative nonmetals, such as oxygen and the halogens. Sulfur is also oxidized by other strong oxidizing agents. For example, concentrated nitric acid oxidizes sulfur to the sulfate ion, with the concurrent formation of nitrogen (IV) oxide.

$$S + 6HNO_3 \longrightarrow 2H^+ + SO_4^{2-} + 2H_2O + 6NO_2$$

The products of the reaction with hot, concentrated sulfuric acid are sulfur dioxide and water.

$$S + 2H_2SO_4 \longrightarrow 3SO_2 + 2H_2O$$

Figure 22.15

Diagram of the Frasch process for mining sulfur.

(Diagram labels: Sulfur and air; Air; Water (hot); Soil; Solid sulfur deposit; Liquid sulfur)

Figure 22.16

Molten sulfur from the Frasch process is pumped into large tanks and allowed to solidify.

The general chemical behavior of sulfur is discussed in Section 20.7. The chemistry of sulfur with an oxidation number of -2 is similar to that of oxygen. Unlike oxygen, however, sulfur forms a variety of compounds in which it exhibits positive oxidation numbers. Some of the general reactions of sulfur and its compounds are summarized in Table 22.3.

Table 22.3   Chemical properties of sulfur and its compounds

General equation	Comments
Reactions with elements	
$nM + mS \longrightarrow M_nS_m$	Most metals combine with sulfur, giving sulfides which contain the S^{2-} ion
$mM + xsS \longrightarrow M_m(S_x)_n$	With excess S, some metals give polysulfides, S_x^{2-}, $x = 2, 3, 4, 5$
$H_2 + S \longrightarrow H_2S$	In low yield; H_2S decomposes at elevated temperatures
$E + 2S \longrightarrow ES_2$	E = C, Si, Ge, the lighter members of Group IVA
$E + S \longrightarrow ES$	E = Sn, Pb, the heavier members of Group IVA
$N_2 + S \longrightarrow$ N.R.	N_2 does not react with S
$P_4 + 10S \longrightarrow P_4S_{10}$	With less S, many other phosphorus sulfides are possible
$2E + 3S \longrightarrow E_2S_3$	E = As, Sb, Bi, the heavier members of Group VA
$S + O_2 \longrightarrow SO_2$	Traces of SO_3 also form when S burns in air
$S + 3F_2 \longrightarrow SF_6$	Other sulfur fluorides can be prepared indirectly
$S + nCl_2 \longrightarrow S_2Cl_2$ or SCl_2	Products depend on the reaction stoichiometry
$2S + Br_2 \longrightarrow S_2Br_2$	I_2 does not react with S
Reactions of compounds	
$S^{2-} + 2H^+ \longrightarrow H_2S$	S^{2-} is a strongly basic anion
$S^{2-} + \text{oxidant} \longrightarrow S$	S^{2-} can be oxidized to S or to higher oxidation states
$2H_2S + O_2 \longrightarrow S + 2H_2O$	Requires a stoichiometric amount of O_2
$2H_2S + 3O_2 \longrightarrow 2SO_2 + 2H_2O$	With excess O_2
$2H_2S + SO_2 \longrightarrow 2H_2O + 3S$	
$CS_2 + 3O_2 \longrightarrow CO_2 + 2SO_2$	
$2SO_2 + O_2 \longrightarrow 2SO_3$	Catalyst required for satisfactory yield
$SO_2 + H_2O \longrightarrow H_2SO_3$	Sulfurous acid
$SO_2 + O^{2-} \longrightarrow SO_3^{2-}$	Sulfite ion
$SO_3 + H_2O \longrightarrow H_2SO_4$	Sulfuric acid
$SO_3 + O^{2-} \longrightarrow SO_4^{2-}$	Sulfate ion
$SO_3 + H_2O_2 \longrightarrow H_2SO_5$	Peroxymonosulfuric acid, $HOS(O)_2OOH$

22.12 Hydrogen Sulfide and Sulfides

Hydrogen sulfide, H_2S, is a colorless gas with an offensive odor. It is responsible for the odor of many hot springs and rotten eggs. Hydrogen sulfide is as toxic as hydrogen cyanide (prussic acid), which has been used in death chambers in some states. Great care must be exercised in handling it. Hydrogen sulfide is particularly deceptive because it paralyzes the olfactory nerves, so after a short exposure one does not smell it.

The production of hydrogen sulfide by the direct reaction of the elements ($H_2 + S$) is unsatisfactory, because the reaction is reversible and hydrogen sulfide decomposes upon heating. A more effective preparation is the reaction of a metal sulfide with a dilute strong acid (Fig. 22.17). For example,

$$FeS + 2H^+ \longrightarrow Fe^{2+} + H_2S$$

An aqueous solution of hydrogen sulfide may be easily prepared by the hydrolysis of the organic sulfur-containing compound **thioacetamide,** CH_3CSNH_2.

$$CH_3CSNH_2 + 2H_2O \longrightarrow CH_3CO_2^- + NH_4^+ + H_2S(aq)$$

This reaction is the most commonly used one to produce hydrogen sulfide as a precipitating agent in solution for metal ion separations in qualitative analysis procedures.

The sulfur in hydrogen sulfide is readily oxidized, making H_2S a good reducing agent (Fig. 22.18). In acidic solutions, hydrogen sulfide reduces Fe^{3+} to Fe^{2+}, Br_2 to Br^-, MnO_4^- to Mn^{2+}, $Cr_2O_7^{2-}$ to Cr^{3+}, and HNO_3 to NO_2. The sulfur in the H_2S is usually oxidized to elemental sulfur, unless a large excess of the oxidizing agent is present. In this case the sulfide may be oxidized to SO_3^{2-} or SO_4^{2-} (or to SO_2 and SO_3 in the absence of water).

Hydrogen sulfide burns in air, forming water and sulfur dioxide. When heated with a limited supply of air or with sulfur dioxide (which can be produced by the combustion of H_2S), free sulfur is formed. In this way sulfur is recovered from the hydrogen sulfide found in many sources of natural gas. The deposits of sulfur in volcanic regions may be the result of the reaction between H_2S and SO_2, since both are constituents of volcanic gases.

Most of the more reactive metals will displace hydrogen from hydrogen sulfide. Lead sulfide is formed by the action of hydrogen sulfide on metallic lead, according to the equation

$$Pb(s) + H_2S(g) \longrightarrow PbS(s) + H_2(g)$$

Silver tableware tarnishes when used with eggs and other foods that contain certain sulfur compounds, due to the formation of black silver sulfide.

$$4Ag(s) + 2H_2S(g) + O_2(g) \longrightarrow 2Ag_2S(s) + 2H_2O(l)$$

Hydrogen sulfide is a weak acid. An aqueous solution of hydrogen sulfide is called **hydrosulfuric acid.** The acid ionizes in two stages, giving hydrogen sulfide ions, HS^-, in the first stage and sulfide ions, S^{2-}, in the second (Section 17.9). Salts of both the HS^- and S^{2-} ions are known. For example, passing an excess of hydrogen sulfide into a solution of sodium hydroxide yields **sodium hydrogen sulfide.**

$$H_2S + [Na^+] + OH^- \rightleftharpoons [Na^+] + HS^- + H_2O$$

Adding a stoichiometric amount of sodium hydroxide to a solution of sodium hydrogen sulfide gives **sodium sulfide,** Na_2S, upon evaporation.

$$[Na^+] + OH^- + [Na^+] + HS^- \rightleftharpoons [2Na^+] + S^{2-} + H_2O$$

Figure 22.17

The reaction of FeS with H_2SO_4 produces gaseous H_2S.

Figure 22.18

Sulfide ion is oxidized to free sulfur by dichromate ion, $Cr_2O_7^{2-}$.

Hydrogen sulfide is a weak acid, so the sulfide ion and the hydrogen sulfide ion are strong bases. Aqueous solutions of soluble sulfides and hydrogen sulfides are basic.

$$S^{2-} + H_2O \rightleftharpoons HS^- + OH^-$$
$$HS^- + H_2O \rightleftharpoons H_2S + OH^-$$

The sulfide ion is slowly oxidized by oxygen, so sulfur precipitates when a solution of hydrogen sulfide or a sulfide salt is exposed to the air for a time.

When elemental sulfur is added to a solution of a soluble metal sulfide, the sulfur dissolves by combining with the sulfide ion to form complex **polysulfide ions,** S_n^{2-} (n = 2 to 5). The sulfur atoms of these complex ions are linked by covalent bonds (Fig. 22.19). Disulfide ions, S_2^{2-}, are analogous to peroxide ions, O_2^{2-}. The polysulfide ions, like the peroxide ion or free sulfur, are oxidizing agents.

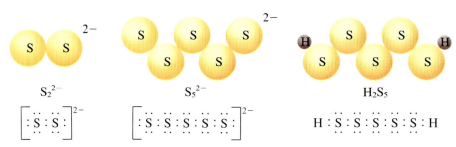

Figure 22.19

Lewis structures and molecular structures of S_2^{2-}, S_5^{2-}, and H_2S_5.

When solutions containing polysulfide ions are acidified, a white, very finely divided form of free sulfur (milk of sulfur) and hydrogen sulfide are produced (Fig. 22.20).

$$S_2^{2-} + 2H^+ \longrightarrow H_2S(g) + S(s)$$

22.13 Oxides of Sulfur

The two common oxides of sulfur are **sulfur dioxide,** SO_2, and **sulfur trioxide,** SO_3. The odor of burning sulfur is due to sulfur dioxide. Sulfur dioxide occurs in volcanic gases and in the atmosphere near industrial plants that burn coal or oil that contains sulfur compounds. The oxide forms when these sulfur compounds react with oxygen during combustion. Air pollution by sulfur dioxide is a major problem, because it combines with water in the atmosphere to form acid rain.

Sulfur dioxide is produced commercially by burning free sulfur and by **roasting** (heating in air) sulfide ores such as ZnS, FeS_2, and Cu_2S. (Roasting, which forms the metal oxide, is the first step in the separation of the metals from the ores.) Sulfur dioxide is prepared conveniently in the laboratory by the action of sulfuric acid upon either sulfite salts, containing the SO_3^{2-} ion, or hydrogen sulfite salts, containing HSO_3^-. Sulfurous acid, H_2SO_3, is formed first, but it quickly decomposes into sulfur dioxide and water. Sulfur dioxide is also formed when many reducing agents react with hot concentrated sulfuric acid.

Sulfur trioxide forms slowly when sulfur dioxide and oxygen are heated together; the reaction is exothermic.

$$2SO_2 + O_2 \longrightarrow 2SO_3 \qquad \Delta H° = -197.8 \text{ kJ}$$

If the temperature of the system is raised to about 400°C, equilibrium is reached more rapidly, but with reduced yield due to a shift in equilibrium to the left with increasing

Figure 22.20

Addition of acid to a solution of a polysulfide produces finely divided sulfur.

temperature (see van't Hoff's law, Section 15.6). However, even at 400°C the rate is too slow for the reaction to be commercially useful unless a catalyst such as finely divided platinum or vanadium(V) oxide is used.

Sulfur dioxide is a gas at room temperatures, and the SO_2 molecule is bent (Fig. 22.21). Sulfur trioxide is a liquid or a solid at room temperatures; it melts at 17°C and boils at 43°C. In the vapor state its molecules are single SO_3 units (Fig. 22.22). There are several solid forms of sulfur trioxide, all of which contain four-coordinate sulfur. Two of these are illustrated in Figure 22.23.

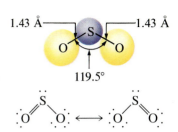

Figure 22.21

The molecular structure and resonance forms of sulfur dioxide.

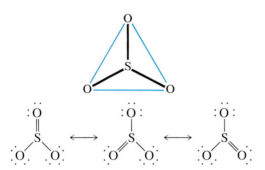

Figure 22.22

The structure of sulfur trioxide in the gas phase and its resonance forms.

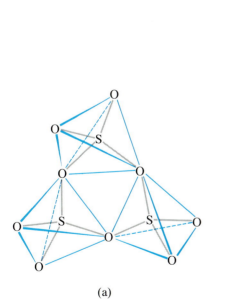

(a)

An icelike form with three tetrahedral SO_4 groupings joined through common oxygen atoms.

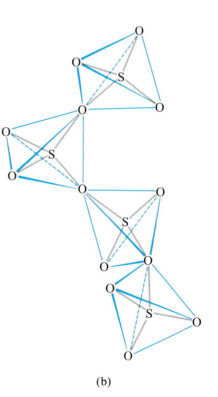

(b)

An asbestoslike form with tetrahedral SO_4 groupings, each joined through a common oxygen atom to form infinite chain molecules.

Figure 22.23

Two of the structures of solid sulfur trioxide.

Sulfur dioxide and sulfur trioxide are acidic oxides. Both react with water to give oxyacids; SO_2 produces the moderately strong acid **sulfurous acid,** H_2SO_3, and SO_3 produces the strong acid **sulfuric acid,** H_2SO_4. Sulfur trioxide also dissolves readily in concentrated sulfuric acid and forms **pyrosulfuric acid,** $H_2S_2O_7$, also known as fuming sulfuric acid or oleum.

The oxides react as Lewis acids with many oxides and hydroxides in Lewis acid-base reactions (Section 16.15), with the formation of **sulfides** or **hydrogen sulfides** and **sulfates** or **hydrogen sulfates,** respectively.

$$BaO + SO_2 \longrightarrow BaSO_3 \text{ (a sulfite)}$$
$$KOH + SO_2 \longrightarrow KHSO_3 \text{ (a hydrogen sulfite)}$$
$$BaO + SO_3 \longrightarrow BaSO_4 \text{ (a sulfate)}$$
$$KOH + SO_3 \longrightarrow KHSO_4 \text{ (a hydrogen sulfate)}$$

The sulfur atom in sulfur trioxide exhibits its maximum oxidation number, so SO_3 cannot be oxidized. It can act as an oxidizing agent, however. Sulfur dioxide can be both oxidized and reduced. It is slowly oxidized by oxygen to sulfur trioxide. Chlorine oxidizes SO_2 to **sulfuryl chloride,** SO_2Cl_2 [Fig. 22.24(a)]. A related compound, **thionyl chloride,** $SOCl_2$ [Fig. 22.24(b)], is formed by the metathetical reaction of sulfur dioxide with phosphorus(V) chloride.

$$SO_2(g) + PCl_5(s) \longrightarrow SOCl_2(l) + POCl_3(l)$$

Like most nonmetal halides, $SOCl_2$ and SO_2Cl_2 react with water, giving hydrogen chloride and sulfur dioxide or sulfur trioxide (or the acids), respectively.

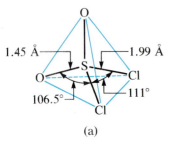

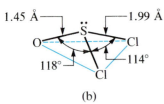

Figure 22.24

The molecular structures (a) of sulfuryl chloride (tetrahedral) and (b) of thionyl chloride (trigonal pyramidal).

22.14 Sulfurous Acid and Sulfites

Sulfur dioxide dissolves in water to form a solution of sulfurous acid, as expected for the oxide of a nonmetal. **Sulfurous acid** is unstable, and anhydrous H_2SO_3 cannot be isolated. Boiling a solution of sulfurous acid expels the sulfur dioxide. Like other diprotic acids, sulfurous acid ionizes in two steps; the **hydrogen sulfite ion,** HSO_3^-, and the **sulfite ion,** SO_3^{2-}, are formed. Sulfurous acid is a moderately strong acid. Ionization in the first stage is about 25% but is much less in the second.

Solid sulfite and hydrogen sulfite salts can be prepared by adding a stoichiometric amount of a base to a sulfurous acid solution and then evaporating the water. These salts are also formed by the reaction of SO_2 with oxides and hydroxides (Section 22.13). Solid sodium hydrogen sulfite forms sodium sulfite, sulfur dioxide, and water when heated.

$$2NaHSO_3 \longrightarrow Na_2SO_3 + SO_2 + H_2O$$

Sulfurous acid can be oxidized by strong oxidizing agents. Oxygen of the air oxidizes it slowly to the more stable sulfuric acid.

$$2H_2SO_3 + O_2 \longrightarrow 4H^+ + 2SO_4^{2-}$$

Solutions containing the purple permanganate ion rapidly turn colorless when sulfurous acid is added; the MnO_4^- ion is reduced to the colorless Mn^{2+} ion as sulfurous acid is oxidized.

$$2MnO_4^- + 5H_2SO_3 \longrightarrow 2Mn^{2+} + 4H^+ + 5SO_4^{2-} + 3H_2O$$

Solutions of sulfites are also very susceptible to air oxidation, and sulfates are formed. Thus solutions of sulfites always contain sulfates after standing in contact with air.

22.15 Sulfuric Acid and Sulfates

Sulfuric acid, H_2SO_4 (Fig. 22.25), is prepared by oxidizing sulfur to sulfur trioxide and then converting the trioxide to sulfuric acid (Section 9.9). Pure sulfuric acid is a colorless, oily liquid that freezes at 10.5°C. It fumes when heated because the acid decomposes to water and sulfur trioxide. More sulfur trioxide than water is lost during the heating, until a concentration of 98.33% acid is reached. Acid of this concentration boils at 338°C without further change in concentration and is sold as concentrated H_2SO_4.

Concentrated sulfuric acid dissolves in water with the evolution of a large amount of heat. **Caution! Adding water to the concentrated acid may cause dangerous spattering of the acid. Do not add water to concentrated sulfuric acid; add the acid slowly to water while stirring the solution to distribute the heat of dilution.** The large heat of dilution of sulfuric acid is caused by hydrate and hydronium ion formation. The hydrates $H_2SO_4 \cdot H_2O$, $H_2SO_4 \cdot 2H_2O$, and $H_2SO_4 \cdot 4H_2O$ are known.

The strong affinity of concentrated sulfuric acid for water makes it a good dehydrating agent. Gases that do not react with the acid may be dried by passing them through it. So great is the affinity of concentrated sulfuric acid for water that it will remove hydrogen and oxygen, in the form of water, from many organic compounds containing these elements in a 2-to-1 ratio. For example, cane sugar, $C_{12}H_{22}O_{11}$, is charred by concentrated sulfuric acid (Fig. 22.26).

$$C_{12}H_{22}O_{11} \longrightarrow 12C + 11H_2O$$

Concentrated sulfuric acid can cause serious burns, because it reacts with organic compounds in the skin.

Sulfuric acid is a strong diprotic acid that ionizes in two stages. In aqueous solution the first stage is essentially complete. The secondary ionization is less complete, but even so HSO_4^- is a moderately strong acid (about 25% ionized in solution).

Being a diprotic acid, sulfuric acid forms both **sulfates,** such as Na_2SO_4, and **hydrogen sulfates,** such as $NaHSO_4$. The sulfates of barium, strontium, calcium, and lead are only slightly soluble in water. These salts occur in nature as the minerals barite, $BaSO_4$; celestite, $SrSO_4$; gypsum, $CaSO_4 \cdot 2H_2O$; and anglesite, $PbSO_4$. They can be prepared in the laboratory by metathetical reactions. For example, adding barium nitrate to a solution of sodium sulfate causes the precipitation of white barium sulfate.

$$Ba^{2+} + [2NO_3^-] + [2Na^+] + SO_4^{2-} \longrightarrow BaSO_4(s) + [2Na^+] + [2NO_3^-]$$

This reaction is the basis of a qualitative and quantitative test for the sulfate ion and the barium ion.

Among the important soluble sulfates are Glauber's salt, $Na_2SO_4 \cdot 10H_2O$; Epsom salt, $MgSO_4 \cdot 7H_2O$; blue vitriol, $CuSO_4 \cdot 5H_2O$; green vitriol, $FeSO_4 \cdot 7H_2O$; and white vitriol, $ZnSO_4 \cdot 7H_2O$. Because the HSO_4^- ion is an acid, hydrogen sulfates, such as $NaHSO_4$, exhibit acidic behavior. Sodium hydrogen sulfate is the primary ingredient in some household cleansers.

Hot, concentrated sulfuric acid is an oxidizing agent. Depending on its concentration, the temperature, and the strength of the reducing agent, sulfuric acid oxidizes many compounds and, in the process, undergoes reduction to either SO_2, HSO_3^-, SO_3^{2-}, S, H_2S, or S^{2-}. The displacement of volatile acids from their salts by concentrated sulfuric acid is described in Section 16.8.

The amount of sulfuric acid used in industry exceeds that of any other manufactured compound. During 1986, 36.8 million tons were produced in the United States alone.

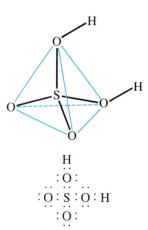

Figure 22.25

The tetrahedral molecular structure and the Lewis structure of sulfuric acid.

Figure 22.26

Concentrated sulfuric acid reacts with sugar (left), removing the elements of water and leaving carbon (right).

The major uses of sulfuric acid are based largely on its strongly acidic character. It is used to produce ammonium sulfate (2.1 million tons in the United States in 1986) and soluble phosphate fertilizers; to refine petroleum and to remove impurities from such products as gasoline and kerosene; to pickle steel (clean its surface of rust); to produce dyes, drugs and disinfectants from coal tar; as an electrolyte itself or to produce metal sulfates for electrolytes; to manufacture other chemicals, such as hydrochloric and nitric acids; and to produce textiles, paints, plastics, explosives, and lead storage batteries.

22.16 Derivatives of Sulfuric Acid and of Sulfates

There are several compounds that may be viewed as sulfuric acid or sulfate derivatives—that is, compounds in which the oxygen or hydroxide group of a sulfuric acid molecule or a sulfate ion has been replaced by some other atom or combination of atoms. For example, a **thiosulfate ion,** $S_2O_3^{2-}$, may be viewed as a sulfate ion in which one oxygen atom has been replaced by a sulfur atom (Fig. 22.27).

Oxygen oxidizes sulfites to sulfates (Section 22.14); sulfur oxidizes sulfites to thiosulfates. For example, when a mixture of sulfur and a solution of sodium sulfite is boiled, **sodium thiosulfate,** $Na_2S_2O_3$, is formed.

$$[2Na^+] + SO_3^{2-} + S \longrightarrow [2Na^+] + S_2O_3^{2-}$$

Crystals of the pentahydrate, $Na_2S_2O_3 \cdot 5H_2O$, separate when the solvent is evaporated.

Sodium thiosulfate, also called **hypo,** is used in the photographic process in the fixing solution. Hypo dissolves from the plate or film any silver halides that have not been reduced to metallic silver by the developer. Thiosulfate ion forms a soluble complex ion with silver ion, even with insoluble silver halides.

$$AgBr(s) + 2S_2O_3^{2-} \longrightarrow [Ag(S_2O_3)_2]^{3-} + Br^-$$

The thiosulfate ion is a reducing agent. It is oxidized by iodine to the tetrathionate ion, $S_4O_6^{2-}$.

$$2\,:\!\ddot{O}\!-\!\underset{\underset{:\ddot{O}:}{|}}{\overset{\overset{:\ddot{O}:}{|}}{S}}\!-\!\ddot{S}\!:^{2-} + :\!\ddot{I}\!-\!\ddot{I}\!: \longrightarrow :\!\ddot{O}\!-\!\underset{\underset{:\ddot{O}:}{|}}{\overset{\overset{:\ddot{O}:}{|}}{S}}\!-\!\ddot{S}\!-\!\ddot{S}\!-\!\underset{\underset{:\ddot{O}:}{|}}{\overset{\overset{:\ddot{O}:}{|}}{S}}\!-\!\ddot{O}\!:^{2-} + 2\,:\!\ddot{I}\!:^-$$

This reaction is used extensively in analytical chemistry. For example, in one quantitative analysis procedure for copper ion, iodine is produced by the following reaction and then titrated with a standard thiosulfate solution.

$$Cu^{2+} + 3I^- \longrightarrow I_2 + CuI(s)$$

When a solution of sodium thiosulfate is acidified, unstable thiosulfuric acid, $H_2S_2O_3$, is formed.

$$S_2O_3^{2-} + 2H^+ \longrightarrow H_2S_2O_3$$

The acid decomposes immediately into sulfurous acid and sulfur.

$$H_2S_2O_3 \longrightarrow H_2SO_3 + S(s)$$

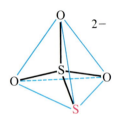

Thiosulfate ion, $S_2O_3^{2-}$

(a)

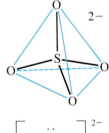

Sulfate ion, SO_4^{2-}

(b)

Figure 22.27

The tetrahedral molecular structures and the Lewis structures of (a) the thiosulfate ion and (b) the sulfate ion.

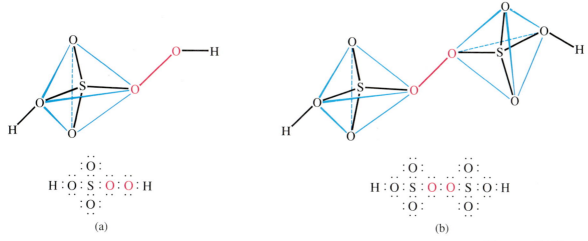

(a)   (b)

Figure 22.28

Structures of (a) peroxymonosulfuric acid and (b) peroxydisulfuric acid. Each acid possesses an oxygen-oxygen (peroxide) linkage; hence the use of the prefix *peroxy-* in the names.

Both **peroxymonosulfuric acid,** H_2SO_5, and **peroxydisulfuric acid,** $H_2S_2O_8$, (Fig. 22.28) may be viewed as peroxide derivatives of sulfuric acid. A hydroxyl group of the sulfuric acid molecule has been replaced by a —OOH group in peroxymonosulfuric acid and by a —OOSO$_3$H group in peroxydisulfuric acid.

Peroxydisulfuric acid is produced commercially by the anodic oxidation of hydrogen sulfate ions in 45–55% sulfuric acid at a low temperature.

$$2HSO_4^- \longrightarrow H_2S_2O_8 + 2e^-$$

Electrolysis of potassium hydrogen sulfate in an aqueous solution gives **potassium peroxydisulfate,** $K_2S_2O_8$. Treating this salt at low temperatures with concentrated sulfuric acid produces peroxymonosulfuric acid, commonly called Caro's acid, H_2SO_5. This acid is also produced by the reaction of sulfur trioxide with H_2O_2. This reaction is analogous to that of sulfur trioxide with water.

The peroxysulfuric acids and their salts are useful as strong oxidizing agents.

Chlorosulfonic acid, HSO$_3$Cl, and **sulfamic acid,** SO$_3$NH$_3$, are related structurally to sulfuric acid, as shown in Fig. 22.29. The formula for sulfamic acid, sometimes written HSO$_3$NH$_2$, is more properly written SO$_3$NH$_3$.

Chlorosulfonic acid is a colorless liquid that fumes in moist air, reacts vigorously with water to give sulfuric acid and hydrogen chloride (like many nonmetal halides), and is used to introduce the sulfonate group, SO$_3$H, into many organic compounds. Sulfamic acid is a white, crystalline, nonhygroscopic solid. It is one of the few strong

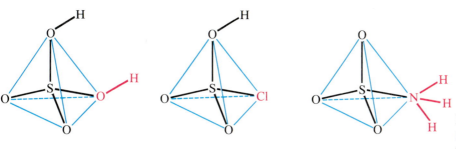

Sulfuric acid   Chorosulfonic acid   Sulfamic acid

Figure 22.29

Structure of a sulfuric acid molecule compared to the structures of two related acids.

monoprotic acids that can be weighed with no special drying precautions required. It is of some importance, therefore, in analytical chemistry. Sulfamates are important ingredients in some weed killers.

The Halogens

The elements of Group VIIA of the Periodic Table are known as the **halogens,** which means ''salt formers.'' Their binary compounds are called **halides.** The first four halogens—fluorine, chlorine, bromine, and iodine—were first isolated between 1774 and 1886. Salts of these elements are common in nature. Astatine, the fifth halogen, was first prepared artificially in 1940. It is a radioactive element of limited stability.

22.17 Occurrence and Preparation of the Halogens

The halogens are too reactive to occur free in nature, but their compounds are widely distributed. Chlorides are the most abundant, and although fluorides, bromides, and iodides are less common, they are reasonably available.

All of the halogens occur in sea water as halide ions. The concentration of the chloride ion is 0.54 molar; that of the other halides is less than 10^{-4} molar. Fluorine is also found in the minerals fluorite, CaF_2, fluoroapatite, $Ca_5(PO_4)_3F$, and cryolite, Na_3AlF_6. Fluorine is also found in teeth, bone, and the blood, in small amounts. Chlorine is found in high concentrations in the Great Salt Lake and the Dead Sea and in extensive salt beds (Fig. 22.30) that contain NaCl, $MgCl_2$, or $CaCl_2$. It is a component of stomach acid, which is hydrochloric acid. Bromine occurs in the Dead Sea and in underground brines. Iodine is found in small quantities in Chile saltpeter, in underground brines, and in sea kelp. Iodine is an essential component of the thyroid gland.

The best sources of the halogens (except iodine) are halide salts. The halide ions may be oxidized to free diatomic halogen molecules by various methods, depending upon the ease of oxidation of the halide ion. This increases with increasing atomic size, in the order $F^- < Cl^- < Br^- < I^-$.

Figure 22.30
An underground salt mine in a bed of essentially pure salt.

Fluorine is the most powerful oxidizing agent of the known elements; it spontaneously oxidizes most other elements. The reverse reaction, the oxidation of fluorides, is very difficult; electrolytic oxidation is needed to prepare elemental **fluorine.** Electrolysis is often used to carry out oxidation-reduction reactions with species that are oxidized or reduced with difficulty. The electrolysis is commonly carried out in a molten mixture of potassium hydrogen fluoride, KHF_2, and anhydrous hydrogen fluoride (melting point 72°C). When electrolysis begins, HF is decomposed to form fluorine gas at the anode and hydrogen at the cathode. The two gases are kept separated to prevent their recombination to form hydrogen fluoride.

Most commercial **chlorine** is produced by electrolysis of the chloride ion in aqueous solutions of sodium chloride. Other products of the electrolysis are hydrogen and sodium hydroxide (Sections 13.5 and 21.3). Chlorine is also a product when metals such as sodium, calcium, and magnesium are produced by the electrolysis of their fused chlorides (Section 13.2 and 21.2).

Chloride, bromide, and iodide ions are easier to oxidize than fluoride ions. Thus chlorine, bromine, and iodine also can be prepared by the chemical oxidation of the respective halides. Small quantities of chlorine are sometimes prepared by oxidation of the chloride ion in acid solution with strong oxidizing agents such as manganese dioxide (MnO_2), potassium permanganate ($KMnO_4$), and sodium dichromate ($Na_2Cr_2O_7$). The reaction with manganese dioxide is

$$MnO_2 + 2Cl^- + 4H^+ \longrightarrow Mn^{2+} + Cl_2(g) + 2H_2O$$

The methods for small-scale oxidation of bromides to bromine are like those used for the oxidation of chlorides.

Bromine is prepared commercially by the oxidation of bromide ion by chlorine.

$$2Br^-(aq) + Cl_2(g) \longrightarrow Br_2(l) + 2Cl^-(aq)$$

Chlorine is a stronger oxidizing agent than bromine, and the equilibrium for this reaction lies well to the right. Essentially all domestic bromine is produced by chlorine oxidation of bromide ions obtained from underground brines found in Arkansas.

Elemental **iodine** is sometimes produced by the oxidation of iodide ion with chlorine. An excess of chlorine must be avoided; it forms iodine monochloride, ICl, and iodic acid, HIO_3. Iodine is produced commercially by the reduction of sodium iodate, $NaIO_3$, an impurity in deposits of Chile saltpeter, with sodium hydrogen sulfite.

$$2IO_3^- + 5HSO_3^- \longrightarrow 3HSO_4^- + 2SO_4^{2-} + H_2O + I_2(s)$$

22.18 Uses of the Halogens

Fluorine gas has been used to fluorinate organic compounds (to replace hydrogen with fluorine) since its initial discovery by Moissan in 1886. The resulting **fluorocarbon** compounds are quite stable and nonflammable. Freon-12, CCl_2F_2, is widely used as a refrigerant. Teflon is a polymer composed of $-CF_2CF_2-$ units. Perfluorodecalin ($C_{10}F_{16}$) is useful as a blood substitute, in part because oxygen is very soluble in this chemically inert substance. Fluorine gas is used in the production of uranium hexafluoride, UF_6, which is used in the separation of isotopes for the production of atomic energy. Fluoride ion is added to water supplies and to some toothpastes (Fig. 22.31) as SnF_2 or NaF to decrease cavities.

Chlorine is used to bleach wood pulp and cotton cloth. The chlorine reacts with water to form hypochlorous acid, which oxidizes colored substances to colorless ones.

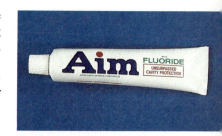

Figure 22.31

Fluorides are used in fighting tooth decay.

Large quantities of chlorine are used in chlorinating hydrocarbons (replacing hydrogen with chlorine) to produce compounds such as carbon tetrachloride (CCl_4), chloroform ($CHCl_3$), and ethyl chloride (C_2H_5Cl) and in the production of polyvinyl chloride (PVC) and other polymers. Chlorine is also used to kill the bacteria in community water supplies.

Bromine is used to produce certain dyes, light-sensitive silver bromide for photographic film, and sodium and potassium bromides for sedatives and soporifics.

Iodine in alcohol solution with potassium iodide is used as an antiseptic (tincture of iodine). Iodine salts are essential for the proper functioning of the thyroid gland; an iodine deficiency may lead to the development of goiter. Iodized table salt contains 0.023% potassium iodide. Silver iodide is used in photographic film and in the seeding of clouds to induce rain. Iodoform, CHI_3, is an antiseptic.

22.19 Properties of the Halogens

Fluorine is a pale yellow gas, chlorine is a greenish-yellow gas, bromine is a deep reddish-brown liquid three times as dense as water, and iodine is a grayish-black crystalline solid with a low melting point (Fig. 22.32). Liquid bromine has a high vapor pressure, and the reddish vapor can easily be seen in a bottle containing the liquid. Iodine crystals have a high vapor pressure. When gently heated, these crystals sublime and form a beautiful deep violet vapor.

Bromine is only slightly soluble in water, but it is miscible in all proportions in less polar (or nonpolar) solvents such as alcohol, ether, chloroform, carbon tetrachloride, and carbon disulfide, forming solutions that vary in color from yellow to reddish-brown, depending upon the concentration.

Iodine is soluble in chloroform, carbon tetrachloride, carbon disulfide, and many hydrocarbons, giving violet solutions of I_2 molecules (Fig. 15.4). The solutions have the same color as I_2 molecules in the gas phase. Iodine dissolves only slightly in water, giving brown solutions. It is quite soluble in alcohol, ether, and aqueous solutions of iodides, with which it also forms brown solutions. These brown solutions (Fig. 15.4) result because iodine molecules are weak Lewis acids (Section 16.15) and combine with solvent molecules that can function as Lewis bases, or with the iodide ion, which

Figure 22.32

Chlorine is a pale yellow-green gas, gaseous bromine is deep orange, and gaseous iodine (produced by warming the solid) is purple. (Fluorine is so reactive that it is too dangerous for the photographer to handle.)

can also act as a Lewis base. The equation for the reversible reaction of iodine with the iodide ion to give the triiodide ion, $I_3{}^-$, is

$$:\!\ddot{I}\!:^- + \; :\!\ddot{I}\!:\!\ddot{I}\!: \; \rightleftharpoons \; :\!\ddot{I}\!:\!\ddot{I}\!:\!\ddot{I}\!:^-$$

The elemental (free) halogens are oxidizing agents with strengths decreasing in the order $F_2 > Cl_2 > Br_2 > I_2$. In general, the heavier a halogen, the less its strength as an oxidizing agent. Fluorine generally oxidizes an element to its highest oxidation number, whereas the heavier halogens may not. For example, when fluorine reacts with sulfur, SF_6 is formed. Chlorine gives SCl_2 and bromine, S_2Br_2. Iodine does not react with sulfur.

The general chemical behavior of the halogens is discussed in Section 20.8. The reactions of elemental halogens with a variety of substances are summarized in Table 22.4.

Fluorine reacts directly and forms binary fluorides with all of the elements except the lighter noble gases (He, Ne, and Ar). Fluorine is such a strong oxidizing agent that

---

Table 22.4   Chemical properties of the elemental halogens

General equation	Comments
Reactions with elements	
$2M + nX_2 \longrightarrow 2MX_n$	With almost all metals
$H_2 + X_2 \longrightarrow 2HX$	With decreasing reactivity in the order $F_2 > Cl_2 > Br_2 > I_2$
$Xe + \dfrac{n}{2} F_2 \longrightarrow XeF_n$	$n = 2, 4,$ or 6
$nX_2 + X'_2 \longrightarrow 2X'X_n$	X′ heavier than X; n an odd integer
$S + 3X_2 \longrightarrow SX_6$	With F_2; Se and Te can replace S
$S + X_2 \longrightarrow SX_2$	With Cl_2
$2S + X_2 \longrightarrow S_2X_2$	With Cl_2 or Br_2
$2P + 3X_2 \longrightarrow 2PX_3$	With excess P; also with As, Sb, or Bi
$2P + 5X_2 \longrightarrow 2PX_5$	With excess halogen, except I_2
Reactions with compounds	
$X_2 + 2X'^- \longrightarrow 2X^- + X'_2$	X′ heavier than X
$F_2 + H_2O \longrightarrow$ $\quad\quad O_2 + O_3 + OF_2 + H_2O_2 + HF$	Unbalanced; actual distribution of products depends on reaction conditions
$X_2 + H_2O \longrightarrow H^+ + X^- + HOX$	Not with F_2
$X_2 + H_2S \longrightarrow 2HX + S$	With Cl_2 or Br_2
$X_2 + CO \longrightarrow COX_2$	With Cl_2 or Br_2
$X_2 + SO_2 \longrightarrow SO_2X_2$	With F_2 or Cl_2
$X_2 + PX_3 \longrightarrow PX_5$	Not with I_2
$-\!\overset{\mid}{\underset{\mid}{C}}\!-H + X_2 \longrightarrow -\!\overset{\mid}{\underset{\mid}{C}}\!-X + HX$	With Cl_2 or Br_2 and many hydrocarbons
$\diagup\!\!\!\!C\!=\!C\diagdown + X_2 \longrightarrow -\overset{X}{\underset{\mid}{C}}-\overset{X}{\underset{\mid}{C}}-$	With Cl_2, Br_2, I_2, and many hydrocarbons containing $C\!=\!C$ double bonds

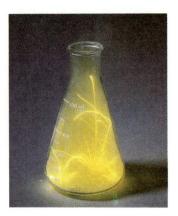

Figure 22.33
Molten sodium (mp 97.8°C) inflames in an atmosphere of chlorine.

many substances ignite upon contact with it. Drops of water inflame in fluorine and form O_2, OF_2, H_2O_2, O_3, and HF. Wood and asbestos ignite and burn in fluorine gas. Most hot metals burn vigorously in fluorine. However, fluorine can be handled in copper, iron, magnesium, or nickel containers, because an adherent film of the fluoride protects their surfaces from further attack.

Fluorine readily displaces chlorine and the other halogens from solid metal halides; in an excess of fluorine, halogen fluorides are formed. Fluorine and hydrogen react explosively. Fluorine is the only element that will react directly with the noble gas xenon.

Although it is a strong oxidizing agent, **chlorine** is less active than fluorine. For example, fluorine and hydrogen react explosively, but when chlorine and hydrogen are mixed in the dark, the reaction between them is so slow as to be imperceptible. When the mixture is exposed to light, the reaction is explosive. Chlorine is less active toward metals than fluorine, and oxidation reactions usually require higher temperatures. Molten sodium ignites in chlorine (Fig. 22.33). Chlorine attacks most nonmetals (C, N_2, and O_2 are notable exceptions), forming covalent molecular compounds. Chlorine generally reacts with compounds that contain only carbon and hydrogen (hydrocarbons) by adding to multiple bonds or by substitution (Table 22.4).

When chlorine is added to water, it is both oxidized and reduced in a **disproportionation** reaction.

$$Cl_2 + H_2O \rightleftharpoons HOCl + H^+ + Cl^-$$

Half the chlorine atoms oxidize to the +1 oxidation state (in hypochlorous acid), and the other half reduce to the −1 oxidation state (in chloride ion). This disproportionation is incomplete, so chlorine water is a solution of chlorine molecules, hypochlorous acid molecules, hydrogen ions, and chlorine ions. When exposed to light, this solution undergoes a photochemical decomposition.

$$2HOCl \xrightarrow{\text{Sunlight}} 2H^+ + 2Cl^- + O_2(g)$$

The chemical properties of **bromine** are similar to those of chlorine, although bromine is the weaker oxidizing agent and its reactivity is less than that of chlorine.

Iodine is the least reactive of the four naturally occurring halogens. It is the weakest oxidizing agent, and its ion is the most easily oxidized. Iodine reacts with metals, but heating is often required. It will not oxidize other halide ions but will oxidize some other nonmetal anions such as sulfide ion, S^{2-}.

$$S^{2-} + I_2 \longrightarrow S + 2I^-$$

Compared with the other halogens, iodine reacts only slightly with water. Traces of iodine in water react with a mixture of starch and iodide ion and form a deep blue color. This reaction is used as a very sensitive test for the presence of iodine in water.

22.20 Interhalogens

Compounds formed from two different halogens are called **interhalogens**. Interhalogen molecules consist of one atom of the heavier halogen bonded by single bonds to an odd number of atoms of the lighter halogen. The structures of IF_3, IF_5, and IF_7 are shown in Fig. 22.34. Formulas for other interhalogens, each of which can be prepared by the reaction of the respective halogens, are given in Table 22.5.

Table 22.5
Interhalogens

XX′	XX′$_3$	XX′$_5$	XX′$_7$
ClF	ClF$_3$	ClF$_5$	
BrF	BrF$_3$	BrF$_5$	
BrCl			
IF	IF$_3$	IF$_5$	IF$_7$
ICl	ICl$_3$		
IBr			

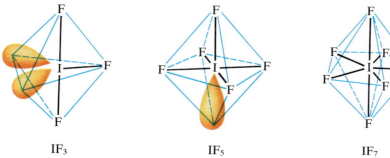

IF₃   IF₅   IF₇

Figure 22.34
Structures of IF₃ (T-shaped).
IF₅ (square pyramidal), and
IF₇ (pentagonal bipyramidal).

Because smaller halogens are grouped about a larger one, the maximum number of smaller atoms possible increases as the radius of the larger atom increases (see Section 11.17). Many of these compounds are unstable, and most are extremely reactive. The interhalogens react like their component halides; halogen fluorides, for example, are stronger oxidizing agents than halogen chlorides.

The ionic **polyhalides** of the alkali metals, compounds such as KI₃, KICl₂, KICl₄, CsIBr₂, and CsBrCl₂ that contain an anion composed of at least three halogen atoms, are closely related to the interhalogens (Fig. 22.35). The formation of the polyhalide anion I₃⁻ is responsible for the solubility of iodine in aqueous solutions containing iodide ion.

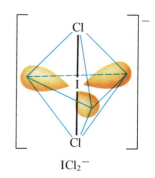

ICl₂⁻

22.21 Hydrogen Halides

Binary compounds containing only hydrogen and a halogen are called **hydrogen halides.** At room temperature the pure hydrogen halides HF, HCl, HBr, and HI are gases.

The anhydrous hydrogen halides are rather inactive chemically and do not attack dry metals at room temperature. However, they will react with many metals at elevated temperatures, forming metal halides and hydrogen. These reactions are sometimes used to prepare anhydrous metal halides.

$$Fe + 2HCl(g) \xrightarrow{300°C} FeCl_2 + H_2(g)$$

The hydrogen halides can be prepared by the general techniques used to prepare other acids (Section 16.8) although each technique is not suitable for every hydrogen halide. As indicated in Table 22.4, fluorine, chlorine, and bromine react directly with hydrogen to form the respective hydrogen halide. This reaction is used to prepare hydrogen chloride and hydrogen bromide commercially. Bromine reacts much less vigorously with hydrogen than chlorine. The direct reaction of hydrogen and iodine is unsatisfactory for the preparation of hydrogen iodide, because the reaction is slow and the equilibrium yield is low. Heat decomposes HI, so the rate of reaction cannot be increased by heating. Hydrogen fluoride can be prepared by the direct reaction of the elements, but elemental fluorine is expensive and the reaction is violent; other routes are used to prepare HF.

Hydrogen halides can be prepared by acid-base reactions between a nonvolatile strong acid and a metal halide. The escape of the gaseous hydrogen halide drives the reaction to completion. For example, hydrogen fluoride is usually prepared by heating a mixture of calcium fluoride, CaF₂, and concentrated sulfuric acid.

$$CaF_2 + H_2SO_4 \longrightarrow CaSO_4 + 2HF(g)$$

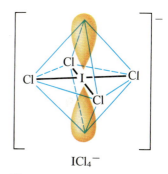

ICl₄⁻

Figure 22.35
Structures of the ions ICl₂⁻
(linear) and ICl₄⁻ (square planar).

Figure 22.36

The reaction of sulfuric acid with NaCl (left) produces gaseous HCl. With NaI, sulfuric acid oxidizes the NaI and produces I₂ (right).

Gaseous hydrogen fluoride is also a by-product in the preparation of phosphate fertilizers by the reaction of fluoroapatite, $Ca_5(PO_4)_3F$, with sulfuric acid (Section 9.9). Hydrogen chloride is prepared, both in the laboratory and commercially, by the reaction of concentrated sulfuric acid with a chloride salt [Fig. 22.36 (left)]. Sodium chloride is usually used, since it is the least expensive chloride. Hydrogen bromide and hydrogen iodide cannot be prepared in similar reactions because sulfuric acid is a strong enough oxidizing agent that it oxidizes both bromide and iodide ions [Fig. 22.36 (right)]. Sulfuric acid is reduced to sulfur dioxide by bromide ion. Iodide ion reduces sulfuric acid to sulfur dioxide, sulfur, or hydrogen sulfide, depending upon the relative amounts of acid and iodide ion used in the reaction.

The reaction of a covalent nonmetal bromide or iodide with water produces hydrogen bromide or hydrogen iodide, respectively.

$$PBr_3 + 3H_2O \longrightarrow 3HBr(g) + H_3PO_3$$
$$PI_3 + 3H_2O \longrightarrow 3HI(g) + H_3PO_3$$

This method is rarely used to prepare hydrogen chloride; less expensive methods are used.

About two-thirds of the hydrogen chloride produced commercially in the United States (2,983,000 tons in 1986) is a by-product of the reactions used to prepare chlorinated hydrocarbons. For example, the reaction used to manufacture ethyl chloride from ethane (C_2H_6) and chlorine also produces hydrogen chloride.

$$C_2H_6(g) + Cl_2(g) \longrightarrow C_2H_5Cl(g) + HCl(g)$$

Hydrogen bromide is a by-product of the manufacture of brominated hydrocarbons by a similar reaction involving bromine and hydrocarbons.

All of the hydrogen halides are very soluble in water. With the exception of hydrogen fluoride, they are strong acids; they ionize completely in dilute aqueous solution. Aqueous solutions of the hydrogen halides are called **hydrofluoric acid, hydrochloric acid, hydrobromic acid,** and **hydriodic acid** (Fig. 22.37). Reactions of the hydrohalic acids with metals or with metal hydroxides, oxides, and carbonates are often used to prepare soluble salts of the halides. Most chloride salts are soluble (AgCl, $PbCl_2$, and Hg_2Cl_2 are the common exceptions).

The halide ions in hydrohalic acids give these substances the properties associated with $X^-(aq)$. The heavier halide ions (Cl^-, Br^-, I^-) can act as reducing agents and are

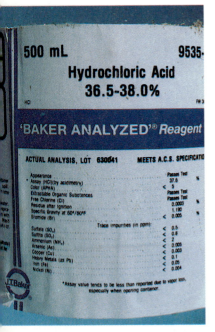

Figure 22.37

Hydrochloric acid is supplied as a 36.5–38% solution of HCl in water.

oxidized by lighter halogens or other oxidizing agents. They also serve as precipitating agents for insoluble metal halides.

Pure hydrogen fluoride differs from the other hydrogen halides because its molecules are associated through hydrogen bonding (Section 11.12). This gives liquid hydrogen fluoride an anomalously high boiling point for a hydrogen halide (Fig. 11.20). Hydrogen-bonded dimers, $(HF)_2$, are observed in the vapor. Only very weak hydrogen bonds can occur in hydrogen chloride, hydrogen bromide, and hydrogen iodide, because the electronegativities of the halogen atoms in these molecules are low, and consequently the polarities of the bonds are small.

Hydrofluoric acid is unique in its reactions with sand (silicon dioxide) and with glass, which is a mixture of silicates (mainly calcium silicate).

$$SiO_2 + 4HF \longrightarrow SiF_4(g) + 2H_2O$$
$$CaSiO_3 + 6HF \longrightarrow CaF_2 + SiF_4(g) + 3H_2O$$

The silicon escapes from these reactions as silicon tetrafluoride, a volatile compound. Because hydrogen fluoride attacks glass, it is used to frost or etch glass. Light bulbs are frosted with hydrogen fluoride, and markings on thermometers, burets, and other glassware are made with it.

The largest use for hydrogen fluoride is in the production of fluorocarbons for refrigerants such as the Freons, plastics, and propellants. The second largest use is in the manufacture of cryolite, K_3AlF_6, which is important in the production of aluminum. The acid is also used in the production of other inorganic fluorides (such as BF_3), which are used as catalysts in the industrial synthesis of certain organic compounds.

Extreme care should be used when handling hydrofluoric acid, for it causes painful and slow-healing burns. It is a local anesthetic, and its presence may not be noticed until much damage has been done.

Hydrochloric acid is inexpensive. It is the most important acid in industry after sulfuric acid and is used in the manufacture of metal chlorides, dyes, glue, glucose, and various other chemicals. A considerable amount is also used in the activation of oil wells and as a pickle liquor, an acid used to remove oxide coatings from iron or steel that is to be galvanized, tinned, or enameled. The amounts of hydrobromic acid and hydroiodic acid used commercially are insignificant compared to hydrochloric acid.

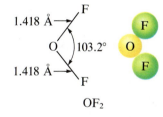

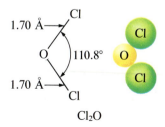

22.22 Binary Oxygen-Halogen Compounds

The halogens do not react directly with oxygen, but binary halogen-oxygen compounds can be prepared by the reactions of the halogens with oxygen-containing compounds. Oxygen compounds with chlorine, bromine, and iodine are called oxides because oxygen is the more electronegative element in these compounds. On the other hand, fluorine compounds with oxygen are called fluorides because fluorine is the more electronegative element. Most binary oxygen-halogen compounds are extremely reactive and unstable. Iodine(V) oxide, I_2O_5, is the only one of these compounds that does not decompose upon warming.

The most stable oxygen fluoride, oxygen difluoride (Fig. 22.38), is prepared by the reaction of fluorine with a solution of a hydroxide.

$$2F_2 + 2OH^- \longrightarrow OF_2(g) + 2F^- + H_2O$$

Although OF_2 is appreciably soluble in water, it is not the anhydride of an acid; it does not react with water.

Figure 22.38

Angular structures of the OF_2 and Cl_2O molecules (drawn approximately to scale). Bond distances are measured between the nuclei of the atoms.

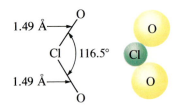

Figure 22.39
Structure of the ClO_2 molecule.

Chlorine(I) oxide, or simply chlorine monoxide, a yellow-red gas that is apt to explode violently, has the same structure as oxygen difluoride (Fig. 22.38). It is prepared by the reaction of chlorine with freshly prepared mercury(II) oxide

$$2Cl_2 + 2HgO \longrightarrow Cl_2O(g) + HgCl_2 \cdot HgO$$

Chlorine monoxide is an active oxidizing agent and reacts with water to give hypochlorous acid, HOCl. Chlorine dioxide, ClO_2 (Fig. 22.39), can be prepared by the reaction of chlorine with silver chlorate:

$$Cl_2 + 2AgClO_3 \longrightarrow 2ClO_2(g) + 2AgCl + O_2$$

On an industrial scale, ClO_2 is prepared by oxidation of sodium chlorite with chlorine:

$$2NaClO_2 + Cl_2 \longrightarrow 2NaCl + 2ClO_2(g)$$

When pure, this yellow gas is apt to explode violently, but it can be handled safely if diluted with carbon dioxide or air. Chlorine dioxide disproportionates slowly in water, forming chloric and hydrochloric acids. Chlorine dioxide is used for the bleaching of flour and in water treatment.

The reaction of perchloric acid, $HClO_4$, with phosphorus(V) oxide produces the violently explosive chlorine heptaoxide, Cl_2O_7.

$$P_4O_{10} + 4HClO_4 \longrightarrow 4HPO_3 + 2Cl_2O_7$$

This reaction is analogous to that used for the production of nitrogen pentaoxide by the reaction of P_4O_{10} with nitric acid (Section 23.12).

The most stable of the halogen oxides, iodine(V) oxide, is prepared by heating iodic acid, HIO_3, which decomposes to form I_2O_5 and water.

22.23 Oxyacids of the Halogens and Their Salts

The compounds HXO, HXO_2, HXO_3, and HXO_4, where X represents one of the heavier halogens (Cl, Br, or I), are called **hypohalous, halous, halic,** and **perhalic** acids, respectively. The strengths of these acids increase from the hypohalous acids, which are very weak acids, to the perhalic acids, which are very strong.

The only known oxyacid of fluorine is the very unstable **hypofluorous acid,** HOF, which is prepared by the reaction of gaseous fluorine with ice:

$$F_2(g) + H_2O(s) \longrightarrow HOF(g) + HF(g)$$

This compound does not ionize in water, and no salts are known.

The reactions of chlorine and bromine with water are analogous to that of fluorine with ice, but these reactions do not go to completion, and mixtures of the halogen and the respective hypohalous and hydrohalic acids result. The reaction of the halogen with mercury(II) oxide, similar to that used for Cl_2O (Section 22.22), is used to prepare a solution of the pure **hypohalous acid** (X = Cl, Br, I).

$$2X_2 + 3HgO + H_2O \longrightarrow HgX_2 \cdot 2HgO(s) + 2HOX(aq)$$

None of the hypohalous acids, except HOF, has been isolated in the free state. Because of their thermal instability, they are stable only in solution. The hypohalous acids are all very weak acids; HOCl is a stronger acid than HOBr, which in turn is stronger than HOI.

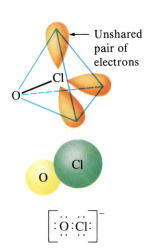

Figure 22.40
Structure of the hypochlorite ion, ClO^-.

Solutions of salts containing the basic **hypohalite** ions, OX^- (Fig. 22.40), may be prepared by adding base to solutions of hypohalous acids. The salts have been isolated as solids. All of the hypohalites are unstable with respect to disproportionation in solution, but the reaction is slow for hypochlorite. Hypobromite and hypoiodite disproportionate rapidly, even in the cold.

$$3XO^- \longrightarrow 2X^- + XO_3^-$$

Sodium hypochlorite is used as an inexpensive bleach (Chlorox) and germicide. It is produced commercially by the electrolysis of cold, dilute aqueous sodium chloride solutions under conditions where the resulting chlorine and hydroxide ion can react. The net reaction is

$$Cl^- + H_2O \xrightarrow{\text{Electrical energy}} ClO^- + H_2$$

The only known halous acid is **chlorous acid,** $HClO_2$, obtained by the reaction of barium chlorate with dilute sulfuric acid.

$$Ba(ClO_2)_2(aq) + H_2SO_4(aq) \longrightarrow BaSO_4(s) + 2HClO_2(aq)$$

A solution of $HClO_2$ is obtained by filtering off the barium sulfate. Chlorous acid is not stable; it slowly decomposes in solution to give chlorine dioxide, hydrochloric acid, and water. Chlorous acid reacts with bases to give salts containing the **chlorite ion** (Fig. 22.41). Metal chlorite salts can also be prepared by the action of chlorine dioxide on a metal peroxide. For example, sodium peroxide reacts as follows:

$$2ClO_2 + Na_2O_2 \longrightarrow 2NaClO_2 + O_2$$

Sodium chlorite is used extensively in the bleaching of paper, because it is a strong oxidizing agent that does not damage the paper.

Chloric acid, $HClO_3$, and **bromic acid,** $HBrO_3$, are stable only in solution, but **iodic acid,** HIO_3, can be isolated as a stable white solid from the reaction of iodine with concentrated nitric acid.

$$I_2 + 10HNO_3 \longrightarrow 2HIO_3 + 10NO_2(g) + 4H_2O$$

The lighter halic acids can be obtained from their barium salts by reaction with dilute sulfuric acid. The reaction is analogous to that used to prepare chlorous acid. All of the halic acids are strong acids and very active oxidizing agents. Salts containing **halate** ions (Fig. 22.42) can be prepared by reaction of the acids with bases. Metal chlorites are also prepared by electrochemical oxidation of a hot solution of a metal halide. Bromates can be produced by the oxidation of bromides with hypochlorite ion; oxidation of iodides with chlorates gives iodates. Sodium chlorate is used as a weed killer; potassium chlorate is used in some matches and to prepare oxygen in the laboratory (Section 22.5).

Perchloric acid, $HClO_4$, may be obtained by treating a perchlorate, such as potassium perchlorate, with sulfuric acid under reduced pressure. The $HClO_4$ distills from the mixture.

$$KClO_4 + H_2SO_4 \longrightarrow HClO_4 + KHSO_4$$

Perchloric acid explodes above 92°C, but it will distill at temperatures below 92° at reduced pressures. Dilute aqueous solutions of perchloric acid are quite stable thermally, but concentrations above 60% are unstable and dangerous. Perchloric acid and its salts are powerful oxidizing agents (Fig. 22.13). Serious explosions have occurred

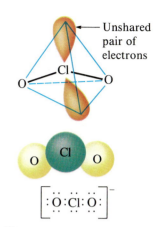

Figure 22.41
Structure of the chlorite ion, ClO_2^-.

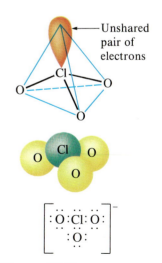

Figure 22.42
Structure of the chlorate ion, ClO_3^-.

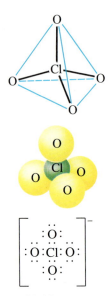

Figure 22.43
Structure of the perchlorate ion, ClO_4^-.

when concentrated solutions were heated with easily oxidized substances. However, its reactions as an oxidizing agent are slow when it is cold and dilute. The acid is among the strongest of all acids. Most salts containing the **perchlorate ion** (Fig. 22.43) are soluble. They are prepared by reactions of bases with perchloric acid and, commercially, by the electrolysis of hot solutions of their chlorides.

Perbromate salts are difficult to prepare. They require the reaction of bromates with fluorine in basic solution. Perbromic acid, $HBrO_4$, is prepared by the acidification of perbromate salts.

Several different acids containing iodine in the +7 oxidation state are known; these include **metaperiodic acid,** HIO_4, and **paraperiodic acid,** H_5IO_6. Salts of these acids can be readily prepared by reactions with bases. In addition, sodium metaperiodate, $NaIO_4$, may be prepared by the oxidation of sodium iodate, $NaIO_3$, by chlorine in a hot alkaline solution. Paraperiodates may also be prepared by the electrolytic oxidation of iodates, a method analogous to the preparation of perchlorates. Free paraperiodic acid may be obtained by evaporating a solution of metaperiodic acid, obtained by the reaction of barium metaperiodate with sulfuric acid. Salts of paraperiodic acid such as $Na_2H_3IO_6$, $Na_3H_2IO_6$, and Ag_5IO_6 have been prepared.

For Review

Summary

Hydrogen has the chemical properties of a nonmetal with a relatively low electronegativity. It forms ionic hydrides with active metals, covalent compounds in which it has an oxidation number of −1 with less electronegative elements, and covalent compounds in which it has an oxidation number of +1 with more electronegative nonmetals. It will react explosively with oxygen, fluorine, and chlorine; less readily with bromine; and much less readily with iodine, sulfur, and nitrogen. Hydrogen will reduce the oxides of those metals lying below chromium in the activity series to form the metal and water.

Oxygen (Group VIA) forms compounds with almost all of the elements. Except in a few compounds with fluorine, oxygen exhibits negative oxidation numbers in its compounds. The most common oxidation number is −2, in **oxides.** The −1 oxidation number is found in compounds that contain O—O bonds, **peroxides.** Oxygen is a strong oxidizing agent. **Ozone,** O_3, is an allotrope of oxygen, O_2. Ozone forms from oxygen in an endothermic reaction and is a stronger oxidizing agent than O_2.

Sulfur (Group VIA) reacts with almost all metals and readily forms the **sulfide ion,** S^{2-}, in which it has an oxidation number of −2, or **polysulfides,** S_n^{2-}. Sulfur also reacts with most nonmetals. It exhibits an oxidation number of −2 in covalent compounds with less electronegative elements. With more electronegative elements it commonly exhibits oxidation numbers of +4 and +6, and occasionally +1 and +2.

Sulfur burns in air and forms sulfur dioxide, which reacts with water to form the weak, unstable **sulfurous acid.** Neutralization of sulfurous acid and reactions of sulfur dioxide with metal oxides produce **sulfites.** The slow reaction of sulfur dioxide with oxygen produces sulfur trioxide. Sulfur trioxide is used to prepare **sulfuric acid,** a strong acid and an oxidizing agent. Neutralization of sulfuric acid and reactions of sulfur trioxide with metal oxides produce **sulfates.** Replacement of an OH group or an oxygen atom in a sulfuric acid molecule or a sulfate ion produces derivatives of sulfuric acid and of the sulfate ion.

The **halogens** are members of Group VIIA. The heavier halogens form compounds in which they have oxidation numbers of -1, $+1$, $+3$, $+5$, and $+7$, although they sometimes exhibit other oxidation numbers. Fluorine always exhibits an oxidation number of -1 in compounds. A halogen exhibits the highest electronegativity of any of the elements in each row of the Periodic Table.

The oxidizing ability of the elemental halogens and the resistance of the halides to oxidation decrease as halogen size increases. This is reflected in the ease of preparation of the elements from their halide salts. The gaseous **hydrogen halides** form from the direct reaction of halogens and hydrogen and from the other general techniques used to make acids. They dissolve in water to give solutions of **hydrohalic acids.**

The halogens form **halides** with less electronegative elements. Halides of the metals vary from ionic to covalent; halides of nonmetals are covalent. **Interhalogens** are formed by the combination of two different halogens. Binary halogen-oxygen compounds, generally of low stability, are known, as are **hypohalous, halous, halic,** and **perhalic** acids and their salts.

Key Terms and Concepts

derivatives of sulfuric acid (22.16)
deuterium (22.3)
disproportionation (22.19)
disulfide (22.12)
Frasch process (22.10)
halic acid (22.23)
halides (22.21)
halous acid (22.23)

hydrides (22.2)
hydrocarbon (22.1)
hydrogen halide (22.21)
hydrohalic acid (22.21)
hydrosulfuric acid (22.12)
hypohalous acid (22.23)
interhalogen (22.20)
oxide (22.6)
perhalic acid (22.23)

peroxide (22.6)
peroxysulfuric acids (2.16)
photosynthesis (22.8)
polyhalide ion (22.20)
polysulfide (22.12)
sulfate (22.15)
sulfide (22.12)
sulfite (22.14)
tritium (22.3)

Exercises

1. Both hydrogen and oxygen are colorless, odorless and tasteless gases. How may we distinguish between them using physical properties? Using chemical properties?

2. Why is it easier and cheaper to prepare commercial quantities of liquid oxygen than to prepare commercial quantities of liquid hydrogen?

3. Describe five chemical properties of sulfur or of compounds of sulfur that characterize it as a nonmetal.

4. Which of the following elements will exhibit a positive oxidation number when combined with sulfur?
   (a) Al   (b) Br   (c) Ca
   (d) Cl   (e) F   (f) O
   (g) Si

5. Arrange the halogens in order of increasing
   (a) atomic radius
   (b) electronegativity
   (c) boiling point
   (d) oxidizing activity

6. From the positions of the elements in the Periodic Table, predict which of the following pairs will
   (a) reduce S; Cl_2 or Mg
   (b) oxidize O_2; F_2 or N_2

   (c) oxidize Br; I_2 or Cl_2
   (d) oxidize S; H_2 or Cl_2
   (e) reduce O_2; H_2 or F_2

Hydrogen

7. Write the equation that describes the preparation of hydrogen when steam is passed through a red-hot gun barrel.

8. Write balanced equations for the preparation of hydrogen from $C + H_2O$; $Fe + H_2SO_4$; H_2O + electrical energy; $H_2O + NaH$.

9. Write balanced equations for and name the compounds formed in the reaction, if any, of each of the following metals with water and with hydrobromic acid: K, Mg, Ga, Bi, Fe, Pt.

10. Write a balanced equation for the reaction of an excess of hydrogen with each of the following: PtO, Na, Fe_2O_3, Cl_2, O_3.

11. Explain: "Metal hydrides are convenient and portable sources of hydrogen."

12. The reaction of NaH with H_2O may be characterized as an acid-base reaction. Identify the acid and the base of the reactants. The reaction is also an oxidation-reduction reac-

tion. Identify the oxidizing agent, the reducing agent, and the changes in oxidation number that occur in the reaction.

13. Hydrogen is not very soluble. At 25°C a saturated solution of H_2 in water is 7.86×10^{-5} M. How many milliliters of H_2 at 25°C and 1.00 atm are dissolved in a liter of solution? *Ans. 1.91 mL*

14. What mass of hydrogen would result from the reaction of 27.2 g of LiH with water? *Ans. 6.90 g*

15. What mass of CaH_2 is required to provide enough hydrogen by reaction with water to fill a balloon at STP with a volume of 3.50 L? *Ans. 3.29 g*

16. How many grams of hydrogen and of zinc acetate, $Zn(CH_3CO_2)_2$, can be prepared by the reaction of 22 g of zinc with acetic acid, CH_3CO_2H? *Ans. 0.68 g H_2; 62 g $Zn(CH_3CO_2)_2$*

17. Deuterium, ^2_1H, usually indicated as D, may be separated from ordinary hydrogen by repeated distillation of water. Ultimately, pure D_2O can be isolated. What is the molecular weight of D_2O to three significant figures? (The atomic weight of D may be found in Table 5.2). *Ans. 20.0*

18. What mass of PD_3 may be prepared by the reaction of 0.498 g of Na_3P with D_2O? (See exercise 17.) *Ans. 0.184 g*

Oxygen

19. Write the reaction describing the preparation of oxygen by using a lens to focus the sun's rays on a sample of mercury(II) oxide.

20. Which of the various laboratory techniques for the preparation of oxygen should be used if a sample of oxygen free of traces of water (dry oxygen) is required for an experiment?

21. Write balanced equations for the preparation of oxygen from
    (a) $Cu(NO_3)_2$ (b) $NaNO_3$
    (c) H_2O (d) BaO_2

22. Write balanced equations showing the release of oxygen gas upon heating the following oxides:
    (a) Au_2O (b) PtO_2 (c) CrO_3

23. Which of the following materials will burn in O_2: $SiH_4(g)$, SiO_2, CO, CO_2, Mg, CaO? Why won't some of these materials burn in oxygen?

24. Why will magnesium ribbon burn more rapidly in pure oxygen than in air?

25. Write a balanced equation for the reaction of an excess of oxygen with each of the following. (Keep in mind that an element tends to reach its highest oxidation number when it combines with an excess of oxygen.)
    (a) Ca (b) Cs (c) As
    (d) Ge (e) Na_2SO_3 (f) AlP
    (g) C_2H_6 (h) CO

26. Determine whether the heat of formation of MgO is positive or negative from Figure 22.8.

27. Assume that when O_2 dissolves in water, the volume of the solution is the same as the volume of water from which the solution was made. What is the molar concentration of O_2 in a solution at 0°C that is saturated with O_2? *Ans. 2.2×10^{-3} M*

28. What mass and volume of $O_2(g)$ (density = 1.429 g/L) are required to prepare 41.3 g of MgO? *Ans. 16.4 g; 11.5 L*

29. What mass of oxygen can be obtained by the thermal decomposition of 48.19 g of $Cu(NO_3)_2$? What is the volume of O_2 produced at STP? *Ans. 4.111 g; 2.878 L*

30. Elements generally exhibit their highest oxidation numbers in compounds containing oxygen or fluorine. What is the oxidation number of osmium in 0.3789 g of an osmium oxide prepared by the reaction of 0.2827 g of Os with O_2? Write the equation for the reaction that gives this compound. *Ans. +8; $Os + 2O_2 \longrightarrow OsO_4$*

Sulfur

31. Based on the location of sulfur in the Periodic Table, predict the products of and write a balanced equation for each of the following. (There may be more than one correct answer, depending on your choice of stoichiometries.)
    (a) $Ca + S \longrightarrow$ (b) $Al + S \longrightarrow$
    (c) $F_2 + S \longrightarrow$ (d) $H_2 + S \longrightarrow$
    (e) $C + S \longrightarrow$

32. Write an equation for the reaction of sulfur with each of the following:
    (a) Cl_2 (b) Li
    (c) Ga (d) O_2
    (e) HNO_3

33. Write chemical equations describing three chemical reactions in which elemental sulfur acts as an oxidizing agent; three in which it acts as a reducing agent.

34. Determine the oxidation number of sulfur in each of the following species:
    (a) H_2S (b) SO_2Cl_2
    (c) SF_4 (d) $NaHS$
    (e) $S_2O_3^{2-}$ (f) $S_4O_6^{2-}$

35. Explain the fact that hydrogen sulfide is a gas at room temperature, whereas water, which has a smaller molecular weight, is a liquid.

36. Why are solutions of sulfides and hydrogen sulfides alkaline? Write equations.

37. Write an equation for the reaction of hydrogen sulfide in acidic solution with each of the following:
    (a) Fe^{3+} (b) Br_2
    (c) MnO_4^- (d) $Cr_2O_7^{2-}$

38. How can sodium sulfite be made in the laboratory?

39. Why do the two sulfur-oxygen bonds in sulfur dioxide have the same length?

40. Describe the hybridization of the sulfur atom in gaseous molecules of SO_2 and SO_3.

41. How does the hybridization of the sulfur atom change when gaseous SO_3 condenses to give solid SO_3?

42. Why is sulfuric acid a stronger acid than sulfurous acid?

43. Which is the stronger acid, $NaHSO_3$ or $NaHSO_4$?

44. Why do solutions of sulfites usually contain sulfate ions?
45. Write the Lewis structure of each of the following:
    (a) S^{2-} (b) H_2S_2 (c) SO_2
    (d) H_2SO_3 (e) SO_3 (f) Na_2SO_4
    (g) H_2SO_5 (h) $H_2S_2O_8$
46. In 1986, 2.086×10^6 tons of ammonium sulfate was produced in the United States. If all of this ammonium sulfate resulted from the reaction of ammonia with sulfuric acid, what percentage of the 3.6822×10^7 tons of sulfuric acid produced in 1986 was used in ammonium sulfate manufacture?   *Ans. 4.205%*
47. A volume of 22.85 mL of a 0.1023 M standard sodium thiosulfate solution is required to titrate a 25.00-mL sample of a solution containing iodine. What is the iodine concentration in the 25.00-mL sample?   *Ans. 0.04675 M*
48. Based on the location of sulfur in the Periodic Table, predict the products of and write a balanced equation for each of the following reactions. (There may be more than one correct answer, depending on your choice of stoichiometries.)
    (a) $F_2 + SO_2 \longrightarrow$ (b) $Na_2O + SO_2 \longrightarrow$
    (c) $KClO + Na_2S \longrightarrow$ (d) $O_2 + H_2S \longrightarrow$
    (e) $Cl_2 + H_2S \longrightarrow$ (f) $I^- + SO_2 \longrightarrow$
49. What volume of hydrogen sulfide at STP can be produced by the reaction of 308 g of aluminum sulfide, Al_2S_3, with an excess of phosphoric acid?   *Ans. 138 L*
50. The following compounds can be considered to be derived from sulfurous acid, sulfuric acid, or their salts, by replacement of one or more atoms in these molecules. Suggest structures for these molecules and write their Lewis structures.
    (a) SOF_2 (b) $Ca(FSO_3)_2$
    (c) $(CH_3)_2SO_4$ (d) $Na_2S_2O_8$
    (e) $H_2S_2O_7$ (f) $S_2O_6F_2$

Halogens

51. Why can fluorine, which reacts with all of the metals, be stored in certain metal cylinders?
52. Write the equation that describes the electrochemical preparation of chlorine (a) from an aqueous solution of sodium chloride and (b) from molten iron(III) chloride.
53. Suggest two reasons why fluorine is not used in the extraction of bromine from brine.
54. Describe the production of iodine from sodium iodate.
55. Write balanced chemical equations that describe the reaction of chlorine with each of the following:
    (a) lithium
    (b) magnesium
    (c) hydrogen
    (d) an excess of phosphorus
    (e) iodine
    (f) sulfur
56. Using the reactions of the halogens with sulfur as examples, show that the oxidizing abilities of the halogens decrease as their size increases.

57. Show that the reaction of bromine with water is a disproportionation reaction by considering the changes in oxidation number that occur.
58. Why is iodine more soluble in a solution of calcium iodide than in pure water?
59. Why is iodine monofluoride more polar than iodine monochloride?
60. Write chemical equations describing the changes that occur in each of the following cases.
    (a) Sulfur is burned in a fluorine atmosphere.
    (b) Molten potassium bromide is electrolyzed.
    (c) Sodium iodide is treated with chlorine.
    (d) Hydrogen sulfide is bubbled into a solution of bromine.
    (e) Hydrogen and fluorine mix.
    (f) Magnesium is heated with iodine.
61. Write the Lewis structure for each of the iodine fluorides.
62. Describe the molecular geometry of each of the iodine fluorides.
63. Write an equation describing a convenient laboratory preparation for each of the hydrogen halides.
64. How do hydrogen fluoride and hydrofluoric acid differ in their chemical behavior from the other hydrogen halides and hydrohalic acids?
65. Write a balanced chemical equation that describes the reaction that occurs in each of the following cases.
    (a) Calcium is added to hydrobromic acid.
    (b) Potassium hydroxide is added to hydrofluoric acid.
    (c) Ammonia is bubbled through hydrofluoric acid.
    (d) Sodium acetate is added to hydroiodic acid.
    (e) Silver oxide is added to hydrobromic acid.
    (f) Chlorine is bubbled through hydrobromic acid.
66. Write chemical equations that show that Cl_2O and I_2O_5 are acid anhydrides.
67. Write the Lewis structure of each of the oxyacids of chlorine. Calculate the oxidation number of chlorine in each of these acids.
68. Write a chemical equation for the reaction that occurs in each of the following cases.
    (a) Sodium hydroxide is added to a solution of chloric acid.
    (b) Metallic zinc is added to a solution of perbromic acid.
    (c) Ammonia is bubbled through a solution of chlorous acid.
    (d) An excess of calcium hydroxide reacts with paraperiodic acid.
    (e) Bromine is added to a solution of lithium hydroxide.
    (f) Solid sodium hypochlorite is added to water.
69. Which is the stronger acid, $HClO_2$ or $HClO_3$? Why?
70. Which is the stronger acid, $HClO_3$ or $HBrO_3$? Why?
71. Predict the products of each of the following.
    (a) Aluminum reacts with oxygen difluoride.
    (b) A solution of hypochlorous acid is heated.
    (c) Hydrochloric acid is added to a solution of hypochlorous acid.

Additional Exercises

72. Which of the following compounds will react with water to give an acid solution? Which will give a basic solution? Write a balanced equation for the reaction of each with water.
    (a) BaH_2 (b) H_2S (c) HBr
    (d) SO_2 (e) Li_2O (f) Na_2S
    (g) Cl_2O_7

73. Write the balanced equations or equations necessary to carry out the following transformations (H_2O, H_2, and/or O_2 may be used as needed):
    (a) Na_2O_2 from Na
    (b) NaOH from Na and O_3
    (c) NaCl from Na_2O_2 and Cl_2
    (d) $ZnSO_4$ from Zn and H_2S
    (e) Fe from Fe_3O_4

74. Write the balanced equation for the chemical reaction that occurs when ICl dissolves in water.

75. Write the chemical reaction for the preparation of ethylene dibromide, $C_2H_4Br_2$, a constituent of leaded gasoline, from ethylene and bromine. Use Lewis structures instead of chemical formulas in your equation.

76. (a) Which is a stronger acid, H_2O or H_2S?
    (b) Which is a stronger base, OH^- or SH^-?
    (c) Which is a stronger base, SH^- or S^{2-}?
    Explain each answer.

77. The reaction of titanium metal with F_2 yields a titanium fluoride that contains 38.7% titanium. Write the chemical equation that describes the reaction.
    Ans. $Ti + 2F_2 \longrightarrow TiF_4$

78. What mass of PCl_3 can be prepared by the reaction of chlorine with 13.55 g of phosphorus?   *Ans. 60.08 g*

79. The reaction of V_2O_3 with Cl_2 gives a yellow liquid that contains 29.42% vanadium, 61.3% chlorine, and the remainder oxygen. At 19°C a sample of the liquid with a mass of 0.433 g vaporized in a 115-mL flask, giving a gas with a pressure of 390 torr. What is the molecular formula of this vanadium oxychloride?   *Ans. $VOCl_3$*

80. Air contains 20.99% oxygen by volume. What volume of air in cubic meters at 27.0°C and 0.9868 atm is required to oxidize 1.00 metric ton (1000 kg) of sulfur to sulfur dioxide, assuming that all of the oxygen reacts with sulfur?
    Ans. $3.71 \times 10^3 \ m^3$

81. The bond length in the O_2 molecule is 1.209 Å, and that in the O_3 molecule is 1.278 Å. Why does ozone have a longer bond?

82. Which of the following compounds or ions have the same structure; $ClICl^-$, Cl_2O, OF_2, BrF_3, ClO_3^-, ClO_4^-, ICl_4^-, IF_4^+, XeF_4?   *Ans. Cl_2O and OF_2; ICl_4^- and XeF_4*

83. The average oxidation number of sulfur is not one of its common ones in Na_2S_2, H_2S_2, K_2S_5, and $Na_2S_2O_3$. Calculate the average oxidation number of sulfur in these com-pounds. What is the common structural feature in these compounds? In view of your answer to the preceding question, write a Lewis structure of S_2F_{10}. What is the oxidation number of sulfur in S_2F_{10}?

84. A sample of cadmium chloride with a mass of 1.766 g gave 1.089 g of cadmium on electrolysis. Determine the formula of cadmium chloride.   *Ans. $CdCl_2$*

85. Iodine reacts with chlorine to give a compound containing 78.2% iodine. Write the balanced equation for the reaction.   *Ans. $I_2 + Cl_2 \longrightarrow 2ICl$*

86. How many grams of sodium iodate are required to prepare 0.100 kg of iodine by reduction with sodium hydrogen sulfite?   *Ans. 156 g*

87. A molten mixture of rubidium fluoride and uranium(IV) fluoride is oxidized with fluorine to produce a uranium compound in which most but not all of the uranium has an oxidation number of +5. The product is found to contain 54.43% uranium. A 1.0357-g sample of the product immersed in 100.0 mL of 0.1007 M acidified potassium iodide solution reacts according to the following equation:

    $$2I^- + 2UF_6^- \longrightarrow 2UF_4 + I_2 + 4F^-$$

    The iodine produced is titrated with 14.80 mL of 0.1494 M sodium thiosulfate solution. What percentage of the original uranium was oxidized to the +5 oxidation number?   *Ans. 93.36%*

88. The density of HF(g), 0.991 g/L at 19°C and 1.00 atm, is unexpectedly high. Calculate the apparent molecular weight of HF and explain this result.
    Ans. 23.7; due to HF dimers

89. The reactions involved in the preparation of sulfuric acid are highly exothermic. Using the data in Appendix I, calculate the enthalpy changes for the following reactions:
    (a) $S(s) + O_2(g) \longrightarrow SO_2(g)$
    (b) $2SO_2(g) + O_2(g) \longrightarrow 2SO_3(g)$
    (c) $SO_3(g) + H_2O(l) \longrightarrow H_2SO_4(l)$
    Ans. (a) −296.83 kJ; (b) −197.8 kJ; (c) −132.5 kJ

90. The combustion of 9.180 g of sodium to sodium oxide is exothermic; 100.74 kJ of heat is evolved. What is the heat of formation of sodium oxide?
    Ans. $−504.6 \ kJ \ mol^{-1}$

91. From the data in Appendix I, calculate the free energy change for the reaction of hydrogen with each of the halogens. Determine which of these reactions is not spontaneous at 25°C.   *Ans. $H_2 + I_2 \longrightarrow 2HI$*

92. A gas mixture at equilibrium at 1375 K exhibits a partial pressure of HBr of 0.998 atm, of H_2 of 3.82×10^{-3} atm, and of Br_2 of 3.82×10^{-3} atm. What is K_p for the decomposition of HBr to H_2 and Br_2?   *Ans. 1.46×10^{-5}*

93. Thallium(I) chloride is one of the few insoluble chlorides. What is the molar solubility of TlCl, whose solubility product is 1.8×10^{-4}?   *Ans. $1.3 \times 10^{-2} \ M$*

The Nonmetals, Part 2: Carbon, Nitrogen, Phosphorus, and the Noble Gases

23

Nitrogen and phosphorus fertilizers are essential for modern agriculture.

This chapter continues the more detailed discussion (begun in Chapter 22) of the chemistry of certain specific nonmetals. In this chapter we will consider the behavior of the elements carbon, nitrogen, phosphorus, and the noble gases (helium, neon, argon, krypton, xenon, and radon).

Carbon

Carbon is nineteenth among the elements in abundance; it constitutes only about 0.027% of the earth's crust. Without carbon, life would be impossible because carbon compounds are essential components of all animal and plant life.

About five million carbon compounds found in or produced by living organisms have been characterized, and thousands of carbon compounds have been produced by chemists. The existence of so many compounds is due primarily to the ability of carbon

Figure 23.1
Samples of diamond and graphite, two forms of carbon.

atoms to combine with other carbon atoms, forming chains of atoms of different lengths and rings of different sizes. Most compounds containing carbon-carbon bonds fall into the realm of organic chemistry; examples are discussed in Chapter 30. The role of carbon in living systems is introduced in Chapter 31.

23.1 Properties of Carbon

Carbon is found in the free state as diamond and graphite (Fig. 23.1). It is combined in both living and dead organic matter and in natural gas, petroleum, coal, limestone, dolomite, coral, and chalk. Carbon is found in the air as carbon dioxide and in natural waters as carbon dioxide, carbonic acid, and carbonates.

All forms of elemental carbon are almost inert toward most reagents at ordinary temperatures. However, graphite is slowly oxidized by a mixture of nitric acid and sodium chlorate. The activity of carbon increases rapidly with rising temperatures, and at elevated temperatures it is very reactive. Hot carbon is a reducing agent. It reduces oxygen, forming either carbon monoxide, CO, or carbon dioxide, CO_2, depending on the amount of oxygen present. With sulfur it forms carbon disulfide, CS_2, and with fluorine it forms carbon tetrafluoride, CF_4. Hot carbon reduces to many metal oxides, forming carbon monoxide or carbon dioxide and the elemental metal (Fig. 23.2). With certain metals it forms carbides, such as iron carbide, Fe_3C, with metallic properties. Some of the general reactions of carbon and its compounds are summarized in Table 23.1.

Because of its resistance to heat and chemical action and its conductivity, graphite is used to make electrodes and crucibles. It is also used in the manufacture of paints, commutator brushes, and lead pencils. Suspensions of colloidal graphite in water or in oil provide excellent lubricants. Diamond is used in grinding, polishing, and cutting operations (Fig. 23.3) because of its hardness.

Table 23.1   Chemical properties of carbon and its compounds

General reaction	Comments
Reactions with elements	
$2C + O_2 \longrightarrow 2CO$	With heat and a stoichiometric amount of O_2
$C + O_2 \longrightarrow CO_2$	With heat and excess O_2
$C + 2S \longrightarrow CS_2$	With heat in an inert atmosphere
$C + 2F_2 \longrightarrow CF_4$	
$C + CuO \longrightarrow Cu + CO$	
$3C + CaO \longrightarrow CaC_2 + CO$	Product contains C_2^{2-} ion
Reactions of compounds	
$2CO + O_2 \longrightarrow 2CO_2$	
$CO_2 + H_2O \rightleftharpoons H_2CO_3$	CO_2 is an acidic oxide
$MO + CO \longrightarrow M + CO_2$	Many metal oxides are reduced by CO
$CH_4 + 2O_2 \longrightarrow CO_2 + 2H_2O$	Most hydrocarbons burn, producing CO_2 and H_2O
$CaCO_3 \longrightarrow CaO + CO_2$	Many carbonates decompose upon heating
$CO_2 + O^{2-} \longrightarrow CO_3^{2-}$	CO_2 is a Lewis acid
$CS_2 + S^{2-} \longrightarrow CS_3^{2-}$	CS_2 is a Lewis acid

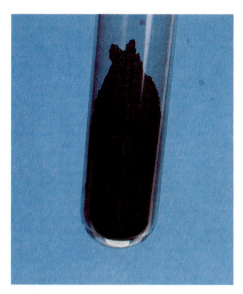

Figure 23.2
A mixture of copper oxide and carbon (left) produces copper metal (right) upon heating. The carbon reduces the CuO to free Cu.

Charcoal, a microcrystalline form of carbon, is produced by heating of wood in the absence of air. It is used as a fuel. Charcoal is also used as a decolorizing agent for liquids such as sugar solutions, alcohol, and petroleum products, because it can adsorb substances that discolor the liquids. **Adsorption** is a surface phenomenon in which the surface forces of the adsorbing agent attract and hold molecules of the substance being adsorbed. Charcoal is used in water purification to remove small amounts of chloroform and other organic molecules that may be present.

23.2 Carbon Monoxide

Carbon monoxide, CO, is produced when carbon is heated or burned without enough oxygen to oxidize it fully. The reaction of steam with red-hot coke also produces carbon monoxide, along with hydrogen in the mixture called water gas, as described in Section 22.1.

Carbon monoxide is prepared in the laboratory by heating crystals of oxalic acid, $H_2C_2O_4$, with concentrated sulfuric acid, which removes the elements of water from the oxalic acid and absorbs the water produced (see Section 22.15).

$$H_2C_2O_4 \xrightarrow{\text{Conc. } H_2SO_4} H_2O + CO + CO_2$$

The carbon dioxide in the resulting mixture may be removed by passing the mixture through solid sodium hydroxide, which absorbs the carbon dioxide, leaving pure carbon monoxide.

Carbon monoxide burns readily in oxygen, forming carbon dioxide.

$$2CO(g) + O_2(g) \longrightarrow 2CO_2(g) \qquad \Delta H° = -566 \text{ kJ}$$

This reaction makes carbon monoxide useful as a gaseous fuel. At high temperatures carbon monoxide will reduce many metal oxides, so it is an important reducing agent in metallurgical processes. Copper(II) oxide, CuO, and iron(III) oxide, Fe_2O_3, are two oxides that are reduced to the metal with formation of carbon dioxide.

Figure 23.3
Industrial diamonds are arranged in a mold used to cast a diamond-studded bit for drilling through very hard rock formations.

Carbon monoxide is a very dangerous poison; it is an odorless and tasteless gas and therefore gives no warning of its presence. It combines with the hemoglobin (Section 26.1) in blood to form a compound that is too stable to be broken down by body processes. Because the hemoglobin is combined with CO, it cannot combine with oxygen; this destroys the blood's ability to carry oxygen.

23.3 Carbon Dioxide

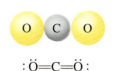

$$: \ddot{O} = C = \ddot{O} :$$

Figure 23.4

The electronic and molecular structures of CO_2.

Carbon dioxide, CO_2 (Fig. 23.4), is produced when any form of carbon or almost any carbon compound is burned in an excess of oxygen. Many carbonates liberate carbon dioxide when they are heated. Heating calcium carbonate, $CaCO_3$, produces carbon dioxide and calcium oxide, CaO (Section 9.9).

Large quantities of industrial carbon dioxide are obtained as a by-product of the fermentation of sugar (glucose) during the preparation of alcohol and alcoholic beverages. The net reaction is

$$\underset{\text{Glucose}}{C_6H_{12}O_6} \xrightarrow{\text{Yeast}} \underset{\text{Ethanol}}{2C_2H_5OH} + 2CO_2(g)$$

Carbon dioxide can be produced in the laboratory by the reaction of acids with carbonates.

$$CaCO_3 + 2H^+ \longrightarrow Ca^{2+} + H_2O + CO_2(g)$$

Carbon dioxide is a colorless and odorless gas that is 1.5 times as heavy as air. It is not toxic, although a large concentration can cause suffocation (lack of oxygen). Carbon dioxide is a component of all carbonated beverages. One liter of water at 20° dissolves 0.9 liter of carbon dioxide at 1 atmosphere, forming carbonic acid, which has a mildly acid taste. The gas is easily liquefied by compression, because its critical temperature is relatively high (31.1°C). Solid carbon dioxide vaporizes without melting; its vapor pressure is 1 atmosphere at −78.5°C. This property makes solid carbon dioxide, **dry ice,** a valuable refrigerant that is always free of the liquid.

Carbon dioxide is used as a fire extinguisher because most substances will not burn in it, it is easily generated, and it is cheap. Air containing as little as 2.5% carbon dioxide will extinguish a flame.

The atmosphere contains about 0.04% by volume of carbon dioxide and serves as a huge reservoir of this compound. With the help of sunlight and chlorophyll (as a catalyst), green plants convert carbon dioxide and water into sugar and oxygen (Section 22.8).

Carbon dioxide is a product of respiration and is returned to the air by green plants and by animals. The organisms (mostly nongreen plants) that cause plant and animal matter to decay and sugars to ferment also produce carbon dioxide, as does the combustion of carbon-containing fuels. In fact, the carbon dioxide content of the atmosphere has increased significantly in the last few years because of the burning of fossil fuels (Fig. 23.5). The gas from volcanoes and other geological formations are other minor sources of carbon dioxide. The solubility of carbon dioxide in water makes oceans and lakes significant reservoirs of this compound.

Figure 23.5

Ice cores from a Yukon glacier contain bubbles of air that was trapped over 130 years ago. Analysis of this trapped air shows that carbon dioxide levels in today's air are 27% higher than before 1850.

23.4 Carbonic Acid and Carbonates

Carbon dioxide is the anhydride of **carbonic acid,** H_2CO_3, which forms in small amounts when carbon dioxide dissolves in water. Carbonic acid is a diprotic acid.

When carbon dioxide dissolves in water, most of it is present as CO_2 molecules rather than H_2CO_3 molecules. The first ionization constant for carbonic acid, taking into account the true concentration of H_2CO_3 molecules in a carbon dioxide-water solution, is about 2×10^{-4}. However, in considering equilibria involving carbonic acid, it is convenient to assume that all of the carbon dioxide in solution is present as carbonic acid, and thus the value of 4.3×10^{-7} is commonly used for K_a in such calculations. This leads to correct values for the concentration of hydronium ion and hydrogen carbonate ion in solution. It does, however, give the impression that carbonic acid is a weaker acid than it actually is.

Carbon dioxide can behave as a Lewis acid (Section 16.15). It reacts with oxide ions to form the carbonate ion and with hydroxide ion to form the hydrogen carbonate ion.

$$CO_2 + O^{2-} \longrightarrow CO_3^{2-}$$
$$CO_2 + OH^- \longrightarrow HCO_3^-$$

When a solution of sodium hydroxide is saturated with carbon dioxide, **sodium hydrogen carbonate,** $NaHCO_3$, is formed.

$$Na^+ + OH^- + CO_2 \longrightarrow Na^+ + HCO_3^-$$

This compound is also called sodium bicarbonate, or baking soda. The action of baking soda is due to the reaction of the basic hydrogen carbonate anion with an acid, producing gaseous carbon dioxide and water (Fig. 23.6).

$$HCO_3^- + H^+ \longrightarrow H_2CO_3 \longrightarrow CO_2 + H_2O$$

Because the hydrogen carbonate anion is stronger as a base than it is as an acid, solutions of salts of hydrogen carbonate are weakly alkaline.

$$HCO_3^- + H_2O \rightleftharpoons H_2CO_3 + OH^-$$

If equivalent amounts of sodium hydroxide and a solution of sodium hydrogen carbonate are mixed, crystals of the 10-hydrate of **sodium carbonate,** $Na_2CO_3 \cdot 10H_2O$, will form upon evaporation. This hydrate, also called washing soda, is commonly used as a water softener in the home. If heated gently, it forms anhydrous sodium carbonate, Na_2CO_3, called soda ash industrially. Solutions of sodium carbonate are basic because the carbonate anion is a moderately weak base.

$$CO_3^{2-} + H_2O \rightleftharpoons HCO_3^- + OH^- \qquad K_b = 1.4 \times 10^{-4}$$

23.5 Carbon Disulfide

At high temperatures carbon is oxidized by sulfur to **carbon disulfide,** CS_2 (Fig. 23.7). Air must be excluded because the volatile carbon disulfide is highly flammable and burns to form CO_2 and SO_2.

Pure carbon disulfide is a colorless, very volatile liquid with a disagreeable odor. The liquid is dense (specific gravity 1.28) and is immiscible with water. The vapor is denser than air, very poisonous, and highly flammable.

Carbon disulfide, like carbon dioxide, is a Lewis acid. Carbon disulfide reacts with the sulfide ion, S^{2-}, giving the thiocarbonate ion, CS_3^{2-} [Fig. 23.8(a)], and with the amide ion, NH_2^-, giving the dithiocarbamate ion, $H_2NCS_2^-$ [Fig. 23.8(b)].

When sulfur replaces part or all of the oxygen in a molecule or ion, the prefix *thio* is used. Thus CO_3^{2-} is the carbonate ion, whereas CS_3^{2-} is the thiocarbonate ion.

Large quantities of carbon disulfide are used in making rayon by the viscose process and in the manufacture of cellophane.

Figure 23.6

The reaction of baking soda, $NaHCO_3$, with an acid produces CO_2, a sodium salt of the acid, and water.

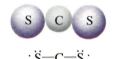

Figure 23.7

The electronic and molecular structures of CS_2.

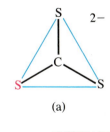

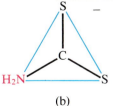

Figure 23.8

The structures of (a) the thiocarbonate ion and (b) the dithiocarbamate ion.

23.6 Carbon Tetrachloride

Carbon tetrachloride, CCl_4, is manufactured by the reaction of chlorine with methane, CH_4.

$$CH_4 + 4Cl_2 \longrightarrow CCl_4 + 4HCl$$

It is an excellent solvent for fats, oils, and greases, and it once was used for dry cleaning fabric. Because its vapor is about five times as heavy as air and it does not burn, it was also used as a fire extinguisher. However, since carbon tetrachloride has been recognized as a carcinogen, its use has been sharply curtailed.

23.7 Calcium Carbide and Cyanides

Calcium carbide, CaC_2, is an important commercial product made by heating calcium oxide with carbon (coke). It reacts with water (Fig. 23.9) forming acetylene, $HC{\equiv}CH$.

$$CaC_2 + 2H_2O \longrightarrow Ca(OH)_2 + C_2H_2(g)$$

The reaction of calcium carbide with nitrogen at about 1100°C gives **calcium cyanamide,** $CaCN_2$.

$$CaC_2 + N_2 \xrightarrow{1100°} CaCN_2 + C$$

The linear cyanamide ion, NCN^{2-}, is isoelectronic and isostructural with carbon dioxide and carbon disulfide (compare Fig. 23.10 with Fig. 23.4 and Fig. 23.7).

The fusion of calcium cyanamide with carbon and sodium carbonate produces **sodium cyanide,** NaCN.

$$CaCN_2 + C + Na_2CO_3 \longrightarrow CaCO_3 + 2NaCN$$

The sodium cyanide is separated from the insoluble calcium carbonate by dissolving it in water. Sodium cyanide can also be prepared by the reaction of sodium amide with carbon at 500° to 600° C.

$$NaNH_2 + C \xrightarrow{500°-600° C} NaCN + H_2$$

The cyanide ion is a basic ion. It combines with acid to form the very weak acid **hydrogen cyanide,** HCN.

$$H^+(aq) + CN^-(aq) \longrightarrow HCN(g)$$

Hydrogen cyanide is a gas that smells like bitter almonds. It is very poisonous; a dose of about 0.05 gram is fatal to humans.

Both cyanamides and cyanides find extensive use in the plastics and polymer industries. Calcium cyanamide is used to prepare melamine, a raw material for the production of plastics such as Melmac. Hydrogen cyanide is used in the preparation of acrylonitrile polymers used in synthetic fibers such as Dynel and Orlon.

Figure 23.9

The reaction of solid calcium carbide with water produces acetylene, C_2H_2, a flammable gas.

$$:\overset{..}{N}{=}C{=}\overset{..}{N}:^{2-}$$

Figure 23.10

The structure of the cyanamide ion. Note the similarity to the isoelectronic molecules CO_2 (Fig. 23.4) and CS_2 (Fig. 23.7).

Nitrogen

Compounds of nitrogen have an extensive history; the first historical record of ammonium chloride is from the fifth century B.C. Ammonium salts, nitric acid, nitrates, and aqua regia were well known to the alchemists. Elemental nitrogen was first recognized as an element by the Scotch botanist Daniel Rutherford in 1772. He demonstrated that this gas supports neither life nor combustion.

23.8 Occurrence and Preparation of Nitrogen

The atmosphere consists of 78% nitrogen by volume and 75% by weight. There are more than 20 million tons of nitrogen over every square mile of the earth's surface. Nitrogen is a component of the proteins and DNA of all plants and animals. The most important mineral sources of nitrogen in combined form are deposits of saltpeter, KNO_3, in India and other countries of the Far East and deposits of Chile saltpeter, $NaNO_3$, in South America.

Nitrogen is obtained industrially by the fractional distillation of liquid air (Section 29.2). Nitrogen is prepared in the laboratory by heating a solution of ammonium nitrite.

$$NH_4NO_2(aq) \longrightarrow 2H_2O(l) + N_2(g) \qquad \Delta H° = -305.1 \text{ kJ}$$

In actual practice, a mixture of sodium nitrite and ammonium chloride is used to furnish the ions of ammonium nitrite, because NH_4NO_2 is so unstable that it cannot be stored. Atmospheric nitrogen (nitrogen plus the noble gases) can be isolated in the laboratory by burning phosphorus in air that is confined over water. The P_4O_{10} formed dissolves in the water, and the residual gas is primarily nitrogen.

Large volumes of atmospheric nitrogen are used for making ammonia, the principal starting material used for preparation of large quantities of other nitrogen-containing compounds. Most other uses of elemental nitrogen are based on its inactivity. It is used when a chemical process requires an inert atmosphere. Canned foods and lunch meats cannot oxidize in a pure nitrogen atmosphere, so they retain a better flavor and color if sealed with nitrogen instead of air. Luncheon meats are often packed in nitrogen to retard spoilage.

23.9 Properties of Nitrogen

Under ordinary conditions, nitrogen is a colorless, odorless, and tasteless gas. It boils at $-195.8°C$ and freezes at $-210.0°C$. It is slightly less dense than air, for air contains the heavier molecules of oxygen and argon as well as those of nitrogen.

Nitrogen molecules are very unreactive. The only common reactions of N_2 at room temperature are with lithium to give Li_3N, with certain transition metal complexes, and with nitrogen-fixing bacteria. Nitrogen forms nitrides upon heating with active metals and low yields of ammonia upon heating with hydrogen. Heating with oxygen followed by rapid cooling (quenching) produces nitric oxide, NO. The general unreactivity of nitrogen makes the remarkable ability of some bacteria to manufacture nitrogen compounds one of the exciting unsolved problems of chemistry.

Compounds of nitrogen in all of the possible oxidation states from -3 to $+5$ are known. Much of the chemistry of nitrogen involves oxidation-reduction reactions, which convert nitrogen from one oxidation number to another.

The general chemical behavior of nitrogen is discussed in Section 20.6. Some of the reactions of nitrogen and its compounds are summarized in Table 23.2 (p. 696).

23.10 Ammonia

Ammonia, NH_3, is produced in nature when any nitrogen-containing organic material decomposes in the absence of air. Its odor is common in decaying organic matter. Ammonia is usually prepared in the laboratory by the reaction of an

Table 23.2   Chemical properties of nitrogen and its compounds

General equation	Comments
Reactions with elements	
$N_2 + 6Li \longrightarrow 2Li_3N$	N^{3-} forms with active metals
$N_2 + 3Mg \xrightarrow{\Delta} Mg_3N_2$	
$N_2 + 3H_2 \underset{\text{High P}}{\overset{\Delta}{\rightleftharpoons}} 2NH_3$	Reversible, slow reaction, low yield at elevated temperatures
$N_2 + O_2 \xrightarrow{\Delta} 2NO$	Low-yield, endothermic reaction; requires heating
Reactions of compounds	
$NH_3 + H^+ \longrightarrow NH_4^+$	NH_3 acts both as a weak base and
$NH_3 + LiCH_3 \longrightarrow LiNH_2 + CH_4$	a very weak acid
$NH_4^+ + OH^- \longrightarrow NH_3 + H_2O$	NH_4^+ is a weak acid
$4NH_3 + 5O_2 \xrightarrow{\Delta} 4NO + 6H_2O$	
$NH_4NO_2 \longrightarrow N_2 + 2H_2O$	The NH_4^+ ion is oxidized by NO_2^-
$NH_4NO_3 \xrightarrow{\Delta} N_2O + 2H_2O$	Potentially explosive reaction
$2NH_3 + 3CuO \xrightarrow{\Delta} N_2 + 3Cu + 3H_2O$	
$NH_3 + OCl^- \longrightarrow NH_2Cl + OH^-$	
$NH_2Cl + NH_3 + OH^- \longrightarrow N_2H_4 + Cl^- + H_2O$	
$Cu + HNO_3 \longrightarrow NO_2$ or NO, $Cu(NO_3)_2$, and H_2O	Conc. HNO_3 gives NO_2; dilute HNO_3 gives NO
$2NO + O_2 \longrightarrow 2NO_2$	
$NO + NO_2 \rightleftharpoons N_2O_3$	
$2NO_2 \rightleftharpoons N_2O_4$	
$2NO_2 + H_2O \longrightarrow HNO_3 + HNO_2$	NO_2 disproportionates in water
$3NO_2 + H_2O \longrightarrow 2HNO_3 + NO$	
$4HNO_3 + P_4O_{10} \longrightarrow 2N_2O_5 + 4HPO_3$	

ammonium salt with a strong base such as sodium hydroxide. The acid-base reaction with the weakly acidic ammonium ion gives ammonia (Fig. 23.11). Ammonia also forms when ionic nitrides react with water. The nitride ion is a much stronger base than the hydroxide ion.

$$Mg_3N_2 + 6H_2O \longrightarrow 3Mg(OH)_2 + 2NH_3(g)$$

Ammonia is produced commercially by the direct combination of the elements in the **Haber process.**

$$N_2(g) + 3H_2(g) \xrightarrow{\text{Catalyst}} 2NH_3(g) \qquad \Delta H° = -92 \text{ kJ}$$

This reaction is very slow at room temperature; it is carried out at an elevated temperature and pressure with a catalyst so that the rate is fast enough for the reaction to be practical. Because the reaction is exothermic, the yield decreases as the temperature is raised. However, four volumes of reactants ($1N_2$ and $3H_2$) give two volumes of product ($2NH_3$), so high pressure increases the yield. Thus the process is carried out at the lowest temperatures and highest pressures practicable; 400–600°C and 200–600 atmospheres. The most efficient catalyst is a mixture of iron oxide and potassium aluminate. Nitrogen for the process is obtained from liquid air, and much of the hydrogen is obtained from water gas (Section 22.1). The mixture of hydrogen and nitrogen is compressed, heated, and passed over the catalyst. The ammonia is removed by liquefaction, and the residual hydrogen and nitrogen are recycled through the process. Fritz

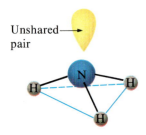

Unshared → pair

Figure 23.11
Structure of an ammonia molecule.

Haber, a German chemist, received the 1918 Nobel prize in chemistry for his success in developing the direct synthesis of ammonia on a commercial scale.

The equilibrium concentrations of ammonia at several temperatures and pressures are shown in Table 23.3.

Ammonia is a colorless gas with an irritating odor. It is a powerful stimulant (Fig. 23.12). Gaseous ammonia is readily liquefied, giving a colorless liquid that boils at $-33°C$. Liquid ammonia has a vapor pressure of only about 10 atmospheres at 25°C; it is readily handled in steel cylinders. Solid ammonia is white and crystalline; it melts at $-78°C$. The heat of vaporization of liquid ammonia is higher than that of any other liquid except water, so ammonia is used as a refrigerant. Ammonia is quite soluble in alcohol, ether, and water (1180 L at STP dissolve in 1 L of H_2O).

The chemical properties of ammonia may be outlined as follows:

1. Ammonia acts as a Brønsted base, since it readily accepts protons, and as a Lewis base, in that it can be an electron-pair donor (Chapter 16). When ammonia dissolves in water, only about 1% reacts to form ammonium and hydroxide ions; the remainder is present as unreacted NH_3 molecules. Although a weak base, ammonia readily accepts protons from acids and hydronium ions, forming salts of the **ammonium ion**, NH_4^+. The ammonium ion is similar in size to the potassium ion, and ionic compounds of the two ions exhibit many similarities in their structures and solubilities.

Ammonia forms **ammines** by sharing electrons with many metal ions (Chapter 26). The diamminesilver ion forms by the reaction

$$Ag^+ + 2\,NH_3 \longrightarrow [H_3N{-}Ag{-}NH_3]^+$$

In these species, ammonia functions as a Lewis base.

2. Ammonia displays acidic behavior, although it is a much weaker acid than water. It will react with very strong bases such as the CH_3^- ion. Like other acids, ammonia reacts with metals, although it is so weak that high temperatures are often required. Hydrogen and (depending upon the stoichiometry) **amides** (salts of NH_2^-), **imides** (salts of NH^{2-}), or **nitrides** (salts of N^{3-}) are formed.

3. The nitrogen atom in ammonia has its lowest possible oxidation number and thus is not susceptible to reduction. However, it can be oxidized. Ammonia will burn in air, giving NO and water. Hot ammonia and the ammonium ion are active reducing agents. Of particular interest are the oxidations of ammonium ion by nitrite ion, NO_2^-, to give pure nitrogen and by nitrate ion to give nitrous oxide, N_2O.

About three-fourths of the 14 million tons of ammonia produced in the United States during 1986 was used in fertilizers, either as the compound or as an ammonium salt such as the sulfate, nitrate, or dihydrogen phosphate. The application of anhydrous ammonia (Fig. 23.13) or of solutions of ammonium compounds is standard farm

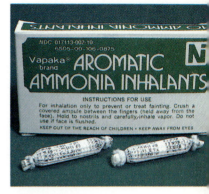

Figure 23.12

Smelling salts contain ammonium carbonate, which is in equilibrium with ammonia, carbon dioxide, and water at room temperature. When a tube of smelling salts is broken, the ammonia is released and may be inhaled as a stimulant.

Figure 23.13

Anhydrous ammonia, stored under pressure as a liquid in the tank, is applied as a fertilizer on this Iowa farm.

Table 23.3 The equilibrium concentrations of ammonia (percent by volume) at several temperatures and pressures for the reaction $N_2 + 3H_2 \rightleftharpoons 2NH_3$

Pressure, atm	Temperature, °C					
	200°	300°	400°	500°	600°	800°
1	15.3	2.18	0.44	0.129	0.05	0.012
100	80.6	52.1	25.1	10.4	4.47	1.15
200	85.8	62.8	36.3	17.6	8.25	2.24
1000	98.3	92.5	80.0	57.5	31.5	—

procedure. Large quantities of ammonia are used in the production of nitric acid, urea, and other nitrogen compounds. Ammonia is a common refrigerant used in refrigerating plants. Household ammonia is an aqueous solution of ammonia that is usually used as a cleanser.

23.11 Derivatives of Ammonia

There are a number of compounds that can be considered as derivatives of ammonia, in that one or more of the hydrogen atoms in the ammonia molecule has been replaced by some other atom or group of atoms, bonded to the nitrogen atom by a single bond. Inorganic derivatives include chloramine, NH_2Cl, hydrazine, N_2H_4, and hydroxylamine NH_2OH.

$$
\begin{array}{cccc}
\overset{\displaystyle H}{\underset{\displaystyle H-N-H}{|}} & \overset{\displaystyle H}{\underset{\displaystyle H-N-Cl}{|}} & \overset{\displaystyle H\ \ H}{\underset{\displaystyle H-N-N-H}{|\ \ \ |}} & \overset{\displaystyle H}{\underset{\displaystyle H-N-O-H}{|}} \\[1em]
\text{Ammonia} & \text{Chloramine} & \text{Hydrazine} & \text{Hydroxylamine}
\end{array}
$$

Organic analogues that contain nitrogen-carbon bonds are called *amines* and are discussed in Chapter 30.

Chloroamine, NH_2Cl, results from the reaction of sodium hypochlorite, $NaOCl$, with ammonia in basic solution. In the presence of a large excess of ammonia at low temperature, the chloramine reacts further to produce **hydrazine,** N_2H_4.

$$NH_3 + OCl^- \longrightarrow NH_2Cl + OH^-$$

$$NH_2Cl + NH_3 + OH^- \longrightarrow N_2H_4 + Cl^- + H_2O$$

For reasons that are not fully understood, the highest yields of hydrazine are obtained when the reaction is run in the presence of gelatin or glue.

Anhydrous hydrazine is thermally stable but very reactive. It burns in air and reacts vigorously with the halogens. Like ammonia, hydrazine is a base, although it is weaker than ammonia. It reacts with strong acids and forms two series of salts that contain the $N_2H_5^+$ and $N_2H_6^{2+}$ ions, respectively.

Hydrazine is readily oxidized by hydrogen peroxide; nitrogen and water are the products. The reaction products are both gases at the temperature of the reaction, and the reaction is highly exothermic. Hydrazine and hydrogen peroxide have therefore been used as rocket propellants.

Pure **hydroxylamine,** NH_2OH, is an unstable white solid that decomposes to ammonia, water, nitrogen, and nitrous oxide at about 15°C. The decomposition can be explosive at elevated temperatures. Aqueous solutions are more stable. Hydroxylamine is a base with a strength somewhat less than that of ammonia. It is an active reducing agent. It is usually prepared and handled as a hydroxylammonium salt, such as the chloride, $[NH_3OH^+]Cl^-$.

23.12 Oxides of Nitrogen

Nitrogen oxides in which nitrogen exhibits each of its positive oxidation numbers from +1 to +5 are well characterized. They are listed in Table 23.4.

When ammonium nitrate is heated, **nitrous oxide,** N_2O (Fig. 23.14), is formed. In this oxidation-reduction reaction the nitrogen in the ammonium ion is oxidized by the nitrogen in the nitrate ion. **Caution! If ammonium nitrate is heated too strongly, an explosion will occur.**

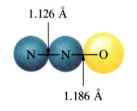

1.126 Å

1.186 Å

$:\ddot{N}\!=\!\!N\!=\!\ddot{O}: \longleftrightarrow :N\!\equiv\!N\!-\!\ddot{\underset{\cdot\cdot}{O}}:$

Figure 23.14

The molecular and resonance structures of a molecule of nitrous oxide, N_2O.

Table 23.4   The oxides of nitrogen

Formula	Oxidation number of nitrogen	Common name
N_2O	I	Nitrous oxide
NO	II	Nitric oxide
N_2O_3	III	Dinitrogen trioxide
NO_2	IV	Nitrogen dioxide
N_2O_4	IV	Dinitrogen tetraoxide
N_2O_5	V	Dinitrogen pentaoxide

Nitrous oxide is a colorless gas possessing a mild, pleasing odor and a sweet taste. It is used as an anesthetic for minor operations, especially in dentistry, under the name laughing gas.

Low yields of **nitric oxide,** NO, are produced when nitrogen and oxygen are heated together. It also forms by direct union of nitrogen and oxygen in the air by lightning during thunderstorms. Nitric oxide is produced commercially by burning ammonia. In the laboratory, nitric oxide is produced by reduction of nitric acid. When copper reacts with dilute nitric acid (Fig. 23.15), nitric oxide is the principal reduction product.

$$3Cu + 8HNO_3 \longrightarrow 2NO + 3Cu(NO_3)_2 + 4H_2O$$

Pure nitric oxide may be obtained by reducing nitric acid with an iron(II) salt in dilute acid solution.

$$3Fe^{2+} + NO_3^- + 4H^+ \longrightarrow 3Fe^{3+} + NO(g) + 2H_2O$$

Gaseous nitric oxide is the most thermally stable of the nitrogen oxides. It is one of the air pollutants from internal combustion engines, due to the reaction of nitrogen and oxygen from the air during the combustion process.

At room temperature nitric oxide is a slightly soluble colorless gas consisting of diatomic molecules. As is often the case with molecules that contain an unpaired electron, two molecules combine to form a dimer by pairing their unpaired electrons. Liquid NO is partially dimerized, and solid NO contains dimers.

$$2NO \longrightarrow \quad \overset{..}{\underset{..}{O}}\!=\!N\!-\!N\!\overset{\overset{..}{O}}{\underset{..}{}}$$

When a mixture of equal parts of nitric oxide and nitrogen dioxide is cooled to −21°C, the gases form **dinitrogen trioxide,** a blue liquid consisting of N_2O_3 molecules (Fig. 23.16). Dinitrogen trioxide exists only in liquid and solid forms. When heated, it forms a mixture of NO and NO_2.

Figure 23.15

The reaction of copper metal with dilute nitric acid produces a solution of $Cu(NO_3)_2$ and colorless NO. When it leaves the flask, the colorless NO is oxidized by the oxygen in the air to NO_2, which is yellow to brown depending on its concentration.

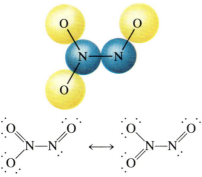

Figure 23.16

The molecular and resonance structures of a molecule of dinitrogen trioxide, N_2O_3.

Nitrogen dioxide is prepared in the laboratory by heating the nitrate of a heavy metal,

$$2Pb(NO_3)_2 \longrightarrow 2PbO + 4NO_2 + O_2$$

or by the reduction of concentrated nitric acid with copper metal (Fig. 23.17). Nitrogen dioxide is prepared commercially by oxidizing nitric oxide with air.

The nitrogen dioxide molecule (Fig. 23.18) contains an unpaired electron, which is responsible for its color and paramagnetism. It is also responsible for the dimerization of NO_2. At low pressures or at high temperatures nitrogen dioxide has a deep brown color due to the presence of the NO_2 molecule. At low temperatures the color almost entirely disappears as **dinitrogen tetraoxide,** N_2O_4, is formed (Fig. 15.6). At room temperature an equilibrium exists.

$$2NO_2(g) \rightleftharpoons N_2O_4(g) \qquad \Delta H_{298}^\circ = -57.20 \text{ kJ}; \quad \Delta G_{298}^\circ = -4.77 \text{ kJ}$$

Figure 23.17

The reaction of copper metal with concentrated HNO_3 produces a solution of $Cu(NO_3)_2$ and brown fumes of NO_2.

Figure 23.18

The molecular structures of molecules of nitrogen dioxide, NO_2, and dinitrogen tetraoxide, N_2O_4, and the resonance structures of nitrogen dioxide.

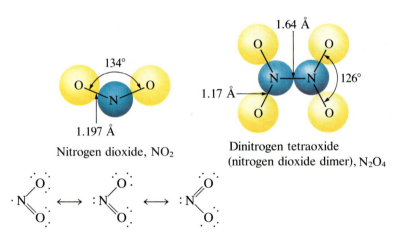

Nitrogen dioxide, NO_2

Dinitrogen tetraoxide
(nitrogen dioxide dimer), N_2O_4

Dinitrogen pentaoxide, N_2O_5 (Fig. 23.19), is a white solid formed by the dehydration of nitric acid by phosphorus(V) oxide.

$$P_4O_{10} + 4HNO_3 \longrightarrow 4HPO_3 + 2N_2O_5$$

The oxides of nitrogen(III), nitrogen(IV), and nitrogen(V) react with water and form nitrogen-containing oxyacids. Nitrogen(III) oxide, N_2O_3, is the anhydride of nitrous acid; HNO_2 forms when N_2O_3 reacts with water. There are no stable oxyacids containing nitrogen with an oxidation number of $+4$. Thus nitrogen(IV) oxide, NO_2, disproportionates in one of two ways when it reacts with water. In cold water a mixture of HNO_2 and HNO_3 is formed. At higher temperatures HNO_3 and NO form. Nitrogen(V) oxide, N_2O_5, is the anhydride of nitric acid; HNO_3 is produced when N_2O_5 reacts with water.

Little is known about the reactions of nitrogen(V) oxide; it decomposes to NO_2 and O_2 above 30°C. However, the other nitrogen oxides are known to exhibit extensive oxidation-reduction behavior. **Nitrous oxide,** nitrogen(I) oxide, resembles oxygen in its behavior when heated with combustible substances. It is a strong oxidizing agent that decomposes when heated to form nitrogen and oxygen. Because one-third of the gas liberated is oxygen, nitrous oxide supports combustion better than air. A glowing splint will burst into flame when thrust into a bottle of this gas. **Nitric oxide** will act both as an oxidizing agent or as a reducing agent. For example,

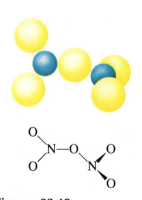

Figure 23.19

The molecular structure of a molecule of dinitrogen pentaoxide, N_2O_5.

Oxidizing agent: $\qquad P_4 + 6NO \longrightarrow P_4O_6 + 3N_2$

Reducing agent: $\qquad Cl_2 + 2NO \longrightarrow 2ClNO$

Nitrogen dioxide (or dinitrogen tetraoxide) is a good oxidizing agent. For example,

$$NO_2 + CO \longrightarrow NO + CO_2$$
$$NO_2 + 2HCl \longrightarrow NO + Cl_2 + H_2O$$

23.13 Nitric Acid and Nitrates

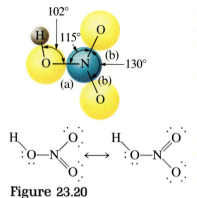

Both N_2O_5 and NO_2 react with water to form **nitric acid**, HNO_3. Nitric acid (Fig. 23.20) was known to the alchemists of the eighth century as *aqua fortis* (meaning "strong water"). It was prepared from KNO_3 and was used in the separation of gold from silver; it dissolves silver but leaves gold. Traces of nitric acid occur in the atmosphere after thunderstorms, and its salts are widely distributed in nature. Chile saltpeter, $NaNO_3$, is found in tremendous deposits (3 by 300 kilometers, and up to 2 meters thick) in the desert region near the boundary of Chile and Peru. Bengal saltpeter, KNO_3, is found in India and other countries of the Far East.

Nitric acid can be prepared in the laboratory by heating a nitrate salt (such as sodium or potassium nitrate) with concentrated sulfuric acid.

$$NaNO_3 + H_2SO_4 \xrightarrow{\triangle} NaHSO_4 + HNO_3(g)$$

Figure 23.20
The molecular and resonance structures of a molecule of nitric acid, HNO_3. N—O distances are (a) 1.41 Å and (b) 1.22 Å.

Nitric acid boils at 86°C, whereas sulfuric acid boils at 338° ($NaNO_3$ and $NaHSO_4$ are nonvolatile), and nitric acid is readily removed by distillation.

Nitric acid is produced commercially by the **Ostwald process:** oxidation of ammonia to nitric oxide, NO; oxidation of nitric oxide to nitrogen dioxide, NO_2; and conversion of nitrogen dioxide to nitric acid (Section 9.9, Part 2). A mixture of ammonia and excess air, heated to 600°–700°C, literally burns with a flame when passed through a platinum-rhodium catalyst (Fig. 23.21). Admission of additional air oxidizes the nitric oxide to nitrogen dioxide. The nitrogen dioxide, excess oxygen, and the unreactive nitrogen from the air are passed through a water spray, and nitric acid and nitric oxide form as the nitrogen dioxide disproportionates. The nitric oxide is combined with more oxygen and recycled through the process. The nitric acid is drawn off and concentrated. Most of the 6.5 million tons of domestic nitric acid produced in 1986 came from the Ostwald process.

Pure nitric acid is a colorless liquid that boils at 86°C and freezes at −42°C. It decomposes in light or when heated, producing a mixture of oxygen, water, and nitrogen oxides, of which nitrogen dioxide is predominant.

$$4HNO_3 \longrightarrow 4NO_2 + 2H_2O + O_2$$

Consequently, pure nitric acid fumes in moist air, forming a cloud of very small droplets of aqueous nitric acid as the NO_2 reacts with water in the air. It is often yellow or brown in color due to the NO_2 formed as it decomposes. Nitric acid is stable in aqueous solution; solutions containing 68% of the acid are sold as concentrated nitric acid. It is both a strong oxidizing agent and a strong acid.

The action of nitric acid on a metal rarely produces H_2 (by reduction of H^+) in more than small amounts. Instead, the acid is reduced. The products formed depend upon the concentration of the acid, the activity of the metal, and the temperature. A mixture of nitrogen oxides, nitrates, and other reduction products is usually produced. Less active metals such as copper, silver, and lead reduce concentrated nitric acid primarily to nitrogen dioxide (Fig. 23.17). The reaction of dilute nitric acid with copper gives NO (Fig. 23.15). The more active metals, such as zinc and iron, give nitrous oxide with dilute nitric oxide. With zinc, we have

$$4Zn + 10H^+ + 2NO_3{}^- \longrightarrow 4Zn^{2+} + N_2O(g) + 5H_2O$$

Figure 23.21
The catalyst used in the combustion of NH_3 to form NO is a platinum gauze.

When the acid is very dilute, either nitrogen or ammonium ions may be formed, depending on the conditions. In each case, the nitrate salts of the metals crystallize when the resulting solutions are evaporated.

Nonmetallic elements, such as sulfur, carbon, iodine, and phosphorus, are oxidized by concentrated nitric acid to their oxides or oxyacids, with the formation of NO_2.

$$S + 6HNO_3 \longrightarrow H_2SO_4 + 6NO_2 + 2H_2O$$

$$C + 4HNO_3 \longrightarrow CO_2 + 4NO_2 + 2H_2O$$

Many compounds are oxidized by nitric acid. Hydrochloric acid is readily oxidized by concentrated nitric acid to chlorine and chlorine dioxide. A mixture of 1 part concentrated nitric acid and 3 parts concentrated hydrochloric acid (called *aqua regia,* meaning "royal water"), reacts vigorously with metals. This mixture is particularly useful in dissolving gold and platinum and other metals that lie below hydrogen in the activity series. The action of aqua regia on gold may be represented, in a somewhat simplified form, by the equation

$$Au + 4HCl + 3HNO_3 \longrightarrow HAuCl_4 + 3NO_2(g) + 3H_2O$$

Nitric acid reacts with proteins, such as those in the skin, to give a yellow material called **xanthoprotein.** You may have noticed that if you get nitric acid on your fingers, they turn yellow.

Nitrates, salts of nitric acid, form when metals or their oxides, hydroxides, or carbonates react with nitric acid. Most nitrates are soluble in water; indeed, one of the significant uses of nitric acid is to prepare soluble metal nitrates.

All nitrates decompose when heated. When sodium or potassium nitrate is heated, a nitrite is formed and oxygen is evolved.

$$2NaNO_3 \longrightarrow 2NaNO_2 + O_2$$

A nitrate of a heavy metal produces the oxide of the metal when heated. With copper(II) nitrate, the equation is

$$2Cu(NO_3)_2 \longrightarrow 2CuO + 4NO_2 + O_2$$

Ammonium nitrate produces nitrous oxide. The nitrates, especially ammonium nitrate, can explode when heated or detonated.

Nitric acid is used extensively in the laboratory and in chemical industries as a strong acid and as an active oxidizing agent. It is used in the manufacture of explosives, dyes, plastics, and drugs. Salts of nitric acid (nitrates) are valuable as fertilizers. **Gunpowder** is a mixture of potassium nitrate, sulfur, and charcoal. **Ammonal,** an explosive, is a mixture of ammonium nitrate and aluminum powder.

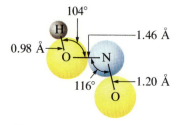

Figure 23.22

The molecular structure of a molecule of nitrous acid, HNO_2.

23.14 Nitrous Acid and Nitrites

The reaction of N_2O_3 with water gives a pale blue solution of **nitrous acid,** HNO_2 (Fig. 23.22). However, it is easier to prepare by addition of an acid to a solution of a nitrite; nitrous acid is a weak acid, so the nitrite ion is basic.

$$NO_2^- + H^+ \longrightarrow HNO_2$$

Nitrous acid is very unstable and exists only in solution. It disproportionates slowly at room temperature (rapidly when heated) into nitric acid and nitric oxide. Nitrous acid is an active oxidizing agent with strong reducing agents, and it is oxidized to nitric acid by active oxidizing agents.

Sodium nitrite is the most important salt of nitrous acid. It is usually made by reducing molten sodium nitrate with lead.

$$NaNO_3 + Pb \longrightarrow NaNO_2 + PbO$$

The nitrites are much more stable than the acid (the salts of all oxyacids are more stable than the acids themselves), but nitrites, like nitrates, can explode. Like the nitrates, nitrites are soluble in water ($AgNO_2$ is only slightly soluble).

When nitrous acid is reduced by hydrazine, **hydrazoic acid,** HN_3 (Fig. 23.23) is formed.

$$N_2H_4(aq) + HNO_2(aq) \longrightarrow HN_3(aq) + 2H_2O(l)$$

It is a weak acid and reacts with both oxidizing and reducing agents. Its salts are called **azides;** like the acid, the azides are unstable. Azides of electropositive metals decompose smoothly to nitrogen and the metal when heated; other metal azides (and the acid itself) explode. Lead azide, $Pb(N_3)_2$, is used as a detonator for explosives.

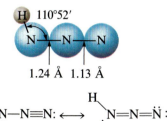

Figure 23.23
The molecular structure and resonance forms of a molecule of hydrazoic acid, HN_3.

23.15 The Nitrogen Cycle

Nitrogen is an essential constituent of all plants and animals; it is present principally in proteins, complex organic materials that contain carbon, hydrogen, and oxygen as well (Chapter 31). Most plants obtain the nitrogen necessary for growth by absorbing nitrogen compounds, primarily ammonium and nitrate salts, through their roots. However, some legumes, such as clover, alfalfa, peas, and beans, are able to obtain their nitrogen from the air by means of nitrogen-fixing bacteria, which live in nodules on their roots. These bacteria convert atmospheric nitrogen into nitrites and nitrates, which are assimilated by the host plant. Generally, the bacteria fix more nitrogen than used by the plant; when the legume dies, the unused nitrogen remains in the soil and is available for other plants. Animals obtain their nitrogenous compounds by eating plants and other animals.

Tremendous quantities of nitrogen are fixed as nitric oxide by lightning (Fig. 20.14). The nitric oxide is then oxidized to nitrogen dioxide by atmospheric oxygen, and reaction with water forms nitrous and nitric acids by the same reactions as in the Ostwald process (Sections 9.9 and 23.13). These acids are carried by rain to the soil, where they react with oxides and carbonates of metals to form nitrites and nitrates, respectively. Some soil bacteria oxidize nitrites to nitrates, a process called **nitrification.** Other bacteria change ammonia into nitrites. Denitrifying bacteria decompose nitrates and other nitrogen compounds to free nitrogen.

The decay of both plant and animal matter returns nitrogen to the soil in the form of nitrates, and either ammonia or free nitrogen is produced. Thus nitrogen passes through a cycle of fundamental importance to all plants and animals.

Phosphorus

Phosphorus is too active a nonmetal to be found free in nature. It is found in many minerals and is a component of both plants and animals. Bones, teeth, and nerve and muscle tissue contain combined phosphorus. The nucleic acids DNA and RNA, molecules that provide an organism's genetic information, contain phosphorus. The metabolism of foods and the production of energy in the body require phosphorus compounds. Foods like eggs, beans, peas, and milk furnish phosphorus for our body requirements. Plants obtain it from soluble phosphates in the soil, so phosphates are important fertilizers. Phosphorus forms compounds that are used in foodstuffs, explosives, matches, agricultural chemicals (including insecticides, herbicides, and fertilizers), soaps and detergents, and special metal alloys.

23.16 Occurrence and Preparation of Phosphorus

Phosphorus is produced commercially by heating calcium phosphate, obtained from phosphate rock (Section 9.9), with sand and coke.

$$2Ca_3(PO_4)_2 + 6SiO_2 + 10C \longrightarrow 6CaSiO_3 + 10CO(g) + P_4(g)$$

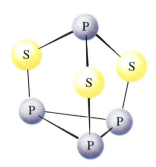

Figure 23.24
The molecular structure of P_4S_3 shows the P—P bonds.

The phosphorus distills out of the furnace and is condensed to a solid, or burned to form P_4O_{10}, from which other phosphorus compounds are manufactured. Elemental phosphorus is shipped to plants where it is converted into phosphoric acid and phosphates. In this way, the freight costs for the oxygen and water used in the manufacture of phosphorus compounds are avoided. The profits of chemical industries often depend on such considerations.

Large quantities of phosphorus compounds are converted into acids and salts to be used in fertilizers, in baking powder, and in the chemical industries. Other uses are in the manufacture of special alloys such as ferrophosphorus and phosphorbronze. Considerable quantities of phosphorus are used in making fireworks, bombs, and rat poisons. Burning phosphorus has been used to produce smoke screens during warfare.

White phosphorus once was used in the manufacture of matches. Because the workers who were exposed to the fumes suffered from necrosis of the bones, it was replaced in the heads of strike-anywhere matches by tetraphosphorus trisulfide, P_4S_3 (Fig. 23.24).

23.17 Properties of Phosphorus

Phosphorus is an active nonmetal. In compounds, phosphorus is most commonly observed with oxidation numbers of -3, $+3$, and $+5$. Phosphorus exhibits oxidation numbers that are unusual for a Group VA element in compounds that contain phosphorus-phosphorus bonds—for example, diphosphorus tetrahydride, $H_2P—PH_2$, and tetraphosphorus trisulfide, P_4S_3 (Fig. 23.24).

The most important chemical property of elemental phosphorus is its reactivity with oxygen. Slow oxidation of white phosphorus causes it to get warm, and it spontaneously inflames when it reaches 35–45°C. Because of this, white phosphorus must be stored under water (Fig. 23.25). Burns caused by phosphorus are very painful and slow to heal. *White phosphorus should be handled only with forceps and put in water immediately after use.* Red phosphorus is less active than the white form; it does not ignite in air unless it is heated to about 250°. However, the products of the reactions of red phosphorus are the same as those of the white form.

Phosphorus is one of the least electronegative of the nonmetals. It reacts with active metals, forming salts that contain the very basic **phosphide ion,** P^{3-}. With transition metals it forms phosphides that are not ionic. Phosphorus is oxidized by the nonmetals oxygen, sulfur, and the halogens. The general chemical behavior of phosphorus is discussed in Section 20.6. Some of the reactions of phosphorus and its compounds are summarized in Table 23.5.

Figure 23.25
White phosphorus is so reactive that it must be stored under water.

Table 23.5 Chemical properties of phosphorus and its compounds

General equation	Comments
Reactions of phosphorus	
$P_4 + 12Na \longrightarrow 4Na_3P$	Active metals reduce P to P^{3-}
$P_4 + 6Mg \longrightarrow 2Mg_3P_2$	
$P_4 + 3O_2 \longrightarrow P_4O_6$	In about 50% yield
$P_4 + 5O_2 \longrightarrow P_4O_{10}$	Very exothermic reaction
$P_4 + S \longrightarrow P_4S_{10}, P_4S_3$, and others	A variety of phosphorus sulfides form, depending on the stoichiometry
$P_4 + 6X_2 \longrightarrow 4PX_3$	X = F, Cl, Br, I
$P_4 + 10X_2 \longrightarrow 4PX_5$	X = F, Cl, Br
$P_4 + 3NaOH + 3H_2O \longrightarrow 3NaPO_2H_2 + PH_3$	P disproportionates in base
Reactions of compounds	
$P^{3-} + 3H^+ \longrightarrow PH_3$	
$PX_3 + 3H_2O \longrightarrow H_3PO_3 + 3HX$	Forms phosphorous acid; X = halogen
$PX_5 + H_2O \longrightarrow POX_3 + 2HX$	X = halogen
$PX_5 + 4H_2O \longrightarrow H_3PO_4 + 5HX$	Forms phosphoric acid; X = halogen
$P_4O_6 + 6H_2O \longrightarrow 4H_3PO_3$	The oxides of phosphorus are acidic
$P_4O_{10} + 6H_2O \longrightarrow 4H_3PO_4$	
$H_3PO_4 + OH^- \longrightarrow H_2PO_4^-, HPO_4^{2-}$, or PO_4^{3-}	Product depends on stoichiometry
$2H_3PO_4 \longrightarrow H_4P_2O_7 + H_2O$	Contains P—O—P bonds
$H_3PO_4 \longrightarrow HPO_3 + H_2O$	Contains P—O—P bonds
$nMH_2PO_4 \longrightarrow (MPO_3)_n + nH_2O$	Contains P—O—P bonds
$2M_2HPO_4 \longrightarrow M_4P_2O_7 + H_2O$	Contains P—O—P bonds

23.18 Phosphine

The most important hydride of phosphorus is **phosphine,** PH_3, a gaseous analogue of ammonia in formula and structure. Unlike ammonia, phosphine cannot be made by the direct union of the elements. It is prepared by the reaction of an ionic phosphide with acid or by the disproportionation of white phosphorus in a hot concentrated solution of sodium hydroxide. The disproportionation reaction produces phosphine and sodium hypophosphate, $NaPO_2H_2$. Pure phosphine is not spontaneously flammable. However, diphosphorus tetrahydride, P_2H_4, a by-product of the disproportionation, is spontaneously flammable and will ignite the phosphine on contact with air. Consequently, phosphine is best prepared and handled under an atmosphere of nitrogen or some other inert gas.

Phosphine is a colorless, very poisonous gas, which has an odor like that of decaying fish. It is easily decomposed by heat ($4PH_3 \longrightarrow P_4 + 6H_2$). Like ammonia, gaseous phosphine unites with gaseous hydrogen halides, forming the phosphonium compounds, PH_4Cl, PH_4Br, and PH_4I. However, phosphine is a much weaker base than ammonia; these compounds decompose in water, and the insoluble PH_3 escapes from solution.

23.19 Phosphorus Halides

Phosphorus reacts directly with the halogens, forming **trihalides,** PX_3, and, with one exception, **pentahalides,** PX_5. PI_5 is not known. The trihalides are much more stable than the corresponding nitrogen trihalides; nitrogen pentahalides are not stable.

The chlorides PCl_3 and PCl_5 (Fig. 23.26) are the most important halides of phosphorus. The colorless, liquid **phosphorus trichloride** is prepared by passing chlorine over molten phosphorus. Solid **phosphorus pentachloride** is prepared by oxidizing the trichloride with excess chlorine. The pentachloride is a straw-colored solid that sublimes when warmed and decomposes reversibly into the trichloride and chlorine when heated.

As most other nonmetal halides, both phosphorus chlorides react with an excess of water and form hydrogen chloride and an oxyacid; PCl_3 gives phosphorous acid, H_3PO_3, and PCl_5 gives phosphoric acid, H_3PO_4. All of the halides of phosphorus fume in moist air because they react with water vapor. Partial reaction produces **phosphorus(V) oxyhalides,** POX_3, which have tetrahedral structures (Fig. 23.27).

The pentahalides of phosphorus are Lewis bases (Section 16.15). They pick up a halide ion to give the anion PX_6^-. X-ray studies show that solid phosphorus pentachloride is an ionic compound, $[PCl_4^+][PCl_6^-]$, with a tetrahedral cation and octahedral anion. PCl_5 also forms these ions in solution in polar solvents.

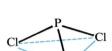
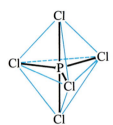

Figure 23.26
The molecular structure of PCl_3 and of PCl_5 in the gas phase.

23.20 Phosphorus Oxides

Phosphorus forms two common oxides, **phosphorus(III) oxide,** P_4O_6, and **phosphorus(V) oxide,** P_4O_{10} (Fig. 23.28). Phosphorus(III) oxide is a white crystalline solid with a garlic-like odor. Its vapor is very poisonous. It oxidizes slowly in air and inflames when heated to 70°, forming P_4O_{10}. Phosphorus(III) oxide dissolves slowly in cold water to form phosphorous acid, H_3PO_3.

Phosphorus(V) oxide, P_4O_{10}, is a white flocculent powder that is prepared by burning phosphorus. Its enthalpy of formation is very high (-2984 kJ), and it is quite stable and a very poor oxidizing agent. When P_4O_{10} is dropped into water, it reacts with a hissing sound, and heat is liberated as orthophosphoric acid is formed.

$$P_4O_{10}(s) + 6H_2O(l) \longrightarrow 4H_3PO_4(aq)$$

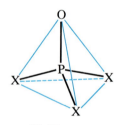

Figure 23.27
The molecular structure of a phosphorus oxyhalide, POX_3.

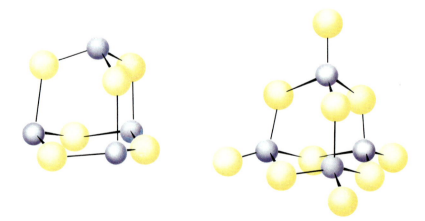

Figure 23.28
The molecular structures of P_4O_6 and P_4O_{10}.

Because of its great affinity for water, phosphorus(V) oxide is used extensively for drying gases and removing water from many compounds.

23.21 Orthophosphoric Acid

Pure **orthophosphoric acid,** H_3PO_4 (Fig. 23.29), forms colorless, deliquescent crystals that melt at 42°C. It is commonly called phosphoric acid and is commercially available as an 82% solution known as syrupy phosphoric acid.

One commercial method of preparing orthophosphoric acid is to treat calcium phosphate with concentrated sulfuric acid.

$$Ca_3(PO_4)_2 + 3H_2SO_4 \longrightarrow 2H_3PO_4 + 3CaSO_4(s)$$

The products are diluted with water, and the calcium sulfate is removed by filtration. This method gives a dilute acid that is contaminated with calcium dihydrogen phosphate, $Ca(H_2PO_4)_2$, and other compounds associated with native calcium phosphate. Pure orthophosphoric acid is manufactured by dissolving P_4O_{10} in water.

Orthophosphoric acid is a triprotic acid and forms three series of salts: **dihydrogen phosphates,** containing the $H_2PO_4^-$ ion; **hydrogen phosphates,** containing the HPO_4^{2-} ion, and **orthophosphates** (or simply phosphates), containing the PO_4^{3-} ion. Soluble dihydrogen phosphate salts form aqueous solutions that are weakly acidic, because $H_2PO_4^-$ is a weak acid. Solutions of hydrogen phosphates are basic, because the monohydrogen phosphate ion is stronger as a base than as an acid. Solutions of orthophosphates are strongly basic, because the PO_4^{3-} ion is a strong base.

Orthophosphoric acid, dihydrogen phosphate salts, and hydrogen phosphate salts decompose when heated, and water is given off. The resulting phosphorus-containing products have P—O—P bonds (Section 23.22).

Large quantities of impure orthophosphoric acid are used in the manufacture of fertilizers. The pure acid is also used in medicine as an astringent, an antipyretic, and a stimulant. It is also used in some cola drinks.

Phosphorus, in the form of soluble orthophosphates, is essential to plant growth. Calcium dihydrogen phosphate is sufficiently soluble in water to be suitable as a fertilizer. This compound is prepared commercially by treating insoluble fluorapatite, $Ca_5(PO_4)_3F$, or calcium orthophosphate, $Ca_3(PO_4)_2$, with sulfuric acid (Section 9.9).

Sodium dihydrogen phosphate is used to prevent the formation of boiler scale and as an acid in some baking powders, disodium hydrogen phosphate as a boiler cleansing compound and in the weighting of silk, and trisodium phosphate as a water softener and boiler cleansing compound. Calcium dihydrogen phosphate, $Ca(H_2PO_4)_2$, is a constituent of baking powders. Calcium hydrogen phosphate, $CaHPO_4$, is added to animal feed to provide phosphorus in the diet and is used as a polishing agent in toothpastes.

23.22 Condensed Phosphoric Acids

Orthophosphoric acid decomposes when heated; water is lost, and acids containing P—O—P bonds are formed. Among these acids are **diphosphoric acid,** $H_2P_2O_7$, **triphosphoric acid,** $H_5P_3O_{10}$, and **metaphosphoric acid,** $(HPO_3)_n$.

Diphosphoric acid, $H_4P_2O_7$ (Fig. 23.30), may be prepared by heating orthophosphoric acid to 250°C.

$$2H_3PO_4 \rightleftharpoons H_4P_2O_7 + H_2O$$

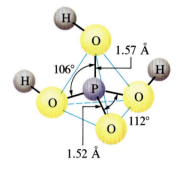

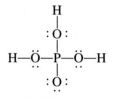

Figure 23.29
The molecular and electronic structures of orthophosphoric acid, H_3PO_4.

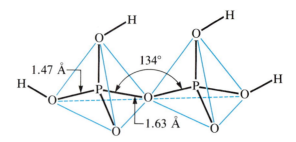

Figure 23.30

The molecular structure of diphosphoric acid, $H_4P_2O_7$, consists of two tetrahedra sharing a corner.

This reaction involves the elimination of a molecule of water from two molecules of orthophosphoric acid, as shown by the equation

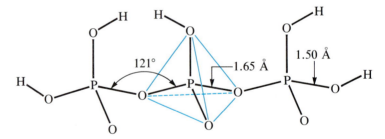

Diphosphoric acid is a white crystalline solid that melts at 61°C. When dissolved in water, it gradually hydrolyzes to the orthophosphoric acid. The hydrogen atoms in diphosphoric acid undergo stepwise neutralization, giving the ions $H_3P_2O_7^-$, $H_2P_2O_7^{2-}$, $HP_2O_7^{3-}$, and $P_2O_7^{4-}$. As you might expect, the acid strengths of the **hydrogen diphosphate anions** ($H_3P_2O_7^-$, $H_2P_2O_7^{2-}$, $HP_2O_7^{3-}$) decrease and the base strengths increase with increasing negative charge.

Triphosphoric acid (Fig. 23.31), $H_5P_3O_{10}$, is formed by the elimination of two molecules of water from three molecules of orthophosphoric acid.

$$3H_3PO_4 \longrightarrow H_5P_3O_{10} + 2H_2O$$

The reaction is reversible, and triphosphoric acid will form orthophosphoric acid upon standing in water.

Figure 23.31

The molecular structure of triphosphoric acid, $H_5P_3O_{10}$.

The sodium salt of triphosphoric acid, $Na_5P_3O_{10}$, is used as a water softener. It may be prepared by fusing equimolar quantities of sodium diphosphate and sodium metaphosphate.

$$Na_4P_2O_7 + NaPO_3 \longrightarrow Na_5P_3O_{10}$$

Polyphosphate linkages like those in pyrophosphates and triphosphates have great biochemical importance. The energy required for the contraction of muscles results from hydrolysis of P—O—P bonds in a complex organic triphosphate known as **adenosine triphosphate** (Section 31.14). The enzyme-catalyzed hydrolysis releases 29 kilojoules of energy per mole of adenosine triphosphate, and this energy is used in muscle contraction. The following equation for the reaction is given (R represents the complex organic portion of the molecule):

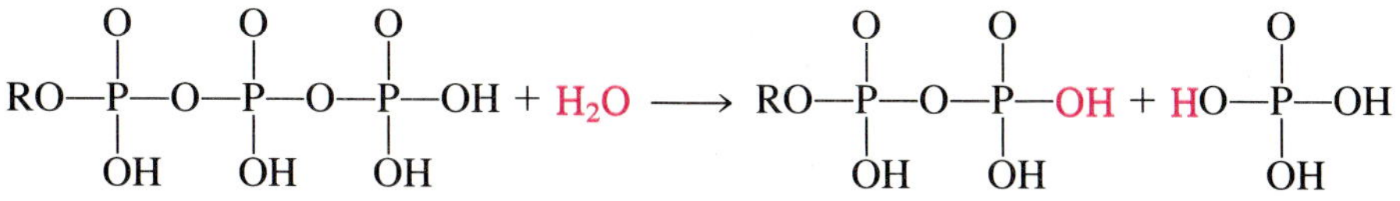

Adenosine triphosphate   Adenosine diphosphate   Phosphoric acid

Glassy **metaphosphoric acid,** $(HPO_3)_n$, solidifies from the melt produced by heating orthophosphoric acid above 400°C.

$$nH_3PO_4 \longrightarrow (HPO_3)_n + nH_2O(g)$$

The various forms of metaphosphoric acid contain rings or chains built up by oxygen bridges between adjacent phosphorus atoms. There are four oxygen atoms linked to every phosphorus atom. The polymeric HPO_3 units are represented by the formula $(HPO_3)_n$. Metaphosphoric acid dissolves readily in water, in which it slowly changes into orthophosphoric acid as the P—O—P bonds hydrolyze, producing two P—OH bonds.

Sodium metaphosphate is made by heating sodium dihydrogen orthophosphate.

$$nNaH_2PO_4 \longrightarrow (NaPO_3)_n + nH_2O(g)$$

If the product is heated to about 700°C and then cooled rapidly, a water-soluble, glassy polymetaphosphate is formed. This is a long-chain polymer that is best formulated as $(NaPO_3)_x$. Under different conditions metaphosphate anions form as rings formulated as $(PO_3^-)_n$, with $n = 3$, 4, 6, or 8. Metaphosphates form soluble complexes with calcium ion and reduce its concentration in solution, so it cannot be precipitated by soaps. For this reason, sodium polymetaphosphate and other forms of phosphate have been used extensively as water softeners in detergents. However, their use has been restricted because they enhance the growth of algae in water (Chapter 29).

23.23 Phosphorous Acid

The action of water upon P_4O_6, PCl_3, PBr_3, or PI_3 forms **phosphorous acid,** H_3PO_3 (Fig. 23.32). Pure phosphorous acid is most readily obtained by hydrolyzing phosphorus trichloride,

$$PCl_3 + 3H_2O \longrightarrow H_3PO_3 + 3HCl(g)$$

heating the resulting solution to expel the hydrogen chloride, and evaporating the water until white crystals of phosphorous acid appear upon cooling. The crystals are deliquescent, very soluble in water, and have an odor like that of garlic. The solid melts at 70.1°C and decomposes at about 200° by disproportionation into phosphine and orthophosphoric acid.

$$4H_3PO_3 \longrightarrow PH_3 + 3H_3PO_4$$

Phosphorous acid and its salts are active reducing agents, because they are readily oxidized to phosphoric acid and phosphates, respectively. Phosphorous acid reduces the silver ion to free silver, mercury(II) salts to mercury(I) salts, and sulfurous acid to sulfur.

Phosphorous acid forms only two series of salts, containing the **dihydrogen phosphite ion,** $H_2PO_3^-$, and the **hydrogen phosphite ion,** HPO_3^{2-}, respectively. The third atom of hydrogen cannot be replaced; it is bonded to the phosphorus atom rather than to an oxygen atom.

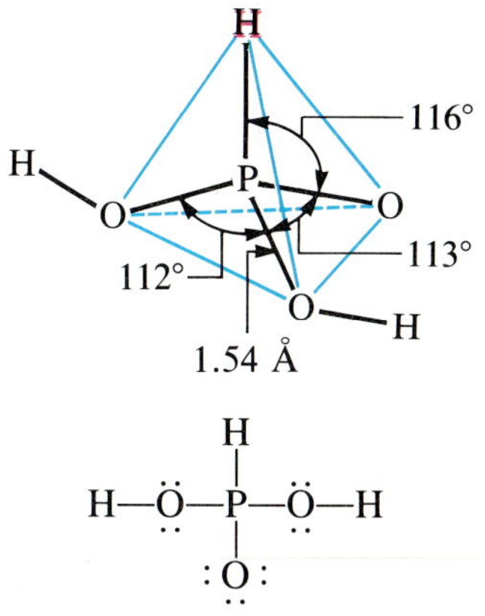

Figure 23.32

Phosphorous acid, H_3PO_3. Note that one hydrogen atom is bonded directly to the phosphorus atom. The other two hydrogen atoms are bonded to oxygen atoms. Only those hydrogen atoms bonded through an oxygen atom are acidic.

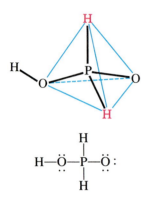

Figure 23.33

Hypophosphorous acid, H_3PO_2. Note that two of the three hydrogen atoms are attached directly to the phosphorus atom. Hence only one hydrogen atom is acidic, the one bonded through an oxygen atom.

23.24 Hypophosphorous Acid

The solution remaining from the preparation of phosphine from white phosphorus and a base contains the **hypophosphite ion,** $H_2PO_2^-$. The barium salt may be obtained by using barium hydroxide in the preparation. When barium hypophosphite is treated with sulfuric acid, barium sulfate precipitates, and **hypophosphorous acid,** H_3PO_2 (Fig. 23.33), forms in solution.

$$Ba^{2+} + 2H_2PO_2^- + 2H^+ + SO_4^{2-} \longrightarrow BaSO_4(s) + 2H_3PO_2$$

The acid is weak and monoprotic, forming only one series of salts. Two nonacidic hydrogen atoms are bonded directly to the phosphorus atom. Hypophosphorous acid and its salts are strong reducing agents because phosphorus is in the unusually low oxidation state of $+1$.

The Noble Gases

No compounds of the elements of Group VIIIA of the Periodic Table were known for many years, so these elements were called the *inert* gases. However, when it was discovered that some of these gases form a limited range of compounds, the name *noble* gases was adopted. They are sometimes called the *rare* gases.

23.25 Discovery and Production of the Noble Gases

Although all of the noble gases are present in the atmosphere in small amounts, the first was discovered in the sun. In 1868, the French astronomer Pierre-Jules-César Janssen went to India to study a total eclipse of the sun. Using a spectroscope, he found one new yellow line in its spectrum. The English chemist Edward Frankland and the English astronomer Sir Norman Lockyer concluded that the sun contained a new element, which they jointly named **helium,** from the Greek word *helios*, meaning "the sun." Twenty-seven years later the Scottish chemist Sir William Ramsay detected helium on the earth for the first time. He showed that the traces of gas present in the uranium mineral cleveite have a spectrum identical to that of helium.

In 1905 Professor Hamilton P. Cady and D. F. McFarland of the University of Kansas discovered that a sample of natural gas contained 1.84% helium by mass. Helium is now isolated from certain natural gases by liquefying the condensable components in a liquid-air machine, leaving only helium as a gas. The United States has most of the world's commercial supply of this element in its helium-bearing gas fields.

In 1894 the British physicist Lord Rayleigh observed that a sample of nitrogen prepared by removing the oxygen, carbon dioxide, and water vapor from air had a density of 1.2572 grams per liter, whereas a liter of nitrogen prepared from ammonia had a density of only 1.2506 grams per liter under the same conditions. This difference caused Rayleigh to suspect a previously undiscovered element in the atmosphere. About the same time, Sir William Ramsay isolated a small amount of gas that would not combine with any other element by passing nitrogen obtained from the air over red-hot magnesium.

$$3Mg + N_2 \longrightarrow Mg_3N_2$$

Rayleigh and Ramsay both found that the residual gases showed spectral lines never before observed. In 1894 they announced the isolation of the first noble gas, which they called **argon,** meaning "the lazy one." It was later found that this residual gas also contained traces of the other noble gases: helium, neon, krypton, and xenon.

In 1898 Ramsay and his assistant Morris W. Travers isolated **neon** (meaning "the new element") by the fractional distillation of impure liquid oxygen. Shortly thereafter they showed that the less volatile fractions from liquid air contain two other new elements, **krypton** ("the hidden element") and **xenon** ("the stranger").

In 1900 the German physicist Friedrich E. Dorn discovered that one of the disintegration products of radium is a gas similar in chemical properties to the noble gases. At first this gas was called *radium emanation,* but later its name was changed to **radon.**

Argon, neon, krypton, and xenon are produced by the fractional distillation of liquid air. Radon is collected from radium salts. More recently it has been observed that this radioactive gas is present in very small amounts in many different soils and minerals. Its accumulation in buildings may constitute a health hazard.

23.26 Properties of the Noble Gases

The boiling points and melting points of the noble gases are extremely low compared to those of other substances of comparable atomic or molecular weights. The reason for this is that no strong chemical bonds hold the atoms together in the liquid or solid states. Only weak London forces (Section 11.11) are present, and these can hold the atoms together only when molecular motion is very slight, as it is at very low temperatures. Helium is the only substance known that does not solidify on cooling. It remains liquid down to absolute zero ($-273.15°C$) at ordinary pressures, but it will solidify under elevated pressure.

Almost from the time of their discovery, chemists tried to combine the noble gases with other elements, but until 1962 no covalently bonded compounds of these nonmetals had been prepared. In 1962 Neil Bartlett, then in Canada, reported the preparation of a yellow compound of xenon, which he formulated as $Xe[PtF_6]$. This discovery led a group of chemists at the Argonne National Laboratory to think that perhaps xenon might be oxidized by the highly electronegative element fluorine, and they prepared XeF_4. Since that time a number of additional compounds have been reported, although the only direct reaction of xenon is with fluorine, and the only reported stable compounds of the noble gases are limited to xenon compounds.

23.27 Noble Gas Compounds

Stable compounds of xenon form when xenon reacts with fluorine (Fig. 23.34). **Xenon difluoride,** XeF_2, is prepared by heating an excess of xenon gas with fluorine gas at 400°C and cooling. The material forms colorless crystals, which are stable at room temperature in a dry atmosphere. **Xenon tetrafluoride,** XeF_4, and **xenon hexafluoride,** XeF_6, are prepared in an analogous manner, with a stoichiometric amount of fluorine and an excess of fluorine, respectively. Compounds with oxygen are prepared by replacing fluorine atoms in the xenon fluorides with oxygen.

Xenon compounds are very strong oxidizing agents. They may disproportionate in water, and they react with strong Lewis acids by donating a fluoride ion. The reaction with AsF_5 is

$$XeF_2 + AsF_5 \longrightarrow (XeF)AsF_6$$

Figure 23.34
Crystals of xenon tetrafluoride.

Xenon difluoride is quite soluble in water. It dissolves as undissociated molecules that soon oxidize the water to produce xenon, hydrogen fluoride, and oxygen.

$$2XeF_2 + 2H_2O \longrightarrow 2Xe + 4HF + O_2$$

The reaction of xenon tetrafluoride is more complicated because it is accompanied by disproportionation of the xenon(IV). With a stoichiometric amount of water, $XeOF_4$ is formed, but with excess water the covalent Xe—F bonds in the oxyfluoride hydrolyze and the oxide, XeO_3, forms.

$$6XeF_4 + 8H_2O \longrightarrow 2XeOF_4 + 16HF + 3O_2$$
$$XeOF_4 + 2H_2O \longrightarrow XeO_3 + 4HF$$

Dry, solid **xenon trioxide,** XeO_3, is an extremely sensitive explosive that must be handled with care. When XeF_6 reacts with water, the +6 oxidation state for the xenon is retained.

$$XeF_6 + 3H_2O \longrightarrow XeO_3 + 6HF$$

Both XeF_6 and XeO_3 disproportionate in basic solution, producing xenon, oxygen, and salts of the **perxenate ion,** XeO_6^{4-}.

Xenon difluoride reacts with acids such as HSO_3F, F_5TeOH, and $HClO_4$, which are resistant to oxidation, with evolution of hydrogen fluoride and formation of one or two Xe—O bonds, depending on the stoichiometry of the reaction. Compounds formed this way include $FXeOSO_2F$, $FXeOTeF_5$, and $Xe(OTeF_5)_2$.

23.28 Uses of the Noble Gases

Helium is used for filling balloons, and lighter-than-air craft; because it does not burn, it is safer to use than hydrogen. Helium at high pressures is not a narcotic; nitrogen is. Thus mixtures of oxygen and helium are used by divers working under high pressures in order to avoid the disoriented mental state known as nitrogen narcosis, the so-called rapture of the deep, which can result from breathing air. Mixtures of helium and oxygen are also beneficial in the treatment of certain respiratory diseases such as asthma. The lightness and rapid diffusion of helium decrease the muscular effort involved in breathing. Helium is used as an inert atmosphere for the melting and welding of easily oxidizable metals and for many chemical processes that are sensitive to air.

Liquid helium is used to reach low temperatures for cryogenic research, and it is essential for the production of the low temperatures necessary to produce superconduction in traditional superconducting materials.

Neon is used in neon lamps and signs. When an electric spark is passed through a tube containing neon at low pressure, a brilliant orange-red glow it emitted (Fig. 23.35). The color of the light given off by a neon tube may be changed by mixing argon or mercury vapor with the neon and by utilizing tubes made of glasses of special color. Neon lamps cost less to operate than ordinary electric lamps, and their light penetrates fog better.

Argon is used in gas-filled electric light bulbs, where its lower heat conductivity and chemical inertness makes it preferable to nitrogen for inhibiting vaporization of the tungsten filament and prolonging the life of the bulb. Fluorescent tubes commonly contain a mixture of argon and mercury vapor. Many Geiger-counter tubes are filled with argon.

Krypton-xenon flash tubes are used for taking high-speed photographs. An electric discharge through the tube gives a very intense light that lasts only $\frac{1}{50,000}$ of a second.

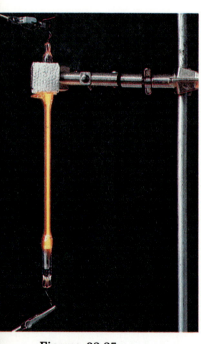

Figure 23.35

When a tube of neon gas is excited by an electric discharge, red-orange light is emitted.

For Review

Summary

Carbon (Group IVA) exhibits oxidation numbers ranging from -4 to $+4$. It is covalently bonded in the vast majority of its compounds. The immense number and variety of carbon compounds are due to the ability of carbon atoms to bond together and form chains and rings of atoms. These compounds are also discussed in Chapters 30 and 31, which describe the organic and biochemical behavior of carbon, respectively. This chapter describes the inorganic chemistry of carbon.

Carbon is relatively nonreactive as an element; the only nonmetals that oxidize it are oxygen, sulfur, and fluorine. **Carbon monoxide** forms when carbon and oxygen react in stoichiometric amounts. **Carbon dioxide** forms when carbon or carbon monoxide react with an excess of oxygen. Carbon dioxide is a weakly acidic nonmetal oxide. It will act as a Lewis acid toward oxide ion (forming the carbonate ion) and other Lewis bases, and it gives a low yield of **carbonic acid** when it dissolves in water. Carbonic acid is a weak acid that is a source of the weakly basic **hydrogen carbonate ion,** HCO_3^-, and the more strongly basic **carbonate ion,** CO_3^{2-}. Other anions that contain carbon include the acetylide ion, C_2^{2-}, the cyanide ion, CN^-, and the cyanamide ion, CN_2^{2-}.

Nitrogen (Group VA) is not very reactive. At room temperature it reacts with lithium. At elevated temperatures it reacts with active metals, forming **nitrides,** N^{3-}, and with hydrogen or oxygen in reversible low-yield reactions, forming ammonia or nitric oxide, respectively.

Ammonia exhibits both Brønsted and Lewis base behavior, and it will undergo oxidation. **Hydrazine, hydroxylamine, nitric oxide,** and **ammonium salts** are prepared from ammonia. Ammonia is prepared by the direct reaction of hydrogen and nitrogen by the **Haber process.** Because the rate of reaction is slow at low temperatures, high pressures and elevated temperatures are required.

Nitrogen exhibits oxidation states ranging from -3 to $+5$. The most important oxides are **nitric oxide** and **nitrogen dioxide,** since they are intermediates in the preparation of nitric acid by oxidation of ammonia. These two oxides can also be prepared by reduction of nitric acid. Dinitrogen trioxide is the anhydride of nitrous acid; dinitrogen pentaoxide, the anhydride of nitric acid.

Nitric acid is one of the important strong acids of commerce. It is also a strong oxidizing agent. Metals dissolve in nitric acid by reduction of the nitrate ion, generally with formation of nitrogen dioxide or nitric oxide, rather than by reduction of hydrogen ion with the production of hydrogen gas. **Nitrous acid,** a weak acid, can be prepared by the acidification of a solution of a nitrite salt.

Phosphorus (Group VA) commonly exhibits an oxidation number of -3 with active metals and $+3$ and $+5$ with more electronegative nonmetals. Hydrolysis of **phosphides** or disproportionation of phosphorus in base produces phosphine, PH_3. Phosphorus is oxidized by halogens and by oxygen. The oxides are phosphorus(V) oxide, P_4O_{10}, and phosphorus(III) oxide, P_4O_6.

Orthophosphoric acid, H_3PO_4, is prepared by the reaction of phosphates with sulfuric acid or of water with phosphorus(V) oxide. Orthophosphoric acid is a triprotic acid that forms three types of salts. Upon heating, orthophosphoric acid loses water and forms condensed phosphoric acids.

The reaction of PCl_3 with water produces **phosphorous acid,** H_3PO_3. One hydrogen atom is bonded directly to the phosphorus atom, so this acid contains only two acidic hydrogen atoms. **Hypophosphorous acid,** H_3PO_2, contains two hydrogen atoms bonded directly to the phosphorus atom and one acidic hydrogen.

The noble gas elements exhibit a very limited chemistry. Only xenon will react directly with fluorine, forming the xenon fluorides XeF_2, XeF_4, and XeF_6. Xenon oxides may be prepared by the reaction of water with the fluorides, and the reaction of certain acids with XeF_2 will form derivatives containing Xe—O bonds.

Key Terms and Concepts

adsorption (23.1)
amide (23.10)
carbonate (23.4)
derivatives of ammonia (23.11)
disproportionation (23.27)
Haber process (23.10)

hypophosphite (23.24)
imide (23.10)
nitrate (23.13)
nitride (23.10)
nitrite (23.14)
nitrogen cycle (23.15)

noble gases (23.25–23.28)
Ostwald process (23.13)
phosphate (23.21)
phosphide (23.17)
phosphite (23.23)

Exercises

1. Compare and contrast the chemical properties of elemental phosphorus and nitrogen; phosphoric acid and nitric acid; phosphine and ammonia.

2. With what elements can nitrogen form binary compounds in which it exhibits a positive oxidation number?

3. With which elements will phosphorus form compounds in which it has a positive oxidation number?

4. Complete and balance the equations for the following oxidation-reduction reactions. In some cases there may be more than one correct answer, depending on the amounts of reactants used.
   (a) $Al + P_4 \longrightarrow$ (b) $Mg + N_2 \longrightarrow$
   (c) $K + P_4 \longrightarrow$ (d) $C + F_2 \longrightarrow$
   (e) $P_4 + xsS \longrightarrow$ (f) $Li + N_2 \longrightarrow$
   (g) $P_4O_6 + O_2 \longrightarrow$ (h) $C + O_2 \longrightarrow$
   (i) $N_2 + O_2 \longrightarrow$

Nitrogen

5. Write electronic structures for the following:
   (a) N_2 (b) NH_3 (c) NH_4^+
   (d) N_2H_4 (e) HN_3 (f) NH_2OH

6. What are the products of the chemical reaction between an ionic nitride and water?

7. Explain the effects of temperature, pressure, and a catalyst on the direct synthesis of ammonia from hydrogen and nitrogen.

8. The salt hydroxylammonium chloride, $[NH_3OH]Cl$, can be formed by the reaction of hydrogen chloride and hydroxylamine. Would you expect the proton from the hydrogen chloride to be found on the oxygen or on the nitrogen of the hydroxylamine? Explain. (*Hint:* Ammonia is more basic than water.)

9. What mass of ammonia is produced by adding water to the product of the reaction of 93.0 g of lithium with elemental nitrogen? *Ans. 76.1 g*

10. What volume of ammonia measured at 1.20 atm and 25°C is required to react with 0.800 L of 0.510 M HCl solution? *Ans. 8.31 L*

11. What is the maximum volume of ammonia that can be collected at 27°C and 750 torr by the treatment of 2.00 g of ammonium chloride with 0.500 L of 0.500 M sodium hydroxide? *Ans. 0.933 L*

12. What is the anhydride of nitrous acid? Of nitric acid?

13. Write equations for the preparation of each of the oxides of nitrogen.

14. Using the data in Appendix I, determine ΔH°_{298} and ΔG°_{298} for the reaction

$$2NO_2(g) \longrightarrow N_2O_4(g)$$

The values are widely different. Discuss possible reasons for this difference, and discuss the significance of each of the two values as completely as you can. (You may wish to refer to Chapter 19 in answering this question.)

15. Outline the chemistry of the production of nitric acid from ammonia.

16. Write equations for the reaction of concentrated HNO_3 with each of the following:
   (a) Na_2CO_3 (b) $Ca(OH)_2$ (c) Cu
   (d) S   (e) C   (f) SO_2
   (g) Na_2SO_3

17. Write electronic structures for the following:
   (a) HNO_3 (b) NO_3^- (c) HNO_2
   (d) N_2O (e) N_2O_4 (f) NO_2^-
   (g) N_2O_3 (h) N_2O_5 (i) NCl_3
   (j) $ClNO$

18. Write equations for the preparation of nitrous acid starting with sodium nitrate.

19. The oxidation of ammonia and of nitric oxide are exothermic processes. Using the data in Appendix I, calculate ΔH°_{298} for these reactions. *Ans. −905.50 kJ; −114.1 kJ*

Phosphorus

20. Contrast and compare the chemical behavior of white and red phosphorus.

21. Complete and balance the following equations (*xs* indicates that the reactant is present in excess):
    (a) $P_4 + xsO_2 \longrightarrow$ (b) $P_4 + K \longrightarrow$
    (c) $P_4 + xsS \longrightarrow$ (d) $P_4 + Ca \longrightarrow$
    (e) $P_4 + xsF_2 \longrightarrow$ (f) $xsP_4 + Cl_2 \longrightarrow$
    (g) $P_4 + I_2 \longrightarrow$

22. How many tons of $Ca_3(PO_4)_2$ would be needed to prepare 1.0 ton of phosphorus if a yield of 94% is obtained?
    Ans. 5.3 tons

23. Write a Lewis structure for each of the following:
    (a) PH_3 (b) P_2H_4
    (c) H_3PO_2 (d) H_3PO_3
    (e) H_3PO_4 (f) $H_4P_2O_7$
    (g) PH_4I (an ionic compound) (h) PCl_3
    (i) P_4O_6 (j) P_4O_{10}

24. Write equations showing the stepwise ionization of phosphoric acid.

25. Write equations for the preparation and hydrolysis of sodium phosphide. Compare the hydrolysis of sodium phosphide to that of sodium nitride.

26. How can the phosphorus in insoluble tricalcium phosphate be made available for plant nutrition?

27. Write the equation to show the effect of heating H_3PO_4. The effect of heating NaH_2PO_4.

28. Why does phosphorous acid form only two series of salts although its molecule contains three hydrogen atoms?

29. Write equations for the preparation of hypophosphorous acid, starting with white phosphorus.

30. Show that the decomposition of phosphorous acid by heat involves a disproportionation reaction.

31. Write equations for each of the following preparations:
    (a) P_4 from $Ca_3(PO_4)_2$
    (b) P_4O_{10} from P_4
    (c) H_3PO_4 from P_4O_{10}
    (d) Na_2HPO_4 from H_3PO_4
    (e) $Na_4P_2O_7$ from H_3PO_4

32. Compare the structures of PCl_4^+, PCl_5, PCl_6^-, and $POCl_3$.

33. Complete and balance the following equations (*xs* indicates that the reactant is present in excess):
    (a) $NaH_2PO_4 + NH_3 \longrightarrow$
    (b) $PF_5 + KF \longrightarrow$
    (c) $P_4O_{10} + K_2O \longrightarrow$
    (d) $xsK_3PO_4 + HCl \longrightarrow$
    (e) $Na_3P + xsH_2O \longrightarrow$
    (f) $Na_4P_2O_7 + xsH_2O \longrightarrow$
    (g) $PCl_3 + CH_3OH \longrightarrow$
    (treat CH_3OH as a derivative of water in which one H is replaced by the —CH_3 group.)

34. How much $POCl_3$ can be produced from 50.0 g of PCl_5 and the appropriate amount of H_2O? *Ans. 36.8 g*

35. What volume of 0.100 M NaOH will be required to neutralize the solution produced by dissolving 1.00 g of PCl_3 in an excess of water? Note that when H_3PO_3 is titrated under these conditions, only one proton of the phosphorous acid molecule reacts. *Ans. 291 mL*

Carbon

36. Write the Lewis structure, including resonance forms where necessary, for each of the following molecules or ions:
    (a) CS_2 (b) CO_3^{2-} (c) CCl_4
    (d) CO (e) H_2C_2 (f) $H_2NCS_2^-$
    Identify the hybridization of the carbon atom in each.

37. Write the equation describing the chemical reaction that occurs when carbon dioxide is bubbled through sodium hydroxide solution.

38. Explain the toxicity of carbon monoxide.

39. Determine the heat of combustion of carbon monoxide and of carbon when carbon dioxide is the product.
    Ans. -282.99 kJ mol^{-1}; -393.51 kJ mol^{-1}

40. Write a balanced equation for the complete combustion of each of the following:
    (a) CH_4 (b) CH_3OH
    (c) HCO_2H (d) C_2H_2

41. Write equations for the production of the following:
    (a) carbon disulfide
    (b) carbon tetrachloride
    (c) calcium carbide
    (d) acetylene
    (e) sodium cyanide
    (f) hydrogen cyanide
    (g) calcium cyanamide

42. How much calcium carbonate would be formed from the addition of 10.0 L of carbon dioxide, measured at 27°C and 770 torr, to an excess of a solution of calcium hydroxide? *Ans. 41.2 g*

The Noble Gases

43. Based on the periodic properties of the elements, write balanced chemical equations for the reactions, if any, between the components of air given in Table 29.1 and hot magnesium.

44. Write the Lewis structures, for XeF_2, XeF_4, XeF_6, F—Xe—O—SO$_2$F, and XeO_3.

45. Write complete and balanced equations for the following reactions:
    (a) $XeO_3 + Ba(OH)_2$ (in water) $\longrightarrow$
    (b) $XeF_2 + HClO_4 \longrightarrow$
    (c) $XeF_2 + 2HClO_4 \longrightarrow$
    (d) $XeF_6 + Na \longrightarrow$
    (e) $xsXeF_6 + P \longrightarrow$

46. A mixture of xenon and fluorine was heated. A sample of the white solid that formed reacted with hydrogen to give 81 mL of xenon at STP and hydrogen fluoride, which was collected in water, giving a solution of hydrofluoric acid.

The hydrofluoric acid solution was titrated, and 68.43 mL of 0.3172 M sodium hydroxide was required to reach the equivalence point. Determine the empirical formula for the white solid and write balanced chemical equations for the reactions involving xenon.

$$Ans.\ Xe + 3F_2 \longrightarrow XeF_6$$
$$XeF_6 + 3H_2 \longrightarrow Xe + 6HF$$

47. Basic solutions of Na_4XeO_6 are powerful oxidants. What mass of $Mn(NO_3)_2 \cdot 6H_2O$ will react with 125.0 mL of a 0.1717 M basic solution of Na_4XeO_6 if the products include Xe and a solution of sodium permanganate?

Ans. 9.857 g

48. (a) Calculate the density of air, of hydrogen, and of helium at STP in g L^{-1}. (Assume that the average molecular weight of air is 29.1.)

Ans. Air, 1.30 g L^{-1}; hydrogen, 0.0900 g L^{-1}; helium, 0.179 g L^{-1}

(b) The lifting power of a gas with a density less than that of air is determined by the difference between its mass and that of an equal volume of air. Calculate the lifting power of hydrogen, the lifting power of helium [which is shown in Part (a) to be approximately twice as heavy as hydrogen for a given volume], and the lifting power of helium as a percentage of that of hydrogen.

Ans. Hydrogen, 1.21 g L^{-1}; helium, 1.12 g L^{-1}; 92.6%

Additional Exercises

49. Which of the following compounds will react with water to give an acid solution? Which will give a basic solution? Write a balanced equation for the reaction of each with water.

(a) Na_3N (b) N_2O_3 (c) Na_2CO_3
(d) CO_2 (e) Ca_3P_2 (f) Na_2NH
(g) CaC_2 (h) $LiNO_2$

50. Write the balanced equation or equations necessary to carry out the following transformations (H_2O, H_2, and/or O_2 may be used as needed):

(a) $Al(NO_3)_3$ from Al and N_2
(b) $Zn_3(PO_4)_2$ from $Zn(OH)_2$ and P
(c) Na_2CO_3 from Na_2O and C
(d) $XeF(OTeF_5)$ from Xe, F_2, and $HOTeF_5$
(e) P from K_3PO_4 and any other necessary reagents

51. Increases in the price of petroleum result in increases in the price of ammonia. Why?

52. (a) Which is a stronger acid, NH_3 or PH_3?

(b) Which is a stronger base, N^{3-} or P^{3-}?
(c) Which is a stronger base, HPO_4^{2-} or PO_4^{3-}?
Explain each answer.

53. What evidence is there that a coordinate-covalent bond between the proton and an ammonia molecule is stronger than that between the proton and water?

54. Which is the stronger oxidizing agent, hydrazine or hydrogen peroxide? Give evidence for your answer.

55. The heat of formation of $CS_2(l)$ is +21.4 kcal mol^{-1}. How much heat is required for the reaction of 10.00 g of S with carbon according to the following equation?

$$C(s) + 2S(s) \longrightarrow CS_2(l)$$

Is the reaction exothermic or endothermic?

Ans. 3.34 kcal; endothermic

56. What quantity of nitric acid could be prepared by the Ostwald process from 300 ft^3 of ammonia, measured at 4.00 atm and 250°C, if a 93.0% yield is obtained based on the original quantity of ammonia.   *Ans. 46.4 kg*

57. What mass of phosphorus is required to prepare 60.08 g of PCl_3 by the reaction of chlorine with phosphorus?

Ans. 13.55 g

58. (a) When phosphine burns to give liquid orthophosphoric acid, an unusually large amount of energy is involved. Calculate ΔH°_{298} for the reaction using the data in Appendix I.   *Ans. −1272 kJ*

(b) Calculate the value ΔH°_{298} for production of *solid* orthophosphoric acid by this reaction.

Ans. −1284 kJ

(c) Calculate ΔG°_{298} for the reaction that gives solid orthophosphoric acid.   *Ans. −1132 kJ*

59. At 25°C and 1.0 atm a mixture of $N_2O_4(g)$ and $NO_2(g)$ contains 30.0% $NO_2(g)$ by volume. Calculate the partial pressures of the two gases when they are at equilibrium at 25°C with a total pressure of 9.0 atm.

Ans. $P_{N_2O_4} = 8.0$ atm
$P_{NO_2} = 1.0$ atm

60. At 422 K the partial pressures of gaseous PCl_5, PCl_3, and Cl_2 in a closed system are 0.453 atm, 6.08×10^{-2} atm, and 6.08×10^{-2} atm, respectively. Determine K for the dissociation reaction.

$$PCl_5(g) \longrightarrow PCl_3(g) + Cl_2(g)$$

Ans. 8.16×10^{-3}

61. The solubility product of $Ca_3(PO_4)_2$ is 1×10^{-25}. What is the concentration of $Ca_3(PO_4)_2$ in a saturated solution of $Ca_3(PO_4)_2$?   *Ans. 4×10^{-6} M*

Nuclear Chemistry

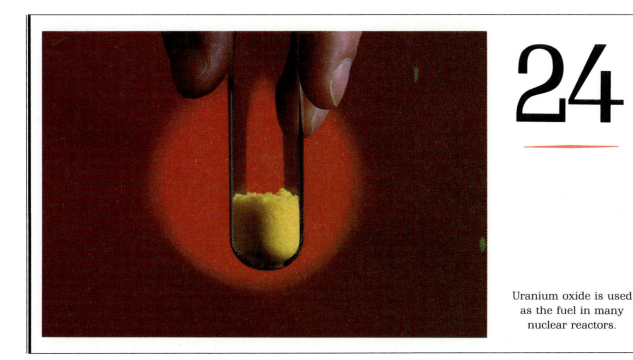

24

Uranium oxide is used as the fuel in many nuclear reactors.

The chemical changes (transformations of one form of matter into another) we have studied thus far have involved only the electrons in atoms. When we discussed changes in atoms during reactions, we focused on changes in the electronic structures of the atoms. However, there is a branch of chemistry called **nuclear chemistry** that considers changes and differences in atomic nuclei. This branch is on the borderline between physics and chemistry. It began with the discovery of radioactivity in 1896 by Antoine Becquerel, a French physicist, and has become increasingly important during the past 90 years.

The Stability of Nuclei

24.1 The Nucleus

The nucleus of an atom is composed of protons and, with the exception of ^1_1H, neutrons (Sections 5.5–5.6). The number of protons in the nucleus is called the **atomic number, Z,** of the element, and the sum of the number of protons and number of neutrons is the **mass number, A.** Atoms with the same atomic number but different mass numbers are **isotopes** of the same element. When talking about a single type of nucleus, the term **nuclide** is often used.

Protons and neutrons, collectively called **nucleons,** are packed together tightly in a nucleus. A nucleus is quite small (about 10^{-13} cm) compared to the entire atom (10^{-8} cm). Nuclei are extremely dense; they average 1.8×10^{14} grams per cubic centimeter. This density is very large compared to the densities of familiar materials. Water, for example, has a density of 1 gram per cubic centimeter. Osmium, the densest element known, has a density of 22.6 grams per cubic centimeter. If the earth's density were equal to the average nuclear density, the earth's radius would be only about 200 meters.

To hold positively charged protons together in the very small volume of a nucleus requires a very strong force, since the positive charges repel one another strongly at very short distances. The **nuclear force** is the force of attraction between nucleons that holds the nucleus together. This force acts between protons, between neutrons, and between protons and neutrons. The nuclear force is very different from and much stronger than the electrostatic force that holds negatively charged electrons around a positively charged nucleus. In fact, the nuclear force is the strongest force known and is about 30–40 times stronger than electrostatic repulsions between protons in a nucleus. Although the exact nature of the nuclear force is unknown, it is known that it is a short-range force, effective only within distances of about 10^{-13} centimeter.

24.2 Nuclear Binding Energy

Although the nature of the nuclear force is unknown, the magnitude of the energy changes associated with its action can be determined. As an example, let us consider the nucleus of the helium atom, ^4_2He, which consists of two protons and two neutrons. If a helium nucleus were to be formed by the combination of two protons and two neutrons without change of mass, the mass of the helium atom (including the mass of the two electrons outside the nucleus) would be

$$(2 \times 1.0073) + (2 \times 1.0087) + (2 \times 0.00055) = 4.0331 \text{ amu}$$

(See Section 5.3 or Appendix C for masses of the proton, neutron, and electron.) However, the mass is only 4.0026 atomic mass units. This difference between the calculated and experimental masses, the **mass defect** of the atom, indicates a loss in mass of 0.0305 atomic mass unit. The loss in mass accompanying the formation of an atom from protons, neutrons, and electrons is due to conversion of that much mass to **nuclear binding energy.** Such conversions are extremely exothermic; thus the reverse reaction, decomposition of an atom to protons, neutrons, and electrons is extremely endothermic, and very difficult to accomplish.

The nuclear binding energy can be calculated from the mass defect by the **Einstein equation**

$$E = mc^2$$

where E represents energy in joules, m stands for mass in kilograms, and c is the speed of light in meters per second. A tremendous quantity of energy results from the conversion of a small quantity of matter to energy.

EXAMPLE 24.1   Calculate the binding energy for the nuclide ^4_2He in joules per mole and electron-volts per nucleus.

The difference in mass between a ^4_2He nucleus and two protons plus two neutrons is 0.0305 amu. In order to use the Einstein equation to convert this mass defect into equivalent energy in joules, the mass defect must be expressed in kilograms (1 amu = 1.6605×10^{-27} kg).

$$0.0305 \text{ amu} \times \frac{1.6605 \times 10^{-27} \text{ kg}}{1 \text{ amu}} = 5.06 \times 10^{-29} \text{ kg}$$

The nuclear binding energy in *one* nucleus is found from the Einstein equation ($E = mc^2$).

$$
\begin{aligned}
E &= 5.06 \times 10^{-29} \text{ kg} \times (3.00 \times 10^8 \text{ m s}^{-1})^2 \\
&= 4.54 \times 10^{-12} \text{ kg m}^2\text{s}^{-2} \\
&= 4.54 \times 10^{-12} \text{ J}
\end{aligned}
$$

Units of kg m^{2}s^{-2} are equivalent to J (Section 4.1). The binding energy in 1 mol of nuclei is

$$
\begin{aligned}
4.54 \times 10^{-12} \text{ J/nucleus} \times 6.022 \times 10^{23} \text{ nuclei/mol} &= 2.73 \times 10^{12} \text{ J mol}^{-1} \\
&= 2.73 \times 10^9 \text{ kJ mol}^{-1}
\end{aligned}
$$

Binding energies are often expressed in millions of electron-volts (MeV) per nucleus rather than in joules. The conversion factor is

$$1 \text{ MeV} = 1.602189 \times 10^{-13} \text{ J}$$

giving

$$4.54 \times 10^{-12} \text{ J/nucleus} \times \frac{1 \text{ MeV}}{1.602189 \times 10^{-13} \text{ J}} = 28.3 \text{ MeV/nucleus}$$

The changes in mass in all ordinary chemical reactions are negligible, because only chemical bonds change—that is, they form or break. On the other hand, the changes in mass in nuclear reactions are very significant. If the nuclear reaction of 2 moles of neutrons with 2 moles of hydrogen atoms to give 1 mole of helium atoms could be made to occur, 2.73×10^9 kilojoules of energy would be released (Example 24.1). For comparison, burning 1 mole of methane releases only 8.9×10^2 kilojoules, about three million times less energy.

The **binding energy per nucleon** (that is, the total binding energy for the nucleus divided by the sum of the numbers of protons and neutrons present in the nucleus) is greatest for the nuclei of elements with mass numbers between 40 and 100 and decreases with mass numbers less than 40 or greater than 100 (Fig. 24.1). The most

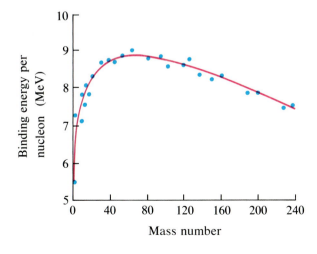

Figure 24.1

Binding energy curve for the elements.

stable nuclei are those with the largest binding energy per nucleon, those in the vicinity of iron, cobalt, and nickel in the Periodic Table. The binding energy per nucleon in helium is

$$\frac{28.3}{4} = 7.08 \text{ MeV/nucleon}$$

The binding energy per nucleon in $^{56}_{26}\text{Fe}$ is almost 25% larger than in $^{4}_{2}\text{He}$.

EXAMPLE 24.2   Calculate the binding energy per nucleon in MeV for the iron isotope with an atomic number of 26 and a mass number of 56, $^{56}_{26}\text{Fe}$, which has an atomic mass of 55.9349 amu.

To determine the binding energy per nucleon we must first determine the nuclear binding energy from the mass defect of the nucleus. The mass defect is the difference between the mass of 26 protons, 30 neutrons, and 26 electrons and the observed mass of a $^{56}_{26}\text{Fe}$ atom.

$$\begin{aligned}
\text{Mass defect} &= [(26 \times 1.0073) + (30 \times 1.0087) + (26 \times 0.00055)] - 55.9349 \\
&= 56.4651 - 55.9349 \\
&= 0.5302 \text{ amu}
\end{aligned}$$

Now we calculate the nuclear binding energy in millions of electron-volts.

$$\begin{aligned}
E &= mc^2 \\
&= 0.5302 \text{ amu} \times \frac{1.6605 \times 10^{-27} \text{ kg}}{1 \text{ amu}} \times (3.00 \times 10^8 \text{ m s}^{-1})^2 \\
&= 7.92 \times 10^{-11} \text{ kg m}^2 \text{ s}^{-2} \\
&= 7.92 \times 10^{-11} \text{ J}
\end{aligned}$$

$$7.92 \times 10^{-11} \text{ J} \times \frac{1 \text{ MeV}}{1.602189 \times 10^{-13} \text{ J}} = 494 \text{ MeV}$$

The binding energy per nucleon is found by dividing the total nuclear binding energy by 56, the number of nucleons in the atom.

$$\begin{aligned}
\text{Binding energy per nucleon} &= \frac{494 \text{ MeV}}{56} \\
&= 8.82 \text{ MeV}
\end{aligned}$$

The iron-56 nucleus is one of the more stable nuclei.

24.3 Nuclear Stability

A nucleus is stable if it cannot be transformed into another configuration without adding energy from the outside. A plot of the number of neutrons versus the number of protons for stable nuclei (the curve shown in Fig. 24.2) shows that the stable isotopes fall into a narrow band, which is called the **band of stability.** The straight line in Fig. 24.2 represents equal numbers of protons and neutrons for comparison. The figure indicates that the lighter, stable nuclei, in general, have equal numbers of protons and neutrons. For example, nitrogen-14 has 7 protons and 7 neutrons. Heavier stable nuclei, however, have slightly more neutrons than protons, as the curve shows.

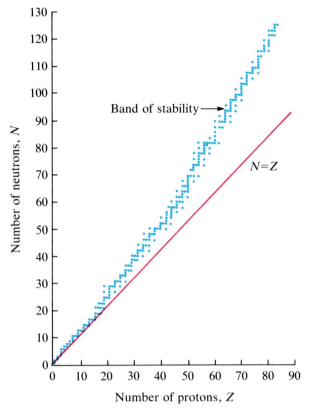

Figure 24.2

A plot of number of neutrons versus number of protons for naturally occurring nuclei. Each dot shows the number of protons and neutrons in a known stable nucleus. All isotopes of elements with atomic numbers greater than 83 are unstable.

For example, iron-56 has 30 neutrons and 26 protons. Lead-207 has 125 neutrons and 82 protons, a neutron-to-proton ratio approximately equal to 1.5.

The isotopes that fall to the left or the right of the band of stability have unstable nuclei and are **radioactive.** Note that all isotopes with atomic numbers above 83 are radioactive. Radioactive nuclei change spontaneously (decay) to other nuclei that fall either in or closer to the band of stability. The exact nature of these changes toward more stable nuclei will be discussed in subsequent sections.

24.4 The Half-Life

The number of atoms of a radioactive element that disintegrate per unit of time is a constant fraction of the total number of atoms present. The time required for $\frac{1}{2}$ of the atoms in a sample to decay is called its **half-life.** The half-lives of radioactive nuclides vary widely. For example, the half-life of $^{142}_{58}Ce$ is 5×10^{15} years, that of $^{226}_{88}Ra$ is 1590 years, that of $^{222}_{86}Rn$ is 3.82 days, and that of $^{216}_{84}Po$ is 0.16 second. Only $\frac{1}{2}$ of a sample of radium-226 will remain unchanged after 1590 years, and at the end of another 1590 years the sample will have decayed to $\frac{1}{4}$ of its initial mass, and so on. The half-lives of a number of nuclides are listed in Appendix K.

Radioactive decay is a first-order process. The rate constant k for a radioactive disintegration can be expressed in terms of the half-life, $t_{1/2}$, by the following equation (which applies to all first-order reactions; see Section 14.7):

$$k = \frac{0.693}{t_{1/2}}$$

The relationship between the initial amount of a radioactive isotope and the amount remaining after a given period of time is expressed by the logarithmic relationship that applies to all other first-order reactions (Section 14.7).

$$\log \frac{c_0}{c_t} = \frac{kt}{2.303}$$

where c_0 is the initial amount and c_t is the amount remaining at time t.

EXAMPLE 24.3   Calculate the rate constant for the radioactive disintegration of cobalt-60, an isotope used in cancer therapy. $^{60}_{27}\text{Co}$ decays with a half-life of 5.2 years to produce $^{60}_{28}\text{Ni}$.

$$k = \frac{0.693}{t_{1/2}}$$

$$= \frac{0.693}{5.2 \text{ yr}} = 0.13 \text{ yr}^{-1}$$

EXAMPLE 24.4   Calculate the fraction and the percentage of a sample of the $^{60}_{27}\text{Co}$ isotope that will remain after 15 years.

$$\log \frac{c_0}{c_t} = \frac{kt}{2.303}$$

$$= \frac{(0.13 \text{ yr}^{-1})(15 \text{ yr})}{2.303} = 0.847$$

$$\frac{c_0}{c_t} = \text{antilog } 0.847 = 7.031$$

The fraction remaining will be the amount at time t divided by the initial amount, or c_t/c_0.

$$\frac{c_t}{c_0} = \frac{1}{7.031} = 0.14$$

The fraction of $^{60}_{27}\text{Co}$ remaining after 15 years is 0.14. Thus 14% of the $^{60}_{27}\text{Co}$ originally present will remain after 15 years.

EXAMPLE 24.5   How long would it take for a sample of $^{60}_{27}\text{Co}$ to disintegrate to the extent that only 2.0% of the original amount remains?

$$c_t = 0.020 \times c_0$$

Therefore

$$\frac{c_0}{c_t} = \frac{1}{0.020} = 50$$

$$\log 50 = \frac{(0.13 \text{ yr}^{-1})t}{2.303}$$

$$t = \frac{2.303 \log 50}{0.13 \text{ yr}^{-1}} = 30 \text{ yr}$$

EXAMPLE 24.6   The half-life of $^{216}_{84}\text{Po}$ is 0.16 s. How long would it take to reduce a sample of it to the negligible amount of 0.000010% of the original amount (1.0×10^{-7} times the original amount)?

$$k = \frac{0.693}{0.16 \text{ s}} = 4.33 \text{ s}^{-1}$$

$$c_t = 1.0 \times 10^{-7} \times c_0$$

Hence

$$\frac{c_0}{c_t} = 1.0 \times 10^7$$

$$\log(1.0 \times 10^7) = \frac{(4.33 \text{ s}^{-1})t}{2.303}$$

$$t = \frac{2.303 \log(1.0 \times 10^7)}{4.33 \text{ s}^{-1}} = 3.7 \text{ s}$$

These examples indicate part of the problem of disposal of radioactive isotopes. In 3.7 seconds any reasonable quantity of $^{216}_{84}\text{Po}$ would be reduced to a negligible quantity and would itself no longer be a problem. However, the burst of radiation during the first second or so would be extremely intense and very hazardous. Furthermore, $^{216}_{84}\text{Po}$ is a part of a series (see Table 24.1, Section 24.8), decaying into $^{212}_{82}\text{Pb}$, which is itself radioactive with a half-life of 10.6 seconds, producing $^{212}_{83}\text{Bi}$, which has a half-life of 60.5 minutes, and so on, until finally the stable isotope $^{208}_{82}\text{Pb}$ is reached. Hence, in coping with radiation effects and isotope disposal, all steps in the decay of the material must be taken into account. In contrast to the few seconds required for the decay of $^{216}_{84}\text{Po}$, it takes 30 years for 98% of a sample of $^{60}_{27}\text{Co}$ to disintegrate. Radium-226, an isotope with a half-life of 1590 years, would take even longer to decay. Even after 200 years, 91.6% of the original quantity of $^{226}_{88}\text{Ra}$ would still be present. Obviously we cannot simply leave radium-226 or other radioactive isotopes lying around on the assumption that the radiation danger will disappear within a short period.

24.5 The Age of the Earth

One of the most interesting applications of the study of radioactivity has been in determining the age of the earth. To estimate the lower limit for the earth's age, scientists determine the age of various rocks and minerals, assuming that the earth must be at least as old as the rocks and minerals in its crust. One method is to measure the relative amounts of $^{87}_{37}\text{Rb}$ and $^{87}_{38}\text{Sr}$ in rock. $^{87}_{37}\text{Rb}$ decays to $^{87}_{38}\text{Sr}$ with a half-life of 4.7×10^{10} years. Thus 1 gram of $^{87}_{37}\text{Rb}$ would produce 0.5000 gram of $^{87}_{38}\text{Sr}$ and leave 0.5000 gram of $^{87}_{37}\text{Rb}$ after decaying for 47 billion years. The $^{87}_{37}\text{Rb}$-to-$^{87}_{38}\text{Sr}$ ratio in a rock formation in southwestern Greenland has shown this formation to have an age of 3.75×10^9 years. This is the oldest known rock on earth.

Nuclear Reactions

The following sections describe reactions of nuclei that result in changes in their atomic numbers, mass numbers, and energy states. Such reactions are called **nuclear reactions.** Nuclear reactions result (1) from the spontaneous decay of naturally occurring or artificially produced radioactive nuclei, (2) from the fission of unstable heavy nuclei, (3) from the fusion of light nuclei, and (4) from the bombardment of nuclei with other nuclei or with other fast-moving particles.

24.6 Equations for Nuclear Reactions

An equation that describes a nuclear reaction identifies the nuclides involved in the reaction, their mass numbers and atomic numbers, and the other particles involved in the reaction. These particles include:

1. **alpha particles ($_2^4$He, or α)**, helium nuclei consisting of two neutrons and two protons
2. **beta particles ($_{-1}^0$e, or β)**, high-speed electrons
3. **positrons ($_{+1}^0$e, or β^+)**, particles with the same mass as an electron but with 1 unit of *positive* charge
4. **protons ($_1^1$H, or p)**, nuclei of hydrogen atoms
5. **neutrons ($_0^1$n, or n)**, particles with a mass approximately equal to that of a proton but with no charge

Nuclear reactions also often involve **gamma rays (γ).** Gamma rays are electromagnetic radiation, somewhat like X rays in character but with higher energies and shorter wavelengths.

Examples of equations for some types of nuclear reactions of historical interest follow.

1. The first radioactive element, naturally occurring polonium, was discovered by the Polish scientist Marie Curie and her husband, Pierre, in 1898. It decays by the reaction

$$_{84}^{212}\text{Po} \longrightarrow {}_{82}^{208}\text{Pb} + {}_2^4\text{He} \qquad (1)$$

2. Technetium, a radioactive element that does not occur naturally on the earth and was first prepared in 1937, decays by the reaction

$$_{43}^{98}\text{Tc} \longrightarrow {}_{44}^{98}\text{Ru} + {}_{-1}^0\text{e} \qquad (2)$$

3. One of the many fission reactions of uranium in the first nuclear reactor (1942) is

$$_{92}^{235}\text{U} + {}_0^1\text{n} \longrightarrow {}_{35}^{87}\text{Br} + {}_{57}^{146}\text{La} + 3\,{}_0^1\text{n} \qquad (3)$$

4. One of the fusion reactions present during the detonation of a hydrogen bomb is

$$_1^3\text{H} + {}_1^2\text{H} \longrightarrow {}_2^4\text{He} + {}_0^1\text{n} \qquad (4)$$

5. The first element prepared by artificial means was prepared in 1919 by bombarding nitrogen atoms with α particles.

$$_7^{14}\text{N} + {}_2^4\text{He} \longrightarrow {}_8^{17}\text{O} + {}_1^1\text{H} \qquad (5)$$

A correctly written nuclear reaction is balanced; that is, the sum of the mass numbers of the reactants equals the sum of the mass numbers of the products, and the sum of the atomic numbers of the reactants equals the sum of the atomic numbers of the products. If the atomic number and the mass number of all but one of the particles in a nuclear reaction are known, the particle can be identified by balancing the reaction.

EXAMPLE 24.7   The reaction of an α particle with magnesium-25 ($_{12}^{25}$Mg) produces a proton ($_1^1$H) and a nuclide of another element. Identify the new nuclide produced.

The nuclear reaction may be written as

$$_{12}^{25}\text{Mg} + {}_2^4\text{He} \longrightarrow {}_Z^A\text{X} + {}_1^1\text{H}$$

where A is the mass number and Z is the atomic number of the new nuclide, X. Since the sum of the mass numbers of the reactants must equal the sum of the mass numbers of the products, $25 + 4 = A + 1$, or $A = 28$. Similarly, the atomic numbers must balance, so $12 + 2 = Z + 1$ and $Z = 13$. The element with atomic number 13 is aluminum. Thus the product is $^{28}_{13}\text{Al}$, an unstable nuclide with a half-life of 2.3 min that decays by β decay (Section 24.7) to $^{28}_{14}\text{Si}$.

24.7 Radioactive Decay

The spontaneous change of an unstable nuclide into another is called **radioactive decay.** The unstable nuclide is often called the **parent nuclide;** the nuclide that results from the decay, the **daughter nuclide.** The daughter nuclide may be stable, or it may decay further. Different types of radioactive decay can be classified by the radiation produced:

1. The loss of an α particle, **α decay,** occurs primarily in heavy nuclei ($A \geq 200$, $Z > 83$). Loss of an α particle gives a daughter nuclide with a mass number 4 units smaller and an atomic number 2 units smaller than those of the parent nuclide. Consequently, the daughter nuclide has a larger neutron-to-proton ratio than the parent nuclide, and, if the parent nuclide lies below the band of stability (Fig. 24.2), the daughter nuclide will lie closer to the band.

2. A nuclide with a large neutron-to-proton ratio (one that lies above the band of stability in Fig. 24.2) can reduce this ratio by electron emission, or **β decay,** in which a neutron in the nucleus decays into a proton that remains in the nucleus and an electron that is emitted.

$$^{1}_{0}\text{n} \longrightarrow {}^{1}_{1}\text{H} + {}^{0}_{-1}\text{e}$$

Electron emission does not change the mass number of the nuclide but does increase the number of its protons and decrease the number of its neutrons. Consequently, the neutron-to-proton ratio is decreased, and the daughter nuclide lies closer to the band of stability than did the parent nuclide.

3. Certain artificially produced nuclides in which the neutron-to-proton ratio is low (nuclides that lie below the band of stability in Fig. 24.2) undergo **β^{+} decay,** the emission of a positron. During the course of β^{+} decay, a proton is converted into a neutron with the emission of a positron. The neutron-to-proton ratio increases, and the daughter nuclide lies closer to the band of stability than did the parent nuclide.

When a positron and an electron interact, they annihilate each other. All of their mass is converted into energy—two 0.511-MeV γ rays are produced.

$$^{0}_{-1}\text{e} + {}^{0}_{+1}\text{e} \longrightarrow 2\gamma \qquad (0.511 \text{ MeV each})$$

4. A proton is converted to a neutron when one of the electrons in an atom is captured by the nucleus (this capture is called orbital **electron capture**).

$$^{1}_{1}\text{H} + {}^{0}_{-1}\text{e} \longrightarrow {}^{1}_{0}\text{n}$$

Like β^{+} decay, electron capture occurs when the neutron-to-proton ratio is low (the nuclide lies below the band of stability). Electron capture has the same effect on the nucleus as that of positron emission; the atomic number is decreased by 1 as a proton is converted into a neutron. This increases the neutron-to-proton ratio, and the daughter nuclide lies closer to the band of stability curve than did the parent nuclide.

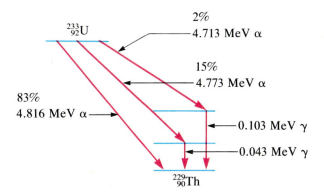

Figure 24.3

The decay of $^{233}_{92}U$ to $^{229}_{90}Th$. The decay can occur by three paths. Two paths give excited states of the $^{229}_{90}Th$ nuclide, which decay to the ground state by emission of γ rays.

5. A daughter nuclide may be formed in an excited state and then decay to its ground state with the emission of a γ ray, a quantum of high-energy electromagnetic radiation. Figure 24.3 illustrates the relationships for the decay of uranium-233. Note that there is no change of mass number or atomic number during emission of a γ ray. **γ-ray emission** is observed only when a nuclear reaction produces a daughter nuclide that is in an excited state.

The naturally occurring radioactive isotopes of the heavier elements fall into chains of successive disintegrations, or decays, and all the species in one chain constitute a radioactive family, or series. Three of these series include most of the naturally radioactive elements of the Periodic Table. They are the **uranium series,** the **actinium series,** and the **thorium series.** Each series is characterized by a parent (first member) with a long half-life and a series of decay processes that ultimately lead to a stable end-product—that is, an isotope on the band of stability (Fig. 24.2). In all three series the end-product is an isotope of lead: $^{206}_{82}Pb$ in the uranium series, $^{207}_{82}Pb$ in the actinium series, and $^{208}_{82}Pb$ in the thorium series.

The steps in the thorium decay series are given in Table 24.1 as an illustration of one natural chain of successive decays.

24.8 Synthesis of Nuclides

Atoms with stable nuclei can be converted to other atoms **(transmuted)** by bombardment with other nuclei or with high-speed particles. The first artificial nucleus was made in Lord Rutherford's laboratory in 1919. He bombarded nitrogen atoms with high-speed α particles emanating from a naturally radioactive isotope of radium and observed the nuclear reaction

$$^{14}_{7}N + {}^{4}_{2}He \longrightarrow {}^{17}_{8}O + {}^{1}_{1}H \tag{1}$$

The $^{17}_{8}O$ nucleus is stable, so this reaction does not lead to further changes.

Many artificially induced nuclear reactions produce unstable nuclei, which then undergo radioactive disintegration. The disintegration in these cases is referred to as **artificial radioactivity.** For example, boron-10 reacts with an α particle to form nitrogen-13 and a neutron.

$$^{10}_{5}B + {}^{4}_{2}He \longrightarrow {}^{13}_{7}N + {}^{1}_{0}n \tag{2}$$

Nitrogen-13 is radioactive and decays by β^{+} decay.

$$^{13}_{7}N \longrightarrow {}^{13}_{6}C + \beta^{+}$$

Table 24.1   The thorium decay series

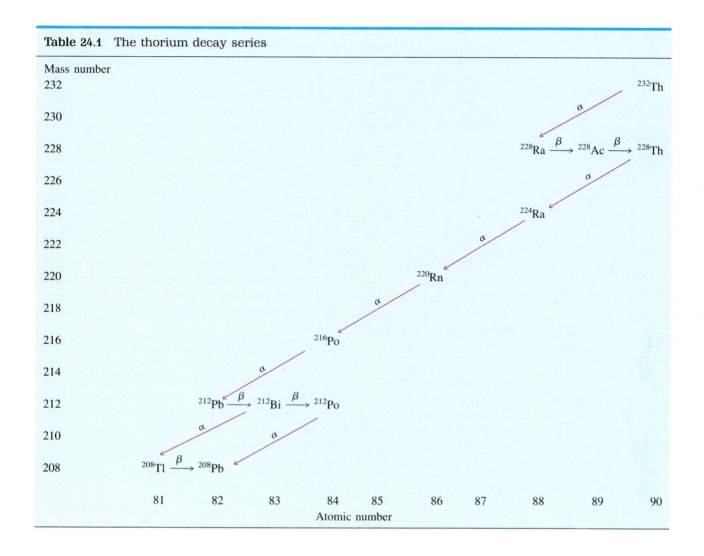

In an abbreviated notation for **transmutation reactions,** the bombarding particle and the product particle are written in parentheses between the symbols for the reactant and product nuclides. Thus Equations (1) and (2) can be written as

$$^{14}_{7}\text{N}(\alpha,\ \text{p})\ ^{17}_{8}\text{O} \qquad \text{and} \qquad ^{10}_{5}\text{B}(\alpha,\ \text{n})\ ^{13}_{7}\text{N}$$

respectively, where α, p, and n are symbols for an α particle, a proton, and a neutron; $^{14}_{7}\text{N}$ and $^{10}_{5}\text{B}$ are symbols for the reactants; and $^{17}_{8}\text{O}$ and $^{13}_{7}\text{N}$ are symbols for the products.

Generally, the charged particles used to cause transmutation reactions are accelerated to the kinetic energies that will produce reactions by machines (**accelerators**), which use magnetic and electric fields to accelerate the particles. In all accelerators the particles move in a vacuum to avoid collisions with gas molecules.

Since neutrons are not charged, they cannot be accelerated to the velocities necessary for nuclear transmutations. Neutrons used in transmutation reactions are therefore obtained from radioactive decay and other nuclear reactions occurring in a nuclear reactor (Section 24.11).

Many known elements have been transmuted into other known elements. Also, a considerable number of new elements have been synthesized. These include element 43, technetium (Tc); element 85, astatine (At); element 87, francium (Fr); and element 61, promethium (Pm).

Prior to 1940 the heaviest known element was uranium, whose atomic number is 92. In 1940 McMillan and Abelson were able to make element 93, neptunium (Np), by bombarding uranium with high-velocity neutrons. The nuclear reactions are

$$^{238}_{92}U + {}^{1}_{0}n \longrightarrow {}^{239}_{92}U$$

$$^{239}_{92}U \xrightarrow{\text{23 min}} {}^{239}_{93}Np + {}^{0}_{-1}e$$

Neptunium-239 is also radioactive, with a half-life of 2.3 days, and converts to plutonium (Pu), whose atomic number is 94.

$$^{239}_{93}Np \longrightarrow {}^{239}_{94}Pu + {}^{0}_{-1}e$$

Elements 95 through 109 have also been prepared artificially. The elements beyond element 92 (uranium) are called **transuranium elements.** Elements 89 through 103 make up the **actinide series.** Element 109 was prepared in the summer of 1982 by a West German group who bombarded a target of bismuth-209 with accelerated nuclei of iron-58.

$$^{209}_{83}Bi + {}^{58}_{26}Fe \longrightarrow {}^{266}_{109}X + {}^{1}_{0}n$$

The researchers identified the element even though they made only one atom of it. The decay products of the atom were characteristic of those expected for element 109. It began to decay about 5×10^{-3} second after its formation, with the emission of an α particle and the formation of the same isotope of element 107 that the research team had first produced the preceding year.

$$^{266}_{109}Une \longrightarrow {}^{262}_{107}Uns + {}^{4}_{2}He$$

$$\searrow {}^{258}_{105}Unp + {}^{4}_{2}He$$

$$\downarrow \text{Electron capture}$$

$$^{258}_{104}Unq \longrightarrow \text{Smaller nuclei}$$

Equations describing the preparation of some other isotopes of the transuranium elements are given in Table 24.2.

The name *rutherfordium,* with the symbol Rf, has been suggested for element 104 by American scientists who synthesized the element. Soviet scientists, who also worked out the synthesis of the element, have suggested the name *kurchatovium.* It has been proposed that element 105 be named *hahnium,* with the symbol Ha, to honor Otto Hahn, the late German scientist who won the Nobel Prize for the discovery of nuclear fission.

Both a Soviet and an American group have announced the synthesis of element 106, by quite different methods. The two groups have agreed not to propose any name for the element until a decision is reached as to who will get credit for being the first to synthesize it. The failure of these two groups to reach agreement is responsible, in part, for the systematic names, suggested by the International Union of Pure and Applied Chemistry (Section 5.13 and Table 24.2), for the elements beyond element 103.

Elements 106, 107, 108, and 109 are very unstable. Whether or not additional elements can be made cannot be predicted at this time. However, their synthesis may

Table 24.2 Preparation of the transuranium elements

Name	Symbol	Atomic number	Reaction
Neptunium	Np	93	$^{238}_{92}U + ^1_0n \longrightarrow ^{239}_{93}Np + ^0_{-1}e$
Plutonium	Pu	94	$^{238}_{92}U + ^2_1H \longrightarrow ^{238}_{93}Np + 2\,^1_0n$
			$^{238}_{93}Np \longrightarrow ^{238}_{94}Pu + ^0_{-1}e$
Americium	Am	95	$^{239}_{94}Pu + ^1_0n \longrightarrow ^{240}_{95}Am + ^0_{-1}e$
Curium	Cm	96	$^{239}_{94}Pu + ^4_2He \longrightarrow ^{242}_{96}Cm + ^1_0n$
Berkelium	Bk	97	$^{241}_{95}Am + ^4_2He \longrightarrow ^{243}_{97}Bk + 2\,^1_0n$
Californium	Cf	98	$^{242}_{96}Cm + ^4_2He \longrightarrow ^{245}_{98}Cf + ^1_0n$
Einsteinium	Es	99	$^{238}_{92}U + ^{14}_7N \longrightarrow ^{253}_{99}Es + 5\,^1_0n$
Fermium	Fm	100	$^{238}_{92}U + ^{16}_8O \longrightarrow ^{254}_{100}Fm + 5\,^1_0n$
Mendelevium	Md	101	$^{253}_{99}Es + ^4_2He \longrightarrow ^{256}_{101}Md + ^1_0n$
Nobelium	No	102	$^{246}_{96}Cm + ^{12}_6C \longrightarrow ^{254}_{102}No + 4\,^1_0n$
Lawrencium	Lr	103	$^{250}_{98}Cf + ^{11}_5B \longrightarrow ^{257}_{103}Lr + 4\,^1_0n$
Unnilquadium	Unq	104	$^{249}_{98}Cf + ^{12}_6C \longrightarrow ^{257}_{104}Unq + 4\,^1_0n$
Unnilpentium	Unp	105	$^{249}_{98}Cf + ^{15}_7N \longrightarrow ^{260}_{105}Unp + 4\,^1_0n$
Unnilhexium	Unh	106	$^{206}_{82}Pb + ^{54}_{24}Cr \longrightarrow ^{257}_{106}Unh + 3\,^1_0n$
			$^{249}_{98}Cf + ^{18}_8O \longrightarrow ^{263}_{106}Unh + 4\,^1_0n$

prove difficult; isotopes containing more than 157 neutrons are exceedingly unstable (an observation for which there is as yet no explanation).

Glenn Seaborg has suggested extending the Periodic Table to include new elements whose synthesis might be possible (Table 24.3). (It may be helpful to review the material in Chapter 5 on filling shells and subshells by electrons.) Element 104 is the first element of the **transactinide series** (that is, the elements beyond the actinide series). The chemical properties of these elements can be predicted by the method of Mendeleev—comparison with their lighter analogues in the table. Element 104 should be an analogue of hafnium; 105, an analogue of tantalum; and so on to element 118, a noble gas analogous to radon.

The most striking feature of Seaborg's extension of the Periodic Table is the addition of another inner transition series of elements starting with atomic number 121 and extending through atomic number 153. He calls this grouping the **superactinide series.** This series, after element 121, would include elements in which the $5g$ orbitals fill (see Table 5.4). Element 121 would receive the first $7d$ electron (the $n = 7$ major shell being the second from the outside) and be analogous to scandium, yttrium, lanthanum, and actinium. The first of eighteen $5g$ electrons could enter at element 122, for which the $n = 5$ shell is the fourth from the outside; the eighteenth $5g$ electron would enter at element 139. This would presumably be followed by addition of fourteen $6f$ electrons (third major shell from the outside) for elements 140 through 153, making this latter series an inner transition series analogous to the lanthanide and actinide series. Following the superactinide inner transition series, elements 154 through 168 would be analogous to elements 104 through 118, with the addition of the remaining ten $7d$ electrons in the second major shell from the outside and the $6p$ electrons in the outermost major shell. Element 168 would be another noble gas—or should we expect a noble liquid?

Table 24.3 Predicted locations of new elements in the Periodic Table

																					s	
																1 H	2 He

s		d												p						
1 H																	
3 Li	4 Be												5 B	6 C	7 N	8 O	9 F	10 Ne
11 Na	12 Mg												13 Al	14 Si	15 P	16 S	17 Cl	18 Ar
19 K	20 Ca	21 Sc	22 Ti	23 V	24 Cr	25 Mn	26 Fe	27 Co	28 Ni	29 Cu	30 Zn	31 Ga	32 Ge	33 As	34 Se	35 Br	36 Kr
37 Rb	38 Sr	39 Y	40 Zr	41 Nb	42 Mo	43 Tc	44 Ru	45 Rh	46 Pd	47 Ag	48 Cd	49 In	50 Sn	51 Sb	52 Te	53 I	54 Xe
55 Cs	56 Ba	[57–71] *	72 Hf	73 Ta	74 W	75 Re	76 Os	77 Ir	78 Pt	79 Au	80 Hg	81 Tl	82 Pb	83 Bi	84 Po	85 At	86 Rn
87 Fr	88 Ra	[89–103] †	104 Unq	105 Unp	106 Unh	107 Uns	108 Uno	109 Une	110	111	112	113	114	115	116	117	118
119	120	[121–153] ‡	154	155	156	157	158	159	160	161	162	163	164	165	166	167	168

	f															
*LANTHANIDE SERIES	57 La	58 Ce	59 Pr	60 Nd	61 Pm	62 Sm	63 Eu	64 Gd	65 Tb	66 Dy	67 Ho	68 Er	69 Tm	70 Yb	71 Lu
†ACTINIDE SERIES	89 Ac	90 Th	91 Pa	92 U	93 Np	94 Pu	95 Am	96 Cm	97 Bk	98 Cf	99 Es	100 Fm	101 Md	102 No	103 Lr

			g				f			
‡SUPER ACTINIDES	121	122	123			139	140			153

24.9 Nuclear Fission

The greater stability of the nuclei of elements with mass numbers of intermediate values suggested the possibility of spontaneous decomposition of the less stable nuclei of the heavy elements into more stable fragments of approximately half their sizes. Two German scientists, Hahn and Strassman, reported in 1939 that uranium-235 atoms bombarded with slow-moving neutrons split into smaller fragments, consisting of elements near the middle of the Periodic Table and several neutrons. The process is called **fission.** Among the fission products were barium, krypton, lanthanum, and cerium, all of which have more stable nuclei than uranium.

$$^{235}_{92}\text{U} + ^{1}_{0}\text{n} \longrightarrow \text{fission fragments} + \text{neutrons} + \text{energy}$$
(Isotopes of Ba, Kr, etc.)

The sum of the atomic numbers of the fission products is 92, the atomic number of the original uranium nucleus. A loss of mass of about 0.2 atomic mass unit per uranium atom occurs in these fission reactions. This mass is converted into a fantastic quantity of energy—fission of 1 pound of uranium-235 produces about 2.5 million times as much energy as is produced by burning 1 pound of coal.

Fission of a uranium-235 nucleus produces, on the average, 2.5 neutrons as well as fission fragments. These neutrons may cause the fission of other uranium-235 atoms, which in turn provide more neutrons, setting up a **chain reaction.** Nuclear fission becomes self-sustaining when the number of neutrons produced by fission equals or exceeds the number of neutrons absorbed by splitting nuclei plus the number lost to the surroundings. The amount of a fissionable material that will support a self-sustaining chain reaction is called a **critical mass.** The critical mass of a fissionable material depends on the shape of the sample as well as on the type of material.

An atomic bomb contains several pounds of fissionable material, $^{235}_{92}\text{U}$ or $^{239}_{94}\text{Pu}$, and an explosive device for compressing it quickly into a small volume. When fissionable material is in small pieces, the proportion of neutrons that escape at the relatively large surface area is great, and a chain reaction does not take place. When the small pieces of fissionable material are brought together quickly to form a body with a mass larger than the critical mass, the relative number of escaping neutrons decreases, and a chain reaction and explosion result. The explosion of an atomic bomb can release more energy than the explosion of thousands of tons of TNT.

Chain reactions of fissionable materials can be controlled in a nuclear reactor (Fig. 24.4 and Section 24.11).

Figure 24.4
The core of a research reactor at Sandia National Laboratories, seen looking down through shielding water. The characteristic blue glow is Cerenkov radiation, which is light produced by rapidly moving charged particles.

24.10 Nuclear Fusion

The process of combining very light nuclei into heavier nuclei is also accompanied by the conversion of mass into large amounts of energy. The process is called **fusion** and is the focus of an intensive research effort to develop a practical thermonuclear reactor (Section 24.13). The principal source of energy in the sun is the fusion of four hydrogen nuclei into one helium nucleus. Four hydrogen nuclei have a mass that is 0.7% greater than that of a helium nucleus; this extra matter is converted into energy during fusion.

It has been found that a deuteron, $_1^2H$, and a triton, $_1^3H$, which are the nuclei of the heavy isotopes of hydrogen, will undergo fusion at extremely high temperatures (**thermonuclear fusion**) to form a helium nucleus and a neutron.

$$_1^2H + _1^3H \longrightarrow _2^4He + _0^1n$$

This change is accompanied by a conversion of a portion of the mass into energy and is the nuclear reaction of the hydrogen bomb. In a hydrogen bomb a fission bomb (uranium or plutonium) is exploded inside a charge of deuterium and tritium to provide the temperature of many millions of degrees required for the fusion of the deuterium and tritium.

If the fusion of heavy isotopes of hydrogen can be controlled, hydrogen from the water of the oceans could supply energy for future generations.

Nuclear Energy

24.11 Nuclear Power Reactors

Any **nuclear reactor** (Fig. 24.5) that produces power by the fission of uranium or plutonium by bombardment with slow neutrons must have at least five components (see Fig. 24.6).

Figure 24.5

A commercial nuclear plant for generating electricity. The two large domed structures are the containment buildings that house the reactors.

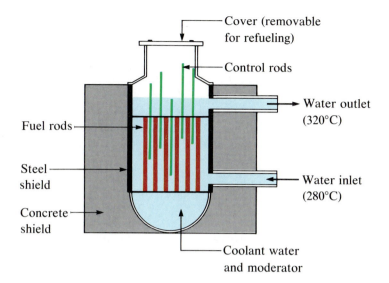

Figure 24.6

A light-water nuclear reactor. This reactor uses pressurized liquid water at 280°C and 150 atm as both a coolant and a moderator.

1. A NUCLEAR FUEL. A fissionable isotope (commonly $^{235}_{92}U$, $^{233}_{92}U$, or $^{239}_{94}Pu$) must be present in sufficient quantity to provide a self-sustaining chain reaction. Most reactors used in the United States use pellets of the uranium oxide U_3O_8, in which the concentration of uranium-235 has been increased (*enriched*) from its natural level to about 3%. The U_3O_8 pellets are contained in a tube (a fuel rod) of a protective material, usually a zirconium alloy. The core of a typical nuclear power reactor in the United States has about 40,000 kilograms of enriched U_3O_8, contained in several hundred fuel rods.

Naturally occurring uranium is a mixture of several isotopes; uranium-238 is the most abundant. About 1 in every 140 uranium atoms is the uranium-235 isotope. To obtain a higher concentration of uranium-235, it is necessary to separate uranium-235. The most successful method separates $^{235}_{92}UF_6$ from $^{238}_{92}UF_6$ by fractional diffusion of large volumes of gaseous UF_6 at low pressure through porous barriers. This method is based on the fact that the lighter $^{235}_{92}UF_6$ molecules diffuse through a porous barrier faster than do the heavier $^{238}_{92}UF_6$ molecules. The enriched UF_6 is chemically converted to U_3O_8.

2. A MODERATOR. Neutrons produced by nuclear reactions move too fast to cause fission. They must be slowed down before they will be absorbed by the fuel and produce additional nuclear reactions. In a reactor neutrons are slowed by collision with the nuclei of a moderator such as heavy water (D_2O; Section 22.3), graphite, carbon dioxide, or light (ordinary) water. These materials are used because they do not react with or absorb neutrons. Most reactors in operation in the United States use light water as the moderator. Graphite is used in reactors in many other countries.

3. A COOLANT. The coolant carries the heat from the fission reaction to an external boiler and turbine where it is transformed into electricity (Fig. 24.7). The coolant is a gas or liquid that is pumped through the reactor core. Some coolants also serve as moderators.

4. A CONTROL SYSTEM. A nuclear reactor is controlled by adjusting the number of slow neutrons present to keep the rate of the chain reaction at a safe level. Control is maintained by control rods that absorb neutrons. Rods containing cadmium or boron-10 are often used. Boron-10, for example, absorbs neutrons by the reaction $^{10}_5B$ (n, α)

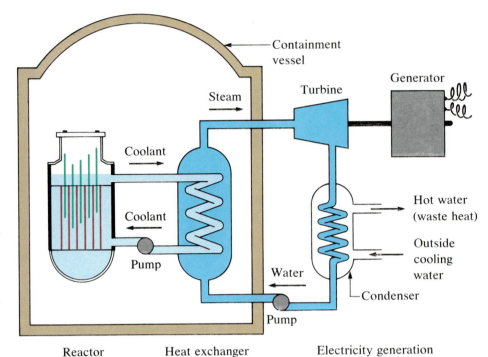

Figure 24.7

Power-generating plant employing a nuclear power reactor. In a coal-fired power plant the steam is generated in a boiler. Note that the reactor coolant is contained in a closed system and does not come in contact with outside cooling water. The reactor shielding has been omitted for clarity.

$^{7}_{3}$Li. The chain reaction can be completely stopped by inserting all of the control rods into the core between the fuel rods.

5. A SHIELD AND CONTAINMENT SYSTEM. A nuclear reactor produces neutrons and other particles by radioactive decay of the products resulting from fusion. In addition, a reactor is very hot, and high pressures result from the circulation of water or another coolant through it. A reactor must withstand high temperatures and pressures and protect operating personnel from the radiation. A reactor container often consists of three parts: (1) the reactor vessel, a steel shell that is 3–20 centimeters thick and absorbs much of the radiation produced by the reactor; (2) a main shield of 1–3 meters of high-density concrete; and (3) a personnel shield of lighter materials to absorb γ rays and X rays. In addition, reactors are often covered with a steel or concrete dome designed to contain any radioactive materials released by a reactor accident.

The importance of a containment vessel is amply illustrated by two recent accidents at nuclear reactors. In March 1979 the cooling system of one of the reactors at the Three Mile Island plant in Pennsylvania failed, and the cooling water spilled from the reactor into the floor of the containment building. The temperature of the core climbed to at least 2200°C, and the upper portion of the core began to melt. In addition, the zirconium alloy cladding of the fuel rods began to react with steam and produced hydrogen.

$$Zr(s) + 2H_2O(g) \longrightarrow ZrO_2(s) + 2H_2(g)$$

The hydrogen accumulated in the confinement building and it was feared that there was danger of an explosion of a mixture of hydrogen and air in the building. Consequently, the hydrogen and other, radioactive, gases (primarily krypton and xenon) were vented from the building. Within a week, cooling water circulation was restored and the core began to cool. The plant is still closed and the process of cleanup is expected to last until about 1989.

Although no discharge of radioactive material is desirable, the discharge of radioactive krypton and xenon from the Three Mile Island plant is among the least intolerable. These gases readily disperse in the atmosphere and thus do not produce highly radioactive areas. Moreover, they are noble gases and are not absorbed into plant and animal matter in the food chain. Effectively none of the heavy elements of the core of the reactor were released into the environment and no cleanup of the area outside of the containment building was necessary.

A second major nuclear accident occurred in April 1986 at the V. I. Lenin power plant near Chernobyl in the Ukraine. While operating at low power during an unauthorized experiment with some of its safety devices shut off, one of the RBMK model reactors at the plant became unstable. Its chain reaction became uncontrollable and increased to a level far beyond that for which the reactor was designed. The steam pressure in the reactor went to between 100 and 500 times the full power pressure and ruptured the reactor. Because the reactor was not enclosed in a containment building, the hot core was exposed to the environment. In the initial burst, a large amount of radioactive material was released, and additional fission products were released as the graphite moderator of the core continued to burn. The fire was controlled, but a total of 203 plant workers and firemen developed acute radiation sickness and 31 died. The reactor has been encapsulated in concrete, and, after a year of decontamination of the reactor site and surrounding countryside, other reactors at the plant have been restarted and residents are returning to towns and villages in the region.

It should be noted that the RBMK reactors used in some places in the Soviet Union are unique. They are the only reactors that are designed with the potential low power instability. These reactors have been modified since the accident to reduce the risk of a recurrence.

The energy produced by a reactor fueled with enriched uranium results from the fission of $^{235}_{92}U$ (Section 24.9) as well as from the fission of $^{239}_{94}Pu$. Plutonium-239 forms from $^{238}_{92}U$ present in the fuel (Section 24.8). In any nuclear reactor only about 0.1% of the mass of the fuel is converted into energy. The other 99.9% remains in the fuel rods as fission products and unused fuel. All of the fission products absorb neutrons, and after a period of several months to a year, depending on the reactor, the fission products must be removed by changing the fuel rods. Otherwise, the concentration of these fission products will increase until the reactor can no longer operate because of their absorption of neutrons.

Spent fuel rods contain a variety of products, consisting of unstable nuclei ranging in atomic number from 25 to 60, some transuranium elements, including $^{239}_{94}Pu$ and $^{241}_{95}Am$, and unreacted $^{235}_{92}U$ and $^{238}_{92}U$. The unstable nuclei and the transuranium elements give the spent fuel a dangerously high level of radioactivity. The long-lived isotopes $^{90}_{38}Sr$ and $^{137}_{55}Cs$ and the shorter-lived isotope $^{131}_{53}I$ are particularly dangerous because they can be incorporated into human bodies if the radioactive material is dispersed in the environment. Consequently, it is absolutely essential that this material not be allowed to be released into the biosphere. It takes about 400 years for the radioactivity of $^{90}_{38}Sr$ and $^{137}_{55}Cs$ to decrease to a reasonably safe level. Other nuclides such as $^{241}_{95}Am$ and $^{239}_{94}Pu$ have much longer half-lives and require thousands of years to decay to a safe level. The ultimate fate of the nuclear reactor as a significant source of energy in the United States probably rests on whether or not a scientifically and politically satisfactory technique for processing and storing the components of spent fuel rods can be developed.

24.12 Breeder Reactors

A **breeder reactor** is a nuclear reactor that produces more fissionable material than it consumes. Because the supply of naturally occurring uranium is limited (some estimates suggest that the known reserves will only last for another 50 years of full-scale use), the conversion of nonfissionable material to nuclear fuel in a breeder reactor may provide a long-term energy supply. The breeder reactor is constructed with a blanket of $^{238}_{92}U$ or $^{232}_{90}Th$ surrounding the core of fissionable material. Extra neutrons produced by the fission in the core are captured by the blanket to form (breed) more fissionable atoms.

$^{238}_{92}U$ produces the fissionable nuclide $^{239}_{94}Pu$, according to the reactions described in Section 24.8. $^{232}_{90}Th$ produces the fissionable nuclide $^{233}_{92}U$ by the following series of reactions:

$$^{232}_{90}Th + ^{1}_{0}n \longrightarrow ^{233}_{90}Th$$

$$^{233}_{90}Th \xrightarrow{23 \text{ min}} ^{233}_{91}Pa + ^{0}_{-1}e$$

$$^{233}_{91}Pa \xrightarrow{27 \text{ day}} ^{233}_{92}U + ^{0}_{-1}e$$

Experimental breeder reactors are in operation in the United States, Great Britain, the Soviet Union, and France.

Breeder reactors present challenges in addition to those outlined in Section 24.11. Not only must spent fuel be stored, but the blanket of fertile material must be processed (by remote control) to separate the radioactive fuel from other radioactive by-products, and the large amounts of plutonium produced must be safely stored. Plutonium-239 decays with a half-life of 24,000 years by emission of an α particle; thus it will remain in the biosphere for a very long time if it is dispersed. Plutonium is one of the most toxic substances known, estimates of a fatal dose being as low as 1 microgram (10^{-6} g). Moreover, plutonium can be used to make bombs, and the existence of quantities of the pure material could be attractive to terrorist groups and to some countries not yet capable of producing their own atomic weapons.

24.13 Fusion Reactors

A fusion reactor is a nuclear reactor in which fusion reactions of light nuclei (Section 24.10) are controlled. At the time of this writing, there are no self-sustaining fusion reactors operating in the world, although small-scale fusion reactions have been run for very brief periods.

Fusion reactions require very high temperatures, about 10^8 K. At these temperatures, all molecules dissociate into atoms, and the atoms ionize, forming a new state of matter called a **plasma.** Since no solid materials are stable at 10^8 K, a plasma cannot be contained by mechanical devices. Two techniques to contain a plasma at the necessary density and temperature for a fusion reaction to occur are currently under study. The techniques involve containment by a magnetic field (Fig. 24.8) and by the use of focused laser beams (Fig. 24.9).

After a plasma of hydrogen isotopes is generated and contained, it will undergo fusion reactions when the temperature exceeds about 10^8 K. One such reaction is

$$^{2}_{1}H + ^{3}_{1}H \longrightarrow ^{4}_{2}He + ^{1}_{0}n$$

which proceeds with a mass loss of 0.0188 atomic mass unit, corresponding to the release of 1.69×10^9 kilojoules per mole of $^{4}_{2}He$ formed. Deuterium ($^{2}_{1}H$) is available

Figure 24.8
The experimental fusion reactor at Princeton. The plasma in this device is contained by a magnetic field.

Figure 24.9
This bank of lasers produces the very intense light necessary to induce fusion.

from heavy water. Tritium ($^3_1$H) can be prepared by reaction of the neutrons from the fusion reaction with lithium.

$$^1_0n + ^6_3Li \longrightarrow ^3_1H + ^4_2He$$

For Review

Summary

Protons and neutrons, collectively called **nucleons,** are held together by a short-range, but very strong, force called the nuclear force. The **nuclear binding energy** may be calculated from the **mass defect** (the difference in mass between a nucleus and the nucleons of which it is composed) by the **Einstein equation,**

$$E = mc^2$$

The binding energy per nucleon is largest for the elements with mass numbers between 40 and 100; these are the most stable nuclei.

Stable nuclei have equal numbers of neutrons and protons or a few more neutrons than protons. Nuclei that deviate from stable neutron-to-proton ratios are radioactive and decay by losing one of several different kinds of particles. These include **α particles,** $^4_2$He; **β particles,** $^0_{-1}$e; **positrons,** $^0_{+1}$e; and **neutrons,** $^1_0$n. Some nuclei decay by **electron capture.** Each of these modes of decay leads to a new nucleus with a stable neutron-to-proton ratio. The kinetics of the decay process are first order. The half-life of a radioactive isotope is the time that is required for $\frac{1}{2}$ of the atoms in a sample to decay. Each radioactive nuclide has its own characteristic half-life.

New atoms can be produced by bombarding other atoms with nuclei or high-speed particles. The products of these **transmutation reactions** can be stable or radioactive. A number of artificial elements, including technetium, astatine, and the transuranium elements, have been produced in this way.

Nuclear power can be generated through **fission,** or reactions in which a heavy nucleus breaks up into two or more lighter nuclei and several neutrons. Since the neutrons may induce additional fission reactions when they combine with other heavy nuclei, a **chain reaction** can result. Useful power is obtained if the fission process is carried out in a **nuclear reactor.** The conversion of light nuclei into heavier nuclei (**fusion**) also produces energy. At present, this energy has not been contained and is used only in nuclear weapons.

Key Terms and Concepts

alpha decay (24.7)
alpha particle (24.6)
artificial radioactivity (24.8)
atomic number (24.1)
band of stability (24.3)
β decay (24.7)
β particle (24.6)
breeder reactor (24.12)
chain reaction (24.9)
daughter nuclide (24.7)
Einstein equation (24.2)

electron capture (24.7)
fission (24.9)
fusion (24.10)
fusion reactor (24.13)
γ ray (24.6)
half-life (24.4)
mass defect (24.2)
mass number (24.1)
neutron (24.6)
nuclear binding energy (24.2)
nuclear force (24.1)

nuclear reactor (24.11)
nucleon (24.1)
nuclide (24.1)
parent nuclide (24.7)
positron (24.6)
proton (24.6)
radioactive decay (24.7)
stability curve (24.3)
transmutation reaction (24.8)

Exercises

1. Write a brief description or definition of each of the following:
   (a) nucleon   (b) α particle
   (c) β particle   (d) positron
   (e) γ ray   (f) mass number
   (g) atomic number
2. Indicate the number of protons and neutrons in each of the following nuclei:
   (a) $^{16}_{8}O$ (b) neon-20   (c) lead-208   (d) $^{233}_{92}U$

Nuclear Stability

3. Which of the following nuclei lie within the band of stability shown in Figure 24.2?
   (a) $^{5}_{3}Li$ (b) beryllium-9
   (c) chlorine-35   (d) $^{60}_{30}Zn$
   (e) radon-210
4. Which of the following nuclei would you expect to be unstable because it does not lie in the band of stability?
   (a) $^{3}_{1}H$ (b) $^{34}_{15}P$ (c) $^{42}_{20}Ca$ (d) $^{238}_{92}U$
5. Define and illustrate the term *half-life*.
6. The mass of the atom $^{19}_{9}F$ is 18.99840 amu. (a) Calculate its binding energy per atom in millions of electron-volts. (b) Calculate its binding energy per nucleon. (See Appendix C.)   *Ans. 148.3 MeV, 7.805 MeV*
7. A $^{7}_{4}Be$ atom (mass = 7.0169 amu) decays to a $^{7}_{3}Li$ atom (mass = 7.0160 amu) by electron capture. How much energy (in millions of electron-volts) is produced by this reaction?   *Ans. 0.8 MeV*

8. The mass of a hydrogen atom ($^{1}_{1}H$) is 1.007825 amu; that of a tritium atom ($^{3}_{1}H$), 3.01605 amu; and that of an α particle, 4.00150 amu. How much energy in kilojoules per mole of $^{4}_{2}He$ produced is released by the following reaction?
$$^{1}_{1}H + {}^{3}_{1}H \longrightarrow {}^{4}_{2}He$$
Ans. 2.011 × 10⁹ kJ
9. The half-life of ^{239}Pu is 24,000 yr. What fraction of the ^{239}Pu present in nuclear wastes generated today will be present in 1000 yr?   *Ans. 0.97 (97%)*
10. What percentage of $^{254}_{102}No$ remains of a 0.100-g sample 5.0 min after it is formed (half-life of 55 s)? 1.0 h after it is formed?   *Ans. 2.3%; 2.0 × 10⁻¹⁸%*
11. The isotope ^{208}Tl undergoes β decay with a half-life of 3.1 min.
    (a) What isotope is the product of the decay?
    (b) Is ^{208}Tl more stable or less stable than an isotope with a half-life of 54.5 s?
    (c) How long will it take for 99.0% of a sample of pure ^{208}Tl to decay?   *Ans. 20.6 min*
    (d) What percentage of a sample of pure ^{208}Tl will remain undecayed after 1.0 h?   *Ans. 1.5 × 10⁻⁴%*
12. Calculate the time required for 99.999% of each of the following radioactive isotopes to decay:
    (a) $^{240}_{94}Pu$ (half-life = 6580 yr)   *Ans. 1.09 × 10⁵ yr*
    (b) $^{13}_{5}B$ (half-life = 1.9 × 10⁻² s)   *Ans. 3.16 s*
    (c) $^{233}_{92}U$ (half-life = 1.62 × 10⁵ yr)   *Ans. 2.69 × 10⁶ yr*
13. The isotope $^{90}_{38}Sr$ is one of the extremely hazardous species

in the fallout from a nuclear fission explosion. A 0.500-g sample diminishes to 0.393 g in 10.0 yr. Calculate the half-life.   *Ans. 28.8 yr*

14. If 1.000 g of $^{226}_{88}$Ra (atomic weight = 226) produces 0.0001 mL of the gas $^{222}_{86}$Rn (atomic weight = 222) at STP in 24 h, what is the half-life of ^{226}Ra in years?
                    Ans. 2 × 10³ yr

Nuclear Decay

15. Describe the types of radiation emitted from the nuclei of naturally radioactive elements.

16. What is the change in the nucleus that gives rise to a β particle? a β^+ particle?

17. The loss of an α particle by a nucleus causes what change in the atomic number and the mass of the nucleus? What is the change in the atomic number and mass when a β particle is emitted?

18. How do nuclear reactions differ from ordinary chemical changes?

19. Many nuclides with atomic numbers greater than 83 decay by processes such as electron emission. Rationalize the observation that the emissions from these unstable nuclides normally include α particles also.

20. Identify the various particles that may be produced in a nuclear reaction.

21. Write a balanced equation for each of the following nuclear reactions.
    (a) Uranium-230 undergoes α decay.
    (b) Bismuth-212 decays to polonium-212.
    (c) Beryllium-8 and a positron are produced by the decay of an unstable nucleus.
    (d) Neptunium-239 forms from the reaction of uranium-238 with a neutron, then spontaneously converts to plutonium-239.
    (e) Strontium-90 decays to yttrium-90.

22. Write a nuclear reaction for each step in the formation of $^{218}_{84}$Po from $^{238}_{92}$U, which proceeds by a series of decay reactions involving stepwise emission of α, β, β, α, α, α, α particles, in that order.

23. Complete the following equations:
    (a) $^{27}_{13}$Al + $^{4}_{2}$He $\longrightarrow$? + $^{1}_{0}$n
    (b) $^{7}_{3}$Li + ? $\longrightarrow$ 2 $^{4}_{2}$He
    (c) $^{239}_{94}$Pu + ? $\longrightarrow$ $^{242}_{96}$Cm + $^{1}_{0}$n
    (d) $^{14}_{6}$C $\longrightarrow$ $^{14}_{7}$N + ?
    (e) $^{14}_{7}$N + $^{4}_{2}$He $\longrightarrow$? + $^{1}_{1}$H

24. Fill in the atomic number of the initial nucleus and write out the complete nuclear symbol for the product of each of the following nuclear reactions:
    (a) ^{9}Be (α, n) (b) ^{23}Na $(^{2}$H, p)
    (c) ^{33}S (n, p) (d) ^{33}S (p, n)
    (e) ^{10}B (α, p) (f) ^{27}Al (α, n)
    (g) ^{63}Cu (p, n)

25. Complete the following notations by filling in the missing parts:
    (a) $(n, 2n)^{66}$Cu   (b) $(\alpha, n)^{30}$P

    (c) $(^{2}$H, n)^{238}Np   (d) ^{10}B $(\alpha, \ )^{13}$C
    (e) ^{232}Th $(\ , n)^{235}$U   (f) ^{2}H $(^{12}$C, $)^{10}$B
    (g) ^{24}Mg (α, n) (h) ^{238}U $(^{12}_{6}$C, 4n)

26. Use the abbreviated system of notation, as in exercise 25, to describe:
    (a) the production of ^{17}O from ^{14}N by α-particle bombardment
    (b) the production of ^{14}C from ^{14}N by neutron bombardment
    (c) the production of ^{233}Th from ^{232}Th by neutron bombardment
    (d) the production of ^{239}U from ^{238}U by $^{2}_{1}$H bombardment

27. For each of the following unstable isotopes, predict by what mode(s) spontaneous radioactive decay might proceed:
    (a) $^{6}_{2}$He (n/p ratio too large)
    (b) $^{60}_{30}$Zn (n/p ratio too small)
    (c) ^{235}Pa (too much mass, n/p ratio too large)
    (d) $^{241}_{94}$Np
    (e) ^{18}F
    (f) ^{129}Ba
    (g) ^{237}Pu

28. Which of the following nuclei is most likely to decay by positron emission: chromium-53, manganese-51, or iron-59? Explain your choice.

29. Explain in terms of Fig. 24.2 how unstable heavy nuclides (atomic number greater than 83) may decompose to form nuclides of greater stability (a) if they are below the band of stability and (b) if they are above the band of stability.

30. Technetium is used in nuclear medicine because it is absorbed by certain damaged tissues. The location of the technetium (and the tissue) can be detected by the γ ray that an excited Tc nucleus emits. Technetium is prepared from ^{98}Mo. Molybdenum-98 combines with a neutron to give molybdenum-99, an unstable isotope which decays by β emission to give an excited form of ^{99}Tc. This excited nucleus relaxes to the ground state by emission of a γ ray. The ground state of ^{99}Tc decays by β emission. Write the equations for each of these nuclear reactions.

Nuclear Power

31. How does nuclear fission differ from nuclear fusion? Why are both of these processes exothermic?

32. How are atomic bombs and hydrogen bombs detonated?

33. Describe the components of a nuclear reactor.

34. Describe how the potential energy of uranium is converted into electrical energy in a nuclear power plant.

35. What is a breeder reactor?

36. List advantages and disadvantages of nuclear energy as a source of electrical power, as compared to coal, fuel oil, natural gas, and water.

37. Discuss and compare the problems of radioactive wastes for radioactive substances of short half-life and those of long half-life.

The Semi-Metals

25

Lava is a mixture of molten silicates.

A series of elements called the **semi-metals,** sometimes referred to as the **metalloids,** separate the metals from the nonmetals in the Periodic Table. These elements look metallic, but they conduct electricity poorly. They are semiconductors. Their chemical behavior is intermediate between that of metals and nonmetals; for example, the pure elements form covalent crystals like the nonmetals, but like metals, they generally do not form monatomic anions.

In this chapter we will discuss the chemical behavior of semi-metals. The chapter begins with an overview of these elements, followed by sections dealing with the chemistry of selected semi-metals.

25.1 The Chemical Behavior of the Semi-Metals

The semi-metals are boron, silicon, germanium, antimony, and tellurium (Fig. 25.1). They include certain members of Groups IIIA, IVA, VA, and VIA. Semi-metals are fairly nonreactive elements, with electronegativities slightly lower than that of hydrogen.

Boron has a large affinity for fluorine and oxygen. It combines with fluorine at room temperature, forming BF_3, and with chlorine, bromine, oxygen, and sulfur at elevated temperatures, forming BCl_3, BBr_3, B_2O_3, and B_2S_3, respectively. It does not combine directly with iodine. Boron reacts with carbon at very high temperatures and forms boron carbide, B_4C, which is second only to diamond in hardness. With the exception of BF_3 and B_4C, boron compounds react with water or oxidize readily to give boric acid, $B(OH)_3$, or boric oxide, B_2O_3, respectively.

Boron is oxidized when heated with water, sulfur dioxide, nitric oxide, carbon dioxide, or many other oxides. It is also oxidized by both concentrated nitric and sulfuric acid, forming boric acid. In most of its chemical properties, boron behaves as a nonmetal.

Boron is covalent in its compounds. In compounds such as BF_3 and B_2O_3 (Fig. 11.2) it forms three single covalent bonds. The geometry about boron in such compounds is trigonal planar (Sections 8.1–8.2), and the boron atom is sp^2 hybridized. The p orbital that is not hybridized is empty; consequently, boron compounds with three covalent bonds are strong Lewis acids (Section 16.15). The addition of a fourth group in a Lewis acid-base reaction gives a four-coordinate boron atom with a tetrahedral geometry and sp^3 hybridization (Sections 8.1–8.2). For example, the reaction of BF_3 with NaF produces the tetrahedral BF_4^- ion by the reaction

$$BF_3(g) + NaF(s) \longrightarrow NaBF_4(s)$$

In addition to compounds in which it exhibits an oxidation number of $+3$, as expected for a member of Group IIIA, boron forms an extensive variety of compounds

Figure 25.1

The location of the semi-metals (shown in blue) in the Periodic Table. Nonmetals are shown in green; the metals, in red.

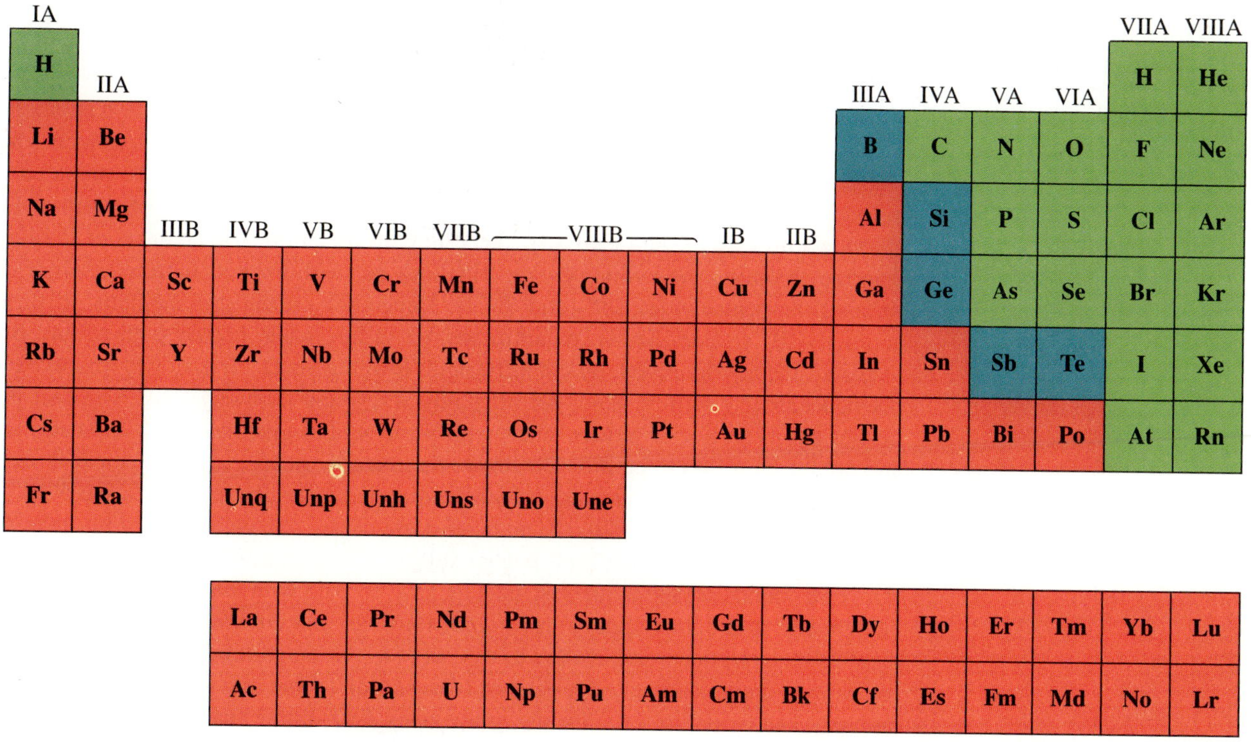

containing boron-boron bonds, in which it has other, sometimes fractional, oxidation numbers. These include the boron hydrides (Section 25.6) and compounds such as BO, B_2Cl_4, and B_4Cl_4.

Silicon is unreactive at low temperatures and resists attack by air, water, and acids. A very thin film of silicon dioxide, SiO_2, protects the surface from attack. This film dissolves in base and exposes the surface, so silicon dissolves in hot sodium hydroxide or potassium hydroxide solutions, forming silicates and hydrogen.

$$Si + 2OH^- + H_2O \longrightarrow SiO_3^{2-} + 2H_2(g)$$

Silicon reacts with the halogens at high temperatures and forms volatile tetrahalides, such as SiF_4. It oxidizes in air at elevated temperatures to give silicon dioxide, SiO_2, a solid. In the majority of its compounds silicon exhibits an oxidation number of +4.

Silicon does not form double or triple bonds. It forms compounds in which it is sp^3 or d^2sp^3 hybridized, with four or six single covalent bonds, respectively. Thus silicon compounds of the general formula SiX_4, where X is a highly electronegative group, can act as Lewis acids and form six-coordinate silicon (six bonds) with d^2sp^3 hybridization. For example, silicon tetrafluoride, SiF_4, reacts with sodium fluoride to give $Na_2[SiF_6]$, which contains the octahedral $[SiF_6]^{2-}$ ion.

Except for silicon tetrafluoride, silicon halides are extremely sensitive to water. For example, when $SiCl_4$ is exposed to water, it reacts rapidly, and all four chlorine atoms are replaced by hydroxide groups. The overall reaction is

$$SiCl_4 + 4H_2O \longrightarrow Si(OH)_4 + 4HCl$$

Silicic acid, $Si(OH)_4$ or H_4SiO_4, is unstable and gradually dehydrates to SiO_2.

$$Si(OH)_4 \longrightarrow SiO_2 + 2H_2O$$

Germanium is very similar to silicon in its chemical behavior. However, it also forms a series of air- and moisture-sensitive compounds, such as GeO and $GeCl_2$, in which it exhibits an oxidation number of +2.

Antimony generally forms compounds in which it exhibits an oxidation number of +3 or +5. The element tarnishes only slightly in dry air but is readily oxidized when warmed, giving antimony(III) oxide, Sb_4O_6. Antimony(V) oxide must be prepared by indirect means, such as the dehydration of the product of the reaction of antimony(V) chloride with water. The metal reacts readily with stoichiometric amounts of fluorine, chlorine, bromine, or iodine, giving trihalides or, with excess fluorine or chlorine, forming the pentahalides SbF_5 and $SbCl_5$. Depending on the stoichiometry, it forms antimony(III) sulfide, Sb_2S_3, or antimony(V) sulfide when heated with sulfur. It is oxidized, but not dissolved, by hot nitric acid, forming Sb_4O_6. It is slowly dissolved by hot concentrated sulfuric acid, forming $Sb_2(SO_4)_3$ and evolving SO_2. The fact that neither the metal nor its oxides reacts with nitric acid to give the nitrate indicates that antimony is not very metallic. However, the metallic nature of the element is more pronounced than that of arsenic, which lies immediately above it in Group VA.

Tellurium combines directly with most elements, although less readily than the lighter members of Group VIA (oxygen, sulfur, and selenium). The most stable tellurium compounds are the tellurides, salts of Te^{2-} formed with the active metals and the lanthanides, and compounds with oxygen, fluorine, and chlorine, in which tellurium exhibits an oxidation number of +2, +4, or +6. Although tellurium(VI) compounds are known, there is a marked resistance to oxidation to this maximum group oxidation number. The oxidation of tellurium in air gives TeO_2 rather than TeO_3, for example. Tellurium(VI) compounds are good oxidizing agents.

25.2 Structures of Semi-Metals

The crystal structures of the semi-metals are characterized by covalent bonding. In this regard, these elements resemble nonmetals in their behavior.

Elemental silicon, germanium, antimony, and tellurium are lustrous metallic-looking solids. Silicon and germanium crystallize with a diamond-like structure (Fig. 20.6). Each atom within the crystal is covalently bonded to four neighboring atoms at the corners of a regular tetrahedron. Single crystals of silicon and germanium are giant three-dimensional molecules. The crystal structure of antimony is layerlike and is similar to that of black phosphorus (Fig. 20.10). Each antimony atom forms covalent bonds to three adjacent atoms in the layer. Tellurium forms crystals that contain chains of tellurium atoms. Each atom in the chain is bonded to two other atoms.

Pure crystalline boron is transparent. The crystals have a rather unusual structure; they consist of icosahedra (Fig. 25.2), with a boron atom at each corner. In the most common form of boron the icosahedra are packed together in a manner similar to cubic closest packing of spheres. All boron-boron bonds within each icosahedron are identical and are 1.76 Å long. However, there are two kinds of boron-boron bonds between the icosahedra. Each of six of the twelve boron atoms of an icosahedron is joined to a boron atom in adjacent icosahedra by a regular covalent bond between the two atoms that is 1.71 Å long. (These bonds are not shown in Fig. 25.3). Each of the other six atoms is bonded to *two* other atoms, one in each of two other icosahedra, by a bond in which two electrons bond three atoms, forming a bond that is 2.03 Å long (Fig. 25.3).

In Chapters 6 and 7, we discussed bonds in which two atoms are bonded together by one pair of electrons in a bond that results from the overlap of two atomic orbitals. These bonds are called **two-electron two-center bonds.** However, in some situations three atoms are bonded together by one pair of electrons in a molecular orbital formed from the overlap of three atomic orbitals. These bonds are called **two-electron three-center bonds.** As was pointed out in Chapter 7, in a two-electron two-center bond, one bonding molecular orbital and one antibonding molecular orbital result from the over-

Figure 25.2

An icosahedron, a symmetrical solid shape with 20 faces, each of which is an equilateral triangle. The faces meet at 12 corners.

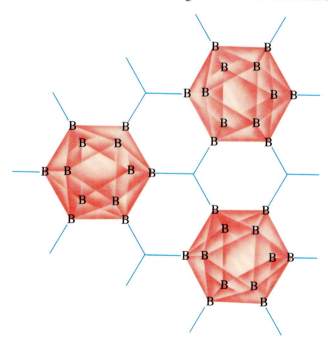

Figure 25.3

The structure of boron. Each icosahedron contains twelve boron atoms. Each of six of these is bonded to a boron atom in another icosahedron by a two-center bond (1.71 Å) and each of the other six is bonded to two boron atoms, each in a separate icosahedron, by a three-center bond (2.03 Å). Only the three-center bonds between icosahedra are shown here.

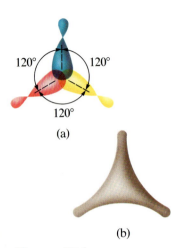

Figure 25.4

The three-center B—B—B bond involving the overlap of three sp^3 hybrid orbitals, which share one pair of electrons. (a) The overlap diagram for the hybrid orbitals. (b) The overall outline of the bonding molecular orbital.

lap of two atomic orbitals (Section 7.1). In a two-electron three-center bond, one bonding molecular orbital and two antibonding molecular orbitals (or one bonding orbital, one nonbonding orbital, and one antibonding orbital) result from the overlap of three atomic orbitals. The two electrons occupy the bonding molecular orbital. The formation of the bonding orbital in elemental boron is shown in Fig. 25.4.

The two-electron three-center bond is found with atoms that do not have sufficient electrons to satisfy ordinary two-center bond requirements. Boron makes the greatest use of the three-center bond; it exhibits this type of bond not only in elemental boron but also in several boron hydrides (Section 25.6).

25.3 Semiconductors, Metals, and Insulators

Semi-metals are intermediate between metals and nonmetals in their chemical properties and in their electrical conductivity. They conduct electricity better than nonmetals (which are insulators), but not as well as metals; they are classified as semiconductors.

Band theory accounts for the variation in electrical conductivity among metals, semiconductors, and insulators. According to this theory, atomic orbitals of the atoms in a crystal combine to yield molecular orbitals that extend through the entire crystal. As the number of atoms (and thus the number of atomic orbitals) increases, the number of molecular orbitals increases (Fig. 25.5). As the number of molecular orbitals, or **energy levels** as they are sometimes called, increases, the difference in energy between them becomes smaller and smaller until there is very little difference in energy between adjacent molecular orbitals. The effective result is a continuous *band* of molecular orbitals, or energy levels, that extends through the entire crystal. There is one energy level in the band for each atomic orbital that participates in forming it. Each of the energy levels in the band can contain two electrons. However, only the electrons in the higher-energy portion of a band are sufficiently free, or mobile, to cause electrical conductivity.

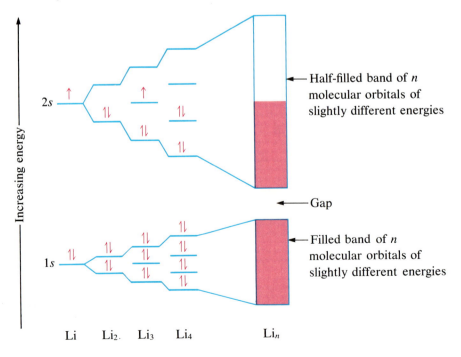

Figure 25.5

Atomic orbitals in a Li atom; molecular orbitals in Li$_2$, Li$_3$, and Li$_4$; and bands in a lithium crystal (Li$_n$).

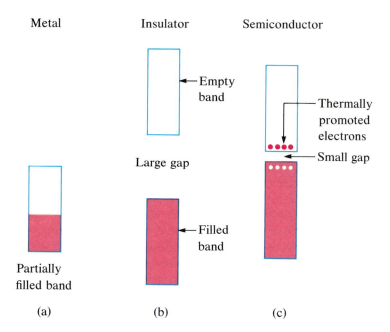

Metal Insulator Semiconductor

Empty band

Thermally promoted electrons

Large gap

Small gap

Filled band

Partially filled band

(a) (b) (c)

Figure 25.6
(a) Partially filled bands are found in metals. (b) Completely filled and completely empty bands separated by a large gap are found in insulators. (c) A completely filled band and a completely empty band separated by a small gap are found in semiconductors at 0 K. When warmed to room temperature, a few electrons are thermally promoted from the previously filled band to the previously empty band, as shown.

To gain a better understanding of band theory, consider the case of lithium, whose electronic structure is $1s^2 2s^1$. A very small crystal of lithium contains about 10^{18} atoms. For this many atoms the energy difference between energy levels is so small that the levels are essentially continuous, and they constitute a band. In a lithium crystal one band arises from the $1s$ atomic orbitals and is fully occupied with electrons. However, the second band, arising from the $2s$ atomic orbitals, is only half-filled with electrons (Fig. 25.5). Between the two bands is a **gap,** a range of energy in which no energy levels, or molecular orbitals, are located. No electrons can be found with these energies, since there are no energy levels for them to occupy.

The spacing of the bands in a substance and their filling determines whether the substance is a conductor, a nonconductor (insulator), or a semiconductor (a poor conductor). A substance such as lithium, which contains partially filled bands [Fig. 25.6(a)], exhibits **metallic conduction.** If the bands are completely filled or completely empty and the energy gap between bands is large [Fig. 25.6(b)], the substance will be an **insulator.** Diamond is an example of such an insulator. A substance that contains a completely filled band and a completely empty band can behave as a **semiconductor** if the energy gap between the filled and empty bands is so small that electrons from the filled band can be promoted to energy levels in the empty band by thermal energy (heat). The previously empty band then contains a few electrons that can conduct an electric current [Fig. 25.6(c)]. Since the number of electrons is much smaller than in a metal, the material is a semiconductor, as is silicon and the other semi-metals. Alternatively, electrons can be provided to the empty band by addition of impurities with extra electrons (for example, arsenic with the valence shell $4s^2 4p^3$, as an impurity in silicon, $3s^3 3p^2$). The extra electrons occupy what would be an empty band in the pure material.

25.4 The Solar Cell

An enormous amount of energy is given off by the sun. The earth receives more energy from sunlight in two days than is stored in all known reserves of fossil fuels. Several devices that convert sunlight into electrical energy are in use or under study. The conventional photoelectric cell, used in electric-eye doors, transforms light into electrical energy (Section 5.1) but delivers as power only about 0.5% of the total light energy it absorbs. Another type of photoelectric device is the **solar cell.** It is about 20 times more efficient than the photoelectric cell and generates electric power from sunlight at the rate of 90 watts per square yard of illuminated surface.

The basic unit of a solar cell is a thin wafer of very pure silicon containing a tiny amount of arsenic. Since arsenic has five valence electrons, compared to four for silicon, replacement of silicon atoms with arsenic atoms creates **n-type silicon** (silicon that contains electrons in what would be an empty band in pure silicon). A thin layer of silicon containing a trace of boron is placed on the surface of the wafer. Since boron has only three valence electrons, replacement of silicon atoms with boron atoms forms **p-type silicon** (silicon that is missing electrons in what would be a filled band in pure silicon). Holes (a few vacant energy levels in an otherwise filled band) thus exist at the surface of the wafer, and a junction, called a p-n junction, exists between the body of the wafer and its thin surface layer. Electrons diffuse through the junction from the wafer to the holes; at the same time holes, or electron vacancies, move to the body of the wafer. This results in a net positive charge within the body of the wafer, which was neutral prior to the diffusion of electrons (negative charges) out of it, and a net negative charge on the surface layer. Thus an electrostatic potential develops between the two regions, until it is just large enough to prevent any further charge migration. When the diffusion of electrons is just balanced by the electrostatic potential, equilibrium is established between the two forces and hence also between electrons and holes. At equilibrium a net difference in potential exists between the two regions.

In a solar cell one electrical lead is attached to the body of the wafer, and one lead to the surface. When the cell is exposed to sunlight, energy from the sunlight causes electrons to be released from their positions in the lattice near the p-n junction, thereby upsetting the equilibrium between electrons and holes and causing additional electrons to move across the junction into the body of the wafer. Thus a sufficient electromotive force is built up to cause electrons to move through the electrical leads, and a current flows through the wire. The device is thus an electric cell, with the positive terminal at the p-contact and the negative terminal at the n-contact (Fig. 25.7).

In actual practice a series of such wafers makes up a solar cell. The first practical application of the solar cell was as a power source for eight telephones on a rural line in Georgia in 1955. Such cells are used to power communication devices in satellites and spacecraft, especially those designed to remain in space for long periods.

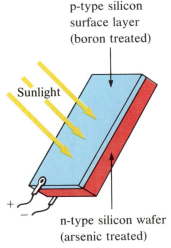

p-type silicon
surface layer
(boron treated)

Sunlight

+
−

n-type silicon wafer
(arsenic treated)

Figure 25.7
A solar cell.

Boron

Boron and its compounds have been known for a long time. The boron-containing compound borax is mentioned in early Latin works on chemistry. Impure elemental boron was first prepared in 1808 by Gay-Lussac and Thénard in France and by Davy in England, by reducing boric acid with potassium.

25.5 Occurrence and Preparation of Boron

Boron constitutes less than 0.001% of the earth's crust. It does not occur in the free state in nature, but is found in compounds with oxygen. Boron is widely distributed in volcanic regions as **boric acid,** $B(OH)_3$ (or H_3BO_3), and in dry lake regions, including the desert areas of California, as **borates,** such as **borax,** $Na_2B_4O_7 \cdot 10H_2O$, **kernite,** $Na_2B_4O_7 \cdot 4H_2O$, and **colemanite,** $Ca_2B_6O_{11} \cdot 5H_2O$.

Reduction of boric oxide with powdered magnesium forms impure boron as a brown amorphous powder.

$$B_2O_3 + 3Mg \longrightarrow 2B + 3MgO$$

The magnesium oxide is removed by dissolving it in hydrochloric acid. Pure boron may be obtained by passing a mixture of the trichloride and hydrogen either through an electric arc or over a hot tungsten filament.

$$2BCl_3(g) + 3H_2(g) \xrightarrow{1500°} 2B(s) + 6HCl(g) \qquad \Delta H° = 253.7 \text{ kJ}$$

25.6 Boron Hydrides

Boron forms a series of volatile hydrides that are quite different from the hydrides of carbon and silicon. The hydride having the molecular formula BH_3 is not stable at room temperature, although Lewis acid-base adducts, such as H_3B—CO, are known. The simplest and most important of the boron hydrides is **diborane, B_2H_6,** the dimer of BH_3. However, an electronic structure cannot be written for B_2H_6 that is consistent with the theory of regular covalent bonds as outlined in Chapter 6 and with the properties of the compound. It would require 14 valence electrons (7 electron pairs) to form the covalent structure

$$
\begin{array}{ccc}
 & H \quad H & \\
 & | \quad\; | & \\
H\!\!-\!\!&B\!\!-\!\!B&\!\!-\!\!H \\
 & | \quad\; | & \\
 & H \quad H &
\end{array}
$$

but only 12 electrons are available for bond formation. (Compounds of this sort are sometimes called electron-deficient.) The structure most in keeping with the properties of diborane is one with two hydrogen bridges (Fig. 25.8). Each boron atom is bonded to two hydrogen atoms by regular two-center single covalent bonds. These four bonds are all in the same plane and use 8 of the 12 valence electrons. A hydrogen above and below the plane is connected to the two boron atoms by a two-electron three-center B—H—B bond. The three atoms in each bond are bonded by a molecular orbital that contains one pair of electrons, analogous to the B—B—B three-center bond in elemental boron. The two three-center bonds hold the remaining 4 electrons.

The reaction of lithium aluminum hydride with boron trifluoride in ether solution yields diborane.

$$4BF_3 + 3LiAlH_4 \longrightarrow 2B_2H_6(g) + 3LiF + 3AlF_3$$

Above 300° diborane rapidly decomposes into boron and hydrogen. It ignites spontaneously in moist air and forms boric oxide and water. Diborane reacts violently with

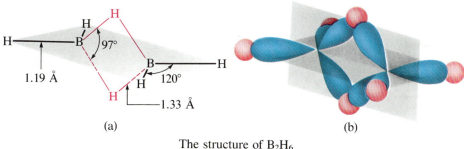

(a)

The structure of B_2H_6

(b)

Figure 25.8

The structure of the diborane molecule, B_2H_6. (a) The spatial arrangement of atoms showing the bridging hydrogens (red). (b) The spatial arrangement of the bonding orbitals.

chlorine to form boron trichloride and hydrogen chloride. Diborane is not an acidic hydride; it reacts with water, forming boric acid and hydrogen.

$$B_2H_6(g) + 6H_2O(l) \longrightarrow 2H_3BO_3(s) + 6H_2(g) \qquad \Delta H° = -509.2 \text{ kJ}$$

Hydrides of higher molecular weight than B_2H_6 are possible, and many have been made. Examples include B_4H_{10}, B_5H_9, B_6H_{10}, and $B_{10}H_{14}$.

When an excess of sodium hydride reacts with boron trifluoride, **sodium borohydride,** $NaBH_4$, is formed. The four hydrogen atoms in the tetrahedral anion $[BH_4]^-$ are covalently bonded to the boron atom.

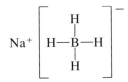

Sodium borohydride

The formal oxidation number of the hydrogen atoms in the BH_4^- ion is -1. Salts containing the borohydride ion or other similar complex hydrides, such as the aluminum hydride ion, AlH_4^-, are very useful reducing agents, particularly for organic compounds.

25.7 Boron Halides

Boron trifluoride, trichloride, and tribromide can be prepared by direct union of the elements. These trigonal planar molecules contain boron with sp^2 hybridization. The fluoride is a colorless gas, the trichloride is a colorless mobile liquid, the bromide is a viscous liquid, and the iodide is a white crystalline solid.

The heavier boron trihalides hydrolyze in water to form boric acid and the corresponding hydrohalic acid. Boron trichloride reacts according to the equation

$$BCl_3(g) + 3H_2O(l) \longrightarrow H_3BO_3(aq) + 3HCl(aq)$$

When boron trifluoride is added to hydrofluoric acid, it reacts and gives **fluoroboric acid,** HBF_4.

$$BF_3(aq) + HF(aq) \longrightarrow H^+(aq) + BF_4^-(aq)$$

In the latter reaction the BF_3 molecule acts as a Lewis acid (electron-pair acceptor) and accepts a pair of electrons from a fluoride ion, as shown by the equation

$$\ddot{:}\!\ddot{F}\!\ddot{:}^- \; + \; \begin{matrix} :\ddot{F}: \\ B:\ddot{F}: \\ :\ddot{F}: \end{matrix} \longrightarrow \begin{bmatrix} :\ddot{F}: \\ :\ddot{F}:B:\ddot{F}: \\ :\ddot{F}: \end{bmatrix}^-$$

Fluoroboric acid is a strong acid that is stable only in solution.

25.8 Boron Oxide and Oxyacids of Boron

Boron burns at 700°C in oxygen, forming **boric oxide,** B_2O_3. Boric oxide can be obtained as crystals, but when melted and allowed to solidify it forms a glass

Figure 25.9
Pyrex is a glass prepared using boron oxide. It is used in laboratory glassware because it does not break when heated.

(Fig. 11.2). Boric oxide is used in the production of Pyrex glass (Fig. 25.9) and certain optical glasses. It dissolves in hot water to form **boric acid,** H_3BO_3.

$$B_2O_3 + 3H_2O \longrightarrow 2H_3BO_3$$

The boron atom in H_3BO_3 is at the center of an equilateral triangle with oxygen atoms at the corners.

H
O
120° B 120°
O 120° O
H H

Boric acid

In the solid acid these triangular units are held together by hydrogen bonding.

When boric acid is heated to 100°C, a molecule of water is split out between a pair of adjacent OH groups to form **metaboric acid,** HBO_2. With further heating at about 150° additional B—O—B linkages form, connecting the BO_3 groups together with shared oxygen atoms to form **tetraboric acid,** $H_2B_4O_7$. At still higher temperatures, boric oxide is formed.

$$H_3BO_3 \longrightarrow HBO_2 + H_2O$$
$$4HBO_2 \longrightarrow H_2B_4O_7 + H_2O$$
$$H_2B_4O_7 \longrightarrow 2B_2O_3 + H_2O$$

The formation of oxides and oxyanions with bridging oxygen atoms is one of the characteristic properties of the chemistry of the semi-metals. Silicon forms an extensive array of bridged oxyanions (Section 25.16).

25.9 Borates

Borates are salts of the oxyacids of boron. Borates result from the reaction of a base with an oxyacid or from the fusion of boric acid or boric oxide with a metal oxide or hydroxide. Borate anions range from the simple trigonal planar BO_3^{3-} ion to complex species containing chains and rings of three- and four-coordinated boron

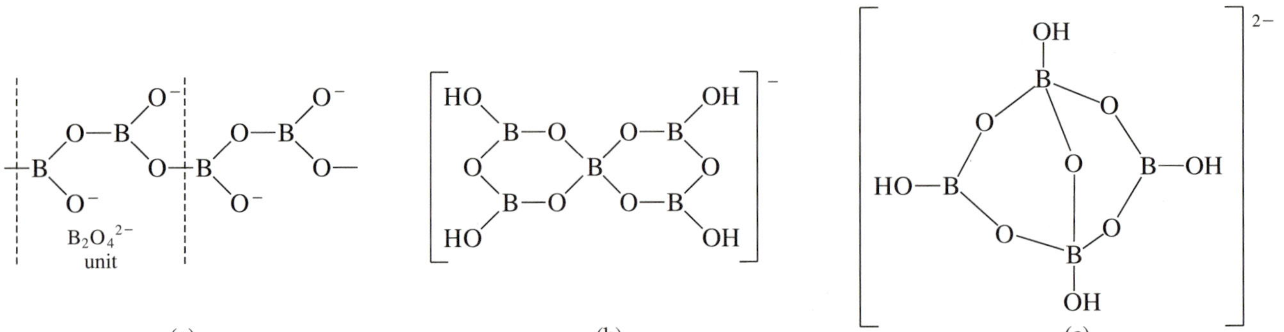

Figure 25.10
The borate anions found in (a) CaB_2O_4, (b) $KB_5O_8 \cdot 4H_2O$, and (c) $Na_2B_4O_7 \cdot 10H_2O$. The anion in CaB_2O_4 is an infinite chain.

atoms. The structures of the anions found in CaB_2O_4, $K[B_5O_6(OH)_4] \cdot 2H_2O$ (commonly written $KB_5O_8 \cdot 4H_2O$), and $Na_2[B_4O_5(OH)_4] \cdot 8H_2O$ (commonly written $Na_2B_4O_7 \cdot 10H_2O$) are shown in Fig. 25.10.

Commercially, the most important borate is **borax, or sodium tetraborate 10-hydrate, $Na_2B_4O_7 \cdot 10H_2O$.** Most of the supply of borax comes directly from dry lakes, such as Searles Lake in California, or is prepared from kernite, $Na_2B_4O_7 \cdot 4H_2O$. Borax is a salt of a strong base and a weak acid, so its aqueous solutions are basic due to the hydrolysis of the tetraborate ion.

$$B_4O_7{}^{2-} + 7H_2O \rightleftharpoons 4H_3BO_3 + 2OH^-$$

Borax is used to soften water and to make washing compounds. These uses depend on the alkaline character of its solutions and the insolubility of the borates of calcium and magnesium.

When heated, borax fuses to form glass that dissolves metal oxides. An example is given by the equation

$$Na_2B_4O_7 + CuO \longrightarrow 2NaBO_2 + Cu(BO_2)_2$$

Because of this property, borax is used as a flux to remove oxides from metal surfaces in soldering and welding. Different metals form borax glasses of different colors, a fact applied to detecting certain metals. Cobalt borax glass is blue, for example.

25.10 Boron-Nitrogen Compounds

A number of compounds containing boron-nitrogen bonds have been prepared and studied. Two of particular interest are **boron nitride, BN,** and **borazine, $B_3N_3H_6$.**

Boron nitride is the final product of the thermal decomposition of many boron-nitrogen compounds, such as $B(NH_2)_3$ and $BF_3 \cdot NH_3$. It can also be prepared by heating boron with nitrogen or ammonia or by heating borax with ammonium chloride. Crystalline boron nitride is a white solid that sublimes somewhat below 3000°C; it melts at this temperature under pressure. It is inert to most reagents but can be decomposed by reaction with molten bases. The crystalline structure of boron nitride is analogous to that of graphite (Section 20.3 and Fig. 20.6)—layers of atoms in both are made up of hexagonal rings.

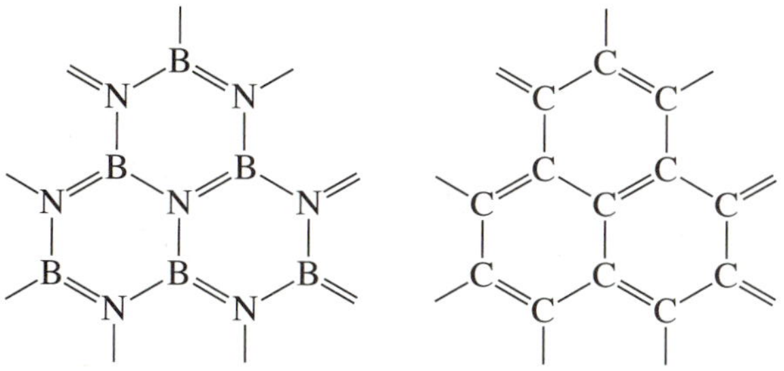

Boron nitride
layer

Graphite
layer

This form of boron nitride is sometimes called **inorganic graphite.**

A second form of boron nitride, **Borazon,** has the same structure as diamond (Section 20.3 and Fig. 20.6) and is almost as hard as diamond. It is now produced commercially.

Borazine, $B_3N_3H_6$, and hydrogen are formed when ammoniates of the boron hydrides, such as $BH_3 \cdot NH_3$, are heated at 180–200°C. Borazine has been called **inorganic benzene,** because its structure and physical properties are similar to those of benzene. However, its chemistry is different, because the B—N bonds are polar and the C—C bonds are not.

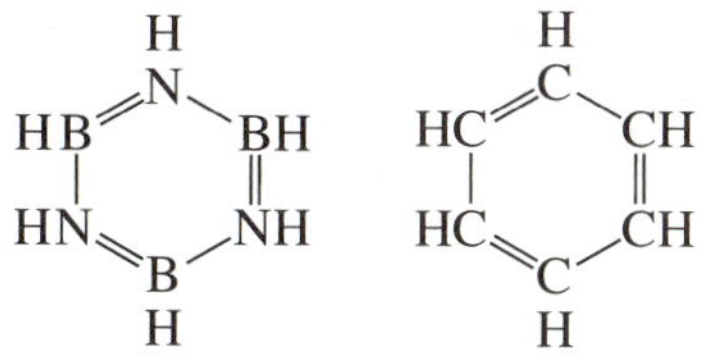

Borazine

Benzene

Since a boron atom and a nitrogen atom bonded together have the same number of electrons as two carbon atoms, and the sizes of these two groups are about the same, it is not surprising that the structures of boron nitride are similar to those of graphite and diamond and that borazine resembles benzene.

Silicon

Both carbon and silicon are members of Group IVA. Carbon, with its ability to form compounds containing carbon-carbon bonds, plays the dominant structural role in the animal and vegetable worlds, and silicon, which readily forms compounds containing Si—O—Si bonds, is of prime importance in the mineral world. The name *silicon* is derived from the Latin word for flint, *silex.*

25.11 Occurrence and Preparation of Silicon

Silicon composes nearly one-fourth of the mass of the earth's crust— second only to oxygen. The crust is composed almost entirely of minerals in which silicon atoms are connected by oxygen atoms in complex structures involving chains, layers, and three-dimensional frameworks. These minerals constitute the bulk of most common rocks (except limestone and dolomite), soils, clays, and sands. Sand and sandstone are forms of impure silicon dioxide, as are quartz, amethyst, agate, and flint. Most rocks are built up of the common metal cations and silicate anions. Materials such as granite, bricks, cement, mortar, ceramics, and glasses are composed of silicon compounds. A great variety of silicates exists in nature, and many of these have important specific uses. However, they are so stable that it is not economical to use them as sources of metals (except for beryl, $Be_3Al_2Si_6O_{18}$, from which beryllium is obtained).

Elemental silicon was first prepared in an impure form by Berzelius in 1823, by heating silicon tetrafluoride with potassium. It can also be obtained by the reduction of silicon dioxide with strong reducing agents at high temperatures. With carbon and magnesium as the reducing agents, the equations are

$$SiO_2(s) + 2C(s) \longrightarrow Si(s) + 2CO(g)$$
$$SiO_2(s) + 2Mg(s) \longrightarrow Si(s) + 2MgO(s)$$

Extremely pure silicon, such as that required for the fabrication of semiconductor electronic devices, is prepared by the decomposition of silicon tetrahalides or silane, SiH_4, at high temperatures.

Highly purified silicon, containing no more than one part impurity per million parts silicon, is the most important element in the computer industry (Fig. 25.11). It is used in semiconductor electronic devices such as transistors, microcomputers, and solar cells (Fig. 25.7). Elemental silicon is also used as a deoxidizer in the production of steel, copper, and bronze and in the manufacture of acid-resistant iron alloys.

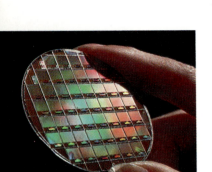

Figure 25.11
Computer chips are manufactured on thin slices of very pure crystalline silicon.

25.12 Silicon Hydrides

Silicon forms a series of hydrides, including SiH_4, Si_2H_6, Si_3H_8, Si_4H_{10}, and Si_6H_{14}. These compounds contain Si—H and Si—Si single bonds. The chemical behavior of these silicon hydrides is decidedly different from that of the hydrocarbons (compounds of carbon and hydrogen) of similar formulas. For example, the silicon hydrides inflame spontaneously in air, whereas the hydrocarbons do not.

Acids react with magnesium silicide to form **silane,** SiH_4, with a structure and formula analogous to methane, CH_4.

$$Mg_2Si + 4H^+ \longrightarrow 2Mg^{2+} + SiH_4(g)$$

Silane, a colorless gas, is thermally stable at ordinary temperatures but spontaneously inflames when exposed to air. The products of its oxidation are silicon dioxide and water.

$$SiH_4(g) + 2O_2(g) \longrightarrow SiO_2(s) + 2H_2O(g) \qquad \Delta H° = -1429 \text{ kJ}$$

The hydrogen atoms in silane can be replaced one at a time by halogen atoms by reaction of silane with hydrogen halides, using the corresponding aluminum halide as a catalyst. With HBr, using $AlBr_3$ as the catalyst, the reactions are given by the equations

$$SiH_4 + HBr \longrightarrow SiH_3Br + H_2$$
$$SiH_3Br + HBr \longrightarrow SiH_2Br_2 + H_2$$
$$SiH_2Br_2 + HBr \longrightarrow SiHBr_3 + H_2$$
$$SiHBr_3 + HBr \longrightarrow SiBr_4 + H_2$$

Silane is extremely sensitive to hydroxides, reacting easily to give silicates and hydrogen.

$$SiH_4 + 2OH^- + H_2O \longrightarrow SiO_3^{2-} + 4H_2(g)$$

In contrast, methane, CH_4, is unreactive to hydroxides.

Perhaps the most important reaction of compounds containing an Si—H bond, at least from a commercial standpoint, is the reaction with hydrocarbons such as propene, $CH_3CH{=}CH_2$, that contain carbon-carbon double bonds.

$$CH_3CH{=}CH_2 + H_2SiCl_2 \longrightarrow CH_3CH_2CH_2SiHCl_2$$

$$CH_3CH{=}CH_2 + CH_3CH_2CH_2SiHCl_2 \longrightarrow (CH_3CH_2CH_2)_2SiCl_2$$

These reactions are used in the preparation of silicones (Section 25.17).

25.13 Silicon Carbide

When a mixture of sand and a large excess of coke is heated in an electric furnace, **silicon carbide (carborundum), SiC,** is produced according to the equation

$$SiO_2(s) + 3C(s) \longrightarrow SiC(s) + 2CO(g) \qquad \Delta H° = 624.7 \text{ kJ}$$

The product is blue-black, iridescent crystals, nearly as hard as diamonds and very stable at high temperatures. The crystals are crushed, and the particles are screened to uniform size, mixed with a binder of clay or sodium silicate, molded into various shapes such as grinding wheels, and fired. Silicon carbide is used as an abrasive for cutting, grinding, and polishing.

Silicon carbide exists in many different crystalline forms; in all of these forms each atom is surrounded tetrahedrally by four of the other kind. One crystalline form has a structure like that of diamond (Section 20.3 and Fig. 20.6). To rupture a crystal of silicon carbide, a number of very strong covalent bonds must be broken. The high decomposition temperature (above 2200°C), the extreme hardness, the brittleness, and the chemical inactivity of silicon carbide are due to these strong covalent bonds.

25.14 Silicon Halides

All the tetrahalides of silicon, SiX_4, and several mixed halides of the type $SiCl_2F_2$ have been prepared. **Silicon tetrachloride** can be prepared by direct chlorination at elevated temperatures or by heating silicon dioxide with chlorine and carbon. The equations are

$$Si(s) + 2Cl_2(g) \longrightarrow SiCl_4(g)$$

$$SiO_2(s) + 2C(s) + 2Cl_2(g) \longrightarrow SiCl_4(g) + 2CO(g)$$

Silicon tetrachloride is a covalent tetrahedral molecule containing four covalent Si—Cl bonds. It is a low-boiling (57°C) colorless liquid that fumes strongly in moist air to produce a dense smoke of finely divided silica as the Si—Cl bonds are replaced by Si—O bonds.

$$SiCl_4(g) + 2H_2O(g) \longrightarrow SiO_2(s) + 4HCl(g)$$

The Si—Cl bonds in silicon tetrachloride can be replaced by Si—C bonds in a stepwise fashion by reaction with a stoichiometric amount of $Na^+CH_3^-$ or other compounds that contain carbon anions.

$$SiCl_4 + Na^+CH_3^- \longrightarrow Cl_3SiCH_3 + NaCl$$

$$SiCl_4 + 2ClMgC_2H_5 \longrightarrow Cl_2Si(CH_2CH_3)_2 + 2MgCl_2$$

Elemental silicon ignites spontaneously in an atmosphere of fluorine, forming gaseous **silicon tetrafluoride,** SiF_4. The reaction of hydrofluoric acid with silica or a silicate also produces SiF_4.

$$SiO_2(s) + 4HF(g) \longrightarrow SiF_4(g) + 2H_2O(l) \qquad \Delta H° = -191.2 \text{ kJ}$$

$$CaSiO_3(s) + 6HF(g) \longrightarrow SiF_4(g) + CaF_2 + 3H_2O$$

Silicon tetrafluoride hydrolyzes in water and produces **fluorosilicic acid** as well as **orthosilicic acid.**

$$3SiF_4 + 4H_2O \longrightarrow \underset{\substack{\text{Orthosilicic} \\ \text{acid}}}{H_4SiO_4(s)} + 4H^+ + \underset{\substack{\text{Fluorosilicic} \\ \text{acid}}}{2SiF_6{}^{2-}}$$

Fluorosilicic acid is a stronger acid than sulfuric acid. It is stable only in solution, however, and upon evaporation decomposes according to the equation

$$H_2SiF_6(aq) \longrightarrow 2HF(g) + SiF_4(g)$$

The difference in the reactivity of SiF_4 and $SiCl_4$ with water can be attributed to the strengths of Si—O, Si—Cl, and Si—F bonds. These bonds are particularly strong. Exposure of $SiCl_4$ and most other silicon compounds to water or oxygen results in their decomposition to compounds containing Si—O bonds, unless the compounds are stabilized by the presence of Si—O, Si—C, or Si—F bonds.

The valence shell of silicon contains d orbitals and can accommodate more than eight electrons. Thus silicon compounds of the general formula SiX_4, where X is a highly electronegative group, can act as Lewis acids giving six-coordinate silicon (six bonds) with d^2sp^3 hybridization. For example, silicon tetrafluoride, SiF_4, reacts with hydrofluoric acid to give a solution of $H_2[SiF_6]$, which contains the octahedral $[SiF_6]^{2-}$ ion.

25.15 Silicon Dioxide (Silica)

Silicon dioxide, silica, is found in both crystalline and amorphous forms. The usual crystalline form of silicon dioxide is **quartz,** a hard, brittle, clear, colorless solid. It is used in many ways—for architectural decorations, semiprecious jewels, optical instruments, and frequency control in radio transmitters. The contrast in structure and physical properties between silicon dioxide and its carbon analog, carbon dioxide, is interesting. Solid carbon dioxide (dry ice) contains single CO_2 molecules, with very weak intermolecular forces holding them together in the crystal. The low melting point and volatility of dry ice reflect these weak forces between molecules. Each of the two oxygen atoms is attached to the central carbon atom by double bonds. In contrast, silicon dioxide does not form double bonds. Hence in silicon dioxide each silicon atom is linked to four oxygen atoms by single bonds directed toward the corners of a regular tetrahedron, and SiO_4 tetrahedra share oxygen atoms. This structure gives a three-dimensional, continuous silicon-oxygen network. A quartz crystal is a macromolecule of silicon dioxide (Fig. 11.22).

At 1600°, quartz melts to give a viscous liquid with a random internal structure. When the liquid is cooled, it does not crystallize readily but usually supercools and forms a glass, also called **silica.** The SiO_4 tetrahedra in glassy silica have the random arrangement characteristic of supercooled liquids, and the glass has some very useful properties. Silica is highly transparent to both visible and ultraviolet light. It is used in lamps that give radiation rich in ultraviolet light and in certain optical instruments that operate with ultraviolet light. The coefficient of expansion of silica glass is very low, so it is not easily fractured by sudden changes in temperature. It is insoluble in water and inert toward all acids except hydrofluoric acid.

$$SiO_2(s) + 4HF(aq) \longrightarrow SiF_4(g) + 2H_2O$$

This reaction is used in the quantitative separation of silica from other oxides, since both products are volatile. Hot solutions of metal hydroxides and fused metal carbonates convert silica into soluble silicates (SiO_4^{4-} and SiO_3^{2-}). Typical examples are

$$SiO_2 + 4OH^- \longrightarrow SiO_4^{4-} + 2H_2O$$
$$SiO_2 + Na_2CO_3 \longrightarrow Na_2SiO_3 + CO_2$$

The latter reaction is employed in the conversion of silicate rocks to soluble forms for analysis and is part of the glassmaking process (Section 25.18).

25.16 Natural Silicates

Silicates are salts containing anions composed of silicon and oxygen. There are many types of silicates, because the silicon-to-oxygen ratio can vary widely. In all silicates, however, sp^3 hybridized silicon atoms are found at the centers of tetrahedra with oxygen at the corners, and silicon is tetravalent. The variation in the silicon-to-oxygen ratio occurs because the silicon-oxygen tetrahedra may exist as discrete, independent units or may share oxygen atoms at corners, edges, or more rarely faces, in a variety of ways. The silicon-to-oxygen ratio varies according to the extent of sharing of oxygen atoms by silicon atoms in the linking together of the tetrahedra.

It is convenient to classify the silicates into groups, based on how the silicon-oxygen tetrahedra are linked.

1. Individual SiO_4^{4-} tetrahedra exist as independent groups in some minerals. Examples include **olivine,** Mg_2SiO_4, and **zircon,** $ZrSiO_4$. Positively charged metallic ions (Mg^{2+}, Zr^{4+}) bind together the negatively charged SiO_4^{4-} ions, which have the tetrahedral structure shown in Fig. 25.12. (Note that only the oxygen atoms are shown in the figure; a silicon atom is in the center of each tetrahedron but is not shown.)

2. Two SiO_4 tetrahedra share one corner oxygen and form discrete $Si_2O_7^{6-}$ ions (Fig. 25.13) in **hardystonite, $Ca_2ZnSi_2O_7$,** and **hemimorphite, $Zn_4(OH)_2Si_2O_7 \cdot H_2O$,** for example. The cations are between the $Si_2O_7^{6-}$ ions and bind them together.

3. Three SiO_4 tetrahedra share corners and form closed rings in **benitoite, $BaTiSi_3O_9$.** The $Si_3O_9^{6-}$ ions (Fig. 25.14) are held together by the metal cations. The rings are arranged in sheets with their planes parallel. Six SiO_4 tetrahedra share corners to form a closed ring in **beryl (emerald), $Be_3Al_2Si_6O_{18}$.**

4. Chains of SiO_4 tetrahedra in which each tetrahedron shares two oxygen atoms are present in **diopside, $CaMg(SiO_3)_2$,** and other minerals. Such a chain (Fig. 25.15) has an empirical formula of SiO_3^{2-}, but no SiO_3^{2-} ions are present as independent

Figure 25.12
The tetrahedral structure of the SiO_4^{4-} ion. (Only the oxygen atoms are shown; the silicon atom, not shown, is at the center.)

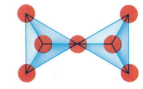

Figure 25.13
The structure of the $Si_2O_7^{6-}$ ion. (A silicon atom, not shown, is at the center of each tetrahedron.)

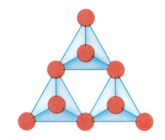

Figure 25.14
The structure of the $Si_3O_9^{6-}$ ion. (A silicon atom, not shown, is at the center of each tetrahedron.)

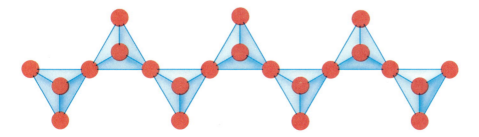

Figure 25.15
A portion of an SiO_3^{2-} chain. (A silicon atom, not shown, is at the center of each tetrahedron.)

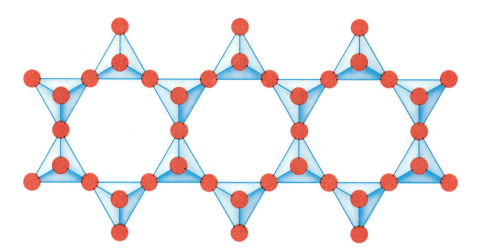

Figure 25.16

A portion of a double silicon-oxygen chain. (A silicon atom, not shown, is at the center of each tetrahedron.)

entities. Parallel chains extend the full length of the crystal and are held together by the positively charged metal ions lying between them.

5. Double silicon-oxygen chains form when SiO_4 tetrahedra in single chains share oxygen atoms (Fig. 25.16). Metal cations link the parallel chains together. The fact that these ionic bonds are not as strong as the silicon-oxygen bonds within the chains accounts for the fibrous nature of many of these minerals. **Asbestos** is a chain-type silicate.

6. Silicon-oxygen sheets are formed by sharing of oxygen atoms between double chains. Metal ions form ionic bonds between the sheets. These ionic bonds are weaker than the silicon-bonds within the sheets; thus minerals with this structure cleave into thin layers. Examples include the micas **talc, $Mg_3Si_4O_{10}(OH)_2$,** and **muscovite, $KAl_2(AlSi_3O_{10})(OH,F)_2$.** Note that aluminum atoms can substitute for silicon atoms, as in muscovite. When this occurs, a larger charge on the cations is required to produce a neutral entity.

7. Three-dimensional silicon-oxygen networks with some of the tetravalent silicon replaced by trivalent aluminum are also observed. The negative charge that results is neutralized by a distribution of positive ions throughout the network. Examples are **feldspar, $K(AlSi_3O_8)$,** and the **zeolites,** such as **$Na_2(Al_2Si_3O_{10}) \cdot 2H_2O$.**

25.17 Silicones

Polymeric organosilicon compounds containing Si—O—Si linkages and Si—C bonds are known as **silicones.** Because these compounds contain strong Si—O and Si—C bonds, they are generally very stable. Silicones have the general formula $(R_2SiO)_x$ and may be linear, cyclic, or cross-linked polymers of the types shown.

Linear silicone

Cyclic silicone

Cross-linked silicone

The R in the formulas represents methyl, $-CH_3$, ethyl, $-C_2H_5$, phenyl, $-C_6H_5$, or some other hydrocarbon group. The linear and cyclic silicones are produced by hydrolyzing organochlorosilanes of the type R_2SiCl_2, followed by polymerization through the elimination of a molecule of water from two hydroxyl groups of adjacent $R_2Si(OH)_2$ molecules.

$$R_2SiCl_2 + 2H_2O \longrightarrow R_2Si(OH)_2 + 2HCl$$

$$\underset{R}{\overset{R}{HO-Si-O}} \; \overline{[H + HO]} \underset{R}{\overset{R}{-Si-OH}} \longrightarrow \underset{R}{\overset{R}{HO-Si-O}}\underset{R}{\overset{R}{-Si-OH}} + H_2O$$

Organosilicon polymers incorporate some of the properties of both hydrocarbons and oxysilicon compounds. They are remarkably stable toward heat and chemical reagents and are water-repellent. Depending on the extent of polymerization and the molecular complexity, silicones may be oils, greases, rubberlike substances, or resins. They are used as lubricants, hydraulic fluids, electrical insulators, and moisture-proofing agents. The viscosity of silicone oils changes so little as the temperature changes that they can be employed as lubricants where there are extreme variations in temperature.

Paper, wool, silk, and other fabrics can be coated with a water-repellent film by exposing them for a second or two to the vapor of **trimethylchlorosilane, $(CH_3)_3SiCl$.** The $-OH$ groups on the surface of the material react, and the surface becomes coated with a thin film (monolayer) of $(CH_3)_3Si-O-$ groups. This layer repels water because it is similar to a hydrocarbon film.

$$(Surface)-OH + Cl-Si(CH_3)_3 \longrightarrow (Surface)-O-Si(CH_3)_3 + HCl$$

25.18 Glass

Glass is a supercooled liquid consisting of a complex mixture of silicates. It is transparent, brittle, and entirely lacking in the ordered internal structure characteristic of crystals. When heated, it does not melt sharply. Instead it gradually softens until it becomes liquid.

The glass used for windowpanes, bottles, dishes, and so forth is a mixture of sodium silicates and calcium silicates with an excess of silica. It is made by heating together sand, sodium carbonate (or sodium sulfate), and calcium carbonate.

$$Na_2CO_3 + SiO_2 \longrightarrow Na_2SiO_3 + CO_2(g)$$
$$Na_2SO_4 + SiO_2 \longrightarrow Na_2SiO_3 + SO_3(g)$$
$$CaCO_3 + SiO_2 \longrightarrow CaSiO_3 + CO_2(g)$$

After the bubbles of gas have been expelled, a clear viscous melt results. This is poured into molds or stamped with dies to produce pressed glassware. Articles such as bottles, flasks, and beakers are formed by taking a lump of molten glass on a hollow tube, inserting it into a mold and blowing with compressed air until the glass assumes the outline of the mold. Plate glass (float glass) is made by pouring molten glass on a layer of very pure molten tin. Since the molten tin surface is perfectly smooth, the glass floating on it is also perfectly smooth and does not need to be ground and polished after hardening.

Glassware is annealed by heating it for a time just below its softening temperature and then cooling slowly. Annealing lessens internal strains and thereby reduces the chances of breakage from shock or temperature change.

If sodium is replaced by potassium in a glass melt, a higher-melting, harder, and less-soluble glass is obtained. If part of the calcium is replaced by lead, a glass of high density and high refractive index is formed. This variety of glass is called **flint glass** and is used in making lenses and cut-glass articles. **Pyrex glass,** which is used for test tubes, flasks, and other laboratory glassware because it is resistant to sudden changes in temperature and to chemical action, is a borosilicate glass (some of the silicon atoms are replaced by boron atoms).

One form of **safety glass** used in automobiles contains a thin layer of plastic held between two pieces of thin plate glass. Adhesion of the glass to the flexible plastic reduces the danger from flying glass and jagged edges if it is broken.

Enamels on kitchen utensils, sinks, and bathtubs and glazes on pottery are made of easily melted glass containing opacifiers (opaquing agents), such as titanium dioxide and tin(IV) oxide.

25.19 Cement

Portland cement is essentially powdered calcium aluminosilicate, which sets to a hard mass when treated with water. It is made by pulverizing a mixture of limestone ($CaCO_3$) and clay (an aluminosilicate) and then roasting the powder in a rotary kiln at about 1500°C. This treatment yields marble-sized lumps called *clinker,* which are ground with a little gypsum, $CaSO_4 \cdot 2H_2O$, to a very fine powder.

Concrete is made by adding water to a mixture of Portland cement, sand, and stone or gravel. The reactions taking place as the portland cement mixture sets (starts to harden to form concrete) are complex and not completely understood. It is known that calcium aluminate hydrolyzes, forming calcium hydroxide and aluminum hydroxide. These compounds then react with calcium silicates, forming calcium aluminosilicate in the form of interlocking crystals. Portland cement sets rapidly (within 24 hours) and then hardens slowly over a period of years.

For Review

Summary

The elements boron, silicon, germanium, antimony, and tellurium separate the metals from the nonmetals in the Periodic Table. These elements, called **semi-metals,** or sometimes **metalloids,** exhibit properties characteristic of both metals and nonmetals. Their electronegativities are less than that of hydrogen, but they do not form positive ions. The structures of these elements are similar to those of nonmetals, but the elements are electrical conductors. The electrical conductivity can be explained by **band theory.**

Boron exhibits some metallic characteristics, but the large majority of its chemical behavior is that of a nonmetal that exhibits an oxidation state of $+3$ in its compounds (although other oxidation states are known). Its valence shell configuration is $2s^2 2p^1$, so it forms trigonal planar compounds with three single covalent bonds. The resulting compounds are Lewis acids, since the unhybridized p orbital does not contain an electron pair. The most stable boron compounds are those that contain oxygen and fluorine. Diborane, the simplest stable boron hydride, contains **two-electron three-center bonds,** as does elemental boron.

Silicon is a semi-metal with a valence shell configuration of $3s^2 3p^2 3d^0$. It commonly forms tetrahedral compounds in which silicon exhibits an oxidation state of $+4$. Although the d orbital is unfilled in four coordinate silicon compounds, its presence makes silicon compounds much more reactive than the corresponding carbon compounds. Silicon forms strong single bonds with carbon, giving rise to the stability of silicon carbide and **silicones.** Silicon also forms strong bonds to oxygen and fluorine. **Silicates** contain oxyanions of silicon and are important components of minerals and glass.

Elemental silicon is a semiconductor that is used in **solar cells.**

Key Terms and Concepts

band (25.3)	metal (25.3)	three-center bond (25.2)
energy level (25.3)	semiconductor (25.3)	two-center bond (25.2)
insulator (25.3)	solar cell (25.4)	

Exercises

1. What are the common oxidation numbers exhibited by each of the semi-metals?
2. Predict the products of the reactions of germanium with an excess of each of the following: F_2, O_2, S_2, and Cl_2.
3. Predict the products of the reaction of antimony with an excess of each of the following: F_2, S, Cl_2, H_2SO_4.
4. Predict the products of the reactions of tellurium with an excess of each of the following: F_2, O_2, Na, and Al.
5. Why are boron compounds generally good Lewis acids?
6. How does the Lewis acid behavior of silicon tetrafluoride differ from that of boron trifluoride?
7. The bonding in crystalline boron is more like that of a metal than a nonmetal, whereas the bonding in crystalline tellurium is more like that of a nonmetal than a metal. Explain this statement.
8. Which of the following crystalline materials contain partially filled bands? Na, S_8, NaCl, Mn, MgO.
9. Explain why the partially filled band in lithium is exactly half filled.
10. The bands in both insulators and some semiconductors are either filled or empty at 0 K. Explain why semiconductors will carry an electrical current as the temperature increases, whereas insulators will not.

Boron

11. What is the electron configuration of a boron atom?
12. Why is boron limited to a maximum coordination number of 4 in its compounds?
13. Describe the molecular structure and the hybridization of boron in each of the following molecules or ions:
    (a) BCl_3 (b) BH_4^- (c) B_2H_6
    (d) $B(OH)_3$ (e) $B(OH)_4^-$ (f) H_3BCO
14. Write a Lewis structure for each of the following molecules or ions:

(a) BCl_3 (b) BH_4^- (c) $B(OH)_3$
(d) $B(OH)_4^-$ (e) H_3BCO

15. Write two equations for reactions in which boron exhibits metallic behavior and two equations for reactions in which boron exhibits nonmetallic behavior.
16. Write balanced equations describing the reactions of boron with elemental fluorine, carbon, oxygen, and nitrogen, respectively.
17. Why is B_2H_6 said to be an electron-deficient compound?
18. Identify the oxidizing agent and the reducing agent in the chemical reaction for the preparation of diborane from boron trifluoride and lithium aluminum hydride.
19. Write equations to show the formation of fluoroboric acid from boron trifluoride.
20. Show by equations the relationship of the boric acids to boric oxide.
21. Why does an aqueous solution of borax turn red litmus blue?
22. Explain the chemistry of the use of borax as a flux in soldering and welding.
23. Why is boron nitride sometimes referred to as inorganic graphite?
24. Why should the physical properties of borazine be similar to those of benzene?
25. Complete and balance the following equations. (In some instances, two or more reactions may be correct, depending on the stoichiometry.)
    (a) $B + Cl_2 \longrightarrow$
    (b) $B + S_8 \longrightarrow$
    (c) $B + SO_2 \longrightarrow$
    (d) $B_2H_6 + NaH \longrightarrow$
    (e) $BCl_3 + C_2H_5OH \longrightarrow$
    (f) $HBO_2 + H_2O \longrightarrow$
    (g) $B_2O_3 + CuO \longrightarrow$

26. A 0.7849-g sample of boron was oxidized in air, giving 2.5274 g of B_2O_3. From this information and the atomic weight of oxygen, calculate the atomic weight of boron.

Ans. 10.81

27. From the data given in Appendix I, determine the standard enthalpy change and standard free energy change for each of the following reactions:
    (a) $BF_3(g) + 3H_2O(l) \longrightarrow B(OH)_3(s) + 3HF(g)$

    Ans. $\Delta H° = 87$ kJ; $\Delta G° = 44$ kJ
    (b) $BCl_3(g) + 3H_2O(l) \longrightarrow B(OH)_3(s) + 3HCl(g)$

    Ans. $\Delta H° = -109.9$ kJ; $\Delta G° = -154.7$ kJ
    (c) $B_2H_6(g) + 6H_2O(l) \longrightarrow 2B(OH)_3(s) + 6H_2(g)$

    Ans. $\Delta H° = -510$ kJ; $\Delta G° = -601.5$ kJ

Silicon

28. What is the electron configuration of a silicon atom? Include the unfilled orbitals in the valence shell.

29. Describe the hybridization and the bonding of a silicon atom in elemental silicon.

30. Describe the molecular structure and the hybridization of silicon in each of the following molecules or ions:
    (a) $SiCl_4$ (b) SiO_4^{4-}
    (c) $(CH_3)_2SiH_2$ (d) SiF_6^{2-}

31. Write a Lewis structure for each of the following molecules or ions:
    (a) $SiCl_4$ (b) SiO_4^{4-} (c) $(CH_3)_2SiH_2$
    (d) SiF_6^{2-} (e) Si_3H_8 (contains Si—Si single bonds)

32. Write two equations for reactions in which silicon exhibits metallic behavior and two equations for reactions in which silicon exhibits nonmetallic behavior.

33. Heating a mixture of sand and coke in an electric furnace produces either silicon or silicon carbide. What determines which will be formed?

34. In terms of molecular and crystal structure, account for the low melting point of dry ice (solid CO_2) and the high melting point of SiO_2.

35. How does the following reaction show the acidic character of SiO_2? Which acid-base theory applies?

$$CaO + SiO_2 \xrightarrow{\Delta} CaSiO_3$$

36. Describe the conversion of silicate rocks to soluble forms for the purpose of analysis.

37. Account for the existence of the great variety of silicates in terms of how SiO_4 tetrahedra are linked.

38. Give equations for the reactions involved in the production of common window glass.

39. How does the internal structure of silica glass differ from that of quartz?

40. How does the structure of acetone, $(CH_3)_2CO$, differ from that of the silicone, $[(CH_3)_2SiO]_3$?

41. Compare and contrast the chemistry of carbon and that of silicon.

42. Write an equation for the reaction of silicon with each of the following (heated if necessary for reaction):
    (a) F_2 (b) NaOH   (c) O_2 (d) N_2

43. Write equations to contrast the hydrolysis of SiF_4, $SiCl_4$, and CCl_4.

44. Write an equation for the combustion of SiH_4; of Si_3H_8.

Additional Exercises

45. The experimentally determined ratio of hydrogen atoms to boron atoms in a gas is 3 to 1. How could it be proved that the gas actually consists of B_2H_6 molecules?

46. A hydride of boron contains 88.5% boron and 11.5% hydrogen by mass. The density of the hydride at 125°C and 450 torr is 2.21 g/L. A 5.0-mL gaseous sample of this hydride at 125°C and 450 torr decomposes to the elements, giving 35 mL of H_2 under the same conditions. Determine the molecular formula of the hydride and show that all data are consistent with that formula.

Ans. $B_{10}H_{14}$

47. Write balanced equations describing at least two different reactions in which solid silica exhibits acidic character.

48. Which would you expect to be a better Lewis base: the carbide ion, C^{4-}, or the silicide ion, Si^{4-}?

49. In what ways does the chemistry of silicates resemble that of phosphates?

50. Carbon forms carbonic acid, H_2CO_3, while silicon forms silicic acid, H_4SiO_4. Explain. (*Hint:* Draw the Lewis structures for the two compounds.)

51. (a) From the table of bond energies given in Section 6.9 calculate the approximate enthalpy change for the following reaction:

$$SiH_4(g) + 4HF(g) \longrightarrow SiF_4(g) + 4H_2(g)$$

Ans. -48 kJ
    (b) Using bond energy data, calculate the approximate enthalpy change for the corresponding reaction with methane:

$$CH_4(g) + 4HF(g) \longrightarrow CF_4(g) + 4H_2(g)$$

Ans. $+436$ kJ
    (c) Why do the two enthalpy changes differ so greatly?
    (d) Which reaction is more likely to proceed spontaneously?

52. Silicon reacts with sulfur at elevated temperatures. If 0.0923 g of silicon reacts with sulfur to give 0.3030 g of silicon sulfide, determine the empirical formula of silicon sulfide.   *Ans. SiS_2*

53. A hydride of silicon prepared by the reaction of Mg_2Si with acid exerted a pressure of 306 torr at 26°C in a bulb with a volume of 57.0 mL. If the mass of the hydride was 0.0861 g, what is its molecular weight? What is the molecular formula of the hydride?   *Ans. 92.0; Si_3H_8*

Coordination Compounds

26

The colors of these copper bearing minerals are due to differing ligand field-splitting of the copper ions' *d*-orbitals.

The hemoglobin in your blood, the blue dye in the ink in your ballpoint pen and in your blue jeans, chlorophyll, vitamin B-12, and the catalyst used in the manufacture of polyethylene all contain **coordination compounds,** or **complexes**—compounds in which ions and/or neutral molecules are attached to metal ions. Examples of simple complexes include $[Ag(NH_3)_2]^+$, $[Cu(NH_3)_4]^{2+}$, $[Fe(CN)_6]^{3-}$, $[Fe(CN)_6]^{4-}$, $[Co(NH_3)_6]^{3+}$, $[Co(NH_3)_3(NO_2)_3]$, $[Pt(NH_3)_2Cl_2]$, $Fe(CO)_5$, $[Al(H_2O)_6]^{3+}$, and $[Zn(OH)_4]^{2-}$. In this chapter we will consider the structures and properties of such complexes.

Formation of a complex requires two kinds of species: (1) an ion or molecule that has at least one pair of electrons available for coordinate covalent bonding, and (2) a metal ion or atom that has a sufficient attraction for electrons to form a coordinate covalent bond with the attaching group. Ions of the transition metals, inner transition metals, and a few metals near these series in the Periodic Table are especially prone to combine in complexes. Many of these complexes are highly colored (Fig. 26.1).

Figure 26.1
Metal ions containing partially filled d subshells usually form colored complex ions; ions with empty d subshells (d^0) or with filled d subshells (d^{10}) usually form colorless complexes. This figure shows, from left to right, solutions containing $[M(H_2O)_6]^{n+}$ ions with M = Sc^{3+} (d^0), Cr^{3+} (d^3), Co^{2+} (d^7), Ni^{2+} (d^8), Cu^{2+} (d^9), and Zn^{2+} (d^{10}).

The enthalpies of formation for coordination compounds vary greatly, but many are very stable. For example, the enthalpy of formation of $[Ni(NH_3)_6]I_3$ is -808 kilojoules per mole, whereas that of $[Co(NH_3)_6](NO_3)_3$ is -1282 kilojoules per mole (Appendix I).

Nomenclature, Structures, and Properties of Coordination Compounds

26.1 Definitions of Terms

The metal ion or atom in a complex is called the **central metal ion** or **atom;** the groups attached to it are called **ligands.** Ligands may be either ions or neutral molecules. Within a ligand, the atom that attaches directly to the metal by a coordinate covalent bond (Section 6.4) is called the **donor atom.**

The **coordination sphere,** usually the atoms within square brackets in a formula, includes the central metal ion plus the attached ligands. The **coordination number** of the central metal ion is the number of donor atoms bonded to it. The coordination number for the silver ion in $[Ag(NH_3)_2]^+$ is 2; that for the copper ion in $[Cu(NH_3)_4]^{2+}$ is 4; and that for the iron(III) ion in $[Fe(CN)_6]^{3-}$ is 6. In each of these examples the coordination number is equal to the number of ligands in the coordination sphere, but such is not always the case. Some ligands, such as ethylenediamine,

$$H-\overset{\overset{\displaystyle H}{|}}{\underset{\overset{\displaystyle \cdot\cdot}{}}{N}}-\overset{\overset{\displaystyle H}{|}}{\underset{\overset{\displaystyle H}{|}}{C}}-\overset{\overset{\displaystyle H}{|}}{\underset{\overset{\displaystyle H}{|}}{C}}-\overset{\overset{\displaystyle H}{|}}{\underset{\overset{\displaystyle \cdot\cdot}{}}{N}}-H$$

contain two donor atoms (shown in color). Thus the coordination number for cobalt in $[Co(H_2NCH_2CH_2NH_2)_3]^{3+}$ is 6. Although the coordination sphere of this complex contains only three ligands, all six nitrogen atoms are bonded to the cobalt. The most common coordination numbers are 2, 4, and 6. Coordination numbers of 3, 5, 7, and 8 do occur, as well as (though much less commonly) 9, 10, 11, and 12.

When a ligand attaches itself to a central metal ion by bonds from two or more donor atoms, it is referred to as a **polydentate ligand,** or a **chelating ligand.** The resulting complex is a **metal chelate;** examples include

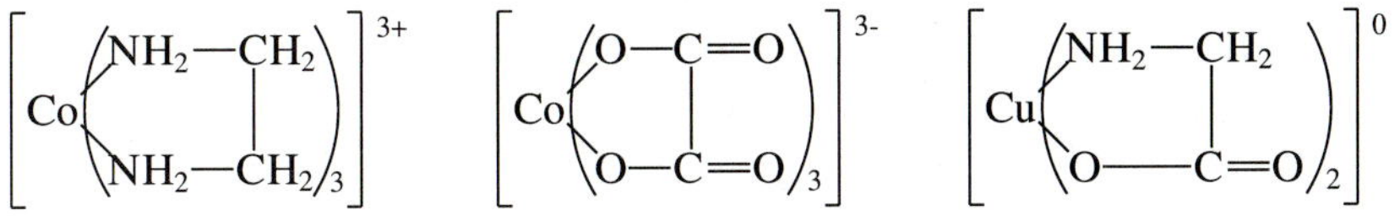

$$\left[\begin{array}{c} Co\left(\genfrac{}{}{0pt}{}{NH_2{-}CH_2}{NH_2{-}CH_2}\right)_3 \end{array}\right]^{3+} \quad \left[\begin{array}{c} Co\left(\genfrac{}{}{0pt}{}{O{-}C{=}O}{O{-}C{=}O}\right)_3 \end{array}\right]^{3-} \quad \left[\begin{array}{c} Cu\left(\genfrac{}{}{0pt}{}{NH_2{-}CH_2}{O{-}C{=}O}\right)_2 \end{array}\right]^{0}$$

The complex heme in hemoglobin (Fig. 26.2) contains a polydentate ligand with four donor atoms.

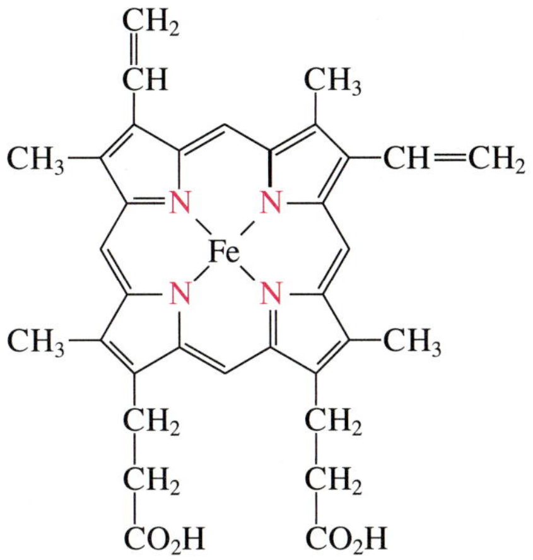

Figure 26.2

Heme, the square planar complex of iron found in hemoglobin. The donor atoms are shown in color.

26.2 The Naming of Complexes

The nomenclature of complexes is patterned after a system suggested by Alfred Werner, a Swiss chemist and Nobel laureate, whose outstanding work more than 90 years ago laid the foundation for a clearer understanding of these compounds. The following rules are used for naming complexes.

1. If a complex is ionic, name the cation first and the anion second, in accordance with usual nomenclature.

2. Name the ligands first, followed by the central metal.

3. Name the ligands alphabetically. (An older system names negative ligands alphabetically, then neutral ligands alphabetically, and finally positive ligands alphabetically.) Negative ligands (anions) have names formed by adding -*o* to the stem name of the group; for example,

F^-	*fluoro*	NO_3^-	*nitrato*
Cl^-	*chloro*	OH^-	*hydroxo*
Br^-	*bromo*	O^{2-}	*oxo*
I^-	*iodo*	NH_2^-	*amido*
CN^-	*cyano*	$C_2O_4^-$	*oxalato*
NO_2^-	*nitro* (M—N bond)	CO_3^{2-}	*carbonato*
ONO^-	*nitrito* (M—O bond)		

For most neutral ligands the name of the molecule is used. The four common exceptions are *aqua* (H_2O), *ammine* (NH_3), *carbonyl* (CO), and *nitrosyl* (NO).

4. If more than one ligand of a given type is present, the number is indicated by the prefixes *di-* (for two), *tri-* (for three), *tetra-* (for four), *penta-* (for five), and *hexa-* (for six). Sometimes the prefixes *bis-* (for two), *tris-* (for three), and *tetrakis-* (for four) are used when the name of the ligand contains numbers, begins with a vowel, is for a polydentate ligand, or includes *di-*, *tri-*, or *tetra-*.

5. When the complex is either a cation or a neutral molecule, the name of the central metal atom is spelled exactly as the name of the element and is followed by a Roman numeral in parentheses to indicate its oxidation number. When the complex is an anion, the suffix *-ate* is added to the stem for the name of the metal (or sometimes to the stem for the Latin name of the metal), followed by the Roman numeral designation of its oxidation number. Examples in which the complex is a cation (shown in color) are as follows:

$[Co(NH_3)_6]Cl_3$	Hexaamminecobalt(III) chloride
$[Pt(NH_3)_4Cl_2]^{2+}$	Tetraamminedichloroplatinum(IV) ion
$[Ag(NH_3)_2]^+$	Diamminesilver(I) ion
$[Cr(H_2O)_4Cl_2]Cl$	Tetraaquadichlorochromium(III) chloride
$[Co(H_2NCH_2CH_2NH_2)_3]_2(SO_4)_3$	Tris(ethylenediamine)cobalt(III) sulfate

Examples in which the complex is neutral:

$[Pt(NH_3)_2Cl_4]$	Diamminetetrachloroplatinum(IV)
$[Co(NH_3)_3(NO_2)_3]$	Triamminetrinitrocobalt(III)
$[Ni(H_2NCH_2CH_2NH_2)_2Cl_2]$	Dichlorobis(ethylenediamine)nickel(II)

Examples in which the complex is an anion (shown in color):

$K_3[Co(NO_2)_6]$	Potassium hexanitrocobaltate(III)
$[PtCl_6]^{2-}$	Hexachloroplatinate(IV) ion
$Na_2[SnCl_6]$	Sodium hexachlorostannate(IV)

26.3 The Structures of Complexes

The structures of many simple compounds and ions were discussed in Chapter 8. Spatial arrangements of atoms include geometries that are linear, trigonal planar, tetrahedral, square planar, square pyramidal, trigonal bipyramidal, and octahedral, among others. Now we can extend our study to the structures of coordination compounds. It may be helpful to reread Chapter 8 to refresh your memory of the structures discussed there. Many of these structures are found in coordination compounds as well as in simple compounds.

In 1893 Alfred Werner suggested that when ions or polar molecules are coordinated to a metal ion they are arranged in a definite geometrical pattern about it. This concept accounted for the properties of hydrates such as $CoCl_2 \cdot 6H_2O$, ammoniates such as $CoCl_3 \cdot 6NH_3$, and various so-called double salts such as $PtCl_4 \cdot 2KCl$. Werner assigned the formulas $[Co(H_2O)_6]Cl_2$, $[Co(NH_3)_6]Cl_3$, and $K_2[PtCl_6]$ to these compounds and pointed out that the properties of the complexes $[Co(H_2O)_6]^{2+}$, $[Co(NH_3)_6]^{3+}$, and $[PtCl_6]^{2-}$ could be explained by postulating that the six ligands are arranged about the central ion at the corners of a regular octahedron (Fig. 26.3).

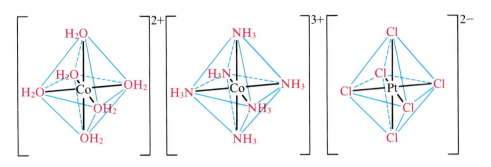

Figure 26.3

Octahedral structures of the $[Co(H_2O)_6]^{2+}$, the $[Co(NH_3)_6]^{3+}$, and the $[PtCl_6]^{2-}$ ions.

Many chemists use an abbreviated drawing of an octahedron [Fig. 26.4(a)] to show the geometry about a metal ion. However, a more realistic octahedron can be drawn by following the steps outlined in Fig. 26.4(b).

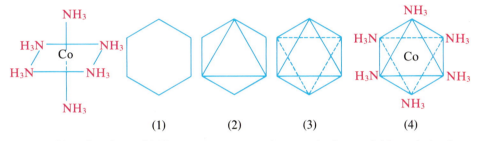

(1)   (2)   (3)   (4)

(a) An abbreviated drawing of an octahedron.

(b) To construct a complete octahedron quickly and simply: (1) Draw a regular hexagon. (2) Draw a triangle inside the hexagon. (3) Draw another triangle with dashed lines upside-down inside the hexagon. (4) Add symbols for central metal (black) and ligands (red).

Figure 26.4

Steps in drawing an octahedron.

Complexes in which the metal shows a coordination number of 4 exist in either of two geometric arrangements, the square planar or the tetrahedral configurations. Examples of four-coordinated complex ions with the square planar configuration are $[Ni(CN)_4]^{2-}$ and $[Cu(NH_3)_4]^{2+}$ and with the tetrahedral configuration are $[Zn(CN)_4]^{2-}$ and $[Zn(NH_3)_4]^{2+}$ (Fig. 26.5).

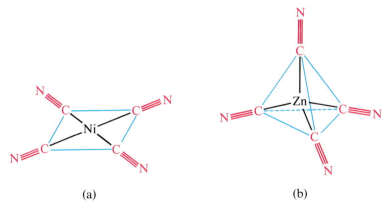

(a)   (b)

Figure 26.5

(a) The square planar configuration of the $[Ni(CN)_4]^{2-}$ ion, and (b) the tetrahedral configuration of the $[Zn(CN)_4]^{2-}$ ion.

Many other geometries are possible. Table 26.1 shows several of the known types.

Table 26.1 Geometric shapes and corresponding hybridization of orbitals for several complexes

Geometric shape	Diagram of shape	Coordination number	Hybridization	Examples
Linear		2	sp	$[Ag(NH_3)_2]^+$, $[Cu(CN)_2]^-$
Trigonal planar	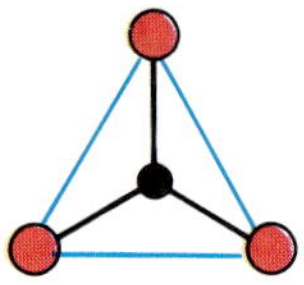	3	sp^2	$[HgCl_3]^-$
Tetrahedral	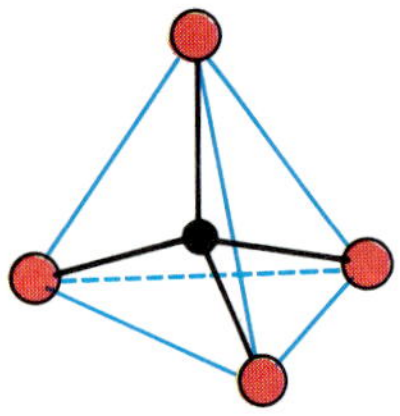	4	sp^3 (or sd^3)	$[Zn(CN)_4]^{2-}$, $[FeCl_4]^-$, $[CoBr_4]^{2-}$
Square planar	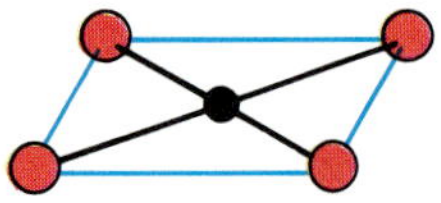	4	dsp^2	$[Ni(CN)_4]^{2-}$, $[Cu(NH_3)_4]^{2+}$
Square pyramidal	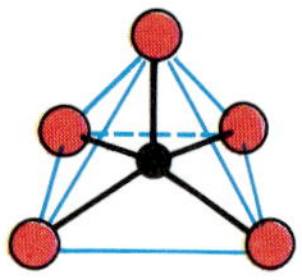	5	d^2sp^2 (or d^4s)	$[VOCl_4]^{2-}$, $[Ni(Br)_3\{(C_2H_5)_3P\}_2]$
Trigonal bipyramidal	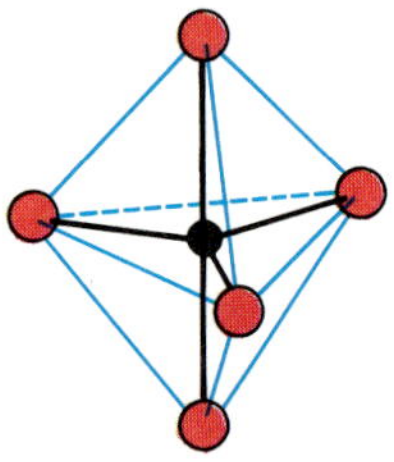	5	dsp^3 (or d^3sp)	$[Fe(CO)_5]$, $[Mn(CO)_4NO]$
Octahedral	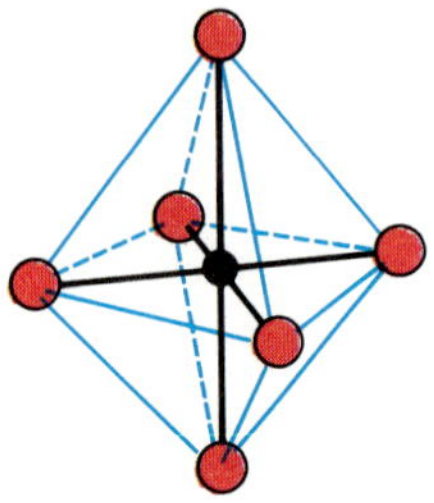	6	d^2sp^3	$[Co(NH_3)_6]^{3+}$, $[PtCl_6]^{2-}$, $[MoF_6]^-$

26.4 Isomerism in Complexes

Certain complexes, such as $[Co(NH_3)_4Cl_2]^+$, the tetraamminedichloro-cobalt(III) ion, have more than one form. These different forms with the same formula are **isomers.** Isomers have the same molecular formula, and hence the same composition, but different physical and chemical properties because the arrangement of the atoms in the structure is different. The $[Co(NH_3)_4Cl_2]^+$ ion has two isomers, one violet and the other green. Crystal structure analysis, using X-ray methods and other techniques, shows that the violet form has the ***cis* configuration** (chloride ions occupy adjacent corners of the octahedron) and the green form has the ***trans* configuration** (chloride ions occupy opposite corners), as shown in Fig. 26.6. Isomers such as these, which differ only in the way that the atoms are oriented in space relative to each other, are called **geometrical isomers,** or **stereoisomers.**

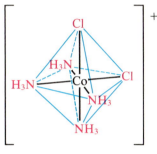

Violet, *cis* form

Green, *trans* form

Figure 26.6
The *cis* and *trans* isomers of $[Co(NH_3)_4Cl_2]^+$.

A complex such as $[Cr(NH_3)_2(H_2O)_2(Br)_2]^+$ has a variety of different geometrical isomers. Each ligand could be *trans* to one like it (as shown in the figure on the left below) or the ammonia molecules could be *trans* to each other while the water molecules and bromine atoms are *cis* to their counterparts (figure on the right).

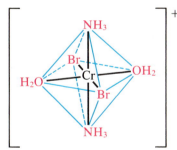

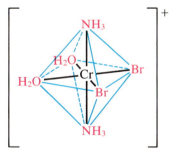

Similarly, the water molecules or the bromine atoms could be *trans* to each other while the other ligands are *cis* to ones like themselves:

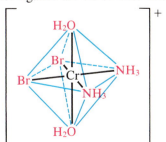

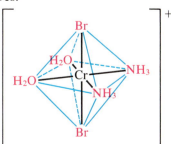

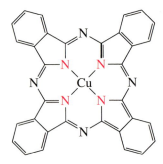

Figure 26.10
Copper phthalocyanine blue, a square planar copper complex found in some blue dyes.

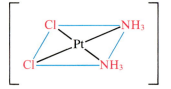

Figure 26.11
The square planar structure of the anticancer drug Platinol, cis-[Pt(NH₃)₂Cl₂].

a large protein molecule (globin) by coordination of the protein to a position above the plane of the heme molecule. Oxygen molecules are transported by hemoglobin in the blood by being bound to the coordination site opposite the binding site of the globin molecule.

Complexing agents are often used for water softening because they tie up such ions as Ca^{2+}, Mg^{2+}, and Fe^{2+}, which make water hard (Section 29.10). Complexing agents that tie up metal ions are also used as drugs. British Anti-Lewisite, $HSCH_2CH(SH)CH_2OH$, a drug developed during World War I as an antidote for the arsenic-based war gas Lewisite, is now used to treat poisoning by heavy metals such as arsenic, mercury, thallium, and chromium. The drug (abbreviated as BAL) is a ligand and functions by making a water-soluble chelate of the metal; this metal chelate is eliminated by the kidneys. Another polydentate ligand, enterobactin, which is isolated from certain bacteria, is used to form complexes of iron and thereby to control the severe iron build-up found in patients suffering from blood diseases such as Cooley's anemia. This disease prevents a patient's own blood from transporting oxygen adequately. Such patients need regular blood transfusions to survive, but as the new blood breaks down, the usual metabolic processes that remove excess iron are overloaded. This excess iron can build up to fatal levels in the heart, kidneys, and liver. Enterobactin forms a water-soluble complex with the excess iron, and this complex can be eliminated by the body.

The anticancer drug Platinol is cis-[Pt(NH₃)₂Cl₂], a square planar platinum(II) complex (Fig. 26.11). This substance, whose biological activity was first observed by scientists at Michigan State University in 1969, is believed to cross-link DNA strands (Section 31.11) and thereby interfere with mitosis (cell division).

Some complexing groups, when coordinated to certain metals, make the metal more easily assimilated by plants; in other cases, they keep the metal from being effectively utilized by the plant. Hence complexes can be effective in soil treatment. An example is the tris(ethylenediamine)iron(III) ion, $[Fe(en)_3]^{3+}$, which is the active agent in a substance that is added to the soil to provide iron to pin oak trees.

In the electroplating industry it has been found that many metals plate out as a smoother, more uniform, better-looking, and more adherent surface when plated from a bath containing the metal as a complex ion. Thus complexes such as $[Ag(CN)_2]^-$ and $[Au(CN)_2]^-$ are used extensively in the electroplating industry.

Bonding in Coordination Compounds

Any theory of the bonding in coordination compounds must explain four important properties: their stabilities, their structures, their colors, and their magnetic properties. The first modern attempt to explain the properties of complex compounds was made by applying the concepts of orbital hybridization (Sections 8.3 through 8.7) and valence bond theory. More recent models use ligand field theory, a model that examines the electrostatic attractions and repulsions between the central metal ion and the ligands, or they use molecular orbital theory.

26.6 Valence Bond Theory

The valence bond theory treats a metal-ligand bond as a coordinate covalent bond (Section 6.4) that forms when an orbital of the donor atom overlaps a hybrid orbital of the central metal atom. Electron pairs from the ligands are shared with the

The [Co(en)$_2$Cl$_2$]$^+$ ion, where en stands for ethylenediamine, has two *cis* isomers, which are a pair of optical isomers, and one *trans* isomer. The *trans* configuration is symmetrical and has no optical isomerism. Its mirror image is superimposable on it and is therefore identical to it. The three isomers are shown in Fig. 26.8.

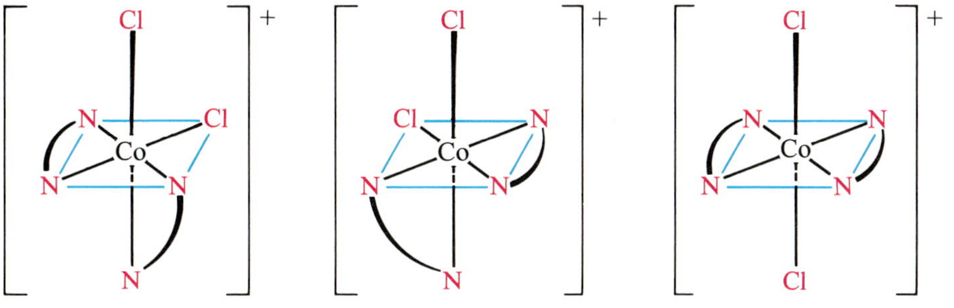

Cis forms (optical isomers)   *Trans* form

Figure 26.8

The three isomeric forms of [Co(en)$_2$Cl$_2$]$^+$. In these abbreviated formulas, N⌒N stands for H$_2$NCH$_2$CH$_2$NH$_2$.

The subject of isomers, including geometrical and optical isomers and a third type called structural isomers, will be discussed in more detail in Chapter 30 (Sections 30.3 and 30.6).

26.5 Uses of Complexes

Many uses of complexes are based on their colors, their solubilities, or the changes in chemical behavior of metal ions and ligands when they form complexes.

Chlorophyll (Fig. 26.9), the green pigment in plants, is a complex that contains magnesium. Plants appear green because chlorophyll absorbs yellow light; the reflected light consequently appears green (Section 26.10). The energy resulting from the absorption of this light is used in photosynthesis (Section 31.14).

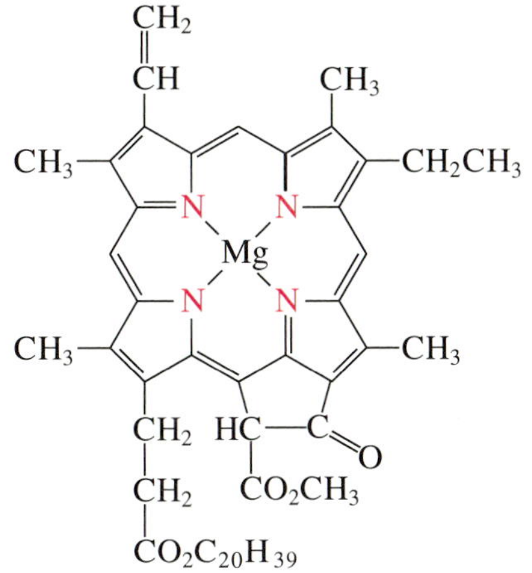

Figure 26.9

Chlorophyll, a square planar magnesium complex found in plants.

The square planar copper(II) complex phthalocyanine blue (Fig. 26.10) is one of many complexes used as pigments or dyes. This complex is used in blue ink, blue jeans, and certain blue paints.

The structure of heme (Fig. 26.2), the iron-containing complex in hemoglobin, is very similar to that of chlorophyll. In hemoglobin the red heme complex is bonded to

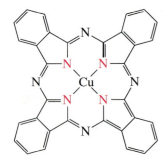

Figure 26.10

Copper phthalocyanine blue, a square planar copper complex found in some blue dyes.

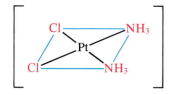

Figure 26.11

The square planar structure of the anticancer drug Platinol, cis-[Pt(NH₃)₂Cl₂].

a large protein molecule (globin) by coordination of the protein to a position above the plane of the heme molecule. Oxygen molecules are transported by hemoglobin in the blood by being bound to the coordination site opposite the binding site of the globin molecule.

Complexing agents are often used for water softening because they tie up such ions as Ca^{2+}, Mg^{2+}, and Fe^{2+}, which make water hard (Section 29.10). Complexing agents that tie up metal ions are also used as drugs. British Anti-Lewisite, $HSCH_2CH(SH)CH_2OH$, a drug developed during World War I as an antidote for the arsenic-based war gas Lewisite, is now used to treat poisoning by heavy metals such as arsenic, mercury, thallium, and chromium. The drug (abbreviated as BAL) is a ligand and functions by making a water-soluble chelate of the metal; this metal chelate is eliminated by the kidneys. Another polydentate ligand, enterobactin, which is isolated from certain bacteria, is used to form complexes of iron and thereby to control the severe iron build-up found in patients suffering from blood diseases such as Cooley's anemia. This disease prevents a patient's own blood from transporting oxygen adequately. Such patients need regular blood transfusions to survive, but as the new blood breaks down, the usual metabolic processes that remove excess iron are overloaded. This excess iron can build up to fatal levels in the heart, kidneys, and liver. Enterobactin forms a water-soluble complex with the excess iron, and this complex can be eliminated by the body.

The anticancer drug Platinol is cis-[Pt(NH₃)₂Cl₂], a square planar platinum(II) complex (Fig. 26.11). This substance, whose biological activity was first observed by scientists at Michigan State University in 1969, is believed to cross-link DNA strands (Section 31.11) and thereby interfere with mitosis (cell division).

Some complexing groups, when coordinated to certain metals, make the metal more easily assimilated by plants; in other cases, they keep the metal from being effectively utilized by the plant. Hence complexes can be effective in soil treatment. An example is the tris(ethylenediamine)iron(III) ion, $[Fe(en)_3]^{3+}$, which is the active agent in a substance that is added to the soil to provide iron to pin oak trees.

In the electroplating industry it has been found that many metals plate out as a smoother, more uniform, better-looking, and more adherent surface when plated from a bath containing the metal as a complex ion. Thus complexes such as $[Ag(CN)_2]^-$ and $[Au(CN)_2]^-$ are used extensively in the electroplating industry.

Bonding in Coordination Compounds

Any theory of the bonding in coordination compounds must explain four important properties: their stabilities, their structures, their colors, and their magnetic properties. The first modern attempt to explain the properties of complex compounds was made by applying the concepts of orbital hybridization (Sections 8.3 through 8.7) and valence bond theory. More recent models use ligand field theory, a model that examines the electrostatic attractions and repulsions between the central metal ion and the ligands, or they use molecular orbital theory.

26.6 Valence Bond Theory

The valence bond theory treats a metal-ligand bond as a coordinate covalent bond (Section 6.4) that forms when an orbital of the donor atom overlaps a hybrid orbital of the central metal atom. Electron pairs from the ligands are shared with the

26.4 Isomerism in Complexes

Certain complexes, such as $[Co(NH_3)_4Cl_2]^+$, the tetraamminedichloro-cobalt(III) ion, have more than one form. These different forms with the same formula are **isomers.** Isomers have the same molecular formula, and hence the same composition, but different physical and chemical properties because the arrangement of the atoms in the structure is different. The $[Co(NH_3)_4Cl_2]^+$ ion has two isomers, one violet and the other green. Crystal structure analysis, using X-ray methods and other techniques, shows that the violet form has the *cis* **configuration** (chloride ions occupy adjacent corners of the octahedron) and the green form has the *trans* **configuration** (chloride ions occupy opposite corners), as shown in Fig. 26.6. Isomers such as these, which differ only in the way that the atoms are oriented in space relative to each other, are called **geometrical isomers,** or **stereoisomers.**

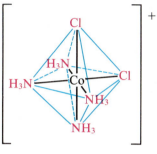

Violet, *cis* form   Green, *trans* form

Figure 26.6

The *cis* and *trans* isomers of $[Co(NH_3)_4Cl_2]^+$.

A complex such as $[Cr(NH_3)_2(H_2O)_2(Br)_2]^+$ has a variety of different geometrical isomers. Each ligand could be *trans* to one like it (as shown in the figure on the left below) or the ammonia molecules could be *trans* to each other while the water molecules and bromine atoms are *cis* to their counterparts (figure on the right).

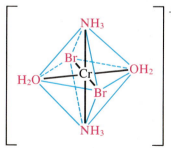
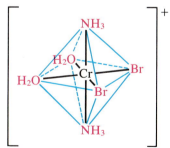

Similarly, the water molecules or the bromine atoms could be *trans* to each other while the other ligands are *cis* to ones like themselves:

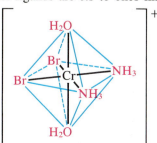
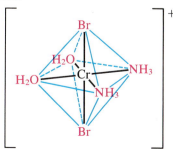

Finally, each group could be *cis* to one like itself; in this case two different arrangements are possible:

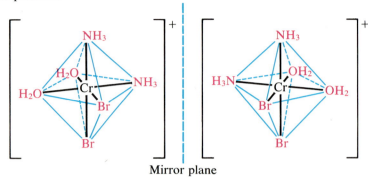

These two geometrical isomers have the same arrangement of ligands in the coordination sphere (*cis*-H_2O, *cis*-Br, *cis*-NH_3), but they are mirror images of each other and are not superimposable. Therefore they are not geometrically identical. Isomers that are mirror images of each other but not identical are called **optical isomers**. Mirror images of the other geometrical isomers of $[Cr(NH_3)_2(H_2O)_2(Br)_2]^+$ can be drawn, but the mirror images can be superimposed and so are identical. For example, the following mirror images are identical to each other; either may be turned 90° on an axis through the two corners occupied by the ammonia molecules and superimposed on the other:

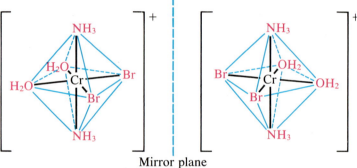

The $[Cr(NH_3)_2(H_2O)_2(Br)_2]^+$ complex thus has a total of six geometrical isomers, two of which are a pair of optical isomers.

The tris(ethylenediamine)cobalt(III) ion, $[Co(H_2NCH_2CH_2NH_2)_3]^{3+}$, has two optical isomers, as shown in Fig. 26.7.

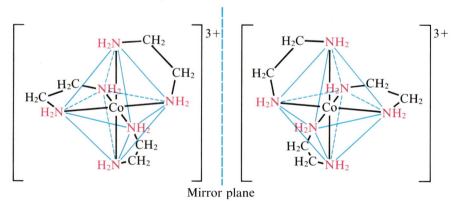

Figure 26.7
Optical isomers of
$[Co(H_2NCH_2CH_2NH_2)_3]^{3+}$.

Colors of various compounds of chromium, with different oxidation states and ligands. From left to right, solutions of K_2CrO_4, $CrCl_3 \cdot 6H_2O$, $Cr(NO_3)_3$, $K_2Cr_2O_7$, and $[Cr(H_2O)_4Cl_2]Cl$.

metal, and each electron pair occupies both an atomic orbital of a ligand and one of several equivalent hybridized orbitals of the metal. The ligand is a Lewis base; the metal, a Lewis acid (Section 16.15).

In an octahedral complex such as $[Co(NH_3)_6]^{3+}$, a total of six electron pairs from the six ligands may be considered to have entered hybrid orbitals of the metal ion. As discussed in Section 8.6, six equivalent hybrid orbitals for the metal atom result from the hybridization of two d, one s, and three p orbitals; d^2sp^3 hybridization. All octahedral complexes, for example, $[Co(NH_3)_6]^{3+}$, $[SnCl_6]^{2-}$, $[Co(H_2O)_6]^{2+}$, $[Co(CN)_6]^{3-}$, $[Fe(CN)_6]^{3-}$, and $[Cr(en)_3]^{3+}$, can be described as being d^2sp^3 hybridized.

Let us look more closely at the octahedral $[Co(NH_3)_6]^{3+}$ ion. An isolated cobalt atom in its ground state has a $1s^2 2s^2 2p^6 3s^2 3p^6 3d^7 4s^2$ electron configuration (Section 5.12). Each atomic orbital can accommodate two electrons of opposing spin. The electrons enter the orbitals of a given type singly before any pairing of electrons occurs within those orbitals—Hund's Rule (Section 5.11). Each orbital in the cobalt atom can be represented by a circle as follows:

Co atom

1s	2s	2p	3s	3p	3d	4s	4p
⑪	⑪	⑪⑪⑪	⑪	⑪⑪⑪	⑪⑪↑↑↑	⑪	○○○

When the cobalt(III) ion, Co^{3+}, is formed, the atom loses three electrons, two from the 4s orbital and one from a completely filled 3d orbital, giving a structure possessing four unpaired electrons.

Co^{3+}

1s	2s	2p	3s	3p	3d	4s	4p
⑪	⑪	⑪⑪⑪	⑪	⑪⑪⑪	⑪↑↑↑↑	○	○○○

Six pairs of electrons are shared with the cobalt when six ammonia molecules bond to a cobalt(III) ion. It is postulated that the unpaired electrons in the 3d orbitals of a cobalt(III) ion pair as the electron pairs in the ligands approach, and the two empty 3d orbitals that result hybridize with the $4s$ and the three $4p$ orbitals to form six d^2sp^3 hybrid orbitals directed toward the corners of an octahedron. These hybrid orbitals accept the incoming electrons from the ammonia (indicated below in color). However, once the structure is formed, it is impossible to distinguish the source of the electrons, since all electrons are identical regardless of their origin. The distribution of electrons on cobalt in the $[Co(NH_3)_6]^{3+}$ ion is

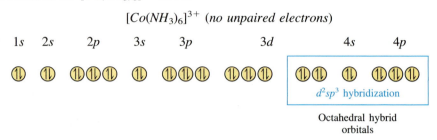

$[Co(NH_3)_6]^{3+}$ (*no unpaired electrons*)

Octahedral complexes of cobalt(II), $[Co(CN)_6]^{4-}$, for example, contain one more electron than do those of cobalt(III). To permit d^2sp^3 bonding this extra electron must be promoted to a $4d$ (or perhaps a $5s$) orbital, where it is loosely held. This electron is readily removed, as indicated by the ease with which most Co(II) complexes are oxidized to Co(III) complexes.

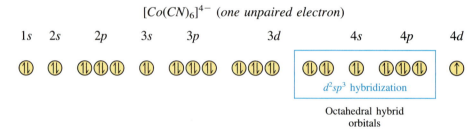

$[Co(CN)_6]^{4-}$ (*one unpaired electron*)

For four-coordinated complexes with a tetrahedral structure, the four bonding orbitals of the central atom come from the hybridization of one s and three p orbitals, or sp^3 hybridization. This bonding is analogous to that in methane (Section 8.3), except that in this case the coordinate covalent bonds are formed with both electrons of each shared pair coming from the ligand.

The distribution of electrons in the tetrahedral complex $[Zn(CN)_4]^{2-}$ is represented diagrammatically as follows:

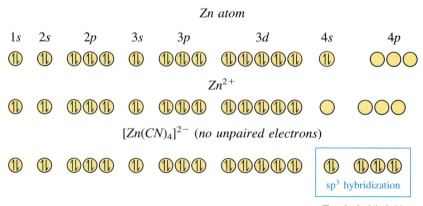

Zn atom

Zn^{2+}

$[Zn(CN)_4]^{2-}$ (*no unpaired electrons*)

In the case of four-coordinated structures with a square planar structure, the four bonds to the central atom arise from dsp^2 hybridization, as illustrated for $[Ni(CN)_4]^{2-}$.

$[Ni(CN)_4]^{2-}$ *(no unpaired electrons)*

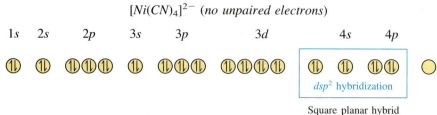

Square planar hybrid
orbitals

Table 26.1 shows other types of hybridization in coordination compounds.

Some properties of complexes cannot be explained satisfactorily by assuming hybridization and coordinate covalent bonding. For example, magnetic moments (Section 26.9) of iron(II) complexes indicate that $[Fe(CN)_6]^{4-}$ has no unpaired electrons and $[Fe(H_2O)_6]^{2+}$ has four. Assuming covalent bonding, no unpaired electrons are expected for either complex, and $[Fe(CN)_6]^{4-}$ meets this expectation.

$[Fe(CN)_6]^{4-}$

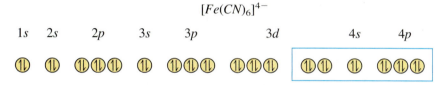

However, the $[Fe(H_2O)_6]^{2+}$ ion with its four unpaired electrons does not meet the expectation.

If $[Fe(H_2O)_6]^{2+}$ is assumed to be "ionically" bonded by the electrostatic attraction of the Fe^{2+} ion for the negative end of the polar water molecule, the four unpaired electrons can be explained. Just as in the simple iron(II) ion, there are four unpaired electrons in the d orbitals of the iron atom since the ligand electrons do not enter these orbitals through sharing because the bonding is ionic.

Fe^{2+} and $[Fe(H_2O)_6]^{2+}$

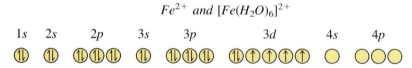

The difference in the bonding in $[Fe(CN)_6]^{4-}$ and in $[Fe(H_2O)_6]^{2+}$ is not surprising, since the cyanide ion makes an electron pair more readily available for sharing with a central metal ion than does the water molecule. Thus cyanide forms more stable complexes. Although the concept of two different types of bonding in complexes satisfactorily explains some facts, such as their magnetic behavior, it does not adequately explain certain other properties, such as their structural configurations and colors.

26.7 Ligand Field Theory

Even though coordination compounds have coordinate covalent bonding between the metal and the ligands, it is possible to understand, interpret, and predict the properties of many of these compounds, especially those of transition metals, on the assumption that there is no covalent bonding, but only simple electrostatic interactions. This totally ionic model of the bonding in complexes is called **ligand field theory.**

According to the assumptions of ligand field theory, the basic reason for the formation of a complex ion or molecule is the electrostatic attraction of a positively charged metal ion for negative ions or for the negative ends (the electron pairs) of polar molecules. The colors of complexes and their magnetic properties result from the electrostatic repulsions between the electron pairs of the ligands and electrons in the d orbitals of the metal ion.

In Section 5.10 the shapes of the s, p, and d orbitals were given. An s orbital is spherical. A p orbital has a dumbbell shape, and the three p orbitals for a given level are oriented at right angles to one another along the x, y, and z axes (Fig. 5.17). The d orbitals, which occur in sets of five, consist of lobe-shaped regions and are arranged in space as shown in Fig. 26.12, reproduced within an octahedral structure.

The lobes in two of the five d orbitals, the d_{z^2} and $d_{x^2-y^2}$ orbitals, point toward the ligands on the corners of the octahedron around the metal (Fig. 26.12). These two orbitals are called the e_g **orbitals** (the symbol actually refers to the symmetry of the orbitals, but we will use it as a convenient name for these two orbitals in an octahedral complex). The other three orbitals, the d_{xy}, d_{xz} and d_{yz} orbitals, whose lobes point between the ligands on the corners of the octahedron, are called the t_{2g} **orbitals** (again the symbol really refers to the symmetry of the orbitals). As six ligands approach the metal ion along the axes of the octahedron, the electron pairs of the ligands repel the electrons in the d orbitals of the metal ion. However, the repulsions between the electrons in the e_g orbitals (the d_{z^2} and $d_{x^2-y^2}$ orbitals) and the electron pairs of the ligands are greater than the repulsions between the electrons in the t_{2g} orbitals (the d_{xy}, d_{xz}, and d_{yz} orbitals) and the electron pairs of the ligands because the lobes of the e_g orbitals point directly at the ligand electron pairs, whereas the lobes of the t_{2g} orbitals point between them. Thus electrons in e_g orbitals of a metal ion in an octahedral

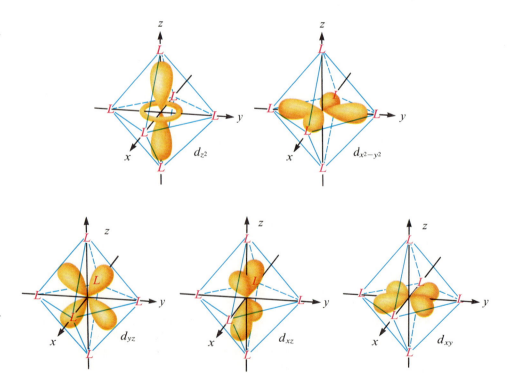

Figure 26.12

Diagrams showing the directional characteristics of the five d orbitals. L indicates a ligand at each corner of the octahedron.

complex have higher potential energies than those of electrons in t_{2g} orbitals. The difference in energy may be represented as follows:

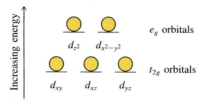

The difference in energy between the e_g and the t_{2g} orbitals is called the **ligand field splitting** and is symbolized by **Δ** or **10Dq.**

The size of the ligand field splitting (10Dq) depends on the nature of the six ligands located around the central metal ion. Different ligands have different concentrations of electrons in the orbital that points toward the central atom. Ligands with a low density of electrons produce a small ligand field splitting (small 10Dq value), while ligands with a high density of electrons produce a large ligand field splitting (large 10Dq value). The increasing ligand field splitting produced by ligands is expressed in the **spectrochemical series,** a short version of which is given here.

$$I^- < Br^- < Cl^- < F^- < H_2O < C_2O_4{}^{2-} < NH_3 < en < NO_2{}^- < CN^-$$

A few ligands of the spectrochemical series, in order of increasing
field strength of the ligand electron pairs

In this series the ligands on the left project a low density of electrons toward the metal (low field strength), and those on the right have a large density of electrons (high field strength). Thus the ligand field splitting produced by an iodide ion (I^-) as a ligand is much smaller than that produced by a cyanide ion (CN^-) as a ligand.

In a simple metal ion in the gas phase, the electrons are distributed among the five $3d$ orbitals in accord with Hund's Rule (Section 5.11), since the orbitals all have the same energy.

Ion	3d Orbitals	Ion	3d Orbitals
Ti^{3+}	⬆〇〇〇〇	Fe^{2+}	⬆⬇ ⬆ ⬆ ⬆ ⬆
V^{3+}	⬆⬆〇〇〇	Co^{3+}	⬆⬇ ⬆ ⬆ ⬆ ⬆
Cr^{3+}	⬆⬆⬆〇〇	Co^{2+}	⬆⬇ ⬆⬇ ⬆ ⬆ ⬆
Cr^{2+}	⬆⬆⬆⬆〇	Ni^{2+}	⬆⬇ ⬆⬇ ⬆⬇ ⬆ ⬆
Mn^{2+}	⬆⬆⬆⬆⬆	Cu^{2+}	⬆⬇ ⬆⬇ ⬆⬇ ⬆⬇ ⬆
Fe^{3+}	⬆⬆⬆⬆⬆	Zn^{2+}	⬆⬇ ⬆⬇ ⬆⬇ ⬆⬇ ⬆⬇

However, if the metal ion lies in an octahedron formed by six ligands, the energies of the d orbitals are no longer the same, and two opposing forces are set up. One force tends to keep the electrons of the metal ion distributed with unpaired spins within all of the d orbitals (it requires energy to pair up electrons in an orbital). The other force tends to reduce the average energy of the d electrons by placing as many of them as possible in the lower-energy t_{2g} orbitals. The d electrons will end up with the lowest possible total energy. If it requires less energy for the d electrons to be excited to the upper e_g orbitals than to pair in the lower t_{2g} orbitals, then they will remain unpaired. If it requires less energy to pair d electrons in the lower t_{2g} orbitals than to put them in the upper e_g orbitals, then they will pair.

In $[Fe(CN)_6]^{4-}$ the strong field of six cyanide ions produces a large ligand field splitting. Under these conditions the electrons require less energy to pair than to be excited to the e_g orbitals. Thus the six $3d$ electrons of the Fe^{2+} ion pair in the three t_{2g} orbitals. The result is in agreement with the experimentally measured magnetic moment of $[Fe(CN)_6]^{4-}$ (no unpaired electrons).

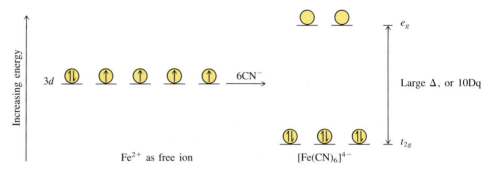

In $[Fe(H_2O)_6]^{2+}$, on the other hand, the weak field of the water molecules produces only a small ligand field splitting. Since it requires less energy to excite electrons to the e_g orbitals than to pair them, they remain distributed in all five $3d$ orbitals.

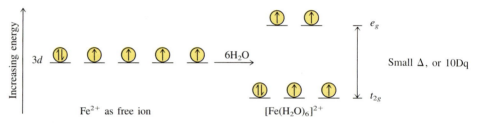

Thus four unpaired electrons should be present, as is verified by measuring the magnetic moment of $[Fe(H_2O)_6]^{2+}$. A similar line of reasoning can be followed for complexes of iron(III) to show why $[Fe(CN)_6]^{3-}$ has only one unpaired electron while $[Fe(H_2O)_6]^{3+}$ and $[FeF_6]^{3-}$ both have five unpaired electrons.

Actually, as was pointed out in Section 6.6, there is no sharp dividing line between covalent and ionic bonding. The molecular orbital theory, discussed in Section 26.11, introduces a covalent component into the ionic viewpoint.

26.8 Ligand Field Stabilization Energy

As indicated earlier, the difference in energy between the e_g and t_{2g} orbitals in an octahedral complex is commonly symbolized by either Δ or 10Dq.

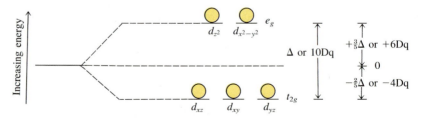

Since the splitting of the five d orbitals into two energy levels does not change the total energy, the zero line on the energy axis is the weighted average energy of these orbitals. Thus the e_g and t_{2g} energies are $+\frac{3}{5}\Delta$ and $-\frac{2}{5}\Delta$, respectively (or $+6$Dq and -4Dq, respectively). Hence

$$\left(\frac{3}{5}\Delta\right)2 + \left(-\frac{2}{5}\Delta\right)3 = 0$$

or

$$\left(\frac{3}{5} \times 10\text{Dq}\right)2 + \left(-\frac{2}{5} \times 10\text{Dq}\right)3 = (+6\text{Dq})2 + (-4\text{Dq})3 = 0$$

A value for each complex, called the **ligand field stabilization energy (LFSE)**, can be calculated and is related to the stability of the complex. Actually the ligand field stabilization energy is a measure of how much more stable a complex is than a hypothetical complex that is identical in every way except showing no ligand field splitting. Consider, for example, an octahedral complex of the vanadium(III) ion, which has two d electrons (referred to as d^2). The two unpaired electrons occupy two of the lower-energy t_{2g} orbitals, whether the vanadium is coordinated to a ligand producing a weak field or to one producing a strong field. In calculating the ligand field stabilization energy, we take into account the fact that there are two electrons, each with an energy of -4Dq relative to the zero-energy line.

$$\text{LFSE} = 2(-4\text{Dq}) = -8\text{Dq}$$

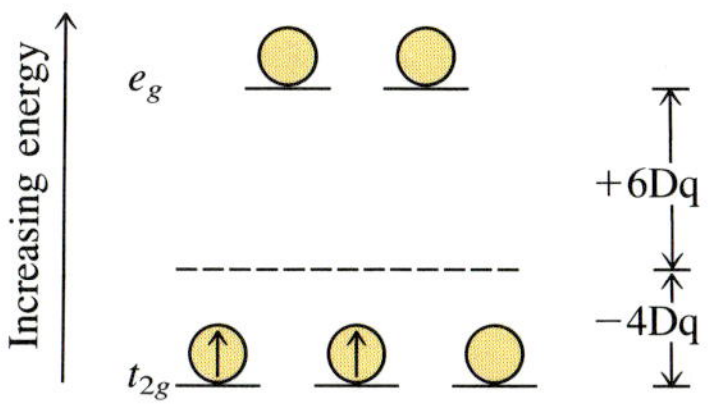

V^{3+} with weak-field or strong-field ligands
LFSE $= 2(-4\text{Dq}) = -8\text{Dq}$

Now consider another metal ion, the manganese(II) ion, which has five d electrons (d^5). In the free ion these electrons are distributed evenly throughout the five d orbitals in accord with Hund's Rule. If the ion is coordinated to ligands that produce a weak field, the five electrons remain evenly distributed; that is, there are three unpaired electrons in the t_{2g} level and two in the e_g level.

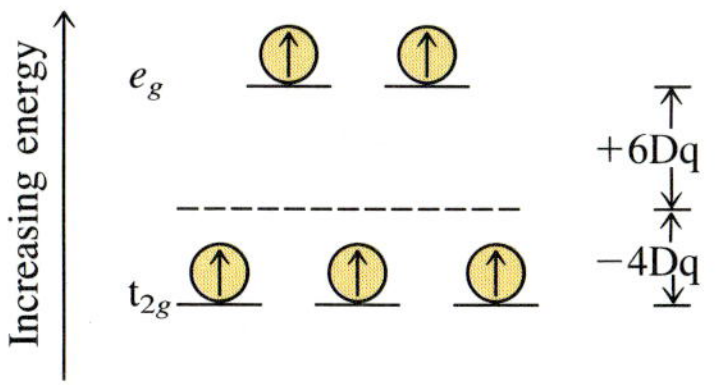

Mn^{2+} with weak-field ligands
LFSE $= 2(+6\text{Dq}) + 3(-4\text{Dq}) = 0$

The two electrons in the e_g orbitals are in the higher energy level, corresponding to $+6$Dq, and the three electrons in the t_{2g} orbitals are in the lower energy level, corresponding to -4Dq. Hence the ligand field stabilization energy is zero in this weak-field case.

$$\text{LFSE} = 2(+6\text{Dq}) + 3(-4\text{Dq}) = 0$$

If, however, the Mn^{2+} ion is coordinated to ligands that produce a strong field, all five electrons fall into the lower-energy nonbonding t_{2g} orbitals, and none are in the higher-energy e_g orbitals.

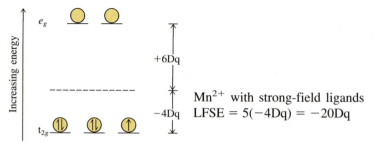

The ligand field stabilization energy for Mn^{2+} in this strong-field case is $-20Dq$.

$$LFSE = 5(-4Dq) = -20Dq$$

In the weak-field case for Mn^{2+}, the splitting of the two energy levels (the 10Dq value) is much smaller than it is for the strong-field case. The possible gain in energy that would result if the fourth and fifth electrons occupied the lower-energy t_{2g} orbitals is not large enough to overcome the repulsion of the electrons for each other and to pair them up. Hence the fourth and fifth electrons go singly into the higher-energy e_g orbitals.

In general, the more negative the value of the ligand field stabilization energy, the more stable is the complex. A complex of Mn^{2+} with ligands of high field strength is more stable than one with ligands of low field strength. Sometimes the term **low-spin complex,** or **spin-paired complex,** is applied to complexes of a given metal with ligands of high field strength; the term **high-spin complex,** or **spin-free complex,** is sometimes used for complexes with ligands of low field strength.

Table 26.2 shows the ligand field stabilization energy for a variety of octahedral complexes with weak and strong fields. The calculated LFSE values in the table have been experimentally verified by spectroscopic methods. A relatively large negative number indicates that more energy is lost in the formation of the complex and that, in general, the complex is more stable for a given ligand. Remember, however, each

Table 26.2 Ligand field stabilization energy (LFSE) of octahedral complexes[a]

| Configuration | Weak field (high spin) | | | | Strong field (low spin) | | | | Examples |
	t_{2g}	e_g	Number of unpaired electrons	LFSE (Dq)	t_{2g}	e_g	Number of unpaired electrons	LFSE (Dq)	
d^0	0	0	0	0	0	0	0	0	Ca^{2+}, Sc^{3+}
d^1	1	0	1	-4	1	0	1	-4	Ti^{3+}
d^2	2	0	2	-8	2	0	2	-8	V^{3+}
d^3	3	0	3	-12	3	0	3	-12	Cr^{3+}, V^{2+}
d^4	3	1	4	-6	4	0	2	-16	Cr^{2+}, Mn^{3+}
d^5	3	2	5	0	5	0	1	-20	Mn^{2+}, Fe^{3+}
d^6	4	2	4	-4	6	0	0	-24	Fe^{2+}, Co^{3+}
d^7	5	2	3	-8	6	1	1	-18	Co^{2+}
d^8	6	2	2	-12	6	2	2	-12	Ni^{2+}
d^9	6	3	1	-6	6	3	1	-6	Cu^{2+}
d^{10}	6	4	0	0	6	4	0	0	Cu^+, Zn^{2+}

[a]Used by permission from *Concepts and Models of Inorganic Chemistry,* B. E. Douglas, D. H. McDaniel, and J. J. Alexander (2nd ed.). John Wiley and Sons Inc., New York, 1983, p. 270.

ligand has its own specific 10Dq value, and this must be considered when the stabilities of complexes with different ligands are calculated. Weak-field ligands cause less splitting between the energy levels for a given metal ion and have smaller 10Dq values, in general, than do strong-field ligands attached to the same metal ion.

The most negative LFSE value in the table for strong-field complexes is for the d^6 configuration, since in this case the lower-energy t_{2g} orbitals are completely occupied and the higher-energy e_g orbitals are empty. The most negative LFSE values for weak-field complexes are for the d^3 and d^8 configurations. In the d^3 case all of the lower-energy t_{2g} orbitals are singly occupied, and the higher-energy e_g orbitals are empty. The d^8 configuration represents the lowest number of electrons for which the lower-energy t_{2g} orbitals are completely filled; however, since two electrons are present in the higher-energy e_g orbitals, the most negative LFSE value for weak-field complexes (-12) is less negative than that for strong-field complexes (-24). The LFSE is zero, indicating very low stability, at two places in the table for strong-field complexes: in the cases when no d electrons are present and when 10 d electrons are present. In the latter case all t_{2g} and e_g orbitals are filled, so that the stability gained in the formation of the complex by electrons going into lower-energy orbitals is exactly counterbalanced by the stability lost by electrons going into higher-energy orbitals. For weak-field complexes the d^0 and d^{10} configurations have LFSE values of zero, as does the d^5 configuration, in which all t_{2g} and e_g orbitals are singly occupied.

26.9 Magnetic Moments of Molecules and Ions

Molecules such as O_2 (Section 7.11) and NO (Section 7.14) and ions such as $[Co(CN)_6]^{4-}$ (Section 26.6) and $[Fe(H_2O)_6]^{2+}$ (Section 26.7) that contain unpaired electrons are **paramagnetic.** Paramagnetic substances tend to move into a magnetic field, such as that between the poles of a magnet (Fig. 22.7). Many transition metal complexes have unpaired electrons and hence are paramagnetic. Molecules such as N_2 (Section 7.10) and C_2H_6 (Section 8.3) and ions such as Na^+, $[Co(NH_3)_6]^{3+}$ (Section 26.6), and $[Fe(CN)_6]^{4-}$ (Section 26.7) that contain no unpaired electrons are **diamagnetic.** Diamagnetic substances have a slight tendency to move *out* of a magnetic field.

An electron in an atom spins about its own axis. Because the electron is electrically charged, this spin gives it the properties of a small magnet, with north and south poles. Two electrons in the same orbital spin in opposite directions, and their magnetic moments cancel because their north and south poles are opposed. When an electron in an atom or ion is unpaired, the magnetic moment due to its spin makes the entire atom or ion paramagnetic. Thus a sample containing such atoms or ions is paramagnetic. The size of the magnetic moment of a system containing unpaired electrons is related directly to the number of such electrons; the greater the number of unpaired electrons, the larger is the magnetic moment. Therefore, the observed magnetic moment is used to determine the number of unpaired electrons present.

26.10 Colors of Transition Metal Complexes

When atoms absorb light of the proper frequency, their electrons are excited to higher energy levels (Chapter 5). The same thing can happen in coordination compounds. Electrons can be excited from the lower-energy t_{2g} to the higher-energy e_g orbitals, provided that the latter are not already filled with paired electrons.

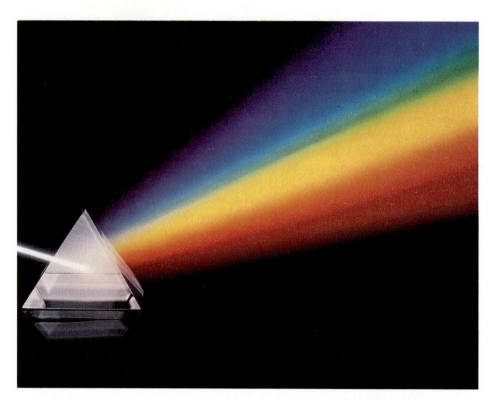

Figure 26.13
Passing white light through a prism shows that it is actually a mixture of all colors of visible light (red, orange, yellow, green, blue, indigo, and violet).

The human eye perceives a mixture of all the colors, in the proportions present in sunlight, as white light (Fig. 26.13) and utilizes complementary colors in color vision. The eye perceives a mixture of two complementary colors, in the proper proportions, as white light. Likewise, when a color is missing from white light, the eye sees its complement (Fig. 26.14). For example, as shown in Table 26.3, if red light is removed from white light, the eye sees the color blue-green; if violet is removed from white light, the eye sees lemon yellow; if green light is removed, the eye sees purple. The

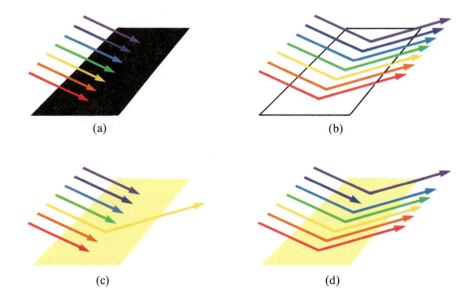

(a)

(b)

(c)

(d)

Figure 26.14
An object is black if it absorbs all colors of light (a); if it reflects all colors of light, it is white (b). An object (such as this yellow strip) has a color if (c) it absorbs all colors except one (yellow in this case) or if (d) it absorbs the complementary color from white light (the complementary color of yellow is indigo).

Table 26.3 Complementary colors

Wavelength, Å	Spectral color	Complementary color[a]
4100	Violet	Lemon yellow
4300	Indigo	Yellow
4800	Blue	Orange
5000	Blue-green	Red
5300	Green	Purple
5600	Lemon yellow	Violet
5800	Yellow	Indigo
6100	Orange	Blue
6800	Red	Blue-green

[a]The complementary color is seen when the spectral color is removed from white light.

Figure 26.15
A solution of $[Cu(NH_3)_4]^{2+}$

blue color of the $[Cu(NH_3)_4]^{2+}$ ion (Fig. 26.15) results because this ion absorbs orange and red light, leaving the complementary colors of blue and blue-green (Fig. 26.16).

Consider $[Fe(CN)_6]^{4-}$ (Section 26.7). The electrons in the t_{2g} orbitals can absorb energy and be excited to the higher energy level. The necessary energy corresponds to photons of violet light. If white light impinges on $[Fe(CN)_6]^{4-}$, violet light is absorbed (to accomplish the excitation), and the eye sees the unabsorbed complement, lemon yellow. $K_4[Fe(CN)_6]$ is lemon yellow. In contrast, if white light strikes $[Fe(H_2O)_6]^{2+}$, red light (longer wavelength, lower energy) is absorbed. The eye sees its complement, blue-green. $[Fe(H_2O)_6]SO_4$, for example, is therefore blue-green.

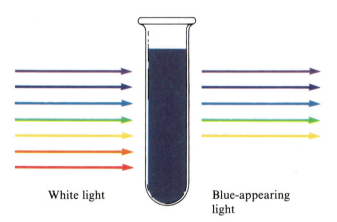

White light   Blue-appearing light

Figure 26.16
A solution of $[Cu(NH_3)_4]^{2+}$ is blue because the ion absorbs orange light and red light, the complementary colors of blue and blue-green, respectively.

As shown in Table 26.2, a coordination compound of the Cu^+ ion has a d^{10} configuration, and all the e_g orbitals are filled. In order to excite an electron to a higher level, such as the $4p$ orbital, photons of very high energy are needed. This energy corresponds to very short wavelengths in the ultraviolet region of the spectrum. Since no visible light is absorbed, the eye sees no change, and the compound appears white or colorless. A solution containing $[Cu(CN)_2]^-$, for example, is colorless (Fig. 26.17). On the other hand, Cu^{2+} complexes have a vacancy in the e_g orbitals, and electrons can be excited to this level. The wavelength (energy) of the light absorbed corresponds to the visible part of the spectrum, and Cu^{2+} complexes are almost always colored—blue, blue-green, violet, or yellow (Fig. 26.17).

Figure 26.17

From left to right, solutions of the colorless Cu^+ complex $[Cu(CN)_2]^-$ and the colored Cu^{2+} complexes $[Cu(H_2O)_4]^{2+}$ $[Cu(en)_2]^{2+}$, and $[CuBr_4]^{2-}$.

As we have noted earlier, strong-field ligands cause a large split in the energies of the *d* orbitals of the central metal atom (significantly negative 10Dq value). Transition metal coordination compounds with these ligands are yellow, orange, or red since they absorb higher-energy violet or blue light. On the other hand, coordination compounds of transition metals with weak-field ligands are blue-green, blue, or indigo since they absorb lower-energy yellow, orange, or red light.

26.11 The Molecular Orbital Theory

We have seen that a theory based primarily on covalent bonding is especially useful in explaining the structures of complexes and that a theory based primarily on ionic bonding is particularly useful in explaining the magnetic properties and the energies involved in the electron transitions that give rise to the colors of complexes. However, neither of these theories can explain certain other properties of complexes. Since most bonds have both ionic and covalent character, the theory of bonding in complexes can be improved by using molecular orbital concepts to introduce some contribution due to covalent bonding into the ligand field theory. In Chapter 7 we discussed the molecular orbital theory, and in later chapters we apply the theory to a variety of substances. Now we can use it to understand the nature of the bonding in coordination compounds.

Molecular orbital energy diagrams for complexes are the same in principle as those for diatomic species (Chapter 7) but are, of course, more complicated since several atoms are involved, instead of just two, and a large number of electrons must be considered. The molecular orbital energy diagrams for $[CoF_6]^{3-}$, possessing weak-field ligands, and for $[Co(NH_3)_6]^{3+}$, possessing strong-field ligands, are shown in Fig. 26.18. Only the six *d* electrons of the Co^{3+} ion and the unshared pair of electrons of each of the six ligands (a total of 18 electrons) are included.

The arrangement and symbols used in Fig. 26.18 are, in general, those described in Chapter 7. In addition, t_{1u}, a_{1g}, e_g, and t_{2g} are used beside the circles (orbitals) to label the various energy levels. In some cases the molecular orbital is also labeled $d_{x^2-y^2}$, d_{z^2}, or d_{xy}, etc., corresponding to the directional characteristics of the *d* orbital (Fig. 26.12). The antibonding and nonbonding energy levels use the e_g^* and t_{2g} designations, respectively, in the molecular orbital energy level diagram. The energy difference between them is Δ or 10Dq (Sections 26.7 and 26.8).

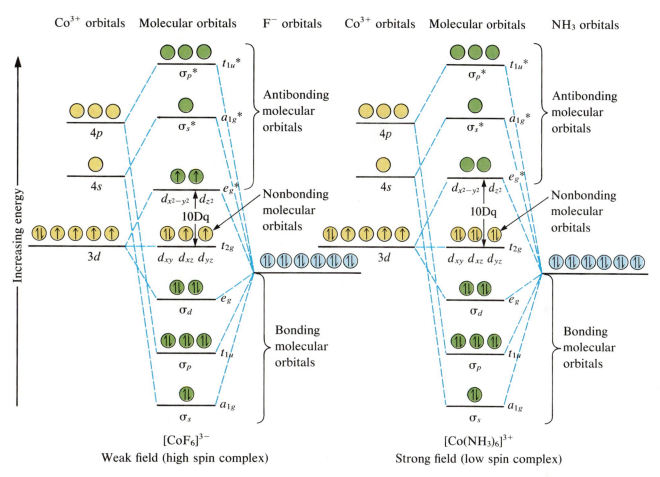

[CoF₆]³⁻
Weak field (high spin complex)

[Co(NH₃)₆]³⁺
Strong field (low spin complex)

Figure 26.18

Molecular orbital energy diagrams for $[CoF_6]^{3-}$ and $[Co(NH_3)_6]^{3+}$

Figure 26.18 shows that the ligand atomic orbitals are lower in energy than the corresponding metal atomic orbitals. This is indicative of some ionic character (Section 7.14); that is, some electronic charge has been transferred from the metal to the ligands.

Two of the d orbitals (the e_g orbitals: $d_{x^2-y^2}$ and d_{z^2}), are directed, as indicated in Section 26.7, toward the corners of the octahedron, where the ligands are located. In addition, the one $4s$ atomic orbital and the three $4p$ atomic orbitals of the metal are also so oriented with respect to the octahedron corners. Hence these metal atomic orbitals overlap with ligand orbitals, and six antibonding and six bonding molecular orbitals are formed. The other three d orbitals (the t_{2g} orbitals: d_{xy}, d_{xz}, and d_{yz}) do not point toward the ligand orbitals and hence are not involved in σ bonding; these molecular orbitals are called nonbonding orbitals. In complexes, therefore, the t_{2g} orbitals are unaffected by σ bonding, whereas each of the e_g orbitals combines with a ligand orbital to give an antibonding molecular orbital and a bonding molecular orbital.

Note in Fig. 26.18 that the 10Dq value is larger for the low-spin complex $[Co(NH_3)_6]^{3+}$. Measurements show the 10Dq value to be approximately 290 kilojoules per mole. Note also that the electrons in the t_{2g} orbitals are paired. For the high-spin complex $[CoF_6]^{3-}$ the 10Dq value is smaller (approximately 150 kilojoules per mole), and two unpaired electrons are in the antibonding molecular orbitals labeled e_g^*.

Analogous molecular orbital energy diagrams can be constructed for complexes that have geometrical structures other than octahedral, such as planar and tetrahedral.

26.12 Metal-to-Metal Bonds

In some compounds and complexes bonds occur between one metal and another. For example, the ions Hg_2^{2+}, $Re_2Br_8^{2-}$, and $Mo_2Cl_8^{4-}$ contain metal-to-metal bonds (Fig. 26.19). Several divalent metal acetates also contain metal-to-metal bonds. One typical example is chromium(II) acetate monohydrate, $[Cr(CH_3CO_2)_2 \cdot H_2O]_2$, which occurs as a dimer (Fig. 26.20). Each chromium atom in this molecule is in the center of an octahedron.

Figure 26.19

The structures of (a) Hg_2^{2+}, (b) $Re_2Br_8^{2-}$, and (c) $Mo_2Cl_8^{4-}$. In both (b) and (c) each metal atom is surrounded by a square plane of halide ions and occupies the apex of a square pyramid with respect to the other metal atom and its halide ions. The two planes are parallel to each other and are perpendicular to the rhenium-rhenium and molybdenum-molybdenum bonds.

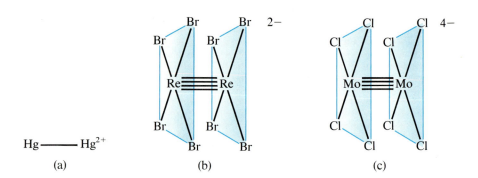

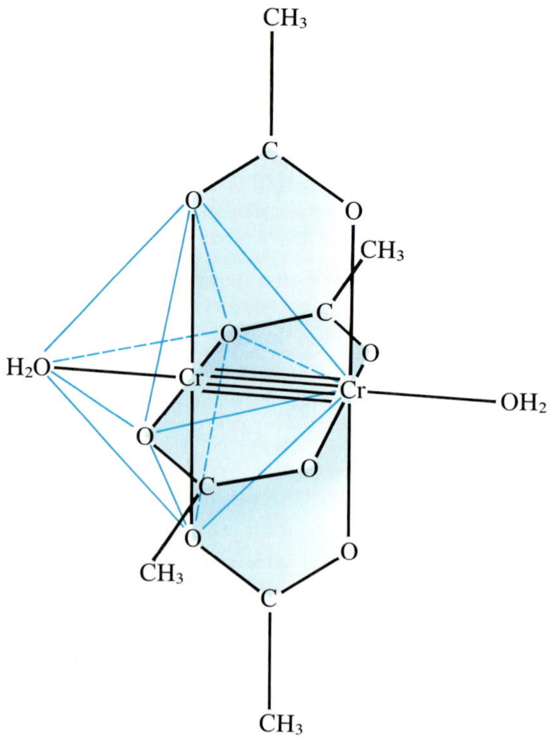

Figure 26.20

The structure of the chromium(II) acetate mono-hydrate dimer, which has a metal-to-metal bond. Each chromium atom is in the center of an octahedron. The octahedron is shown in color for one of the two chromium atoms.

Bonds between metal atoms in molecules and in simple ions are covalent bonds and can be described by the molecular orbital theory.

For Review

Summary

Metal ions can form **coordination compounds,** or **complexes,** in which a **central metal atom or ion** is bonded to two or more **ligands** by coordinate covalent bonds. Ligands with more than one donor atom are called **polydentate ligands** and form **chelates.** The common geometries found in complexes are tetrahedral and square planar (both with a **coordination number** of 4) and octahedral (with a coordination number of 6). *Cis* and *trans* **configurations** are possible in some octahedral and square planar complexes. In addition to these geometrical isomers, **optical isomers** (molecules or ions that are mirror images but not identical) are possible in certain octahedral complexes.

The bonding in coordination compounds can be described in terms of valence bonds, although this model of bonding does not explain the magnetic properties and colors very well. In the valence bond model the bonds are regarded as normal coordinate covalent bonds between a ligand, which acts as a Lewis base, and an empty hybrid metal atomic orbital.

The **ligand field theory** treats bonding between the metal and the ligands as a simple electrostatic attraction. However, the presence of the ligands near the metal ion changes the energies of the metal *d* orbitals relative to their energies in the free ion. Both the color and the magnetic properties of a complex can be attributed to this **ligand field splitting. Ligand field stabilization energy,** which also results from the ligand field splitting, can enhance the stability of a complex. The magnitude of the splitting (**10Dq** or **Δ**) depends on the nature of the ligands bonded to the metal. Strong-field ligands produce large splittings and favor formation of **low-spin complexes** in which the t_{2g} **orbitals** are completely filled before any electrons occupy the e_g **orbitals.** Weak-field ligands favor formation of **high-spin complexes.** Both the t_{2g} and e_g orbitals are singly occupied before any are doubly occupied.

The **molecular orbital theory** can describe the bonding in coordination compounds, and it can also explain the colors and magnetic properties of such complexes.

Key Terms and Concepts

bonding in complexes (26.6–26.11)
central metal atom or ion (26.1)
chelating ligand (26.1)
cis configuration (26.4)
colors of complexes (26.10)
coordination number (26.1)
coordination sphere (26.1)
diamagnetic (26.9)
donor atom (26.1)

e_g orbitals (26.7)
geometric isomers (26.4)
high-spin complex (26.8)
hybridization of orbitals (26.3)
isomerism (26.4)
ligand field splitting (26.7)
ligand field stabilization energy, LFSE (26.8)
ligand field theory (26.7)

ligands (26.1)
low-spin complex (26.8)
metal-to-metal bonding (26.11)
optical isomers (26.4)
paramagnetic (26.9)
polydentate ligand (26.1)
structures of complexes (26.3)
t_{2g} orbitals (26.7)
trans configuration (26.4)

Exercises

Structure and Nomenclature of Complexes

1. Indicate the coordination number for the central metal atom in each of the following compounds or ions:
   (a) $[Cu(NH_3)_4]^{2+}$
   (b) $[Cr(NH_3)_2(H_2O)_2Br_2]^+$
   (c) $[Co(NH_3)_4Br_2]_2SO_4$
   (d) $[Co(CO_3)_3]^{3-}$ (CO_3^{2-} is bidentate in this complex.)
   (e) $[Cr(en)_3]^{3+}$ (en = ethylenediamine)
   (f) $[Co(en)_2(NO_2)Cl]^+$
   (g) $[Pt(NH_3)(py)(Cl)(Br)]$ (py = pyridine, C_5H_5N)
   (h) $[Zn(NH_3)_2Cl_2]$
   (i) $[Co(C_2O_4)_2Cl_2]^{3-}$
   (j) $[Zn(NH_3)(py)(Cl)(Br)]$

2. Indicate the coordination number for the central metal atom in each of the following coordination compounds or ions:
   (a) $[Pd(NH_3)_2Br_2]$
   (b) $[Cr(en)_3](NO_3)_3$ (en = ethylenediamine)
   (c) $[Pt(NH_3)_2Cl_4]$
   (d) $[Pt(NH_3)_2Cl_2]$
   (e) $K_2[Ni(CN)_4]$
   (f) $[Co(en)_2Cl_2]^+$
   (g) $[Pt(H_2O)_2Br_2]$
   (h) $K_3[Fe(CN)_6]$
   (i) $[Ni(H_2O)_4Cl_2]$
   (j) $[PtNH_3)_4][PtCl_4]$

3. Explain what is meant by (a) coordination sphere, (b) ligand, (c) donor atom, (d) coordination number, (e) coordination compound, (f) complex, and (g) chelating ligand.

4. Using the metal ions Co^{3+}, Pt^{2+}, and Zn^{2+} and the ligands Cl^- and NH_3, construct two examples each of square planar complexes, tetrahedral complexes, and octahedral complexes. Do not use isomers.

5. Give the coordination numbers and write the formulas for each of the following, including all isomers where appropriate:
   (a) diamminedibromoplatinum(II)
   (b) dichlorobis(ethylenediamine)chromium(III) nitrate
   (c) hexacyanopalladate(IV) ion
   (d) potassium diamminetetrabromocobaltate(III)
   (e) hexaamminecobalt(III) hexacyanochromate(III)
   (f) tetrahydroxonickelate(II) ion
   (g) dibromoaurate(I) ion   (*aurum* is Latin for gold)

6. Name each of the compounds given in exercises 1 and 2 (including any isomers).

7. Name and sketch the structures of the following complexes. Indicate any *cis, trans,* and optical isomers.
   (a) $[Zn(NH_3)_2Cl_2]$
   (b) $[Zn(NH_3)(py)(Cl)(Br)]$ (py = pyridine, C_5H_5N)
   (c) $[Pt(H_2O)_2Cl_2]$

(d) $[Pt(NH_3)(py)(Cl)(Br)]$
(e) $[Co(C_2O_4)_2Cl_2]^{3-}$
(f) $[Ni(H_2O)_4Cl_2]$
(g) $[Pt(en)(NH_3)_2Br_2]^{2+}$

8. Draw diagrams for any *cis, trans,* and optical isomers that could exist for each of the following:
   (a) $[Pt(NH_3)_2Br_2]$ (b) $[Pt(NH_3)_2Br_4]$
   (c) $[Cr(en)_2Cl_2]^+$ (d) $[Co(H_2O)_2(NH_3)_2Cl_2]^+$
   (e) $[Co(en)_2(NO_2)Cl]^+$ (f) $[Co(en)_3]^{3+}$

9. Using NH_3 and NO_2^- as ligands, (a) give the formula of a six-coordinated cobalt(III) complex that would not give ions when dissolved in water; (b) give the formula of a complex containing one six-coordinated cobalt(III) ion that would give two ions when dissolved in water; and (c) sketch the isomers of the compounds formulated in (a) and (b).

10. In this and other chapters we have discussed substances whose structures encompass a considerable variety of geometric shapes. (a) Name each shape; (b) draw a diagram of each; (c) give a specific example of each; (d) indicate the relationship of atomic orbitals to each of the geometric shapes; (e) discuss the relationship of the geometric shape to the number of groups attached to the central element. (*Hint:* You will find it useful to make use of coordination compounds in your answer, but do not confine your answer to coordination compounds nor to the material in this chapter alone.)

Bonding in Complexes

11. Explain how the diphosphate ion, $P_2O_7^{4-}$ (Section 23.23) can function as a water softener by complexing Fe^{2+}.

12. Draw orbital diagrams and indicate the type of hybridization you would expect for each of the following:
    (a) $[Co(NH_3)_6]^{3+}$ (b) $[Cd(CN)_4]^{2-}$
    (c) $[Ni(CN)_4]^{2-}$ (d) $[Pd(NH_3)_2Br_2]$
    (e) $[Zn(NH_3)_2Br_2]$ (f) $[Cu(NH_3)_4]^{2+}$
    (g) $[Co(en)_2Cl_2]^-$ (h) $[CoCl_4]^{2-}$
    (i) $[PtCl_4]^{2-}$

13. Show by means of orbital diagrams the hybridization for each of the examples given in Table 26.1.

14. (a) Verify, by calculation, the ligand field stabilization energies listed in Table 26.2; (b) discuss the relationship of ligand field stabilization energies to the stabilities of complexes, and the limitations to the concept.

15. How many unpaired electrons will be present in each of the following?
    (a) $[MnCl_6]^{4-}$ (high-spin)
    (b) $[Mn(CN)_6]^{3-}$ (low-spin)
    (c) $[Mn(CN)_6]^{4-}$ (low-spin)
    (d) $[CoF_6]^{3-}$ (high-spin)
    (e) $[RhCl_6]^{3-}$ (low-spin)
    (f) $[Co(en)_3]^{3-}$ (low-spin)

16. The ligand field splitting in a tetrahedral complex is

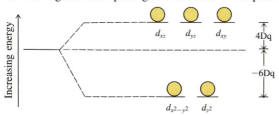

(a) Show that the ligand field stabilization energy for a tetrahedral complex such as VCl_4 with a d^1 configuration is $-6Dq$.

(b) All known tetrahedral complexes are high-spin. For what configuration will the value of the ligand field stabilization energy be greatest?

(c) Calculate the LFSE value for each of the following tetrahedral complexes: $[NiCl_4]^{2-}$, $[FeCl_4]^-$, $[CoBr_4]^{2-}$, $[Zn(OH)_4]^{2-}$, $[MnCl_4]^{2-}$.

17. Assume that you have a complex of a transition metal ion with a d^6 configuration. Can you tell whether the complex is octahedral or tetrahedral if a measurement of the magnetic moment establishes that it has no unpaired electrons? (See exercise 16 for the ligand field splitting in a tetrahedral complex.)

18. Is it possible for a complex of a metal of the first transition series to have six or seven unpaired electrons? Explain.

19. For complexes of the same metal ion with no change in oxidation number, the stability increases as the LFSE of the complex increases. Which complex in each of the following pairs of complexes is the more stable? $[Fe(H_2O)_6]^{2+}$ or $[Fe(CN)_6]^{4-}$; $[Co(NH_3)_6]^{3+}$ or $[CoF_6]^{3-}$; $[Mn(H_2O)_6]^{2+}$ or $[MnCl_6]^{4-}$; $[Co(NH_3)_6]^{3+}$ or $[Co(en)_3]^{3+}$.

20. What is the significance of the relative energy values of ligand atomic orbitals, metal atomic orbitals, and the corresponding molecular orbitals for $[CoF_6]^{3-}$ and $[Co(NH_3)_6]^{3+}$?

Additional Exercises

21. What type of tetrahedral complexes can form optical isomers?

22. The formation constant for $[Fe(CN)_6]^{3-}$ is 1×10^{44}; that for $[Fe(CN)_6]^{4-}$ is 1×10^{37}. Is this consistent with the

difference in LFSE between the two complexes? CN^- is a strong-field ligand in both complexes. What other feature of the ligand field model can explain the difference in formation constants?

23. Determine the ligand field splitting of the d orbitals in a two-coordinated complex as the two ligands approach the metal along the z axis.

24. The standard reduction potential for the reaction

$$[Co(H_2O)_6]^{3+} + e^- \longrightarrow [Co(H_2O)_6]^{2+}$$

is about 1.8 V. The standard reduction potential for the reaction

$$[Co(NH_3)_6]^{3+} + e^- \longrightarrow [Co(NH_3)_6]^{2+}$$

is $+ 0.1$ V. Show by calculation of the cell potentials which of the complex ions, $[Co(H_2O)_6]^{2+}$ or $[Co(NH_3)_6]^{2+}$, can be oxidized to the corresponding cobalt(III) complex by oxygen.   *Ans. $[Co(NH_3)_6]^{2+}$*

25. Calculate the concentration of free nickel ion that is present in equilibrium with 1.5×10^{-3} M $[Ni(NH_3)_6]^{2+}$ and 0.10 M NH_3.   *Ans. 8.3×10^{-6} M*

26. What is the ligand field splitting of the Ni^{2+} ion in NiO? Of the Fe^{3+} ion in $FeCl_3$? (See Section 11.17.)

27. Trimethylphosphine, $:P(CH_3)_3$, can act as a ligand by donating the free pair of electrons on the phosphorus atom. If trimethylphosphine is added to a solution of nickel(II) chloride in acetone, a blue compound with a molecular weight of approximately 270 and containing 21.5% Ni, 26.0% Cl, and 52.5% $P(CH_3)_3$ can be isolated. This blue compound does not have any isomeric forms. What is the geometry and molecular formula of the blue compound?   *Ans. Tetrahedral; $NiCl_2[P(CH_3)_3]_2$*

28. The complex ion $[Co(en)_3]^{3+}$ is diamagnetic. Would you expect the $[Co(en)_3]^{2+}$ ion to be diamagnetic or paramagnetic? The $[Co(CN)_6]^{3-}$ ion? Explain your reasoning in each case.

29. Using the radius ratio rule and the ionic radii given inside the back cover, predict whether complexes of Ni^{2+}, Mn^{2+}, and Sc^{3+} with Cl^- will be tetrahedral or octahedral. Write the formulas for these complexes.
    Ans. $[NiCl_4]^{2-}$, tetrahedral; $[MnCl_6]^{4-}$, octahedral; $[ScCl_6]^{3-}$, octahedral

The Transition Elements

27

Because it is malleable, ductile, and relatively unreactive, copper is used to make flexible pipe.

We have daily contact with several of the transition elements: the metals used for decoration and to manufacture many of the items we use regularly. Iron, copper, nickel, chromium, silver, and gold are the transition metals that are most often encountered. Iron is used in the fabrication of many metal items, ranging from the rings in your notebook and the cutlery in your kitchen to automobiles, ships, and buildings. Copper is used in electrical wiring and coins, nickel in coins, chromium as a protective plating on automobile bumpers and plumbing fixtures, and silver and gold in jewelry.

All of the transition elements are metals. Their physical properties are typical of metals, but they vary in reactivity from moderately active to almost inert.

The outermost *s* orbital and the first inner *d* orbitals counting from the outside comprise the valence shells of the transition elements. Some or all of the electrons in these two subshells are used when these elements form compounds. Because a given element can often exhibit several different oxidation numbers and because the presence of a partially filled *d* subshell leads to colored compounds, these elements exhibit a rich and fascinating chemistry with a variety of oxidation numbers.

27.1 The Transition Metals

Two sets of metals that include a total of 62 elements may be identified as transition metals (Fig. 27.1): the **d-block elements,** which are sometimes called the transition elements, and the **f-block elements,** sometimes called the inner transition elements. The *d*-block elements are those elements in which the second shell, counting in from the outside is filling from 8 to 18 electrons as its *d* orbitals fill. The *f*-block elements are those elements in which the third shell, counting in from the outside, is building from 18 to 32 electrons as its *f* orbitals fill. Table 5.5 lists the electron configurations of the atoms.

The *d*-block elements are sometimes classified as the **first transition series** (the elements Sc through Cu), the **second transition series** (the elements Y through Ag), and the **third transition series** (the elements La and Hf through Au). Actinium, Ac, is the first member of the **fourth transition series,** which is incomplete at present. The elements Ce through Lu constitute the **lanthanide series** or **rare earth elements;** and the elements Th through Lr are the **actinide series.** The rare earth elements and the actinide elements are the *f*-block elements.

Because lanthanum behaves very much like the lanthanide elements, it is often considered to be a lanthanide element even though its electron configuration makes it the first member of the third transition series. Similarly, the behavior of actinium often causes it to be considered with the actinide series, although its electron configuration makes it the first member of the fourth transition series.

All of the transition elements have common properties. They are almost all hard, strong, high-melting metals (Fig. 27.2) that conduct heat and electricity well. They readily form alloys with one another and with other metallic elements. The elements of the *f*-block, of Group IIIB, and of the first transition series are sufficiently active that

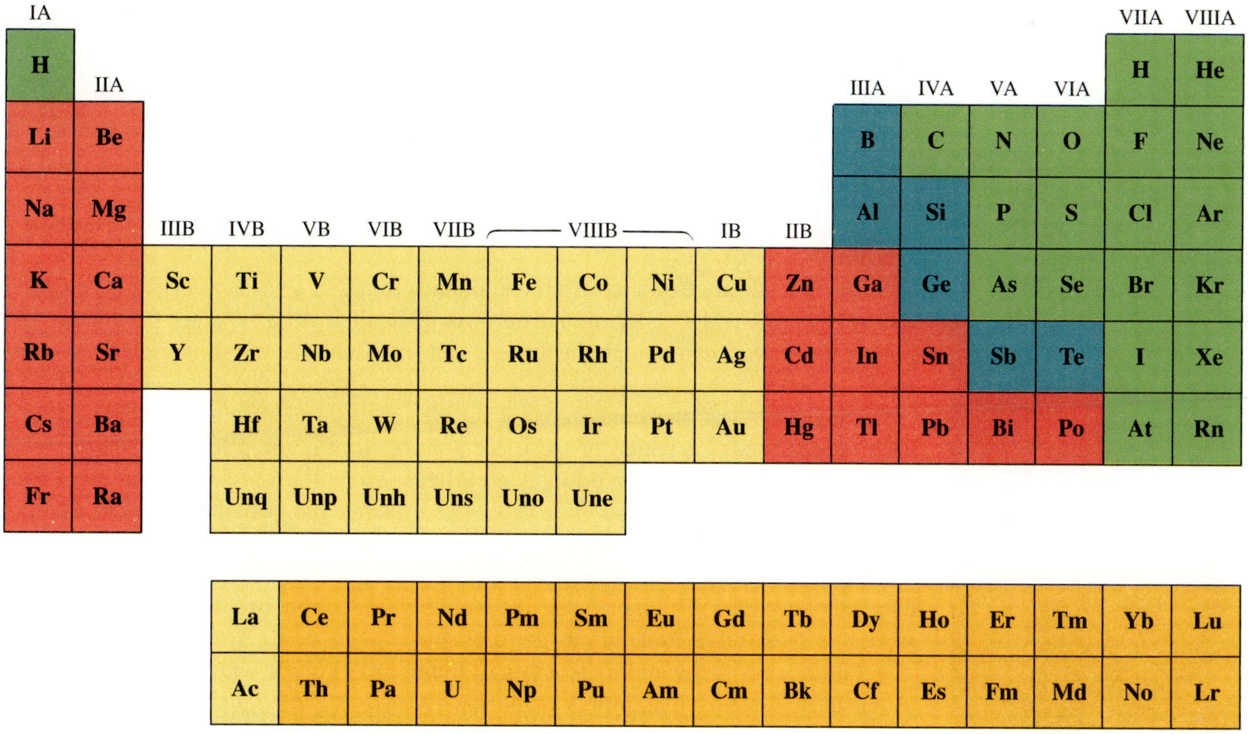

Figure 27.1

The location of the *d*-block elements, (shown in yellow) and the *f*-block elements, the inner transition metals, (shown in orange) in the Periodic Table. Nonmetals are shown in green; semi-metals, in blue; and the remaining metals in red.

Figure 27.2
Several transition metals are used in the fabrication of durable items.

they react with acids. With a few exceptions they form compounds that are colored and exhibit variable oxidation numbers. Both the *d*-block and the *f*-block elements form a vast array of coordination compounds (Chapter 26).

In this chapter we will focus primarily on the chemical behavior of the elements of the first transition series, although occasional comparisons of these elements with their heavier congeners will appear.

27.2 Preparation of the Transition Metals

Figure 27.3
A naturally occurring nugget of copper.

Iron, copper, silver, and gold were known to the ancients because these metals occur free in nature (Fig. 27.3). Elemental iron resulting from meteorite falls is found in small quantities; free copper, silver, and gold occur in larger amounts. The largest known deposit of free copper is in the Upper Peninsula of Michigan, near the border between Canada and the United States. The largest single mass of elemental copper yet found, weighing 420 tons, is on display in the Smithsonian Institution in Washington, D.C.

Most metals are extracted from compounds found in a variety of ores. The ease of recovery varies widely, depending on how difficult it is to separate the metal compound from other compounds in the ore and to reduce the combined element to the metal.

The majority of ores contain metal sulfides or metal oxides, and the processing of such ores depends on which of these are present. Iron is often found as oxides in its ores, and copper and silver are found as sulfides. We will discuss the isolation of these metals in detail because these three processes illustrate the principal routes to isolation of metals. First, however, let us look at the general procedures for refining metals.

Most ores of titanium, vanadium, chromium, manganese, and iron contain these elements as oxides. The metals are recovered by reduction after the oxides have been separated from other minerals in the ores.

Titanium is recovered by reaction of titanium(IV) oxide with carbon and chlorine at red heat, forming titanium(IV) chloride.

$$TiO_2 + C + 2Cl_2 \longrightarrow TiCl_4 + CO_2$$

The $TiCl_4$ is then reduced with magnesium metal at about 1000°C. At the temperatures of the reaction, the magnesium chloride is volatile and sublimes away leaving titanium metal.

$$TiCl_4(g) + 2Mg \longrightarrow Ti + 2MgCl_2(g)$$

An inert atmosphere is used to prevent oxidation of the titanium metal. **Vanadium** is produced in a similar way, using calcium metal to reduce vanadium(V) oxide to vanadium metal.

Metallic **chromium** and **manganese** are produced by the reduction of the oxides Cr_2O_3 and MnO_2, respectively, with aluminum metal. Chromium is also produced by the electrolytic reduction of its compounds from aqueous solution, a process used in chromium plating.

Iron is produced in a blast furnace by the reduction of iron oxides with carbon at elevated temperatures, as described later in this section.

The common ores of cobalt, nickel, and copper contain these elements as mixtures of sulfides that often contain iron sulfides. These elements are more difficult to recover than iron because of the need to separate the various metal sulfides and the difficulty of reducing some of the metal ions. Seventeenth-century metallurgists thought that nickel ores were copper ores because of the similarity in appearance, but when the ores failed to yield copper, they were called *kupfernickel* (German). *Kupfer* means copper, and *nickel* means a devil, which was thought to prevent the extraction of copper from the ores. The name cobalt is derived from the German word *kobold*, which means goblin. Early metallurgists believed that goblins carefully guarded the cobalt ores to prevent human beings from liberating the metal.

The metallurgy of **cobalt** involves the separation of cobalt sulfide from the nickel, copper, and iron sulfides, conversion to Co_3O_4, and the reduction of this oxide with aluminum or hydrogen.

$$3Co_3O_4 + 8Al \longrightarrow 9Co + 4Al_2O_3$$
$$Co_3O_4 + 4H_2 \longrightarrow 3Co + 4H_2O$$

The impure metal is purified by electrolytic deposition on a rotating stainless steel cathode.

Nickel is also recovered by conversion of the sulfide to the oxide by roasting. The oxide is then reduced by carbon, giving a metal that is approximately 96% pure. It is purified electrolytically to 99.98% nickel. Any gold, silver, or platinum that may be present is not affected by the electrolysis and is recovered from the residues at the anode.

When copper(I) sulfide has been separated from the sulfides of the other metals, **copper** is recovered by heating with air. The heating converts part of the Cu_2S to Cu_2O, and as soon as copper(I) oxide is formed, it is reduced by the remaining copper(I) sulfide to metallic copper. The equations are

$$2Cu_2S + 3O_2 \longrightarrow 2Cu_2O + 2SO_2$$
$$2Cu_2O + Cu_2S \longrightarrow 6Cu + SO_2$$

The impure copper is purified electrolytically.

The precious metals are recovered by **hydrometallurgy,** the separation of an element from other elements using aqueous solutions. For example, silver and gold are recovered by removing the metals from their ores as the complex ions $[Ag(CN)_2]^-$ and $[Au(CN)_2]^-$ followed by reduction to the free metal by the use of metallic zinc.

PRODUCTION OF IRON. The early application of iron to the manufacture of tools and weapons was possible because of the wide distribution of iron ores and the ease

A photomicrograph of crystalline gold.

with which the iron compounds in the ores could be reduced by carbon. For a long time charcoal was the form of carbon used in the reduction process. The production and use of iron on a large scale began about 1620, when coal was introduced as the reducing agent.

The important iron oxide ores are **hematite, Fe₂O₃**, and **magnetite, Fe₃O₄.** The presence of Fe_2O_3 accounts largely for the red coloration of many rocks and soils. Hematite often occurs as the reddish-brown hydrate $2Fe_2O_3 \cdot 3H_2O$, which is called **limonite. Siderite** is a white or brownish carbonate, **FeCO₃. Pyrite, FeS₂**, occurs as pale yellow crystals with a metallic luster, which are easily mistaken for gold and are therefore called fool's gold.

Taconite, a hard rock composed of crystals of magnetite interlocked with crystals of silica, makes up some 95% of the iron-bearing deposits of the famous Mesabi Range in upper Minnesota. Once regarded as useless for steel production, the taconite ores are now used extensively as a result of the development of satisfactory techniques for mining and concentrating them.

The first step in the metallurgy of iron is usually roasting the ore to remove water, decompose carbonates, and oxidize sulfides. The oxides are then reduced in a blast furnace 80–100 feet high and 25 feet in diameter. The charge of roasted ore, coke, and limestone is introduced continuously into the top of the furnace. Molten iron and slag are withdrawn at the bottom. The entire stock in a furnace may weigh several hundred tons.

Near the bottom of a furnace are nozzles through which preheated air (500°C) is blown into the furnace. As soon as the air enters, the coke in the region of the nozzles is oxidized to carbon dioxide with the liberation of a great deal of heat, which raises the temperature to about 1500°C. As the carbon dioxide passes upward through the overlying layer of white-hot coke, it is reduced to carbon monoxide.

$$CO_2 + C \longrightarrow 2CO$$

The carbon monoxide serves as the reducing agent in the upper regions of the furnace. The individual reactions are indicated in Figure 27.4.

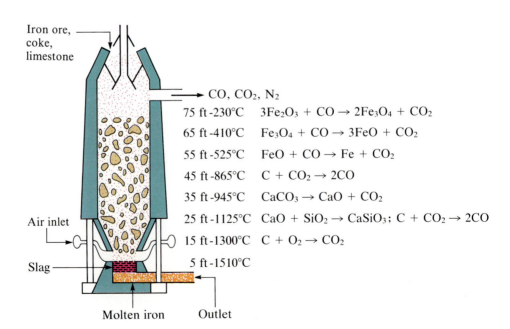

Iron ore, coke, limestone

CO, CO₂, N₂

75 ft -230°C	$3Fe_2O_3 + CO \rightarrow 2Fe_3O_4 + CO_2$
65 ft -410°C	$Fe_3O_4 + CO \rightarrow 3FeO + CO_2$
55 ft -525°C	$FeO + CO \rightarrow Fe + CO_2$
45 ft -865°C	$C + CO_2 \rightarrow 2CO$
35 ft -945°C	$CaCO_3 \rightarrow CaO + CO_2$
25 ft -1125°C	$CaO + SiO_2 \rightarrow CaSiO_3$; $C + CO_2 \rightarrow 2CO$
15 ft -1300°C	$C + O_2 \rightarrow CO_2$
5 ft -1510°C	

Air inlet

Slag

Molten iron   Outlet

Figure 27.4

Reactions that occur in a blast furnace.

The iron oxides are completely reduced in the middle region of the furnace. Also in this region the limestone (calcium carbonate) decomposes, and the resulting calcium oxide combines with silica and silicates in the ore to form slag.

$$CaO + SiO_2 \longrightarrow CaSiO_3$$

Just below the middle of the furnace, the temperature is high enough to melt both the iron and the slag. They collect in layers at the bottom of the furnace; the less dense slag floats on the iron and protects it from oxidation. Several times a day, the slag and molten iron are withdrawn from the furnace. The iron is transferred to casting machines or to a steelmaking plant (Fig. 27.5). The slag is often used in the manufacture of Portland cement. A modern blast furnace can produce as much as 3000 tons of iron daily.

The hot exhaust gases, which issue from the top of the blast furnace, contain some unoxidized carbon monoxide and are mixed with air and burned to preheat the air used in the operation of the blast furnace.

PREPARATION OF COPPER. The most important copper ores are sulfides such as **chalcocite, Cu_2S,** and **chalcopyrite, $CuFeS_2$.** Other ores are **cuprite, Cu_2O, melaconite, CuO,** and **malachite, $Cu_2(OH)_2CO_3$.**

Ores that contain elemental (free) copper are pulverized, the unwanted rock material is washed away, and the copper is melted and poured into molds to cool. Oxide and carbonate ores are often leached with sulfuric acid to produce copper(II) sulfate solutions, from which copper metal may be obtained by electrodeposition. High-grade oxide or carbonate ores are reduced by heating them with coke mixed with a suitable flux, which removes impurities by combining with them.

Sulfide ores usually contain less than 10% copper. These ores are first concentrated by **flotation.** In this process the pulverized ore is mixed with water to which a carefully selected oil has been added. A froth is produced by blowing air through the mixture. The metal sulfide particles have little or no attraction for water, but they do have an attraction for oil. They are preferentially coated by the oil, and the oily particles adhere to the air bubbles in the froth, which floats on the surface. The particles of sand, rock, clay, etc., are wetted by water and sink to the bottom of the flotation vat. The froth containing the metal-bearing particles is removed, and the ore is recovered in a highly concentrated form.

The concentrate is then roasted in a furnace, at a temperature below its melting point, to drive off the moisture and remove part of the sulfur as sulfur dioxide. The remaining mixture, which consists of Cu_2S, FeS, FeO, and SiO_2 is **smelted** (extracted in the molten state) by mixing it with limestone, which serves as a flux, and heating the mixture so that it melts. The reactions that take place during the formation of the slag are

$$CaCO_3 + SiO_2 \longrightarrow CaSiO_3 + CO_2$$
$$FeO + SiO_2 \longrightarrow FeSiO_3$$

Reduction of the Cu_2S that remains after smelting is accomplished by blowing air through the molten material. The air converts part of the Cu_2S to Cu_2O. As soon as copper(I) oxide is formed, it is reduced by the remaining copper(I) sulfide to metallic copper. The last traces of copper oxide are removed by reduction with H_2 and CO produced from natural gas. The copper obtained in this way is called **blister copper** because of its characteristic appearance due to the air blisters it contains.

The impure copper is cast into large plates, which are used as anodes in the electrolytic purification of the metal (Section 21.4). Thin sheets of pure copper serve as the cathodes; copper(II) sulfate, in a solution of sulfuric acid, is used as the electrolyte.

Figure 27.5
Casting molten iron.

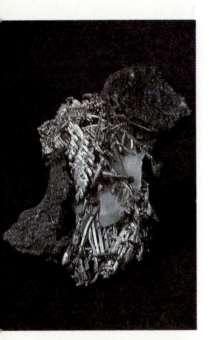

Naturally occurring free silver may be found as nuggets or in veins.

The impure copper passes into solution from the anodes, and pure copper plates out on the cathodes as electrolysis proceeds. Inert metals in the anodes do not oxidize but fall to the bottom of the electrolytic cell as anode mud, along with bits of slag and Cu_2O. Silver, gold, and the platinum metals are recovered from the anode mud. Metals more active than copper, such as zinc and iron, are oxidized at the anode, and their cations pass into solution, where they remain. The copper deposited on the cathode has an average purity of 99.95%. The value of the precious metals recovered from the anode mud is often sufficient to pay the cost of the electrolytic refining.

PREPARATION OF SILVER. Silver is sometimes found in large nuggets but more frequently in veins and related deposits. It is frequently found alloyed with gold, copper, or mercury. It occurs combined as the chloride, AgCl (horn silver), and as the sulfide, Ag_2S, which is usually mixed with the sulfides of lead, copper, nickel, arsenic, and antimony. Most silver produced is a by-product of the mining of other metals such as lead and copper.

When lead is obtained from lead sulfide ore, it is usually mixed with some silver. The silver is extracted by the use of zinc, in which silver is about 3000 times more soluble than it is in lead. The lead is melted and thoroughly mixed with a small quantity of zinc. Lead and zinc are immiscible. Most of the silver leaves the lead and dissolves in the zinc. When mixing is stopped, the zinc rises to the surface of the lead and solidifies. The zinc-silver alloy is removed from the lead and the more volatile zinc separated from the silver by distillation. This extraction procedure is known as the **Parkes process.**

The extraction of silver from its ores is dependent on the formation of the complex dicyanoargentate ion, $[Ag(CN)_2]^-$. Silver metal and all of its compounds are readily dissolved by alkali metal cyanides in the presence of air. Representative equations for the extraction of silver are

$$4Ag + 8CN^- + O_2 + 2H_2O \longrightarrow 4[Ag(CN)_2]^- + 4OH^-$$
$$2Ag_2S + 8CN^- + O_2 + 2H_2O \longrightarrow 4[Ag(CN)_2]^- + 2S + 4OH^-$$
$$AgCl + 2CN^- \rightleftharpoons [Ag(CN)_2]^- + Cl^-$$

The silver is precipitated from the cyanide solution by addition of either zinc or aluminum.

$$2[Ag(CN)_2]^- + Zn \longrightarrow 2Ag + [Zn(CN)_4]^{2-}$$

Silver is also obtained from the anode mud formed during the electrolytic refining of copper.

27.3 Properties of the Transition Elements

Both the d-block and f-block elements are metals and generally have the usual physical properties of metals. However, these elements vary from active metals to very inactive metals.

The elements of the first transition series (Sc–Cu) form ionic compounds that dissolve in water giving stable solvated cations. The heavier d-block elements usually do not form simple positive ions that are stable in water. Thus, the Cr^{3+}, Fe^{3+}, and Co^{2+} ions are stable in aqueous solutions (Fig. 27.6) whereas the Mo^{3+}, Ru^{2+}, and Ir^{2+} ions are not. The majority of simple water-stable ions formed by the heavier d-block elements are oxyanions such as MoO_4^{2-}, and ReO_4^-. In general, higher oxidation numbers are more stable for members of the second and third transition series than for the first transition series.

The heavier elements of Group VIIIB (ruthenium, osmium, rhodium, iridium, palladium, and platinum) are sometimes called the **platinum metals.** These elements and gold are particularly nonreactive. They do not form simple cations that are stable in water and, unlike the earlier elements in the second and third transition series, they do not form stable oxyanions.

The inner transition elements, the f-block elements, are among the most active of the transition elements.

The transition elements, using the term broadly to refer to both the d-block elements and the f-block elements, react with nonmetals, forming binary compounds; heating is usually required. Both the d-block and the f-block elements react with halogens to form a variety of halides, ranging in oxidation number from $+1$ to $+6$. With the exception of palladium, platinum, and gold, they react with sulfur to form sulfides. Oxygen reacts with all of the transition elements except for palladium, platinum, silver, and gold, whose oxides decompose upon heating. The f-block elements, the elements of Group IIIB, and the elements of the first transition series (except copper) are sufficiently active to react with aqueous solutions of strong acids forming hydrogen and solutions of the corresponding salts.

In crossing the three series of the d-block elements, each successive element can form compounds with a wider range of oxidation numbers. The common oxidation numbers of the elements of the first transition series are listed in Table 27.1.

The elements at the beginning of the first transition series exhibit a highest oxidation number that corresponds to the loss of all of the electrons in both the s and d orbitals of their valence shell. The titanium(IV) ion, for example, is formed when the titanium atom loses its two $3d$ and two $4s$ electrons. These group oxidation numbers are the most stable oxidation numbers for scandium, titanium, and vanadium. Moving across the series, it becomes progressively more difficult to form these highest oxidation numbers. Scandium, like aluminum, exhibits no lower oxidation numbers in its compounds. Compounds of titanium(IV) are harder to reduce than those of vanadium(V), which in turn are harder to reduce than those of chromium(VI) or manganese(VII). No compounds of iron(VIII) are known. Beyond manganese, the elements are stable with oxidation numbers of $+1$, $+2$, or $+3$. All of the elements of the first transition series form ions with a charge of $+2$ or $+3$ that are stable in water, although those of the early members of the series can be readily oxidized by air.

Figure 27.6
Solutions containing $Cr^{3+}(aq)$, $Fe^{3+}(aq)$, and $Co^{2+}(aq)$ (left, center, and right, respectively) are stable.

Table 27.1   Common oxidation numbers of the elements of the first transition series. Numbers in boldface indicate oxidation numbers that are stable in water.

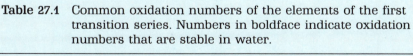

Sc	Ti	V	Cr	Mn	Fe	Co	Ni	Cu
								$+1$
	$+2$	$+2$	$+2$	$+2$	$+2$	$+2$	$+2$	$+2$
$+3$	$+3$	$+3$	$+3$	$+3$	$+3$	$+3$		
	$+4$	$+4$	$+4$	$+4$				
		$+5$						
			$+6$					
				$+7$				

The elements of the second and third transition series generally are more stable with higher oxidation numbers than are the elements of the first series. For example, the simple chemistry of molybdenum and tungsten, members of Group VIB, is limited to an oxidation number of +6 in aqueous solution. Chromium, the lightest member of the group, forms stable Cr^{3+} ions in water and, in the absence of air, stable Cr^{2+} ions in water. The sulfide with highest oxidation number for chromium is Cr_2S_3, which contains the Cr^{3+}ion. Molybdenum and tungsten form sulfides in which the metal exhibits oxidation numbers of +4 and +6.

27.4 The Lanthanide Contraction

Much of the similarity of the chemistry of the elements of the second and third transition series can be attributed to the similar sizes of these elements within each group, in addition to the similarity in valence electron structure. As may be seen from the table of atomic and ionic radii inside the back cover of this book, there is the expected increase in radii of the atoms or ions in going down a group from the first d-block transition series to the second series. However, within a given group of d-block transition elements, the radii of the elements of the second and third series are almost the same. The two series of d-block elements have similar radii, because the lanthanide (f-block) elements show a decreasing size with increasing atomic number, as is shown by the ionic radii (+3 ion) listed in Table 27.2. This size decrease, referred to as the **lanthanide contraction,** results in an expected regular gradation in properties through the series, including a decrease in the basicity of the hydroxides, $M(OH)_3$, with increasing atomic number of the elements.

Also as a result of the lanthanide contraction and illustrative of the statement above regarding similarities in sizes of corresponding d-block transition elements of the second and third series in a given group, zirconium and hafnium are the same size (atomic radius, 1.57 Å); hafnium is smaller than expected because of the contraction in size

Table 27.2   Electron configurations and ionic radii of the rare earth elements

Atomic number	Element	Electron configuration	Ionic (M^{3+}) radius, Å
57	La	$1s^2 2s^2 2p^6 3s^2 3p^6 3d^{10} 4s^2 4p^6 4d^{10} 4f^0\ 5s^2 5p^6 5d^1 6s^2$	1.15
58	Ce	$4f^2\ 5s^2 5p^6 5d^0 6s^2$	1.034
59	Pr	$4f^3\ 5s^2 5p^6 5d^0 6s^2$	1.013
60	Nd	$4f^4\ 5s^2 5p^6 5d^0 6s^2$	0.995
61	Pm	$4f^5\ 5s^2 5p^6 5d^0 6s^2$	0.979
62	Sm	$4f^6\ 5s^2 5p^6 5d^0 6s^2$	0.964
63	Eu	$4f^7\ 5s^2 5p^6 5d^0 6s^2$	0.950
64	Gd	$4f^7\ 5s^2 5p^6 5d^1 6s^2$	0.938
65	Tb	$4f^9\ 5s^2 5p^6 5d^0 6s^2$	0.923
66	Dy	$4f^{10} 5s^2 5p^6 5d^0 6s^2$	0.908
67	Ho	$4f^{11} 5s^2 5p^6 5d^0 6s^2$	0.894
68	Er	$4f^{12} 5s^2 5p^6 5d^0 6s^2$	0.881
69	Tm	$4f^{13} 5s^2 5p^6 5d^0 6s^2$	0.869
70	Yb	$4f^{14} 5s^2 5p^6 5d^0 6s^2$	0.858
71	Lu	$4f^{14} 5s^2 5p^6 5d^1 6s^2$	0.848

through the lanthanide elements immediately preceding it. Zirconium and hafnium, since they are not only in the same family but also of the same size, are similar in chemical properties, occur together in nature, and are difficult to separate (Section 27.10). Similarly, the pairs niobium and tantalum and molybdenum and tungsten are more similar in size than expected, with a resulting enhanced similarity in properties for each pair. The lanthanide contraction also accounts for the fact that yttrium has nearly the same radius as gadolinium, which is about midway through the lanthanide series. With very similar charge and size characteristics as compared to the lanthanide elements, yttrium has very similar properties and occurs with the rare earth elements in nature.

27.5 Compounds of the Transition Metals

The simple compounds of the transition metals range from ionic to covalent. In their lower oxidation states the transition metals form ionic compounds; in their higher oxidation states, they form covalent compounds. Unlike the active metals, the variation in oxidation states of the transition metals gives a metal-based oxidation-reduction chemistry to these compounds.

1. HALIDES. Anhydrous halides of each of the transition elements can be prepared by the direct reaction of the metal with halogens. For example,

$$2Fe + 3Cl_2 \longrightarrow 2FeCl_3$$

Heating a metal halide with additional metal can be used to form a halide of the metal with a lower oxidation number.

$$Fe + 2FeCl_3 \longrightarrow 3FeCl_2$$

The stoichiometry of the metal halide that results from the reaction of a metal with a halogen is determined by the relative amounts of metal and halogen and by the strength of the halogen as an oxidizing agent. Fluorine forms halides containing metals with the highest oxidation numbers. The other halogens may not form analogous compounds. In fact, the iodide ion is a sufficiently good reducing agent that the iodides of some metals are limited to compounds containing the metal with its lowest oxidation number. The oxidation numbers of the first transition series metals in their binary halides are given in Table 27.3.

The nature of the bonding in the anhydrous halides of the elements of the first transition series varies with the oxidation number of the metal. Halides of metals with lower oxidation numbers are ionic; halides of metals with higher oxidation numbers are covalent. For example, the titanium chlorides $TiCl_2$ and $TiCl_3$ are ionic compounds

Table 27.3 Oxidation numbers of some transition metals in their binary halides

	Sc	Ti	V	Cr	Mn	Fe	Co	Ni	Cu
F	3	3,4	3,4,5	2,3,4,5	2,3,4	2,3	2,3	2	2
Cl	3	2,3,4	2,3,4	2,3	2	2,3	2	2	1,2
Br	3	2,3,4	2,3,4	2,3	2	2,3	2	2	1,2
I	3	2,3,4	2,3	2,3	2	2	2	2	1

with high melting points, whereas $TiCl_4$ is a volatile liquid with covalent titanium-chlorine bonds. The halides of the heavier d-block elements all have a significant covalent component.

Stable water solutions of the halides of the metals of the first transition series, the rare earth elements, and of the actinides are generally prepared by addition of a hydrohalic acid to carbonates, hydroxides, oxides, or other compounds containing basic anions. Sample reactions are indicated.

$$NiCO_3(s) + 2H^+(aq) + 2F^-(aq) \longrightarrow Ni^{2+}(aq) + 2F^-(aq) + H_2O(l) + CO_2(g)$$
$$Co(OH)_2(s) + 2H^+(aq) + 2Br^-(aq) \longrightarrow Co^{2+}(aq) + 2Br^-(aq) + 2H_2O(l)$$
$$La_2O_3(s) + 6H^+(aq) + 6I^-(aq) \longrightarrow 2La^{3+}(aq) + 6I^-(aq) + 3H_2O(l)$$

Many of these metals also dissolve in these acids, forming a solution of the salt and hydrogen gas. Evaporation of the water gives hydrated halides such as $NiF_2 \cdot 6H_2O$, $CoBr_2 \cdot 7H_2O$, and $LaI_3 \cdot 9H_2O$, where the center dots in the formulas indicate that the water is present in the solids as **water of hydration** (that is, as identifiable molecules). The metals of the second and third transition series are generally not soluble in acids.

The reaction of water with the anhydrous halides of the elements of the first transition series depends on the oxidation number of the transition metal. Compounds with metal oxidation numbers of $+1$, $+2$, and $+3$, if soluble, dissolve forming solutions containing metal ions and the respective halide ions. Those transition metal ions that are not stable in water react with it or disproportionate; Ti^{2+} is oxidized by water, Co^{3+} is reduced by water, and Cu^+ and Mn^{3+} disproportionate. The majority of the halides of the elements of the second and third transition series are either not soluble or decompose in water.

The covalent behavior of the transition metals with higher oxidation numbers is exemplified by the reaction of the metal tetrahalides with water. Both the covalent titanium and vanadium tetrahalides react with water to give solutions containing the corresponding hydrohalic acids and the covalent oxyions TiO^{2+} and VO^{2+}, respectively.

$$TiCl_4(l) + H_2O(l) \longrightarrow TiO^{2+}(aq) + 2H^+(aq) + 4Cl^-(aq)$$
$$VBr_4(l) + H_2O(l) \longrightarrow VO^{2+}(aq) + 2H^+(aq) + 4Br^-(aq)$$

The tetrafluorides of chromium and manganese react with water forming hydrofluoric acid and precipitates of the dioxides. Vanadium(V) fluoride reacts with water and forms hydrofluoric acid and vanadium(V) oxide.

2. OXIDES. Oxides of the transition metals with oxidation numbers of $+1$, $+2$, $+3$, and $+4$ behave as ionic compounds that contain metal ions and oxide ions, whereas those with oxidation numbers of $+5$, $+6$, and $+7$ contain covalent metal-oxygen bonds. Oxides of transition metals with the lowest oxidation numbers are basic, the intermediate ones are amphoteric, and the highest ones are primarily acidic. Table 27.4 lists the common oxides of the metals of the first transition series.

The oxides indicated in boldface in Table 27.4 are stable when heated in air and form when the metals are oxidized by heating in air. They also form when hydroxides, carbonates, oxalates, and certain other oxides of these metals are heated in air.

Oxides of the metals of the first transition series with lower metal oxidation numbers may be prepared by heating an oxide containing the metal in a higher oxidation number with a reducing agent such as hydrogen or the metal. For example, **Vanadium(III) oxide, V_2O_3,** can be prepared by heating vanadium(V) oxide in an

Table 27.4   Oxides of the metals of the first transition series

								Cu$_2$O,
	TiO	VO		MnO	FeO	CoO	NiO	CuO
				Mn$_3$O$_4$	**Fe$_3$O$_4$**	**Co$_3$O$_4$**		
Sc$_2$O$_3$	Ti$_2$O$_3$	V$_2$O$_3$	**Cr$_2$O$_3$**	Mn$_2$O$_3$	Fe$_2$O$_3$	Co$_2$O$_3$		
	TiO$_2$	VO$_2$	CrO$_2$	MnO$_2$				
		V$_2$O$_5$						
			CrO$_3$					
				Mn$_2$O$_7$				

The oxides in boldface form when the metal is heated in air.

atmosphere of hydrogen; **titanium(II) oxide, TiO,** can be prepared from titanium(IV) oxide and titanium metal.

$$V_2O_5 + 2H_2(g) \xrightarrow{800°C} V_2O_3 + 2H_2O(g)$$

$$TiO_2 + Ti \xrightarrow{1200°C} 2TiO$$

Alternatively, many of these compounds may be produced by heating the corresponding hydroxides, carbonates, or oxalates in an inert atmosphere. **Iron(II) oxide** may be produced by heating iron(II) oxalate, and **cobalt(II) oxide** by heating cobalt(II) hydroxide.

$$FeC_2O_4 \longrightarrow FeO + CO + CO_2$$

$$Co(OH)_2 \longrightarrow CoO + H_2O(g)$$

Oxides of the transition metals with higher oxidation numbers are formed by precipitation from acidic solution or by oxidation or reduction under special conditions. For example, **chromium(VI) oxide, CrO$_3$,** is produced as scarlet crystals when concentrated sulfuric acid is added to a concentrated solution of potassium dichromate. **Cobalt(III) oxide, Co$_2$O$_3$,** may be produced by gently heating cobalt(II) nitrate.

The elements of the early portion of the second and third transition series are generally oxidized to their highest oxidation numbers when heated in air. The oxides of the platinum metals are unstable at elevated temperatures and must be prepared by indirect methods, such as the careful dehydration of the hydroxides or precipitation.

Ionic oxides crystallize with the metal ions occupying holes in an array of oxide ions. In such structures one metal ion can substitute for another of about the same size. As can be seen in the table of ionic radii inside the back cover of this text, the radii of the transition metal ions are similar, so substitution of one transition metal ion for another is easy. Thus oxides that contain mixtures of the same metal ions with different oxidation numbers or mixtures of two or more types of metal ions are well known. For example, Mn$_3$O$_4$, Fe$_3$O$_4$, and Co$_3$O$_4$ each contain one M^{2+} ion and two M^{3+} ions for every four O^{2-} ions. Some of the metal ions can be replaced by other divalent metal ions. For example, the Mn^{2+} ion in Mn$_3$O$_4$ can be replaced by Mg^{2+} or by Fe^{2+}, giving MgMn$_2$O$_4$ or FeMn$_2$O$_4$, respectively. The trivalent ions can also be replaced; MnAl$_2$O$_4$, MnCo$_2$O$_4$, and MnCr$_2$O$_4$ are examples.

With the exception of CrO$_3$ and Mn$_2$O$_7$, transition metal oxides are not soluble in water. They exhibit their acid-base properties by reacting with acids or bases. Trends in acidity are as expected from the principles enumerated in Section 16.6 regarding size and oxidation number of the central element. The oxides of metals with oxidation numbers of $+1$, $+2$, and $+3$ are basic; they react with aqueous acids to form solutions of salts and water. Examples include

$$CoO(s) + 2H^+(aq) + [2NO_3{}^-(aq)] \longrightarrow Co^{2+}(aq) + [2NO_3{}^-(aq)] + H_2O(l)$$

$$NiO(s) + 2H^+(aq) + [SO_4{}^{2-}(aq)] \longrightarrow Ni^{2+}(aq) + [SO_4{}^{2-}(aq)] + H_2O(l)$$

$$Sc_2O_3(s) + 6H^+(aq) + [6Cl(aq)] \longrightarrow 2Sc^{3+}(aq) + [6Cl^-(aq)] + 3H_2O(l)$$

The oxides of metals with oxidation numbers of $+4$ are amphoteric and, in general, are not soluble in either acids or bases. **Vanadium(V) oxide, chromium(VI) oxide,** and **manganese(VII) oxide** are acidic. They react with solutions of hydroxides to form salts of the oxyanions $VO_4{}^{3-}$, $CrO_4{}^{2-}$, and $MnO_4{}^-$. For example,

$$CrO_3(s) + [2Na^{2+}(aq)] + 2OH^-(aq) \longrightarrow [2Na^+(aq)] + CrO_4{}^{2-}(aq) + H_2O(l)$$

Chromium(VI) oxide and manganese(VII) oxide react with water to form the acids H_2CrO_4 and $HMnO_4$, respectively.

3. HYDROXIDES. When a solution of a hdyroxide base is added to an aqueous solution of a salt of a transition metal of the first transition series, a gelatinous precipitate forms. For example, addition of a solution of sodium hydroxide to a solution of cobalt sulfate produces a gelatinous blue precipitate of **cobalt(II) hydroxide.**

$$Co^{2+}(aq) + [SO_4{}^{2-}(aq)] + [2Na^+(aq)] + 2OH^-(aq) \longrightarrow$$
$$Co(OH)_2(s) + [2Na^+(aq)] + [SO_4{}^{2-}(aq)]$$

In this and many other cases these precipitates are hydroxides containing the transition metal ion, hydroxide ions, and water coordinated to the transition metal. Upon gentle warming the anhydrous hydroxides can be obtained from these precipitates, but if the precipitates are heated too strongly, they may decompose to oxides. In other cases the precipitates are hydrated oxides composed of the metal ion, oxide ions, and water of hydration. These substances do not contain hydroxide ions. Both the hydroxides and the hydrated oxides react with acids to form salts and water or react with many of the oxides of the nonmetals to form salts with oxyanions such as those shown in Table 13.4.

4. CARBONATES. Many of the elements of the first transition series form insoluble carbonates. Thus these carbonates can be prepared by the addition of a soluble carbonate salt to a solution of a transition metal salt. For example, **nickel carbonate** can be prepared from solutions of nickel nitrate and sodium carbonate.

$$Ni^{2+}(aq) + [2NO_3{}^-(aq)] + [2Na^+(aq)] + CO_3{}^{2-}(aq) \longrightarrow$$
$$NiCO_3(s) + [2Na^+(aq)] + [2NO_3{}^-(aq)]$$

Solutions of the carbonate ion contain hydroxide ion, due to the basic character of the carbonate ion. If a transition metal forms a hydroxide or a hydrated oxide that is less soluble than the transition metal carbonate, the hydroxide or hydrated oxide will form in preference to the carbonate.

The reactions of the transition metal carbonates are similar to those of the active metal carbonates (Section 13.6). They react with acids to form metal salts, carbon dioxide, and water. Upon heating, they decompose forming the transition metal oxides.

5. SULFIDES. Transition metal sulfides can be prepared by heating the metals with sulfur, by heating gaseous hydrogen sulfide with the metals or with certain metal oxides, hydroxides, or carbonates, and, since most transition metal sulfides are insolu-

ble, by addition of a solution of a soluble sulfide to aqueous solutions of transition metal salts.

$$Fe + S \longrightarrow FeS$$

$$Mo + 2H_2S \longrightarrow MoS_2 + 2H_2$$

$$Cu^{2+}(aq) + S^{2-}(aq) \longrightarrow CuS(s)$$

Precipitation reactions and the reactions with hydrogen sulfide usually give compounds in which the metal exhibits a single oxidation number. The stoichiometry of the direct reaction of sulfur with a metal can be adjusted so that the product contains mixed oxidation states, as observed in V_5S_8, for example, which contains V^{3+} and V^{4+} ions.

The sulfide ion is a stronger reducing agent than the oxide ion, so many sulfides in which the metal exhibits a high oxidation number are not stable; the metal is reduced by the sulfide ion. With the exception of TiS_2 and VS_2, sulfides of the first transition series are limited to those in which the metals have oxidation numbers of $+1$, $+2$, or $+3$. Metals of the second and third transition series form sulfides with higher oxidation numbers, because these elements are more stable with higher oxidation numbers.

6. OTHER SALTS. A wide variety of salts containing transition metals has been prepared. Simple salts of oxyanions such as the nitrate, sulfate, and perchlorate ions are usually stable only with metals of the first transition series. Soluble salts of the heavier transition metals are generally found only as the oxyanions of these metals. A variety of salts can be prepared from the metals or from the common compounds that we have just discussed. Examples of these preparations include the following:

1. *Reactions of acids with basic metal compounds.* Oxides, hydroxides, and carbonates of the elements of the first transition series react with solutions of most acids to form solutions of salts, which can be isolated by evaporation of the solvent. Like the halides, the solid compounds usually contain water of hydration. The reaction of nickel carbonate with nitric acid is an example; hydrated nickel nitrate forms when the resulting solution is evaporated.

$$NiCO_3(s) + 2H^+(aq) + [2NO_3{}^-(aq)] \longrightarrow$$
$$Ni^{2+}(aq) + [2NO_3{}^-(aq)] + CO_2(g) + H_2O$$

$$Ni^{2+}(aq) + 2NO_3{}^-(aq) + 6H_2O \xrightarrow{\text{Evaporation}} Ni(NO_3)_2 \cdot 6H_2O$$

2. *Oxidation-reduction reactions.* With the exception of copper, the elements of the first transition series are more active than hydrogen and will react with acids liberating hydrogen gas. The reaction of these metals with acids is thus a second source for their salts. A solution of **iron(II) perchlorate,** for example, can be prepared by the reaction of metallic iron with an aqueous solution of perchloric acid.

$$Fe(s) + 2H^+(aq) + [2ClO_4{}^-(aq)] \longrightarrow Fe^{2+}(aq) + [2ClO_4{}^-(aq)] + H_2(g)$$

Copper and the other elements will react with nitric acid and concentrated sulfuric acid by oxidation-reduction reactions involving the oxyanion of the acid.

3. *Metathetical reactions.* Insoluble salts can be prepared by mixing solutions of two soluble compounds. **Barium chromate** is prepared in this way from the soluble salts barium chloride and potassium chromate.

$$Ba^{2+}(aq) + [2Cl^-(aq)] + [2K^+(aq)] + CrO_4{}^{2-}(aq) \longrightarrow$$
$$BaCrO_4(s) + [2K^+(aq)] + [2Cl^-(aq)]$$

4. *Preparation of oxyanions.* Salts containing transition metal oxyanions such as $VO_4{}^{3-}$, $CrO_4{}^{2-}$, $MoO_4{}^{2-}$, and $WO_4{}^{2-}$ can be prepared by heating active metal oxides or hydroxides with the corresponding oxides of the transition metals, or by the reaction of solutions of hydroxides with the oxides. These reactions generally occur without a change in the oxidation number of the transition metal. For example,

$$V_2O_5 + 3Li_2O \xrightarrow{\Delta} 2Li_3VO_4$$

$$MoO_3(s) + 2NaOH(s) \xrightarrow{\Delta} Na_2MoO_4(s) + H_2O(g)$$

$$CrO_3(s) + [2K^+(aq)] + 2OH^-(aq) \longrightarrow [2K^+(aq)] + CrO_4^{2-}(aq) + H_2O(l)$$

The Elements of Group IIIB

The elements of Group IIIB include scandium, yttrium, lanthanum, and actinium. Because the rare earth elements and the actinide elements resemble lanthanum and actinium, respectively, in many of their properties due to similar sizes and valence subshell electron configurations, we will also consider the *f*-block elements in this section.

27.6 Scandium, Yttrium, Lanthanum, and Actinium

Scandium, yttrium, lanthanum, and actinium are active metals. The chemistry of scandium resembles that of aluminum. These elements form ionic compounds containing trivalent ions, M^{3+}. The free metals are sufficiently active that they react with the oxygen of the air at room temperature. All burn readily, giving the oxides M_2O_3. They react with halogens at room temperature and with most nonmetals upon warming. The metals react readily with acids, producing salts and hydrogen gas.

Very little scandium is produced, because it is of little technological importance. Yttrium and lanthanum are isolated during the separation of the rare earths (Section 27.7). Actinium is a product of the radioactive decay of ^{235}U, and it can also be produced artificially. Only trace amounts of actinium are available.

Yttrium is used in the electronics industry. Yttrium compounds are used in the phosphors that produce the red color on color television tubes. **Yttrium iron garnet, $Y_3Fe_5O_{12}$,** is used in microwave devices, and **yttrium aluminum garnet (YAG), $Y_3Al_5O_{12}$,** is used in fabricating lasers and as an artificial gemstone. Yttrium is also a component of a new high-temperature superconductor (Section 27.28).

Lanthanum oxide is used in some kinds of glass. The metal is a component of misch metal (Section 27.7).

27.7 The Lanthanide Series (Rare Earth Elements)

The **lanthanide series (rare earth elements),** beginning with cerium (element 58) and extending through lutetium (element 71), are very similar in chemical and physical properties to each other and to lanthanum. This similarity is attributed to the similarity in sizes and to the fact that the electron configurations of the atoms differ principally in the number of electrons in the 4*f* and 5*d* subshells rather than in their outer $6s$ subshell (see Table 27.2). The electron configurations of the trivalent ions differ only in the $4f$ subshell. These ions have the configuration $[Xe]4f^n5d^06s^0$.

The chemical behavior of all of the rare earths is generally similar to that of lanthanum (Section 27.6). They are active metals that form oxides with the general formula M_2O_3 upon exposure to air. Their usual oxidation number is $+3$; except for cerium, this is the only oxidation number of these elements that is stable in water. Cerium also forms salts in which it exhibits an oxidation number of $+4$. These are quite strong oxidizing agents.

The rare earths are found in many minerals, and some of the rare earths are actually more common than several of the most familiar elements. **Monazite sand** is the most important source of these minerals; it contains the various elements of the group, in the form of phosphates, except promethium (atomic number 61), which does not occur in nature at all.

The rare earth elements resemble each other so closely that their separation has been an exceedingly difficult task. Formerly the only effective method of separation was fractional crystallization of certain of their salts, involving many hundreds of crystallizations. Now ion-exchange liquid chromatography techniques provide a much more rapid and effective separation. This technique employs the selective adsorption of the ions of the various rare earths on a chromatographic column containing an ion-exchange resin (Section 29.11), followed by fractional elution of the adsorbed material by a suitable solvent.

An alloy of iron with a mixture of the rare earth metals, known commercially as **misch metal,** is used in making the flints in cigarette lighters. A mixture of the fluorides of these metals forms the core of the carbon arcs used in motion picture projectors. Recently the rare earths have been found to have great catalytic power in some important industrial reactions such as petroleum cracking and reforming. Some of the rare earth elements are used in color television tubes to improve the color rendition (especially of reds).

27.8 The Actinide Series

The series of elements beginning with thorium (element 90) and extending through lawrencium (element 103) is called the **actinide series.** All the elements of this series are radioactive. They are analogous to the rare earths in that the electron configurations of succeeding elements in the series differ by one electron, in this case in either the $5f$ or $6d$ subshells. The properties of the actinide elements are similar to those of the rare earths. The actinide elements have a characteristic oxidation number of $+3$ but show a much wider variation in oxidation number than do the elements of the lanthanide series. Thorium resembles cerium in that it has an oxidation number of $+4$.

Elements 93 through 103 have been prepared synthetically by nuclear reactions (Section 24.8). The most important elements in the actinide series are thorium, uranium, and plutonium, the importance of the latter two being related to their use in atomic energy (Section 24.11).

Thorium occurs as the dioxide in monazite sand, in some cases to the extent of 10%, along with the rare earths. Metallic thorium is obtained by reducing **thorium(IV) chloride, ThCl$_4$,** with sodium; thorium(IV) chloride is prepared by the reaction of chlorine and a heated mixture of the dioxide and carbon. Thorium is very similar to the elements of Group IVB. It exhibits an oxidation number of $+4$ in its compounds. **Thorium hydroxide, Th(OH)$_4$,** is highly basic, and its salts react to a limited extent with water. Tungsten filaments for electric lamps are thoriated to increase their efficiency and retard their disintegration.

Uranium, the first radioactive element to be discovered, is found in the minerals pitchblende, U_3O_8, and carnotite, $K_2(UO_2)_2(VO_4)_2 \cdot 8H_2O$. The principal sources of uranium minerals are in Colorado, Utah, the Congo, and the Great Bear Lake region of Canada. The metal may be obtained by reduction of the oxide with either carbon, calcium, or aluminum. It is a white, lustrous metal of high density (18.9 g/cm^3), and it melts at 1132°C. The metal is moderately active and forms compounds in which it shows oxidation numbers of +2, +3, +4, +5, and +6. The "uranium" salts of commerce contain the dipositive uranyl ion, UO_2^{2+}. These include **uranyl nitrate, $UO_2(NO_3)_2 \cdot 6H_2O$,** and **uranyl acetate, $UO_2(CH_3CO_2)_2 \cdot 2H_2O$.** There are several series of uranates, the most important of which are the diuranates, such as **sodium diuranate, $Na_2U_2O_7 \cdot 6H_2O$.** Uranium compounds are used for producing yellow glazes on ceramic ware, as mordants, and in the dye industry. The chief interest in uranium at present is in its release of nuclear energy and its conversion to elements that will release nuclear energy (see Section 24.11).

The actinide elements exhibit a contraction in size (the **actinide contraction**) with increasing atomic number analogous to the lanthanide contraction (Section 27.4).

The Elements of Group IVB

The elements of Group IVB [titanium, zirconium, hafnium, and element 104 (unnilquadium)] commonly exhibit the oxidation number of +4; the two d electrons of the incomplete inner subshell and the two s electrons of the outer shell are used in bonding. These elements, however, also show oxidation numbers of +2, and +3. Zirconium and hafnium are even more similar in chemical properties than expected from their being in the same family. This similarity results from their similarity in size, which arises because of the lanthanide contraction (Section 27.4).

27.9 Titanium

Titanium is one of the more abundant elements. It is tenth among the elements in abundance in the earth's crust, ranking above such useful metals as nickel, copper, zinc, lead, tin, and mercury.

The most important ores of titanium from the commercial standpoint are ilmenite, $FeTiO_3$, and rutile, TiO_2. The dioxide is an excellent white pigment for paints and is used in the preparation of white rubber and white leather. In the production of chemicals in the United States, titanium dioxide ranks 44th with 917,000 tons produced in 1986.

Titanium metal is very strong, light (specific gravity = 4.49), high-melting, and corrosion-resistant. These properties make the metal and its alloys valuable in the production of jet engines and high speed aircraft. An alloy, ferrotitanium, is used in making special steels of great strength and toughness, in which the titanium acts as a "scavenger" to remove nitrogen and other undesirable impurities.

Titanium forms three oxides, TiO, Ti_2O_3, and TiO_2, and their corresponding hydroxides. The dioxide may be prepared from ilmenite by the reaction

$$2FeTiO_3 + 4HCl + Cl_2 \xrightarrow{\triangle} 2FeCl_3 + 2TiO_2 + 2H_2O$$

The temperature is high enough to volatilize the iron(III) chloride that is formed.

Titanium(IV) chloride, TiCl₄, is a liquid boiling at 136.4°C. It is produced by passing chlorine over a heated mixture of carbon and titanium dioxide.

$$TiO_2(s) + C(s) + 2Cl_2(g) \longrightarrow TiCl_4(l) + CO_2(g) \qquad \Delta H° = -252.9 \text{ kJ}$$

The tetrachloride reacts quickly with the water in moist air, producing a dense white smoke consisting of finely divided TiO_2. Titanium(IV) chloride is used in producing smoke screens and in skywriting.

Titanium(III) chloride, TiCl₃, and **titanium(III) sulfate, Ti₂(SO₄)₃,** are powerful reducing agents and are used in this capacity in the chemical laboratory. Molten alkali metal hydroxides react with titanium dioxide, forming **titanates** such as sodium titanate, Na_2TiO_3.

27.10 Zirconium and Hafnium

Zirconium is not as abundant as titanium but is far more abundant than such familiar metals as lead, copper, nickel, zinc, mercury, and tin. Its chief ores are zircon, $ZrSiO_4$, and baddeleyite, ZrO_2. The metal is obtained by heating K_2ZrF_6 with sodium, potassium, or aluminum. Ferrozirconium has been used with some success in the steel industry in the production of a tough steel.

Zirconium dioxide, ZrO₂, which is called **zirconia,** is the most important compound of the element because of its excellent refractory qualities. These include a high melting point (about 2700°C), a low coefficient of expansion, and a high resistance to corrosion.

Although zirconia is attacked by scarcely any acid except hydrofluoric, it reacts with molten alkali metal hydroxides, forming **zirconates** such as Na_2ZrO_3. Insoluble zirconates are formed when the dioxide is used in the production of enamels and opaque glass.

Zirconium(IV) chloride, ZrCl₄, does not react with water as readily as titanium tetrachloride. Reaction with water results in the formation of derivatives such as **zirconium oxychloride, ZrOCl₂,** rather than the oxide as is the case with $TiCl_4$.

Zircon, the naturally occurring silicate, $ZrSiO_4$, is found in a variety of colors and, because of its beauty and hardness, is used as a semiprecious stone in jewelry.

Hafnium was discovered in 1923 by Coster and Hevesy in zircon from Norway by means of spectroscopic analysis. It is found in nearly all zirconium minerals, most of which contain about 5% hafnium. The fact that hafnium has chemical properties very similar to those of zirconium makes the separation of the two elements difficult.

The Elements of Group VIB

The elements that constitute Group VIB of the Periodic Table are chromium, molybdenum, tungsten, and element 106 (unnilhexium). These metals are typical transition metals; they show a variety of oxidation numbers, form highly colored compounds, and form stable complex ions. Chromium exhibits behavior characteristic of the more active metals of the first transition series. The behavior of molybdenum and tungsten is characteristic of the early members of the second and third transition series. Simple salts containing chromium with oxidation numbers of $+2$, $+3$, and $+6$ are stable in water. The stable water-soluble salts of molybdenum and tungsten are limited to those containing oxyanions in which the metals exhibit oxidation numbers of $+6$ (MoO_4^{2-} and WO_4^{2-}).

Colors of various compounds of chromium with different oxidation states.

27.11 Properties of Chromium

Chromium is the lightest member of Group VIB. Elemental chromium does not occur in nature; its most important ore is **chromite, FeCr$_2$O$_4$.** Practically all of the chromium used in the United States is imported.

Reduction of chromite by carbon yields carbon monoxide and an alloy of iron and chromium called ferrochrome, which is used in making chromium steels. Pure chromium is prepared by reducing chromium(III) oxide with aluminum. It can also be produced by the electrolytic reduction of its compounds from aqueous solution, a process used in chromium plating.

Chromium is a very hard, silvery white, crystalline metal. It becomes passive when coated with a thin layer of oxide, which protects it against further corrosive attack. This property, along with its metallic luster, accounts for its extensive use in the plating of iron and copper objects. Passivity is produced by treatment with concentrated nitric acid or chromic acid or by exposure of the metal to air. When not in the passive state, chromium dissolves readily in dilute acids, with the evolution of hydrogen.

Chromium exhibits several oxidation numbers; the principal ones are +2, +3, and +6. The compounds of chromium are highly colored. The particular color exhibited depends on the oxidation number and on the structure of the compound.

27.12 Compounds of Chromium

Most compounds containing chromium(II) are oxidized by air, so they are prepared under nitrogen or some other inert atmosphere. **Chromium(II) chloride, CrCl$_2$,** can be prepared (under nitrogen) by dissolving chromium in hydrochloric acid or by reducing a solution of chromium(III) chloride with zinc in the presence of an acid. The anhydrous salt is colorless, but its solutions have the bright blue color of the hydrated ion $[Cr(H_2O)_6]^{2+}$. This ion is rapidly oxidized in air to the chromium(III) ion, $[Cr(H_2O)_6]^{3+}$; for this reason solutions of chromium(II) salts are frequently employed to remove the last traces of oxygen from gases. **Chromium(II) sulfate 7-hydrate, CrSO$_4$ · 7H$_2$O,** is prepared by dissolving the metal in sulfuric acid. **Chromium(II) acetate, Cr(CH$_3$CO$_2$)$_2$,** forms as a red precipitate when a saturated solution of sodium acetate is added to a solution of chromium(II) chloride. Chromium(II) acetate is one of few exceptions to the rule that acetates are soluble. **Chromium(II) hydroxide** is basic.

Green **chromium(III) oxide, Cr$_2$O$_3$,** is formed by heating the metal in air, by igniting ammonium dichromate, or by reducing a dichromate with sulfur or another similar reducing agent.

$$(NH_4)_2Cr_2O_7(s) \xrightarrow{\triangle} N_2(g) + 4H_2O(g) + Cr_2O_3(s) \qquad \Delta H° = -300 \text{ kJ}$$
$$Na_2Cr_2O_7 + S \longrightarrow Na_2SO_4 + Cr_2O_3$$

This oxide is used as a pigment called **chrome green.** Bluish-green, gelatinous **chromium(III) hydroxide** is precipitated ($K_{sp} = 6.7 \times 10^{-31}$) when a soluble hydroxide, sulfide, or carbonate is added to a solution of a chromium(III) compound. Chromium(III) hydroxide is amphoteric, dissolving in both acids and bases.

$$Cr(OH)_3 + 3H^+ \longrightarrow Cr^{3+} + 3H_2O$$
$$Cr(OH)_3 + OH^- \longrightarrow [Cr(OH)_4]^- \quad \text{(chromite ion)}$$

Chromium(III) chloride 6-hydrate, CrCl$_3$ · 6H$_2$O, and **chromium(III) sulfate 18-hydrate, Cr$_2$(SO$_4$)$_3$ · 18H$_2$O,** are the best known chromium(III) salts. Dilute solutions of the chloride are violet; the hexaaquachromium(III) ion, $[Cr(H_2O)_6]^{3+}$, is responsible for the color. In more concentrated solutions of the chloride, green $[Cr(H_2O)_4Cl_2]^+$ is formed. The hydrates of chromium(III) sulfate also exist in violet and green modifications. Chromium(III) forms alums of the type $KCr(SO_4)_2 \cdot 12H_2O$, solutions of which are bluish-violet when cold and green when hot. This alum is used in the tanning of leather, the printing of calico, and the waterproofing of fabrics, and as a mordant. It is the most important of the soluble chromium salts.

Chromium with a +6 oxidation number is combined covalently with oxygen in the oxide **chromium(VI) oxide, CrO_3**, and in the oxyanions chromate, CrO_4^{2-}, and dichromate, $Cr_2O_7^{2-}$. **Potassium chromate, K_2CrO_4**, is produced commercially by the oxidation of chromite ore mixed and heated in air with potassium carbonate.

$$4FeCr_2O_4 + 8K_2CO_3 + 7O_2 \xrightarrow{\Delta} 2Fe_2O_3 + 8K_2CrO_4 + 8CO_2$$

Lead chromate, $PbCrO_4$ ($K_{sp} = 1.8 \times 10^{-14}$), and **barium chromate, $BaCrO_4$** ($K_{sp} = 2 \times 10^{-10}$), are both insoluble in water and are used as yellow pigments.

When an acid is added to a solution containing chromate ions, the solution changes from yellow to orange-red due to the formation of the **dichromate ion, $Cr_2O_7^{2-}$**.

$$2CrO_4^{2-} + 2H^+ \rightleftharpoons Cr_2O_7^{2-} + H_2O$$

The addition of a base or dilution with water reverses the reaction.

$$Cr_2O_7^{2-} + 2OH^- \rightleftharpoons 2CrO_4^{2-} + H_2O$$

The conversion of chromate to dichromate involves the formation of an oxygen bridge between two chromium atoms. The Lewis structure of the dichromate ion is

Dichromates are used as oxidizing agents in many reactions. The dichromate ion is readily reduced to the chromium(III) ion by reducing agents such as sulfurous acid.

$$Cr_2O_7^{2-} + 3H_2SO_3 + 2H^+ \longrightarrow 2Cr^{3+} + 3SO_4^{2-} + 4H_2O$$

As the reduction progresses, a change from the orange color of the dichromate to the violet color of the chromium(III) ion occurs.

Chromium trioxide, CrO_3, reacts with water in different proportions, forming **chromic acid, H_2CrO_4**, or **dichromic acid, $H_2Cr_2O_7$**.

$$2CrO_3 + H_2O \longrightarrow 2H^+ + Cr_2O_7^{2-}$$
$$CrO_3 + H_2O \longrightarrow 2H^+ + CrO_4^{2-}$$

These solutions are active oxidizing agents. When heated, the trioxide decomposes, liberating oxygen and forming green chromium(III) oxide.

$$4CrO_3(s) \longrightarrow 2Cr_2O_3(s) + 3O_2(g) \qquad \Delta H° = 497.1 \text{ kJ}$$

27.13 Uses of Chromium

A large portion of the chromium produced goes into steel alloys, which are very hard and strong. **Stainless steel**, which usually contains chromium and some nickel, is used because of its corrosion resistance. Nonferrous chromium alloys include nichrome and chromel (Ni and Cr), which are used in wires in heating elements because of their electrical resistance. Chromium is widely used as a protective and decorative coating for other metals, such as plumbing fixtures and automobile trim.

The compounds of chromium have many uses, including uses as paint pigments and mordants and in the tanning of leather. Certain refractories used as linings for high-temperature furnaces are made by mixing pulverized chromite ore with clay or magnesia (MgO).

Manganese (A Group VIIB Element)

Manganese is the lightest member of Group VIIB. It is found principally as **pyrolusite, MnO_2.** Nearly pure manganese may be produced by the high-temperature reduction of the dioxide by aluminum. However, since alloys of manganese and iron are used extensively in the production of steel, such alloys are usually produced instead of pure manganese.

Manganese is a gray-white metal with a slightly reddish tinge. It is brittle and looks like cast iron. It is readily oxidized by moist air and reacts slowly with water, forming manganese(II) hydroxide and hydrogen. It dissolves readily in dilute acids, forming manganese(II) salts. Manganese forms five oxides and five corresponding series of salts (Table 27.5). These compounds are considered in the following sections.

Table 27.5   Classes of manganese compounds

Oxidation number	Oxide	Hydroxide	Character	Derivative	Name	Color
+2	MnO	$Mn(OH)_2$	Moderately basic	$MnCl_2$	Manganese(II) chloride	Pink
+3	Mn_2O_3	$Mn(OH)_3$	Weakly basic	$MnCl_3$	Manganese(III) chloride	Violet
+4	MnO_2	H_2MnO_3	Weakly acidic	$CaMnO_3$	Calcium manganite	Brown
+6	MnO_3	H_2MnO_4	Moderately acidic	K_2MnO_4	Potassium manganate	Green
+7	Mn_2O_7	$HMnO_4$	Strongly acidic	$KMnO_4$	Potassium permanganate	Purple

27.14 Manganese(II) Compounds

Although manganese forms compounds in which it exhibits oxidation numbers of +2, +3, +4, +6, and +7, the only stable cation is the dipositive manganese ion, Mn^{2+}. The common soluble manganese(II) salts are the chloride, sulfate, and nitrate. Each imparts a faint pink color to its solutions.

Alkali metal hydroxides precipitate pale pink **manganese(II) hydroxide, $Mn(OH)_2$,** which is oxidized when exposed to air to the dark brown **manganese(III) oxyhydroxide, MnO(OH).** Manganese(II) hydroxide is only partially precipitated by aqueous ammonia and is dissolved by solutions of the ammonium salts of strong acids, according to the equation

$$Mn(OH)_2 + 2NH_4^+ \rightleftharpoons Mn^{2+} + 2NH_3 + 2H_2O$$

Manganese(II) hydroxide is entirely basic in character; it dissolves in acids but not in bases such as sodium hydroxide. **Manganese(II) oxide, MnO,** may be produced by heating the corresponding hydroxide in the absence of air.

From solutions of manganese(II) compounds, alkali metal sulfides precipitate pink **manganese(II) sulfide, MnS** ($K_{sp} = 4.3 \times 10^{-22}$), which is readily soluble in dilute acids.

27.15 Manganese(III) Compounds

Dimanganese trioxide, Mn$_2$O$_3$, and **manganese oxyhydroxide, MnO(OH),** occur naturally, but the manganese(III) ion is unstable in aqueous solution and is readily reduced to the manganese(II) ion. **Manganese(III) chloride** is formed in solution by the reaction of hydrochloric acid with manganese dioxide at a low temperature. When the solution is warmed, the manganese(III) chloride decomposes with the formation of manganese(II) chloride and chlorine.

$$MnO_2 + 4H^+ + 4Cl^- \longrightarrow MnCl_4 + 2H_2O$$

$$2MnCl_4 \longrightarrow 2MnCl_3 + Cl_2(g)$$

$$2MnCl_3 \xrightarrow{\Delta} 2Mn^{2+} + 4Cl^- + Cl_2(g)$$

27.16 Manganese(IV) Compounds

The most important manganese(IV) compound is the dioxide, MnO_2, which is amphoteric but relatively inert toward acids and bases. Cold, concentrated hydrochloric acid reacts with the dioxide, giving a green solution of the unstable **manganese(IV) chloride.** The sulfate $Mn(SO_4)_2$ is also unstable, but the complex salt $K_2[MnF_6]$ is not readily decomposed. When the dioxide is fused with calcium oxide, **calcium manganite, CaMnO$_3$,** is formed.

When manganese dioxide is heated to 535°C, it is transformed to **trimanganese tetraoxide, Mn$_3$O$_4$,** and oxygen, according to the equation

$$3MnO_2(s) \xrightarrow{\Delta} Mn_3O_4(s) + O_2(g) \qquad \Delta H^\circ = 172 \text{ kJ}$$

A historical note is that Scheele, independently of Priestley and Lavoisier, discovered oxygen by the reaction of concentrated sulfuric acid with manganese dioxide.

$$MnO_2 + 2H_2SO_4 \longrightarrow Mn(SO_4)_2 + 2H_2O$$

$$2Mn(SO_4)_2 + 2H_2O \longrightarrow 2MnSO_4 + 2H_2SO_4 + O_2$$

27.17 Manganates, Manganese(VI) Compounds

When an oxide of manganese is melted with an alkali metal hydroxide or carbonate in the presence of air or some other oxidizing agent (such as potassium chlorate or potassium nitrate), a **manganate** is formed.

$$2MnO_2 + 4KOH + O_2 \longrightarrow 2K_2MnO_4 + 2H_2O$$

Manganates are green and stable only in alkaline solution. The addition of water to solutions of manganates may bring about disproportionation, with the precipitation of manganese dioxide and the formation of the purple **permanganate ion, MnO$_4^-$.**

$$3MnO_4^{2-} + 2H_2O \longrightarrow 2MnO_4^- + 4OH^- + MnO_2(s)$$

The hypothetical manganic acid, H_2MnO_4, is too unstable to be prepared and isolated. Solutions containing the manganate ion become active oxidizing agents when acidified. The half-reaction is

$$MnO_4^{2-} + 8H^+ + 4e^- \longrightarrow Mn^{2+} + 4H_2O$$

27.18 Permanganates, Manganese(VII) Compounds

An important compound of manganese is **potassium permanganate, $KMnO_4$.** It is prepared commercially by oxidizing potassium manganate in alkaline solution with chlorine.

$$2MnO_4^{2-} + Cl_2 \longrightarrow 2MnO_4^- + 2Cl^-$$

The resultant purple solution deposits crystals when sufficiently concentrated. Potassium permanganate is a valuable laboratory reagent because it acts as a strong oxidizing agent. A solution of the free **permanganic acid, $HMnO_4$,** may be prepared by the reaction of dilute sulfuric acid and barium permanganate.

$$Ba^{2+} + 2MnO_4^- + 2H^+ + SO_4^{2-} \longrightarrow BaSO_4(s) + 2H^+ + 2MnO_4^-$$

Mn_2O_7, a dark brown, highly explosive liquid, is the anhydride of permanganic acid.

27.19 Uses of Manganese and Its Compounds

Manganese forms a number of alloys used in making very hard steels for the manufacture of rails, safes, and heavy machinery. About 12.5 pounds of manganese metal are used in the production of every ton of steel to remove oxygen, nitrogen, and sulfur. An alloy called **manganin** (Cu, 84%; Mn, 12%; Ni, 4%) is used in instruments for making electrical measurements, because its electrical resistance does not change significantly with changes in temperature.

Manganese dioxide is used to correct the green color produced in glass by iron(II) compounds, to color glass and enamels black, as an oxidizing agent or dryer in black paints, and as a depolarizer in dry cells. Potassium permanganate is used as a disinfectant, a deodorant, and a germicide.

The oxidation of glycerol by potassium permanganate is sufficiently exothermic that the glycerol inflames.

Iron (A Group VIIIB Element)

Iron was known at least as early as 4000 B.C. and was probably used to some extent during prehistoric times. Since free iron is not commonly found in nature, that used in prehistoric times may well have been of meteoric origin. The early use of iron to manufacture tools and weapons was possible because of the wide distribution of iron ores and the ease with which the iron compounds in the ores can be reduced by charcoal. The production and use of iron on a large scale began when coal was introduced as the reducing agent.

Iron is the most widely used of all metals. The amount used is 14 times the amount of all other metals combined. The reasons for this are that its ores are abundant and widely distributed, the metal is easily and cheaply produced from the ores, and its properties can be varied over a wide range by the addition of other substances and by different methods of treatment, such as tempering and annealing.

Iron ranks second in abundance among the metals (aluminum is first) and fourth among all elements in the earth's crust. The central core of the earth is largely iron. Metallic meteorites are usually about 90% iron, and the remainder is principally nickel. Practically all rocks, minerals, and soils, as well as plants, contain some iron. Iron is present in the hemoglobin of the blood, which acts as a carrier of oxygen (Fig. 26.2).

27.20 Properties of Iron

Pure iron is silvery white, capable of taking a high polish, ductile, relatively soft, and high in tensile strength. These properties are greatly modified by impurities, especially carbon. Pure iron is attracted by a magnet but does not retain magnetism. Iron dissolves in hydrochloric acid and dilute sulfuric acid, forming hydrogen and iron(II) ions, Fe^{2+}. Hot concentrated sulfuric acid forms Fe^{3+}, and SO_2 is evolved. Cold concentrated nitric acid induces **passivity** in iron. When passive, a metal does not react with dilute acids that otherwise would readily dissolve it. The passivity is easily destroyed by a scratch or by shock. Hot dilute nitric acid oxidizes iron to Fe^{3+} with the evolution of nitric oxide. When heated to redness, iron reduces steam to produce hydrogen and iron(II,III) oxide, Fe_3O_4, which is the same oxide that is produced when iron burns in oxygen. When exposed to moist air at ordinary temperatures, iron becomes oxidized, with a loose coating of partially hydrated iron(III) oxide, $Fe_2O_3 \cdot nH_2O$, rust (Section 21.16).

27.21 Compounds of Iron

Iron forms two principal series of compounds in which its oxidation numbers are +2 and +3. The +2 and +3 series are called iron(II) or ferrous compounds and iron(III) or ferric compounds, respectively. The oxides and hydroxides of iron(II) and (III) are basic, with little or no acidic character.

Black **iron(II) oxide, FeO,** may be obtained by the thermal decomposition of **iron(II) oxalate, FeC_2O_4,** or iron(II) hydroxide. **Iron(II) hydroxide, $Fe(OH)_2$,** precipitates when alkali metal hydroxides are added to solutions of iron(II) salts. Ammonium salts and many organic acids greatly increase the solubility of iron(II) hydroxide.

In the air the white hydroxide quickly turns green and then reddish-brown, due to oxidation to **iron(III) hydroxide, Fe(OH)$_3$.**

$$4Fe(OH)_2(s) + O_2(g) + 2H_2O(g) \longrightarrow 4Fe(OH)_3(s) \qquad \Delta H° = -532.2 \text{ kJ}$$

Iron(III) oxide, Fe$_2$O$_3$, occurs in nature as the mineral **hematite**, one of the principal iron ores. It may also be prepared by heating Fe(OH)$_3$ or **pyrite, FeS$_2$,** in the air. It is a red powder, which is used as a paint pigment under the names rouge and Venetian red. When treated with either alkali metal hydroxides or aqueous ammonia, solutions of Fe^{3+} yield Fe(OH)$_3$, which is reddish-brown and insoluble in an excess of either reagent.

Iron(II,III) oxide, Fe$_3$O$_4$, sometimes called **magnetic oxide of iron,** or **lodestone,** is also a valuable ore of iron.

Pale green **iron(II) chloride 4-hydrate, FeCl$_2$ · 4H$_2$O,** is produced by the reaction of dilute hydrochloric acid with iron. Sulfuric acid reacts with iron or iron(II) oxide, forming **green vitriol,** or **copperas,** which crystallizes as the 7-hydrate FeSO$_4$ · 7H$_2$O.

Iron(II) ammonium sulfate 6-hydrate, (NH$_4$)$_2$Fe(SO$_4$)$_2$ · 6H$_2$O, crystallizes from an equimolar solution of ammonium sulfate and iron(II) sulfate. It is readily obtained in a pure state and is used extensively in laboratory work as a reducing agent; the iron(II) is readily oxidized to iron(III).

Iron(II) carbonate, FeCO$_3$, occurs in nature as the mineral **siderite.** It is also produced as a white precipitate ($K_{sp} = 2.11 \times 10^{-11}$) by the reaction of alkali metal carbonates with solutions of iron(II) salts in the absence of air. The carbonate dissolves in carbonic acid and forms **iron(II) hydrogen carbonate, Fe(HCO$_3$)$_2$,** a common constituent of hard waters.

Iron(II) sulfide, FeS ($K_{sp} = 8 \times 10^{-26}$), is precipitated when an alkali metal sulfide is added to a solution of an iron(II) salt. It is prepared commercially by heating a mixture of iron and sulfur. It reacts with acids, forming hydrogen sulfide.

An important iron(III) compound is **iron(III) chloride, FeCl$_3$,** which may be obtained by heating iron with chlorine. It is appreciably covalent, as indicated by its volatility and its solubility in nonpolar solvents. When the anhydrous chloride is dissolved in water, the hydration of the ions is highly exothermic. Iron(III) chloride crystallizes from water as the brownish-yellow hydrate [Fe(H$_2$O)$_6$]Cl$_3$. This compound, like all soluble iron(III) salts, gives an acid solution. The first ionization step is

$$[Fe(H_2O)_6]^{3+} + H_2O \longrightarrow [Fe(H_2O)_5(OH)]^{2+} + H_3O^+$$

Several additional steps also occur. Evidence for this is the appearance of the reddish-brown gelatinous [Fe(H$_2$O)$_3$(OH)$_3$], which is simply hydrated Fe(OH)$_3$.

Hydrated iron(III) chloride, FeCl$_3$ · 6H$_2$O, is used to coagulate blood and thus stop bleeding, and in the treatment of anemia.

When a solution of an iron(III) salt is added to a solution containing thiocyanate ions, a blood-red color appears. An iron(II) salt yields a colorless solution. These facts are applied to the detection of iron(III) ions in the presence of iron(II) ions. It appears that the species responsible for the red color is the complex cation [Fe(H$_2$O)$_5$NCS]$^{2+}$.

Other iron(III) salts include the sulfate Fe$_2$(SO$_4$)$_3$ and the iron **alums** formed from this salt, such as NH$_4$Fe(SO$_4$)$_2$ · 12H$_2$O.

When an excess of potassium cyanide is added to a solution of an iron(II) salt, **potassium hexacyanoferrate(II), K$_4$[Fe(CN)$_6$],** also called potassium ferrocyanide, is formed in solution.

$$Fe^{2+} + 6CN^- \longrightarrow [Fe(CN)_6]^{4-}$$

The complex hexacyanoferrate(II) ion is so stable that it does not give positive results in any of the common qualitative tests for either iron(II) or cyanide ions. When the hexacyanoferrate(II) ion is oxidized by chlorine, **hexacyanoferrate(III) ion, $[Fe(CN)_6]^{3-}$,** also called ferricyanide, is produced.

$$2[Fe(CN)_6]^{4-} + Cl_2 \longrightarrow 2[Fe(CN)_6]^{3-} + 2Cl^-$$

Solutions of potassium hexacyanoferrate(III) do not give positive results in the usual test for either iron(III) or cyanide ions.

The addition of iron(II) ions to hexacyanoferrate(III) solutions produces a blue precipitate known as **Turnbull's blue.** Iron(III) ions react with hexacyanoferrate(II) ions to form **Prussian blue.** These precipitates have a deep blue color and are thought to be the same, with the approximate formula $KFe[Fe(CN)_6] \cdot H_2O$. When solutions of iron(III) ions and hexacyanoferrate(III) ions are mixed, no precipitate forms, but the solution turns brown. Iron(II) ions and hexacyanoferrate(II) ions produce a white precipitate, $K_2FeFe(CN)_6$. These reactions serve as sensitive tests to distinguish between Fe^{2+} and Fe^{3+} in solution.

Blueprint paper is coated with a mixture of iron(III) ammonium citrate and potassium hexacyanoferrate(III); this gives the paper the bronze-green color of iron(III) hexacyanoferrate(III). When a drawing in black ink on tracing cloth is placed over the blueprint paper and the paper is exposed to light, the iron(III) ions are reduced by the citrate ions to iron(II) wherever light transmission occurs. The iron(II) ion immediately reacts with the hexacyanoferrate(III) ion to form an insoluble blue precipitate. After exposure, the paper is washed with water to remove the unchanged iron(III) hexacyanoferrate(III), leaving the body of the paper blue and the lines produced by the tracing white.

27.22 The Manufacture of Steel

The term steel is applied to many widely different alloys of iron. **Steel** is made from iron by removing impurities and adding substances such as manganese, chromium, nickel, tungsten, molybdenum, and vanadium to produce alloys with properties that make the material suitable for specific uses. Most steels also contain small but definite percentages of carbon (0.04–2.5%). Thus a large part of the carbon contained in iron must be removed in the manufacture of steel.

Steels can be classified in three main categories: (1) **carbon steels,** which are primarily iron and carbon; (2) **stainless steels,** low-carbon steels containing about 12% chromium; and (3) **alloy steels,** specialty steels that contain large amounts of other elements to impart special properties for specific uses. Of the 141 million tons of steel produced in the United States in a recent year, 124.6 million tons (88.4%) were carbon steels, 14.8 million tons (10.5%) were stainless steels, and 1.6 million tons (1.1%) were alloy steels.

The principal process used in the production of steel is the basic oxygen process, which utilizes a cylindrical furnace with a basic lining, such as magnesium oxides and calcium oxides. A typical charge is 80 tons of scrap iron, 200 tons of molten iron, and 18 tons of limestone (to form slag). A jet of high-purity (99.5%) oxygen is directed into the white-hot molten charge through a water-cooled lance. The oxygen produces a vigorous reaction that oxidizes the impurities in the charge. In the central reaction zone, temperatures reach a level close to the boiling point of iron. The entire steelmaking cycle is completed in one hour or less. Electrostatic precipitators (see Section 12.30) clean the gases resulting from the furnace reactions, making the furnaces virtually smokeless. The steel produced is of extremely high and uniform quality.

Carbon is the most important alloying element in steel. It may be present in combination with iron as **cementite, Fe$_3$C,** or as crystals of graphite. The reaction of iron with carbon is reversible, and cementite is stable only at elevated temperatures. If cementite is cooled slowly, it decomposes to iron and graphite. Thus steel containing cementite in solid solution in iron must be made by quenching (quickly cooling) the hot metal in water or oil. The rate of decomposition of cementite is very slow at lower temperatures, so quick cooling does not provide enough time for the decomposition to occur. This steel is hard, brittle, and light-colored. If the metal is cooled slowly, the cementite decomposes and the carbon is deposited largely as separate crystals of graphite; the product is softer and more pliable and has a much higher tensile strength than does steel that has been cooled rapidly.

Certain materials called **scavengers** are added to iron in the manufacture of steel to remove impurities, especially oxygen and nitrogen, and thus improve the quality of the product. The most important scavengers are aluminum, ferrosilicon, ferromanganese, and ferrotitanium. They react with dissolved oxygen and nitrogen, forming oxides and nitrides, respectively, which are removed with the slag.

For every ton of steel produced, about 25–30 pounds of nonferrous metals are added or used as coatings. By the appropriate choice of the number and percentages of these elements, alloy steel of widely varying properties can be manufactured. Some of the important alloy steels and their features are given in Table 27.6.

Table 27.6   Alloy steels

Name	Composition	Characteristic properties	Uses
Manganese steel	10–18% Mn	Hard, tough, resistant to wear	Railroad rails, safes, armor plate, rock-crushing machinery
Silicon steel	1–5% Si	Hard, strong, highly magnetic	Magnets
Duriron	12–15% Si	Resistant to corrosion, acids	Pipes, kettles, condensers, etc.
Invar	36% Ni	Low coefficient of expansion	Meter scales, measuring tapes, pendulum rods
Chrome-vanadium	1–10% Cr, 0.15% V	Strong, resistant to strains	Axles
Stainless steel	14–18% Cr, 7–9% Ni	Resistant to corrosion	Cutlery, instruments
Permalloy	78% Ni	High magnetic susceptibility	Ocean cables
High-speed steels	14–20% W or 6–12% Mo	Retain temper at high temperatures	High-speed cutting tools
Nickel steel	2–4% Ni	Hard and elastic, resistant to corrosion	Drive shafts, gears, cables

Cobalt (A Group VIIIB Element)

Cobalt is similar to iron in appearance but has a faint tinge of pink. Like iron and nickel, it is magnetic. Almost all cobalt produced is a by-product of the extraction of nickel, copper, silver, and other metals. Cobalt dissolves slowly in warm dilute hydrochloric or sulfuric acid and more rapidly in dilute nitric acid, forming salts of cobalt(II). The Co^{2+} ion is the only simple hydrated ion of cobalt that is stable in aqueous solution. Like iron and nickel, cobalt is rendered passive (unreactive) by contact with concentrated nitric acid. It is not oxidized when exposed to air, but at red heat it reduces hydrogen in steam, with the evolution of hydrogen. The halogens, except fluorine, convert the metal to cobalt(II) halides. When cobalt is heated with fluorine, **cobalt(III) fluorine, CoF_3,** is formed.

Cobalt is alloyed with iron and small percentages of other metals for use in high-speed cutting tools and surgical instruments. Permanent magnets are made from the alloys **Alnico** (Al, Ni, Co, and Fe), **Hiperco** (Co, Fe, and Cr), and **Vicalloy** (Co, Fe, and V).

27.23 Compounds of Cobalt

Cobalt(II) hydroxide, $Co(OH)_2$, is formed as a blue flocculent precipitate ($K_{sp} = 2 \times 10^{-16}$) when an alkali metal hydroxide is added to a solution of a cobalt(II) salt. The blue color of the precipitate changes to violet and then to pink, probably as a result of hydration. Cobalt(II) hydroxide is readily soluble in aqueous ammonia to form **hexaamminecobalt(II) hydroxide, $[Co(NH_3)_6](OH)_2$.** Solutions of the latter compound are oxidized by oxygen in air to various cobalt(III) compounds; the oxidation is accompanied by a darkening of the solution.

When cobalt(II) hydroxide is heated in the absence of air, **cobalt(II) oxide, CoO,** results. This oxide is a black substance, but when dissolved in molten glass, it gives the glass a blue color. Such glass is called cobalt glass and contains **cobalt(II) silicate.** Ignition of cobalt(II) hydroxide or oxide in air gives rise to **cobalt(II, III) oxide, Co_3O_4. Cobalt(III) oxide, Co_2O_3,** may be produced by gently heating **cobalt(II) nitrate, $Co(NO_3)_2$.**

$$4Co(NO_3)_2 \xrightarrow{\triangle} 2Co_2O_3 + 8NO_2 + O_2$$

Note that three elements in this reaction change their oxidation numbers. (You will find balancing this equation by an oxidation-reduction method a challenging review.)

Cobalt(II) oxide and hydroxide readily dissolve in hydrochloric acid. Concentration of the resulting solution causes crystallization of **cobalt(II) chloride 6-hydrate, $CoCl_2 \cdot 6H_2O$,** which is red. The hydrated cobalt(II) ion, $[Co(H_2O)_6]^{2+}$, exhibits a pink color in solution. When partially dehydrated, cobalt(II) chloride changes to a deep blue color. This is believed to result from a change in the coordination number of the cobalt(II) ion from 6 to 4.

$$\underset{\text{Pink}}{[Co(H_2O)_6]Cl_2} \longrightarrow \underset{\text{Blue}}{[Co(H_2O)_4]Cl_2} + 2H_2O$$

The same changes in color is effected by dissolving $CoCl_2 \cdot 6H_2O$ in alcohol. Writing made on paper with a dilute solution of the hydrated salt is almost invisible, but when

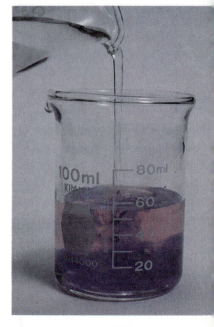

A precipitate of $Co(OH)_2$ forms when NaOH is added to a solution of a salt of cobalt(II).

the paper is warmed, dehydration of the salt occurs and the writing becomes blue. It fades again as hydration takes place from moisture of the air. This is the chemistry of one kind of "invisible" ink.

Black **cobalt(II) sulfide, CoS,** can be completely precipitated only from basic solutions. However, the precipitated sulfide is but slightly soluble in hydrochloric acid. Aqua regia readily dissolves it.

In addition to the simple salts, oxides, and hydroxides, cobalt forms a large number of coordination compounds. Cobalt(II) simple salts are more stable than are those of cobalt(III); cobalt(III) coordination compounds, in contrast, are much more stable than the corresponding cobalt(II) ones. Thus the majority of the more stable coordination compounds of cobalt contain the metal with the +3 oxidation number. Some of the more important cobalt(III) coordination compounds are **hexaamminecobalt(III) chloride, [Co(NH₃)₆]Cl₃, potassium hexacyanocobaltate(III), K₃[Co(CN)₆],** and **sodium hexanitrocobaltate(III), Na₃[Co(NO₂)₆].** In these coordination compounds the six ligands, such as NH_3, CN^-, and NO_2^-, occupy the corners of a regular octahedron with the cobalt at the center (see Section 26.3).

Nickel (A Group VIIIB Element)

Nickel is a silvery white metal that is hard, malleable, and ductile. Like iron and cobalt, it is highly magnetic. It is not oxidized by air under ordinary conditions, and it is resistant to the action of bases. Dilute acids slowly dissolve nickel; hydrogen is evolved. Nickel is made passive by treatment with concentrated nitric acid; nickel so treated will no longer displace hydrogen from dilute acids.

Because of its hardness, resistance to corrosion, and high reflectivity when polished, nickel is widely used in the plating of iron, steel, and copper. It is also a constituent of many important alloys, such as **Monel metal** (Ni, Cu, and a little Fe), and **Permalloy** (Ni and Fe), used in instruments for electrical transmission and reproduction of sound. **German silver** is a nickel-copper alloy. **Nichrome** and **chromel** contain nickel, iron, and chromium. **Alnico** contains aluminum, nickel, iron, and cobalt; it is highly magnetic, able to lift 4000 times its own weight of iron. **Platinite** and **invar** are nickel alloys that have the same coefficient of expansion as that of glass and are therefore used for "seal-in" wires through glass, such as those in electric light bulbs. Finely divided nickel is used as a catalyst in the hydrogenation of oils.

27.24 Compounds of Nickel

In its compounds, nickel exhibits an oxidation number of +2 almost exclusively. One notable exception is the +4 it shows in **nickel(IV) oxide, NiO₂,** a black hydrous substance formed by the oxidation of nickel(II) salts in alkaline solution. This oxide forms one of the electrodes in the nickel-cadmium cell (Section 21.14).

Nickel(II) oxide, NiO, is prepared by heating nickel(II) hydroxide, carbonate, or nitrate. When alkali metal hydroxides are added to solutions of nickel(II) salts, pale green **nickel(II) hydroxide, Ni(OH)₂,** precipitates ($K_{sp} = 1.6 \times 10^{-14}$). Both the oxide and the hydroxide are dissolved by aqueous ammonia with the formation of deep blue **hexaamminenickel(II) hydroxide, [Ni(NH₃)₆](OH)₂.**

$$Ni(OH)_2(s) + 6NH_3 \longrightarrow [Ni(NH_3)_6]^{2+} + 2OH^-$$

The soluble salts of nickel are prepared from nickel(II) oxide, hydroxide, or carbonate by treatment with the proper acid. The important soluble salts of the metal are the acetate, chloride, nitrate, and sulfate, and also hexaamminenickel(II) sulfate. Ammoniacal solutions of the latter compound are used in nickel-plating baths. The hydrated nickel(II) ion, $[Ni(H_2O)_6]^{2+}$, imparts a pale green color to the solutions and crystallized salts of the nickel(II) ion.

Nickel(II) sulfide, NiS, is produced as a black precipitate by the action of ammoniacal sulfide solutions on nickel(II) salts.

Nickel carbonyl, Ni(CO)₄, is formed as a colorless volatile liquid (boiling at 43°C) when carbon monoxide is passed over finely divided nickel. Nickel carbonyl readily decomposes at higher temperatures and deposits pure nickel metal. The **Mond process** for separating nickel from other metals is based on the formation and decomposition of nickel carbonyl. Carbon monoxide is passed over a mixture of nickel and other metals; nickel carbonyl is formed and carried along with the excess carbon monoxide, leaving other metals behind. When heated to 200°C, the nickel carbonyl decomposes, and the metal deposits as a fine dust. As is true for most metal carbonyl compounds, nickel carbonyl is highly toxic and must be handled with extreme care.

The Elements of Group IB

Copper, silver, and gold, which are sometimes called the **coinage metals,** are members of Group IB of the Periodic Table. Copper exhibits behavior similar to that of the other elements of the latter half of the first transition series. Silver and gold are relatively inactive transition metals with behavior that is characteristic of the platinum metals. Each element exhibits multiple oxidation numbers of +1, +2, and +3. However, the most common oxidation number for copper is +2; for silver, +1; and for gold, +3.

These metals were the first to be used by primitive races, because they are found in the free state, they tarnish slowly, and their appearance is pleasing to the eye. Ornaments of gold have been found in Egyptian tombs constructed in the prehistoric Stone Age. Gold and silver coins were used as a medium of exchange in India and Egypt thousands of years ago. It is believed that copper was the first metal to be fashioned into utensils, instruments, and weapons; its use for such purposes began sometime during the Stone Age. The practice of alloying copper with tin and the use of the resulting bronze in place of stone introduced the Bronze Age.

27.25 Properties of Copper

Copper is a reddish-yellow metal. It is ductile and malleable, so it is readily fashioned into wire, tubing, and sheets. Copper is the best electrical conductor among the cheaper metals, but when used for this purpose it must be quite pure, since small amounts of impurities greatly reduce its conductivity.

The extensive uses of copper make it the metal second in commercial importance to iron. The chief use of copper is in the production of all types of electrical wiring. Copper is also used in the production of a great many alloys. Among the more important ones are **brass** (Cu, 60–82%; Zn, 18–40%), **bronze** (Cu, 70–95%; Zn, 1–25%; Sn, 1–18%), **aluminum bronze** (Cu, 90–98%; Al, 2–10%), and **German silver** (Cu, 50–60%; Zn, 20%; Ni, 20–25%).

Figure 27.7

The green color of the Statue of Liberty is due to a coating of $Cu_2(CO_3)(OH)_2$, produced by weathering of its copper covering.

Copper is relatively inactive chemically. In moist air it first turns brown, due to the formation of a very thin, adherent film of either copper oxide or copper sulfide. Prolonged weathering causes copper to become coated with a green film of the hydroxycarbonate $Cu_2(OH)_2CO_3$. This compound or the hydroxysulfate is responsible for the green color of well-weathered copper (Fig. 27.7).

When heated in air, copper oxidizes, forming copper(II) oxide along with some copper(I) oxide. Oxidizing acids, and nonoxidizing acids in the presence of air, convert copper to the corresponding copper(II) salts. Aqueous ammonia in the presence of air dissolves copper, producing a blue solution containing $[Cu(NH_3)_4]^{2+}$. Alkali metal hydroxides have no effect on the metal. Sulfur vapor reacts with hot copper to give both Cu_2S and CuS. Hot copper burns in chlorine to give $CuCl_2$.

Copper forms two principal series of compounds, in which it has the oxidation numbers of $+1$ and $+2$. Copper also exhibits an oxidation number of $+3$, but copper(III) compounds are rare. Copper(I) forms a complete series of binary compounds, such as the halides and the oxide, but oxysalts of copper(I) are few in number and readily decomposed by water. In contrast, copper(II) forms a complete series of oxysalts, while some of its binary compounds are unstable and decompose spontaneously.

27.26 Copper(I) Compounds

Solid salts of copper(I) show a variety of colors but in solution are generally colorless.

Copper(I) oxide, Cu_2O, may be prepared as a reddish-brown precipitate by boiling copper(I) chloride with sodium hydroxide.

$$2CuCl(s) + 2OH^- \xrightarrow{\Delta} Cu_2O(s) + 2Cl^- + H_2O$$

When basic solutions of copper(II) salts are heated with reducing agents, copper(I) oxide precipitates.

$$2Cu^{2+} + 2OH^- + 2e^- \text{ (reducing agent)} \xrightarrow{\Delta} Cu_2O(s) + H_2O$$

This reaction is the basis for the test for the presence of reducing sugars in the urine in the diagnosis of diabetes. The reagent most often used for this purpose is Benedict's solution, which is a solution of copper(II) sulfate, sodium carbonate, and sodium citrate. The addition of a reducing sugar, such as glucose, causes the precipitation of the reddish-brown copper(I) oxide.

Copper(I) hydroxide, CuOH, is formed as a yellow precipitate when a cold solution of copper(I) chloride in hydrochloric acid is treated with sodium hydroxide. The hydroxide is unstable and decomposes into the oxide and water when heated.

When copper metal is heated with a solution of copper(II) chloride acidified with hydrochloric acid, an oxidation-reduction reaction takes place, with the precipitation of **copper(I) chloride.**

$$Cu^{2+} + 2Cl^- + Cu \longrightarrow 2CuCl(s)$$

This white crystalline compound is readily soluble in concentrated hydrochloric acid, forming the complex ion $[CuCl_2]^-$.

$$CuCl + Cl^- \rightleftharpoons [CuCl_2]^-$$

The two chlorides of the $[CuCl_2]^-$ ion are bonded to the copper by coordinate-covalent bonds. Dilution with water brings about reprecipitation of the copper(I) chloride ($K_{sp} = 1.85 \times 10^{-7}$).

Among other stable and insoluble copper(I) binary salts are Cu_2S (black), $CuCN$ (white), $CuBr$ (white), and CuI (white).

Copper(I) forms linear complex ions with ammonia, $[Cu(NH_3)_2]^+$, with the halides, $[CuX_2]^-$, and with cyanide ions, $[Cu(CN)_2]^-$. Unlike copper(II) complexes, most copper(I) complexes are colorless.

27.27 Copper(II) Compounds

The more common compounds of copper are those in which the metal has an oxidation number of $+2$. Solid copper(II) compounds may be blue, black, green, yellow, or white. However, dilute solutions of all soluble copper(II) salts have the blue color of the hydrated copper(II) ion, $[Cu(H_2O)_4]^{2+}$.

Copper(II) oxide, CuO, is a black, insoluble compound that may be obtained by heating copper(II) carbonate or copper(II) nitrate or finely divided copper in oxygen. Because of the ease with which copper can be reduced either to an oxidation number $+1$ or to the metal, hot copper(II) oxide is a good oxidizing agent. It is used as one in the analysis of organic compounds to determine the percentage of carbon they contain. Copper(II) oxide oxidizes the carbon to carbon dioxide, which is absorbed by sodium hydroxide. The gain in weight of the sodium hydroxide is a measure of the quantity of carbon dioxide produced and absorbed. The copper oxide is continuously regenerated by passing a stream of air through the tube in which the organic matter and the copper oxide are reacting.

When hydroxide ions are added to cold solutions of copper(II) salts, a bluish-green gelatinous precipitate of **copper(II) hydroxide, Cu(OH)$_2$** ($K_{sp} = 5.6 \times 10^{-20}$), is formed. With hot solutions, CuO is obtained. Copper(II) hydroxide is soluble in aqueous ammonia, giving **tetraamminecopper(II) hydroxide, [Cu(NH$_3$)$_4$](OH)$_2$,** which is a strong base with a deep blue color due to the color of the $[Cu(NH_3)_4]^{2+}$ ion.

$$Cu(OH)_2(s) + 4NH_3 \longrightarrow [Cu(NH_3)_4]^{2+} + 2OH^-$$

Copper(II) sulfate, CuSO$_4$, is the most important of the copper salts. The anhydrous salt is colorless, but when it is crystallized from aqueous solution the blue pentahydrate $[Cu(H_2O)_4]SO_4 \cdot H_2O$, known as **blue vitriol,** is formed. As indicated by the formula, four of the water molecules are coordinated to the copper(II) ion and the fifth is attached to the sulfate ion and two of the coordinated water molecules by hydrogen bonds. The structure of the pentahydrate may be represented as shown below.

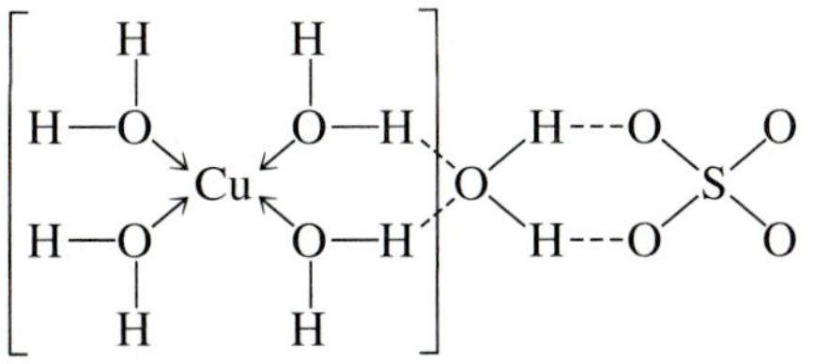

Copper(II) sulfate is produced commercially by oxidizing copper(II) sulfide either directly to the sulfate or to the oxide, which is then converted to the sulfate by reaction with sulfuric acid. In the laboratory the anhydrous sulfate may be prepared by oxidizing copper with hot concentrated sulfuric acid.

$$Cu(s) + 2H_2SO_4(l) \longrightarrow CuSO_4(s) + SO_2(g) + 2H_2O(l) \qquad \Delta H° = -11.9 \text{ kJ}$$

Copper(II) sulfate is used in the electrolytic refining of copper, in electroplating, in the Daniell cell (Section 21.13), in the manufacture of pigments, as a color-fixing agent in the textile industry, and in the prevention of algae in reservoirs and swimming pools. Because anhydrous copper(II) sulfate is insoluble in alcohol and ether and readily takes up water, turning blue, this salt is used for detecting water in and removing it from liquids such as alcohol and ether. Of course, copper(II) sulfate cannot be used to remove or detect water in liquids with which it reacts.

Anhydrous **copper(II) chloride, $CuCl_2$,** may be prepared as a yellow crystalline salt by direct union of the elements. The blue-green hydrated salt $Cu(H_2O)_2Cl_2$ may be obtained by treating either the carbonate or the hydroxide of copper(II) with hydrochloric acid and evaporating the solution. Concentrated solutions of copper(II) chloride are green, because they contain both the hydrated copper(II) ion $[Cu(H_2O)_4]^{2+}$, which is blue, and the tetrachlorocuprate(II) ion $[CuCl_4]^{2-}$, which is yellow. (The combination of blue and yellow looks green to the eye.) When such a solution is diluted, water molecules replace the chloride ions of the $[CuCl_4]^{2-}$ ion, and the solution turns blue.

Copper(II) bromide, $CuBr_2$, is a black solid that may be obtained by direct union of the elements or by the action of hydrobromic acid on copper(II) oxide or carbonate.

It is interesting that copper(II) iodide, CuI_2, does not exist. When a solution containing the copper(II) ion is treated with an excess of iodide, an oxidation-reduction reaction occurs, with precipitation of **copper(I) iodide** and formation of elementary iodine.

$$2Cu^{2+} + 4I^- \longrightarrow 2CuI(s) + I_2$$

This reaction is used in the quantitative determination of copper by titration of the liberated iodine with a standard solution of sodium thiosulfate. Once the amount of elemental iodine is known, the amount of copper can be calculated.

Copper(II) sulfide, CuS, black, is precipitated ($K_{sp} = 6.7 \times 10^{-42}$) by passing hydrogen sulfide into acidic, basic, or neutral solutions of copper(II) salts. The sulfide readily dissolves in warm dilute nitric acid with the formation of elemental sulfur and nitric oxide.

$$3CuS + 8H^+ + 2NO_3^- \rightleftharpoons 3Cu^{2+} + 3S(s) + 2NO(g) + 4H_2O$$

A sample of hydrated copper(II) chloride.

Copper possesses the property of forming both complex cations and anions that are quite stable. Copper(II) ions coordinate with four neutral ammonia molecules to form the deep blue **tetraamminecopper(II) ion.**

$$[Cu(H_2O)_4]^{2+} + 4NH_3 \rightleftharpoons [Cu(NH_3)_4]^{2+} + 4H_2O$$

The fact that the ammonia molecules displace the coordinated water molecules indicates the greater stability of the ammine complex. In fact, the ammine complex is so stable that most slightly soluble copper(II) compounds are dissolved by aqueous ammonia. The copper ion is located at the center of a square formed by the four attached groups, whether they are water, $[Cu(H_2O)_4]^{2+}$, or ammonia, $[Cu(NH_3)_4]^{2+}$. An example of a negatively charged copper(II) complex ion is $[CuCl_4]^{2-}$.

27.28 Copper Oxide Superconductors

One of the most exciting scientific discoveries of the 1980s is the characterization of new materials that exhibit superconductivity at temperatures above 90 K. **Superconductors** are conductors that conduct electricity with no resistance. They also possess unusual magnetic properties.

These new superconducting materials are ternary oxides containing yttrium or one of several rare earth elements, barium, and copper in a 1:2:3 ratio. The formula of the yttrium compound is $YBa_2Cu_3O_x$ where x ranges from about 6.5 to 7 depending upon the conditions used to prepare the compound. The yttrium(III) ion can be replaced by samarium (Sm), europium (Eu), gadolinium (Gd), dysprosium (Dy), holmium (Ho), or ytterbium (Yb). The variable amount of oxygen reflects the fact that some of the copper in these compounds can be oxidized from copper(II) to copper(III). For $x = 6.5$, all of the copper is present as copper(II); for $x = 7$, one-third of the copper is oxidized to copper(III).

Preparation of these materials is straightforward. A stoichiometric mixture of the oxides and carbonates is finely ground together and then heated to a temperature between 900 and 1100°C to decompose the carbonates and form the black ternary oxide.

$$\frac{1}{2}Y_2O_3(s) + 2BaCO_3(s) + 3CuCO_3(s) \longrightarrow YBa_2Cu_3O_{6.5}(s) + 5CO_2(g)$$

After regrinding and reheating, the material is pressed into a pellet and sintered (heated so it bonds without melting) at high temperature. The sintered pellets are annealed in a pure oxygen atmosphere at 400 to 900°C in order to oxidize some of the copper to copper(III) and then allowed to cool slowly.

The superconducting copper oxides are ionic compounds. The structure of a unit cell of $YBa_2Cu_3O_7$ is illustrated in Fig. 27.8. The copper(II) and copper(III) ions cannot be distinguished in the structure because the compound is a solid solution (Section 12.8) containing copper(II) and copper(III) ions.

Because these oxides are ionic compounds (ceramics), they are brittle and fragile and cannot be drawn out to form wires, as can a metal such as copper itself. However ceramic wires and other shapes can be formed by mixing the ceramic powder or its ingredients with an organic binder (a glue), and then extruding a thin strip through a press. (This works just like squeezing a fine strip of toothpaste out of a tube.) The sample is then heated to burn off the binder and annealed, giving a current-carrying ceramic strip.

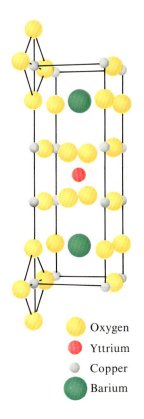

Oxygen

Yttrium

Copper

Barium

Figure 27.8
The unit cell of $YBa_2Cu_3O_7$.

60. In terms of electronic structure, explain the three oxidation numbers of copper.
61. Explain the use of copper(II) sulfate in urine analysis.
62. Compare the stability of copper(I) halides to that of copper(II) halides.
63. A white precipitate forms when copper metal is added to a solution of copper(II) chloride and hydrochloric acid. The white precipitate dissolves in excess concentrated hydrochloric acid. Dilution with water causes the white precipitate to reappear. Write equations for the reactions involved.
64. How many grams of $CuCl_2$ contain the same mass of copper as in 100 g of CuCl? *Ans. 136 g*
65. Why are most slightly soluble copper salts readily dissolved by aqueous ammonia?
66. The formation constant of $[Cu(NH_3)_4]^{2+}$ is 1.2×10^{12}. What will be the equilibrium concentration of Cu^{2+} if 1.0 g of Cu is oxidized and put into 1.0 L of 0.25 M NH_3 solution? *Ans. 1.1×10^{-11} M*
67. Sketch the structure of $Cu(H_2O)_4SO_4 \cdot H_2O$.
68. Would you expect CuS to dissolve in 1 M NH_3 solution?
69. A 1.008-g sample of a silver-copper alloy is dissolved and treated with excess iodide ion. The liberated I_2 is titrated with a 0.1052 M $S_2O_3^{2-}$ solution. If 29.84 mL of this solution is required, what is the percentage of Cu in the alloy? *Ans. 19.79%*

Silver

70. What properties of silver have made it valuable as a coinage metal?
71. Explain the tarnishing of silver in the presence of sulfur-containing substances.
72. Compare the solubilities of silver halides in water and in aqueous ammonia.

73. Why does silver dissolve in nitric acid but not in hydrochloric acid?
74. Write equations for the electrode reactions when $Na[Ag(CN)_2]$ is used as the electrolyte in silver plating.
75. Dilute sodium cyanide solution is slowly dripped into a slowly stirred silver nitrate solution. A white precipitate forms temporarily but dissolves as the addition of sodium cyanide continues. Use chemical equations to explain this observation.
76. Show several ways in which coordination compounds play an important part in the chemistry of silver.
77. The formation constant of $[Ag(CN)_2]^-$ is 1.0×10^{20}. What will be the equilibrium concentration of Ag^+ if 1.0 g of Ag is oxidized and put into 1 L of 1.0×10^{-1} M CN^- solution? *Ans. 1.4×10^{-20} M*
78. A 1.4820-g sample of a pure solid alkali metal chloride is dissolved in water and treated with excess silver nitrate. The resulting precipitate, filtered and dried, weighs 2.849 g. What percent by mass of chloride ion is in the original compound? What is the identity of the salt? *Ans. 47.55%; KCl*

Gold

79. Account for the variable oxidation number of gold in terms of electronic structure.
80. Would you expect salts of the gold(I) ion, Au^+, to be colored? Explain.
81. Write balanced equations for the following changes, which occur during the recovery of gold from an ore:

$$Au \longrightarrow [Au(CN)_2]^- \longrightarrow Au$$

82. How many gold atoms are there in 5.0 g of 20-carat gold? *Ans. 1.3×10^{22} atoms*

(b) $CrO + 2HNO_3$ in water.
(c) $Cr_2(SO_4)_3 + Zn$ in acid.
(d) $CrCl_3$ is added to an aqueous solution of NaOH.
(e) $TiCl_2$ is added to a solution containing an excess of $CrO_4{}^{2-}$ ion.
(f) Mn is heated with CrO_3.

Manganese

28. How are the ferromanganese alloys produced?
29. Write a balanced equation for the reaction of manganese dioxide and hydrochloric acid.
30. Explain the dissolution of manganese(II) hydroxide in aqueous ammonium chloride in terms of an acid-base reaction.
31. Identify two uses of manganese dioxide.
32. Write the names and formulas for compounds in which manganese exhibits each of the following oxidation numbers: $+2$, $+3$, $+4$, $+6$, and $+7$.
33. How is potassium permanganate prepared commercially?
34. Would you expect a manganese(VII) oxide solution to have a pH greater or less than 7.0? Justify your answer.
35. How can the insolubility of barium sulfate be applied to the preparation of free permanganic acid?
36. Predict the products of the following reactions.
    (a) MnO is heated in air.
    (b) $MnCO_3$ is added to a solution of HNO_3.
    (c) MnO_2 is warmed in a solution of HBr.
    (d) A mixture of MnO_2 and BaO is heated in air.
    (e) VCl_3 is added to a solution of $KMnO_4$.
    (f) An excess of Mn metal is added to a solution of $KMnO_4$.

Iron

37. Write the names and formulas of the principal ores of iron.
38. Identify each of the following: hematite, magnetite, rouge, Mohr's salt, and Prussian blue.
39. What is the composition of iron rust?
40. What compound of iron is a common constituent of hard water?
41. What is the gas produced when iron(II) sulfide is treated with a nonoxidizing acid?
42. Give the general equation for the reaction of iron(III) salts with water. Write equations showing the steps in the reactions of iron(III) chloride to produce a precipitate of hydrated $Fe(OH)_3$, $[Fe(H_2O)_3(OH)_3]$.
43. Iron(II) can be titrated to iron(III) by dichromate ion, which is reduced to chromium(III) in acid solution. A 2.500-g sample of iron ore is dissolved and the iron converted to iron(II). Exactly 19.17 mL of 0.0100 M $Na_2Cr_2O_7$ is required in the titration. What percentage of iron was the ore sample?   *Ans. 2.57%*
44. How many cubic feet of air at STP is required per ton of Fe_2O_3 to convert it into iron in a blast furnace? Assume air is 19% oxygen by volume.   *Ans. 3.5×10^4 ft^3*

45. Predict the products of the following reactions.
    (a) Fe is heated in air.
    (b) Fe is added to a dilute solution of H_2SO_4.
    (c) A solution of $Fe(NO_3)_2$ and HNO_3 is allowed to stand in air.
    (d) Fe is heated in an atmosphere of steam.
    (e) $FeCO_3$ is added to a solution of $HClO_4$.
    (f) NaOH is added to a solution of $Fe(NO_3)_3$.
    (g) $FeSO_4$ is added to an acidic solution of K_2CrO_4.

Cobalt

46. Sketch the geometrical structure of $[Co(CN)_6]^{3-}$.
47. Balance the following equation by oxidation-reduction methods:

$$Co(NO_3)_2 \longrightarrow Co_2O_3 + NO_2 + O_2$$

48. On the basis of the reactions of elemental fluorine, chlorine, and oxygen with metallic cobalt, predict which is the strongest oxidizing agent.
49. What is the percent by mass of cobalt in sodium hexanitrocobaltate(III)?   *Ans. 14.5898%*
50. Why does cobalt precede nickel in the Periodic Table when cobalt has a higher atomic weight than that of nickel?
51. Predict the products of the following reactions.
    (a) Co is heated in air.
    (b) Co is added to a dilute solution of H_2SO_4.
    (c) NaOH is added to a solution of $Co(NO_3)_2$.
    (d) Co is heated with I_2.
    (e) $CoCO_3$ is added to a solution of $HClO_4$.
    (f) NaOH and Na_2S are added to a solution of $Co(NO_3)_2$.
    (g) $CoSO_4$ is added to solution of NH_3.

Nickel

52. What is meant by *passivity*?
53. How is nickel rendered passive?
54. What is the composition of Alnico? What use is made of this alloy?
55. Write the equation for the dissolution of nickel(II) hydroxide in aqueous ammonia.
56. What is the physical nature of nickel carbonyl?
57. What is the potential of the following electrochemical cell:

$$Cd|Cd^{2+}, M = 0.10\|Ni^{2+}, M = 0.50|Ni$$

Ans. 0.174 V

58. Predict the products of the following reactions.
    (a) Ni is heated in air.
    (b) Ni is added to a dilute solution of H_2SO_4.
    (c) KOH is added to a solution of $NiCl_2$.
    (d) Ni is heated with Cl_2.
    (e) $NiCO_3$ is added to a solution of $HClO_4$.
    (f) NH_3 and Na_2S are added to a solution of $Ni(NO_3)_2$.
    (g) $NiSO_4$ is added to solution of NH_3.

Copper

59. Copper plate makes up the surface of the Statue of Liberty in New York harbor. Why does it appear green?

60. In terms of electronic structure, explain the three oxidation numbers of copper.
61. Explain the use of copper(II) sulfate in urine analysis.
62. Compare the stability of copper(I) halides to that of copper(II) halides.
63. A white precipitate forms when copper metal is added to a solution of copper(II) chloride and hydrochloric acid. The white precipitate dissolves in excess concentrated hydrochloric acid. Dilution with water causes the white precipitate to reappear. Write equations for the reactions involved.
64. How many grams of $CuCl_2$ contain the same mass of copper as in 100 g of CuCl?   *Ans. 136 g*
65. Why are most slightly soluble copper salts readily dissolved by aqueous ammonia?
66. The formation constant of $[Cu(NH_3)_4]^{2+}$ is 1.2×10^{12}. What will be the equilibrium concentration of Cu^{2+} if 1.0 g of Cu is oxidized and put into 1.0 L of 0.25 M NH_3 solution?   *Ans. 1.1×10^{-11} M*
67. Sketch the structure of $Cu(H_2O)_4SO_4 \cdot H_2O$.
68. Would you expect CuS to dissolve in 1 M NH_3 solution?
69. A 1.008-g sample of a silver-copper alloy is dissolved and treated with excess iodide ion. The liberated I_2 is titrated with a 0.1052 M $S_2O_3^{2-}$ solution. If 29.84 mL of this solution is required, what is the percentage of Cu in the alloy?   *Ans. 19.79%*

Silver

70. What properties of silver have made it valuable as a coinage metal?
71. Explain the tarnishing of silver in the presence of sulfur-containing substances.
72. Compare the solubilities of silver halides in water and in aqueous ammonia.

73. Why does silver dissolve in nitric acid but not in hydrochloric acid?
74. Write equations for the electrode reactions when $Na[Ag(CN)_2]$ is used as the electrolyte in silver plating.
75. Dilute sodium cyanide solution is slowly dripped into a slowly stirred silver nitrate solution. A white precipitate forms temporarily but dissolves as the addition of sodium cyanide continues. Use chemical equations to explain this observation.
76. Show several ways in which coordination compounds play an important part in the chemistry of silver.
77. The formation constant of $[Ag(CN)_2]^-$ is 1.0×10^{20}. What will be the equilibrium concentration of Ag^+ if 1.0 g of Ag is oxidized and put into 1 L of 1.0×10^{-1} M CN^- solution?   *Ans. 1.4×10^{-20} M*
78. A 1.4820-g sample of a pure solid alkali metal chloride is dissolved in water and treated with excess silver nitrate. The resulting precipitate, filtered and dried, weighs 2.849 g. What percent by mass of chloride ion is in the original compound? What is the identity of the salt?   *Ans. 47.55%; KCl*

Gold

79. Account for the variable oxidation number of gold in terms of electronic structure.
80. Would you expect salts of the gold(I) ion, Au^+, to be colored? Explain.
81. Write balanced equations for the following changes, which occur during the recovery of gold from an ore:

$$Au \longrightarrow [Au(CN)_2]^- \longrightarrow Au$$

82. How many gold atoms are there in 5.0 g of 20-carat gold?   *Ans. 1.3×10^{22} atoms*

Key Terms and Concepts

actinide contraction (27.8)
actinide series (27.1, 27.3, 27.8)
basic oxygen process (27.22)
blast furnace (27.2)
blister copper (27.2)
carbonates (27.5)
cementite (27.22)
d-block elements (27.1, 27.3)
f-block elements (27.1, 27.3)
flotation process (27.2)
halides (27.5)

hydrometallurgy (27.2)
hydroxides (27.5)
inner transition elements
    (27.1, 27.3)
lanthanide contraction (27.4)
lanthanide series (27.1, 27.3,
    27.7)
Mond process (27.24)
oxides (27.5)
Parkes process (27.2)
platinum metals (27.3)

rare earth elements
    (27.1, 27.3, 27.7)
roasting process (27.2)
scavenger (27.22)
smelting process (27.2)
stainless steel (27.13)
steel (27.22)
sulfides (27.5)
superconductor (27.28)
transition elements (27.1, 27.3)

Exercises

1. Write the electron configurations for the following elements: Sc, Ti, Cr, Fe, Mo, Ru.
2. Write the electron configurations for Ti and the Ti^{2+}, Ti^{3+}, and Ti^{4+} ions.
3. Write the electron configurations for the following elements and their +3 ions: La, Sm, and Lu.
4. Which of the elements of the first transition series are reduced to the elemental state by carbon during their isolation?
5. What are the various reducing agents used to isolate the elements of the first transition series?
6. What are the reactions that occur in a blast furnace? Which of these are oxidation-reduction reactions?
7. Describe the electrolytic process for refining copper.
8. Describe the Parks process for extracting silver from lead.
9. Why are the rare earths found associated with one another in nature?
10. Why are the elemental (free) rare earth elements not found in nature?
11. Which of the following elements is most likely to be used to prepare La by the reduction of La_2O_3: Al, C, or Fe? Why?
12. Which of the following elements is most likely to form an oxide with the formula MO_3: Zr, Nb, Mo?
13. Which of the following ions are not stable in water: Sc^{3+}, Ti^{2+}, Cr^{2+}, Fe^{4+}, Co^{3+}, Ni^{2+}, Cu^+?
14. Which of the following is the strongest oxidizing agent: VO_4^{3-}, CrO_4^{2-}, or WO_4^{2-}?
15. Describe the lanthanide contraction.
16. What is the effect of the lanthanide contraction on the chemistry of the elements of the second and third transition series?
17. Predict the products of the following reactions.

    (a) $V + VCl_3 \xrightarrow{\triangle}$

    (b) $Ti + xsF_2 \xrightarrow{\triangle}$

    (c) $Co + xsF_2 \longrightarrow$

    (d) $Mn(OH)_2 + HBr(aq) \longrightarrow$
    (e) $CuCO_3 + HI(aq) \longrightarrow$
    (f) $Mn_2O_3 + HCl(aq) \longrightarrow$
    (g) $TiBr_4 + H_2O \longrightarrow$

    (h) $Cr + O_2 \xrightarrow{\triangle}$

    (i) $CoO + O_2 \xrightarrow{\triangle}$

    (j) $La + O_2 \xrightarrow{\triangle}$

    (k) $W + O_2 \xrightarrow{\triangle}$

    (l) $CrO_3 + CsOH(aq) \longrightarrow$
    $Fe + H_2SO_4 \longrightarrow$
    $LaCl_3(aq) + NaOH(aq) \longrightarrow$

18. In what way are the Group VIB metals typical of transition metals in general?
19. In what ways are the Group VIIIB metals Fe, Co, and Ni typical of transition metals in general?

Chromium

20. Write a balanced chemical equation describing the oxidation in air of aqueous chromium(II) to chromium(III).
21. Write an equation for the decomposition of ammonium dichromate by ignition.
22. Account for the extensive use of chromium in the plating of iron and copper objects, such as the trim on automobiles.
23. Reduction of chromite ore by carbon yields an alloy of iron and chromium. How would you separate the two components of this alloy into solutions of their respective cations, if necessary?
24. What is the reaction between hydrogen ions and chromate ions? This is called a condensation reaction. Explain.
25. Write the Lewis structure of chromic acid.
26. Write an equation to illustrate the action of the dichromate ion as an oxidizing agent.
27. Predict the products of the following reactions.
    (a) $Cr^{2+} + CrO_4^{2-}$ in acid solution.

metal, while a 10-carat alloy is $\frac{10}{24}$ gold by weight. Red or yellow gold alloys contain copper, and white gold alloys are made using palladium, nickel, or zinc.

Gold is a very inactive metal. It neither combines directly with oxygen nor corrodes in the atmosphere. The metal is not affected by any single common acid or by bases. It does, however, dissolve readily in aqua regia.

$$Au + 6H^+ + 3NO_3^- + 4Cl^- \longrightarrow [AuCl_4]^- + 3NO_2(g) + 3H_2O$$

It is also dissolved by a solution of chlorine and by cyanide in the presence of air; the latter reaction is used to separate gold from its ores.

In most of its compounds gold has an oxidation number of $+1$ or $+3$. When gold is dissolved in aqua regia and the solution is evaporated, yellow crystals of **chlorauric acid, $HAuCl_4 \cdot 4H_2O$,** are formed. When this coordination compound is heated, hydrogen chloride is evolved, and red crystalline **gold(III) chloride, $AuCl_3$,** remains. When this compound is heated to 175°C it decomposes to **gold(I) chloride, AuCl.** At higher temperatures the metal is obtained. The thermal instability of these compounds is typical, for all compounds of gold are decomposed by heat. Gold(I) chloride undergoes an auto-oxidation-reduction reaction in water, forming gold(III) chloride and the metal. Gold forms two oxides, Au_2O and Au_2O_3, and the corresponding hydroxides, AuOH (a weak base), and $Au(OH)_3$. The trihydroxide is a weak acid capable of reacting with strong bases to form **aurates,** such as $NaAuO_2$. Potassium cyanide reacts with gold(I) and gold(III) compounds, giving the complex soluble salts $Na[Au(CN)_2]$ and $Na[Au(CN)_4]$, which are important in the extraction of gold from its ores and in gold plating operations.

For Review

Summary

The transition elements are the elements with partially filled d orbitals (d-block elements) or f orbitals (f-block elements) in their valence shells. The d-block elements are further classified into the first, second, third, and fourth transition series; the f-block is divided into the lanthanide and the actinide series.

All of the transition elements are metallic; all react with halogens, and most react with sulfur and oxygen. The f-block elements, the elements of Group IIIB, and the first transition series (except copper) are sufficiently active to react with aqueous solutions of acids. The second and third transition (d-block) series are less active, and the elements near platinum are particularly nonreactive. The d-block elements exhibit a variety of oxidation states. The elements of the first half of the d-block exhibit oxidation numbers that correspond to loss of all of their valence electrons. The common oxidation numbers of the elements in the second half of the first transition series are limited to $+1$, $+2$, and $+3$. The chemical behavior of the elements of the first transition series is similar to that of the active metals in those compounds that contain the metal with an oxidation number that results in compounds that are stable in water. However, unlike the active metals, the metals in these compounds can also undergo oxidation-reduction reactions.

oxide is only slightly soluble in water, it dissolves enough to give a distinctly basic solution. The equilibrium is described by the equation

$$Ag_2O(s) + H_2O \rightleftharpoons 2Ag^+ + 2OH^-$$

Silver oxide is a convenient reagent, both in inorganic and organic chemistry, for preparing soluble hydroxides from the corresponding halides because the silver halide formed at the same time may be removed conveniently by filtration. Cesium hydroxide, for example, may be prepared by the reaction

$$[2Cs^+] + 2Cl^- + Ag_2O + H_2O \longrightarrow [2Cs^+] + 2OH^- + 2AgCl(s)$$

This reaction proceeds to the right because AgCl is much less soluble than Ag_2O. Silver oxide dissolves readily in aqueous ammonia to form the strong base **diamminesilver hydroxide.**

$$Ag_2O + 4NH_3 + H_2O \longrightarrow 2[Ag(NH_3)_2]^+ + 2OH^-$$

When silver oxide is heated at atmospheric pressure in air, it readily decomposes, a behavior characteristic of the oxides of the platinum metals.

$$2Ag_2O(s) \xrightarrow{\triangle} 4Ag(s) + O_2(g) \qquad \Delta H° = 62.09 \text{ kJ}$$

Silver fluoride, AgF, is very soluble in water, but the chloride, bromide, and iodide are relatively insoluble—the higher the atomic weight of the halogen, the lower is the solubility of the silver halide. The insoluble silver halides are formed as curdy precipitates when halide ions are added to solutions of silver salts. Silver chloride ($K_{sp} = 1.8 \times 10^{-10}$) is white; silver bromide ($K_{sp} = 3.3 \times 10^{-13}$) is pale yellow; silver iodide ($K_{sp} = 1.5 \times 10^{-16}$) is yellow. When exposed to light, the halides of silver turn violet at first and finally black; during this process, they are decomposed into the elements.

$$2AgX + light \longrightarrow 2Ag + X_2 \qquad (X = halogen)$$

Silver chloride, AgCl, is readily soluble in an excess of dilute aqueous ammonia to form the diamminesilver complex.

$$AgCl(s) + 2NH_3 \longrightarrow [Ag(NH_3)_2]^+ + Cl^-$$

Silver nitrate, AgNO₃, is the only simple silver salt that is usefully soluble. It is obtained by dissolving silver in nitric acid and evaporating the solution. Organic materials, such as the skin, readily reduce silver nitrate to give free silver, which forms a black stain. Silver nitrate is heavily used in the production of photographic materials, as a laboratory reagent, and in the manufacture of other silver compounds.

27.31 Gold

Gold is a soft yellow metal and is the most malleable and ductile of all metals. Gold foil can be made by hammering the metal into sheets so thin that 300,000 of them make a pile only 1 inch thick; 1 gram of gold can be drawn into a wire more than a mile and a half in length. Pure gold is too soft to be used for jewelry and coinage; for such purposes it is always alloyed with copper, silver, or some other metal. The purity of gold is expressed in **carats,** a designation that indicates the number of parts by weight of gold in 24 parts of alloy. Thus 24-carat gold is the pure

such as eggs and mustard. Its reaction with hydrogen sulfide in the presence of air is

$$4Ag(s) + 2H_2S(g) + O_2(g) \longrightarrow 2Ag_2S(s) + 2H_2O(l) \qquad \Delta H° = -595.59 \text{ kJ}$$

Silver tarnish is a thin film of silver sulfide.

The halogens react with silver, forming the halides, and the metal is soluble in oxidizing acids such as nitric acid and hot sulfuric acid.

$$3Ag + 4H^+ + NO_3^- \longrightarrow 3Ag^+ + NO(g) + 2H_2O$$

Since silver lies below hydrogen in the activity series, it is not soluble in nonoxidizing acids such as hydrochloric acid. The common oxidation number exhibited by silver is +1.

A large amount of silver is used for the making of coins, silverware, and ornaments. For most of its uses silver is too soft to wear well, and it is therefore hardened by alloying it with other metals, particularly copper. **Sterling silver** contains 7.5% copper, and jewelry silver contains 20% copper. Large amounts of silver are used in the preparation of dental alloys, photographic films, and mirrors.

The electroplating industry uses a large percentage of the silver produced. The object to be plated with silver is made the cathode in an electrolytic cell containing a solution of $Na[Ag(CN)_2]$ as the electrolyte. The anode is a bar of pure silver, which dissolves to replace the silver ions removed from the solution as plating proceeds. The electrode reactions are

Cathodic reduction: $\quad [Ag(CN)_2]^- + e^- \longrightarrow Ag + 2CN^-$

Anodic oxidation: $\qquad Ag + 2CN^- \longrightarrow [Ag(CN)_2]^- + e^-$

The film of deposited silver has a flat white appearance but can be given a brilliant luster by burnishing.

The characteristic luster can be restored to tarnished silverware by the following application of the activity series. The tarnished silver object is immersed in a solution of a teaspoonful of baking soda and a teaspoonful of salt in a liter of water, held in an aluminum vessel. (A piece of aluminum metal foil in a glass vessel will serve equally well.) The aluminum and the silver must be in contact. After the tarnish is removed, the silver object should be thoroughly washed. The chemistry involved may be explained as follows: The more active aluminum displaces the silver from the tarnish, causing it to plate out on the silver object, according to the equation

$$3Ag^+ + Al \longrightarrow 3Ag + Al^{3+}$$

The tarnish is removed with little loss of silver. However, the "antiquing" (actually silver sulfide tarnish), which often gives a desirable appearance by defining the crevices of the design, will be removed also. Silver polish can be used to remove tarnish, but the silver in the tarnish is lost.

Silver mirrors are formed by depositing a thin layer of silver on glass. This is accomplished by reducing a solution of silver nitrate in ammonia with some mild reducing agent, such as glucose or formaldehyde.

27.30 Compounds of Silver

Silver oxide, Ag_2O, is formed when silver is exposed to ozone or when finely divided silver is heated in oxygen under pressure. Hydroxide bases react with silver nitrate to give a dark brown amorphous precipitate of silver oxide. Although the

Copper possesses the property of forming both complex cations and anions that are quite stable. Copper(II) ions coordinate with four neutral ammonia molecules to form the deep blue **tetraamminecopper(II) ion.**

$$[Cu(H_2O)_4]^{2+} + 4NH_3 \rightleftharpoons [Cu(NH_3)_4]^{2+} + 4H_2O$$

The fact that the ammonia molecules displace the coordinated water molecules indicates the greater stability of the ammine complex. In fact, the ammine complex is so stable that most slightly soluble copper(II) compounds are dissolved by aqueous ammonia. The copper ion is located at the center of a square formed by the four attached groups, whether they are water, $[Cu(H_2O)_4]^{2+}$, or ammonia, $[Cu(NH_3)_4]^{2+}$. An example of a negatively charged copper(II) complex ion is $[CuCl_4]^{2-}$.

27.28 Copper Oxide Superconductors

One of the most exciting scientific discoveries of the 1980s is the characterization of new materials that exhibit superconductivity at temperatures above 90 K. **Superconductors** are conductors that conduct electricity with no resistance. They also possess unusual magnetic properties.

These new superconducting materials are ternary oxides containing yttrium or one of several rare earth elements, barium, and copper in a 1:2:3 ratio. The formula of the yttrium compound is $YBa_2Cu_3O_x$ where x ranges from about 6.5 to 7 depending upon the conditions used to prepare the compound. The yttrium(III) ion can be replaced by samarium (Sm), europium (Eu), gadolinium (Gd), dysprosium (Dy), holmium (Ho), or ytterbium (Yb). The variable amount of oxygen reflects the fact that some of the copper in these compounds can be oxidized from copper(II) to copper(III). For $x = 6.5$, all of the copper is present as copper(II); for $x = 7$, one-third of the copper is oxidized to copper(III).

Preparation of these materials is straightforward. A stoichiometric mixture of the oxides and carbonates is finely ground together and then heated to a temperature between 900 and 1100°C to decompose the carbonates and form the black ternary oxide.

$$\frac{1}{2}Y_2O_3(s) + 2BaCO_3(s) + 3CuCO_3(s) \longrightarrow YBa_2Cu_3O_{6.5}(s) + 5CO_2(g)$$

After regrinding and reheating, the material is pressed into a pellet and sintered (heated so it bonds without melting) at high temperature. The sintered pellets are annealed in a pure oxygen atmosphere at 400 to 900°C in order to oxidize some of the copper to copper(III) and then allowed to cool slowly.

The superconducting copper oxides are ionic compounds. The structure of a unit cell of $YBa_2Cu_3O_7$ is illustrated in Fig. 27.8. The copper(II) and copper(III) ions cannot be distinguished in the structure because the compound is a solid solution (Section 12.8) containing copper(II) and copper(III) ions.

Because these oxides are ionic compounds (ceramics), they are brittle and fragile and cannot be drawn out to form wires, as can a metal such as copper itself. However ceramic wires and other shapes can be formed by mixing the ceramic powder or its ingredients with an organic binder (a glue), and then extruding a thin strip through a press. (This works just like squeezing a fine strip of toothpaste out of a tube.) The sample is then heated to burn off the binder and annealed, giving a current-carrying ceramic strip.

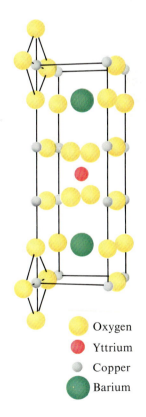

Oxygen

Yttrium

Copper

Barium

Figure 27.8
The unit cell of $YBa_2Cu_3O_7$.

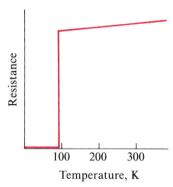

Figure 27.9

A graph of resistance versus temperature for $YBa_2Cu_3O_{6.5}$ (a superconductor below 92 K). Notice how the resistance falls to zero upon cooling.

Most commercial superconducting materials, niobium alloys such as NbTi and Nb_3Sn, do not become superconducting until they are cooled below 23 K ($-250°C$). This requires the use of liquid helium, which has a boiling temperature of 4 K and is expensive and difficult to handle. The new materials become superconducting at temperatures close to 90 K (Fig. 27.9), temperatures that can be reached by cooling with liquid nitrogen (boiling temperature 77 K). A sample immersed in boiling liquid nitrogen is cooled to 77 K. Not only are liquid-nitrogen-cooled materials easier to handle, but the cooling costs are about 1000 times less than for liquid helium.

Liquid-nitrogen-cooled superconductors could revolutionize society. Power companies, for example, could use underground superconducting transmission lines that would carry current for hundreds of miles with no loss of power due to resistance in the wires. This would allow generating stations to be located in areas remote from population centers and near to the natural resources necessary for power production.

Superconducting microchips could lead to more powerful supercomputers. Present chips produce waste heat due to the resistance of the electric current as it flows through the chip, so the chips have to be spaced to allow for a coolant to flow through them. Superconducting chips would produce no waste heat because they would have no resistance. Thus, chips could be packed closer together and the signals would require less time to travel from chip to chip. This would lead to smaller computers that work faster.

Superconducting materials are particularly useful for generating very strong magnetic fields. Because they have no resistance, superconducting coils can carry very large currents without melting, and these large currents can be used to generate very strong magnetic fields.

The Japanese National Railways has already built an experimental train that is levitated above the tracks by frictionless magnetic force generated by superconducting magnets cooled by liquid helium. This train can travel up to 300 miles an hour, but it is not practical for routine use because of the expense and difficulty of carrying liquid helium. The new higher-temperature superconductors could make such trains common.

The medical use of nuclear magnetic resonance (NMR) imaging devices requires the use of superconducting magnets to image tissues inside the body. Today's superconducting magnets are cooled by liquid helium making existing scanners complex and expensive. The new superconductors could lower the price of these instruments and increase their power. The development of nuclear fusion (Section 24.13) would be enhanced if these materials can be used to produce magnetic fields strong enough to contain the hot plasma in which the nuclear reactions take place.

The conversion from laboratory to practical technology will require a tremendous amount of research and development. In particular, shaping the superconducting ceramics into useful shapes with the right physical and electrical properties presents a challenge. However, work in this area has already started; experimental superconducting rods, rings, wires, tapes, and thin films have been fabricated.

27.29 Properties of Silver

Silver is a white, lustrous metal whose polished surface is an excellent reflector of light. As a conductor of heat and electricity it is second only to gold. Silver is noted for its ductility and malleability.

Silver is not attacked by oxygen in the air under ordinary conditions but tarnishes quickly in the presence of hydrogen sulfide or on contact with sulfur-containing foods

The Post-Transition Metals

28

The bronze horse was cast in the third century B.C.

Ten metals in the Periodic Table form ions with pseudo noble gas electron configurations. These elements are known as the post-transition metals. They are zinc, cadmium, and mercury from Group IIB, gallium, indium, and thallium from Group IIIA, tin and lead from Group IVA, bismuth from Group VA, and polonium from Group VIA. The reactivities of these elements range from those of fairly active metals (zinc, cadmium, and gallium) to those of much less active metals (lead and mercury). All of the post-transition metals form ionic compounds, as expected for metallic elements, but they also form a variety of covalent compounds.

Several of these elements were known in ancient times, because they are easily extracted from their ores. Mercury was known to the Greek philosopher Aristotle who referred to it as quicksilver (living silver), because of its appearance and liquid state. The alchemists named the element after Mercury, the swift messenger of the gods in Roman mythology. One of the early uses of mercury was in the alchemists' attempts to transmute common metals into gold. They assumed that mercury had unusual powers because it is a liquid and it dissolves other metals, forming amalgams.

Lead was known to the early Egyptians and Babylonians, and it is mentioned in the Old Testament. Lead pipes were commonly used by the Romans for conveying water, and in the Middle Ages lead was used as a roofing material. The stained glass windows of the great cathedrals of that age are set in lead.

Tin was an essential component of the alloy that gave its name to the Bronze Age.

28.1 The Chemical Behavior of the Post-Transition Metals

The distribution of the post-transition metals in the Periodic Table is shown in Figure 28.1.

Zinc, cadmium, and **mercury** (Fig. 28.2) are the three elements in Group IIB in the Periodic Table. Each of these elements has two electrons in its outer shell and 18 in the underlying shell $[(n-1)s^2,(n-1)p^6,(n-1)d^{10},ns^2]$ (Section 5.12). When these metals form ions with a charge of +2, the two outer s electrons are lost, giving pseudo noble gas electron configurations $[(n-1)s^2,(n-1)p^6,(n-1)d^{10}]$ (Section 6.2).

Zinc, cadmium, and mercury are representative metals that commonly exhibit the group oxidation number of +2, although mercury also exhibits an oxidation number of +1 in compounds that contain the Hg_2^{2+} group. The melting points, boiling points,

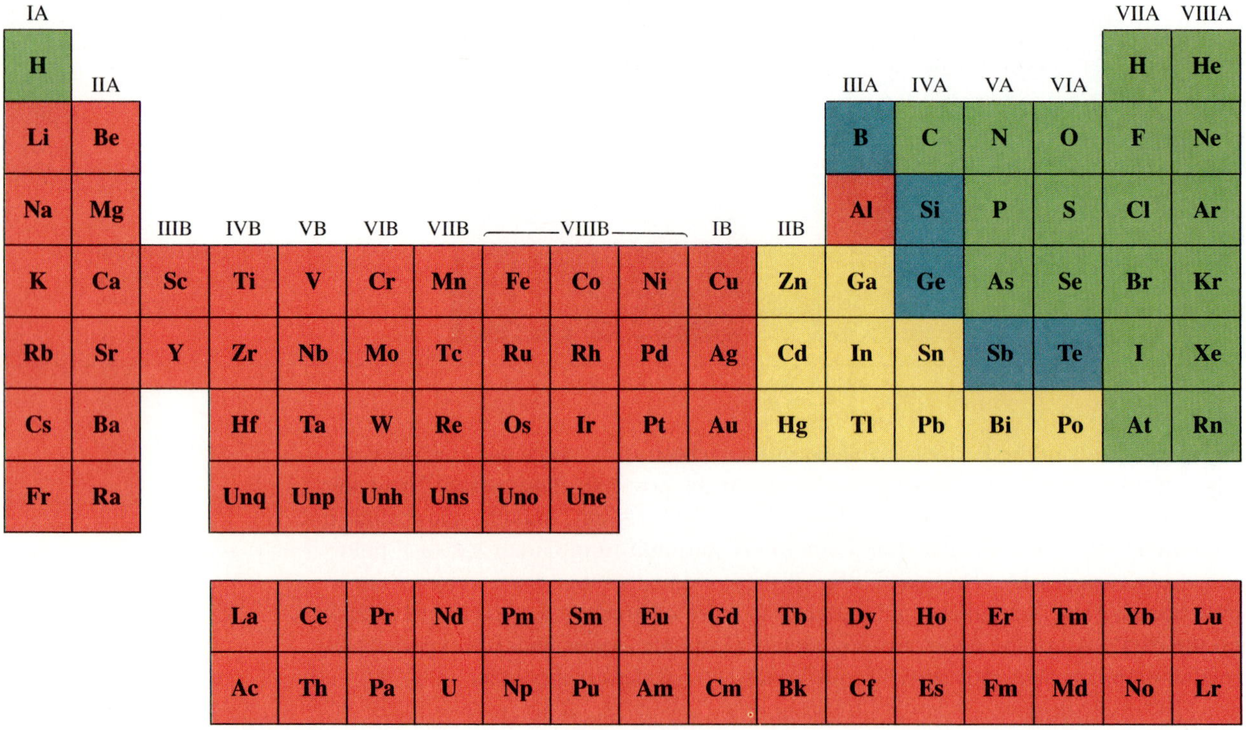

Figure 28.1

The location of the post-transition metals (shown in yellow) in the Periodic Table. Nonmetals are shown in green, semi-metals in blue, and the remaining metals in red.

Figure 28.2

Zinc, cadmium, and mercury.

and heats of vaporization for the Group IIB metals are lower than those for any other group of metals except the alkali metals (Group IA). Of the three elements, zinc is the most reactive and mercury the least. Both cadmium and mercury compounds are toxic.

The chemical behaviors of zinc and cadmium are quite similar. Both elements lie above hydrogen in the activity series (Section 9.7). They react with oxygen, sulfur, phosphorus, and the halogens to form compounds containing the metals with the group oxidation number of +2. For example,

$$Zn + Cl_2 \longrightarrow ZnCl_2$$
$$Cd + I_2 \longrightarrow CdI_2$$
$$2Zn + O_2 \longrightarrow 2ZnO$$
$$Cd + S \longrightarrow CdS$$
$$6Zn + P_4 \longrightarrow 2Zn_3P_2$$

They also react with solutions of acids, resulting in the liberation of hydrogen and the formation of salts that generally are soluble. The reaction of zinc with hydrochloric acid (Fig. 28.3) is

$$Zn(s) + 2H_3O^+(aq) + [2Cl^-(aq)] \longrightarrow H_2(g) + Zn^{2+}(aq) + [2Cl^-(aq)] + 2H_2O(l)$$

Mercury is very different from zinc and cadmium. It is a liquid at 25°C, and many metals dissolve in it, forming **amalgams.** Mercury is a nonreactive element that lies well below hydrogen in the activity series. Thus it does not displace hydrogen from acids, and it dissolves only in oxidizing acids such as nitric acid (Fig. 28.4).

$$Hg + HCl(aq) \longrightarrow \text{no reaction}$$
$$3Hg + 8HNO_3(aq) \longrightarrow 3Hg(NO_3)_2 + 4H_2O + 2NO$$

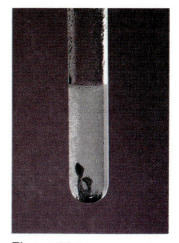

Figure 28.3

Zinc is an active post-transition metal. It dissolves in hydrochloric acid, forming a solution of colorless $ZnCl_2$ and hydrogen gas.

Figure 28.4

Mercury is an inactive post-transition metal. It does not react with hydrochloric acid (left). It dissolves in nitric acid (right) because this acid is a strong oxidizing agent. (The brown fumes result from atmospheric oxidation of the NO produced from the reaction of Hg with HNO_3.)

Mercury compounds decompose when they are heated, so elemental mercury reacts to form compounds only at relatively low temperatures—and then only with oxygen, sulfur, and the halogens.

$$2Hg + O_2 \xrightarrow{\text{Warm}} 2HgO$$
$$Hg + S \longrightarrow HgS$$
$$Hg + I_2 \longrightarrow HgI_2$$

Mercury exhibits an oxidation number of +2 in these compounds.

Mercury exhibits an oxidation number of +1 in covalent compounds containing X—Hg—Hg—X groups, such as Hg_2Cl_2, and in solutions containing salts of the Hg_2^{2+} ion. The Hg_2^{2+} ion is a diatomic species in which one valence electron of each mercury atom serves to form a covalent single bond between the two mercury atoms.

The +2 ions of all three of the elements of Group IIB are small in relation to their charges, so they show strong tendencies toward the formation of covalent compounds and complex ions. The simplest complex ions containing these elements are those with halides, cyanide, or ammonia. Coordination numbers of 4 and 6 are usual in the complexes of these elements. The four-coordinated complexes are tetrahedral. The six-coordinated complexes have octahedral structures (Section 26.3).

Gallium, indium, and **thallium** (Fig. 28.5) are members of Group IIIA. Each of these elements has three electrons in its outer shell and 18 in the underlying shell $[(n-1)s^2,(n-1)p^6,(n-1)d^{10},ns^2,np^1]$ (Section 5.12). As described in Section 13.1, these elements form compounds in which they exhibit the group oxidation number of +3. Their +3 ions exhibit a pseudo noble gas electron configuration $[(n-1)s^2,(n-1)p^6,(n-1)d^{10}]$. In addition, gallium and indium exhibit oxidation numbers of +1 in certain unstable compounds, and thallium readily forms compounds with an oxidation number of +1. All of these elements lie above hydrogen in the activity series and react with acids, liberating hydrogen. Gallium and indium are oxidized to the +3 ion, but thallium is oxidized only to the Tl+ ion, the stable oxidation number for thallium in water. Gallium, indium, and thallium react with nonmetals to form salts. Thallium(III) salts form with strong oxidizing agents.

Thallium is an example of an element that exhibits the so-called **inert pair effect.** The stability of an oxidation number that is two below the group oxidation number is a result of the resistance of the pair of s electrons in the valence shell to being lost during formation of an ion or to participation in the formation of a covalent bond. The s electrons are, in effect, an inert pair of electrons. The electron configuration of the thallium atom is $[Xe]4f^{14}5d^{10}6s^26p^1$, and that of the Tl+ ion is $[Xe]4f^{14}5d^{10}6s^2$, showing that the 6s electrons are not lost when this ion forms. The stability of the s electrons in the valence shells of the post-transition metals increases down a group in the Periodic Table, so the heaviest member of a group exhibits the most pronounced inert pair effect.

Figure 28.5

Gallium, indium, and thallium. Gallium has a low melting temperature (29.8°C). The dull color of the thallium is due to a layer of thallium(I) oxide and hydroxide resulting from the reaction of the metal with the atmosphere.

Figure 28.6
Tin and lead.

Tin and **lead** (Fig. 28.6) are the heavy members of Group IVA of the Periodic Table. The light elements in this group, carbon, silicon, and germanium, are primarily nonmetallic in character, whereas tin and lead become increasingly metallic in their properties with increasing atomic weight. Tin and lead form stable dipositive cations, Sn^{2+} and Pb^{2+}, with oxidation numbers two below the group oxidation number of $+4$. The stability of this oxidation state can also be attributed to the inert pair effect. The fact that the hydroxides of these ions are amphoteric is an indication of some nonmetallic character. Tin and lead also form covalent compounds in which the $+4$ oxidation number is exhibited; for example, $SnCl_4$ and $PbCl_4$ are low-boiling covalent liquids (Fig. 28.7).

Tin is a moderately active metal. It reacts with acids to form tin(II) compounds and with nonmetals to form either tin(II) or tin(IV) compounds, depending on the stoichiometry. Lead is less reactive. It lies just above hydrogen in the activity series and is attacked only by hot concentrated acids.

Bismuth, the heaviest member of Group VA, is a typical metal (Fig. 28.8). It readily gives up three of its five valence electrons to active nonmetals to form the tripositive ion, Bi^{3+}. It forms compounds with the group oxidation number of $+5$ only when treated with strong oxidizing agents.

Polonium is a member of Group VIA. It is an intensively radioactive element that results from the radioactive decay of uranium and thorium (Section 24.7). Its chemistry is similar to the metallic behavior of tellurium (Section 25.1). We will not consider polonium further in this chapter.

Figure 28.7
Tin(IV) chloride is a covalent liquid; tin(II) chloride is an ionic solid.

28.2 Occurrence and Isolation of the Post-Transition Metals

Ores of the post-transition elements usually contain mixtures of these elements as oxides or sulfides. Some ores also contain certain transition elements. Thus isolation of a post-transition metal often requires reduction of a mixture of metal oxides or sulfides to the metals, followed by separation of two or more metals.

Because the activities of the post-transition metals are lower than those of the transition metals and much lower than those of the active metals, compounds of the post-transition metals are readily reduced. Carbon is sufficient to reduce most of their ores. The post-transition metals have low melting points and low boiling points, compared to those of the transition metals. Thus, unlike the transition metals, they can be separated by fractional distillation. Electrolysis is sometimes used for separation; careful control of the applied voltage allows deposition of one metal at a potential that is too low to deposit a second.

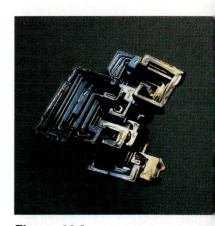

Figure 28.8
Bismuth.

Zinc is usually obtained from **sphalerite,** or **zinc blende, ZnS,** the principal ore of zinc. Less important ores include **zincite, ZnO, smithsonite, ZnCO₃, franklinite,** a mixture of oxides of zinc, iron, and manganese, and **willemite, Zn₂SiO₄.** Sulfide ores are first concentrated by flotation (Section 27.2) and then roasted to convert zinc sulfide to zinc oxide. Carbonate ores that contain zinc are converted to the oxide simply by heating.

$$2ZnS + 3O_2 \xrightarrow{\triangle} 2ZnO + 2SO_2 \qquad \Delta H° = -878.3 \text{ kJ}$$

$$ZnCO_3 \xrightarrow{\triangle} ZnO + CO_2 \qquad \Delta H° = 71.0 \text{ kJ}$$

Zinc oxide is reduced by heating it with coal in a fire-clay retort. As rapidly as zinc is produced, it distills out and is condensed. It contains impurities of cadmium, iron, lead, and arsenic, but it can be purified by careful redistillation.

An electrolytic process is also employed to produce zinc. Roasted ore is treated with dilute sulfuric acid, giving a solution of zinc sulfate and the sulfates of other metals. Silver sulfate and lead sulfate remain in the residue because they are insoluble. The resulting solution is treated with powdered zinc to reduce the less active metals, such as copper, arsenic, and antimony, to the elemental state. Any iron(II) present is oxidized by passing air through the solution. Manganese and iron are precipitated as hydroxides by adding a calculated amount of slaked lime (calcium hydroxide). After the solution is thus purified, it is electrolyzed, and zinc is deposited on aluminum cathodes. Sulfuric acid is regenerated as electrolysis proceeds; it is used in the leaching of additional roasted ore. At intervals the cathodes are removed, and the zinc is scraped off, melted, and cast into ingots. Electrolytic zinc has a purity of 99.95%.

Cadmium is found in the rare mineral **greenockite,** CdS, and in small amounts (less than 1%) in several zinc ores. Most cadmium comes from zinc smelters and from the sludge obtained from the electrolytic refining of zinc.

In the smelting of cadmium-containing zinc ores, the two metals are reduced together. Because cadmium is more volatile (bp = 767°C) than zinc (bp = 907°C), their separation can be accomplished by fractional distillation. Separation of the two metals is also possible by selective electrolytic deposition. Cadmium is less active than zinc and is deposited at a lower voltage.

Cadmium is a silvery, crystalline metal that is only slightly tarnished by air or water at ordinary temperatures. A large part of the cadmium produced is used in electroplating metals, such as iron and steel, to protect them from corrosion. The electrolyte solution contains tetracyanocadmate ions, $[Cd(CN)_4]^{2-}$, made by mixing cadmium chloride and sodium cyanide. The equation for the cathodic reduction is

$$[Cd(CN)_4]^{2-} + 2e^- \longrightarrow Cd + 4CN^-$$

The most important commercial source of **mercury** is the dark red sulfide, HgS, known as **cinnabar** (Fig. 28.9). When cinnabar is roasted, metallic mercury distills from the furnace and is condensed to the liquid. Ordinarily the roasting of a mineral in air results in the oxidation of the metal to an oxide. However, the mercury oxides are thermally unstable and decompose to the elements at the temperature of the furnace. Mercury may be purified by washing it with nitric acid to oxidize the metallic impurities or by distilling it in an atmosphere of oxygen to remove the more active metals.

No ores contain large fractions of **gallium, indium,** or **thallium.** Gallium is present in aluminum ores in small amounts and is recovered as a by-product during the refining of aluminum. Because of the tremendous scale of the aluminum industry, gallium is readily available. Indium occurs in small amounts in zinc ores, particularly zinc sulfide ores. Indium is recovered from the flue dust produced when ZnS is roasted

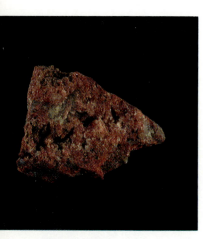

Figure 28.9
A sample of cinnabar, red HgS.

during the metallurgy of zinc. Thallium occurs along with lead. It is recovered during the smelting of lead.

Tin is obtained from **cassiterite,** or **tinstone,** SnO_2, the most abundant tin ore. Roasting of the ore removes arsenic and sulfur as volatile oxides. Oxides of other metals are extracted by means of hydrochloric acid. The purified ore is reduced by carbon.

$$SnO_2(s) + 2C(s) \xrightarrow{\triangle} Sn(l) + 2CO(g) \qquad \Delta H° = 360 \text{ kJ}$$

The molten tin, which collects on the bottom of the furnace, is drawn off and cast into blocks (**block tin**). The crude tin is remelted and permitted to flow away from the higher-melting impurities (chiefly compounds of iron and arsenic). Further purification of the tin is accomplished by electrolysis, using impure tin anodes, pure tin cathodes, and an electrolyte of fluorosilicic acid (H_2SiF_6) and sulfuric acid.

Lead is obtained principally from lead sulfide, PbS, commonly called **galena.** Other common ores are lead carbonate, $PbCO_3$, called **cerrusite,** and lead sulfate, $PbSO_4$, called **anglesite.** Both of the latter may have been formed by the weathering of sulfide ores.

Lead ores are first concentrated by flotation to remove the unwanted materials and a large part of the zinc sulfide that is usually associated with lead ores. The concentrated ore is then roasted in air to convert most of the sulfide to the oxide.

$$2PbS(s) + 3O_2(g) \xrightarrow{\triangle} 2PbO(s) + 2SO_2(g) \qquad \Delta H° = -710.4 \text{ kJ}$$

The roasted product is reduced in a blast furnace (Section 27.2) with coke and scrap iron.

$$PbO(s) + C(s) \xrightarrow{\triangle} Pb(s) + CO(g) \qquad \Delta H° = 106.8 \text{ kJ}$$
$$PbO(s) + CO(g) \xrightarrow{\triangle} Pb(s) + CO_2(g) \qquad \Delta H° = -65.7 \text{ kJ}$$
$$PbS(s) + Fe(s) \xrightarrow{\triangle} Pb(s) + FeS(s) \qquad \Delta H° = 0.4 \text{ kJ}$$

The lead obtained from a blast furnace generally contains copper, antimony, arsenic, bismuth, gold, and silver. If this crude lead is melted and stirred, antimony, arsenic, and bismuth are oxidized. The oxides of these metals rise to the surface, and the molten lead may be drained off for further refining. Gold and silver may be extracted from the lead by the Parkes process (Section 27.2) or by the electrolytic **Betts process.** In the Betts process thin sheets of pure lead are used as cathodes, and plates of impure lead are used as anodes. The electrolyte is a solution containing lead hexafluorosilicate, $PbSiF_6$, and hexafluorosilicic acid, H_2SiF_6.

Bismuth is most often found in nature in the free state. It sometimes occurs in compounds such as bismuth(III) oxide, Bi_2O_3 (called **bismuth ocher**), or bismuth(III) sulfide, Bi_2S_3 (called **bismuth glance**). Ores containing the free metal are treated by heating them in inclined iron pipes. The metal melts and flows away from the remaining materials. The oxide and sulfide ores are roasted and then heated with charcoal. As the elemental bismuth forms, it melts and collects beneath the less dense material. Bismuth is produced in the United States as a by-product of the refining of other metals, particularly lead.

28.3 Uses of the Post-Transition Metals

Zinc is a silvery metal that quickly tarnishes to a blue-gray appearance. This color is due to an adherent coating of a basic carbonate, $Zn_2(OH)_2CO_3$, which protects the underlying metal from further corrosion. The metal is hard and brittle at ordinary temperatures but ductile and malleable at 100–150°C. When molten zinc is poured into cold water, it solidifies in irregular masses called **granulated,** or **mossy, zinc.**

A large amount of zinc is used in the manufacture of dry cells (Section 21.13) and in the production of alloys such as brass and bronze. About half of the zinc metal produced is used to protect iron and other metals from corrosion by air and water. Its protective action results from the basic carbonate film on its surface. Furthermore, it protects a less active metal such as iron, because it forms an electrochemical cell in which zinc serves as the anode and iron as the cathode (see Section 21.16). The zinc coating on iron may be applied in several ways, and the product is called **galvanized iron.**

Cadmium is a silvery, crystalline metal that resembles zinc. It is only lightly tarnished by air or water at ordinary temperatures. Cadmium can be used in electroplating metals, such as iron and steel, to protect them from corrosion. Cadmium-plated metals are more resistant to corrosion, more easily soldered, and more attractive in appearance than galvanized (zinc-coated) metals, but their use has been restricted due to the toxicity of cadmium (Section 28.4).

Several alloys of cadmium are easily fusible; these include **Wood's metal** (12.5% Cd), melting at 65.5°C, and **Lipowitz alloy** (10% cadmium), melting at 70°C. Because these alloys will melt from the heat produced by the first stages of a fire, they are used in fire-protection devices. Some antifriction alloys used for bearings contain cadmium. Rods of cadmium are used in nuclear reactors (Section 24.11) to absorb neutrons and thus control the chain reaction.

Mercury is the only metal that is a liquid at room temperature. Its low freezing point ($-38.87°C$), its high boiling point (356.6°C), its uniform coefficient of expansion, and the fact that it does not wet glass make mercury an excellent thermometric substance. Because of its chemical inactivity, mobility, high density, and electrical conductivity, it is used extensively in barometers, vacuum pumps, and liquid seals, and for electrical contacts. Except for iron and platinum, all metals readily dissolve in (or are wet by) mercury to form amalgams. Sodium amalgam is used as a reducing agent, because it is less active than sodium metal alone. Amalgams of tin, silver, and gold are used in dentistry.

Tin exists as three allotropic solids. They are **gray tin** (cubic crystals), **white tin** (tetragonal crystals), and **brittle tin** (rhombic crystals). The white form is malleable. When a rod of it is bent, the crystals slip over one another, producing a sound described as tin cry. When white tin is heated, it changes to the brittle form. At temperatures below 13.2°C, white tin slowly changes into gray tin, which is powdery. Consequently, articles made of tin are likely to disintegrate in cold weather, particularly if the cold spell is lengthy. The change progresses slowly from the spot of origin, and the gray tin that is first formed catalyzes further change. This effect is, in a way, similar to the spread of an infection in a plant or animal body. For this reason, it is called tin disease or tin pest.

The principal use of tin is in the electrolytic production of tin plate, sheet iron coated with tin to protect it from corrosion. However, when a scratch through the tin exposes the iron, corrosion sets in rapidly. Iron is more active than tin and corrodes more rapidly when in contact with tin than otherwise. This is because iron and tin in contact with each other, when both are in contact with moist air, comprise an electrolytic cell that promotes corrosion (Section 21.16). Tin is also used in making alloys, such as **bronze** (Cu, Sn, and Zn), **solder** (Sn and Pb), and **type metal** (Sn, Pb, Sb, and Cu.)

Lead is a soft metal that has little tensile strength and the greatest density of the common metals (except for gold and mercury). Lead has a metallic luster when freshly cut but quickly acquires a dull gray color when exposed to moist air. It becomes oxidized on its surface, forming a protective layer that is both compact and adherent; this film is probably lead hydroxycarbonate, $Pb_3(OH)_2(CO_3)_2$.

The major uses of lead depend on the ease with which it is worked, its low melting point, its great density, and its resistance to corrosion. Lead is a principal constituent of the lead storage battery (Section 21.14). It is an important component of solder (Sn and Pb), and type metal (Sn, Pb, Sb, and Cu.)

Bismuth is a lustrous, hard, and brittle metal with a reddish tint. Like antimony, melted bismuth expands when it solidifies—a most unusual property. Because of this, it is used in the formation of alloys to prevent them from shrinking on solidification. This is particularly important in alloys used in casting; these must expand to fill all of the space in the mold.

Alloys of bismuth, tin, and lead have low melting points, which makes them useful for electrical fuses, safety plugs for boilers, and automatic sprinkler systems. Some of these alloys melt even in hot water. For example, **Rose's metal** (Bi, 50%; Pb, 25%; Sn, 25%) melts at 94°C; and **Wood's metal** (Bi, 50%; Pb, 25%; Sn, 12.5%; Cd, 12.5%) melts at 65.5°C.

28.4 Physiological Action of the Post-Transition Metals

Zinc is one of the most important metals in biological processes, and it appears to be necessary to all forms of life. It is a key component of many enzymes (Section 31.5), which are the catalysts of biochemical reactions in living organisms. The body of an adult human contains about 2 grams of zinc. Interestingly, cadmium and mercury, which are members of the same group as zinc, have no known biological function and are, in fact, among the most toxic of elements.

Metallic mercury is a cumulative poison; breathing air containing mercury vapor or contact with the liquid metal should be avoided. Mercury compounds are also poisonous. The fatal dose of mercury(II) chloride is 0.2–0.4 gram. The mercury(II) ions combine with the proteins in the kidneys and destroy their ability to remove waste products from the blood. Mercury also has a permanent effect on the brain and central nervous system. Antidotes for mercury poisoning are egg white and milk; the proteins in these substances precipitate the mercury in the stomach before it can be absorbed into the body. Mercury poisoning is treated with chelates like BAL, British Anti-Lewisite (Section 26.5).

Cadmium is extremely toxic. Its effects on the liver and nervous system are similar to those of mercury. As a consequence, the use of cadmium is significantly limited.

Thallium, the heaviest member of Group IIIA, is dangerous. Both thallium metal and its compounds are extremely toxic.

Tin is nontoxic, but lead is a heavy-metal poison. It interferes with the catalytic ability of enzymes in various metabolic processes. Typical symptoms of lead poisoning are anemia, headaches, convulsions, and kidney, brain, and nervous-system disorders. Lead poisoning, like mercury poisoning, is treated with chelating agents like BAL, British Anti-Lewisite.

At one time, lead hydroxycarbonate, $Pb_3(OH)_2(CO_3)_2$, was used extensively as the paint pigment **white lead.** Many people have suffered lead poisoning because of its extensive use as a house paint. Because of the danger of poisoning it has been largely replaced as a paint pigment by titanium dioxide, zinc oxide, and lithopone (a mixture of barium sulfate and zinc sulfide, especially on objects used by children.

Compounds of the Post-Transition Metals

28.5 Oxides of the Post-Transition Metals

The common oxides of the post-transition metals are tabulated in Table 28.1.

Table 28.1	Common oxides of the post-transition metals. Oxides for which formulas are printed in red are formed when the corresponding metal burns in air.		
Group IIB	Group IIIA	Group IVA	Group VA
ZnO	Ga_2O_3		
CdO	In_2O_3	SnO	
		SnO_2	
HgO	Tl_2O	PbO	Bi_2O_3
	Tl_2O_3	Pb_3O_4	Bi_2O_5
		PbO_2	

1. PREPARATION OF THE OXIDES. The oxides of the post-transition metals can be prepared by reactions of the metals with oxygen, by thermal decomposition of hydroxides or other salts, or by oxidation of oxides or other compounds.

1. Zinc oxide, cadmium oxide, gallium oxide, indium oxide, thallium(I) oxide, tin(IV) oxide, lead(II) oxide, and bismuth(III) oxide form when the corresponding metals burn in air. Examples of the reactions involved are

$$4Tl + O_2 \longrightarrow 2Tl_2O$$

$$2Cd + O_2 \longrightarrow 2CdO$$

$$4In + 3O_2 \longrightarrow 2In_2O_3$$

$$Sn + O_2 \longrightarrow SnO_2$$

Under these conditions, thallium, lead, and bismuth, the heaviest elements in their respective groups and the elements for which the inert pair effect is most pronounced, form oxides in which they exhibit oxidation numbers two less than the group oxidation number.

Mercury(II) oxide forms slowly when mercury is warmed below 500°C in oxygen; it decomposes above this temperature.

2. Tin(II) oxide and the oxides that result from burning the metal can also be prepared by the thermal decomposition of the corresponding metal hydroxides and carbonates, as well as certain of their other salts. In some cases the thermal decomposition occurs at room temperature. For example, **mercury(II) oxide** may be prepared by addition of a solution of a strong base to a solution of a mercury(II) salt; $Hg(OH)_2$ is unstable, and HgO precipitates from the mixture.

Tin(II) oxide, SnO, may be prepared by treating a hot solution of a tin(II) compound with an alkali metal carbonate or by heating tin(II) oxalate in the absence of air.

The equations are

$$Sn^{2+}(aq) + CO_3^{2-}(aq) \longrightarrow SnO(s) + CO_2(g)$$

$$SnC_2O_4(s) \xrightarrow[\text{Inert atm.}]{\text{Heat}} SnO(s) + CO_2(g) + CO(g)$$

3. Formation of the oxides that contain thallium(III), lead(IV), or bismuth(V) may require stronger oxidizing agents than the oxygen present in air. Oxides containing the metal with a lower oxidation number are often used as starting materials in these reactions. **Lead dioxide, PbO₂,** is a chocolate-brown powder (Fig. 28.10) formed by oxidizing lead(II) compounds in alkaline solution. The equation for the oxidation with sodium hypochlorite as the oxidizing agent is

$$PbO(s) + [Na^+(aq)] + ClO^-(aq) \longrightarrow PbO_2(s) + [Na^+(aq)] + Cl^-(aq)$$

Bismuth(V) oxide, Bi₂O₅, is produced by the action of very strong oxidizing agents on the trioxide.

Trilead tetraoxide, Pb₃O₄, called red lead (Fig. 28.10), is prepared by carefully heating the monoxide in air at temperatures of 400–500°C. Red lead contains lead with two oxidation numbers; when it is treated with nitric acid, two-thirds of the lead dissolves as lead nitrate (oxidation number = +2), and the other third remains as insoluble lead dioxide (oxidation number = +4). The equation for the reaction is

$$Pb_3O_4 + 4HNO_3 \longrightarrow PbO_2 + 2Pb(NO_3)_2 + 2H_2O$$

2. REACTIONS AND USES OF THE OXIDES. **Zinc oxide, ZnO,** is used as a paint pigment called **zinc white,** or **Chinese white.** Unlike white lead, it does not blacken with hydrogen sulfide, because zinc sulfide is white. Zinc oxide is also used in the manufacture of automobile tires and other rubber goods and in the preparation of medicinal ointments. It is used in zinc-oxide-based sun screens to prevent sunburn.

When a strong base is added to a cold solution of a mercury(II) compound, **mercury(II) oxide** forms as a yellow precipitate, but when precipitated from hot solutions, it is red. Since the crystal structure is the same in both cases, this difference in color has been attributed to a difference in the size of the HgO particles.

Lead dioxide is the principal constituent of the cathode of the charged lead storage battery (Section 21.14). Since lead(IV) tends to revert to the more stable lead(II) ion by gaining two electrons, lead dioxide is a powerful oxidizing agent.

Figure 28.10
Yellow PbO, red Pb₃O₄, and brown PbO₂.

28.6 Hydroxides of the Post-Transition Metals

With the exception of thallium(I) hydroxide, the known hydroxides of the post-transition metals are insoluble. They can be prepared by adding soluble hydroxides, such as the alkali metal hydroxides, to solutions of their salts. For example, one reaction for the formation of zinc hydroxide (Fig. 28.11) is

$$Zn^{2+}(aq) + [2NO_3^-(aq)] + [2Na^+(aq)] + 2OH^-(aq) \longrightarrow$$
$$Zn(OH)_2(s) + [2Na^+(aq)] + [2NO_3^-(aq)]$$

An excess of the soluble hydroxide should be avoided, because many of the hydroxides of the post-transition elements are amphoteric and dissolve in an excess of base. Zinc hydroxide, for example, dissolves in a solution of sodium hydroxide (Fig. 28.11).

$$Zn(OH)_2(s) + [2Na^+(aq)] + 2OH^-(aq) \longrightarrow [2Na^+(aq)] + Zn(OH)_4^{2-}(aq)$$

An excess of base causes **tin(II) hydroxide** to dissolve, with the formation of the soluble **stannite ion.**

$$Sn(OH)_2(s) + OH^-(aq) \longrightarrow [Sn(OH)_3]^-(aq)$$

The stannite ion is an active reducing agent. It is used in qualitative analysis to test for bismuth because it reduces white bismuth(III) hydroxide to black metallic bismuth.

$$2Bi(OH)_3(s) + 3[Sn(OH)_3]^- + 3OH^- \longrightarrow 2Bi(s) + 3[Sn(OH)_6]^{2-}$$

Thallium(I) hydroxide is soluble. It forms when thallium(I) oxide is added to water.

$$Tl_2O + H_2O \longrightarrow 2TlOH$$

Mercury(I) hydroxide is unstable. Addition of a strong base to a mercury(I) salt produces a mixture of mercury metal and mercury(II) oxide. **Mercury(II) hydroxide** is also unstable; it immediately decomposes to mercury(II) oxide and water.

$$Hg(OH)_2 \longrightarrow HgO + H_2O$$

Addition of a strong base to a solution of a salt of thallium(III), tin(IV), lead(IV), or bismuth(V) does not produce hydroxides. Instead, oxides containing water of hydration (Section 13.7), called hydrated oxides, are formed.

The hydrated oxide of tin(IV) has the formula $SnO_2 \cdot H_2O$. When this hydrated oxide is melted with sodium hydroxide, **sodium metastannate, Na_2SnO_3,** and water are formed. Because sodium metastannate can be regarded as a salt, $SnO_2 \cdot H_2O$ has been called **metastannic acid,** or sometimes **alpha-stannic acid** to distinguish it from **beta-stannic acid,** which is the hydrated oxide formed when hot concentrated nitric acid reacts with tin. Beta-stannic acid has the same composition as the alpha form, but it is insoluble in acids and only very slightly soluble in strongly alkaline solutions.

Some other alkali metal stannates appear to be derived from **orthostannic acid,** which is unknown but whose formula may be written as $H_2[Sn(OH)_6]$. The sodium salt, for example, is known and is $Na_2[Sn(OH)_6]$.

When a suspension of **bismuth hydroxide** is boiled, the hydroxide decomposes to **bismuth oxyhydroxide,** BiO(OH), and water.

Figure 28.11

Mixing solutions of NaOH and $Zn(NO_3)_2$ produces a white precipitate of $Zn(OH)_2$ (top). Addition of an excess of NaOH results in dissolution of the precipitate (bottom).

28.7 Halides of the Post-Transition Metals

Halides of the post-transition metals can be prepared by the techniques used to prepare halides of other metals (Sections 13.7 and 27.5). However, there are complications resulting from the varying stabilities of the oxidation numbers of these elements and the relative nonreactivity of mercury and lead.

1. PREPARATION OF THE HALIDES. These compounds can be prepared by a variety of methods.

1. The direct reaction of a post-transition metal and a halogen will produce the anhydrous halide of the metal. If the post-transition metal exhibits two oxidation numbers, it may be necessary to control the stoichiometry in order to obtain the halide with the lower oxidation number. For example, preparation of tin(II) chloride requires a one-to-one mole ratio of Sn to Cl_2, whereas preparation of tin(IV) chloride requires a one-to-two ratio.

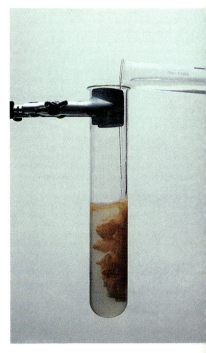

$$Sn + Cl_2 \longrightarrow SnCl_2$$
$$Sn + 2Cl_2 \longrightarrow SnCl_4$$

2. The active post-transition metals, zinc, cadmium, gallium, indium, thallium, tin, and bismuth, dissolve in solutions of the hydrogen halides producing hydrogen and generally forming solutions of halide salts that can be evaporated to produce hydrated metal halides. Zinc, cadmium, gallium, and indium exhibit their respective group oxidation numbers; thallium forms thallium(I) salts, tin forms tin(II) salts, and bismuth forms bismuth(III) salts. The heavier halides of thallium(I) (the chloride, bromide, and iodide) are insoluble and precipitate. Sample reactions include

$$2Tl + 2HCl(aq) \longrightarrow 2TlCl(s) + H_2(g)$$
$$Cd + 2HBr(aq) \longrightarrow CdBr_2(aq) + H_2(g)$$
$$Sn + 2HCl(aq) \longrightarrow SnCl_2(aq) + H_2(g)$$
$$2Ga + 6HI(aq) \longrightarrow 2GaI_3(aq) + 3H_2(g)$$

3. Hydroxides, carbonates, and some oxides dissolve in solutions of the hydrohalic acids, forming solutions of halide salts.

4. Several of the post-transition metals form insoluble halides. The heavier mercury(I) halides, mercury(II) iodide, the heavier thallium(I) halides, and the heavier lead(II) halides are insoluble. Hence halides such as Hg_2Cl_2, HgI_2, TlBr, and PbI_2 can be prepared by metathetical reactions between soluble salts of these metals and soluble halide salts (Fig. 28.12). For example,

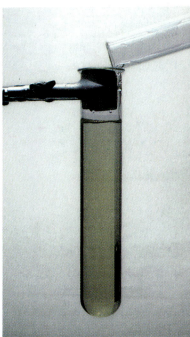

$$Hg^{2+}(aq) + 2I^-(aq) \longrightarrow HgI_2(s)$$

2. REACTIONS AND USES OF THE HALIDES. Anhydrous **zinc chloride, $ZnCl_2$,** is a white, very deliquescent solid. It is made by the reaction of zinc with chlorine or by fusing hydrated zinc chloride to drive off the water of hydration. The product, which melts at 262°C, is cast into sticks.

The anhydrous salt is used as a caustic in surgery and as a dehydrating agent in certain organic reactions. Aqueous solutions of zinc chloride are acidic because the **hexaaquozinc(II) ion** is acidic (Section 17.15).

$$[Zn(H_2O)_6]^{2+} + H_2O \longrightarrow [Zn(H_2O)_5OH]^+ + H_3O^+$$

Figure 28.12

HgI_2 forms when solutions of KI and $Hg(NO_3)_2$ are mixed (top). The precipitate redissolves in excess KI (bottom) due to the formation of $HgI_4{}^{2-}$.

Because of its acidic nature, a solution of zinc chloride (with ammonium chloride) is used as a **flux** to dissolve the oxides on metal surfaces before soldering.

Concentrated solutions of zinc chloride dissolve cellulose, forming a gelatinous mass, which may be molded into various shapes. Fiberboard is made of this material. Zinc oxide dissolves in concentrated solutions of zinc chloride, producing **zinc oxychloride,** Zn_2OCl_2, which sets to a hard mass and is used as a cement.

Cadmium chloride, $CdCl_2$, is converted almost completely into $[CdCl_4]^{2-}$ by high concentrations of the chloride ion. Most cadmium compounds are soluble in an excess of sodium iodide, due to the formation of the tetraiodocadmate(II) ion $[CdI_4]^{2-}$.

$$Cd(OH)_2(s) + 4I^-(aq) \longrightarrow [CdI_4]^{2-}(aq) + 2OH^-(aq)$$

Mercury(I) chloride, Hg_2Cl_2 (commonly called mercurous chloride), is a white, insoluble crystalline compound ($K_{sp} = 1.1 \times 10^{-18}$), which is of analytical importance because its insolubility serves to precipitate the mercury(I) ion. Mercury(I) chloride is manufactured on a commercial scale by heating a mixture of mercury(II) sulfate, mercury, and sodium chloride.

$$HgSO_4 + Hg + 2NaCl \longrightarrow Hg_2Cl_2 + Na_2SO_4$$

Mercury(I) chloride is volatile and sublimes from the reaction mixture. It is used in the field of medicine under the name calomel as a cathartic and a diuretic (a stimulant for the organs of secretion).

When exposed to light, mercury(I) chloride slowly decomposes.

$$Hg_2Cl_2 \longrightarrow HgCl_2 + Hg$$

Hence it is usually stored in amber-colored bottles.

Reducing agents, such as tetrachlorostannate(II) ion, readily reduce mercury(I) compounds to metallic mercury. This is used in qualitative tests for the ion.

Mercury(II) chloride, $HgCl_2$, is prepared commercially by heating mercury(II) sulfate with sodium chloride; the $HgCl_2$ sublimes from the reaction mixture.

$$HgSO_4 + 2NaCl \longrightarrow HgCl_2 + Na_2SO_4$$

The compound $HgCl_2$ is also called **corrosive sublimate.** In dilute solutions it is used as an antiseptic. It is moderately soluble in water but only slightly ionized, as indicated by the low electrical conductivity of its aqueous solutions. Its solubility can be increased by adding an excess of chloride ions, which cause the formation of the complex **tetrachloromercurate(II) ion** according to the equation

$$HgCl_2 + 2Cl^- \rightleftharpoons [HgCl_4]^{2-}$$

Iodide ions precipitate mercury(II) ions from solution as HgI_2, which is red. **Mercury(II) iodide** is soluble in a solution containing an excess of iodide ions (Fig. 28.12); **tetraiodomercurate(II) ions** are formed according to the equation

$$HgI_2 + 2I^- \rightleftharpoons [HgI_4]^{2-}$$

Tin(II) chloride is obtained as the dihydrate ($SnCl_2 \cdot 2H_2O$) by evaporating a solution of $SnCl_2$ in hydrochloric acid. When dissolved in neutral water, tin(II) chloride will react with the water, forming **tin hydroxychloride, Sn(OH)Cl.**

$$SnCl_2 + 2H_2O \rightleftharpoons Sn(OH)Cl + H_3O^+ + Cl^-$$

Tin(II) chloride is widely used as a reducing agent, because of the ease with which it is oxidized to tin(IV) compounds. Aqueous solutions of tin(II) compounds must be protected against oxidation by air.

Tin(IV) chloride is a colorless liquid that is soluble in organic solvents such as carbon tetrachloride and is a nonconductor of electricity. These properties are typical of covalent compounds. When dissolved in water, the tetrachloride reacts with water, but in hydrochloric acid, hexachlorostannic acid is produced.

$$SnCl_4(l) + [2H_3O^+(aq)] + 2Cl^-(aq) \longrightarrow [2H_3O^+(aq)] + SnCl_6^{2-}(aq)$$

28.8 Oxysalts of the Post-Transition Metals

A wide variety of salts of post-transition metals with oxyacids have been characterized.

1. PREPARATION OF OXYSALTS. Post-transition metal salts of the oxyacids can be prepared by methods that are similar to those used to prepare halides of these metals.

1. Active post-transition metals dissolve in solutions of oxyacids, producing hydrogen and solutions of salts of the acids. For exampie, the reaction of zinc with sulfuric acid is

$$Zn + H_2SO_4 \longrightarrow ZnSO_4 + H_2$$

These salts can be recovered by evaporation of the solutions. The oxidation numbers of the metals are the same as those produced when the metals dissolve in hydrohalic acids (Section 28.7).

2. Hydroxides and some oxides dissolve in solutions of oxyacids, forming solutions of salts of these acids. For example,

$$Cd(OH)_2 + 2HClO_4 \longrightarrow Cd(ClO_4)_2 + 2H_2O$$

$$Tl_2O + 2HNO_3 \longrightarrow 2TlNO_3 + H_2O$$

The oxides of lead(IV) and bismuth(V) are not soluble in acids.

3. The less active post-transition metals dissolve in hot concentrated oxidizing acids such as nitric acid and sulfuric acid. For example, the reaction of lead and hot sulfuric acid produces lead(II) sulfate.

$$Pb + 2H_2SO_4 \longrightarrow PbSO_4(s) + SO_2(g) + 2H_2O$$

Mercury(II) nitrate is formed by the reaction of mercury with an excess of concentrated nitric acid. With an excess of mercury, mercury(I) nitrate is formed.

$$3Hg + 8HNO_3 \longrightarrow 3Hg(NO_3)_2 + 2NO(g) + 4H_2O$$

$$6Hg + 8HNO_3 \longrightarrow 3Hg_2(NO_3)_2 + 2NO(g) + 4H_2O$$

2. REACTIONS AND USES OF THE OXYSALTS. **Zinc sulfate, ZnSO₄,** is produced commercially in large quantities by roasting **zinc blende, ZnS,** at low red heat.

$$ZnS(s) + 2O_2(g) \xrightarrow{\triangle} ZnSO_4(s) \qquad \Delta H° = -777.0 \text{ kJ}$$

The product is extracted with water and recovered by recrystallization to form crystals of **zinc sulfate 7-hydrate, ZnSO₄ · 7H₂O,** which is called **white vitriol.** It is used principally in the production of the white paint pigment **lithopone,** which is a mixture of barium sulfate and zinc sulfide formed by the reaction

$$Ba^{2+}(aq) + S^{2-}(aq) + Zn^{2+}(aq) + SO_4^{2-}(aq) \longrightarrow BaSO_4(s) + ZnS(s)$$

When either metallic lead or lead monoxide, PbO, is dissolved in nitric acid, **lead(II) nitrate, Pb(NO₃)₂,** is formed. This salt is readily soluble in water, but unless the solution is made slightly acidic with nitric acid, reaction with water occurs and **lead(II) hydroxynitrate** precipitates.

$$Pb^{2+} + 2NO_3^- + 2H_2O \longrightarrow Pb(OH)NO_3(s) + H_3O^+ + NO_3^-$$

Lead nitrate is unstable at moderately high temperatures and decomposes in the same manner as the nitrates of other heavy metals.

$$2Pb(NO_3)_2 \longrightarrow 2PbO(s) + 4NO_2 + O_2$$

Lead carbonate, PbCO$_3$, may be prepared by the action of sodium hydrogen carbonate on lead chloride. **Lead hydroxycarbonate, Pb$_3$(OH)$_2$(CO$_3$)$_2$,** is formed when alkali metal carbonates are added to solutions containing the lead ion. This compound is the paint pigment white lead; because of the danger of lead poisoning, its use has been largely curtailed.

Lead sulfate, PbSO$_4$, forms when solutions of a soluble lead salt and a soluble sulfate are mixed. It is insoluble in water ($K_{sp} = 1.8 \times 10^{-8}$) but readily dissolves in solutions containing an excess of alkali metal halide or acetate ions. **Lead acetate, Pb(CH$_3$CO$_2$)$_2$,** is one of the few soluble compounds of lead; it is a weak electrolyte, indicating that it dissolves mainly as a covalent compound rather than as an ionic one. It is extremely toxic. **Lead chromate, PbCrO$_4$,** is used as a yellow art pigment. It is insoluble in water but dissolves readily in acids and alkali metal hydroxides.

Bismuth(III) nitrate, Bi(NO$_3$)$_3$, reacts with water, forming **bismuth hydroxynitrate, Bi(OH)$_2$NO$_3$.**

$$Bi(NO_3)_3 + 4H_2O \longrightarrow Bi(OH)_2NO_3 + 2H_3O^+ + 2NO_3^-$$

When dried, bismuth hydroxynitrate forms **bismuth oxynitrate, BiONO$_3$.** Bismuth oxynitrate and bismuth oxycarbonate, (BiO)$_2$CO$_3$ (under the names bismuth subnitrate and bismuth subcarbonate), are used in medicine for the treatment of stomach disorders, such as gastritis and ulcers, and skin diseases, such as eczema.

28.9 Sulfides of the Post-Transition Metals

Sulfides of the post-transition metals occur commonly in nature. The sources of many of these metals are sulfide ores.

There are a wide variety of post-transition metal sulfides that contain metal ions with two different oxidation numbers. We will not consider these compounds; instead, we will consider only those sulfides that contain the metal with a single oxidation number.

1. PREPARATION OF THE SULFIDES. Post-transition metal sulfides can be prepared by the direct reaction of sulfur with a metal or by precipitation from a solution of a metal salt by addition of hydrogen sulfide or a sulfide salt.

1. The reaction of sulfur and a metal usually requires warming in order to form the sulfide. Mercury(II) sulfide slowly forms at room temperature, however.

$$2Tl + S \xrightarrow{\triangle} Tl_2S$$

$$Zn + S \xrightarrow{\triangle} ZnS$$

$$Hg + S \longrightarrow HgS$$

$$2Ga + 3S \xrightarrow{\triangle} Ga_2S_3$$

Tin forms two sulfides, depending on the stoichiometry of the reaction.

$$Sn + S \xrightarrow{\Delta} SnS$$
$$Sn + 2S \xrightarrow{\Delta} SnS_2$$

The sulfides of the heavy elements thallium, lead, and bismuth (Tl_2S, PbS, and Bi_2S_3) form with the metals exhibiting oxidation numbers two less than the group oxidation number, even if an excess of sulfur is used in their preparation. Sulfide ions are good enough reducing agents to reduce these metal ions when they exhibit their group oxidation numbers.

2. The sulfides of cadmium(II), mercury(II), tin(II), tin(IV), lead(II), and bismuth(III) precipitate when hydrogen sulfide or a soluble sulfide salt is added to solutions of the metal salts. Examples include the following reactions

$$Cd^{2+} + H_2S \longrightarrow CdS(s) + 2H^+$$
$$Sn^{2+} + S^{2-} \longrightarrow SnS(s)$$
$$Sn^{4+} + 2H_2S \longrightarrow SnS_2(s) + 4H^+$$
$$2Bi^{3+} + 3S^{2-} \longrightarrow Bi_2S_3(s)$$

Zinc sulfide is soluble in acids and precipitates only from basic solution. The sulfides of the post-transition metals of Group IIIA react with water and form the metal hydroxides and hydrogen sulfide; they cannot be prepared from aqueous solution.

2. REACTIONS AND USES OF THE SULFIDES. **Zinc sulfide, ZnS,** is insoluble in water ($K_{sp} = 1 \times 10^{-27}$) and in acetic acid but dissolves readily in stronger acids such as hydrochloric acid or sulfuric acid.

$$ZnS + 2H_3O^+ \longrightarrow Zn^{2+} + H_2S + 2H_2O$$

Zinc sulfide is used as a white paint pigment, either alone or mixed with zinc oxide or barium sulfate.

Cadmium sulfide, CdS, is an insoluble bright yellow solid ($K_{sp} = 2.8 \times 10^{-35}$) (Fig. 28.13). It is used as a paint pigment called **cadmium yellow.**

Mercury(I) sulfide, Hg₂S, is unstable and immediately decomposes into mercury and mercury(II) sulfide when it is formed by passing hydrogen sulfide into a solution containing mercury(I) ions.

$$Hg_2^{2+} + H_2S \longrightarrow HgS + Hg + 2H^+$$

Hydrogen sulfide precipitates black **mercury(II) sulfide, HgS** ($K_{sp} = 2 \times 10^{-59}$), from solutions of mercury(II) salts, even from strongly acidic solutions. The precipitate formed by the interaction of $HgCl_2$ in solution with H_2S is first white, then yellow, then red, and finally black. The white compound has the formula $HgCl_2 \cdot 3HgS$. When heated, HgS becomes bright red; the red sulfide is isomeric with the black and has in the past been used as the pigment **vermilion.**

Mercury(II) sulfide dissolves in solutions of soluble sulfides in the presence of an excess of hydroxide ions and forms the **thiomercurate ion.**

$$HgS + S^{2-} \rightleftharpoons [HgS_2]^{2-}$$

Dark brown **tin(II) sulfide, SnS,** ($K_{sp} = 6 \times 10^{-35}$) is not dissolved by alkali metal sulfides or by alkaline polysulfides. However, **tin(IV) sulfide** dissolves in alkali metal sulfides, with the formation of **thiostannate ions,** $[SnS_3]^{2-}$. The addition of acid reprecipitates SnS_2.

$$[SnS_3]^{2-} + 2H_3O^+ \longrightarrow SnS_2(s) + H_2S + 2H_2O$$

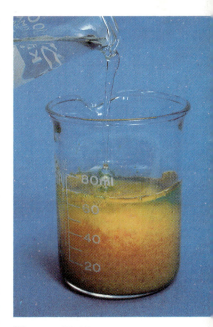

Figure 28.13

CdS forms when solutions of $CdCl_2$ and Na_2S are mixed.

Concentrated hydrochloric acid dissolves tin(IV) sulfide, because the complex ion $[SnCl_6]^{2-}$ is formed.

$$SnS_2 + 4H_3O^+(aq) + 6Cl^-(aq) \longrightarrow SnCl_6^{2-}(aq) + 2H_2S + 4H_2O$$

Lead sulfide, PbS, is a black insoluble solid ($K_{sp} = 6.5 \times 10^{-34}$), which is insoluble in solutions of dilute acids and soluble metal sulfides.

For Review

Summary

The post-transition metals are zinc, cadmium, and mercury from Group IIB, gallium, indium, and thallium from Group IIIA, tin and lead from Group IVA, bismuth from Group VA, and polonium from Group VIA. The reactivities of these elements range from those of fairly active metals (zinc, cadmium, and gallium) to those of much less active metals (lead and mercury). All of the post-transition metals form ionic compounds as expected for metallic elements, but they also form a variety of covalent compounds.

Each of the post-transition elements can potentially form an ion with a pseudo noble gas electron configuration $[(n-1)s^2, (n-1)p^6, (n-1)d^{10}]$ by the loss of all of its valence electrons. However, ions with this configuration become more difficult to form down a group and from left to right across a period. Thus the heaviest post-transition elements exhibit the group oxidation number only under strongly oxidizing conditions, and generally only in covalent compounds.

Zinc, cadmium, and mercury commonly exhibit the group oxidation number of +2, although mercury also exhibits an oxidation number of +1 in compounds that contain the Hg_2^{2+} group. The chemical behaviors of zinc and cadmium are quite similar. Both elements lie above hydrogen in the activity series. Mercury is very different from zinc and cadmium. It is a nonreactive element that lies well below hydrogen in the active series.

Gallium, indium, and thallium lie above hydrogen in the activity series and react with acids, liberating hydrogen. Gallium and indium are commonly found with an oxidation number of +3, but thallium is commonly found also as the Tl^+ ion, whose stability is attributed to the **inert pair effect.**

Tin and lead form stable dipositive cations, Sn^{2+} and Pb^{2+}, with oxidation numbers two below the group oxidation number of +4. This is another example of stability that can be attributed to the inert pair effect. Tin and lead also form covalent compounds in which the +4 oxidation number is exhibited. Tin is a moderately active metal. Lead is less reactive; it lies just above hydrogen in the activity series and is attacked only by hot concentrated acids.

Bismuth readily gives up three of its five valence electrons to active nonmetals to form the tripositive ion, Bi^{3+}. It forms compounds with the group oxidation number of +5 only when treated with strong oxidizing agents.

Polonium is a member of Group VIA. It is an intensively radioactive element that results from the radioactive decay of uranium and thorium.

Techniques for the preparation of compounds of the post-transition metals are similar to those used for the preparation of compounds of other metals of varying activity.

Key Terms and Concepts

active post-transition metals (28.1)

Betts process (28.2)

halides (28.7)

hydroxides (28.6)

inactive post-transition metals (28.1)

inert pair effect (28.1)

nitrates (28.8)

ores (28.2)

oxidation numbers (28.1)

oxides (28.5)

pseudo noble gas electron configuration (28.1)

sulfates (28.8)

sulfides (28.9)

Exercises

Periodic Properties

1. How do the properties of the post-transition metals tin and lead differ from the properties of the elements at the top of Group IVA in the Periodic Table?

2. Identify the group oxidation numbers possible for the post-transition metals in their compounds.

3. Which post-transition metals form common oxides and halides in which the metal does not exhibit the group oxidation number?

4. Why is $+2$ found to be a stable oxidation number for lead when the group oxidation number is $+4$?

5. Describe the inert pair effect.

6. Which of the post-transition metals have positive reduction potentials for the reduction of the metal ion to the metal?

7. Write the electron configurations for an Sn atom, a Sn^{2+} ion, and a Sn^{4+} ion.

8. Which of the post-transition metals is used to make amalgams?

9. Which of the post-transition metals do not dissolve readily in acids, with the evolution of hydrogen?

10. Write the chemical formulas for the fluorides of each of the post-transition metals with their common oxidation numbers.

Zinc

11. What is the common ore of zinc and how is zinc extracted from it?

12. How is zinc metal separated from the impurities cadmium, iron, lead, and arsenic during the refining process?

13. What is galvanized iron, how is it produced, and what useful properties does it have?

14. A dilute solution of perchloric acid is dripped into a solution of sodium zincate, $Na_2[Zn(OH)_4]$. A white gelatinous precipitate is formed; upon analysis it proves to be a hydroxide. Upon addition of more acid, a clear solution results. Use chemical equations to explain these observations.

15. What is lithopone and how is it produced? In what way is lithopone superior to white lead as a paint pigment?

16. Write balanced chemical equations for the following reactions:

    (a) Zinc is burned in air.

    (b) An aqueous solution of ammonia is added dropwise to a solution of zinc chloride until the mixture becomes clear again.

    (c) Zinc is heated with sulfur.

    (d) Mossy zinc is added to a solution of hydrobromic acid.

    (e) Zinc carbonate is added to a solution of acetic acid, CH_3CO_2H.

    (f) Zinc carbonate is heated until weight loss stops.

    (g) Zinc metal is added to a solution of cadmium nitrate.

    (h) Zinc metal is added to a solution of lead(II) nitrate.

    (i) Zinc sulfide is heated in a stream of oxygen gas.

Cadmium

17. What is the common source of cadmium and how is cadmium extracted from it?

18. How is cadmium metal separated from impurities during the refining process?

19. What properties make cadmium effective as a protective coating for other metals?

20. What properties make cadmium unsuitable for general use as a protective coating on car door handles?

21. Write balanced chemical equations for the following reactions:

    (a) Cadmium is burned in air.

    (b) An aqueous solution of ammonia is added dropwise to a solution of cadmium chloride until the mixture becomes clear again.

    (c) Cadmium is heated with sulfur.

    (d) Cadmium is added to a solution of hydrochloric acid.

    (e) Cadmium hydroxide is added to a solution of acetic acid, CH_3CO_2H.

    (f) Cadmium hydroxide is heated until weight loss stops.

    (g) Cadmium metal is added to a solution of mercury(II) nitrate.

    (h) Cadmium metal is added to a solution of lead(II) nitrate.

    (i) Cadmium sulfide is heated in a stream of oxygen gas.

22. How many pounds of cadmium can be obtained from 10.0 tons of ore that is 5.0% greenockite? *Ans. 7.8×10^2 lb*

23. Calculate the emf of the following electrochemical cell:

$$Cd|Cd^{2+}, M = 0.10\|Ni^{2+}, M = 0.50|Ni$$

Ans. 0.174 V

24. How many grams of CdS will precipitate when a solution obtained by dissolving 2.5 g of Wood's metal is treated with excess sulfide ion?   *Ans. 0.40 g*

Mercury

25. What is the common ore of mercury and how is mercury separated from it?
26. How is mercury metal separated from impurities?
27. The roasting of an ore of a metal usually results in the conversion of the metal to the oxide. Why does the roasting of cinnabar produce metallic mercury rather than an oxide of mercury?
28. Why does mercury dissolve readily in nitric acid or hot sulfuric acid even though it is not attacked by hydrochloric acid?
29. What does it mean to say that mercury(II) halides are weak electrolytes?
30. What properties make mercury valuable as a thermometric substance?
31. Write Lewis structures for $HgCl_2$ and Hg_2Cl_2.
32. Hydrogen sulfide gas is bubbled through a solution marked mercury(I) nitrate. A precipitate forms. Of what substance(s) is the precipitate composed?
33. How many moles of ionic species would be present in a solution marked 1.0 M mercury(I) nitrate? How would you demonstrate the accuracy of your prediction?
34. What is the mass of fish that one would have to consume to obtain a fatal dose of mercury, if the fish contains 30 parts per million of mercury by weight? (Assume all the mercury from the fish ends up as mercury(II) chloride in the body and that a fatal dose is 0.20 g of $HgCl_2$.) How many pounds of fish would this be?   *Ans. 4.9 kg; 11 lb*
35. How many pounds of mercury can be obtained from 10 tons of ore that is 7.4% cinnabar?   *Ans. 1.3×10^3 lb.*
36. Write balanced chemical equations for the following reactions:
    (a) Mercury(II) oxide is added to a solution of nitric acid.
    (b) Mercury is heated with sulfur.
    (c) Cadmium metal is added to a solution of mercury(II) nitrate.
    (d) Mercury(II) sulfide is heated in a stream of oxygen gas.
    (e) A solution of mercury(II) nitrate is stirred with liquid mercury.
    (f) A solution mercury(I) nitrate is added to a solution of potassium sulfide.
    (g) An excess of zinc is added to a solution of mercury(II) nitrate.
    (h) Zinc is added to a solution containing an excess of mercury(II) nitrate.

Tin

37. What is the common ore of tin and how is tin separated from it?

38. How is tin metal separated from impurities during the refining process?
39. Mercury(I) solutions can be protected from air oxidation to mercury(II) by keeping them in contact with metallic mercury. Can tin(II) solutions be similarly stabilized against oxidation to tin(IV) by keeping them in contact with metallic tin? [The standard reduction potential for tin(IV) to tin(II) is $+0.15$ V.]
40. How and why is tin plate applied to the surface of iron?
41. What is tin pest, also known as tin disease?
42. Does metallic tin react with HCl? with HNO_3?
43. Write balanced chemical equations describing:
    (a) the burning of tin metal
    (b) the dissolution of tin in a solution of hydrochloric acid
    (c) the preparation of sodium metastannate
    (d) the purification of tin by electrolysis
    (e) the reaction of dry bromine with tin (write equations for both possible products)
    (f) the formation of thiostannate ion
    (g) the dissolution of tin oxides by basic solution (two equations)
    (h) the reactions involved when an excess of aqueous NaOH is slowly added to a solution of tin(II) chloride
    (i) the thermal decomposition of tin(II) oxalate
    (j) the reaction of tin with sulfur (write equations for both possible products)
44. Why is $SnCl_4$ not classified as a salt?
45. Why must aqueous solutions of $SnCl_2$ be protected from the air?
46. A 1.497-g sample of type metal is dissolved in nitric acid, and metastannic acid precipitates. This is dehydrated by heating to stannic oxide, which is found to weigh 0.4909 g. What percent tin was in the original type metal sample?   *Ans. 25.83%*

Lead

47. What is the common ore of lead and how is lead separated from it?
48. How is lead metal separated from impurities during the refining process?
49. Describe the Betts process for the refining of lead.
50. Compare the nature of the bonds in $PbCl_2$ to that of the bonds in $PbCl_4$. Would you expect the existence of Pb^{4+} ions? Explain.
51. Show by suitable equations that lead(II) hydroxide is amphoteric.
52. What is the composition of the tarnish on lead?
53. Why should water to be used for human consumption not be conveyed in lead pipes?
54. Why is $PbCl_2$ somewhat soluble in solutions of high chloride ion concentration?
55. When elements in the first transition metal series form dipositive ions, they usually give up their s electrons. Does this apply to lead when it forms Pb^{2+} ion?

56. A 100-mL volume of a saturated solution of lead(II) iodide contains 0.41 g of solute at 100°C. Calculate the solubility product at this temperature.   *Ans. 2.8×10^{-6}*

Bismuth

57. What are the properties of metallic bismuth that make it commercially useful?
58. Compare the basicity and acidity of Bi_2O_3 and Bi_2O_5.
59. Write equations for the reaction of $BiCl_3$ with water.
60. Write balanced equations for the following reactions:
    (a) Bismuth is heated in air.
    (b) An aqueous solution of sodium hydroxide is added dropwise to a solution of bismuth(III) chloride.
    (c) Bismuth is heated with an excess of sulfur.
    (d) Bismuth is added to a solution of hydrobromic acid.
    (e) Bismuth(III) oxide is added to a solution of nitric acid.
    (f) Bismuth hydroxide is heated until weight loss stops.
    (g) Gallium metal is added to a solution of bismuth nitrate.
    (h) Bismuth metal is added to a solution of lead(II) nitrate.
    (i) Bismuth sulfide is heated in a stream of oxygen gas.

Additional Exercises

61. Write balanced equations for the following reactions:
    (a) Gallium is heated in air.
    (b) An aqueous solution of sodium hydroxide is added dropwise to a solution of gallium chloride until the solution becomes clear again.
    (c) Indium is heated with an excess of sulfur.
    (d) Indium is added to a solution of hydrobromic acid.
    (e) Gallium hydroxide is added to a solution of nitric acid.
    (f) Indium hydroxide is heated until weight loss stops.

(g) Gallium metal is added to a solution of indium nitrate.
(h) Indium metal is added to a solution of lead(II) nitrate.
(i) Indium sulfide is heated in a stream of oxygen gas.

62. Write balanced equations for the following reactions:
    (a) Thallium is heated in air.
    (b) An aqueous solution of sodium hydroxide is added dropwise to a solution of thallium chloride.
    (c) Thallium is heated with an excess of sulfur.
    (d) Thallium is added to a solution of nitric acid.
    (e) Thallium(I) hydroxide is added to a solution of nitric acid.
    (f) Thallium(I) carbonate is heated until weight loss stops.
    (g) A solution of thallium(I) hydroxide is added to a solution of gallium nitrate.
    (h) A solution of thallium(I) nitrate is added to a solution of gallium chloride.
    (i) Thallium sulfide is heated in air.

63. Calculate the molar solubilities of $AgBr$, Hg_2Cl_2, and $PbSO_4$.
    Ans. AgBr, 5.7×10^{-7} M; Hg_2Cl_2, 6.5×10^{-7} M; $PbSO_4$, 1.3×10^{-4} M.

64. Calculate the concentration of $[HgCl_4]^{2-}$ in a solution prepared by adding 4.0×10^{-3} mol of NaCl to 50.0 mL of a 0.040 M $HgCl_2$ solution.   *Ans. 0.040 M*

65. What is the maximum $[S^{2-}]$ possible in a solution that is 0.0010 M in Cd^{2+}?   *Ans. 2.8×10^{-32} M*

66. In 1774 Joseph Priestly prepared O_2 by heating red HgO with sunlight focused through a lens. How much heat is required to decompose exactly 1 mol of red HgO(s) to Hg(l) and $O_2(g)$ under standard state conditions?
    Ans. 90.83 kJ

The Atmosphere and Natural Waters

29

The Hurricane River near Grand Marais, Michigan.

Both air and water are essential to our existence. We could live for about five minutes without air and about seven days without fresh water. The surface of the earth is covered with these two substances. The atmosphere is a fluid consisting of a mixture of gases; oceans, rivers, and lakes are fluids consisting of water of varying purity.

Air and natural waters serve as sources of raw materials for many chemical processes. Oxygen, nitrogen, and certain noble gases are extracted from the air. The oxygen in air is an essential part of almost all combustion reactions. Water is widely used as a solvent, to produce steam in a variety of different types of power plants, in cooling, and in cleaning. It is a source of hydrogen, and it is used in the manufacture of both acids and bases from oxides. The oceans serve as a source of a variety of minerals. Both the air and many bodies of water are used as dumps for a variety of wastes.

Although air and water are absolutely essential to the existence of life on the earth, the quality of both has deteriorated as a result of human activities. In this chapter we will examine the properties of the atmosphere and of naturally occurring waters, some of the sources of their pollution, its effects, and some ways in which this pollution can be reduced.

The Atmosphere

29.1 The Composition of the Atmosphere

The **atmosphere** is the mixture of gaseous substances, the air, that surrounds the earth. For the most part, it consists of almost constant proportions of uncombined nitrogen, oxygen, and the noble gases. The percentage of carbon dioxide varies somewhat, and that of dust and water vapor is widely variable. There are also varying trace amounts of ammonia, hydrogen sulfide, nitrogen oxides, sulfur dioxide, and other gases that are commonly regarded as pollutants. The average composition of unpolluted dry air at sea level is given in Table 29.1.

Table 29.1 Composition of dry air

Component	Percent by volume	Component	Percent by volume
Nitrogen	78.03	Helium	0.0005
Oxygen	20.99	Krypton	0.0001
Argon	0.94	Ozone	0.00006
Carbon dioxide	0.035–0.04	Hydrogen	0.00005
Neon	0.0012	Xenon	0.000009

The chemical behavior of these gases has been described in Chapters 20, 22, and 23.

The percent composition of dry air does not vary much with location or altitude. However, the density of air, as reflected by its pressure, decreases with increasing altitude. The average pressure at sea level is 1.0 atmosphere; at 15,000 feet (almost 3 miles up), 0.53 atmosphere; at 10 miles, about 0.05 atmosphere; and at 30 miles, only about 1×10^{-4} atmosphere.

29.2 Liquid Air

Before air is liquefied, carbon dioxide and water vapor are removed, because these substances freeze when cooled to liquid air temperature and would clog the pipes of a liquid-air machine (Fig. 29.1). The air is then compressed to about 200 atmospheres and cooled to remove the heat produced by compression. The cold, compressed air is passed into the liquefier, where it is further cooled by air that has already undergone expansion. It then is allowed to expand to a pressure of about 20 atmospheres. At this point part of the air liquefies, and the remainder passes back through the large outer coil of the liquefier and cools more incoming air in the small inner coil.

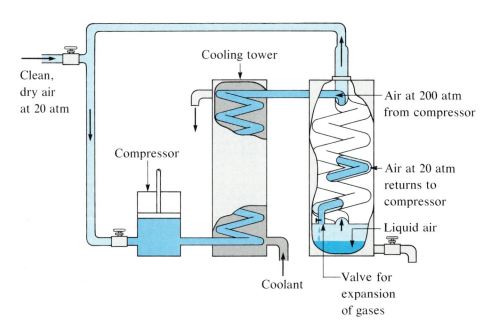

Figure 29.1

Diagram of the liquefier used in the commercial preparation of liquid air.

Liquid air (Fig. 29.2) is a mobile liquid with a faint blue color. It absorbs heat and evaporates rapidly in an open container. To reduce the rate of evaporation, liquid air is stored in Dewar flasks (Fig. 29.3). These flasks have an evacuated space between their inner and outer walls, because a vacuum is an excellent insulator. Most Dewar flasks are silvered inside the evacuated space to reflect heat. Thermos bottles (Fig. 29.4) are Dewar flasks.

Liquid air is a mixture; thus it has no definite boiling point. Nitrogen, oxygen, and argon are obtained by the fractional distillation of liquid air. The low temperature of liquid air makes it very useful in research that requires low temperatures.

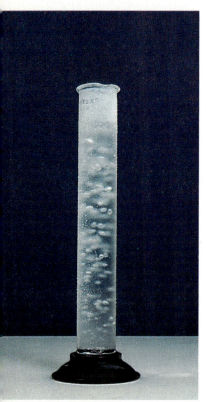

Figure 29.2

Liquid air. Note that the liquid boils when exposed to room temperature.

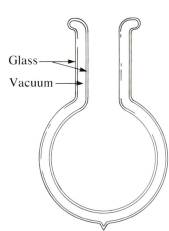

Figure 29.3

Cross section of a Dewar flask.

Figure 29.4

The familiar thermos bottle is a Dewar flask.

29.3 The Air Pollution Problem

Air pollution is not a new problem. William Shakespeare described the problem in *Hamlet* in about 1601:

> This most excellent canopy, the air, look you, this brave o'erhanging firmament, this majestical roof fretted with golden fire, why, it appears no other thing to me than a foul and pestilent congregation of vapors.*

In 1272 Edward I banned the use of smoke-producing coal in an effort to clear the smoky air of London, and the British Parliament ordered a man tortured and hanged for illegally burning the banned coal. Later, Richard III put a high tax on the use of coal. However, these efforts accomplished little, and London remained smoky. A combination of smoke and fog killed almost 4000 people there during a four-day period in 1952. In Donora, Pennsylvania, in 1948, 20 died and 5900 became ill during a period of serious air pollution. Less dramatic, but no less tragic for those involved, are the deaths attributable to present-day air pollution (Fig. 29.5). Air pollution alerts and warnings are not uncommon in many cities. Polluted air may ultimately result in shortened lives for all of us, but it is a special problem for those suffering from respiratory diseases such as emphysema and bronchitis. Lung cancer is found to be twice as prevalent among people living in air-polluted cities as among those living in rural areas with cleaner air.

The problems due to air pollution reach well beyond those related to human health (Fig. 29.6). It can damage plants, discolor paint, damage textiles, discolor dyes, damage limestone and marble buildings and statues, corrode metals, and damage tires, to name but a few of its harmful effects. In the United States alone, it is estimated that well over 200 million tons of pollutants are emitted into the air annually. Strictly on an economic basis, it has been estimated that air pollution costs the people in the United States over $12 billion per year. Directly or indirectly, air pollution is a major problem. It will be advantageous for us to learn more about it and take steps to solve it.

29.4 The Causes

The six most common air pollutants, with a partial list of the effects attributed to each, are as follows.

1. *Sulfur oxides*—cause acute leaf injury and attack trees; irritate the human respiratory tract; corrode metals and other building materials; ruin textiles, disintegrate book pages and leather; destroy paint pigments; erode statues. Sulfur dioxide is suspected to be the principal cause of acid rain (Section 29.8).

2. *Particulates (solids)*—obscure vision; aggravate lung illnesses; deposit grime on buildings and personal belongings; corrode metals.

3. *Carbon monoxide*—causes headaches, dizziness, and nausea; reduces oxygen content in the blood, impairing mental processes; in sufficient quantity, causes death.

4. *Nitrogen oxides*—cause leaf damage; stunt plant growth; irritate eyes and nasal membranes; cause brown pungent haze; corrode metals and other building materials; damage rubber. Smog (Fig. 29.5) results from the reactions induced by sunlight in a mixture of nitrogen oxides, ozone, and hydrocarbons.

*Hamlet, Prince of Denmark, William Shakespeare (1601); Act II, Scene 2.

Figure 29.5
Breathing the air in some cities is equivalent to smoking two packages of cigarettes a day.

Figure 29.6
The marble of this sculpture has been attacked by sulfur oxides.

5. *Hydrocarbons*—may be carcinogenic (cancer-producing); retard plant growth and cause abnormal leaf and bud development; produce smog with nitrogen oxides and ozone.

6. *Photochemical oxidants:* (a) Ozone and resultant chemical products—discolor leaves of many crops, trees, and shrubs; damage and fade textiles; reduce physical ability, including athletic performance; speed deterioration of rubber; disturb lung function; irritate eyes, nose, and throat; induce coughing; produce smog with nitrogen oxides and hydrocarbons. (b) Peroxyacetyl nitrate (PAN)—discolors leaves; irritates eyes; disturbs lung function.

Over 100 specific polluting compounds fall into these six categories. In addition to these compounds, many other materials are contaminants in specific locations. These include, to list only a few, fluorides, lead compounds, beryllium, mercury, pesticides, and natural radioactive materials such as radon gas.

Most air pollution arises from combustion. Coal is largely carbon; fuel oils, gasoline, and natural gas are largely hydrocarbons (compounds containing only carbon and hydrogen). If combustion is complete, carbon burns and produces only carbon dioxide; hydrocarbons (for example, CH_4, methane) produce only carbon dioxide and water.

$$C + O_2 \longrightarrow CO_2$$
$$CH_4 + 2O_2 \longrightarrow CO_2 + 2H_2O$$

However, if combustion is incomplete because of a deficiency of oxygen, carbon monoxide forms.

$$2C + O_2 \longrightarrow 2CO$$
$$2CH_4 + 3O_2 \longrightarrow 2CO + 4H_2O$$

Sulfur dioxide is produced when generating plants burn coal containing impurities (Fig. 29.7) such as elemental sulfur or pyrite (FeS_2).

$$S + O_2 \longrightarrow SO_2$$
$$4FeS_2 + 11O_2 \longrightarrow 2Fe_2O_3 + 8SO_2$$

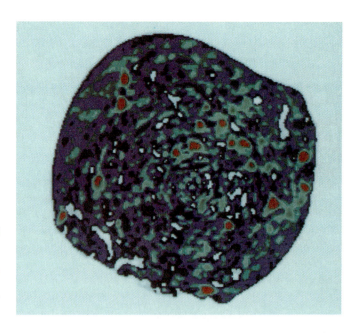

Figure 29.7

Reddish spots in this piece of coal are minerals; white spots are voids. (Microtomography was used to make this image of a piece of coal approximately 0.5 mm in diameter.)

The SO_2 either reacts with water to form sulfurous acid (which is readily oxidized by oxygen to sulfuric acid) or is oxidized by nitrogen oxides or other oxidizing agents to SO_3 (which unites with water to form sulfuric acid; Section 9.9). These acids are among the main substances responsible for acid rain. The catalytic converters of automobile exhaust systems produce small quantities of sulfur dioxide, sulfur trioxide, and sulfuric acid from reactions of the sulfur compounds present in gasoline.

Nitrogen oxides are produced in combustion processes because the nitrogen and oxygen in the air will react at high temperatures (Section 23.9).

$$N_2 + O_2 \longrightarrow 2NO \qquad \Delta H° = 180.5 \text{ kJ}$$
$$2NO + O_2 \longrightarrow 2NO_2 \qquad \Delta H° = -114.1 \text{ kJ}$$

The reaction of nitrogen dioxide with water is also believed to be partially responsible for the increased acidity of rain in many parts of the world.

$$2NO_2 + H_2O \longrightarrow HNO_3 + HNO_2$$

Many of the solids in polluted air are noncombustible or unburned material (ash) from fuels.

Other materials that contribute to air pollution result from vaporization of liquids (or sometimes sublimation of solids). Solid particles are introduced into the air by abrasion (Fig. 29.8) or grinding of materials. Examples include metal filings from polishing or grinding metals, asbestos loosened by procedures used in its removal from buildings, dust from a dust storm or automobile wheels on a dusty road, and (what is said to be the most common particulate matter in human lungs) finely divided rubber from wear of automobile tires.

Secondary reactions produce additional pollution problems. The oxidation of sulfur dioxide to sulfur trioxide and the subsequent reaction with moisture in the atmosphere to produce sulfuric acid has been mentioned. Another example is the photochemical reaction of unburned hydrocarbons and nitrogen oxides from automobile exhausts to form peroxyacetyl nitrate (PAN) and ozone. These compounds are primary constituents of smog, a problem in many of our cities.

The problem of smog is intensified by air inversion. Normally, the layer of air near the earth's surface becomes warmer than the air above and rises, allowing fresh air to take its place. Sometimes, however, a warm layer of air develops above the ground layer and traps the ground layer at the earth's surface. Pollutants from auto exhausts and other sources are also trapped, sometimes for days at a time.

Tetraethyl lead is used in some gasoline to provide greater anti-knock capability. Its use has been reduced, however, with the introduction of catalytic converters, because lead compounds poison the catalysts used in the converters and render them ineffective. When leaded gasoline is burned in an engine, volatile lead compounds are emitted through the exhaust. Lead compounds are known to have serious physiological effects on humans.

Figure 29.8
Air borne particles of asbestos can be carried into the lungs, increasing a person's chances of developing lung cancer. Workers use extreme caution as they wet, scrape, and bag asbestos for disposal.

29.5 Some Solutions to the Air Pollution Problem

Removal of particulate (solid) pollutants from smokestack gases is easily accomplished by the use of electrostatic Cottrell precipitators (see Section 12.30). Processes for removing sulfur oxides, however, are not well developed, though several methods are being studied. The wet limestone process is one that shows promise:

$$SO_2 + CaCO_3 \xrightarrow{H_2O} CaSO_3 + CO_2$$

$$SO_3 + CaCO_3 \xrightarrow{H_2O} CaSO_4 + CO_2$$

The use of low-sulfur coal, natural gas, or fuel oil, instead of high-sulfur coal, reduces the problem, but natural gas and fuel oil are becoming more expensive, and reserves of low-sulfur coal are distant from where they are needed.

Removing the sulfur is one solution. Fuel oil can be fairly readily desulfurized, and only 12% of the sulfur oxides currently emitted by combustion originate in fuel oil, whereas 65% originate in coal. Coal contains 2.5–3% sulfur on the average. Of the coal mined in the United States, 25% is such that its sulfur can be reduced to 1% by existing processes. For the other 75% of coal mined, it is now possible to remove only about 40% of its sulfur.

Over 95% of nitrogen oxide emissions result from combustion; 25% of the emissions come from combustion in steam-electric power plants. Some reduction in nitrogen oxide emissions can be achieved by modifying the combustion process, though with some loss of efficiency, but the technology is not well defined yet. The technology for removal of nitrogen oxides from stack gases is also still in the laboratory stage.

Carbon monoxide emission can be lessened by finer control of combustion, better tuning of auto engines, and use of catalytic converters. Catalytic converters (Fig. 29.9) decrease carbon monoxide, nitrogen oxides, and unburned hydrocarbon pollutants in automobile exhausts to low levels. One system contains two converters, each containing a different catalytic material. In the first, the pollutants react with oxygen from additional added air on the surface of the solid catalyst. The catalyst in the second converter promotes the spontaneous decomposition of nitrogen oxides to the elements.

$$2NO \longrightarrow N_2 + O_2 \qquad \Delta G° = -173.1 \text{ kJ}$$

$$2NO_2 \longrightarrow N_2 + 2O_2 \qquad \Delta G° = -102.6 \text{ kJ}$$

Unfortunately, catalytic converters also catalyze reactions involving sulfur compounds in the gasoline. Most serious is the catalysis of the reaction with oxygen to

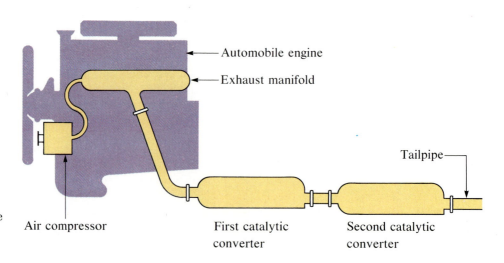

Figure 29.9
Catalytic converters decrease pollutants in the exhaust of automobiles.

Automobile engine

Exhaust manifold

Tailpipe

Air compressor

First catalytic converter

Second catalytic converter

produce sulfur dioxide and sulfur trioxide. These react with water and ultimately form small quantities of sulfuric acid mist. Fortunately, technology is available for removing sulfur compounds from gasoline during refining.

As pointed out in Section 29.4, lead compounds poison the catalyst in converters, so unleaded gasolines must be used in automobiles equipped with converters (Fig. 29.10). The undesirable emission of toxic lead compounds in the exhaust is also eliminated by the use of unleaded gasolines.

Evaporation of gasoline from tanks during filling and during use contributes surprising amounts of hydrocarbon vapors to the air. It has been estimated that in a single large U.S. city, before the introduction of vapor tight fuel-tank caps, 16 tons of hydrocarbons evaporated into the air each day. Automobile engines in the same city at one time put over 1700 tons of hydrocarbons and approximately 650 tons of nitrogen oxides into the air each day.

Alternatives to the internal combustion engine, as well as modifications of existing internal combustion engines, are under study. Alternatives include electric automobiles with rechargeable batteries. It should be noted, however, that recharging batteries requires electrical power, which in turn requires additional generating facilities. These may cause a subsequent increase in pollution. A change to avoid one type of pollution often introduces another type. The complete solution to the pollution dilemma involves solving exceedingly complex, intertwined problems.

To many people, nuclear energy is the answer to our energy crisis, and nuclear energy is used in many ways in many different countries. With it come problems of radioactive emissions (Chapter 24) and of thermal discharge, which raises the temperature of lakes, streams, and the surrounding air. Thermal pollution by both nuclear and conventional power plants is a serious problem. There is also the possibility that radioactive materials may escape, as demonstrated by the serious accident at the Chernobyl nuclear power station in the U.S.S.R. and the escape of lesser amounts of radioactive materials at other locations. Radioactive wastes must be handled very carefully to prevent such a loss. (Nuclear power is discussed in Chapter 24.)

The use of solar energy (energy from the sun) is in a very rudimentary state of development, but holds promise as a clean method for solving some of our power needs. Windmills are used to generate small quantities of electrical power in areas where wind velocities are sufficient and reasonably consistent. Meanwhile, we must solve, at least on a short-term basis, the problems involving combustion of fossil fuels.

In 1971 the United States government announced strict air quality standards, to go into effect in 1975. In part, these standards have been postponed several times, and at the time this text was written (1987) they were still not fully in effect. The standards establish limits for the amounts of sulfur oxides, particulate matter, carbon monoxide, photochemical oxidants, nitrogen oxides, and hydrocarbons that can safely exist in the air. Full enforcement of the limits would undoubtedly cause a large number of changes, such as closing some sections of large cities to automobile traffic at certain hours, a greater use of public transportation and car pools, and very significant changes in the fuels used by electric generating plants and other industries. Hence there has been considerable controversy.

In summary, air pollution involves the loss of natural resources, degrades the environment, and destroys health. Methods presently known to control pollution cost money, time, and effort. Development of new technology necessary for really adequate control of pollution carries a high price. Nevertheless, it appears that we really have no choice but to pay the necessary price in dollars and inconvenience.

Figure 29.10
Unleaded fuel became a necessity when cars were equipped with catalytic converters.

Natural Waters

Almost everyone knows ''aitch-two-oh'' (H_2O), the formula of water. This is appropriate, because none of the hundreds of thousands of other chemical substances is more important than water.

Water covers nearly three-fourths of the earth's surface. It is present in the atmosphere and the earth's crust and composes a large part of all living plant and animal matter. Nutrients are transported into the roots of plants as solutions in water. Biological reactions occur in solution in the water in cells. Many industrial reactions are run in water, and much of the chemistry that is discussed in a general chemistry course occurs in water. Water has its own chemical behavior, and it has been used as a standard for many physical constants and units.

Pure water is an odorless, tasteless, and colorless liquid. Large bodies of natural water appear bluish-green. The taste of drinking water is due to dissolved gases and salts.

The yearly use of water in the United States averages about 730,000 gallons per person. It is estimated that in the year 2000 the world's consumption of fresh water will be 9700 cubic kilometers (2300 cubic miles) per year.

29.6 Water, an Important Natural Resource

Although there is enough water in the oceans to cover the entire earth to a depth of 2 miles, human uses generally require fresh water. About a half-gallon of water per day is required to satisfy the biological needs of a human being. About 90 gallons per day per person are used in maintaining cleanliness, cooking food, and heating and air-conditioning homes. These quantities are dwarfed by the 2000 gallons of water used per person per day by industries in the United States (Fig. 29.11). To produce 1 egg requires 120 gallons of water, 10 gallons are required to produce 1 gallon of gasoline, 80 gallons for 1 kilowatt of electricity, 65,000 gallons for 1 ton of steel, and 3 million gallons for 1 ton of nylon. More water is used for irrigation in the United States than for any other single purpose; the average consumption is over 750 gallons per person per day for this purpose.

Unlike most mineral resources, fresh water is renewable, because it is part of a gigantic cycle. Water evaporates from the earth into the atmosphere, is transported in the atmosphere by the winds, and finally falls back to the earth in some form of precipitation. A large part of the water that precipitates is lost to our daily use, however, through evaporation and transpiration by plants.

At the present time the oceans are becoming increasingly important as potential sources of fresh water, because of recent advances in devising economical and practical desalting processes. It has been estimated that about 2 billion gallons of pure drinking water per day are produced world-wide by desalting water.

29.7 Naturally Occurring Water

All natural waters, even when not polluted by people, are impure, since they contain many dissolved substances. Rain water is relatively pure; its chief impurities are dust and dissolved gases. After rain has fallen for some time, the air will have been

Figure 29.11
Water is used in many facets of the steelmaking industry.

washed free of polluting gases, dust, and bacteria, and any subsequent rain will be quite free of such impurities. Sea water contains about 3.6% dissolved solids, principally sodium chloride (Table 29.2). Sea water is known to contain at least 72 elements, and it is possible that all naturally occurring elements exist in the sea.

The impurities in the fresh water on the earth's surface vary with the nature of the soil and rocks that it has passed over or through. Increasingly, fresh water also contains impurities added by people. The natural impurities may include dissolved gases such as oxygen, nitrogen, and carbon dioxide (from the air) and ammonia and hydrogen sulfide; dissolved salts; dissolved organic substances from the decay of plant and animal matter; and suspended solids such as sand, clay, silt, and organic material; and microorganisms.

During the normal life cycle of a body of fresh water such as a lake, organic sediment slowly accumulates from algae, bacteria, aquatic plants, and animal by-products. Bacteria and other organisms in the lake feed on this sediment and, in the presence of oxygen, decompose the sediment to various molecules and ions, including nitrates and phosphates. This decomposition of organic matter by bacteria in the presence of oxygen is called **aerobic decomposition.** If the supply of oxygen is reduced or if the amount of organic sediment increases to the point where aerobic decomposition cannot keep up with it, the aerobic bacteria die, and other bacteria that live in the absence of oxygen thrive. They too decompose the sediment, but the products are much less pleasant. Decomposition of organic matter in the absence of oxygen is called **anaerobic decomposition.** Anaerobic decomposition is accompanied by the smell of rotten eggs due to hydrogen sulfide (which in fact is the substance responsible for the unpleasant odor of rotten eggs); bubbles of methane are usually visible; and the water is black and often filled with slime.

The process by which a lake grows rich in nutrients (phosphates and nitrates, primarily) and gradually fills with organic sediment and aquatic plants is called **eutrophication.** As a lake ages naturally, it becomes shallower because it fills with sediment, and plant life abounds on the banks and within the lake itself. The lake is slowly transformed into a marsh and ultimately into dry land by the processes accompanying eutrophication (Fig. 29.12).

Table 29.2 The most abundant elements dissolved in sea water (excluding dissolved gases)

Element	Concentration, g/L
Cl	18.98
Na	10.56
Mg	1.272
S	0.884
Ca	0.400
K	0.380
Br	0.065
C[a]	0.028
Sr	0.013
B[b]	0.0045
F	0.0014

[a] In CO_3^{2-} and HCO_3^-
[b] In H_3BO_3

Figure 29.12
This lake in the Pocono Mountains is well along in the process of natural eutrophication; it is near the end of its conversion to a marsh.

29.8 Water Pollution

In addition to the natural contaminants in water, many other pollutants are added to our water supplies by people. The very existence of this problem went unrecognized for many years, and its magnitude may not be realized yet. Unfortunately, very little is known of the long-term effects on human health of the many chemical compounds that enter water supplies. A whole new field of chemistry is developing rapidly in which pollutants in our *overall* environment, not just in our water supplies, are studied in connection with their short-range and long-range effects on human health and with their efficient removal.

Several water pollutants are currently causing serious concern, and their effects are worthy of our attention.

1. PHOSPHATES AND OTHER NUTRIENTS.   Phosphates have been used widely in laundry detergents and hence have been released in large quantities in municipal wastes into rivers and streams. Furthermore, many fertilizers contain phosphates (Section 9.9). A number of phosphate compounds are also present in certain rocks. Thus additional phosphates reach streams through natural runoff. A high concentration of phosphates in streams and lakes presents a problem, because such substances are significant nutrients for algae, bacteria, and aquatic plants. Other nutrients include nitrogen (in the form of ammonia, ammonium salts such as ammonium sulfate, and nitrate salts such as calcium nitrate) and potassium (in potassium salts such as potassium nitrate and potassium sulfate), which are released in municipal wastes, in the runoff from fertilizer application, and in the wastes from animals.

The growth of algae and other aquatic plants is usually controlled by the amount of nutrients available. Excess nutrients lead to rapid plant growth, with a consequent rapid increase in the rate of eutrophication of natural waters. In the worst cases the deposition of organic sediment can become so extensive that anaerobic bacteria replace the aerobic bacteria and the lake is no longer able to support life.

Development of effective long-term controls of eutrophication will require a better understanding of the significant nutrients in streams and lakes, the natural plant and animal populations in them, and how these are affected by externally applied factors such as the added nutrients. In the short range, substitutes for phosphates in detergents are now in use, and methods of removing phosphates from and controlling their undesirable effects in municipal wastes are increasingly being used.

2. SEWAGE AND OTHER WASTE MATERIALS.   In addition to nutrients, municipal sewage contains microorganisms as well as the by-products from many industrial processes. In some communities raw sewage is dumped directly into natural waters that serve as sources of public water supplies. Contamination may also result from breakdowns in sewage treatment plants or from flooding of such plants. Chlorination of municipal water supplies has largely eliminated the danger from bacteria and viruses in drinking water. However, many bodies of natural water are still contaminated with microorganisms and certain chemical by-products, which may pose hazards in domestic water supplies.

The presence of organic matter, nutrients, and microorganisms in water is measured by three tests: **coliform count, algal count,** and **biological oxygen demand (BOD).** A coliform count determines the number of *E. coli* (the characteristic bacteria in animal wastes) present. The algal count is a test for microorganisms other than bacteria and viruses. The BOD measures the volume of oxygen gas taken up by a given amount of water in five days at 20°C. Since the oxygen is consumed by microorganisms in the water, the BOD is a measure of the total quantity of microorganisms and the nutrients available to them. Pure water has a BOD of 1 part per million (equivalent to 1 milligram of O_2 per liter of water) or less. A BOD of 20 parts per million or less is acceptable for municipal or industrial sewage. Raw sewage exhibits a BOD of 100–400 parts per million.

The coliform count, the algal count, and the BOD do not determine the presence of small amounts of organic compounds (often resulting from industrial wastes), but specialized tests for compounds such as benzene, acetone, toluene, chloroform, and carbon tetrachloride have shown small amounts (1–40 parts per billion) of these materials in the drinking water of cities in industrial areas. The normal water purification procedures do not remove these compounds. In fact, the presence of chloroform, $CHCl_3$, in domestic water supplies may be due to the reaction of chlorine (used in water purification) with naturally occurring organic compounds that result from the decay of plant matter in the soil. Some of these organic pollutants can be removed by filtering domestic water through activated charcoal.

Other sources of industrial chemical wastes include improperly constructed waste storage or treatment facilities and the random (and often illegal) discharge or dumping of such wastes.

3. TOXIC POLLUTANTS. Many waters contain pesticides (organic compounds that are used to kill insects, weeds, and fungi). Pesticides can concentrate in the food chain of fish and wildlife and kill these species (Fig. 29.13) or decrease their ability to reproduce. Some of these compounds, such as Aldrin, DDT, Dieldrin, and Kepone, have proved so toxic that their manufacture and use have been prohibited.

The very poisonous compound methylmercury is produced in inland waters by the action of microorganisms in the bottom mud on mercury, which may remain from prior discharge of waste mercury and mercury compounds. (The discharge of mercury and mercury compounds is now prohibited.) Fish take in methylmercury both through their food and through their gills to the extent that the concentration in their flesh may be several thousand times that in the surrounding water. Hence the possibility of mercury poisoning in humans from eating some fish (or other animals) is greater than might be expected from the actual concentration of the element in water, air, or soil.

Figure 29.13
A fish kill resulting from a runoff of toxic pollutants.

Considerable attention has recently been devoted to pollution by cadmium, an element in the same periodic family as mercury. The discharge of cadmium as well as other toxic pollutants such as cyanide (Fig. 29.13), benzidine, and polybromobiphenyl (PBB) compounds, has also been prohibited, but it may be some time before these materials are absent from natural waters.

4. ACID RAIN. Rain with an acid concentration of 0.0001–0.001 M is falling over the eastern United States and in Minnesota, Colorado, Oregon and Washington, southeastern Canada, Scandinavia, Japan, and China. Sulfuric acid appears to be the principal component of this acid rain. Such rain increases the acid content of many lakes to the point where plants and animals cannot live in the water. It may also be stripping the soil of elements such as calcium, magnesium, and potassium, which are nutrients necessary for plant life. Some of this loss, however, appears to be made up by the sulfates, nitrates, and ammonium salts present in the rain.

Natural sources of acid in rain include lightning, volcanic eruptions, and anaerobic decomposition of organic matter. However, many scientists believe that most acid rain is a result of the combustion of fossil fuels (Section 29.4).

It has been suggested that acid rain could be significantly reduced if emission control technology and/or low-sulfur coal were used. However, a switch to low-sulfur coal could depress the midwestern coal mining industry, and consumers of electricity in many states would face substantial rate hikes to pay for the emission control equipment. Some scientists claim that such controls would have very little effect on the acidity of rainfall. The acid rain problem illustrates the conflict between the need to respond promptly to a problem and the need to gather enough information to act effectively.

5. THERMAL POLLUTION. Discharges that raise the temperature of a body of water may harm fish and increase the growth of algae and other microorganisms. A few degrees rise in water temperature due to thermal pollution can decrease the solubility of oxygen in the water to a level too low for many forms of marine life to survive. Cooling towers and other facilities to reduce the heat transferred to natural waters are increasingly employed in industrial situations.

It is conceivable that thermal pollution could be beneficial under certain conditions. For example, the Ralston Purina Company and the Florida Power Company, which has replaced a conventional power plant with a nuclear power plant, cooperated in a study to determine if selected saltwater fish and shellfish might thrive in warmer water, thereby improving the food supply from the sea.

29.9 Purification of Water

Water is purified for city water supplies by first allowing it to stand in large reservoirs where most of the mud, clay, and silt settle out, a process called **sedimentation,** and then filtering it through beds of sand and gravel. Prior to filtration, lime and aluminum sulfate are often added to the water. These chemicals react to form aluminum hydroxide as a gelatinous precipitate that settles slowly, carrying with it much of the suspended matter, including most of the bacteria. The bacteria that remain in the water after filtration are killed by adding chlorine.

Relatively pure water for laboratory use is commonly prepared by distillation (Section 11.4). Gases from the air, particularly carbon dioxide, often contaminate distilled water and must be removed for some laboratory operations.

29.10 Hard Water

Water containing dissolved calcium, magnesium, and iron salts is known as **hard water.** The negative ions present in hard water are usually chloride, sulfate, and hydrogen carbonate. Hardness in water is objectionable for two reasons:

1. The calcium, magnesium, and iron ions form insoluble soaps such as calcium stearate, $(C_{17}H_{35}CO_2)_2Ca$, by reaction with soluble soaps such as sodium stearate, $C_{17}H_{35}CO_2Na$. Insoluble soaps have no cleansing power, and they adhere to fabrics, giving them a dingy appearance. They also make up the ring in the bathtub. Soap in excess of that needed to precipitate the calcium and magnesium must be added to hard water in order to obtain cleansing action.

2. Hard water is responsible for the formation of boiler scale. At high temperatures much of the mineral matter precipitates as scale [substances such as calcium carbonate, magnesium carbonate, iron(II) carbonate, and calcium sulfate]. A major component of some scale is calcium sulfate, which is *less* soluble in hot water than in cold and precipitates partly because of that fact. The scale is a poor conductor of heat and thus causes a waste of fuel. Furthermore, scale may cause boiler explosions. Since it is not a good heat conductor, the metal under it must be very hot (often red hot) in order to boil the water in the boiler. If the scale cracks, the water seeps through and comes in contact with the hot metal.

$$4H_2O + 3Fe(hot) \longrightarrow Fe_3O_4 + 4H_2$$

As it is formed, the hydrogen gas breaks the scale loose and more water gets in, producing still more hydrogen and setting the stage for a violent explosion.

29.11 Water-Softening

It is important that the substances responsible for hardness in water be removed before the water is used for washing or in boilers. The removal of the metallic ions responsible for the hardness is **water-softening.**

Water that contains Ca^{2+}, Mg^{2+}, or Fe^{2+} as hydrogen carbonate salts can be softened by boiling. This drives off carbon dioxide, and the hardness is removed as the metal carbonates precipitate.

$$Ca^{2+} + 2HCO_3^- \xrightarrow{\Delta} CaCO_3(s) + CO_2 + H_2O$$

Carbonates of calcium, magnesium, and iron form some of the deposits found in teakettles and boilers (Fig. 29.14). The hydrogen carbonate ion may also be converted to the carbonate ion by the addition of a base. Commercially, calcium hydroxide is added in the quantity needed to react with the hydrogen carbonate ions to form insoluble calcium carbonate.

$$Ca^{2+} + 2OH^- + Ca^{2+} + 2HCO_3^- \longrightarrow 2CaCO_3(s) + 2H_2O$$

Similar reactions remove Mg^{2+} or Fe^{2+} as insoluble magnesium or iron(II) carbonates. The precipitated carbonates are readily removed from the water by filtration.

Unlike hydrogen carbonate salts, chloride and sulfate salts are not removed by boiling. Crude sodium hydroxide (caustic soda) is often used in large installations to remove both carbonate and noncarbonate salts.

$$Ca^{2+} + 2HCO_3^- + [2Na^+] + 2OH^- \longrightarrow CaCO_3(s) + [2Na^+] + CO_3^{2-} + 2H_2O$$

Figure 29.14
A hard-water deposit in a hot water pipe.

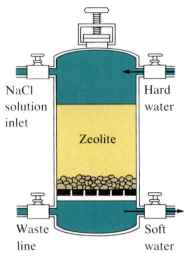

Figure 29.15

Industrial water-softening by ion-exchange. When regeneration is necessary, the soft-water outlet valve is closed, and a concentrated NaCl solution is run through the calcium zeolite and out the waste line. Regeneration is complete when the calcium ions are replaced by sodium ions.

The sodium carbonate produced then reacts to remove any remaining noncarbonate salts. For example,

$$Ca^{2+} + [SO_4^{2-}] + [2Na^+] + CO_3^{2-} \longrightarrow CaCO_3(s) + [2Na^+] + [SO_4^{2-}]$$

The sodium sulfate or sodium chloride produced is relatively soluble, and does not interfere with the cleansing action of soaps or form boiler scale.

Ion exchange provides another useful and highly important method of softening water. When water containing calcium, magnesium, and iron(II) ions filters slowly through layers of sodium aluminosilicates, those ions are replaced by sodium ions and the water is softened. Aluminosilicates are known as **zeolites.**

$$\underset{\substack{\text{Sodium aluminosilicate} \\ \text{(insoluble)}}}{2NaAlSi_2O_6} + Ca^{2+} \longrightarrow \underset{\substack{\text{Calcium aluminosilicate} \\ \text{(insoluble)}}}{Ca(AlSi_2O_6)_2} + 2Na^+$$

Calcium aluminosilicate can be reconverted into the sodium compound using a concentrated solution of sodium chloride, thus regenerating the system to remove more calcium ions.

$$Ca(AlSi_2O_6)_2 + 2Na^+ \longrightarrow 2NaAlSi_2O_6 + Ca^{2+}$$

This reaction is the reverse of the preceding one, and the direction the reaction takes is controlled by an excess of either Ca^{2+} or Na^+.

$$2Na(AlSi_2O_6) + Ca^{2+} \rightleftharpoons Ca(AlSi_2O_6)_2 + 2Na^+$$

After the sodium zeolite is regenerated, it is ready for use again. The ion-exchange method of water-softening is effective for the removal of both carbonate and noncarbonate salts (Fig. 29.15).

Synthetic **ion-exchange resins** have also been developed to soften hard water. These insoluble resins remove both cations (positive ions) and anions (negative ions) from hard water; that is, they demineralize the water completely. Examples of such resins are Amberlite and Zeo-Carb. Metal ions in the hard water displace hydrogen ions from one type of ion-exchange resin, RCO_2H. (R represents a complex hydrocarbon portion of the resin that does not enter into the exchange reaction.)

$$2RCO_2H + Ca^{2+} + [SO_4^{2-}] + 2H_2O \longrightarrow (RCO_2)_2Ca + 2H_3O^+ + [SO_4^{2-}]$$

Another type of ion-exchange resin then removes the ions of the acid from the water.

$$2RNH_2 + 2H_3O^+ + SO_4^{2-} \longrightarrow (RNH_3^+)_2SO_4^{2-}(s) + 2H_2O$$

Hence if hard water (or a salt solution) is passed first through a resin that replaces all metal ions with H_3O^+ and then through a resin that combines with the resulting acid, it can be completely demineralized. The result is essentially the same as that achieved through distillation of water. In recent years the use of ion-exchange resins in the treatment of water for industrial, laboratory, and domestic purposes has grown tremendously.

For Review

Summary

The **atmosphere** surrounding the earth is a mixture of gases consisting principally of nitrogen (78%) and oxygen (21%) with smaller amounts of water vapor, carbon dioxide, ozone, the noble gases (helium, neon, argon, krypton, xenon, radon), and gases that are usually regarded as air pollutants. The principal classes of pollutants include sulfur oxides, carbon monoxide, nitrogen oxides, hydrocarbons, photochemical oxidants, and particulates (which are not gases but suspended solids).

Carbon monoxide and hydrocarbons in the atmosphere generally result from incomplete combustion. Sulfur oxides result from combustion of fuels containing sulfur or from the use of sulfide-containing ores in metallurgical processing. Nitrogen oxides are formed as nitrogen and oxygen in the atmosphere combine under the high temperature conditions that accompany combustion. Nitrogen oxides and hydrocarbons combine in sunlight to produce photochemical oxidants such as peroxyacetyl nitrate (PAN) and ozone, components of photochemical smog. Each of these air pollutants can produce harmful effects on the health of plants and animals and can damage metals, fibers, and building materials.

A number of techniques have been used to reduce the amounts of pollutants released into the atmosphere. These range from relatively simple devices such as gastight fuel tank caps, which prevent the evaporation of hydrocarbons, to catalytic converters, desulfurization of coal, and washing of stack gas. Present pollution control methods cost money, time, effort, and, in some cases, inconvenience. However, air pollution can result in loss of natural resources, degradation of the environment, and destruction of health, so it may prove necessary to pay the price for its control.

Water is a vital natural resource, satisfying both important biological and industrial requirements. Fresh water is available from rivers, lakes, and wells, but it may require purification to remove sediment and microorganisms. It may also require softening in order to remove the Ca^{2+}, Mg^{2+}, and Fe^{2+} ions that produce **hard water.** Pollution of natural waters is of great concern. This pollution may be due to excess nutrients finding their way into natural waters (leading to increased rates of **eutrophication**), toxic chemicals from sewage or chemical waste dumps, acid rain, or simply from the discharge of heat into the waters.

Key Terms and Concepts

aerobic decomposition (29.7)
air pollution (29.3–29.5)
anaerobic decomposition (29.7)
atmosphere (29.1)
biological oxygen demand, BOD
   (29.8)

catalytic converter (29.5)
Dewar flask (29.2)
eutrophication (29.7)
hard water (29.10, 29.11)
ion exchange (29.11)
liquid air (29.2)

purification of water (29.9)
water pollution (29.8)
water-softening (29.11)
zeolite (29.11)

Exercises

The Atmosphere

1. What evidence can you cite to show that air is a mixture rather than a compound?
2. Name the six principal air pollutants and identify a source of each. Briefly describe one technique for controlling each of these pollutants.
3. Write balanced equations for each of the following processes:
   (a) the formation of sulfur dioxide from zinc sulfide present in burning coal
   (b) the formation of nitrogen dioxide in an internal combustion engine
   (c) the formation of sulfuric acid mist from sulfur dioxide

   (d) the removal of carbon monoxide from an automobile's exhaust using a catalytic converter
4. A concentration of 750 parts per million by volume of carbon monoxide is lethal. What mass of carbon monoxide is required to provide a lethal uniform concentration in a room 6.8 m long, 5.3 m wide, and 2.2 m high (about $22 \times 17 \times 7$ ft) if the temperature is 22°C and the atmospheric pressure is 740 torr?   *Ans. 67 g*
5. Based on the periodic properties of the elements, write balanced chemical equations for the reactions, if any, between the components of air given in Table 29.1 and hot magnesium.
6. (a) Calculate the density of air, of hydrogen, and of he-

lium at STP in g L^{-1}. (Assume the average molecular weight of air is 29.10.)

   Ans. Air, 1.30 g L^{-1}; hydrogen, 0.0900 g L^{-1}; helium, 0.179 g L^{-1}

(b) The lifting power of a gas with a density less than that of air is determined by the difference between its mass and that of an equal volume of air. Calculate the lifting power of hydrogen, the lifting power of helium [which by calculation in part (a) is shown to be approximately twice as heavy as hydrogen for a given volume], and the lifting power of helium as a percentage of that of hydrogen.

   Ans. Hydrogen, 1.21; helium, 1.12; 92.6%

Natural Waters

7. What naturally occurring impurities are typically present in unpolluted rain water? In unpolluted river water?
8. What is the effect of excess nutrients in a lake?
9. Define the following terms:
   (a) BOD
   (b) eutrophication
   (c) aerobic decomposition
   (d) anaerobic decomposition
   (e) hard water
   (f) zeolites

10. What are some of the primary substances responsible for water pollution?
11. What is thermal pollution of water? Is it necessarily undesirable?
12. Can the purification of water lead to its pollution? Can the pollution of water produce any beneficial results?
13. What metal ions are commonly responsible for the hardness of water?
14. Water that is hard due to the presence of hydrogen carbonate salts may be softened by boiling. Explain, using an equation.
15. Distinguish between softening and demineralizing of water. Does demineralization soften water?
16. One of the reasons that so much slaked lime (calcium hydroxide) and aluminum sulfate are produced industrially (Section 9.9) is that these compounds are used in water treatment. Explain how they are used.
17. What are the molarities of Na^+ and Cl^- in sea water (see Table 29.2)?

   Ans. Na^+, 0.4593 M; Cl^-, 0.5354 M

18. Assuming 95% recovery, what volume in liters of sea water must be processed to isolate 1.000 metric ton (1000 kg) of magnesium?   *Ans. 8.3×10^5 L*

Organic Chemistry

30

Photomicrograph of urea crystals with polarized light.

T he term *organic* was first used about 1777 to describe compounds occurring in or derived from living organisms. Such substances as starch, alcohol, and urea were classified as organic, for starch is produced by living plants, alcohol is a product of fermentation caused by microorganisms, and urea is contained in urine. In 1828, however, the German chemist Friedrich Wohler synthesized urea from materials obtained from inanimate sources, and the original meaning of organic was no longer applicable. **Organic compounds,** in the modern sense, are compounds of carbon that contain either carbon-carbon bonds or carbon-hydrogen bonds or both. Thousands of carbon compounds not found in or derived from living organisms have been produced by chemists, and about five million organic compounds have been characterized.

That so many organic compounds exist is due primarily to the ability of carbon atoms to combine with other carbon atoms, forming chains of different lengths and rings of different sizes. As pointed out in Chapter 23, carbon forms covalent bonds in compounds, with all of its four valence electrons usually used in the bonding. It exhibits sp^3, sp^2, or sp hybridization of its atomic orbitals. Carbon thereby displays a coordination number of 4, 3, or 2 with a tetrahedral, trigonal planar, or linear geometry. Examples are C_2H_6 (ethane), C_2H_4 (ethylene), or C_2H_2 (acetylene), respectively.

Other than carbon, the elements most frequently found in organic compounds are hydrogen, oxygen, nitrogen, sulfur, the halogens, phosphorus, and some of the metals.

The simplest organic compounds contain only carbon and hydrogen and are called **hydrocarbons.** Many hydrocarbons are found in plants, animals, and their fossils; other hydrocarbons have been prepared in the laboratory. Several types of hydrocarbons have been characterized. They are distinguished by the type of bonding and the hybridization of the orbitals (Sections 8.3 and 8.8).

30.1 Saturated Hydrocarbons

Saturated hydrocarbons contain only single covalent bonds. All of the carbon atoms in a saturated hydrocarbon have sp^3 hybridization and are bonded to four other carbon or hydrogen atoms. Saturated hydrocarbons of one series have the general molecular formula C_nH_{2n+2}, where n is an integer, and are called **alkanes,** or **paraffins** (from the Latin for having little affinity, or being not very reactive).

A few alkanes and some of their properties are listed in Table 30.1. With the exception of methane, each of the alkanes listed in the table contains a chain of carbon atoms. All of the carbon atoms are sp^3 hybridized and bonded either to other carbon atoms or to hydrogen atoms by single bonds. All of the C—C bond lengths are about 1.54 Å, all of the C—H lengths are about 1.09 Å, and the bond angles are close to 109.5° (the tetrahedral angle). Thus chains of carbon atoms have a staggered, or zig-zag, configuration. The Lewis structures and molecular structures of methane, ethane, and pentane are illustrated in Fig. 30.1. In the molecular structures the bonds shown as striped wedges extend behind the page, and the bonds shown as solid wedges extend in

Table 30.1   Some alkanes[a]

	Molecular formula	Melting point, °C	Boiling point, °C	Usual form	Number of structural isomers
Methane	CH_4	−182.5	−161.5	Gas	1
Ethane	C_2H_6	−183.2	−88.6	Gas	1
Propane	C_3H_8	−187.7	−42.1	Gas	1
Butane	C_4H_{10}	−138.3	−0.5	Gas	2
Pentane	C_5H_{12}	−129.7	36.1	Liquid	3
Hexane	C_6H_{14}	−95.3	68.7	Liquid	5
Heptane	C_7H_{16}	−90.6	98.4	Liquid	9
Octane	C_8H_{18}	−56.8	125.7	Liquid	18
Nonane	C_9H_{20}	−53.6	150.8	Liquid	35
Decane	$C_{10}H_{22}$	−29.7	174.0	Liquid	75
Undecane	$C_{11}H_{24}$	−25.6	195.8	Liquid	159
Dodecane	$C_{12}H_{26}$	−9.6	216.3	Liquid	355
Tridecane	$C_{13}H_{28}$	−5.4	235.4	Liquid	802
Tetradecane	$C_{14}H_{30}$	5.9	253.5	Liquid	1858
Octadecane	$C_{18}H_{38}$	28.2	316.1	Solid	60,523

[a]Physical properties for C_4H_{10} and the heavier molecules are those of the *normal* isomer, *n*-butane, *n*-pentane, etc.

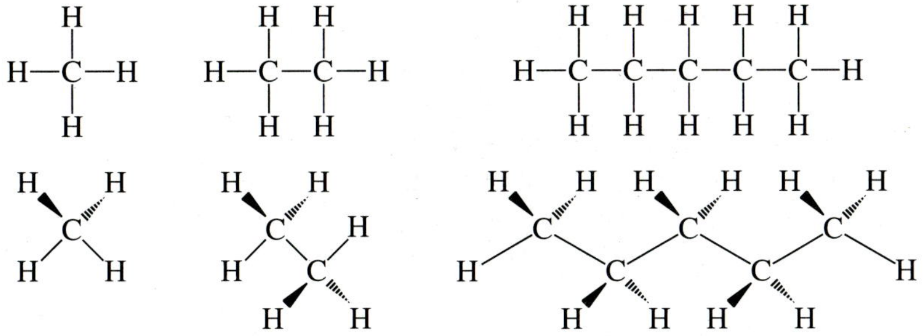

Figure 30.1

The structures of methane, ethane, and pentane. Striped wedges indicate bonds that extend behind the plane of the paper, and solid wedges indicate bonds that extend in front of the plane of the paper.

front of the page. Each carbon atom can rotate about its carbon-carbon single bonds; hence the chain of carbon atoms need not lie in a plane. A long chain can twist itself into many different shapes. For clarity, carbon atoms are usually written in straight lines in Lewis structures, but remember that Lewis structures usually show only electron distribution and bonds, not the geometry about each atom.

Two hydrocarbons have the formula C_4H_{10}; their common names are normal butane (or *n*-butane) and isobutane. These two butanes are **structural isomers.** They have the same molecular formula and hence the same composition, but different physical and chemical properties because the arrangement of the atoms in their molecules is different. Normal butane, or *n*-butane, is a straight-chain molecule, and isobutane is a branched-chain molecule.

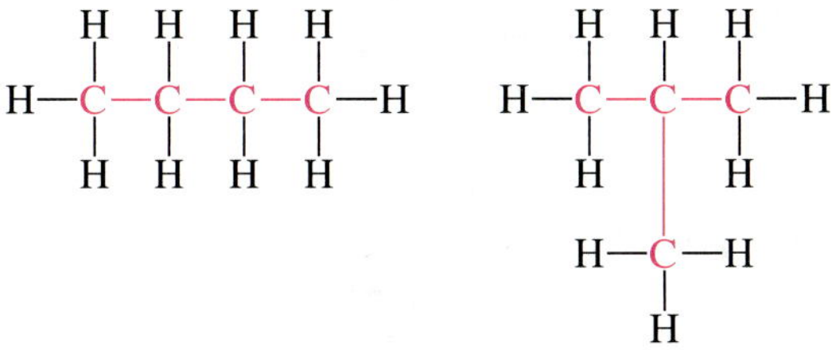

n-Butane   Isobutane

The number of possible isomers increases with the number of carbon atoms in the hydrocarbon. Hydrocarbons with a large number of carbon atoms have a great many isomers (see Table 30.1). The three structural isomers of the hydrocarbon C_5H_{12} have the common names *n*-pentane, isopentane, and neopentane.

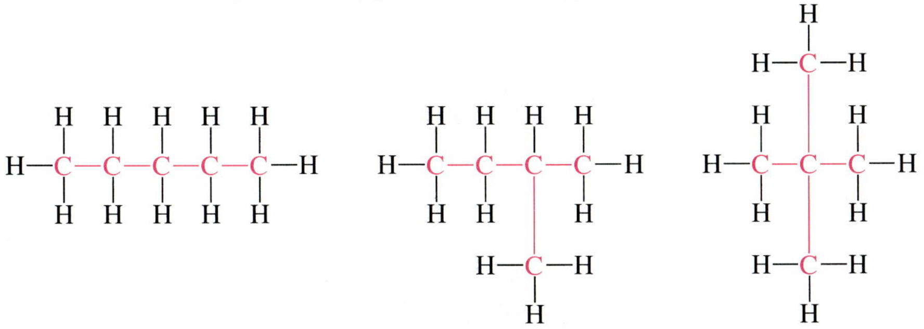

n-Pentane, or
pentane
$CH_3CH_2CH_2CH_2CH_3$

Isopentane, or
2-methylbutane
$CH_3CH_2CH(CH_3)CH_3$

Neopentane, or
2,2-dimethylpropane
$CH_3C(CH_3)_2CH_3$

Note that the following three structures, however, all represent the same molecule, *n*-pentane, and hence are not separate isomers. They are identical—they each contain an unbranched chain of five carbon atoms.

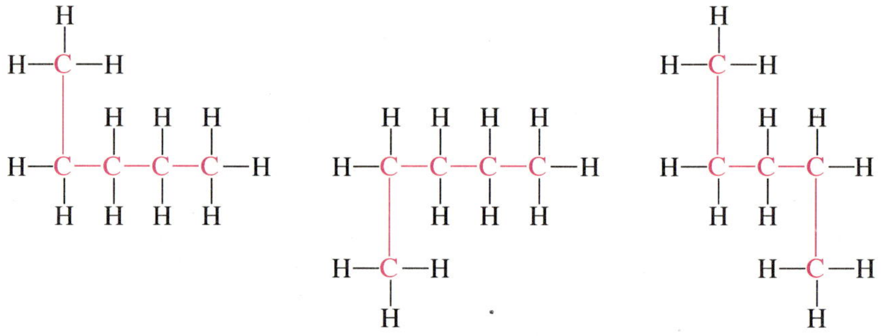

All of the hydrogen atoms in the —CH$_3$ groups at the ends of a hydrocarbon chain (like that in Fig. 30.1, for example) are equivalent, because a carbon atom can rotate freely around a C—C bond.

The members of a second series of saturated hydrocarbons have the general molecular formula C$_n$H$_{2n}$, where *n* is an integer, and are called **cycloalkanes.** As the name implies, the molecules of these substances are cyclic (possessing rings). The smallest member of the series is cyclopropane; next are cyclobutane, cyclopentane, etc.

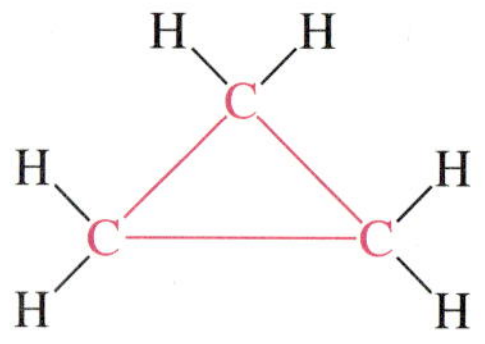

Cyclopropane 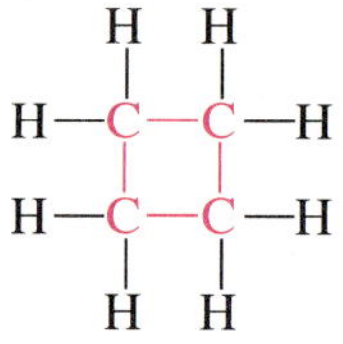   Cyclobutane 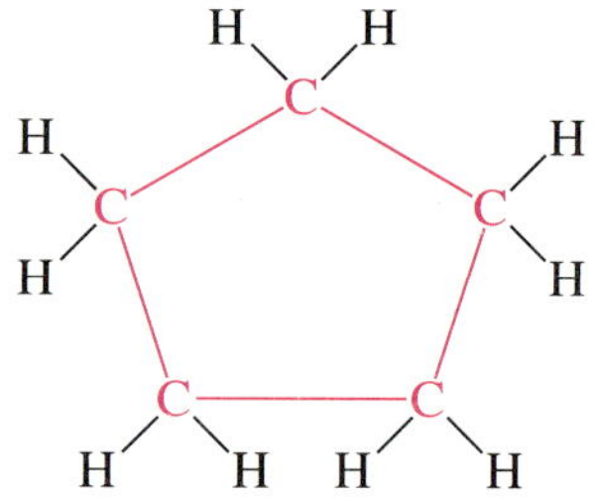   Cyclopentane

The reactions of alkanes all involve the breaking of C—H or C—C single bonds. In a **substitution reaction,** a typical reaction of alkanes, a second type of atom is substituted for a hydrogen atom.

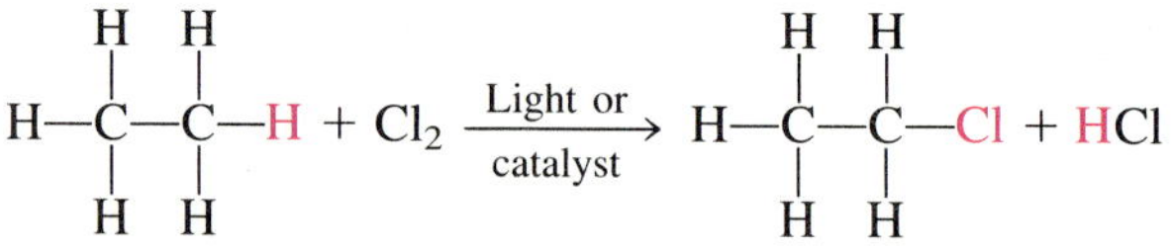

Ethane   Ethyl chloride

This reaction transforms an alkane molecule into one that contains a more reactive species or group, a **functional group.** The functional group, a halogen in this case, makes it possible for the molecule to take part in many different kinds of reactions.

Alkanes burn in air; the reaction is a highly exothermic oxidation-reduction reaction. Thus they are excellent fuels. A typical combustion reaction is

$$2C_2H_6(g) + 7O_2(g) \longrightarrow 4CO_2(g) + 6H_2O(g) \qquad \Delta H° = -2856 \text{ kJ}$$

Methane is the principal component of natural gas. The butane used in camping stoves and lighters is an alkane. Gasoline is a mixture of straight and branched chain alkanes containing from five to nine carbon atoms plus various additives. Kerosene, diesel oil, and fuel oil are primarily mixtures of alkanes with higher molecular weights.

30.2 The Nomenclature of Saturated Hydrocarbons

The International Union of Pure and Applied Chemistry (IUPAC) has devised a system of nomenclature for the hydrocarbons, their isomers, and their **derivatives** (compounds in which one or more hydrogen atoms have been replaced by other atoms or groups of atoms). The nomenclature for saturated hydrocarbons is based on two rules.

1. The longest *continuous* chain of more than four carbon atoms in a saturated hydrocarbon is indicated by one of the following prefixes: five carbons, *penta-;* six carbons, *hexa-;* seven carbons, *hepta-;* eight carbons, *octa-;* nine carbons, *nona-;* and ten carbons, *deca-.* The suffix *-ane* is added at the end (one *a* is dropped when two occur in succession). Thus these names are pentane, hexane, heptane, octane, nonane, and decane, respectively. A two-carbon chain is called ethane; a three-carbon chain, propane; and a four-carbon chain, butane. Thus the names of the hydrocarbons listed in Table 30.1 indicate that these compounds, which are the normal isomers when more than one form exists, contain a single chain of carbon atoms consisting of the same number of carbon atoms as the number of carbon atoms in the molecule.

2. The positions and names of branches from the longest chain or of atoms that replace hydrogen atoms on the chain (**substituents**) are added as prefixes to the name of the chain. The position of attachment is identified by the number of the carbon atom to which a branch or substituent is attached; the number of the carbon atom is found by counting from the end of the chain nearer to the substituent.

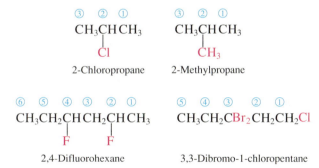

Notice that *-o* replaces *-ide* at the end of the name of an electronegative substituent and that the number of substituents of the same type is indicated by the prefixes *di-* (two), *tri-* (three), *tetra-* (four), etc. (for example, *difluoro-* for two fluoride substituents).

The longest continuous chain of carbon atoms in *n*-butane is four, but in isobutane the longest continuous chain of carbon atoms is only three even though there are a total of four carbon atoms in the molecule. Hence, in IUPAC nomenclature, *n*-butane is called butane, but isobutane is 2-methylpropane.

A substituent that contains one less hydrogen than the corresponding alkane is called an **alkyl group.** The name of an alkyl group is obtained by dropping the suffix

-ane of the alkane name and adding *-yl*. For example, methane becomes *methyl*, and ethane becomes *ethyl*.

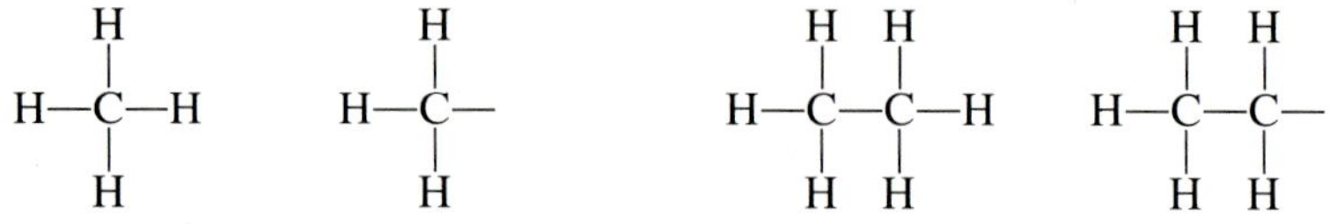

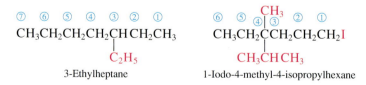

| Methane | A methyl group | Ethane | An ethyl group |

The open bonds in the methyl and ethyl groups indicate that these groups are bonded to another atom.

Removal of any of the four hydrogen atoms from methane forms a methyl group. These four hydrogen atoms are equivalent. Likewise, removing any one of the six equivalent hydrogen atoms in ethane gives an ethyl group. However, in both propane and 2-methylpropane there are two different types of hydrogen atoms, distinguished by the adjacent atoms or groups of atoms.

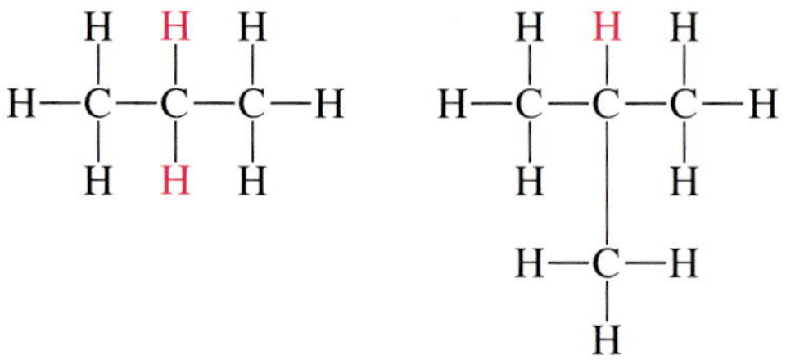

Propane   2-Methylpropane

The six equivalent hydrogen atoms of the first type in propane and the nine equivalent hydrogen atoms of that type in 2-methylpropane (all shown in black) are each bonded to a **primary carbon,** a carbon atom bonded to only one other carbon atom. The two colored hydrogen atoms in propane are of a second type. They differ from the six hydrogen atoms of the first type in that they are bonded to a **secondary carbon,** a carbon atom bonded to two other carbon atoms. The colored hydrogen atom in 2-methylpropane differs from the other nine hydrogen atoms in that molecule; it is bonded to a **tertiary carbon,** a carbon atom bonded to three other carbon atoms. Two different alkyl groups can be formed from each of these molecules, depending on which hydrogen atom is removed. The names and structures of these and several other alkyl groups are listed in Table 30.2. Note that alkyl groups do not exist as stable independent entities. They are always part of some larger molecule.

The location of an alkyl group on a hydrocarbon chain is indicated in the same way as any other substituent. If more than one substituent is present on the same carbon atom, they are listed either alphabetically or in order of increasing complexity.

⑦ ⑥ ⑤ ④ ③ ② ①
$CH_3CH_2CH_2CH_2CHCH_2CH_3$
C_2H_5
3-Ethylheptane

⑥ ⑤ ④ ③ ② ①
CH_3
$CH_3CH_2CCH_2CH_2CH_2I$
CH_3CHCH_3
1-Iodo-4-methyl-4-isopropylhexane

According to IUPAC nomenclature, the longest continuous chain of carbon atoms is numbered to produce the lowest number or sum of numbers for the substituted atoms(s) and/or group(s). For example, the six carbon atoms of the chain in 1-iodo-4-methyl-4-isopropylhexane (above) could be numbered from left to right, resulting in

Table 30.2   Some alkyl groups

Alkyl group	Structure
Methyl	CH_3-
Ethyl	CH_3CH_2-
n-Propyl	$CH_3CH_2CH_2-$
Isopropyl	$CH_3\overset{\mid}{C}HCH_3$
n-Butyl	$CH_3CH_2CH_2CH_2-$
sec-Butyl (where *sec* stands for secondary)	$CH_3CH_2\overset{\mid}{C}HCH_3$
Isobutyl	$CH_3\underset{\underset{CH_3}{\mid}}{C}HCH_2-$
t-Butyl (where t stands for tertiary)	$CH_3\overset{\overset{CH_3}{\mid}}{\underset{\underset{CH_3}{\mid}}{C}}CH_3$

the name 3-methyl-3-isopropyl-6-iodohexane. However, the latter gives a sum of 12 for the numbers of the substituents, whereas the preferred numbering from right to left gives a sum of 9.

Cycloalkanes are named in the same way as alkanes. The carbons in the ring are numbered, with a substituted carbon being the first.

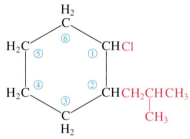

1-Chloro-2-isobutylcyclohexane

30.3 Alkenes

Hydrocarbon molecules that contain a double bond are members of another series called **alkenes.** The two carbon atoms linked by a double bond are bound together by two bonds, one σ bond and one π bond (Section 8.8). The alkenes have the general molecular formula C_nH_{2n}.

Ethene, C_2H_4, commonly called **ethylene,** is the simplest alkene. Each carbon atom in ethene has a trigonal planar structure, as illustrated in Fig. 8.19 in Section 8.8. The second member of the series is **propene (propylene);** the next members are the butene isomers. The series is analogous to the alkane series.

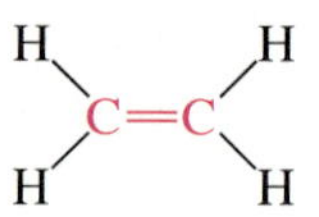

Ethene
(ethylene)

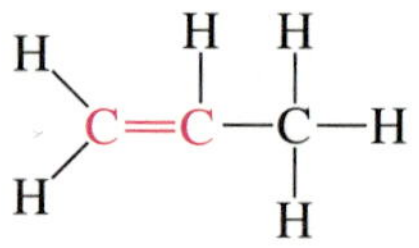

Propene
(propylene)

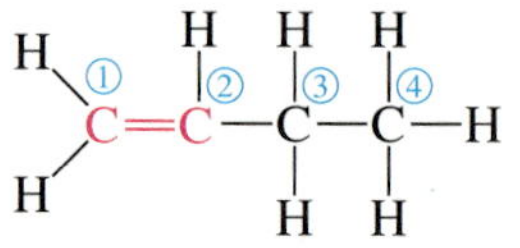

1-Butene

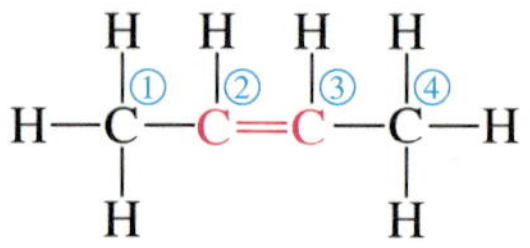

2-Butene

The IUPAC name of an alkene is derived from that of the alkane with the same number of carbon atoms. The presence of the double bond in the chain is indicated by replacing the suffix -ane with the suffix -ene. The location of the double bond in the chain is indicated by a number that specifies the position of the first carbon atom in the double bond.

The carbon atoms involved in single bonds in alkenes have sp^3 hybridization, and those in double bonds have sp^2 hybridization. Thus the geometry around the carbon atoms at the double bonds is trigonal planar (Section 8.8). The molecular structure of propene is illustrated in Fig. 30.2.

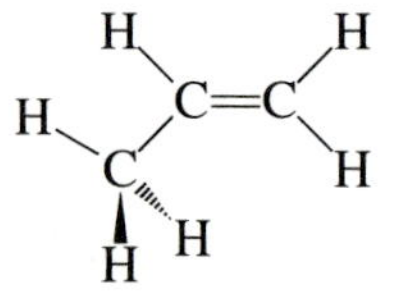

Figure 30.2
The molecular structure of propene.

Carbon atoms are free to rotate around a single bond but not around a double bond; a double bond is quite rigid. This makes it possible to have two separate isomers of 2-butene, one with both methyl groups on the same side of the double bond and one with the methyl groups on opposite sides. When formulas of butene are written with 120° bond angles around the doubly bonded carbon atoms, the isomers are apparent.

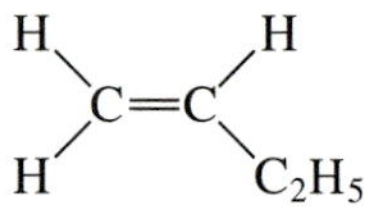

1-Butene

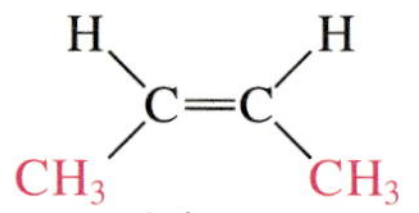

cis isomer 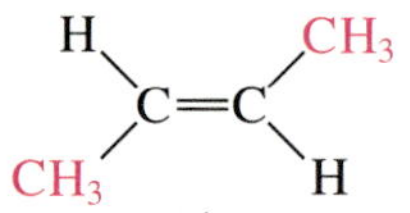   *trans* isomer

2-Butene

The 2-butene isomer in which the two methyl groups are on the same side is called the *cis* isomer; the one in which the two methyl groups are opposite is called the *trans* isomer. In these **geometrical isomers,** the same types of atoms are attached to each other, but the geometries of the two molecules differ. (Recall that in Section 26.4 we discussed geometrical isomers for coordination compounds. It would be a good idea to review that section to see the analogy.) The different geometries produce different properties that make separation of the isomers possible.

Alkenes are much more reactive than alkanes. A π bond, being a relatively weaker bond, is disrupted much more easily than a σ bond. Thus the reaction characteristic of alkenes is a type in which the π bond is broken and replaced by two σ bonds. Such a reaction is called an **addition reaction.**

In the presence of a suitable catalyst (Pt, Pd, or Ni, for example), alkenes undergo an addition reaction with hydrogen to form the corresponding alkanes.

$$\underset{\text{Ethene}}{\begin{array}{c} H \\ \diagdown \\ H \end{array}\!\!C\!\!=\!\!C\!\!\begin{array}{c} H \\ \diagup \\ H \end{array}} + H_2 \xrightarrow{\text{Pt, pressure}} \underset{\text{Ethane}}{H\!-\!\overset{\displaystyle H}{\underset{\displaystyle H}{C}}\!-\!\overset{\displaystyle H}{\underset{\displaystyle H}{C}}\!-\!H}$$

Chlorine adds to the double bond in an alkene, instead of replacing a hydrogen as it does in an alkane.

$$\underset{\text{Ethene}}{\begin{array}{c} H \\ \diagdown \\ H \end{array}\!\!C\!\!=\!\!C\!\!\begin{array}{c} H \\ \diagup \\ H \end{array}} + Cl_2 \longrightarrow \underset{\text{1,2-Dichloroethane}}{H\!-\!\overset{\displaystyle H}{\underset{\displaystyle Cl}{C}}\!-\!\overset{\displaystyle H}{\underset{\displaystyle Cl}{C}}\!-\!H}$$

Many other reagents react with the double bond in alkenes. An example is the acid-catalyzed addition of water to an alkene.

$$\underset{\text{Ethene}}{\begin{array}{c} H \\ \diagdown \\ H \end{array}\!\!C\!\!=\!\!C\!\!\begin{array}{c} H \\ \diagup \\ H \end{array}} + H_2O \xrightarrow[\text{catalyst}]{\text{Acid,}} \underset{\text{Ethanol}}{H\!-\!\overset{\displaystyle H}{\underset{\displaystyle H}{C}}\!-\!\overset{\displaystyle H}{\underset{\displaystyle OH}{C}}\!-\!H}$$

Alkenes can add to themselves. This very important addition reaction gives polymers.

$$n\left(\!\!\begin{array}{c} H \\ \diagdown \\ H \end{array}\!\!C\!\!=\!\!C\!\!\begin{array}{c} H \\ \diagup \\ H \end{array}\!\!\right) \xrightarrow{\text{Catalyst}} \underset{\text{Polyethylene}}{\left(\!\!-\!\overset{\displaystyle H}{\underset{\displaystyle H}{C}}\!-\!\overset{\displaystyle H}{\underset{\displaystyle H}{C}}\!-\!\right)_n} \qquad \text{(where } n \text{ is large)}$$

Most of the ethylene (ethene) used to make polyethylene and in other industrial reactions is produced from alkanes by cracking operations. The preparation of ethylene from ethane is described in Section 14.14. Ethylene is a basic raw material used in the polymer, petrochemical, and plastics industries. Over 16.5 million tons of ethylene were produced in the United States in 1986. Only three chemicals were produced in larger amounts.

30.4 Alkynes

Hydrocarbon molecules with a triple bond are called **alkynes;** they make up another series of unsaturated hydrocarbons. Two carbon atoms joined by a triple bond are bound together by one σ bond and two π bonds (Section 8.8). The alkynes have the general molecular formula C_nH_{2n-2}.

The simplest and most important member of the alkyne series is **ethyne, C_2H_2,** commonly called **acetylene.** The Lewis structure for ethyne is

$$H\!-\!C\!\equiv\!C\!-\!H$$
Ethyne (acetylene)

The bonding in the ethyne molecule, which has a linear structure, is illustrated in Fig. 8.22 in Section 8.8.

The IUPAC nomenclature for alkynes is similar to that for alkenes except that the suffix *-yne* is used to indicate a triple bond in the chain.

$$\overset{④}{CH_3}\overset{③}{CH_2}\overset{②}{C}\equiv\overset{①}{CH}$$

$$\overset{⑥}{CH_3}\overset{⑤}{CH_2}\overset{④}{C}\equiv\overset{③}{C}\overset{②}{CH}\overset{①}{CH_3}$$
$$\underset{CH_3}{|}$$

1-Butyne　　　　　　　　　　　2-Methyl-3-hexyne

Chemically, the alkynes are similar to the alkenes, except, having two π bonds, they react even more readily, adding twice as much reagent in addition reactions. The reaction of acetylene with bromine is a typical example.

$$H\!-\!C\equiv\!C\!-\!H + 2Br_2 \longrightarrow H\!-\!\overset{\overset{\textstyle Br}{|}}{C}\!-\!\overset{\overset{\textstyle Br}{|}}{C}\!-\!H$$
$$\qquad\qquad\qquad\qquad\qquad\quad\underset{Br}{|}\;\;\underset{Br}{|}$$

Tetrabromoethane

Acetylene and all the other alkynes burn very easily.

$$2H\!-\!C\equiv\!C\!-\!H(g) + 5O_2(g) \longrightarrow 4CO_2(g) + 2H_2O(g) \qquad \Delta H° = -2511.2 \text{ kJ}$$

The flame from burning acetylene is very hot; it is used in welding and cutting metal. When acetylene burns, some of it breaks down a second way according to the equation

$$2C_2H_2 + O_2 \longrightarrow 4C + 2H_2O$$

At the temperature of the flame, the particles of carbon give off a brilliant white light. At one time acetylene was used to light homes and in bicycle and automobile lamps; acetylene lamps are still used by some cavers.

30.5 Aromatic Hydrocarbons

Benzene, C_6H_6, is the simplest member of a large family of hydrocarbons, called **aromatic hydrocarbons,** that contain ring structures. The benzene molecule contains a hexagonal ring of sp^2 hybridized carbon atoms with the unhybridized p orbital of each one perpendicular to the ring. The electronic structure of benzene can be described as a resonance hybrid of two Lewis structures (Section 6.8).

Three valence electrons in the sp^2 hybrid orbitals of each carbon atom and the valence electron of each hydrogen atom form the framework of sigma bonds in the benzene molecule (Fig. 30.3). The unhybridized carbon p orbitals, each with one electron, combine to form π bonds. As shown in Figs. 30.4 and 30.5, the electrons in the π bonds are delocalized around the ring, because the electronic structure is the average of the resonance forms. Since the π electrons are delocalized, benzene does not have alkene character. Each bond between two carbon atoms is neither a single nor a double bond but is intermediate in character (a hybrid). Each bond is equivalent to the others and has a bond order of $1\frac{1}{2}$. Although benzene is often written as one of its resonance

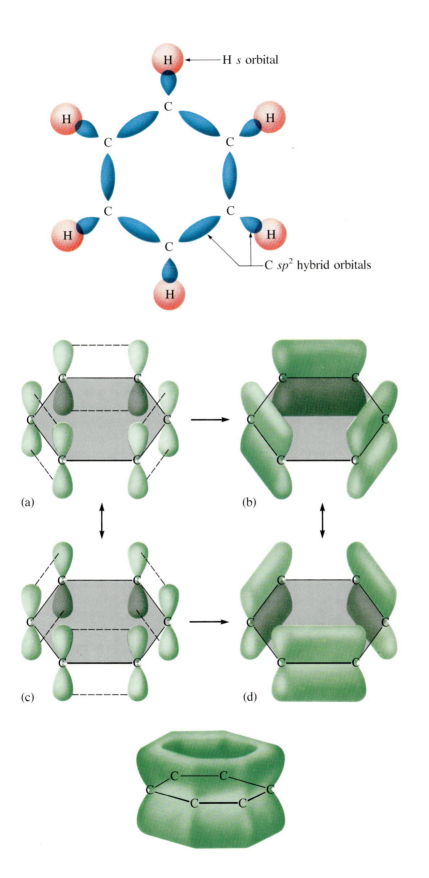

Figure 30.3
The overlap sp^2 hybrid orbitals of the carbon atoms and s orbitals of the hydrogen atoms that form σ bonds in a benzene ring.

Figure 30.4
The π overlap in a benzene molecule. Hydrogen atoms have been omitted for clarity. The six sp^2 hybridized carbon atoms lie in a plane. The six unhybridized p orbitals, shown in color in (a) and (c), extend above and below the plane and are perpendicular to it. The dashed lines between lobes of p orbitals indicate side-by-side overlap. The overlap can occur in either direction between adjacent lobes, as shown in (a) and (c). The resulting π orbitals are shown in (b) and (d), with the orbital portion of each (in color) lying above and below the hexagonal plane. The resonance forms in (a) and (c) and in (b) and (d) are exactly equivalent. The true structure is an average (hybrid) of the two (see Fig. 30.5).

Figure 30.5
The distribution of π electrons in a benzene molecule.

Figure 30.6

The three filled π bonding molecular orbitals in benzene.

forms in chemical formulas, it is important for you to remember that all of the carbon-carbon bonds in benzene are equivalent. The structure of benzene is sometimes written with a circle within the ring (right-hand structure above) to emphasize the fact that the bonds are equivalent and the electrons are delocalized.

The π bonding in benzene can be described by the molecular orbital model. The six unhybridized p orbitals can be combined to give three π bonding molecular orbitals and three π antibonding molecular orbitals (Section 7.1). The six electrons (as three pairs) from the unhybridized p orbitals occupy the three π bonding molecular orbitals. These π bonding molecular orbitals are shown in Fig. 30.6.

There are many derivatives of benzene. The hydrogen atoms can be replaced by many different species of groups. The following are typical examples.

Toluene   Styrene   Bromobenzene

Xylene is a derivative of benzene in which two hydrogen atoms have been replaced by methyl groups. Xylene has three isomers since the two methyl groups can occupy three different relative positions on the ring.

o-Xylene   m-Xylene   p-Xylene

Any disubstituted benzene can take one of these three structurally different (isomeric) forms. The relative positions of the two substituents are indicated by the prefixes *o-* *(ortho-)*, *m-* *(meta-)*, and *p-* *(para-)*.

Some important aromatic hydrocarbons and their derivatives contain more than one ring. Examples are naphthalene, $C_{10}H_8$, found in moth balls, anthracene, $C_{14}H_{10}$, and benzpyrene, $C_{20}H_{12}$.

Naphthalene   Anthracene   Benzpyrene

Aromatic hydrocarbons are widely used in manufacturing dyes, synthetic drugs, explosives, and plastics. For example, the important explosive trinitrotoluene (TNT), $C_6H_2(CH_3)(NO_2)_3$, is synthesized from toluene by replacing three hydrogen atoms with nitro groups, $-NO_2$.

Trinitrotoluene

30.6 Isomerism

Isomers are groups of atoms with identical chemical formulas but different arrangements of the atoms. Thus far we have seen two kinds of isomerism: structural and geometrical. Structural isomers (Section 30.1) are molecules with different arrangements of bonds, such as *n*-butane and 2-methylpropane or the three isomers of xylene (Section 30.5). Geometrical isomers have the same atoms bonded in the same order, but some parts of the two molecules have different relative spatial arrangements. In *cis* and *trans* isomers of organic compounds, for example, groups are arranged either on the same side (*cis-*) or on different sides (*trans-*) of bonds that cannot rotate (Section 30.3). Recall, too, that we pointed out in Section 26.4 the existence of geometrical *cis-* and *trans-* isomers for certain coordination compounds.

A third type of isomers, called **optical isomers,** is a category of isomerism also characteristic of some coordination compounds (Section 26.4). Optical isomers are detected by their effect on polarized light. Ordinary light is composed of rays that vibrate in all planes parallel to the direction in which the light travels. When ordinary light is passed through a polarizer, such as is used in Polaroid sunglasses or a Nicol prism, only light that vibrates in one plane is transmitted; such light is called **plane-polarized light.** The direction of the plane in which the light vibrates can be detected using another polarizer. When a beam of plane-polarized light passes through a solution containing a single form of an optical isomer, the plane of vibration of the light is rotated. Molecules that rotate the plane of vibration of plane-polarized light are **optically active molecules.** The amount of rotation and thus the optical activity vary from compound to compound.

A compound that forms optical isomers exists as at least two different forms. One of these isomers rotates plane-polarized light clockwise; such an isomer is called a **dextrorotatory isomer,** or a + isomer. The other isomer rotates plane-polarized light counterclockwise by an amount exactly equal to the clockwise rotation of the + isomer. The second isomer is called a **levorotatory isomer,** or a − isomer. A mixture of equal amounts of a dextrorotatory isomer and a levorotatory isomer of the same substance shows no rotation of plane-polarized light because the effects of the two isomers cancel. In most properties, such as melting point, boiling point, and solubility, optical isomers exhibit identical characteristics. However, optical isomers may exhibit different degrees of reactivity in reactions with other optically active compounds.

Optical isomers of a compound have the same relationship to each other as that of a right hand to a left hand. Both have the same shape and the same order of attachment of the various parts to one another; they are almost alike. However, they differ in that

Figure 30.7
Mirror images of a hand, the molecule CHClBrI, and the molecule alanine. Although these objects have the same composition and are the same size and shape, their mirror images cannot be superimposed.

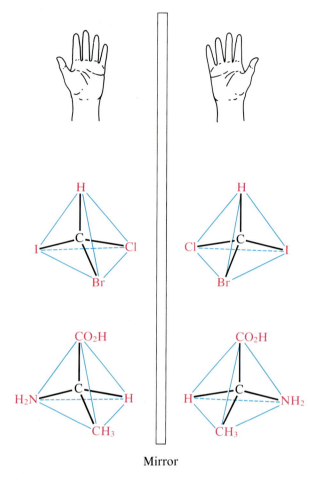

Mirror

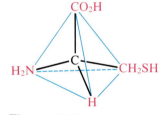

Figure 30.8
The structure of L-cysteine.

they are mirror images of each other. Just as your right and left hands cannot be superimposed, two optical isomers of a molecule cannot be superimposed, no matter how they are rotated. Figure 30.7 illustrates the mirror relationship for a hand, for the molecule CHClBrI, and for the amino acid alanine. Note that simple molecules that contain sp^3 hybridized carbon atoms form optical isomers only when a carbon atom is bonded to four different groups; such a carbon atom within a molecule is referred to as an **asymmetric carbon.**

Practically all optically active molecules found in nature are found as only one of the two possible isomers. For example, the amino acids found in proteins (Chapter 31) are all of the type designated as L. The structure of L-cysteine, one of these amino acids, is shown in Fig. 30.8.

A striking example of the effect of optical activity is found in the oils responsible for the scents of spearmint and caraway seeds. Both oils are composed of carvone (Fig. 30.9). However, caraway seed oil contains D-carvone, and spearmint oil contains the other optical isomer, L-carvone.

30.7 Petroleum

Petroleum is composed chiefly of saturated hydrocarbons, but may also contain unsaturated hydrocarbons, aromatic hydrocarbons, and their derivatives. The main step in refining petroleum is distilling it into a number of fractions (fractional distillation; Section 12.19, each of which is a complex mixture of hydrocarbons with proper-

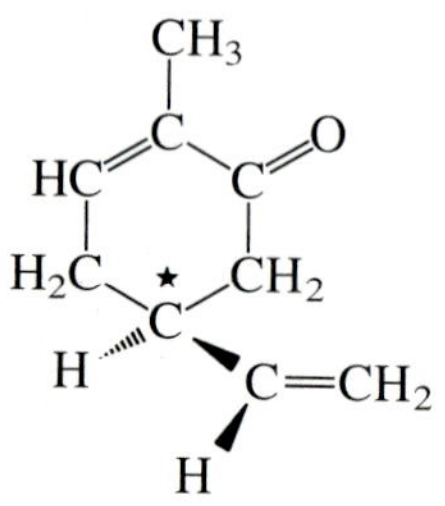

Figure 30.9
The molecule of carvone. The asymmetric carbon is indicated by the asterisk.

Table 30.3   Some hydrocarbon products from petroleum

	Approximate composition	Boiling point range, °C	Uses
Petroleum ether	C_4 to C_{10}	35–80	Solvent
Gasoline	C_4 to C_{13}	40–225	Motor fuel
Kerosene	C_{10} to C_{16}	175–300	Fuel, lighting
Lubricating oils	C_{20} up	350 up	Lubrication
Paraffin	C_{23} to C_{29}	50–60 (m.p.)	Candles, waxed paper
Asphalt		Viscous liquids	Paving, roofing
Coke		Solid	Fuel

ties that make it valuable. Additional purification is usually necessary before these fractions can be used. Some of the more important hydrocarbon products from the refining of petroleum are listed in Table 30.3.

Gasoline, a mixture of *n*-hexane, *n*-heptane, *n*-octane, and their isomers, is rated on an arbitrary scale in which isooctane (2,2,4-trimethylpentane), once thought to be the ideal fuel for gasoline engines, is given a rating, or octane number, of 100. Gasolines with octane numbers less than 100 are less efficient than isooctane, and those with higher numbers, more efficient. Until recently, gasoline was a blend of aliphatic (open-chain) hydrocarbons and aromatic hydrocarbons, and tetraethyl lead. However, manufacturers are now making most gasoline without the tetraethyl lead because of the toxic effects of lead compounds and their poisoning effect on catalytic converters (Section 29.4). Gasoline with sufficient octane number can be produced without tetraethyl lead by blending in more highly branched hydrocarbon isomers, methanol or ethanol, or a combination of these. At the same time, engines are being designed to run on lower-octane gasolines.

Drilling for petroleum in southeast Utah.

Derivatives of Hydrocarbons

30.8 Introduction

Hydrocarbon derivatives such as chloroethane, ethanol, and acetic acid are formed by replacing one or more hydrogen atoms of a hydrocarbon by other atoms or groups of atoms, called functional groups. An alcohol is a derivative of a hydrocarbon that contains an —OH group in place of a hydrogen atom. Thus methanol, CH_3OH [Fig. 30.10(a)], is a derivative of methane; ethanol, C_2H_5OH [Fig. 30.10(b)], is a derivative of ethane.

Methane   Methanol   Ethane   Ethanol

(a)   (b)

Figure 30.10
(a) The molecular structures of methane and methanol.
(b) The molecular structures of ethane and ethanol.

Table 30.4 lists several important types of compounds derived from hydrocarbons, the corresponding functional groups, and the general formulas. In the general formulas, R stands for the portion of the molecule other than the functional group; R might represent an alkyl group (Section 30.2), or a phenyl group, C_6H_5—, for example.

Table 30.4   Several types of derivatives of hydrocarbons

Derivative type	Functional group	General formula	Typical examples
Alcohols	—OH	R—OH	CH_3CH_2OH Ethanol (a primary alcohol) $\overset{\displaystyle OH}{\underset{\displaystyle \vert}{CH_3CHCH_3}}$ 2-Propanol (a secondary alcohol) $\overset{\displaystyle OH}{\underset{\displaystyle \vert}{CH_3\underset{\underset{\displaystyle CH_3}{\vert}}{C}CH_3}}$ 2-Methyl-2-propanol (a tertiary alcohol)
Ethers	—O—	R—O—R	CH_3—O—C_2H_5 C_2H_5—O—C_2H_5 Methyl ethyl ether Diethyl ether
Amines	$-\overset{\displaystyle \vert}{N}-$	$R-\overset{\displaystyle H}{\underset{\displaystyle H}{N}}$	CH_3NH_2 Methylamine (a primary amine)
		$R-\overset{\displaystyle R}{\underset{}{N}}-H$	$(C_2H_5)_2NH$ Diethylamine (a secondary amine)
		$R-\overset{\displaystyle R}{\underset{}{N}}-R$	$(C_3H_7)_3N$ Tri-n-propylamine (a tertiary amine)
Thiols	—SH	R—SH	C_2H_5SH Ethanethiol
Aldehydes	$-\overset{\displaystyle O}{\overset{\displaystyle \parallel}{C}}H$	$R-\overset{\displaystyle O}{\overset{\displaystyle \parallel}{C}}H$	$H\overset{\displaystyle O}{\overset{\displaystyle \parallel}{C}}H$ $CH_3\overset{\displaystyle O}{\overset{\displaystyle \parallel}{C}}H$ Formaldehyde Acetaldehyde
Ketones	$-\overset{\displaystyle O}{\overset{\displaystyle \parallel}{C}}-$	$R-\overset{\displaystyle O}{\overset{\displaystyle \parallel}{C}}-R$	$CH_3-\overset{\displaystyle O}{\overset{\displaystyle \parallel}{C}}-CH_3$ $CH_3-\overset{\displaystyle O}{\overset{\displaystyle \parallel}{C}}-C_6H_5$ Dimethyl ketone Methyl phenyl ketone
Acids	$-\overset{\displaystyle O}{\overset{\displaystyle \parallel}{C}}OH$	$R\overset{\displaystyle O}{\overset{\displaystyle \parallel}{C}}OH$	$CH_3\overset{\displaystyle O}{\overset{\displaystyle \parallel}{C}}OH$ $C_6H_5\overset{\displaystyle O}{\overset{\displaystyle \parallel}{C}}OH$ Acetic acid Benzoic acid

Table 30.4 (continued)

Derivative type	Functional group	General formula	Typical examples	
Esters	$\overset{\displaystyle O}{\overset{\|}{-C-OR}}$	$R\overset{\displaystyle O}{\overset{\|}{C}}-OR$	$CH_3\overset{\displaystyle O}{\overset{\|}{C}}OC_2H_5$ — Ethyl acetate	$C_3H_7\overset{\displaystyle O}{\overset{\|}{C}}OC_2H_5$ — Ethyl butyrate
Salts	$\overset{\displaystyle O}{\overset{\|}{-C-O^-M^+}}$	$R\overset{\displaystyle O}{\overset{\|}{C}}O^-M^+$	$CH_3\overset{\displaystyle O}{\overset{\|}{C}}O^-Na^+$ — Sodium acetate	$C_2H_5\overset{\displaystyle O}{\overset{\|}{C}}O^-K^+$ — Potassium pro-pionate
Amides	$\overset{\displaystyle O}{\overset{\|}{-C-NH_2}}$	$R\overset{\displaystyle O}{\overset{\|}{C}}NH_2$	$CH_3\overset{\displaystyle O}{\overset{\|}{C}}NH_2$ — Acetamide	$C_6H_5\overset{\displaystyle O}{\overset{\|}{C}}NH_2$ — Benzamide

The following sections describe the properties of some specific hydrocarbon derivatives. Their functional groups should be noted carefully.

30.9 Alcohols, R—OH

All **alcohols** have one or more —OH functional groups, yet they do not behave like bases such as sodium hydroxide and potassium hydroxide. They do not form hydroxide ions in water, nor do they have the other usual properties of bases.

Methanol, or **methyl alcohol, CH₃OH,** is produced commercially from either carbon monoxide or carbon dioxide.

$$CO + 2H_2 \xrightarrow{\text{Ag or Cu}} CH_3OH$$
$$CO_2 + 3H_2 \longrightarrow CH_3OH + H_2O$$

It is a colorless liquid, which boils at 65°C. It smells and tastes like ethyl alcohol. However, methanol is poisonous—breathing the vapor or drinking the liquid may cause blindness or death. Methanol is used in the manufacture of formaldehyde and other organic products; as a solvent for resins, gums, and shellac; and to denature ethyl alcohol (make it unsafe to drink).

Ethanol, C₂H₅OH, called **ethyl alcohol, grain alcohol,** or simply **alcohol,** is the most important of the alcohols. It has long been prepared by fermentation of starch, cellulose, and various sugars.

$$\underset{\text{Glucose}}{C_6H_{12}O_6} \xrightarrow[\text{yeast}]{\text{Enzymes in}} \underset{\text{Ethanol}}{2C_2H_5OH} + 2CO_2(g)$$

Large quantities of ethanol are synthesized from ethylene or acetylene. The synthesis from ethylene is summarized by the equation

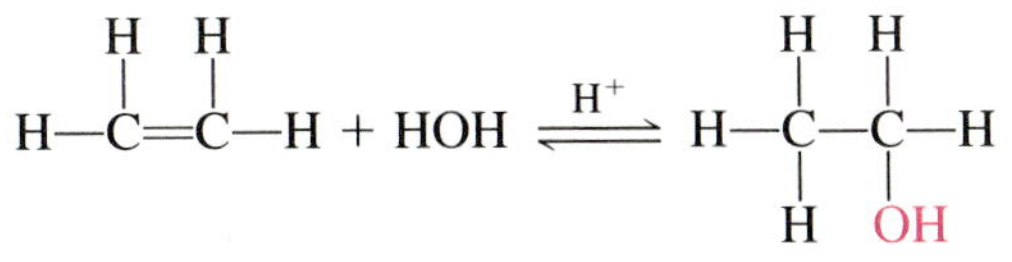

$$\underset{H}{\overset{H}{\underset{|}{C}}}=\underset{H}{\overset{H}{\underset{|}{C}}}-H + HOH \underset{}{\overset{H^+}{\rightleftharpoons}} H-\underset{H}{\overset{H}{\underset{|}{C}}}-\underset{OH}{\overset{H}{\underset{|}{C}}}-H$$

The conversion of sugar in grape juice to alcohol by the enzymes in certain yeasts produces wine.

Ethanol is a colorless liquid with a characteristic and somewhat pleasant odor. It is miscible with water in all proportions. The boiling point of the pure alcohol is 78.37°C. It forms a constant-boiling mixture with water that contains 95.57% alcohol by weight and boils at 78.15°C. Ethanol is the least toxic of all the alcohols and is found in all alcoholic beverages. It is used as a solvent in tinctures, essences, and extracts; in the preparation of iodoform, ether, dyes, perfumes, and collodion; as a solvent in the lacquer industry; and to some extent as a motor fuel. It is currently used as an additive in some brands of unleaded gasoline.

Alcohols containing two or more hydroxyl groups can be made. Important examples are 1,2-ethanediol (ethylene glycol), $C_2H_4(OH)_2$, and 1,2,3-propanetriol (glycerol), $C_3H_5(OH)_3$.

$$
\begin{array}{cc}
 & H \\
 & | \\
 & H-C-OH \\
H & | \\
| & H-C-OH \\
H-C-OH & | \\
| & H-C-OH \\
H-C-OH & | \\
| & H \\
H & \\
\text{Ethylene glycol} & \text{Glycerol}
\end{array}
$$

Ethylene glycol is used as a solvent and as an antifreeze in automobile radiators. Glycerol (also called glycerin) is produced, along with soap, when either fats or oils are heated with an alkali metal hydroxide.

An alcohol's name comes from the hydrocarbon from which it is derived. The hydrocarbon's final -*e* is replaced by -*ol*, and the carbon atom to which the —OH group is bonded is indicated by a number placed before the name.

30.10 Ethers, R—O—R

Ethers are compounds obtained from alcohols by the elimination of a molecule of water from two molecules of the alcohol. For example, when ethanol is treated with a limited amount of sulfuric acid and heated to 140°C, **diethyl ether** (ordinary ether) and water are formed.

$$
\begin{array}{c}
H \ H \\
| \ | \\
H-C-C-O-[H+HO]-C-C-H \xrightarrow{H_2SO_4} H-C-C-O-C-C-H + HOH \\
| \ | \\
H \ H
\end{array}
$$

Diethyl ether

In the general formula for ethers, R—O—R, the hydrocarbon groups (R) may be the same or different. Diethyl ether is the most important compound of this class. It is a colorless volatile liquid (boiling point = 35°C), which is highly flammable. It has been used since 1846 as an anesthetic, though other better anesthetics have largely taken its place. Diethyl ether and other ethers are valuable solvents for gums, fats, waxes, and resins.

30.11 Aldehydes, R—$\overset{\overset{\textstyle O}{\|}}{C}$—H

Alcohols represent the first stage of oxidation of hydrocarbons. Further oxidation produces **aldehydes,** compounds containing the group —CHO. Aldehydes are generally made by the oxidation of primary alcohols. When a mixture of methanol and air is passed through a heated tube containing either silver or a mixture of iron powder and molybdenum oxide, the simplest aldehyde, called **formaldehyde, HCHO,** is formed.

$$2H-\overset{\displaystyle H}{\underset{\displaystyle H}{\overset{|}{\underset{|}{C}}}}-OH + O_2 \xrightarrow{\text{Catalyst}} 2H-C\overset{\displaystyle H}{\underset{\displaystyle O}{\diagdown}} + 2H_2O$$

Methanol   Formaldehyde

Formaldehyde is a colorless gas with a pungent and irritating odor; it is soluble in water in all proportions. Formaldehyde is sold in an aqueous solution called **formalin,** which contains about 37% formaldehyde by weight. The solution also contains 7% methanol, added to inhibit the formation of an insoluble polymer (a compound of high molecular weight) that occurs when formaldehyde molecules react with one another. In fact, one of the early polymers, Bakelite, was a formaldehyde polymer that made formaldehyde industrially important. Formaldehyde causes coagulation of proteins, so it is used to preserve tissue specimens and to embalm. It is also useful as a disinfectant and as a reducing agent in the production of silvered mirrors.

Acetaldehyde, CH$_3$CHO, boils at 20.2°C. It is colorless, water-soluble, and smells like freshly cut green apples. It is used in the manufacture of aniline dyes, synthetic rubber, and other organic materials.

The systematic names of aldehydes are formed by replacing the *-e* ending of the parent alkane by *-al*. Acetaldehyde is the common name for ethanal. The chain is numbered with the —CHO group as the first carbon. Thus 3-methyl-butanal is

$$\overset{④}{CH_3}-\overset{③}{\underset{\displaystyle CH_3}{\overset{|}{CH}}}-\overset{②}{CH_2}-\overset{①}{C}\overset{\displaystyle O}{\underset{\displaystyle H}{\diagup}}$$

30.12 Ketones, $R-\overset{\displaystyle O}{\overset{\|}{C}}-R$

Ketones, like aldehydes, contain the **carbonyl group,** $\diagdown C=O.$ In fact, a ketone may be regarded as an aldehyde in which the hydrogen in the aldehyde group is replaced by a hydrocarbon group. In the general structural formula for ketones (see section heading), the R groups may be the same or different. Ketones are often prepared by the oxidation of secondary alcohols.

Dimethyl ketone, CH$_3$COCH$_3$, commonly called **acetone,** is the simplest and most important ketone. It is made commercially by fermenting corn or molasses or by dehydrogenation of 2-propanol (isopropyl alcohol).

$$CH_3\overset{\displaystyle OH}{\overset{|}{C}}HCH_3 \xrightarrow[\text{Cu catalyst}]{350°C} CH_3\overset{\displaystyle O}{\overset{\|}{C}}CH_3 + H_2$$

2-Propanol   Dimethyl ketone

It is also a product of the destructive distillation of wood. Acetone is a colorless liquid, boiling at 56.5°C and possessing a characteristic pungent odor and a sweet taste. Among the many uses of acetone are as a solvent for cellulose acetate, cellulose nitrate, acetylene, plastics, and varnishes; as a paint, varnish, and fingernail polish remover; and as a solvent in the manufacture of drugs, chemicals, smokeless powder, and the explosive cordite.

30.13 Carboxylic Acids, R—C—OH

Carboxylic acids contain the **carboxyl group, —CO₂H.** They are usually weak acids, yet they readily form metal salts. Carboxylic acids may be prepared by the oxidation of primary alcohols or of aldehydes.

$$RCH_2OH \xrightarrow{\text{Oxidation}} R-C\begin{smallmatrix}H\\ \\O\end{smallmatrix} \xrightarrow{\text{Oxidation}} R-C\begin{smallmatrix}OH\\ \\O\end{smallmatrix}$$

Primary alcohol   Aldehyde   Carboxylic acid

The simplest carboxylic acid is **formic acid, HCO₂H,** known since 1670. Its name comes from the Latin word *formicus,* for ant; it was first isolated by the distillation of red ants. It is partially responsible for the irritation of ant bites and bee stings.

Acetic acid, CH₃CO₂H, constitutes 3–6% of vinegar. Cider vinegar is produced by allowing apple juice to ferment, which changes the sugar present to ethanol. Then bacteria produce an enzyme (Chapter 31) that catalyzes the air oxidation of the alcohol to acetic acid. Pure, anhydrous acetic acid, sometimes called **glacial acetic acid,** boils at 118.1°C and freezes at 16.6°C. Pure acetic acid has a penetrating odor and produces painful burns. It is an excellent solvent for many organic and some inorganic compounds. Acetic acid is essential in the production of cellulose acetate, a component of many synthetic fibers such as rayon.

Oxalic acid is a dicarboxylic acid with the formula (CO₂H)₂, or H₂C₂O₄. Its molecule consists of two carboxyl groups (dicarboxylic) bonded together.

Oxalic acid

Oxalic acid is a colorless crystalline solid, found as a metal hydrogen salt in sorrel, rhubarb, and other plants, to which it imparts a sour taste. In concentrated form it is poisonous.

Benzoic acid, C₆H₅CO₂H, is a colorless crystalline solid with a cyclic structure. It is a monocarboxylic acid that occurs in cranberries and coal tar. The sodium salt of benzoic acid, **sodium benzoate,** is used to preserve foods such as tomato ketchup and fruit juices.

Benzoic acid

Chemists carry out sniff tests on perfumes produced by reactions of organic compounds.

30.14 Esters, R—C—OR

Esters are produced by the reaction of acids with alcohols. For example, the ester **ethyl acetate, CH₃CO₂C₂H₅,** is formed when acetic acid reacts with ethanol.

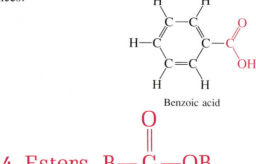

The distinctive and attractive odors and flavors of many flowers, perfumes, and ripe fruits are due to the presence of one or more esters. Among the most important of the natural esters are fats (such as lard, tallow, and butter) and oils (such as linseed, cottonseed, and olive oils), which are esters of the trihydroxyl alcohol (triol) glycerol, $C_3H_5(OH)_3$, with large acids such as palmitic acid, stearic acid, and oleic acid.

Polymers

30.15 Introduction

Polymers and **plastics,** compounds of very high molecular weights, are built up of a large number of simple molecules, or **monomers,** which have reacted with one another. Cellulose, starch, proteins, and rubber are natural polymers (although most rubber in use is a synthetic polymer). Nylon, rayon, polyethylene, and Dacron are synthetic polymers. Polymers are used in clothing, fibers, insulation, and construction materials; and recently, in cars to save weight and thus save gasoline. However, polymers are themselves made from petroleum, so part of the advantage of the lower gasoline use is offset.

30.16 Rubber

Natural rubber comes mainly from latex, the sap of the rubber tree. Rubber consists of very long molecules, which are polymers formed by the union of isoprene units, C_5H_8.

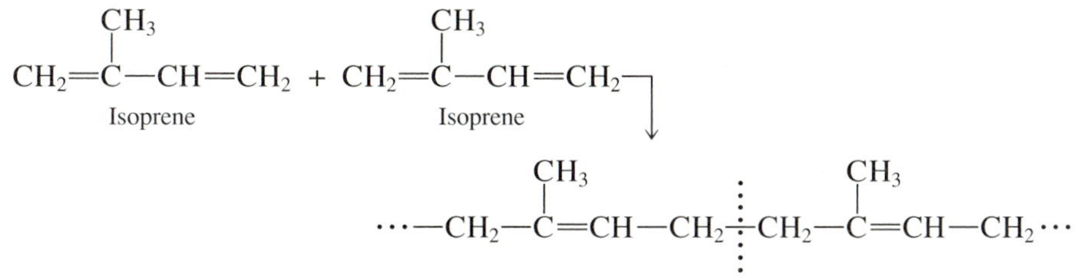

The number of isoprene units in a rubber molecule is about 2000, giving it a molecular weight of approximately 136,000.

Rubber has the undesirable property of becoming sticky when warmed, but this can be eliminated by **vulcanization.** Rubber is vulcanized by heating it with sulfur to about 140°C. During the process sulfur atoms add at the double bonds in the linear polymer and form bridges that bind one rubber molecule to another. In this way, a linear polymer is converted into a three-dimensional polymer. During vulcanization, fillers are added to increase the wearing qualities of the rubber and to color it. Among the substances used as fillers are carbon black (black), zinc oxide (white), antimony(V) sulfide (orange), and titanium dioxide (white).

Synthetic rubbers resemble natural rubber and are often superior to it in certain respects. For example, neoprene is a synthetic elastomer with rubberlike properties.

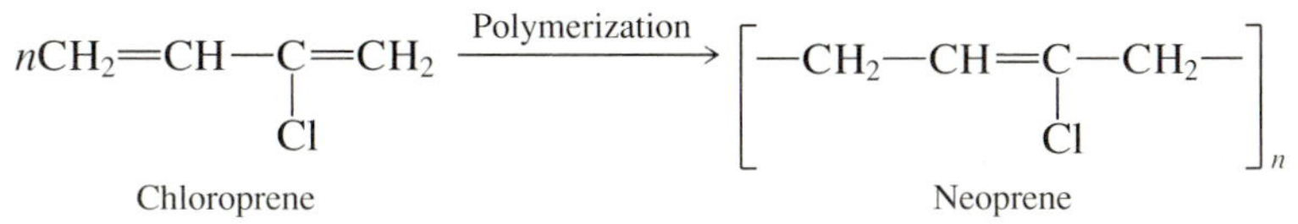

A strand of raw nylon may be drawn from the film of nylon that forms at the interface between solutions of adipoyl chloride and hexamethylenediamine.

The monomer chloroprene is similar to isoprene, except that a chlorine atom replaces the methyl group. Neoprene is more elastic than natural rubber, resists abrasion well, and is less affected by oil and gasoline. It is used for making gasoline and oil hoses, automobile and refrigerator parts, and electrical insulation. There are many other synthetic elastomers of this type.

Both natural rubber and neoprene are examples of **addition polymers,** polymers that form by addition reactions (Section 30.3).

30.17 Synthetic Fibers

The synthetic fiber nylon is a **condensation polymer.** Nylon is made from hexamethylenediamine, which contains an amine group ($-NH_2$) at both ends, and adipic acid, which contains a carboxylic acid group ($-CO_2H$) at both ends. During condensation, linkages of the type $R-NH-CO-R$ are formed and water is eliminated. The part shown in color is an **amide linkage.**

$$NH_2-(CH_2)_6-\overset{H}{N}+H + HO+\overset{O}{\overset{\|}{C}}-(CH_2)_4-\overset{O}{\overset{\|}{C}}OH \longrightarrow$$

Hexamethylenediamine   Adipic acid

$$NH_2-(CH_2)_6-NH-\overset{O}{\overset{\|}{C}}-(CH_2)_4-\overset{O}{\overset{\|}{C}}OH + H_2O$$

Because the molecule resulting from the condensation has an $-NH_2$ group at one end and a $-CO_2H$ group at the other, the condensation process can be repeated many times to form a linear polymer of great length.

$$(n + 1)H_2N(CH_2)_6NH_2 + (n + 1)HO_2C(CH_2)_4CO_2H \longrightarrow$$

Hexamethylenediamine   Adipic acid

$$H_2N(CH_2)_6NH-\left[\overset{O}{\overset{\|}{C}}(CH_2)_4\overset{O}{\overset{\|}{C}}NH(CH_2)_6NH\right]-\overset{O}{\overset{\|}{C}}(CH_2)_4CO_2H + 2nH_2O$$

Nylon

Fine threads are produced by extruding melted nylon through a spinneret. Nylon is used in hosiery and other clothing, bristles for toothbrushes, surgical sutures, strings for tennis rackets, fishing line leaders, and many other products.

Dacron is made from ethylene glycol and terephthalic acid by a condensation process that forms an ester linkage. Dacron is one of the family of polyesters.

$$(n + 1)HO\overset{H\ H}{\underset{H\ H}{C-C}}OH + (n + 1)HO_2C-C\underset{\underset{H\ H}{C=C}}{\overset{\overset{H\ H}{C-C}}{}}C-CO_2H \longrightarrow$$

Ethylene glycol   Terephthalic acid

$$HO\overset{H\ H}{\underset{H\ H}{C-C}}O\left[\overset{O}{\overset{\|}{C}}-C\underset{\underset{H\ H}{C=C}}{\overset{\overset{H\ H}{C-C}}{}}C-\overset{O}{\overset{\|}{C}}-O-\overset{H\ H}{\underset{H\ H}{C-C}}-O\right]\overset{O}{\overset{\|}{C}}-C\underset{\underset{H\ H}{C=C}}{\overset{\overset{H\ H}{C-C}}{}}C-CO_2H + 2nH_2O$$

Dacron

30.18 Polyethylene

Polyethylene is widely used. Polyethylene results from the polymerization of ethylene (Section 30.3). It is a flexible, tough polymer that is very water-resistant and has excellent insulating properties. It is used in plastic bottles, bags for fruits and vegetables, and many other items.

If the ethylene molecules contain substituents, the polymer will also contain the substituents.

$$n\text{CH}_2\!=\!\text{CHR} \xrightarrow{\text{Catalysts}} \left[\!\!\begin{array}{c} \text{CH}_2-\!\!\underset{\underset{\text{R}}{|}}{\text{CH}} \end{array}\!\!\right]_n$$

Polypropylene (R = CH$_3$) is the polymer of propylene, CH$_3$CH=CH$_2$. **Polyvinylchloride,** or **PVC** (R = Cl), is the polymer of vinyl chloride, CH$_2$=CHCl. **Teflon** results from the polymerization of tetrafluoroethylene, CF$_2$=CF$_2$; and **polystyrene** (R = C$_6$H$_5$) is the polymer of styrene, C$_6$H$_5$CH=CH$_2$.

Plastic bags and strips of polymer films such as Saran Wrap are formed from large continuous tubes blown from the molten polymer.

For Review

Summary

With the valence shell electron configuration $2s^2 2p^2$, **carbon** is the first member of Group IVA of the Periodic Table. Carbon generally completes its valence shell by sharing electrons in covalent bonds. Since the valence shell of a carbon atom can contain a maximum of eight electrons, hybridization is limited to *sp*, *sp*^{2}, and *sp*^{3}. Strong, stable bonds between carbon atoms produce complex molecules containing chains and rings. The chemistry of these compounds is called **organic chemistry.**

Hydrocarbons are organic compounds composed only of carbon and hydrogen. The **alkanes** are **saturated hydrocarbons;** that is, hydrocarbons that contain only single bonds. **Alkenes** are hydrocarbons that contain one or more carbon-carbon double bonds. **Alkynes** contain one or more carbon-carbon triple bonds. **Aromatic hydrocarbons** contain ring structures with delocalized π-electron systems.

Most hydrocarbons have **structural isomers,** compounds with the same chemical formula but different arrangements of atoms. For example, *n*-butane and isobutane are structural isomers. Many aromatic hydrocarbons also have structural isomers; *o*-xylene, *m*-xylene, and *p*-xylene are examples. In addition, molecules of the alkenes and alkynes exhibit isomerism based on the position of the multiple bond in the molecule. Most alkenes also exhibit *cis-trans* **isomerism,** which is the result of the lack of rotation about a carbon-carbon double bond. Certain organic compounds form **optical isomers,** which contain one or more **asymmetric carbons,** carbon atoms bonded to four different groups.

Alkanes undergo **substitution reactions** in which one of the hydrogen atoms of the alkane is replaced by another type of atom. Alkenes and alkynes undergo **addition reactions** in which a reactant adds to the molecule and converts the carbon-carbon π bond into σ bonds. All hydrocarbons react with oxygen and burn.

Organic compounds that are not hydrocarbons can be considered to be derivatives of hydrocarbons. A hydrocarbon derivative can be formed by replacing one or more hydrogen atoms of a hydrocarbon by a **functional group,** which contains at least one atom of an element other than carbon or hydrogen. The properties of hydrocarbon derivatives are determined largely by the functional group. The —OH group is the functional group of an **alcohol.** The —O— group is the functional group of an **ether.** Other functional groups include the —CHO group of an **aldehyde,** the —CO— group of a **ketone,** and the —CO$_2$H group of a **carboxylic acid.**

Polymers, compounds of very high molecular weights, form when a large number of smaller molecules **(monomers)** react with one another. **Addition polymers** such as rubber and polyethylene form by addition reactions, while **condensation polymers** such as nylon and the polyesters form by condensation reactions.

Key Terms and Concepts

addition polymer (30.16)
addition reaction (30.3)
alcohols (30.9)
aldehydes (30.11)
alkanes (30.1)
alkenes (30.3)
alkyl group (30.2)
alkynes (30.4)
aromatic hydrocarbons (30.5)
asymmetric carbon (30.6)
carbonyl group (30.12)

carboxyl group (30.13)
carboxylic acid (30.13)
condensation polymer (30.17)
cycloalkanes (30.1)
derivatives (30.8–30.14)
esters (30.14)
ethers (30.10)
functional group (30.1)
geometrical isomers (30.3)
ketones (30.12)
monomers (30.15)

optical isomers (30.6)
optically active molecules (30.6)
polymers (30.15)
primary carbon (30.2)
secondary carbon (30.2)
structural isomers (30.1)
substituents (30.12)
substitution reaction (30.1)
tertiary carbon (30.2)

Exercises

Hydrocarbons

1. Write the chemical formula and Lewis structure of an alkane, an alkene, an alkyne, and an aromatic hydrocarbon, each of which contains six carbon atoms.

2. Write a Lewis structure for each of the following molecules:
   (a) propane
   (b) propanethiol
   (c) isopropyl alcohol
   (d) methyl ethyl ether
   (e) *trans*-4-chloro-2-octene
   (f) 2,3,4-trimethylpentene
   (g) *o*-dichlorobenzene
   (h) methylcyclohexane
   (i) 3-ethyl-1-heptyne

3. What is the difference between the electronic structures of saturated and unsaturated hydrocarbons?

4. Give a complete name for each of the following compounds:
   (a) $(CH_3)_3CH$
   (b) $CH_3CH_2CHBrCHICH_3$
   (c) $CH_3CH_2CHCH_3$
       CH_3CHCH_3
   (d) $CH_3CH_2CH_2CH{=}CHCH_3$
   (e) $CH_3CH_2CHClCHCH_2CH_2CH_3$
       $CH{=}CH_2$
   (f)
       CH_3 H
          $C{=}C$
       H C_2H_5
   (g) H_2C $CH_2{-}CH_2$ $CHCH_2CH(CH_3)CH_3$
              $CH_2{-}CH_2$

(h) $(CH_3)_2CHCH_2C{\equiv}CH$

5. Draw a three-dimensional structure for each of the following, using solid and striped wedge-shaped bonds where appropriate:
   (a) CH_4
   (b) C_2H_6
   (c) *n*-C_4H_{10}
   (d) C_3H_6
   (e) C_3H_4

6. Write Lewis structures for all of the isomers of the alkene C_5H_{10}.

7. Write Lewis structures for all of the isomers of the alkyne C_5H_8.

8. Write Lewis structures for all of the isomers of C_4H_9Br. Do any of these isomers exist as optical isomers?

9. Write the Lewis structures and the names of all isomers of the alkyl groups —C_3H_7 and —C_4H_9.

10. Write Lewis structures for all of the isomers of the alkyl group —C_5H_{11}.

11. Indicate which of the following molecules can form optical isomers:
    (a) *trans*-2-pentene
    (b) dichlorodifluoromethane
    (c) 3-ethyloctane
    (d) 4-ethyloctane
    (e) 2-butanethiol
    (f) 1-cyanoethanol, [$CH_3CH(CN)OH$]

12. Draw three-dimensional structures, using solid and striped wedge-shaped bonds, for the optical isomers of phenylalanine, $H_2NCH(R)CO_2H$, where R is $C_6H_5CH_2$.

13. Does the molecule, shown at the top of the next page, exist as *cis-trans* isomers? Optical isomers?

$$\begin{array}{c}
\text{CH}_2 \\
\text{H}_2\text{C} \qquad \text{CHCl} \\
\text{H}_2\text{C} \qquad \text{CHBr} \\
\text{CH}_2
\end{array}$$

14. (a) How much heat will be released by the combustion of exactly 1 g of C_2H_6? Of C_2H_2?

    Ans. 47.483 kJ g^{-1}; 48.218 kJ g^{-1}

    (b) How much heat will be released by the combustion of 1 mol of each of the compounds in part (a)?

    Ans. 1427.80 kJ mol^{-1}; 1255.5 kJ mol^{-1}

Hydrocarbon Derivatives

15. Identify each of the following hydrocarbon derivatives, in which R is an alkyl group:

    (a) ROH   (b) RNH_2
    (c) RCO_2H (d) ROR
    (e) RCHO   (f) RC(O)R

16. Write the Lewis structure, indicating resonance structures where appropriate, for each of the following molecules or ions:

    (a) methanol   (b) ethanol
    (c) acetone   (d) formaldehyde
    (e) bromobenzene   (f) acetate ion

17. Draw a three-dimensional structure, using solid and striped wedge-shaped bonds where appropriate, for each of the following molecules:

    (a) CH_3OCH_3 (b) CH_3CO_2H
    (c) C_2H_5CHO (d) CH_3OH
    (e) $C_2H_5NH_2$

18. Identify the functional groups present in the molecule carvone (Fig. 30.9).

19. What are the products formed by the stepwise oxidation of methanol? Of ethanol?

20. Write a complete balanced equation for each of the following reactions:

    (a) Butene is burned in air.
    (b) Propanol is oxidized to an aldehyde by oxygen.
    (c) Acetylene (1 mol) + bromine (1 mol) $\longrightarrow$
    (d) 2-butene + water $\longrightarrow$
    (e) Benzoic acid reacts with propanol.

21. What alcohol can be oxidized to give the acid $(CH_3)_2CHCH_2CO_2H$? Which aldehyde?

Additional Exercises

22. Why is CH_3CH_2OH soluble in water, whereas its isomer CH_3OCH_3 is not?

23. Write a complete balanced equation for each of the following reactions:

    (a) $CH_3Li + H_2O \longrightarrow$
    (b) $(CH_3)_2NH + Li \longrightarrow$
    (c) $(CH_3)_2CHCH_2OH + Na \longrightarrow$
    (d) $(CH_3)_3CCH_2SH + NaOH \longrightarrow$
    (e) $CH_3OLi + CS_2 \longrightarrow$

24. Indicate the types of hybridized orbitals used and the geometry about each carbon atom in each of the following molecules:

    (a) CH_3CONH_2 (b) C_2H_6
    (c) C_2H_4 (d) $H_2C{=}O$
    (e) $CH_3C{\equiv}CH$

25. If all of the 30.55 billion lb of ethylene produced in the United States during 1985 was produced by the pyrolysis of ethane

    $$C_2H_6 \longrightarrow C_2H_4 + H_2$$

    what mass of hydrogen, in pounds, would have been produced?   *Ans. 2.195 billion lb*

26. (a) What volume of $C_2H_2(g)$ at STP would result from the hydrolysis of 15.5 g of CaC_2?   *Ans. 5.42 L*

    (b) How many grams of magnesium would have to react with HCl to provide enough hydrogen to reduce the C_2H_2 of part (a) to ethane?   *Ans. 11.8 g*

27. (a) From the reactions

    $$C_2H_4(g) + 2H_2(g) \longrightarrow 2CH_4(g)$$
    $$C(g) + 2H_2(g) \longrightarrow CH_4(g)$$

    and by considering the bonds that must be broken and formed to get from reactants to products, show that the C=C bond energy, $D_{C{=}C}$, is given by

    $$D_{C{=}C} = \Delta H^\circ_{f_{C(g)}} + \Delta H^\circ_{f_{CH_4(g)}} - \Delta H^\circ_{f_{C_2H_4(g)}}$$

    Calculate, using data in Appendix I, a value for $D_{C{=}C}$.   *Ans. 590 kJ*

    (b) By considerations similar to those in part (a), show that the reaction

    $$C_2H_4(g) + H_2(g) \longrightarrow C_2H_6(g)$$

    leads to the conclusion that

    $$D_{C{=}C} = D_{C{-}C} + 2D_{C{-}H}$$
    $$- D_{H{-}H} + \Delta H^\circ_{f_{C_2H_6(g)}} - \Delta H^\circ_{f_{C_2H_4(g)}}$$

    and calculate, using data in Table 6.3 and Appendix I, a value for $D_{C{=}C}$.   *Ans. 602 kJ*

28. How much acetic acid, in grams, is in exactly 1 qt of vinegar if the vinegar contains 3.00% acetic acid by volume? The density of acetic acid is 1.049 g/mL. (See Appendix B for conversion factors.)   *Ans. 29.8 g*

29. On the basis of hybridization of orbitals of the carbon atom, explain why cyclohexane exists in the form of a "puckered" (bent) ring and benzene in the form of a planar ring.

30. Assuming a value of (n) equal to six in the formula for nylon, calculate how many pounds of hexamethylenediamine would be needed to produce 1.00 ton of nylon. (See Appendix B for conversion factors.)

    Ans. 1.02 × 10^3 lb

Biochemistry

Narrow beech fern

Biochemistry is the study of the chemical composition and structure of living organisms and the reactions that take place within them. Considering the diverse and complex functions of a cell, it is not surprising to find that its molecules are among the most complicated substances known.

Biomolecules are often divided into four major classes—proteins, carbohydrates, nucleic acids, and lipids—according to their structures or functions. Proteins, carbohydrates, and nucleic acids are found in the form of macromolecules with molecular weights in the range 10^4–10^6, or larger, which form by the polymerization of monomers of relatively low molecular weights. Molecules of the lipids are considerably smaller than those of the other three classes. However, they can associate to form clusters of molecules of enormous size.

31.1 The Cell

All cells are basically similar. They contain many of the same structures, the same or similar systems of enzymes, and the same kind of genetic material. The principal constituent of living matter is water. Most plants and animals are 60–90% water, most of which is the solvent in which both organic and inorganic substances are dissolved and transported in the cell or through the organism.

An understanding of the role played by water in biological systems requires that a distinction be made between substances that are **hydrophilic** (literally water-loving) and those that are **hydrophobic** (water-hating). Water is a polar solvent that readily dissolves either polar or ionic substances. In general, the less polar a substance is, the more hydrophobic it is. The nonpolar hydrocarbons are excellent examples of hydrophobic compounds; they are almost totally immiscible with water. Some biomolecules contain polar or ionic substituents that impart hydrophilic character, and they are therefore water-soluble. Those biomolecules that do not contain polar or ionic substituents are hydrophobic; they are effectively insoluble in water and tend to escape from an aqueous environment.

Cells are commonly divided into two classes: **eukaryotic cells** such as that represented in Fig. 31.1 contain a nucleus and include most organisms. **Prokaryotic cells** do not contain a nucleus and are restricted to certain microorganisms such as blue-green algae and some bacteria. The nucleus contains the chromosomes, which are a mixture of two substances: deoxyribonucleic acid (DNA) and protein. The nucleus stores the genetic information in DNA that enables the cell to replicate itself and to synthesize the various proteins that it needs for biological activity.

All cells are surrounded by a cell membrane, which is the ''skin'' or envelope containing the cell contents. The cell membrane insulates the contents of the cell from surrounding fluids.

The **cytoplasm** is the entire contents of the cell, except the nucleus. It can be divided into the **cytosol** (the water-soluble contents of the cell) and a variety of small organs, referred to as **cellular organelles,** such as the ribosomes, mitochondria, and chloroplasts.

Ribosomes, the site of the directed polymerization (genetic coding) of amino acids to form enzymes and other proteins, are small spherical bodies that are present in all cells. **Mitochondria,** a cell's power plants, are present in eukaryotic cells. A variety of biomolecules (including carbohydrates, amino acids, and fatty acids) are oxidized to carbon dioxide and water in the mitochondria. Some of the energy released during this

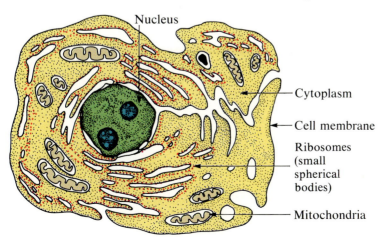

Nucleus

Cytoplasm

Cell membrane

Ribosomes
(small
spherical
bodies)

Mitochondria

Figure 31.1

Simplified drawing of a eukaryotic cell.

oxidation is stored as chemical energy through the synthesis of compounds called **nucleoside triphosphates,** for example, adenosine triphosphate (ATP) and guanosine triphosphate (GTP). **Chloroplasts** are present in all plants and algae that are capable of converting light energy into chemical energy in photosynthesis. Light-induced synthesis of carbohydrates from carbon dioxide and water takes place in the chloroplasts.

Proteins

31.2 Amino Acids

Proteins are substances that are composed of more than 40 small compounds called **amino acids.** When proteins are heated with aqueous acid or base or are treated with certain hydrolytic enzymes, they release these amino acids.

Figure 31.2

The 20 α-amino acids commonly found as constituents of proteins.

Nonpolar R Group

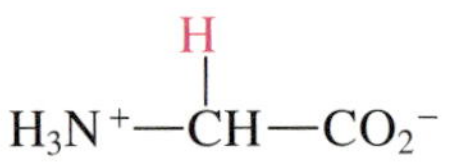

Glycine (GLY)

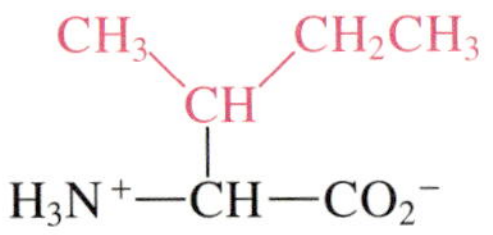

Isoleucine (ILE)

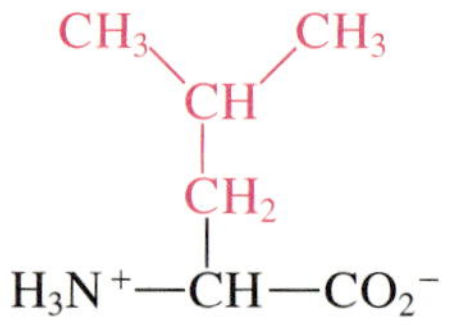

Leucine (LEU)

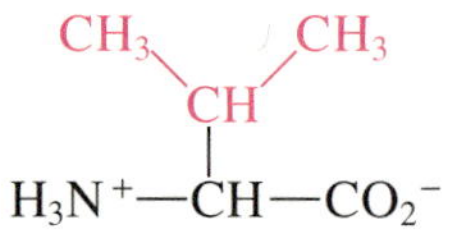

Valine (VAL)

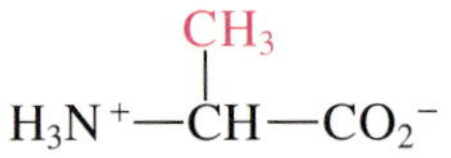

Alanine (ALA)

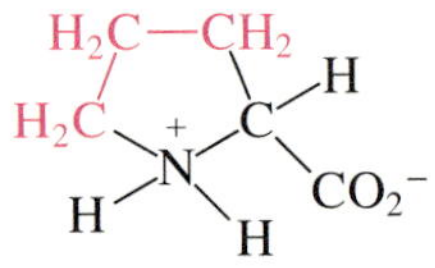

Proline (PRO)

Aromatic R Group

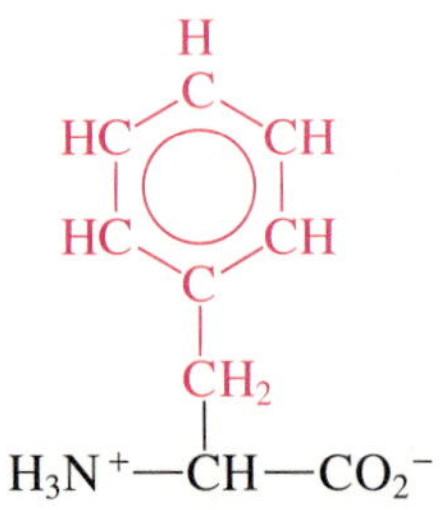

Phenylalanine (PHE)

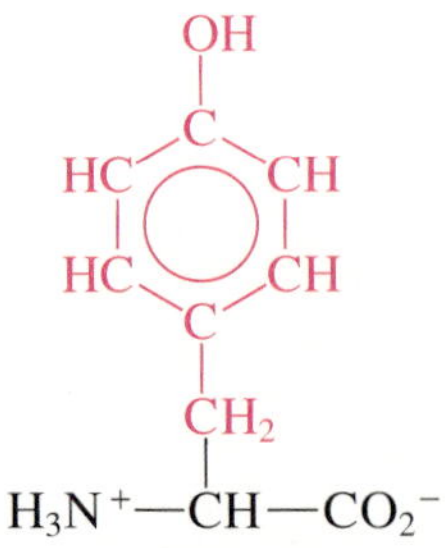

Tyrosine (TYR)

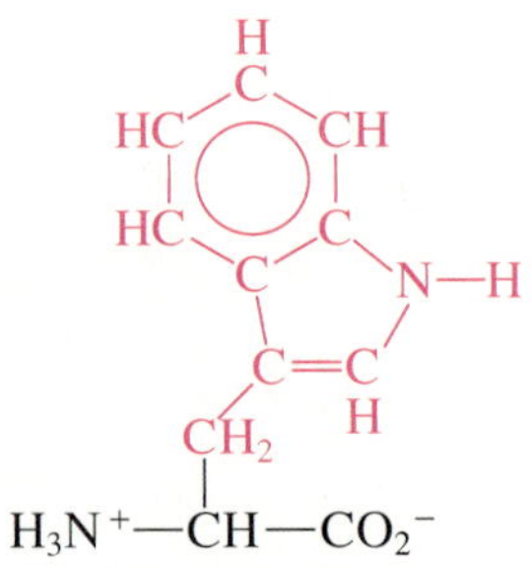

Tryptophan (TRP)

All amino acids contain both an amine group, $-NH_2$, and a carboxylic acid group, $-CO_2H$ (Section 30.13). At least 150 naturally occurring amino acids are known; the most important, however, are the 20 amino acids used in the synthesis of proteins. The amino acids are all **α-amino acids,** which means that the amino group is attached to the carbon atom adjacent, or α, to the carbon of the carboxylic acid group. The primary α-amino acids can be represented by the following structure:

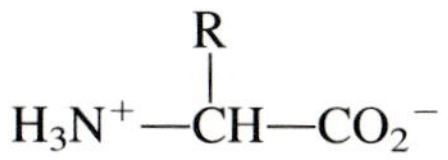

$$\underset{\displaystyle \text{H}_3\text{N}^+-\text{CH}-\text{CO}_2{}^-}{\overset{\displaystyle \text{R}}{|}}$$

The structures of these amino acids differ only in the nature of the substituent (R). Thus they are usually grouped on the basis of similarities in the structure of this side chain. The names and structures of the 20 most important α-amino acids are given in Fig. 31.2.

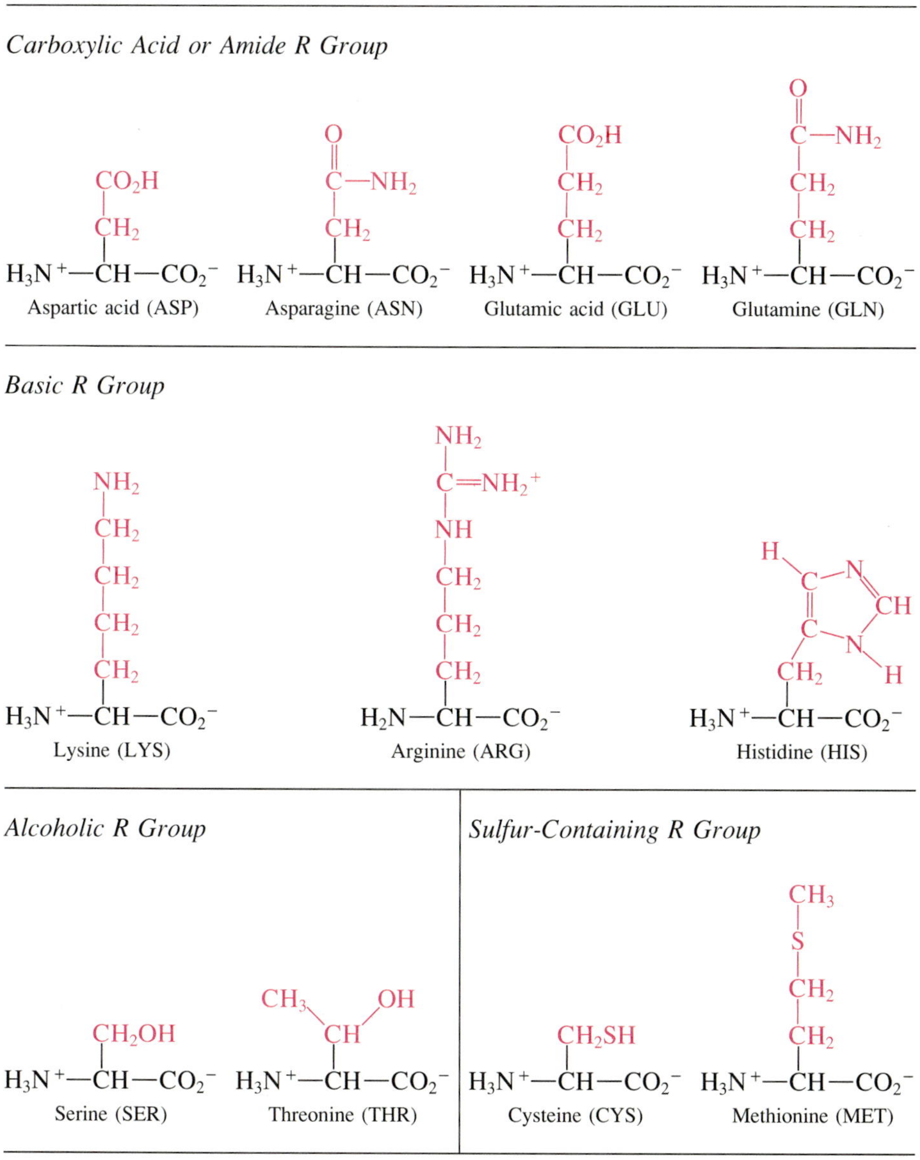

Carboxylic Acid or Amide R Group

Aspartic acid (ASP)   Asparagine (ASN)   Glutamic acid (GLU)   Glutamine (GLN)

Basic R Group

Lysine (LYS)   Arginine (ARG)   Histidine (HIS)

Alcoholic R Group

Serine (SER)   Threonine (THR)

Sulfur-Containing R Group

Cysteine (CYS)   Methionine (MET)

The formulas of the amino acids are often written in the form

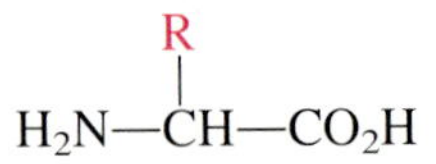

$$\underset{\displaystyle H_2N-CH-CO_2H}{\overset{\displaystyle R}{|}}$$

However, it is important to recognize that the basic H_2N- end of this molecule acts as a proton acceptor, and the $-CO_2H$ end functions as a proton donor. The proton transfer forms a species called a **zwitterion** (from German, meaning hybrid ion or dipolar ion). Zwitterions are molecules that contain ionic charges but that are electrically neutral overall. For example, the amino acid alanine forms a zwitterion.

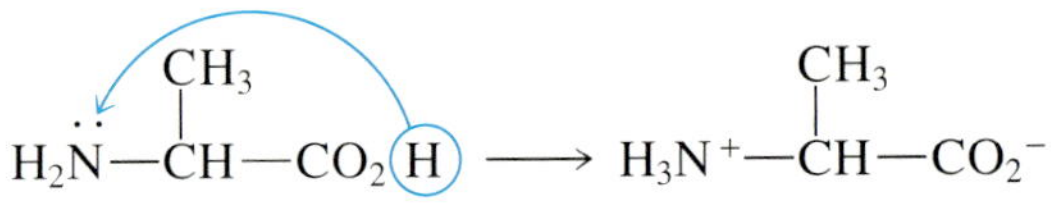

$$\overset{\displaystyle CH_3}{\underset{\displaystyle H_2N-CH-CO_2(H)}{|}} \longrightarrow \overset{\displaystyle CH_3}{\underset{\displaystyle H_3N^+-CH-CO_2^-}{|}}$$

Zwitterion

A zwitterion is the dominant form of an amino acid in aqueous solution at or near neutral pH (pH = 7). In an acidic solution, however, the amino acid becomes positively charged as the carboxylate ion is protonated. Conversely, in a basic solution, the amino acid becomes negatively charged as the amino group is deprotonated.

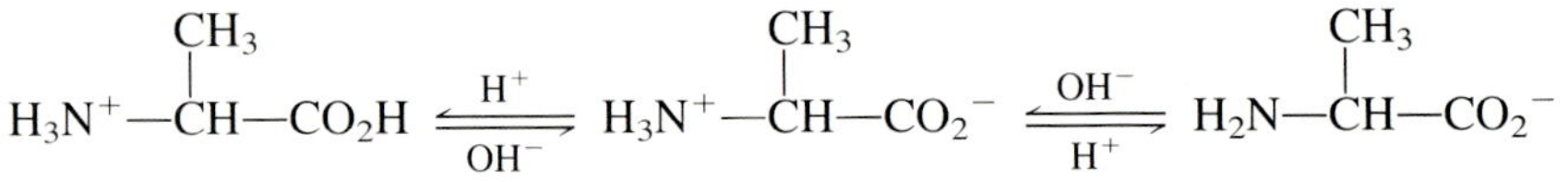

$$\overset{\displaystyle CH_3}{\underset{\displaystyle H_3N^+-CH-CO_2H}{|}} \underset{OH^-}{\overset{H^+}{\rightleftharpoons}} \overset{\displaystyle CH_3}{\underset{\displaystyle H_3N^+-CH-CO_2^-}{|}} \underset{H^+}{\overset{OH^-}{\rightleftharpoons}} \overset{\displaystyle CH_3}{\underset{\displaystyle H_2N-CH-CO_2^-}{|}}$$

     Acidic   Neutral   Basic
     solution   solution   solution

Cysteine and methionine, as shown in Fig. 31.2, both have a sulfur-containing side chain. Cysteine often links to another cysteine through formation of a disulfide bond ($-S-S-$), which is a covalent bond between the side chains, to form the amino acid cystine.

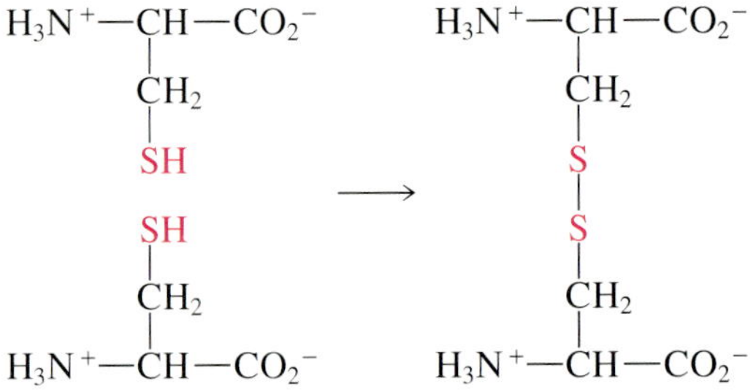

$$\begin{array}{ccc} H_3N^+-CH-CO_2^- & & H_3N^+-CH-CO_2^- \\ | & & | \\ CH_2 & & CH_2 \\ | & & | \\ SH & & S \\ | & \longrightarrow & | \\ SH & & S \\ | & & | \\ CH_2 & & CH_2 \\ | & & | \\ H_3N^+-CH-CO_2^- & & H_3N^+-CH-CO_2^- \end{array}$$

    Two cysteine   Cystine
    molecules

With the exception of glycine the α-amino acids are optically active (Section 30.6). Two optical isomers, which differ in their interaction with plane-polarized light, are possible for every optically active substance. The dextrorotatory (+) isomer rotates the plane of plane-polarized light in a clockwise direction, whereas the levorotatory (−) isomer rotates the plane of such light in a counterclockwise direction (Section 30.6).

Unfortunately, there is no direct relationship between the molecular structure of an isomer and its classification as either (+) or (−). Optical isomers whose molecular structures are related to the structure of the (+) isomer of glyceraldehyde, $HOCH_2CH(OH)CHO$, are labeled D isomers, and their mirror images, which are re-

lated structurally to the (−) isomer of glyceraldehyde, are labeled L isomers. Every amino acid isolated from a naturally occurring protein is an L amino acid. Looking down the H—C bond of an L amino acid, we see that the —CO_2^-, —R, and —NH_3^+ substituents are arranged in a clockwise fashion. In a D amino acid these substituents are arranged in a counterclockwise manner (Fig. 31.3). Though occasionally found in nature, D amino acids are not genetically coded into proteins.

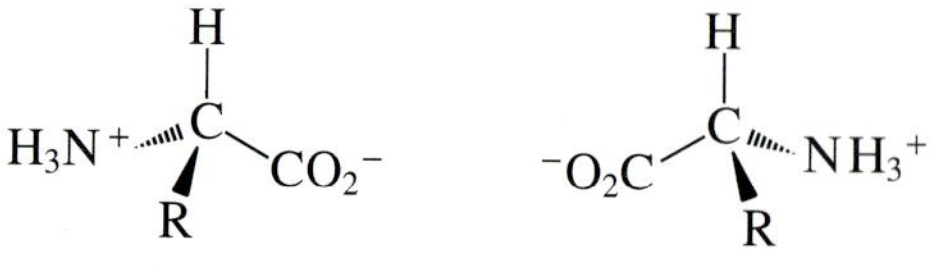

L isomer   D isomer

Figure 31.3
The molecular structures of the D and L isomers of an amino acid.

31.3 Peptides and Proteins

The structures of peptides and proteins were first determined by the German chemist Emil Fischer between 1900 and 1910. **Peptides** are polymers formed from two or more amino acids linked by covalent amide linkages given the special name **peptide bonds.** The nature of a peptide bond (in color) can be seen in the following structure for a dipeptide.

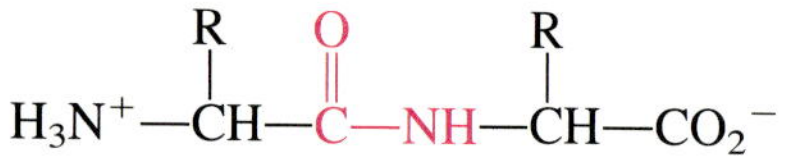

$$H_3N^+ —CH—C—NH—CH—CO_2^-$$

The name *peptide* is usually restricted to polymers made up of only a small number of amino acids. Such polymers are often called **oligopeptides** (from the Greek word *oligo,* meaning few). Polymers containing approximately 10–40 amino acid units are called **polypeptides.** Relatively large polymers containing 40–10,000 (or more) amino acid units are called **proteins.** Aspartame, the artificial sweetener with the brand name NutraSweet, is the methyl ester of the dipeptide aspartylphenylalanine.

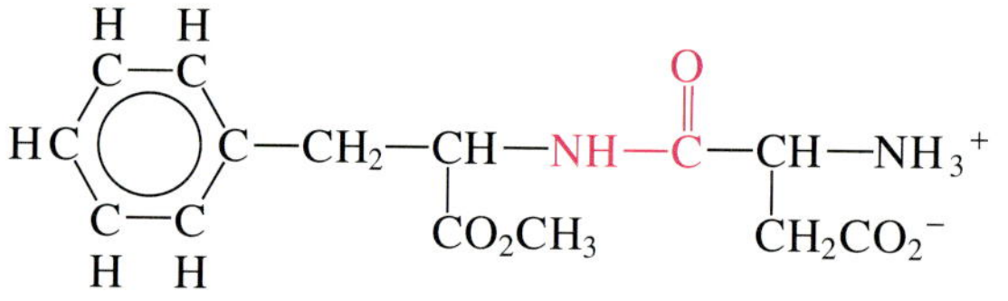

Aspartylphenylalanine methyl ester
(aspartame)

The peptide bond results from combining the α-carboxyl end of one amino acid with the α-amine end of another. Starting with two different amino acids, for example, glycine (GLY) and alanine (ALA), two different dipeptides can be formed.

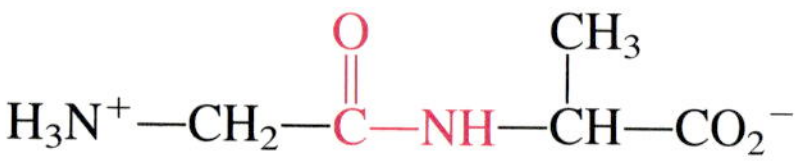

Glycylalanine
(GLY-ALA)

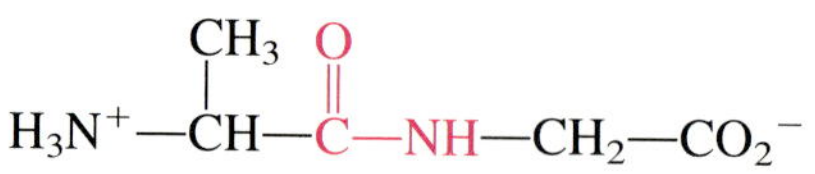

Alanylglycine
(ALA-GLY)

The glycine unit in the dipeptide GLY-ALA has a free amine, and this unit is therefore called the **N-terminal amino acid.** In the dipeptide ALA-GLY, however, the glycine

unit is the **C-terminal amino acid.** Peptides and proteins are written and named starting with the N-terminal amino acid. The tetrapeptide

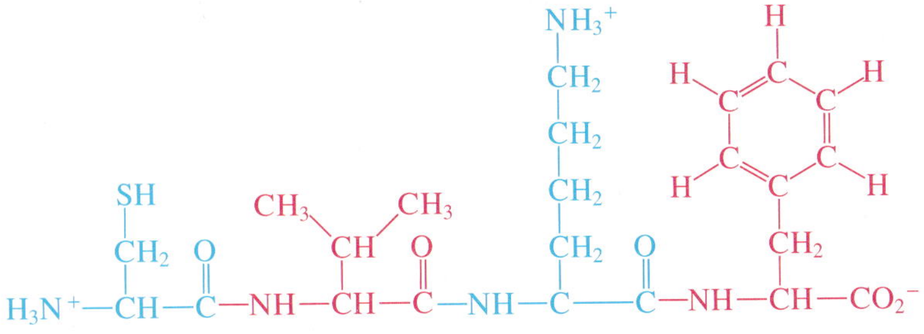

is made up of cysteine, valine, lycine, and phenylalanine bonded together by peptide bonds and is therefore called cysteinylvalyllysylphenylalanine, or CYS-VAL-LYS-PHE. The four amino acids in this tetrapeptide are shown in different colors. This is an example of an **amino acid sequence.** Most protein molecules contain hundreds of amino acids linked one to another in a specific order. Any changes, even small ones, in the amino acid sequence produce differences in the chemical and physiological properties of the peptide.

It would be very useful to be able to synthesize any given protein, since the body's inability to synthesize one or more proteins is the cause of such diseases as hemophilia, sickle-cell anemia, and diabetes. Unfortunately, there are three major obstacles to the chemical synthesis of peptides or proteins. First, the order in which the amino acids are incorporated must be rigorously controlled. Second, reactions must be prevented from occurring at the amino acid side chains. Finally, the free energy for the reaction of the peptide with water is negative, so proteins decompose spontaneously.

$$H_3N^+\!-\!CH\!-\!C\!-\!NH\!-\!CH\!-\!CO_2^- + H_2O \longrightarrow H_3N^+\!-\!CH\!-\!CO_2^- + H_3N^+\!-\!CH\!-\!CO_2^-$$
$$\Delta G^\circ = -17 \text{ kJ/mol}$$

Each of these problems has been solved for many syntheses. Peptides and proteins can be produced by chemical syntheses, but not in large amounts at the present time. However, genetic engineering techniques, which use bacterial cells to synthesize mammalian proteins, now make it possible to synthesize human insulin, interferon, and other important proteins.

31.4 The Structures of Proteins

The extraordinary variations possible in the structures of proteins allow for remarkable differences in their types and functions. Among the types of proteins and their physiological functions are the following:

1. *Enzymes.* Enzymes catalyze biochemical reactions with high catalytic efficiency and a high degree of specificity (Section 31.5).

2. *Hormones.* A number of oligopeptides and proteins, such as insulin and human growth hormone, bind to specific sites in or on the cell to stimulate or control specific physiological functions.

3. *Structural proteins.* A wide variety of proteins are involved in the formation of various kinds of connective tissue, skin, hair, horn, and hoof.

4. *Transport proteins*. Proteins are involved in the transport of both organic and inorganic nutrients across cell membranes. The metal-containing proteins, or metalloproteins, such as the cytochromes, are involved in electron transport.

5. *Antibodies*. The antibodies produced by the immune response system are proteins that neutralize viruses, bacteria, and other "foreign" substances.

6. *Muscle tissue*. Muscles are composed of fibrous bundles of the proteins actin and myosin, which are involved in the contraction and relaxation of muscle tissue.

Proteins can function in many different ways because their amino acid sequences and three-dimensional structures vary widely. Their structures can vary so much (even though they are formed from only 20 different amino acids) because the amino acids can be used in many different orders in a peptide chain of given length. For example, there are 20^2 or 400 possible dipeptides, 20^3 or 8000 tripeptides, and 20^4 or 160,000 tetrapeptides. Even for a protein as small as insulin, with only 51 amino acid units, there are 2.3×10^{66} possible amino acid sequences. What is truly remarkable is that the number of proteins found in living systems is limited to the tens of thousands instead of the astronomically large numbers that are mathematically possible. This is so because only certain amino acid sequences are capable of forming a stable three-dimensional structure that has a biological function.

The **primary structure** of a protein, the organic chemical structure, is simply the sequence of amino acids, listed as though printed on ticker tape, starting with the N-terminal amino acid and proceeding through the C-terminal amino acid.

A polypeptide chain can fold upon itself to produce a **secondary structure,** in which the number of hydrogen bonds between peptide linkages is maximized.

$$\text{>N—H} \cdots \text{O=C<}$$

There are only a limited number of secondary structures that a polypeptide chain can adopt that maximize the number of hydrogen bonds between peptide linkages. Two of the most common forms of secondary structure are the **α-helix** (Fig. 31.4) and the **β-sheet,** or **pleated sheet** (Fig. 31.5). In the α-helix the polypeptide backbone forms a helical coil with approximately 3.6 amino acids per 360° turn. The α-helix is the dominant secondary structure in fibrous proteins such as collagen or the α-keratins of

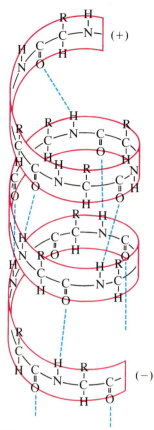

Figure 31.4

The α-helix stucture of a peptide chain, one form of secondary structure of protein molecules. The hydrogen bonds that stabilize this structure are indicated by blue dashed lines.

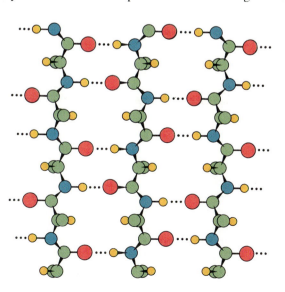

Figure 31.5

The β-sheet, or pleated sheet, secondary structure of protein molecules. Carbon atoms are green; nitrogen, blue; oxygen, red; and hydrogen, yellow. [*From H. D. Springall, The Structural Chemistry of Proteins (Butterworth Scientific Publishers, London, 1954.)*]

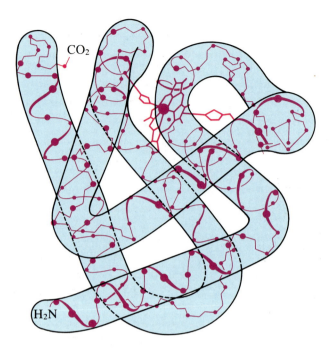

Figure 31.6

The tertiary structure of myoglobin, an oxygen-storing protein found in muscle tissue. The folded tertiary structure is shown, as well as the helical secondary structure (cylindrical regions). The primary structure is not shown in detail but as a series of red dots.

hair and wool. The β-sheet structure results from hydrogen bonds between adjacent strands of a polypeptide backbone. Extended arrays of the β-sheet secondary structure are found in the β-keratins, such as silk.

Proteins often contain regions where a helical or sheet structure is folded into a complex conformation known as the **tertiary structure** (Fig. 31.6). Interactions between amino acid side chains produce and stabilize this structure.

Many larger proteins contain more than one polypeptide chain. The **quaternary structure** of a protein describes how these independent molecules are related to each other. The oxygen-carrying protein hemoglobin (Section 26.5) consists of four separate polypeptide chains, two α subunits and two β subunits, held together in a quaternary structure.

Factors that **denature** a protein disturb its natural conformation, without chemically altering the primary structure, and lead to the loss of biological activity. A denatured protein has lost some or all of its secondary, tertiary, or quaternary structure.

31.5 Enzymes

Enzymes are proteins that catalyze chemical reactions in living systems with remarkable efficiency. The enzyme carbonic anhydrase, for example, which catalyzes the reaction

$$CO_2 + H_2O \rightleftharpoons H^+ + HCO_3^-$$

is such a good catalyst that a single molecule can transform over 36 million carbon dioxide molecules per minute into product. This corresponds to an increase of 10^7–10^8 in the rate of this reaction. One explanation for this efficiency is that the binding of the substrate (the molecule undergoing the reaction) to the enzyme may stretch or strain the substrate until its conformation more closely resembles the transition state for the reaction (Section 14.8), thereby reducing the activation energy. Enzymes are not only efficient; they can also be remarkably selective. The ability of some enzymes to be specific for a very few substrates in a variety of substances with related structures suggests that the enzymes and substrate fit together somewhat like a lock and key do. The enzyme activates only those substrates that fit, just as a lock is opened only by keys that fit.

In 1913 L. Michaelis and M. L. Menten suggested that enzyme-catalyzed reactions proceed through at least two steps. First, the substrate binds to the enzyme to form an enzyme-substrate complex. Second, within the complex the enzyme acts on the substrate to convert it into product. Many substances may bind to an enzyme, but only a limited number of them can react. The enzyme succinate dehydrogenase, for example, not only binds its normal substrate, succinate, but also binds other dicarboxylates, such as malonate, which do not react. By occupying a portion of the available binding sites on the enzyme, malonate slows down the rate of reaction (fewer molecules of enzyme are available to catalyze it). In effect, the activity of the enzyme is **inhibited.** Since the substrate (in this case, succinate) and the inhibitor (malonate) compete for the same binding sites on the enzyme, this form of enzyme inhibition is called **competitive inhibition.** Enzymes can also be inhibited by substances that have no resemblance to the substrate, for example, heavy metal ions such as Ag^+, Pb^{2+}, and Hg^{2+}, which bind either in the presence or in the absence of substrate binding. This form of enzyme inhibition is called **noncompetitive inhibition.**

For a cell to function at its maximum efficiency, the rates of its chemical reactions must be adjusted to its needs. Enzymes alone are not enough; the cell must also have control mechanisms that can turn on enzyme activity when the product of an enzyme-catalyzed reaction is needed and switch off enzyme activity when the product is present in excess. The cell can control the rate of enzyme-catalyzed reactions in four ways. First, it can influence the rate of a reaction by changing the concentration of the substrate. In the absence of substrate, the enzyme is effectively switched off. Second, the cell can produce natural inhibitors of enzyme activity, which operate by either competitive or noncompetitive inhibition mechanisms. Third, the cell can use **allosteric enzymes,** which contain two or more binding sites. Allosteric enzymes bind both a substrate molecule and a small molecule known as the **effector.** Binding of the effector is thought to alter the structure of the enzyme, thereby either switching on or switching off its activity. Binding of the effector usually does not influence the binding of the substrate; it either facilitates or prevents conversion of the substrate into product. There are several forms of control of allosteric enzymes. In so-called **feedback inhibition** an enzyme involved in the synthesis of an end-product is inhibited by binding the end-product as an effector. Alternatively, when the synthesis of a particular product should occur only when the substrate is present in excess, an enzyme may exhibit **substrate activation,** where more than one substrate must bind to the enzyme before it is activated. Fourth, the cell can control the rate of an enzyme-catalyzed reaction by controlling the population of enzyme molecules. This can be done by controlling either the rate at which the enzyme is synthesized or the rate at which the enzyme is degraded or destroyed.

Carbohydrates

31.6 Introduction

The name *carbohydrate* originally meant a compound with the empirical formula CH_2O, literally, a hydrate of carbon. This definition has since been broadened to include all aldehydes and ketones having two or more hydroxyl groups. Among the compounds that fall into the class of **carbohydrates** are the starches, cellulose, glycogen, and the sugars. The carbohydrates are an important source of energy for all organisms and form the supporting tissue of plants and some animals. There are three important classes of carbohydrates: the monosaccharides, the disaccharides, and the polysaccharides.

31.7 Monosaccharides

The simplest carbohydrates contain only a single aldehyde or ketone functional group (Sections 30.11 and 30.12) and are known as **monosaccharides.** They are colorless, crystalline, and frequently possess a sweet taste. A monosaccharide is classified as either an **aldose** or a **ketose,** depending on whether it contains an aldehyde or a ketone functional group, respectively. With the exception of the simplest ketose, the three-carbon monosaccharide dihydroxyacetone, all of the monosaccharides are optically active. Although both D and L isomers are possible, the dominant form of carbohydrates isolated from natural sources is the D isomer. Structures of the D and L forms of the simplest aldose, glyceraldehyde, are

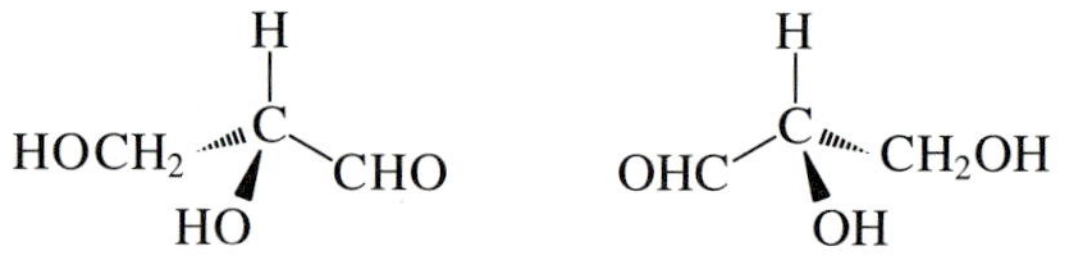

D-Glyceraldehyde   L-Glyceraldehyde

Monosaccharides can undergo a reversible intramolecular reaction to change from a linear structure to a cyclic structure. An aldose, such as glucose, forms a six-membered **pyranose ring** (Fig. 31.7), whereas a ketose, such as fructose, forms a five-membered **furanose ring** (Fig. 31.8). In each case two isomers known as the α and β forms are produced; they differ only in orientation at the first carbon atom, or C(1).

Glucose, $C_6H_{12}O_6$ (Fig. 31.7), also called **dextrose,** is the most abundant organic compound in nature. It is the sole component of the polysaccharides cellulose, starch, and glycogen. In medicine, glucose is often called blood sugar, since it is by far the most abundant carbohydrate in the bloodstream. Human blood normally contains about 1 gram of glucose per liter. Individuals suffering from diabetes are unable to assimilate glucose; they eliminate it through the kidneys. In 100 milliliters of a diabetic's urine, there may be as much as 8–10 grams of glucose, and its presence is one symptom of the disease. **Fructose** (Fig. 31.8), also called **levulose** or **fruit sugar,** has the same

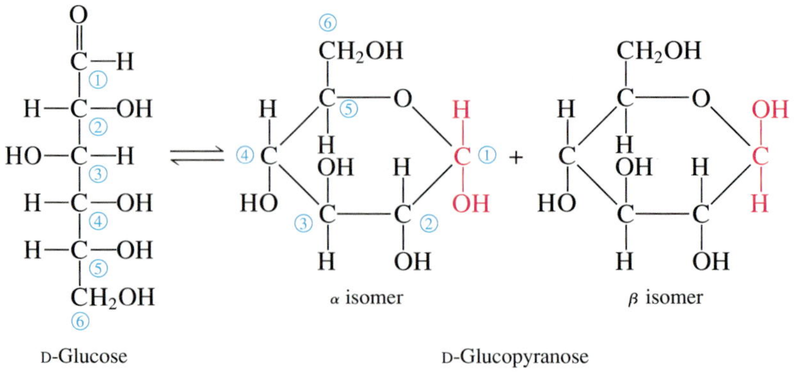

Figure 31.7

Glucose undergoes a reversible reaction leading to the formation of a stable six-membered glucopyranose ring. Two isomers, differing only in orientation at the C(1) atom, are formed.

D-Glucose   D-Glucopyranose

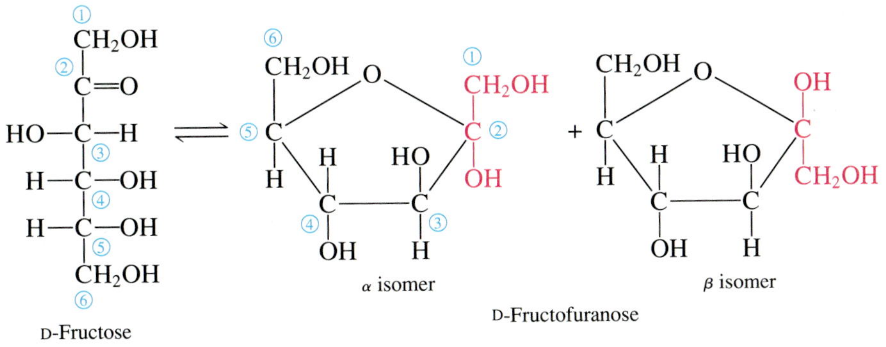

Figure 31.8
Fructose undergoes a reversible reaction leading to the formation of α and β isomers of a five-membered fructofuranose ring.

molecular formula as glucose but contains a ketone rather than an aldehyde functional group. Fructose occurs naturally in both fruits and honey and is found combined with glucose in the disaccharide sucrose, called cane sugar. Fructose is the sweetest of all sugars; a given mass of fructose is nearly twice as sweet as the same amount of sucrose.

31.8 Disaccharides and Polysaccharides

In addition to undergoing reversible reactions that lead to pyranose and furanose rings, monosaccharides can also polymerize to form chains. However, the polymerization reaction can be reversed only in the presence of acid or a suitable enzyme catalyst, so polysaccharides are relatively stable toward reaction with water.

The most important **disaccharide** is **sucrose,** or **cane sugar** (Fig. 31.9). The reaction of water and sucrose, $C_{12}H_{22}O_{11}$, yields a 50:50 mixture of the monosaccharides fructose and glucose (both $C_6H_{12}O_6$), a mixture often called **invert sugar,** which is a major component of honey. **Lactose,** or **milk sugar,** is a disaccharide that contains an α-D-glucopyranose ring and a β-D-galactopyranose ring (Fig. 31.10). Many individuals of African or Asiatic origin cannot digest milk because they lack enzymes that hydrolyze the β-linkage (the ether linkage between the rings) in lactose.

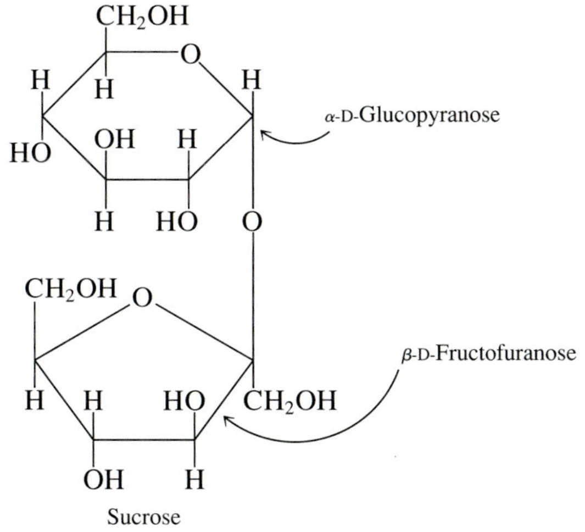

Figure 31.9
The structure of the disaccharide sucrose, formed from reaction of α-D-glucopyranose with β-D-fructofuranose. (Carbon atoms in the rings are not shown.)

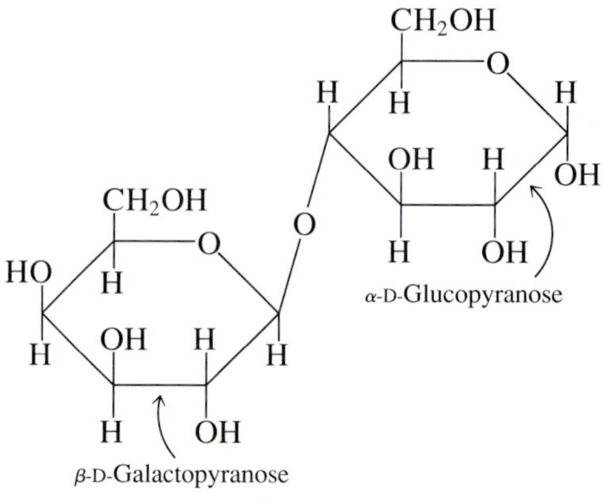

Figure 31.10

The structure of the disaccharide lactose, formed by reaction of β-D-galactopyranose with α-D-glucopyranose. (Carbon atoms in the rings are not shown.)

α-D-Glucopyranose

β-D-Galactopyranose

Lactose

Three major classes of **polysaccharides** are formed by the polymerization of D-glucopyranose rings: starch, glycogen, and cellulose. **Starch** serves primarily as a long-term storage medium for food energy in plants. It accumulates in seeds, tubers, and fruits. Starch is often the main food supply for the young plant until the development of a leaf system enables the plant to manufacture its own food through photosynthesis.

Glycogen is a polysaccharide that serves the same function in animals as starch serves in plants. Glycogen is found primarily in liver, heart, and muscle tissue.

Cellulose is a linear polysaccharide of β-D-glucopyranose units (Fig. 31.11) found in plants. It is the principal component of the rigid cell wall of plant cells and therefore provides structure rather than energy. The lack of an enzyme capable of cleaving the β-linkage in cellulose is the only thing that prevents humans from being able to digest the paper of this text or cotton (which is at least 90% cellulose by weight). Most animals cannot digest cellulose because they lack suitable enzymes to catalyze its cleavage. Termites and the ruminant animals, such as cattle, sheep, and goats, are able to use cellulose as food because bacteria in their digestive tracts contain enzymes that can cleave the β-linkages. Termites are commonly destroyed by killing these bacteria— they then starve to death.

Acid mucopolysaccharides are polysaccharides that contain N-acetylglucosamine and glucuronic acid. Hyaluronic acid is an example of an acid mucopolysaccharide, consisting of repeating units of N-acetylglucosamine linked to glucuronic acid (Fig. 31.12).

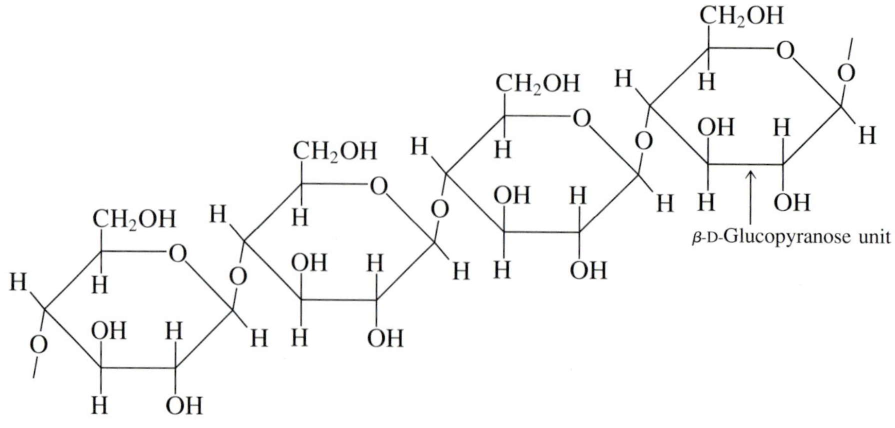

Figure 31.11

The structure of cellulose, a linear polysaccharide formed by the polymerization of β-D-glucopyranose. (Carbon atoms in the rings are not shown.)

β-D-Glucopyranose unit

Cellulose

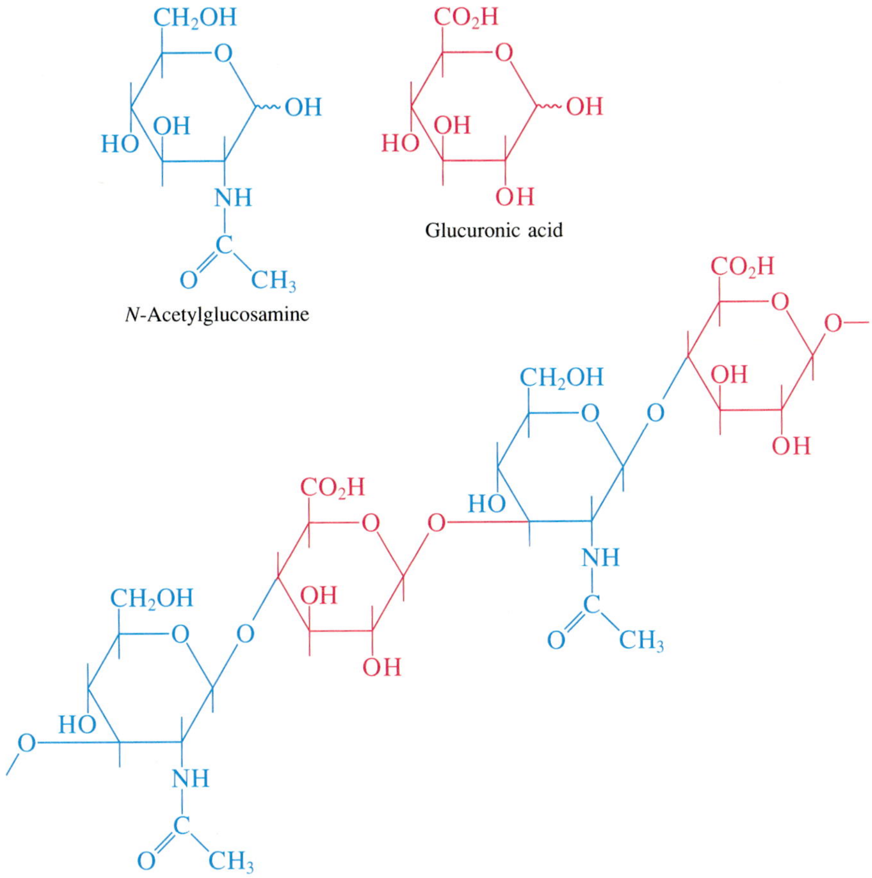

N-Acetylglucosamine

Glucuronic acid

Hyaluronic acid

Figure 31.12
The structures of *N*-acetylglucosamine (shown in blue), glucuronic acid (shown in red), and hyaluronic acid. (Carbon atoms in the rings are not shown.)

Hyaluronic acid is a part of the connective tissue in the body and helps to hold cells together. Another acid mucopolysaccharide is heparin, an important natural blood coagulant. It contains a glucuronic acid sulfate unit and a glucosamine sulfate unit (Fig. 31.13).

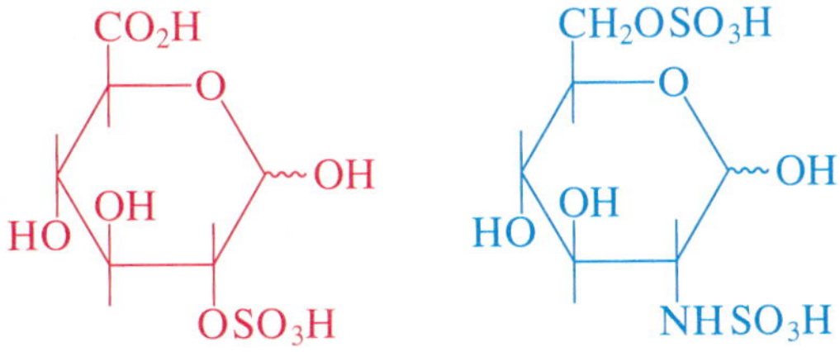

Glucuronic acid sulfate unit

Glucosamine sulfate unit

Glucuronic acid sulfate unit

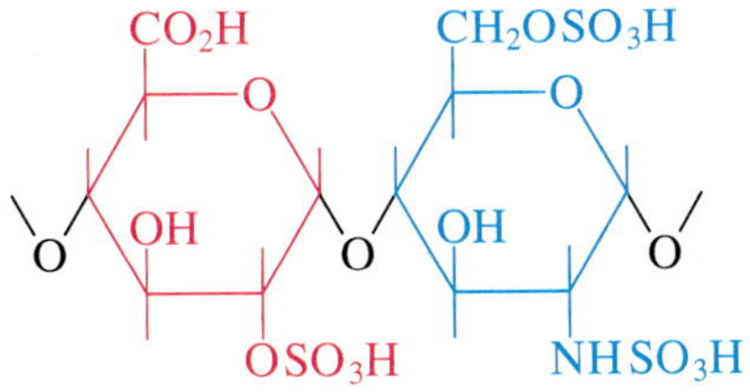

Glucosamine sulfate unit

Heparin

Figure 31.13
The structures of a glucuronic acid sulfate unit (shown in red), a glucosamine sulfate unit (shown in blue), and heparin. (Carbon atoms in the rings are not shown.)

Glycoproteins are protein molecules that have carbohydrate units attached to side chains of certain amino acid residues. Glycoproteins include protein blood-clotting materials and many of the membrane proteins. A very important glycoprotein is collagen, which is the most abundant protein in the human body, comprising about one-third of the protein there. It occurs in all organs, imparting strength and stiffness to them. Collagen is present in such diverse parts of the body as teeth, bones, cartilage, tendons, skin, and the lens of the eye.

Lipids

31.9 Complex Lipids

Lipids are more diverse than either proteins or carbohydrates; this class of biomolecules includes compounds of widely differing structures. They consist of a hydrophobic hydrocarbon backbone with a hydrophilic head, which may or may not be highly charged. The word *lipid* is derived from the Greek word *lipos* (meaning fat), although not all lipids are fats.

Lipids are often divided into two classes, complex and simple, on the basis of the presence or absence of a fatty acid structure. The **complex lipids** are derivatives of long-chain carboxylic acids, or **fatty acids,** such as stearic acid.

$$CH_3CH_2CH_2CH_2CH_2CH_2CH_2CH_2CH_2CH_2CH_2CH_2CH_2CH_2CH_2CH_2CH_2CO_2H$$

Stearic acid

Although fatty acids are available to a cell from a wide variety of sources, the concentration of free fatty acid in the cell is negligible. Most fatty acids occur in plant and animal cells in the form of triesters (Section 30.14) of the alcohol 1,2,3-propanetriol (glycerol), $HOCH_2CH(OH)CH_2OH$

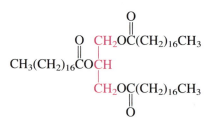

Tristearylglycerol

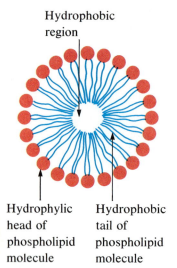

Hydrophobic region

Hydrophylic head of phospholipid molecule   **Hydrophobic tail of phospholipid molecule**

Figure 31.14

The phosphatidyl cholines, which possess a charged hydrophilic head and a non-polar hydrophobic tail, can associate to form micelles in which the hydrophilic heads are oriented toward the surrounding water molecules and the hydrophobic tails share a region of the solution where water molecules are partially excluded.

Such lipids are commonly known as **triglycerides,** although the proper name is **triacylglycerols.** The primary function of these triglycerides is the long-term storage of food energy. Triglycerides formed from saturated fatty acids, such as tristearylglycerol, are typically solids and are known as **fats.** Triglycerides formed from unsaturated fatty acids, which contain one or more carbon-carbon double bonds, are more frequently liquids and are known as **oils;** examples include cottonseed oil, linseed oil, and palm oil.

One of the characteristic reactions of fats and oils is **saponification,** which is the reaction of the triester in the presence of aqueous base to form glycerol and a salt of the fatty acid.

$$\text{CH}_3(\text{CH}_2)_{16}\overset{\displaystyle O}{\overset{\displaystyle \|}{\text{C}}}\text{O}\overset{\displaystyle\text{CH}_2\text{O}\overset{O}{\overset{\|}{\text{C}}}(\text{CH}_2)_{16}\text{CH}_3}{\underset{\displaystyle\text{CH}_2\text{O}\underset{O}{\underset{\|}{\text{C}}}(\text{CH}_2)_{16}\text{CH}_3}{\text{CH}}} \quad + \text{ 3NaOH}(aq) \longrightarrow$$

$$\overset{\text{CH}_2\text{OH}}{\underset{\text{CH}_2\text{OH}}{\text{CHOH}}} \; + \; 3\text{CH}_3(\text{CH}_2)_{16}\text{CO}_2{}^-\text{Na}^+(aq)$$

The salts of the fatty acids thus produced are soaps (Section 12.28).

A second major class of complex lipids are the **phosphoglycerides,** or **glycerophosphatides,** which are major components of cell membranes. These lipids can be thought of as derivatives of phosphoric acid and glycerol, with the following structure:

$$\begin{array}{c}\text{R}-\overset{\displaystyle O}{\overset{\displaystyle \|}{\text{C}}}-\text{O}-\text{CH}_2\\[4pt]\text{R}-\overset{\displaystyle \|}{\underset{\displaystyle O}{\text{C}}}-\text{O}-\text{CH}\qquad\overset{O}{}\\[6pt]\text{CH}_2-\text{O}-\overset{\displaystyle}{\text{P}}-\text{OX}\\[4pt]\text{OH}\end{array}$$

(where X is choline, ethanolamine, or serine). Among the most important of the phosphoglycerides are the **phosphatidyl cholines,** or **lecithins,** of which the following is a representative example:

$$\begin{array}{c}\text{CH}_3(\text{CH}_2)_{16}\overset{\displaystyle O}{\overset{\displaystyle \|}{\text{C}}}\text{O}\text{CH}_2\\[4pt]\text{CH}_3(\text{CH}_2)_{16}\overset{}{\text{C}}\text{O}\text{CH}\qquad O\\[2pt]\underset{\displaystyle O}{\overset{\displaystyle \|}{}}\;\;\text{CH}_2\text{OPOCH}_2\text{CH}_2\text{N}(\text{CH}_3)_3{}^+\\[4pt]\underset{\displaystyle O^-}{}\end{array}$$

Like soaps, phosphatidyl cholines contain a highly charged hydrophilic head and a nonpolar hydrophobic tail. These lipids can spontaneously group to form **micelles,** in which the ionic heads face the surrounding water and the nonpolar tails form a hydrophobic pocket (Fig. 31.14, p. 906). Alternatively, they can form bilayers, which serve as the basis for membranes that separate the cell from its surroundings or segment the interior of the cell (Fig. 31.15).

31.10 Simple Lipids

Lipids that do not contain a fatty acid are given the rather misleading name **simple lipids.** The simple lipids include substances such as the **fatty alcohols** (cetyl alcohol and myricyl alcohol, for example), which are long-chain alcohols.

$$\underset{\text{Cetyl alcohol}}{\text{CH}_3(\text{CH}_2)_{14}\text{CH}_2\text{OH}} \qquad \underset{\text{Myricyl alcohol}}{\text{CH}_3(\text{CH}_2)_{29}\text{CH}_2\text{OH}}$$

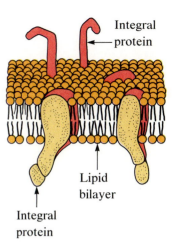

Figure 31.15
The phosphatidyl cholines can associate to form bilayers that serve as the basis for cell membranes. The nonpolar tails associate to form hydrophobic pockets, and the hydrophilic heads point toward the polar solvent, water. Membranes may also contain other lipids, such as cholesterol, and proteins that are involved in the transport of molecules across the cell membrane. *(From "The Assembly of Cell Membranes," by Harvey F. Lodish and James E. Rothman. Copyright © 1979 by Scientific American, Inc. All rights reserved.)*

With fatty acids these fatty alcohols form esters (Section 30.14) known as **waxes.** The family of simple lipids also includes the **terpenes,** which include pigments such as β-carotene, fat-soluble vitamins such as vitamin A, and forms of natural rubber (Section 30.16).

A remarkably diverse group of lipids, **steroids,** can be synthesized from terpenes by a complex set of reactions. Among the steroids are such compounds as cholesterol, an important component of cell membranes, and the sex hormones testosterone and estradiol-17 (Fig. 31.16).

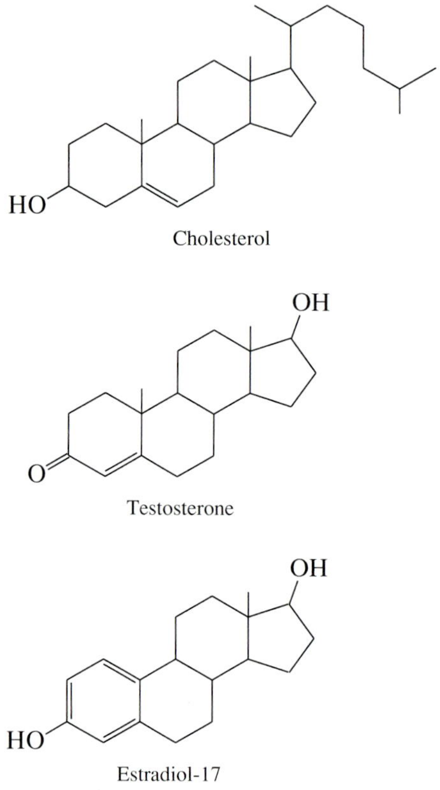

Figure 31.16

The steroids represent a class of compounds that have the same essential fused-ring structure. The steroids function as important components of cell membranes, such as cholesterol; as precursors in the synthesis of vitamins, such as vitamin D_2; as hormones, such as the male and female sex hormones testosterone and estradiol-17; and as ingredients of oral contraceptives.

Cholesterol

Testosterone

Estradiol-17

Nucleic Acids and the Genetic Code

31.11 Nucleic Acids

Understanding the chemical nature of nucleic acids is the basis for understanding biochemical genetics. The most important **nucleic acids** are macromolecules of very high molecular weights whose sole function is the storage and transfer of genetic information. There are two types of such nucleic acids: **ribonucleic acids (RNA),** located primarily in the cytoplasm of the cell, and **deoxyribonucleic acids (DNA),** concentrated in the nucleus of the cell.

The ribonucleic acids (RNA) are built on the framework of a β-D-ribofuranose ring, and the deoxyribonucleic acids (DNA) contain a modified ribofuranose ring in which the —OH group on the second, or 2′, carbon is replaced by a hydrogen atom.

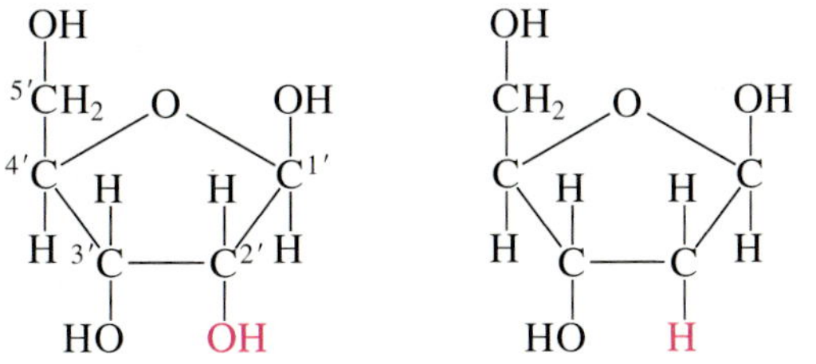

β-D-Ribofuranose 2-Deoxy-β-D-ribofuranose

Nucleosides are formed by combining one of these monosaccharides with one of five nitrogen-containing bases (Fig. 31.17). **Nucleotides** are phosphate esters of nucleosides. For example, the base uracil, which is found only in RNA, not in DNA, can be

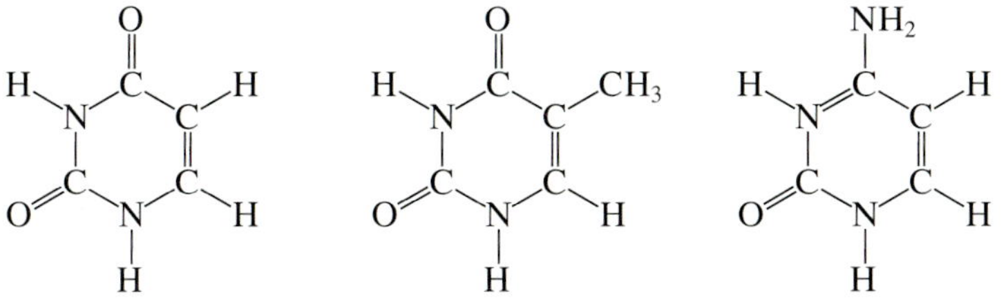

Uracil (U) Thymine (T) Cytosine (C)

(a)

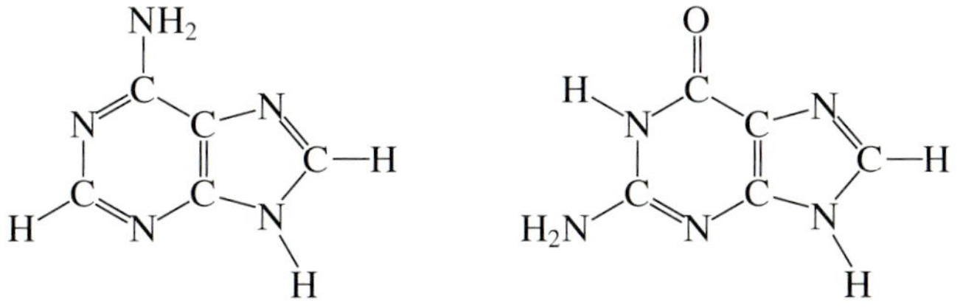

Adenine (A) Guanine (G)

(b)

Figure 31.17

The five most common nitrogen-containing bases found in nucleic acids are divided into two families: (a) the pyrimidines uracil (U), thymine (T), and cytosine (C); and (b) the purines adenine (A) and guanine (G).

joined to a ribofuranose ring at the 1′ carbon to form the nucleoside **uridine.** This nucleoside can then combine with the phosphate ion, PO_4^{3-}, to form the nucleotide **uridine monophosphate.**

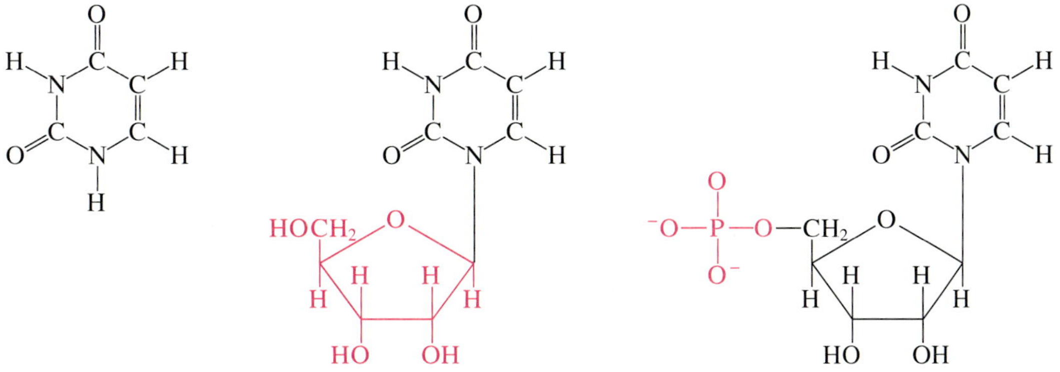

Uracil Uridine Uridine monophosphate

Nucleic acids are formed by polymerization of nucleotides. Polymerization occurs by reaction of the 3′ hydroxyl group of one nucleotide with the acidic phosphate that is attached to the 5′ carbon of another to form an ester. A segment of a DNA chain containing the four nucleotides that are found in deoxyribonucleic acids is shown in Fig. 31.18. The sequence of nucleotides in Fig. 31.18, reading from the 5′ end of this chain to the 3′ end, can be symbolized as T-A-C-G. This figure shows the tetranucleotide as it would exist at or near neutral pH. The phosphodiester linkages

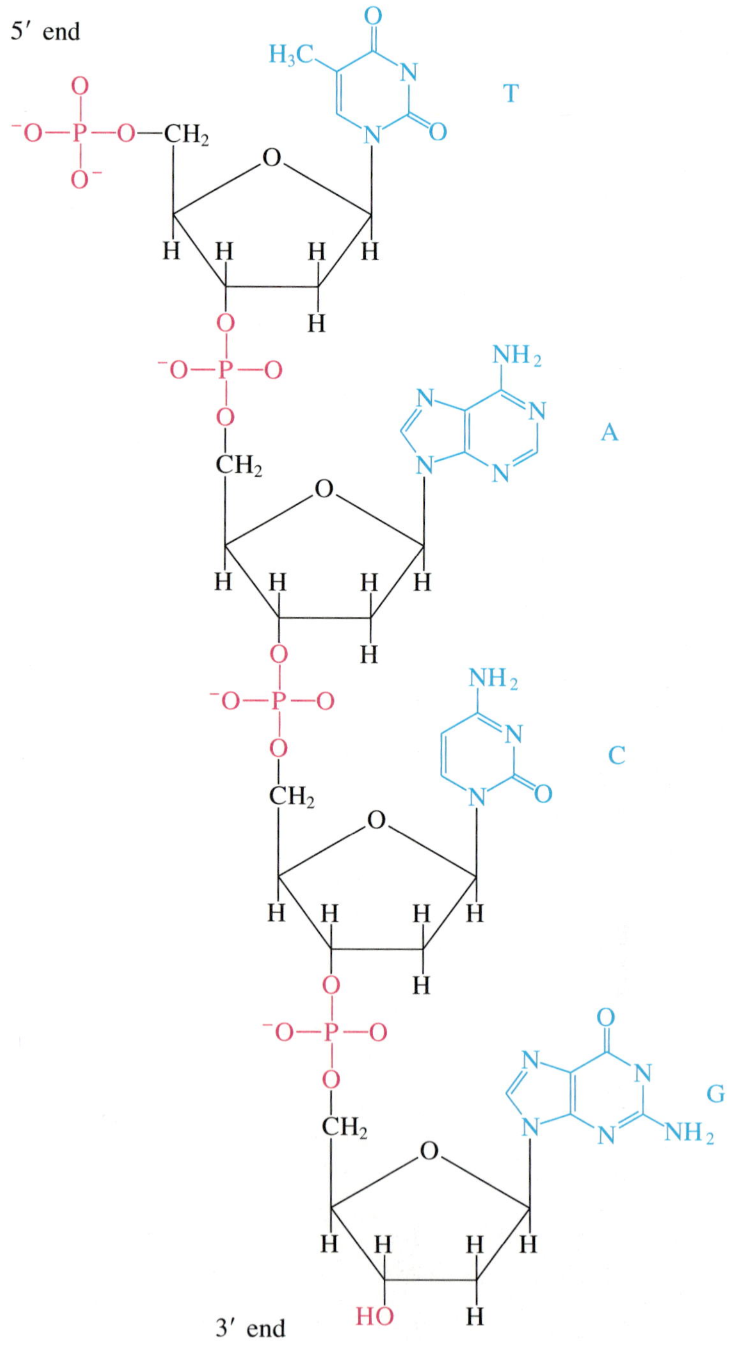

Figure 31.18

The structure of a tetranucleotide segment of a single strand of deoxyribonucleic acid (DNA). Reading from the 5′ to the 3′ end, this tetranucleotide can be symbolized as T-A-C-G. (Carbon atoms in the rings are not shown.)

between adjacent nucleosides in nucleic acids are strong acids. The phosphate groups should therefore be expected to carry a negative charge in aqueous solution at or near neutral pH. The nucleic acids are therefore normally found associated either with highly charged positive proteins known as **histones** or with cations such as the Mg^{2+} ion.

There are several important structural differences between the nucleic acids DNA and RNA. Each RNA nucleotide contains an —OH group on the $2'$ carbon of the furanose ring, but DNA nucleotides do not. Furthermore, the nitrogen-containing bases used to form nucleotides are different. DNA contains exclusively the two purines adenine (A) and guanine (G) and the two pyrimidines thymine (T) and cytosine (C). RNA contains the pyrimidine uracil (U) instead of thymine and often a number of other nucleotides. Evidence suggests, however, that these additional nucleotides are formed by modification of adenine, guanine, cytosine, or uracil after the RNA polymer is produced.

In 1953 J. D. Watson and F. H. C. Crick proposed a mechanism for the storage of genetic information based on the structure of DNA. For this work they subsequently received the Nobel Prize. According to Watson and Crick, DNA is composed of two strands, running in opposite directions, which are bridged by hydrogen bonds between specific pyrimidine groups on one strand and purine groups on the other (Fig. 31.19). These two strands are twisted into a double α-helix structure with approximately 10 nucleotide pairs per 360° turn (Fig. 31.20). Information is stored in this structure because the strongest hydrogen bonds form between adenine and thymine and between guanine and cytosine (Fig. 31.21). Wherever adenine is found on one strand, thymine will be found on the other, and wherever guanine is found on one strand, cytosine will be found on the other. The DNA helix can only accommodate a purine base opposite a pyrimidine; this results exclusively in A-T or G-C hydrogen-bonded interactions. The sequence of nucleotides on one strand of the double helix determines the sequence on the other.

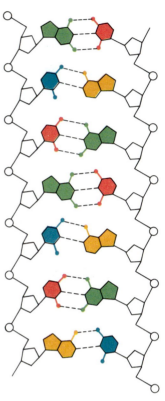

Figure 31.19

Schematic diagram of the structure of DNA. The sugar-phosphate backbone is shown in black, whereas the purine and pyrimidine bases are shown in various colors. (*After A. Kornberg. "The Synthesis of DNA." Copyright © 1968 by Scientific American, Inc. All rights reserved.*)

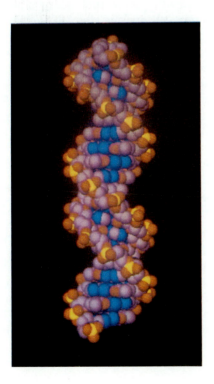

Figure 31.20

The double helix structure of deoxyribonucleic acid (DNA). Schematic drawing (right) and model of the double-stranded α-helix (left).

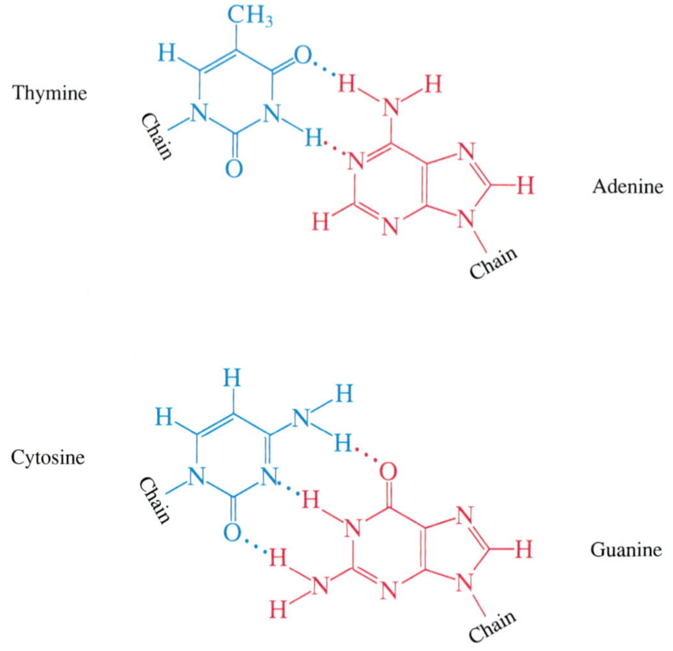

Figure 31.21
The hydrogen bonds in A-T and G-C base pairs. (Carbon atoms in the ring are not shown.)

31.12 The Genetic Code and Protein Synthesis

Modern genetics, and much of modern biochemistry, rests on the theory that the double helix of DNA serves as the medium for storing the information necessary to control the sequence of incorporation of amino acids during protein biosynthesis. To transmit this information during cell division, the DNA molecule must be capable of faultless **replication,** meaning "copying" or "reproduction," so that each of the new cells gets a perfect copy of the information. To be useful to the cell, the information coded within the structure of the DNA double helix must be capable of being **transcribed,** meaning "rewritten," thereby passing the information contained in DNA to another nucleic acid called **messenger RNA (m-RNA).** Finally, a mechanism must exist to **translate** this code from one "language" to another during the assembly of the protein—specifically from the language of nucleic acids to a completely different language, the amino acid sequence of a protein.

We can summarize the three processes of replication, transcription, and translation as follows:

$$DNA \xrightarrow{\text{Replication}} DNA \xrightarrow{\text{Transcription}} \text{m-RNA} \xrightarrow{\text{Translation}} \text{Protein}$$

These three processes are very important in understanding the relationship between heredity and protein synthesis and thus are considered central to the subject of molecular biology. We will now consider each of the three processes in greater detail.

The successful replication of the DNA double helix is based on the preference of adenine (A) on one strand of a double helix for thymine (T) on the complementary strand and of guanine (G) for cytosine (C) on the complementary strand. The first step in the replication of DNA during cell division involves the binding of certain proteins that separate a portion of the double helix into two single strands of DNA. Each strand is then copied by the formation of short segments with A in the strand bonded to T in

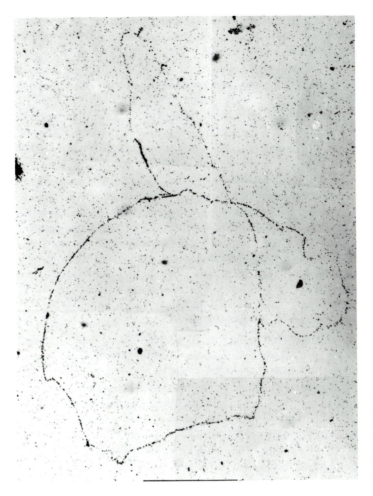

Autoradiograph of replicating DNA from *E. coli*. Replication started at the Y at 12 o'clock and produced the two strands that converge at the Y at 3 o'clock on the large loop.

the new segment, G to C, T to A, and C to G. These segments finally join to form the complement of the strand being copied. Replication proceeds until two identical copies of the original DNA molecule are eventually formed. Replication is a semiconservative process, since each copy retains one strand from the parent DNA molecule (Fig. 31.22).

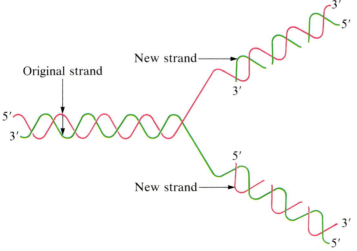

Figure 31.22

The replication of the DNA double helix. Both strands are copied simultaneously, and each strand is copied in small segments that are then joined together. The overall process is semiconservative, since each copy retains one strand from the parent molecule.

DNA double helix

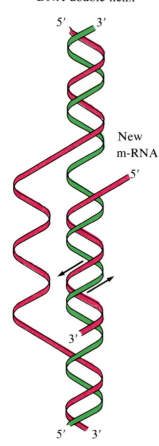

New m-RNA

Figure 31.23

The transcription of a sequence of messenger RNA on the template of one strand of a DNA double helix. Note that the messenger RNA chain is the complement of the DNA strand being copied.

DNA stores the information, in its sequence of bases, that enables a cell to synthesize any of several thousand different proteins. This information is processed in a two-step mechanism. In **transcription,** an RNA complement of one strand of the DNA double helix is synthesized under the control of certain enzymes. Since this segment of RNA carries the message contained in the DNA molecule, it is called **messenger RNA,** or **m-RNA.** Just as the two strands of the DNA double helix run in opposite directions, the m-RNA segment synthesized during transcription is a reversed copy of the original DNA strand (Fig. 31.23), with one major difference—RNA contains the nucleotide uracil, whereas DNA would contain thymine. Thus the base pairs used during transcription are A-U and G-C.

Now that the basic information from DNA is transcribed onto a messenger RNA segment, we must have a system for translating that information into an amino acid sequence. The second step in the mechanism of processing the information contained within the DNA double helix therefore is called **translation.** During translation the m-RNA segment serves as the template on which the protein is synthesized. The synthesis of a protein takes place in a ribosome (Section 31.1), which contains both proteins and a second form of RNA known as **ribosomal RNA.** The first step in translation, therefore, is binding of the m-RNA segment to a ribosome. The amino acids are transported to the ribosome by a third form of RNA known as **transfer RNA,** or **t-RNA.** The t-RNA molecules are the smallest of all the nucleic acid structures. Individual t-RNA molecules, each carrying an amino acid, enter the ribosome and bind to the m-RNA chain at specific positions through hydrogen bonds analogous to those linking the base pairs in the DNA double helix. The amino acids are then joined to form a polypeptide chain. The process is repeated, increasing the size of the polypeptide chain. The elongation of the chain continues until a ''chain termination'' group is encountered, whereupon the ribosome releases both the finished polypeptide chain and the m-RNA segment.

The sequence of nucleotides in a DNA strand is the basis for coding the sequence of amino acids in proteins. Each sequence of three adjacent nucleotides (a code word or **codon**) is specific for a single amino acid. There are 64 possible codons and only 20 amino acids to be coded. Among the codons are three that signal chain termination (Table 31.1).

The identity of the amino acid coded by any three-nucleotide sequence, or codon, on a m-RNA segment can be determined from the genetic code dictionary in Table 31.1.

EXAMPLE 31.1

What is the sequence of amino acids coded by the following sequence of nucleotides on a strand of DNA, reading from the 5′ to the 3′ end?

A-T-C-G-C-T-A-C-G-A-A-T

The m-RNA segment that forms on this DNA template is the complement of this sequence and runs in the opposite direction. Thus the m-RNA chain, reading from the 3′ to the 5′ end, would be U-A-G-C-G-A-U-G-C-U-U-A. Reading this message in the other direction, from the 5′ to the 3′ end, yields the triplets A-U-U, C-G-U, A-G-C, and G-A-U, which code for the sequence ILE-ARG-SER-ASP, reading from the N- to the C-terminal amino acid.

Table 31.1 The amino acids coded by specific triplets (codons) of nucleotide residues on messenger RNA

First position (5' end)	Second position				Third position (3' end)
	U	C	A	G	
U	PHE	SER	TYR	CYS	U
	PHE	SER	TYR	CYS	C
	LEU	SER	Term[a]	Term[a]	A
	LEU	SER	Term[a]	TRP	G
C	LEU	PRO	HIS	ARG	U
	LEU	PRO	HIS	ARG	C
	LEU	PRO	GLN	ARG	A
	LEU	PRO	GLN	ARG	G
A	ILE	THR	ASN	SER	U
	ILE	THR	ASN	SER	C
	ILE	THR	LYS	ARG	A
	MET	THR	LYS	ARG	G
G	VAL	ALA	ASP	GLY	U
	VAL	ALA	ASP	GLY	C
	VAL	ALA	GLU	GLY	A
	VAL	ALA	GLU	GLY	G

[a]The three codons UAA, UGA, and UAG code for the termination of the polypeptide chain.

All of the information necessary for the synthesis of the 2000–3000 different proteins found in the simplest cell is contained in the DNA molecules of the cell's **chromosomes.** The portion of the DNA molecule that codes for a single polypeptide chain is known as a **gene.** Each gene is specific for the synthesis of a single kind of protein, although multiple copies of a gene may be found on a chromosome.

31.13 Gene Mutations

Gene mutations are substitutions, deletions, or additions of one or more nucleotides in the DNA molecule. **Mutations** may be either beneficial or harmful to a living species.

An example of gene mutation occurs with **sickle-cell anemia,** which affects approximately three of every thousand Americans of African descent (Fig. 31.24). In sickle-cell anemia the red blood cells contain an abnormal hemoglobin in which a negatively charged glutamic acid residue at the sixth position of the beta chain of normal adult hemoglobin is replaced by an uncharged valine residue.

Normal adult hemoglobin: VAL-HIS-LEU-THR-PRO-GLU-GLU-LYS- . . .

Abnormal adult hemoglobin: VAL-HIS-LEU-THR-PRO-VAL-GLU-LYS- . . .
         1 2 3 4 5 6 7 8

The difference in the electrical charge causes the abnormal hemoglobin to have a hydrophobic section (neutral VAL), compared to the normal hemoglobin with a

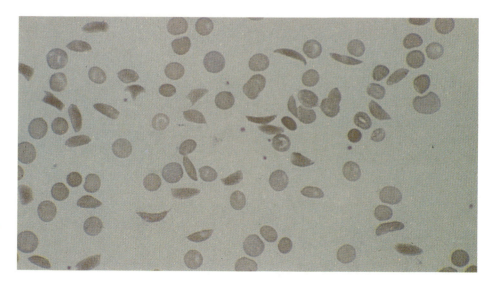

Figure 31.24

Photo of normal (circular) red blood cells and (crescent shaped) sickled cells from sickle-cell anemia.

corresponding hydrophilic section (negative GLU). This difference causes the abnormal hemoglobin, particularly in its deoxygenated form, to be considerably less soluble in water than normal hemoglobin is. The abnormal hemoglobin thus tends to precipitate in the red blood cells, causing the cells to "sickle" and sometimes burst. The sickled cells block the flow of blood to the capillaries, causing the painful effects of sickle-cell anemia.

Many persons that do not actually develop sickle-cell anemia carry the sickle-cell trait; half their hemoglobin is normal and half abnormal. These people usually lead normal lives, but exposure to low oxygen levels, as at high altitudes in mountains or on airplane flights, can increase the risk by increasing the deoxygenated form of the abnormal hemoglobin. One treatment for the symptoms of sickle-cell anemia, therefore, is administration of oxygen at higher than atmospheric pressure.

Persons with sickle-cell trait or sickle-cell anemia, for reasons not well understood, have a high resistance to malaria, illustrating that even the same gene mutation can have both beneficial and harmful effects. It is particularly striking that both these effects of the sickle-cell trait arise from the substitution of a *single* nucleotide base in the gene that controls the synthesis of the beta chain of a hemoglobin molecule that contains a total of 574 amino acid residues.

31.14 Metabolism

The term **metabolism** is used to describe the chemical reactions that take place in living organisms. Metabolism includes the reactions the cell uses to produce energy, to capture nutrients for the synthesis of biomolecules, to degrade macromolecules to produce the basic building blocks for future construction, and to shuffle intermediates between more abundant and less abundant species. Metabolism is divided into pathways that result in either the synthesis or the degradation of a compound. **Catabolic pathways** lead to the degradation of a metabolic intermediate, and **anabolic pathways** result in the synthesis of a compound.

The energy required for the growth and maintenance of cells is provided by metabolic reactions of two general types: photosynthesis and respiration (fermentation). **Photosynthesis** is an endothermic process by which the chloroplasts of green plants and algae convert light into chemical energy. Photosynthetic reactions lead to the reaction of carbon dioxide and water producing oxygen and a carbohydrate, such as glucose.

$$6CO_2 + 6H_2O \xrightarrow{\text{Light}} C_6H_{12}O_6 + 6O_2 \qquad \Delta G^\circ = 2870 \text{ kJ mol}^{-1}$$

Respiration, or **fermentation,** is a set of reactions that provide energy through the metabolic degradation, or catabolism, of organic compounds. The reactions involved in respiration can be divided into two categories: (1) those that can occur in the absence of molecular oxygen **(anaerobic reactions),** and (2) those that require oxygen **(aerobic reactions).** Anaerobic pathways typically do not lead to a net oxidation or reduction of a metabolic intermediate; they liberate significantly smaller amounts of energy than do the oxidation-reduction reactions that accompany aerobic pathways.

The sequence of reactions involved in the anaerobic degradation of glucose is known collectively as **glycolysis.** Glycolysis ultimately degrades a single molecule of glucose into two pyruvate ions, with the release of more than 200 kilojoules of energy per mole of glucose consumed.

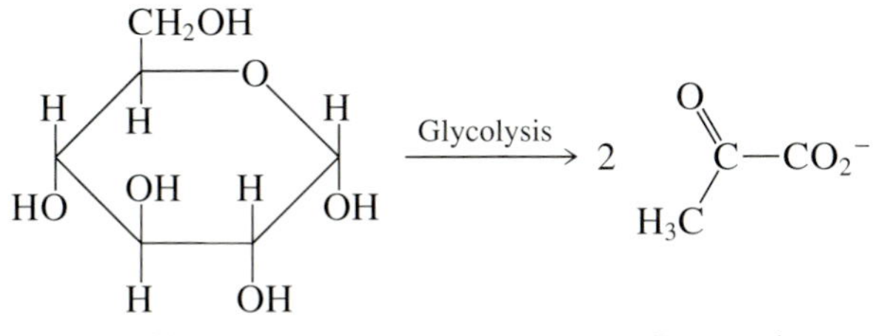

Glucose   Pyruvate ion

Anaerobic organisms, such as the yeast cells used in the production of alcohol by the fermentation of sugars, convert the pyruvate produced in glycolysis to acetaldehyde and then to ethanol.

Aerobic organisms ultimately convert pyruvate into carbon dioxide and water through a sequence of reactions known as the **citric acid cycle,** or the **Krebs cycle.** The net reaction in aerobic organisms is thus the opposite of the photosynthetic reaction and liberates approximately 2870 kilojoules per mole of glucose consumed.

$$C_6H_{12}O_6 + 6O_2 \longrightarrow 6CO_2 + 6H_2O \qquad \Delta G^\circ = -2870 \text{ kJ mol}^{-1}$$

About 50% of the energy released during the degradation of glucose is lost in the form of heat. The remainder is stored as chemical energy through the synthesis of nucleoside triphosphates such as **adenosine triphosphate, ATP.**

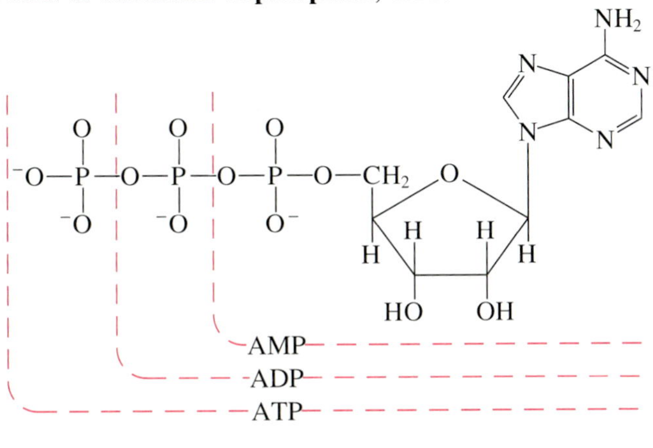

ATP, and the related triphosphates GTP, CTP, and UTP, can react with water to produce adenosine diphosphate, ADP (or GDP, CDP, or UDP), and the phosphate ion, with the release of approximately 31 kilojoules per mole of ATP. The nucleoside diphosphates, such as ADP, can undergo further reaction to form nucleoside mono-phosphates, such as AMP, with the release of an additional 30 kilojoules per mole. The nucleoside triphosphates, and to some extent the diphosphates, serve as the cell's short-term energy-storage mechanism. ATP, for example, is synthesized from AMP and ADP during the degradation of carbohydrates, proteins, and lipids. When the cell needs energy either to drive a chemical reaction, such as the synthesis of proteins from amino acids, or for a mechanical process, such as the contraction of a muscle, ATP or its equivalent is consumed.

Glycolysis and the citric acid cycle are important means of producing biological energy in the form of ATP, but these pathways have other functions as well. There are several points where intermediates are produced that can be used in the synthesis of amino acids, or where intermediates from degradation of amino acids and proteins can be inserted for eventual conversion to carbohydrates. The metabolism of carbohydrates and the metabolism of proteins are therefore intimately related as are those of carbohydrates and the complex lipids.

The complete degradation of either carbohydrates or lipids leads eventually to the same products: carbon dioxide and water. However, $2\frac{1}{2}$ times more energy per gram is released during the oxidation of fats. In times of plenty, that is, when biological energy is abundant, the cell first stores this energy by synthesizing ATP from ADP and/or AMP. Once this short-term storage capacity is exceeded, the cell may store additional energy by synthesizing carbohydrates. These carbohydrates may then be linked to form polysaccharides such as glycogen and starch, which serve as a more efficient means of storing energy for times of shortage. In times of extreme plenty, the cell stores excess energy in the form of complex lipids, or fats. Each of these forms of storage—ATP, carbohydrate, polysaccharide, and lipid—represents a progressively more efficient means of storing excess energy during times of plenty to provide for times of shortage that could come. It is an unfortunate fact, in modern times, that for some humans these times of shortage never come, while for others, the times of plenty are all too infrequent.

31.15 Recombinant DNA Technology

We have noted that DNA contains subunits, repeated many times, that are used to store the information necessary for life. DNA specifies all our physical characteristics. DNA molecules are divided into regions of genes, each of which is responsible for a specific component of the life processes. For example, certain genes are responsible for determining eye color and others for determining blood type.

In the mid-1970s, scientists discovered how to cut a DNA chain and to transfer particular pieces, containing specific information, from one organism to another, thereby changing the characteristics of the second organism. This procedure has come to be known as **recombinant DNA technology.**

The recipient organisms are most often bacteria. In the process, the fragment of transferred DNA and the recipient organism are both duplicated **(cloned)** many times, producing millions of identical cells called a **clone of cells.** Thus it is possible to obtain millions of copies of a particular piece of a DNA chain by placing it in a bacterial cell and allowing the cell to multiply millions of times. This gene cloning is part of a process known as **genetic engineering.** Many possible processes arise, such as making new breeds of plants or improving the characteristics of present breeds, replacement of defective genes with normal genes in persons with genetic diseases, and changing personal characteristics from ones considered to be undesirable to ones thought to be more desirable.

Several steps are necessary in carrying out these procedures:

1. Break open living cells.
2. Remove genetic information from the cells.
3. Cut specific genes of interest away from the other parts of the DNA chain.
4. Splice the cut-away DNA sections into small DNA molecules, called **cloning vehicles,** that can penetrate the walls of a living cell.
5. Transfer the cloning vehicle and the genes spliced into it into a host cell, such as a single-celled bacterial organism or yeast.
6. Allow the host cell to multiply to a clone that has millions of identical cells.

Through these steps, a gene can be transferred into a cell where it would never occur naturally. Gene cloning is a procedure for transferring small pieces of genetic information from one organism to another, thereby altering the characteristics of the recipient organism in predictable and (it is hoped) useful ways.

There have been notable successes in recombinant DNA technology. Improvement in the manufacture of insulin is one example. The insulin gene in human DNA contains information for producing insulin. Before gene cloning was developed, insulin could only be obtained by an expensive process of extracting it from the hog pancreas. Now, through gene cloning techniques, large quantities of insulin are produced much more easily inside bacteria, using human insulin genes. This has the added advantage that the product is human insulin, which is important to the many diabetic persons who are allergic to hog insulin.

Another example is the use of genetically altered bacteria to inhibit frost damage in plants, which causes an estimated $1.6 billion in annual damage to agriculture in the United States. The first two authorized outdoor tests of the procedure, in April, 1987, involved spraying strawberries and potatoes with bacteria altered to eliminate frost-inducing characteristics.

Research is proceeding with great caution. Gene transplantation provides the potential for control of the genetic makeup of plants and animals, including human beings. This necessitates choices that can have profound ethical and moral consequences.

31.16 Viruses

Viruses consist of a protein coat, called a **capsid,** around a nucleic acid core. Viruses are much smaller than bacteria, ranging in size from 0.02 to 0.04 micrometers (μm) compared to about 1 μm for a prokaryotic cell and 10 to 20 μm for a eukaryotic cell (Section 31.1).

The overall shapes of viruses vary, depending largely on the arrangement of the coat proteins. Common shapes include rods, filaments, and spheres (Fig. 31.25).

When infecting an organism, a virus invades certain of the organism's cells and interrupts the transcription and translation function of the cell (Section 31.12). Instead of producing cellular DNA and proteins, the cell produces copies of the virus's DNA and the proteins of its coat; thus, many copies of the virus are replicated. Cells reproduce the virus until they become filled with the virus and burst, releasing new virus to

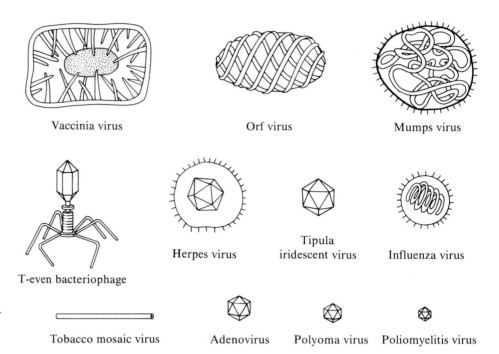

Figure 31.25

Some typical shapes of viruses. (*From an illustration by James Egleson in "The Structure of Viruses" by R. W. Horne, Scientific American, January, 1963. Redrawn by permission*)

continue the infection. Infection by viruses is responsible for many diseases in plants and animals—for example, measles, smallpox, mumps, influenza, and rabies.

31.17 Antibody Immune Response

Defenses against infections of the body result when white blood cells ingest and destroy foreign matter, or when special cells in the body produce antibodies (an **antibody immune response**). **Antibodies** are large protein molecules manufactured by cells called **lymphocytes,** which are produced in bone marrow and in the lymphatic tissues (the lymph nodes, the spleen, and the thymus). The antibodies function by reacting with the infecting substance, called the antigen. **Antigens** are materials that the body considers to be foreign, including bacteria, viruses, and other substances.

The bloodstream carries the antibodies to the body tissues. The antibody proteins released into the bloodstream are called **immunoglobulins,** the most common of which are the **gamma globulins.** Gamma globulin antibodies, which have been studied extensively, have been found to consist of pairs of high- and low-molecular-weight peptide chains, arranged in a Y shape.

Many of the amino acid sequences are similar for various antibodies, but short variable sections exist in both the heavy and the light chains. The antigens bond to these variable regions. The formation of the antibody-antigen complexes is responsible for the destruction of the antigen. A single kind of antibody binds only to one kind of antigen. Huge numbers of kinds of antigens exist, but, fortunately, the body manufactures enough antibodies to bind to just about any antigen. The number of antibodies synthesized by the body is enormous—trillions of antibodies and millions of different types each minute or two. When a particular antigen shows up, the call goes out to the appropriate lymphocytes to synthesize many additional copies of the proper antibody.

The invasion of a certain class of human white blood cells, the T-lymphocytes, by the virus HIV-1 (and perhaps other HIV viruses) leads to the condition called **acquired immune deficiency syndrome, A.I.D.S.** The virus destroys these cells and, thus, the body's ability to produce certain antibodies. Death from A.I.D.S. results from the body's inability to defend itself from other infections because it cannot produce these antibodies.

For Review

Summary

Molecules present in living organisms are called biomolecules and are often divided into four classes—proteins, carbohydrates, lipids, and nucleic acids—each having its own structure and functions.

Proteins are polymers composed of **amino acids,** molecules with the general formula H_3N^+—CHR—CO_2^-. The 20 amino acids that are genetically coded into proteins are distinguished by the nature of the R side chain on the α-carbon. Most amino acids are optically active. The combination of two or more amino acids by formation of **peptide bonds,**

$$\begin{array}{c} O \\ \parallel \\ -C-NH- \end{array}$$

produces a **polypeptide.** Proteins are large polypeptides that can fulfill a variety of functions because their structures vary widely. The **primary structure** of a protein is the order in which its amino acids are connected. The **secondary structure** is produced by the folding of the chain of amino acids to maximize hydrogen bonding between peptide bonds; an α-**helix** or a β-**sheet (pleated sheet)** is a common secondary structure. One of the most remarkable classes of proteins is the **enzymes,** which are catalysts of biochemical reactions.

Carbohydrates are aldehydes and ketones that have two or more —OH groups and that can be energy sources or energy stores in plants and animals. The simplest carbohydrates are the **monosaccharides,** such as the sugars glucose and fructose. **Disaccharides** are dimers composed of two monosaccharides; sucrose, for example, consists of glucose and fructose. **Starch,** a food storage carbohydrate found in plants; **glycogen** (a food storage carbohydrate found in animals) and **cellulose,** the carbohydrate that is the structural component of plants, are all **polysaccharides,** large polymers formed from glucose.

Fats and **oils** are members of the class of molecules known as **complex lipids,** esters formed from long-chain carboxylic acids (**fatty acids**) and glycerol. The closely related **phosphoglycerides** serve as the basis for cell membranes. **Simple lipids,** which include vitamin A and the steroids, do not contain fatty acids.

Nucleic acids are polymers of very high molecular weights that store and transfer genetic information. They also determine which proteins are synthesized by a cell. A nucleic acid consists of a polymeric chain of alternating phosphate groups and ribofuranose rings with nitrogen-containing bases bonded to the rings. The **DNA** polymer is a **double helix** that is held together by hydrogen bonds between pairs of specific bases, one on each strand of the helix. This base pairing makes each strand a complement of the other. The sequence of groups of three adjacent nucleotides (**codons**) in a DNA strand is the basis for coding the sequence of amino acids in proteins.

Protein synthesis is carried out in the body through a series of steps referred to as **replication, transcription,** and **translation.**

Gene mutations, such as those that occur in sickle-cell anemia, are substitutions, deletions, or additions of one or more nucleotides in the DNA molecule. Mutations can have both harmful and beneficial effects.

Research with **recombinant DNA** is a relatively new field that seeks to alter the nature of the protein being synthesized by cleaving the DNA chain, inserting a new piece of DNA in the opening, and resealing the chain.

Viruses (consisting of a protein coat, called a capsid, around a nucleic acid core) are responsible for a number of diseases that result from infection of cells by the viruses. The body sets up a defense against chronic infections through an **antibody immune response.**

The possible benefits to society from research in such fields as recombinant DNA are virtually without limit, but important moral and ethical choices must be made. The work must proceed with caution and good judgment.

Key Terms and Concepts

acid mucopolysaccharide (31.8)

acquired immune deficiency syndrome, A.I.D.S. (31.17)

adenosine triphosphate, ATP (31.14)

aerobic reaction (31.14)

aldose (31.7)

amino acid (31.2)

amino acid sequence (31.3)

anabolic pathway (31.14)

anaerobic reaction (31.14)

antibody (31.4, 31.17)

antibody immune response (31.17)

antigen (31.17)

carbohydrate (31.6)

cell (31.1)

cellulose (31.8)

chromosome (31.12)

cloning (31.15)

codon (31.12)

complex lipid (31.9)

deoxyribonucleic acid, DNA (31.11)

disaccharide (31.8)

double helix (31.11)

enzyme (31.4, 31.5)

gene (31.12)

gene cloning (31.15)

gene mutation (31.13)

genetic code (31.12)

genetic engineering (31.15)

glycolysis (31.14)

glycoprotein (31.8)

α-helix (31.4)

hormone (31.4)

hydrophilic (31.1)

hydrophobic (31.1)

immunoglobulin (31.17)

inhibition (31.5)

ketose (31.7)

lipid (31.9, 31.10)

metabolism (31.14)

monosaccharide (31.7)

nucleic acid (31.11)

nucleoside (31.11)

nucleotide (31.11)

oligopeptide (31.3)

peptide (31.3)

photosynthesis (31.14)

polypeptide (31.3)

polysaccharide (31.8)

primary structure (31.4)

protein (31.2, 31.3, 31.4)

protein structure (31.4)

protein synthesis (31.12)

quaternary structure (31.4)

recombinant DNA (31.15)

replication (31.12)

ribonucleic acid, RNA (31.11, 31.12)

saponification (31.9)

secondary structure (31.4)

β-sheet (31.4)

simple lipid (31.10)

tertiary structure (31.4)

transcription (31.12)

translation (31.12)

virus (31.16)

zwitterion (31.2)

Exercises

1. List the four principal chemical constituents of living cells. Identify the specific functions of each of these constituents.
2. Define *hydrophilic* and *hydrophobic*. What factor or factors determine whether a compound is hydrophilic or hydrophobic? Give an example of each class of compound.

Proteins

3. Define the following: amino acid, peptide, and protein.
4. Distinguish between the primary, secondary, tertiary, and quaternary structures of a protein.
5. Define what is meant by *denaturation*.

6. Define the following: enzyme, substrate, and product.
7. Describe what happens to the rate of an enzyme-catalyzed reaction from the instant that the enzyme and substrate are mixed until the substrate is consumed.
8. Into what simpler components may a protein be reduced?
9. Write the structure of the tripeptide CYS-LYS-GLU at neutral pH.
10. Write symbolic formulas for all of the possible tripeptides that could be formed from the amino acids GLY, ALA, and VAL.
11. Describe the process of protein synthesis by the body and indicate its significance.

12. Fungal lactase, a blue protein found in wood-rotting fungi, contains approximately 0.397% copper by mass. If the molecular weight of fungal lactase is approximately 64,000, how many copper atoms are there in each protein molecule?   *Ans. 4 Cu atoms/protein molecule*

Carbohydrates

13. What is the literal meaning of *carbohydrate,* and how did this name arise?
14. Which carbon atoms in glucose are asymmetric (Section 30.6)?
15. How does a pyranose differ from a furanose?
16. Write the name and formula for one example each of a monosaccharide, a disaccharide, and a polysaccharide.
17. What is the difference between the α and β isomers of glucopyranose?
18. How many grams of invert sugar can be produced by the reaction of 1.00 g of sucrose, $C_{12}H_{22}O_{11}$, with water?   *Ans. 1.05 g of invert sugar*

Lipids

19. Write the general formula for a neutral fat. How do fats and oils differ in composition?
20. Describe the common feature or features of complex and simple lipids, and provide an example of each class of compound. In what way do complex and simple lipids differ?
21. Define saponification and write an equation for the saponification of tripalmitylglycerol; the triester of palmitic acid, $CH_3(CH_2)_{14}CO_2H$.
22. Explain why the triacylglycerols do not form micelles.
23. The molecular weight of an unknown triglyceride can be estimated by determining the mass of KOH required to saponify a known mass of lipid. What is the molecular weight of a triglyceride if 3.12 mg of the triglyceride requires 0.590 mg of KOH for saponification?   *Ans. 890 g/mol (tripalmitylglycerol)*

Nucleic Acids

24. What is the connection between nucleic acids, amino acids, and proteins?
25. Write the complete structure of the following segment of DNA, reading from the 5′ end to the 3′ end: G-T-C-A.
26. If a single strand of DNA contains the nucleotide sequence A-G-G-C-T-C-A-G-C-T-A-G, reading from the 5′ to the 3′ carbon, what would be the sequence of nucleotides on the complementary strand of the DNA double helix, reading in the same direction?
27. What would be (a) the product of replication and (b) the product of transcription of the following fragment of a DNA molecule?

    5′   G-T-C-A-A-T-G-G-A   3′

28. What is the difference between the nucleotides adenosine monophosphate and deoxyadenosine monophosphate?

29. Complete the following table.

	DNA	RNA
Sugar unit		
Purine bases		
Pyrimidine bases		

30. What is a virus? What is its effect within the body?
31. Explain how sickle-cell anemia occurs.
32. Contrast cysteine and cystine.
33. How does the body build a defense against chronic infection?

Additional Exercises

34. Discuss the manner in which the energy requirements of a cell are satisfied.
35. How would you explain the observation that proteins can contain additional amino acids that are structurally related to the 20 genetically coded amino acids?
36. What sequence of amino acids would be specified by a single strand of DNA containing the nucleotide sequence A-G-G-C-T-C-A-G-C-T-A-G, reading from the 5′ to the 3′ carbon?
37. What evidence do we have that a twenty-first genetically coded amino acid will not be found?
38. Would the alanine in a solution with a pH of 5.0 migrate toward a positive electrode, a negative electrode, or show no preference for either electrode on electrolysis of the solution?
39. The synthesis of organic molecules in nature from carbon dioxide and water is an endothermic process. What is the source of the required energy?
40. Distinguish between aerobic and anaerobic processes.
41. What is recombinant DNA technology, and what are some of its consequences for society? What ethical and moral concerns arise? Cite examples both of benefits and problems.
42. In what sense can sucrose be considered an anhydride?
43. The degradation of both fatty acids, such as palmitic acid, and carbohydrates, such as glucose, leads to the formation of carbon dioxide and water. However, the degradation of palmitic acid produces $2\frac{1}{2}$ times as much energy as that of glucose, per gram of substance consumed. Explain this observation in terms of the average oxidation numbers of the carbon atoms in fatty acids, carbohydrates, and carbon dioxide.
44. If the degradation of 1 mol of glucose and 1 mol of palmitic acid yields 38 and 129 mol of ATP, respectively, calculate the moles of ATP produced per gram of lipid or carbohydrate consumed. The formula of palmitic acid is $CH_3(CH_2)_{14}CO_2H$.   *Ans. 0.21 mol ATP/g glucose; 0.504 mol ATP/g palmitic acid*

Semimicro
Qualitative Analysis

Chemistry of the Qualitative Analysis Scheme

Analytical samples

Q**ualitative analysis** is the determination of the identities, but not the concentrations, of the constituents present in a substance, a mixture of substances, or a solution. The process of determining the concentrations of the species present or the percent composition of a mixture is called **quantitative analysis.**

In this chapter we will discuss a method of analyzing a mixture of common metal ions that is based on the solubilities of their chlorides, sulfides, and carbonates. This scheme is not widely used by professional analytical chemists, because modern instrumental methods are more efficient for analysis of large numbers of samples. However, chemists occasionally use some of the test described in this chapter when they wish to determine the presence of a particular ion in a single sample.

The qualitative analysis scheme is included in general chemistry laboratory programs because it illustrates the applications of a variety of chemical phenomena, such as solution equilibria, solubility-product equilibria, buffers, acid-base reactions, acid-base equilibria, and oxidation-reduction reactions. It also illustrates many of the different types of chemical behavior that have been discussed in the preceding chapters.

32.1 Qualitative Analysis: An Example

During the course of a qualitative analysis, a cation is usually detected by formation of a solution with a characteristic color or by formation of a characteristic precipitate. The barium ion, for example, gives a characteristic yellow precipitate of barium chromate when it reacts with chromate ion (Fig. 32.1). However, certain other ions also give yellow precipitates with chromate, and these must be removed in a systematic manner before testing for barium. As an example, let us consider the analysis for barium ion in a sample that might also contain lead ion.

One cannot test for the presence of barium by adding potassium chromate to a solution that could contain lead, because both of these ions form an insoluble yellow chromate. However, it is possible to remove the interfering ion (lead) and then test for the presence of barium. The following steps will also separate the lead ion so we can test for it as well:

STEP 1. Precipitate any lead present as lead sulfide by adding an excess of hydrogen sulfide to an acidic solution of the mixture.

$$Pb^{2+} + H_2S \longrightarrow \underline{PbS} + 2H^+$$

(When describing the reactions of qualitative analysis it is customary to indicate the presence of a precipitate in a reaction by underlining the formula of the solid in the chemical equation.) Filtration or centrifugation of the mixture (Fig. 32.2) will remove any solid lead sulfide, leaving a solution that may contain barium ions.

STEP 2. Test for barium by addition of potassium chromate to the remaining solution. If Step 1 has been performed correctly, any yellow precipitate that forms at this point must be due to the presence of barium ion in the solution.

$$Ba^{2+} + 2K^+ + CrO_4^{2-} \longrightarrow \underline{BaCrO_4} + 2K^+$$

STEP 3. If no black precipitate forms in Step 1, we may assume that lead is not present in the solution. If a precipitate does form, we can confirm that it contains lead by dissolving the precipitate in nitric acid, evaporating off the nitric acid, and adding sulfuric acid. The formation of a white precipitate, lead sulfate, would confirm the presence of lead. (Note that we must separate lead ion from barium ion before testing for lead by adding sulfuric acid. Barium ion also gives a white precipitate with sulfuric acid.) The underline indicates a precipitate. We could alternatively use $BaCrO_4(s)$.

Flow charts are often used in qualitative analysis to illustrate the various steps in separation and identification. Steps 1 through 3 can be summarized in a flow chart like the one illustrated in Table 32.1.

In the qualitative analysis of a solution that may contain any or all of the common metal ions, the first step is that of separating the mixture of cations into groups that

Figure 32.1

A precipitate of barium chromate.

Figure 32.2

A precipitate of lead sulfide (left) settles rapidly to the bottom of the test tube (center left) when it is centrifuged. It can be separated easily by pouring off (center right) the centrifugate from (right) the precipitate.

Table 32.1 Flow chart for the separation and identification of lead and barium

$$Pb^{2+} \Big\} \xrightarrow{H_2S} \quad \underline{PbS} \xrightarrow{HNO_3} Pb^{2+} \xrightarrow{H_2SO_4} \underline{PbSO_4}\ (\text{white})$$

$$Ba^{2+} \qquad\qquad Ba^{2+} \xrightarrow{K_2CrO_4} \underline{PbCrO_4}\ (\text{yellow})$$

each contain only a small number of ions. This separation is usually accomplished by adding reagents that precipitate some, but not the majority, of the ions in solution.

We will discuss these separations in the next section and then discuss the separation and identification of the components of each group in subsequent sections. Finally, we will survey some of the principles employed in the separation and identification of the common metal ions.

32.2 Separation of the Metals into Analytical Groups

One series of steps for the separation of the common metal ions into groups is outlined in Table 32.2 and described briefly here.

Table 32.2 Separation of Groups I–V (Formulas for precipitates are underlined.)

Flow chart for the separation of metal ions into Groups I–V:

Starting mixture: Hg_2^{2+}, Ag^+, Pb^{2+}, Bi^{3+}, Cu^{2+}, Cd^{2+}, Hg^{2+}, As^{3+}, Sb^{3+}, Sn^{4+}, Co^{2+}, Ni^{2+}, Mn^{2+}, Fe^{3+}, Al^{3+}, Cr^{3+}, Zn^{2+}, Ba^{2+}, Sr^{2+}, Ca^{2+}, Mg^{2+}, NH_4^+, Na^+, K^+

Add HCl →

Group I — Chlorides: $\underline{Hg_2Cl_2}$, $\underline{AgCl}$, $\underline{PbCl_2}$

Remaining: Pb^{2+}, Bi^{3+}, Cu^{2+}, Cd^{2+}, Hg^{2+}, As^{3+}, Sb^{3+}, Sn^{4+}, Co^{2+}, Ni^{2+}, Mn^{2+}, Fe^{3+}, Al^{3+}, Cr^{3+}, Zn^{2+}, Ba^{2+}, Sr^{2+}, Ca^{2+}, Mg^{2+}, NH_4^+, Na^+, K^+

Add 0.3 M HCl, H_2S →

Group II — Sulfides: $\underline{PbS}$, $\underline{Bi_2S_3}$, $\underline{CuS}$, $\underline{CdS}$, $\underline{HgS}$, $\underline{As_2S_3}$, $\underline{Sb_2S_3}$, $\underline{SnS_2}$

Remaining: Co^{2+}, Ni^{2+}, Mn^{2+}, Fe^{2+}, Al^{3+}, Cr^{3+}, Zn^{2+}, Ba^{2+}, Sr^{2+}, Ca^{2+}, Mg^{2+}, NH_4^+, Na^+, K^+

Add NH_3, H_2S →

Group III — Sulfides and Hydroxides: $\underline{CoS}$, $\underline{NiS}$, $\underline{MnS}$, $\underline{FeS}$, $\underline{Al(OH)_3}$, $\underline{Cr(OH)_3}$, $\underline{ZnS}$

Remaining: Ba^{2+}, Sr^{2+}, Ca^{2+}, Mg^{2+}, NH_4^+, Na^+, K^+

Add NH_3, NH_4Cl, $(NH_4)_2CO_3$ →

Group IV — Carbonates: $\underline{BaCO_3}$, $\underline{SrCO_3}$, $\underline{CaCO_3}$

Group V — Soluble Ions: Mg^{2+}, NH_4^+, Na^+, K^+

1. THE METALS OF ANALYTICAL GROUP I. When dilute hydrochloric acid is added to a solution of the common metal ions (and ammonium ion), mercury(I) chloride, silver chloride, and lead chloride precipitate. The chlorides of all the other common metal ions are soluble in this acid solution and can be separated from those of Group I by filtration or centrifugation. The solubilities of Hg_2Cl_2 and AgCl are low enough that the mercury(I) ion and the silver ion are completely precipitated. However, lead chloride is somewhat soluble, and lead is found in Analytical Group II as well as in Group I.

2. THE METALS OF ANALYTICAL GROUP II. After the Group I chlorides have been separated, the solution is made acidic, and the Group II metal ions are precipitated as sulfides by the addition of hydrogen sulfide. The precipitate formed consists of the sulfides of lead, bismuth, copper, cadmium, mercury(II), arsenic, antimony, and tin. Mercury occurs in both Groups I and II because mercury(I) chloride is insoluble, whereas mercury(II) chloride is soluble in dilute hydrochloric acid.

3. THE METALS OF ANALYTICAL GROUP III. After the Group II sulfides have been separated, the solution is saturated with hydrogen sulfide, and then an excess of aqueous ammonia is added. Under these basic conditions the sulfides of cobalt, nickel, manganese, iron, and zinc precipitate, and the insoluble hydroxides of aluminum and chromium precipitate. The sulfides of Group III do not precipitate with the sulfides of Group II because the high acidity used during the precipitation of the Group II ions reduces the concentration of sulfide ion to such an extent that the solubility products of the Group III sulfides are not exceeded.

4. THE METALS OF ANALYTICAL GROUP IV. Ions of the Group IV metals (barium, strontium, and calcium) are precipitated as their carbonates by ammonium carbonate in a buffer of aqueous ammonia and ammonium chloride.

5. THE METALS OF ANALYTICAL GROUP V. The filtrate from the Group IV separation contains the sodium, potassium, magnesium, and ammonium ions. These ions constitute Group V.

32.3 Separation and Identification of the Ions of Group I

After the ions of Group I have been isolated from an unknown as the insoluble chlorides AgCl, Hg_2Cl_2, and $PbCl_2$ (Section 32.2), they can be separated and identified by the series of steps outlined in Table 32.3.

Table 32.3 Separation and identification of the ions of Group I

$$\left.\begin{array}{l}Hg_2Cl_2 \\ \overline{AgCl} \\ \overline{PbCl_2}\end{array}\right\} \xrightarrow{hot\ H_2O} \left.\begin{array}{l}Pb^{2+} \xrightarrow{CrO_4^{2-}} PbCrO_4\ (yellow) \\ \\ \left.\begin{array}{l}Hg_2Cl_2 \\ \overline{AgCl}\end{array}\right\} \xrightarrow{NH_3 + H_2O} Hg\ (black) + \underline{HgNH_2Cl}\ (white) \\ \\ \qquad\qquad\qquad Ag(NH_3)_2^+ \xrightarrow{HNO_3} \underline{AgCl}\ (white)\end{array}\right.$$

Lead chloride is about three times as soluble in hot water as in cold water and can be separated from silver chloride and mercury(I) chloride by extraction of the mixed chlorides with hot water. The $PbCl_2$ dissolves and the presence of the lead ion in the hot water can be confirmed by precipitation of yellow lead chromate.

$$Pb^{2+} + CrO_4^{2-} \longrightarrow \underline{PbCrO_4}$$

When aqueous ammonia is added to the remaining mixture of silver chloride and mercury(I) chloride, the silver chloride dissolves by forming soluble diamminesilver(I) chloride.

$$\underline{AgCl} + 2NH_3 \longrightarrow [Ag(NH_3)_2]^+ + Cl^-$$

In the presence of ammonia, mercury(I) chloride undergoes disproportionation, with the formation of finely divided metallic mercury, which is black, and mercury(II) amidochloride, $HgNH_2Cl$, which is white. The equation is

$$\underline{Hg_2Cl_2} + 2NH_3 \longrightarrow \underline{Hg} + \underline{HgNH_2Cl} + NH_4^+ + Cl^-$$

The formation of this black (or gray) precipitate proves the presence of mercury(I) ions.

The presence of the silver ion may be confirmed by adding nitric acid to the solution of $[Ag(NH_3)_2]^{2+}$ and Cl^-. This reprecipitates white silver chloride.

$$[Ag(NH_3)_2]^+ + Cl^- + 2H^+ + 2NO_3^- \longrightarrow \underline{AgCl} + 2NH_4^+ + 2NO_3^-$$

32.4 Separation of the Group II Ions into Divisions A and B

The Group II ions are separated from an unknown by precipitation of their sulfides (Section 32.2). HgS, As_2S_3, Sb_2S_3, and SnS_2 form soluble complex sulfides and dissolve in alkaline sulfide solutions, but PbS, Bi_2S_3, CuS, and CdS do not form these complex ions and remain as solids (Table 32.4). The cations whose sulfides do not dissolve in a solution of sodium sulfide constitute Division A of Group II. They are Pb^{2+}, Bi^{3+}, Cu^{2+}, and Cd^{2+}. Those cations that dissolve (Hg^{2+}, As^{3+}, Sb^{3+}, and Sn^{4+}) are Division B.

Table 32.4 Separation of Group II into Divisions A and B

PbS		
Bi_2S_3		
CuS		PbS
CdS	$\xrightarrow{Na_2S}$	Bi_2S_3
HgS		CuS
As_2S_3		CdS $\Big\}$ Division A
Sb_2S_3		HgS_2^{2-}
SnS_2		AsS_3^{3-}
		SbS_3^{3-} $\Big\}$ Division B
		SnS_3^{2-}

32.5 Separation and Identification of the Ions of Division A of Group II

The separation and identification of the ions of Division A of Group II (Section 32.4) are outlined in Table 32.5.

Table 32.5 Separation and identification of the ions of Group II, Division A

$$
\begin{array}{c}
\left.\begin{array}{l} PbS \\ Bi_2S_3 \\ CuS \\ CdS \end{array}\right\} \xrightarrow{HNO_3}
\left.\begin{array}{l} Pb^{2+} \\ Bi^{3+} \\ Cu^{2+} \\ Cd^{2+} \end{array}\right| \xrightarrow[H_2O]{H_2SO_4}
\begin{array}{l} PbSO_4 \\ \hline Bi^{3+} \\ Cu^{2+} \\ Cd^{2+} \end{array}
\end{array}
$$

$PbSO_4 \xrightarrow{NH_4CH_3CO_2} Pb(CH_3CO_2)_2 \xrightarrow{K_2CrO_4} PbCrO_4 \text{ (yellow)}$

$\underline{Bi(OH)_3} \text{ (white)} \xrightarrow{NaSn(OH)_3} \underline{Bi} \text{ (black)}$

$\left.\begin{array}{l} Cu(NH_3)_4{}^{2+} \\ Cd(NH_3)_4{}^{2+} \end{array}\right\} \xrightarrow{KCN} \begin{array}{l} Cu(CN)_2{}^- \\ Cd(CN)_4{}^{2-} \end{array} \xrightarrow{H_2S} \begin{array}{l} \underline{CdS} \text{ (yellow)} \\ Cu(CN)_2{}^- \end{array}$

$\left\{\begin{array}{l} \\ \end{array}\right. \xrightarrow{NH_3 + H_2O}$

$\left| \xrightarrow{K_4Fe(CN)_6} \underline{Cu_2Fe(CN)_6} \text{ (pink)} \right.$

$\underline{Cd_2Fe(CN)_6} \text{ (white)}$

The sulfides of Division A dissolve in hot dilute nitric acid. A sample reaction (for lead sulfide) is

$$3\underline{PbS} + 8H^+ + 2NO_3{}^- \longrightarrow 3Pb^{2+} + 3S + 2NO(g) + 4H_2O$$

The ions can be separated and identified as follows.

1. SEPARATION AND IDENTIFICATION OF LEAD. Lead is separated from bismuth, copper and cadmium by precipitating lead sulfate, $PbSO_4$, which is white. The lead sulfate is then dissolved in a solution of ammonium acetate, with the formation of the weak electrolyte lead acetate. The equation is

$$\underline{PbSO_4} + 2CH_3CO_2{}^- \longrightarrow Pb(CH_3CO_2)_2(aq) + SO_4{}^{2-}$$

When chromate ions are added to this solution, yellow lead chromate, $PbCrO_4$, precipitates and serves to confirm the presence of lead.

2. SEPARATION AND IDENTIFICATION OF BISMUTH. When an excess of aqueous ammonia is added to the solution of Bi^{3+}, Cu^{2+}, and Cd^{2+} from the lead sulfate separation, bismuth(III) hydroxide precipitates and soluble tetraammine complexes of Cu^{2+} and Cd^{2+} form. After $Bi(OH)_3$ is separated from the complexes, the presence of bismuth is confirmed by reducing it to the elemental state with sodium stannite, a strong reducing agent. The equation is

$$2\underline{Bi(OH)_3} + 3[Sn(OH)_3]^- + 3OH^- \longrightarrow 2\underline{Bi} + 3[Sn(OH)_6]^{2-}$$
$$\qquad\quad \text{Stannite ion} \qquad\qquad\qquad\qquad\quad \text{Stannate ion}$$

3. IDENTIFICATION OF COPPER AND CADMIUM. The presence of copper in the solution of tetraammine complexes may be apparent from the deep blue color of the complex ion $[Cu(NH_3)_4]^{2+}$. The formation of pink copper(II) hexacyanoferrate(II) serves as a more sensitive test. The equation for the reaction is

$$2[Cu(NH_3)_4]^{2+} + [Fe(CN)_6]^{4-} \longrightarrow \underline{Cu_2Fe(CN)_6} + 8NH_3$$

If cadmium is present, a white precipitate of cadmium hexacyanoferrate(II), $Cd_2Fe(CN)_6$, will also form.

The presence of cadmium ions may be confirmed by forming yellow cadmium sulfide. However, if we attempt to precipitate CdS in the presence of copper, the black color of CuS obscures the yellow color of CdS. By first adding cyanide ions to a solution containing Cu^{2+} and Cd^{2+}, the very stable dicyanocopper(I) complex will be formed; it will not react with sulfide ions.

$$2[Cu(NH_3)_4]^{2+} + 6CN^- \longrightarrow 2[Cu(CN)_2]^- + (CN)_2(g) + 8NH_3$$

The tetracyanocadmium complex also forms, but it reacts with the sulfide ion to form yellow CdS.

$$[Cd(CN)_4]^{2-} + S^{2-} \longrightarrow \underline{CdS} + 4CN^-$$

32.6 Separation and Identification of the Ions of Division B of Group II

The separation and identification of the ions of Division B, which are present as soluble complex sulfides (Section 32.4), are outlined in Table 32.6.

Table 32.6   Separation and identification of the ions of Group II, Division B

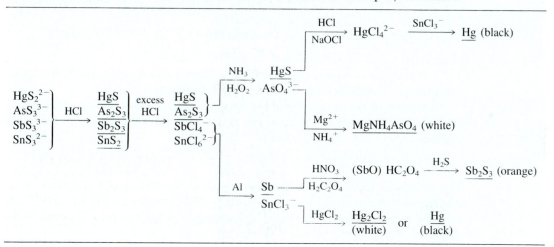

The sulfides of mercury, arsenic, antimony, and tin reprecipitate upon addition of hydrochloric acid to the solution of their soluble complex sulfides. These sulfides are separated and identified as described below.

1. SEPARATION AND IDENTIFICATION OF ANTIMONY AND TIN. Antimony and tin are separated by addition of an excess of concentrated hydrochloric acid, which dissolves their sulfides as complex ions.

$$\underline{Sb_2S_3} + 6H^+ + 8Cl^- \longrightarrow 2[SbCl_4]^- + 3H_2S(g)$$
$$\underline{SnS_2} + 4H^+ + 6Cl^- \longrightarrow [SnCl_6]^{2-} + 2H_2S(g)$$

The sulfides of mercury and arsenic remain undissolved.

Aluminum metal will reduce the antimony in this solution of concentrated hydrochloric acid to the metallic state and will reduce the tin(IV) ion to the tin(II) ion, thus effecting a separation of the two metals.

$$[SbCl_4]^- + Al \longrightarrow \underline{Sb} + Al^{3+} + 4Cl^-$$

$$3[SnCl_6]^{2-} + 2Al \longrightarrow 3[SnCl_3]^- + 2Al^{3+} + 9Cl^-$$

The tin(II) ion is a strong reducing agent, and the reduction of mercury(II) is used to confirm the presence of tin. Two reactions could occur, depending upon the concentration of tin in the solution.

$$[SnCl_3]^- + 2HgCl_4^{2-} \longrightarrow [SnCl_6]^{2-} + \underline{Hg_2Cl_2} + 3Cl^-$$

$$[SnCl_3]^- + Hg_2Cl_2 + Cl^- \longrightarrow [SnCl_6]^{2-} + \underline{2Hg}$$

Antimony metal reacts with nitric acid to form the insoluble oxide Sb_4O_6, which dissolves in oxalic acid, $H_2C_2O_4$.

$$4\underline{Sb} + 4H^+ + 4NO_3^- \longrightarrow \underline{Sb_4O_6} + 4NO + 2H_2O$$

$$\underline{Sb_4O_6} + 4H_2C_2O_4 \longrightarrow 4(SbO)HC_2O_4(aq) + 2H_2O$$

The presence of antimony is confirmed by precipitating the orange-red sulfide, Sb_2S_3.

2. IDENTIFICATION OF ARSENIC AND MERCURY. The remaining ions of Division B, As^{3+} and Hg^{2+}, are separated with a mixture of ammonia and hydrogen peroxide. In basic solution, hydrogen peroxide oxidizes As_2S_3 to the arsenate ion, AsO_4^{3-}.

$$\underline{As_2S_3} + 12OH^- + 14H_2O_2 \longrightarrow 2[AsO_4]^{3-} + 3SO_4^{2-} + 20H_2O$$

Mercury(II) sulfide is not dissolved by this treatment.

The presence of arsenic (in the form of arsenate ions) is confirmed by the formation of white crystalline magnesium ammonium arsenate.

A mixture of hydrochloric acid and hypochlorus acid (from adding sodium hypochlorite to HCl) is used to dissolve the precipitate of HgS in order to confirm the presence of mercury. The equation is

$$\underline{HgS} + 3H^+ + 3Cl^- + HClO \longrightarrow 2H^+ + [HgCl_4]^{2-} + S + H_2O$$

The solution is then boiled to decompose excess hypochlorous acid, which would interfere with the confirmatory test for mercury [HOCl oxidizes tin(II) to tin(IV)]. The presence of mercury(II) is confirmed by reducing $[HgCl_4]^{2-}$ to Hg_2Cl_2 (white) or Hg (black) by Sn(II), which is added as $[SnCl_3]^-$ ions.

32.7 Separation and Identification of the Ions of Group III

Some indication of the cations present in a solution containing the Group III ions can be obtained from the colors of the ions: Co^{2+}, pale red; Ni^{2+}, pale green; Zn^{2+}, colorless; Mn^{2+}, colorless in low concentrations; Fe^{2+}, pale green; Fe^{3+}, reddish brown; Al^{3+}, colorless; Cr^{3+}, dark green or blue. Due to the phenomenon of complementary colors (Section 26.10), a solution containing certain combinations of colored ions may appear colorless.

The sulfides and hydroxides of Group III (Section 32.2) can be separated into two small groups by addition of hydrochloric acid (Table 32.7). The sulfides of cobalt and nickel will precipitate only from basic solutions, but once formed these sulfides are only slightly soluble in dilute HCl. The other precipitates dissolve and can be separated from CoS and NiS.

Table 32.7 Subdivision of Group III

$\underline{CoS}$		$\underline{CoS}$
$\underline{NiS}$		$\underline{NiS}$
$\underline{MnS}$	HCl $\longrightarrow$	Mn^{2+}
$\underline{FeS}$		Fe^{2+}
$\underline{Al(OH)_3}$		Al^{3+}
$\underline{Cr(OH)_3}$		Cr^{3+}
$\underline{ZnS}$		Zn^{2+}

1. IDENTIFICATION OF COBALT AND NICKEL. The identification of cobalt and nickel is outlined in Table 32.8.

Table 32.8 Identification of cobalt and nickel

$$\left.\begin{array}{c}\underline{CoS}\\ \underline{NiS}\end{array}\right\} \xrightarrow[HNO_3]{HCl} \left.\begin{array}{c}Co^{2+}\\ Ni^{2+}\end{array}\right\} \xrightarrow[H_2O]{NH_3} \begin{array}{c}Co(NH_3)_6{}^{2+}\\ Ni(NH_3)_6{}^{2+}\end{array}$$

$$\xrightarrow[HCl]{NH_4SCN} Co(NCS)_4{}^{2-}\ (blue)$$

$$\xrightarrow{DMG} \underline{Ni(DMG)_2}\ (red)$$

The sulfides of cobalt and nickel readily dissolve in a mixture of nitric and hydrochloric acids.

$$3\underline{CoS} + 8H^+ + 2NO_3{}^- \longrightarrow 3Co^{2+} + 2NO(g) + 3S + 4H_2O$$
$$3\underline{NiS} + 8H^+ + 2NO_3{}^- \longrightarrow 3Ni^{2+} + 2NO(g) + 3S + 4H_2O$$

After removing the oxides of nitrogen, an excess of aqueous ammonia is added, forming the hexaammine complexes $[Co(NH_3)_6]^{2+}$ (pink) and $[Ni(NH_3)_6]^{2+}$ (blue).

In the ammoniacal solution, $[Ni(NH_3)_6]^{2+}$ will react with dimethylglyoxime to form a very insoluble, bright red coordination compound that confirms the presence of nickel.

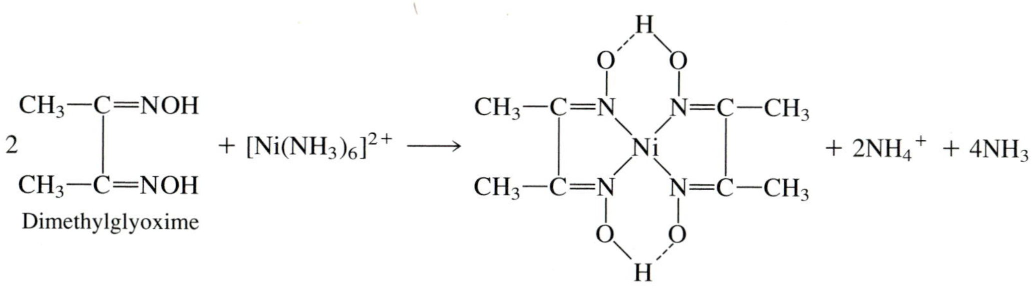

$$2\ \begin{array}{c}CH_3{-}C{=}NOH\\ |\\ CH_3{-}C{=}NOH\end{array} + [Ni(NH_3)_6]^{2+} \longrightarrow$$

Dimethylglyoxime

$+ 2NH_4{}^+ + 4NH_3$

A concentrated solution of ammonium thiocyanate is used to identify the cobalt ion; cobalt(II) and thiocyanate ions form a deep blue complex ion, $[Co(NCS)_4]^{2-}$.

$$[Co(NH_3)_6]^{2+} + 4NCS^- \longrightarrow [Co(NCS)_4]^{2-} + 6NH_3$$

The separation and identification of the remaining ions of Group III are outlined in Table 32.9.

Table 32.9 Separation and identification of manganese, iron, aluminum, chromium, and zinc

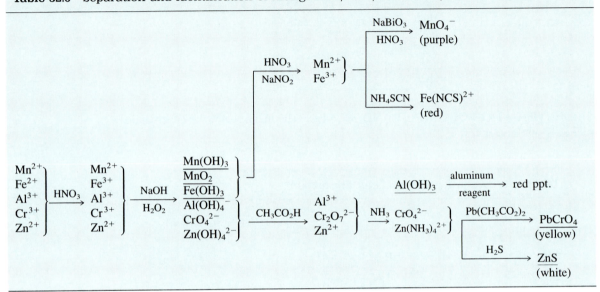

2. SEPARATION AND IDENTIFICATION OF IRON AND MANGANESE. The solution containing Mn^{2+}, Fe^{2+}, Al^{3+}, Cr^{3+}, and Zn^{2+} (Table 32.7) is treated with nitric acid to remove sulfide ion by oxidizing it to sulfur and to oxidize Fe^{2+} to Fe^{3+}. The equation for the oxidation of the iron is

$$3Fe^{2+} + 4H^+ + NO_3^- \longrightarrow 3Fe^{3+} + NO(g) + 2H_2O$$

In order to separate iron and manganese from aluminum, chromium, and zinc, the solution is treated with an excess of sodium hydroxide and with hydrogen peroxide. Iron and manganese hydroxides precipitate while aluminum and zinc form soluble hydroxide complexes.

$$Al^{3+} + 4OH^- \longrightarrow [Al(OH)_4]^- \text{ (colorless)}$$
$$Zn^{2+} + 4OH^- \longrightarrow [Zn(OH)_4]^{2-} \text{ (colorless)}$$

In base hydrogen peroxide oxidizes chromium(III) to the soluble chromate ion, CrO_4^{2-}. The hydrogen peroxide also oxidizes manganese(II) hydroxide to a mixture of manganese dioxide and manganese(III) hydroxide.

Following its separation, the precipitate of MnO_2, $Mn(OH)_3$, and $Fe(OH)_3$ is dissolved by a mixture of HNO_3 and $NaNO_2$. The nitrite ion reduces manganese(III) and manganese(IV) to manganese(II).

The presence of Fe^{3+} is confirmed by its reaction with the thiocyanate ion, SCN^-, to produce the blood-red complex ion $[Fe(NCS)]^{2+}$.

Manganese may be identified by oxidizing it to the purple permanganate ion with sodium bismuthate in nitric acid.

$$2Mn^{2+} + 5BiO_3^- + 14H^+ \longrightarrow 2MnO_4^- + 5Bi^{3+} + 7H_2O$$

3. IDENTIFICATION OF ALUMINUM, CHROMIUM, AND ZINC. When acetic acid is added to the solution containing $[Al(OH)_4]^-$, CrO_4^{2-}, and $[Zn(OH)_4]^{2-}$, aluminum and zinc are converted to their cations and the chromate ion to the dichromate ion.

$$2CrO_4^{2-} + 2H^+ \rightleftharpoons Cr_2O_7^{2-} + H_2O$$

The addition of excess aqueous ammonia precipitates aluminum as the hydroxide, complexes the zinc as the soluble $[Zn(NH_3)_4]^{2+}$ ion, and converts the dichromate ion to the chromate ion.

The presence of aluminum is confirmed by dissolving aluminum hydroxide in acetic acid and adding aluminon and $(NH_4)_2CO_3$. Aurin-tricarboxylic acid (aluminon) imparts a red color to the $Al(OH)_3$ formed.

The presence of the chromate ion is confirmed by precipitating yellow lead chromate, and zinc is confirmed by precipitation of white zinc sulfide.

32.8 Separation and Identification of the Ions of Group IV

The separation and identification of the ions of Group IV, which are precipitated as carbonates from an unknown (Section 32.2), are outlined in Table 32.10.

Table 32.10 Separation and identification of the ions of Group IV

$$\left.\begin{matrix}\underline{BaCO_3}\\ \underline{SrCO_3}\\ \underline{CaCO_3}\end{matrix}\right\} \xrightarrow{CH_3CO_2H} \left.\begin{matrix}Ba^{2+}\\ Sr^{2+}\\ Ca^{2+}\end{matrix}\right\} \xrightarrow[CH_3CO_2H,\ NH_4CH_3CO_2]{K_2CrO_4} \begin{cases}\underline{BaCrO_4} \xrightarrow{HCl} Ba^{2+} \xrightarrow{H_2SO_4} \underline{BaSO_4}\text{ (white)}\\ \left.\begin{matrix}Sr^{2+}\\ Ca^{2+}\end{matrix}\right\} \xrightarrow[K_2CrO_4,\ \text{alcohol}]{NH_3 + H_2O} \begin{cases}\underline{SrCrO_4}\text{ (yellow)}\\ Ca^{2+} \xrightarrow{(NH_4)_2C_2O_4} \underline{CaC_2O_4}\text{ (white)}\end{cases}\end{cases}$$

The carbonates of barium, strontium, and calcium dissolve in acetic acid. The reaction for barium carbonate is

$$\underline{BaCO_3} + 2CH_3CO_2H \longrightarrow Ba^{2+} + 2CH_3CO_2^- + H_2O + CO_2(g)$$

Barium is separated from strontium and calcium in a buffered acidic solution by precipitation of barium chromate, redissolved in hydrochloric acid, and confirmed by precipitation of barium sulfate upon addition of sulfuric acid. Additional evidence of the presence of the barium ion may be obtained by a flame test. When heated in the Bunsen burner flame, barium salts give it a yellow-green color.

Although the chromate ion concentration in a solution buffered with acetic acid and ammonium acetate is too low to precipitate strontium chromate, this compound will precipitate when aqueous ammonia and ethyl alcohol are added. The formation of a fine yellow precipitate of strontium chromate confirms the presence of strontium. Calcium chromate is soluble in this solution. A crimson color in a flame test also is characteristic of the strontium ion.

Following removal of strontium, the addition of ammonium oxalate precipitates calcium oxalate, CaC_2O_4, a white crystalline salt. Calcium ions give a brick-red color in a flame test.

32.9 Analysis of Group V

The analysis of the ions of Group V is outlined in Table 32.11. These are the ions that remain in solution after the ions of the other four groups have been removed (Section 32.2).

Table 32.11   Separation and identification of the ions of Group V

The sodium ion may be detected by the formation of sodium zinc uranyl acetate 6-hydrate, a yellow crystalline compound.

$$Na^+ + Zn^{2+} + 3UO_2^{2+} + 9CH_3CO_2^- + 6H_2O \longrightarrow$$
$$NaZn(UO_2)_3(CH_3CO_2)_9 \cdot 6H_2O$$

The sodium ion gives a yellow color in a flame test.

The formation of a white crystalline precipitate, $MgNH_4PO_4$, is evidence of the presence of the magnesium ion. The equation is

$$Mg^{2+} + NH_4^+ + HPO_4^{2-} \longrightarrow \underline{MgNH_4PO_4} + H^+$$

After the magnesium ammonium phosphate has been separated, it is dissolved in acetic acid, and magnesium reagent and sodium hydroxide are added to the solution. The magnesium hydroxide that is formed adsorbs the magnesium reagent [a dye, 4-nitrophenylazo-1-naphthol] and forms a blue precipitate that confirms the presence of magnesium.

Ammonium ions must be removed prior to the test for the potassium ion, because the precipitant for potassium ions will also precipitate ammonium ions. Concentrated nitric acid is added to the solution, it is evaporated to dryness, and then it is ignited to decompose the ammonium salts.

$$NH_4Cl \xrightarrow{\Delta} NH_3(g) + HCl(g)$$

$$NH_4NO_3 \xrightarrow{\Delta} N_2O(g) + 2H_2O(g)$$

When a solution of $Na_3[Co(NO_2)_6]$ is added to an acidic solution of the residue, the potassium ion precipitates in the form of the yellow salt $K_2Na[Co(NO_2)_6]$.

Potassium gives a characteristic violet flame test. By observing the flame through cobalt glass, which filters out the yellow color of the sodium ion, we may detect the potassium ion in the presence of the sodium ion.

Since ammonium ions are added in the course of the analysis at various points, it is necessary to test the original unknown solution for this ion. The presence of the ammonium ion may be detected by the addition of sodium hydroxide, with subsequent heating of the solution and liberation of gaseous ammonia.

$$NH_4^+ + OH^- \text{ (from NaOH)} \longrightarrow NH_3(g) + H_2O$$

The gaseous ammonia may be detected by moist red litmus paper, which turns blue in the presence of a base, and by its characteristic odor.

The Chemical Basis of Qualitative Analysis

The qualitative analysis scheme described in the preceding sections involves three principal components:

1. Systematic separation of the ions in a solution into groups of ions by formation of groups of insoluble precipitates.
2. Separation of the components of each group of precipitates by solubility or by reactivity.
3. Identification of each component by converting it to a compound whose solubility or reactivity is unique at that step in the scheme, if the scheme has been followed properly.

Control of chemical equilibrium plays a central role in qualitative analysis. Normally, equilibrium is controlled by adding an excess of a reactant. However, there are circumstances where more careful control is important. In the following sections we will consider how the separations and identifications in the qualitative analysis scheme are accomplished and how the equilibria in these steps are controlled.

32.10 Separation of the Groups

The ions of Analytical Groups I, II, III, and IV are separated by selective precipitation of chlorides, acid-insoluble sulfides, acid-soluble sulfides, and carbonates, respectively (Section 32.2). Group V ions remain in solution after the ions of the other groups have been removed. Each of these precipitations requires that the concentrations of the precipitating agents be carefully controlled to obtain a clean separation of the ions of each group. The conditions necessary are discussed below.

1. PRECIPITATION OF GROUP I. Hydrochloric acid is used to precipitate the chlorides of the ions of Group I: Ag^+, Hg_2^{2+}, and Pb^{2+}. The chlorides of all the other ions considered in this scheme are soluble under these conditions.

A slight excess of hydrochloric acid is used to prevent the precipitation of bismuth oxychloride, BiOCl, and antimony oxychloride, AsOCl, which form by the following reactions:

$$Bi^{3+} + H_2O + Cl^- \rightleftharpoons BiOCl + 2H^+$$
$$As^{3+} + H_2O + Cl^- \rightleftharpoons AsOCl + 2H^+$$

The hydrogen ions of the hydrochloric acid keep the equilibrium shifted to the left. The slight excess of chloride ions ensures more complete precipitation of Hg_2Cl_2, AgCl, and $PbCl_2$, due to the common ion effect (Section 17.7). For example, the common ion effect will reduce the concentration of Ag^+ from 1.3×10^{-5} M in a saturated solution of AgCl to 6.0×10^{-10} M in a solution that is 0.30 M in Cl^-. On the other hand, a large excess of chloride ions must be avoided, since silver chloride and lead chloride tend to react with chloride ions, forming the soluble complex ions $AgCl_2^-$ and $PbCl_4^{2-}$.

Lead chloride is about 2800 times more soluble than silver chloride and about 14,500 times more soluble than mercury(I) chloride. Thus lead is not completely precipitated, and some is carried through to Group II, although the mercury(I) ion and the silver ion are completely removed from solution.

2. PRECIPITATION OF GROUP II. The sulfides of the ions of Group II are insoluble in acid and will precipitate from an acidic solution when hydrogen sulfide is added (Section 32.2). The sulfides of most of the ions of Group III are insoluble in neutral solution, so the pH of the solution must be carefully controlled in order to separate the two groups.

Cadmium sulfide is the most soluble of the sulfides of Group II, and zinc sulfide is the least soluble of the sulfides of Groups III. Therefore, the separation of the sulfides of Group II from those of Group III is ensured if the sulfide ion concentration of the solution is controlled in such a way that the cadmium is completely precipitated as the sulfide without exceeding the solubility product of zinc sulfide. Hydrogen sulfide ionizes in aqueous solution according to the equation

$$H_2S \rightleftharpoons 2H^+ + S^{2-}$$

The addition of hydrogen ions represses the ionization of hydrogen sulfide and thus causes a decrease in the sulfide ion concentration of the solution. (The ionization of hydrogen sulfide and its control are discussed in detail in Section 17.9.) In a solution that is saturated with hydrogen sulfide and is 0.3 M in hydrogen ion, the sulfide ion concentration is just right to effect the separation of cadmium from zinc and, hence, of Group II from Group III. The theory for these separations is discussed in Section 18.9.

3. PRECIPITATION OF GROUP III. The ions of Group III form soluble chlorides, and their sulfides do not precipitate from acidic solution. However, the sulfides will precipitate from a basic solution saturated with hydrogen sulfide.

Analytical Group III contains the metallic ions Ni^{2+}, Co^{2+}, Mn^{2+}, Fe^{3+}, Al^{3+}, Cr^{3+}, and Zn^{2+}. The concentration of sulfide ions in the 0.3 M hydrogen ion solution of the Group II precipitation is so small that none of the solubility products of the Group III sulfides is exceeded. If the sulfide ion concentration is increased by addition of a base such as ammonia, the sulfides of Co^{2+}, Ni^{2+}, Mn^{2+}, Fe^{2+}, and Zn^{2+} precipitate. In addition, the hydroxides of Al^{3+} and Cr^{3+} precipitate from the basic solution. The increase in sulfide ion concentration is due to the reaction

$$2NH_3 + H_2S \rightleftharpoons 2NH_4^+ + S^{2-}$$

A strong base such as sodium hydroxide is not used to form the sulfide ion, and the solution is buffered to a pH of 9 with a mixture of ammonia and ammonium chloride in order to prevent the precipitation of magnesium hydroxide. Of the cations remaining in solution, only the magnesium ion (Group V) forms an insoluble hydroxide. The hydroxide ion concentration is kept just low enough to prevent the precipitation of magnesium hydroxide by buffering. This point is considered quantitatively in Example 18.9 of Section 18.9.

4. PRECIPITATION OF GROUP IV. The ions of Group IV are separated by adding ammonium carbonate to precipitate them as carbonates. The carbonate ion is basic, and its concentration must be controlled by adjusting the pH of the solution.

The concentration of the carbonate ion in a solution of ammonium carbonate is too low for complete precipitation of barium, strontium, and calcium; the carbonate ion reacts with water and forms the hydrogen carbonate ion.

$$CO_3^{2-} + H_2O \rightleftharpoons HCO_3^- + OH^-$$

This reaction is repressed by adding a basic buffer consisting of aqueous ammonia and ammonium chloride. This is the same buffer used in the precipitation of the sulfides and hydroxides of the Group III ions; it is used to prevent the precipitation of magnesium hydroxide. These conditions give a carbonate concentration high enough to precipitate $BaCO_3$, $SrCO_3$, and $CaCO_3$, but not $MgCO_3$. At the same time the hydroxide ion concentration is not high enough to precipitate $Mg(OH)_2$.

5. ISOLATION OF GROUP V. After the ions of Groups I–IV have been removed from the solution to be analyzed, the Group V ions remain in solution and can be identified. Prior to this point, ammonium chloride has been added to the solution, so the solution will give a positive test for the ammonium ion. The test for the presence of ammonium ion in the original sample must be run on a portion of the original solution.

32.11 Separation of Individual Ions

After the ions of the analysis scheme have been separated into five groups, each group of ions is further subdivided. In some instances individual ions are obtained; in other cases, small subdivisions consisting of several ions result. The techniques used in these separations include selective dissolution of a portion of a group precipitate, complete dissolution of a group precipitate followed by selective precipitation of one or more members of the group, and dissolution by formation of a complex ion. Examples of each type of separation will be discussed.

1. SELECTIVE DISSOLUTION. The separation of lead chloride from silver chloride and mercury(I) chloride (Section 32.3) during the analysis of Group I is an interesting example of selective dissolution. The solubility product (Section 18.1) of lead chloride increases with temperature, and the molar solubility of lead chloride is about three times greater in hot water than in cold water. The solubilities of silver chloride and mercury(I) chloride do not change significantly from hot water to cold water. Thus lead chloride dissolves selectively in hot water and can be separated from a mixture of the Group I chlorides.

The sulfides and hydroxides of Group III can be separated into two subdivisions by selective dissolution in 1 M hydrochloric acid (Section 32.7). The sulfides of cobalt(II) and nickel(II) will not precipitate from acidic solution, but once they precipitate from basic solution, the initial crystalline form of these sulfides (the α form) changes rapidly into a crystalline modification (the β form) that is not soluble in acid. Thus MnS, FeS, ZnS, $Al(OH)_3$, and $Cr(OH)_3$, each of which is soluble in acid, can be separated from CoS and NiS, which are insoluble in hydrochloric acid.

2. DISSOLUTION FOLLOWED BY SELECTIVE PRECIPITATION. The separation of lead in Group II (Section 32.5) is an example of dissolution of a set of precipitates followed by selective precipitation. Lead occurs in a set of four very insoluble sulfides, PbS, Bi_2S_3, CuS, and CdS. In order to dissolve these sulfides it is necessary to reduce the sulfide ion concentration in solution to zero by oxidation of the sulfide ion to free sulfur. An excess of nitric acid is used as the oxidizing agent; when this set of sulfides is heated with nitric acid, free sulfur and soluble nitrate salts are formed. A sample equation (for the reaction of copper sulfide) is

$$3\underline{CuS} + 8HNO_3 \longrightarrow 3Cu(NO_3)_2 + 3\underline{S} + 2NO + 4H_2O$$

Lead can be separated from the resulting solution of nitrate salts by precipitation of lead sulfate. However, before lead sulfate will precipitate, the excess nitric acid in the solution must be removed, because the sulfate ion is converted largely to hydrogen sulfate ion by hydrogen ions in high concentration.

$$SO_4{}^{2-} + H^+ \rightleftharpoons HSO_4{}^-$$

Under these conditions the concentration of sulfate ion is too low for the solubility product of lead sulfate to be exceeded. The removal of the nitric acid is accomplished by adding sulfuric acid and evaporating the solution to the point where sulfur trioxide fumes appear.

$$H_2SO_4 \longrightarrow H_2O + SO_3$$

The more volatile nitric acid will have been expelled before the SO_3 fumes appear. When the sulfuric acid solution is diluted with water, sulfate ions are formed, and the lead precipitates as lead sulfate.

$$Pb^{2+} + SO_4{}^{2-} \longrightarrow \underline{PbSO_4}$$

A second example of complete dissolution followed by selective precipitation is the isolation of barium ion from the Group IV precipitate (Section 32.8). In this case very careful control of the equilibrium concentration of the precipitating agent (the chromate ion) is required.

The Group IV precipitate consists of the carbonates of barium, strontium, and calcium. These carbonates dissolve readily in a solution of acetic acid. Barium chromate is less soluble than strontium chromate and calcium chromate, so barium can be separated from strontium and calcium in the solution.

The solubility products are 2×10^{-10} for $BaCrO_4$ and 3.6×10^{-5} for $SrCrO_4$. Calcium chromate is relatively very soluble. The problem in this separation, then, is to control the concentration of the chromate ion so that the concentration of the barium ion will be reduced to at least 0.0001 M and strontium chromate will not be precipitated from a solution that is 0.1 M in the strontium ion. Let us first calculate the concentration of chromate ion required to reduce the barium ion concentration to 0.0001 M.

$$[Ba^{2+}][CrO_4{}^{2-}] = 2 \times 10^{-10}$$

$$[CrO_4{}^{2-}] = \frac{2 \times 10^{-10}}{[Ba^{2+}]} = \frac{2 \times 10^{-10}}{1 \times 10^{-4}}$$

$$= 2 \times 10^{-6} \text{ M}$$

Now let us determine the maximum concentration of chromate ion that can be present without precipitating any strontium chromate when the concentration of the strontium ion is 0.1 M.

$$[Sr^{2+}][CrO_4^{2-}] = 3.6 \times 10^{-5}$$

$$[CrO_4^{2-}] = \frac{3.6 \times 10^{-5}}{[Sr^{2+}]} = \frac{3.6 \times 10^{-5}}{0.1}$$

$$= 4 \times 10^{-4} \text{ M}$$

Thus we find that the concentration of the chromate ion must be held between 2×10^{-6} M and 4×10^{-4} M to cause a nearly complete precipitation of $BaCrO_4$ without precipitation of any $SrCrO_4$.

The following equilibrium occurs in a solution containing chromate ions:

$$Cr_2O_7^{2-} + H_2O \rightleftharpoons 2CrO_4^{2-} + 2H^+$$

The concentration of chromate ions in a solution made by dissolving a dichromate salt in water can be controlled by adjusting the hydrogen ion concentration. The equation for the equilibrium constant is

$$\frac{[CrO_4^{2-}]^2[H^+]^2}{[Cr_2O_7^{2-}]} = 2.4 \times 10^{-15}$$

The chromate ion concentration is a function of the hydrogen ion concentration. The more acidic the solution, the lower will be the chromate ion concentration; the more basic the solution, the higher will be the chromate ion concentration. By substituting in the above expression, it can be shown that for a 0.01 M solution of dichromate ions, the hydrogen ion concentration must be adjusted to about 1×10^{-5} M (pH = 5) to maintain the chromate ion concentration at about 5×10^{-4} M. This concentration of chromate ions is within the range (calculated above) necessary for the separation of Ba^{2+} from Sr^{2+}. A buffer composed of acetic acid and ammonium acetate is used to maintain a pH of 5.

As chromate ions are removed from the system through precipitation of barium chromate, the concentration of chromate does not change appreciably. This is because the excess of dichromate ions serves as a reservoir for chromate ions, and as CrO_4^{2-} is used in precipitating $BaCrO_4$, the equilibrium

$$Cr_2O_7^{2-} + H_2O \rightleftharpoons 2CrO_4^{2-} + 2H^+$$

shifts to the right and generates more CrO_4^{2-}. The hydrogen ions that are formed unite with the acetate ions of the buffer, and a constant pH is maintained.

3. DISSOLUTION BY COMPLEX ION FORMATION. Many precipitates in the qualitative analysis scheme dissolve, and the metal ions they contain can be separated, because they form stable complex ions. Silver ion, for example, is separated from a mixture of silver chloride and mercury(I) chloride by addition of ammonia (Section 32.3). The soluble diamminesilver(I) complex forms.

$$\underline{AgCl} + 2NH_3 \rightleftharpoons Ag(NH_3)^+ + Cl^-$$

This equilibrium can be reversed, and silver chloride will reprecipitate if nitric acid is added to a solution of the complex. The acid reacts with ammonia and reduces the concentration of free ammonia to such an extent that the complex is no longer stable.

A second example of the use of complex ion formation to separate ions may be found in Group II. Antimony and tin are separated from mercury(II) sulfide and arsenic(III) sulfide because antimony(III) sulfide and tin(IV) sulfide react with hydrochloric acid and form the complex ions $SbCl_4^-$ and $SnCl_6^{2-}$, respectively, but the mercury and arsenic compounds do not form complex ions under these conditions (Section 32.6).

32.12 Confirmation of Ions

As the final step in the qualitative analysis of an ion, it is necessary to identify the ion by the solubility or reactivity of a compound that contains the ion and whose properties are unique at that step in the scheme. The final steps involve identification by complex ion formation, by oxidation-reduction, by precipitation of a colored compound, or by an acid-base reaction.

1. IDENTIFICATION INVOLVING COMPLEX IONS. The identification of cobalt, nickel, copper, and iron is based on the formation of colored complexes of these ions (Sections 32.5 and 32.7). The complexes and their colors are:

Group II:	$Cu(NH_3)_4^{2+}$	(blue)
Group III:	$Co(NCS)_4^{2-}$	(blue)
Group III:	$Ni(DMG)_2$	(red)
Group III:	$Fe(NCS)^{2+}$	(red)

In some cases complexes are formed in order to reduce the concentration of certain metal ions so as to prevent their interference with the test for other metal ions. The formation of $[Cu(CN)_2]^-$ (Section 32.5) is an example of such a use.

In general, the formation of complex ions is limited to ions of the transition metals and the post-transition metals. Thus the use of complex ions is limited to the members of Analytical Groups I–III, the groups that contain such metal ions.

2. IDENTIFICATION USING OXIDATION-REDUCTION REACTIONS. Oxidation-reduction reactions are used to confirm the presence of mercury(I), mercury(II), bismuth, manganese, chromium, and tin. The reaction in Group I involving mercury(I) occurs when aqueous ammonia is added to solid Hg_2Cl_2 (Section 32.3).

$$\underline{Hg_2Cl_2} + NH_3(aq) \longrightarrow \underline{Hg} + \underline{Hg(NH_2)Cl} + NH_4Cl$$

Mercury is both oxidized and reduced in this disproportionation reaction; elemental mercury (oxidation number = 0) and mercury(II) are formed.

The other reactions involve the oxidation or reduction of the ion that is to be identified with a second oxidizing or reducing agent. Bismuth is identified in Group II (Section 32.5) by the reduction of bismuth(III) to black elemental bismuth in basic solution with a complex ion containing the tin(II) ion.

$$2\underline{Bi(OH)_3} + 3[Sn(OH)_3]^- + 3OH^- \longrightarrow 2\underline{Bi} + 3[Sn(OH)_6]^{2-}$$

Tin is oxidized to tin(IV) during this reaction.

Chromium and manganese are identified in Group III by oxidation (Section 32.7). The chromium(III) ion is oxidized to the yellow chromate ion by hydrogen peroxide, whereas the manganese(II) ion requires BiO_3^-, a much stronger oxidizing agent than hydrogen peroxide, to form the purple permanganate ion.

There is an interesting relationship between mercury and tin in Group II (Section

32.5). If $SnCl_3^-$ is added to a solution of Hg^{2+} in hydrochloric acid, Hg (black) or Hg_2Cl_2 (white) is formed.

$$2[HgCl_4]^{2-} + [SnCl_3]^- \longrightarrow \underline{Hg_2Cl_2} + [SnCl_6]^{2-} + 3Cl^-$$

$$\underline{Hg_2Cl_2} + [SnCl_3]^- + Cl^- \longrightarrow 2\underline{Hg} + [SnCl_6]^{2-}$$

Thus a solution of $HgCl_4^{2-}$ can be used to test for the presence of tin(II) in an unknown, and a solution of $SnCl_3^-$ can be used to test for mercury(II) in an unknown.

Inasmuch as the active metal ions of Groups IA and IIA in the Periodic Table exhibit only one oxidation number in their compounds, the metal ions of Analytical Groups IV and V (the Analytical Groups that consist primarily of active metal ions) exhibit no oxidation-reduction chemistry.

3. IDENTIFICATION BY PRECIPITATION. Many of the final confirmatory tests for the presence of the various ions in the qualitative analysis scheme involve precipitation as the final test. Most of these tests are conclusive because the precipitate is the only one that will form under conditions of the test or because only the ion being considered will give the color of precipitate expected. The following example illustrates these considerations.

The confirmation of lead in Group I (Section 32.3) involves precipitation of yellow lead chromate. If the test solution contains no lead ion and no contamination, no precipitate will form. If the solution contains no lead ion but is contaminated with either silver or mercury(I) ions, a precipitate will form, but it will not be yellow. A solution containing lead ion contaminated with silver or mercury(I) ion will give a precipitate that is not yellow. A clean separation must be accomplished; the lead ion must be the only Group I ion in the solution before a yellow precipitate can be obtained.

4. IDENTIFICATION BY ACID-BASE REACTIONS. Although many acid-base reactions are used to dissolve precipitates and to adjust conditions during the course of an analysis, there is only one confirmatory test that can be considered to be a Brønsted acid-base reaction; it is the identification of the ammonium ion. The ammonium ion is identified when it reacts with a strong base and liberates ammonia.

$$NH_4^+ + OH^- \text{ (from a strong base)} \longrightarrow NH_3(g) + H_2O$$

If the reaction is warmed, gaseous ammonia is liberated and can be detected by its odor or by its reaction with damp red litmus paper (red litmus turns blue in base).

For Review

Summary

Twenty-four common cations can be separated and detected by the **qualitative analysis** scheme described in this chapter. The scheme involves the systematic separation of the ions into groups of precipitates, separation of the components of each group by differences in solubility or reactivity of the precipitates, and identification of each component by its characteristic solubility or reactivity.

Control of chemical equilibria is central in the separation and identification of these metal ions. Precipitation reactions, dissolution reactions, complex ion formation, acid-base reactions, and oxidation-reduction reactions are all employed during the course of separation and analysis.

Exercises

1. What are the three principal steps that are used to separate and identify the cations of the qualitative analysis scheme?
2. Distinguish between qualitative analysis and quantitative analysis.
3. A student has a sample that might contain lead and barium. She adds some hydrochloric acid, centrifuges to remove a white precipitate, and adds potassium chromate to the centrifugate. Is the formation of a yellow precipitate clear evidence for the presence of barium in the sample? Why? Is the absence of a yellow precipitate clear evidence for the absence of barium in the precipitate? Why?

Group Separations

4. Why is a slight excess of HCl added during the precipitation of Group I, but not a large excess?
5. Why is it necessary to add HCl during the precipitation of the sulfides of Group II when the sulfides of Group II are even less soluble in neutral solution?
6. Suggest a reason why $Al(OH)_3$ is formed during the precipitation of Group III instead of Al_2S_3.
7. Why is a buffer used during the precipitation of the carbonates of Group IV?
8. If the ions of Group I were not precipitated by the addition of hydrochloric acid at the beginning of an analysis, in which groups would these ions appear? (Check in the table of solubility products in Appendix D for useful information.)
9. If thallium were present in an unknown, in what two groups would the thallium ion appear?
10. Calculate the concentration of Hg_2^{2+} ion in a solution of a Group I sample containing 0.20 M HCl in contact with solid Hg_2Cl_2.   *Ans. 2.8×10^{-17} M*
11. What fraction of the cadmium(II) ion from a 0.10 M solution (of a Group II sample) remains in solution after precipitation of CdS if the hydrogen ion concentration in the final solution is 0.30 M and the solution is saturated with H_2S?   *Ans. 2.5×10^{-6}%*
12. What is the sulfide ion concentration in a solution used to separate the ions of Group III if the solution is saturated with hydrogen sulfide and is buffered to a pH of 9.0.   *Ans. 1×10^{-9} M*
13. What is the maximum aluminum concentration remaining in the solution resulting from the separation of the ions of Group III if the solution is saturated with hydrogen sulfide and is buffered to a pH of 9.0?   *Ans. 2×10^{-18} M*
14. Calculate the highest pH that can be used in the precipitation of the Group IV carbonates if the precipitation of $Mg(OH)_2$ is to be avoided. (Assume $[Mg^{2+}]$ is 0.10 M.)   *Ans. 9.09*

Group I

15. What metals are members of Group I? What oxidation numbers do they exhibit as they are carried through separation and identification?
16. Identify the oxidation-reduction reactions used in the separation and identification of the ions of Group I. Write chemical equations for these reactions.
17. Identify the precipitates used in the separation and identification of the ions of Group I. Write chemical equations for their formation.
18. Identify the complex ions formed during the separation and identification of the ions of Group I. Write chemical equations for their formation.
19. Identify any buffers used in the separation and identification of the ions of Group I.
20. Write chemical equations for the acid-base reactions that occur during the analysis of Group I.
21. Which metal ion (or metal ions) of Group I gives
    (a) a chloride that is soluble in hot H_2O?
    (b) a chloride that does not turn dark or dissolve in aqueous NH_3?
    (c) a white amide?
    (d) a chloride that dissolves in aqueous ammonia?
    (e) a chloride that dissolves in concentrated HCl?
22. In the reaction of Hg_2Cl_2 with NH_3, half the mercury is converted to Hg and the other half to mercury(II). Could the reaction have given 40% of one and 60% of the other? Explain your answer.
23. A solution of sodium chloride is used to recharge a water softener. What chemical tests from Group I could you use to see if all of the salt had been washed out of a softener before it was placed back into use? Which of these would be the most sensitive? The least sensitive?
24. Prepare a flow sheet for analysis of a mixture that might contain only mercury(I) and lead, leaving out all unnecessary steps.

Group II

25. What metals are members of Group II? What oxidation numbers do they exhibit as they are carried through separation and identification?
26. Identify the oxidation-reduction reactions used in the separation and identification of the ions of Group II. Write chemical equations for these reactions.
27. Identify the precipitates used in the separation and identification of the ions of Group II. Write chemical equations for their formation.
28. Identify the complex ions formed during the separation and identification of the ions of Group II. Write chemical equations for their formation.
29. Identify any buffers used in the separation and identification of the ions of Group II. Do these buffers give an acidic or a basic solution?

30. Write chemical equations for the acid-base reactions that occur during the analysis of Group II.

31. What ions are members of Division A of Group II? Of Division B?

32. How are the ions of Division A of Group II separated from the ions of Division B?

33. An unknown is suspected of containing only Bi^{3+} and Cu^{2+}. Will addition of aqueous NH_3 provide enough evidence to determine which of these ions is present?

34. How can one differentiate between samples of each of the following using only a single reaction?
    (a) HgS and CuS
    (b) $SnCl_4^-$ and $SnCl_6^{2-}$
    (c) CdS and SnS_2
    (d) $AsCl_3$ and $CuCl_2$
    (e) $BiCl_3$ and $CdCl_2$
    (f) a solution of NH_4Cl and a solution of NH_3, using an ion of Group II

35. For each of the following pairs identify a reagent that will dissolve one member but not the other.
    (a) HgS and As_2S_3
    (b) $[Cu(NH_3)_4](OH)_2$ and CuS
    (c) HgS and Bi_2S_3
    (d) As_2S_3 and Sb_2S_3
    (e) $PbSO_4$ and $CuSO_4$

36. Lead sulfate is insoluble in water or in sulfuric acid. Explain how the formation of $Pb(CH_3CO_2)_2$, a weak electrolyte, gives rise to the dissolution of lead sulfate in a solution of acetic acid.

37. Prepare a flow sheet for analysis of a mixture that might contain only lead, copper, and cadmium, leaving out all unnecessary steps.

38. Prepare a flow sheet for analysis of a mixture that might contain only mercury, arsenic, and tin, leaving out all unnecessary steps.

39. Prepare a flow sheet for analysis of a mixture that might contain only bismuth, cadmium, and antimony, leaving out all unnecessary steps.

40. Prepare a flow sheet for analysis of a mixture that might contain only copper, arsenic, and mercury, leaving out all unnecessary steps.

Group III

41. What ions are members of Group III? What oxidation numbers do they exhibit as they are carried through separation and identification?

42. Identify the oxidation-reduction reactions used in the separation and identification of the ions of Group III. Write chemical equations for these reactions.

43. Identify the precipitates used in the separation and identification of the ions of Group III. Write chemical equations for their formation.

44. Identify the complex ions formed during the separation

and identification of the ions of Group III. Write chemical equations for their formation.

45. Identify any buffers used in the separation and identification of the ions of Group III. Do these buffers give an acidic or a basic solution?

46. Write chemical equations for the acid-base reactions that occur during the analysis of Group III.

47. Outline a separation of Fe^{3+} from Cr^{3+}.

48. For each of the following pairs of ions, outline brief but effective tests that will distinguish between them.
    (a) Fe^{3+} and Zn^{2+} (b) Ni^{2+} and Cr^{3+}
    (c) Ni^{2+} and Al^{3+} (d) Mn^{2+} and Co^{2+}
    (e) Al^{3+} and Fe^{3+}

49. A solution of a single Group III ion is colorless. What ions might be present?

50. Prepare a flow sheet for analysis of a mixture that might contain only iron, cobalt, and aluminum, leaving out all unnecessary steps.

51. Prepare a flow sheet for analysis of a mixture that might contain only zinc, chromium, and manganese, leaving out all unnecessary steps.

Group IV

52. What ions are members of Group IV? What oxidation numbers do they exhibit as they are carried through separation and identification?

53. Identify any oxidation-reduction reactions used in the separation and identification of the ions of Group IV.

54. Identify the precipitates used in the separation and identification of the ions of Group IV. Write chemical equations for their formation.

55. Identify the complex ions formed during the separation and identification of the ions of Group IV. Write chemical equations for their formation.

56. Identify any buffers used in the separation and identification of the ions of Group IV. Do these buffers give an acidic or a basic solution?

57. Write chemical equations for any acid-base reactions that occur during the analysis of Group IV.

58. For each of the following pairs, indicate a single reagent that could be used to distinguish between the members.
    (a) $CaCl_2$ and $BaCl_2$
    (b) $Sr(NO_3)_2$ and $Ca(NO_3)_2$
    (c) $BaCrO_4$ and $CaCrO_4$

59. Why won't $BaCrO_4$ precipitate when the pH is 1?

60. Write the chemical equation for the reaction that causes the dissolution of the Group IV carbonates.

61. Prepare a flow sheet for analysis of a mixture that might contain only barium and calcium, leaving out all unnecessary steps.

62. Prepare a flow sheet for analysis of a mixture that might contain only strontium and calcium, leaving out all unnecessary steps.

Group V

63. What ions are members of Group V? What oxidation numbers do they exhibit as they are carried through separation and identification?
64. Identify any oxidation-reduction reactions used in the separation and identification of the ions of Group V.
65. Identify the precipitates used in the separation and identification of the ions of Group V. Write chemical equations for their formation.
66. Identify any complex ions formed during the separation and identification of the ions of Group V.
67. Identify any buffers used in the separation and identification of the ions of Group V. Do these buffers give an acidic or a basic solution?
68. Write chemical equations for the acid-base reactions that occur during the analysis of Group V.
69. Prepare a flow sheet for analysis of a mixture that might contain only sodium and potassium, leaving out all unnecessary steps.

Additional Exercises

70. Calculate the approximate value of the solubility product of $PbCl_2$ in hot water.   *Ans. 5×10^{-4}*
71. Aluminum hydroxide precipitates in a saturated solution of H_2S in a solution with a pH of 9.0 because the K_{sp} of $Al(OH)_3$ is smaller than that of Al_2S_3. Calculate the minimum possible value of the K_{sp} of aluminum sulfide.
    Ans. 4×10^{-63}
72. Would CuS be expected to dissolve in a 1.0 M solution of NH_3? Explain.
73. Calculate the molar solubility of Hg_2Cl_2
    Ans. 6.5×10^{-7} M
74. What is the solubility of $SrSO_4$ in g/100 mL of solution?   *Ans. 9.7×10^{-3} g/100 mL*
75. What concentration of phosphate ion, PO_4^{3-}, is required for formation of the first trace of $MgNH_4PO_4$ during the analysis of Group V if the concentrations of Mg^{2+} and NH_4^+ are 0.10 M.   *Ans. 2.5×10^{-11} M*

General Laboratory Directions

33

This course in qualitative analysis has two principal objectives. One of these is to give you the reasons for the analytical procedures and results in terms of the theory of ionic equilibria, especially that relating to weak electrolytes, solubility products, complex ions, and oxidation-reduction. The other objective is the practical one of introducing the sights and smells of chemical reactions and of teaching you careful laboratory manipulation, critical observation, and logical interpretation of observed results.

33.1 Classification of the Metals into Analytical Groups

Qualitative analysis pertains to the identification of the constituents present in a sample of a substance, a mixture of substances, or a solution. In the qualitative analysis of a solution that may contain any or all of the common metal ions, the first step is that of separating the ions into several groups, each of which contains ions exhibiting a common chemical property that is the basis of the separation. The separation of the common metal ions into groups is usually done as outlined in Section 32.2.

THE METALS OF ANALYTICAL GROUP I. When dilute hydrochloric acid is added to a solution containing all of the common metal ions (and ammonium ion), mercury(I) chloride, silver chloride, and lead chloride precipitate and can be separated by filtration or centrifugation.

THE METALS OF ANALYTICAL GROUP II. The Group II metals are precipitated as sulfides by the addition of hydrogen sulfide to a 0.3 M acidic solution, then separated. The precipitate formed consists of the sulfides of lead, bismuth, copper, cadmium, mercury(II), arsenic, antimony, and tin.

THE METALS OF ANALYTICAL GROUP III. After the Group II sulfides have been separated, the solution is saturated with hydrogen sulfide, and then an excess of aqueous ammonia is added to it. Under these basic conditions the sulfides of cobalt, nickel, manganese, iron, and zinc and the hydroxides of aluminum and chromium are precipitated, then separated.

THE METALS OF ANALYTICAL GROUP IV. Barium, strontium, and calcium are precipitated as carbonates by ammonium carbonate in the presence of aqueous ammonia and ammonium chloride.

THE METALS OF ANALYTICAL GROUP V. The filtrate from the Group IV separation contains the sodium, potassium, magnesium, and ammonium ions, which constitute Group V.

The flow sheet diagramming the separations of the metal ions into the various analytical groups is given in Table 32.2. Specific details for these separations are given in Chapters 34 through 38. Chapters 34 through 38 cover the analysis procedures for the metals. Chapter 39 contains procedures for the qualitative analysis of a sample of anions, and Chapter 40 provides the procedures to analyze solid materials.

Even though rather definite directions are given for each analysis to be carried out, no two analyses will be exactly alike. For this reason directions should never be followed blindly, to the letter, but with careful thought; procedures should be adapted to the particular problem at hand.

33.2 Equipment

In semimicro qualitative analysis volumes of solutions from 1 drop to about 5 mL are employed, and small test tubes, centrifuge tubes, capillary syringes, and medicine droppers are used to carry out the separations and identifying tests.

Appendixes L and M are lists of apparatus and of solutions and solid reagents that each student will need in carrying out the laboratory work of this course. Some of the necessary equipment will be stocked in the laboratory desk or in the stockroom. Other equipment will have to be fabricated. Wash *all* of the apparatus before beginning your laboratory work.

33.3 Wash Bottle and Stirring Rods

Wash bottles made of plastic are supplied to students in some laboratories. Alternatively, you may construct one using a 250-mL Florence flask and 6-mm glass tubing, as shown in Fig. 33.1. Make uniform bends with no constrictions, and fire-polish all ends of the glass tubing.

Keep the wash bottle filled with distilled or deionized water for use in the analytical procedures. Ordinary tap water may contain such ions as Ca^{2+}, Mg^{2+}, Fe^{3+}, Al^{3+}, Cl^-, SO_4^{2-}, and HCO_3^-. Since these ions are among those being tested for in the unknown solutions, all water used in the procedures and for cleaning glassware must be distilled or deionized.

In addition, make at least five glass stirring rods, approximately 15 cm in length and 3 mm in diameter. Fire-polish both ends of each rod.

33.4 Capillary Syringes

Standard medicine droppers of approximately 1-mL capacity are used for measuring and transferring solutions of reagents. These droppers deliver approximately 20 drops per milliliter. A second type of dropper, called a capillary syringe, is

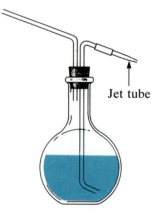

Jet tube

Figure 33.1

(top) A commercially available wash bottle. (bottom) A laboratory-constructed wash bottle.

used for the removal of liquids from precipitates held in small test tubes or centrifuge tubes. These may be supplied. If not, capillary syringes may be made from glass tubing by the following method. Heat the middle portion of a 7-in section of 8-mm glass tubing over a Bunsen burner flame, rotating until the glass softens. Remove the tube from the flame and slowly draw it out until the bore is about 1 mm. When the tube has cooled, cut the capillary at the midpoint and fire-polish the capillary ends. Flare the wide ends of the tubes by heating until soft and quickly pressing down against a flat metal surface. When the syringes are cold, attach medicine dropper bulbs to the flared ends. These syringes deliver approximately 40 drops per milliliter.

33.5 Reagents

The solids and solutions called for in the analytical procedures will be available in small (typically 10-mL) reagent bottles (Fig. 33.2). See Appendix M for a list of these reagents. Only a small quantity of the starred reagents will be needed during the course.

If you are to fill reagent bottles from stock bottles for your own use, be sure that your bottles are completely clean before filling them. To avoid mistakes, it is advisable to label each bottle before filling it.

33.6 Precipitation

Practically all of the precipitations are carried out in either 4-mL Pyrex test tubes or 2-mL conical test tubes. Check for completeness of precipitation by adding a drop of reagent to the solution (centrifugate) obtained from the separation of the precipitate. If the addition of more reagent to the solution shows that precipitation is incomplete, separate the mixture and test the second solution for completeness of precipitation.

The precipitating agent should be added slowly, preferably from a medicine dropper, and with vigorous shaking or stirring of the reaction mixture. The formation of larger crystals of the precipitate is favored by warming the solution, and separation of the precipitate should not be attempted before the crystals become large enough to settle.

A slight excess of the precipitating agent is added to reduce the solubility of the precipitate by the common ion effect (Section 17.7). However, a very large excess of the precipitating agent should be avoided, since it may actually increase the solubility of the precipitate. For example, in precipitating silver chloride a large excess of Cl^- will bring about the formation of $AgCl_2^-$ and thereby increase the solubility of AgCl. Many precipitates are dissolved, at least partially, by the formation of related complex ions.

33.7 Centrifugation of Precipitates

A precipitate may be separated from a liquid in a centrifuge (Fig. 33.3). Spinning a mixture of solid and liquid at high speed in a centrifuge forces the denser precipitate to the bottom of the containing tube by a centrifugal force that is many times the force of gravity. This accounts for the much shorter time required for settling a precipitate when centrifugation is employed. Colloidal precipitates require longer centrifugation than do crystalline precipitates because of the small size of colloidal particles.

Before centrifuging a test tube or centrifuge tube, prepare another tube to balance the first by filling an empty tube to the same level with water. Insert the tubes in

Figure 33.2
Reagent bottle equipped with dropper, used in semimicro analytical work.

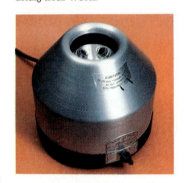

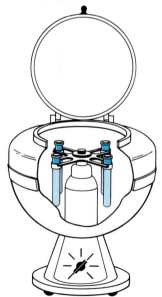

Figure 33.3
(top) A centrifuge. (bottom) Drawing of a centrifuge, with cutaway showing arrangement for holding test tubes used in centrifuging a precipitate.

opposite positions in the centrifuge, and turn the machine on. Allow the centrifugation to continue for at least 30 s. Turn the machine off and, after the rotation has stopped completely, remove the tubes.

33.8 Transfer of the Centrifugate

After centrifugation the precipitate should be found packed in the bottom of the tube. The supernatant liquid, or centrifugate, is separated from the precipitate by holding the tube at an angle of about 30° (Fig. 33.4) and removing the liquid by slowly

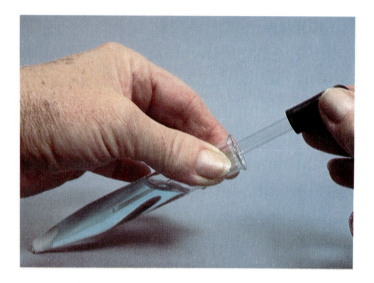

Figure 33.4

Capillary syringe used to separate centrifugate from precipitate.

drawing it into a capillary syringe. The tip of the syringe is held just below the surface of the liquid. As the pressure on the bulb is slowly released, causing the liquid to rise in the syringe, the capillary is lowered into the tube until all of the liquid is removed. As the capillary approaches the bottom of the tube, the tip must not be allowed to stir up the mixture by touching the precipitate.

33.9 Washing of the Precipitate

The precipitate left in the tube after the removal of a supernatant liquid is still wet with a solution containing the ions of this liquid. The precipitate must be washed, usually with water, to dilute the solution adhering to the precipitate. The wash liquid is added to the precipitate, and the mixture is stirred thoroughly. The mixture is then centrifuged to cause the precipitate to settle again. After centrifugation, the washings are removed by a capillary syringe as described in Section 33.8. A precipitate is usually washed at least twice. The first wash liquid is ordinarily saved and added to the first centrifugate. If the precipitate must be transferred to another container, the reagent to be used is added, the mixture is well stirred, and then it is poured into the other container. After the precipitate has settled, the supernatant liquid may be employed to remove any precipitate remaining in the centrifuge tube.

Failure to wash precipitates thoroughly is one of the principal sources of error in qualitative analyses.

33.10 Dissolution and Extraction of Precipitates

When all or a part of a precipitate is to be dissolved by a reagent, the solvent is added to the precipitate that is in the centrifuge tube and the mixture is stirred. The mixture is then separated by centrifugation, and the operation is repeated using fresh solvent. Often the extraction of a precipitate is more efficient at an elevated temperature.

33.11 Heating of Mixtures or Solutions

Whenever it is necessary to heat a mixture in order to bring about a precipitation or to dissolve or extract a precipitate, the test tube or centrifuge tube is placed in a water bath (Fig. 33.5) maintained at a suitable temperature. The water in the water bath should be kept hot throughout the work period.

33.12 Evaporation

It is often necessary to heat a solution to boiling and hold it at the boiling temperature in order to concentrate it or to remove a volatile acid or base, or even to evaporate it to dryness. Evaporation should be carried out in a small casserole or porcelain evaporating dish. The contents of the container should be agitated constantly while the heating continues. The evaporation of solutions contained in small test tubes should be avoided because the contents of the tube may be lost due to overheating.

Figure 33.5
Individual water bath for heating reaction mixtures.

33.13 Cleaning Glassware

Because small amounts of contaminants may give rise to erroneous results, all glassware used in the analytical procedures should be thoroughly cleaned before it is used. The cleaning should be done with a brush and some cleansing powder such as a synthetic detergent. The apparatus should then be rinsed first with tap water and finally with distilled water. Test tube brushes and centrifuge tube brushes are available. Medicine droppers, capillary syringes, and stirring rods should be cleaned, rinsed, and stored in a beaker of distilled water.

33.14 Flame Tests

Flame tests are made using a platinum wire sealed in the end of a piece of glass rod. The wire is cleaned by first dipping the end of it in 6 M HCl and then heating it in the burner flame (Fig. 33.6). Rather than use the hottest portion of the flame, one should bring the looped end of the wire slowly up to the edge of the flame. The platinum wire should never come into the reducing part of the flame (the darker blue inner cone). The sequence of dipping in acid and heating in the flame should be repeated until the wire no longer imparts a color to the flame. The wire loop is dipped in the solution to be tested and then heated in the flame. Do not rely on memory to judge the color imparted to the flame; a known solution of the ion should be tested and its color compared to that given by the unknown.

Figure 33.6
Platinum wire used in making flame tests for unknown ions.

33.15 Known and Unknown Solutions

You should know the details of the analytical procedures before attempting to analyze a solution of unknown composition. Therefore, known solutions containing all the ions of a given group are provided. You should become familiar with the separations and confirmatory tests for the ions of each group by practicing on a known solution before trying to determine the ions present in an unknown solution. You should especially note the quantities of precipitates obtained and the colors of precipitates and solutions as you proceed with the analysis of a known solution. These observations and the equations for the reactions involved should be recorded systematically in a notebook at the time the observations are made.

A typical set of laboratory assignments for knowns and unknowns is given in Appendix O.

33.16 Report Notebook

A notebook should be obtained (of the type designated by your instructor) in which to record in a systematic way the results of the analyses that are performed. All observations that are made should be recorded *as soon as they have been completed*. An example of one method of keeping the systematic record is shown below.

Sample Unknown Report

Name: _____ Section: _____ Date: _____

Instructor's Approval: _____ Ions Found: _____ Grade: _____

Cation Unknown Report

No.	Substance	Reagent	Result	Inference or Conclusion	Precipitate or Residue	Centrifugate or Solution
1	Group I	HCl	White ppt	Group I present	One or more of $AgCl$, $PbCl_2$, Hg_2Cl_2	Possibly Pb^{2+} or Hg^{2+}
2	Ppt from 1	Hot water	No visible action	Hg_2^{2+} and/or Ag^+ present; Pb^{2+} uncertain	Hg_2Cl_2 and/or $AgCl$	Possibly Pb^{2+}
3	Filtrate from 2 or 1	K_2CrO_4	Yellow ppt	Pb^{2+} present	$PbCrO_4$	
4	Residue from 2	$NH_3 + H_2O$	Residue dissolves completely	Hg_2^{2+} absent; Ag^+ probable		$[Ag(NH_3)_2]^+$
5	Filtrate from 4	HNO_3	White ppt	Ag^+ present	$AgCl$	
6						

The Analysis of Group I

34

34.1 Introduction

The procedures in this chapter are the first in the analysis of a solution of a known or unknown sample containing metal ions. Depending on the nature of the known or unknown, the solution may contain only the ions of Group I, or it may contain the ions of any or all of Groups I through V. If your sample is a general unknown, reserve a part of the original solution for the test for the ammonium ion in Group V. Review Section 32.2 for the classification of the metal ions into analytical groups before proceeding with the analysis of Group I. A schematic outline of the analysis for the Group I metals is given in the flow sheet in Table 34.1.

A discussion of the chemistry involved in each step of the analysis of each group will be given just prior to the laboratory procedure for the step. Careful study of these discussions will help you learn the chemistry involved in the analysis.

Table 34.1   Group I flow sheet

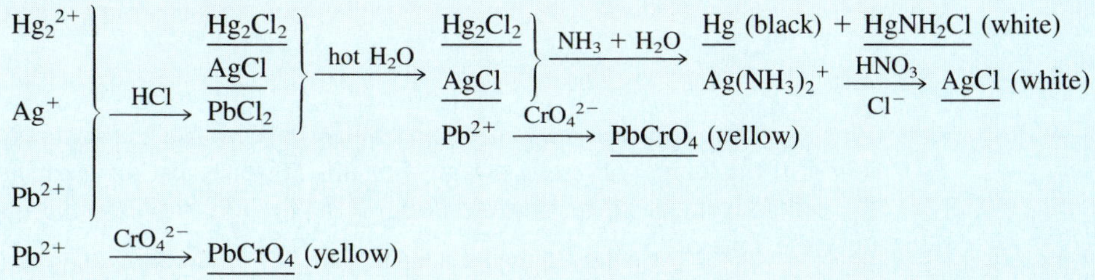

$$
\begin{array}{c}
Hg_2^{2+} \\
Ag^+ \\
Pb^{2+}
\end{array}
\left.\begin{array}{c}\\\\\\\end{array}\right\}
\xrightarrow{HCl}
\begin{array}{c}
Hg_2Cl_2 \\
AgCl \\
PbCl_2
\end{array}
\left.\begin{array}{c}\\\\\\\end{array}\right\}
\xrightarrow{hot\ H_2O}
\begin{array}{c}
Hg_2Cl_2 \\
AgCl \\
Pb^{2+}
\end{array}
$$

Hg_2Cl_2, $AgCl$ $\xrightarrow{NH_3 + H_2O}$ Hg (black) + $HgNH_2Cl$ (white)

$Ag(NH_3)_2^+ \xrightarrow[Cl^-]{HNO_3} AgCl$ (white)

$Pb^{2+} \xrightarrow{CrO_4^{2-}} PbCrO_4$ (yellow)

$Pb^{2+} \xrightarrow{CrO_4^{2-}} PbCrO_4$ (yellow)

34.2 Precipitation of Group I

Hydrochloric acid is the reagent used to precipitate the metal ions of Group I. The chlorides of mercury(I), silver, and lead are the only ones of the cations under consideration in this qualitative scheme that are insoluble in acid solutions. The equations for the precipitation of the metal ions of Group I are

$$Hg_2^{2+} + 2Cl^- \rightleftharpoons \underline{Hg_2Cl_2}$$
$$Ag^+ + Cl^- \rightleftharpoons \underline{AgCl}$$
$$Pb^{2+} + 2Cl^- \rightleftharpoons \underline{PbCl_2}$$

The mercury(I) ion and the silver ion are in effect completely removed from a solution but lead is not completely precipitated by chloride ion, and some of it is carried through to Group II.

A slight excess of hydrochloric acid is used in the precipitation of the Group I ions to prevent the precipitation of bismuth oxychloride and antimony oxychloride (Section 32.10) and to ensure more complete precipitation of Hg_2Cl_2, $AgCl$, and $PbCl_2$, due to the common ion effect (Section 17.7). A large excess of chloride ions must be avoided, since silver chloride and lead chloride tend to react with chloride ions, forming the soluble complex ions $[AgCl_2]^-$ and $[PbCl_4]^{2-}$.

PROCEDURE 1

Precipitation of Group I: Ag^+, Hg_2^{2+}, and Pb^{2+}

To 10 drops of the solution to be analyzed, add enough water to make a total volume of 1 mL. Add 2 drops of 6 M HCl to this solution (avoid using a large excess of HCl to eliminate the possibility of formation of $[AgCl_2]^-$ or $[PbCl_4]^{2-}$). Separate the precipitate by centrifugation. If your sample is a general unknown, reserve the solution (centrifugate) for the analysis of Groups II through V. If your sample is a Group I known or unknown, test the solution for lead as directed in Procedure 2 below. Treat the precipitate according to Procedure 2.

Precipitate of Group I chlorides: AgCl, Hg_2Cl_2, and $PbCl_2$

34.3 Separation and Identification of the Lead Ion

Lead chloride may fail to precipitate at all if the lead ion concentration is too low or if the temperature is too high. For this reason a test for lead should be made on the centrifugate from the original Group I separation provided the unknown contains only the ions of Group I.

Lead chloride is separated from silver chloride and mercury(I) chloride by extraction of the mixed chlorides with hot water. The presence of the lead ion may be confirmed by precipitation of the slightly soluble yellow lead chromate.

$$Pb^{2+} + CrO_4^{2-} \rightleftharpoons \underline{PbCrO_4}$$

Precipitate from Procedure 1:
Hg_2Cl_2, $AgCl$, and $PbCl_2$

Extract the precipitate twice with 10 drops of hot water to dissolve the $PbCl_2$. Reserve the residue for Procedure 3. Add 2 drops of 1 M K_2CrO_4 to the solution (hot water extract). A yellow precipitate of $PbCrO_4$ confirms the presence of lead.

PROCEDURE 2

Precipitate of $PbCrO_4$.

34.4 Separation and Identification of the Mercury(I) Ion

When aqueous ammonia is added to a mixture of silver chloride and mercury(I) chloride, the silver chloride dissolves by forming soluble diamminesilver chloride.

$$\underline{AgCl} + 2NH_3 \rightleftharpoons [Ag(NH_3)_2]^+ + Cl^-$$

In the presence of ammonia, mercury(I) chloride undergoes disproportionation, with the formation of finely divided metallic mercury, which is black, and mercury(II) amidochloride, $HgNH_2Cl$, which is white. The equation is

$$\underline{Hg_2Cl_2} + 2NH_3 \longrightarrow \underline{Hg} + \underline{HgNH_2Cl} + NH_4^+ + Cl^-$$

The formation of this black (or gray) precipitate is sufficient proof of the presence of mercury(I) ions.

Residue from Procedure 2:
Hg_2Cl_2 and $AgCl$

Extract the residue twice with 5 drops of 4 M aqueous ammonia. A black or gray residue confirms the presence of mercury(I). Reserve the aqueous ammonia extract for Procedure 4.

PROCEDURE 3

Mixture of precipitates of Hg (black) and Hg_2Cl_2 (white).

34.5 Identification of the Silver Ion

The presence of the silver ion may be confirmed by treating the aqueous ammonia extract of mercury(I) chloride and silver chloride with nitric acid. This causes the reprecipitation of white silver chloride.

$$[Ag(NH_3)_2]^+ + Cl^- + 2H^+ \rightleftharpoons \underline{AgCl} + 2NH_4^+$$

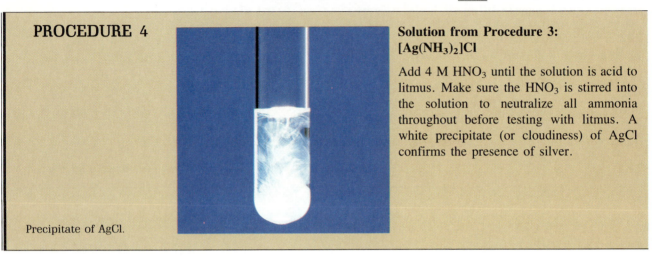

PROCEDURE 4

Solution from Procedure 3: [Ag(NH₃)₂]Cl

Add 4 M HNO_3 until the solution is acid to litmus. Make sure the HNO_3 is stirred into the solution to neutralize all ammonia throughout before testing with litmus. A white precipitate (or cloudiness) of AgCl confirms the presence of silver.

Precipitate of AgCl.

For Review

Exercises

1. What general statement can be made concerning the solubility of common chloride salts other than those of the cations of Analytical Group I?

2. Write out the flow sheet for the Group I analysis.

3. Give the color of each of the following: AgCl, $PbCl_2$, Hg_2Cl_2, $PbCrO_4$, Hg, $HgNH_2Cl$, and $[Ag(NH_3)_2]Cl$ (in solution).

4. Why must a large excess of chloride ions be avoided in the precipitation of the Group I chlorides?

5. In the case of an unknown containing only the cations of Group I, why is it advisable to make a confirmatory test for lead on the filtrate from the Group I precipitation?

6. Select a reagent used in the analysis of Group I that will in one step separate each of the following pairs:
   (a) Hg_2Cl_2, $PbCl_2$ (b) Hg_2Cl_2, AgCl
   (c) AgCl, $CuCl_2$ (d) Hg_2^{2+}, Hg^{2+}
   (e) AgCl, $PbCl_2$

7. In terms of ionic equilibria and solubility product theory, explain the dissolution of silver chloride in aqueous ammonia and its reprecipitation with nitric acid.

8. Show that the reaction of ammonia with mercury(I) chloride is an oxidation-reduction reaction.

9. What prevents the slightly soluble oxychlorides of bismuth and antimony from precipitating with the Group I chlorides?

10. A solution is 0.020 M in both Pb^{2+} and Ag^+. If Cl^- is added to this solution, what is the concentration of Ag^+ when $PbCl_2$ begins to precipitate? (K_{sp} for AgCl is 1.8×10^{-10}, and K_{sp} for $PbCl_2$ is 1.7×10^{-5}.)
    Ans. 6.2×10^{-9} M

11. Calculate the concentration of Ag^+ in a 0.0010 M solution of $[Ag(NH_3)_2]Cl$ that is 0.25 M in aqueous ammonia. (K_f for $[Ag(NH_3)_2]^+$ is 1.6×10^7.)   *Ans. 1.0×10^{-9} M*

12. From the K_{sp} data given in Appendix D, calculate the number of cations present in 1.0 mL of saturated solutions of Hg_2Cl_2, AgCl, and $PbCl_2$.
    Ans. 3.9×10^{14} Hg_2^{2+} ions;
    8.1×10^{15} Ag^+ ions; 9.8×10^{18} Pb^{2+} ions

13. Calculate the solubility of Hg_2Cl_2 in 0.020 M NaCl. Compare this solubility with that of Hg_2Cl_2 in water. (K_{sp} for Hg_2Cl_2 is 1.1×10^{-18}.)
    Ans. 2.8×10^{-15} M in NaCl; 6.5×10^{-7} M in water

14. Calculate the solubility of AgCl in 6.0 M HCl. (The formation constant for $Ag^+ + 2Cl^- \rightarrow [AgCl_2]^-$ is 2.5×10^5, and K_{sp} of AgCl is 1.8×10^{-10}.)
    Ans. 2.7×10^{-4} M

The Analysis of Group II

35

T he solution to be analyzed may be a Group II known or unknown, or it may be the solution from the Group I separation. In either case proceed according to Procedure 1 below. See Table 35.1 (p. 960) for the Group II flow sheet.

35.1 Precipitation of Group II Sulfides

1. SEPARATION OF GROUP II FROM GROUPS III–V. Ions of lead, bismuth, copper, cadmium, mercury(II), arsenic, antimony, and tin form sulfides that are insoluble in solutions that are 0.3 M in hydrogen ion. Cadmium sulfide is the most soluble of the sulfides of Group II, and zinc sulfide is the least soluble of the sulfides of Group III. Therefore, the separation of the sulfides of Group II from those of Group III is ensured if the sulfide ion concentration of the solution is controlled in such a way that the cadmium is completely precipitated as the sulfide without exceeding the solubility product of zinc sulfide (Section 32.10). In a solution that is saturated with hydrogen sulfide and 0.3 M in hydrogen ion, the sulfide ion concentration is just right to effect the separation of cadmium from zinc and, hence, of Group II from Group III.

The hydrogen sulfide for the precipitation of the Group II cations is supplied by thioacetamide, CH_3CSNH_2, which reacts with water in hot acidic solutions, with the formation of ammonium acetate and hydrogen sulfide.

$$CH_3CSNH_2 + 2H_2O \xrightarrow{\triangle} CH_3CO_2^- + NH_4^+ + H_2S$$

Table 35.1 Group II flow sheet

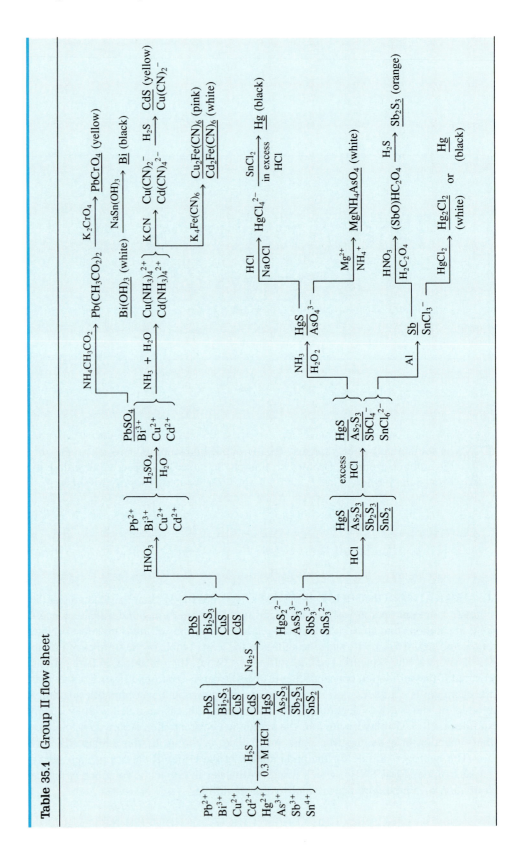

The precipitation of the Group II sulfides may be represented by the following equations:

$$Pb^{2+} + H_2S \rightleftharpoons \underline{PbS} + 2H^+$$
$$2Bi^{3+} + 3H_2S \rightleftharpoons \underline{Bi_2S_3} + 6H^+$$
$$Cu^{2+} + H_2S \rightleftharpoons \underline{CuS} + 2H^+$$
$$Cd^{2+} + H_2S \rightleftharpoons \underline{CdS} + 2H^+$$
$$Hg^{2+} + H_2S \rightleftharpoons \underline{HgS} + 2H^+$$
$$2As^{3+} + 3H_2S \rightleftharpoons \underline{As_2S_3} + 6H^+$$
$$2Sb^{3+} + 3H_2S \rightleftharpoons \underline{Sb_2S_3} + 6H^+$$
$$Sn^{4+} + 2H_2S \rightleftharpoons \underline{SnS_2} + 4H^+$$

2. COLORS OF THE SULFIDES OF GROUP II. The sulfides of lead(II), copper(II), and mercury(II) are all black. The sulfides of tin(IV), cadmium, and arsenic(III) are yellow, that of bismuth is dark brown or black, and that of antimony(III) is orange-red (sometimes black).

Tin(II) sulfide, SnS, forms a gelatinous precipitate that does not dissolve in sodium sulfide solution, while SnS_2 is readily soluble in this reagent. For this reason tin(II) is oxidized to tin(IV) by adding nitric acid and heating the original solution before the sulfides of the Group II ions are precipitated.

Precipitation of Group II: Pb^{2+}, Bi^{3+}, Cu^{2+}, Cd^{2+}, Hg^{2+}, As^{3+}, Sb^{3+}, Sn^{2+}, Sn^{4+}   **PROCEDURE 1**

Add 5 drops of 4 M HNO_3 to 10 drops of the known or unknown (or to the centrifugate from the Group I separation) and evaporate the solution not quite to dryness, ending with a moist residue (Section 33.12). Cool and add 10 drops of water. Add 1 M aqueous ammonia until the solution (or mixture) is just basic to litmus. (Dip a stirring rod into the solution and then touch the moist end of the rod to a piece of litmus paper to test for acidity or basicity of solutions.) Add 1.0 M HCl until the solution is just acidic to litmus. Now add 2 drops of 6 M HCl and dilute the solution to 1.5 mL. Add 10 drops of 5% thioacetamide solution. The solution should now be 0.3 M in hydrogen ion if the directions have been followed carefully. Heat the solution contained in a test tube in a hot water bath for 10 min. Sufficient reaction of the thioacetamide to get complete precipitation of the Group II sulfides is essential. Add 1 mL of water and heat the mixture for another 5 min. Separate the precipitate. If working on a general unknown, reserve the centrifugate for the analysis of Groups III through V. Wash the precipitate twice with 10 drops of 0.1 M HCl. Add the first 10 drops of wash solution to the original centrifugate and discard the rest.

35.2 Separation of the Group II Ions into Divisions A and B

The separation of the Group II ions into two divisions is based on the fact that HgS, As_2S_3, Sb_2S_3, and SnS_2 form soluble complex sulfides in alkaline sulfide solutions, whereas PbS, Bi_2S_3, CuS, and CdS do not form these complex ions (Section 32.4).

The reagent used in this separation is sodium sulfide, which is produced by the reaction of hydrogen sulfide (formed from thioacetamide) with sodium hydroxide. The equation is

$$2Na^+ + 2OH^- + H_2S \rightleftharpoons 2Na^+ + S^{2-} + 2H_2O$$

The presence of an excess of sodium hydroxide increases the concentration of the sulfide ion. This provides the relatively high concentration of sulfide ion required for the dissolution of the Division B sulfides, particularly HgS. The equations are

$$\underline{HgS} + S^{2-} \rightleftharpoons [HgS_2]^{2-} \quad [\text{dithiomercurate(II)}]$$
$$\underline{As_2S_3} + 3S^{2-} \rightleftharpoons 2[AsS_3]^{3-} \quad [\text{trithioarsenate(III)}]$$
$$\underline{Sb_2S_3} + 3S^{2-} \rightleftharpoons 2[SbS_3]^{3-} \quad [\text{trithioantimonate(III)}]$$
$$\underline{SnS_2} + S^{2-} \rightleftharpoons [SnS_3]^{2-} \quad [\text{trithiostannate(IV)}]$$

The cations whose sulfides fail to dissolve in the presence of a high concentration of sulfide ion constitute Division A of Group II. They are Pb^{2+}, Bi^{3+}, Cu^{2+}, and Cd^{2+}.

PROCEDURE 2

Precipitate from Procedure 1: PbS, Bi$_2$S$_3$, CuS, CdS, HgS, As$_2$S$_3$, Sb$_2$S$_3$, and SnS$_2$

Add 12 drops of 4 M NaOH and 4 drops of thioacetamide solution to the precipitate. Heat the mixture in a hot water bath for 5 min. Separate the mixture and reserve the solution. Treat the precipitate a second time with 12 drops of 4 M NaOH and 4 drops of thioacetamide solution. Heat the mixture in a hot water bath for 5 min. Separate the mixture and add the solution to that obtained from the first treatment. Save the combined solutions (complex sulfides of the Division B ions) for Procedure 6. The residue consists of the sulfides of the ions of Division A.

35.3 Separation and Identification of Lead

The undissolved sulfides of Division A (PbS, Bi$_2$S$_3$, CuS, and CdS) are washed with a solution of ammonium nitrate. This is used rather than pure water to decrease the tendency of the sulfides to become colloidal. In a colloidal condition these sulfides would pass into the centrifugate and be partly lost.

The sulfides are then dissolved by treatment with hot dilute nitric acid. The acid oxidizes the sulfide ion to elemental sulfur and causes the metal sulfides to dissolve (Section 32.11). In the case of lead sulfide, the net reaction is

$$3\underline{PbS} + 8H^+ + 2NO_3^- \longrightarrow 3Pb^{2+} + 3\underline{S} + 2NO(g) + 4H_2O$$

Lead is separated from bismuth, copper, and cadmium by precipitating lead sulfate, $PbSO_4$, which is white. The sulfates of the other three metals are soluble. The nitric acid in the solution must first be removed because the sulfate ion is converted largely to hydrogen sulfate ion by hydrogen ions in high concentration.

$$SO_4^{2-} + H^+ \rightleftharpoons HSO_4^-$$

Under these conditions the concentration of sulfate ion is too low for the solubility product of lead sulfate to be exceeded. The removal of the nitric acid is accomplished by adding sulfuric acid and evaporating the solution to the point where sulfur trioxide fumes appear.

$$H_2SO_4 \longrightarrow H_2O + SO_3$$

The more volatile nitric acid will have been expelled before the SO_3 fumes appear. When the sulfuric acid solution is diluted with water, sulfate ions are formed and the lead precipitates as lead sulfate. The equations are

$$H_2SO_4 \rightleftharpoons H^+ + HSO_4^-$$
$$HSO_4^- \rightleftharpoons H^+ + SO_4^{2-}$$
$$Pb^{2+} + SO_4^{2-} \rightleftharpoons \underline{PbSO_4}$$

The lead sulfate is then dissolved in a solution of ammonium acetate, with the formation of slightly ionized lead acetate. The equation is

$$PbSO_4 + 2CH_3CO_2^- \rightleftharpoons Pb(CH_3CO_2)_2 + SO_4^{2-}$$

When chromate ions are added to this acetate solution, yellow lead chromate, $PbCrO_4$, precipitates.

$$Pb(CH_3CO_2)_2 + CrO_4^{2-} \rightleftharpoons \underline{PbCrO_4} + 2CH_3CO_2^-$$

Residue from Procedure 2: PbS, CuS, CdS, and Bi$_2$S$_3$ **PROCEDURE 3**

Wash the residue with 1 mL of water to which 2 drops of 1 M NH_4NO_3 have been added. Discard the wash solution. Add 15 drops of 6 M HNO_3 to the residue and heat the mixture in a hot water bath for several minutes to dissolve the sulfides. Separate and discard any sulfur that is formed by the oxidation of sulfide ions, and transfer the solution to a casserole or evaporating dish. Add 5 drops of 4 M H_2SO_4 and evaporate the solution under the hood (very important) until white SO_3 fumes appear. (The SO_3 fumes are dense, not at all like steam. Do not prolong the heating after SO_3 fumes appear.) Cool and add 1 mL of water to the mixture. Warm the mixture and stir up the precipitate. (The precipitate, $PbSO_4$, may appear slowly, especially if the solution is too acidic.) Separate the mixture. Reserve the solution for Procedure 4. Wash the residue ($PbSO_4$) with a few drops of water and discard the wash solution. Extract the residue (Section 33.10) with a mixture of 5 drops of 1 M ammonium acetate ($NH_4CH_3CO_2$) and 1 drop of 1 M acetic acid (CH_3CO_2H). Add 1 drop of 1 M K_2CrO_4 to the extract. Scratch the inside wall of the test tube with a glass rod to initiate precipitation. The formation of a yellow precipitate confirms the presence of lead as the chromate, $PbCrO_4$.

Precipitate of $PbSO_4$

Precipitate of $PbCrO_4$.

35.4 Separation and Identification of Bismuth

When a small amount of aqueous ammonia is added to the solution from the lead sulfate separation, the hydroxides of bismuth (white), copper (pale blue), and cadmium (white) are precipitated. The equations are

$$Bi^{3+} + 3NH_3 + 3H_2O \longrightarrow \underline{Bi(OH)_3} + 3NH_4^+$$
$$Cu^{2+} + 2NH_3 + 2H_2O \longrightarrow \underline{Cu(OH)_2} + 2NH_4^+$$
$$Cd^{2+} + 2NH_3 + 2H_2O \longrightarrow \underline{Cd(OH)_2} + 2NH_4^+$$

When an excess of aqueous ammonia is added, the hydroxides of copper and cadmium dissolve by forming tetraammine complexes, but bismuth(III) hydroxide does not dissolve.

$$Bi(OH)_3 + NH_3 \longrightarrow \text{no reaction}$$
$$\underline{Cu(OH)_2} + 4NH_3 \longrightarrow [Cu(NH_3)_4]^{2+} + 2OH^-$$
$$\underline{Cd(OH)_2} + 4NH_3 \longrightarrow [Cd(NH_3)_4]^{2+} + 2OH^-$$

After it is separated from the complexes of copper and cadmium, bismuth(III) hydroxide is treated with sodium stannite, a strong reducing agent. The bismuth is reduced to the elemental state, in which condition it appears black. The equation is

$$\underline{2Bi(OH)_3} + 3[Sn(OH)_3]^- + 3OH^- \longrightarrow 2\underline{Bi} + 3[Sn(OH)_6]^{2-}$$
$$\qquad\qquad\qquad \text{Stannite ion} \qquad\qquad\qquad\qquad\qquad \text{Stannate ion}$$

Because sodium stannite is unstable and darkens on standing, it must be prepared just prior to being used, and the test for bismuth performed immediately. Sodium stannite is prepared by treating a solution of tin(II) chloride with an excess of sodium hydroxide. The equations are

$$Sn^{2+} + 2OH^- \rightleftharpoons \underline{Sn(OH)_2} \text{ (white)}$$
$$\underline{Sn(OH)_2} + \text{excess } OH^- \rightleftharpoons [Sn(OH)_3]^- \text{ (colorless)}$$

The darkening of the stannite solution on standing is due to the formation of black tin(II) oxide. The equation is

$$[Sn(OH)_3]^- \longrightarrow \underline{SnO} + H_2O + OH^-$$

PROCEDURE 4 **Solution from Procedure 3: Bi^{3+}, Cu^{2+}, and Cd^{2+}**

Add 15 M aqueous ammonia dropwise to the solution until it is distinctly basic to litmus (about 5 drops). The development of a deep blue color in the solution indicates the presence of the tetraamminecopper(II) ion, $[Cu(NH_3)_4]^{2+}$. The formation of a white precipitate indicates the presence of Bi^{3+} as $Bi(OH)_3$. Separate the mixture and reserve the solution for Procedure 5. Wash the precipitate with 15 drops of hot water, and then add several drops of fresh sodium stannite solution to the precipitate. The immediate formation of a black residue (finely divided metallic bismuth) confirms the presence of Bi^{3+}. (Preparation of sodium stannite: In a separate test tube, place 3 drops of 0.4 M $SnCl_2$ and add sufficient 4 M NaOH to dissolve the white precipitate that first forms. Be sure that you are not fooled by an initial precipitate that forms because of a temporary local excess of reagent and dissolves just with stirring.)

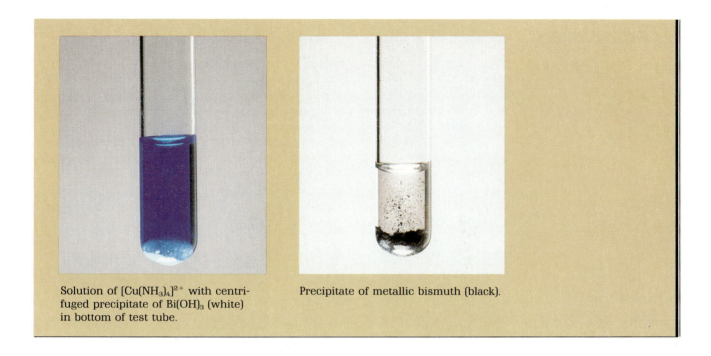

Solution of $[Cu(NH_3)_4]^{2+}$ with centrifuged precipitate of $Bi(OH)_3$ (white) in bottom of test tube.

Precipitate of metallic bismuth (black).

35.5 Identification of Copper and Cadmium

The presence of copper is apparent from the deep blue color exhibited by the complex ion $[Cu(NH_3)_4]^{2+}$. However, if the blue color is too faint to discern, as is the case with a trace of copper, the formation of pink copper(II) hexacyanoferrate(II) serves as a more sensitive test. The equation for the reaction is

$$2[Cu(NH_3)_4]^{2+} + [Fe(CN)_6]^{4-} \longrightarrow \underline{Cu_2[Fe(CN)_6]} + 8NH_3$$

If cadmium is present, a white precipitate of cadmium hexacyanoferrate(II), $Cd_2[Fe(CN)_6]$, will form.

The presence of cadmium ions may be confirmed by forming yellow cadmium sulfide with hydrogen sulfide. However, if we attempt to precipitate CdS in the presence of copper, the black color of CuS obscures the yellow color of CdS. By first adding cyanide ions to a solution containing Cu^{2+} and Cd^{2+}, the very stable dicyanocopper(I) complex will be formed, which will not react with sulfide ions. The copper(II) ion is reduced to copper(I) by the cyanide ion, and the cyanide ion is oxidized to cyanogen, C_2N_2.

$$2[Cu(NH_3)_4]^{2+} + 6CN^- \longrightarrow 2[Cu(CN)_2]^- + C_2N_2(g) + 8NH_3$$
$$[Cu(CN)_2]^- + S^{2-} \longrightarrow \text{no reaction}$$

The corresponding tetracyanocadmium complex is much less stable than the dicyanocopper complex, and it will react with the sulfide ion to form yellow CdS.

$$[Cd(NH_3)_4]^{2+} + 4CN^- \rightleftharpoons [Cd(CN)_4]^{2-} + 4NH_3$$
$$[Cd(CN)_4]^{2-} + S^{2-} \longrightarrow \underline{CdS} + 4CN^-$$

PROCEDURE 5   **Solution from Procedure 4: $[Cu(NH_3)_4]^{2+}$ (blue) and $[Cd(NH_3)_4]^{2+}$ (colorless)**

If the solution is colorless, a trace of copper may be present. Place 10 drops of the solution in a test tube, acidify with acetic acid, and add several drops of 0.1 M $K_4Fe(CN)_6$. The formation of a pink precipitate, $Cu_2Fe(CN)_6$, indicates the presence of a small concentration of Cu^{2+}; the formation of a white precipitate indicates the absence of Cu^{2+} and the probable presence of Cd^{2+} as $Cd_2Fe(CN)_6$.

If copper(II) ions are present, add 2 drops of 1 M KCN to another 10 drops of the solution from Procedure 4, and then continue to add 1 M KCN dropwise until the solution becomes colorless. Add 2 drops of thioacetamide solution and heat the mixture in a hot water bath. The formation of either a yellow or an olive-green precipitate (CdS) confirms the presence of cadmium. If Cu^{2+} is found to be absent, test the second sample for Cd^{2+} as described above but leave out the KCN.

Mixture of precipitates of $Cu_2[Fe(CN)_6]$ (pink) and $Cd_2[Fe(CN)_6]$ (white).

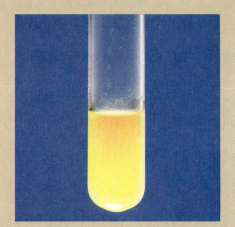

Precipitate of CdS. Pure CdS is yellow, but mixed with a small amount of impurity it may appear olive-green.

35.6 Reprecipitation of the Sulfides of the Division B Ions

To reprecipitate the sulfides of mercury, arsenic, antimony, and tin from the solution containing their complex thiosalts, it is necessary to increase the concentration of each cation to a value such that the ion product exceeds the solubility product. For example,

$$[Hg^{2+}][S^{2-}] > K_{sp}$$

This is accomplished by the addition of hydrochloric acid.

$$[HgS_2]^{2-} + 4H^+ \rightleftharpoons Hg^{2+} + 2H_2S(g)$$

The added acid displaces the equilibria in the direction of an increased concentration of Hg^{2+} as a result of the formation and escape of the weak electrolyte H_2S as a gas. This causes an increase in the concentration of the cation to such an extent that the product of the concentration of the mercury(II) ion and the sulfide ion becomes larger than the solubility product, and reprecipitation of the sulfide occurs. The net reactions for these reprecipitations are

$$[HgS_2]^{2-} + 2H^+ \rightleftharpoons \underline{HgS} + H_2S(g)$$
$$2[AsS_3]^{3-} + 6H^+ \rightleftharpoons \underline{As_2S_3} + 3H_2S(g)$$
$$2[SbS_3]^{3-} + 6H^+ \rightleftharpoons \underline{Sb_2S_3} + 3H_2S(g)$$
$$[SnS_3]^{2-} + 2H^+ \rightleftharpoons \underline{SnS_2} + H_2S(g)$$

Solution from Procedure 2: Thiosalts of the Division B Ions, $[HgS_2]^{2-}$, $[AsS_3]^{3-}$, $[SbS_3]^{3-}$, and $[SnS_3]^{2-}$ **PROCEDURE 6**

Add 1 M HCl until the solution is just acidic to litmus. Heat the mixture in a hot water bath for several minutes. Separate the mixture and discard the solution. The precipitate consists of HgS, As_2S_3, Sb_2S_3, and SnS_2.

35.7 Separation of Mercury and Arsenic from Antimony and Tin

To dissolve Sb_2S_3 and SnS_2, and leave HgS and As_2S_3 undissolved, the concentrations of the cations and sulfide ion must be made such that

$$[Sb^{3+}]^2[S^{2-}]^3 < K_{Sb_2S_3} \quad \text{and} \quad [Hg^{2+}][S^{2-}] > K_{HgS}$$

The addition of 6 M HCl reduces the sulfide ion concentration

$$S^{2-} + 2H^+ \rightleftharpoons 2H_2S$$

and the concentrations of antimony ion and tin ion

$$Sb^{3+} + 4Cl^- \rightleftharpoons [SbCl_4]^-$$
$$Sn^{4+} + 6Cl^- \rightleftharpoons [SnCl_6]^{2-}$$

to values such that these sulfides dissolve. The net reactions are

$$\underline{Sb_2S_3} + 6H^+ + 8Cl^- \longrightarrow 2[SbCl_4]^- + 3H_2S(g)$$
$$\underline{SnS_2} + 4H^+ + 6Cl^- \longrightarrow [SnCl_6]^{2-} + 2H_2S(g)$$

The sulfides of mercury and arsenic remain undissolved in the presence of 6 M HCl due to the very small values of their solubility products.

Precipitate from Procedure 6: HgS, As_2S_3, Sb_2S_3, and SnS_2   **PROCEDURE 7**

Add 1 mL of 6 M HCl to the precipitate and stir the mixture. Heat the mixture in a hot water bath and then separate the residue. Add 15 drops of 6 M HCl to the residue and heat the mixture. Separate the residue and reserve the combined centrifugates for Procedure 11. Treat the residue according to Procedure 8.

35.8 Separation of Arsenic from Mercury

In order to dissolve HgS and As_2S_3, it is necessary to reduce either the concentration of the sulfide ion by oxidation or the concentration of the cation by the formation of a complex ion. A mixture of aqueous ammonia and hydrogen peroxide is used to separate arsenic from mercury. In basic solution, hydrogen peroxide oxidizes As_2S_3.

$$As_2S_3 + 12OH^- + 14H_2O_2 \longrightarrow 2AsO_4^{3-} + 3SO_4^{2-} + 20H_2O$$

The arsenic(III) ion is oxidized to arsenate, AsO_4^{3-}, and the sulfide ion, to sulfate. Mercury(II) sulfide is not dissolved by this treatment.

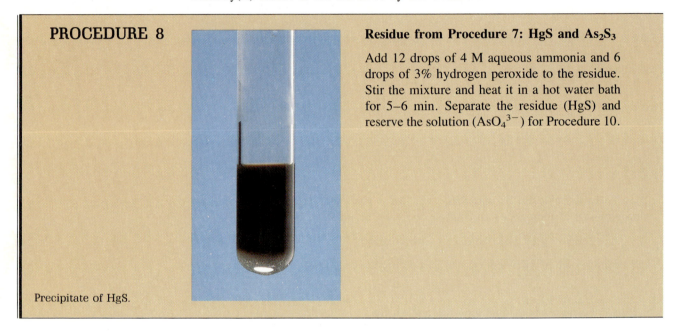

PROCEDURE 8

Residue from Procedure 7: HgS and As_2S_3

Add 12 drops of 4 M aqueous ammonia and 6 drops of 3% hydrogen peroxide to the residue. Stir the mixture and heat it in a hot water bath for 5–6 min. Separate the residue (HgS) and reserve the solution (AsO_4^{3-}) for Procedure 10.

Precipitate of HgS.

35.9 Identification of Mercury

In order to dissolve the extremely insoluble mercury(II) sulfide, it is necessary to reduce the mercury(II) and sulfide ion concentrations to the extent that $[Hg^{2+}][S^{2-}]$ will be less than the solubility product of HgS. This may be done by forming the complex ion $[HgCl_4]^{2-}$ and oxidizing the sulfide ion to sulfur. A mixture of hydrochloric acid and sodium hypochlorite is used to dissolve HgS. The chloride ions from hydrochloric acid combine with mercury(II) ions and form $[HgCl_4]^{2-}$; hypochlorite ions in acid solution oxidize the sulfide ions to sulfur. The equation is

$$\underline{HgS} + 2H^+ + 3Cl^- + ClO^- \longrightarrow [HgCl_4]^{2-} + \underline{S} + H_2O$$

The solution is then boiled to decompose excess hypochlorite ions, which would otherwise interfere with the confirmatory test for mercury by oxidizing tin(II) to tin(IV), thus destroying its reducing power. The decomposition of the hypochlorite ion is according to the equation

$$2H^+ + Cl^- + ClO^- \xrightarrow{\triangle} H_2O + Cl_2(g)$$

The presence of mercury(II) is confirmed by reducing $[HgCl_4]^{2-}$ to Hg_2Cl_2 (white) or Hg (black) by means of $[SnCl_3]^-$ ions.

$$2[HgCl_4]^{2-} + [SnCl_3]^- \longrightarrow \underline{Hg_2Cl_2} + [SnCl_6]^{2-} + 3Cl^-$$

$$\underline{Hg_2Cl_2} + [SnCl_3]^- + Cl^- \longrightarrow 2\underline{Hg} + [SnCl_6]^{2-}$$

The trichlorostannate(II) ion, $[SnCl_3]^-$, is formed when tin(II) chloride is added to the hydrochloric acid solution.

$$SnCl_2 + Cl^- \longrightarrow [SnCl_3]^-$$

PROCEDURE 9

Residue from Procedure 8: HgS

To the black residue add 6 drops of 5% NaClO and 2 drops of 6 M HCl. Stir the mixture, add 1 mL of water, and separate the sulfur from the solution. Heat the solution to boiling. Add 2 drops of 1 M $SnCl_2$ to the solution. The formation of a white, gray, or black precipitate confirms the presence of mercury.

Mixture of precipitates of Hg_2Cl_2 (white) and Hg (black). Precipitate may be white, grey, or black, depending upon the relative amounts of the two substances.

35.10 Identification of Arsenic

The presence of arsenic (in the form of arsenate ions) is confirmed by the formation of white crystalline magnesium ammonium arsenate when magnesia mixture ($MgCl_2$, NH_4Cl, and aqueous ammonia) is added to a solution containing arsenate ions.

$$Mg^{2+} + NH_4^+ + AsO_4^{3-} \longrightarrow \underline{MgNH_4AsO_4}$$

PROCEDURE 10

Solution from Procedure 8: $AsO_4{}^{3-}$

Add 2 drops of 15 M aqueous ammonia and 5 drops of magnesia mixture to the solution. The formation of a white precipitate ($MgNH_4AsO_4$), frequently slow in forming, indicates the presence of arsenate ions.

Precipitate of $MgNH_4AsO_4$.

35.11 Identification of Antimony and Tin

Antimony and tin in hydrochloric acid solutions are in the form of the complex ions $[SbCl_4]^-$ and $[SnCl_6]^{2-}$. Aluminum metal will reduce antimony to the metallic state and tin(IV) to tin(II) in hydrochloric acid solution, thus effecting a separation of the two metals.

$$[SbCl_4]^- + Al \longrightarrow \underline{Sb} + Al^{3+} + 4Cl^-$$

$$3[SnCl_6]^{2-} + 2Al \longrightarrow 3[SnCl_3]^- + 2Al^{3+} + 9Cl^-$$

To be exact, aluminum actually reduces tin to the metallic state. When all of the aluminum is gone, the metallic tin dissolves in the hydrochloric acid, forming $[SnCl_3]^-$ ions and liberating hydrogen.

The fact that the tin(II) ion is a strong reducing agent is used in the confirmatory test for the ion. $[SnCl_3]^-$ will reduce mercury(II) to mercury(I), or metallic mercury, depending on the amount of the tin(II) present in solution.

$$[SnCl_3]^- + 2HgCl_4^{2-} \longrightarrow [SnCl_6]^{2-} + \underline{Hg_2Cl_2} + 3Cl^-$$

$$[SnCl_3]^- + \underline{Hg_2Cl_2} + Cl^- \longrightarrow [SnCl_6]^{2-} + \underline{2Hg}$$

Antimony metal reacts with nitric acid to form the insoluble oxide Sb_4O_6. This oxide is soluble in oxalic acid, $H_2C_2O_4$, forming oxyantimony(III) hydrogen oxalate, $(SbO)HC_2O_4$.

$$4Sb + 4H^+ + 4NO_3^- \longrightarrow \underline{Sb_4O_6} + 4NO + 2H_2O$$

$$\underline{Sb_4O_6} + 4H_2C_2O_4 \longrightarrow 4(SbO)HC_2O_4 + 2H_2O$$

The presence of antimony is then confirmed by precipitating it as the orange-red sulfide, Sb_2S_3.

$$2(SbO)HC_2O_4 + 3H_2S \longrightarrow \underline{Sb_2S_3} + 2H_2C_2O_4 + 2H_2O$$

PROCEDURE 11   **Solution from Procedure 7: $[SnCl_6]^{2-}$ and $[SbCl_4]^-$**

Precipitate of metallic Sb.

Boil the solution until all of the H_2S has been expelled. Add a volume of water equal to that of the solution and add 2 drops of 6 M HCl. Place a piece of aluminum wire about $\frac{1}{8}$ inch long in the solution, and heat the mixture until the aluminum has completely dissolved. Add 1 drop of 6 M HCl to the mixture and heat it for a few minutes. If antimony is present, black flakes of the metal will appear. Separate the mixture and save the solution. Treat the black flakes with 3 drops of 4 M HNO$_3$ and several drops of 1 M oxalic acid. Add 2 drops of thioacetamide to the solution and place the test tube in a hot water bath. The formation of an

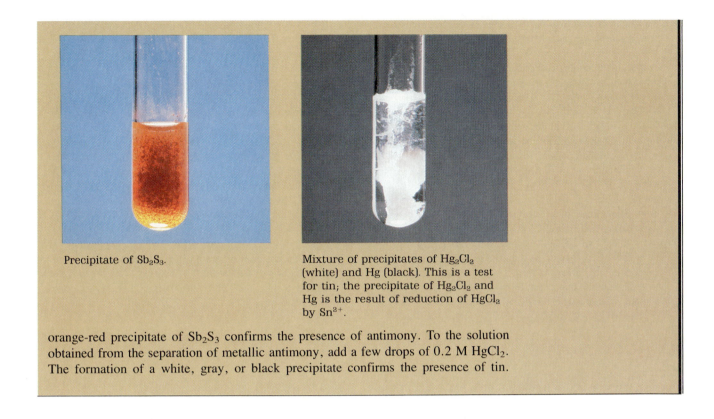

Precipitate of Sb_2S_3.

Mixture of precipitates of Hg_2Cl_2 (white) and Hg (black). This is a test for tin; the precipitate of Hg_2Cl_2 and Hg is the result of reduction of $HgCl_2$ by Sn^{2+}.

orange-red precipitate of Sb_2S_3 confirms the presence of antimony. To the solution obtained from the separation of metallic antimony, add a few drops of 0.2 M $HgCl_2$. The formation of a white, gray, or black precipitate confirms the presence of tin.

For Review

Exercises

1. Give the color of each of the following: CuS, HgS, $[Cu(H_2O)_4]^{2+}$, $PbSO_4$, $[Cu(NH_3)_4]^{2+}$, $[Cu(CN)_2]^-$, $MgNH_4AsO_4$, PbS, $Bi(OH)_3$, Sb_2S_3, As_2S_3, SnS_2, $[Cd(NH_3)_4]^{2+}$, Bi (finely divided), Sb, and Bi_2S_3.

2. Why are the ions of lead and mercury found in both Groups I and II?

3. Explain in terms of ionic equilibria and solubility product theory why Group II can be separated from Group III by hydrogen sulfide in the presence of hydrogen ions at a concentration of 0.3 M.

4. In terms of ionic equilibria theory, discuss the effect of added hydrochloric acid on the concentration of sulfide ion in a solution of hydrogen sulfide. How would adding ammonia or hydroxide ions influence the concentration of sulfide ions?

5. Explain in terms of ionic equilibria and solubility product theory the dissolution of the sulfides of copper, bismuth, cadmium, and lead in nitric acid.

6. Write the equation for the hydrogen-ion catalyzed reaction of water with thioacetamide. Hydrogen sulfide, as one of the products, is a diprotic acid; illustrate this property of hydrogen sulfide by suitable equations.

7. Explain the dissolution of lead sulfate in ammonium acetate and the reprecipitation of the lead as lead chromate in terms of ionic equilibria and solubility product theory.

8. Why is it necessary to remove the nitric acid present before attempting to precipitate lead as lead sulfate?

9. If a yellow precipitate is obtained when an unknown solution for Group II is treated with hydrogen sulfide, what ions are probably absent?

10. The Division A sulfides are washed with water containing ammonium nitrate. What is the function of the ammonium nitrate?

11. Why must sodium stannite, which is used in the identification of bismuth, be prepared just prior to its use?

12. How will the separation of Groups II and III be affected if the concentration of hydronium ion in the solution saturated with hydrogen sulfide is 0.1 M? 1 M?

13. If a Group II unknown contains copper in an appreciable concentration, this fact may be evident from an inspection of the unknown solution. Why?

14. Outline the separation of the following groups of ions, leaving out all unnecessary steps:
    (a) Bi^{3+}, As^{3+}, Sb^{3+} (b) Ag^+, Hg^{2+}, Co^{2+}
    (c) Pb^{2+}, Cu^{2+}, Cd^{2+}

15. Select a reagent used in the analysis of Group I or Group II that will separate each of the following pairs in one step:
    (a) As_2S_3, SnS_2 (b) $[SbCl_4]^-$, $[SnCl_6]^{2-}$
    (c) CdS, HgS   (d) CuS, CdS
    (e) Ag^+, Bi^{3+} (f) Ag^+, Fe^{3+}
    (g) Bi^{3+}, Cd^{2+}

16. How many drops of 1 M aqueous ammonia will be required to neutralize the HCl in 2 mL of a Group II solution that is 0.5 M in HCl? (1 drop is 0.05 mL.) How many drops of 6 M HCl would be required to make the resulting solution 0.3 M in HCl?   *Ans. 20 drops; 3 drops*

17. (a) Calculate the concentration of S^{2-} in an acid solution that is saturated with H_2S and has a pH of 0.52. (Saturated H_2S is 0.10 M.)   *Ans. 1.1×10^{-26}*

(b) Calculate the number of S^{2-} ions that would be present in 1.0 mL of solution.   *Ans. 6.6×10^{-6}*

(c) Calculate the number of HS^- ions that would be present in 1.0 mL of solution.   *Ans. 2.0×10^{13}*

(d) Calculate the concentration of H^+ in the solution.   *Ans. 0.30 M*

18. How many grams of thioacetamide is required to precipitate quantitatively the bismuth and copper from 10 mL of a solution that is 0.010 M in each ion?   *Ans. 0.019 g*

19. Calculate the solubility of ZnS in a solution saturated with H_2S and 0.30 M in HCl.   *Ans. 9×10^{-2} M*

20. Calculate the concentration of H^+, H_2S, HS^-, and S^{2-} in a 0.052 M solution of hydrogen sulfide.
    Ans. $[H^+] = [HS^-] = 7.2 \times 10^{-5}$ M;
    $[S^{2-}] = 1.0 \times 10^{-19}$ M;
    $[H_2S] = 0.052$ M

21. Calculate the concentration of H^+ required to prevent the precipitation of CdS from 0.010 M Cd^{2+} that is saturated with H_2S. (K_{sp} for CdS is 2.8×10^{-35}.) Is this an attainable concentration?   *Ans. 6.0×10^2 M; No*

The Analysis of Group III

The solution to be analyzed may be a Group III known or unknown, or it may be the solution from the Group II separation. Table 36.1 (p. 974) is the Group III flow sheet.

36.1 Precipitation of the Group III Ions

Analytical Group III contains the metallic ions Ni^{2+}, Co^{2+}, Mn^{2+}, Fe^{3+}, Al^{3+}, Cr^{3+}, and Zn^{2+}. These ions are not precipitated by hydrochloric acid (the Group I precipitant) or by sulfide ions in solutions that are 0.3 M in hydrogen ion (the Group II precipitant). However, an ammonium sulfide solution precipitates Ni^{2+}, Co^{2+}, Mn^{2+}, Fe^{2+}, and Zn^{2+} as sulfides and Al^{3+} and Cr^{3+} as hydroxides.

The concentration of sulfide ions in the 0.3 M hydrogen ion solution of the Group II precipitation is so small that the solubility products of the Group III sulfides are not exceeded. Hydrogen sulfide in an aqueous solution of ammonia has a much higher sulfide ion concentration, due to the formation of ammonium sulfide, according to the equation

$$2NH_3 + H_2S \rightleftharpoons 2NH_4^+ + S^{2-}$$

The resultant sulfide ion concentration is sufficiently large that the solubility products of the sulfides of cobalt, nickel, manganese, iron, and zinc are exceeded and precipitation occurs. Similarly, the hydroxide ion concentration of the ammonium sulfide solution is great enough to precipitate the hydroxides of aluminum and chromium. The solubility products of aluminum hydroxide and chromium hydroxide are lower than those of the corresponding sulfides; thus the hydroxides form, not the sulfides.

Table 36.1   Group III flow sheet

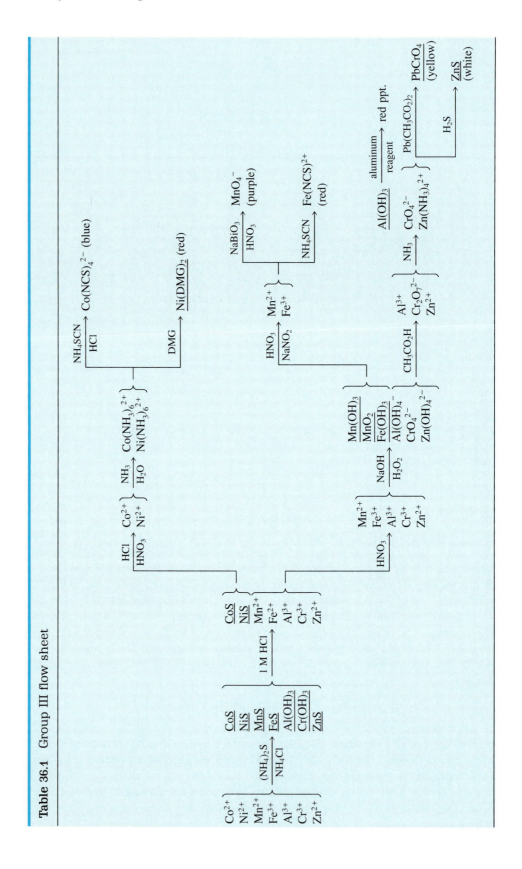

Hydrogen sulfide is produced for the precipitation of this group by the reaction of water with thioacetamide in an acidic solution. When aqueous ammonia is added, the following precipitations occur:

$$Co^{2+} + S^{2-} \rightleftharpoons \underline{CoS} \text{ (brown or black)}$$
$$Ni^{2+} + S^{2-} \rightleftharpoons \underline{NiS} \text{ (brown or black)}$$
$$Mn^{2+} + S^{2-} \rightleftharpoons \underline{MnS} \text{ (light pink)}$$
$$Fe^{2+} + S^{2-} \rightleftharpoons \underline{FeS} \text{ (black)}$$
$$Zn^{2+} + S^{2-} \rightleftharpoons \underline{ZnS} \text{ (white)}$$
$$Al^{3+} + 3OH^- \rightleftharpoons \underline{Al(OH)_3} \text{ (white)}$$
$$Cr^{3+} + 3OH^- \rightleftharpoons \underline{Cr(OH)_3} \text{ (blue-green)}$$

In the event that iron is present as Fe^{3+} in the original solution, it will be reduced by hydrogen sulfide in the acidic solution according to the equation

$$2Fe^{3+} + H_2S \longrightarrow 2Fe^{2+} + S + 2H^+$$

Of the cations remaining in solution, only the magnesium ion, of Group V, forms an insoluble hydroxide. The hydroxide ion concentration is kept just low enough in the aqueous ammonia to prevent the precipitation of magnesium hydroxide by buffering with ammonium chloride.

Some evidence regarding the cations present in a solution containing the Group III ions can be obtained by noting the colors of the ions: Co^{2+}, pale red; Ni^{2+}, pale green; Zn^{2+}, colorless; Mn^{2+}, colorless in low concentrations; Fe^{2+}, pale green; Fe^{3+}, reddish brown; Al^{3+}, colorless; Cr^{3+}, dark green or blue. Due to the phenomenon of complementary colors (Section 26.10), a solution containing certain combinations of colored ions may appear colorless.

Precipitation of Group III: Co^{2+}, Ni^{2+}, Mn^{2+}, Fe^{3+}, Al^{3+}, Cr^{3+}, and Zn^{2+}   **PROCEDURE 1**

(a) If the solution to be analyzed is a known or unknown for Group III only, take 10 drops of the solution, add 1 drop of 6 M HCl, dilute to 1 mL, and add 5 drops of 5% thioacetamide solution. Heat the solution in a hot water bath for at least 5 min. (b) If the solution to be analyzed is that from the Group II separation, add 5 drops of 5% thioacetamide solution and heat the mixture in a hot water bath for at least 5 min.

To the solution resulting from either procedure (a) or (b) above, add 5 drops of 15 M aqueous ammonia and stir up the precipitate. Heat the mixture for 5 min in the hot water bath. Separate the precipitate and wash it with a few drops of water. Reserve the solution for the analysis of Groups IV and V.

36.2 Separation of Cobalt and Nickel

Although the sulfides of cobalt and nickel will completely precipitate only from basic solutions, once formed, these sulfides are only slightly soluble in dilute HCl. It appears that these sulfides precipitate in a form (the α form) that is soluble in acid but change rapidly into other crystalline modifications (the β forms) that have much lower solubility products. The K_{sp} of α CoS is 4.5×10^{-27}, of β CoS is 6.7×10^{-29}, of α NiS is 2×10^{-27}, and of β NiS is 8×10^{-33}. These facts are used in separating CoS and NiS from MnS, FeS, $Al(OH)_3$, $Cr(OH)_3$, and ZnS, the latter five precipitates being soluble in 1 M HCl.

$$\underline{MnS} + 2H^+ \rightleftharpoons Mn^{2+} + H_2S(g) \qquad \underline{Al(OH)_3} + 3H^+ \rightleftharpoons Al^{3+} + 3H_2O$$
$$\underline{FeS} + 2H^+ \rightleftharpoons Fe^{2+} + H_2S(g) \qquad \underline{Cr(OH)_3} + 3H^+ \rightleftharpoons Cr^{3+} + 3H_2O$$
$$\underline{ZnS} + 2H^+ \rightleftharpoons Zn^{2+} + H_2S(g)$$

PROCEDURE 2   **Precipitate from Procedure 1: CoS, NiS, FeS, MnS, $Al(OH)_3$, $Cr(OH)_3$, and ZnS**

Add 10 drops of 1 M HCl to the precipitate and stir the mixture. Separate the mixture immediately since prolonged contact with the acid causes some dissolution of CoS and NiS. Wash the sulfides that remain (CoS and NiS) with 4 drops of 1 M HCl. Reserve the combined centrifugates for Procedure 4.

36.3 Identification of Cobalt and Nickel

The sulfides of cobalt and nickel readily dissolve in a mixture of nitric and hydrochloric acids due to the oxidation of the sulfide ion to sulfur.

$$3\underline{CoS} + 8H^+ + 2NO_3^- \longrightarrow 3Co^{2+} + 2NO(g) + 3\underline{S} + 4H_2O$$
$$3\underline{NiS} + 8H^+ + 2NO_3^- \longrightarrow 3Ni^{2+} + 2NO(g) + 3\underline{S} + 4H_2O$$

After boiling the solution containing the cobalt and nickel ions to remove oxides of nitrogen, which would destroy the reagents used in the confirmation of these ions, an excess of aqueous ammonia is added. Ammonia in excess reacts with cobalt ions and nickel ions to form complex ions.

$$Co^{2+} + 6NH_3 \rightleftharpoons [Co(NH_3)_6]^{2+} \text{ (pink)}$$
$$Ni^{2+} + 6NH_3 \rightleftharpoons [Ni(NH_3)_6]^{2+} \text{ (blue)}$$

In the ammoniacal solution, Ni^{2+} will react with dimethylglyoxime to form a very insoluble, bright red coordination compound, $[Ni(DMG)_2]$ (Section 32.7). Co^{2+} forms a brown-colored soluble complex with dimethylglyoxime, which does not interfere with the test.

A concentrated solution of ammonium thiocyanate is used in the identification of the cobalt ion, with which the thiocyanate ion forms a complex ion, $[Co(NCS)_4]^{2-}$, that has a characteristic blue color.

$$[Co(NH_3)_6]^{2+} + 4NCS^- \rightleftharpoons [Co(NCS)_4]^{2-} + 6NH_3$$

When iron(III) ions are present as a contaminant, they interfere and form a bright red complex ion of the formula $[Fe(NCS)]^{2+}$. This interference may be avoided by converting the iron to the colorless and very stable hexafluoroferrate(III) ion, $[FeF_6]^{3-}$. The equation is

$$Fe^{3+} + 6F^- \rightleftharpoons [FeF_6]^{3-}$$

A second confirmatory test for cobalt involves its oxidation by the nitrite ion from

Co^{2+} to Co^{3+} and the precipitation of the complex yellow salt $K_3[Co(NO_2)_6]$, tripotassium hexanitrocobaltate(III).

$$Co^{2+} + 6NO_2^- \longrightarrow [Co(NO_2)_6]^{4-}$$

$$[Co(NO_2)_6]^{4-} + NO_2^- + 2H^+ \longrightarrow [Co(NO_2)_6]^{3-} + NO(g) + H_2O$$

$$3K^+ + [Co(NO_2)_6]^{3-} \longrightarrow \underline{K_3[Co(NO_2)_6]}$$

**Residue from Procedure 2:
CoS and NiS**

Add 3 drops of 12 M HCl and 1 drop of 14 M HNO_3 to the residue and heat the mixture in a hot water bath. Separate any sulfur that forms and boil the solution to remove any excess nitric acid or oxides of nitrogen. Add sufficient 4 M aqueous ammonia to the solution to make it *slightly* basic to litmus. (Excess alkalinity favors the formation of the brown cobalt complex when testing for nickel with dimethylglyoxime.) Dilute the solution to 1 mL and divide it into three parts.

(a) *Test for Nickel.* Add 1 drop of dimethylglyoxime to one portion of the solution. The formation of a pink or red precipitate confirms the presence of nickel.

(b) *Test for Cobalt.* Acidify a second portion of the solution with 1 M HCl and add several crystals of NH_4NCS to it. Now add an equal volume of acetone and agitate the mixture. The development of a blue color proves the presence of cobalt. If the solution turns red when the NH_4NCS is added, iron(III) ions are present. Add 1 drop of 1 M NaF to the solution. Now, if the solution is bluish-green to green, the presence of cobalt is confirmed.

(c) *Test for Cobalt.* Acidify the third portion of the solution with 4 M CH_3CO_2H, and add several large crystals of KNO_2. Warm the mixture. The formation of a yellow precipitate confirms the presence of cobalt.

PROCEDURE 3

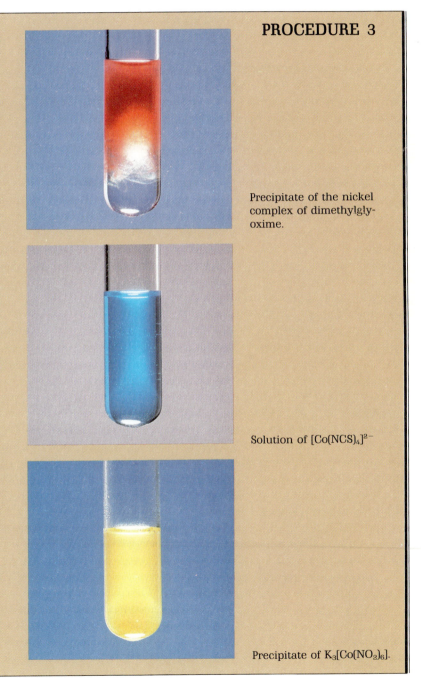

Precipitate of the nickel complex of dimethylglyoxime.

Solution of $[Co(NCS)_4]^{2-}$

Precipitate of $K_3[Co(NO_2)_6]$.

36.4 Separation of Iron and Manganese from Aluminum, Chromium, and Zinc

The solution containing Mn^{2+}, Fe^{2+}, Al^{3+}, and Zn^{2+} is treated with nitric acid to remove the sulfide ion by oxidizing it to sulfur and to oxidize Fe^{2+} to Fe^{3+}. The equation for the oxidation of the iron is

$$3Fe^{2+} + 4H^+ + NO_3^- \longrightarrow 3Fe^{3+} + NO(g) + 2H_2O$$

It is desirable to have the iron as Fe^{3+} because $Fe(OH)_3$ is less soluble and less gelatinous in character than is $Fe(OH)_2$.

In order to separate iron and manganese from aluminum, chromium, and zinc, the solution is first treated with sodium hydroxide and then with hydrogen peroxide. The precipitations that first occur, as sodium hydroxide is added to the solution, are described by the following:

$$Mn^{2+} + 2OH^- \rightleftharpoons \underline{Mn(OH)_2} \text{ (white)}$$
$$Fe^{3+} + 3OH^- \rightleftharpoons \underline{Fe(OH)_3} \text{ (reddish-brown)}$$
$$Al^{3+} + 3OH^- \rightleftharpoons \underline{Al(OH)_3} \text{ (white)}$$
$$Cr^{3+} + 3OH^- \rightleftharpoons \underline{Cr(OH)_3} \text{ (dark green or blue)}$$
$$Zn^{2+} + 2OH^- \rightleftharpoons \underline{Zn(OH)_2} \text{ (white)}$$

The addition of an excess of NaOH dissolves the amphoteric hydroxides of aluminum, chromium, and zinc, but not the hydroxides of iron and manganese.

$$Al(OH)_3 + OH^- \rightleftharpoons [Al(OH)_4]^- \text{ (aluminate) (colorless)}$$
$$Cr(OH)_3 + OH^- \rightleftharpoons [Cr(OH)_4]^- \text{ (chromite) (green)}$$
$$Zn(OH)_2 + 2OH^- \rightleftharpoons [Zn(OH)_4]^{2-} \text{ (zincate) (colorless)}$$

Hydrogen peroxide is added to the resulting mixture to oxidize manganese(II) hydroxide to a mixture of manganese dioxide and manganese(III) hydroxide, both of which are much less soluble than $Mn(OH)_2$.

$$\underline{Mn(OH)_2} + H_2O_2 \longrightarrow \underline{MnO_2} + 2H_2O$$
$$2\underline{Mn(OH)_2} + H_2O_2 \longrightarrow 2\underline{Mn(OH)_3}$$

Hydrogen peroxide oxidizes the chromite ion, $[Cr(OH)_4]^-$, to the chromate ion, CrO_4^{2-}. This is desirable because the reactions of the chromate ion permit better identification of chromium than do those of either Cr^{3+} or $[Cr(OH)_4]^-$.

$$2[Cr(OH)_4]^- + 3H_2O_2 + 2OH^- \longrightarrow 2CrO_4^{2-} + 8H_2O$$

The mixture containing MnO_2 and $Mn(OH)_3$ is brown, and solutions of CrO_4^{2-} are yellow.

PROCEDURE 4 **Solution from Procedure 2: Mn^{2+}, Fe^{2+}, Al^{3+}, Cr^{3+}, and Zn^{2+}**

Transfer the solution to a casserole, add 1 mL of 4 M HNO_3, and evaporate the solution to a moist residue. Take up the residue in 1 mL of water and transfer the solution to a test tube. Add 10 drops of 4 M NaOH beyond the amount of this reagent that is required to initiate precipitation. Now add 6 drops of 3% hydrogen peroxide to the mixture and heat it in a hot water bath for 5 min. Separate the mixture and wash the residue with 10 drops of water to which has been added 1 drop of 4 M NaOH. Save the combined centrifugates for Procedure 6.

36.5 Separation and Identification of Manganese and Iron

The residue containing MnO_2, $Mn(OH)_3$, and $Fe(OH)_3$ is dissolved by a mixture of HNO_3 and $NaNO_2$. The nitrite ion in acidic solution reduces manganese to the Mn(II) ion. The iron(III) hydroxide is dissolved by nitric acid. The equations for these reactions are

$$2Mn(OH)_3 + 4H^+ + NO_2^- \longrightarrow 2Mn^{2+} + NO_3^- + 5H_2O$$

$$MnO_2 + 2H^+ + NO_2^- \longrightarrow Mn^{2+} + NO_3^- + H_2O$$

$$Fe(OH)_3 + 3H^+ \longrightarrow Fe^{3+} + 3H_2O$$

The presence of Fe^{3+} is confirmed by its reaction with the thiocyanate ion, SCN^-, to produce the blood-red complex ion $[Fe(NCS)]^{2+}$ or one of the possible complexes containing from two to six SCN^- groups.

$$Fe^{3+} + SCN^- \rightleftharpoons [Fe(NCS)]^{2+}$$

This test for iron may be conducted on a solution containing Mn^{2+} without interference by this ion. It should be noted that there are traces of Fe^{3+} existing in many reagents. Therefore a light pink color in this test may be due to iron as an impurity. Blank tests can be run on the reagents used in the test to establish this point.

Manganese may be identified in the presence of iron by oxidizing it to the permanganate ion by means of sodium bismuthate in nitric acid. The MnO_4^- ion is purple, but in dilute solutions it may appear pink.

$$2Mn^{2+} + 5BiO_3^- + 14H^+ \longrightarrow 2MnO_4^- + 5Bi^{3+} + 7H_2O$$

PROCEDURE 5

Residue from Procedure 4:
Mn(OH)$_3$, MnO$_2$, and Fe(OH)$_3$

Treat the residue with 1 mL of 4 M HNO_3 and 2 drops of 1 M $NaNO_2$. Stir the mixture and heat it in a hot water bath. Separate any residue that remains. Heat the solution to boiling, then cool it, and divide it into two parts.

(a) *Test for Iron*. Dilute one portion of the solution to 1 mL and add 2 or 3 crystals of NH$_4$NCS. If iron is present, a dark blood-red color will develop in the solu-

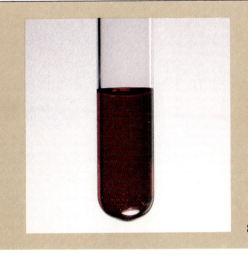

Solution of $[Fe(NCS)]^{2+}$.

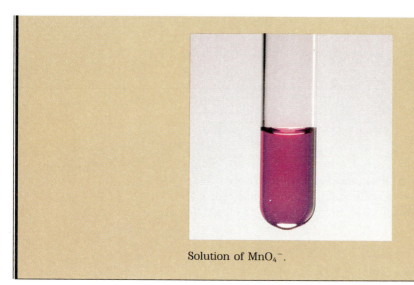

Solution of MnO_4^-.

tion. (Faint or pale redness indicates a trace of iron due to impurities; a red color that does not persist indicates the presence of reducing agents but does not invalidate the test.)

(b) *Test for Manganese.* To the second portion of the solution, add a small quantity of solid $NaBiO_3$ and a few drops of 4 M HNO_3. The formation of a pink or purple color that persists confirms the presence of manganese. (Pink or purple that does not persist indicates the presence of reducing agents but does not invalidate the test.)

36.6 Separation and Identification of Aluminum

When acetic acid is added to the solution containing $[Al(OH)_4]^-$, CrO_4^{2-}, and $[Zn(OH)_4]^{2-}$, aluminum and zinc are converted to their cations.

$$[Al(OH)_4]^- + 4H^+ \longrightarrow Al^{3+} + 4H_2O$$
$$[Zn(OH)_4]^{2-} + 4H^+ \longrightarrow Zn^{2+} + 4H_2O$$

Acids react with yellow chromate, CrO_4^{2-}, and convert some of it to dichromate ion, $Cr_2O_7^{2-}$. That extent of the conversion depends on the concentration of the hydrogen ion.

$$2CrO_4^{2-} + 2H^+ \rightleftharpoons Cr_2O_7^{2-} + H_2O$$

The addition of excess aqueous ammonia precipitates aluminum as the hydroxide, converts $Cr_2O_7^{2-}$ to CrO_4^{2-}, and complexes the zinc as the soluble $[Zn(NH_3)_4]^{2+}$ ion.

$$Al^{3+} + 3NH_3 + 3H_2O \rightleftharpoons Al(OH)_3 + 3NH_4^+$$
$$Cr_2O_7^{2-} + 2OH^- \rightleftharpoons 2CrO_4^{2-} + H_2O$$
$$Zn^{2+} + 4NH_3 \rightleftharpoons [Zn(NH_3)_4]^{2+}$$

The presence of aluminum is confirmed by dissolving aluminum hydroxide in acetic acid and adding the aluminum reagent and $(NH_4)_2CO_3$. Aurin-tricarboxylic acid (aluminon) imparts a red color to the $Al(OH)_3$ formed.

PROCEDURE 6   **Solution from Procedure 4: $[Al(OH)_4]^-$, CrO_4^{2-}, and $[Zn(OH)_4]^{2-}$**

Add 4 M CH_3CO_2H to the solution until it is acid to litmus and then add 2 or 3 drops of the acid in excess. Now add 4 M aqueous ammonia until the solution is distinctly alkaline to litmus. If a white gelatinous precipitate forms, it is probably aluminum hydroxide. Separate the precipitate and reserve the solution for Procedure 7. Confirm the presence of aluminum by dissolving the precipitate in 4 M CH_3CO_2H and adding 2 drops of the aluminum reagent and enough 1 M $(NH_4)_2CO_3$ to make the solution basic. The formation of a reddish-colored precipitate confirms the presence of aluminum.

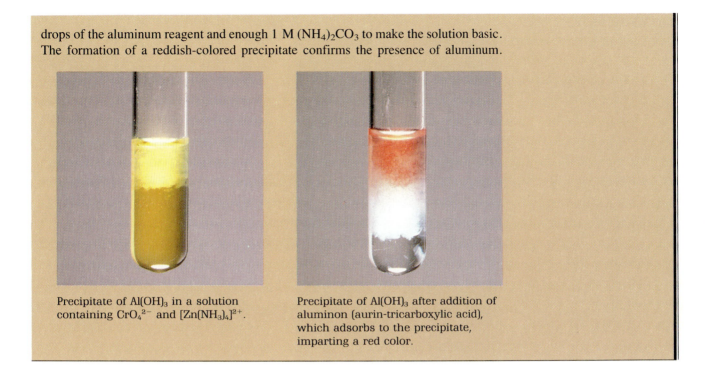

Precipitate of $Al(OH)_3$ in a solution containing CrO_4^{2-} and $[Zn(NH_3)_4]^{2+}$.

Precipitate of $Al(OH)_3$ after addition of aluminon (aurin-tricarboxylic acid), which adsorbs to the precipitate, imparting a red color.

36.7 Identification of Chromium and Zinc

The presence of the chromate ion is confirmed by precipitating yellow lead chromate. Zinc ions do not interfere with the chromate test.

The zinc ions are precipitated by sulfide ions from a thioacetamide solution, white zinc sulfide being formed. The chromate ion may oxidize some of the sulfide ions to elemental sulfur, but ZnS is soluble in hydrochloric acid and sulfur is not. A second confirmatory test for zinc may be made by boiling all of the hydrogen sulfide out of the solution and precipitating the zinc as zinc hexacyanoferrate(II), which is white.

$$2K^+ + Zn^{2+} + Fe(CN)_6^{4-} \longrightarrow \underline{K_2Zn[Fe(CN)_6]}$$

PROCEDURE 7

Solution from Procedure 6:
CrO_4^{2-} and $[Zn(NH_3)_4]^{2+}$

Divide the solution into two parts.

(a) *Test for Chromium.* To one portion of the solution, add 1 M CH_3CO_2H until the solution is acid to litmus. Then add 2 drops of 0.1 M $Pb(CH_3CO_2)_2$. The formation of a yellow precipitate, $PbCrO_4$, confirms the presence of chromium.

Precipitate of $PbCrO_4$.

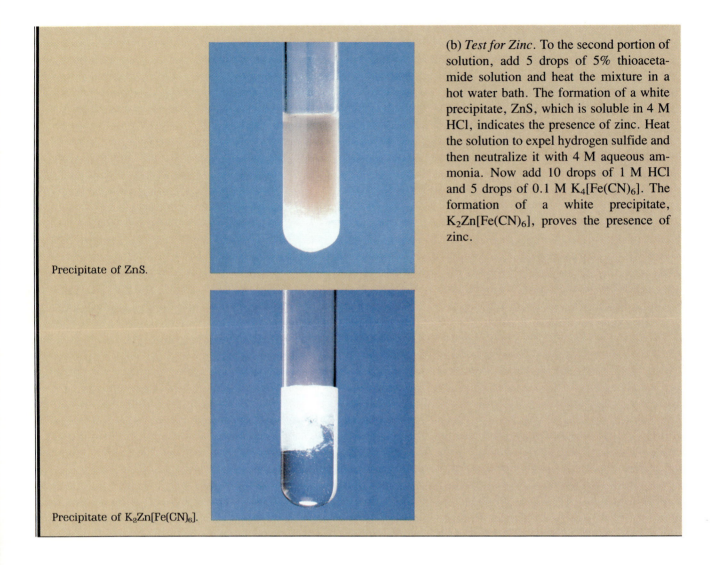

Precipitate of ZnS.

Precipitate of $K_2Zn[Fe(CN)_6]$.

(b) *Test for Zinc.* To the second portion of solution, add 5 drops of 5% thioacetamide solution and heat the mixture in a hot water bath. The formation of a white precipitate, ZnS, which is soluble in 4 M HCl, indicates the presence of zinc. Heat the solution to expel hydrogen sulfide and then neutralize it with 4 M aqueous ammonia. Now add 10 drops of 1 M HCl and 5 drops of 0.1 M $K_4[Fe(CN)_6]$. The formation of a white precipitate, $K_2Zn[Fe(CN)_6]$, proves the presence of zinc.

For Review

Exercises

1. A Group III unknown is colorless. What cations are probably absent? Why should one not rely definitely on such an observation?
2. Why do aluminum and chromium precipitate as hydroxides rather than as sulfides in Group III?
3. What is the function of the ammonium chloride used in the Group III precipitant?
4. Why do the sulfides of Co^{2+}, Ni^{2+}, Mn^{2+}, Fe^{2+}, and Zn^{2+} precipitate in an ammonium sulfide solution but not in a 0.3 M HCl solution of hydrogen sulfide?
5. Why does iron precipitate as iron(II) sulfide rather than iron(III) sulfide in an ammonium sulfide solution?
6. Cite two examples of the use of complex ions in the analysis of Group III.
7. Account for the fact that CoS and NiS fail to precipitate in the 0.3 M HCl solution of Group II, yet they dissolve only very slowly in 1 M HCl.
8. Give the color of each of the following: $Fe(OH)_3$, Fe^{3+}, $Fe(OH)_2$, Fe^{2+}, $[Fe(NCS)]^{2+}$, $[FeF_6]^{3-}$.
9. Write equations showing the amphoteric nature of the hydroxides of zinc, chromium, and aluminum.
10. Outline the separation of the following groups of ions

leaving out all unnecessary steps.

(a) Ni^{2+}, Mn^{2+}, Zn^{2+}

(b) Hg_2^{2+}, Hg^{2+}, Cu^{2+}, Fe^{2+}

(c) Cd^{2+}, Co^{2+}, Ca^{2+}

11. Select a reagent used in Group III that will separate each of the following pairs.

(a) Al^{3+}, Zn^{2+} (b) $[Zn(NH_3)_4]^{2+}$, CrO_4^{2-}

(c) CoS, ZnS   (d) Mn^{2+}, Mg^{2+}

(e) Fe^{3+}, Al^{3+}

12. Why will $PbCrO_4$ precipitate when Pb^{2+} is added to a solution made up from $K_2Cr_2O_7$?

13. When and why is fluoride added in the test for cobalt using thiocyanate?

14. Show by the proper formulas that manganese can act as either a metal or a nonmetal, depending on its oxidation number.

15. What concentration of NH_4^+ must be present in 0.20 M aqueous ammonia to prevent the precipitation of $Mg(OH)_2$ if the solution contains 1.0×10^{-3} mol of Mg^{2+} per liter? (K_{sp} for $Mg(OH)_2$ is 1.5×10^{-11}.)   *Ans. 0.029 M*

16. A solution from the Group II separation is 0.3 M in HCl and has a volume of 5.0 mL. Calculate the number of drops of 15 M aqueous ammonia required to react with the HCl (assume that 1 drop is equal to 0.05 mL).   *Ans. 2 drops*

17. Calculate the concentration of Pb^{2+} in a saturated solution of $PbCrO_4$. (K_{sp} for $PbCrO_4$ is 1.8×10^{-14}.)   *Ans. 1.3×10^{-7} M*

18. Calculate the concentration of Pb^{2+} in a saturated solution of $PbCrO_4$ in the presence of 1.0×10^{-3} M Na_2CrO_4.   *Ans. 1.8×10^{-11} M*

19. Using 8×10^{-26} for the solubility product of FeS, show that 0.3 M HCl should dissolve or prevent the precipitation of FeS in a saturated H_2S solution.

Ans. These conditions permit an iron(II) ion concentration of 7 M before precipitation would occur.

37

The Analysis of Group IV

T he solution to be analyzed may be a Group IV known or unknown, or it may be the solution from the Group III separation. The Group IV flow sheet appears in Table 37.1.

Table 37.1   Group IV flow sheet

$\left.\begin{array}{l} Ba^{2+} \\ Sr^{2+} \\ Ca^{2+} \end{array}\right\} \xrightarrow[\substack{NH_4Cl \\ NH_3 + H_2O}]{(NH_4)_2CO_3} \left.\begin{array}{l} \overline{BaCO_3} \\ \overline{SrCO_3} \\ \overline{CaCO_3} \end{array}\right\} \xrightarrow{CH_3CO_2H} \left.\begin{array}{l} Ba^{2+} \\ Sr^{2+} \\ Ca^{2+} \end{array}\right\} \xrightarrow[\substack{CH_3CO_2H \\ NH_4CH_3CO_2}]{K_2CrO_4}$

$\begin{array}{l} BaCrO_4 \xrightarrow{HCl} Ba^{2+} \xrightarrow{H_2SO_4} BaSO_4 \ (white) \\ \left.\begin{array}{l} Sr^{2+} \\ Ca^{2+} \end{array}\right\} \xrightarrow[\substack{K_2CrO_4 \\ alcohol}]{NH_3 + H_2O} \begin{array}{l} SrCrO_4 \ (yellow) \\ Ca^{2+} \xrightarrow{(NH_4)_2C_2O_4} CaC_2O_4 \ (white) \end{array} \end{array}$

37.1 Precipitation of the Group IV Ions

Analytical Group IV contains the metallic ions Ba^{2+}, Sr^{2+}, and Ca^{2+}. These metals form chlorides, sulfides, and hydroxides that are soluble under the conditions that prevail in the precipitations of Groups I, II, and III. The carbonates of barium, strontium, and calcium precipitate in aqueous ammonia solutions containing ammonium carbonate. These conditions give a carbonate concentration high enough to precipitate $BaCO_3$, $SrCO_3$, and $CaCO_3$, but not $MgCO_3$. At the same time the hydroxide ion concentration in this buffered solution is not high enough to precipitate $Mg(OH)_2$ (Section 32.10).

To bring the solution to the required ammonium ion concentration, the centrifugate from the Group III separation, which contains ammonium salts, is first evaporated to dryness and then heated strongly to expel the ammonium salts present. The equations are

$$NH_4Cl \xrightarrow{\triangle} NH_3(g) + HCl(g)$$

$$NH_4NO_3 \xrightarrow{\triangle} N_2O(g) + 2H_2O(g)$$

The necessary ammonium ion concentration is then obtained by adding the Group IV precipitant, which consists of ammonium carbonate, ammonium chloride, and aqueous ammonia in the required concentrations. The equations for the precipitations of the Group IV carbonates are

$$Ba^{2+} + CO_3{}^{2-} \rightleftharpoons \underline{BaCO_3} \text{ (white)}$$
$$Sr^{2+} + CO_3{}^{2-} \rightleftharpoons \underline{SrCO_3} \text{ (white)}$$
$$Ca^{2+} + CO_3{}^{2-} \rightleftharpoons \underline{CaCO_3} \text{ (white)}$$

Precipitation of Group IV: Ba²⁺, Sr²⁺, and Ca²⁺   **PROCEDURE 1**

Evaporate the solution (10 drops of a Group IV known or unknown or the solution from the Group III separation) to dryness and ignite in a casserole to expel ammonium salts. Dissolve the residue in a mixture of 1 drop of 12 M HCl and 12 drops of water. Make the solution alkaline by adding 4 M aqueous ammonia. Add just 1 drop of aqueous ammonia in excess. Add 2 drops of 1 M $(NH_4)_2CO_3$, or more if necessary, to effect complete precipitation, and warm the mixture in a hot water bath. Allow the mixture to cool; separate the precipitate and reserve the solution for the Group V analysis.

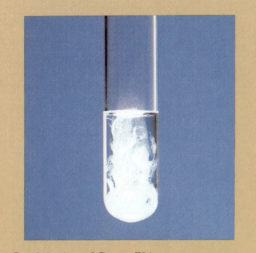

Precipitation of Group IV ions as carbonates, $BaCO_3$, $SrCO_3$, and $CaCO_3$.

37.2 Separation and Identification of Barium

1. DISSOLUTION OF THE CARBONATES. The carbonates of barium, strontium, and calcium are dissolved readily by strong acids such as hydrochloric acid or by weak acids such as acetic acid. The latter acid is used in these procedures. The equation, uses $BaCO_3$ as an example; similar equations may be written for the dissolution of the carbonates of strontium and calcium in acetic acid.

$$BaCO_3 + 2CH_3CO_2H \rightleftharpoons Ba^{2+} + 2CH_3CO_2{}^- + H_2O + CO_2(g)$$

2. SEPARATION OF BARIUM FROM STRONTIUM AND CALCIUM. By taking advantage of the fact that barium chromate is less soluble than strontium chromate and calcium chromate, barium can be separated from strontium and calcium (Section 32.11). However, the concentration of the chromate ion must be controlled so that barium chromate will precipitate (Section 32.11) and strontium chromate will not.

In a solution containing chromate ions, we have the following equilibrium:

$$Cr_2O_7{}^{2-} + H_2O \rightleftharpoons 2CrO_4{}^{2-} + 2H^+$$

The concentration of chromate ions in a solution made by dissolving a dichromate salt in water can be controlled by adjusting the hydrogen ion concentration. A buffer composed of acetic acid and ammonium acetate is used to maintain the hydrogen ion concentration required to give the appropriate chromate ion concentration.

3. DISSOLUTION OF BARIUM CHROMATE. Even though barium chromate is precipitated from a weakly acidic solution, it is readily dissolved by strong acids. The high hydrogen ion concentration converts the chromate ion to dichromate. A solution of 12 M HCl is used in this analysis to bring about the dissolution of $BaCrO_4$.

4. IDENTIFICATION OF BARIUM. The presence of the barium ion in the solution is confirmed by precipitating it as barium sulfate, using sulfuric acid as the precipitant.

Additional evidence for the presence of the barium ion is obtained by a flame test. When heated in the Bunsen burner flame, barium salts give it a yellow-green color.

PROCEDURE 2   Precipitate from Procedure 1: $BaCO_3$, $SrCO_3$, and $CaCO_3$

Dissolve the precipitate in a mixture of 2 drops of 4 M CH_3CO_2H and 4 drops of 1 M $NH_4CH_3CO_2$. Add 1 drop of 1 M K_2CrO_4 to the solution. The formation of a yellow precipitate indicates the presence of barium. Separate the mixture and reserve the solution for Procedure 3. Dissolve the precipitate ($BaCrO_4$) in 2 drops of 12 M HCl. Make a flame test (Section 33.14) with this solution. The barium ion imparts a weak and fleeting greenish-yellow color to the Bunsen burner flame. Add 1 drop of 4 M H_2SO_4 to the remainder of the solution. A white precipitate ($BaSO_4$) confirms the presence of barium.

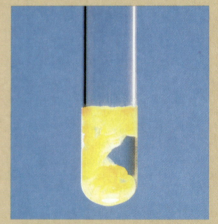

Precipitate of $BaCrO_4$.

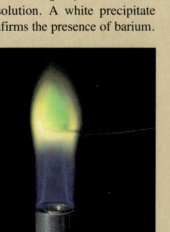

Flame test for barium.

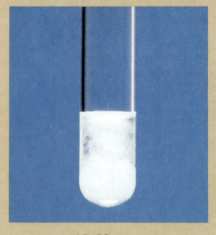

Precipitate of $BaSO_4$.

37.3 Separation and Identification of Strontium

1. SEPARATION OF STRONTIUM FROM CALCIUM. Although the chromate ion concentration in an acetic acid-ammonium acetate buffered solution is too low to exceed the solubility product of strontium chromate, this compound will precipitate when the solution is made basic with aqueous ammonia. The equation for the formation of additional chromate ion is

$$Cr_2O_7^{2-} + 2NH_3 + H_2O \rightleftharpoons 2CrO_4^{2-} + 2NH_4^+$$

The resulting concentration of chromate ions is large enough that the solubility product of strontium chromate is exceeded and precipitation occurs. The complete precipitation of strontium chromate is ensured by further reducing its solubility through the addition of enough ethyl alcohol to the solution to make it 50% alcohol by volume. Calcium chromate is soluble in this solution.

2. IDENTIFICATION OF STRONTIUM. The formation of a fine yellow crystalline precipitate of strontium chromate confirms the presence of strontium. A crimson color imparted to the Bunsen burner flame is characteristic of the strontium ion.

Solution from Procedure 2: Sr^{2+} and Ca^{2+}   **PROCEDURE 3**

Add 4 M aqueous ammonia to the solution until the color changes from orange to yellow. Now add a volume of ethyl alcohol equal to the volume of the solution. The formation of a fine yellow precipitate indicates $SrCrO_4$. Separate the mixture and reserve the solution for Procedure 4. Dissolve the precipitate in 2 drops of 12 M HCl and make a flame test with this solution. A crimson color imparted to the flame is characteristic of the strontium ion.

Precipitate of $SrCrO_4$.

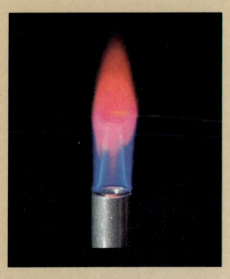

Flame test for strontium.

37.4 Identification of Calcium

The solution from the strontium chromate separation contains the calcium ion. The addition of ammonium oxalate precipitates calcium oxalate, CaC_2O_4, a white crystalline salt.

$$Ca^{2+} + C_2O_4^{2-} \rightleftharpoons \underline{CaC_2O_4}$$

Calcium ions give a brick-red color to the Bunsen burner flame.

PROCEDURE 4   **Solution from Procedure 3: Ca^{2+}**

Heat the solution to boiling and add 2 drops of 0.4 M $(NH_4)_2C_2O_4$. The formation of a white precipitate, which may form slowly, confirms the presence of calcium. Dissolve the precipitate in 12 M HCl and make a flame test with the solution. Calcium ions give a brick-red color to the flame.

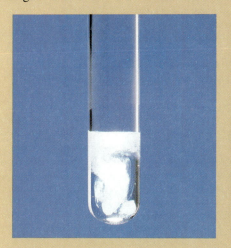

Precipitate of CaC_2O_4.

Flame test for calcium.

PROCEDURE 5   **Original Solution: Ba^{2+}, Sr^{2+}, and Ca^{2+}**

Make flame tests with the original solution and compare with tests of known solutions that you make up yourself.

For Review

Exercises

1. Why are ammonium salts expelled prior to the precipitation of Group IV?
2. What reagents compose the Group IV precipitant? What is the function of each of these chemicals in the precipitation?
3. In terms of ionic equilibria and solubility product theory, explain the dissolution of $BaCrO_4$ in HCl.
4. Barium chromate is precipitated from a solution of potassium chromate containing acetic acid. Why is the acetic acid present?

5. Write equations representing the various ionic equilibria involved in the dissolution of barium carbonate in acetic acid.

6. What are the characteristic colors imparted to the Bunsen burner flame by Ca^{2+}, Sr^{2+}, and Ba^{2+}?

7. Account for the color change from orange to yellow when aqueous ammonia is added to an acetic acid solution of potassium dichromate.

8. If the magnesium concentration in the unknown is quite high or if too large an amount of carbonate is added in precipitating Group IV, some magnesium carbonate will also precipitate. What effect would such an error have on the rest of the analytical scheme for Group IV?

9. Explain why ammonium carbonate rather than sodium carbonate is used in the precipitation of Group IV.

10. What characteristic properties cause Ca^{2+}, Sr^{2+}, and Ba^{2+} to be members of the same group of the analytical scheme?

11. Outline the separation of the following groups of ions, leaving out all unnecessary steps:
    (a) Ag^+, Cu^{2+}, Zn^{2+}, Ca^{2+}, Na^+
    (b) Fe^{3+}, Ba^{2+}, Mg^{2+}
    (c) Pb^{2+}, Sn^{4+}, Al^{3+}, Sr^{2+}

12. What is the total weight of oxalates that can be obtained from a solution containing 3.00 mg each of Ba^{2+}, Sr^{2+}, and Ca^{2+}?   *Ans. 20.5 mg*

13. Calculate the hydroxide ion concentration of a solution that is 0.40 M in NH_3 and 0.30 M in NH_4^+.
                                            Ans. 2.4×10^{-5} M

14. If sodium sulfate is added to a mixture originally 0.010 M in Sr^{2+} and 0.010 M in Ba^{2+}, what percentage of the barium remains unprecipitated before any $SrSO_4$ precipitates?   *Ans. 0.039%*

15. What concentration of carbonate ion is required to initiate precipitation of each of the following:
    (a) $CaCO_3$ from 0.010 M Ca^{2+} ($K_{CaCO_3} = 4.8 \times 10^{-9}$)
    (b) $SrCO_3$ from 0.010 M Sr^{2+} ($K_{SrCO_3} = 9.42 \times 10^{-10}$)
                        Ans. (a) 4.8×10^{-7} M; (b) 9.4×10^{-8} M

16. How many drops (assume 1 drop = 0.050 mL) of 1.0 M acetic acid would be required to dissolve 5 mg of $SrCO_3$?   *Ans. 2 drops*

17. Will the concentration of OH^- in exercise 13 be large enough to cause $Mg(OH)_2$ to precipitate in a 0.010 M solution of Mg^{2+}? (K_{sp} for $Mg(OH)_2$ is 1.5×10^{-11}.)
                                            Ans. No

38

The Analysis of Group V

The solution to be analyzed may be a Group V known or unknown, or it may be the solution from the Group IV separation. The Group V flow sheet is given in Table 38.1.

Table 38.1 Group V flow sheet

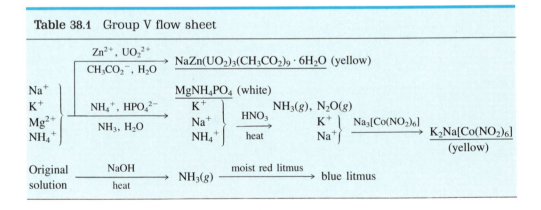

38.1 Removal of Traces of Calcium and Barium

The cations in Group V (Na^+, K^+, Mg^{2+}, and NH_4^+) do not form precipitates with the reagents used to separate Groups I, II, III, and IV. In fact, the four ions of Analytical Group V have no precipitant in common.

If ammonium salts were present in excessive quantities during the precipitation of the carbonates of the Group IV ions, then trace amounts of barium and calcium ions may have come through into the Group V solution. Because these ions will interfere with the identification of the Group V ions, they must be removed. Ammonium oxalate is added to precipitate the calcium ions, and ammonium sulfate is added to precipitate the barium ions.

Solution from the Group IV Separation: Na^+, K^+, Mg^{2+}, and NH_4^+ (Ca^{2+} and Ba^{2+})   **PROCEDURE 1**

Add to this solution 1 drop of 0.4 M $(NH_4)_2C_2O_4$ and 1 drop of 1 M $(NH_4)_2SO_4$. Separate and discard any precipitate that may form and use the solution for Procedure 2.

38.2 Identification of Sodium

After the removal of the traces of barium and calcium, the sodium ion may be detected by the formation of the triple salt sodium zinc uranyl acetate 6-hydrate, which is a yellow crystalline compound.

$$Na^+ + Zn^{2+} + 3UO_2^{2+} + 9CH_3CO_2^- + 6H_2O \rightleftharpoons$$
$$NaZn(UO_2)_3(CH_3CO_2)_9 \cdot 6H_2O$$

The sodium reagent is a saturated solution of zinc acetate and uranyl acetate in acetic acid. Ammonium, magnesium, and potassium ions do not interfere with the test. The sodium ion imparts a yellow color to the Bunsen burner flame.

Solution from Procedure 1: K^+, Na^+, Mg^{2+}, and NH_4^+   **PROCEDURE 2**

To 2 drops of the solution from Procedure 1 or a Group V known or unknown, add 1 M CH_3CO_2H until the solution is acid to litmus. Add 1 drop of this acidified solution to 5 drops of the sodium reagent. Shake the mixture and set it aside for an hour. The formation of a yellow crystalline precipitate indicates the presence of sodium.

Precipitate of $NaZn(UO_2)_3(CH_3CO_2)_9 \cdot 6H_2O$.

38.3 Separation and Identification of Magnesium

To a second portion of the solution for analysis is added an excess of aqueous ammonia and disodium hydrogen phosphate. The formation of a white crystalline precipitate, $MgNH_4PO_4$, is evidence of the presence of the magnesium ion. The equation is

$$Mg^{2+} + NH_4^+ + HPO_4^{2-} \rightleftharpoons MgNH_4PO_4 + H^+$$

Note that the hydrogen ion is one product of the reaction. Completeness of precipitation of the magnesium ammonium phosphate is ensured by making the solution basic with aqueous ammonia, which combines with the hydrogen ion and causes a shift of the equilibrium to the right.

After the magnesium ammonium phosphate is separated from the solution containing sodium ion and ammonium ion, it is dissolved in acetic acid. This treatment with acid causes the following equilibrium to be shifted to the right:

$$\underline{MgNH_4PO_4} + H^+ \rightleftharpoons Mg^{2+} + NH_4^+ + HPO_4^{2-}$$

The reason is that the phosphate ion in a saturated solution of $MgNH_4PO_4$ is removed from the solution by combining with hydrogen ions, making the product of the concentrations of the Mg^{2+}, NH_4^+, and PO_4^{3-} ions less than the solubility product. The magnesium reagent and sodium hydroxide are added to the solution containing magnesium ions. The magnesium hydroxide that is formed adsorbs the magnesium reagent [a dye, 4-nitrophenylazo-1-naphthol] and forms a blue precipitate, which confirms the presence of magnesium.

PROCEDURE 3

Use 10 drops of the solution from Procedure 1 or 10 drops of a Group V known or unknown: K^+, Na^+, Mg^{2+}, and NH_4^+

Add 4 M aqueous ammonia to the solution until it is alkaline to litmus and then add 1 drop in excess. Now add 2 drops of 1 M Na_2HPO_4 to the solution. The formation of a white crystalline precipitate, often slow in forming, confirms the presence of magnesium. Separate the mixture and save the solution for Procedure 4. Dissolve the precipitate in a mixture of 2 drops of 1 M CH_3CO_2H and 3 drops of water. Add 1 drop of the magnesium reagent and an excess of 4 M NaOH to the solution. The formation of a blue precipitate confirms the presence of magnesium. (Do not confuse the sky-blue color of the precipitate with the purple color of the magnesium reagent. Be sure a precipitate exists.)

Precipitate of $MgNH_4PO_4$ (left).

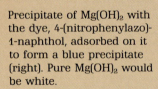

Precipitate of $Mg(OH)_2$ with the dye, 4-(nitrophenylazo)-1-naphthol, adsorbed on it to form a blue precipitate (right). Pure $Mg(OH)_2$ would be white.

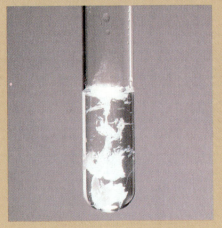

38.4 Identification of Potassium

The identification of potassium is made using the solution obtained from the separation of magnesium ammonium phosphate. Ammonium ions must be removed prior to the test for the potassium ion, because the precipitant for potassium ions will also precipitate ammonium ions. Concentrated nitric acid is added to the solution, it is evaporated to dryness, and then it is ignited to expel volatile ammonium salts.

$$NH_4Cl \xrightarrow{\triangle} NH_3(g) + HCl(g)$$
$$NH_4NO_3 \xrightarrow{\triangle} N_2O(g) + 2H_2O(g)$$

When a solution of $Na_3[Co(NO_2)_6]$ is added to an acidic solution of the residue, the potassium ion precipitates in the form of the yellow complex salt $K_2Na[Co(NO_2)_6]$.

$$2K^+ + Na^+ + [Co(NO_2)_6]^{3-} \longrightarrow \underline{K_2Na[Co(NO_2)_6]}$$

Potassium gives a characteristic violet flame test, which is masked by the intense yellow color produced by the sodium ion. By observing the flame through cobalt glass, which filters out the yellow color of the sodium ion, we may detect the potassium ion in the presence of the sodium ion.

Solution from Procedure 3: K^+, Na^+, and NH_4^+

PROCEDURE 4

Add 4 drops of concentrated HNO_3 to the solution held in a casserole, evaporate it to dryness, and then heat the dry residue for several minutes. After the casserole is cool, add 1 drop of 1 M HCl and 5 drops of 4 M CH_3CO_2H to the residue and boil the resultant solution. Now add 2 drops of this solution to 5 drops of a saturated solution of $Na_3[Co(NO_2)_6]$. The formation of a yellow precipitate confirms the presence of potassium.

Precipitate of $K_2Na[Co(NO_2)_6]$.

Original Known or Unknown Solution for Group V or Solution from Procedure 1

PROCEDURE 5

Make flame tests with this solution and compare the results with known solutions of the ions in question. When testing for potassium, observe the flame through a cobalt glass.

Flame test for sodium.

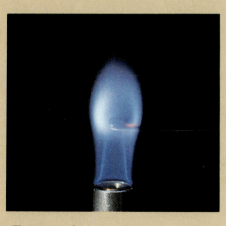

Flame test for potassium.

Sodium ions impart an intense yellow color to the flame. A trace of the sodium ion as contaminant will give a yellow coloration to the flame, so do not rely entirely on the flame test in reporting the presence of sodium ions. Potassium ions give a violet color to the flame.

38.5 Identification of Ammonium Ion

Since ammonium ions are added in the course of the analysis at various points, it is necessary to test the original solution of a known or unknown for this ion. The presence of the ammonium ion in a solution may be detected by the addition of a strong, nonvolatile base such as sodium hydroxide, with subsequent heating of the solution and liberation of gaseous ammonia.

$$NH_4^+ + OH^- \text{ (from NaOH)} \rightleftharpoons NH_3(g) + H_2O$$

The gaseous ammonia may be detected by moist red litmus paper and by its characteristic odor.

$$NH_3 + H_2O \rightleftharpoons NH_4^+ + OH^-$$

During the heating of the solution one must not permit the sodium hydroxide solution to come in contact with the litmus paper, because this would void the test for ammonia.

PROCEDURE 6   **Original Solution of the Known or Unknown**

Add 4 M NaOH to the solution until it is basic. Moisten a piece of red litmus with distilled water and place it on the convex side of a watch glass. Place the watch glass on a beaker containing the solution and warm the solution gently. Avoid spattering of the solution by overheating and do not permit the litmus to come into contact with the solution. If the litmus turns blue within a short time, the presence of the ammonium ion is confirmed.

Test for ammonium ion. (The beaker on the right exhibits a positive test.)

For Review

Exercises

1. Why must the ammonium ion be removed before making the chemical test for the potassium ion?
2. Explain why the test for the ammonium ion must be made on a portion of the original sample during the analysis of either a Group V known or unknown or a general unknown.
3. Why is Group V often referred to as the soluble group?
4. Why are $(NH_4)_2SO_4$ and $(NH_4)_2C_2O_4$ added prior to the analysis of Group V in a general unknown?
5. If magnesium carbonate had, through an error in carrying out the procedure, precipitated with Group IV, how would you determine its presence? What error or errors might have caused its precipitation with Group IV?
6. When $MgNH_4PO_4$ is heated to a high temperature (1000°C) it is converted to magnesium diphosphate, $Mg_2P_2O_7$. Write a balanced equation for this reaction.
7. Write the equation for the dissolution of $MgNH_4PO_4$ in an acid. Explain this reaction in terms of solubility product and ionic equilibria theory.
8. Why may red litmus turn blue after prolonged exposure to the air of the laboratory?
9. What is the chemistry of the chemical test for potassium?
10. Explain the use of cobalt glass in the flame test for potassium.
11. What weight of $K_2Na[Co(NO_2)_6]$ can be formed from 2.5 mg of K^+?   *Ans. 14 mg*
12. How many mL of NH_3 gas measured at 20°C and 750 torr will be evolved when a solution containing 10 mg of NH_4Cl is treated with NaOH and heated?
                                                        Ans. 4.6 mL

39 The Analysis of Anions

39.1 Introduction

The analytical scheme for anions includes 13 of the more common and important negative ions. The anions considered are carbonate, sulfide, sulfite, nitrite, sulfate, nitrate, phosphate, metaborate, oxalate, fluoride, chloride, bromide, and iodide. Among the anions that were covered by the analytical scheme for cations are arsenite, arsenate, stannite, stannate, permanganate, aluminate, chromate, dichromate, and zincate.

Although several schemes for anion analysis involving the separation of the ions into groups have been developed, the procedures are in general more complicated and unreliable than are those for cation analysis. Instead of using a systematic scheme of analysis involving the same solution throughout, the procedures outlined in this chapter involve a series of elimination tests that prove the absence of certain anions. Tests are then made on different samples of the unknown solution for the presence of the anions whose absence was not indicated in the preliminary elimination tests. (A flow sheet of the anion elimination tests is given in Table 39.1.)

The properties of the acids and their anions have been considered in the first part of the text in connection with the descriptive chemistry of the nonmetals. You will find it necessary to refer to earlier chapters in order to answer some of the questions that arise concerning the chemistry of the anions.

The anion knowns and unknowns are usually furnished in the form of mixtures of dry salts due to the fact that some of the anions react with one another in solution. Thus certain anions may be detected in freshly prepared solutions, whereas, after standing, these same anions may have undergone decomposition. Certain combinations of anions react almost immediately in alkaline solutions, and still other anions cannot exist

Table 39.1   Anion elimination flow sheet

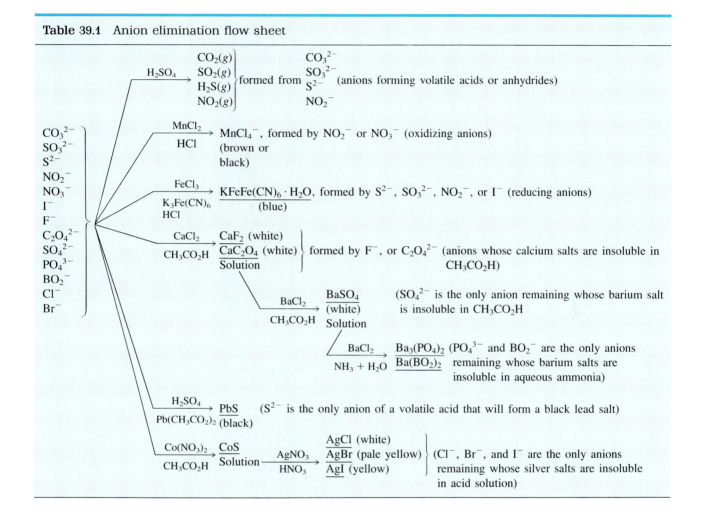

together in acidic solutions. As an example of anion incompatibility, anions that are strong oxidizing agents usually react on acidification with anions that are strong reducing agents.

Preliminary Elimination Tests for the Anions

39.2 Anion Elimination Chart

Prepare an anion elimination chart like the one shown in Table 39.2. As you perform each elimination test, check off in the appropriate spaces those anions that you have found to be absent. When all the elimination tests have been made, it will be apparent which anions may be present. Tests can then be made for the anions that have not been eliminated.

Table 39.2 Anion elimination chart

Name: _____ Section: _____ Date: _____

Instructor's Approval: _____ Ions Found: _____

Anion Elimination Chart

Anion	Test I	Test II	Test III	Test IV	Test V	Test VI	Test VII	Test VIII	Make Confirmatory Tests for
CO_3^{2-}									
SO_3^{2-}									
S^{2-}									
NO_2^-									
NO_3^-									
I^-									
F^-									
$C_2O_4^{2-}$									
SO_4^{2-}									
PO_4^{3-}									
BO_2^-									
Cl^-									
Br^-									

39.3 Elimination of Anions That Form Volatile Acid Anhydrides

The anions CO_3^{2-}, SO_3^{2-}, S^{2-}, and NO_2^- are derived from the weak acids H_2CO_3, H_2SO_3, H_2S, and HNO_2, respectively. As acids in solution, H_2CO_3, H_2SO_3, and HNO_2 are unstable and decompose, producing gases. H_2S does not decompose but has a limited solubility in water. When salts of these anions are treated with strong acids, the anions, being strong Brønsted bases, readily combine with the hydrogen ions from the strong acids. The equilibria are

$$CO_3^{2-} + H^+ \rightleftharpoons HCO_3^-$$
$$HCO_3^- + H^+ \rightleftharpoons H_2CO_3$$
$$H_2CO_3 \rightleftharpoons H_2O + CO_2(g)$$

$$SO_3^{2-} + H^+ \rightleftharpoons HSO_3^-$$
$$HSO_3^- + H^+ \rightleftharpoons H_2SO_3$$
$$H_2SO_3 \rightleftharpoons H_2O + SO_2(g)$$

$$S^{2-} + H^+ \rightleftharpoons HS^-$$
$$HS^- + H^+ \rightleftharpoons H_2S(g)$$
$$NO_2^- + H^+ \rightleftharpoons HNO_2$$
$$3HNO_2 \longrightarrow H^+ + NO_3^- + 2NO(g) + H_2O$$
$$2NO + O_2 \longrightarrow 2NO_2(g)$$

The high concentration of hydrogen ion furnished by the sulfuric acid used in this elimination test serves to shift the equilibrium to the right in each case. When the solubility of each gas is exceeded, it will escape from solution in the form of bubbles. Gentle heating decreases the solubilities of the gases and aids in the test. Boiling of the solution should be avoided, because bubbles of steam may be mistaken for bubbles of gaseous reaction products.

If no gas is evolved during acidification and gentle heating of an unknown sample, CO_3^{2-}, SO_3^{2-}, S^{2-}, and NO_2^- are absent. If a gas is evolved, then one or more of these anions is present, and it is necessary to make tests for the presence of each of these ions unless they are proved absent by other elimination tests.

Elimination of Anions That Form Volatile Acid Anhydrides: CO_3^{2-}, SO_3^{2-}, S^{2-}, and NO_2^-

TEST I

Put 25 mg of the solid unknown in a test tube and treat it with 2 drops of 1.5 M H_2SO_4. Examine the mixture for the formation of gas bubbles. If no gas is evolved, heat the mixture in the hot water bath. If no gas is evolved, then CO_3^{2-}, SO_3^{2-}, S^{2-}, and NO_2^- are proved to be absent. If these anions are absent, place a check mark after the formula for each of these anions in the Test I column of the anion elimination chart; leave these spaces blank if a gas is evolved.

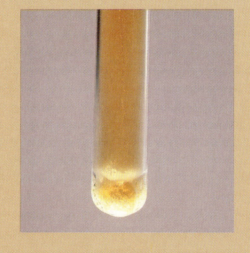

Evolution of nitrogen dioxide by oxidation of NO_2^-.

If a gas is evolved, note its odor and color. Carbon dioxide is colorless and odorless. Sulfur dioxide is colorless and has the odor of burning sulfur. Hydrogen sulfide is colorless and has a characteristic odor. If the nitrate ion is present, nitrogen dioxide, with a reddish-brown color and characteristic odor, is evolved. The solution may be pale blue in color due to the presence of HNO_2 if the nitrite ion is present.

39.4 Preparation of the Sample for Analysis

The procedures outlined for the detection of the anions are applicable, in general, to salts containing only the alkali metals as cations. It is obvious that complications in the anion analysis will arise if certain metal ions are present to form precipitates, or possibly colored solutions, with the anions being detected.

Most of the metal ions may be removed from solution as insoluble carbonates, hydroxycarbonates, hydroxides, or oxides by heating a solution of the unknown with sodium carbonate. The ammonium ion, NH_4^+, is converted to NH_3, which is evolved from the hot solution. Since the carbonate ion is added in excess the identification of this ion must be made on the original sample. The CO_3^{2-} must be removed before certain other tests are made.

PROCEDURE 1   **Preparation of the Sample for Analysis: Removal of Heavy Metal Ions by Na_2CO_3**

Put 100 mg of the powdered unknown in a test tube and add 2 mL of 1.5 M Na_2CO_3 solution. Heat the mixture for 10 min in the hot water bath. If ammonia is given off, continue the heating until no more gas is evolved. Separate any precipitate that forms. The remaining solution contains the anions in the form of sodium salts and will be referred to as the *prepared solution*.

39.5 Elimination of Oxidizing Anions

Note that the nitrite ion is included in the elimination tests for both oxidizing anions and reducing anions. The nitrite ion in acid solution is an oxidizing agent in the presence of strong reducing agents and a reducing agent in the presence of strong oxidizing agents.

The prepared solution is tested for the oxidizing anions NO_2^- and NO_3^- with a saturated solution of manganese(II) chloride in concentrated hydrochloric acid. The development of a dark-brown or black color is the result of the oxidation, in strongly acidic solution, of the manganese(II) ion, Mn^{2+}, to the manganese(III) ion, Mn^{3+}. The manganese(III) ion then unites with chloride ions and forms the dark-colored complex $[MnCl_4]^-$.

$$Mn^{2+} + HNO_2 + H^+ + 4Cl^- \longrightarrow [MnCl_4]^- + NO(g) + H_2O$$
$$3Mn^{2+} + 4H^+ + NO_3^- + 12Cl^- \longrightarrow 3[MnCl_4]^- + NO(g) + 2H_2O$$

TEST II

Solution of $MnCl_4^-$, produced by the action of oxidizing anions on $MnCl_2$ in HCl.

Elimination of Oxidizing Anions: NO_3^- and NO_2^-

Add 6 drops of a saturated solution of $MnCl_2$ in 12 M HCl to 4 drops of the prepared solution and heat the mixture to boiling. The formation of a dark-brown or black color, due to $MnCl_4^-$, indicates the presence of NO_3^-, NO_2^-, or a mixture of these ions. Record the results of the test on the anion elimination chart.

39.6 Elimination of Reducing Anions

A sample of the prepared solution is acidified with hydrochloric acid and then iron(III) chloride and potassium hexacyanoferrate(III) are added. The prompt appearance of a blue to blue-green color or precipitate proves the presence of one or more of the reducing anions S^{2-}, SO_3^{2-}, NO_2^-, and I^-.

Iron(III), Fe^{3+}, forms a brown solution when fresh, and a green solution when old, with $Fe(CN)_6^{3-}$. In the presence of a strong reducing agent, the Fe^{3+} may be reduced to Fe^{2+}, or the $[Fe(CN)_6]^{3-}$ reduced to $[Fe(CN)_6]^{4-}$, with the formation of $KFe[Fe(CN)_6] \cdot H_2O$, a blue pigment. The equations are

$$2Fe^{3+} + S^{2-} \longrightarrow 2Fe^{2+} + S$$
$$2Fe^{3+} + SO_3^{2-} + H_2O \longrightarrow 2Fe^{2+} + SO_4^{2-} + 2H^+$$
$$2Fe^{3+} + 2I^- \longrightarrow 2Fe^{2+} + I_2$$
$$2[Fe(CN)_6]^{3-} + NO_2^- + H_2O \longrightarrow 2[Fe(CN)_6]^{4-} + 2H^+ + NO_3^-$$
$$K^+ + Fe^{2+} + [Fe(CN)_6]^{3-} + H_2O \longrightarrow KFe[Fe(CN)_6] \cdot H_2O$$
$$K^+ + Fe^{3+} + [Fe(CN)_6]^{4-} + H_2O \longrightarrow KFe[Fe(CN)_6] \cdot H_2O$$

Elimination of Reducing Anions: S^{2-}, SO_3^{2-}, I^-, and NO_2^-   **TEST III**

Mix 2 drops of a recently prepared saturated solution of $K_3[Fe(CN)_6]$ with 1 drop of 0.1 M $FeCl_3$, and add 2 drops of 6 M HCl. Add 2 drops of the prepared solution to this mixture and let the mixture stand for a few minutes. The development of a blue color or the formation of a blue precipitate indicates the presence of a reducing agent (S^{2-}, SO_3^{2-}, I^-, NO_2^-). Run a blank test with the same reagents, leaving out the prepared solution, and compare the colors and intensities of the unknown and blank test solutions.

Solution of $[Fe(CN)_6]^{4-}$ and precipitate of $KFe[Fe(CN)_6] \cdot H_2O$, produced by the reduction of $[Fe(CN)_6]^{3-}$ by reducing anions.

39.7 Elimination of Fluoride and Oxalate

Of the 13 anions that are included in this analytical scheme, only fluoride and oxalate form calcium salts that are insoluble in 4 M acetic acid.

$$Ca^{2+} + 2F^- \rightleftharpoons \underline{CaF_2} \text{ (white)}$$
$$Ca^{2+} + C_2O_4^{2-} \rightleftharpoons \underline{CaC_2O_4} \text{ (white)}$$

If no precipitate forms when calcium chloride is added to an acetic acid solution of the unknown, then fluoride and oxalate are absent.

TEST IV

Elimination of Fluoride and Oxalate: F^- and $C_2O_4^{2-}$

To 20 drops of the prepared solution, add 4 M CH_3CO_2H (count the drops) until the solution is just acid to litmus. Now add an equal number of drops of 4 M CH_3CO_2H in excess. Tap the test tube repeatedly until the excess of CO_2 from the Na_2CO_3 present is expelled. Add 8 drops of 0.1 M $CaCl_2$ and shake the contents of the tube. If a precipitate forms after a few minutes, it may be CaF_2, CaC_2O_4, or a mixture of these compounds. If no precipitate forms, F^- and $C_2O_4^{2-}$ are absent. Separate the precipitate and use the solution for Test V. Save the precipitate for the confirmatory test for $C_2O_4^{2-}$ and F^-.

39.8 Elimination of Sulfate

After the separation of fluoride and oxalate as insoluble calcium salts, barium chloride is added to precipitate the sulfate ion as white crystalline barium sulfate from the solution that is still acidic with acetic acid. If the sulfite ion is present in relatively large amounts, barium sulfite may precipitate. $BaSO_4$ is insoluble in dilute hydrochloric acid, while $BaSO_3$ will dissolve and evolve SO_2 when treated with hydrochloric acid.

Owing to the ease with which the sulfite ion is oxidized to the sulfate ion by oxygen in the air, if sulfite is present in the original sample, you may obtain a test for sulfate although none was present at the outset.

TEST V

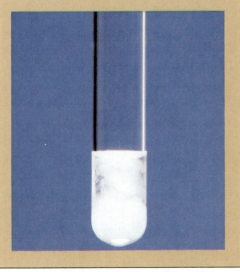

Precipitate of $BaSO_4$.

Elimination of Sulfate: SO_4^{2-}

Test for completeness of precipitation in the solution from Test IV by adding 1 drop of 0.1 M $CaCl_2$. Now add 8 drops of 0.1 M $BaCl_2$. Shake the mixture and let it stand for 5 min. If SO_4^{2-} is present, a precipitate of $BaSO_4$ will form. Separate the mixture and reserve the solution for Test VI. If SO_3^{2-} was present in the original solution, you may obtain a test for SO_4^{2-} because sulfite is oxidized to sulfate by oxygen in the air.

39.9 Elimination of Phosphate and Metaborate

The solution from the barium sulfate separation is acidified with hydrochloric acid and heated to expel CO_2 from CO_3^{2-} and SO_2 from SO_3^{2-}, if present. If they were not first removed, these ions would be precipitated as barium salts in the ammoniacal solution used to precipitate barium phosphate and barium metaborate.

$$3Ba^{2+} + 2PO_4^{3-} \rightleftharpoons \underline{Ba_3(PO_4)_2} \text{ (white)}$$
$$Ba^{2+} + 2BO_2^{-} \rightleftharpoons \underline{Ba(BO_2)_2} \text{ (white)}$$

Even though most phosphates and metaborates are insoluble in water, these ions did not precipitate as their calcium or barium salts in the acetic acid solutions in which CaF_2, CaC_2O_4, and $BaSO_4$ were formed. Barium phosphate does not precipitate in an acetic acid solution, due to the low concentration of free phosphate ions. The phosphate ion is a strong base.

$$PO_4^{3-} + H^+ \rightleftharpoons HPO_4^{2-}$$

In an acid solution, its concentration is so low that the solubility product of $Ba_3(PO_4)_2$ is not exceeded. Raising the pH of the solution by adding aqueous ammonia, on the other hand, increases the concentration of the phosphate ion

$$HPO_4^{2-} + NH_3 \rightleftharpoons PO_4^{3-} + NH_4^+$$

and permits the solubility product of $Ba_3(PO_4)_2$ to be exceeded.

Elimination of Phosphate and Metaborate: PO_4^{3-} and BO_2^{-}   **TEST VI**

Transfer the solution from Test V to a casserole, add 6 drops of 12 M HCl, and heat to expel SO_2 and CO_2. Make the solution alkaline with aqueous ammonia and then add 5 drops in excess. A white precipitate may be $Ba_3(PO_4)_2$, $Ba(BO_2)_2$, or a mixture of these. Because barium metaborate is prone to form supersaturated solutions, it is necessary to carry out the confirmatory test for BO_2^- even though no precipitate forms in this elimination test.

39.10 Elimination of Sulfide

The addition of dilute sulfuric acid to a sample of the prepared solution causes hydrogen sulfide gas to be evolved if the sulfide ion is present. Hydrogen sulfide reacts with lead acetate and forms a black precipitate of lead sulfide.

$$Pb^{2+} + H_2S \rightleftharpoons \underline{PbS} + 2H^+$$

This elimination test also serves to confirm the sulfide ion if it is present.

Elimination of Sulfide: S^{2-}   **TEST VII**

To 4 drops of the prepared solution in a small beaker, add sufficient 1.5 M H_2SO_4 to make the solution acidic. Cover the beaker with a watch glass to the underside of which is attached a small piece of moist lead acetate paper. The appearance of a black stain of PbS on the paper after a few minutes indicates the presence of S^{2-}.

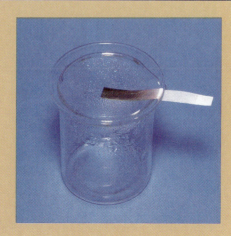

Test for S^{2-} using moist lead acetate paper, on which PbS forms if S^{2-} is present in the solution.

39.11 Elimination of Chloride, Bromide, and Iodide

If the sulfide ion is present, it is first removed as the insoluble CoS before attempting to eliminate chloride, bromide, and iodide. Silver nitrate is then added to precipitate the halides as AgCl (white), AgBr (pale yellow), and AgI (yellow). Silver chloride is completely soluble in the mixture of aqueous ammonia and silver nitrate solutions used in Test VIII, silver bromide is partially soluble, and silver iodide is insoluble in this reagent.

$$AgCl + 2NH_3 \rightleftharpoons [Ag(NH_3)_2]^+ + Cl^-$$

$$AgBr + 2NH_3 \rightleftharpoons [Ag(NH_3)_2]^+ + Br^- \text{ (partially soluble)}$$

$$AgI + NH_3 \rightleftharpoons \text{ no reaction}$$

These facts reflect the decreasing solubility of the silver halides with increasing atomic weight of the halogen. The solubility products are 1.8×10^{-10} for AgCl, 3.3×10^{-13} for AgBr, and 1.5×10^{-16} for AgI.

If a white precipitate forms when the aqueous ammonia extract of the silver halide precipitate is acidified, then the presence of the chloride ion is confirmed.

$$[Ag(NH_3)_2]^+ + Cl^- + 2H^+ \longrightarrow AgCl + 2NH_4^+$$

A slight precipitate, yellow in color, may be silver bromide.

TEST VIII **Elimination of Chloride, Bromide, and Iodide: Cl^-, Br^-, and I^-**

If the sulfide ion is present, treat 10 drops of the prepared solution with 4 M CH_3CO_2H until the solution is acid to litmus. Now add, dropwise, with shaking, 10 drops of 1 M $Co(NO_3)_2$. Heat in the hot water bath for 5 min and separate the precipitate of CoS. The solution may contain Cl^-, Br^-, and I^-. To the solution (or, if sulfide is absent, to 10 drops of the prepared solution acidified with 1 M HNO_3), add 5 drops of 1 M HNO_3. Now add 0.1 M $AgNO_3$ to this solution until precipitation is complete. Separate and wash the precipitate with 10 drops of water. Discard the solution. Mix 10

(left to right)
Precipitates of (a) AgCl,
(b) AgBr, and (c) AgI.

drops of 0.1 M $AgNO_3$, 6 drops of 4 M aqueous ammonia, and 4 mL of water. Add this mixture to the precipitate and stir thoroughly. If the precipitate dissolves completely, Br^- and I^- are absent, and Cl^- is indicated. Acidify the solution with 4 M HNO_3. A white precipitate confirms chloride. If the precipitate did not dissolve completely in the aqueous ammonia and silver nitrate mixture, it may have partially dissolved. Separate and acidify the solution with 4 M HNO_3. If no precipitate forms, Cl^- is absent. The formation of a heavy white precipitate confirms the presence of Cl^-. A slight yellow precipitate or yellowish turbidity may be silver bromide.

Confirmatory Tests for the Anions

39.12 Identification of Carbonate

A sample of the original unknown is treated with hydrogen peroxide to oxidize any sulfite or nitrite that might be present. Then dilute sulfuric acid is added to decompose the carbonate and liberate carbon dioxide gas. The carbon dioxide gas evolved is brought in contact with a drop of barium hydroxide solution, which is held in a platinum wire loop suspended above the reaction mixture. The formation of a definite turbidity in the drop of barium hydroxide indicates the presence of carbonate.

$$CO_3^{2-} + 2H^+ \rightleftharpoons H_2O + CO_2(g)$$
$$CO_2(g) + Ba^{2+} + 2OH^- \rightleftharpoons \underline{BaCO_3} + H_2O$$

If the sulfite and nitrite were not removed by oxidation to sulfate and nitrate, respectively, sulfur dioxide and nitrogen dioxide would escape following acidification, enter the drop of barium hydroxide solution, and either prevent the formation of barium carbonate or mask its presence.

Identification of CO_3^{2-}	PROCEDURE 2

To 25 mg of the powdered solid unknown, add 10 drops of hydrogen peroxide and heat the mixture in the hot water bath. Place 1 drop of $Ba(OH)_2$ solution in a loop of platinum wire. Add 3 drops of 4 M H_2SO_4 to the unknown mixture and immediately hold the drop of $Ba(OH)_2$ over the reaction mixture. The formation of a definite turbidity in the $Ba(OH)_2$ solution indicates the presence of CO_3^{2-}.

39.13 Identification of Sulfate and Sulfite

The formation of a white precipitate of barium sulfate when barium chloride is added to a hydrochloric acid solution of the unknown serves to confirm the sulfate ion. Barium salts of the other anions are soluble in 6 M hydrochloric acid.

Bromine water is added to the solution obtained from the separation of barium sulfate for the purpose of oxidizing the sulfite ion to sulfate.

$$SO_3^{2-} + Br_2 + H_2O \longrightarrow SO_4^{2-} + 2Br^- + 2H^+$$

The appearance of a white precipitate (barium sulfate) indicates the presence of the sulfite ion in the original solution.

PROCEDURE 3

Precipitate of $BaSO_4$.

Identification of SO_4^{2-}

Acidify 5 drops of the prepared solution with 6 M HCl and add 2 drops in excess. Heat in the hot water bath to expel excess CO_2. Add 0.1 M $BaCl_2$ until precipitation is complete; then allow the mixture to stand. A white, finely divided precipitate of $BaSO_4$ confirms the presence of SO_4^{2-}. Separate the mixture and save the solution for Procedure 4.

PROCEDURE 4   Identification of SO_3^{2-}

Add 5 drops of bromine water and 2 drops of 0.1 M $BaCl_2$ to the solution from Procedure 2. Heat in the hot water bath and allow to stand for 5 min. A white precipitate of $BaSO_4$ shows the presence of sulfite in the original solution.

PROCEDURE 5   Identification of S^{2-}

The sulfide ion is confirmed in its elimination test (Test VII).

39.14 Identification of Nitrite

The oxidizing ability of the nitrite ion in acid solution is applied in the confirmatory test for this ion. The nitrite ion oxidizes iron(II) and iron(III), and nitric oxide is formed from the reduction of the nitrite ion.

$$NO_2^- + Fe^{2+} + 2H^+ \longrightarrow Fe^{3+} + NO + H_2O$$

The nitric oxide produced combines with some of the excess iron(II) ions in the reaction mixture, forming the brown complex $[Fe(NO)]^{2+}$.

$$Fe^{2+} + NO \longrightarrow [Fe(NO)]^{2+}$$

PROCEDURE 6

Solution of $[Fe(NO)]^{2+}$.

Identification of NO_2^-

To 5 drops of the prepared solution, add 4 M H_2SO_4 dropwise until the solution is acidic. Now add 5 drops of freshly prepared 0.1 M $FeSO_4$ solution. If NO_2^- is present, the solution will assume a dark brown color.

39.15 Identification of Nitrate

If nitrite, bromide, and iodide ions are absent, we may identify the nitrate ion by the brown ring test. Iron(II) sulfate is added to a sample of the prepared solution in a test tube and then acidified with dilute sulfuric acid. The test tube containing the solution is tilted slightly and concentrated H_2SO_4 is poured down the inside wall in such a way as to form a layer of acid in the bottom of the tube. If the nitrate ion is present, a brown ring will soon form at the interface of the two layers. The reduction of the nitrate ion to nitric oxide by iron(II) takes place rapidly only in a solution of very high hydrogen ion concentration and at relatively high temperatures, conditions that prevail at the interface of the two liquids.

$$3Fe^{2+} + NO_3^- + 4H^+ \longrightarrow 3Fe^{3+} + NO + 2H_2O$$
$$Fe^{2+} + NO \longrightarrow [Fe(NO)]^{2+}$$

If the nitrite ion is present, it must be removed prior to the nitrate test; otherwise the entire solution will turn brown and the development of the brown ring will be obscured. The removal of the nitrite ion is accomplished by adding ammonium sulfate and evaporating nearly to dryness over a flame. The added ammonium ion reacts with the nitrite ion and forms nitrogen and water.

$$NH_4^+ + NO_2^- + heat \longrightarrow N_2(g) + 2H_2O$$

The residue is taken up in water, and the brown ring test for nitrate is conducted on the solution.

If bromide or iodide ions are present, they will be oxidized by the concentrated sulfuric acid used in the test for nitrate, forming bromine and iodine, respectively, both of which form colored layers. Therefore these ions must be removed before making the brown ring test for nitrate. This is accomplished by adding solid nitrate-free silver sulfate to an acidified sample of the prepared solution to precipitate the bromide and iodide ions as their silver salts.

$$Ag_2SO_4 + 2Br^- \longrightarrow \underline{2AgBr} + SO_4^{2-}$$
$$Ag_2SO_4 + 2I^- \longrightarrow \underline{2AgI} + SO_4^{2-}$$

Identification of NO_3^- in the Absence of NO_2^-, Br^-, and I^-　　　　　　**PROCEDURE 7**

To 5 drops of the prepared solution in a test tube, add 4 M H_2SO_4 dropwise until the solution is acidic. Now add 5 drops of freshly prepared 0.1 M $FeSO_4$ solution. Add 5 drops of 18 M H_2SO_4, holding the test tube in an inclined position so that the sulfuric acid runs down the side of the test tube and forms a separate layer on the bottom. If NO_3^- is present, a brown ring will form at the interface of the two liquids, within a few minutes.

Identification of NO_3^- in the Presence of NO_2^-. Removal of NO_2^-.　　　　**PROCEDURE 8**

To 6 drops of the prepared solution add 4 M H_2SO_4 until the solution is acidic, and then add 4 drops of 1 M $(NH_4)_2SO_4$ solution. Place the mixture in a casserole and slowly evaporate the solution until only a moist residue remains (do not evaporate to dryness). Add 4 drops of water and evaporate to a moist residue a second time. Dissolve the residue in 10 drops of water and transfer the mixture to a small test tube. Perform the brown ring test for NO_3^- as described in Procedure 7 on the mixture resulting from the removal of NO_2^-.

PROCEDURE 9

Brown ring test for NO_3^-.

Identification of NO_3^- in the Presence of Br^- and I^-. Removal of Br^- and I^-

To 6 drops of the prepared solution in a test tube, add 10 drops of water. Acidify the solution with 4 M CH_3CO_2H and then add 80 mg of powdered Ag_2SO_4 (nitrate-free). Stir and grind the mixture in the test tube for 2-3 min. Separate the precipitate and transfer the solution to a test tube. Perform the brown ring test for NO_3^- as described in Procedure 7 on the solution obtained from the removal of Br^- and I^-.

39.16 Identification of Oxalate

The precipitate formed during the elimination test for oxalate and fluoride ions (Test IV) is used for the identification of the oxalate ion. It consists of CaC_2O_4 and CaF_2 if both ions were present in the original solution. Treatment of the precipitate with sulfuric acid dissolves the calcium oxalate and forms oxalic acid.

$$CaC_2O_4 + 2H^+ \rightleftharpoons Ca^{2+} + H_2C_2O_4$$

Oxalic acid is a weak reducing agent that is readily oxidized to carbon dioxide and water by potassium permanganate in sulfuric acid solution. A dilute solution of the pink permanganate is employed, and the permanganate ion is reduced to the colorless manganese(II) ion by the oxalic acid. This bleaching of the colored permanganate solution shows the presence of the weak reducing agent oxalic acid.

$$5H_2C_2O_4 + 2MnO_4^- + 6H^+ \longrightarrow 10CO_2(g) + 2Mn^{2+} + 8H_2O$$

PROCEDURE 10   **Identification of $C_2O_4^{2-}$**

Wash the precipitate obtained in Test IV twice. Use 10 drops of water each time and discard the washings. To the residue add 10 drops of water and 10 drops of 4 M H_2SO_4, and shake the mixture. Add 0.002 M $KMnO_4$ dropwise until 1 drop imparts a permanent pink color to the solution. If $C_2O_4^{2-}$ is present, several drops of $KMnO_4$ should be required to give a permanent pink coloration to the solution. Run a blank by repeating the test but leaving out the precipitate from Test IV.

39.17 Identification of Fluoride

In the confirmatory test for the fluoride ion, a sample of the solid unknown is treated with concentrated sulfuric acid in the presence of silica, SiO_2.

$$2NaF + H_2SO_4 \longrightarrow Na_2SO_4 + 2HF(g)$$

The hydrofluoric acid produced reacts with the silica present and forms gaseous silicon tetrafluoride, SiF_4.

$$SiO_2 + 4HF \longrightarrow SiF_4(g) + 2H_2O$$

The gaseous SiF_4 is brought in contact with a drop of water contained in a loop of platinum wire suspended over the reaction mixture. The development of a white precipitate, H_4SiO_4, in the water confirms the presence of the fluoride ion.

$$3SiF_4 + 4H_2O \longrightarrow H_4SiO_4 + 4H^+ + 2SiF_6^{2-}$$

Identification of F⁻

PROCEDURE 11

Place 25 mg of the powdered solid unknown in a test tube and add an equal volume of powdered silica. Select a cork to fit the test tube. Cut a notch on the side of the cork and bore a hole through its center. Through the hole insert a glass rod to the end of which is sealed a platinum wire with a looped end. Add 2 drops of 18 M H_2SO_4 to the mixture in the tube. Now place the platinum wire loop with a drop of water in it in the test tube directly above the reaction mixture. Warm the test tube in the hot water bath and then set it aside to cool. The formation of a white precipitate or cloudiness in the drop of water confirms the presence of F⁻.

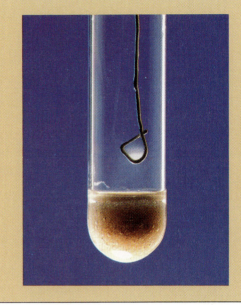

Precipitate of H_4SiO_4, produced by the reaction of water with SiF_4.

39.18 Identification of Phosphate

Magnesia mixture ($MgCl_2$, NH_4Cl, and aqueous ammonia), when added to a slightly acidic solution of the unknown, will precipitate the phosphate ion as white crystalline magnesium ammonium phosphate. Since this compound tends to form supersaturated solutions, the precipitate may be slow in appearing.

$$Mg^{2+} + NH_4^+ + HPO_4^{2-} \rightleftharpoons \underline{MgNH_4PO_4} + H^+$$

Identification of PO₄³⁻

PROCEDURE 12

Dilute 4 drops of the prepared solution with 10 drops of water. Make the solution just acidic with 4 M HNO_3 and add 10 drops of magnesia mixture. A white precipitate, often slow in forming, confirms the presence of PO_4^{3-}.

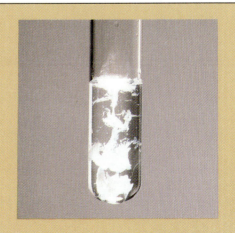

Precipitate of $MgNH_4PO_4$.

39.19 Identification of Metaborate

A sample of the prepared solution is evaporated to a small volume and then treated with methyl alcohol and sulfuric acid. The sulfuric acid converts the metaborate ion to orthoboric acid.

$$BO_2^- + H^+ + H_2O \longrightarrow H_3BO_3$$

In the presence of the dehydrating agent, concentrated sulfuric acid, orthoboric acid reacts with methyl alcohol and forms the volatile methyl orthoborate.

$$H_3BO_3 + 3CH_3OH \rightleftharpoons B(OCH_3)_3(g) + 3H_2O$$

The sulfuric acid present takes up the water produced in the reaction and shifts the equilibrium to the right. Methyl orthoborate burns with a characteristic green flame when ignited.

$$2B(OCH_3)_3 + 9O_2 \longrightarrow B_2O_3 + 6CO_2(g) + 9H_2O$$

PROCEDURE 13

Flame containing vapors of methyl orthoborate in the test for BO_2^-.

Identification of BO_2^-

Evaporate 6 drops of the prepared solution to a small volume in a casserole. Add 1 mL of methyl alcohol and 5 drops of 18 M H_2SO_4. Transfer the mixture to a test tube and place the tube in the hot water bath. When the alcohol begins to boil, ignite the vapors. A green tinge to the flame confirms the presence of the metaborate ion.

39.20 Identification of Iodide, Bromide, and Chloride

The confirmatory test for chloride was outlined in the elimination test for chloride, bromide, and iodide (Test VIII); no further test for this ion is necessary.

In general the methods used for the detection of the halide ions in the presence of one another depend on the differences in the ease of oxidation of the ions. Of the three ions chloride, bromide, and iodide, the iodide ion is the most easily oxidized, the bromide ion is second, and the chloride ion is the most difficult to oxidize.

The iodide ion may be oxidized to iodine by iron(III) in acid solution without affecting bromide or chloride ions.

$$2I^- + 2Fe^{3+} \longrightarrow I_2 + 2Fe^{2+}$$

The elemental iodine is then extracted from the aqueous solution by dichloromethane, in which it exhibits a characteristic violet color.

After the complete removal of the iodide ion, a stronger oxidizing agent than iron(III) is used to oxidize the bromide to bromine. A dilute solution of potassium permanganate in the presence of nitric acid will oxidize bromide to bromine without interference from chloride.

$$10Br^- + 2MnO_4^- + 16H^+ \longrightarrow 5Br_2 + 2Mn^{2+} + 8H_2O$$

The free bromine is extracted with dichloromethane, to which it imparts a yellow or orange color, depending on its concentration.

Identification of I⁻   **PROCEDURE 14**

Dilute 6 drops of the prepared solution with 12 drops of water. Add 4 M HNO_3 until the solution is just acid to litmus, and then add 2 drops of the acid in excess. Treat the solution with 1 mL of 0.1 M $Fe(NO_3)_3$ and 10 drops of CH_2Cl_2 and shake the test tube. The development of a violet color in the organic layer (the lower layer) proves the presence of I⁻. Remove and discard the organic layer by means of a capillary syringe. Add 10 more drops of CH_2Cl_2, shake, and remove the organic layer. Repeat the extraction until the organic layer remains colorless, and use the aqueous solution in Procedure 15.

Identification of Br⁻   **PROCEDURE 15**

To the aqueous solution from Procedure 14, add 2 drops of 4 M HNO_3, then 0.1 M $KMnO_4$ solution dropwise until the solution remains pink. Extract the solution with CH_2Cl_2. A yellow or orange color in the organic layer indicates the presence of Br⁻.

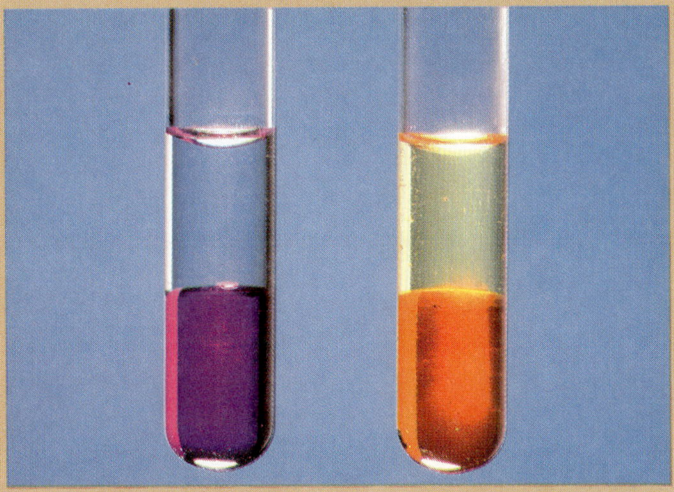

Extraction of I_2 (left) and of Br_2 (right) into an organic liquid (bottom layers).

For Review

Exercises

1. What is meant by incompatibility of anions? Give an example.
2. What principle is involved in the tests for the halide ions in the presence of one another?
3. Explain why the nitrite ion responds to both the test for oxidizing anions and that for reducing anions.
4. In terms of solubility product and ionic equilibria theory, explain the following:
   (a) $Ca_3(PO_4)_2$ is soluble in HCl, whereas $BaSO_4$ is not.
   (b) $CaCO_3$ is soluble in CH_3CO_2H, whereas CaC_2O_4 is not.
5. Make a list of anions that are reducing agents and list their oxidation products.
6. Make a list of anions that are oxidizing agents and list their reduction products.
7. If an anion unknown contains the sulfite ion, it may give a positive test for the sulfate ion even though sulfate was not used in making up the unknown. Explain.
8. An acidic solution contains silver ions. What anions are probably absent?
9. A neutral solution contains calcium ions. What anions are probably absent?
10. Explain how the nitrite ion interferes with the test for nitrate. How is the nitrite ion removed prior to the test for nitrate?
11. A solution is slightly acidic and contains sulfide ions. What cations are probably absent?
12. A solution known to contain either Na_2CO_3 or Na_2SO_4 turns red litmus blue. Which salt is present?
13. Why is it necessary to extract the iodine in the confirmatory test for the iodide ion?
14. Outline a simple test to distinguish between the ions of each pair in the following:
    (a) CO_3^{2-} and SO_4^{2-} (b) Cl^- and I^-
    (c) F^- and BO_2^- (d) PO_4^{3-} and Br^-
    (e) S^{2-} and SO_3^{2-} (f) $C_2O_4^{2-}$ and Cl^-
15. If a solution is strongly acidic, what anions cannot be present in appreciable concentrations?
16. Would you expect the test for F^- to work if no powdered silica were added to the reaction mixture?
17. How many mL of a solution that is 0.010 M in Fe^{3+} ion would be needed to oxidize 1.0 mg of the iodide ion?
    Ans. 0.79 mL
18. How many drops of 0.020 M $KMnO_4$ solution (assume 0.05 mL per drop) would be needed to react with 1.00 mg of oxalic acid and impart a pink color to the resulting solution?
    Ans. 5 drops

The Analysis of Solid Materials

40

The analytical schemes that have been outlined for cations and anions in the preceding chapters are usually applicable to solutions of the substances being analyzed. However, many substances are not readily soluble in water or even in acids. Thus special procedures for effecting the dissolution of certain solid substances must be used before proceeding with the analyses.

40.1 Dissolution of Nonmetallic Solids

The procedures outlined in this section are applicable to solid inorganic materials other than metals and alloys. The dissolution of metals and alloys is described in Section 40.2.

For Samples Soluble in Water   **PROCEDURE 1**

Grind the solid sample to a fine powder. Treat 100 mg of the powder with 1 mL of water. If no dissolution is apparent, heat the mixture in the hot water bath for a few minutes. If none of the sample appears to dissolve, evaporate a few drops of the supernatant liquid on a watch glass to determine whether partial dissolution has occurred. If a solid residue remains on the watch glass, heat the sample with fresh portions of water until complete dissolution has been effected. Concentrate the combined extracts by evaporation, and proceed with the analyses for cations and anions as outlined in the preceding chapters. Certain solid substances, such as $BiCl_3$ and $SbCl_3$, hydrolyze when treated with water to form new solid substances. In such cases Procedure 2 should be followed.

PROCEDURE 2   For Samples Not Soluble in Water

Warm a small sample of the solid substance with 6 M HCl or, if this has no effect, with 12 M HCl. If dissolution does not occur, or if it is incomplete, try 6 M HNO_3, 15 M HNO_3, and aqua regia (1 part 15 M HNO_3 to 3 parts 12 M HCl) in succession on small samples of the solid until a suitable solvent is found. If dissolution has been effected by any of these reagents, dissolve 100 mg of the sample in the effective solvent. Evaporate the solution nearly to dryness, and then take up the residue in 2 mL of water. Use this solution for the analytical schemes for cations and anions.

PROCEDURE 3   Treatment of the Residue with Na_2CO_3 Solution

If HCl, HNO_3, or aqua regia does not completely dissolve the solid, the residue remaining after the acid treatment may be treated with a Na_2CO_3 solution. Treat the residue with 2 mL of 1.5 M Na_2CO_3 in a casserole and boil the mixture for 10 min, replacing the water that is lost during evaporation. This procedure will convert many insoluble salts, such as $BaSO_4$, $CaSO_4$, and $PbSO_4$, and many oxides into acid-soluble carbonates. Separate the mixture and discard the solution unless it is to be used for anion analysis. Wash the precipitate with water and then dissolve it in a few drops of 6 M HNO_3. Dilute the solution with 1 mL of water and add it to the original acid solution that is to be evaporated. If dissolution of the residue was not complete, repeat the Na_2CO_3 treatment. The halides of silver, silicate salts, some oxides, and calcined salts are not dissolved by this procedure.

PROCEDURE 4   Reduction of Silver Halides with Zinc

Suspend the residue containing the insoluble silver halides in 5 drops of water, add 1 mL of 1 M H_2SO_4 and a few granules of zinc metal. Warm and stir the mixture for several minutes. Add more zinc if the evolution of hydrogen ceases. Separate the residue of precipitated silver. Wash the residue with water, dissolve it in 6 M HNO_3, and test the solution for Ag^+. Conduct tests for the halide ions on the solution from the silver separation.

PROCEDURE 5   Fusion with Sodium Carbonate

Silicates and certain oxides and calcined salts may not be taken into solution by treatment with acid or Na_2CO_3 solution or by reduction with zinc. Fusion with Na_2CO_3 is effective with many of these substances. Transfer the residue remaining after the zinc reduction of silver halides to a small nickel crucible. Add to the residue 100 mg of anhydrous Na_2CO_3, about half as much K_2CO_3, and a few milligrams of $NaNO_3$. Place the crucible in a small clay triangle and heat it in the hot flame of a Meker burner until the mixture fuses. Cool the crucible, add 1 mL of

Fusion with sodium carbonate. The mixture is heated until it melts.

water, and warm the mixture until the solid mass has disintegrated. Separate the mixture and reserve the solution. Treat the residue with 8 drops of 6 M HNO_3 and warm in the hot water bath for a few minutes. Separate and add this nitric acid solution to the original solution. Analyze the solution for its cations.

40.2 Dissolution of Metals and Alloys

All of the common metals except Al, Cr, Mn, Fe, Sb, and Sn may be taken into solution with dilute nitric acid. Al, Cr, Mn, and Fe react superficially with nitric acid, forming an oxide film, which makes dissolution of these metals too slow to be practical. Metastannic acid, H_2SnO_3, and the insoluble oxides of antimony, Sb_4O_6, Sb_2O_4, and Sb_2O_5, are formed when nitric acid reacts directly with tin and antimony, respectively. Although these compounds are insoluble in nitric acid, they may be readily converted to the corresponding sulfides, which are soluble in nitric acid. Hydrochloric acid is in general not suitable as a solvent for alloys of unknown composition since the volatile hydrides of sulfur, arsenic, phosphorus, and antimony may be formed; consequently, these elements could escape detection. On the other hand, nitric acid oxidizes the first three of these elements to sulfate, arsenate, and phosphate, respectively, and antimony to oxides, all of which are nonvolatile. Aqua regia will generally dissolve alloys that resist the action of nitric acid alone.

Preparation of the Sample for Dissolution   **PROCEDURE 6**

Convert the metal sample into a finely divided state with a large surface for interaction with the solvent. This may be accomplished with a steel file, a mortar and pestle for brittle metals, a hammer for malleable metals, or a knife for soft metals.

Selection of a Suitable Solvent   **PROCEDURE 7**

Treat a small sample of the metal with 6 M HNO_3 in a test tube and warm the mixture if necessary. If the sample reacts completely, with or without the formation of a white precipitate, carry out Procedure 3. If the sample does not dissolve readily, try aqua regia (1 part of 15 M HNO_3 and 3 parts of 12 M HCl). If aqua regia fails to dissolve the sample, it may be treated with a mixture of concentrated HCl and Br_2 or fused with solid NaOH in a silver crucible.

Dissolution by HNO_3   **PROCEDURE 8**

Place 20 mg of the sample in a test tube, add 1 mL of 6 M HNO_3 and heat the mixture in the hot water bath. If a white residue forms, stir the mixture to remove the coating from the surface of the undissolved metallic particles. Add more HNO_3 if necessary to complete the reaction. After the metal has dissolved, transfer the solution or mixture to a casserole and evaporate it nearly to dryness. Add 5 drops of 15 M HNO_3 and again evaporate the solution nearly to dryness. Now add 5 drops of 6 M HNO_3 and 1 mL of water and transfer the solution to a test tube. If a clear solution is obtained, analyze it according to the procedures for the cations. If there is a residue (H_2SnO_3, Sb_2O_5, or a mixture of these), separate and analyze the solution according to the procedures for the cations, omitting the tests for tin and antimony. Add 10 drops of 4 M NaOH and 5 drops of 5% thioacetamide solution to the residue. Heat the mixture in the water bath for 5 min. Separate any residue and analyze the solution for tin and antimony.

PROCEDURE 9   **Dissolution by Aqua Regia**

Place 20 mg of the sample in a test tube, and add 10 drops of 15 M HNO_3 and 30 drops of 12 M HCl. Heat the mixture in the hot water bath until the reaction is complete. Transfer the reaction mixture to a casserole and evaporate it to a small volume (not to dryness). When it is cool, add 2 mL of water and 4 drops of 6 M HCl. Separate any precipitate that forms. Analyze the centrifugate, or the clear solution if no precipitate forms, by the procedures for Groups II, III, IV, and V. The precipitate may consist of AgCl, $PbCl_2$, or SiO_2. Analyze the precipitate for Group I.

Appendixes

Appendix A: Chemical Arithmetic

A.1 Exponential Arithmetic

Exponential notation is used to express very large and very small numbers as a product of two numbers. The first number of the product, the *digit term,* is usually a number not less than 1 and not greater than 10. The second number of the product, the *exponential term,* is written as 10 with an exponent. Some examples of exponential notation are

$$1000 = 1 \times 10^3 \qquad 0.01 = 1 \times 10^{-2}$$
$$100 = 1 \times 10^2 \qquad 0.001 = 1 \times 10^{-3}$$
$$10 = 1 \times 10^1 \qquad 2386 = 2.386 \times 1000 = 2.386 \times 10^3$$
$$1 = 1 \times 10^0 \qquad 0.123 = 1.23 \times 0.1 = 1.23 \times 10^{-1}$$
$$0.1 = 1 \times 10^{-1}$$

The power (exponent) of 10 is equal to the number of places the decimal is shifted to give the digit number. The exponential method is a particularly useful notation for very large and very small numbers. For example, $1{,}230{,}000{,}000 = 1.23 \times 10^9$; and $0.00000000036 = 3.6 \times 10^{-10}$.

1. ADDITION OF EXPONENTIALS. Convert all numbers to the same power of 10 and add the digit terms of the numbers.

Add 5.00×10^{-5} and 3.00×10^{-3}.

$$3.00 \times 10^{-3} = 300 \times 10^{-5}$$
$$(5.00 \times 10^{-5}) + (300 \times 10^{-5}) = 305 \times 10^{-5} = 3.05 \times 10^{-3}$$

EXAMPLE A.1

---

2. SUBTRACTION OF EXPONENTIALS. Convert all numbers to the same power of 10 and take the difference of the digit terms.

A1

EXAMPLE A.2 Subtract 4.0×10^{-7} from 5.0×10^{-6}.

$$4.0 \times 10^{-7} = 0.40 \times 10^{-6}$$
$$(5.0 \times 10^{-6}) - (0.40 \times 10^{-6}) = 4.6 \times 10^{-6}$$

3. MULTIPLICATION OF EXPONENTIALS. Multiply the digit terms in the usual way and add the exponents of the exponential terms.

EXAMPLE A.3 Multiply 4.2×10^{-8} by 2×10^{3}.

$$(4.2 \times 10^{-8}) \times (2 \times 10^{3}) = (4.2 \times 2) \times 10^{(-8)+(+3)} = 8.4 \times 10^{-5}$$

4. DIVISION OF EXPONENTIALS. Divide the digit term of the numerator by the digit term of the denominator and subtract algebraically the exponents of the exponential terms.

EXAMPLE A.4 Divide 3.6×10^{-5} by 6×10^{-4}.

$$\frac{3.6 \times 10^{-5}}{6 \times 10^{-4}} = \left(\frac{3.6}{6}\right) \times 10^{(-5)-(-4)} = 0.6 \times 10^{-1} = 6 \times 10^{-2}$$

5. SQUARING OF EXPONENTIALS. Square the digit term in the usual way and multiply the exponent of the exponential term by 2.

EXAMPLE A.5 Square the number 4.0×10^{-6}.

$$(4.0 \times 10^{-6})^2 = 4 \times 4 \times 10^{2 \times (-6)} = 16 \times 10^{-12} = 1.6 \times 10^{-11}$$

6. CUBING OF EXPONENTIALS. Cube the digit term in the usual way and multiply the exponent of the exponential term by 3.

EXAMPLE A.6 Cube the number 2×10^{4}.

$$(2 \times 10^{4})^3 = 2 \times 2 \times 2 \times 10^{3 \times 4} = 8 \times 10^{12}$$

7. TAKING SQUARE ROOTS OF EXPONENTIALS. Decrease or increase the exponential term so that the power of 10 is evenly divisible by 2. Extract the square root of the digit term and divide the exponential term by 2.

EXAMPLE A.7 Find the square root of 1.6×10^{-7}.

$$1.6 \times 10^{-7} = 16 \times 10^{-8}$$
$$\sqrt{16 \times 10^{-8}} = \sqrt{16} \times \sqrt{10^{-8}} = 4.0 \times 10^{-4}$$

A.2 Significant Figures

A beekeeper reports that he has 525,341 bees. The last three figures of the number are obviously inaccurate, for during the time the keeper was counting the bees, some of them would have died and others would have hatched; this would have made the exact number of bees quite difficult to determine. It would have been more accurate if he had reported the number 525,000. In other words, the last three figures are not significant, except to set the position of the decimal point. Their exact values have no meaning. In reporting any information as numbers, only as many significant figures should be used as are warranted by the accuracy of the measurement.

The importance of significant figures lies in their application to fundamental computation. When adding or subtracting, the last digit that is retained in the sum or difference should correspond to the first doubtful decimal place (indicated by underscoring in the following example).

Add 4.383 g and 0.0023 g.

$$
\begin{array}{r}
4.38\underline{3}\text{ g} \\
0.002\underline{3} \\
\hline
4.38\underline{5}\text{ g}
\end{array}
$$

EXAMPLE A.8

When multiplying or dividing, the product or quotient should contain no more digits than the least number of significant figures in the numbers involved in the computation.

Multiply 0.6238 by 6.6.

$$0.623\underline{8} \times 6.\underline{6} = 4.\underline{1}$$

EXAMPLE A.9

In rounding off numbers, increase the last digit retained by 1 if it is followed by a number larger than 5 or by a 5 followed by other nonzero digits. Do not change the last digit retained if the following digits are less than 5. If the last digit retained is followed by 5, increase the last digit retained by 1 if it is odd or leave the last digit retained unchanged if it is even.

A.3 The Use of Logarithms and Exponential Numbers

The common logarithm of a number (log) is the power to which 10 must be raised to equal that number. For example, the logarithm of 100 is 2 because 10 must be raised to the second power to equal 100. Additional examples follow.

Number	Number expressed exponentially	Logarithm
10,000	10^4	4
1000	10^3	3
10	10^1	1
1	10^0	0
0.1	10^{-1}	-1
0.01	10^{-2}	-2
0.001	10^{-3}	-3
0.0001	10^{-4}	-4

What is the logarithm of 60? Because 60 lies between 10 and 100, which have logarithms of 1 and 2, respectively, the logarithm of 60 must lie between 1 and 2. The logarithm of 60 is 1.7782; or

$$60 = 10^{1.7782}$$

The logarithm of a number less than 1 has a negative value. The logarithm of 0.03918 is -1.4069; or

$$0.03918 = 10^{-1.4069} = \frac{1}{10^{1.4069}}$$

To obtain the logarithm of a number, use the log button on your calculator. To calculate a number from its logarithm, enter the logarithm into your calculator and either push the antilog button, take the "inverse" log of the logarithm, or calculate 10^x (where x is the logarithm of the number).

The natural logarithm of a number (1n) is the power to which e must be raised to equal the number; e is the constant 2.7182818. For example, the natural logarithm of 10 is 2.303; or

$$10 = e^{2.303} = 2.7182818^{2.303}$$

We have not used natural logarithms in this text.

Logarithms are exponents; thus operations involving logarithms follow the same rules as operations involving exponents.

1. The logarithm of a product of two numbers is the sum of the logarithms of the two numbers: $\log xy = \log x + \log y$.
2. The logarithm of the number resulting from the division of two numbers is the difference of the logarithms of the two numbers: $\log x/y = \log x - \log y$.
3. The logarithm of the square root of a number is one-half of the logarithm of the number: $\log x^{1/2} = 1/2 \log x$.
4. The logarithm of the cube root of a number is one-third of the logarithm of the number: $\log x^{1/3} = 1/3 \log x$.

A.4 The Solution of Quadratic Equations

Any quadratic equation can be expressed in the following form:

$$ax^2 + bx + c = 0$$

In order to solve a quadratic equation, the following formula is used.

$$x = \frac{-b \pm \sqrt{b^2 - 4ac}}{2a}$$

Solve the quadratic equation $3x^2 + 13x - 10 = 0$.

Substituting the values $a = 3$, $b = 13$, and $c = -10$ in the formula, we obtain

$$x = \frac{-13 \pm \sqrt{(13)^2 - 4 \times 3 \times (-10)}}{2 \times 3}$$

$$x = \frac{-13 \pm \sqrt{169 + 120}}{6} = \frac{-13 \pm \sqrt{289}}{6} = \frac{-13 \pm 17}{6}$$

The two roots are therefore

$$x = \frac{-13 + 17}{6} = 0.67 \qquad \text{and} \qquad x = \frac{-13 - 17}{6} = -5$$

Equations constructed on physical data always have real roots, and of these real roots only those having positive values are usually of any significance.

Appendix B: Units and Conversion Factors

Base units of International System of Units (SI)

Physical property	Name of unit	Symbol
Length	Meter	m
Mass	Kilogram	kg
Time	Second	s
Electric current	Ampere	A
Thermodynamic temperature	Kelvin	K
Luminous intensity	Candela	cd
Amount of substance	Mole	mol

Units of length

Meter (m) = 39.37 inches (in) = 1.094 yards (yd)
Centimeter (cm) = 0.01 m
Millimeter (mm) = 0.001 m
Kilometer (km) = 1000 m
Angstrom unit (Å) = 10^{-8} cm = 10^{-10} m

Yard = 0.9144 m (exact)
Inch = 2.54 cm (exact)
Mile (U.S.) = 1.60934 km

Units of volume

Liter (L) = 0.001 m^3 = 1000 cm^3
Milliliter (mL) = 0.001 L = 1 cm^3

Liquid quart (U.S.) = 0.9463 L
$\qquad\qquad\qquad$ = 32 (U.S.) liquid ounces
$\qquad\qquad\qquad$ = $\frac{1}{4}$ (U.S.) gallon
Dry quart = 1.1012 L
Cubic foot (U.S.) = 28.316 L

Units and Conversion Factors (continued)

Units of mass

Gram (g) = 0.001 kg	Ounce (oz) (avoirdupois) = 28.35 g
Milligram (mg) = 0.001 g	Pound (lb) (avoirdupois) = 0.45359237 kg
Kilogram (kg) = 1000 g	Ton (short) = 2000 lb = 907.185 kg
Ton (metric) = 1000 kg = 2204.62 lb	Ton (long) = 2240 lb = 1.016 metric ton

Units of energy

4.184 joule (J) = 1 thermochemical calorie (cal) = 4.184×10^7 erg
Erg = 10^{-7} J
Electron-volt (eV) = $1.60217733 \times 10^{-19}$ J = 23.061 kcal mol^{-1}
Liter atmosphere = 24.217 cal = 101.32 J

Unit of force

Newton (N) = 1 kg m s^{-2} (force that when applied for 1 second will give to a 1-kilogram mass a speed of 1 meter per second)

Units of pressure

Torr = 1 mmHg
Atmosphere (atm) = 760 mm Hg = 760 torr = 101,325 N m^{-2} = 101,325 Pa
Pascal (Pa) = kg m^{-1} s^{-2} = N m^{-2}

Appendix C: General Physical Constants
(1986 Values)

Avogadro's number	6.0221367×10^{23} mol^{-1}
Electron charge, e	$1.60217733 \times 10^{-19}$ coulomb (C)
Electron rest mass, m_e	9.109390×10^{-31} kg
Proton rest mass, m_p	$1.6726231 \times 10^{-27}$ kg
Neutron rest mass, m_n	$1.6749286 \times 10^{-27}$ kg
Charge-to-mass ratio for electron, e/m_e	$1.75881962 \times 10^{11}$ coulomb kg^{-1}
Faraday constant, F	9.6485309×10^4 coulomb/equivalent
Planck constant, h	$6.6260755 \times 10^{-34}$ J s
Boltzmann constant, k	1.380658×10^{-23} J K^{-1}
Gas constant, R	8.205784×10^{-2} L atm mol^{-1} K^{-1}
	= 8.314510 J mol^{-1} K^{-1}
Speed of light (in vacuum), c	2.99792458×10^8 m s^{-1}
Atomic mass unit (= $\frac{1}{12}$ the mass of an atom of the ^{12}C nuclide), amu	$1.6605402 \times 10^{-27}$ kg
Rydberg constant, R$_\infty$	1.0973731534×10^7 m^{-1}

Appendix D: Solubility Products

Substance	K_{sp} at 25°C	Substance	K_{sp} at 25°C
Aluminum		**Lead**	
$Al(OH)_3$	1.9×10^{-33}	$Pb(OH)_2$	2.8×10^{-16}
Barium		PbF_2	3.7×10^{-8}
$BaCO_3$	8.1×10^{-9}	$PbCl_2$	1.7×10^{-5}
$BaC_2O_4 \cdot 2H_2O$	1.1×10^{-7}	$PbBr_2$	6.3×10^{-6}
$BaSO_4$	1.08×10^{-10}	PbI_2	8.7×10^{-9}
$BaCrO_4$	2×10^{-10}	$PbCO_3$	1.5×10^{-13}
BaF_2	1.7×10^{-6}	PbS	6.5×10^{-34}
$Ba(OH)_2 \cdot 8H_2O$	5.0×10^{-3}	$PbCrO_4$	1.8×10^{-14}
$Ba_3(PO_4)_2$	1.3×10^{-29}	$PbSO_4$	1.8×10^{-8}
$Ba_3(AsO_4)_2$	1.1×10^{-13}	$Pb_3(PO_4)_2$	3×10^{-44}
Bismuth		**Magnesium**	
$BiO(OH)$	1×10^{-12}	$Mg(OH)_2$	1.5×10^{-11}
$BiOCl$	7×10^{-9}	$MgCO_3 \cdot 3H_2O$	$ca\ 1 \times 10^{-5}$
Bi_2S_3	7.3×10^{-91}	$MgNH_4PO_4$	2.5×10^{-13}
Cadmium		MgF_2	6.4×10^{-9}
$Cd(OH)_2$	1.2×10^{-14}	MgC_2O_4	8.6×10^{-5}
CdS	2.8×10^{-35}	**Manganese**	
$CdCO_3$	2.5×10^{-14}	$Mn(OH)_2$	4.5×10^{-14}
Calcium		$MnCO_3$	8.8×10^{-11}
$Ca(OH)_2$	7.9×10^{-6}	MnS	4.3×10^{-22}
$CaCO_3$	4.8×10^{-9}	**Mercury**	
$CaSO_4 \cdot 2H_2O$	2.4×10^{-5}	$Hg_2O \cdot H_2O$	1.6×10^{-23}
$CaC_2O_4 \cdot H_2O$	2.27×10^{-9}	Hg_2Cl_2	1.1×10^{-18}
$Ca_3(PO_4)_2$	1×10^{-25}	Hg_2Br_2	1.26×10^{-22}
$CaHPO_4$	5×10^{-6}	Hg_2I_2	4.5×10^{-29}
CaF_2	3.9×10^{-11}	Hg_2CO_3	9×10^{-17}
Chromium		Hg_2SO_4	6.2×10^{-7}
$Cr(OH)_3$	6.7×10^{-31}	Hg_2S	8×10^{-52}
Cobalt		Hg_2CrO_4	2×10^{-9}
$Co(OH)_2$	2×10^{-16}	HgS	2×10^{-59}
$CoS(\alpha)$	4.5×10^{-27}	**Nickel**	
$CoS(\beta)$	6.7×10^{-29}	$Ni(OH)_2$	1.6×10^{-14}
$CoCO_3$	1.0×10^{-12}	$NiCO_3$	1.36×10^{-7}
$Co(OH)_3$	2.5×10^{-43}	$NiS(\alpha)$	2×10^{-27}
Copper		$NiS(\beta)$	8×10^{-33}
$CuCl$	1.85×10^{-7}	**Potassium**	
$CuBr$	5.3×10^{-9}	$KClO_4$	1.07×10^{-2}
CuI	5.1×10^{-12}	K_2PtCl_6	1.1×10^{-5}
$CuSCN$	4×10^{-14}	$KHC_4H_4O_6$	3×10^{-4}
Cu_2S	1.2×10^{-54}	**Silver**	
$Cu(OH)_2$	5.6×10^{-20}	$\frac{1}{2}Ag_2O\ (Ag^+ + OH^-)$	2×10^{-8}
CuS	6.7×10^{-42}	$AgCl$	1.8×10^{-10}
$CuCO_3$	1.37×10^{-10}	$AgBr$	3.3×10^{-13}
Iron		AgI	1.5×10^{-16}
$Fe(OH)_2$	7.9×10^{-15}	$AgCN$	1.2×10^{-16}
$FeCO_3$	2.11×10^{-11}	$AgSCN$	1.0×10^{-12}
FeS	8×10^{-26}	Ag_2S	8×10^{-58}
$Fe(OH)_3$	1.1×10^{-36}	Ag_2CO_3	8.2×10^{-12}

Solubility Products (continued)

Substance	K_{sp} at 25°C	Substance	K_{sp} at 25°C
Ag_2CrO_4	9×10^{-12}	TlSCN	5.8×10^{-4}
$Ag_4Fe(CN)_6$	1.55×10^{-41}	Tl_2S	9.2×10^{-31}
Ag_2SO_4	1.18×10^{-5}	$Tl(OH)_3$	1.5×10^{-44}
Ag_3PO_4	1.8×10^{-18}	Tin	
Strontium		$Sn(OH)_2$	5×10^{-26}
$Sr(OH)_2 \cdot 8H_2O$	3.2×10^{-4}	SnS	6×10^{-35}
$SrCO_3$	9.42×10^{-10}	$Sn(OH)_4$	1×10^{-56}
$SrCrO_4$	3.6×10^{-5}	Zinc	
$SrSO_4$	2.8×10^{-7}	$ZnCO_3$	6×10^{-11}
$SrC_2O_4 \cdot H_2O$	5.61×10^{-8}	$Zn(OH)_2$	4.5×10^{-17}
Thallium		ZnS	1×10^{-27}
TlCl	1.9×10^{-4}		

Appendix E:
Formation Constants for Complex Ions

Equilibrium	K_f
$Al^{3+} + 6F^- \rightleftharpoons [AlF_6]^{3-}$	5×10^{23}
$Cd^{2+} + 4NH_3 \rightleftharpoons [Cd(NH_3)_4]^{2+}$	4.0×10^6
$Cd^{2+} + 4CN^- \rightleftharpoons [Cd(CN)_4]^{2-}$	1.3×10^{17}
$Co^{2+} + 6NH_3 \rightleftharpoons [Co(NH_3)_6]^{2+}$	8.3×10^4
$Co^{3+} + 6NH_3 \rightleftharpoons [Co(NH_3)_6]^{3+}$	4.5×10^{33}
$Cu^+ + 2CN^- \rightleftharpoons [Cu(CN)_2]^-$	1×10^{16}
$Cu^{2+} + 4NH_3 \rightleftharpoons [Cu(NH_3)_4]^{2+}$	1.2×10^{12}
$Fe^{2+} + 6CN^- \rightleftharpoons [Fe(CN)_6]^{4-}$	1×10^{37}
$Fe^{3+} + 6CN^- \rightleftharpoons [Fe(CN)_6]^{3-}$	1×10^{44}
$Fe^{3+} + 6SCN^- \rightleftharpoons [Fe(NCS)_6]^{3-}$	3.2×10^3
$Hg^{2+} + 4Cl^- \rightleftharpoons [HgCl_4]^{2-}$	1.2×10^{15}
$Ni^{2+} + 6NH_3 \rightleftharpoons [Ni(NH_3)_6]^{2+}$	1.8×10^8
$Ag^+ + 2Cl^- \rightleftharpoons [AgCl_2]^-$	2.5×10^5
$Ag^+ + 2CN^- \rightleftharpoons [Ag(CN)_2]^-$	1×10^{20}
$Ag^+ + 2NH_3 \rightleftharpoons [Ag(NH_3)_2]^+$	1.6×10^7
$Zn^{2+} + 4CN^- \rightleftharpoons [Zn(CN)_4]^{2-}$	1×10^{19}
$Zn^{2+} + 4OH^- \rightleftharpoons [Zn(OH)_4]^{2-}$	2.9×10^{15}

Appendix F: Ionization Constants of Weak Acids

Acid	Formula	K_a at 25°C
Acetic	CH_3CO_2H	1.8×10^{-5}
Arsenic	H_3AsO_4	4.8×10^{-3}
	$H_2AsO_4^-$	1×10^{-7}
	$HAsO_4^{2-}$	1×10^{-13}
Arsenous	H_3AsO_3	5.8×10^{-10}
Boric	H_3BO_3	5.8×10^{-10}
Carbonic	H_2CO_3	4.3×10^{-7}
	HCO_3^-	7×10^{-11}
Cyanic	$HCNO$	3.46×10^{-4}
Formic	HCO_2H	1.8×10^{-4}
Hydrazoic	HN_3	1×10^{-4}
Hydrocyanic	HCN	4×10^{-10}
Hydrofluoric	HF	7.2×10^{-4}
Hydrogen peroxide	H_2O_2	2.4×10^{-12}
Hydrogen selenide	H_2Se	1.7×10^{-4}
	HSe^-	1×10^{-10}
Hydrogen sulfate ion	HSO_4^-	1.2×10^{-2}
Hydrogen sulfide	H_2S	1.0×10^{-7}
	HS^-	1.0×10^{-19}
Hydrogen telluride	H_2Te	2.3×10^{-3}
	HTe^-	1×10^{-5}
Hypobromous	$HBrO$	2×10^{-9}
Hypochlorous	$HClO$	3.5×10^{-8}
Nitrous	HNO_2	4.5×10^{-4}
Oxalic	$H_2C_2O_4$	5.9×10^{-2}
	$HC_2O_4^-$	6.4×10^{-5}
Phosphoric	H_3PO_4	7.5×10^{-3}
	$H_2PO_4^-$	6.3×10^{-8}
	HPO_4^{2-}	3.6×10^{-13}
Phosphorous	H_3PO_3	1.6×10^{-2}
	$H_2PO_3^-$	7×10^{-7}
Sulfurous	H_2SO_3	1.2×10^{-2}
	HSO_3^-	6.2×10^{-8}

Appendix G: Ionization Constants of Weak Bases

Base	Ionization equation	K_b at 25°C
Ammonia	$NH_3 + H_2O \rightleftharpoons NH_4^+ + OH^-$	1.8×10^{-5}
Dimethylamine	$(CH_3)_2NH + H_2O \rightleftharpoons (CH_3)_2NH_2^+ + OH^-$	7.4×10^{-4}
Methylamine	$CH_3NH_2 + H_2O \rightleftharpoons CH_3NH_3^+ + OH^-$	4.4×10^{-4}
Phenylamine (aniline)	$C_6H_5NH_2 + H_2O \rightleftharpoons C_6H_5NH_3^+ + OH^-$	4.6×10^{-10}
Trimethylamine	$(CH_3)_3N + H_2O \rightleftharpoons (CH_3)_3NH^+ + OH^-$	7.4×10^{-5}

Appendix H: Standard Electrode (Reduction) Potentials

Half-reaction	$E°$, V	Half-reaction	$E°$, V
$Li^+ + e^- \longrightarrow Li$	-3.09	$Zn^{2+} + 2e^- \longrightarrow Zn$	-0.763
$K^+ + e^- \longrightarrow K$	-2.925	$Cr^{3+} + 3e^- \longrightarrow Cr$	-0.74
$Rb^+ + e^- \longrightarrow Rb$	-2.925	$HgS + 2e^- \longrightarrow Hg + S^{2-}$	-0.72
$Ra^{2+} + 2e^- \longrightarrow Ra$	-2.92	$[Cd(NH_3)_4]^{2+} + 2e^- \longrightarrow Cd + 4NH_3$	-0.597
$Ba^{2+} + 2e^- \longrightarrow Ba$	-2.90	$Ga^{3+} + 3e^- \longrightarrow Ga$	-0.53
$Sr^{2+} + 2e^- \longrightarrow Sr$	-2.89	$S + 2e^- \longrightarrow S^{2-}$	-0.48
$Ca^{2+} + 2e^- \longrightarrow Ca$	-2.87	$[Ni(NH_3)_6]^{2+} + 2e^- \longrightarrow Ni + 6NH_3$	-0.47
$Na^+ + e^- \longrightarrow Na$	-2.714	$Fe^{2+} + 2e^- \longrightarrow Fe$	-0.440
$La^{3+} + 3e^- \longrightarrow La$	-2.52	$[Cu(CN)_2]^- + e^- \longrightarrow Cu + 2CN^-$	-0.43
$Ce^{3+} + 3e^- \longrightarrow Ce$	-2.48	$Cr^{3+} + e^- \longrightarrow Cr^{2+}$	-0.41
$Nd^{3+} + 3e^- \longrightarrow Nd$	-2.44	$Cd^{2+} + 2e^- \longrightarrow Cd$	-0.403
$Sm^{3+} + 3e^- \longrightarrow Sm$	-2.41	$Se + 2H^+ + 2e^- \longrightarrow H_2Se$	-0.40
$Gd^{3+} + 3e^- \longrightarrow Gd$	-2.40	$[Hg(CN)_4]^{2-} + 2e^- \longrightarrow Hg + 4CN^-$	-0.37
$Mg^{2+} + 2e^- \longrightarrow Mg$	-2.37	$ClO_4^- + H_2O + 2e^- \longrightarrow ClO_3^- + 2OH^-$	-0.36
$Y^{3+} + 3e^- \longrightarrow Y$	-2.37	$PbSO_4 + 2e^- \longrightarrow Pb + SO_4^{2-}$	-0.356
$Am^{3+} + 3e^- \longrightarrow Am$	-2.32	$In^{3+} + 3e^- \longrightarrow In$	-0.342
$Lu^{3+} + 3e^- \longrightarrow Lu$	-2.25	$[Ag(CN)_2]^- + e^- \longrightarrow Ag + 2CN^-$	-0.31
$\frac{1}{2}H_2 + e^- \longrightarrow H^-$	-2.25	$Co^{2+} + 2e^- \longrightarrow Co$	-0.277
$Sc^{3+} + 3e^- \longrightarrow Sc$	-2.08	$[SnF_6]^{2-} + 4e^- \longrightarrow Sn + 6F^-$	-0.25
$[AlF_6]^{3-} + 3e^- \longrightarrow Al + 6F^-$	-2.07	$Ni^{2+} + 2e^- \longrightarrow Ni$	-0.250
$Pu^{3+} + 3e^- \longrightarrow Pu$	-2.07	$Sn^{2+} + 2e^- \longrightarrow Sn$	-0.136
$Th^{4+} + 4e^- \longrightarrow Th$	-1.90	$CrO_4^{2-} + 4H_2O + 3e^- \longrightarrow Cr(OH)_3 + 5OH^-$	-0.13
$Np^{3+} + 3e^- \longrightarrow Np$	-1.86	$Pb^{2+} + 2e^- \longrightarrow Pb$	-0.126
$Be^{2+} + 2e^- \longrightarrow Be$	-1.85	$MnO_2 + 2H_2O + 2e^- \longrightarrow Mn(OH)_2 + 2OH^-$	-0.05
$U^{3+} + 3e^- \longrightarrow U$	-1.80	$[HgI_4]^{2-} + 2e^- \longrightarrow Hg + 4I^-$	-0.04
$Hf^{4+} + 4e^- \longrightarrow Hf$	-1.70	$2H^+ + 2e^- \longrightarrow H_2$	0.00
$SiO_3^{2-} + 3H_2O + 4e^- \longrightarrow Si + 6OH^-$	-1.70	$NO_3^- + H_2O + 2e^- \longrightarrow NO_2^- + 2OH^-$	$+0.01$
$Al^{3+} + 3e^- \longrightarrow Al$	-1.66	$[Ag(S_2O_3)_2]^{3-} + e^- \longrightarrow Ag^+ + 2S_2O_3^{2-}$	$+0.01$
$Ti^{2+} + 2e^- \longrightarrow Ti$	-1.63	$[Co(NH_3)_6]^{3+} + e^- \longrightarrow [Co(NH_3)_6]^{2+}$	$+0.1$
$Zr^{4+} + 4e^- \longrightarrow Zr$	-1.53	$S + 2H^+ + 2e^- \longrightarrow H_2S$	$+0.141$
$ZnS + 2e^- \longrightarrow Zn + S^{2-}$	-1.44	$Sn^{4+} + 2e^- \longrightarrow Sn^{2+}$	$+0.15$
$Cr(OH)_3 + 3e^- \longrightarrow Cr + 30H^-$	-1.3	$Cu^{2+} + e^- \longrightarrow Cu^+$	$+0.153$
$[Zn(CN)_4]^{2-} + 2e^- \longrightarrow Zn + 4CN^-$	-1.26	$Co(OH)_3 + e^- \longrightarrow Co(OH)_2 + OH^-$	$+0.17$
$Zn(OH)_2 + 2e^- \longrightarrow Zn + 2OH^-$	-1.245	$[HgBr_4]^{2-} + 2e^- \longrightarrow Hg + 4Br^-$	$+0.21$
$[Zn(OH)_4]^{2-} + 2e^- \longrightarrow Zn + 4OH^-$	-1.216	$AgCl + e^- \longrightarrow Ag + Cl^-$	$+0.222$
$CdS + 2e^- \longrightarrow Cd + S^{2-}$	-1.21	$Hg_2Cl_2 + 2e^- \longrightarrow 2Hg + 2Cl^-$	$+0.27$
$[Cr(OH)_4]^- + 3e^- \longrightarrow Cr + 4OH^-$	-1.2	$ClO_3^- + H_2O + 2e^- \longrightarrow ClO_2^- + 2OH^-$	$+0.33$
$[SiF_6]^{2-} + 4e^- \longrightarrow Si + 6F^-$	-1.2	$Cu^{2+} + 2e^- \longrightarrow Cu$	$+0.337$
$V^{2+} + 2e^- \longrightarrow V$	ca -1.18	$[Fe(CN)_6]^{3-} + e^- \longrightarrow [Fe(CN)_6]^{4-}$	$+0.36$
$Mn^{2+} + 2e^- \longrightarrow Mn$	-1.18	$[Ag(NH_3)_2]^+ + e^- \longrightarrow Ag + 2NH_3$	$+0.373$
$[Cd(CN)_4]^{2-} + 2e^- \longrightarrow Cd + 4CN^-$	-1.03	$O_2 + 2H_2O + 4e^- \longrightarrow 4OH^-$	$+0.401$
$[Zn(NH_3)_4]^{2+} + 2e^- \longrightarrow Zn + 4NH_3$	-1.03	$[RhCl_6]^{3-} + 3e^- \longrightarrow Rh + 6Cl^-$	$+0.44$
$FeS + 2e^- \longrightarrow Fe + S^{2-}$	-1.01	$Ag_2CrO_4 + 2e^- \longrightarrow 2Ag + CrO_4^{2-}$	$+0.446$
$PbS + 2e^- \longrightarrow Pb + S^{2-}$	-0.95	$NiO_2 + 2H_2O + 2e^- \longrightarrow Ni(OH)_2 + 2OH^-$	$+0.49$
$SnS + 2e^- \longrightarrow Sn + S^{2-}$	-0.94	$Cu^+ + e^- \longrightarrow Cu$	$+0.521$
$Cr^{2+} + 2e^- \longrightarrow Cr$	-0.91	$TeO_2 + 4H^+ + 4e^- \longrightarrow Te + 2H_2O$	$+0.529$
$Fe(OH)_2 + 2e^- \longrightarrow Fe + 2OH^-$	-0.877	$I_2 + 2e^- \longrightarrow 2I^-$	$+0.5355$
$SiO_2 + 4H^+ + 4e^- \longrightarrow Si + 2H_2O$	-0.86	$[PtBr_4]^{2-} + 2e^- \longrightarrow Pt + 4Br^-$	$+0.58$
$NiS + 2e^- \longrightarrow Ni + S^{2-}$	-0.83	$MnO_4^- + 2H_2O + 3e^- \longrightarrow MnO_2 + 4OH^-$	$+0.588$
$2H_2O + 2e^- \longrightarrow H_2 + 2OH^-$	-0.828	$[PdCl_4]^{2-} + 2e^- \longrightarrow Pd + 4Cl^-$	$+0.62$

Standard Electrode (Reduction) Potentials (continued)

Half-reaction	$E°$, V	Half-reaction	$E°$, V
$ClO_2^- + H_2O + 2e^- \longrightarrow ClO^- + 2OH^-$	+0.66	$O_2 + 4H^+ + 4e^- \longrightarrow 2H_2O$	+1.23
$[PtCl_6]^{2-} + 2e^- \longrightarrow [PtCl_4]^{2-} + 2Cl^-$	+0.68	$MnO_2 + 4H^+ + 2e^- \longrightarrow Mn^{2+} + 2H_2O$	+1.23
$O_2 + 2H^+ + 2e^- \longrightarrow H_2O_2$	+0.682	$Cr_2O_7^{2-} + 14H^+ + 6e^- \longrightarrow 2Cr^{3+} + 7H_2O$	+1.33
$[PtCl_4]^{2-} + 2e^- \longrightarrow Pt + 4Cl^-$	+0.73	$Cl_2 + 2e^- \longrightarrow 2Cl^-$	+1.3595
$Fe^{3+} + e^- \longrightarrow Fe^{2+}$	+0.771	$HClO + H^+ + 2e^- \longrightarrow Cl^- + H_2O$	+1.49
$Hg_2^{2+} + 2e^- \longrightarrow 2Hg$	+0.789	$Au^{3+} + 3e^- \longrightarrow Au$	+1.50
$Ag^+ + e^- \longrightarrow Ag$	+0.7991	$MnO_4^- + 8H^+ + 5e^- \longrightarrow Mn^{2+} + 4H_2O$	+1.51
$Hg^{2+} + 2e^- \longrightarrow Hg$	+0.854	$Ce^{4+} + e^- \longrightarrow Ce^{3+}$	+1.61
$HO_2^- + H_2O + 2e^- \longrightarrow 3OH^-$	+0.88	$HClO + H^+ + e^- \longrightarrow \frac{1}{2}Cl_2 + H_2O$	+1.63
$ClO^- + H_2O + 2e^- \longrightarrow Cl^- + 2OH^-$	+0.89	$HClO_2 + 2H^+ + 2e^- \longrightarrow HClO + H_2O$	+1.64
$2Hg^{2+} + 2e^- \longrightarrow Hg_2^{2+}$	+0.920	$Au^+ + e^- \longrightarrow Au$	ca +1.68
$NO_3^- + 3H^+ + 2e^- \longrightarrow HNO_2 + H_2O$	+0.94	$NiO_2 + 4H^+ + 2e^- \longrightarrow Ni^{2+} + 2H_2O$	+1.68
$NO_3^- + 4H^+ + 3e^- \longrightarrow NO + H_2O$	+0.96	$PbO_2 + SO_4^{2-} + 4H^+ + 2e^- \longrightarrow$	
$Pd^{2+} + 2e^- \longrightarrow Pd$	+0.987	$\phantom{PbO_2 + SO_4^{2-} + 4H^+ + 2e^- \longrightarrow} PbSO_4 + 2H_2O$	+1.685
$Br_2(l) + 2e^- \longrightarrow 2Br^-$	+1.0652	$H_2O_2 + 2H^+ + 2e^- \longrightarrow 2H_2O$	+1.77
$ClO_4^- + 2H^+ + 2e^- \longrightarrow ClO_3^- + H_2O$	+1.19	$Co^{3+} + e^- \longrightarrow Co^{2+}$	+1.82
$Pt^{2+} + 2e^- \longrightarrow Pt$	ca +1.2	$F_2 + 2e^- \longrightarrow 2F^-$	+2.87
$ClO_3^- + 3H^+ + 2e^- \longrightarrow HClO_2 + H_2O$	+1.21		

Appendix I: Standard Molar Enthalpies of Formation, Standard Molar Free Energies of Formation, and Absolute Standard Entropies [298.15 K (25°C), 1 atm]

Substance	$\Delta H_f°$, kJ mol^{-1}	$\Delta G_f°$, kJ mol^{-1}	$S_{298}°$, J K^{-1} mol^{-1}
Aluminum			
$Al(s)$	0	0	28.3
$Al(g)$	326	286	164.4
$Al_2O_3(s)$	−1676	−1582	50.92
$AlF_3(s)$	−1504	−1425	66.44
$AlCl_3(s)$	−704.2	−628.9	110.7
$AlCl_3 \cdot 6H_2O(s)$	−2692	—	—
$Al_2S_3(s)$	−724	−492.4	—
$Al_2(SO_4)_3(s)$	−3440.8	−3100.1	239
Antimony			
$Sb(s)$	0	0	45.69
$Sb(g)$	262	222	180.2
$Sb_4O_6(s)$	−1441	−1268	221
$SbCl_3(g)$	−314	−301	337.7
$SbCl_5(g)$	−394.3	−334.3	401.8
$Sb_2S_3(s)$	−175	−174	182
$SbCl_3(s)$	−382.2	−323.7	184
$SbOCl(s)$	−374	—	—

Standard Molar Enthalpies of Formation, Standard Molar Free Energies of
Formation, and Absolute Standard Entropies [298.15 K (25°C), 1 atm]
(continued)

Substance	ΔH_f°, kJ mol^{-1}	ΔG_f°, kJ mol^{-1}	S_{298}°, J K^{-1} mol^{-1}
Arsenic			
As(s)	0	0	35
As(g)	303	261	174.1
As$_4$(g)	144	92.5	314
As$_4$O$_6$(s)	−1313.9	−1152.5	214
As$_2$O$_5$(s)	−924.87	−782.4	105
AsCl$_3$(g)	−258.6	−245.9	327.1
As$_2$S$_3$(s)	−169	−169	164
AsH$_3$(g)	66.44	68.91	222.7
H$_3$AsO$_4$(s)	−906.3	—	—
Barium			
Ba(s)	0	0	66.9
Ba(g)	175.6	144.8	170.3
BaO(s)	−558.1	−528.4	70.3
BaCl$_2$(s)	−860.06	−810.9	126
BaSO$_4$(s)	−1465	−1353	132
Beryllium			
Be(s)	0	0	9.54
Be(g)	320.6	282.8	136.17
BeO(s)	−610.9	−581.6	14.1
Bismuth			
Bi(s)	0	0	56.74
Bi(g)	207	168	186.90
Bi$_2$O$_3$(s)	−573.88	−493.7	151
BiCl$_3$(s)	−379	−315	177
Bi$_2$S$_3$(s)	−143	−141	200
Boron			
B(s)	0	0	5.86
B(g)	562.7	518.8	153.3
B$_2$O$_3$(s)	−1272.8	−1193.7	53.97
B$_2$H$_6$(g)	36	86.6	232.0
B(OH)$_3$(s)	−1094.3	−969.01	88.83
BF$_3$(g)	−1137.3	−1120.3	254.0
BCl$_3$(g)	−403.8	−388.7	290.0
B$_3$N$_3$H$_6$(l)	−541.0	−392.8	200
HBO$_2$(s)	−794.25	−723.4	40
Bromine			
Br$_2$(l)	0	0	152.23
Br$_2$(g)	30.91	3.142	245.35
Br(g)	111.88	82.429	174.91
BrF$_3$(g)	−255.6	−229.5	292.4
HBr(g)	−36.4	−53.43	198.59
Cadmium			
Cd(s)	0	0	51.76
Cd(g)	112.0	77.45	167.64
CdO(s)	−258	−228	54.8
CdCl$_2$(s)	−391.5	−344.0	115.3

Standard Molar Enthalpies of Formation, Standard Molar Free Energies of
Formation, and Absolute Standard Entropies [298.15 K (25°C), 1 atm]
(continued)

Substance	ΔH_f°, kJ mol^{-1}	ΔG_f°, kJ mol^{-1}	S_{298}°, J K^{-1} mol^{-1}
CdSO$_4$(s)	−933.28	−822.78	123.04
CdS(s)	−162	−156	64.9
Calcium			
Ca(s)	0	0	41.6
Ca(g)	192.6	158.9	154.78
CaO(s)	−635.5	−604.2	40
Ca(OH)$_2$(s)	−986.59	−896.76	76.1
CaSO$_4$(s)	−1432.7	−1320.3	107
CaSO$_4 \cdot$ 2H$_2$O(s)	−2021.1	−1795.7	194.0
CaCO$_3$(s) (calcite)	−1206.9	−1128.8	92.9
CaSO$_3 \cdot$ 2H$_2$O(s)	−1762	−1565	184
Carbon			
C(s) (graphite)	0	0	5.740
C(s) (diamond)	1.897	2.900	2.38
C(g)	716.681	671.289	157.987
CO(g)	−110.52	−137.15	197.56
CO$_2$(g)	−393.51	−394.36	213.6
CH$_4$(g)	−74.81	−50.75	186.15
CH$_3$OH(l)	−238.7	−166.4	127
CH$_3$OH(g)	−200.7	−162.0	239.7
CCl$_4$(l)	−135.4	−65.27	216.4
CCl$_4$(g)	−102.9	−60.63	309.7
CHCl$_3$(l)	−134.5	−73.72	202
CHCl$_3$(g)	−103.1	−70.37	295.6
CS$_2$(l)	89.70	65.27	151.3
CS$_2$(g)	117.4	67.15	237.7
C$_2$H$_2$(g)	226.7	209.2	200.8
C$_2$H$_4$(g)	52.26	68.12	219.5
C$_2$H$_6$(g)	−84.68	−32.9	229.5
CH$_3$COOH(l)	−484.5	−390	160
CH$_3$COOH(g)	−432.25	−374	282
C$_2$H$_5$OH(l)	−277.7	−174.9	161
C$_2$H$_5$OH(g)	−235.1	−168.6	282.6
C$_3$H$_8$(g)	−103.85	−23.49	269.9
C$_6$H$_6$(g)	82.927	129.66	269.2
C$_6$H$_6$(l)	49.028	124.50	172.8
CH$_2$Cl$_2$(l)	−121.5	−67.32	178
CH$_2$Cl$_2$(g)	−92.47	−65.90	270.1
CH$_3$Cl(g)	−80.83	−57.40	234.5
C$_2$H$_5$Cl(l)	−136.5	−59.41	190.8
C$_2$H$_5$Cl(g)	−112.2	−60.46	275.9
C$_2$N$_2$(g)	308.9	297.4	241.8
HCN(l)	108.9	124.9	112.8
HCN(g)	135	124.7	201.7
Chlorine			
Cl$_2$(g)	0	0	222.96
Cl(g)	121.68	105.70	165.09

Standard Molar Enthalpies of Formation, Standard Molar Free Energies of
Formation, and Absolute Standard Entropies [298.15 K (25°C), 1 atm]
(continued)

Substance	ΔH_f°, kJ mol^{-1}	ΔG_f°, kJ mol^{-1}	S_{298}°, J K^{-1} mol^{-1}
ClF(g)	−54.48	−55.94	217.8
ClF$_3$(g)	−163	−123	281.5
Cl$_2$O(g)	80.3	97.9	266.1
Cl$_2$O$_7$(l)	238	—	—
Cl$_2$O$_7$(g)	272	—	—
HCl(g)	−92.307	−95.299	186.80
HClO$_4$(l)	−40.6	—	—
Chromium			
Cr(s)	0	0	23.8
Cr(g)	397	352	174.4
Cr$_2$O$_3$(s)	−1140	−1058	81.2
CrO$_3$(s)	−589.5	—	—
(NH$_4$)$_2$Cr$_2$O$_7$(s)	−1807	—	—
Cobalt			
Co(s)	0	0	30.0
CoO(s)	−237.9	−214.2	52.97
Co$_3$O$_4$(s)	−891.2	−774.0	103
Co(NO$_3$)$_2$(s)	−420.5	—	—
Copper			
Cu(s)	0	0	33.15
Cu(g)	338.3	298.5	166.3
CuO(s)	−157	−130	42.63
Cu$_2$O(s)	−169	−146	93.14
CuS(s)	−53.1	−53.6	66.5
Cu$_2$S(s)	−79.5	−86.2	121
CuSO$_4$(s)	−771.36	−661.9	109
Cu(NO$_3$)$_2$(s)	−303	—	—
Fluorine			
F$_2$(g)	0	0	202.7
F(g)	78.99	61.92	158.64
F$_2$O(g)	−22	−4.6	247.3
HF(g)	−271	−273	173.67
Hydrogen			
H$_2$(g)	0	0	130.57
H(g)	217.97	203.26	114.60
H$_2$O(l)	−285.83	−237.18	69.91
H$_2$O(g)	−241.82	−228.59	188.71
H$_2$O$_2$(l)	−187.8	−120.4	110
H$_2$O$_2$(g)	−136.3	−105.6	233
HF(g)	−271	−273	173.67
HCl(g)	−92.307	−95.299	186.80
HBr(g)	−36.4	−53.43	198.59
HI(g)	26.5	1.7	206.48
H$_2$S(g)	−20.6	−33.6	205.7
H$_2$Se(g)	30	16	218.9
Iodine			
I$_2$(s)	0	0	116.14

Standard Molar Enthalpies of Formation, Standard Molar Free Energies of Formation, and Absolute Standard Entropies [298.15 K (25°C), 1 atm] (continued)

Substance	ΔH_f°, kJ mol^{-1}	ΔG_f°, kJ mol^{-1}	S_{298}°, J K^{-1} mol^{-1}
$I_2(g)$	62.438	19.36	260.6
$I(g)$	106.84	70.283	180.68
$IF(g)$	95.65	−118.5	236.1
$ICl(g)$	17.8	−5.44	247.44
$IBr(g)$	40.8	3.7	258.66
$IF_7(g)$	−943.9	−818.4	346
$HI(g)$	26.5	1.7	206.48
Iron			
$Fe(s)$	0	0	27.3
$Fe(g)$	416	371	180.38
$Fe_2O_3(s)$	−824.2	−742.2	87.40
$Fe_3O_4(s)$	−1118	−1015	146
$Fe(CO)_5(l)$	−774.0	−705.4	338
$Fe(CO)_5(g)$	−733.9	−697.26	445.2
$FeSeO_3(s)$	−1200	—	—
$FeO(s)$	−272	—	—
$FeAsS(s)$	−42	−50	120
$Fe(OH)_2(s)$	−569.0	−486.6	88
$Fe(OH)_3(s)$	−823.0	−696.6	107
$FeS(s)$	−100	−100	60.29
$Fe_3C(s)$	25	20	105
Lead			
$Pb(s)$	0	0	64.81
$Pb(g)$	195	162	175.26
$PbO(s)$ (yellow)	−217.3	−187.9	68.70
$PbO(s)$ (red)	−219.0	−188.9	66.5
$Pb(OH)_2(s)$	−515.9	—	—
$PbS(s)$	−100	−98.7	91.2
$Pb(NO_3)_2(s)$	−451.9	—	—
$PbO_2(s)$	−277	−217.4	68.6
$PbCl_2(s)$	−359.4	−314.1	136
Lithium			
$Li(s)$	0	0	28.0
$Li(g)$	155.1	122.1	138.67
$LiH(s)$	−90.42	−69.96	25
$Li(OH)(s)$	−487.23	−443.9	50.2
$LiF(s)$	−612.1	−584.1	35.9
$Li_2CO_3(s)$	−1215.6	−1132.4	90.4
Manganese			
$Mn(s)$	0	0	32.0
$Mn(g)$	281	238	173.6
$MnO(s)$	−385.2	−362.9	59.71
$MnO_2(s)$	−520.03	−465.18	53.05
$Mn_2O_3(s)$	−959.0	−881.2	110
$Mn_3O_4(s)$	−1388	−1283	156
Mercury			
$Hg(l)$	0	0	76.02

Standard Molar Enthalpies of Formation, Standard Molar Free Energies of Formation, and Absolute Standard Entropies [298.15 K (25°C), 1 atm] (continued)

Substance	ΔH_f°, kJ mol^{-1}	ΔG_f°, kJ mol^{-1}	S_{298}°, J K^{-1} mol^{-1}
Hg(g)	61.317	31.85	174.8
HgO(s) (red)	−90.83	−58.555	70.29
HgO(g) (yellow)	−90.46	−57.296	71.1
HgCl$_2$(s)	−224	−179	146
Hg$_2$Cl$_2$(s)	−265.2	−210.78	192
HgS(s) (red)	−58.16	−50.6	82.4
HgS(s) (black)	−53.6	−47.7	88.3
HgSO$_4$(s)	−707.5	—	—
Nitrogen			
N$_2$(g)	0	0	191.5
N(g)	472.704	455.579	153.19
NO(g)	90.25	86.57	210.65
NO$_2$(g)	33.2	51.30	239.9
N$_2$O(g)	82.05	104.2	219.7
N$_2$O$_3$(g)	83.72	139.4	312.2
N$_2$O$_4$(g)	9.16	97.82	304.2
N$_2$O$_5$(g)	11	115	356
NH$_3$(g)	−46.11	−16.5	192.3
N$_2$H$_4$(l)	50.63	149.2	121.2
N$_2$H$_4$(g)	95.4	159.3	238.4
NH$_4$NO$_3$(s)	−365.6	−184.0	151.1
NH$_4$Cl(s)	−314.4	−201.5	94.6
NH$_4$Br(s)	−270.8	−175	113
NH$_4$I(s)	−201.4	−113	117
NH$_4$NO$_2$(s)	−256	—	—
HNO$_3$(l)	−174.1	−80.79	155.6
HNO$_3$(g)	−135.1	−74.77	266.2
Oxygen			
O$_2$(g)	0	0	205.03
O(g)	249.17	231.75	160.95
O$_3$(g)	143	163	238.8
Phosphorus			
P(s)	0	0	41.1
P(g)	58.91	24.5	280.0
P$_4$(g)	314.6	278.3	163.08
PH$_3$(g)	5.4	13	210.1
PCl$_3$(g)	−287	−268	311.7
PCl$_5$(g)	−375	−305	364.5
P$_4$O$_6$(s)	−1640	—	—
P$_4$O$_{10}$(s)	−2984	−2698	228.9
HPO$_3$(s)	−948.5	—	—
H$_3$PO$_2$(s)	−604.6	—	—
H$_3$PO$_3$(s)	−964.4	—	—
H$_3$PO$_4$(s)	−1279	−1119	110.5
H$_3$PO$_4$(l)	−1267	—	—
H$_4$P$_2$O$_7$(s)	−2241	—	—
POCl$_3$(l)	−597.1	−520.9	222.5
POCl$_3$(g)	−558.48	−512.96	325.3

Standard Molar Enthalpies of Formation, Standard Molar Free Energies of Formation, and Absolute Standard Entropies [298.15 K (25°C), 1 atm] (continued)

Substance	ΔH_f°, kJ mol^{-1}	ΔG_f°, kJ mol^{-1}	S_{298}°, J K^{-1} mol^{-1}
Potassium			
K(s)	0	0	63.6
K(g)	90.00	61.17	160.23
KF(s)	−562.58	−533.12	66.57
KCl(s)	−435.868	−408.32	82.68
Silicon			
Si(s)	0	0	18.8
Si(g)	455.6	411	167.9
SiO$_2$(s)	−910.94	−856.67	41.84
SiH$_4$(g)	34	56.9	204.5
H$_2$SiO$_3$(s)	−1189	−1092	130
H$_4$SiO$_4$(s)	−1481	−1333	190
SiF$_4$(g)	−1614.9	−1572.7	282.4
SiCl$_4$(l)	−687.0	−619.90	240
SiCl$_4$(g)	−657.01	−617.01	330.6
SiC(s)	−65.3	−62.8	16.6
Silver			
Ag(s)	0	0	42.55
Ag(g)	284.6	245.7	172.89
Ag$_2$O(s)	−31.0	−11.2	121
AgCl(s)	−127.1	−109.8	96.2
Ag$_2$S(s)	−32.6	−40.7	144.0
Sodium			
Na(s)	0	0	51.0
Na(g)	108.7	78.11	153.62
Na$_2$O(s)	−415.9	−377	72.8
NaCl(s)	−411.00	−384.03	72.38
Sulfur			
S(s) (rhombic)	0	0	31.8
S(g)	278.80	238.27	167.75
SO$_2$(g)	−296.83	−300.19	248.1
SO$_3$(g)	−395.7	−371.1	256.6
H$_2$S(g)	−20.6	−33.6	205.7
H$_2$SO$_4$(l)	−813.989	690.101	156.90
H$_2$S$_2$O$_7$(s)	−1274	—	—
SF$_4$(g)	−774.9	−731.4	291.9
SF$_6$(g)	−1210	−1105	291.7
SCl$_2$(l)	−50	—	—
SCl$_2$(g)	−20	—	—
S$_2$Cl$_2$(l)	−59.4	—	—
S$_2$Cl$_2$(g)	−18	−32	331.4
SOCl$_2$(l)	−246	—	—
SOCl$_2$(g)	−213	−198	309.7
SO$_2$Cl$_2$(l)	−394	—	—
SO$_2$Cl$_2$(g)	−364	−320	311.8
Tin			
Sn(s)	0	0	51.55
Sn(g)	302	267	168.38

Standard Molar Enthalpies of Formation, Standard Molar Free Energies of Formation, and Absolute Standard Entropies [298.15 K (25°C), 1 atm] (continued)

Substance	ΔH_f°, kJ mol^{-1}	ΔG_f°, kJ mol^{-1}	S_{298}°, J K^{-1} mol^{-1}
SnO(s)	−286	−257	56.5
SnO$_2$(s)	−580.7	−519.7	52.3
SnCl$_4$(l)	−511.2	−440.2	259
SnCl$_4$(g)	−471.5	−432.2	366
Titanium			
Ti(s)	0	0	30.6
Ti(g)	469.9	425.1	180.19
TiO$_2$(s)	−944.7	−889.5	50.33
TiCl$_4$(l)	−804.2	−737.2	252.3
TiCl$_4$(g)	−763.2	−726.8	354.8
Tungsten			
W(s)	0	0	32.6
W(g)	849.4	807.1	173.84
WO$_3$(s)	−842.87	−764.08	75.90
Zinc			
Zn(s)	0	0	41.6
Zn(g)	130.73	95.178	160.87
ZnO(s)	−348.3	−318.3	43.64
ZnCl$_2$(s)	−415.1	−369.43	111.5
ZnS(s)	−206.0	−201.3	57.7
ZnSO$_4$(s)	−982.8	−874.5	120
ZnCO$_3$(s)	−812.78	−731.57	82.4
Complexes			
[Co(NH$_3$)$_4$(NO$_2$)$_2$]NO$_3$, *cis*	−898.7	—	—
[Co(NH$_3$)$_4$(NO$_2$)$_2$]NO$_3$, *trans*	−896.2	—	—
NH$_4$[Co(NH$_3$)$_2$(NO$_2$)$_4$]	−837.6	—	—
[Co(NH$_3$)$_6$][Co(NH$_3$)$_2$(NO$_2$)$_4$]$_3$	−2733	—	—
[Co(NH$_3$)$_4$Cl$_2$]Cl, *cis*	−997.0	—	—
[Co(NH$_3$)$_4$Cl$_2$]Cl, *trans*	−999.6	—	—
[Co(en)$_2$(NO$_2$)$_2$]NO$_3$, *cis*	−689.5	—	—
[Co(en)$_2$Cl$_2$]Cl, *cis*	−681.1	—	—
[Co(en)$_2$Cl$_2$]Cl, *trans*	−677.4	—	—
[Co(en)$_3$](ClO$_4$)$_3$	−762.7	—	—
[Co(en)$_3$]Br$_2$	−595.8	—	—
[Co(en)$_3$]I$_2$	−475.3	—	—
[Co(en)$_3$]I$_3$	−519.2	—	—
[Co(NH$_3$)$_6$](ClO$_4$)$_3$	−1035	−227	636
[Co(NH$_3$)$_5$NO$_2$](NO$_3$)$_2$	−1089	−418.4	350
[Co(NH$_3$)$_6$](NO$_3$)$_3$	−1282	−530.5	469
[Co(NH$_3$)$_5$Cl]Cl$_2$	−1017	−582.8	366
[Pt(NH$_3$)$_4$]Cl$_2$	−728.0	—	—
[Ni(NH$_3$)$_6$]Cl$_2$	−994.1	—	—
[Ni(NH$_3$)$_6$]Br$_2$	−923.8	—	—
[Ni(NH$_3$)$_6$]I$_2$	−808.3	—	—

Appendix J:
Composition of Commercial Acids and Bases

Acid or base	Specific gravity	Percentage by mass	Molarity	Normality
Hydrochloric acid	1.19	38%	12.4	12.4
Nitric acid	1.42	70%	15.8	15.8
Sulfuric acid	1.84	95%	17.8	35.6
Acetic acid	1.05	99%	17.3	17.3
Aqueous ammonia	0.90	28%	14.8	14.8

Appendix K:
Half-Life Times for Several Radioactive Isotopes

(Symbol in parentheses indicates type of emission; $E.C.$ = K-electron capture, $S.F.$ = spontaneous fission; y = years, d = days, h = hours, m = minutes, s = seconds.)

$^{14}_{6}C$	5770 y	(β^-)		$^{226}_{88}Ra$	1590 y	(α)
$^{13}_{7}N$	10.0 m	(β^+)		$^{228}_{88}Ra$	6.7 y	(β^-)
$^{24}_{11}Na$	15.0 h	(β^-)		$^{228}_{89}Ac$	6.13 h	(β^-)
$^{32}_{15}P$	14.3 d	(β^-)		$^{228}_{90}Th$	1.90 y	(α)
$^{40}_{19}K$	1.3×10^9 y	$(\beta^-$ or $E.C.)$		$^{232}_{90}Th$	1.39×10^{10} y	$(\alpha, \beta^-,$ or $S.F.)$
$^{60}_{27}Co$	5.2 y	(β^-)		$^{233}_{90}Th$	23 m	(β^-)
$^{87}_{37}Rb$	4.7×10^{10} y	(β^-)		$^{234}_{90}Th$	24.1 d	(β^-)
$^{90}_{38}Sr$	28 y	(β^-)		$^{223}_{91}Pa$	27 d	(β^-)
$^{115}_{49}In$	6×10^{14} y	(β^-)		$^{233}_{92}U$	1.62×10^5 y	(α)
$^{131}_{53}I$	8.05 d	(β^-)		$^{234}_{92}U$	2.4×10^5 y	$(\alpha$ or $S.F.)$
$^{142}_{58}Ce$	5×10^{15} y	(α)		$^{235}_{92}U$	7.3×10^8 y	$(\alpha$ or $S.F.)$
$^{198}_{79}Au$	64.8 h	(β^-)		$^{238}_{92}U$	4.5×10^9 y	$(\alpha$ or $S.F.)$
$^{208}_{81}Tl$	3.1 m	(β^-)		$^{239}_{92}U$	23 m	(β^-)
$^{210}_{82}Pb$	21 y	(β^-)		$^{239}_{93}Np$	2.3 d	(β^-)
$^{212}_{82}Pb$	10.6 h	(β^-)		$^{239}_{94}Pu$	24,360 y	$(\alpha$ or $S.F.)$
$^{214}_{82}Pb$	26.8 m	(β^-)		$^{240}_{94}Pu$	6.58×10^3 y	$(\alpha$ or $S.F.)$
$^{206}_{83}Bi$	6.3 d	$(\beta^+$ or $E.C.)$		$^{241}_{94}Pu$	13 y	$(\alpha$ or $\beta^-)$
$^{210}_{83}Bi$	5.0 d	(β^-)		$^{241}_{95}Am$	458 y	(α)
$^{212}_{83}Bi$	60.5 m	$(\alpha$ or $\beta^-)$		$^{242}_{96}Cm$	163 d	$(\alpha$ or $S.F.)$
$^{207}_{84}Po$	5.7 h	$(\alpha, \beta^+,$ or $E.C.)$		$^{243}_{97}Bk$	4.5 h	$(\alpha$ or $E.C.)$
$^{210}_{84}Po$	138.4 d	(α)		$^{245}_{98}Cf$	350 d	$(\alpha$ or $E.C.)$
$^{212}_{84}Po$	3×10^{-7} s	(α)		$^{253}_{99}Es$	20.0 d	$(\alpha$ or $S.F.)$
$^{216}_{84}Po$	0.16 s	(α)		$^{254}_{100}Fm$	3.24 h	$(S.F.)$
$^{218}_{84}Po$	3.0 m	$(\alpha$ or $\beta^-)$		$^{255}_{100}Fm$	22 h	(α)
$^{215}_{85}At$	10^{-4} s	(α)		$^{256}_{101}Md$	1.5 h	$(E.C.)$
$^{218}_{85}At$	1.3 s	(α)		$^{254}_{102}No$	3 s	(α)
$^{220}_{86}Rn$	54.5 s	(α)		$^{257}_{103}Lr$	8 s	(α)
$^{222}_{86}Rn$	3.82 d	(α)		$^{263}_{106}Unh$	0.9 s	(α)
$^{224}_{88}Ra$	3.64 d	(α)				

Appendix L: Apparatus for Qualitative Analysis (one student)

50 Reagent bottles, dropper type, 10-mL
1 Rack for 50 reagent bottles
1 Test tube block
1 Flask, Florence, 250-mL
2 Beakers, 250-mL
2 Beakers, 50-mL
2 Beakers, 20-mL
1 Graduate, 10-mL
1 Casserole, 15-mL
6 Centrifuge tubes
6 Test tubes, 65 × 10 mm
6 Test tubes, 75 × 10 mm
1 Metal rack for a water bath

1 Micro burner
1 Bunsen burner
2 Watch glasses, 2.5-cm
1 File
1 Box of labels
1 Bottle of litmus paper, red
1 Bottle of litmus paper, blue
1 Test tube holder, small
1 Test tube brush, small, tapered
1 Test tube brush, small, not tapered
1 Wire, platinum, 2-in
1 Wire gauze
1 Ring stand (with ring) small

1 Spatula, micro (Monel metal)
1 Forceps
1 Two-hole rubber stopper to fit 250-mL
    flask
1 Wing top
1 Box of matches
1 Towel
1 Box of detergent, small
6 Medicine droppers, 1-mL
6 Capillary syringes
100-cm Glass tubing, 6-mm
100-cm Glass rod, 3-mm
1 Cobalt glass

Appendix M: Reagents for Cation and Anion Analysis

Reagents for Cation Analysis

Group I   Reagents

Hydrochloric acid, HCl, 6 M
Nitric acid, HNO_3, 4 M
Aqueous ammonia, $NH_3 + H_2O$, 4 M
*Potassium chromate, K_2CrO_4, 1 M

Group II   Reagents (in addition to those listed for Group I)

Aqueous ammonia, $NH_3 + H_2O$, 6 M
Hydrochloric acid, HCl, 1.0 M
Thioacetamide, CH_3CSNH_2, 5% solution
Sodium hydroxide, NaOH, 4 M
Ammonium nitrate, NH_4NO_3, 1 M
Sulfuric acid, H_2SO_4, 4 M
Acetic acid, CH_3CO_2H, 1 M
Oxalic acid, $H_2C_2O_4$, 1 M
Ammonium acetate, $NH_4CH_3CO_2$, 1 M
Aqueous ammonia, $NH_3 + H_2O$, 15 M
*Potassium hexacyanoferrate(II), $K_4Fe(CN)_6$, 0.1 M
Potassium cyanide, KCN, 1 M
Hydrogen peroxide, H_2O_2, 3% solution
*Magnesia mixture: Dissolve 50 g of $MgCl_2 \cdot 6H_2O$ and 70 g of NH_4Cl in 400 mL of water. Add 100 mL of 15 M aqueous ammonia and dilute to 1 L. Filter.
*Aluminum wire, Al
Mercury(II) chloride, $HgCl_2$, 0.2 M
Sodium hypochlorite, NaOCl, 5% solution

Group III   Reagents (in addition to those listed for Groups I and II)

Hydrochloric acid, HCl, 12 M
Hydrochloric acid, HCl, 1 M
Nitric acid, HNO_3, 14 M
*Dimethylglyoxime, 1% solution: Dissolve 10 g in 1 L of alcohol
Acetic acid, CH_3CO_2H, 4 M
*Ammonium thiocyanate, NH_4SCN, solid
Acetone, $(CH_3)_2CO$
*Sodium fluoride, NaF, 1 M
*Potassium nitrite, KNO_2, solid
*Sodium nitrite, $NaNO_2$, 1 M
*Sodium bismuthate, $NaBiO_3$, solid
*Aluminum reagent: Dissolve 1 g of the ammonium salt of aurin-tricarboxylic acid in 1 L of water
Lead acetate, $Pb(CH_3CO_2)_2$, 0.1 M
Ammonium carbonate, $(NH_4)_2CO_3$, 1 M

Group IV   Reagents (in addition to those listed for Groups I–III)

Ethyl alcohol, C_2H_5OH
*Ammonium oxalate, $(NH_4)_2C_2O_4$, 0.4 M

*Fill reagent bottles only about one-fourth full of starred reagents.

Reagents for Cation and Anion Analysis (continued)

Group V   Reagents (in addition to those listed for Groups I–IV)

*Ammonium sulfate, $(NH_4)_2SO_4$, 1 M

*Sodium reagent: Mix 30 g of $UO_2(CH_3CO_2)_2 \cdot 2H_2O$ with 80 g of $Zn(CH_3CO_2)_2 \cdot 2H_2O$ and 10 mL of glacial acetic acid. Dilute the solution to 250 mL, let it stand for several hours, and filter. Use the clear solution.

*Sodium hexanitrocobaltate(III), $Na_3Co(NO_2)_6$: Dissolve 30 g of $NaNO_2$ in 97 mL of water; add 3 mL of glacial CH_3CO_2H and 3.3 g of $Co(NO_3)_2 \cdot 6H_2O$. Filter and use the clear solution.

*Disodium hydrogen phosphate, Na_2HPO_4, 1 M

*Magnesium reagent, 4-(nitrophenylazo)-1-naphthol: Dissolve 0.25 g of this reagent and 2.5 g of NaOH in sufficient water to make 250 mL of solution.

Reagents for Anion Analysis (in addition to those listed for cation analysis)

*Sulfuric acid, H_2SO_4, 1.5 M

Sodium carbonate, Na_2CO_3, 1.5 M

*Manganese(II) chloride, $MnCl_2$: Saturate 12 M HCl with $MnCl_2$.

*Potassium hexacyanoferrate(III), $K_3Fe(CN)_6$, a freshly prepared saturated solution

*Iron(III) chloride, $FeCl_3$, 0.1 M

Calcium chloride, $CaCl_2$, 0.1 M

Barium chloride, $BaCl_2$, 0.1 M

*Cobalt(II) nitrate, $Co(NO_3)_2$, 1 M

*Silver nitrate, $AgNO_3$, 0.1 M

*Bromine water, $Br_2 + H_2O$, saturated solution

*Barium hydroxide, $Ba(OH)_2$, saturated solution

*Iron(II) sulfate, $FeSO_4$, 0.1 M

*Sulfuric acid, H_2SO_4, 18 M

*Silver sulfate, Ag_2SO_4, solid (nitrate-free)

*Potassium permanganate, $KMnO_4$, 0.002 M

*Silica, SiO_2, powdered

*Methyl alcohol, CH_3OH

*Iron(III) nitrate, $Fe(NO_3)_3$, 0.1 M

*Dichloromethane, CH_2Cl_2

*Potassium permanganate, $KMnO_4$, 0.1 M

*Lead acetate paper, filter paper moist with 0.1 M $Pb(CH_3CO_2)_2$

*Fill reagent bottles only about one-fourth full of starred reagents.

Appendix N: Preparation of Solutions of Cations

Stock Solutions

Stock solutions should contain the cations in question at a concentration of 50 mg per milliliter. These stock solutions may be prepared by grinding to a powder the weight of salt given below and adding enough water (or acid if specified) to make the volume 100 mL. Dissolution of the salts may be hastened by heating.

Known and Unknown Solutions

To prepare known or unknown solutions of cations, mix 20 mL of the stock solutions (40 mL of $AsCl_3$) of the cations desired and dilute the solution to 100 mL with water. This solution will contain 10 mg of cations per milliliter. This procedure allows for a maximum of five cations at a concentration of 10 mg of cations per milliliter. Dilution to 200 mL will allow a maximum of ten cations at a concentration of 5 mg per milliliter. Give each student about 1 mL of solution.

Stock Solutions of Cations (50 mg of cations per milliliter)

Group	Ion	Formula of salt	Grams per 100 mL of solution
I	Ag^+	$AgNO_3$	8.0
	Pb^{2+}	$Pb(NO_3)_2$	8.0
	Hg_2^{2+}	$Hg_2(NO_3)_2$	7.0 (dissolve in 0.6 M HNO_3)
II	Pb^{2+}	$Pb(NO_3)_2$	8.0
	Bi^{3+}	$Bi(NO_3)_3 \cdot 5H_2O$	11.5 (dissolve in 3 M HNO_3)
	Cu^{2+}	$Cu(NO_3)_2 \cdot 3H_2O$	19.0
	Cd^{2+}	$Cd(NO_3)_2 \cdot 4H_2O$	13.8
	Hg^{2+}	$HgCl_2$	6.8
	As^{3+}	As_4O_6	3.3 (heat in 50 mL of 12 M HCl, then add 50 mL of water)
	Sb^{3+}	$SbCl_3$	9.5 (dissolve in 6 M HCl and dilute with 2 M HCl)
	Sn^{2+}	$SnCl_2 \cdot 2H_2O$	9.5 (dissolve in 50 mL of 12 M HCl. Dilute to 100 mL with water. Add a piece of tin metal)
	Sn^{4+}	$SnCl_4 \cdot 3H_2O$	13.3 (dissolve in 6 M HCl)
III	Co^{2+}	$Co(NO_3)_2 \cdot 6H_2O$	24.7
	Ni^{2+}	$Ni(NO_3)_2 \cdot 6H_2O$	24.8
	Mn^{2+}	$Mn(NO_3)_2 \cdot 6H_2O$	26.2
	Fe^{3+}	$Fe(NO_3)_3 \cdot 9H_2O$	36.2
	Al^{3+}	$Al(NO_3)_3 \cdot 9H_2O$	69.5
	Cr^{3+}	$Cr(NO_3)_3$	23.0
	Zn^{2+}	$Zn(NO_3)_2$	14.5
IV	Ba^{2+}	$BaCl_2 \cdot 2H_2O$	8.9
	Sr^{2+}	$Sr(NO_3)_2$	12.0
	Ca^{2+}	$Ca(NO_3)_2 \cdot 4H_2O$	29.5
V	Mg^{2+}	$Mg(NO_3)_2 \cdot 6H_2O$	52.8
	NH_4^+	NH_4NO_3	22.2
	Na^+	$NaNO_3$	18.5
	K^+	KNO_3	13.0

Appendix O: Laboratory Assignments

The following list suggests a set of laboratory assignments.

1. Construct a wash bottle (if necessary), stirring rods, and capillary syringes.
2. Analyze a known solution containing all the cations of Group I.
3. Analyze an unknown solution containing Group I cations.
4. Analyze a known solution containing all the cations of Group II.
5. Analyze an unknown solution containing Group II cations.
6. Analyze a known solution containing all the cations of Group III.
7. Analyze an unknown solution containing Group III cations.
8. Analyze a known solution containing all the cations of Group IV.
9. Analyze a known solution containing all the cations of Group V.
10. Analyze an unknown solution containing cations from Groups IV and V.
11. Analyze a general unknown solution containing cations of all the groups.
12. Analyze a known salt mixture containing no oxidizing anions.
13. Analyze an unknown salt mixture containing no oxidizing anions.
14. Analyze a known salt mixture containing no reducing anions.
15. Analyze an unknown salt mixture containing no reducing anions.
16. Analyze a salt mixture for both cations and anions.
17. Analyze an alloy.

Photo Credits

Chapter 1 p. 1, E. R. Degginger; p. 2, Gary Milburn/Tom Stack & Associates; p. 3, Ken O'Donoghue; p. 4 (Fig. 1.1), E. R. Degginger; p. 6 (Fig. 1.3, both), E. R. Degginger; p. 6 (Fig. 1.4), E. R. Degginger; p. 8 (Fig. 1.6), E. R. Degginger; p. 8 (Fig. 1.7), Yoav Levy/Phototake; p. 9, Ken O'Donoghue; p. 11, Eric Kroll/Taurus Photos; p. 14, Yoav/Phototake; p. 16 (Fig. 1.9), National Bureau of Standards; p. 16 (Fig. 1.11), E. R. Degginger; p. 18 (Fig. 1.13), Tom Pantages.

Chapter 2 p. 26, Dennis Kunkel/Phototake; p. 27 (all), Ken O'Donoghue; p. 31, Ken O'Donoghue; p. 32 (Fig. 2.2), E. R. Degginger; p. 34 (Fig. 2.3), E. R. Degginger; p. 35, Russ Kinne/Photo Researchers, Inc.; p. 37, E. R. Degginger; p. 39 (Fig. 2.5), Yoav Levy/Phototake; p. 40, Tom Pantages; p. 44 (Fig. 2.6, both), Ken O'Donoghue.

Chapter 3 p. 52, Dario Perla/After-Image; p. 53, Ken O'Donoghue; p. 55 (Fig. 3.1), E. R. Degginger; p. 55 (Fig. 3.2), E. R. Degginger; p. 55 (Fig. 3.3, both), E. R. Degginger; p. 57, Tom Pantages; p. 61, E. R. Degginger.

Chapter 4 p. 72, Malcolm Lockwood; p. 73 (Fig. 4.1), E. R. Degginger; p. 73 (Fig. 4.2), E. R. Degginger; p. 79, Ken O'Donoghue; p. 81, E. R. Degginger; p. 88 (Fig. 4.6), E. R. Degginger.

Chapter 5 p. 93, Chuck O'Rear/West Light; p. 94 (Fig. 5.1), L. S. Stepanowicz/Panographics; p. 104 (Fig. 5.11), David Parker/Photo Researchers, Inc.; p. 104 (Fig. 5.12), from *General College Chemistry, Fifth edition* (1976), by Charles W. Keenan, Jesse H. Wood and Donald Kleinfelter. By permission of the authors. p. 118 (upper left clockwise), Manfred Kage/Peter Arnold, Inc.; Ken O'Donoghue; Ken O'Donoghue; E. R. Degginger; Paul Silverman/Fundamental Photographs; Paul Silverman/Fundamental Photographs; p. 119 (upper left clockwise), E. R. Degginger; Tom Pantages; E. R. Degginger; E. R. Degginger; Yoav/Phototake.

Chapter 6 p. 137, Phil Degginger; p. 147 (Fig. 6.5, both), Tom Pantages.

Chapter 7 p. 165, Michael Siegel/Phototake.

Chapter 8 p. 183, Langridge/McCoy/Rainbow; p. 188 (Fig. 8.5), Ken O'Donoghue; p. 188 (Fig. 8.6), Ken O'Donoghue; p. 195 (Fig. 8.17), Ken O'Donoghue; p. 195 (Fig. 8.18), Ken O'Donoghue.

Chapter 9 p. 203, Barry L. Runk/Grant Heilman Photography; p. 211 (Fig. 9.3), E. R. Degginger; p. 215 (Fig. 9.5), E. R. Degginger; p. 218 (Fig. 9.6, both), L. S. Stepanowicz/Panographics; p. 220 (both), E. R. Degginger; p. 223, E. R. Degginger; p. 225 (Fig. 9.8), E. R. Degginger; p. 225 (Fig. 9.9), E. R. Degginger; p. 227 (Fig. 9.12), Grant Heilman/Grant Heilman Photography; p. 228 (Fig. 9.14), B. J. Spenceley/Bruce Coleman, Inc.; p. 230 (Fig. 9.16), E. R. Degginger. •

Chapter 10 p. 236, Bob Sell/Tom Stack & Associates; p. 241 (Fig. 10.6, all), Ken O'Donoghue; p. 249, UPI/Bettmann Newsphotos.

Chapter 11 p. 274, L. S. Stepanowicz/Panographics; p. 275 (Fig. 11.1), L. S. Stepanowicz/Panographics; p. 279 (Fig. 11.8), E. R. Degginger; p. 281, E. R. Degginger; p. 286, E. R. Degginger/Bruce Coleman, Inc.; p. 287, E. R. Degginger; p. 292 (Fig. 11.23), Joseph Nettis/Photo Researchers, Inc.; p. 304, E. R. Degginger/Bruce Coleman, Inc.; p. 306, E. R. Degginger.

Chapter 12 p. 313, John Koivula/Photo Researchers, Inc.; p. 314 (Fig. 12.1), L. S. Stepanowicz/Panographics; p. 315 (Fig. 12.2), Diane Schiumo/Fundamental Photographs; p. 316 (Fig. 12.3), E. R. Degginger; p. 317 (Fig. 12.4), L. S. Stepanowicz/Panographics; p. 317 (Fig. 12.5), Kip Petikolas/Fundamental Photographs; p. 318 (Fig. 12.6), E. R. Degginger; p. 319 (Fig. 12.8, all), Fundamental Photographs; p. 324 (Fig. 12.13), Tom Pantages; p. 329 (Fig. 12.15, both), L. S. Stepanowicz/Panographics; p. 332, Tom Pantages; p. 340 (Fig. 12.17), Tom Pantages.

Chapter 13 p. 357, Steve Dunwell/The Image Bank; p. 359 (Fig. 13.2), Ken O'Donoghue; p. 359 (Fig. 13.3), Tom Pantages; p. 361 (Fig. 13.4), Tom Pantages; p. 361 (Fig. 13.5), E. R. Degginger; p. 364 (Fig. 13.7), Ken O'Donoghue; p. 367 (Fig. 13.10), E. R. Degginger; p. 367 (Fig. 13.11), E. R. Degginger; p. 368 (Fig. 13.12), E. R. Degginger; p. 369 (Fig. 13.13), E. R. Degginger; p. 372 (Fig. 13.14), Tom Pantages; p. 374 (Fig. 13.16), E. R. Degginger; p. 374 (bottom), New York State Department of Environmental Conservation; p. 376 (Fig. 13.17), E. R. Degginger; p. 377 (Fig. 13.18), Bausch & Lomb Inc.; p. 378 (Fig. 13.19), Ronald F. Thomas/Taurus Photos; p. 379 (Fig. 13.20), International Salt Company; p. 380 (Fig. 13.21), Judi Benvenuti/Taurus Photos; p. 384 (Fig. 13.23), Susan Leavines/Photo Researchers, Inc.

Chapter 14 p. 391, Hans Reinhard/Bruce Coleman Inc.; p. 395 (Fig. 14.3), Wide World Photos, Inc.; p. 417 (Fig. 14.15), A. C. Spark Plug/General Motors Corp.

Chapter 15 p. 422, NASA; p. 423 (Fig. 15.1), Gregg Mancuso/After-Image; p. 436 (Fig. 15.4, a–e), E. R. Degginger; p. 437 (Fig. 15.5, a–c), Tom Pantages; p. 442 (Fig. 15.6, both), Ken O'Donoghue; p. 444 (Fig. 15.8), Tom Pantages.

Chapter 16 p. 451, Bruce Coleman Inc.; p. 460 (Fig. 16.2, a & b), E. R. Degginger; p. 470 (Fig. 16.3), Paul Silverman/Fundamental Photographs.

Chapter 17 p. 477, Science Photo Library/Photo Researchers, Inc.; p. 481 (Fig. 17.1), E. R. Degginger; p. 481 (Fig. 17.3), Tom Pantages; p. 485 (Fig. 17.4), E. R. Degginger; p. 490 (Fig. 17.5), E. R. Degginger; p. 495 (Fig. 17.7, a & b), E. R. Degginger; p. 496, E. R. Degginger; p. 497, E. R. Degginger; p. 498, E. R. Degginger; p. 500 (Fig. 17.8), E. R. Degginger; p. 510 (Fig. 17.10), E. R. Degginger; p. 512, E. R. Degginger; p. 515 (Fig. 17.11), E. R. Degginger.

Chapter 18 p. 529, E. R. Degginger; p. 531 (top), E. R. Degginger; p. 531 (bottom), Tom Pantages; p. 532, E. R. Degginger; p. 533, Tom Pantages; p. 535, E. R. Degginger; p. 536, E. R. Degginger; p. 539, E. R. Degginger; p. 541, Tom Pantages; p. 546, Leonard Lessin/Peter Arnold, Inc.; p. 549 (Fig. 18.14), Ken O'Donoghue.

Chapter 19 p. 557, E. R. Degginger; p. 558 (Fig. 19.1), E. R. Degginger; p. 558 (Fig. 19.2), E. R. Degginger; p. 560 (Fig. 19.3), Erika Stone/Peter Arnold, Inc.; p. 564 (Fig. 19.4), E. R. Degginger; p. 568 (Fig. 19.5), E. R. Degginger; p. 575 (Fig. 19.6), E. R. Degginger; p. 577 (Fig. 19.7), E. R. Degginger.

Chapter 20 p. 589, C. B. Jones/Taurus Photos; p. 591 (Fig. 20.2), E. R. Degginger; p. 592 (Fig. 20.3, both), E. R. Degginger; p. 593 (Fig. 20.5), Paul Silverman/Fundamental Photographs; p. 596 (Fig. 20.11) (a), Paul Silverman/Fundamental Photographs, (b), E. R. Degginger; p. 598 (Fig. 20.13, a & b), Ken O'Donoghue; p. 599 (Fig.

Glossary

Absolute entropy   Represents the entropy change of a substance taken from absolute zero to a given temperature (T). (19.13)

Absolute zero   The temperature at which all possible heat has been removed from an object. (10.5)

Acid   A compound that donates a hydrogen ion (H^+) to another compound. (9.2)

Acid-base reaction   Reaction occurring when a hydrogen atom is transferred from a Brønsted acid to a Brønsted base. (9.3)

Acquired immune deficiency syndrome (AIDS)   The invasion of a certain class of human white blood cells, the T-lymphocytes, by the HIV virus, leading to a fatal inability to produce antibodies against disease. (31.17)

Activated complex   An unstable combination of reacting molecules that is intermediate between reactants and products. (14.8)

Activation energy (E_a)   The minimum energy necessary to form an activated complex in a reaction. (14.9)

Addition polymer   A polymer formed by an addition reaction. (30.16)

Addition reaction   Reaction of two or more substances to give another substance. (9.3, 30.3)

Aerobic decomposition   The decomposition of organic matter in the presence of oxygen. (29.7)

Alkyl group   A substituent that contains one less hydrogen than the corresponding alkane. (30.2)

Alpha decay   The loss of an alpha particle during radioactive decay. (24.7)

Alpha (α) particle   A helium nucleus; i.e., a helium atom that has lost two electrons. (5.4)

Amino acid   A substance containing both an amine group and a carboxylic acid group. Proteins are composed of amino acids. (31.2)

Amorphous solid   A solid that lacks a crystalline structure. (11.1)

Amphiprotic   A species that may either gain or lose a proton in reaction. (16.2)

Amphoteric behavior   The behavior of elements that can exhibit the properties of either a metal or a nonmetal. (9.5)

Anaerobic decomposition   The decomposition of organic matter in the absence of oxygen. (29.7)

Anion   A negative ion. (6.1)

Anode   The electrode at which oxidation takes place in an electrochemical cell. (21.2)

Antibonding orbital   Molecular orbital located outside of the region between two nuclei. Electrons in an antibonding orbital destabilize the molecule. (7.1)

Arrhenius equation   ($k = A \times 10^{-E_a/2.303RT}$)   Expresses the relationship between the rate constant and the activation energy of a reaction. (14.9)

Asymmetric carbon   A carbon atom that is bonded to four different groups within a molecule. (30.6)

Atom   The smallest particle of an element that can enter into a chemical combination. (1.6)

Atomic mass unit (amu)   A unit of mass equal to $\frac{1}{12}$ of the mass of a ^{12}C atom. (2.2)

Atomic number (Z)   The number of protons in the nucleus of an atom. (5.5)

Atomic weight   The average atomic mass of an atom of an element, numerically equal to its mass in amu but expressed without a unit. (2.2)

Aufbau process   Process by which chemists illustrate the electronic structures of the elements by ''building'' them in atomic order, adding one proton to the nucleus and one electron to the proper subshell at a time. (5.12)

Avogadro's law   Equal volumes of all gases, measured under the same conditions of temperature and pressure, contain the same number of molecules. (10.7)

Avogadro's number   The number of atoms contained in exactly 12 grams of ^{12}C, equal to 6.022×10^{23} atoms. (2.2)

Azeotropic mixture   A solution that forms a vapor with the same concentration as the solution, distilling without a change in concentration. (12.19)

Azimuthal quantum number (*l*) A quantum number distinguishing the different shapes of orbitals. (5.10)

Band The orbitals, or energy levels, that extend through a crystal. The way in which these bands are filled or not filled with electrons determines whether the substance is a metal, a semiconductor or an insulator. (25.3)

Barometer A device used to measure air pressure. (10.2)

Base A compound that accepts a hydrogen ion (H^+). (9.2)

Beta decay The breakdown of a neutron into a proton, which remains in the nucleus, and an electron, which is emitted as a beta particle. (24.7)

Beta (*β*) particle An electron emitted during radioactive decay. (5.4)

Bimolecular reaction The collision and combination of two reactants to give an activated complex in an elementary reaction. (14.12)

Binary compound A compound containing two different elements. (6.13)

Biological oxygen demand (BOD) A measure of the volume of oxygen gas taken up by a given amount of water in five days at 20°C, indicating the quantity of microorganisms in the sample. (29.8)

Body-centered cubic structure A crystalline structure that has a cubic unit cell with lattice points at the corners and in the center of the cell. (11.18)

Boiling point The temperature at which the vapor pressure of a liquid equals the pressure of the gas above it. (11.3)

Boiling-point elevation ($\Delta T = K_b m$) The elevation of the boiling point of a liquid by addition of a solute. (12.18)

Bond angle The angle between any two covalent bonds that include a common atom. (8.1)

Bond distance The distance between the nuclei of two bonded atoms. (8.1)

Bond energy The energy required to break a covalent bond in a gaseous substance. (6.9)

Bond order In a Lewis formula, the number of bonding pairs of electrons between two atoms. The molecular-orbital bond order is the net number of pairs of bonding electrons, or the difference between the number of bonding and antibonding electrons divided by two. (7.3)

Bonding orbital Molecular orbital located between two nuclei. Electrons in a bonding orbital stabilize a molecule. (7.1)

Born-Haber cycle Cyclic process used to relate the enthalpy of formation (ΔH_f) of a compound to its lattice energy *(U)*, the ionization energy *(I)*, the electron affinity *(E.A.)*, the enthalpy of sublimation (ΔH_s), and the bond dissociation energy *(D)* of its constituents. (11.21)

Boyle's law The volume of a given mass of gas held at constant temperature is inversely proportional to the pressure under which it is measured. *PV = k*. (10.3)

Bragg equation ($n\lambda = 2d \sin\theta$) An equation which relates the angles (θ) at which X rays of wavelength λ are scattered by planes with a separation *d*.

Brønsted acid A compound that donates a hydrogen ion (H^+) to another compound. (9.2)

Brønsted base A compound that accepts a hydrogen ion (H^+). (9.2)

Buffer capacity The amount of an acid or base that can be added to a volume of a buffer solution before its pH changes significantly. (17.8)

Buffer solution A mixture of a weak acid or weak base and its salt. The pH of a buffer resists change when small amounts of acid or base are added. (17.8)

Calorie A non-SI unit representing the amount of heat or other energy necessary to raise the temperature of one gram of water one degree Celsius. 1 cal = 4.184 J. (4.1)

Calorimetry The process of measuring the amount of heat involved in a chemical or physical change. (4.2)

Catalysis The increase in the speed of a chemical reaction caused by a catalyst. (14.15)

Catalyst A substance that changes the speed of a chemical reaction without affecting the yield or undergoing permanent chemical change. (9.9)

Cathode The electrode at which reduction takes place in an electrochemical cell. (21.2)

Cathode ray Stream of particles (electrons) emanating from the negative electrode in an evacuated glass tube. (5.1)

Cathodic protection Method of preventing corrosion of iron or steel by suspending stainless steel anodes in the tank and passing a small current continuously through the system. (21.16)

Cation A positive ion. (6.1)

Cell potential The difference in potential between the two electrodes of a cell. (21.7)

Chain mechanism A series of elementary reactions that repeat over and over to produce a product. (14.14)

Chain reaction Repeated fission caused when the neutrons released in fission bombard other atoms. (24.9)

Charles's law The volume of a given mass of gas is directly proportional to its Kelvin temperature when the pressure is held constant. *V/T = k*. (10.4)

Chelating ligand A ligand that is attached to a central metal ion by bonds from two or more donor atoms. (26.1)

Chemical change Change producing one or more different kinds of matter from an original kind of matter. (1.4)

Chemical property A chemical property describes how one kind of matter is changed into another kind. (1.4)

Chemical thermodynamics The chemical science that deals with the energy transfers and transformations that accompany chemical and physical changes. (4.3, 19.1)

***Cis* configuration** Configuration of a geometrical isomer in which two groups are on the same side of an imaginary reference line on the molecule. (26.4)

Cloning The exact duplication of a cell's DNA and, therefore, of an organism's characteristics. (31.15)

Colligative properties Properties of a solution that depend only on the concentration of a solute species. (12.17)

Colloid Insoluble particles (larger than single molecules) in a stable suspension. (12.26)

Combination reaction Reaction of two or more substances to give another substance. (9.3)

Common ion effect The shift in equilibrium caused by the addition of a substance with an ion in common with the substances in equilibrium. (17.7)

Compound A pure substance with an invariant composition that can be decomposed, producing either elements or other compounds, by chemical change. (1.5)

Concentration The relative amounts of solute and solvent present in a solution. (3.1)

Condensation The change from a vapor to a condensed state (solid or liquid). (11.2)

Condensation polymer A polymer formed by linking together molecules in a reaction that eliminates small molecules such as water. (30.17)

Conjugate acid Substance formed when a base gains a hydrogen ion. Considered an acid because it can lose a hydrogen ion to reform the base. (16.1)

Conjugate base Substance formed when an acid loses a hydrogen ion. Considered a base because it can gain a hydrogen ion to reform the acid. (16.1)

Constant boiling solution A solution that forms a vapor with the same concentration as the solution, distilling without a change in concentration. (12.19)

Coordinate covalent bond A bond formed when one atom provides both electrons in a shared pair. (6.4)

Coordination compound A molecule or ion formed by the bonding of a metal atom or ion to two or more ligands by coordinate covalent bonds. (26.1)

Coordination number The number of atoms closest to any given atom in a crystal. (11.15)

Coordination sphere The central metal ion plus the attached ligands of a coordination compound. (26.1)

Coulomb The quantity of electricity involved when a current of one ampere flows for one second. (21.5)

Covalent bond Bond formed when pairs of electrons are shared between atoms. (6.3)

Covalent radius Half the distance between the nuclei of two identical atoms when they are joined by a single covalent bond. (5.15)

Critical pressure The pressure required to liquefy a gas at its critical temperature. (11.8)

Critical temperature The temperature above which a gas cannot be liquefied, no matter how much pressure is applied. (11.8)

Crystal defect A variation in the regular arrangement of the atoms or molecules of a crystal. (11.14)

Crystalline solid A homogeneous solid in which the atoms, ions or molecules assume ordered positions. (11.1)

Cubic closest packed structure A crystalline structure in which planes of closest packed atoms or ions are stacked ABCABC. (11.15)

Dalton's law The total pressure of a mixture of ideal gases is equal to the sum of the partial pressures of the component gases. (10.12)

Daughter nuclide A nuclide produced by the radioactive decay of another nuclide. May be stable or may decay further. (24.7)

Decomposition reaction Reaction in which one compound breaks down into two or more substances. (9.3)

Degenerate orbitals Orbitals having the same energy. (5.10)

Density Mass per unit volume of a substance. (1.13)

Diamagnetic substance A substance that contains no unpaired electrons. Diamagnetic substances tend to move out of a magnetic field. (26.9)

Diffusion The movement of gas molecules through the gas. (10.1)

Dipole The separation of charge in a bond or a molecule with a positively charged end and a negatively charged end. (6.6)

Dipole-dipole attraction The intermolecular attraction of two dipoles. (11.11)

Diprotic acid An acid containing two ionizable hydrogen atoms per molecule. A diprotic acid ionizes in two steps. (16.9)

Disproportionation Oxidation-reduction reaction in which the same element is both oxidized and reduced. (20.2)

Dissociation constant (K_d) The equilibrium constant for the decomposition of a complex ion into its components in solution. (18.13)

Double bond A bond in which two pairs of electrons are shared between two atoms. (6.3)

Einstein equation Equation for determining the amount of energy resulting from the conversion of matter to energy. $E = mc^2$. (24.2)

Electrode A conductor that delivers electricity into a cell without necessarily entering into the cell reaction. Also, a system in which a conductor is in contact with a mixture of oxidized and reduced forms of some chemical species. (21.6)

Electrode potential The difference between the charge on an electrode and the charge in the solution. (21.6)

Electrolysis The input of electrical energy as a direct current to force a nonspontaneous reaction to occur. (21.2)

Electrolyte Ionic or covalent compounds that melt to give liquids that contain ions or that dissolve to give solutions that contain ions. (9.2, 12.6)

Electrolytic conduction The movement of ions through a molten substance, a solution, or occasionally, a solid. (21.1)

Electromotive force (emf) The difference in potential between two half-cells. (21.7)

Electron A small, negatively charged subatomic particle. (5.1)

Electron affinity A measure of the energy involved when an electron is added to a gaseous atom to form a negative ion. (5.15)

Electron capture Capture of an electron by an unstable nucleus. The electron converts a proton to a neutron in the nucleus. (24.7)

Electron configuration Electronic structure of an atom. (5.12)

Electronegativity The attraction of an atom for the electrons in a covalent bond. (6.5)

Element A pure substance that cannot be decomposed by a chemical change. (1.5)

Elementary reaction A reaction that cannot be broken down into smaller steps. (14.10)

Empirical formula A formula showing the composition of a compound given as the simplest whole-number ratio of atoms. (2.1)

End point The point during a titration when an indicator shows that the amount of reactant necessary for a complete reaction has been added to a solution. (3.2)

Endothermic reaction A chemical reaction or physical change that occurs with the absorption of heat. (4.1)

Energy The capacity to do work. (1.2)

Enthalpy change (ΔH) The heat lost or absorbed by a system under constant pressure during a reaction. (4.4, 19.5)

Entropy (S) The randomness, or amount of disorder, of a system. (19.7)

Entropy change (ΔS) The change in entropy that accompanies a chemical or physical change. ΔS is given by the sum of the entropies of the products of a chemical change minus the sum of the entropies of the reactants. $\Delta S = \Sigma S_{products} - \Sigma S_{reactants}$. (19.7)

Equilibrium The state at which the conversion of reactants into products and the conversion of products back into reactants occur simultaneously at the same rate. (15.1)

Equilibrium constant (K) The value of the reaction quotient for a system at equilibrium. (15.2)

Equivalent An equivalent of an acid is the mass of acid required to provide one mole of hydrogen ions in a reaction; an equivalent of a base is the mass of base required to react with one mole of hydrogen ions. (16.14) An equivalent of an oxidizing agent or of a reducing agent is the mass of the agent that combines with or releases one mole of electrons, respectively. (21.5)

Eutrophication The process by which a lake grows rich in nutrients and gradually fills with organic sediment and plants. (29.7)

Evaporation The change of a liquid into a gas. (11.2)

Excited state State in which an atom or molecule picks up outside energy, causing an electron to move into a higher-energy orbital. (5.7)

Exothermic reaction A chemical reaction or physical change that produces heat. (4.1)

Expansion work Work transferred between a system and its surroundings as the system expands or contracts against a constant pressure. (19.2)

Face-centered cubic structure A crystalline structure with a cubic unit cell with lattice points on the corners and in the center of each face. (11.18)

Faraday (F) The charge on one mole of electrons. $1 F = 96,487$ coulombs. (21.5)

Faraday's law The amount of a substance undergoing a chemical change at each electrode during electrolysis is directly proportional to the quantity of electricity that passes through the electrolytic cell. (21.5)

First ionization energy The energy required to remove the most loosely bound electron from a gaseous atom. (5.15)

First law of thermodynamics The total amount of energy in the universe is constant. (19.3)

Fission The splitting of a heavier nucleus into two or more lighter nuclei, usually accompanied by the conversion of mass into large amounts of energy. (24.9)

Formation constant (K_f) The equilibrium constant for the formation of a complex ion from its components in solution. (18.13)

Formula weight The sum of the atomic weights of the atoms found in one formula unit of an ionic compound. (2.3)

Free energy change (ΔG) A predictor of the spontaneity of a chemical reaction at constant temperature. $\Delta G = \Delta H - T\Delta S$. (19.9)

Freezing point The temperature at which the solid and liquid phases of a substance are in equilibrium. (11.5)

Freezing-point depression ($\Delta T = K_f m$) The lowering of the freezing point of a liquid by addition of a solute. (12.20)

Frequency factor (A) In the Arrhenius equation, a constant indicating how many collisions have the correct orientation to lead to products. (14.9)

Functional group A part of an organic molecule responsible for chemical behaviors of the molecule. (30.1)

Fusion The combining of very light nuclei into heavier nuclei, accompanied by the conversion of mass into large amounts of energy. (24.10)

Gamma (γ) ray High energy electromagnetic radiation. (5.4)

Gas The state in which matter has neither a definite volume nor shape. (1.2)

Gas constant (R) Constant derived from the ideal gas equation, $PV = nRT$. $R = 0.08206$ L atm/mol K; $R = 8.314$ L kPa/mol K. (10.8)

Gay-Lussac's law The volume of gases involved in a reaction, at constant temperature and pressure, can be expressed as a ratio of small whole numbers. (10.6)

Gene cloning A procedure for transferring small pieces of genetic information from one organism to another, thereby altering the characteristics of the second organism. (31.15)

Genetic code The information coded in a DNA molecule that is replicated when new cells are produced. (31.12)

Geometrical isomers   Isomers that differ only in the way that atoms are oriented in space relative to each other. (26.4, 30.3)

Glass   An amorphous solid. (11.1)

Graham's law   The rates of diffusion of gases are inversely proportional to the square roots of their densities (or their molecular weights). (10.13)

Gram (g)   A unit of measure for mass. $1 \text{ g} = 1 \times 10^{-3} \text{ kg}$. (1.12)

Ground state   State in which the electrons in an atom, ion, or molecule are in the lowest energy orbitals possible. (5.7)

Half-cell   An electrode containing both an oxidized and a reduced species. (21.6)

Half-life   The time required for half of the atoms in a radioactive sample to decay. (24.4)

Half-life of a reaction ($t_{1/2}$)   The time required for half of the original concentration of the limiting reactant to be consumed. (14.7)

Half-reaction   One of the two parts (oxidation or reduction) of an oxidation-reduction reaction. (21.2)

Hard water   Water containing dissolved calcium, magnesium, and/or iron salts. (29.10)

Heat   The form of energy that moves spontaneously from a warmer object to a cooler one. (4.1, 19.2)

Heat capacity   A property of a body of matter that represents the quantity of heat required to increase its temperature by one degree Celsius. (4.1)

Heat of fusion (ΔH_{fus})   The energy needed to change a given quantity of a substance from the solid state to the liquid state at a constant temperature. (11.7)

Heat of reaction   The heat lost or absorbed by a system in a reaction. (4.4)

Heat of vaporization (ΔH_{vap})   The energy needed to evaporate a given quantity of liquid at a constant specified temperature. (11.6)

Heisenberg uncertainty principle   It is impossible to determine accurately both the momentum and the position of a particle simultaneously. (5.9)

Henry's law   The mass of a gas that dissolves in a definite volume of liquid is directly proportional to the pressure of the gas provided the gas does not react with the solvent. (12.2)

Hess's law   If a process can be written as the sum of several stepwise processes, the enthalpy change of the total process equals the sum of the enthalpy changes of the various steps. (4.5)

Heterogeneous catalyst   A catalyst present in a different phase than the reactants, furnishing a surface at which a reaction can occur. (14.15)

Heterogeneous equilibrium   An equilibrium between two or more different phases, involving a boundary surface between the two phases. (15.8)

Heteronuclear diatomic molecule   Diatomic molecules composed of two different elements. (7.14)

Hexagonal closest packed structure   A crystalline structure in which close packed layers of atoms or ions are stacked ABABAB; the unit cell is hexagonal. (11.15)

Homogeneous catalyst   A catalyst present in the same phase as the reactants. (14.15)

Homogeneous equilibrium   An equilibrium within a single phase. (15.8)

Homonuclear diatomic molecule   Molecules composed of two identical atoms. (7.1)

Hund's rule   Every orbital in a subshell is singly occupied with one electron before any one orbital is doubly occupied, and all electrons in singly occupied orbitals have the same spin. (5.11)

Hybridization   A model that describes the changes in the atomic orbitals of an atom when it forms a covalent compound. (8.3)

Hydrocarbon   A compound composed only of hydrogen and carbon. The major component of fossil fuels. (4.6)

Hydrogen bond   The strong electrostatic attraction that occurs between molecules in which hydrogen is in a covalent bond with a highly electronegative element (i.e., fluorine, oxygen, nitrogen, chlorine). (11.12)

Hydrogen salt   A salt in which only part of the acidic hydrogens of a polyprotic acid are replaced by a cation. (16.12)

Hydronium ion (H_3O^+)   A water molecule with an added hydrogen ion. (9.2)

Ideal gas   A gas that follows Boyle's law, Charles's law, and Avogadro's law perfectly. (10.8, 10.14)

Ideal gas equation $PV = nRT$.   (10.8)

Ideal solution   A solution formed with no accompanying energy change, when the intermolecular attractive forces between the molecules of the solvent are the same as those between the molecules in the separate components. (12.9)

Insulator   A crystal that has bands that are completely filled or completely empty, with large energy gaps between them. An insulator does not conduct electricity. (25.3)

Intermolecular force   The attractive force between two molecules. (11.11)

Internal energy (*E*)   The total of all possible kinds of energy present in a substance or substances. (4.3, 19.2)

International System of Units (SI Units)   An updated version of the metric system used by scientists, adopted by the U.S. National Bureau of Standards in 1964. (1.8)

Interstitial site   A position between the regular positions in an array of atoms or ions that can be occupied by other atoms or ions. (11.14)

Intramolecular force   The attractive force between the atoms making up a molecule. (11.11)

Ion   Charged particle resulting from the loss or gain of one or more electrons from an atom or a molecule. (5.2)

Ion activity   The effective concentration of any particular kind of ion in solution. It is less than indicated by the actual concentration of a solution. (12.24)

Ion exchange   A method of softening water by filtering it through layers of sodium aluminosilicates (zeolites), which exchange sodium or hydrogen ions for Ca^{2+}, Mg^{2+}, or Fe^{2+} in the water. (29.11)

Ion product for water (K_w)   The equilibrium constant for the autoionization of water. At 25°C, $K_w = 1.00 \times 10^{-14}$ $(mol/L)^2$. (17.1)

Ion-dipole attraction   The electrostatic attraction between an ion and the dipole of a molecule. (12.10)

Ionic bond   The electrostatic attraction between positive and negative ions in an ionic compound. (6.1)

Ionic compound   A compound composed of ions. (6.1)

Ionic radius   The radius of an ion. (5.15)

Ionization constant   The equilibrium constant for the ionization of a weak acid or base. (17.4)

Ionization energy   The amount of energy required to remove an electron from the valence shell of a gaseous atom. (5.15)

Isoelectronic species   A group of ions, atoms, or molecules that have the same arrangement of electrons. (5.15, 6.3)

Isomorphous structures   Two elements or compounds that crystallize with the same structure. (11.15)

Isotopes   Atoms with the same atomic number and different numbers of neutrons. (5.6)

Joule   The SI unit of energy. One joule is the kinetic energy of an object with a mass of 2 kilograms moving with a velocity of one meter per second. $1 J = 1$ kg m^2/s^2. $4.184 J = 1$ cal. (4.1)

K_a   The equilibrium constant for the reaction of an acid with water. (17.4)

K_b   The equilibrium constant for the reaction of a base with water. (17.5)

Kelvin (K)   The SI unit of temperature. 273.15 K $= 0°$ Celsius. (1.14)

Kilogram   Standard SI unit of mass. Approximately 2.2 pounds. (1.12)

Kinetic energy (KE)   The kinetic energy of a moving body, in joules, is equal to $\frac{1}{2}mu^2$ ($m =$ mass; $u =$ speed in meters/second). (10.15)

Kinetic-molecular theory   Theory which explains the properties of an ideal gas and assumes that they consist of continuously moving molecules of negligible size. (10.14)

Lattice energy (U)   The energy required to separate the ions or molecules in a mole of a compound by infinite distances. (11.20, 11.21)

Lattice point   A point in a space lattice. (11.18)

Law of definite proportion   All samples of a pure compound contain the same elements in the same proportion by mass. (2.4)

Law of mass action   When a reversible reaction has attained equilibrium at a given temperature, the reaction quotient remains constant. (15.2)

Le Châtelier's principle   If a stress is applied to a system in equilibrium, the equilibrium shifts in a way that tends to undo the effect of the stress. (15.4)

Lewis acid   Any species that can accept a pair of electrons and form a coordinate covalent bond. (16.15)

Lewis base   Any species that can donate a pair of electrons and form a coordinate covalent bond. (16.15)

Lewis structure   Diagram showing shared and unshared pairs of electrons in an atom, molecule, or ion. (6.3)

Lewis symbol   The symbol for an element showing a dot for each valence electron in the element or ion. (6.1)

Ligand   An ion or neutral molecule attached to the central metal ion in a coordination compound. (26.1)

Ligand field splitting (10Dq)   The difference in energy between the metal's e_g and t_{2g} orbitals in a coordination complex. (26.7)

Ligand field stabilization energy (LFSE)   A measure of the increased stability of a complex showing ligand field splitting. (26.8)

Limiting reagent   The reactant completely consumed by a chemical reaction. The amount of the limiting reagent limits the amount of product that can be formed. (2.9)

Liquid   The state in which matter takes the shape of its container, assumes a horizontal upper surface and has a fairly definite volume. (1.2)

Liter (L)   Unit of volume. 1 L $= 1000$ cm^3. (1.11)

London force   The attraction between two rapidly fluctuating, temporary dipoles. Of significance only if atoms are very close together. (11.11)

Magnetic quantum number (m)   Quantum number signifying the orientation of an orbital around the nucleus. (5.10)

Mass   The quantity of matter contained by an object. Mass is measured by the force required to change the speed or direction of its movement. (1.2)

Mass defect   The difference between the calculated and experimental mass of a nucleus. (24.2)

Mass number (A)   The sum of the number of neutrons and protons in the nucleus of an atom. (5.6, 24.1)

Matter   Anything that occupies space and possesses mass. (1.2)

Melting point   The temperature at which the solid and liquid phases of a substance are in equilibrium. (11.5)

Metal   A substance that is malleable and ductile, has a characteristic luster, and is generally a good conductor of heat and electricity. The bands of a metal are partially filled. (9.4, 25.3)

Metallic conduction   The movement of electrons through a metal, with no changes in the metal and no movement of the metal atoms. (21.1)

Metathetical reaction A reaction in which two or more compounds exchange parts. (9.3)

Meter Standard metric and SI unit of length. Approximately 1.094 yards. (1.11)

Miscibility The ability of a liquid to mix with another liquid. (12.3)

Mixture Matter that can be separated into its components by physical means. (1.5)

Molality (m) The number of moles of solute dissolved in exactly one kilogram of solvent. (12.14)

Molar mass Mass in grams of one mole of an element or compound. Numerically equal to the molecular weight of a molecule or the atomic weight of an atom. (2.3)

Molar volume The volume of one mole of an ideal gas (22.4 liters at STP). (10.9)

Molarity (M) The number of moles of solute dissolved in one liter of solution. (3.1)

Mole (mol) The number of atoms contained in exactly 12 grams of ^{12}C (Avogadro's number). 1 mol = 6.022 × 10^{23} atoms, molecules, or ions. (2.2)

Mole fraction (X) The number of moles of a component of a solution divided by the total number of moles of all components. (12.15)

Molecular formula A formula indicating the composition of a molecule of a compound and giving the actual number of atoms of each element in a molecule of the compound. (2.1)

Molecular orbital The discrete energy and region of space around the nuclei of the atoms in a molecule in which an electron can be found. (7.1)

Molecular structure The three-dimensional, geometrical arrangement of the atoms in a molecule. (8.1)

Molecular weight The sum of the atomic weights of each atom in a particular molecule. (2.3)

Molecule The smallest particle of an element or compound that exists independently, retaining its identity. (1.6)

Monoprotic acid An acid containing one ionizable hydrogen atom per molecule. (16.9)

Nernst equation Used to calculate emf values at other than standard conditions. At 25°C, $E = E° - \dfrac{0.05915}{n} \log Q$. (21.11)

Neutralization A reaction that occurs when stoichiometrically equivalent quantities of an acid and a base are mixed. (16.4)

Neutron An uncharged subatomic particle with a mass of 1.0087 amu. (5.3)

Nonelectrolyte A compound that does not ionize when dissolved in water. (9.2)

Nonmetal An element that is brittle and nonductile as a solid, has no luster, and is a poor conductor of heat and electricity. (9.4)

Normal boiling point The temperature at which a liquid's vapor pressure equals one atm (760 torr). (11.3)

Normality (N) The number of equivalents of solute dissolved in one liter of solution. (16.14, 21.5)

Nuclear binding energy The energy produced by the loss of mass accompanying the formation of an atom from protons, electrons, and neutrons. (24.2)

Nuclear force The force of attraction between nucleons that holds a nucleus together. (24.1)

Nucleon Collective term for protons and neutrons. (24.1)

Nucleus The very heavy, positively charged body located at the center of an atom. (5.5)

Nuclide The nucleus of a particular isotope. (24.1)

Octahedral hole An octahedral space between six atoms or ions in a crystal. (11.16)

Optical isomers Isomers that are mirror images of each other and that are optically active. (26.4, 30.6)

Optically active molecules Molecules that rotate the plane of vibration of plane-polarized light. (30.6)

Orbital A three-dimensional region around the nucleus in which an electron moves; can hold up to two electrons. (5.9)

Order of a reaction With respect to one of the reactants, the order of a reaction is equal to the power to which the concentration of that reactant is raised in the rate equation. (14.6)

Osmosis The tendency of a solvent to diffuse through a semipermeable membrane from the less concentrated to the more concentrated solution. (12.22)

Osmotic pressure (π) The pressure required to stop the osmosis from a pure solvent into a solution. $\pi = MRT$. (12.22)

Oxidation The loss of electrons or an increase in oxidation number. (9.3)

Oxidation number Number used to keep track of the redistribution of electrons during chemical reactions. The sum of the oxidation numbers of all atoms in a neutral compound is zero. (6.10)

Oxidation-reduction reaction Reaction in which oxidation numbers change as electrons are lost by one atom and gained by another. (9.3)

Oxidizing agent The substance in an oxidation-reduction reaction that gains electrons and whose oxidation number is reduced. (9.3)

Oxyacid A hydroxide of a nonmetal. (9.4)

Paramagnetic substance A substance containing unpaired electrons. Paramagnetic substances tend to move into a magnetic field. (26.9)

Parent nuclide An unstable nuclide which changes spontaneously into another (daughter) nuclide. (24.7)

Partial pressure The pressure exerted by an individual gas in a mixture of gases. (10.12)

Pauli exclusion principle No two electrons in the same atom can have the same set of four quantum numbers. (5.11)

Percent yield The actual yield of an experiment divided by the theoretical yield and multiplied by 100. (2.10)

Periodic law The properties of the elements are periodic functions of their atomic numbers. (5.13)

pH The negative logarithm of the concentration of hydrogen ions in a solution. (16.5)

Photon A quantum of light or other electromagnetic radiation. The energy of a photon equals the product of Planck's constant and its frequency. (5.8)

Physical change A change in the state or properties of a particular kind of matter that does not involve a chemical change. (1.4)

Physical property Characteristics of a particular kind of matter that do not involve a chemical change (i.e., color, hardness, physical state). (1.4)

Pi (π) orbital A molecular orbital formed by side-by-side overlap of atomic orbitals, in which the electron density is found above and below the bond axis. (7.1)

pOH The negative logarithm of the concentration of hydroxide ions in a solution. (16.5)

Polar covalent bond Covalent bond between atoms of different electronegativities; a covalent bond with a positive end and a negative end. (6.6)

Polydentate ligand A ligand that is attached to a central metal ion by bonds from two or more donor atoms. (26.1)

Polymers Compounds of high molecular weight that are built up of a large number of simple molecules, or monomers. (30.15)

Polymorphism The assumption of two or more crystal structures by the same substance. (11.18)

Polyprotic acid An acid with more than one proton per molecule that can be given up in a reaction. (16.9)

Positron An atomic particle with the same mass as an electron but with one unit of positive charge. (24.6)

Pressure Force exerted on a unit area. The SI unit of pressure is the pascal (Pa). Pa = 1 newton/m^2. (10.2)

Principal quantum number *(n)* Quantum number specifying the shell of an electron in an atom or a monatomic ion. (5.10)

Proton A nuclear particle with a mass of 1.0073 amu and carrying a charge of +1. (5.2)

Quantized A description of the discrete, or individual, values by which the energy of an electron can vary. (5.7)

Radioactive decay The spontaneous change of an unstable nuclide (parent) into another nuclide (daughter). (24.7)

Radioactivity The spontaneous decomposition of an unstable element, with the simultaneous emission of rays of particles. (5.4)

Radius ratio In an ionic compound, the radius of the positive ion, $r+$, divided by the radius of the negative ion, $r-$. (11.17)

Raoult's law The vapor pressure of the solvent in an ideal solution (P_{solv}) is equal to the mole fraction of the solvent (X_{solv}) times the vapor pressure of the pure solvent (P°_{solv}). (12.17)

Rate constant The proportionality constant in the relationship between reaction rate and concentrations of reactants. (14.5)

Rate equations Equations giving the relationship between reaction rate and concentrations of reactants. (14.5)

Rate-determining step The slowest elementary reaction in a reaction path, which determines the maximum rate of the overall reaction. (14.14)

Reaction mechanism The stepwise sequence of elementary reactions in an overall reaction. (14.10, 14.14)

Reaction quotient (Q) A ratio of molar concentrations of the reactants to those of the products, each concentration being raised to the power equal to the coefficient in the equation. For the reaction $aA + bB + \cdots \rightarrow cC + dD + \cdots$, $Q = [C]^c[D]^d \ldots /[A]^a[B]^b \ldots$ (15.2, 19.11)

Recombinant DNA DNA produced by cutting a DNA chain and inserting particular pieces from another organism, changing the characteristics of the original DNA. (31.15)

Reducing agent The substance in an oxidation-reduction reaction that gives up electrons, and whose oxidation number is increased. (9.3)

Reduction The gain of electrons or decrease in oxidation number. (9.3)

Resonance forms Two or more Lewis structures having the same arrangement of atoms but different arrangements of electrons. (6.8)

Resonance hybrid The average of the resonance forms shown by the individual Lewis structures. (6.8)

Reversible reaction A chemical reaction that can proceed in either direction. (9.3)

Salt An ionic compound composed of cations and anions other than hydroxide or oxide ions. (9.2, 16.12)

Saturated solution A solution in which no more solute can be dissolved. (12.1)

Second law of thermodynamics Any spontaneous change that occurs in the universe must be accompanied by an increase in the entropy of the universe. (19.10)

Semiconductor A substance that contains a full band and an empty band, with small energy gaps between the bands. It is a poor conductor. (25.3)

Semi-metal Substances possessing some of the properties of both metals and nonmetals. (9.4)

Shell All of the orbitals in an atom or monatomic ion with the same value of n. (5.10)

Sigma (σ) orbital A molecular orbital in which the electron density is found along the axis of the bond. (7.1)

Simple cubic structure A crystalline structure with a cubic unit cell with lattice points only on the corners. (11.18)

Single bond A bond in which a single pair of electrons is shared between a pair of atoms. (6.3)

Solid The state in which matter is rigid, has a definite shape, and has a fairly constant volume. (1.2)

Solid solution A homogeneous and stable solution of one solid substance in another. (12.8)

Solubility product (K_{sp}) The equilibrium constant for the dissolution of a slightly soluble electrolyte. (18.1)

Solution A homogeneous mixture of a solute in a solvent. (12.1)

Space lattice All points within a crystal that have identical environments. (11.18)

Specific gravity The ratio of the density of a substance to the density of a reference substance—for solids and liquids, usually water; for gases, usually air or hydrogen. (1.13)

Specific heat A property of a substance that represents the quantity of heat required to raise the temperature of one gram of the substance one degree Celsius (or kelvin). (4.1)

Spectrum The component colors, or wavelengths of light. (5.8)

Spin quantum number *(s)* Number specifying the direction of the spin of an electron around its own axis. (5.10)

Spontaneous process A physical or chemical change that occurs without the addition of energy. (12.9) $\Delta G < 0$ for a spontaneous process. (19.8)

Stability curve A plot of the number of neutrons versus the number of protons for stable nuclei. (24.3)

Standard conditions (STP) 273.15 K (0°C) and one atmosphere of pressure (760 torr or 101.325 kilopascals). (10.9)

Standard electrode potential ($E°$) Potential measured with respect to a standard hydrogen electrode at 25°C with one M concentration of each ion in solution and one atm of pressure of each gas involved. (21.7)

Standard hydrogen electrode Assigned an electrode potential of exactly zero. The potential of all other electrodes is reported relative to the standard hydrogen electrode. (21.6)

Standard molar enthalpy of formation ($\Delta H_f°$) The enthalpy change of a chemical reaction in which one mole of a pure substance is formed from the free elements in their most stable states under standard state conditions. (4.4)

Standard molar entropy ($S°_{298}$) Actual entropy content of one mole of a substance in a standard state. (19.7)

Standard state 298.15 K (25 degrees Celsius) and one atmosphere of pressure. (4.4)

State function A property of a system that is not dependent on the way in which the system gets to the state in which it exhibits that property. (19.4)

Stoichiometry The calculation of both material balances and energy balances in a chemical system. (3.3)

Strong acid An acid that gives a 100% yield of hydronium ions when dissolved in water. (9.2)

Strong base A base that gives a 100% yield of hydroxide ions when dissolved in water. (9.2)

Strong electrolyte An electrolyte that gives a 100% yield of ions when dissolved in water. (9.2)

Structural isomers Two substances having the same molecular formula but different physical and chemical properties because the arrangement of their component atoms is different. (30.1)

Sublimation The passing of a solid directly to the vapor state without first melting. (11.2)

Subshell A set of degenerate orbitals with the same values of *n* and *l*. (5.10)

Substitution reaction A reaction in which one atom replaces another in a molecule. (30.1)

Supercooling A liquid that is cooled below its freezing point. (11.1)

Supersaturated solution A solution that contains more solute than it would if the dissolved solute were in equilibrium with the undissolved solute. (12.1)

Surface tension The force that causes the surface of a liquid to contract, reducing its surface area to a minimum. (11.10)

Surroundings The universe outside a thermodynamic system. (4.3)

System The substance or substances involved in a reaction that is being studied. (4.3)

Termolecular reaction An elementary reaction involving the simultaneous collision of any combination of three molecules, ions, or atoms. (14.13)

Ternary compound A compound containing three different elements. (6.13)

Tetrahedral hole A tetrahedral space formed by four atoms or ions in a crystal. (11.16)

Theoretical yield The calculated yield of a reaction based on the assumptions that there is only one reaction involved, that all the reactant is converted into product, and all the product is collected. (2.10)

Third law of thermodynamics The entropy of any pure, perfect crystalline element or compound at absolute zero (0 K) is equal to zero. (19.13)

Three-center bond The bonding of three atoms by one pair of electrons in a molecular orbital formed from the overlap of three atomic orbitals. (25.2)

Titration Method of determining the concentration of a solution by adding a solution of a reactant to a solution of sample until an indicator changes color. (3.2)

***Trans* configuration** Configuration of a geometrical isomer in which two groups are on opposite sides of an imaginary reference line on the molecule. (26.4)

Transition state A combination of reacting molecules that is intermediate between reactants and products. (14.8)

Triple bond A bond in which three pairs of electrons are shared between two atoms. (6.3)

Triple point The point at which an equilibrium exists among the vapor, liquid, and solid phases of a substance. (11.9)

Triprotic acid An acid containing three ionizable hydrogen atoms per molecule. Ionization of triprotic acids occurs in three stages. (16.9)

Unimolecular reaction An elementary reaction in which the rearrangement of a single molecule produces one or more molecules of product. (14.11)

Unit cell The portion of a space lattice that is repeated in order to form the entire lattice. (11.18)

Unsaturated solution A solution in which more solute can be dissolved. (12.1)

Unshared pair Electrons not used in bonding to form a molecule. (6.3)

Valence electrons Electrons in the valence shell of an atom. The number of valence electrons determines how an element reacts. (5.14)

Valence shell The outermost shell of electrons of a representative element; the outermost shell of electrons and the d electrons in the next inner shell of a d-block element; or the outermost shell of electrons, the d electrons in the next inner shell, and the f electrons in the next inner shell of an f-block element. (5.14)

Valence shell electron pair repulsion (VSEPR) theory A theory used to predict the bond angles in a molecule, based on positioning regions of high electron density as far apart as possible to minimize electrostatic repulsion. (8.1)

Van der Waals equation A quantitative expression of the deviations of real gases from the ideal gas laws. (10.18)

Van der Waals force Intermolecular attractive force. (11.11)

Vapor pressure The pressure exerted by a vapor in equilibrium with a solid or a liquid at a given temperature. (11.2)

Vapor-pressure lowering The lowering of the vapor pressure of a liquid by addition of a solute. (12.17)

Volt (V) Difference in electrical potential when one joule of energy is required to move 1/96,485 mole of electrons (one coulomb of charge) from a lower potential to a higher potential. (21.6)

Voltaic cell Any cell that generates an electric current by an oxidation-reduction reaction. (21.7)

Wave function (ψ) A mathematical function that describes the shape of the orbital that an electron occupies, the energy of the electron in the orbital, and the probability of finding the electron at any location in the orbital. (5.9)

Weak acid An acid that gives a 10% or less yield of hydronium ions when dissolved in water. (9.2)

Weak base A base that gives a 10% or less yield of hydroxide ions when dissolved in water. (9.2)

Weak electrolyte An electrolyte that gives a low percentage yield of ions when dissolved in water. (9.2)

Work (w) One process for removing energy from a system or adding energy to it. (4.3, 19.2)

Yield The quantity of a product of a chemical reaction. (2.10)

Index